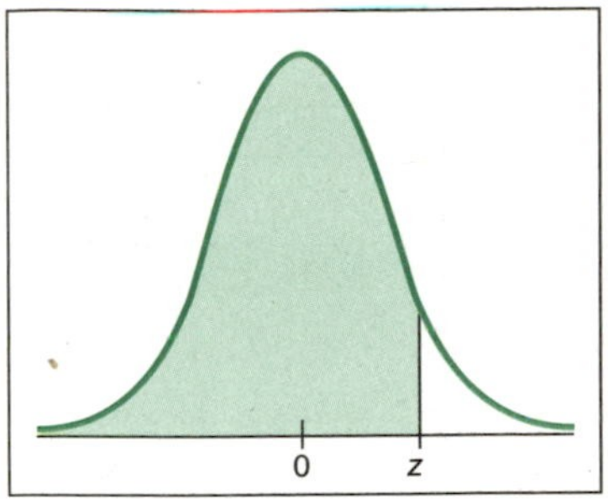

The table entry for *z* is the area to the left of *z*.

Areas of a Standard Normal Distribution *continued*

z	.00	.01	.02	.03	.04	.05	.06	.07	.08	.09
0.0	.5000	.5040	.5080	.5120	.5160	.5199	.5239	.5279	.5319	.5359
0.1	.5398	.5438	.5478	.5517	.5557	.5596	.5636	.5675	.5714	.5753
0.2	.5793	.5832	.5871	.5910	.5948	.5987	.6026	.6064	.6103	.6141
0.3	.6179	.6217	.6255	.6293	.6331	.6368	.6406	.6443	.6480	.6517
0.4	.6554	.6591	.6628	.6664	.6700	.6736	.6772	.6808	.6844	.6879
0.5	.6915	.6950	.6985	.7019	.7054	.7088	.7123	.7157	.7190	.7224
0.6	.7257	.7291	.7324	.7357	.7389	.7422	.7454	.7486	.7517	.7549
0.7	.7580	.7611	.7642	.7673	.7704	.7734	.7764	.7794	.7823	.7852
0.8	.7881	.7910	.7939	.7967	.7995	.8023	.8051	.8078	.8106	.8133
0.9	.8159	.8186	.8212	.8238	.8264	.8289	.8315	.8340	.8365	.8389
1.0	.8413	.8438	.8461	.8485	.8508	.8531	.8554	.8577	.8599	.8621
1.1	.8643	.8665	.8686	.8708	.8729	.8749	.8770	.8790	.8810	.8830
1.2	.8849	.8869	.8888	.8907	.8925	.8944	.8962	.8980	.8997	.9015
1.3	.9032	.9049	.9066	.9082	.9099	.9115	.9131	.9147	.9162	.9177
1.4	.9192	.9207	.9222	.9236	.9251	.9265	.9279	.9292	.9306	.9319
1.5	.9332	.9345	.9357	.9370	.9382	.9394	.9406	.9418	.9429	.9441
1.6	.9452	.9463	.9474	.9484	.9495	.9505	.9515	.9525	.9535	.9545
1.7	.9554	.9564	.9573	.9582	.9591	.9599	.9608	.9616	.9625	.9633
1.8	.9641	.9649	.9656	.9664	.9671	.9678	.9686	.9693	.9699	.9706
1.9	.9713	.9719	.9726	.9732	.9738	.9744	.9750	.9756	.9761	.9767
2.0	.9772	.9778	.9783	.9788	.9793	.9798	.9803	.9808	.9812	.9817
2.1	.9821	.9826	.9830	.9834	.9838	.9842	.9846	.9850	.9854	.9857
2.2	.9861	.9864	.9868	.9871	.9875	.9878	.9881	.9884	.9887	.9890
2.3	.9893	.9896	.9898	.9901	.9904	.9906	.9909	.9911	.9913	.9916
2.4	.9918	.9920	.9922	.9925	.9927	.9929	.9931	.9932	.9934	.9936
2.5	.9938	.9940	.9941	.9943	.9945	.9946	.9948	.9949	.9951	.9952
2.6	.9953	.9955	.9956	.9957	.9959	.9960	.9961	.9962	.9963	.9964
2.7	.9965	.9966	.9967	.9968	.9969	.9970	.9971	.9972	.9973	.9974
2.8	.9974	.9975	.9976	.9977	.9977	.9978	.9979	.9979	.9980	.9981
2.9	.9981	.9982	.9982	.9983	.9984	.9984	.9985	.9985	.9986	.9986
3.0	.9987	.9987	.9987	.9988	.9988	.9989	.9989	.9989	.9990	.9990
3.1	.9990	.9991	.9991	.9991	.9992	.9992	.9992	.9992	.9993	.9993
3.2	.9993	.9993	.9994	.9994	.9994	.9994	.9994	.9995	.9995	.9995
3.3	.9995	.9995	.9995	.9996	.9996	.9996	.9996	.9996	.9996	.9997
3.4	.9997	.9997	.9997	.9997	.9997	.9997	.9997	.9997	.9997	.9998

For *z* values greater than 3.49, use 1.000 to approximate the area.

Areas of a Standard Normal Distribution *continued*

(b) Confidence Interval Critical Values z_c	
Level of Confidence *c*	**Critical Value z_c**
0.70, or 70%	1.04
0.75, or 75%	1.15
0.80, or 80%	1.28
0.85, or 85%	1.44
0.90, or 90%	1.645
0.95, or 95%	1.96
0.98, or 98%	2.33
0.99, or 99%	2.58

Areas of a Standard Normal Distribution *continued*

(c) Hypothesis Testing, Critical Values z_0		
Level of Significance	$\alpha = 0.05$	$\alpha = 0.01$
Critical value z_0 for a left-tailed test	−1.645	−2.33
Critical value z_0 for a right-tailed test	1.645	2.33
Critical values $\pm z_0$ for a two-tailed test	±1.96	±2.58

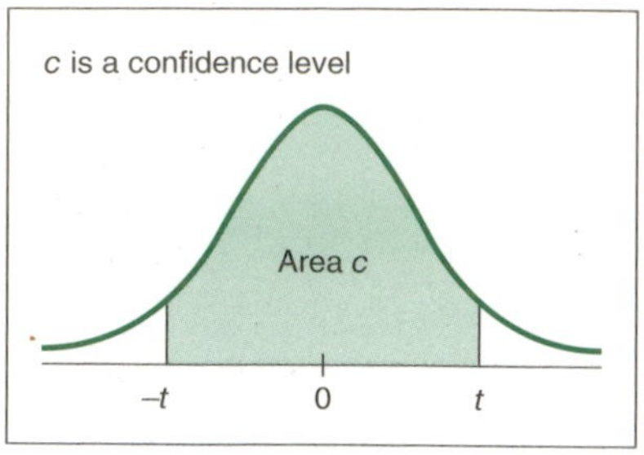

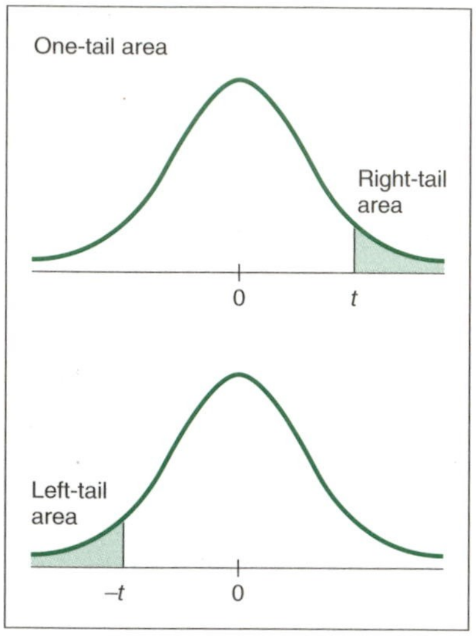

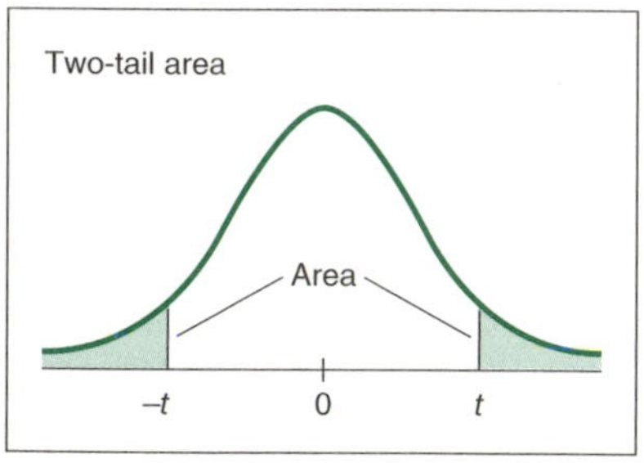

Critical Values for Student's *t* Distribution

one-tail area	0.250	0.125	0.100	0.075	0.050	0.025	0.010	0.005	0.0005
two-tail area	**0.500**	**0.250**	**0.200**	**0.150**	**0.100**	**0.050**	**0.020**	**0.010**	**0.0010**
d.f. *c*	**0.500**	**0.750**	**0.800**	**0.850**	**0.900**	**0.950**	**0.980**	**0.990**	**0.999**
1	1.000	2.414	3.078	4.165	6.314	12.706	31.821	63.657	636.619
2	0.816	1.604	1.886	2.282	2.920	4.303	6.965	9.925	31.599
3	0.765	1.423	1.638	1.924	2.353	3.182	4.541	5.841	12.924
4	0.741	1.344	1.533	1.778	2.132	2.776	3.747	4.604	8.610
5	0.727	1.301	1.476	1.699	2.015	2.571	3.365	4.032	6.869
6	0.718	1.273	1.440	1.650	1.943	2.447	3.143	3.707	5.959
7	0.711	1.254	1.415	1.617	1.895	2.365	2.998	3.499	5.408
8	0.706	1.240	1.397	1.592	1.860	2.306	2.896	3.355	5.041
9	0.703	1.230	1.383	1.574	1.833	2.262	2.821	3.250	4.781
10	0.700	1.221	1.372	1.559	1.812	2.228	2.764	3.169	4.587
11	0.697	1.214	1.363	1.548	1.796	2.201	2.718	3.106	4.437
12	0.695	1.209	1.356	1.538	1.782	2.179	2.681	3.055	4.318
13	0.694	1.204	1.350	1.530	1.771	2.160	2.650	3.012	4.221
14	0.692	1.200	1.345	1.523	1.761	2.145	2.624	2.977	4.140
15	0.691	1.197	1.341	1.517	1.753	2.131	2.602	2.947	4.073
16	0.690	1.194	1.337	1.512	1.746	2.120	2.583	2.921	4.015
17	0.689	1.191	1.333	1.508	1.740	2.110	2.567	2.898	3.965
18	0.688	1.189	1.330	1.504	1.734	2.101	2.552	2.878	3.922
19	0.688	1.187	1.328	1.500	1.729	2.093	2.539	2.861	3.883
20	0.687	1.185	1.325	1.497	1.725	2.086	2.528	2.845	3.850
21	0.686	1.183	1.323	1.494	1.721	2.080	2.518	2.831	3.819
22	0.686	1.182	1.321	1.492	1.717	2.074	2.508	2.819	3.792
23	0.685	1.180	1.319	1.489	1.714	2.069	2.500	2.807	3.768
24	0.685	1.179	1.318	1.487	1.711	2.064	2.492	2.797	3.745
25	0.684	1.178	1.316	1.485	1.708	2.060	2.485	2.787	3.725
26	0.684	1.177	1.315	1.483	1.706	2.056	2.479	2.779	3.707
27	0.684	1.176	1.314	1.482	1.703	2.052	2.473	2.771	3.690
28	0.683	1.175	1.313	1.480	1.701	2.048	2.467	2.763	3.674
29	0.683	1.174	1.311	1.479	1.699	2.045	2.462	2.756	3.659
30	0.683	1.173	1.310	1.477	1.697	2.042	2.457	2.750	3.646
35	0.682	1.170	1.306	1.472	1.690	2.030	2.438	2.724	3.591
40	0.681	1.167	1.303	1.468	1.684	2.021	2.423	2.704	3.551
45	0.680	1.165	1.301	1.465	1.679	2.014	2.412	2.690	3.520
50	0.679	1.164	1.299	1.462	1.676	2.009	2.403	2.678	3.496
60	0.679	1.162	1.296	1.458	1.671	2.000	2.390	2.660	3.460
70	0.678	1.160	1.294	1.456	1.667	1.994	2.381	2.648	3.435
80	0.678	1.159	1.292	1.453	1.664	1.990	2.374	2.639	3.416
100	0.677	1.157	1.290	1.451	1.660	1.984	2.364	2.626	3.390
500	0.675	1.152	1.283	1.442	1.648	1.965	2.334	2.586	3.310
1000	0.675	1.151	1.282	1.441	1.646	1.962	2.330	2.581	3.300
∞	0.674	1.150	1.282	1.440	1.645	1.960	2.326	2.576	3.291

For degrees of freedom *d.f.* not in the table, use the closest *d.f.* that is *smaller*.

STATISTICS FORMULA CARD
for Brase/Brase. *Understandable Statistics*

ELEVENTH EDITION

UNDERSTANDABLE STATISTICS

Concepts and Methods

Charles Henry Brase
Regis University

Corrinne Pellillo Brase
Arapahoe Community College

CENGAGE Learning

Australia • Brazil • Mexico • Singapore • United Kingdom • United States

This book is dedicated to the memory of
a great teacher, mathematician, and friend

Burton W. Jones
Professor Emeritus, University of Colorado

CENGAGE Learning®

Understandable Statistics: Concepts and Methods, Eleventh Edition
Charles Henry Brase and Corrinne Pellillo Brase

Product Director: Liz Covello

Senior Product Manager: Molly Taylor

Senior Content Developer: Jay Campbell

Product Assistant: Danielle Hallock

Media Developer: Andrew Coppola

Market Development Manager: Ryan Ahern

Marketing Brand Manager: Gordon Lee

Content Project Manager: Jill Quinn

Senior Art Director: Linda May

Manufacturing Planner: Sandee Milewski

Rights Acquisition Specialist: Shalice Shah-Caldwell

Production Service: Graphic World Inc.

Text and Cover Designer: Rokusek Design

Cover Image: Shutterstock, © Jan Martin Will

Compositor: Graphic World Inc.

Interior Chapter Opener Photo: Shutterstock, © Jan Martin Will

Library of Congress Control Number: 2013942408

Student Edition:

ISBN-13: 978-1-285-46091-8

ISBN-10: 1-285-46091-X

Annotated Instructor's Edition:

ISBN-13: 978-1-285-46282-0

ISBN-10: 1-285-46282-3

Cengage Learning
200 First Stamford Place, 4th Floor
Stamford, CT 06902
USA

Cengage Learning is a leading provider of customized learning solutions with office locations around the globe, including Singapore, the United Kingdom, Australia, Mexico, Brazil, and Japan. Locate your local office at **www.cengage.com/global**

Cengage Learning products are represented in Canada by Nelson Education, Ltd.

To learn more about Cengage Learning, visit **www.cengage.com**

Purchase any of our products at your local college store or at our preferred online store **www.cengagebrain.com**

Instructors: Please visit **login.cengage.com** and log in to access instructor-specific resources.

Printed in the United States of America
2 3 4 5 6 7 8 18 17 16 15 14

CONTENTS

Students need to develop critical thinking skills in order to understand and evaluate the limitations of statistical methods. *Understandable Statistics: Concepts and Methods* makes students aware of method appropriateness, assumptions, biases, and justifiable conclusions.

CRITICAL THINKING

UNUSUAL VALUES

Chebyshev's theorem tells us that no matter what the data distribution looks like, at least 75% of the data will fall within 2 standard deviations of the mean. As we will see in Chapter 6, when the distribution is mound-shaped and symmetrical, about 95% of the data are within 2 standard deviations of the mean. Data values beyond 2 standard deviations from the mean are less common than those closer to the mean.

In fact, one indicator that a data value might be an outlier is that it is more than 2.5 standard deviations from the mean (Source: *Statistics*, by G. Upton and I. Cook, Oxford University Press).

UNUSUAL VALUES

For a binomial distribution, it is unusual for the number of successes r to be higher than $\mu + 2.5\sigma$ or lower than $\mu - 2.5\sigma$.

We can use this indicator to determine whether a specified number of successes out of n trials in a binomial experiment are unusual.

For instance, consider a binomial experiment with 20 trials for which probability of success on a single trial is $p = 0.70$. The expected number of successes is $\mu = 14$, with a standard deviation of $\sigma \approx 2$. A number of successes above 19 or below 9 would be considered unusual. However, such numbers of successes are possible.

◀ Critical Thinking

Critical thinking is an important skill for students to develop in order to avoid reaching misleading conclusions. The Critical Thinking feature provides additional clarification on specific concepts as a safeguard against incorrect evaluation of information.

Interpretation ▶

Increasingly, calculators and computers are used to generate the numeric results of a statistical process. However, the student still needs to correctly interpret those results in the context of a particular application. The Interpretation feature calls attention to this important step. Interpretation is stressed in examples, in guided exercises, and in the problem sets.

SOLUTION: Since we want to know the number of standard deviations from the mean, we want to convert 6.9 to standard z units.

$$z = \frac{x - \mu}{\sigma} = \frac{6.9 - 8}{0.5} = -2.20$$

Interpretation The amount of cheese on the selected pizza is only 2.20 standard deviations below the mean. The fact that z is negative indicates that the amount of cheese is 2.20 standard deviations *below* the mean. The parlor will not lose its franchise based on this sample.

6. *Interpretation* A campus performance series features plays, music groups, dance troops, and stand-up comedy. The committee responsible for selecting the performance groups include three students chosen at random from a pool of volunteers. This year the 30 volunteers came from a variety of majors. However, the three students for the committee were all music majors. Does this fact indicate there was bias in the selection process and that the selection process was not random? Explain.

7. *Critical Thinking* Greg took a random sample of size 100 from the population of current season ticket holders to State College men's basketball games. Then he took a random sample of size 100 from the population of current season ticket holders to State College women's basketball games.
   (a) What sampling technique (stratified, systematic, cluster, multistage, convenience, random) did Greg use to sample from the population of current season ticket holders to all State College basketball games played by either men or women?
   (b) Is it appropriate to pool the samples and claim to have a random sample of size 200 from the population of current season ticket holders to all State College home basketball games played by either men or women? Explain.

◀ *NEW!* Critical Thinking and Interpretation Exercises

In every section and chapter problem set, Critical Thinking problems provide students with the opportunity to test their understanding of the application of statistical methods and their interpretation of their results. Interpretation problems ask students to apply statistical results to the particular application.

No language, including statistics, can be spoken without learning the vocabulary. *Understandable Statistics: Concepts and Methods* introduces statistical terms with deliberate care.

WHAT DOES THE LEVEL OF MEASUREMENT TELL US?

The level of measurement tells us which arithmetic processes are appropriate for the data. This is important because different statistical processes require various kinds of arithmetic. In some instances all we need to do is count the number of data that meet specified criteria. In such cases nominal (and higher) data levels are all appropriate. In other cases we need to order the data, so nominal data would not be suitable. Many other statistical processes require division, so data need to be at the ratio level. Just keep the nature of the data in mind before beginning statistical computations.

◄ *NEW!* What Does (concept, method, statistical result) Tell Us?

This feature gives a brief summary of the information we obtain from the named concept, method, or statistical result.

NEW! Important Features of a (concept, method, or result) ►

In statistics we use many different types of graphs, samples, data, and analytical methods. The features of each such tool help us select the most appropriate ones to use and help us interpret the information we receive from applications of the tools.

IMPORTANT FEATURES OF A SIMPLE RANDOM SAMPLE

For a simple random sample

- Every sample of specified size *n* from the population has an equal chance of being selected.
- No researcher bias occurs in the items selected for the sample.
- A random sample may not always reflect the diversity of the population. For instance, from a population of 10 cats and 10 dogs, a random sample of size 6 could consist of all cats.

SECTION 6.1 PROBLEMS

1. *Statistical Literacy* Which, if any, of the curves in Figure 6-10 look(s) like a normal curve? If a curve is not a normal curve, tell why.

2. *Statistical Literacy* Look at the normal curve in Figure 6-11, and find μ, $\mu + \sigma$, and σ.

FIGURE 6-10

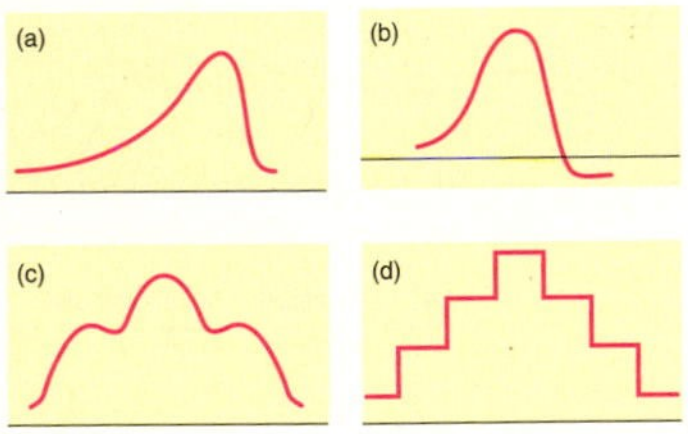

FIGURE 6-11

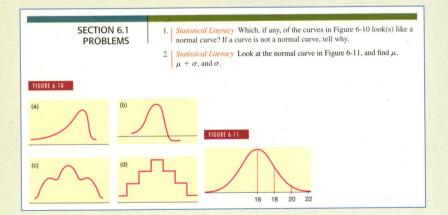

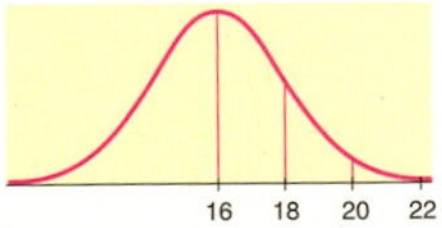

◄ Statistical Literacy Problems

In every section and chapter problem set, Statistical Literacy problems test student understanding of terminology, statistical methods, and the appropriate conditions for use of the different processes.

Definition Boxes ►

Whenever important terms are introduced in text, tan definition boxes appear within the discussions. These boxes make it easy to reference or review terms as they are used further.

BOX-AND-WHISKER PLOTS

The quartiles together with the low and high data values give us a very useful *five-number summary* of the data and their spread.

FIVE-NUMBER SUMMARY

Lowest value, Q_1, median, Q_3, highest value

We will use these five numbers to create a graphic sketch of the data called a *box-and-whisker plot.* Box-and-whisker plots provide another useful technique from exploratory data analysis (EDA) for describing data.

◀ Important Words & Symbols

The Important Words & Symbols within the Chapter Review feature at the end of each chapter summarizes the terms introduced in the Definition Boxes for student review at a glance. Page numbers for first occurrence of term are given for easy reference.

IMPORTANT WORDS & SYMBOLS

Section 4.1
Probability of an event A, $P(A)$ 144
Intuition 144
Relative frequency 144
Equally likely outcomes 144
Law of large numbers 146
Statistical experiment 146
Event 146
Simple event 146
Sample space 146
Complement of event A^c 148

Section 4.2
Independent events 156
Dependent events 156
Probability of *A and B* 156
Event $A \mid B$
Conditional probability 156
$P(A \mid B)$ 156
Multiplication rules of probability (for independent and dependent events) 156
More than two independent events 161
Probability of *A or B* 161
Event *A and B* 161
Event *A or B* 161
Mutually exclusive events 163
Addition rules (for mutually exclusive and general events) 163
More than two mutually exclusive events 165
Basic probability rules 168

Section 4.3
Multiplication rule of counting 177
Tree diagram 178
Factorial notation 181
Permutations rule 181
Combinations rule 183

Linking Concepts: Writing Projects ▶

Much of statistical literacy is the ability to communicate concepts effectively. The Linking Concepts: Writing Projects feature at the end of each chapter tests both statistical literacy and critical thinking by asking the student to express their understanding in words.

LINKING CONCEPTS: WRITING PROJECTS

Discuss each of the following topics in class or review the topics on your own. Then write a brief but complete essay in which you summarize the main points. Please include formulas and graphs as appropriate.

1. What does it mean to say that we are going to use a sample to draw an inference about a population? Why is a random sample so important for this process? If we wanted a random sample of students in the cafeteria, why couldn't we just choose the students who order Diet Pepsi with their lunch? Comment on the statement, "A random sample is like a miniature population, whereas samples that are not random are likely to be biased." Why would the students who order Diet Pepsi with lunch not be a random sample of students in the cafeteria?
2. In your own words, explain the differences among the following sampling techniques: simple random sample, stratified sample, systematic sample, cluster sample, multistage sample, and convenience sample. Describe situations in which each type might be useful.

◀ Basic Computation Problems

These problems focus student attention on relevant formulas, requirements, and computational procedures. After practicing these skills, students are more confident as they approach real-world applications.

5. *Basic Computation: Central Limit Theorem* Suppose x has a distribution with a mean of 8 and a standard deviation of 16. Random samples of size $n = 64$ are drawn.
   (a) Describe the $\bar{x}$ distribution and compute the mean and standard deviation of the distribution.
   (b) Find the z value corresponding to $\bar{x} = 9$.
   (c) Find $P(\bar{x} > 9)$.
   (d) *Interpretation* Would it be unusual for a random sample of size 64 from the x distribution to have a sample mean greater than 9? Explain.

Expand Your Knowledge Problems ▶

Expand Your Knowledge problems present optional enrichment topics that go beyond the material introduced in a section. Vocabulary and concepts needed to solve the problems are included at point-of-use, expanding students' statistical literacy.

30. *Expand Your Knowledge: Geometric Mean* When data consist of percentages, ratios, growth rates, or other rates of change, the *geometric mean* is a useful measure of central tendency. For n data values,

$$\text{Geometric mean} = \sqrt[n]{\text{product of the } n \text{ data values}}, \quad \text{assuming all data values are positive}$$

To find the *average growth factor* over 5 years of an investment in a mutual fund with growth rates of 10% the first year, 12% the second year, 14.8% the third year, 3.8% the fourth year, and 6% the fifth year, take the geometric mean of 1.10, 1.12, 1.148, 1.038, and 1.16. Find the average growth factor of this investment.

Note that for the same data, the relationships among the harmonic, geometric, and arithmetic means are harmonic mean ≤ geometric mean ≤ arithmetic mean (Source: *Oxford Dictionary of Statistics*).

Real knowledge is delivered through direction, not just facts. *Understandable Statistics: Concepts and Methods* ensures the student knows what is being covered and why at every step along the way to statistical literacy.

Chapter Preview ► Questions

Preview Questions at the beginning of each chapter give the student a taste of what types of questions can be answered with an understanding of the knowledge to come.

NORMAL CURVES AND SAMPLING DISTRIBUTIONS

PREVIEW QUESTIONS

What are some characteristics of a normal distribution? What does the empirical rule tell you about data spread around the mean? How can this information be used in quality control? (SECTION 6.1)

Can you compare apples and oranges, or maybe elephants and butterflies? In most cases, the answer is no—unless you first standardize your measurements. What are a standard normal distribution and a standard *z* score? (SECTION 6.2)

How do you convert any normal distribution to a standard normal distribution? How do you find probabilities of "standardized events"? (SECTION 6.3)

As humans, our experiences are finite and limited. Consequently, most of the important decisions in our lives are based on sample (incomplete) information. What is a probability sampling distribution? How will sampling distributions help us make good decisions based on incomplete information? (SECTION 6.4)

There is an old saying: All roads lead to Rome. In statistics, we could recast this saying: All probability distributions average out to be normal distributions (as the sample size increases). How can we take advantage of this in our study of sampling distributions? (SECTION 6.5)

The binomial and normal distributions are two of the most important probability distributions in statistics. Under certain limiting conditions, the binomial can

otoEdit

FOCUS PROBLEM

Benford's Law: The Importance of Being Number 1

Benford's Law states that in a wide variety of circumstances, numbers have "1" as their first nonzero digit disproportionately often. Benford's Law applies to such diverse topics as the drainage areas of rivers; properties of chemicals; populations of towns; figures in newspapers, magazines, and government reports; and the half-lives of radioactive atoms!

Specifically, such diverse measurements begin with "1" about 30% of the time, with "2" about 18% of time, and with "3" about 12.5% of the time. Larger digits occur less often. For example, less than 5% of the numbers in circumstances such as these begin with the digit 9. This is in dramatic contrast to a random sampling situation, in which each of the digits 1 through 9 has an equal chance of appearing.

The first nonzero digits of numbers taken from large bodies of numerical records such

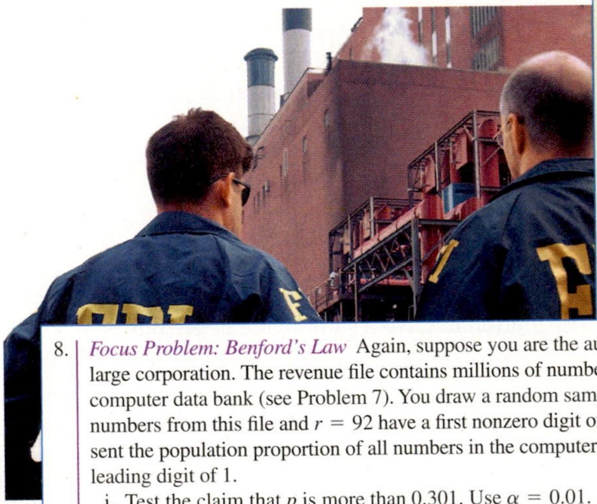

Corbis

8. *Focus Problem: Benford's Law* Again, suppose you are the auditor for a very large corporation. The revenue file contains millions of numbers in a large computer data bank (see Problem 7). You draw a random sample of $n = 228$ numbers from this file and $r = 92$ have a first nonzero digit of 1. Let p represent the population proportion of all numbers in the computer file that have a leading digit of 1.
   i. Test the claim that p is more than 0.301. Use $\alpha = 0.01$.
   ii. If p is in fact larger than 0.301, it would seem there are too many numbers in the file with leading 1's. Could this indicate that the books have been "cooked" by artificially lowering numbers in the file? Comment from the point of view of the Internal Revenue Service. Comment from the perspective of the Federal Bureau of Investigation as it looks for "profit skimming" by unscrupulous employees.
   iii. Comment on the following statement: "If we reject the null hypothesis at level of significance α, we have not *proved* H_0 to be false. We can say that the probability is α that we made a mistake in rejecting H_0." Based on the outcome of the test, would you recommend further investigation before accusing the company of fraud?

▲ Chapter Focus Problems

The Preview Questions in each chapter are followed by a Focus Problem, which serves as a more specific example of what questions the student will soon be able to answer. The Focus Problems are set within appropriate applications and are incorporated into the end-of-section exercises, giving students the opportunity to test their understanding.

DIRECTION AND PURPOSE

Focus Points ▶

Each section opens with bulleted Focus Points describing the primary learning objectives of the section.

SECTION **3.1**

Measures of Central Tendency: Mode, Median, and Mean

FOCUS POINTS

- Compute mean, median, and mode from raw data.
- Interpret what mean, median, and mode tell you.
- Explain how mean, median, and mode can be affected by extreme data values.
- What is a trimmed mean? How do you compute it?
- Compute a weighted average.

This section can be covered quickly. Good discussion topics include *The Story of Old Faithful* in Data Highlights, Problem 1; Linking Concepts, Problem 1; and the trade winds of Hawaii (Using Technology).

The average price of an ounce of gold is $1350. The Zippy car averages 39 miles per gallon on the highway. A survey showed the average shoe size for women is size 9.

Average

In each of the preceding statements, *one* number is used to describe the entire sample or population. Such a number is called an *average*. There are many ways to compute averages, but we will study only three of the major ones.

The easiest average to compute is the *mode*.

Mode

The **mode** of a data set is the value that occurs most frequently. *Note:* If a data set has no single value that occurs more frequently than any other, then that data set has no mode.

EXAMPLE 1

Mode

Count the letters in each word of this sentence and give the mode. The numbers of letters in the words of the sentence are

5 3 7 2 4 4 2 4 8 3 4 3 4

LOOKING FORWARD

In later chapters we will use information based on a sample and sample statistics to estimate population parameters (Chapter 7) or make decisions about the value of population parameters (Chapter 8).

◀ Looking Forward

This feature shows students where the presented material will be used later. It helps motivate students to pay a little extra attention to key topics.

CHAPTER REVIEW

SUMMARY

In this chapter, you've seen that statistics is the study of how to collect, organize, analyze, and interpret numerical information from populations or samples. This chapter discussed some of the features of data and ways to collect data. In particular, the chapter discussed

- Individuals or subjects of a study and the variables associated with those individuals
- Data classification as qualitative or quantitative, and levels of measurement of data
- Sample and population data. Summary measurements from sample data are called statistics, and those from populations are called parameters.
- Sampling strategies, including simple random, stratified, systematic, multistage, and convenience. Inferential techniques presented in this text are based on simple random samples.
- Methods of obtaining data: Use of a census, simulation, observational studies, experiments, and surveys
- Concerns: Undercoverage of a population, nonresponse, bias in data from surveys and other factors, effects of confounding or lurking variables on other variables, generalization of study results beyond the population of the study, and study sponsorship

▲ Chapter Summaries

The Summary within each Chapter Review feature now also appears in bulleted form, so students can see what they need to know at a glance.

Statistics is not done in a vacuum. *Understandable Statistics: Concepts and Methods* gives students valuable skills for the real world with technology instruction, genuine applications, actual data, and group projects.

REVISED! Tech Notes ▶

Tech Notes appearing throughout the text give students helpful hints on using TI-84Plus and TI-nspire (with TI-84Plus keypad) and TI-83Plus calculators, Microsoft Excel 2010, and Minitab to solve a problem. They include display screens to help students visualize and better understand the solution.

TECH NOTES *Box-and-Whisker Plot*

Both Minitab and the TI-84Plus/TI-83Plus/TI-*n*spire calculators support box-and-whisker plots. On the TI-84Plus/TI-83Plus/TI-*n*spire, the quartiles Q_1 and Q_3 are calculated as we calculate them in this text. In Minitab and Excel 2010, they are calculated using a slightly different process.

TI-84Plus/TI-83Plus/TI-*n*spire (with TI-84Plus Keypad) Press **STATPLOT ➤On.** Highlight box plot. Use **Trace** and the arrow keys to display the values of the five-number summary. The display shows the plot for calories in ice cream bars.

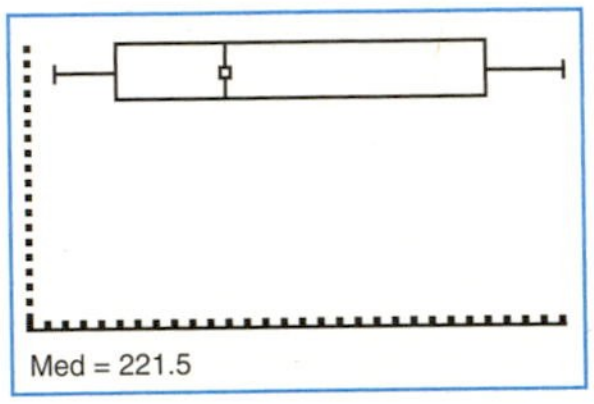

box-and-whisker plot. However, each value of the
und. On the **Home** ribbon, click the **Insert Function**
tatistical as the category and scroll to **Quartile**. In
location and then enter the number of the value you
e quartile box for the first quartile.
t. In the dialogue box, set Display to IQRange Box.

USING TECHNOLOGY

Binomial Distributions

Although tables of binomial probabilities can be found in most libraries, such tables are often inadequate. Either the value of p (the probability of success on a trial) you are looking for is not in the table, or the value of n (the number of trials) you are looking for is too large for the table. In Chapter 6, we will study the normal approximation to the binomial. This approximation is a great help in many practical applications. Even so, we sometimes use the formula for the binomial probability distribution on a computer or graphing calculator to compute the probability we want.

Applications

The following percentages were obtained over many years of observation by the U.S. Weather Bureau. All data listed are for the month of December.

Location	Long-Term Mean % of Clear Days in Dec.
Juneau, Alaska	18%
Seattle, Washington	24%
Hilo, Hawaii	36%
Honolulu, Hawaii	60%
Las Vegas, Nevada	75%
Phoenix, Arizona	77%

Adapted from *Local Climatological Data*, U.S. Weather Bureau publication, "Normals, Means, and Extremes" Table.

In the locations listed, the month of December is a relatively stable month with respect to weather. Since weather patterns from one day to the next are more or less the same, it is reasonable to use a binomial probability model.

1. Let r be the number of clear days in December. Since December has 31 days, $0 \le r \le 31$. Using appropriate computer software or calculators available to you, find the probability $P(r)$ for each of the listed locations when $r = 0, 1, 2, \ldots, 31$.
2. For each location, what is the expected value of the probability distribution? What is the standard deviation?

You may find that using cumulative probabilities and appropriate subtraction of probabilities, rather than addition of probabilities, will make finding the solutions to Applications 3 to 7 easier.

3. Estimate the probability that Juneau will have at most 7 clear days in December.
4. Estimate the probability that Seattle will have from 5 to 10 (including 5 and 10) clear days in December.
5. Estimate the probability that Hilo will have at least 12 clear days in December.
6. Estimate the probability that Phoenix will have 20 or more clear days in December.
7. Estimate the probability that Las Vegas will have from 20 to 25 (including 20 and 25) clear days in December.

Technology Hints

TI-84Plus/TI-83Plus/TI-*n*spire (with TI-84 Plus keypad), Excel 2010, Minitab

The Tech Note in Section 5.2 gives specific instructions for binomial distribution functions on the TI-84Plus/TI-83Plus/TI-*n*spire (with TI-84Plus keypad) calculators, Excel 2010, and Minitab.

SPSS

In SPSS, the function **PDF.BINOM(q,n,p)** gives the probability of q successes out of n trials, where p is the probability of success on a single trial. In the data editor, name a variable r and enter values 0 through n. Name another variable Prob_r. Then use the menu choices **Transform ➤ Compute**. In the dialogue box, use Prob_r for the target variable. In the function group, select **PDF and Noncentral PDF**. In the function box, select **PDF.BINOM(q,n,p)**. Use the variable r for q and appropriate values for n and p. Note that the function **CDF.BINOM(q,n,p)**, from the **CDF and Noncentral CDF** group, gives the cumulative probability of 0 through q successes.

◀ *REVISED!* Using Technology

Further technology instruction is available at the end of each chapter in the Using Technology section. Problems are presented with real-world data from a variety of disciplines that can be solved by using TI-84Plus and TI-*n*spire (with TI-84Plus keypad) and TI-83Plus calculators, Microsoft Excel 2010, and Minitab.

EXAMPLE 13

CENTRAL LIMIT THEOREM

A certain strain of bacteria occurs in all raw milk. Let x be the bacteria count per milliliter of milk. The health department has found that if the milk is not contaminated, then x has a distribution that is more or less mound-shaped and symmetrical. The mean of the x distribution is $\mu = 2500$, and the standard deviation is $\sigma = 300$. In a large commercial dairy, the health inspector takes 42 random samples of the milk produced each day. At the end of the day, the bacteria count in each of the 42 samples is averaged to obtain the sample mean bacteria count $\bar{x}$.

(a) Assuming the milk is not contaminated, what is the distribution of $\bar{x}$?

SOLUTION: The sample size is $n = 42$. Since this value exceeds 30, the central limit theorem applies, and we know
mean and standard deviation

◀ *UPDATED!* Applications

Real-world applications are used from the beginning to introduce each statistical process. Rather than just crunching numbers, students come to appreciate the value of statistics through relevant examples.

Most exercises in each section ▶ are applications problems.

11. *Pain Management: Laser Therapy* "Effect of Helium-Neon Laser Auriculotherapy on Experimental Pain Threshold" is the title of an article in the journal *Physical Therapy* (Vol. 70, No. 1, pp. 24–30). In this article, laser therapy was discussed as a useful alternative to drugs in pain management of chronically ill patients. To measure pain threshold, a machine was used that delivered low-voltage direct current to different parts of the body (wrist, neck, and back). The machine measured current in milliamperes (mA). The pretreatment experimental group in the study had an average threshold of pain (pain was first detectable) at $\mu = 3.15$ mA with standard deviation $\sigma = 1.45$ mA. Assume that the distribution of threshold pain, measured in milliamperes, is symmetrical and more or less mound-shaped. Use the empirical rule to
    (a) estimate a range of milliamperes centered about the mean in which about 68% of the experimental group had a threshold of pain.
    (b) estimate a range of milliamperes centered about the mean in which about 95% of the experimental group had a threshold of pain.

12. *Control Charts: Yellowstone National Park* Yellowstone Park Medical Services (YPMS) provides emergency health care for park visitors. Such health care includes treatment for everything from indigestion and sunburn to more serious injuries. A recent issue of *Yellowstone Today* (National Park

DATA HIGHLIGHTS: GROUP PROJECTS

Break into small groups and discuss the following topics. Organize a brief outline in which you summarize the main points of your group discussion.

1. Examine Figure 2-20, "Everyone Agrees: Slobs Make Worst Roommates." This is a clustered bar graph because two percentages are given for each response category: responses from men and responses from women. Comment about how the artistic rendition has slightly changed the format of a bar graph. Do the bars seem to have lengths that accurately reflect the relative percentages of the responses? In your own opinion, does the artistic rendition enhance or confuse the information? Explain. Which characteristic of "worst roommates" does the graphic seem to illustrate? Can this graph be considered a Pareto chart for men? for women? Why or why not? From the information given in the figure, do you think the survey just listed the four given annoying characteristics? Do you think a respondent could choose more than one characteristic? Explain

FIGURE 2-20

Source: Advantage Business Research for Mattel *Compatibility*

◀ Data Highlights: Group Projects

Using Group Projects, students gain experience working with others by discussing a topic, analyzing data, and collaborating to formulate their response to the questions posed in the exercise.

Get to the "Aha!" moment faster. *Understandable Statistics: Concepts and Methods* provides the push students need to get there through guidance and example.

◀ Procedures and Requirements

Procedure display boxes summarize simple step-by-step strategies for carrying out statistical procedures and methods as they are introduced. Requirements for using the procedures are also stated. Students can refer back to these boxes as they practice using the procedures.

PROCEDURE

HOW TO TEST μ WHEN σ IS KNOWN

Requirements

Let x be a random variable appropriate to your application. Obtain a simple random sample (of size n) of x values from which you compute the sample mean $\bar{x}$. The value of σ is already known (perhaps from a previous study). If you can assume that x has a normal distribution, then any sample size n will work. If you cannot assume this, then use a sample size $n \geq 30$.

Procedure

1. In the context of the application, state the *null and alternate hypotheses* and set the *level of significance* α.
2. Use the known σ, the sample size n, the value of $\bar{x}$ from the sample, and μ from the null hypothesis to compute the standardized *sample test statistic.*

$$z = \frac{\bar{x} - \mu}{\frac{\sigma}{\sqrt{n}}}$$

3. Use the standard normal ... or two-tailed, to find the ...
4. *Conclude* the test. If *P*-va... do not reject H_0.
5. *Interpret your conclusion* ...

GUIDED EXERCISE 11 | **PROBABILITY REGARDING $\bar{x}$**

In mountain country, major highways sometimes use tunnels instead of long, winding roads over high passes. However, too many vehicles in a tunnel at the same time can cause a hazardous situation. Traffic engineers are studying a long tunnel in Colorado. If x represents the time for a vehicle to go through the tunnel, it is known that the x distribution has mean $\mu = 12.1$ minutes and standard deviation $\sigma = 3.8$ minutes under ordinary traffic conditions. From a histogram of x values, it was found that the x distribution is mound-shaped with some symmetry about the mean.

Engineers have calculated that, *on average,* vehicles should spend from 11 to 13 minutes in the tunnel. If the time is less than 11 minutes, traffic is moving too fast for safe travel in the tunnel. If the time is more than 13 minutes, there is a problem of bad air quality (too much carbon monoxide and other pollutants).

Under ordinary conditions, there are about 50 vehicles in the tunnel at one time. What is the probability that the mean time for 50 vehicles in the tunnel will be from 11 to 13 minutes? We will answer this question in steps.

(a) Let $\bar{x}$ represent the sample mean based on samples of size 50. Describe the $\bar{x}$ distribution.

➡ From the central limit theorem, we expect the $\bar{x}$ distribution to be approximately normal, with mean and standard deviation

$$\mu_{\bar{x}} = \mu = 12.1 \quad \sigma_{\bar{x}} = \frac{\sigma}{\sqrt{n}} = \frac{3.8}{\sqrt{50}} \approx 0.54$$

(b) Find $P(11 < \bar{x} < 13)$.

Jupiter Images

➡ We convert the interval

$11 < \bar{x} < 13$

to a standard z interval and use the standard normal probability table to find our answer. Since

$$z = \frac{\bar{x} - \mu}{\sigma/\sqrt{n}} \approx \frac{\bar{x} - 12.1}{0.54}$$

$$\bar{x} = 11 \text{ converts to } z \approx \frac{11 - 12.1}{0.54} = -2.04$$

$$\text{and } \bar{x} = 13 \text{ converts to } z \approx \frac{13 - 12.1}{0.54} = 1.67$$

Therefore,

$$P(11 < \bar{x} < 13) = P(-2.04 < z < 1.67) = 0.9525 - 0.0207 = 0.9318$$

(c) ***Interpret*** your answer to part (b).

➡ It seems that about 93% of the time, there should be no safety hazard for average traffic flow.

Guided Exercises ▶

Students gain experience with new procedures and methods through Guided Exercises. Beside each problem in a Guided Exercise, a completely worked-out solution appears for immediate reinforcement.

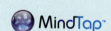

Preface

Welcome to the exciting world of statistics! We have written this text to make statistics accessible to everyone, including those with a limited mathematics background. Statistics affects all aspects of our lives. Whether we are testing new medical devices or determining what will entertain us, applications of statistics are so numerous that, in a sense, we are limited only by our own imagination in discovering new uses for statistics.

OVERVIEW

The eleventh edition of *Understandable Statistics: Concepts and Methods* continues to emphasize concepts of statistics. Statistical methods are carefully presented with a focus on understanding both the *suitability of the method* and the *meaning of the result.* Statistical methods and measurements are developed in the context of applications.

Critical thinking and interpretation are essential in understanding and evaluating information. Statistical literacy is fundamental for applying and comprehending statistical results. In this edition we have expanded and highlighted the treatment of statistical literacy, critical thinking, and interpretation.

We have retained and expanded features that made the first 10 editions of the text very readable. Definition boxes highlight important terms. Procedure displays summarize steps for analyzing data. Examples, exercises, and problems touch on applications appropriate to a broad range of interests.

The eleventh edition continues to have extensive online support. Online homework powered by a choice of Enhanced WebAssign or Aplia™ is now available through CengageBrain.com. Instructional videos are available on DVD. The companion web site at http://www.cengage.com/statistics/brase11e contains more than 100 data sets (in Microsoft Excel, Minitab, SPSS, and TI-84Plus/TI-83Plus/TI-*n*spire with TI-84Plus keypad ASCII file formats), technology guides, lecture aids, a glossary, and statistical tables.

MindTap™

New for the eleventh edition, available via Aplia™, is MindTap™ Reader, Cengage Learning's next-generation eBook. MindTap Reader provides robust opportunities for students to annotate, take notes, navigate, and interact with the text (e.g., ReadSpeaker). Annotations captured in MindTap Reader are automatically tied to the Notepad app, where they can be viewed chronologically and in a cogent, linear fashion. Instructors also can edit the text and assets in the Reader as well as add videos or URLs. Go to http://www.cengage.com/mindtap for more information.

MAJOR CHANGES IN THE ELEVENTH EDITION

With each new edition, the authors reevaluate the scope, appropriateness, and effectiveness of the text's presentation and reflect on extensive user feedback. Revisions have been made throughout the text to clarify explanations of important concepts and to update problems.

Critical Thinking, Interpretation, and Statistical Literacy

The eleventh edition of this text continues and expands the emphasis on critical thinking, interpretation, and statistical literacy. Calculators and computers are very good at providing numerical results of statistical processes. However, numbers from a computer or calculator display are meaningless unless the user knows how to interpret the results and if the statistical process is appropriate. This text helps students determine whether or not a statistical method or process is appropriate. It helps students understand what a statistic measures. It helps students interpret the results of a confidence interval, hypothesis test, or liner regression model.

New Interpretation Features

To further understanding and interpretation of statistical concepts, methods, and results, we have included two new special features: **What Does (a concept, method, or result) Tell Us?** and **Important Features of a (concept, method, or result).** These features summarize the information we obtain from concepts and statistical processes and give additional insights for further application.

New Expand Your Knowledge Problems and Quick Overview Topics With Additional Applications

Expand Your Knowledge problems do just that! These are optional but contain very useful information taken from the vast literature of statistics. These topics are not included in the main text but are easily learned using material from the section or previous sections. Although these topics are optional, the authors feel they add depth and enrich a student's learning experience. Each topic was chosen for its relatively straightforward presentation and useful applications. All such problems and their applications are flagged with a sun logo.

New *Expand Your Knowledge problems* in the eleventh edition involve donut graphs; stratified sampling and the best estimate for the population mean μ; the process of using minimal variance for linear combinations of independent random variables; and serial correlation (also called autocorrelation).

Some of the other topics in *Expand Your Knowledge problems* or quick overviews include graphs such as dotplots and variations on stem-and-leaf plots; outliers in stem-and-leaf plots; harmonic and geometric means; moving averages; calculating odds in favor and odds against; extension of conditional probability to various distributions such as the Poisson distribution and the normal distribution; Bayes's theorem; additional probability distributions such as the multinomial distribution, negative binomial distribution, hypergeometric distribution, continuous uniform distribution, and exponential distribution; waiting time between Poisson events; quick estimate of the standard deviation using the Empirical rule; plus four confidence intervals for proportions; Satterthwaite's approximation for degrees of freedom in confidence intervals and hypothesis tests; relationship between confidence intervals and two-tailed hypothesis testing; pooled two-sample procedures for confidence intervals and hypothesis tests; resampling (also known as bootstrap); simulations of confidence intervals and hypothesis tests using different samples of the same size; mean and standard deviation for linear combinations of dependent random variables; logarithmic transformations with the exponential growth model and the power law model; and polynomial (curvilinear) regression.

For location of these optional topics in the text, please see the index.

Revised Examples and New Section Problems

Examples and guided exercises have been updated and revised. Additional section problems emphasize critical thinking and interpretation of statistical results.

Excel 2010 and Most Recent Operating System for the TI-84Plus/TI-83Plus Calculators

Excel 2010 instructions are included in the *Tech Notes* and *Using Technology*. The latest operating system (v2.55MP) for the TI-84Plus/TI-83Plus calculators is also discussed, with new functions such as the inverse t distribution and the chi-square goodness of fit test described. One convenient feature of the operating system is that it provides on-screen prompts for inputs required for many probability and statistical functions. This operating system is already on new TI-84Plus/TI-83Plus calculators and is available for download to older calculators at the Texas Instruments web site.

Revised Electronic Student Resources

Digital student resources and online tools that accompany *Understandable Statistics* have been revised in accordance with recommendations from both student and faculty users. Online interactive learning solutions, such as Aplia™ for Statistics—featuring the new MindTap™ Reader—and Enhanced WebAssign, are both available.

CONTINUING CONTENT

Introduction of Hypothesis Testing Using *P*-Values

In keeping with the use of computer technology and standard practice in research, hypothesis testing is introduced using *P*-values. The critical region method is still supported but not given primary emphasis.

Use of Student's *t* Distribution in Confidence Intervals and Testing of Means

If the normal distribution is used in confidence intervals and testing of means, then the *population standard deviation must be known.* If the population standard deviation is not known, then under conditions described in the text, the Student's t distribution is used. This is the most commonly used procedure in statistical research. It is also used in statistical software packages such as Microsoft Excel, Minitab, SPSS, and TI-84Plus/TI-83Plus/TI-*n*spire calculators.

Confidence Intervals and Hypothesis Tests of Difference of Means

If the normal distribution is used, then both population standard deviations must be known. When this is not the case, the Student's t distribution incorporates an approximation for t, with a commonly used conservative choice for the degrees of freedom. Satterthwaite's approximation for the degrees of freedom as used in computer software is also discussed. The pooled standard deviation is presented for appropriate applications ($\sigma_1 \approx \sigma_2$).

FEATURES IN THE ELEVENTH EDITION

Chapter and Section Lead-ins

- *Preview Questions* at the beginning of each chapter are keyed to the sections.
- *Focus Problems* at the beginning of each chapter demonstrate types of questions students can answer once they master the concepts and skills presented in the chapter.
- *Focus Points* at the beginning of each section describe the primary learning objectives of the section.

Carefully Developed Pedagogy

- *Examples* show students how to select and use appropriate procedures.
- *Guided Exercises* within the sections give students an opportunity to work with a new concept. Completely worked-out solutions appear beside each exercise to give immediate reinforcement.

- *Definition boxes* highlight important definitions throughout the text.
- *Procedure displays* summarize key strategies for carrying out statistical procedures and methods. Conditions required for using the procedure are also stated.
- NEW! *What Does (a concept method or result) Tell Us?* summarizes information we obtain from the named concepts and statistical processes and gives insight for additional application.
- NEW! *Important Features of a (concept, method, or result)* summarizes the features of the listed item.
- *Looking Forward* features give a brief preview of how a current topic is used later.
- *Labels* for each example or guided exercise highlight the technique, concept, or process illustrated by the example or guided exercise. In addition, labels for section and chapter problems describe the field of application and show the wide variety of subjects in which statistics is used.
- *Section and chapter problems* require the student to use all the new concepts mastered in the section or chapter. Problem sets include a variety of real-world applications with data or settings from identifiable sources. Key steps and solutions to odd-numbered problems appear at the end of the book.
- *Basic Computation problems* ask students to practice using formulas and statistical methods on very small data sets. Such practice helps students understand what a statistic measures.
- *Statistical Literacy problems* ask students to focus on correct terminology and processes of appropriate statistical methods. Such problems occur in every section and chapter problem set.
- *Interpretation problems* ask students to explain the meaning of the statistical results in the context of the application.
- *Critical Thinking problems* ask students to analyze and comment on various issues that arise in the application of statistical methods and in the interpretation of results. These problems occur in every section and chapter problem set.
- *Expand Your Knowledge problems* present enrichment topics such as negative binomial distribution; conditional probability utilizing binomial, Poisson, and normal distributions; estimation of standard deviation from a range of data values; and more.
- *Cumulative review problem sets* occur after every third chapter and include key topics from previous chapters. Answers to *all* cumulative review problems are given at the end of the book.
- *Data Highlights and Linking Concepts* provide group projects and writing projects.
- *Viewpoints* are brief essays presenting diverse situations in which statistics is used.
- *Design and photos* are appealing and enhance readability.

Technology Within the Text

- *Tech Notes* within sections provide brief point-of-use instructions for the TI-84Plus, TI-83Plus, and TI-*n*spire (with 84Plus keypad) calculators, Microsoft Excel 2010, and Minitab.
- *Using Technology* sections show the use of SPSS as well as the TI-84Plus, TI-83Plus, and TI-*n*spire (with TI-84Plus keypad) calculators, Microsoft Excel, and Minitab.

ALTERNATE ROUTES THROUGH THE TEXT

Understandable Statistics: Concepts and Methods, Eleventh Edition, is designed to be flexible. It offers the professor a choice of teaching possibilities. In most one-semester courses, it is not practical to cover all the material in depth. However, depending on the emphasis of the course, the professor may choose to cover various topics. For help in topic selection, refer to the Table of Prerequisite Material on page 1.

- *Introducing linear regression early.* For courses requiring an early presentation of linear regression, the descriptive components of linear regression (Sections 9.1

and 9.2) can be presented any time after Chapter 3. However, inference topics involving predictions, the correlation coefficient ρ, and the slope of the least-squares line β require an introduction to confidence intervals (Sections 7.1 and 7.2) and hypothesis testing (Sections 8.1 and 8.2).

- *Probability.* For courses requiring minimal probability, Section 4.1 (What Is Probability?) and the first part of Section 4.2 (Some Probability Rules—Compound Events) will be sufficient.

ACKNOWLEDGMENTS

It is our pleasure to acknowledge the prepublication reviewers of this text. All of their insights and comments have been very valuable to us. Reviewers of this text include:

Jorge Baca, Cosumnes River College
Wayne Barber, Chemeketa Community College
Molly Beauchman, Yavapai College
Nick Belloit, Florida State College at Jacksonville
Kimberly Benien, Wharton County Junior College
Abraham Biggs, Broward Community College
Dexter Cahoy, Louisiana Tech University
Maggy Carney, Burlington County College
Christopher Donnelly, Macomb Community College
Meike Niederhausen, University of Portland
Deanna Payton, Northern Oklahoma College in Stillwater

We also would like to acknowledge reviewers of previous editions.

Reza Abbasian, Texas Lutheran University
Paul Ache, Kutztown University
Kathleen Almy, Rock Valley College
Polly Amstutz, University of Nebraska at Kearney
Delores Anderson, Truett-McConnell College
Robert J. Astalos, Feather River College
Lynda L. Ballou, Kansas State University
Mary Benson, Pensacola Junior College
Larry Bernett, Benedictine University
Kiran Bhutani, The Catholic University of America
Kristy E. Bland, Valdosta State University
John Bray, Broward Community College
Bill Burgin, Gaston College
Toni Carroll, Siena Heights University
Pinyuen Chen, Syracuse University
Emmanuel des-Bordes, James A. Rhodes State College
Jennifer M. Dollar, Grand Rapids Community College
Larry E. Dunham, Wor-Wic Community College
Andrew Ellett, Indiana University
Ruby Evans, Keiser University
Mary Fine, Moberly Area Community College
Rebecca Fouguette, Santa Rosa Junior College
Rene Garcia, Miami-Dade Community College
Larry Green, Lake Tahoe Community College
Shari Harris, John Wood Community College
Janice Hector, DeAnza College
Jane Keller, Metropolitan Community College
Raja Khoury, Collin County Community College
Diane Koenig, Rock Valley College
Charles G. Laws, Cleveland State Community College

Michael R. Lloyd, Henderson State University
Beth Long, Pellissippi State Technical and Community College
Lewis Lum, University of Portland
Darcy P. Mays, Virginia Commonwealth University
Charles C. Okeke, College of Southern Nevada, Las Vegas
Peg Pankowski, Community College of Allegheny County
Ram Polepeddi, Westwood College, Denver North Campus
Azar Raiszadeh, Chattanooga State Technical Community College
Traei Reed, St. Johns River Community College
Michael L. Russo, Suffolk County Community College
Janel Schultz, Saint Mary's University of Minnesota
Sankara Sethuraman, Augusta State University
Stephen Soltys, West Chester University of Pennsylvania
Ron Spicer, Colorado Technical University
Winson Taam, Oakland University
Jennifer L. Taggart, Rockford College
William Truman, University of North Carolina at Pembroke
Bill White, University of South Carolina Upstate
Jim Wienckowski, State University of New York at Buffalo
Stephen M. Wilkerson, Susquehanna University
Hongkai Zhang, East Central University
Shunpu Zhang, University of Alaska, Fairbanks
Cathy Zucco-Teveloff, Trinity College

We would especially like to thank Roger Lipsett for his careful accuracy review of this text. We are especially appreciative of the excellent work by the editorial and production professionals at Cengage Learning. In particular we thank Molly Taylor, Jay Campbell, Jill Quinn, and Amy Simpson.

Without their creative insight and attention to detail, a project of this quality and magnitude would not be possible. Finally, we acknowledge the cooperation of Minitab, Inc., SPSS, Texas Instruments, and Microsoft.

Charles Henry Brase

Corrinne Pellillo Brase

Additional Resources — Get More from Your Textbook!

MindTap™

Available via Aplia™ is MindTap™ Reader, Cengage Learning's next-generation eBook. An integral component of the MindTap environment, MindTap Reader replaces the traditional "e-book" and provides robust opportunities for students to annotate, take notes, navigate, and interact with the text (e.g., ReadSpeaker). Annotations captured in MindTap Reader are automatically tied to the Notepad app, where they can be viewed chronologically and in a cogent, linear fashion. Instructors also can edit the text and assets in the Reader, as well as add videos or URLs.

Go to http://www.cengage.com/mindtap for more information.

Instructor Resources

Annotated Instructor's Edition (AIE) Answers to all exercises, teaching comments, and pedagogical suggestions appear in the margin, or at the end of the text in the case of large graphs.

Solution Builder Contains complete solutions to all exercises in the text, including those in the Chapter Review and Cumulative Review Problems in online format. Solution Builder allows instructors to create customized, secure PDF printouts of solutions matched exactly to the exercises assigned for class. Available to adoptions by signing up at www.cengage.com/solutionbuilder.

Cengage Learning Testing Powered by Cognero A flexible, online system that allows you to:
- author, edit, and manage test bank content from multiple Cengage Learning solutions
- create multiple test versions in an instant
- deliver tests from your LMS, your classroom or wherever you want

Companion Website The companion website at http://www.cengage.com/brase contains a variety of resources.
- Microsoft® PowerPoint® lecture slides
- Figures from the book
- More than 100 data sets in a variety of formats, including
  - Microsoft Excel
  - Minitab
  - SPSS
  - TI-84Plus/TI-83Plus/TI-nspire with 84plus keypad ASCII file formats
- Technology guides for the following programs
  - TI-84Plus, TI-83Plus, and TI-nspire graphing calculators
  - Minitab software (version 14)
  - Microsoft Excel (2010/2007)
  - SPSS Statistics software
- Lecture aids like Teaching Hints and Frequently Used Formulas
- Statistical tables and a glossary

Student Resources

Student Solutions Manual Provides solutions to the odd-numbered section and chapter exercises and to all the Cumulative Review exercises in the student textbook.

Instructional DVDs Hosted by Dana Mosely, these text-specific DVDs cover all sections of the text and provide explanations of key concepts, examples, exercises, and applications in a lecture-based format. DVDs are close-captioned for the hearing-impaired.

aplia

Aplia is an online interactive learning solution that helps students improve comprehension—and their grade—by integrating a variety of mediums and tools such as video, tutorials, practice tests, and an interactive eBook. Created by a professor to enhance his own courses, Aplia provides automatically graded assignments with detailed, immediate feedback on every question, and innovative teaching materials. More than 1,000,000 students have used Aplia at over 1,800 institutions.

jmp

JMP is a statistics software for Windows and Macintosh computers from SAS, the market leader in analytics software and services for industry. JMP Student Edition is a streamlined, easy-to-use version that provides all the statistical analysis and graphics covered in this textbook. Once data is imported, students will find that most procedures require just two or three mouse clicks. JMP can import data from a variety of formats, including Excel and other statistical packages, and you can easily copy and paste graphs and output into documents.

JMP also provides an interface to explore data visually and interactively, which will help your students develop a healthy relationship with their data, work more efficiently with data, and tackle difficult statistical problems more easily. Because its output provides both statistics and graphs together, the student will better see and understand the application of concepts covered in this book as well. JMP Student Edition also contains some unique platforms for student projects, such as mapping and scripting. JMP functions in the same way on both Windows and Macintosh platforms and instructions contained with this book apply to both platforms.

Access to this software is available with new copies of the book. Students can purchase JMP standalone via CengageBrain.com or www.jmp.com/getse.

Minitab® and IBM SPSS These statistical software packages manipulate and interpret data to produce textual, graphical, and tabular results. Minitab® and/or SPSS may be packaged with the textbook. Student versions are available.

The companion website at http://www.cengage.com/statistics/brase11e contains useful assets for students.

- **Technology Guides** Separate guides exist with information and examples for each of four technology tools. Guides are available for the TI-84Plus, TI-83Plus, and TI-*n*spire graphing calculators, Minitab software (version 14) Microsoft Excel (2010/2007), and SPSS Statistics software.
- **Interactive Teaching and Learning Tools** include glossary flashcards, online datasets (in Microsoft Excel, Minitab, SPSS, and TI-84Plus/TI-83Plus/TI-*n*spire with TI-84Plus keypad ASCII file formats), statistical tables and formulas, and more.

Enhanced WebAssign Offers an extensive online program for Statistics to encourage the practice that's so critical for concept mastery. The meticulously crafted pedagogy and exercises in Brase and Brase's text become even more effective in Enhanced WebAssign.

CengageBrain.com Provides the freedom to purchase online homework and other materials à la carte exactly what you need, when you need it.

For more information, visit **http://www.cengage.com/statistics/brase11e** or contact your local Cengage Learning sales representative.

Table of Prerequisite Material

Chapter	Prerequisite Sections
1 Getting Started	None
2 Organizing Data	1.1, 1.2
3 Averages and Variation	1.1, 1.2, 2.1
4 Elementary Probability Theory	1.1, 1.2, 2.1, 3.1, 3.2
5 The Binomial Probability Distribution and Related Topics	1.1, 1.2, 2.1, 3.1, 3.2, 4.1, 4.2 4.3 useful but not essential
6 Normal Curves and Sampling Distributions (omit 6.6)	1.1, 1.2, 2.1, 3.1, 3.2, 4.1, 4.2, 5.1
(include 6.6)	also 5.2, 5.3
7 Estimation (omit 7.3 and parts of 7.4)	1.1, 1.2, 2.1, 3.1, 3.2, 4.1, 4.2, 5.1, 6.1, 6.2, 6.3, 6.4, 6.5
(include 7.3 and all of 7.4)	also 5.2, 5.3, 6.6
8 Hypothesis Testing (omit 8.3 and part of 8.5)	1.1, 1.2, 2.1, 3.1, 3.2, 4.1, 4.2, 5.1, 6.1, 6.2, 6.3, 6.4, 6.5
(include 8.3 and all of 8.5)	also 5.2, 5.3, 6.6
9 Correlation and Regression (9.1 and 9.2)	1.1, 1.2, 3.1, 3.2
(9.3 and 9.4)	also 4.1, 4.2, 5.1, 6.1, 6.2, 6.3, 6.4, 6.5, 7.1, 7.2, 8.1, 8.2
10 Chi-Square and *F* Distributions (omit 10.3)	1.1, 1.2, 2.1, 3.1, 3.2, 4.1, 4.2, 5.1, 6.1, 6.2, 6.3, 6.4, 6.5, 8.1 also 7.1
(include 10.3)	
11 Nonparametric Statistics	1.1, 1.2, 2.1, 3.1, 3.2, 4.1, 4.2, 5.1, 6.1, 6.2, 6.3, 6.4, 6.5, 8.1, 8.3

1

MLB Photos/Getty Images Sport/Getty Images

Bettmann/Corbis

Chance favors the prepared mind.

—Louis Pasteur

Statistical techniques are tools of thought . . . not substitutes for thought.

—Abraham Kaplan

Louis Pasteur (1822–1895) is the founder of modern bacteriology. At age 57, Pasteur was studying cholera. He accidentally left some bacillus culture unattended in his laboratory during the summer. In the fall, he injected laboratory animals with this bacilli. To his surprise, the animals did not die—in fact, they thrived and were resistant to cholera.

When the final results were examined, it is said that Pasteur remained silent for a minute and then exclaimed, as if he had seen a vision, "Don't you see they have been vaccinated!" Pasteur's work ultimately saved many human lives.

Most of the important decisions in life involve incomplete information. Such decisions often involve so many complicated factors that a complete analysis is not practical or even possible. We are often forced into the position of making a guess based on limited information.

As the first quote reminds us, our chances of success are greatly improved if we have a "prepared mind." The statistical methods you will learn in this book will help you achieve a prepared mind for the study of many different fields. The second quote reminds us that statistics is an important tool, but it is not a replacement for an in-depth knowledge of the field to which it is being applied.

The authors of this book want you to understand and enjoy statistics. The reading material will *tell you* about the subject. The examples will *show you* how it works. To understand, however, you must *get involved*. Guided exercises, calculator and computer applications, section and chapter problems, and writing exercises are all designed to get you involved in the subject. As you grow in your understanding of statistics, we believe you will enjoy learning a subject that has a world full of interesting applications.

GETTING STARTED

STUDIO SATO/amanaimages/Corbis

PREVIEW QUESTIONS

Why is statistics important? (SECTION 1.1)

What is the nature of data? (SECTION 1.1)

How can you draw a random sample? (SECTION 1.2)

What are other sampling techniques? (SECTION 1.2)

How can you design ways to collect data? (SECTION 1.3)

FOCUS PROBLEM

Where Have All the Fireflies Gone?

A feature article in *The Wall Street Journal* discusses the disappearance of fireflies. In the article, Professor Sara Lewis of Tufts University and other scholars express concern about the decline in the worldwide population of fireflies.

There are a number of possible explanations for the decline, including habitat reduction of woodlands, wetlands, and open fields; pesticides; and pollution. Artificial nighttime lighting might interfere with the Morse-code-like mating ritual of the fireflies. Some chemical companies pay a bounty for fireflies because the insects contain two rare chemicals used in medical research and electronic detection systems used in spacecraft.

Courtesy of Corrinne and Charles Brase

Adapted from Ohio State University Firefly Files logo

What does any of this have to do with statistics?

The truth, at this time, is that no one really knows (a) how much the world firefly population has declined or (b) how to explain the decline. The population of all fireflies is simply too large to study in its entirety.

In any study of fireflies, we must rely on incomplete information from samples. Furthermore, from these samples we must draw realistic conclusions that have statistical integrity. This is the kind of work that makes use of statistical methods to determine ways to collect, analyze, and investigate data.

Suppose you are conducting a study to compare firefly populations exposed to normal daylight/darkness conditions with firefly populations exposed to continuous light (24 hours a day). You set up two firefly colonies in a laboratory environment. The two colonies are identical except that one colony is exposed to normal daylight/darkness conditions and the other is exposed to continuous light. Each colony is populated with the same number of mature fireflies. After 72 hours, you count the number of living fireflies in each colony.

After completing this chapter, you will be able to answer the following questions.

(a) Is this an experiment or an observation study? Explain.

(b) Is there a control group? Is there a treatment group?

For online student resources, visit the Brase/Brase, Understandable Statistics, 11th edition web site at http://www.cengage.com/statistics/brase11e

(c) What is the variable in this study?
(d) What is the level of measurement (nominal, interval, ordinal, or ratio) of the variable? (See Problem 11 of the Chapter 1 Review Problems.)

SECTION **1.1**

What Is Statistics?

FOCUS POINTS

- Identify variables in a statistical study.
- Distinguish between quantitative and qualitative variables.
- Identify populations and samples.
- Distinguish between parameters and statistics.
- Determine the level of measurement.
- Compare descriptive and inferential statistics.

INTRODUCTION

Decision making is an important aspect of our lives. We make decisions based on the information we have, our attitudes, and our values. Statistical methods help us examine information. Moreover, statistics can be used for making decisions when we are faced with uncertainties. For instance, if we wish to estimate the proportion of people who will have a severe reaction to a flu shot without giving the shot to everyone who wants it, statistics provides appropriate methods. Statistical methods enable us to look at information from a small collection of people or items and make inferences about a larger collection of people or items.

Procedures for analyzing data, together with rules of inference, are central topics in the study of statistics.

Statistics

Statistics is the study of how to collect, organize, analyze, and interpret numerical information from data.

The subject of statistics is multifaceted. The following definition of statistics is found in the *International Encyclopedia of Statistical Science*, edited by Miodrag Lovric. Professor David Hand of Imperial College London—the president of the Royal Statistical Society—presents the definition in his article "Statistics: An Overview."

Statistics is both the science of uncertainty and the technology of extracting information from data.

The statistical procedures you will learn in this book should supplement your built-in system of inference—that is, the results from statistical procedures and good sense should dovetail. Of course, statistical methods themselves have no power to work miracles. These methods can help us make some decisions, but not all conceivable decisions. Remember, even a properly applied statistical procedure is no more accurate than the data, or facts, on which it is based. Finally, statistical results should be interpreted by one who understands not only the methods, but also the subject matter to which they have been applied.

The general prerequisite for statistical decision making is the gathering of data. First, we need to identify the individuals or objects to be included in the study and the characteristics or features of the individuals that are of interest.

Individuals
Variable

Individuals are the people or objects included in the study.
A **variable** is a characteristic of the individual to be measured or observed.

For instance, if we want to do a study about the people who have climbed Mt. Everest, then the individuals in the study are all people who have actually made it to the summit. One variable might be the height of such individuals. Other variables might be age, weight, gender, nationality, income, and so on. Regardless of the variables we use, we would not include measurements or observations from people who have not climbed the mountain.

The variables in a study may be *quantitative* or *qualitative* in nature.

Quantitative variable
Qualitative variable

A **quantitative variable** has a value or numerical measurement for which operations such as addition or averaging make sense. A **qualitative variable** describes an individual by placing the individual into a category or group, such as male or female.

For the Mt. Everest climbers, variables such as height, weight, age, or income are *quantitative* variables. *Qualitative variables* involve nonnumerical observations such as gender or nationality. Sometimes qualitative variables are referred to as *categorical variables.*

Categorical variable

Another important issue regarding data is their source. Do the data comprise information from *all* individuals of interest, or from just *some* of the individuals?

Population data
Sample data

In **population data,** the data are from *every* individual of interest.
In **sample data,** the data are from *only some* of the individuals of interest.

It is important to know whether the data are population data or sample data. Data from a specific population are fixed and complete. Data from a sample may vary from sample to sample and are *not* complete.

Population parameter

Sample statistic

A **population parameter** is a numerical measure that describes an aspect of a population.
A **sample statistic** is a numerical measure that describes an aspect of a sample.

For instance, if we have data from *all* the individuals who have climbed Mt. Everest, then we have population data. The proportion of males in the *population* of all climbers who have conquered Mt. Everest is an example of a *parameter.*

On the other hand, if our data come from just some of the climbers, we have sample data. The proportion of male climbers in the *sample* is an example of a *statistic.* Note that different samples may have different values for the proportion of male climbers. One of the important features of sample statistics is that they can vary from sample to sample, whereas population parameters are fixed for a given population.

LOOKING FORWARD

In later chapters we will use information based on a sample and sample statistics to estimate population parameters (Chapter 7) or make decisions about the value of population parameters (Chapter 8).

EXAMPLE 1

USING BASIC TERMINOLOGY

The Hawaii Department of Tropical Agriculture is conducting a study of ready-to-harvest pineapples in an experimental field.

(a) The pineapples are the *objects* (individuals) of the study. If the researchers are interested in the individual weights of pineapples in the field, then the *variable* consists of weights. At this point, it is important to specify units of measurement

Gallo Images/Landbouweekblad/Getty Images

and degrees of accuracy of measurement. The weights could be measured to the nearest ounce or gram. Weight is a *quantitative* variable because it is a numerical measure. If weights of *all* the ready-to-harvest pineapples in the field are included in the data, then we have a *population*. The average weight of all ready-to-harvest pineapples in the field is a *parameter.*

(b) Suppose the researchers also want data on taste. A panel of tasters rates the pineapples according to the categories "poor," "acceptable," and "good." Only some of the pineapples are included in the taste test. In this case, the *variable* is taste. This is a *qualitative* or *categorical* variable. Because only some of the pineapples in the field are included in the study, we have a *sample*. The proportion of pineapples in the sample with a taste rating of "good" is a *statistic.*

Throughout this text, you will encounter *guided exercises* embedded in the reading material. These exercises are included to give you an opportunity to work immediately with new ideas. The questions guide you through appropriate analysis. Cover the answers on the right side (an index card will fit this purpose). After you have thought about or written down *your own response,* check the answers. If there are several parts to an exercise, check each part before you continue. You should be able to answer most of these exercise questions, but don't skip them—they are important.

GUIDED EXERCISE 1 | Using basic terminology

Television station QUE wants to know the proportion of TV owners in Virginia who watch the station's new program at least once a week. The station asks a group of 1000 TV owners in Virginia if they watch the program at least once a week.

(a) Identify the individuals of the study and the variable.

→ The individuals are the 1000 TV owners surveyed. The variable is the response does, or does not, watch the new program at least once a week.

(b) Do the data comprise a sample? If so, what is the underlying population?

→ The data comprise a sample of the population of responses from all TV owners in Virginia.

(c) Is the variable qualitative or quantitative?

→ Qualitative—the categories are the two possible responses, does or does not watch the program.

(d) Identify a quantitative variable that might be of interest.

→ Age or income might be of interest.

(e) Is the proportion of viewers in the sample who watch the new program at least once a week a statistic or a parameter?

→ Statistic—the proportion is computed from sample data.

LEVELS OF MEASUREMENT: NOMINAL, ORDINAL, INTERVAL, RATIO

We have categorized data as either qualitative or quantitative. Another way to classify data is according to one of the four *levels of measurement.* These levels indicate the type of arithmetic that is appropriate for the data, such as ordering, taking differences, or taking ratios.

Levels of Measurement

LEVELS OF MEASUREMENT

Nominal level

The **nominal level of measurement** applies to data that consist of names, labels, or categories. There are no implied criteria by which the data can be ordered from smallest to largest.

Ordinal level

The **ordinal level of measurement** applies to data that can be arranged in order. However, differences between data values either cannot be determined or are meaningless.

Interval level

The **interval level of measurement** applies to data that can be arranged in order. In addition, differences between data values are meaningful.

Ratio level

The **ratio level of measurement** applies to data that can be arranged in order. In addition, both differences between data values and ratios of data values are meaningful. Data at the ratio level have a true zero.

EXAMPLE 2

LEVELS OF MEASUREMENT

Michelle Dulieu/Shutterstock.com

Identify the type of data.

(a) Taos, Acoma, Zuni, and Cochiti are the names of four Native American pueblos from the population of names of all Native American pueblos in Arizona and New Mexico.

SOLUTION: These data are at the *nominal* level. Notice that these data values are simply names. By looking at the name alone, we cannot determine if one name is "greater than or less than" another. Any ordering of the names would be numerically meaningless.

(b) In a high school graduating class of 319 students, Jim ranked 25th, June ranked 19th, Walter ranked 10th, and Julia ranked 4th, where 1 is the highest rank.

SOLUTION: These data are at the *ordinal* level. Ordering the data clearly makes sense. Walter ranked higher than June. Jim had the lowest rank, and Julia the highest. However, numerical differences in ranks do not have meaning. The difference between June's and Jim's ranks is 6, and this is the same difference that exists between Walter's and Julia's ranks. However, this difference doesn't really mean anything significant. For instance, if you looked at grade point average, Walter and Julia may have had a large gap between their grade point averages, whereas June and Jim may have had closer grade point averages. In any ranking system, it is only the relative standing that matters. Differences between ranks are meaningless.

Korban Schwab/iStockphoto.com

(c) Body temperatures (in degrees Celsius) of trout in the Yellowstone River.

SOLUTION: These data are at the *interval* level. We can certainly order the data, and we can compute meaningful differences. However, for Celsius-scale temperatures, there is not an inherent starting point. The value 0°C may seem to be a starting point, but this value does not indicate the state of "no heat." Furthermore, it is not correct to say that 20°C is twice as hot as 10°C.

(d) Length of trout swimming in the Yellowstone River.
SOLUTION: These data are at the *ratio* level. An 18-inch trout is three times as long as a 6-inch trout. Observe that we can divide 6 into 18 to determine a meaningful *ratio* of trout lengths.

In summary, there are four levels of measurement. The nominal level is considered the lowest, and in ascending order we have the ordinal, interval, and ratio levels. In general, calculations based on a particular level of measurement may not be appropriate for a lower level.

PROCEDURE

HOW TO DETERMINE THE LEVEL OF MEASUREMENT

The levels of measurement, listed from lowest to highest, are nominal, ordinal, interval, and ratio. To determine the level of measurement of data, state the *highest level* that can be justified for the entire collection of data. Consider which calculations are suitable for the data.

Level of Measurement	Suitable Calculation
Nominal	We can put the data into categories.
Ordinal	We can order the data from smallest to largest or "worst" to "best." Each data value can be *compared* with another data value.
Interval	We can order the data and also take the differences between data values. At this level, it makes sense to compare the differences between data values. For instance, we can say that one data value is 5 more than or 12 less than another data value.
Ratio	We can order the data, take differences, and also find the ratio between data values. For instance, it makes sense to say that one data value is twice as large as another.

WHAT DOES THE LEVEL OF MEASUREMENT TELL US?

The level of measurement tells us which arithmetic processes are appropriate for the data. This is important because different statistical processes require various kinds of arithmetic. In some instances all we need to do is count the number of data that meet specified criteria. In such cases nominal (and higher) data levels are all appropriate. In other cases we need to order the data, so nominal data would not be suitable. Many other statistical processes require division, so data need to be at the ratio level. Just keep the nature of the data in mind before beginning statistical computations.

GUIDED EXERCISE 2 | LEVELS OF MEASUREMENT

The following describe different data associated with a state senator. For each data entry, indicate the corresponding *level of measurement.*

(a) The senator's name is Sam Wilson. → Nominal level

(b) The senator is 58 years old. → Ratio level. Notice that age has a meaningful zero. It makes sense to give age ratios. For instance, Sam is twice as old as someone who is 29.

Continued

GUIDED EXERCISE 2 *continued*

(c) The years in which the senator was elected to the Senate are 2000, 2006, and 2012.

→ Interval level. Dates can be ordered, and the difference between dates has meaning. For instance, 2006 is 6 years later than 2000. However, ratios do not make sense. The year 2000 is not twice as large as the year 1000. In addition, the year 0 does not mean "no time."

(d) The senator's total taxable income last year was $878,314.

→ Ratio level. It makes sense to say that the senator's income is 10 times that of someone earning $87,831.40.

(e) The senator surveyed his constituents regarding his proposed water protection bill. The choices for response were strong support, support, neutral, against, or strongly against.

→ Ordinal level. The choices can be ordered, but there is no meaningful numerical difference between two choices.

(f) The senator's marital status is "married."

→ Nominal level

(g) A leading news magazine claims the senator is ranked seventh for his voting record on bills regarding public education.

→ Ordinal level. Ranks can be ordered, but differences between ranks may vary in meaning.

CRITICAL THINKING

"Data! Data! Data!" he cried impatiently. "I can't make bricks without clay." Sherlock Holmes said these words in *The Adventure of the Copper Beeches* by Sir Arthur Conan Doyle.

Reliable statistical conclusions require reliable data. This section has provided some of the vocabulary used in discussing data. As you read a statistical study or conduct one, pay attention to the nature of the data and the ways they were collected.

When you select a variable to measure, be sure to specify the process and requirements for measurement. For example, if the variable is the weight of ready-to-harvest pineapples, specify the unit of weight, the accuracy of measurement, and maybe even the particular scale to be used. If some weights are in ounces and others in grams, the data are fairly useless.

Another concern is whether or not your measurement instrument truly measures the variable. Just asking people if they know the geographic location of the island nation of Fiji may not provide accurate results. The answers may reflect the fact that the respondents want you to think they are knowledgeable. Asking people to locate Fiji on a map may give more reliable results.

The level of measurement is also an issue. You can put numbers into a calculator or computer and do all kinds of arithmetic. However, you need to judge whether the operations are meaningful. For ordinal data such as restaurant rankings, you can't conclude that a 4-star restaurant is "twice as good" as a 2-star restaurant, even though the number 4 is twice 2.

Continued

CRITICAL THINKING *continued*

Are the data from a sample, or do they comprise the entire population? Sample data can vary from one sample to another! This means that if you are studying the same statistic from two different samples of the same size, the data values may be different. In fact, the ways in which sample statistics vary among different samples of the same size will be the focus of our study from Section 6.4 on.

INTERPRETATION **When you work with sample data, carefully consider the population from which they are drawn. Observations and analysis of the sample are applicable to only the population from which the sample is drawn.**

Descriptive statistics

Inferential statistics

Descriptive statistics involves methods of organizing, picturing, and summarizing information from samples or populations.

Inferential statistics involves methods of using information from a sample to draw conclusions regarding the population.

LOOKING FORWARD

The purpose of collecting and analyzing data is to obtain information. Statistical methods provide us tools to obtain information from data. These methods break into two branches.

We will look at methods of descriptive statistics in Chapters 2, 3, and 9. These methods may be applied to data from samples or populations.

Sometimes we do not have access to an entire population. At other times, the difficulties or expense of working with the entire population is prohibitive. In such cases, we will use inferential statistics together with probability. These are the topics of Chapters 4 through 11.

VIEWPOINT The First Measured Century

The 20th century saw measurements of aspects of American life that had never been systematically studied before. Social conditions involving crime, sex, food, fun, religion, and work were numerically investigated. The measurements and survey responses taken over the entire century reveal unsuspected statistical trends. The First Measured Century *is a book by Caplow, Hicks, and Wattenberg. It is also a PBS documentary available on video. For more information, visit the Brase/Brase statistics site at* ***http://www.cengage.com/statistics/brase11e*** *and find the link to the PBS* First Measured Century *documentary.*

SECTION 1.1 PROBLEMS

1. *Statistical Literacy* In a statistical study what is the difference between an individual and a variable?

2. *Statistical Literacy* Are data at the nominal level of measurement quantitative or qualitative?

3. *Statistical Literacy* What is the difference between a parameter and a statistic?

4. *Statistical Literacy* For a set population, does a parameter ever change? If there are three different samples of the same size from a set population, is it possible to get three different values for the same statistic?

5. *Critical Thinking* Numbers are often assigned to data that are categorical in nature.
   (a) Consider these number assignments for category items describing electronic ways of expressing personal opinions:

   1 = Twitter; 2 = e-mail; 3 = text message; 4 = Facebook; 5 = blog

   Are these numerical assignments at the ordinal data level or higher? Explain.
   (b) Consider these number assignments for category items describing usefulness of customer service:

   1 = not helpful; 2 = somewhat helpful; 3 = very helpful;
   4 = extremely helpful

   Are these numerical assignments at the ordinal data level? Explain. What about at the interval level or higher? Explain.

6. *Interpretation* Lucy conducted a survey asking some of her friends to specify their favorite type of TV entertainment from the following list of choices:

   sitcom; reality; documentary; drama; cartoon; other

   Do Lucy's observations apply to *all* adults? Explain. From the description of the survey group, can we draw any conclusions regarding age of participants, gender of participants, or education level of participants?

7. *Marketing: Fast Food* A national survey asked 1261 U.S. adult fast-food customers which meal (breakfast, lunch, dinner, snack) they ordered.
   (a) Identify the variable.
   (b) Is the variable quantitative or qualitative?
   (c) What is the implied population?

8. *Advertising: Auto Mileage* What is the average miles per gallon (mpg) for all new cars? Using *Consumer Reports,* a random sample of 35 new cars gave an average of 21.1 mpg.
   (a) Identify the variable.
   (b) Is the variable quantitative or qualitative?
   (c) What is the implied population?

9. *Ecology: Wetlands* Government agencies carefully monitor water quality and its effect on wetlands (Reference: *Environmental Protection Agency Wetland Report* EPA 832-R-93-005). Of particular concern is the concentration of nitrogen in water draining from fertilized lands. Too much nitrogen can kill fish and wildlife. Twenty-eight samples of water were taken at random from a lake. The nitrogen concentration (milligrams of nitrogen per liter of water) was determined for each sample.
   (a) Identify the variable.
   (b) Is the variable quantitative or qualitative?
   (c) What is the implied population?

10. *Archaeology: Ireland* The archaeological site of Tara is more than 4000 years old. Tradition states that Tara was the seat of the high kings of Ireland. Because of its archaeological importance, Tara has received extensive study (Reference: *Tara: An Archaeological Survey* by Conor Newman, Royal Irish Academy, Dublin). Suppose an archaeologist wants to estimate the density of ferromagnetic artifacts in the Tara region. For this purpose, a random sample

of 55 plots, each of size 100 square meters, is used. The number of ferromagnetic artifacts for each plot is determined.
(a) Identify the variable.
(b) Is the variable quantitative or qualitative?
(c) What is the implied population?

11. *Student Life: Levels of Measurement* Categorize these measurements associated with student life according to level: nominal, ordinal, interval, or ratio.
(a) Length of time to complete an exam
(b) Time of first class
(c) Major field of study
(d) Course evaluation scale: poor, acceptable, good
(e) Score on last exam (based on 100 possible points)
(f) Age of student

12. *Business: Levels of Measurement* Categorize these measurements associated with a robotics company according to level: nominal, ordinal, interval, or ratio.
(a) Salesperson's performance: below average, average, above average
(b) Price of company's stock
(c) Names of new products
(d) Temperature (°F) in CEO's private office
(e) Gross income for each of the past 5 years
(f) Color of product packaging

13. *Fishing: Levels of Measurement* Categorize these measurements associated with fishing according to level: nominal, ordinal, interval, or ratio.
(a) Species of fish caught: perch, bass, pike, trout
(b) Cost of rod and reel
(c) Time of return home
(d) Guidebook rating of fishing area: poor, fair, good
(e) Number of fish caught
(f) Temperature of water

14. *Education: Teacher Evaluation* If you were going to apply *statistical methods* to analyze teacher evaluations, which question form, A or B, would be better?

Form A: In your own words, tell how this teacher compares with other teachers you have had.

Form B: Use the following scale to rank your teacher as compared with other teachers you have had.

1	2	3	4	5
worst	**below average**	**average**	**above average**	**best**

15. *Critical Thinking* You are interested in the weights of backpacks students carry to class and decide to conduct a study using the backpacks carried by 30 students.
(a) Give some instructions for weighing the backpacks. Include unit of measure, accuracy of measure, and type of scale.
(b) Do you think each student asked will allow you to weigh his or her backpack?
(c) Do you think telling students ahead of time that you are going to weigh their backpacks will make a difference in the weights?

SECTION **1.2**

Random Samples

FOCUS POINTS

- Explain the importance of random samples.
- Construct a simple random sample using random numbers.
- Simulate a random process.
- Describe stratified sampling, cluster sampling, systematic sampling, multistage sampling, and convenience sampling.

SIMPLE RANDOM SAMPLES

Eat lamb—20,000 coyotes can't be wrong!

This slogan is sometimes found on bumper stickers in the western United States. The slogan indicates the trouble that ranchers have experienced in protecting their flocks from predators. Based on their experience with this sample of the coyote population, the ranchers concluded that *all* coyotes are dangerous to their flocks and should be eliminated! The ranchers used a special poison bait to get rid of the coyotes. Not only was this poison distributed on ranch land, but with government cooperation, it also was distributed widely on public lands.

The ranchers found that the results of the widespread poisoning were not very beneficial. The sheep-eating coyotes continued to thrive while the general population of coyotes and other predators declined. What was the problem? The sheep-eating coyotes that the ranchers had observed were not a representative sample of all coyotes. Modern methods of predator control, however, target the sheep-eating coyotes. To a certain extent, the new methods have come about through a closer examination of the sampling techniques used.

In this section, we will examine several widely used sampling techniques. One of the most important sampling techniques is a *simple random sample.*

Simple random sample

A **simple random sample** of n measurements from a population is a subset of the population selected in such a manner that every sample of size n from the population has an equal chance of being selected.

In a simple random sample, not only does every sample of the specified size have an equal chance of being selected, but every individual of the population also has an equal chance of being selected. However, the fact that each individual has an equal chance of being selected does not necessarily imply a simple random sample. Remember, for a simple random sample, every sample of the given size must also have an equal chance of being selected.

IMPORTANT FEATURES OF A SIMPLE RANDOM SAMPLE

For a simple random sample

- Every sample of specified size n from the population has an equal chance of being selected.
- No researcher bias occurs in the items selected for the sample.
- A random sample may not always reflect the diversity of the population. For instance, from a population of 10 cats and 10 dogs, a random sample of size 6 could consist of all cats.

GUIDED EXERCISE 3 | SIMPLE RANDOM SAMPLE

Is open space around metropolitan areas important? Players of the Colorado Lottery might think so, since some of the proceeds of the game go to fund open space and outdoor recreational space. To play the game, you pay $1 and choose any six different numbers from the group of numbers 1 through 42. If your group of six numbers matches the winning group of six numbers selected by simple random sampling, then you are a winner of a grand prize of at least $1.5 million.

(a) Is the number 25 as likely to be selected in the winning group of six numbers as the number 5?

→ Yes. Because the winning numbers constitute a simple random sample, each number from 1 through 42 has an equal chance of being selected.

(b) Could all the winning numbers be even?

→ Yes, since six even numbers is one of the possible groups of six numbers.

(c) Your friend always plays the numbers

1 2 3 4 5 6

Could she ever win?

→ Yes. In a simple random sample, the listed group of six numbers is *as likely as any* of the 5,245,786 possible groups of six numbers to be selected as the winner. (See Section 4.3 to learn how to compute the number of possible groups of six numbers that can be selected from 42 numbers.)

How do we get random samples? Suppose you need to know if the emission systems of the latest shipment of Toyotas satisfy pollution-control standards. You want to pick a random sample of 30 cars from this shipment of 500 cars and test them. One way to pick a random sample is to number the cars 1 through 500. Write these numbers on cards, mix up the cards, and then draw 30 numbers. The sample will consist of the cars with the chosen numbers. If you mix the cards sufficiently, this procedure produces a random sample.

Random-number table

An easier way to select the numbers is to use a *random-number table.* You can make one yourself by writing the digits 0 through 9 on separate cards and mixing up these cards in a hat. Then draw a card, record the digit, return the card, and mix up the cards again. Draw another card, record the digit, and so on. Table 1 in the Appendix is a ready-made random-number table (adapted from Rand Corporation, *A Million Random Digits with 100,000 Normal Deviates*). Let's see how to pick our random sample of 30 Toyotas by using this random-number table.

EXAMPLE 3

RANDOM-NUMBER TABLE

Use a random-number table to pick a random sample of 30 cars from a population of 500 cars.

Vibrant Image Studio/Shutterstock.com

SOLUTION: Again, we assign each car a different number between 1 and 500, inclusive. Then we use the random-number table to choose the sample. Table 1 in the Appendix has 50 rows and 10 blocks of five digits each; it can be thought of as a solid mass of digits that has been broken up into rows and blocks for user convenience.

You read the digits by beginning anywhere in the table. We dropped a pin on the table, and the head of the pin landed in row 15, block 5. We'll begin there and list all the digits in that row. If we need more digits, we'll move on to row 16, and so on. The digits we begin with are

99281 59640 15221 96079 09961 05371

Since the highest number assigned to a car is 500, and this number has three digits, we regroup our digits into blocks of 3:

992 815 964 015 221 960 790 996 105 371

To construct our random sample, we use the first 30 car numbers we encounter in the random-number table when we start at row 15, block 5. We skip the first three groups—992, 815, and 964—because these numbers are all too large. The next group of three digits is 015, which corresponds to 15. Car number 15 is the first car included in our sample, and the next is car number 221. We skip the next three groups and then include car numbers 105 and 371. To get the rest of the cars in the sample, we continue to the next line and use the random-number table in the same fashion. If we encounter a number we've used before, we skip it.

COMMENT When we use the term *(simple) random sample,* we have very specific criteria in mind for selecting the sample. One proper method for selecting a simple random sample is to use a computer- or calculator-based random-number generator or a table of random numbers as we have done in the example. The term *random* should not be confused with *haphazard*!

LOOKING FORWARD

The runs test for randomness discussed in Section 11.4 shows how to determine if two symbols are randomly mixed in an ordered list of symbols.

PROCEDURE

HOW TO DRAW A RANDOM SAMPLE

1. Number all members of the population sequentially.
2. Use a table, calculator, or computer to select random numbers from the numbers assigned to the population members.
3. Create the sample by using population members with numbers corresponding to those randomly selected.

LOOKING FORWARD

Simple random samples are key components in methods of inferential statistics that we will study in Chapters 7–11. In fact, in order to draw conclusions about a population, the methods we will study *require* that we have simple random samples from the populations of interest.

Another important use of random-number tables is in *simulation.* We use the word *simulation* to refer to the process of providing numerical imitations of "real" phenomena. Simulation methods have been productive in studying a diverse array of subjects such as nuclear reactors, cloud formation, cardiology (and medical science in general), highway design, production control, shipbuilding, airplane design, war games, economics, and electronics. A complete list would probably include something from every aspect of modern life. In Guided Exercise 4 we'll perform a brief simulation.

Simulation

A **simulation** is a numerical facsimile or representation of a real-world phenomenon.

GUIDED EXERCISE 4 | SIMULATION

Use a random-number table to simulate the outcomes of tossing a balanced (that is, fair) penny 10 times.

(a) How many outcomes are possible when you toss a coin once? → Two—heads or tails

Continued

GUIDED EXERCISE 4 *continued*

(b) There are several ways to assign numbers to the two outcomes. Because we assume a fair coin, we can assign an even digit to the outcome "heads" and an odd digit to the outcome "tails." Then, starting at block 3 of row 2 of Table 1 in the Appendix, list the first 10 single digits.	7 1 5 4 9 4 4 8 4 3
(c) What are the outcomes associated with the 10 digits?	T T T H T H H H H T
(d) If you start in a different block and row of Table 1 in the Appendix, will you get the same sequence of outcomes?	It is possible, but not very likely. (In Section 4.3 you will learn how to determine that there are 1024 possible sequences of outcomes for 10 tosses of a coin.)

TECH NOTES

Sampling with replacement

Most statistical software packages, spreadsheet programs, and statistical calculators generate random numbers. In general, these devices sample with replacement. *Sampling with replacement* means that although a number is selected for the sample, it is *not removed* from the population. Therefore, the same number may be selected for the sample more than once. If you need to sample without replacement, generate more items than you need for the sample. Then sort the sample and remove duplicate values. Specific procedures for generating random samples using the TI-84Plus/TI-83Plus/TI-*n*spire (with TI-84Plus keypad) calculator, Excel 2010, Minitab, and SPSS are shown in Using Technology at the end of this chapter. More details are given in the separate *Technology Guides* for each of these technologies.

OTHER SAMPLING TECHNIQUES

Although we will assume throughout this text that (simple) random samples are used, other methods of sampling are also widely used. Appropriate statistical techniques exist for these sampling methods, but they are beyond the scope of this text.

Stratified sampling

One of these sampling methods is called *stratified sampling*. Groups or classes inside a population that share a common characteristic are called *strata* (plural of *stratum*). For example, in the population of all undergraduate college students, some strata might be freshmen, sophomores, juniors, or seniors. Other strata might be men or women, in-state students or out-of-state students, and so on. In the method of stratified sampling, the population is divided into at least two distinct strata. Then a (simple) random sample of a certain size is drawn from each stratum, and the information obtained is carefully adjusted or weighted in all resulting calculations.

The groups or strata are often sampled in proportion to their actual percentages of occurrence in the overall population. However, other (more sophisticated) ways to determine the optimal sample size in each stratum may give the best results. In general, statistical analysis and tests based on data obtained from stratified samples are somewhat different from techniques discussed in an introductory course in statistics. Such methods for stratified sampling will not be discussed in this text.

Systematic sampling

Another popular method of sampling is called *systematic sampling*. In this method, it is assumed that the elements of the population are arranged in some natural sequential order. Then we select a (random) starting point and select every *k*th element for our sample. For example, people lining up to buy rock concert tickets are "in order." To generate a systematic sample of these people (and ask questions regarding topics such as age, smoking habits, income level, etc.), we could include every fifth person in line. The "starting" person is selected at random from the first five.

The advantage of a systematic sample is that it is easy to get. However, there are dangers in using systematic sampling. When the population is repetitive or cyclic in nature, systematic sampling should not be used. For example, consider a fabric mill that produces dress material. Suppose the loom that produces the material makes a mistake every 17th yard, but we check only every 16th yard with an automated electronic scanner. In this case, a random starting point may or may not result in detection of fabric flaws before a large amount of fabric is produced.

Cluster sampling

Cluster sampling is a method used extensively by government agencies and certain private research organizations. In cluster sampling, we begin by dividing the demographic area into sections. Then we randomly select sections or clusters. Every member of the cluster is included in the sample. For example, in conducting a survey of school children in a large city, we could first randomly select five schools and then include all the children from each selected school.

Multistage samples

Often a population is very large or geographically spread out. In such cases, samples are constructed through a *multistage sample design* of several stages, with the final stage consisting of clusters. For instance, the government Current Population Survey interviews about 60,000 households across the United States each month by means of a multistage sample design.

For the Current Population Survey, the first stage consists of selecting samples of large geographic areas that do not cross state lines. These areas are further broken down into smaller blocks, which are stratified according to ethnic and other factors. Stratified samples of the blocks are then taken. Finally, housing units in each chosen block are broken into clusters of nearby housing units. A random sample of these clusters of housing units is selected, and each household in the final cluster is interviewed.

Convenience sampling

Convenience sampling simply uses results or data that are conveniently and readily obtained. In some cases, this may be all that is available, and in many cases, it is better than no information at all. However, convenience sampling does run the risk of being severely biased. For instance, consider a newsperson who wishes to get the "opinions of the people" about a proposed seat tax to be imposed on tickets to all sporting events. The revenues from the seat tax will then be used to support the local symphony. The newsperson stands in front of a concert hall and surveys the first five people exiting after a symphony performance who will cooperate. This method of choosing a sample will produce some opinions, and perhaps some human interest stories, but it certainly has bias. It is hoped that the city council will not use these opinions as the sole basis for a decision about the proposed tax. It is good advice to be very cautious indeed when the data come from the method of convenience sampling.

SAMPLING TECHNIQUES

Random sampling: Use a simple random sample from the entire population.

Stratified sampling: Divide the entire population into distinct subgroups called strata. The strata are based on a specific characteristic such as age, income, education level, and so on. All members of a stratum share the specific characteristic. Draw random samples from each stratum.

Systematic sampling: Number all members of the population sequentially. Then, from a starting point selected at random, include every *k*th member of the population in the sample.

Cluster sampling: Divide the entire population into pre-existing segments or clusters. The clusters are often geographic. Make a random selection of clusters. Include every member of each selected cluster in the sample.

Multistage sampling: Use a variety of sampling methods to create successively smaller groups at each stage. The final sample consists of clusters.

Convenience sampling: Create a sample by using data from population members that are readily available.

CRITICAL THINKING

Sampling frame

We call the list of individuals from which a sample is actually selected the *sampling frame*. Ideally, the sampling frame is the entire population. However, from a practical perspective, not all members of a population may be accessible. For instance, using a telephone directory as the sample frame for residential telephone contacts would not include unlisted numbers.

Undercoverage

When the sample frame does not match the population, we have what is called *undercoverage*. In demographic studies, undercoverage could result if the homeless, fugitives from the law, and so forth, are not included in the study.

A **sampling frame** is a list of individuals from which a sample is actually selected.

Undercoverage results from omitting population members from the sample frame.

Sampling error

In general, even when the sampling frame and the population match, a sample is not a perfect representation of a population. Therefore, information drawn from a sample may not exactly match corresponding information from the population. To the extent that sample information does not match the corresponding population information, we have an error, called a *sampling error.*

Nonsampling error

A **sampling error** is the difference between measurements from a sample and corresponding measurements from the respective population. It is caused by the fact that the sample does not perfectly represent the population.

A **nonsampling error** is the result of poor sample design, sloppy data collection, faulty measuring instruments, bias in questionnaires, and so on.

Sampling errors do not represent mistakes! They are simply the consequences of using samples instead of populations. However, be alert to nonsampling errors, which may sometimes occur inadvertently.

VIEWPOINT Extraterrestrial Life?

Do you believe intelligent life exists on other planets? Using methods of random sampling, a Fox News opinion poll found that about 54% of all U.S. men do believe in intelligent life on other planets, whereas only 47% of U.S. women believe there is such life. How could you conduct a random survey of students on your campus regarding belief in extraterrestrial life?

SECTION 1.2 PROBLEMS

1. *Statistical Literacy* Explain the difference between a stratified sample and a cluster sample.

2. *Statistical Literacy* Explain the difference between a simple random sample and a systematic sample.

3. *Statistical Literacy* Marcie conducted a study of the cost of breakfast cereal. She recorded the costs of several boxes of cereal. However, she neglected to take into account the number of servings in each box. Someone told her not to worry because she just had some sampling error. Comment on that advice.

4. *Statistical Literacy* A random sample of students who use the college recreation center were asked if they approved increasing student fees for all students in order to add a climbing wall to the recreation center. Describe the sample frame. Does the sample frame include all students enrolled in the college? Explain.

5. *Interpretation* In a random sample of 50 students from a large university, all the students were between 18 and 20 years old. Can we conclude that the entire population of students at the university is between 18 and 20 years old? Explain.

6. *Interpretation* A campus performance series features plays, music groups, dance troops, and stand-up comedy. The committee responsible for selecting the performance groups include three students chosen at random from a pool of volunteers. This year the 30 volunteers came from a variety of majors. However, the three students for the committee were all music majors. Does this fact indicate there was bias in the selection process and that the selection process was not random? Explain.

7 *Critical Thinking* Greg took a random sample of size 100 from the population of current season ticket holders to State College men's basketball games. Then he took a random sample of size 100 from the population of current season ticket holders to State College women's basketball games.
   (a) What sampling technique (stratified, systematic, cluster, multistage, convenience, random) did Greg use to sample from the population of current season ticket holders to all State College basketball games played by either men or women?
   (b) Is it appropriate to pool the samples and claim to have a random sample of size 200 from the population of current season ticket holders to all State College home basketball games played by either men or women? Explain.

8. *Critical Thinking* Consider the students in your statistics class as the population and suppose they are seated in four rows of 10 students each. To select a sample, you toss a coin. If it comes up heads, you use the 20 students sitting in the first two rows as your sample. If it comes up tails, you use the 20 students sitting in the last two rows as your sample.
   (a) Does every student have an equal chance of being selected for the sample? Explain.
   (b) Is it possible to include students sitting in row 3 with students sitting in row 2 in your sample? Is your sample a simple random sample? Explain.
   (c) Describe a process you could use to get a simple random sample of size 20 from a class of size 40.

9. *Critical Thinking* Suppose you are assigned the number 1, and the other students in your statistics class call out consecutive numbers until each person in the class has his or her own number. Explain how you could get a random sample of four students from your statistics class.
   (a) Explain why the first four students walking into the classroom would not necessarily form a random sample.
   (b) Explain why four students coming in late would not necessarily form a random sample.
   (c) Explain why four students sitting in the back row would not necessarily form a random sample.
   (d) Explain why the four tallest students would not necessarily form a random sample.

10. *Critical Thinking* In each of the following situations, the sampling frame does not match the population, resulting in undercoverage. Give examples of population members that might have been omitted.
    (a) The population consists of all 250 students in your large statistics class. You plan to obtain a simple random sample of 30 students by using the sampling frame of students present next Monday.
    (b) The population consists of all 15-year-olds living in the attendance district of a local high school. You plan to obtain a simple random sample of 200 such residents by using the student roster of the high school as the sampling frame.

11. *Sampling: Random* Use a random-number table to generate a list of 10 random numbers between 1 and 99. Explain your work.

12. *Sampling: Random* Use a random-number table to generate a list of eight random numbers from 1 to 976. Explain your work.

13. *Sampling: Random* Use a random-number table to generate a list of six random numbers from 1 to 8615. Explain your work.

14. *Simulation: Coin Toss* Use a random-number table to simulate the outcomes of tossing a quarter 25 times. Assume that the quarter is balanced (i.e., fair).

15. *Computer Simulation: Roll of a Die* A die is a cube with dots on each face. The faces have 1, 2, 3, 4, 5, or 6 dots. The table below is a computer simulation (from the software package Minitab) of the results of rolling a fair die 20 times.

DATA DISPLAY

ROW	C1	C2	C3	C4	C5	C6	C7	C8	C9	C10
1	5	2	2	2	5	3	2	3	1	4
2	3	2	4	5	4	5	3	5	3	4

    (a) Assume that each number in the table corresponds to the number of dots on the upward face of the die. Is it appropriate that the same number appears more than once? Why? What is the outcome of the fourth roll?
    (b) If we simulate more rolls of the die, do you expect to get the same sequence of outcomes? Why or why not?

16. *Simulation: Birthday Problem* Suppose there are 30 people at a party. Do you think any two share the same birthday? Let's use the random-number table to simulate the birthdays of the 30 people at the party. Ignoring leap year, let's assume that the year has 365 days. Number the days, with 1 representing January 1, 2 representing January 2, and so forth, with 365 representing December 31. Draw a random sample of 30 days (with replacement). These days represent the birthdays of the people at the party. Were any two of the birthdays the same? Compare your results with those obtained by other students in the class. Would you expect the results to be the same or different?

17. *Education: Test Construction* Professor Gill is designing a multiple-choice test. There are to be 10 questions. Each question is to have five choices for answers. The choices are to be designated by the letters *a, b, c, d,* and *e.* Professor Gill wishes to use a random-number table to determine which letter choice should correspond to the correct answer for a question. Using the number correspondence 1 for *a,* 2 for *b,* 3 for *c,* 4 for *d,* and 5 for *e,* use a random-number table to determine the letter choice for the correct answer for each of the 10 questions.

18. *Education: Test Construction* Professor Gill uses true–false questions. She wishes to place 20 such questions on the next test. To decide whether to place a true statement or a false statement in each of the 20 questions, she uses a random-number table. She selects 20 digits from the table. An even digit tells her to use a true statement. An odd digit tells her to use a false statement. Use a

random-number table to pick a sequence of 20 digits, and describe the corresponding sequence of 20 true–false questions. What would the test key for your sequence look like?

19. *Sampling Methods: Benefits Package* An important part of employee compensation is a benefits package, which might include health insurance, life insurance, child care, vacation days, retirement plan, parental leave, bonuses, etc. Suppose you want to conduct a survey of benefits packages available in private businesses in Hawaii. You want a sample size of 100. Some sampling techniques are described below. Categorize each technique as *simple random sample, stratified sample, systematic sample, cluster sample,* or *convenience sample.*
    (a) Assign each business in the Island Business Directory a number, and then use a random-number table to select the businesses to be included in the sample.
    (b) Use postal ZIP Codes to divide the state into regions. Pick a random sample of 10 ZIP Code areas and then include all the businesses in each selected ZIP Code area.
    (c) Send a team of five research assistants to Bishop Street in downtown Honolulu. Let each assistant select a block or building and interview an employee from each business found. Each researcher can have the rest of the day off after getting responses from 20 different businesses.
    (d) Use the Island Business Directory. Number all the businesses. Select a starting place at random, and then use every 50th business listed until you have 100 businesses.
    (e) Group the businesses according to type: medical, shipping, retail, manufacturing, financial, construction, restaurant, hotel, tourism, other. Then select a random sample of 10 businesses from each business type.

20. *Sampling Methods: Health Care* Modern Managed Hospitals (MMH) is a national for-profit chain of hospitals. Management wants to survey patients discharged this past year to obtain patient satisfaction profiles. They wish to use a sample of such patients. Several sampling techniques are described below. Categorize each technique as *simple random sample, stratified sample, systematic sample, cluster sample,* or *convenience sample.*
    (a) Obtain a list of patients discharged from all MMH facilities. Divide the patients according to length of hospital stay (2 days or less, 3–7 days, 8–14 days, more than 14 days). Draw simple random samples from each group.
    (b) Obtain lists of patients discharged from all MMH facilities. Number these patients, and then use a random-number table to obtain the sample.
    (c) Randomly select some MMH facilities from each of five geographic regions, and then include all the patients on the discharge lists of the selected hospitals.
    (d) At the beginning of the year, instruct each MMH facility to survey every 500th patient discharged.
    (e) Instruct each MMH facility to survey 10 discharged patients this week and send in the results.

Blend Images/Ariel Skelley/Getty Images

SECTION **1.3**

Introduction to Experimental Design

FOCUS POINTS

- Discuss what it means to take a census.
- Describe simulations, observational studies, and experiments.
- Identify control groups, placebo effects, completely randomized experiments, and randomized block experiments.
- Discuss potential pitfalls that might make your data unreliable.

PLANNING A STATISTICAL STUDY

Planning a statistical study and gathering data are essential components of obtaining reliable information. Depending on the nature of the statistical study, a great deal of expertise and resources may be required during the planning stage. In this section, we look at some of the basics of planning a statistical study.

PROCEDURE

BASIC GUIDELINES FOR PLANNING A STATISTICAL STUDY

1. First, identify the individuals or objects of interest.
2. Specify the variables as well as the protocols for taking measurements or making observations.
3. Determine if you will use an entire population or a representative sample. If using a sample, decide on a viable sampling method.
4. In your data collection plan, address issues of ethics, subject confidentiality, and privacy. If you are collecting data at a business, store, college, or other institution, be sure to be courteous and to obtain permission as necessary.
5. Collect the data.
6. Use appropriate descriptive statistics methods (Chapters 2, 3, and 9) and make decisions using appropriate inferential statistics methods (Chapters 7–11).
7. Finally, note any concerns you might have about your data collection methods and list any recommendations for future studies.

Census

One issue to consider is whether to use the entire population in a study or a representative sample. If we use data from the entire population, we have a *census*.

In a **census,** measurements or observations from the *entire* population are used.

When the population is small and easily accessible, a census is very useful because it gives complete information about the population. However, obtaining a census can be both expensive and difficult. Every 10 years, the U.S. Department of Commerce Census Bureau is required to conduct a census of the United States. However, contacting some members of the population—such as the homeless—is almost impossible. Sometimes members of the population will not respond. In such cases, statistical estimates for the missing responses are often supplied.

Overcounting, that is, counting the same person more than once, is also a problem the Census Bureau is addressing. In fact, in 2000, slightly more people were counted twice than the estimated number of people missed. For instance, a college student living on campus might be counted on a parent's census form as well as on his or her own census form.

Sample

If we use data from only part of the population of interest, we have a *sample.*

In a **sample,** measurements or observations from *part* of the population are used.

In the previous section, we examined several sampling strategies: simple random, stratified, cluster, systematic, multistage, and convenience. In this text, we will study methods of inferential statistics based on simple random samples.

Simulation

As discussed in Section 1.2, *simulation* is a numerical facsimile of real-world phenomena. Sometimes simulation is called a "dry lab" approach, in the sense that it is a mathematical imitation of a real situation. Advantages of simulation are that numerical and statistical simulations can fit real-world problems extremely well. The researcher can also explore procedures through simulation that might be very dangerous in real life.

EXPERIMENTS AND OBSERVATION

When gathering data for a statistical study, we want to distinguish between observational studies and experiments.

Observational study

Experiment

In an **observational study,** observations and measurements of individuals are conducted in a way that doesn't change the response or the variable being measured.

In an **experiment,** a *treatment* is deliberately imposed on the individuals in order to observe a possible change in the response or variable being measured.

EXAMPLE 4

EXPERIMENT

In 1778, Captain James Cook landed in what we now call the Hawaiian Islands. He gave the islanders a present of several goats, and over the years these animals multiplied into wild herds totaling several thousand. They eat almost anything, including the famous silver sword plant, which was once unique to Hawaii. At one time, the silver sword grew abundantly on the island of Maui (in Haleakala, a national park on that island, the silver sword can still be found), but each year there seemed to be fewer and fewer plants. Biologists suspected that the goats were partially responsible for the decline in the number of plants and conducted a statistical study that verified their theory.

Tim Davis/Photo Researchers

Silver sword plant, Haleakala National Park

(a) To test the theory, park biologists set up stations in remote areas of Haleakala. At each station two plots of land similar in soil conditions, climate, and plant count were selected. One plot was fenced to keep out the goats, while the other was not. At regular intervals a plant count was made in each plot. This study involved an *experiment* because a *treatment* (the fence) was imposed on one plot.

(b) The experiment involved two plots at each station. The plot that was not fenced represented the *control* plot. This was the plot on which a treatment was specifically not imposed, although the plot was similar to the fenced plot in every other way.

Placebo effect

Statistical experiments are commonly used to determine the effect of a treatment. However, the design of the experiment needs to *control* for other possible causes of the effect. For instance, in medical experiments, the *placebo effect* is the improvement or change that is the result of patients just believing in the treatment, whether or not the treatment itself is effective.

The **placebo effect** occurs when a subject receives no treatment but (incorrectly) believes he or she is in fact receiving treatment and responds favorably.

Treatment group

To account for the placebo effect, patients are divided into two groups. One group receives the prescribed treatment. The other group, called the *control group,* receives a dummy or placebo treatment that is disguised to look like the real treatment. Finally, after the treatment cycle, the medical condition of the patients in the *treatment group* is compared to that of the patients in the control group.

Completely randomized experiment

A common way to assign patients to treatment and control groups is by using a random process. This is the essence of a *completely randomized experiment.*

A **completely randomized experiment** is one in which a random process is used to assign each individual to one of the treatments.

EXAMPLE 5

COMPLETELY RANDOMIZED EXPERIMENT

Can chest pain be relieved by drilling holes in the heart? For more than a decade, surgeons have been using a laser procedure to drill holes in the heart. Many patients report a lasting and dramatic decrease in angina (chest pain) symptoms. Is the relief due to the procedure, or is it a placebo effect? A recent research project at Lenox Hill Hospital in New York City provided some information about this issue by using a completely randomized experiment. The laser treatment was applied through a less invasive (catheter laser) process. A group of 298 volunteers with severe, untreatable chest pain were randomly assigned to get the laser or not. The patients were sedated but awake. They could hear the doctors discuss the laser process. Each patient thought he or she was receiving the treatment.

The experimental design can be pictured as

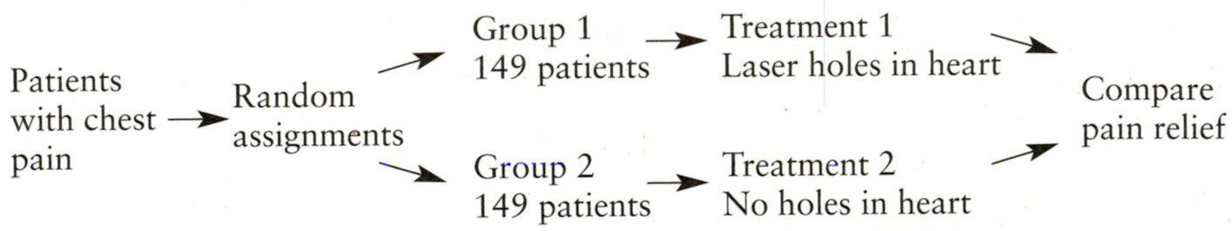

The laser patients did well. But shockingly, the placebo group showed more improvement in pain relief. The medical impacts of this study are still being investigated.

It is difficult to control all the variables that might influence the response to a treatment. One way to control some of the variables is through *blocking.*

Block

Randomized block experiment

A **block** is a group of individuals sharing some common features that might affect the treatment.

In a **randomized block experiment,** individuals are first sorted into blocks, and then a random process is used to assign each individual in the block to one of the treatments.

A randomized block design utilizing gender for blocks in the experiment involving laser holes in the heart would be

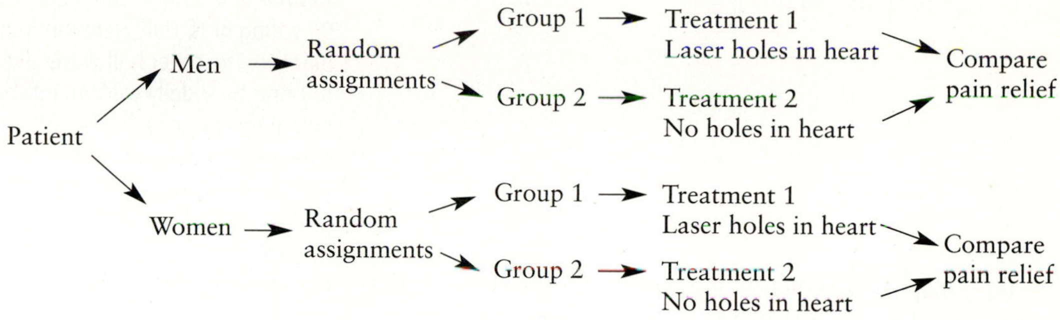

The study cited in Example 5 has many features of good experimental design.

Control group

There is a **control group.** This group receives a dummy treatment, enabling the researchers to control for the placebo effect. In general, a control group is used to account for the influence of other known or unknown variables that might be an underlying cause of a change in response in the experimental group. Such variables are called **lurking** or **confounding variables.**

Randomization

Randomization is used to assign individuals to the two treatment groups. This helps prevent bias in selecting members for each group.

Replication

Replication of the experiment on many patients reduces the possibility that the differences in pain relief for the two groups occurred by chance alone.

Double-blind experiment

Many experiments are also *double-blind*. This means that neither the individuals in the study nor the observers know which subjects are receiving the treatment. Double-blind experiments help control for subtle biases that a doctor might pass on to a patient.

LOOKING FORWARD

One-way and two-way ANOVA (Sections 10.5 and 10.6) are analysis techniques used to study results from completely randomized experiments with several treatments or several blocks with multiple treatments.

GUIDED EXERCISE 5 | Collecting Data

Which technique for gathering data (sampling, experiment, simulation, or census) do you think might be the most appropriate for the following studies?

(a) Study of the effect of stopping the cooling process of a nuclear reactor.

→ Simulation, since you probably do not want to risk a nuclear meltdown.

(b) Study of the amount of time college students taking a full course load spend watching television.

→ Sampling and using an observational study would work well. Notice that obtaining the information from a student will probably not change the amount of time the student spends watching television.

Continued

GUIDED EXERCISE 5 *continued*

(c) Study of the effect on bone mass of a calcium supplement given to young girls.

→ Experimentation. A study by Tom Lloyd reported in the *Journal of the American Medical Association* utilized 94 young girls. Half were randomly selected and given a placebo. The other half were given calcium supplements to bring their daily calcium intake up to about 1400 milligrams per day. The group getting the experimental treatment of calcium gained 1.3% more bone mass in a year than the girls getting the placebo.

(d) Study of the credit hour load of *each* student enrolled at your college at the end of the drop/add period this semester.

→ Census. The registrar can obtain records for *every* student.

Spencer Grant/PhotoEdit

SURVEYS

Once you decide whether you are going to use sampling, census, observation, or experiments, a common means to gather data about people is to ask them questions. This process is the essence of *surveying*. Sometimes the possible responses are simply yes or no. Other times the respondents choose a number on a scale that represents their feelings from, say, strongly disagree to strongly agree. Such a scale is called a *Likert scale*. In the case of an open-ended, discussion-type response, the researcher must determine a way to convert the response to a category or number.

A number of issues can arise when using a survey.

Survey

Likert scale
Nonresponse

Hidden bias

Voluntary response

SOME POTENTIAL PITFALLS OF A SURVEY

Nonresponse: Individuals either cannot be contacted or refuse to participate. Nonresponse can result in significant undercoverage of a population.

Truthfulness of response: Respondents may lie intentionally or inadvertently.

Faulty recall: Respondents may not accurately remember when or whether an event took place.

Hidden bias: The question may be worded in such a way as to elicit a specific response. The order of questions might lead to biased responses. Also, the number of responses on a Likert scale may force responses that do not reflect the respondent's feelings or experience.

Vague wording: Words such as "often," "seldom," and "occasionally" mean different things to different people.

Interviewer influence: Factors such as tone of voice, body language, dress, gender, authority, and ethnicity of the interviewer might influence responses.

Voluntary response: Individuals with strong feelings about a subject are more likely than others to respond. Such a study is interesting but not reflective of the population.

Lurking variable
Confounding variable

Sometimes our goal is to understand the cause-and-effect relationships between two or more variables. Such studies can be complicated by *lurking variables* or *confounding variables*.

A **lurking variable** is one for which no data have been collected but that nevertheless has influence on other variables in the study.

Two variables are **confounded** when the effects of one cannot be distinguished from the effects of the other. Confounding variables may be part of the study, or they may be outside lurking variables.

For instance, consider a study involving just two variables, amount of gasoline used to commute to work and time to commute to work. Level of traffic congestion is a likely lurking variable that increases both of the study variables. In a study involving several variables such as grade point average, difficulty of courses, IQ, and available study time, some of the variables might be confounded. For instance, students with less study time might opt for easier courses.

Generalizing results

Some researchers want to generalize their findings to a situation of wider scope than that of the actual data setting. The true scope of a new discovery must be determined by repeated studies in various real-world settings. Statistical experiments showing that a drug had a certain effect on a collection of laboratory rats do not guarantee that the drug will have a similar effect on a herd of wild horses in Montana.

Study sponsor

The sponsorship of a study is another area of concern. Subtle bias may be introduced. For instance, if a pharmaceutical company is paying freelance researchers to work on a study, the researchers may dismiss rare negative findings about a drug or treatment.

GUIDED EXERCISE 6 | Cautions about Data

Comment on the usefulness of the data collected as described.

(a) A uniformed police officer interviews a group of 20 college freshmen. She asks each one his or her name and then if he or she has used an illegal drug in the last month.

→ Respondents may not answer truthfully. Some may refuse to participate.

(b) Jessica saw some data that show that cities with more low-income housing have more homeless people. Does building low-income housing cause homelessness?

→ There may be some other confounding or lurking variables, such as the size of the city. Larger cities may have more low-income housing and more homeless.

(c) A survey about food in the student cafeteria was conducted by having forms available for customers to pick up at the cash register. A drop box for completed forms was available outside the cafeteria.

→ The voluntary response likely produced more negative comments.

(d) Extensive studies on coronary problems were conducted using men over age 50 as the subjects.

→ Conclusions for men over age 50 may or may not generalize to other age and gender groups. These results may be useful for women or younger people, but studies specifically involving these groups may need to be performed.

CHOOSING DATA COLLECTION TECHNIQUES

We've briefly discussed three common techniques for gathering data: observational studies, experiments, and surveys. Which technique is best? The answer depends on the number of variables of interest and the level of confidence needed regarding statements of relationships among the variables.

- Surveys may be the best choice for gathering information across a wide range of many variables. Many questions can be included in a survey. However, great care must be taken in the construction of the survey instrument and in the administration of the survey. Nonresponse and other issues discussed earlier can introduce bias.
- Observational studies are the next most convenient technique for gathering information on many variables. Protocols for taking measurements or recording observations need to be specified carefully.
- Experiments are the most stringent and restrictive data-gathering technique. They can be time-consuming, expensive, and difficult to administer. In experiments, the goal is often to study the effects of changing only one variable at a time. Because of the requirements, the number of variables may be more limited. Experiments must be designed carefully to ensure that the resulting data are relevant to the research questions.

COMMENT An experiment is the best technique for reaching valid conclusions. By carefully controlling for other variables, the effect of changing one variable in a treatment group and comparing it to a control group yields results carrying high confidence.

The next most effective technique for obtaining results that have high confidence is the use of observational studies. Care must be taken that the act of observation does not change the behavior being measured or observed.

The least effective technique for drawing conclusions is the survey. Surveys have many pitfalls and by their nature cannot give exceedingly precise results. A medical study utilizing a survey asking patients if they feel better after taking a specific drug gives some information, but not precise information about the drug's effects. However, surveys are widely used to gauge attitudes, gather demographic information, study social and political trends, and so on.

IMPORTANT FEATURES OF A DATA COLLECTION PLAN

A data collection plan identifies

- The population
- The variable or variables
- Whether the data are observational or experimental
- Whether there is a control group, use of placebos, double-blind treatment, etc.
- The sampling technique to be used, including whether a block design is to be used
- The method used to collect the data for the variables: survey, method of measurement, count, etc.

VIEWPOINT Is the Placebo Effect a Myth?

Henry Beecher, former Chief of Anesthesiology at Massachusetts General Hospital, published a paper in the Journal of the American Medical Association (1955) in which he claimed that the placebo effect is so powerful that about 35% of patients would improve simply if they believed a dummy treatment (placebo) was real. However, two Danish medical researchers refute this widely accepted claim in the New England Journal of Medicine. They say the placebo effect is nothing more than a "regression effect," referring to a well-known statistical observation that patients who feel especially bad one day will almost always feel better the next day, no matter what is done for them. However, other respected statisticians question the findings of the Danish researchers. Regardless of the new controversy surrounding the placebo effect, medical researchers agree that placebos are still needed in clinical research. Double-blind research using placebos prevents researchers from inadvertently biasing results.

SECTION 1.3 PROBLEMS

1. *Statistical Literacy* A study involves three variables: income level, hours spent watching TV per week, and hours spent at home on the Internet per week. List some ways the variables might be confounded.

2. *Statistical Literacy* Consider a completely randomized experiment in which a control group is given a placebo for congestion relief and a treatment group is given a new drug for congestion relief. Describe a double-blind procedure for this experiment and discuss some benefits of such a procedure.

3. *Critical Thinking* A brief survey regarding opinions about recycling was carefully designed so that the wording of the questions would not influence the responses. Jill administered the survey at a farmer's market. She approached adults and asked if they would fill out the survey, explaining that the results might be used to set trash collection and recycling policy in the city. She stood by silently while the form was filled out. Jill was wearing a green T-shirt with the slogan "fight global warming." Are the respondents a random sample of people in the community? Are there any concerns that Jill might have influenced the respondents?

4. *Critical Thinking* A randomized block design was used to study the amount of grants awarded to students at a large university. One block consisted of undergraduate students and the other block consisted of graduate students. Samples of size 30 were taken from each block. Could the combined sample of 60 be considered a simple random sample from the population of all students, undergraduate and all graduate, at the university? Explain.

5. *Interpretation* Zane is examining two studies involving how different generations classify specified items as either luxuries or necessities. In the first study, the Echo generation is defined to be people ages 18–29. The second study defined the Echo generation to be people ages 20–31. Zane notices that the first study was conducted in 2006 while the second one was conducted in 2008.
   (a) Are the two studies inconsistent in their description of the Echo generation?
   (b) What are the birth years of the Echo generation?

6. *Interpretation* Suppose you are looking at the 2006 results of how the Echo generation classified specified items as either luxuries or necessities. Do you expect the results to reflect how the Echo generation would classify items in 2016? Explain.

7. *Ecology: Gathering Data* Which technique for gathering data (observational study or experiment) do you think was used in the following studies?
(a) The Colorado Division of Wildlife netted and released 774 fish at Quincy Reservoir. There were 219 perch, 315 blue gill, 83 pike, and 157 rainbow trout.
(b) The Colorado Division of Wildlife caught 41 bighorn sheep on Mt. Evans and gave each one an injection to prevent heartworm. A year later, 38 of these sheep did not have heartworm, while the other three did.
(c) The Colorado Division of Wildlife imposed special fishing regulations on the Deckers section of the South Platte River. All trout under 15 inches had to be released. A study of trout before and after the regulation went into effect showed that the average length of a trout increased by 4.2 inches after the new regulation.
(d) An ecology class used binoculars to watch 23 turtles at Lowell Ponds. It was found that 18 were box turtles and 5 were snapping turtles.

8. *General: Gathering Data* Which technique for gathering data (sampling, experiment, simulation, or census) do you think was used in the following studies?
(a) An analysis of a sample of 31,000 patients from New York hospitals suggests that the poor and the elderly sue for malpractice at one-fifth the rate of wealthier patients (*Journal of the American Medical Association*).
(b) The effects of wind shear on airplanes during both landing and takeoff were studied by using complex computer programs that mimic actual flight.
(c) A study of all league football scores attained through touchdowns and field goals was conducted by the National Football League to determine whether field goals account for more scoring events than touchdowns (*USA Today*).
(d) An Australian study included 588 men and women who already had some precancerous skin lesions. Half got a skin cream containing a sunscreen with a sun protection factor of 17; half got an inactive cream. After 7 months, those using the sunscreen with the sun protection had fewer new precancerous skin lesions (*New England Journal of Medicine*).

9. *General: Completely Randomized Experiment* How would you use a completely randomized experiment in each of the following settings? Is a placebo being used or not? Be specific and give details.
(a) A veterinarian wants to test a strain of antibiotic on calves to determine their resistance to common infection. In a pasture are 22 newborn calves. There is enough vaccine for 10 calves. However, blood tests to determine resistance to infection can be done on all calves.
(b) The Denver Police Department wants to improve its image with teenagers. A uniformed officer is sent to a school one day a week for 10 weeks. Each day the officer visits with students, eats lunch with students, attends pep rallies, and so on. There are 18 schools, but the police department can visit only half of these schools this semester. A survey regarding how teenagers view police is sent to all 18 schools at the end of the semester.
(c) A skin patch contains a new drug to help people quit smoking. A group of 75 cigarette smokers have volunteered as subjects to test the new skin patch. For one month, 40 of the volunteers receive skin patches with the new drug. The other volunteers receive skin patches with no drugs. At the end of two months, each subject is surveyed regarding his or her current smoking habits.

10. *Surveys: Manipulation* The *New York Times* did a special report on polling that was carried in papers across the nation. The article pointed out how readily the results of a survey can be manipulated. Some features that can

influence the results of a poll include the following: the number of possible responses, the phrasing of the questions, the sampling techniques used (voluntary response or sample designed to be representative), the fact that words may mean different things to different people, the questions that precede the question of interest, and finally, the fact that respondents can offer opinions on issues they know nothing about.

(a) Consider the expression "over the last few years." Do you think that this expression means the same time span to everyone? What would be a more precise phrase?

(b) Consider this question: "Do you think fines for running stop signs should be doubled?" Do you think the response would be different if the question "Have you ever run a stop sign?" preceded the question about fines?

(c) Consider this question: "Do you watch too much television?" What do you think the responses would be if the only responses possible were yes or no? What do you think the responses would be if the possible responses were "rarely," "sometimes," or "frequently"?

11. *Critical Thinking* An agricultural study is comparing the harvest volume of two types of barley. The site for the experiment is bordered by a river. The field is divided into eight plots of approximately the same size. The experiment calls for the plots to be blocked into four plots per block. Then, two plots of each block will be randomly assigned to one of the two barley types.

Two blocking schemes are shown below, with one block indicated by the white region and the other by the grey region. Which blocking scheme, A or B, would be better? Explain.

Scheme A

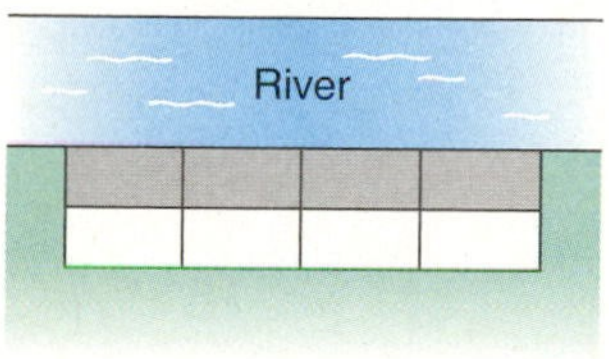

Scheme B

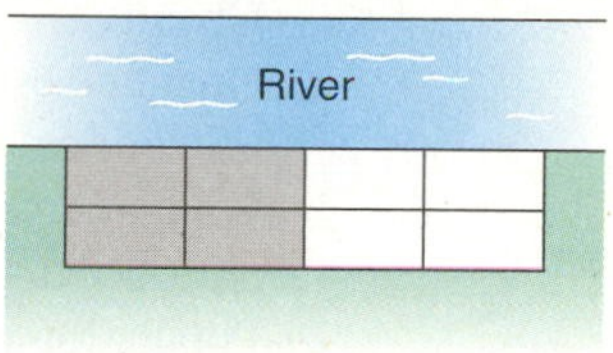

CHAPTER REVIEW

SUMMARY

In this chapter, you've seen that statistics is the study of how to collect, organize, analyze, and interpret numerical information from populations or samples. This chapter discussed some of the features of data and ways to collect data. In particular, the chapter discussed

- Individuals or subjects of a study and the variables associated with those individuals
- Data classification as qualitative or quantitative, and levels of measurement of data
- Sample and population data. Summary measurements from sample data are called statistics, and those from populations are called parameters.
- Sampling strategies, including simple random, stratified, systematic, multistage, and convenience. Inferential techniques presented in this text are based on simple random samples.
- Methods of obtaining data: Use of a census, simulation, observational studies, experiments, and surveys
- Concerns: Undercoverage of a population, nonresponse, bias in data from surveys and other factors, effects of confounding or lurking variables on other variables, generalization of study results beyond the population of the study, and study sponsorship

IMPORTANT WORDS & SYMBOLS

Section 1.1*

Statistics 4
Individuals 5
Variable 5
Quantitative variable 5
Qualitative variable 5
Categorical variable 5
Population data 5
Sample data 5
Population parameter 5
Sample statistic 5
Levels of measurement 7
  Nominal 7
  Ordinal 7
  Interval 7
  Ratio 7
Descriptive statistics 10
Inferential statistics 10

Section 1.2

Simple random sample 13
Random-number table 14
Simulation 15
Sampling with replacement 16
Stratified sampling 16
Systematic sampling 16
Cluster sampling 17
Multistage sample 17
Convenience sampling 17
Sampling frame 18
Undercoverage 18
Sampling error 18
Nonsampling error 18

Section 1.3

Census 22
Sample 23
Observational study 23
Experiment 23
Placebo effect 23
Treatment group 24
Completely randomized experiment 24
Block 24
Randomized block experiment 24
Control group 25
Randomization 25
Replication 25
Double-blind experiment 25
Survey 26
Likert scale 26
Nonresponse 26
Hidden bias 26
Voluntary response 26
Lurking variable 26
Confounding variable 26
Generalizing results 27
Study sponsor 27

*Indicates section of first appearance.

VIEWPOINT Is Chocolate Good for Your Heart?

A study of 7841 Harvard alumni showed that the death rate was 30% lower in those who ate candy compared with those who abstained. It turns out that candy, especially chocolate, contains antioxidants that help slow the aging process. Also, chocolate, like aspirin, reduces the activity of blood platelets that contribute to plaque and blood clotting. Furthermore, chocolate seems to raise levels of high-density lipoprotein (HDL), the good cholesterol. However, these results are all preliminary. The investigation is far from complete. A wealth of information on this topic was published in the August 2000 issue of the Journal of Nutrition. *Statistical studies and reliable experimental design are indispensable in this type of research.*

CHAPTER REVIEW PROBLEMS

Joe McDaniel/iStockphoto.com

1. *Critical Thinking* Sudoku is a puzzle consisting of squares arranged in 9 rows and 9 columns. The 81 squares are further divided into nine 3 × 3 square boxes. The object is to fill in the squares with numerals 1 through 9 so that each column, row, and box contains all nine numbers. However, there is a requirement that each number appear only once in any row, column, or box. Each puzzle already has numbers in some of the squares. Would it be appropriate to use a random-number table to select a digit for each blank square? Explain.

2. *Critical Thinking* Alisha wants to do a statistical study to determine how long it takes people to complete a Sudoku puzzle (see Problem 1 for a description of the puzzle). Her plan is as follows:

   Download 10 different puzzles from the Internet.
   Find 10 friends willing to participate.
   Ask each friend to complete one of the puzzles and time him- or herself.
   Gather the completion times from each friend.

   Describe some of the problems with Alisha's plan for the study. (*Note:* Puzzles differ in difficulty, ranging from beginner to very difficult.) Are the results from Alisha's study anecdotal, or do they apply to the general population?

3. *Statistical Literacy* You are conducting a study of students doing work-study jobs on your campus. Among the questions on the survey instrument are:

   A. How many hours are you scheduled to work each week? Answer to the nearest hour.
   B. How applicable is this work experience to your future employment goals?
   Respond using the following scale: 1 = not at all, 2 = somewhat, 3 = very

   (a) Suppose you take random samples from the following groups: freshmen, sophomores, juniors, and seniors. What kind of sampling technique are you using (simple random, stratified, systematic, cluster, multistage, convenience)?
   (b) Describe the individuals of this study.
   (c) What is the variable for question A? Classify the variable as qualitative or quantitative. What is the level of the measurement?
   (d) What is the variable for question B? Classify the variable as qualitative or quantitative. What is the level of the measurement?
   (e) Is the proportion of responses "3 = very" to question B a statistic or a parameter?
   (f) Suppose only 40% of the students you selected for the sample respond. What is the nonresponse rate? Do you think the nonresponse rate might introduce bias into the study? Explain.
   (g) Would it be appropriate to generalize the results of your study to all work-study students in the nation? Explain.

4. *Radio Talk Show: Sample Bias* A radio talk show host asked listeners to respond either yes or no to the question, "Is the candidate who spends the most on a campaign the most likely to win?" Fifteen people called in and nine said yes. What is the implied population? What is the variable? Can you detect any bias in the selection of the sample?

5. *Simulation: TV Habits* One cable station knows that approximately 30% of its viewers have TIVO and can easily skip over advertising breaks. You are to design a simulation of how a random sample of seven station viewers would respond to the question, "Do you have TIVO?" How would you assign the random digits 0 through 9 to the responses "Yes" or "No" to the TIVO question? Use your random-digit assignment and the random-number table to generate the responses from a random sample of seven station viewers.

6. *General: Type of Sampling* Categorize the type of sampling (simple random, stratified, systematic, cluster, or convenience) used in each of the following situations.
   (a) To conduct a preelection opinion poll on a proposed amendment to the state constitution, a random sample of 10 telephone prefixes (first three digits of the phone number) was selected, and all households from the phone prefixes selected were called.
   (b) To conduct a study on depression among the elderly, a sample of 30 patients in one nursing home was used.
   (c) To maintain quality control in a brewery, every 20th bottle of beer coming off the production line was opened and tested.
   (d) Subscribers to the magazine *Sound Alive* were assigned numbers. Then a sample of 30 subscribers was selected by using a random-number table. The subscribers in the sample were invited to rate new compact disc players for a "What the Subscribers Think" column.
   (e) To judge the appeal of a proposed television sitcom, a random sample of 10 people from each of three different age categories was selected and those chosen were asked to rate a pilot show.

7. *General: Gathering Data* Which technique for gathering data (observational study or experiment) do you think was used in the following studies? Explain.
   (a) The U.S. Census Bureau tracks population age. In 1900, the percentage of the population that was 19 years old or younger was 44.4%. In 1930, the percentage was 38.8%; in 1970, the percentage was 37.9%; and in 2000, the percentage in that age group was down to 28.5% (Reference: *The First Measured Century,* T. Caplow, L. Hicks, and B. J. Wattenberg).
   (b) After receiving the same lessons, a class of 100 students was randomly divided into two groups of 50 each. One group was given a multiple-choice exam covering the material in the lessons. The other group was given an essay exam. The average test scores for the two groups were then compared.

8. *General: Experiment* How would you use a completely randomized experiment in each of the following settings? Is a placebo being used or not? Be specific and give details.
   (a) A charitable nonprofit organization wants to test two methods of fundraising. From a list of 1000 past donors, half will be sent literature about the successful activities of the charity and asked to make another donation. The other 500 donors will be contacted by phone and asked to make another donation. The percentage of people from each group who make a new donation will be compared.
   (b) A tooth-whitening gel is to be tested for effectiveness. A group of 85 adults have volunteered to participate in the study. Of these, 43 are to be given a gel that contains the tooth-whitening chemicals. The remaining 42 are to be given a similar-looking package of gel that does not contain the tooth-whitening chemicals. A standard method will be used to evaluate the whiteness of teeth

for all participants. Then the results for the two groups will be compared. How could this experiment be designed to be double-blind?

(c) Consider the experiment described in part (a). Describe how you would use a randomized block experiment with blocks based on age. Use three blocks: donors under 30 years old, donors 30 to 59 years old, donors 60 and over.

9. *Student Life: Data Collection Project* Make a statistical profile of your own statistics class. Items of interest might be

(a) Height, age, gender, pulse, number of siblings, marital status
(b) Number of college credit hours completed (as of beginning of term); grade point average
(c) Major; number of credit hours enrolled in this term
(d) Number of scheduled work hours per week
(e) Distance from residence to first class; time it takes to travel from residence to first class
(f) Year, make, and color of car usually driven

What directions would you give to people answering these questions? For instance, how accurate should the measurements be? Should age be recorded as of last birthday?

10. *Census: Web Site* *Census and You,* a publication of the Census Bureau, indicates that "Wherever your Web journey ends up, it should start at the Census Bureau's site." Visit the Brase/Brase statistics site at **http://www.cengage.com/statistics/brase11e** and find a link to the Census Bureau's site, as well as to Fedstats, another extensive site offering links to federal data. The Census Bureau site touts itself as the source of "official statistics." But it is willing to share the spotlight. The web site now has links to other "official" sources: other federal agencies, foreign statistical agencies, and state data centers. If you have access to the Internet, try the Census Bureau's site.

11. *Focus Problem: Fireflies* Suppose you are conducting a study to compare firefly populations exposed to normal daylight/darkness conditions with firefly populations exposed to continuous light (24 hours a day). You set up two firefly colonies in a laboratory environment. The two colonies are identical except that one colony is exposed to normal daylight/darkness conditions and the other is exposed to continuous light. Each colony is populated with the same number of mature fireflies. After 72 hours, you count the number of living fireflies in each colony.

(a) Is this an experiment or an observation study? Explain.
(b) Is there a control group? Is there a treatment group?
(c) What is the variable in this study?
(d) What is the level of measurement (nominal, interval, ordinal, or ratio) of the variable?

DATA HIGHLIGHTS: GROUP PROJECTS

1. Use a random-number table or random-number generator to simulate tossing a fair coin 10 times. Generate 20 such simulations of 10 coin tosses. Compare the simulations. Are there any strings of 10 heads? of 4 heads? Does it seem that in most of the simulations, half the outcomes are heads? half are tails? In Chapter 5, we will study the probabilities of getting from 0 to 10 heads in such a simulation.
2. Use a random-number table or random-number generator to generate a random sample of 30 distinct values from the set of integers 1 to 100. Instructions for doing this using the TI-84Plus/TI-83Plus/TI-*n*spire (with TI-84Plus keypad), Excel 2010, Minitab, or SPSS are given in Using Technology at the end of this chapter. Generate five such samples. How many of the samples include the number 1? the number 100? Comment about the differences among the samples. How well do the samples seem to represent the numbers between 1 and 100?

LINKING CONCEPTS: WRITING PROJECTS

Discuss each of the following topics in class or review the topics on your own. Then write a brief but complete essay in which you summarize the main points. Please include formulas and graphs as appropriate.

1. What does it mean to say that we are going to use a sample to draw an inference about a population? Why is a random sample so important for this process? If we wanted a random sample of students in the cafeteria, why couldn't we just choose the students who order Diet Pepsi with their lunch? Comment on the statement, "A random sample is like a miniature population, whereas samples that are not random are likely to be biased." Why would the students who order Diet Pepsi with lunch not be a random sample of students in the cafeteria?
2. In your own words, explain the differences among the following sampling techniques: simple random sample, stratified sample, systematic sample, cluster sample, multistage sample, and convenience sample. Describe situations in which each type might be useful.

USING TECHNOLOGY

General spreadsheet programs such as Microsoft's Excel, specific statistical software packages such as Minitab or SPSS, and graphing calculators such as the TI-84Plus/TI-83Plus/TI-*n*spire all offer computing support for statistical methods. Applications in this section may be completed using software or calculators with statistical functions. Select keystroke or menu choices are shown for the TI-84Plus/TI-83Plus/TI-*n*spire (with TI-84Plus keypad) calculators, Minitab, Excel 2010, and SPSS in the Technology Hints portion of this section. More details can be found in the software-specific *Technology Guide* that accompanies this text.

Applications

Most software packages sample *with replacement.* That is, the same number may be used more than once in the sample. If your applications require sampling without replacement, draw more items than you need. Then use sort commands in the software to put the data in order, and delete repeated data.

1. Simulate the results of tossing a fair die 18 times. Repeat the simulation. Are the results the same? Did you expect them to be the same? Why or why not? Do there appear to be equal numbers of outcomes 1 through 6 in each simulation? In Chapter 4, we will encounter the law of large numbers, which tells us that we would expect equal numbers of outcomes only when the simulation is very large.
2. A college has 5000 students, and the registrar wishes to use a random sample of 50 students to examine credit hour enrollment for this semester. Write a brief description of how a random sample can be drawn. Draw a random sample of 50 students. Are you sampling with or without replacement?

Technology Hints: Random Numbers

TI-84Plus/TI-83Plus/TI-*n*spire (with TI-84Plus keypad)

The TI-*n*spire calculator with the TI-84Plus keypad installed works exactly like other TI-84Plus calculators. Instructions for the TI-84Plus, TI-83Plus, and the TI-*n*spire (with the TI-84Plus keypad installed) calculators are included directly in this text as well as in a separate *Technology Guide*. When the *n*spire keypad is installed, the required keystrokes and screen displays are different from those with the TI-84Plus keypad. A separate *Technology Guide* for this text provides instructions for using the *n*spire keypad to perform statistical operations, create statistical graphs, and apply statistical tests.

The instructions that follow apply to the TI-84Plus, TI-83Plus, and TI-*n*spire (with the TI-84 keypad installed) calculators.

To select a random set of integers between two specified values, press the **MATH** key and highlight **PRB** with **5:randInt** (low value, high value, sample size). Press Enter and fill in the low value, high value, and sample size. To store the sample in list L1, press the **STO➡** key and then L1. The screen display shows two random samples of size 5 drawn from the integers between 1 and 100.

```
randInt(1,100,5)
{63 89 13 46 47}
randInt(1,100,5)
{29 82 99 50 41}
```

Cengage Learning

Excel 2010

Many statistical processes that we will use throughout this text are included in the **Analysis ToolPak** add-in of Excel. This is an add-in that is included with the standard versions of Excel 2010. To see if you have the add-in installed, click the File Tab in the upper-left corner of the spreadsheet, click on the **Excel Options** button, and select **Add-ins**. Check that **Analysis ToolPak** is in the Active Application Add-ins list. If it is not, select it for an active application.

To select a random number between two specified integer values, first select a cell in the active worksheet. Then type the command=**RANDBETWEEN(bottom number, top number)** in the formula bar, where the bottom number is the lower specified value and the top number is the higher specified value.

Alternatively, access a dialogue box for the command by clicking the **Insert Function** button $\boxed{f_x}$ on the ribbon. You will see an insert function dialogue box. Select the category **All,** and then scroll down until you reach **RANDBETWEEN** and click OK. Fill in the bottom and top numbers. In the display shown, the bottom number is 1 and the top number is 100.

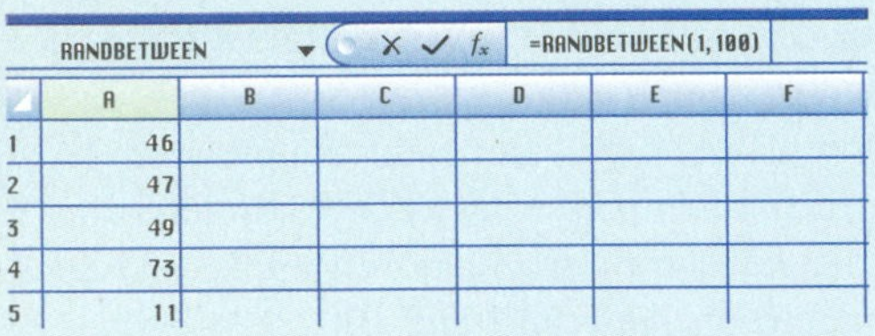

Minitab

To generate random integers between specified values, use the menu selection **Calc ➤ Random Data ➤ Integer.** Fill in the dialogue box to get five random numbers between 1 and 100.

Worksheet 2 ***	
	C1
1	8
2	35
3	33
4	9
5	15

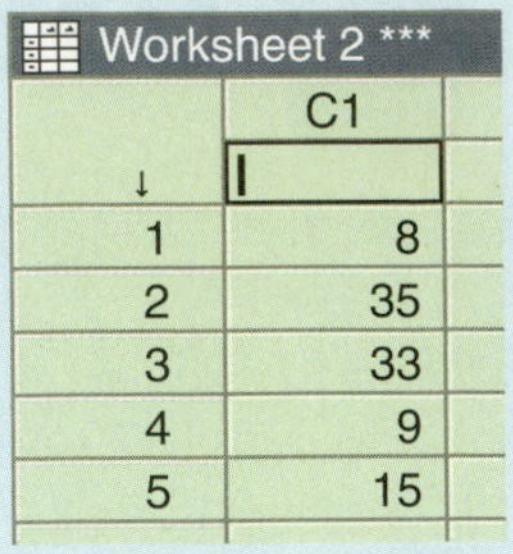

Cengage Learning

SPSS

SPSS is a research statistical package for the social sciences. Data are entered in the data editor, which has a spreadsheet format. In the data editor window, you have a choice of data view (default) or variable view. In the variable view, you name variables, declare type (numeric for measurements, string for category), determine format, and declare measurement type. The choices for measurement type are scale (for ratio or interval data), ordinal, or nominal. Once you have entered data, you can use the menu bar at the top of the screen to select activities, graphs, or analysis appropriate to the data.

SPSS supports several random sample activities. In particular, you can select a random sample from an existing data set or from a variety of probability distributions.

Selecting a random integer between two specified values involves several steps. First, in the data editor, enter the sample numbers in the first column. For instance, to generate five random numbers, list the values 1 through 5 in the first column. Notice that the label for the first column is now var00001. SPSS does not have a direct function for selecting a random sample of integers. However, there is a function for sampling values from the uniform distribution of all real numbers between two specified values. We will use that function and then truncate the values to obtain a random sample of integers between two specified values.

RANDBETWEEN	▼	✕ ✓ f_x	=RANDBETWEEN(1,100)		
	A	B	C	D	E	F
1	46					
2	47					
3	49					
4	73					
5	11					

Cengage Learning

Use the menu options **Transform ➤ Compute.** In the dialogue box, type in var00002 as the target variable. In the Function group select **Random Number**. Then under Functions and Special Variables, select **Rv.Uniform**. In the Numeric Expression box, replace the two question marks by the minimum and maximum. Use 1 as the minimum and 101 as the maximum. The maximum is 101 because numbers between 100 and 101 truncate to 100.

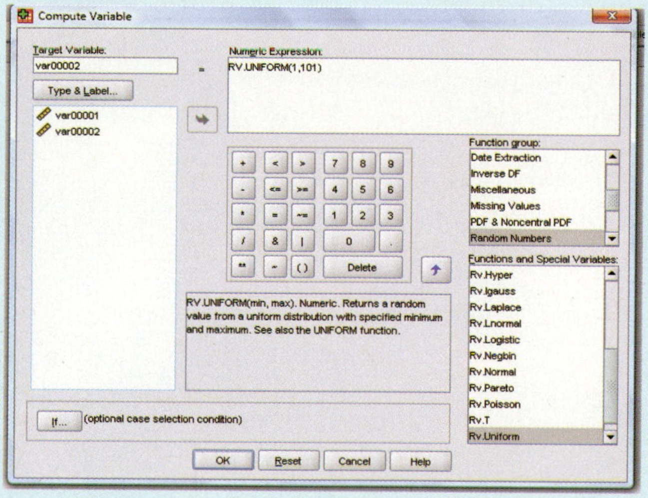

Cengage Learning

The random numbers from the uniform distribution now appear in the second column under var00002. You can visually truncate the values to obtain random integers. However, if you want SPSS to truncate the values for you, you can again use the menu choices **Transform ➤ Compute.** In the dialogue box, enter var00003 for the target variable. In the Function Group select **Arithmetic**. Then in the Functions and Special Variables box select **Trunc(1)**. In the Numeric Expression box use var00002 in place of the question mark representing numexpr. The random integers between 1 and 100 appear in the third column under var00003.

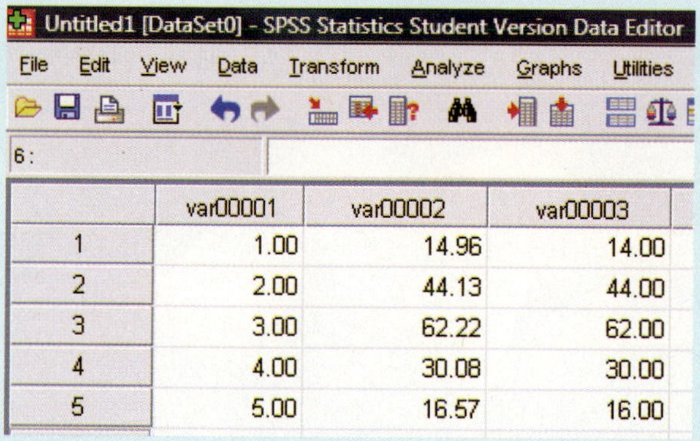

Cengage Learning

2

World History Archive/Alamy

In dwelling upon the vital importance of sound observation, it must never be lost sight of what observation is for. It is not for the sake of piling up miscellaneous information or curious facts, but for the sake of saving life and increasing health and comfort.

—Florence Nightingale, *Notes on Nursing*

Florence Nightingale (1820–1910) has been described as a "passionate statistician" and a "relevant statistician." She viewed statistics as a science that allowed one to transcend his or her narrow individual experience and aspire to the broader service of humanity. She was one of the first nurses to use graphic representation of statistics, illustrating with charts and diagrams how improved sanitation decreased the rate of mortality. Her statistical reports about the appalling sanitary conditions at Scutari (the main British hospital during the Crimean War) were taken very seriously by the English Secretary at War, Sidney Herbert. When sanitary reforms recommended by Nightingale were instituted in military hospitals, the mortality rate dropped from an incredible 42.7% to only 2.2%.

ORGANIZING DATA

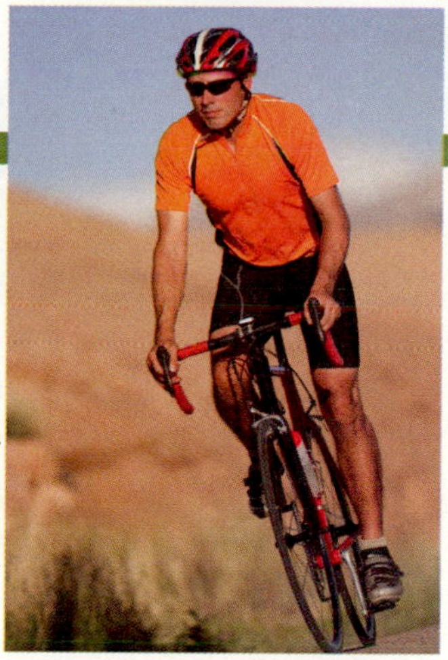

PREVIEW QUESTIONS

What are histograms? When are they used? (SECTION 2.1)

What are common distribution shapes? (SECTION 2.1)

How can you select graphs appropriate for given data sets? (SECTION 2.2)

How can you quickly order data and, at the same time, reveal the distribution shape? (SECTION 2.3)

FOCUS PROBLEM

Say It with Pictures

Edward R. Tufte, in his book *The Visual Display of Quantitative Information*, presents a number of guidelines for producing good graphics. According to the criteria, a graphical display should

- show the data;
- induce the viewer to think about the substance of the graphic rather than about the methodology, the design, the technology, or other production devices;
- avoid distorting what the data have to say.

As an example of a graph that violates some of the criteria, Tufte includes a graphic that appeared in a well-known newspaper. Figure 2-1(a), on the next page, shows a figure similar to the problem graphic, whereas part (b) of the figure shows a better rendition of the data display.

After completing this chapter, you will be able to answer the following questions.

(a) Look at the graph in Figure 2-1(a). Is it essentially a bar graph? Explain. What are some of the flaws of Figure 2-1(a) as a bar graph?

(b) Examine Figure 2-1(b), which shows the same information. Is it essentially a time-series graph? Explain. In what ways does the second graph seem to display the information in a clearer manner?

(See Problem 5 of the Chapter 2 Review Problems.)

FIGURE 2-1

(a) Fuel Economy Standards for Autos

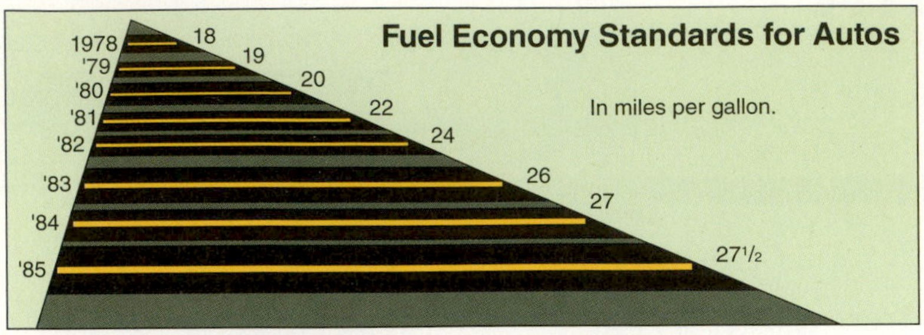

Set by Congress and supplemented by the Transportation Department.

Source: Redrawn from The Visual Display of Quantitative Information by Edward R. Tufte, p. 58. Copyright © 1983. Reprinted by permission of Graphics Press.

Note: Fuel-economy standards for passenger cars remained at 27.5 miles per gallon through 2010. Mileage standards are set to increase to 54.5 miles per gallon by 2025.

(b) Required Fuel Economy Standards: New Cars Built From 1978 To 1985

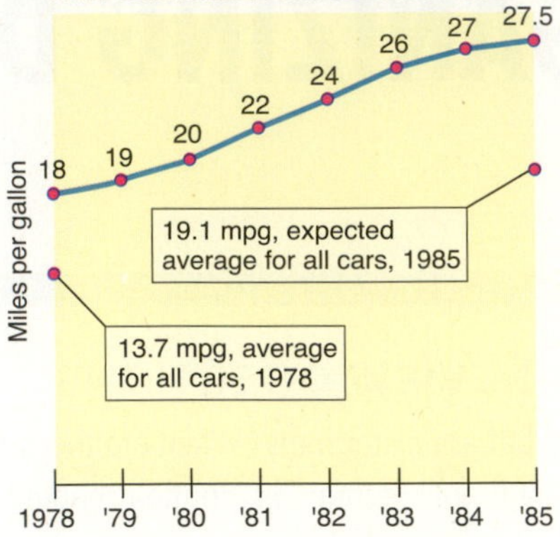

Source: Redrawn from *The Visual Display of Quantitative Information* by Edward R. Tufte, p. 57. Copyright © 1983. Reprinted by permission of Graphics Press.

SECTION **2.1**

Frequency Distributions, Histograms, and Related Topics

FOCUS POINTS

- Organize raw data using a frequency table.
- Construct histograms, relative-frequency histograms, and ogives.
- Recognize basic distribution shapes: uniform, symmetric, skewed, and bimodal.
- Interpret graphs in the context of the data setting.

FREQUENCY TABLES

When we have a large set of quantitative data, it's useful to organize it into smaller intervals or *classes* and count how many data values fall into each class. A frequency table does just that.

Frequency table

A **frequency table** partitions data into classes or intervals of equal width and shows how many data values are in each class. The classes or intervals are constructed so that each data value falls into exactly one class.

Constructing a frequency table involves a number of steps. Example 1 demonstrates the steps.

EXAMPLE 1

FREQUENCY TABLE

A task force to encourage car pooling did a study of one-way commuting distances of workers in the downtown Dallas area. A random sample of 60 of these workers was taken. The commuting distances of the workers in the sample are given in Table 2-1. Make a frequency table for these data.

SOLUTION:

(a) First decide how many classes you want. Five to 15 classes are usually used. If you use fewer than five classes, you risk losing too much information. If you use more than 15 classes, the data may not be sufficiently summarized. Let the

Art Directors & TRIP/Alamy

TABLE 2-1 One-Way Commuting Distances (in Miles) for 60 Workers in Downtown Dallas

13	47	10	3	16	20	17	40	4	2
7	25	8	21	19	15	3	17	14	6
12	45	1	8	4	16	11	18	23	12
6	2	14	13	7	15	46	12	9	18
34	13	41	28	36	17	24	27	29	9
14	26	10	24	37	31	8	16	12	16

spread of the data and the purpose of the frequency table be your guides when selecting the number of classes. In the case of the commuting data, let's use *six* classes.

Class width

(b) Next, find the *class width* for the six classes.

PROCEDURE

HOW TO FIND THE CLASS WIDTH (INTEGER DATA)

1. Compute $\dfrac{\text{Largest data value} - \text{smallest data value}}{\text{Desired number of classes}}$
2. Increase the computed value to the next highest whole number.

Note: To ensure that all the classes taken together cover the data, we need to increase the result of Step 1 to the *next whole number,* even if Step 1 produced a whole number. For instance, if the calculation in Step 1 produces the value 4, we make the class width 5.

To find the class width for the commuting data, we observe that the largest distance commuted is 47 miles and the smallest is 1 mile. Using six classes, the class width is 8, since

$$\text{Class width} = \frac{47 - 1}{6} \approx 7.7 \text{ (increase to 8)}$$

(c) Now we determine the data range for each class.

Class limits

The **lower class limit** is the lowest data value that can fit in a class. The **upper class limit** is the highest data value that can fit in a class. The **class width** is the difference between the *lower* class limit of one class and the *lower* class limit of the next class.

The smallest commuting distance in our sample is 1 mile. We use this *smallest* data value as the lower class limit of the *first* class. Since the class width is 8, we add 8 to 1 to find that the *lower* class limit for the *second* class is 9. Following this pattern, we establish *all* the *lower class limits*. Then we fill in the *upper class limits* so that the classes span the entire range of data. Table 2-2, on the next page, shows the upper and lower class limits for the commuting distance data.

(d) Now we are ready to tally the commuting distance data into the six classes and find the frequency for each class.

TABLE 2-2 Frequency Table of One-Way Commuting Distances for 60 Downtown Dallas Workers (Data in Miles)

Class Limits Lower–Upper	Class Boundaries Lower–Upper	Tally	Frequency	Class Midpoint
1–8	0.5–8.5	𝍸 𝍸 ||||	14	4.5
9–16	8.5–16.5	𝍸 𝍸 𝍸 𝍸 |	21	12.5
17–24	16.5–24.5	𝍸 𝍸 |	11	20.5
25–32	24.5–32.5	𝍸 |	6	28.5
33–40	32.5–40.5	||||	4	36.5
41–48	40.5–48.5	||||	4	44.5

PROCEDURE

HOW TO TALLY DATA

Tallying data is a method of counting data values that fall into a particular class or category.

To tally data into classes of a frequency table, examine each data value. Determine which class contains the data value and make a tally mark or vertical stroke (|) beside that class. For ease of counting, each fifth tally mark of a class is placed diagonally across the prior four marks (𝍸).

Class frequency

The *class frequency* for a class is the number of tally marks corresponding to that class.

Table 2-2 shows the tally and frequency of each class.

Class midpoint or class mark

(e) The center of each class is called the *midpoint* (or *class mark*). The midpoint is often used as a representative value of the entire class. The midpoint is found by adding the lower and upper class limits of one class and dividing by 2.

$$\text{Midpoint} = \frac{\text{Lower class limit} + \text{upper class limit}}{2}$$

Table 2-2 shows the class midpoints.

Class boundaries

(f) There is a space between the upper limit of one class and the lower limit of the next class. The halfway points of these intervals are called *class boundaries*. These are shown in Table 2-2.

Problems 21 and 22 show the procedure for finding the class width, class limits, and class boundaries for noninteger decimal data.

PROCEDURE

HOW TO FIND CLASS BOUNDARIES (INTEGER DATA)

To find **upper class boundaries,** add 0.5 unit to the upper class limits.

To find **lower class boundaries,** subtract 0.5 unit from the lower class limits.

Relative frequency

Basic frequency tables show how many data values fall into each class. It's also useful to know the *relative frequency* of a class. The relative frequency of a class is the proportion of all data values that fall into that class. To find the relative

TABLE 2-3 Relative Frequencies of One-Way Commuting Distances

Class	Frequency f	Relative Frequency f/n
1–8	14	$14/60 \approx 0.23$
9–16	21	$21/60 \approx 0.35$
17–24	11	$11/60 \approx 0.18$
25–32	6	$6/60 \approx 0.10$
33–40	4	$4/60 \approx 0.07$
41–48	4	$4/60 \approx 0.07$

frequency of a particular class, divide the class frequency f by the total of all frequencies n (sample size).

$$\text{Relative frequency} = \frac{f}{n} = \frac{\text{Class frequency}}{\text{Total of all frequencies}}$$

Table 2-3 shows the relative frequencies for the commuter data of Table 2-1. Since we already have the frequency table (Table 2-2), the relative-frequency table is obtained easily. The sample size is $n = 60$. Notice that the sample size is the total of all the frequencies. Therefore, the relative frequency for the first class (the class from 1 to 8) is

$$\text{Relative frequency} = \frac{f}{n} = \frac{14}{60} \approx 0.23$$

The symbol $\approx$ means "approximately equal to." We use the symbol because we rounded the relative frequency. Relative frequencies for the other classes are computed in a similar way.

The total of the relative frequencies should be 1. However, rounded results may make the total slightly higher or lower than 1.

Let's summarize the procedure for making a frequency table that includes relative frequencies.

LOOKING FORWARD

Sorting data into classes can be somewhat tedious. It is an easier task if the data are first ordered from smallest to largest. Stem-and-leaf diagrams, presented in Section 2.3, provide a convenient way to order data by hand.

PROCEDURE

HOW TO MAKE A FREQUENCY TABLE

1. Determine the number of classes and the corresponding class width.
2. Create the distinct classes. We use the convention that the *lower class limit* of the first class is the smallest data value. Add the class width to this number to get the *lower class limit* of the next class.
3. Fill in *upper class limits* to create distinct classes that accommodate all possible data values from the data set.
4. Tally the data into classes. Each data value should fall into exactly one class. Total the tallies to obtain each *class frequency.*
5. Compute the *midpoint* (class mark) for each class.
6. Determine the *class boundaries.*

Relative-frequency table

PROCEDURE

HOW TO MAKE A RELATIVE-FREQUENCY TABLE

First make a frequency table. Then, for each class, compute the *relative frequency, f/n,* where f is the class frequency and n is the total sample size.

HISTOGRAMS AND RELATIVE-FREQUENCY HISTOGRAMS

Histogram
Relative-frequency histogram

Histograms and *relative-frequency histograms* provide effective visual displays of data organized into frequency tables. In these graphs, we use bars to represent each class, where the width of the bar is the class width. For histograms, the height of the bar is the class frequency, whereas for relative-frequency histograms, the height of the bar is the relative frequency of that class.

Problems 23, 24, and 25 introduce dotplots, which are diagrams that show the frequency with which *each* data value occurs.

PROCEDURE

HOW TO MAKE A HISTOGRAM OR A RELATIVE-FREQUENCY HISTOGRAM

1. Make a frequency table (including relative frequencies) with the designated number of classes.
2. Place class boundaries on the horizontal axis and frequencies or relative frequencies on the vertical axis.
3. For each class of the frequency table, draw a bar whose width extends between corresponding class boundaries. For histograms, the height of each bar is the corresponding class frequency. For relative-frequency histograms, the height of each bar is the corresponding class relative frequency.

EXAMPLE 2

HISTOGRAM AND RELATIVE-FREQUENCY HISTOGRAM

Make a histogram and a relative-frequency histogram with six bars for the data in Table 2-1 showing one-way commuting distances.

SOLUTION: The first step is to make a frequency table and a relative-frequency table with six classes. We'll use Table 2-2 and Table 2-3. Figures 2-2 and 2-3 show the histogram and relative-frequency histogram. In both graphs, class boundaries are marked on the horizontal axis. For each class of the frequency table, make a corresponding bar with horizontal width extending from the lower boundary to the upper boundary of the respective class. For a histogram, the height of each bar is the corresponding class frequency. For a relative-frequency histogram, the height of

FIGURE 2-2

Histogram for Dallas Commuters: One-Way Commuting Distances

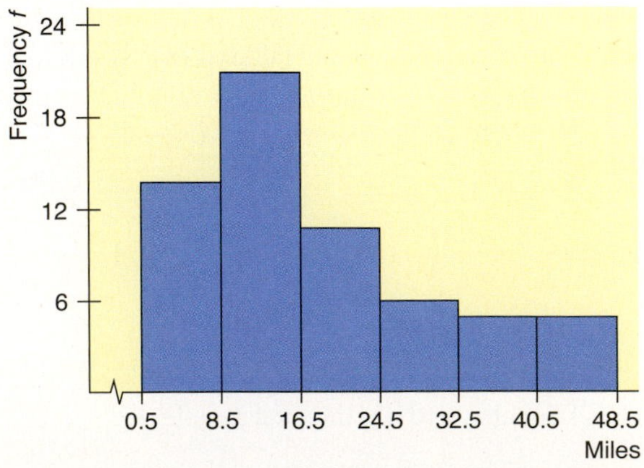

FIGURE 2-3

Relative-Frequency Histogram for Dallas Commuters: One-Way Commuting Distances

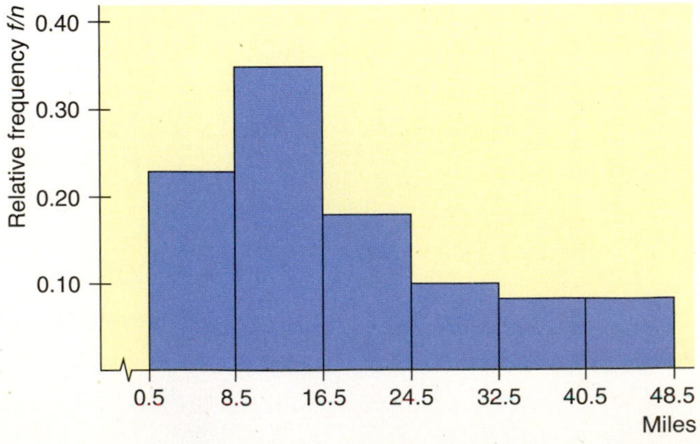

each bar is the corresponding relative frequency. Notice that the basic shapes of the graphs are the same. The only difference involves the vertical axis. The vertical axis of the histogram shows frequencies, whereas that of the relative-frequency histogram shows relative frequencies.

COMMENT The use of class boundaries in histograms assures us that the bars of the histogram touch and that no data fall on the boundaries. Both of these features are important. But a histogram displaying class boundaries may look awkward. For instance, the mileage range of 8.5 to 16.5 miles shown in Figure 2-2 isn't as natural a choice as a mileage range of 8 to 16 miles. For this reason, many magazines and newspapers do not use class boundaries as labels on a histogram. Instead, some use lower class limits as labels, with the convention that *a data value falling on the class limit is included in the next higher class (class to the right of the limit)*. Another convention is to label midpoints instead of class boundaries. Determine the default convention being used before creating frequency tables and histograms on a computer.

GUIDED EXERCISE 1 — HISTOGRAM AND RELATIVE-FREQUENCY HISTOGRAM

An irate customer called Dollar Day Mail Order Company 40 times during the last two weeks to see why his order had not arrived. Each time he called, he recorded the length of time he was put "on hold" before being allowed to talk to a customer service representative. See Table 2-4.

TABLE 2-4 **Length of Time on Hold, in Minutes**

1	5	5	6	7	4	8	7	6	5
5	6	7	6	6	5	8	9	9	10
7	8	11	2	4	6	5	12	13	6
3	7	8	8	9	9	10	9	8	9

(a) What are the largest and smallest values in Table 2-4? If we want five classes in a frequency table, what should the class width be?

→ The largest value is 13; the smallest value is 1. The class width is

$$\frac{13-1}{5} = 2.4 \approx 3$$

Note: Increase the value to 3.

(b) Complete the following frequency table.

TABLE 2-5 **Time on Hold**

Class Limits Lower-Upper	Tally	Frequency	Midpoint
1–3	____	____	____
4–____	____	____	____
____–9	____	____	____
____–____	____	____	____
____–____	____	____	____

→ TABLE 2-6 **Completion of Table 2-5**

Class Limits Lower-Upper	Tally	Frequency	Midpoint
1–3	|||	3	2
4–6	𝍸 𝍸 𝍸	15	5
7–9	𝍸 𝍸 𝍸 ||	17	8
10–12	||||	4	11
13–15			1	14

(c) Recall that the class boundary is halfway between the upper limit of one class and the lower limit of the next. Use this fact to find the class boundaries in Table 2-7 and to complete the partial histogram in Figure 2-4.

Continued

GUIDED EXERCISE 1 *continued*

TABLE 2-7 **Class Boundaries**

Class Limits	Class Boundaries
1–3	0.5–3.5
4–6	3.5–6.5
7–9	6.5–____
10–12	____–____
13–15	____–____

TABLE 2-8 **Completion of Table 2-7**

Class Limits	Class Boundaries
1–3	0.5–3.5
4–6	3.5–6.5
7–9	6.5–9.5
10–12	9.5–12.5
13–15	12.5–15.5

FIGURE 2-4

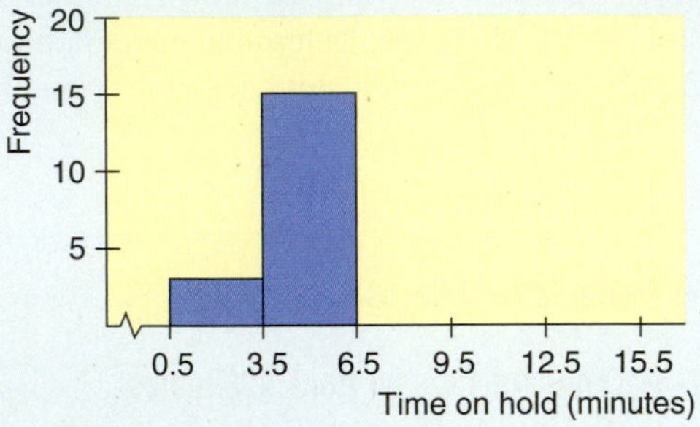

FIGURE 2-5 Completion of Figure 2-4

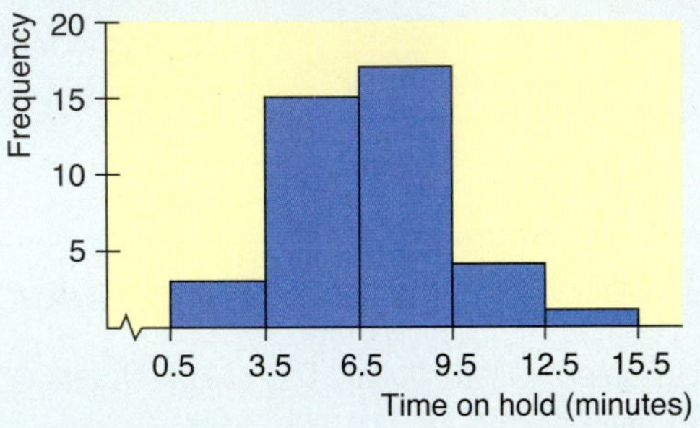

(d) Compute the relative class frequency *f/n* for each class in Table 2-9 and complete the partial relative-frequency histogram in Figure 2-6.

TABLE 2-9 **Relative Class Frequency**

Class	*f/n*
1–3	3/40 = 0.075
4–6	15/40 = 0.375
7–9	____
10–12	____
13–15	____

TABLE 2-10 **Completion of Table 2-9**

Class	*f/n*
1–3	0.075
4–6	0.375
7–9	0.425
10–12	0.100
13–15	0.025

FIGURE 2-6

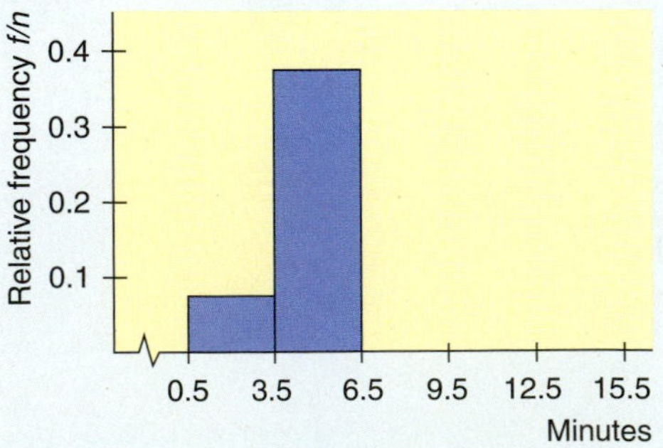

FIGURE 2-7 Completion of Figure 2-6

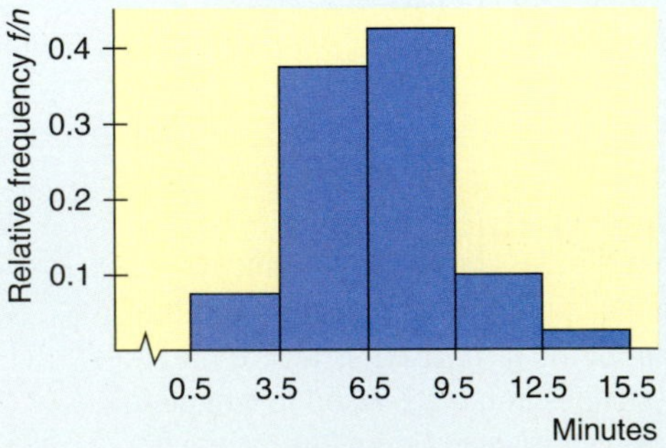

FIGURE 2-8

Types of Histograms

(a) Typical mound-shaped symmetrical histogram
(b) Typical uniform or rectangular histogram
(c) Typical skewed histogram
Skewed left

Skewed right

(d) Typical bimodal histogram
LOOKING FORWARD

We will see relative-frequency distributions again when we study probability in Chapter 4. There we will see that if a random sample is large enough, then we can estimate the probability of an event by the relative frequency of the event. The relative-frequency distribution then can be interpreted as a *probability distribution.* Such distributions will form the basis of our work in inferential statistics.

Mound-shaped symmetrical distribution

Uniform distribution

Skewed left

Skewed right

Bimodal distribution

LOOKING FORWARD

Normal and Student's *t* distributions are two important probability distributions used extensively in inferential statistics (Chapters 6, 7, 8). Both of these distributions are mound-shaped symmetrical. Two other useful probability distributions are the Chi-square and *F* distributions (Chapter 10), which are skewed right.

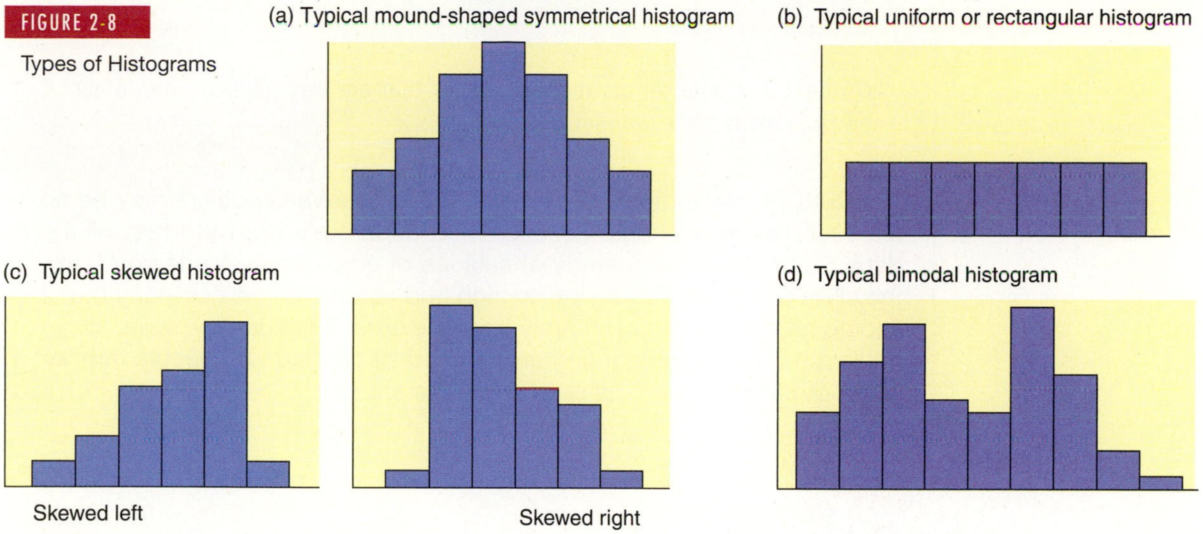

DISTRIBUTION SHAPES

Histograms are valuable and useful tools. If the raw data came from a random sample of population values, the histogram constructed from the sample values should have a distribution shape that is reasonably similar to that of the population.

Several terms are commonly used to describe histograms and their associated population distributions.

(a) ***Mound-shaped symmetrical:*** This term refers to a histogram in which both sides are (more or less) the same when the graph is folded vertically down the middle. Figure 2-8(a) shows a typical mound-shaped symmetrical histogram.

(b) ***Uniform or rectangular:*** These terms refer to a histogram in which every class has equal frequency. From one point of view, a uniform distribution is symmetrical with the added property that the bars are of the same height. Figure 2-8(b) illustrates a typical histogram with a uniform shape.

(c) ***Skewed left or skewed right:*** These terms refer to a histogram in which one tail is stretched out longer than the other. The direction of skewness is on the side of the *longer* tail. So, if the longer tail is on the left, we say the histogram is skewed to the left. Figure 2-8(c) shows a typical histogram skewed to the left and another skewed to the right.

(d) ***Bimodal:*** This term refers to a histogram in which the two classes with the largest frequencies are separated by at least one class. The top two frequencies of these classes may have slightly different values. This type of situation sometimes indicates that we are sampling from two different populations. Figure 2-8(d) illustrates a typical histogram with a bimodal shape.

CRITICAL THINKING

A bimodal distribution shape might indicate that the data are from two different populations. For instance, a histogram showing the heights of a random sample of adults is likely to be bimodal because two populations, male and female, were combined.

If there are gaps in the histogram between bars at either end of the graph, the data set might include *outliers*.

Continued

CRITICAL THINKING *continued*

Outliers

Outliers in a data set are the data values that are very different from other measurements in the data set.

Outliers may indicate data-recording errors. Valid outliers may be so unusual that they should be examined separately from the rest of the data. For instance, in a study of salaries of employees at one company, the chief CEO's salary may be so high and unique for the company that it should be considered separately from the other salaries. Decisions about outliers that are not recording errors need to be made by people familiar with both the field and the purpose of the study.

WHAT DO HISTOGRAMS AND RELATIVE FREQUENCY HISTOGRAMS TELL US?

Histograms and relative frequency histograms show us how the data are distributed. By looking at such graphs, we can tell

- if the data distribution is more symmetric, skewed, or bimodal;
- if there are possible outliers;
- which data intervals contain the most data;
- how spread out the data are.

In the next chapter we will look at measures of center and spread of data. Histograms help us visualize such measures.

TECH NOTES The TI-84Plus/TI-83Plus/TI-*n*spire calculators, Excel 2010, and Minitab all create histograms. However, each technology automatically selects the number of classes to use. In Using Technology at the end of this chapter, you will see instructions for specifying the number of classes yourself and for generating histograms such as those we create "by hand."

CUMULATIVE-FREQUENCY TABLES AND OGIVES

Sometimes we want to study cumulative totals instead of frequencies. Cumulative frequencies tell us how many data values are smaller than an upper class boundary. Once we have a frequency table, it is a fairly straightforward matter to add a column of cumulative frequencies.

Cumulative frequency

The **cumulative frequency** for a class is the sum of the frequencies for *that class* and *all previous classes.*

Ogive

An *ogive* (pronounced "oh-jīve") is a graph that displays cumulative frequencies.

PROCEDURE

HOW TO MAKE AN OGIVE

1. Make a frequency table showing class boundaries and cumulative frequencies.
2. For each class, make a dot over the *upper class boundary* at the height of the cumulative class frequency. The coordinates of the dots are (upper class boundary, cumulative class frequency). Connect these dots with line segments.
3. By convention, an ogive begins on the horizontal axis at the lower class boundary of the first class.

EXAMPLE 3

CUMULATIVE-FREQUENCY TABLE AND OGIVE

Aspen, Colorado, is a world-famous ski area. If the daily high temperature is above 40°F, the surface of the snow tends to melt. It then freezes again at night. This can result in a snow crust that is icy. It also can increase avalanche danger.

Table 2-11 gives a summary of daily high temperatures (°F) in Aspen during the 151-day ski season.

Ph. Royer/Photo Researchers, Inc.

TABLE 2-11 **High Temperatures During the Aspen Ski Season (°F)**

Class Boundaries			
Lower	Upper	Frequency	Cumulative Frequency
10.5	20.5	23	23
20.5	30.5	43	66 (sum 23 + 43)
30.5	40.5	51	117 (sum 66 + 51)
40.5	50.5	27	144 (sum 117 + 27)
50.5	60.5	7	151 (sum 144 + 7)

(a) The cumulative frequency for a class is computed by adding the frequency of that class to the frequencies of previous classes. Table 2-11 shows the cumulative frequencies.

(b) To draw the corresponding ogive, we place a dot at cumulative frequency 0 on the lower class boundary of the first class. Then we place dots over the *upper class boundaries* at the height of the cumulative class frequency for the corresponding class. Finally, we connect the dots. Figure 2-9 shows the corresponding ogive.

(c) Looking at the ogive, estimate the total number of days with a high temperature lower than or equal to 40°F.

FIGURE 2-9

Ogive for Daily High Temperatures (°F) During Aspen Ski Season

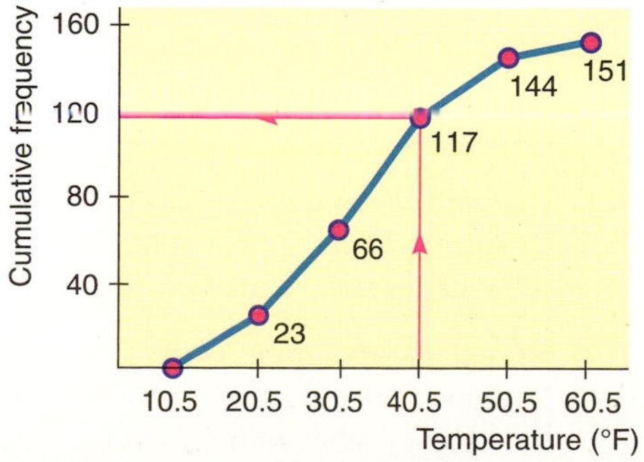

SOLUTION: Following the red lines on the ogive in Figure 2-9, we see that 117 days have had high temperatures of no more than 40°F.

WHAT DOES AN OGIVE TELL US?

An ogive (also known as a cumulative-frequency diagram) tells us

- how many data are less than the indicated value on the horizontal axis;
- how slowly or rapidly the data values accumulate over the range of the data.

In addition, the vertical scale can be changed to cumulative percentages by dividing the cumulative frequencies by the total number of data. Then we can tell what percentage of data are below values specified on the horizontal axis.

VIEWPOINT Mush, You Huskies!

In 1925, the village of Nome, Alaska, had a terrible diphtheria epidemic. Serum was available in Anchorage but had to be brought to Nome by dogsled over the 1161-mile Iditarod Trail. Since 1973, the Iditarod Dog Sled Race from Anchorage to Nome has been an annual sporting event, with a current purse of more than $600,000. Winning times range from more than 20 days to a little over 9 days.

To collect data on winning times, visit the Brase/Brase statistics site at ***http://www.cengage.com/statistics/brase*** *and find the link to the Iditarod. Make a frequency distribution for these times.*

SECTION 2.1 PROBLEMS

1. *Statistical Literacy* What is the difference between a class boundary and a class limit?

2. *Statistical Literacy* A data set has values ranging from a low of 10 to a high of 52. What's wrong with using the class limits 10–19, 20–29, 30–39, 40–49 for a frequency table?

3. *Statistical Literacy* A data set has values ranging from a low of 10 to a high of 50. What's wrong with using the class limits 10–20, 20–30, 30–40, 40–50 for a frequency table?

4. *Statistical Literacy* A data set has values ranging from a low of 10 to a high of 50. The class width is to be 10. What's wrong with using the class limits 10–20, 21–31, 32–42, 43–53 for a frequency table with a class width of 10?

5. *Basic Computation: Class Limits* A data set with whole numbers has a low value of 20 and a high value of 82. Find the class width and class limits for a frequency table with 7 classes.

6. *Basic Computation: Class Limits* A data set with whole numbers has a low value of 10 and a high value of 120. Find the class width and class limits for a frequency table with 5 classes.

7. *Interpretation* You are manager of a specialty coffee shop and collect data throughout a full day regarding waiting time for customers from the time they enter the shop until the time they pick up their order.
   (a) What type of distribution do you think would be most desirable for the waiting times: skewed right, skewed left, mound-shaped symmetrical? Explain.
   (b) What if the distribution for waiting times were bimodal? What might be some explanations?

8. *Critical Thinking* A web site rated 100 colleges and ranked the colleges from 1 to 100, with a rank of 1 being the best. Each college was ranked, and there were no ties. If the ranks were displayed in a histogram, what would be the shape of the histogram: skewed, uniform, mound-shaped?

9. *Critical Thinking* Look at the histogram in Figure 2-10(a), which shows mileage, in miles per gallon (mpg), for a random selection of passenger cars (Reference: *Consumer Reports*).
   (a) Is the shape of the histogram essentially bimodal?
   (b) Jose looked at the raw data and discovered that the 54 data values included both the city and the highway mileages for 27 cars. He used the city mileages for the 27 cars to make the histogram in Figure 2-10(b). Using this information and Figure 2-10, parts (a) and (b), construct a histogram for the highway mileages of the same cars. Use class boundaries 16.5, 20.5, 24.5, 28.5, 32.5, 36.5, and 40.5.

FIGURE 2-10

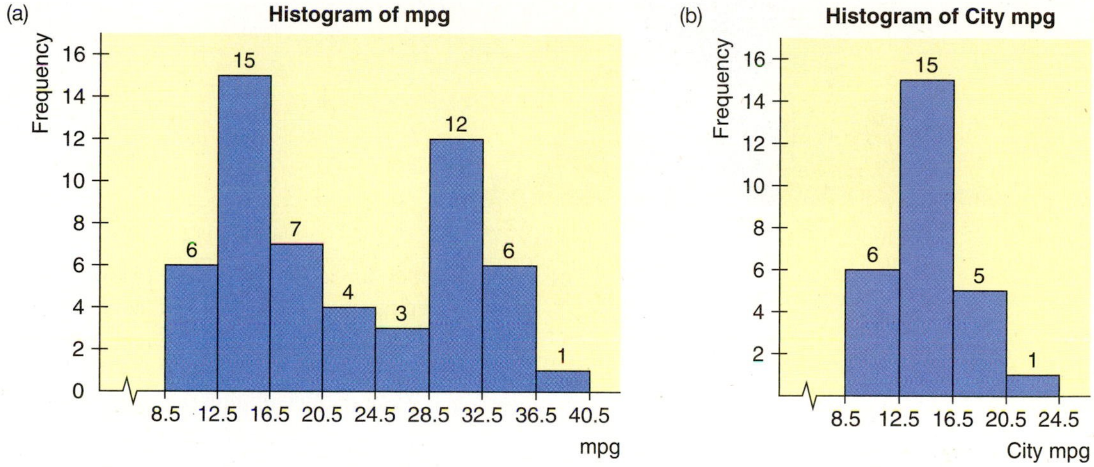

10. *Critical Thinking* The following data represent salaries, in thousands of dollars, for employees of a small company. Notice that the data have been sorted in increasing order.

54	55	55	57	57	59	60	65	65	65	66	68	68
69	69	70	70	70	75	75	75	75	77	82	82	82
88	89	89	91	91	97	98	98	98	280			

   (a) Make a histogram using the class boundaries 53.5, 99.5, 145.5, 191.5, 237.5, 283.5.
   (b) Look at the last data value. Does it appear to be an outlier? Could this be the owner's salary?
   (c) Eliminate the high salary of 280 thousand dollars. Make a new histogram using the class boundaries 53.5, 62.5, 71.5, 80.5, 89.5, 98.5. Does this histogram reflect the salary distribution of most of the employees better than the histogram in part (a)?

11. *Interpretation* Histograms of random sample data are often used as an indication of the shape of the underlying population distribution. The histograms on the next page are based on random samples of size 30, 50, and 100 from the same population.

(i) Sample of size 30

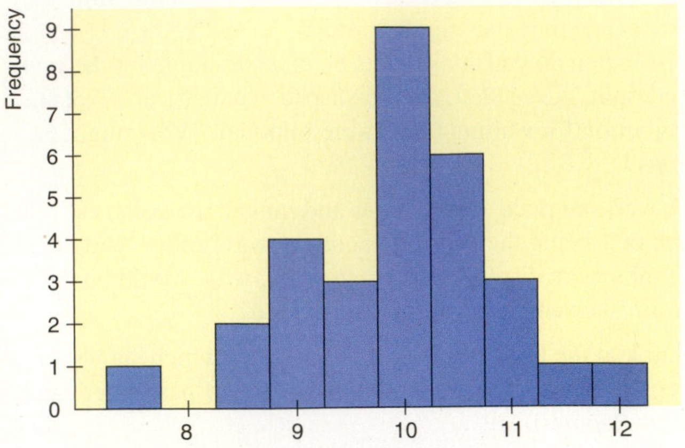

(ii) Sample of size 50

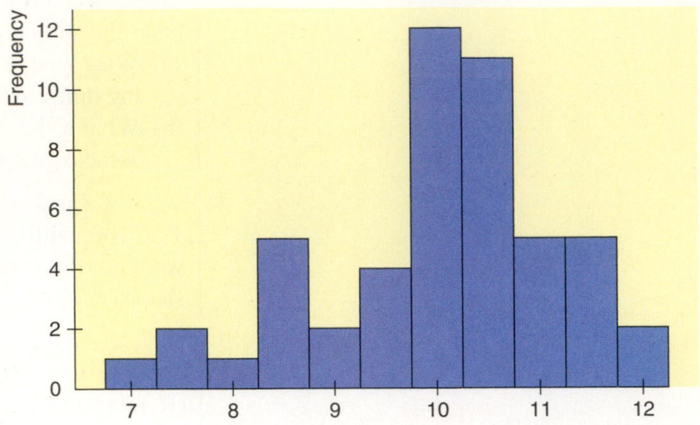

(iii) Sample of size 100

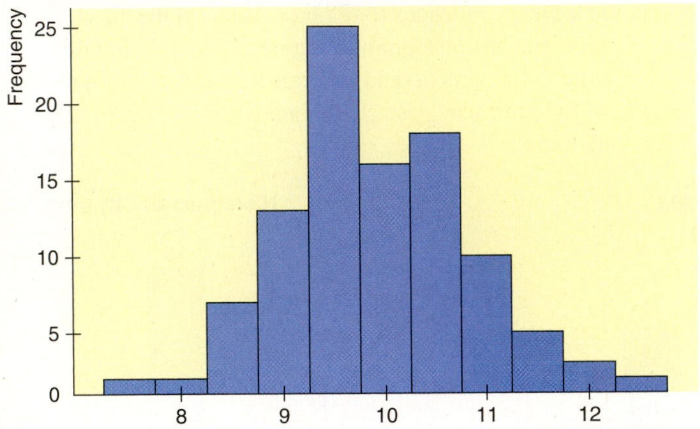

(a) Using the midpoint labels of the three histograms, what would you say about the estimated range of the population data from smallest to largest? Does the bulk of the data seem to be between 8 and 12 in all three histograms?

(b) The population distribution from which the samples were drawn is symmetric and mound-shaped, with the top of the mound at 10, 95% of the data between 8 and 12, and 99.7% of data between 7 and 13. How well does each histogram reflect these characteristics?

12. *Interpretation* The following histograms are based on different random samples of size 100 drawn from the same population.

(i)

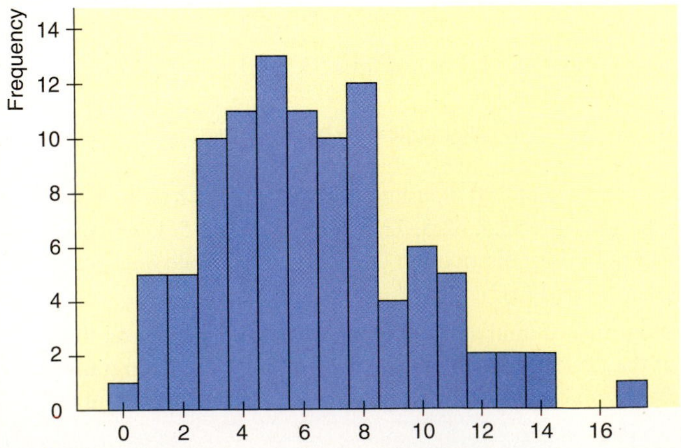

(ii)

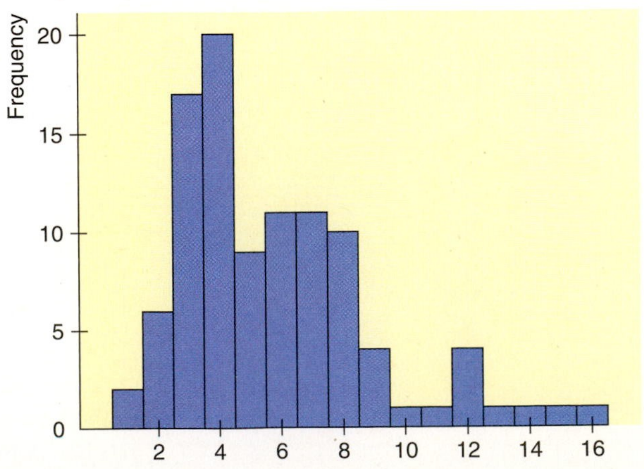

(iii)

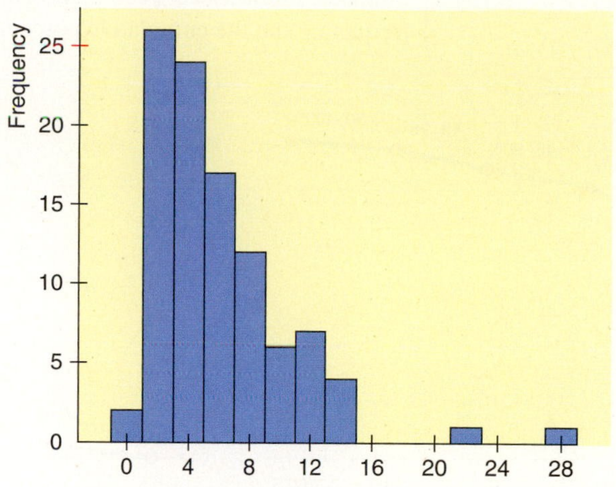

(a) Identify the midpoint of the class with the highest frequency in each of the three histograms.
(b) Using the class midpoints, what is the range of data shown in each histogram?
(c) Which of the histograms are more clearly skewed right?
(d) Based on your study of random sample in Chapter 1, is it surprising to see the variations in the samples as displayed in the histograms? The original population from which the samples were drawn is skewed right with a high frequency near 4. Do all three random samples seem to reflect these properties equally well?

13. *Interpretation* The ogives shown are based on U.S. Census data and show the average personal income per capita for each of the 50 states. The data are rounded to the nearest thousand dollars.

(i) Ogive

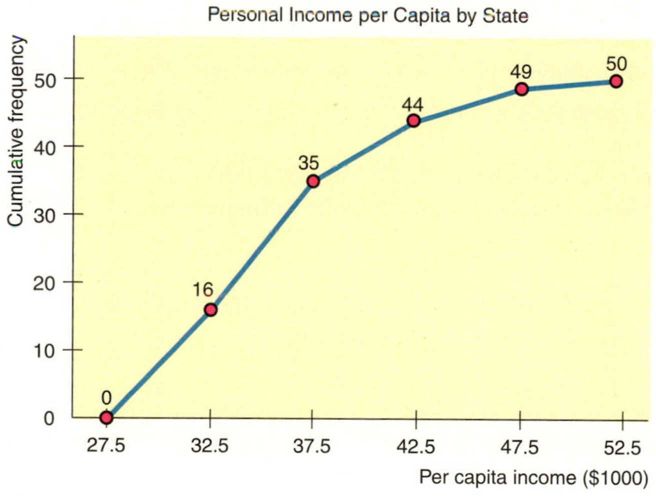

(ii) Ogive Showing Cumulative Percentage of Data

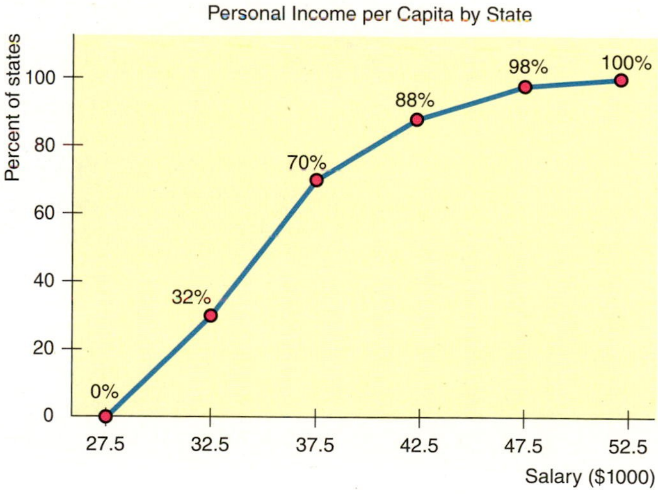

(a) How were the percentages shown in graph (ii) computed?
(b) How many states have average per capita income less than 37.5 thousand dollars?
(c) How many states have average per capita income between 42.5 and 52.5 thousand dollars?
(d) What percentage of the states have average per capita income more than 47.5 thousand dollars?

14. *Critical Thinking* The following ogives come from different distributions of 50 whole numbers between 1 and 60. Labels on each point give the cumulative frequency and the cumulative percentage of data.

(i)

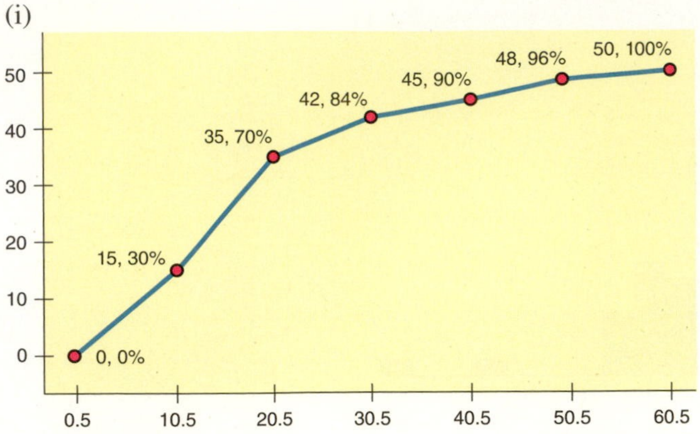

(ii)

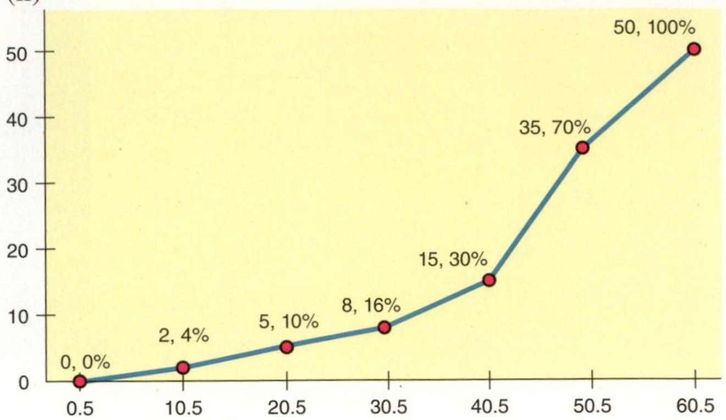

(iii)

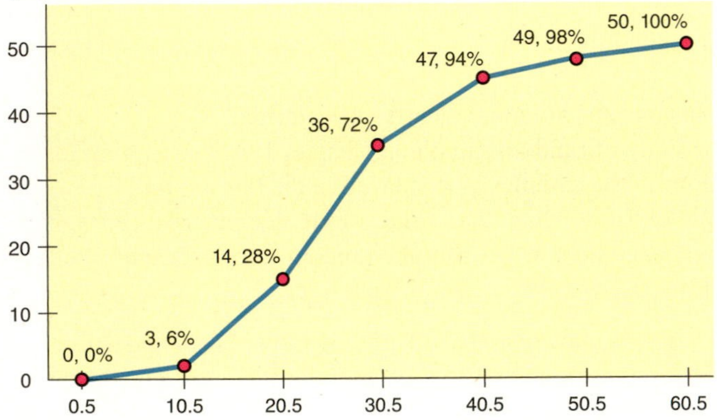

(a) In which distribution does the most data fall below 20.5?
(b) In which distribution does the most data fall below 40.5?
(c) In which distribution does the amount of data below 20.5 most closely match that above 30.5?
(d) Which distribution seems to be skewed right? Skewed left? Mound-shaped?

For Problems 15–20, use the specified number of classes to do the following.
(a) Find the class width.
(b) Make a frequency table showing class limits, class boundaries, midpoints, frequencies, relative frequencies, and cumulative frequencies.
(c) Draw a histogram.
(d) Draw a relative-frequency histogram.
(e) Categorize the basic distribution shape as uniform, mound-shaped symmetrical, bimodal, skewed left, or skewed right.
(f) Draw an ogive.

15. *Sports: Dog Sled Racing* How long does it take to finish the 1161-mile Iditarod Dog Sled Race from Anchorage to Nome, Alaska (*see* Viewpoint)? Finish times (to the nearest hour) for 57 dogsled teams are shown below.

261	271	236	244	279	296	284	299	288	288	247	256
338	360	341	333	261	266	287	296	313	311	307	307
299	303	277	283	304	305	288	290	288	289	297	299
332	330	309	328	307	328	285	291	295	298	306	315
310	318	318	320	333	321	323	324	327			

Use five classes.

16. *Medical: Glucose Testing* The following data represent glucose blood levels (mg/100 ml) after a 12-hour fast for a random sample of 70 women (Reference: *American Journal of Clinical Nutrition,* Vol. 19, pp. 345–351). *Note:* These data are also available for download at the Online Study Center.

45	66	83	71	76	64	59	59
76	82	80	81	85	77	82	90
87	72	79	69	83	71	87	69
81	76	96	83	67	94	101	94
89	94	73	99	93	85	83	80
78	80	85	83	84	74	81	70
65	89	70	80	84	77	65	46
80	70	75	45	101	71	109	73
73	80	72	81	63	74		

Use six classes.

17. *Medical: Tumor Recurrence* Certain kinds of tumors tend to recur. The following data represent the lengths of time, in months, for a tumor to recur after chemotherapy (Reference: D. P. Byar, *Journal of Urology,* Vol. 10, pp. 556–561). *Note:* These data are also available for download at the Online Study Center.

19	18	17	1	21	22	54	46	25	49
50	1	59	39	43	39	5	9	38	18
14	45	54	59	46	50	29	12	19	36
38	40	43	41	10	50	41	25	19	39
27	20								

Use five classes.

18. *Archaeology: New Mexico* The Wind Mountain excavation site in New Mexico is an important archaeological location of the ancient Native American Anasazi culture. The following data represent depths (in cm) below surface grade at which significant artifacts were discovered at this site (Reference: A. I. Woosley and A. J. McIntyre, *Mimbres Mogollon Archaeology,* University of New Mexico Press). *Note:* These data are also available for download at the Online Study Center.

85	45	75	60	90	90	115	30	55	58
78	120	80	65	65	140	65	50	30	125
75	137	80	120	15	45	70	65	50	45
95	70	70	28	40	125	105	75	80	70
90	68	73	75	55	70	95	65	200	75
15	90	46	33	100	65	60	55	85	50
10	68	99	145	45	75	45	95	85	65
65	52	82							

Use seven classes.

19. *Environment: Gasoline Consumption* The following data represent highway fuel consumption in miles per gallon (mpg) for a random sample of 55 models of passenger cars (Source: Environmental Protection Agency). *Note:* These data are also available for download at the Online Study Center.

30	27	22	25	24	25	24	15
35	35	33	52	49	10	27	18
20	23	24	25	30	24	24	24
18	20	25	27	24	32	29	27
24	27	26	25	24	28	33	30
13	13	21	28	37	35	32	33
29	31	28	28	25	29	31	

Use five classes.

20. *Advertising: Readability* "Readability Levels of Magazine Ads," by F. K. Shuptrine and D. D. McVicker, is an article in the *Journal of Advertising Research.* (For more information, visit the Brase/Brase statistics site at **http://www.cengage.com/statistics/brase11e** and find the link to DASL, the Carnegie Mellon University Data and Story Library. Look in Data Subjects under Consumer and then Magazine Ads Readability file.) The following is a list of the number of three-syllable (or longer) words in advertising copy of randomly selected magazine advertisements.

34	21	37	31	10	24	39	10	17	18	32
17	3	10	6	5	6	6	13	22	25	3
5	2	9	3	0	4	29	26	5	5	24
15	3	8	16	9	10	3	12	10	10	10
11	12	13	1	9	43	13	14	32	24	15

Use eight classes.

21. *Expand Your Knowledge: Decimal Data* The following data represent tonnes of wheat harvested each year (1894–1925) from Plot 19 at the Rothamsted Agricultural Experiment Stations, England.

2.71	1.62	2.60	1.64	2.20	2.02	1.67	1.99	2.34	1.26	1.31
1.80	2.82	2.15	2.07	1.62	1.47	2.19	0.59	1.48	0.77	2.04
1.32	0.89	1.35	0.95	0.94	1.39	1.19	1.18	0.46	0.70	

(a) Multiply each data value by 100 to "clear" the decimals.
(b) Use the standard procedures of this section to make a frequency table and histogram with your whole-number data. Use six classes.
(c) Divide class limits, class boundaries, and class midpoints by 100 to get back to your original data values.

22. *Decimal Data: Batting Averages* The following data represent baseball batting averages for a random sample of National League players near the end of the baseball season. The data are from the baseball statistics section of the *Denver Post.*

0.194	0.258	0.190	0.291	0.158	0.295	0.261	0.250	0.181
0.125	0.107	0.260	0.309	0.309	0.276	0.287	0.317	0.252
0.215	0.250	0.246	0.260	0.265	0.182	0.113	0.200	

(a) Multiply each data value by 1000 to "clear" the decimals.

(b) Use the standard procedures of this section to make a frequency table and histogram with your whole-number data. Use five classes.
(c) Divide class limits, class boundaries, and class midpoints by 1000 to get back to your original data.

23. *Expand Your Knowledge: Dotplot* Another display technique that is somewhat similar to a histogram is a *dotplot*. In a dotplot, the data values are displayed along the horizontal axis. A dot is then plotted over each data value in the data set.

PROCEDURE

HOW TO MAKE A DOTPLOT

Display the data along a horizontal axis. Then plot each data value with a dot or point above the corresponding value on the horizontal axis. For repeated data values, stack the dots.

Dotplot

The next display shows a *dotplot* generated by Minitab (**➤Graph ➤Dotplot**) for the number of licensed drivers per 1000 residents by state, including the District of Columbia (Source: U.S. Department of Transportation).

Dotplot for Licensed Drivers per 1000 Residents

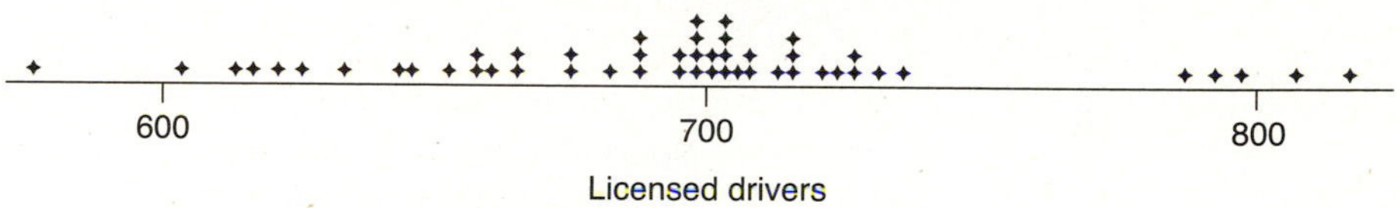

(a) From the dotplot, how many states have 600 or fewer licensed drivers per 1000 residents?
(b) About what percentage of the states (out of 51) seem to have close to 800 licensed drivers per 1000 residents?
(c) Consider the intervals 550 to 650, 650 to 750, and 750 to 850 licensed drivers per 1000 residents. In which interval do most of the states fall?

24. *Dotplot: Dog Sled Racing* Make a dotplot for the data in Problem 15 regarding the finish time (number of hours) for the Iditarod Dog Sled Race. Compare the dotplot to the histogram of Problem 15.

25. *Dotplot: Tumor Recurrence* Make a dotplot for the data in Problem 17 regarding the recurrence of tumors after chemotherapy. Compare the dotplot to the histogram of Problem 17.

SECTION **2.2**

Bar Graphs, Circle Graphs, and Time-Series Graphs

FOCUS POINTS

- Determine types of graphs appropriate for specific data.
- Construct bar graphs, Pareto charts, circle graphs, and time-series graphs.
- Interpret information displayed in graphs.

Histograms provide a useful visual display of the distribution of data. However, the data must be quantitative. In this section, we examine other types of graphs, some of which are suitable for qualitative or category data as well.

Bar graph

Let's start with *bar graphs*. These are graphs that can be used to display quantitative or qualitative data.

FEATURES OF A BAR GRAPH

1. Bars can be vertical or horizontal.
2. Bars are of uniform width and uniformly spaced.
3. The lengths of the bars represent values of the variable being displayed, the frequency of occurrence, or the percentage of occurrence. The same measurement scale is used for the length of each bar.
4. The graph is well annotated with title, labels for each bar, and vertical scale or actual value for the length of each bar.

EXAMPLE 4

BAR GRAPH

Figure 2-11 shows two bar graphs depicting the life expectancies for men and women born in the designated year. Let's analyze the features of these graphs.

Cluster bar graph

SOLUTION: The graphs are called *cluster bar graphs* because there are two bars for each year of birth. One bar represents the life expectancy for men, and the other represents the life expectancy for women. The height of each bar represents the life expectancy (in years).

Changing scale

An important feature illustrated in Figure 2-11(b) is that of a *changing scale.* Notice that the scale between 0 and 65 is compressed. The changing scale amplifies the apparent difference between life spans for men and women, as well as the increase in life spans from those born in 1980 to the projected span of those born in 2010.

CHANGING SCALE

Whenever you use a change in scale in a graphic, warn the viewer by using a squiggle -⁄\/- on the changed axis. Sometimes, if a single bar is unusually long, the bar length is compressed with a squiggle in the bar itself.

FIGURE 2-11

Life Expectancy

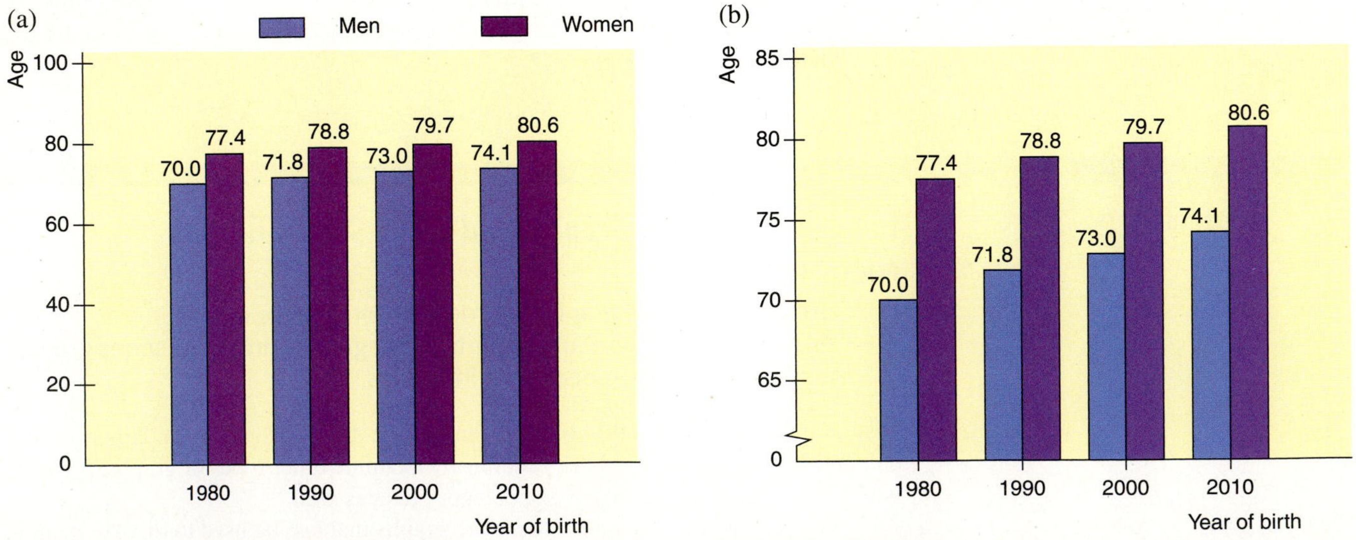

Source: U.S. Census Bureau

LOOKING FORWARD

Cluster bar graphs displaying percentages are particularly useful visuals for showing how different populations fit different categories. For instance, in Example 2 in Chapter 10, we see a cluster graph showing preferences of males and females for dogs, cats, other pets, or no pets. This graph displays the sample data for an inferential test called a *test of homogeneity*.

Quality control is an important aspect of today's production and service industries. Dr. W. Edwards Deming was one of the developers of total quality management (TQM). In his book *Out of Crisis*, he outlines many strategies for monitoring and improving service and production industries. In particular, Dr. Deming recommends the use of some statistical methods to organize and analyze data from industries so that sources of problems can be identified and then corrected. *Pareto* (pronounced "Pah-rāy-tō") *charts* are among the many techniques used in quality-control programs.

Pareto charts

A **Pareto chart** is a bar graph in which the bar height represents frequency of an event. In addition, the bars are arranged from left to right according to decreasing height.

GUIDED EXERCISE 2 | PARETO CHARTS

This exercise is adapted from *The Deming Management Method* by Mary Walton. Suppose you want to arrive at college 15 minutes before your first class so that you can feel relaxed when you walk into class. An early arrival time also allows room for unexpected delays. However, you always find yourself arriving "just in time" or slightly late. What causes you to be late? Charlotte made a list of possible causes and then kept a checklist for 2 months (Table 2-12). On some days more than one item was checked because several events occurred that caused her to be late.

TABLE 2-12 **Causes for Lateness (September–October)**

Cause	Frequency
Snoozing after alarm goes off	15
Car trouble	5
Too long over breakfast	13
Last-minute studying	20
Finding something to wear	8
Talking too long with roommate	9
Other	3

(a) Make a Pareto chart showing the causes for lateness. Be sure to label the causes, and draw the bars using the same vertical scale.

FIGURE 2-12 Pareto Chart: Conditions That Might Cause Lateness.

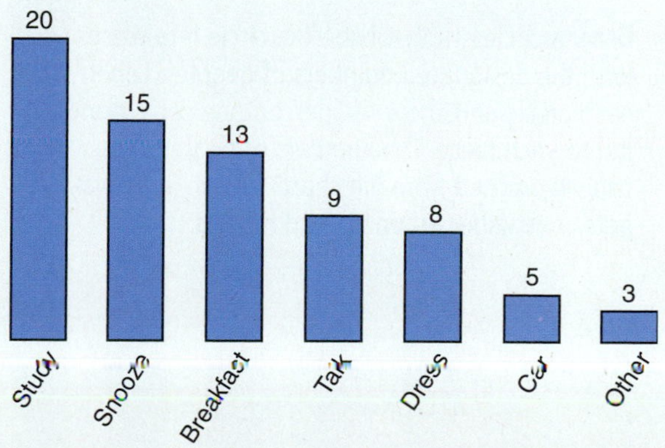

(b) Looking at the Pareto chart, what recommendations do you have for Charlotte?

According to the chart, rearranging study time, or getting up earlier to allow for studying, would cure her most frequent cause for lateness. Repairing the car might be important, but for getting to campus early, it would not be as effective as adjusting study time.

Circle graphs or pie charts

Donut pie charts are a popular variation of pie charts. Problem 15 of this section discusses donut pie charts.

Another popular pictorial representation of data is the *circle graph* or *pie chart.* It is relatively safe from misinterpretation and is especially useful for showing the division of a total quantity into its component parts. The total quantity, or 100%, is represented by the entire circle. Each wedge of the circle represents a component part of the total. These proportional segments are usually labeled with corresponding percentages of the total. Guided Exercise 3 shows how to make a circle graph.

In a **circle graph** or **pie chart,** wedges of a circle visually display proportional parts of the total population that share a common characteristic.

GUIDED EXERCISE 3 | CIRCLE GRAPH

How long do we spend talking on the telephone after hours (at home after 5 P.M.)? The results from a recent survey of 500 people (as reported in *USA Today*) are shown in Table 2-13. We'll make a circle graph to display these data.

TABLE 2-13 **Time Spent on Home Telephone After 5 P.M.**

Time	Number	Fractional Part	Percentage	Number of Degrees
Less than ½ hour	296	296/500	59.2	$59.2\% \times 360° \approx 213°$
½ hour to 1 hour	83	83/500	16.6	$16.6\% \times 360° \approx 60°$
More than 1 hour	121	_____	_____	_____
Total	_____		_____	_____

(a) Fill in the missing parts in Table 2-13 for "More than 1 hour." Remember that the central angle of a circle is 360°. Round to the nearest degree.

For "More than 1 hour," Fractional Part = 121/500; Percentage = 24.2%; Number of Degrees = $24.2\% \times 360° \approx 87°$. The symbol ≈ means "approximately equal."

(b) Fill in the totals. What is the total number of responses? Do the percentages total 100% (within rounding error)? Do the numbers of degrees total 360° (within rounding error)?

The total number of responses is 500. The percentages total 100%. You must have such a total in order to create a circle graph. The numbers of degrees total 360°.

(c) Draw a circle graph. Divide the circle into pieces with the designated numbers of degrees. Label each piece, and show the percentage corresponding to each piece. The numbers of degrees are usually omitted from pie charts shown in newspapers, magazines, journals, and reports.

FIGURE 2-13 Hours on Home Telephone After 5 P.M.

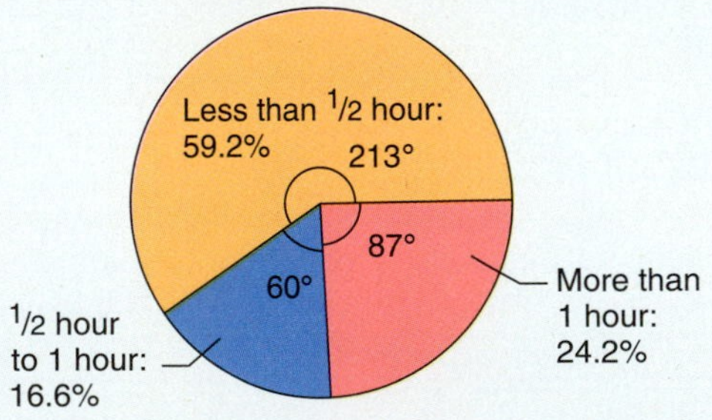

Suppose you begin an exercise program that involves walking or jogging for 30 minutes. You exercise several times a week but monitor yourself by logging the distance you cover in 30 minutes each Saturday. How do you display these data in a meaningful way? Making a bar chart showing the frequency of distances you cover might be interesting, but it does not really show how the distance you cover in 30 minutes has changed over time. A graph showing the distance covered on each date will let you track your performance over time.

Time-series graph

We will use a *time-series graph.* A time-series graph is a graph showing data measurements in chronological order. To make a time-series graph, we put time on the horizontal scale and the variable being measured on the vertical scale. In a basic time-series graph, we connect the data points by line segments.

In a **time-series graph,** data are plotted in order of occurrence at regular intervals over a period of time.

EXAMPLE 5

TIME-SERIES GRAPH

Suppose you have been in the walking/jogging exercise program for 20 weeks, and for each week you have recorded the distance you covered in 30 minutes. Your data log is shown in Table 2-14.

Myrleen Ferguson Cate/PhotoEdit

TABLE 2-14 **Distance (in Miles) Walked/Jogged in 30 Minutes**

Week	1	2	3	4	5	6	7	8	9	10
Distance	1.5	1.4	1.7	1.6	1.9	2.0	1.8	2.0	1.9	2.0
Week	11	12	13	14	15	16	17	18	19	20
Distance	2.1	2.1	2.3	2.3	2.2	2.4	2.5	2.6	2.4	2.7

(a) Make a time-series graph.

SOLUTION: The data are appropriate for a time-series graph because they represent the same measurement (distance covered in a 30-minute period) taken at different times. The measurements are also recorded at equal time intervals (every week). To make our time-series graph, we list the weeks in order on the horizontal scale. Above each week, plot the distance covered that week on the vertical scale. Then connect the dots. Figure 2-14 shows the time-series graph. Be sure the scales are labeled.

FIGURE 2-14

Time-Series Graph of Distance (in miles) Jogged in 30 Minutes

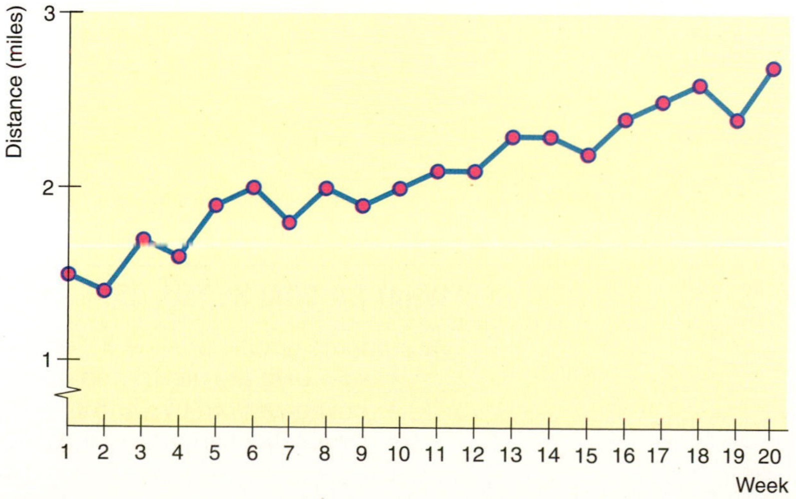

(b) From looking at Figure 2-14, can you detect any patterns?

SOLUTION: There seems to be an upward trend in distance covered. The distances covered in the last few weeks are about a mile farther than those for the first few weeks. However, we cannot conclude that this trend will continue. Perhaps you have reached your goal for this training activity and now wish to maintain a distance of about 2.5 miles in 30 minutes.

Time series

Data sets composed of similar measurements taken at regular intervals over time are called *time series.* Time series are often used in economics, finance, sociology, medicine, and any other situation in which we want to study or monitor a similar measure over a period of time. A time-series graph can reveal some of the main features of a time series.

Time-series data consist of measurements of the same variable for the same subject taken at regular intervals over a period of time.

CRITICAL THINKING

We've seen several styles of graphs. Which kinds are suitable for a specific data collection?

PROCEDURE

HOW TO DECIDE WHICH TYPE OF GRAPH TO USE

Bar graphs are useful for quantitative or qualitative data. With qualitative data, the frequency or percentage of occurrence can be displayed. With quantitative data, the measurement itself can be displayed, as was done in the bar graph showing life expectancy. Watch that the measurement scale is consistent or that a jump scale squiggle is used.

Pareto charts identify the frequency of events or categories in decreasing order of frequency of occurrence.

Circle graphs display how a *total* is dispersed into several categories. The circle graph is very appropriate for qualitative data, or any data for which percentage of occurrence makes sense. Circle graphs are most effective when the number of categories or wedges is 10 or fewer.

Time-series graphs display how data change over time. It is best if the units of time are consistent in a given graph. For instance, measurements taken every day should not be mixed on the same graph with data taken every week.

For any graph: Provide a title, label the axes, and identify units of measure. As Edward Tufte suggests in his book *The Visual Display of Quantitative Information,* don't let artwork or skewed perspective cloud the clarity of the information displayed.

WHAT DO GRAPHS TELL US?

Appropriate graphs provide a visual summary of data that tells us

- how data are distributed over several categories or data intervals;
- how data from two or more data sets compare;
- how data change over time.

TECH NOTES *Bar graphs, circle graphs, and time-series graphs*

TI-84Plus/TI-83Plus/TI-*n*spire (with TI-84Plus keypad) Only time-series. Place consecutive values 1 through the number of time segments in list L1 and corresponding data in L2. Press **Stat Plot** and highlight an *xy* line plot.

Excel 2010 First enter the data into the spreadsheet. Then click the **Insert Tab** and select the type of chart you want to create. A variety of bar graphs, pie charts, and line graphs that can be used as time-series graphs are available. Use the **Design Tab** and the **Layout Tab** on the **Chart Tools** menu to access options such as title, axis labels, etc., for your chart.

Minitab Use the menu selection **Graph.** Select the desired option and follow the instructions in the dialogue boxes.

VIEWPOINT | Do Ethical Standards Vary by the Situation?

The Lutheran Brotherhood conducted a national survey and found that nearly 60% of all U.S. adults claim that ethics vary by the situation; 33% claim that there is only one ethical standard; and 7% are not sure. How could you draw a circle graph to make a visual impression of Americans' views on ethical standards?

SECTION 2.2 PROBLEMS

1. *Interpretation* Consider graph (a) of Reasons People Like Texting on Cell Phones, based on a GfK Roper survey of 1000 adults.

Reasons People Like Texting on Cell Phones

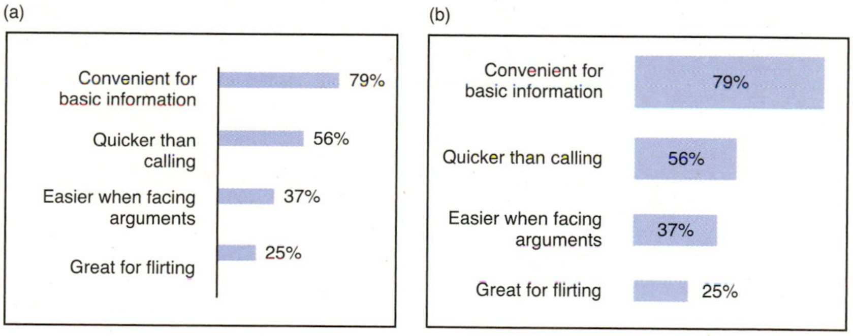

(a) Do you think respondents could select more than one response? Explain.
(b) Could the same information be displayed in a circle graph? Explain.
(c) Is graph (a) a Pareto chart?

2. *Interpretation* Look at graph (b) of Reasons People Like Texting on Cell Phones. Is this a proper bar graph? Explain.

3. *Critical Thinking* A personnel office is gathering data regarding working conditions. Employees are given a list of five conditions that they might want to see improved. They are asked to select the one item that is most critical to them. Which type of graph, circle graph or Pareto chart, would be most useful for displaying the results of the survey? Why?

4. *Critical Thinking* Your friend is thinking about buying shares of stock in a company. You have been tracking the closing prices of the stock shares for the past 90 trading days. Which type of graph for the data, histogram or time-series, would be best to show your friend? Why?

5. *Education: Does College Pay Off?* It is costly in both time and money to go to college. Does it pay off? According to the Bureau of the Census, the answer is yes. The average annual income (in thousands of dollars) of a *household* headed by a person with the stated education level is as follows: 21.6 if ninth grade is the highest level achieved, 39.6 for high school graduates, 56.8 for those holding associate degrees, 75.6 for those with bachelor's degrees, 91.7 for those with master's degrees, and 120.9 for those with doctoral degrees. Make a bar graph showing household income for each education level.

6. *Interpretation* Consider the two graphs depicting the influence of advertisements on making large purchases for two different age groups, those 18–34 years old and those 45–54 years old (based on a Harris Poll of about 2500 adults aged 18 or older). *Note:* Other responses such as "not sure" and "not applicable" were also possible.

Influence of Advertising on Most Recent Large Purchase

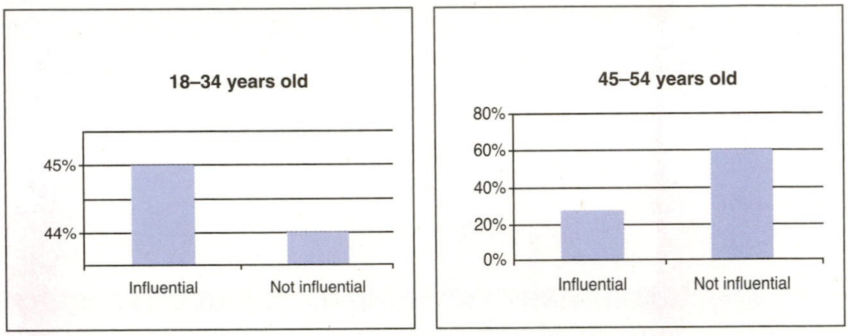

(a) Taking a quick glance at the graphs, Jenna thought that there was very little difference (maybe less than 1%) in the percentage of the two age groups who said that ads were influential. How would you change the graphs so that Jenna would not be misled so easily? *Hint:* Look at the vertical scales of the two graphs.

(b) Take the information from the two graphs and make a cluster bar graph showing the percentage by age group reporting to be influenced by ads and those reporting they were not influenced by ads.

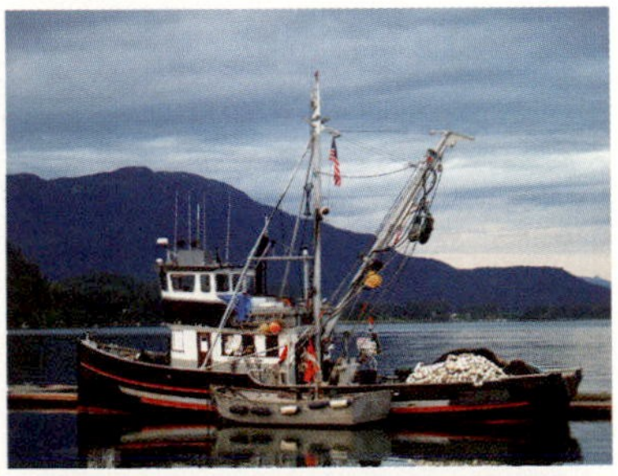

© iStockphoto.com/BergmannD

7. *Commercial Fishing: Gulf of Alaska* It's not an easy life, but it's a good life! Suppose you decide to take the summer off and sign on as a deck hand for a commercial fishing boat in Alaska that specializes in deep-water fishing for groundfish. What kind of fish can you expect to catch? One way to answer this question is to examine government reports on groundfish caught in the Gulf of Alaska. The following list indicates the types of fish caught annually in thousands of metric tons (Source: *Report on the Status of U.S. Living Marine Resources,* National Oceanic and Atmospheric Administration): flatfish, 36.3; Pacific cod, 68.6; sablefish, 16.0; Walleye pollock, 71.2; rockfish, 18.9. Make a Pareto chart showing the annual harvest for commercial fishing in the Gulf of Alaska.

O. Louis Mazzatenta/National Geographic/Getty Images

8. *Archaeology: Ireland* Commercial dredging operations in ancient rivers occasionally uncover archaeological artifacts of great importance. One such artifact is Bronze Age spearheads recovered from ancient rivers in Ireland. A recent study gave the following information regarding discoveries of ancient bronze spearheads in Irish rivers.

River	Bann	Blackwater	Erne	Shannon	Barrow
No. of spearheads	19	8	15	33	14

(Based on information from *Crossing the Rubicon, Bronze Age Studies 5,* Lorraine Bourke, Department of Archaeology, National University of Ireland, Galway.)

(a) Make a Pareto chart for these data.

(b) Make a circle graph for these data.

9. *Lifestyle: Hide the Mess!* A survey of 1000 adults (reported in *USA Today*) uncovered some interesting housekeeping secrets. When unexpected company comes, where do we hide the mess? The survey showed that 68% of the respondents toss their mess into the closet, 23% shove things under the bed, 6% put things into the bathtub, and 3% put the mess into the freezer. Make a circle graph to display this information.

10. *Education: College Professors' Time* How do college professors spend their time? *The National Education Association Almanac of Higher Education* gives the following average distribution of professional time allocation: teaching, 51%; research, 16%; professional growth, 5%; community service, 11%; service to the college, 11%; and consulting outside the college, 6%. Make a pie chart showing the allocation of professional time for college professors.

11. *FBI Report: Hawaii* In the Aloha state, you are very unlikely to be murdered! However, it is considerably more likely that your house might be burgled, your car might be stolen, or you might be punched in the nose. That said, Hawaii is still a great place to vacation or, if you are very lucky, to live. The following numbers represent the crime rates per 100,000 population in Hawaii: murder, 2.6; rape, 33.4; robbery, 93.3; house burglary, 911.6; motor vehicle theft, 550.7; assault, 125.3 (Source: *Crime in the United States,* U.S. Department of Justice, Federal Bureau of Investigation).
    (a) Display this information in a Pareto chart, showing the crime rate for each category.
    (b) Could the information as reported be displayed as a circle graph? Explain. *Hint:* Other forms of crime, such as arson, are not included in the information. In addition, some crimes might occur together.

12. *Driving: Bad Habits* Driving would be more pleasant if we didn't have to put up with the bad habits of other drivers. *USA Today* reported the results of a Valvoline Oil Company survey of 500 drivers, in which the drivers marked their complaints about other drivers. The top complaints turned out to be tailgating, marked by 22% of the respondents; not using turn signals, marked by 19%; being cut off, marked by 16%; other drivers driving too slowly, marked by 11%; and other drivers being inconsiderate, marked by 8%. Make a Pareto chart showing percentage of drivers listing each stated complaint. Could this information as reported be put in a circle graph? Why or why not?

13. *Ecology: Lakes* Pyramid Lake, Nevada, is described as the pride of the Paiute Indian Nation. It is a beautiful desert lake famous for very large trout. The elevation of the lake surface (feet above sea level) varies according to the annual flow of the Truckee River from Lake Tahoe. The U.S. Geological Survey provided the following data:

Year	Elevation
1986	3817
1987	3815
1988	3810
1989	3812
1990	3808
1991	3803

Year	Elevation
1992	3798
1993	3797
1994	3795
1995	3797
1996	3802
1997	3807

Year	Elevation
1998	3811
1999	3816
2000	3817

Make a time-series graph displaying the data. For more information, visit the Brase/Brase statistics site at **http://www.cengage.com/statistics/brase11e** and find the link to the Pyramid Lake Fisheries.

Ryan McVay/Photodisc/Jupiter Images

14. *Vital Statistics: Height* How does average height for boys change as boys get older? According to *Physician's Handbook,* the average heights at different ages are as follows:

Age (years)	0.5	1	2	3	4	5	6	7
Height (inches)	26	29	33	36	39	42	45	47

Age (years)	8	9	10	11	12	13	14
Height (inches)	50	52	54	56	58	60	62

Make a time-series graph for average height for ages 0.5 through 14 years.

15. *Expand Your Knowledge Donut Pie Charts* The book *The Wall Street Journal. Guide to Information Graphics* by Dona M. Wong gives strategies for using graphs and charts to display information effectively. One popular graph discussed is the *donut pie chart*. The donut pie chart is simply a pie chart with the center removed. A recent Harris Poll asked adults about their opinions regarding whether books should be banned from libraries because of social, language, violent, sexual, or religious content. The responses by education level to the question "Do you think that there are any books which should be banned completely?" are shown in the following donut pie charts.

(i)

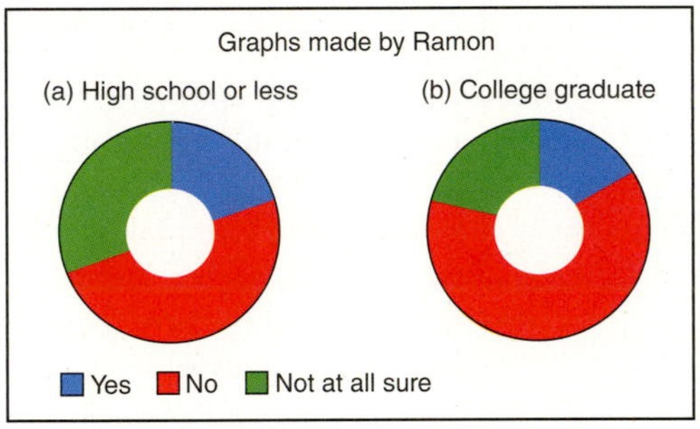

(ii)

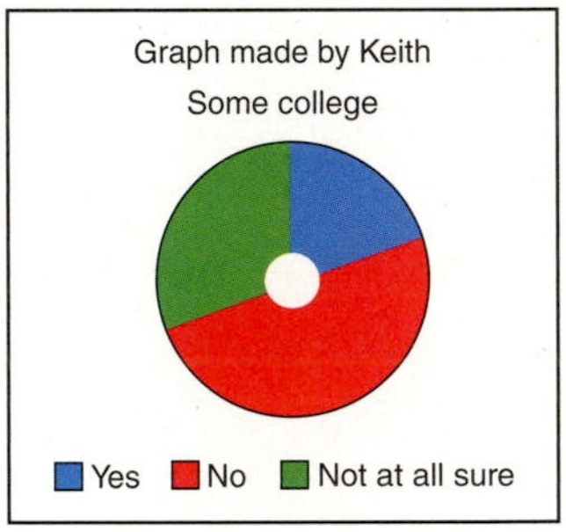

(a) What feature of Keith's graph makes it difficult to visually compare the responses of those with some college to those shown in the other graphs? How would you change Keith's graph for easier comparison?
(b) *Interpretation* Compare graphs (i) and (ii). At which of the two education levels is the "no" response more frequent?

16 *Technology: Cars* The following cluster bar graph shows responses from different age groups to questions regarding connectivity and tracking technology found in new cars. A recent Harris Poll asked respondents how much they agreed or disagreed with statements that they
(1) worry that the technologies cause too much distraction and are dangerous;
(2) worry about letting companies know too much about location and driving habits;
(3) worry that insurance rates could increase because of knowledge of driving habits;
(4) think the technologies make driving more enjoyable;
(5) feel safer with the technologies;
(6) feel it is important to stay connected when in vehicle.

The graph shows the percentage of respondents in each age category who agree strongly or somewhat agree to each of the six statements.

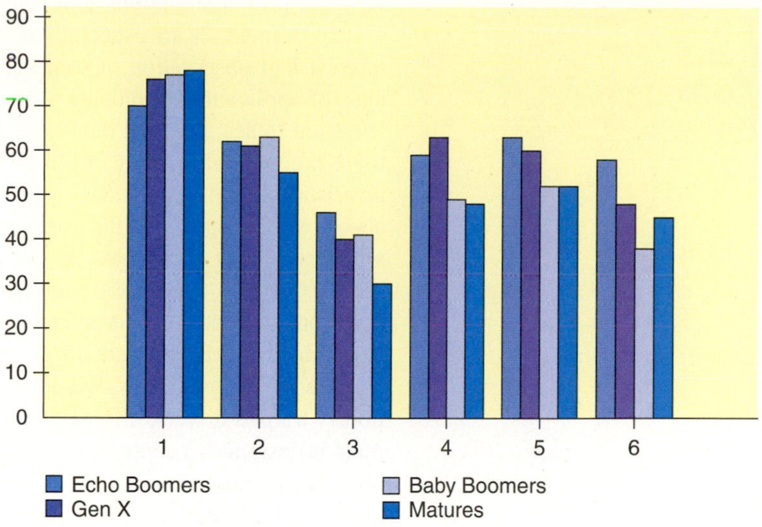

(a) *Interpretation* Which statement has the highest rate of agreement for all four age groups?
(b) *Interpretation* Which age group expresses least worry about insurance companies raising their rates because of the driving habit information collected by the technologies?
(c) *Interpretation* Which age group finds the technologies make driving more enjoyable?

SECTION **2.3**

Stem-and-Leaf Displays

FOCUS POINTS

- Construct a stem-and-leaf display from raw data.
- Use a stem-and-leaf display to visualize data distribution.
- Compare a stem-and-leaf display to a histogram.

EXPLORATORY DATA ANALYSIS

EDA

Together with histograms and other graphics techniques, the stem-and-leaf display is one of many useful ways of studying data in a field called *exploratory data analysis* (often abbreviated as *EDA*). John W. Tukey wrote one of the definitive books on the subject, *Exploratory Data Analysis* (Addison-Wesley). Another very useful reference for EDA techniques is the book *Applications, Basics, and Computing of Exploratory Data Analysis* by Paul F. Velleman and David C. Hoaglin (Duxbury Press). Exploratory data analysis techniques are particularly useful for detecting patterns and extreme data values. They are designed to help us explore a data set, to ask questions we had not thought of before, or to pursue leads in many directions.

EDA techniques are similar to those of an explorer. An explorer has a general idea of destination but is always alert for the unexpected. An explorer needs to assess situations quickly and often simplify and clarify them. An explorer makes pictures—that is, maps showing the relationships of landscape features. The aspects of rapid implementation, visual displays such as graphs and charts, data simplification, and robustness (that is, analysis that is not influenced much by extreme data values) are key ingredients of EDA techniques. In addition, these techniques are good for exploration because they require very few prior assumptions about the data.

EDA methods are especially useful when our data have been gathered for general interest and observation of subjects. For example, we may have data regarding the ages of applicants to graduate programs. We don't have a specific question in mind. We want to see what the data reveal. Are the ages fairly uniform or spread out? Are there exceptionally young or old applicants? If there are, we might look at other characteristics of these applicants, such as field of study. EDA methods help us quickly absorb some aspects of the data and then may lead us to ask specific questions to which we might apply methods of traditional statistics.

In contrast, when we design an experiment to produce data to answer a specific question, we focus on particular aspects of the data that are useful to us. If we want to determine the average highway gas mileage of a specific sports car, we use that model car in well-designed tests. We don't need to worry about unexpected road conditions, poorly trained drivers, different fuel grades, sudden stops and starts, etc. Our experiment is designed to control outside factors. Consequently, we do not need to "explore" our data as much. We can often make valid assumptions about the data. Methods of traditional statistics will be very useful to analyze such data and answer our specific questions.

STEM-AND-LEAF DISPLAY

In this text, we will introduce two EDA techniques: stem-and-leaf displays and, in Section 3.3, box-and-whisker plots. Let's first look at a stem-and-leaf display.

Stem-and-leaf display

A **stem-and-leaf display** is a method of exploratory data analysis that is used to rank-order and arrange data into groups.

We know that frequency distributions and histograms provide a useful organization and summary of data. However, in a histogram, we lose most of the specific data values. A stem-and-leaf display is a device that organizes and groups data but allows us to recover the original data if desired. In the next example, we will make a stem-and-leaf display.

EXAMPLE 6

STEM-AND-LEAF DISPLAY

Many airline passengers seem weighted down by their carry-on luggage. Just how much weight are they carrying? The carry-on luggage weights in pounds for a random sample of 40 passengers returning from a vacation to Hawaii were recorded (see Table 2-15).

TABLE 2-15 **Weights of Carry-On Luggage in Pounds**

30	27	12	42	35	47	38	36	27	35
22	17	29	3	21	0	38	32	41	33
26	45	18	43	18	32	31	32	19	21
33	31	28	29	51	12	32	18	21	26

FIGURE 2-15

Stem-and-Leaf Displays of Airline Carry-On Luggage Weights

(a) Leaves Not Ordered

3 | 2 represents 32 lb

Stem	Leaves
0	3 0
1	2 7 8 8 9 2 8
2	7 7 2 9 1 6 1 8 9 1 6
3	0 5 8 6 5 8 2 3 2 1 2 3 1 2
4	2 7 1 5 3
5	1

(b) Final Display with Leaves Ordered

3 | 2 represents 32 lb

Stem	Leaves
0	0 3
1	2 2 7 8 8 8 9
2	1 1 1 2 6 6 7 7 8 9 9
3	0 1 1 2 2 2 2 3 3 5 5 6 8 8
4	1 2 3 5 7
5	1

Stem

Leaf

To make a stem-and-leaf display, we break the digits of each data value into *two* parts. The left group of digits is called a *stem,* and the remaining group of digits on the right is called a *leaf.* We are free to choose the number of digits to be included in the stem.

The weights in our example consist of two-digit numbers. For a two-digit number, the stem selection is obviously the left digit. In our case, the tens digits will form the stems, and the units digits will form the leaves. For example, for the weight 12, the stem is 1 and the leaf is 2. For the weight 18, the stem is again 1, but the leaf is 8. In the stem-and-leaf display, we list each possible stem once on the left and all its leaves in the same row on the right, as in Figure 2-15(a). Finally, we order the leaves as shown in Figure 2-15(b).

Figure 2-15 shows a stem-and-leaf display for the weights of carry-on luggage. From the stem-and-leaf display in Figure 2-15, we see that two bags weighed 27 lb, one weighed 3 lb, one weighed 51 lb, and so on. We see that most of the weights were in the 30-lb range, only two were less than 10 lb, and six were over 40 lb. Note that the lengths of the lines containing the leaves give the visual impression that a sideways histogram would present.

As a final step, we need to indicate the scale. This is usually done by indicating the value represented by the stem and one leaf.

There are no firm rules for selecting the group of digits for the stem. But whichever group you select, you must list all the possible stems from smallest to largest in the data collection.

PROCEDURE

HOW TO MAKE A STEM-AND-LEAF DISPLAY

1. Divide the digits of each data value into two parts. The leftmost part is called the *stem* and the rightmost part is called the *leaf.*
2. Align all the stems in a vertical column from smallest to largest. Draw a vertical line to the right of all the stems.
3. Place all the leaves with the same stem in the same row as the stem, and arrange the leaves in increasing order.
4. Use a label to indicate the magnitude of the numbers in the display. We include the decimal position in the label rather than with the stems or leaves.

LOOKING FORWARD

Ordering or sorting data is an essential first step to finding the median or center value of the data distribution. We will discuss the median in Section 3.1. The data also need to be ordered before we can construct a box-and-whisker plot (Section 3.3). After a little practice, you will find that stem-and-leaf diagrams provide a very efficient way of ordering data by hand.

GUIDED EXERCISE 4 | Stem-and-Leaf Display

What does it take to win at sports? If you're talking about basketball, one sportswriter gave the answer. He listed the winning scores of the conference championship games over the last 35 years. The scores for those games follow below.

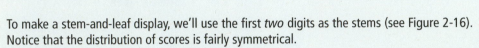

Michael J. Doolittle/The Image Works

132	118	124	109	104	101	125	83	99
131	98	125	97	106	112	92	120	103
111	117	135	143	112	112	116	106	117
119	110	105	128	112	126	105	102	

To make a stem-and-leaf display, we'll use the first *two* digits as the stems (see Figure 2-16). Notice that the distribution of scores is fairly symmetrical.

(a) Use the first *two* digits as the stem. Then order the leaves. Provide a label that shows the meaning and units of the first stem and first leaf.

FIGURE 2-16 Winning Scores

08 | 3 represents 083 or 83 points

08	3
09	2 7 8 9
10	1 2 3 4 5 5 6 6 9
11	0 1 2 2 2 2 6 7 7 8 9
12	0 4 5 5 6 8
13	1 2 5
14	3

(b) Looking at the distribution, would you say that it is fairly symmetrical?

Yes. Notice that stem 11 has the most data.

WHAT DO STEM-AND-LEAF DISPLAYS TELL US?

Stem-and-leaf displays give a visual display that

- shows us all the data (or truncated data) in order from smallest to largest;
- helps us spot extreme data values or clusters of data values;
- displays the shape of the data distribution.

COMMENT Stem-and-leaf displays organize the data, let the data analyst spot extreme values, and are easy to create. In fact, they can be used to organize data so that frequency tables are easier to make. However, at this time, histograms are used more often in formal data presentations, whereas stem-and-leaf displays are used by data analysts to gain initial insights about the data.

TECH NOTES *Stem-and-leaf display*

TI-84Plus/TI-83Plus/TI-*n*spire Does not support stem-and-leaf displays. You can sort the data by using keys **Stat ➤ Edit ➤ 2:SortA.**

Excel 2010 Enter your data and select the data you want to sort. On the **Home** ribbon, click the **Sort and Filter** button in the **Editing** group of the ribbon and select the desired sorting option.

Minitab Use the menu selections **Graph ➤ Stem-and-Leaf** and fill in the dialogue box.

Minitab Stem-and-Leaf Display (for Data in Guided Exercise 4)

```
Stem-and-Leaf of Scores     N=35
Leaf Unit=1.0

   1        8   3
   5        9   2789
  14       10   123455669
 (11)      11   01222267789
  10       12   045568
   4       13   125
   1       14   3
```

The values shown in the left column represent depth. Numbers above the value in parentheses show the cumulative number of values from the top to the stem of the middle value. Numbers below the value in parentheses show the cumulative number of values from the bottom to the stem of the middle value. The number in parentheses shows how many values are on the same line as the middle value.

CRITICAL THINKING

Problems 5 and 6 show how to split a stem.

Problem 10 discusses back-to-back stem-and-leaf displays.

Stem-and-leaf displays show each of the original or truncated data values. By looking at the display "sideways," you can see the distribution shape of the data. If there are large gaps between stems containing leaves, especially at the top or bottom of the display, the data values in the first or last lines may be outliers. Outliers should be examined carefully to see if they are data errors or simply unusual data values. Someone very familiar with the field of study as well as the purpose of the study should decide how to treat outliers.

VIEWPOINT What Does It Take to Win?

Scores for NFL Super Bowl games can be found at the NFL web site. Visit the Brase/Brase statistics site at ***http://www.cengage.com/statistics/brase11e*** *and find the link to the NFL. Once at the NFL web site, follow the links to the Super Bowl. Of special interest in football statistics is the spread, or difference, between scores of the winning and losing teams. If the spread is too large, the game can appear to be lopsided, and TV viewers become less interested in the game (and accompanying commercial ads). Make a stem-and-leaf display of the spread for the NFL Super Bowl games and analyze the results.*

SECTION 2.3 PROBLEMS

1. *Cowboys: Longevity* How long did *real* cowboys live? One answer may be found in the book *The Last Cowboys* by Connie Brooks (University of New Mexico Press). This delightful book presents a thoughtful sociological study of cowboys in west Texas and southeastern New Mexico around the year 1890. A sample of 32 cowboys gave the following years of longevity:

58	52	68	86	72	66	97	89	84	91	91
92	66	68	87	86	73	61	70	75	72	73
85	84	90	57	77	76	84	93	58	47	

(a) Make a stem-and-leaf display for these data.
(b) *Interpretation* Consider the following quote from Baron von Richthofen in his *Cattle Raising on the Plains of North America:* "Cowboys are to be found among the sons of the best families. The truth is probably that most were not a drunken, gambling lot, quick to draw and fire their pistols." Does the data distribution of longevity lend credence to this quote?

2. *Ecology: Habitat* Wetlands offer a diversity of benefits. They provide a habitat for wildlife, spawning grounds for U.S. commercial fish, and renewable timber resources. In the last 200 years, the United States has lost more than half its wetlands. *Environmental Almanac* gives the percentage of wetlands lost in each state in the last 200 years. For the lower 48 states, the percentage loss of wetlands per state is as follows:

46	37	36	42	81	20	73	59	35	50
87	52	24	27	38	56	39	74	56	31
27	91	46	9	54	52	30	33	28	35
35	23	90	72	85	42	59	50	49	
48	38	60	46	87	50	89	49	67	

Make a stem-and-leaf display of these data. Be sure to indicate the scale. How are the percentages distributed? Is the distribution skewed? Are there any gaps?

3. *Health Care: Hospitals* The American Medical Association Center for Health Policy Research, in its publication *State Health Care Data: Utilization, Spending, and Characteristics*, included data, by state, on the number of community hospitals and the average patient stay (in days). The data are shown in the table. Make a stem-and-leaf display of the data for the average length of stay in days. Comment about the general shape of the distribution.

State	No. of Hospitals	Average Length of Stay	State	No. of Hospitals	Average Length of Stay	State	No. of Hospitals	Average Length of Stay
Alabama	119	7.0	Kentucky	107	6.9	New Mexico	37	5.5
Alaska	16	5.7	Louisiana	136	6.7	Oregon	66	5.3
Arizona	61	5.5	Maine	38	7.2	Ohio	193	6.6
Arkansas	88	7.0	Maryland	51	6.8	Oklahoma	113	6.7
California	440	6.0	Massachusetts	101	7.0	Pennsylvania	236	7.5
Colorado	71	6.8	Michigan	175	7.3	Rhode Island	12	6.9
Connecticut	35	7.4	Minnesota	148	8.7	S. Carolina	68	7.1
Delaware	8	6.8	Mississippi	102	7.2	S. Dakota	52	10.3
Dist. of Columbia	11	7.5	Missouri	133	7.4	Tennessee	122	6.8
Florida	227	7.0	Montana	53	10.0	Texas	421	6.2
Georgia	162	7.2	N. Carolina	117	7.3	Utah	42	5.2
Hawaii	19	9.4	N. Dakota	47	11.1	Vermont	15	7.6
Idaho	41	7.1	Nebraska	90	9.6	Virginia	98	7.0
Indiana	113	6.6	Nevada	21	6.4	Washington	92	5.6
Illinois	209	7.3	New Hampshire	27	7.0	W. Virginia	59	7.1
Iowa	123	8.4	New York	231	9.9	Wisconsin	129	7.3
Kansas	133	7.8	New Jersey	96	7.6	Wyoming	27	8.5

4. *Health Care: Hospitals* Using the number of hospitals per state listed in the table in Problem 3, make a stem-and-leaf display for the number of community hospitals per state. Which states have an unusually high number of hospitals?

5. *Expand Your Knowledge: Split Stem* The Boston Marathon is the oldest and best-known U.S. marathon. It covers a route from Hopkinton, Massachusetts, to downtown Boston. The distance is approximately 26 miles. Visit the Brase/Brase statistics site at **http://www.cengage.com/statistics/brase11e** and find the link to the Boston Marathon. Search the marathon site to find a wealth of information about the history of the race. In particular, the site gives the winning times for the Boston Marathon. They are all over 2 hours. The following data are the minutes over 2 hours for the winning male runners:

1961–1980

23	23	18	19	16	17	15	22	13	10
18	15	16	13	9	20	14	10	9	12

1981–2000

9	8	9	10	14	7	11	8	9	8
11	8	9	7	9	9	10	7	9	9

(a) Make a stem-and-leaf display for the minutes over 2 hours of the winning times for the years 1961 to 1980. Use two lines per stem.

PROCEDURE

HOW TO SPLIT A STEM

When a stem has many leaves, it is useful to split the stem into two lines or more. For two lines per stem, place leaves 0 to 4 on the first line and leaves 5 to 9 on the next line.

(b) Make a stem-and-leaf display for the minutes over 2 hours of the winning times for the years 1981 to 2000. Use two lines per stem.

(c) *Interpretation* Compare the two distributions. How many times under 15 minutes are in each distribution?

6. *Split Stem: Golf* The U.S. Open Golf Tournament was played at Congressional Country Club, Bethesda, Maryland, with prizes ranging from $465,000 for first place to $5000. Par for the course was 70. The tournament consisted of four rounds played on different days. The scores for each round of the 32 players who placed in the money (more than $17,000) were given on a web site. For more information, visit the Brase/Brase statistics site at **http://www.cengage.com/statistics/brase11e** and find the link to golf. The scores for the first round were as follows:

71	65	67	73	74	73	71	71	74	73	71
70	75	71	72	71	75	75	71	71	74	75
66	75	75	75	71	72	72	73	71	67	

The scores for the fourth round for these players were as follows:

69	69	73	74	72	72	70	71	71	70	72
73	73	72	71	71	71	69	70	71	72	73
74	72	71	68	69	70	69	71	73	74	

(a) Make a stem-and-leaf display for the first-round scores. Use two lines per stem. (See Problem 5.)
(b) Make a stem-and-leaf display for the fourth-round scores. Use two lines per stem.
(c) *Interpretation* Compare the two distributions. How do the highest scores compare? How do the lowest scores compare?

Are cigarettes bad for people? Cigarette smoking involves tar, carbon monoxide, and nicotine. The first two are definitely not good for a person's health, and the last ingredient can cause addiction. Problems 7, 8, and 9 refer to Table 2-16, which was taken from the web site maintained by the *Journal of Statistics Education.* For more information, visit the Brase/Brase statistics site at **http://www.cengage.com/statistics/brase11e** and find the link to the *Journal of Statistics Education.* Follow the links to the cigarette data.

TABLE 2-16 **Milligrams of Tar, Nicotine, and Carbon Monoxide (CO) per One Cigarette**

Brand	Tar	Nicotine	CO	Brand	Tar	Nicotine	CO
Alpine	14.1	0.86	13.6	MultiFilter	11.4	0.78	10.2
Benson & Hedges	16.0	1.06	16.6	Newport Lights	9.0	0.74	9.5
Bull Durham	29.8	2.03	23.5	Now	1.0	0.13	1.5
Camel Lights	8.0	0.67	10.2	Old Gold	17.0	1.26	18.5
Carlton	4.1	0.40	5.4	Pall Mall Lights	12.8	1.08	12.6
Chesterfield	15.0	1.04	15.0	Raleigh	15.8	0.96	17.5
Golden Lights	8.8	0.76	9.0	Salem Ultra	4.5	0.42	4.9
Kent	12.4	0.95	12.3	Tareyton	14.5	1.01	15.9
Kool	16.6	1.12	16.3	True	7.3	0.61	8.5
L&M	14.9	1.02	15.4	Viceroy Rich Light	8.6	0.69	10.6
Lark Lights	13.7	1.01	13.0	Virginia Slims	15.2	1.02	13.9
Marlboro	15.1	0.90	14.4	Winston Lights	12.0	0.82	14.9
Merit	7.8	0.57	10.0				

Source: Federal Trade Commission, USA (public domain).

7. *Health: Cigarette Smoke* Use the data in Table 2-16 to make a stem-and-leaf display for milligrams of tar per cigarette smoked. Are there any outliers?

8. *Health: Cigarette Smoke* Use the data in Table 2-16 to make a stem-and-leaf display for milligrams of carbon monoxide per cigarette smoked. Are there any outliers?

9. *Health: Cigarette Smoke* Use the data in Table 2-16 to make a stem-and-leaf display for milligrams of nicotine per cigarette smoked. In this case, truncate the measurements at the tenths position and use two lines per stem (see Problem 5, part a).

Back-to-back stem plot

10. *Expand Your Knowledge: Back-to-Back Stem Plot* In archaeology, the depth (below surface grade) at which artifacts are found is very important. Greater depths sometimes indicate older artifacts, perhaps from a different archaeological period. Figure 2-17 is a *back-to-back stem plot* showing the depths of artifact locations at two different archaeological sites. These sites are from similar geographic locations. Notice that the stems are in the center of the diagram. The leaves for Site I artifact depths are shown to the left of the

stem, while the leaves for Site II are to the right of the stem (see *Mimbres Mogollon Archaeology* by A. I. Woosley and A. J. McIntyre, University of New Mexico Press).

(a) What are the least and greatest depths of artifact finds at Site I? at Site II?
(b) Describe the data distribution of depths of artifact finds at Site I and at Site II.
(c) *Interpretation* At Site II, there is a gap in the depths at which artifacts were found. Does the Site II data distribution suggest that there might have been a period of no occupation?

FIGURE 2-17

Depth (in cm) of Artifact Location

5 | 2 | 0 = 25 cm at Site I and 20 cm at Site II

Site I		Site II
5	2	0 5 5
5 0	3	0 0 0 0 5 5
5 5 5 5	4	0 0 0 5
5 0	5	0 0 5 5
5 5 5 5 5 0	6	0 0 0 5 5 5
5 5 5 5 5 5 0 0	7	
5 5 0 0 0	8	
5 5 0 0	9	
5 5	10	
0	11	0 0 5 5 5 5
	12	0 0 0 0 5

CHAPTER REVIEW

SUMMARY

Organizing and presenting data are the main purposes of the branch of statistics called descriptive statistics. Graphs provide an important way to show how the data are distributed.

- Frequency tables show how the data are distributed within set classes. The classes are chosen so that they cover all data values and so that each data value falls within only one class. The number of classes and the class width determine the class limits and class boundaries. The number of data values falling within a class is the class frequency.
- A histogram is a graphical display of the information in a frequency table. Classes are shown on the horizontal axis, with corresponding frequencies on the vertical axis. Relative-frequency histograms show relative frequencies on the vertical axis. Ogives show cumulative frequencies on the vertical axis. Dotplots are like histograms, except that the classes are individual data values.
- Bar graphs, Pareto charts, and pie charts are useful to show how quantitative or qualitative data are distributed over chosen categories.
- Time-series graphs show how data change over set intervals of time.
- Stem-and-leaf displays are effective means of ordering data and showing important features of the distribution.

Graphs aren't just pretty pictures. They help reveal important properties of the data distribution, including the shape and whether or not there are any outliers.

IMPORTANT WORDS & SYMBOLS

Section 2.1

Frequency table 42
Frequency distribution 42
Class width 43
Class, lower limit, upper limit 43
Class frequency 44
Class midpoint 44
Class mark 44
Class boundaries 44
Relative frequency 44
Relative-frequency table 45
Histogram 46
Relative frequency histogram 46
Mound-shaped symmetrical distribution 49
Uniform distribution 49
Skewed left 49
Skewed right 49
Bimodal distribution 49
Outlier 50
Cumulative frequency 50
Ogive 50
Dotplot 59

Section 2.2

Bar graph 59
Cluster bar graph 60
Changing scale 60
Pareto chart 61
Circle graphs or pie charts 62
Time-series graph 63
Time series 64
Donut pie chart 68

Section 2.3

EDA 69
Stem-and-leaf display 70
Stem 71
Leaf 71
Split stem 75
Back-to-back stem plot 76

VIEWPOINT

Personality Clash!

Karl Pearson and Sir Ronald Fisher are two very famous mathematicians who contributed a great deal to the understanding and practice of modern statistics. Each was an outstanding person in his own right; however, the men had a terrific personality clash in the world of statistics. In Chapter 9 we will study the Pearson product moment correlation coefficient. Fisher's F *distribution is the central probability distribution in the field known as* analysis of variance. *In Chapter 8 we will examine two interpretations of statistical testing: the classical method of critical regions, favored by Karl Pearson, and the* P*-value method, favored by Sir Ronald Fisher. For a fixed level of significance, the methods can be shown to be equivalent. However, the* P*-value method is favored strongly by research and technology and is therefore emphasized in this text. The personality battles of Pearson and Fisher are chronicled in the popular book* The Lady Tasting Tea: How Statistics Revolutionized Science in the Twentieth Century *by David Salsburg.*

CHAPTER REVIEW PROBLEMS

1. *Critical Thinking* Consider these types of graphs: histogram, bar graph, Pareto chart, pie chart, stem-and-leaf display.
   (a) Which are suitable for qualitative data?
   (b) Which are suitable for quantitative data?

2. *Critical Thinking* A consumer interest group is tracking the percentage of household income spent on gasoline over the past 30 years. Which graphical display would be more useful, a histogram or a time-series graph? Why?

3. *Critical Thinking* Describe how data outliers might be revealed in histograms and stem-and-leaf plots.

4. *Expand Your Knowledge* How are dotplots and stem-and-leaf displays similar? How are they different?

5. *Focus Problem: Fuel Economy* Solve the focus problem at the beginning of this chapter.

6. *Criminal Justice: Prisoners* The time plot in Figure 2-18 gives the number of state and federal prisoners per 100,000 population (Source: *Statistical Abstract of the United States,* 120th Edition).
   (a) Estimate the number of prisoners per 100,000 people for 1980 and for 1997.
   (b) *Interpretation* During the time period shown, there was increased prosecution of drug offenses, longer sentences for common crimes, and reduced access to parole. What does the time-series graph say about the prison population change per 100,000 people?
   (c) In 1997, the U.S. population was approximately 266,574,000 people. At the rate of 444 prisoners per 100,000 population, about how many prisoners were in the system? The projected U.S. population for the year 2020 is 323,724,000. If the rate of prisoners per 100,000 stays the same as in 1997, about how many prisoners do you expect will be in the system in 2020? To obtain the most recent information, visit the Brase/Brase statistics site at **http://www.cengage.com/statistics/brase11e** and find the link to the Census Bureau.

FIGURE 2-18

Number of State and Federal Prisoners per 100,000 Population

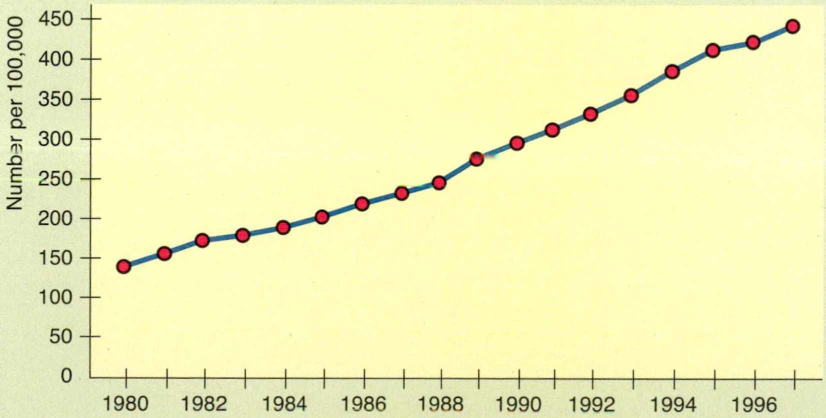

7. *IRS: Tax Returns* Almost everyone files (or will someday file) a federal income tax return. A research poll for TurboTax (a computer software package to aid in tax-return preparation) asked what aspect of filing a return people thought to be the most difficult. The results showed that 43% of the respondents said understanding the IRS jargon, 28% said knowing deductions, 10% said getting the right form, 8% said calculating the numbers, and 10% didn't know. Make a circle graph to display this information. *Note:* Percentages will not total 100% because of rounding.

8. *Law Enforcement: DUI* Driving under the influence of alcohol (DUI) is a serious offense. The following data give the ages of a random sample of 50 drivers arrested while driving under the influence of alcohol. This distribution is based on the age distribution of DUI arrests given in the *Statistical Abstract of the United States* (112th Edition).

46	16	41	26	22	33	30	22	36	34
63	21	26	18	27	24	31	38	26	55
31	47	27	43	35	22	64	40	58	20
49	37	53	25	29	32	23	49	39	40
24	56	30	51	21	45	27	34	47	35

(a) Make a stem-and-leaf display of the age distribution.
(b) Make a frequency table using seven classes.
(c) Make a histogram showing class boundaries.
(d) Identify the shape of the distribution.

9. *Agriculture: Apple Trees* The following data represent trunk circumferences (in mm) for a random sample of 60 four-year-old apple trees at East Malling Agriculture Research Station in England (Reference: S. C. Pearce, University of Kent at Canterbury). *Note:* These data are also available for download at the Online Study Center.

108	99	106	102	115	120	120	117	122	142
106	111	119	109	125	108	116	105	117	123
103	114	101	99	112	120	108	91	115	109
114	105	99	122	106	113	114	75	96	124
91	102	108	110	83	90	69	117	84	142
122	113	105	112	117	122	129	100	138	117

Lisa Valder/Getty Images

(a) Make a frequency table with seven classes showing class limits, class boundaries, midpoints, frequencies, and relative frequencies.
(b) Draw a histogram.
(c) Draw a relative-frequency histogram.
(d) Identify the shape of the distribution.
(e) Draw an ogive.

10. *Law: Corporation Lawsuits* Many people say the civil justice system is overburdened. Many cases center on suits involving businesses. The following data are based on a *Wall Street Journal* report. Researchers conducted a study of lawsuits involving 1908 businesses ranked in the Fortune 1000 over

a 20-year period. They found the following distribution of civil justice caseloads brought before the federal courts involving the businesses:

Case Type	Number of Filings (in thousands)
Contracts	107
General torts (personal injury)	191
Asbestos liability	49
Other product liability	38
All other	21

Note: Contracts cases involve disputes over contracts between businesses.

(a) Make a Pareto chart of the caseloads. Which type of cases occur most frequently?

(b) Make a pie chart showing the percentage of cases of each type.

11. *Archaeology: Tree-Ring Data* *The Sand Canyon Archaeological Project,* edited by W. D. Lipe and published by Crow Canyon Archaeological Center, contains the stem-and-leaf diagram shown in Figure 2-19. The study uses tree rings to accurately determine the year in which a tree was cut. The figure gives the tree-ring-cutting dates for samples of timbers found in the architectural units at Sand Canyon Pueblo. The text referring to the figure says, "The three-digit numbers in the left column represent centuries and decades A.D. The numbers to the right represent individual years, with each number derived from an individual sample. Thus, **124 | 2 2 2** represents three samples dated to A.D. 1242." Use Figure 2-19 and the verbal description to answer the following questions.

(a) Which decade contained the most samples?

(b) How many samples had a tree-ring-cutting date between 1200 A.D. and 1239 A.D., inclusive?

(c) What are the dates of the longest interval during which no tree-cutting samples occurred? What might this indicate about new construction or renovation of the pueblo structures during this period?

FIGURE 2-19

Tree-Ring-Cutting Dates from Architectural Units of Sand Canyon Pueblo: *The Sand Canyon Archaeological Project*

119	5 6
120	0 0 1 2 3 3 3 3 3 3 3 3 3 3 3 3 3 3 3 3 3 3 3 3 3 3 3 3
120	
121	2
121	5 5
122	0 0 1 1 1 1 2 2 3 4 4 4 4 4 4 4
122	5 8 9
123	0 1 2 3 3 4
123	5 5 5 5 5 5 5 5 5 5 5 5 5 5 5 6 8 8 9
124	1 2 2 2 2 2 2 2 2 2 2 2 2 2 2 2 2 2 2 2 2 2 2 2 3 4 4
124	5 6 8 9 9 9 9 9 9 9 9 9 9
125	0 0 0 0 0 0 0 0 0 0 0 0 0 0 0 1 1 1 1 1 1 1 2 2 2
125	
126	0 0 0 1 2 2 2 2 2 2 2 2 2 2 2 2 2 4 4 4 4 4 4 4
126	5 5 5 6 6 7
127	0 1 1 1 1 4 4

12. *Interpretation* A Harris Poll surveyed 2085 U.S. adults regarding use of cell phones while driving. All the adults were asked their opinion regarding how dangerous it is for a driver to use a cell phone while driving. Graph (a) shows the percentage responding "very dangerous" by age group. Only the adults who drive and who have a cell phone were asked how often they talk on the cell phone while driving. Graph (b) shows the percentage responding "never" by age group.

Cell Phone Use and Driving

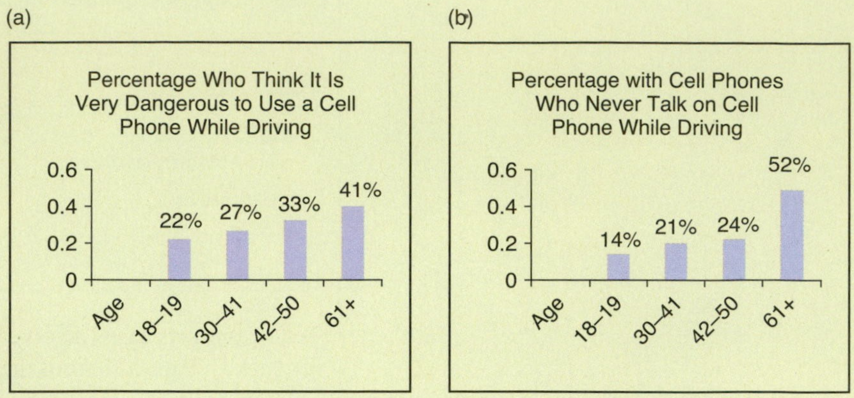

(a) What trend does this survey portray regarding age group and the opinion that using a cell phone while driving is very dangerous?

(b) How does the behavior of never using a cell phone while driving compare to the opinion that using a cell phone while driving is dangerous? Do you think that some of the differences in the behavior (never use a cell phone while driving) and the opinion (using a cell phone while driving is very dangerous) can be attributed to the differences in the survey population? Explain.

DATA HIGHLIGHTS: GROUP PROJECTS

Break into small groups and discuss the following topics. Organize a brief outline in which you summarize the main points of your group discussion.

1. Examine Figure 2-20, "Everyone Agrees: Slobs Make Worst Roommates." This is a clustered bar graph because two percentages are given for each response category: responses from men and responses from women. Comment about how the artistic rendition has slightly changed the format of a bar graph. Do the bars seem to have lengths that accurately reflect the relative percentages of the responses? In your own opinion, does the artistic rendition enhance or confuse the information? Explain. Which characteristic of "worst roommates" does the graphic seem to illustrate? Can this graph be considered a Pareto chart for men? for women? Why or why not? From the information given in the figure, do you think the survey just listed the four given annoying characteristics? Do you think a respondent could choose more than one characteristic? Explain

FIGURE 2-20

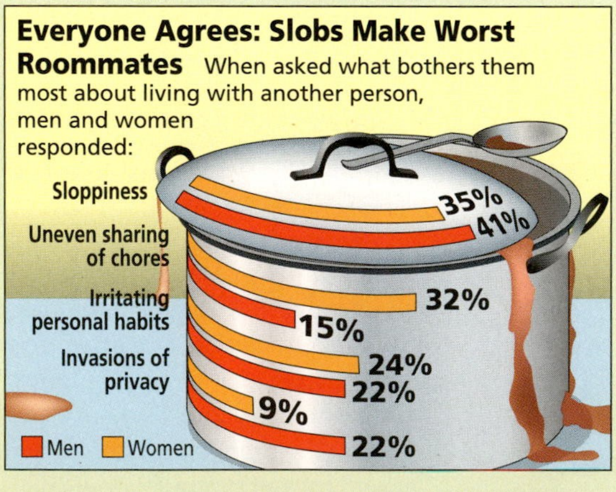

Source: Advantage Business Research for Mattel *Compatibility*

your answer in terms of the percentages given and in terms of the explanation given in the graphic. Could this information also be displayed in one circle graph for men and another for women? Explain.

2. Examine Figure 2-21, "Global Teen Worries." How many countries were contained in the sample? The graph contains bars and a circle. Which bar is the longest? Which bar represents the greatest percentage? Is this a bar graph or not? If not, what changes would need to be made to put the information in a bar graph? Could the graph be made into a Pareto chart? Could it be made into a circle graph? Explain.

FIGURE 2-21

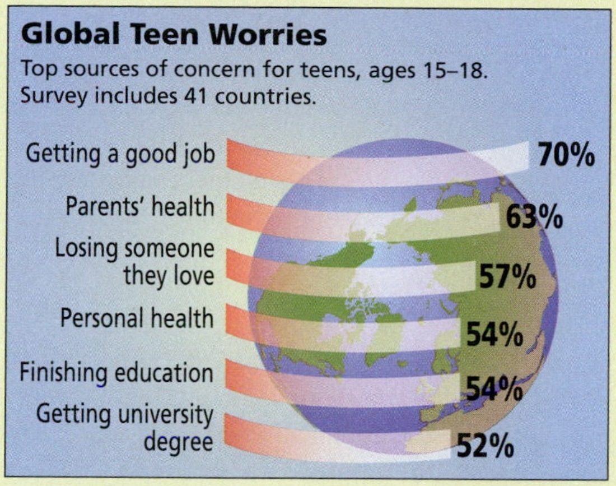

Source: BrainWaves Group's New World Teen Study

LINKING CONCEPTS: WRITING PROJECTS

Discuss each of the following topics in class or review the topics on your own. Then write a brief but complete essay in which you summarize the main points. Please include formulas and graphs as appropriate.

1. In your own words, explain the differences among histograms, relative-frequency histograms, bar graphs, circle graphs, time-series graphs, Pareto charts, and stem-and-leaf displays. If you have nominal data, which graphic displays might be useful? What if you have ordinal, interval, or ratio data?
2. What do we mean when we say a histogram is skewed to the left? to the right? What is a bimodal histogram? Discuss the following statement: "A bimodal histogram usually results if we draw a sample from two populations at once." Suppose you took a sample of weights of college football players and with this sample you included weights of cheerleaders. Do you think a histogram made from the combined weights would be bimodal? Explain.
3. Discuss the statement that stem-and-leaf displays are quick and easy to construct. How can we use a stem-and-leaf display to make the construction of a frequency table easier? How does a stem-and-leaf display help you spot extreme values quickly?
4. Go to the library and pick up a current issue of *The Wall Street Journal, Newsweek, Time, USA Today,* or other news medium. Examine each newspaper or magazine for graphs of the types discussed in this chapter. List the variables used, method of data collection, and general types of conclusions drawn from the graphs. Another source for information is the Internet. Explore several web sites, and categorize the graphs you find as you did for the print media. For interesting web sites, visit the Brase/Brase statistics site at **http://www.cengage.com/statistics/brase11e** and find links to the Social Statistics Briefing Room, to law enforcement, and to golf.

USING TECHNOLOGY

Applications

The following tables show the first-round winning scores of the NCAA men's and women's basketball teams.

TABLE 2-17 **Men's Winning First-Round NCAA Tournament Scores**

95	70	79	99	83	72	79	101
69	82	86	70	79	69	69	70
95	70	77	61	69	68	69	72
89	66	84	77	50	83	63	58

TABLE 2-18 **Women's Winning First-Round NCAA Tournament Scores**

80	68	51	80	83	75	77	100
96	68	89	80	67	84	76	70
98	81	79	89	98	83	72	100
101	83	66	76	77	84	71	77

1. Use the software or method of your choice to construct separate histograms for the men's and women's winning scores. Try 5, 7, and 10 classes for each. Which number of classes seems to be the best choice? Why?
2. Use the same class boundaries for histograms of men's and of women's scores. How do the scores for the two groups compare? What general shape do the histograms follow?
3. Use the software or method of your choice to make stem-and-leaf displays for each set of scores. If your software does not make stem-and-leaf displays, sort the data first and then make a back-to-back display by hand. Do there seem to be any extreme values in either set? How do the data sets compare?

Technology Hints: Creating Histograms

The default histograms produced by the TI-84Plus/TI-83Plus/TI-*n*spire calculators, Minitab, and Excel all determine automatically the number of classes to use. To control the number of classes the technology uses, follow the key steps as indicated. The display screens are generated for data found in Table 2-1, One-Way Commuting Distances (in Miles) for 60 Workers in Downtown Dallas.

TI-84Plus/TI-83Plus/TI-*n*spire (with TI-84Plus keypad)

Determine the class width for the number of classes you want and the lower class boundary for the first class. Enter the data in list L1.

Press **STATPLOT** and highlight On and the histogram plot.

Press **WINDOW** and set X min = lowest class boundary, Xscl = class width. Use appropriate values for the other settings.

Press **GRAPH. TRACE** gives boundaries and frequency.

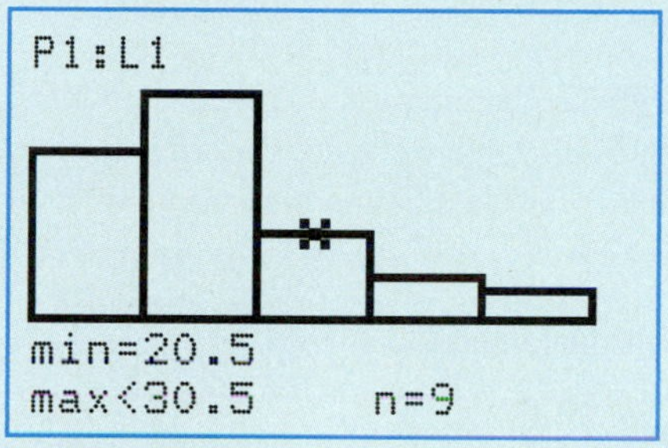

Excel 2010

Determine the upper class boundaries for the five classes. Enter the data in one column. In a separate column, enter the upper class boundaries. Click the **Data** tab, and then click the **Data Analysis** button on the ribbon in the **Analysis** group. Select **Histogram** and click **OK**.

In the Histogram dialogue box, put the data range in the Input Range. Put the upper class boundaries range in the Bin Range. Then check **New Workbook** and **Chart Output** and click **OK**.

To make the bars touch, right click on a bar and select **Format Data Series...** Then under **Series Options** move the **Gap Width** slider to 0% (No Gap) and click **Close**.

Further adjustments to the histogram can be made by clicking the **Layout** tab and selecting options in the labels or axes section of the ribbon.

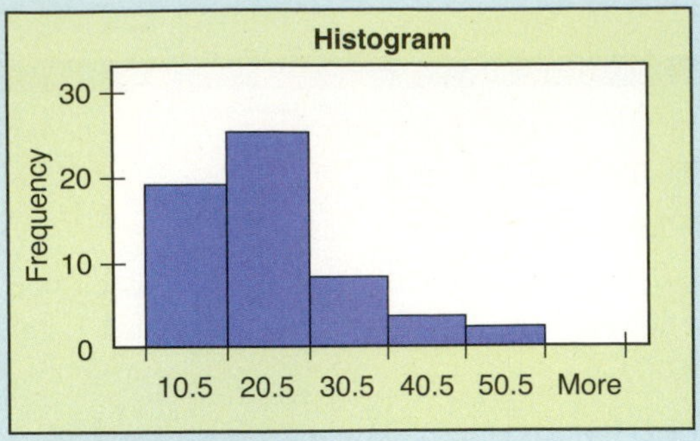

Minitab

Determine the class boundaries. Enter the data. Use the menu selection **Graph ➤ Histogram.**

Choose **Simple** and click **OK.** Select the graph variable and click **OK** to obtain a histogram with automatically selected classes. To set your own class boundaries, double click the displayed histogram. In the dialogue box, select **Binning.** Then choose **Cutpoint** and enter the class boundaries as cutpoint positions.

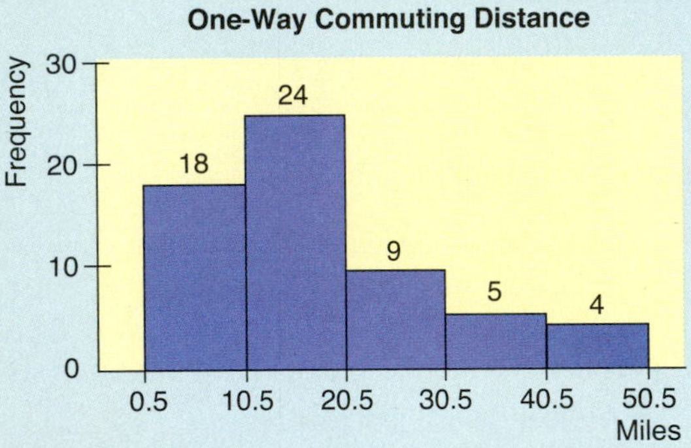

SPSS

The SPSS screen shot shows the default histogram created by the menu choices **Analyze ➤ Descriptive Statistics ➤ Frequencies.** In the dialogue box, move the variable containing the data into the variables window. Click **Charts** and select **Histograms.** Click the Continue button and then the OK button.

In SPSS Student Version 17 there are a number of additional ways to create histograms and other graphs. These options utilize the **Chart Builder** or **Graphboard Template Chooser** items in the **Graphs** menu. Under these options you can set the number of classes and the boundaries of the histogram. Specific instructions for using these options are provided in the *Technology Guide* for SPSS that accompanies this text.

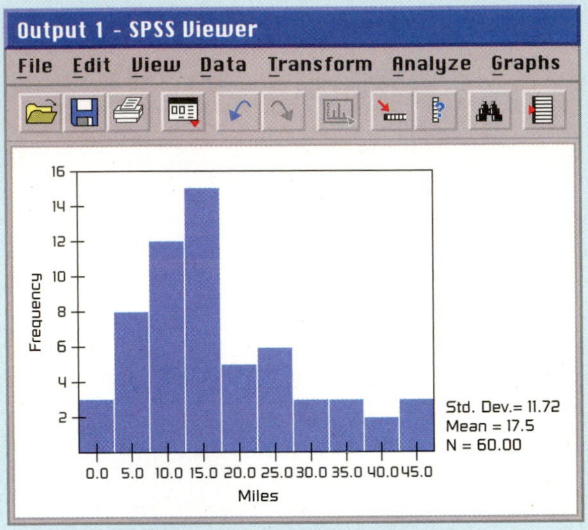

Note that SPSS requires that the measurement level of the data be set before using the **Chart Builder**. In SPSS, the **scale level** includes both interval and ratio levels of measurement. The **nominal** and **ordinal** levels are as described in this text.

Specific instructions for setting class boundaries (cutpoints) of a histogram are provided in the *Technology Guide* that accompanies this text.

3

While the individual man is an insolvable puzzle, in the aggregate he becomes a mathematical certainty. You can, for example, never foretell what any one man will do, but you can say with precision what an average number will be up to.

—Arthur Conan Doyle, *The Sign of Four*

Sherlock Holmes spoke the words quoted here to his colleague, Dr. Watson, as the two were unraveling a mystery. The detective was implying that if a single member is drawn at random from a population, we cannot predict *exactly* what that member will look like. However, there are some "average" features of the entire population that an individual is likely to possess. The degree of certainty with which we would expect to observe such average features in any individual depends on our knowledge of the variations among individuals in the population. Sherlock Holmes has led us to two of the most important statistical concepts: average and variation.

AVERAGES AND VARIATION

PREVIEW QUESTIONS

What are commonly used measures of central tendency? What do they tell you? (SECTION 3.1)

How do variance and standard deviation measure data spread? Why is this important? (SECTION 3.2)

How do you make a box-and-whisker plot, and what does it tell about the spread of the data? (SECTION 3.3)

Nancy Sheehan/PhotoEdit

FOCUS PROBLEM

The Educational Advantage

Is it really worth all the effort to get a college degree? From a philosophical point of view, the love of learning is sufficient reason to get a college degree. However, the U.S. Census Bureau also makes another relevant point. Annually, college graduates (bachelor's degree) earn on average $23,788 more than high school graduates. This means college graduates earn about 78.5% more than high school graduates, and according to "Education Pays" on the next page, the gap in earnings is increasing. Furthermore, as the College Board indicates, for most Americans college remains relatively affordable.

After completing this chapter, you will be able to answer the following questions.

(a) Does a college degree *guarantee* someone an 78.5% increase in earnings over a high school degree? Remember, we are using only *averages* from census data.

(b) Using census data (not shown in "Education Pays"), it is estimated that the standard deviation of college-graduate earnings is about $8500. Compute a 75% Chebyshev confidence interval centered on the mean ($54,091) for bachelor's degree earnings.

(c) How much does college tuition cost? That depends, of course, on where you go to college. The 10th percentile, 25th percentile, median (50th percentile), 75th percentile, and 90th percentile for annual tuition and required fees at public 4-year (in-state) and private 4-year not-for-profit institutions are shown next to "Education pays." This information is from the National Center for Education Statistics. Using these data, what is the range of annual tuition and fees for the group between the 25th percentile and the 75th

Myrleen Ferguson Cate/PhotoEdit

For online student resources, visit the Brase/Brase, *Understandable Statistics*, 11th edition web site at http://www.cengage.com/statistics/brase11e

percentile for each type of institution? About what percent of private not-for-profit 4-year colleges have annual tuition and fees between \$34,417 and \$40,082? What percent of public 4-year colleges have annual tuition and fees at or below \$6,780?

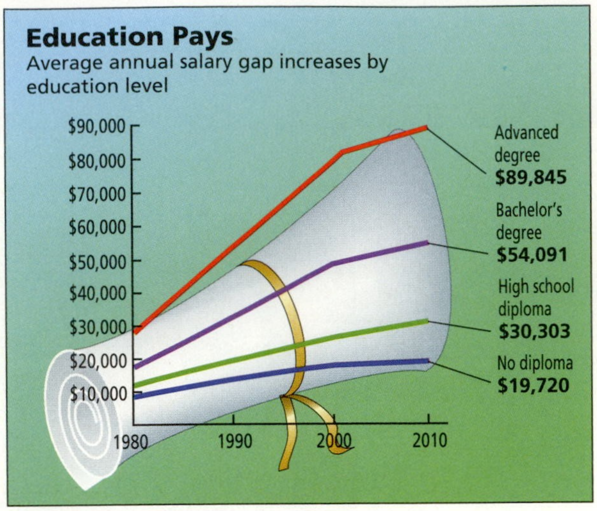

Source: Census Bureau

Annual Undergraduate Tuition and Fees (2010–2011)
Percentiles for Public 4-Year Colleges (for In-State Students)

10th	25th	50th	75th	90th
\$4,336	\$5,105	\$6,780	\$8,689	\$11,029

Percentiles for Private Not-for-Profit 4-Year Colleges

10th	25th	50th	75th	90th
\$12,220	\$19,854	\$27,100	\$34,417	\$40,082

SECTION **3.1**

Measures of Central Tendency: Mode, Median, and Mean

FOCUS POINTS

- Compute mean, median, and mode from raw data.
- Interpret what mean, median, and mode tell you.
- Explain how mean, median, and mode can be affected by extreme data values.
- What is a trimmed mean? How do you compute it?
- Compute a weighted average.

The average price of an ounce of gold is \$1350. The Zippy car averages 39 miles per gallon on the highway. A survey showed the average shoe size for women is size 9.

Average

In each of the preceding statements, *one* number is used to describe the entire sample or population. Such a number is called an *average.* There are many ways to compute averages, but we will study only three of the major ones.

The easiest average to compute is the *mode.*

Mode

The **mode** of a data set is the value that occurs most frequently. *Note:* If a data set has no single value that occurs more frequently than any other, then that data set has no mode.

EXAMPLE 1

MODE

Count the letters in each word of this sentence and give the mode. The numbers of letters in the words of the sentence are

5 3 7 2 4 4 2 4 8 3 4 3 4

Scanning the data, we see that 4 is the mode because more words have 4 letters than any other number. For larger data sets, it is useful to order—or sort—the data before scanning them for the mode.

Not every data set has a mode. For example, if Professor Fair gives equal numbers of A's, B's, C's, D's, and F's, then there is no modal grade. In addition, the mode is not very stable. Changing just one number in a data set can change the mode dramatically. However, the mode is a useful average when we want to know the most frequently occurring data value, such as the most frequently requested shoe size.

Median

Another average that is useful is the *median,* or central value, of an ordered distribution. When you are given the median, you know there are an equal number of data values in the ordered distribution that are above it and below it.

PROCEDURE

HOW TO FIND THE MEDIAN

The **median** is the central value of an ordered distribution. To find it,

1. Order the data from smallest to largest.
2. For an *odd* number of data values in the distribution,

$$\text{Median} = \text{Middle data value}$$

3. For an *even* number of data values in the distribution,

$$\text{Median} = \frac{\text{Sum of middle two values}}{2}$$

EXAMPLE 2

MEDIAN

What do barbecue-flavored potato chips cost? According to *Consumer Reports,* Vol. 66, No. 5, the prices per ounce in cents of the rated chips are

19 19 27 28 18 35

(a) To find the median, we first order the data, and then note that there are an even number of entries. So the median is constructed using the two middle values.

18 19 19 27 28 35

(19 and 27: middle values)

$$\text{Median} = \frac{19 + 27}{2} = 23 \text{ cents}$$

(b) According to *Consumer Reports,* the brand with the lowest overall taste rating costs 35 cents per ounce. Eliminate that brand, and find the median price per ounce for the remaining barbecue-flavored chips. Again order the data. Note that there are an odd number of entries, so the median is simply the middle value.

18 19 19 27 28

(19 ↑ middle values)

$$\text{Median} = \text{middle value} = 19 \text{ cents}$$

(c) One ounce of potato chips is considered a small serving. Is it reasonable to budget about $10.45 to serve the barbecue-flavored chips to 55 people?

Yes, since the median price of the chips is 19 cents per small serving. This budget for chips assumes that there is plenty of other food!

The median uses the *position* rather than the specific value of each data entry. If the extreme values of a data set change, the median usually does not change. This is why the median is often used as the average for house prices. If one mansion costing several million dollars sells in a community of much-lower-priced homes, the median selling price for houses in the community would be affected very little, if at all.

GUIDED EXERCISE 1 — MEDIAN AND MODE

Belleview College must make a report to the budget committee about the average credit hour load a full-time student carries. (A 12-credit-hour load is the minimum requirement for full-time status. For the same tuition, students may take up to 20 credit hours.) A random sample of 40 students yielded the following information (in credit hours):

17	12	14	17	13	16	18	20	13	12
12	17	16	15	14	12	12	13	17	14
15	12	15	16	12	18	20	19	12	15
18	14	16	17	15	19	12	13	12	15

(a) Organize the data from smallest to largest number of credit hours.

→

12	12	12	12	12	12	12	12	12	12
13	13	13	13	14	14	14	14	15	(15)
(15)	15	15	15	16	16	16	16	17	17
17	17	17	18	18	18	19	19	20	20

(b) Since there are an _____ (odd, even) number of values, we add the two middle values and divide by 2 to get the median. What is the median credit hour load?

→ There are an even number of entries. The two middle values are circled in part (a).

$$\text{Median} = \frac{15 + 15}{2} = 15$$

(c) What is the mode of this distribution? Is it different from the median? *Interpretation* If the budget committee is going to fund the college according to the average student credit hour load (more money for higher loads), which of these two averages do you think the college will report?

→ The mode is 12. It is different from the median. Since the median is higher, the college will probably report it and indicate that the average being used is the median.

Note: For small ordered data sets, we can easily scan the set to find the *location* of the median. However, for large ordered data sets of size *n*, it is convenient to have a formula to find the middle of the data set.

> For an ordered data set of size n,
>
> $$\text{Position of the middle value} = \frac{n + 1}{2}$$

For instance, if $n = 99$ then the middle value is the $(99 + 1)/2$ or 50th data value in the ordered data. If $n = 100$, then $(100 + 1)/2 = 50.5$ tells us that the two middle values are in the 50th and 51st positions.

Mean

An average that uses the exact value of each entry is the *mean* (sometimes called the *arithmetic mean*). To compute the mean, we add the values of all the entries and then divide by the number of entries.

$$\text{Mean} = \frac{\text{Sum of all entries}}{\text{Number of entries}}$$

The mean is the average usually used to compute a test average.

EXAMPLE 3

MEAN

To graduate, Linda needs at least a B in biology. She did not do very well on her first three tests; however, she did well on the last four. Here are her scores:

58 67 60 84 93 98 100

Compute the mean and determine if Linda's grade will be a B (80 to 89 average) or a C (70 to 79 average).

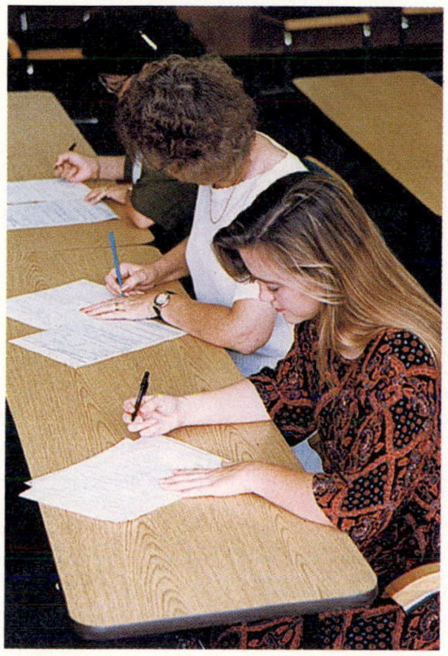
Jose Carillo/PhotoEdit

SOLUTION:

$$\text{Mean} = \frac{\text{Sum of scores}}{\text{Number of scores}} = \frac{58 + 67 + 60 + 84 + 93 + 98 + 100}{7}$$

$$= \frac{560}{7} = 80$$

Since the average is 80, Linda will get the needed B.

COMMENT When we compute the mean, we sum the given data. There is a convenient notation to indicate the sum. Let x represent any value in the data set. Then the notation

Σx (read "the sum of all given x values")

Summation symbol, Σ

means that we are to sum all the data values. In other words, we are to sum all the entries in the distribution. The *summation symbol* Σ means *sum the following* and is capital sigma, the *S* of the Greek alphabet.

Sample mean, $\bar{x}$
Population mean, μ

The symbol for the mean of a *sample* distribution of x values is denoted by $\bar{x}$ (read "x bar"). If your data comprise the entire *population,* we use the symbol μ (lowercase Greek letter mu, pronounced "mew") to represent the mean.

PROCEDURE

HOW TO FIND THE MEAN

1. Compute Σx; that is, find the sum of all the data values.
2. Divide the sum total by the number of data values.

Sample statistic $\bar{x}$	Population parameter μ
$\bar{x} = \frac{\Sigma x}{n}$	$\mu = \frac{\Sigma x}{N}$

where n = number of data values in the sample
N = number of data values in the population

When data consist of rates of change, the *harmonic mean* is the appropriate average to use. Problem 29 discusses the harmonic mean.

Problem 30 discusses the *geometric mean*, which is a useful average for percentages, ratios, or growth rates.

LOOKING FORWARD

In our future work with inferential statistics, we will use the mean $\bar{x}$ from a random sample to estimate the population parameter μ (Chapter 7) or to make decisions regarding the value of μ (Chapter 8).

CALCULATOR NOTE It is very easy to compute the mean on *any* calculator: Simply add the data values and divide the total by the number of data. However, on calculators with a statistics mode, you place the calculator in that mode, *enter* the data, and then press the key for the mean. The key is usually designated $\bar{x}$. Because the formula for the population mean is the same as that for the sample mean, the same key gives the value for μ.

WHAT DO AVERAGES TELL US?

An average provides a one-number summary of a data set.

- The **mode** tells us the single data value that occurs most frequently in the data set. The value of the mode is completely determined by the data value that occurs most frequently. If no data value occurs more frequently than all the other data values, there is no mode. The specific values of the less frequently occurring data do not change the mode.
- The **median** tells us the middle value of a data set that has been arranged in order from smallest to largest. The median is affected by only the relative position of the data values. For instance, if a data value above the median (or above the middle two values of a data set with an even number of data) is changed to another value above the median, the median itself does not change.
- The **mean** tells us the value obtained by adding up *all* the data and dividing by the number of data. As such, the mean can change if just one data value changes. On the other hand, if data values change, but the sum of the data remains the same, the mean will not change.

We have seen three averages: the mode, the median, and the mean. For later work, the mean is the most important. A disadvantage of the mean, however, is that it can be affected by exceptional values.

Resistant measure

A *resistant measure* is one that is not influenced by extremely high or low data values. The mean is not a resistant measure of center because we can make the mean as large as we want by changing the size of only one data value. The median, on the other hand, is more resistant. However, a disadvantage of the median is that it is not sensitive to the specific size of a data value.

Trimmed mean

A measure of center that is more resistant than the mean but still sensitive to specific data values is the *trimmed mean*. A trimmed mean is the mean of the data values left after "trimming" a specified percentage of the smallest and largest data values from the data set. Usually a 5% trimmed mean is used. This implies that we trim the lowest 5% of the data as well as the highest 5% of the data. A similar procedure is used for a 10% trimmed mean.

PROCEDURE

HOW TO COMPUTE A 5% TRIMMED MEAN

1. Order the data from smallest to largest.
2. Delete the bottom 5% of the data and the top 5% of the data. *Note:* If the calculation of 5% of the number of data values does not produce a whole number, *round* to the nearest integer.
3. Compute the mean of the remaining 90% of the data.

GUIDED EXERCISE 2 | MEAN AND TRIMMED MEAN

Barron's Profiles of American Colleges, 19th Edition, lists average class size for introductory lecture courses at each of the profiled institutions. A sample of 20 colleges and universities in California showed class sizes for introductory lecture courses to be

(14)	20	20	20	20	23	25	30	30	30
35	35	35	40	40	42	50	50	80	(80)

(a) Compute the mean for the entire sample.

→ Add all the values and divide by 20:

$$\bar{x} = \frac{\Sigma x}{n} = \frac{719}{20} \approx 36.0$$

(b) Compute a 5% trimmed mean for the sample.

→ The data are already ordered. Since 5% of 20 is 1, we eliminate one data value from the bottom of the list and one from the top. These values are circled in the data set. Then we take the mean of the remaining 18 entries.

$$5\% \text{ trimmed mean} = \frac{\Sigma x}{n} = \frac{625}{18} \approx 34.7$$

(c) Find the median of the original data set.

→ Note that the data are already ordered.

$$\text{Median} = \frac{30 + 35}{2} = 32.5$$

(d) Find the median of the 5% trimmed data set. Does the median change when you trim the data?

→ The median is still 32.5. Notice that trimming the same number of entries from both ends leaves the middle position of the data set unchanged.

(e) Is the trimmed mean or the original mean closer to the median?

→ The trimmed mean is closer to the median.

TECH NOTES

Minitab, Excel 2010, and TI-84Plus/TI-83Plus/TI-*n*spire calculators all provide the mean and median of a data set. Minitab and Excel 2010 also provide the mode. The TI-84Plus/TI-83Plus/TI-*n*spire calculators sort data, so you can easily scan the sorted data for the mode. Minitab provides the 5% trimmed mean, as does Excel 2010.

All this technology is a wonderful aid for analyzing data. However, *a measurement is worthless if you do not know what it represents or how a change in data values might affect the measurement.* The defining formulas and procedures for computing the measures tell you a great deal about the measures. Even if you use a calculator to evaluate all the statistical measures, pay attention to the information the formulas and procedures give you about the components or features of the measurement.

CRITICAL THINKING

Data types and averages

In Chapter 1, we examined four levels of data: nominal, ordinal, interval, and ratio. The mode (if it exists) can be used with all four levels, including nominal. For instance, the modal color of all passenger cars sold last year might be blue. The median may be used with data at the ordinal level or

Continued

CRITICAL THINKING *continued*

above. If we ranked the passenger cars in order of customer satisfaction level, we could identify the median satisfaction level. For the mean, our data need to be at the interval or ratio level (although there are exceptions in which the mean of ordinal-level data is computed). We can certainly find the mean model year of used passenger cars sold or the mean price of new passenger cars.

Another issue of concern is that of taking the average of averages. For instance, if the values \$520, \$640, \$730, \$890, and \$920 represent the mean monthly rents for five different apartment complexes, we can't say that \$740 (the mean of the five numbers) is the mean monthly rent of all the apartments. We need to know the number of apartments in each complex before we can determine an average based on the number of apartments renting at each designated amount.

Distribution shapes and averages

In general, when a data distribution is mound-shaped symmetrical, the values for the mean, median, and mode are the same or almost the same. For skewed-left distributions, the mean is less than the median and the median is less than the mode. For skewed-right distributions, the mode is the smallest value, the median is the next largest, and the mean is the largest. Figure 3-1 shows the general relationships among the mean, median, and mode for different types of distributions.

WEIGHTED AVERAGE

Sometimes we wish to average numbers, but we want to assign more importance, or weight, to some of the numbers. For instance, suppose your professor tells you that your grade will be based on a midterm and a final exam, each of which is based on 100 possible points. However, the final exam will be worth 60% of the grade and the midterm only 40%. How could you determine an average score that would reflect these different weights? The average you need is the *weighted average*.

Weighted average

$$\text{Weighted average} = \frac{\Sigma xw}{\Sigma w}$$

where x is a data value and w is the weight assigned to that data value. The sum is taken over all data values.

FIGURE 3-1

Distribution Types and Averages

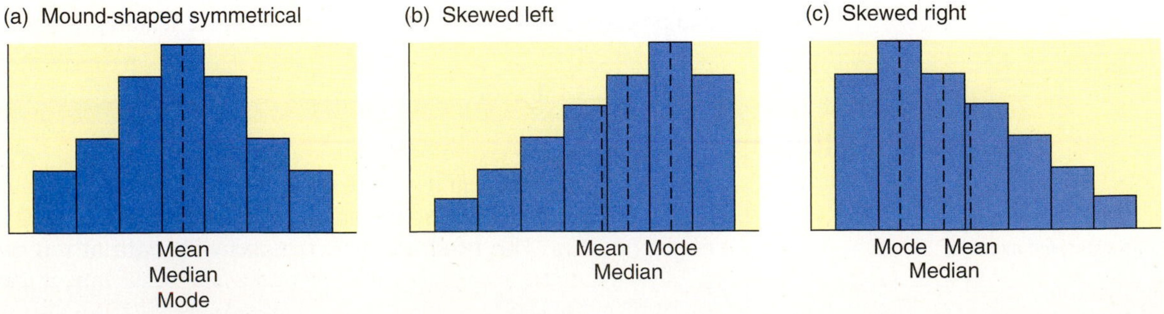

EXAMPLE 4

WEIGHTED AVERAGE

Suppose your midterm test score is 83 and your final exam score is 95. Using weights of 40% for the midterm and 60% for the final exam, compute the weighted average of your scores. If the minimum average for an A is 90, will you earn an A?

SOLUTION: By the formula, we multiply each score by its weight and add the results together. Then we divide by the sum of all the weights. Converting the percentages to decimal notation, we get

$$\text{Weighted average} = \frac{83(0.40) + 95(0.60)}{0.40 + 0.60}$$

$$= \frac{33.2 + 57}{1} = 90.2$$

Your average is high enough to earn an A.

TECH NOTES The TI-84Plus/TI-83Plus/TI-*n*spire calculators directly support weighted averages. Both Excel 2010 and Minitab can be programmed to provide the averages.

TI-84Plus/TI-83Plus/TI-*n*spire (with TI-84 Plus keypad) Enter the data into one list, such as L1, and the corresponding weights into another list, such as L2. Then press **Stat ➤ Calc ➤ 1: 1-Var Stats**. Enter the list containing the data, followed by a comma and the list containing the weights.

VIEWPOINT What's Wrong with Pitching Today?

One way to answer this question is to look at averages. Batting averages and average hits per game are shown for selected years from 1901 to 2000 (Source: The Wall Street Journal*).*

Year	1901	1920	1930	1941	1951	1961	1968	1976	1986	2000
B.A.	0.277	0.284	0.288	0.267	0.263	0.256	0.231	0.256	0.262	0.276
Hits	19.2	19.2	20.0	18.4	17.9	17.3	15.2	17.3	17.8	19.1

A quick scan of the averages shows that batting averages and average hits per game are virtually the same as they were almost 100 years ago. It seems there is nothing wrong with today's pitching! So what's changed? For one thing, the rules have changed! The strike zone is considerably smaller than it once was, and the pitching mound is lower. Both give the hitter an advantage over the pitcher. Even so, pitchers don't give up hits with any greater frequency than they did a century ago (look at the averages). However, modern hits go much farther, which is something a pitcher can't control.

SECTION 3.1 PROBLEMS

1. *Statistical Literacy* Consider the mode, median, and mean. Which average represents the middle value of a data distribution? Which average represents the most frequent value of a distribution? Which average takes all the specific values into account?

2. *Statistical Literacy* What symbol is used for the arithmetic mean when it is a sample statistic? What symbol is used when the arithmetic mean is a population parameter?

3. *Statistical Literacy* Look at the formula for the mean. List the two arithmetic procedures that are used to compute the mean.

4. *Statistical Literacy* In order to find the median of a data set, what do we do first with the data?

5. *Basic Computation: Mean, Median, Mode* Find the mean, median, and mode of the data set

   8 2 7 2 6

6. *Basic Computation: Mean, Median, Mode* Find the mean, median, and mode of the data set

   10 12 20 15 20

7. *Basic Computations Mean, Median, Mode* Find the mean, median, and mode of the data set

   8 2 7 2 6 5

8. *Critical Thinking* Consider a data set with at least three data values. Suppose the highest value is increased by 10 and the lowest is decreased by 5.
   (a) Does the mean change? Explain
   (b) Does the median change? Explain
   (c) Is it possible for the mode to change? Explain.

9. *Critical Thinking* Consider a data set with at least three data values. Suppose the highest value is increased by 10 and the lowest is decreased by 10.
   (a) Does the mean change? Explain
   (b) Does the median change? Explain
   (c) Is it possible for the mode to change? Explain.

10. *Critical Thinking* If a data set has an even number of data, is it true or false that the median is never equal to a value in the data set? Explain.

11. *Critical Thinking* When a distribution is mound-shaped symmetrical, what is the general relationship among the values of the mean, median, and mode?

12. *Critical Thinking* Consider the following types of data that were obtained from a random sample of 49 credit card accounts. Identify all the averages (mean, median, or mode) that can be used to summarize the data.
    (a) Outstanding balance on each account
    (b) Name of credit card (e.g., MasterCard, Visa, American Express, etc.)
    (c) Dollar amount due on next payment

13. *Critical Thinking* Consider the numbers

    2 3 4 5 5

    (a) Compute the mode, median, and mean.
    (b) If the numbers represent codes for the colors of T-shirts ordered from a catalog, which average(s) would make sense?
    (c) If the numbers represent one-way mileages for trails to different lakes, which average(s) would make sense?
    (d) Suppose the numbers represent survey responses from 1 to 5, with 1 = disagree strongly, 2 = disagree, 3 = agree, 4 = agree strongly, and 5 = agree very strongly. Which averages make sense?

14. *Critical Thinking* Consider two data sets.

    Set A: $n = 5; \bar{x} = 10$ Set B: $n = 50; \bar{x} = 10$

    (a) Suppose the number 20 is included as an additional data value in Set A. Compute $\bar{x}$ for the new data set. *Hint:* $\Sigma x = n\bar{x}$. To compute $\bar{x}$ for the new data set, add 20 to Σx of the original data set and divide by 6.

(b) Suppose the number 20 is included as an additional data value in Set B. Compute $\bar{x}$ for the new data set.
(c) Why does the addition of the number 20 to each data set change the mean for Set A more than it does for Set B?

15. *Interpretation* A job-performance evaluation form has these categories:

1 = excellent; 2 = good; 3 = satisfactory; 4 = poor; 5 = unacceptable

Based on 15 client reviews, one employee had

median rating of 4; mode rating of 1

The employee was pleased that most clients had rated her as excellent. The supervisor said improvement was needed because at least half the clients had rated the employee at the poor or unacceptable level. Comment on the different perspectives.

16. *Critical Thinking: Data Transformation* In this problem, we explore the effect on the mean, median, and mode of adding the same number to each data value. Consider the data set 2, 2, 3, 6, 10.
(a) Compute the mode, median, and mean.
(b) Add 5 to each of the data values. Compute the mode, median, and mean.
(c) Compare the results of parts (a) and (b). In general, how do you think the mode, median, and mean are affected when the same constant is added to each data value in a set?

17. *Critical Thinking: Data Transformation* In this problem, we explore the effect on the mean, median, and mode of multiplying each data value by the same number. Consider the data set 2, 2, 3, 6, 10.
(a) Compute the mode, median, and mean.
(b) Multiply each data value by 5. Compute the mode, median, and mean.
(c) Compare the results of parts (a) and (b). In general, how do you think the mode, median, and mean are affected when each data value in a set is multiplied by the same constant?
(d) Suppose you have information about average heights of a random sample of airplane passengers. The mode is 70 inches, the median is 68 inches, and the mean is 71 inches. To convert the data into centimeters, multiply each data value by 2.54. What are the values of the mode, median, and mean in centimeters?

18. *Critical Thinking* Consider a data set of 15 distinct measurements with mean *A* and median *B*.
(a) If the highest number were increased, what would be the effect on the median and mean? Explain.
(b) If the highest number were decreased to a value still larger than *B*, what would be the effect on the median and mean?
(c) If the highest number were decreased to a value smaller than *B*, what would be the effect on the median and mean?

19. *Environmental Studies: Death Valley* How hot does it get in Death Valley? The following data are taken from a study conducted by the National Park System, of which Death Valley is a unit. The ground temperatures (°F) were taken from May to November in the vicinity of Furnace Creek.

146	152	168	174	180	178	179
180	178	178	168	165	152	144

Compute the mean, median, and mode for these ground temperatures.

20. *Ecology: Wolf Packs* How large is a wolf pack? The following information is from a random sample of winter wolf packs in regions of Alaska, Minnesota, Michigan, Wisconsin, Canada, and Finland (Source: *The Wolf*, by L. D. Mech, University of Minnesota Press). Winter pack size:

13	10	7	5	7	7	2	4	3
2	3	15	4	4	2	8	7	8

Compute the mean, median, and mode for the size of winter wolf packs.

21. *Medical: Injuries* The Grand Canyon and the Colorado River are beautiful, rugged, and sometimes dangerous. Thomas Myers is a physician at the park clinic in Grand Canyon Village. Dr. Myers has recorded (for a 5-year period) the number of visitor injuries at different landing points for commercial boat trips down the Colorado River in both the Upper and Lower Grand Canyon (Source: *Fateful Journey* by Myers, Becker, and Stevens).

Upper Canyon: Number of Injuries per Landing Point Between North Canyon and Phantom Ranch

2	3	1	1	3	4	6	9	3	1	3

Lower Canyon: Number of Injuries per Landing Point Between Bright Angel and Lava Falls

8	1	1	0	6	7	2	14	3	0	1	13	2	1

Fuse/Getty Images

(a) Compute the mean, median, and mode for injuries per landing point in the Upper Canyon.
(b) Compute the mean, median, and mode for injuries per landing point in the Lower Canyon.
(c) Compare the results of parts (a) and (b).
(d) The Lower Canyon stretch had some extreme data values. Compute a 5% trimmed mean for this region, and compare this result to the mean for the Upper Canyon computed in part (a).

22. *Football: Age of Professional Players* How old are professional football players? The 11th Edition of *The Pro Football Encyclopedia* gave the following information. Random sample of pro football player ages in years:

24	23	25	23	30	29	28	26	33	29
24	37	25	23	22	27	28	25	31	29
25	22	31	29	22	28	27	26	23	21
25	21	25	24	22	26	25	32	26	29

(a) Compute the mean, median, and mode of the ages.
(b) *Interpretation* Compare the averages. Does one seem to represent the age of the pro football players most accurately? Explain.

© iStockphoto.com/Dale Walsh

23. *Leisure: Maui Vacation* How expensive is Maui? If you want a vacation rental condominium (up to four people), visit the Brase/Brase statistics site at **http://www.cengage.com/statistics/brase11e,** find the link to Maui, and then search for accommodations. The *Maui News* gave the following costs in dollars per day for a random sample of condominiums located throughout the island of Maui.

89	50	68	60	375	55	500	71	40	350
60	50	250	45	45	125	235	65	60	130

(a) Compute the mean, median, and mode for the data.
(b) Compute a 5% trimmed mean for the data, and compare it with the mean computed in part (a). Does the trimmed mean more accurately reflect the general level of the daily rental costs?
(c) *Interpretation* If you were a travel agent and a client asked about the daily cost of renting a condominium on Maui, what average would you use? Explain. Is there any other information about the costs that you think might be useful, such as the spread of the costs?

24. *Basic Computation: Weighted Average* Find the weighted average of a data set where

10 has a weight of 5; 20 has a weight of 3; 30 has a weight of 2

25. *Basic Computation: Weighted Average* Find the weighted average of a data set where

10 has a weight of 2; 20 has a weight of 3; 30 has a weight of 5

26. *Grades: Weighted Average* In your biology class, your final grade is based on several things: a lab score, scores on two major tests, and your score on the final exam. There are 100 points available for each score. However, the lab score is worth 25% of your total grade, each major test is worth 22.5%, and the final exam is worth 30%. Compute the weighted average for the following scores: 92 on the lab, 81 on the first major test, 93 on the second major test, and 85 on the final exam.

27. *Merit Pay Scale: Weighted Average* At General Hospital, nurses are given performance evaluations to determine eligibility for merit pay raises. The supervisor rates the nurses on a scale of 1 to 10 (10 being the highest rating) for several activities: promptness, record keeping, appearance, and bedside manner with patients. Then an average is determined by giving a weight of 2 for promptness, 3 for record keeping, 1 for appearance, and 4 for bedside manner with patients. What is the average rating for a nurse with ratings of 9 for promptness, 7 for record keeping, 6 for appearance, and 10 for bedside manner?

28. *EPA: Wetlands* Where does all the water go? According to the Environmental Protection Agency (EPA), in a typical wetland environment, 38% of the water is outflow; 47% is seepage; 7% evaporates; and 8% remains as water volume in the ecosystem (Reference: U.S. Environmental Protection Agency Case Studies Report 832-R-93-005). Chloride compounds as residuals from residential areas are a problem for wetlands. Suppose that in a particular wetland environment the following concentrations (mg/L) of chloride compounds were found: outflow, 64.1; seepage, 75.8; remaining due to evaporation, 23.9; in the water volume, 68.2.
(a) Compute the weighted average of chlorine compound concentration (mg/L) for this ecological system.
(b) Suppose the EPA has established an average chlorine compound concentration target of no more than 58 mg/L. Comment on whether this wetlands system meets the target standard for chlorine compound concentration.

29. *Expand Your Knowledge: Harmonic Mean* When data consist of rates of change, such as speeds, the *harmonic mean* is an appropriate measure of central tendency. For *n* data values,

Harmonic mean

$$\text{Harmonic mean} = \frac{n}{\Sigma \frac{1}{x}}, \quad \text{assuming no data value is 0}$$

Suppose you drive 60 miles per hour for 100 miles, then 75 miles per hour for 100 miles. Use the harmonic mean to find your average speed.

30. *Expand Your Knowledge: Geometric Mean* When data consist of percentages, ratios, growth rates, or other rates of change, the *geometric mean* is a useful measure of central tendency. For n data values,

Geometric mean

$$\text{Geometric mean} = \sqrt[n]{\text{product of the } n \text{ data values}}, \quad \text{assuming all data values are positive}$$

To find the *average growth factor* over 5 years of an investment in a mutual fund with growth rates of 10% the first year, 12% the second year, 14.8% the third year, 3.8% the fourth year, and 6% the fifth year, take the geometric mean of 1.10, 1.12, 1.148, 1.038, and 1.16. Find the average growth factor of this investment.

Note that for the same data, the relationships among the harmonic, geometric, and arithmetic means are harmonic mean ≤ geometric mean ≤ arithmetic mean (Source: *Oxford Dictionary of Statistics*).

SECTION 3.2 Measures of Variation

FOCUS POINTS

- Find the range, variance, and standard deviation.
- Compute the coefficient of variation from raw data. Why is the coefficient of variation important?
- Apply Chebyshev's theorem to raw data. What does a Chebyshev interval tell us?

An average is an attempt to summarize a set of data using just one number. As some of our examples have shown, an average taken by itself may not always be very meaningful. We need a statistical cross-reference that measures the spread of the data.

The *range* is one such measure of variation.

Range

The **range** is the difference between the largest and smallest values of a data distribution.

EXAMPLE 5

RANGE

A large bakery regularly orders cartons of Maine blueberries. The average weight of the cartons is supposed to be 22 ounces. Random samples of cartons from two suppliers were weighed. The weights in ounces of the cartons were

Supplier I:	17	22	22	22	27
Supplier II:	17	19	20	27	27

(a) Compute the range of carton weights from each supplier.

$$\text{Range} = \text{Largest value} - \text{Smallest value}$$
$$\text{Supplier I range} = 27 - 17 = 10 \text{ ounces}$$
$$\text{Supplier II range} = 27 - 17 = 10 \text{ ounces}$$

(b) Compute the mean weight of cartons from each supplier. In both cases the mean is 22 ounces.

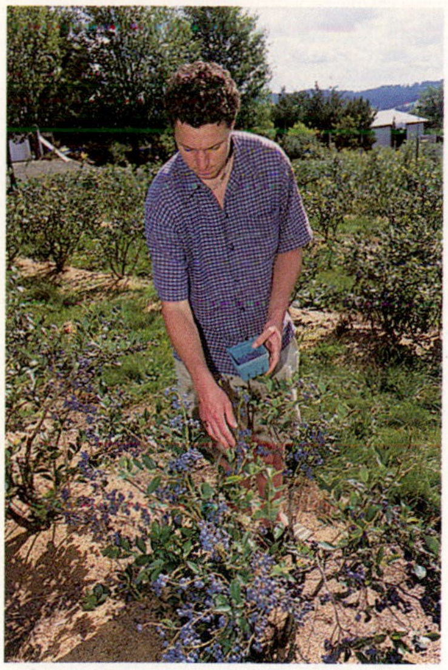

David Young-Wolff/PhotoEdit

Blueberry patch

(c) Look at the two samples again. The samples have the same range and mean. How do they differ? The bakery uses one carton of blueberries in each blueberry muffin recipe. It is important that the cartons be of consistent weight so that the muffins turn out right.

Supplier I provides more cartons that have weights closer to the mean. Or, put another way, the weights of cartons from Supplier I are more clustered around the mean. The bakery might find Supplier I more satisfactory.

As we see in Example 5, although the range tells the difference between the largest and smallest values in a distribution, it does not tell us how much other values vary from one another or from the mean.

VARIANCE AND STANDARD DEVIATION

Variance
Standard deviation

We need a measure of the distribution or spread of data around an expected value (either $\bar{x}$ or μ). *Variance* and *standard deviation* provide such measures. Formulas and rationale for these measures are described in the next Procedure display. Then, examples and guided exercises show how to compute and interpret these measures.

As we will see later, the formulas for variance and standard deviation differ slightly, depending on whether we are using a sample or the entire population.

Sum of squares

Sample variance, s^2

PROCEDURE

HOW TO COMPUTE THE SAMPLE VARIANCE AND SAMPLE STANDARD DEVIATION

Quantity	Description
x	The variable x represents a **data value** or outcome.
Mean $\bar{x} = \dfrac{\Sigma x}{n}$	This is the **average of the data values,** or what you "expect" to happen the next time you conduct the statistical experiment. Note that n is the sample size.
$x - \bar{x}$	This is the **difference** between what happened and what you expected to happen. This represents a "deviation" away from what you "expect" and is a measure of risk.
$\Sigma(x - \bar{x})^2$	The expression $\Sigma(x - \bar{x})^2$ is called the **sum of squares.** The $(x - \bar{x})$ quantity is squared to make it nonnegative. The sum is over all the data. If you don't square $(x - \bar{x})$, then the sum $\Sigma(x - \bar{x})$ is equal to 0 because the negative values cancel the positive values. This occurs even if some $(x - \bar{x})$ values are large, indicating a large deviation or risk.
Sum of squares $\Sigma(x - \bar{x})^2$ or $\Sigma x^2 - \dfrac{(\Sigma x)^2}{n}$	This is an **algebraic simplification of the sum of squares** that is easier to compute. The **defining formula** for the sum of squares is the upper one. The **computation formula** for the sum of squares is the lower one. Both formulas give the same result.
Sample variance $s^2 = \dfrac{\Sigma(x - \bar{x})^2}{n - 1}$ or $s^2 = \dfrac{\Sigma x^2 - (\Sigma x)^2/n}{n - 1}$	The **sample variance is s^2.** The variance can be thought of as a kind of average of the $(x - \bar{x})^2$ values. However, for technical reasons, we divide the sum by the quantity $n - 1$ rather than n. This gives us the best mathematical estimate for the sample variance. The **defining formula** for the variance is the upper one. The **computation formula** for the variance is the lower one. Both formulas give the same result.

Continued

Sample standard deviation, *s*

Sample standard deviation	
$s = \sqrt{\dfrac{\Sigma(x - \bar{x})^2}{n - 1}}$ or $s = \sqrt{\dfrac{\Sigma x^2 - (\Sigma x)^2/n}{n - 1}}$	This is the **sample standard deviation, *s*.** Why do we take the square root? Well, if the original *x* units were, say, days or dollars, then the s^2 units would be days squared or dollars squared (wow, what's that?). We take the square root to return to the original units of the data measurements. The standard deviation can be thought of as a measure of variability or risk. Larger values of *s* imply greater variability in the data. The **defining formula** for the standard deviation is the upper one. The **computation formula** for the standard deviation is the lower one. Both formulas give the same result.

COMMENT Why is *s* called a *sample standard* deviation? First, it is computed from sample data. Then why do we use the word *standard* in the name? We know *s* is a measure of deviation or risk. You should be aware that there are other statistical measures of risk that we have not yet mentioned. However, *s* is the one that everyone uses, so it is called the "standard" (like standard time).

In statistics, the sample standard deviation and sample variance are used to describe the spread of data about the mean $\bar{x}$. The next example shows how to find these quantities by using the defining formulas. Guided Exercise 3 shows how to use the computation formulas.

As you will discover, for "hand" calculations, the computation formulas for s^2 and *s* are much easier to use. However, the defining formulas for s^2 and *s* emphasize the fact that the variance and standard deviation are based on the differences between each data value and the mean.

DEFINING FORMULAS (SAMPLE STATISTIC)

$$\text{Sample variance} = s^2 = \frac{\Sigma(x - \bar{x})^2}{n - 1} \tag{1}$$

$$\text{Sample standard deviation} = s = \sqrt{\frac{\Sigma(x - \bar{x})^2}{n - 1}} \tag{2}$$

where *x* is a member of the data set, $\bar{x}$ is the mean, and *n* is the number of data values. The sum is taken over all data values.

COMPUTATION FORMULAS (SAMPLE STATISTIC)

$$\text{Sample variance} = s^2 = \frac{\Sigma x^2 - (\Sigma x)^2/n}{n - 1} \tag{3}$$

$$\text{Sample standard deviation} = s = \sqrt{\frac{\Sigma x^2 - (\Sigma x)^2/n}{n - 1}} \tag{4}$$

where *x* is a member of the data set, $\bar{x}$ is the mean, and *n* is the number of data values. The sum is taken over all data values.

EXAMPLE 6

SAMPLE STANDARD DEVIATION (DEFINING FORMULA)

Big Blossom Greenhouse was commissioned to develop an extra large rose for the Rose Bowl Parade. A random sample of blossoms from Hybrid A bushes yielded the following diameters (in inches) for mature peak blooms.

2 3 3 8 10 10

Use the defining formula to find the sample variance and standard deviation.

SOLUTION: Several steps are involved in computing the variance and standard deviation. A table will be helpful (see Table 3-1). Since $n = 6$, we take the sum of the entries in the first column of Table 3-1 and divide by 6 to find the mean $\bar{x}$.

$$\bar{x} = \frac{\Sigma x}{n} = \frac{36}{6} = 6.0\text{ inches}$$

TABLE 3-1 Diameters of Rose Blossoms (in inches)

Column I x	Column II $x - \bar{x}$	Column III $(x - \bar{x})^2$
2	$2 - 6 = -4$	$(-4)^2 = 16$
3	$3 - 6 = -3$	$(-3)^2 = 9$
3	$3 - 6 = -3$	$(-3)^2 = 9$
8	$8 - 6 = 2$	$(2)^2 = 4$
10	$10 - 6 = 4$	$(4)^2 = 16$
10	$10 - 6 = 4$	$(4)^2 = 16$
$\Sigma x = 36$		$\Sigma(x - \bar{x})^2 = 70$

Tony Freeman/PhotoEdit

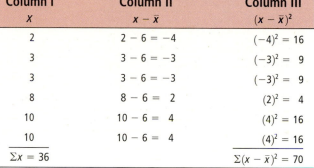

Using this value for $\bar{x}$, we obtain Column II. Square each value in the second column to obtain Column III, and then add the values in Column III. To get the sample variance, divide the sum of Column III by $n - 1$. Since $n = 6$, $n - 1 = 5$.

$$s^2 = \frac{\Sigma(x - \bar{x})^2}{n - 1} = \frac{70}{5} = 14$$

Now obtain the sample standard deviation by taking the square root of the variance.

$$s = \sqrt{s^2} = \sqrt{14} \approx 3.74$$

(Use a calculator to compute the square root. Because of rounding, we use the approximately equal symbol, ≈.)

GUIDED EXERCISE 3

SAMPLE STANDARD DEVIATION (COMPUTATION FORMULA)

Big Blossom Greenhouse gathered another random sample of mature peak blooms from Hybrid B. The six blossoms had the following widths (in inches):

5 5 5 6 7 8

(a) Again, we will construct a table so that we can find the mean, variance, and standard deviation more easily. In this case, what is the value of n? Find the sum of Column I in Table 3-2, and compute the mean.

→ $n = 6$. The sum of Column I is $\Sigma x = 36$, so the mean is

$$\bar{x} = \frac{36}{6} = 6$$

Continued

GUIDED EXERCISE 3 *continued*

TABLE 3-2 **Complete Columns I and II**

I x	II x^2
5	____
5	____
5	____
6	____
7	____
8	____
$\Sigma x =$ ____	$\Sigma x^2 =$ ____

TABLE 3-3 **Completion of Table 3-2**

I x	II x^2
5	25
5	25
5	25
6	36
7	49
8	64
$\Sigma x = 36$	$\Sigma x^2 = 224$

(b) What is the value of n? of $n - 1$? Use the computation formula to find the sample variance s^2. *Note:* Be sure to distinguish between Σx^2 and $(\Sigma x)^2$. For Σx^2, you square the x values first and then sum them. For $(\Sigma x)^2$, you sum the x values first and then square the result.

$n = 6; n - 1 = 5.$

$$s^2 = \frac{\Sigma x^2 - (\Sigma x)^2/n}{n - 1}$$

$$= \frac{224 - (36)^2/6}{5} = \frac{8}{5} = 1.6$$

(c) Use a calculator to find the square root of the variance. Is this the standard deviation?

$s = \sqrt{s^2} = \sqrt{1.6} \approx 1.26$

Yes.

Let's summarize and compare the results of Guided Exercise 3 and Example 6. The greenhouse found the following blossom diameters for Hybrid A and Hybrid B:

Hybrid A: Mean, 6.0 inches; standard deviation, 3.74 inches

Hybrid B: Mean, 6.0 inches; standard deviation, 1.26 inches

Interpretation In both cases, the means are the same: 6 inches. But the first hybrid has a larger standard deviation. This means that the blossoms of Hybrid A are less consistent than those of Hybrid B. If you want a rosebush that occasionally has 10-inch blooms and occasionally has 2-inch blooms, use the first hybrid. But if you want a bush that consistently produces roses close to 6 inches across, use Hybrid B.

ROUNDING NOTE Rounding errors cannot be completely eliminated, even if a computer or calculator does all the computations. However, software and calculator routines are designed to minimize the errors. If the mean is rounded, the value of the standard deviation will change slightly, depending on how much the mean is rounded. If you do your calculations "by hand" or reenter intermediate values into a calculator, try to carry one or two more digits than occur in the original data. If your resulting answers vary slightly from those in this text, do not be overly concerned. The text answers are computer- or calculator-generated.

WHAT DO MEASURES OF VARIATION TELL US?

Measures of variation give information about the spread of the data.

- The **range** tells us the difference between the highest data value and the lowest. It tells us about the spread of data but does not tell us if most of the data is or is not closer to the mean.

- The **sample standard deviation** is based on the difference between *each* data value and the mean of the data set. The magnitude of each data value enters into the calculation. The formula tells us to compute the difference between each data value and the mean, square each difference, add up all the squares, divide by $n-1$, and then take the square root of the result. The standard deviation gives an average of data spread around the mean. The larger the standard deviation, the more spread out the data are around the mean. A smaller standard deviation indicates that the data tend to be closer to the mean.
- The **variance** tells us the square of standard deviation. As such, it is also a measure of data spread around the mean.

Population mean, μ
Population variance, σ^2
Population standard deviation, σ

In most applications of statistics, we work with a random sample of data rather than the entire population of *all* possible data values. However, if we have data for the entire population, we can compute the *population mean* μ, *population variance* σ^2, and *population standard deviation* σ (lowercase Greek letter sigma) using the following formulas:

POPULATION PARAMETERS

$$\text{Population mean} = \mu = \frac{\Sigma x}{N}$$

$$\text{Population variance} = \sigma^2 = \frac{\Sigma(x-\mu)^2}{N}$$

$$\text{Population standard deviation} = \sigma = \sqrt{\frac{\Sigma(x-\mu)^2}{N}}$$

where N is the number of data values in the population and x represents the individual data values of the population.

What if you do not have access to raw data, but you do have a histogram or frequency table of the data and you want to estimate the mean and standard deviation? Problems 22–25 address grouped data, with the formulas shown before Problem 22.

Problem 26 discusses moving averages, and explores the difference in the standard deviations of raw data compared to the standard deviation of the moving average of the raw data.

Population size, N

When we use stratified sampling, how do we determine the sample size from each stratum so that we get a more accurate estimate for the total population mean μ of a specified measurement? Problems 28, 29, and 30 show how the sample standard deviation of the measurement from each stratum help determine optimal samples sizes for the estimate of the total population mean μ.

We note that the formula for μ is the same as the formula for $\bar{x}$ (the sample mean) and that the formulas for σ^2 and σ are the same as those for s^2 and s (sample variance and sample standard deviation), except that the population size N is used instead of $n-1$. Also, μ is used instead of $\bar{x}$ in the formulas for σ^2 and σ.

In the formulas for s and σ, we use $n-1$ to compute s and N to compute σ. Why? The reason is that N (capital letter) represents the *population size,* whereas n (lowercase letter) represents the sample size. Since a random sample usually will not contain extreme data values (large or small), we divide by $n-1$ in the formula for s to make s a little larger than it would have been had we divided by n. Courses in advanced theoretical statistics show that this procedure will give us the best possible estimate for the standard deviation σ. In fact, s is called the *unbiased estimate* for σ. If we have the population of all data values, then extreme data values are, of course, present, so we divide by N instead of $N-1$.

COMMENT The computation formula for the population standard deviation is

$$\sigma = \sqrt{\frac{\Sigma x^2 - (\Sigma x)^2/N}{N}}$$

We've seen that the standard deviation (sample or population) is a measure of data spread. We will use the standard deviation extensively in later chapters.

TECH NOTES Most scientific or business calculators have a statistics mode and provide the mean and sample standard deviation directly. The TI-84Plus/TI-83Plus/TI-*n*spire calculators, Excel 2010, and Minitab provide the median and several other measures as well.

Many technologies display only the sample standard deviation s. You can quickly compute σ if you know s by using the formula

$$\sigma = s\sqrt{\frac{n-1}{n}}$$

The mean given in displays can be interpreted as the sample mean $\bar{x}$ or the population mean μ as appropriate.

The three displays, show output for the hybrid rose data of Guided Exercise 3.

TI-84Plus/TI-83Plus/TI-*n*spire (with TI-84Plus Keypad) Display Press **STAT ➤ CALC ➤ 1:1-Var Stats.** S_x is the sample standard deviation. σ_x is the population standard deviation.

```
1-Var Stats
 x̄=6
 Σx=36
 Σx²=224
 Sx=1.264911064
 σx=1.154700538
↓n=6
```


Excel 2010 Display Click the **Data Tab**. Then in the **Analysis Group** click **Data Analysis**. Select **Descriptive Statistics**. In the dialogue box, fill in the input range with the column and row numbers containing your data. Then check the Summary Statistics box.

Column 1	
Mean	6
Standard Error	0.516398
Median	5.5
Mode	5
Standard Deviation	1.264911
Sample Variance	1.6
Kurtosis	–0.78125
Skewness	0.889391
Range	3
Minimum	5
Maximum	8
Sum	36
Count	6

Minitab Display Menu choices: **Stat ➤ Basic Statistics ➤ Display Descriptive Statistics.** StDev is the sample standard deviation. TrMean is a 5% trimmed mean.

```
N        Mean      Median     TrMean     StDev    SEMean
6        6.000     5.500      6.000      1.265    0.516
Minimum        Maximum        Q1         Q3
5.000          8.000          5.000      7.250
```

Now let's look at two immediate applications of the standard deviation. The first is the coefficient of variation, and the second is Chebyshev's theorem.

COEFFICIENT OF VARIATION

Coefficient of variation, CV

A disadvantage of the standard deviation as a comparative measure of variation is that it depends on the units of measurement. This means that it is difficult to use the standard deviation to compare measurements from different populations. For this reason, statisticians have defined the *coefficient of variation,* which expresses the standard deviation as a percentage of the sample or population mean.

If $\bar{x}$ and s represent the sample mean and sample standard deviation, respectively, then the sample **coefficient of variation** CV is defined to be

$$CV = \frac{s}{\bar{x}} \cdot 100\%$$

If μ and σ represent the population mean and population standard deviation, respectively, then the population coefficient of variation CV is defined to be

$$CV = \frac{\sigma}{\mu} \cdot 100\%$$

Notice that the numerator and denominator in the definition of CV have the same units, so CV itself has no units of measurement. This gives us the advantage of being able to directly compare the variability of two different populations using the coefficient of variation.

In the next example and guided exercise, we will compute the CV of a population and of a sample and then compare the results.

EXAMPLE 7

COEFFICIENT OF VARIATION

The Trading Post on Grand Mesa is a small, family-run store in a remote part of Colorado. The Grand Mesa region contains many good fishing lakes, so the Trading Post sells spinners (a type of fishing lure). The store has a very limited selection of spinners. In fact, the Trading Post has only eight different types of spinners for sale. The prices (in dollars) are

2.10	1.95	2.60	2.00	1.85	2.25	2.15	2.25

Since the Trading Post has only eight different kinds of spinners for sale, we consider the eight data values to be the *population*.

Ron Watts/CORBIS

(a) Use a calculator with appropriate statistics keys to verify that for the Trading Post data, $\mu \approx \$2.14$ and $\sigma \approx \$0.22$.

SOLUTION: Since the computation formulas for $\bar{x}$ and μ are identical, most calculators provide the value of $\bar{x}$ only. Use the output of this key for μ. The computation formulas for the sample standard deviation s and the population standard deviation σ are slightly different. Be sure that you use the key for σ (sometimes designated as σ_n or σ_x).

(b) Compute the CV of prices for the Trading Post and comment on the meaning of the result.

SOLUTION:

$$CV = \frac{\sigma}{\mu} \times 100\% = \frac{0.22}{2.14} \times 100\% = 10.28\%$$

Interpretation The coefficient of variation can be thought of as a measure of the spread of the data relative to the average of the data. Since the Trading Post is very small, it carries a small selection of spinners that are all priced similarly. The *CV* tells us that the standard deviation of the spinner prices is only 10.28% of the mean.

GUIDED EXERCISE 4 — COEFFICIENT OF VARIATION

Cabela's in Sidney, Nebraska, is a very large outfitter that carries a broad selection of fishing tackle. It markets its products nationwide through a catalog service. A random sample of 10 spinners from Cabela's extensive spring catalog gave the following prices (in dollars):

1.69 1.49 3.09 1.79 1.39 2.89 1.49 1.39 1.49 1.99

(a) Use a calculator with sample mean and sample standard deviation keys to compute $\bar{x}$ and s.

→ $\bar{x} = \$1.87$ and $s \approx \$0.62$.

(b) Compute the *CV* for the spinner prices at Cabela's.

→ $$CV = \frac{s}{\bar{x}} \times 100\% = \frac{0.62}{1.87} \times 100\% = 33.16\%$$

(c) *Interpretation* Compare the mean, standard deviation, and *CV* for the spinner prices at the Grand Mesa Trading Post (Example 7) and Cabela's. Comment on the differences.

→ The *CV* for Cabela's is more than three times the *CV* for the Trading Post. Why? First, because of the remote location, the Trading Post tends to have somewhat higher prices (larger μ). Second, the Trading Post is very small, so it has a rather limited selection of spinners with a smaller variation in price.

CHEBYSHEV'S THEOREM

From our earlier discussion about standard deviation, we recall that the spread or dispersion of a set of data about the mean will be small if the standard deviation is small, and it will be large if the standard deviation is large. If we are dealing with a symmetrical, bell-shaped distribution, then we can make very definite statements about the proportion of the data that must lie within a certain number of standard deviations on either side of the mean. This will be discussed in detail in Chapter 6 when we talk about normal distributions.

However, the concept of data spread about the mean can be expressed quite generally for *all data distributions* (skewed, symmetric, or other shapes) by using the remarkable theorem of Chebyshev.

Chebyshev's theorem

CHEBYSHEV'S THEOREM

For *any* set of data (either population or sample) and for any constant k greater than 1, the proportion of the data that must lie within k standard deviations on either side of the mean is *at least*

$$1 - \frac{1}{k^2}$$

Put another way: For sample data with mean $\bar{x}$ and standard deviation s, at least $1 - 1/k^2$ (fractional part) of data must fall between $\bar{x} - ks$ and $\bar{x} + ks$.

RESULTS OF CHEBYSHEV'S THEOREM

For *any* set of data:

- *at least* 75% of the data fall in the interval from $\mu - 2\sigma$ to $\mu + 2\sigma$.
- *at least* 88.9% of the data fall in the interval from $\mu - 3\sigma$ to $\mu + 3\sigma$.
- *at least* 93.8% of the data fall in the interval from $\mu - 4\sigma$ to $\mu + 4\sigma$.

The results of Chebyshev's theorem can be derived by using the theorem and a little arithmetic. For instance, if we create an interval $k = 2$ standard deviations on either side of the mean, Chebyshev's theorem tells us that

$$1 - \frac{1}{2^2} = 1 - \frac{1}{4} = \frac{3}{4} \text{ or } 75\%$$

is the minimum percentage of data in the $\mu - 2\sigma$ to $\mu + 2\sigma$ interval.

Notice that Chebyshev's theorem refers to the *minimum* percentage of data that must fall within the specified number of standard deviations of the mean. If the distribution is mound-shaped, an even *greater* percentage of data will fall into the specified intervals (see the Empirical Rule in Section 6.1).

WHAT DOES CHEBYSHEV'S THEOREM TELL US?

Chebyshev's theorem applies to *any* distribution. It tells us

- the *minimum* percentage of data that falls between the mean and any specified number of standard deviations on either side of the mean.
- a minimum of 88.9% of the data falls between the values 3 standard deviations below the mean and 3 standard deviations above the mean. This implies that a maximum of 11.1% of data fall beyond 3 standard deviations of the mean. Such values might be suspect outliers, particularly for a mound-shaped symmetric distribution.

EXAMPLE 8

CHEBYSHEV'S THEOREM

Bob Daemmrich/The Image Works

Students Who Care is a student volunteer program in which college students donate work time to various community projects such as planting trees. Professor Gill is the faculty sponsor for this student volunteer program. For several years, Dr. Gill has kept a careful record of $x =$ total number of work hours volunteered by a student in the program each semester. For a random sample of students in the program, the mean number of hours was $\bar{x} = 29.1$ hours each semester, with a standard deviation of $s = 1.7$ hours each semester. Find an interval A to B for the number of hours volunteered into which at least 75% of the students in this program would fit.

SOLUTION: According to results of Chebyshev's theorem, at least 75% of the data must fall within 2 standard deviations of the mean. Because the mean is $\bar{x} = 29.1$ and the standard deviation is $s = 1.7$, the interval is

$$\bar{x} - 2s \text{ to } \bar{x} + 2s$$
$$29.1 - 2(1.7) \text{ to } 29.1 + 2(1.7)$$
$$25.7 \text{ to } 32.5$$

At least 75% of the students would fit into the group that volunteered from 25.7 to 32.5 hours each semester.

GUIDED EXERCISE 5 | CHEBYSHEV INTERVAL

The *East Coast Independent News* periodically runs ads in its classified section offering a month's free subscription to those who respond. In this way, management can get a sense about the number of subscribers who read the classified section each day. Over a period of 2 years, careful records have been kept. The mean number of responses per ad is $\bar{x} = 525$ with standard deviation $s = 30$.

Determine a Chebyshev interval about the mean in which at least 88.9% of the data fall.

→ By Chebyshev's theorem, at least 88.9% of the data fall into the interval

$\bar{x} - 3s$ to $\bar{x} + 3s$

Because $\bar{x} = 525$ and $s = 30$, the interval is

$525 - 3(30)$ to $525 + 3(30)$

or from 435 to 615 responses per ad.

LOOKING FORWARD

In Chapter 6 we will study normal distributions. For normal distributions we can make much more precise statements about the percentage of data falling within 1, 2, or 3 standard deviations of the mean. In fact, the empirical rule (see Section 6.1) tells us that for normal distributions, about 68% of the data fall within 1 standard deviation of the mean, about 95% fall within 2 standard deviations of the mean, and almost all (99.7%) fall within 3 standard deviations of the mean.

CRITICAL THINKING

Averages such as the mean are often referred to in the media. However, an average by itself does not tell much about the way data are distributed about the mean. Knowledge about the standard deviation or variance, along with the mean, gives a much better picture of the data distribution.

Chebyshev's theorem tells us that no matter what the data distribution looks like, at least 75% of the data will fall within 2 standard deviations of the mean. As we will see in Chapter 6, when the distribution is mound-shaped and symmetrical, about 95% of the data are within 2 standard deviations of the mean. Data values beyond 2 standard deviations from the mean are less common than those closer to the mean.

Outlier

In fact, one indicator that a data value might be an *outlier* is that it is more than 2.5 standard deviations from the mean (Source: *Oxford Dictionary of Statistics*, Oxford University Press).

VIEWPOINT | Socially Responsible Investing

Make a difference and make money! Socially responsible mutual funds tend to screen out corporations that sell tobacco, weapons, and alcohol, as well as companies that are environmentally unfriendly. In addition, these funds screen out companies that use child labor in sweatshops. There are 68 socially responsible funds tracked by the Social Investment Forum. For more information, visit the Brase/Brase statistics site at **http://www.cengage.com/statistics/brase11e** *and find the link to social investing.*

How do these funds rate compared to other funds? One way to answer this question is to study the annual percent returns of the funds using both the mean *and* standard deviation. *(See Problem 20 of this section.)*

SECTION 3.2 PROBLEMS

1. *Statistical Literacy* Which average—mean, median, or mode—is associated with the standard deviation?

2. *Statistical Literacy* What is the relationship between the variance and the standard deviation for a sample data set?

3. *Statistical Literacy* When computing the standard deviation, does it matter whether the data are sample data or data comprising the entire population? Explain.

4. *Statistical Literacy* What symbol is used for the standard deviation when it is a sample statistic? What symbol is used for the standard deviation when it is a population parameter?

5. *Basic Computation: Range, Standard Deviation* Consider the data set

   2 3 4 5 6

   (a) Find the range.
   (b) Use the defining formula to compute the sample standard deviation s.
   (c) Use the defining formula to compute the population standard deviation σ.

6. *Basic Computation: Range, Standard Deviation* Consider the data set

   1 2 3 4 5

   (a) Find the range.
   (b) Use the defining formula to compute the sample standard deviation s.
   (c) Use the defining formula to compute the population standard deviation σ.

7. *Critical Thinking* For a given data set in which not all data values are equal, which value is smaller, s or σ? Explain.

8. *Critical Thinking* Consider two data sets with equal sample standard deviations. The first data set has 20 data values that are not all equal, and the second has 50 data values that are not all equal. For which data set is the difference between s and σ greater? Explain. *Hint:* Consider the relationship $\sigma = s\sqrt{(n-1)/n}$.

9. *Critical Thinking* Each of the following data sets has a mean of $\bar{x} = 10$.

   (i) 8 9 10 11 12 (ii) 7 9 10 11 13 (iii) 7 8 10 12 13

   (a) Without doing any computations, order the data sets according to increasing value of standard deviations.
   (b) Why do you expect the difference in standard deviations between data sets (i) and (ii) to be greater than the difference in standard deviations between data sets (ii) and (iii)? *Hint:* Consider how much the data in the respective sets differ from the mean.

10. *Critical Thinking: Data Transformation* In this problem, we explore the effect on the standard deviation of adding the same constant to each data value in a data set. Consider the data set 5, 9, 10, 11, 15.

    (a) Use the defining formula, the computation formula, or a calculator to compute s.
    (b) Add 5 to each data value to get the new data set 10, 14, 15, 16, 20. Compute s.

(c) Compare the results of parts (a) and (b). In general, how do you think the standard deviation of a data set changes if the same constant is added to each data value?

11. *Critical Thinking: Data Transformation* In this problem, we explore the effect on the standard deviation of multiplying each data value in a data set by the same constant. Consider the data set 5, 9, 10, 11, 15.
(a) Use the defining formula, the computation formula, or a calculator to compute s.
(b) Multiply each data value by 5 to obtain the new data set 25, 45, 50, 55, 75. Compute s.
(c) Compare the results of parts (a) and (b). In general, how does the standard deviation change if each data value is multiplied by a constant c?
(d) You recorded the weekly distances you bicycled in miles and computed the standard deviation to be $s = 3.1$ miles. Your friend wants to know the standard deviation in kilometers. Do you need to redo all the calculations? Given 1 mile $\approx$ 1.6 kilometers, what is the standard deviation in kilometers?

12. *Critical Thinking: Outliers* One indicator of an outlier is that an observation is more than 2.5 standard deviations from the mean. Consider the data value 80.
(a) If a data set has mean 70 and standard deviation 5, is 80 a suspect outlier?
(b) If a data set has mean 70 and standard deviation 3, is 80 a suspect outlier?

13. *Basic Computation: Variance, Standard Deviation* Given the sample data

x:	23	17	15	30	25

(a) Find the range.
(b) Verify that $\Sigma x = 110$ and $\Sigma x^2 = 2568$.
(c) Use the results of part (b) and appropriate computation formulas to compute the sample variance s^2 and sample standard deviation s.
(d) Use the defining formulas to compute the sample variance s^2 and sample standard deviation s.
(e) Suppose the given data comprise the entire population of all x values. Compute the population variance σ^2 and population standard deviation σ.

14. *Basic Computation: Coefficient of Variation, Chebyshev Interval* Consider sample data with $\bar{x} = 15$ and $s = 3$.
(a) Compute the coefficient of variation.
(b) Compute a 75% Chebyshev interval around the sample mean.

15. *Basic Computation: Coefficient of Variation, Chebyshev Interval* Consider population data with $\mu = 20$ and $\sigma = 2$.
(a) Compute the coefficient of variation.
(b) Compute an 88.9% Chebyshev interval around the population mean.

16. *Investing: Stocks and Bonds* Do bonds reduce the overall risk of an investment portfolio? Let x be a random variable representing annual percent return for Vanguard Total Stock Index (all stocks). Let y be a random variable representing annual return for Vanguard Balanced Index (60% stock and 40% bond). For the past several years, we have the following data (Reference: Morningstar Research Group, Chicago).

x:	11	0	36	21	31	23	24	−11	−11	−21
y:	10	−2	29	14	22	18	14	−2	−3	−10

(a) Compute Σx, Σx^2, Σy, and Σy^2.
(b) Use the results of part (a) to compute the sample mean, variance, and standard deviation for x and for y.

NASA Images

(c) Compute a 75% Chebyshev interval around the mean for x values and also for y values. Use the intervals to compare the two funds.

(d) ***Interpretation:*** Compute the coefficient of variation for each fund. Use the coefficients of variation to compare the two funds. If s represents risks and $\bar{x}$ represents expected return, then $s/\bar{x}$ can be thought of as a measure of risk per unit of expected return. In this case, why is a smaller *CV* better? Explain.

17. *Space Shuttle: Epoxy* Kevlar epoxy is a material used on the NASA space shuttles. Strands of this epoxy were tested at the 90% breaking strength. The following data represent time to failure (in hours) for a random sample of 50 epoxy strands (Reference: R. E. Barlow, University of California, Berkeley). Let x be a random variable representing time to failure (in hours) at 90% breaking strength. *Note:* These data are also available for download at the Online Study Center.

0.54	1.80	1.52	2.05	1.03	1.18	0.80	1.33	1.29	1.11
3.34	1.54	0.08	0.12	0.60	0.72	0.92	1.05	1.43	3.03
1.81	2.17	0.63	0.56	0.03	0.09	0.18	0.34	1.51	1.45
1.52	0.19	1.55	0.02	0.07	0.65	0.40	0.24	1.51	1.45
1.60	1.80	4.69	0.08	7.89	1.58	1.64	0.03	0.23	0.72

(a) Find the range.

(b) Use a calculator to verify that $\Sigma x = 62.11$ and $\Sigma x^2 \approx 164.23$.

(c) Use the results of part (b) to compute the sample mean, variance, and standard deviation for the time to failure.

(d) ***Interpretation*** Use the results of part (c) to compute the coefficient of variation. What does this number say about time to failure? Why does a small *CV* indicate more consistent data, whereas a larger *CV* indicates less consistent data? Explain.

18. *Archaeology: Ireland* The Hill of Tara in Ireland is a place of great archaeological importance. This region has been occupied by people for more than 4000 years. Geomagnetic surveys detect subsurface anomalies in the earth's magnetic field. These surveys have led to many significant archaeological discoveries. After collecting data, the next step is to begin a statistical study. The following data measure magnetic susceptibility (centimeter-gram-second $\times 10^{-6}$) on two of the main grids of the Hill of Tara (Reference: *Tara: An Archaeological Survey* by Conor Newman, Royal Irish Academy, Dublin).

Grid E: x variable

13.20	5.60	19.80	15.05	21.40	17.25	27.45
16.95	23.90	32.40	40.75	5.10	17.75	28.35

Grid H: y variable

11.85	15.25	21.30	17.30	27.50	10.35	14.90
48.70	25.40	25.95	57.60	34.35	38.80	41.00
31.25						

(a) Compute Σx, Σx^2, Σy, and Σy^2.

(b) Use the results of part (a) to compute the sample mean, variance, and standard deviation for x and for y.

(c) Compute a 75% Chebyshev interval around the mean for x values and also for y values. Use the intervals to compare the magnetic susceptibility on the two grids. Higher numbers indicate higher magnetic susceptibility. However, extreme values, high or low, could mean an anomaly and possible archaeological treasure.

(d) ***Interpretation*** Compute the sample coefficient of variation for each grid. Use the *CV*'s to compare the two grids. If s represents variability in the signal (magnetic susceptibility) and $\bar{x}$ represents the expected level of the signal, then $s/\bar{x}$ can be thought of as a measure of the variability per unit of expected signal. Remember, a considerable variability in the signal (above or below average) might indicate buried artifacts. Why, in this case, would a large *CV* be better, or at least more exciting? Explain.

19. *Wildlife: Mallard Ducks and Canada Geese* For mallard ducks and Canada geese, what percentage of nests are successful (at least one offspring survives)? Studies in Montana, Illinois, Wyoming, Utah, and California gave the following percentages of successful nests (Reference: *The Wildlife Society Press,* Washington, D.C.).

x: Percentage success for mallard duck nests

56	85	52	13	39

y: Percentage success for Canada goose nests

24	53	60	69	18

(a) Use a calculator to verify that $\Sigma x = 245$; $\Sigma x^2 = 14{,}755$; $\Sigma y = 224$; and $\Sigma y^2 = 12{,}070$.
(b) Use the results of part (a) to compute the sample mean, variance, and standard deviation for x, the percent of successful mallard nests.
(c) Use the results of part (a) to compute the sample mean, variance, and standard deviation for y, the percent of successful Canada goose nests.
(d) ***Interpretation*** Use the results of parts (b) and (c) to compute the coefficient of variation for successful mallard nests and Canada goose nests. Write a brief explanation of the meaning of these numbers. What do these results say about the nesting success rates for mallards compared to those of Canada geese? Would you say one group of data is more or less consistent than the other? Explain.

20. *Investing: Socially Responsible Mutual Funds* Pax World Balanced is a highly respected, socially responsible mutual fund of stocks and bonds (see Viewpoint). Vanguard Balanced Index is another highly regarded fund that represents the entire U.S. stock and bond market (an index fund). The mean and standard deviation of annualized percent returns are shown below. The annualized mean and standard deviation are based on the years 1993 through 2002 (Source: Morningstar).

Pax World Balanced: $\bar{x} = 9.58\%$; $s = 14.05\%$

Vanguard Balanced Index: $\bar{x} = 9.02\%$; $s = 12.50\%$

(a) *Interpretation* Compute the coefficient of variation for each fund. If $\bar{x}$ represents return and s represents risk, then explain why the coefficient of variation can be taken to represent risk per unit of return. From this point of view, which fund appears to be better? Explain.
(b) *Interpretation* Compute a 75% Chebyshev interval around the mean for each fund. Use the intervals to compare the two funds. As usual, past performance does not guarantee future performance.

21. *Medical: Physician Visits* In some reports, the mean and coefficient of variation are given. For instance, in *Statistical Abstract of the United States,* 116th Edition, one report gives the average number of physician visits by males per year. The average reported is 2.2, and the reported coefficient of variation is 1.5%. Use this information to determine the standard deviation of the annual number of visits to physicians made by males.

Grouped data

EXPAND YOUR KNOWLEDGE: GROUPED DATA

When data are grouped, such as in a frequency table or histogram, we can estimate the mean and standard deviation by using the following formulas. Notice that all data values in a given class are treated as though each of them equals the midpoint x of the class.

Approximating $\bar{x}$ and s from grouped data

Sample Mean for a Frequency Distribution

$$\bar{x} = \frac{\Sigma xf}{n} \qquad (5)$$

Sample Standard Deviation for a Frequency Distribution

$$s = \sqrt{\frac{\Sigma(x - \bar{x})^2 f}{n - 1}} \qquad (6)$$

Computation Formula for the Sample Standard Deviation

$$s = \sqrt{\frac{\Sigma x^2 f - (\Sigma xf)^2/n}{n - 1}} \qquad (7)$$

where

x is the midpoint of a class,
f is the number of entries in that class,
n is the total number of entries in the distribution, and $n = \Sigma f$.
The summation Σ is over all classes in the distribution.

Use formulas (5) and (6) or (5) and (7) to solve Problems 22–25. To use formulas (5) and (6) to evaluate the sample mean and standard deviation, use the following column heads:

Midpoint x	Frequency f	xf	$(x - \bar{x})$	$(x - \bar{x})^2$	$(x - \bar{x})^2 f$

For formulas (5) and (7), use these column heads:

Midpoint x	Frequency f	xf	x^2	$x^2 f$

Note: On the TI-84Plus/TI-83Plus/TI-*n*spire (with TI-84Plus keypad) calculators, enter the midpoints in column L_1 and the frequencies in column $\mathbf{L_2}$. Then use **1-VarStats** $\mathbf{L_1}$, L_2.

22. *Grouped Data: Anthropology* What was the age distribution of prehistoric Native Americans? Extensive anthropologic studies in the southwestern United States gave the following information about a prehistoric extended family group of 80 members on what is now the Navajo Reservation in northwestern New Mexico (Source: Based on information taken from *Prehistory in the Navajo Reservation District*, by F. W. Eddy, Museum of New Mexico Press).

Age range (years)	1–10*	11–20	21–30	31 and over
Number of individuals	34	18	17	11

*Includes infants.

For this community, estimate the mean age expressed in years, the sample variance, and the sample standard deviation. For the class 31 and over, use 35.5 as the class midpoint.

23. *Grouped Data: Shoplifting* What is the age distribution of adult shoplifters (21 years of age or older) in supermarkets? The following is based on information taken from the National Retail Federation. A random sample of 895 incidents of shoplifting gave the following age distribution:

Age range (years)	21–30	31–40	41 and over
Number of shoplifters	260	348	287

Estimate the mean age, sample variance, and sample standard deviation for the shoplifters. For the class 41 and over, use 45.5 as the class midpoint.

24. *Grouped Data: Hours of Sleep per Day* Alexander Borbely is a professor at the University of Zurich Medical School, where he is director of the sleep laboratory. The histogram in Figure 3-2 is based on information from his book *Secrets of Sleep*. The histogram displays hours of sleep per day for a random sample of 200 subjects. Estimate the mean hours of sleep, standard deviation of hours of sleep, and coefficient of variation.

FIGURE 3-2

Hours of Sleep Each Day (24-hour period)

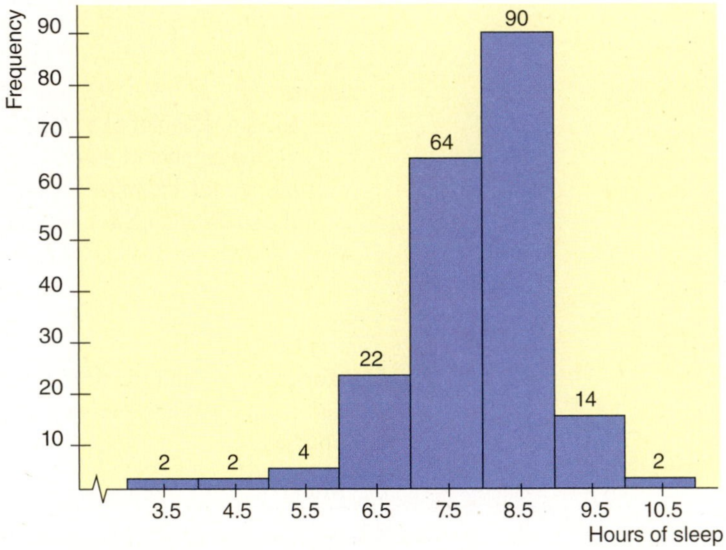

25. *Grouped Data: Business Administration* What are the big corporations doing with their wealth? One way to answer this question is to examine profits as percentage of assets. A random sample of 50 Fortune 500 companies gave the following information (Source: Based on information from *Fortune 500*, Vol. 135, No. 8).

Profit as percentage of assets	8.6–12.5	12.6–16.5	16.6–20.5	20.6–24.5	24.6–28.5
Number of companies	15	20	5	7	3

Estimate the sample mean, sample variance, and sample standard deviation for profit as percentage of assets.

Moving average

26. *Expand Your Knowledge: Moving Averages* You do not need a lot of money to invest in a mutual fund. However, if you decide to put some money into an investment, you are usually advised to leave it in for (at least) several years. Why? Because good years tend to cancel out bad years, giving you a better overall return with less risk. To see what we mean, let's use a 3-year *moving average* on the Calvert Social Balanced Fund (a socially responsible fund).

Year	1990	1991	1992	1993	1994	1995	1996	1997	1998	1999	2000
% Return	1.78	17.79	7.46	5.95	−4.74	25.85	9.03	18.92	17.49	6.80	−2.38

Source: Morningstar

(a) Use a calculator with mean and standard deviation keys to verify that the mean annual return for all 11 years is approximately 9.45%, with standard deviation 9.57%.

(b) To compute a 3-year moving average for 1992, we take the data values for 1992 and the prior two years and average them. To compute a 3-year moving average for 1993, we take the data values for 1993 and the prior two years and average them. Verify that the following 3-year moving averages are correct.

Year	1992	1993	1994	1995	1996	1997	1998	1999	2000
3-year moving average	9.01	10.40	2.89	9.02	10.05	17.93	15.15	14.40	7.30

(c) Use a calculator with mean and standard deviation keys to verify that for the 3-year moving average, the mean is 10.68% with sample standard deviation 4.53%.

(d) ***Interpretation*** Compare the results of parts (a) and (c). Suppose we take the point of view that risk is measured by standard deviation. Is the risk (standard deviation) of the 3-year moving average considerably smaller? This is an example of a general phenomenon that will be studied in more detail in Chapter 6.

27. *Brain Teaser: Sum of Squares* If you like mathematical puzzles or love algebra, try this! Otherwise, just trust that the computational formula for the sum of squares is correct. We have a sample of x values. The sample size is n. Fill in the details for the following steps.

$$\begin{aligned}\Sigma(x - \bar{x})^2 &= \Sigma x^2 - 2\bar{x}\,\Sigma x + n\bar{x}^2 \\ &= \Sigma x^2 - 2n\bar{x}^2 + n\bar{x}^2 \\ &= \Sigma x^2 - \frac{(\Sigma x)^2}{n}\end{aligned}$$

28. *Expand Your Knowledge: Stratified Sampling and Students in Sixth Grade* This is a technique to break down the variation of a random variable into useful components (called stratum) in order to decrease experimental variation and increase accuracy of results. It has been found that a more accurate estimate of population mean μ can often be obtained by taking measurements from naturally occurring subpopulations and combining the results using weighted averages. For example, suppose an accurate estimate of the mean weight of sixth grade students is desired for a large school system. Suppose (for cost reasons) we can only take a random sample of $m = 100$ students. Instead of taking a simple random sample of 100 students from the entire population of all sixth grade students we use stratified sampling as follows. The school system under study consists three large schools. School A has $N_1 = 310$ sixth grade students, School B has $N_2 = 420$ sixth grade students, and School C has $N_3 = 516$ sixth grade students. This is a total population of 1246 sixth grade students in our study and we have strata consisting of the 3 schools. A preliminary study in each school with relatively small sample size has given estimates for the sample standard deviation s of sixth grade student weights in each school. These are shown in the following table

School A	School B	School C
$N_1 = 310$	$N_2 = 420$	$N_3 = 516$
$s_1 = 3$ lb	$s_2 = 12$ lb	$s_3 = 6$ lb

How many students should we randomly choose from each school for a best estimate μ for the population mean weight? A lot of mathematics goes into the answer. Forunately, Bill Williams of Bell Laboratories wrote a book called *A Sampler on Sampling* (John Wiley and Sons, publisher) which provides an answer. Let n_1 be the number of students randomly chosen from School A, n_2 be the number chosen from School B and n_3 be the number chosen from School C. This means our total sample size will be $m = n_1 + n_2 + n_3$. What is the formula for n_i? A popular and widely used technique is the following.

$$n_i = \left[\frac{N_i s_i}{N_1 s_1 + N_2 s_2 + N_3 s_3}\right] m$$

The n_i are usually not whole numbers, so we need to round to the nearest whole number. This formula allocates more studentes to schools that have a larger population of sixth graders and/or have larger sample standard deviations. Remember, this is a popular and widely used technique for stratified sampling. It is not an absolute rule. There are other methods of stratified sampling also in use. In general practice, according to Bill Williams, the use of naturally occurring strata seems to reduce overall variability in measurements by about 20% compared to simple random samples taken from the entire (unstratified) population.

Now suppose you have taken a random sample size n_i from each appropriate school and you got a sample mean weight $\overline{x_i}$ from each school. How do you get the best estimate for population mean weight μ of the all 1246 students? The answer is, we use a weighted average.

$$\mu \approx \frac{n_1}{m}\overline{x_1} + \frac{n_2}{m}\overline{x_2} + \frac{n_3}{m}\overline{x_3}$$

COMMENT: This is an example with three strata. Applications with any number of strata can be solved in a similar way with obvious extensions of formulas.

(a) Compute the size of the random samples n_1, n_2, n_3 to be taken from each school. Round each sample size to the nearest whole number and make sure they add up to $m = 100$.

(b) Suppose you took the appropriate random sample from each school and you got the following average student weights: $\overline{x_1} = 82\ lb$, $\overline{x_2} = 115\ lb$, $\overline{x_3} = 90\ lb$. Compute your best estimate for the population mean weight μ.

29. *Expand Your Knowledge: Stratified Sampling and Politics*
Three local districts are "swing" districts for an upcoming election on a contentious political issue. A survey will be conducted in which voters will be asked to rate their opinion regarding this issue on a scale of 0 (strongly oppose) to 10 (strongly support). A small preliminary random sample from each district was used to estimate the sample standard deviation s of responses for the district. The following table shows the number of voters N in each district and the sample standard deviation s of strength of support in each district.

District 1	District 2	District 3
$N_1 = 1525$	$N_2 = 917$	$N_3 = 2890$
$s_1 = 2.2$	$s_2 = 1.4$	$s_3 = 3.3$

We have a total population of 5332 voters and 3 strata (districts.)

The group doing the survey has enough funding to obtain a random sample of $m = 250$ total responses from all the districts.

(a) Compute the size of the random samples n_1, n_2, n_3 to taken from each district. Round each sample size to the nearest whole number and make sure they add up to $m = 250$. *Hint*: see problem 28.

(b) Suppose you took the appropriate random samples from each district and you got the following average political support measures (scale 0 to 10): $\overline{x_1} = 6.2, \overline{x_2} = 3.1\ \overline{x_3} = 8.5$. Compute your best estimate for the population mean μ (scale 0 to 10) of voter support for the issue from the total population of these three districts. *Hint*: see problem 28.

30. *Expand Your Knowledge: Stratified Sampling, No Preliminary Study, Police Identification* Police are tested for their ability to correctly recognize and identify a suspect based on a witness or victim's verbal description of the suspect. Scores on the identification test range from 0 to 100 (perfect score). Three cities in Massachusetts are under study. The mean score for all police in the three cities is requested. However, funding will only permit $m = 150$ police to be tested. There is no preliminary study to estimate sample standard deviation of scores in each city. City A has $N_1 = 183$ police, City B has $N_2 = 371$ police, and City C has $N_3 = 255$ police.

If we have a preliminary study for sample standard deviations, we can use a method called *proportional sampling* (also called *representative sampling*). The sample size n_1 for each stratum (city police department) is given by

$$n_i = \left[\frac{N_i}{N_1 + N_2 + N_3}\right]m$$

In most cases n_i is not a whole number so we round to the nearest whole number.

Remember our total sample size is $m = n_1 + n_2 + n_3$.

(a) Use the method of proportional sampling to compute n_1, n_2, and n_3, the size of the random sample to be taken from each of the three cities. Round each n_i to the nearest whole number and make sure $m = n_1 + n_2 + n_3$.

(b) Suppose you actually conducted the specified number of tests in each city and obtained the following result: $\overline{x_1} = 96$ is the mean test score from city A, $\overline{x_2} = 85$ is the mean test score from City B, and $\overline{x_3} = 88$ is the mean test score from City C. Use the weighted average

$$\mu \approx \frac{n_1}{m}\overline{x_1} + \frac{n_2}{m}\overline{x_2} + \frac{n_3}{m}\overline{x_3}$$

to get your best estimate for the population mean test score of all police in all three cities.

COMMENT: Estimates for μ are usually better if we have a preliminary study with reasonably accurate estimates for the sample standard deviation s_i of each stratum (*see* Problem 28 and 29). However, in the case where there is no preliminary study we can use proportional sampling and still obtain good results.

SECTION **3.3**

Percentiles and Box-and-Whisker Plots

FOCUS POINTS

- Interpret the meaning of percentile scores.
- Compute the median, quartiles, and five-number summary from raw data.
- Make a box-and-whisker plot. Interpret the results.
- Describe how a box-and-whisker plot indicates spread of data about the median.

We've seen measures of central tendency and spread for a set of data. The arithmetic mean $\bar{x}$ and the standard deviation s will be very useful in later work. However, because they each utilize every data value, they can be heavily influenced by one or two extreme data values. In cases where our data distributions are heavily skewed or even bimodal, we often get a better summary of the distribution by utilizing relative position of data rather than exact values.

Percentile

Recall that the median is an average computed by using relative position of the data. If we are told that 81 is the median score on a biology test, we know that after the data have been ordered, 50% of the data fall at or below the median value of 81. The median is an example of a *percentile;* in fact, it is the 50th percentile. The general definition of the *P*th percentile follows.

> For whole numbers *P* (where $1 \le P \le 99$), the *P*th **percentile** of a distribution is a value such that *P*% of the data fall at or below it and $(100 - P)\%$ of the data fall at or above it.

In Figure 3-3, we see the 60th percentile marked on a histogram. We see that 60% of the data lie below the mark and 40% lie above it.

FIGURE 3-3

A Histogram with the 60th Percentile Shown

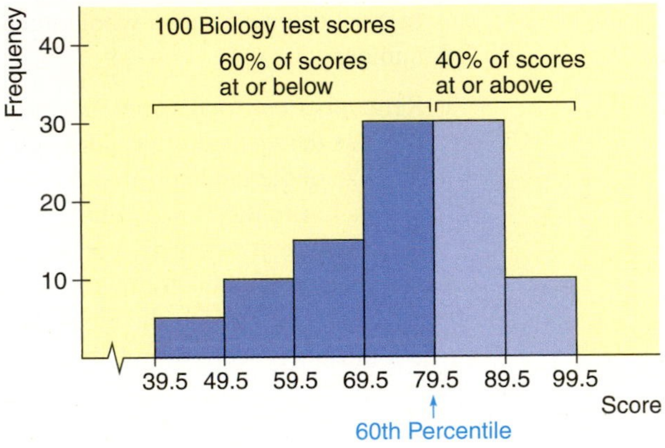

GUIDED EXERCISE 6 | PERCENTILES

You took the English achievement test to obtain college credit in freshman English by examination.

(a) If your score is at the 89th percentile, what percentage of scores are at or below yours?

→ The percentile means that 89% of the scores are at or below yours.

(b) If the scores ranged from 1 to 100 and your raw score is 95, does this necessarily mean that your score is at the 95th percentile?

→ No, the percentile gives an indication of relative position of the scores. The determination of your percentile has to do with the number of scores at or below yours. If everyone did very well and only 80% of the scores fell at or below yours, you would be at the 80th percentile even though you got 95 out of 100 points on the exam.

There are 99 percentiles, and in an ideal situation, the 99 percentiles divide the data set into 100 equal parts. (See Figure 3-4.) However, if the number of data elements is not exactly divisible by 100, the percentiles will not divide the data into equal parts.

FIGURE 3-4

Percentiles

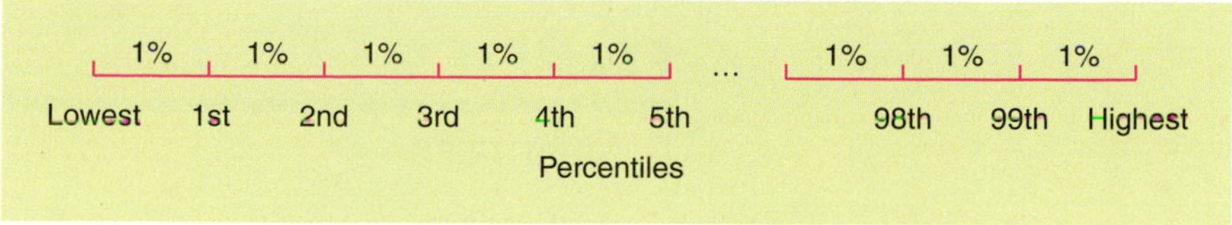

There are several widely used conventions for finding percentiles. They lead to slightly different values for different situations, but these values are close together. For all conventions, the data are first *ranked* or ordered from smallest to largest. A natural way to find the Pth percentile is to then find a value such that $P\%$ of the data fall at or below it. This will not always be possible, so we take the nearest value satisfying the criterion. It is at this point that there is a variety of processes to determine the exact value of the percentile.

Quartiles

We will not be very concerned about exact procedures for evaluating percentiles in general. However, *quartiles* are special percentiles used so frequently that we want to adopt a specific procedure for their computation.

Quartiles are those percentiles that divide the data into fourths. The *first quartile* Q_1 is the 25th percentile, the *second quartile* Q_2 is the median, and the *third quartile* Q_3 is the 75th percentile. (See Figure 3-5.)

FIGURE 3-5

Quartiles

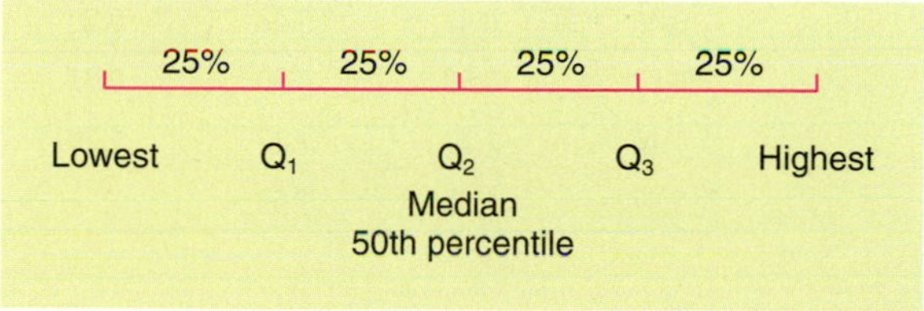

Again, several conventions are used for computing quartiles, but the following convention utilizes the median and is widely adopted.

PROCEDURE

HOW TO COMPUTE QUARTILES

1. Order the data from smallest to largest.
2. Find the median. This is the second quartile.
3. The first quartile Q_1 is then the median of the lower half of the data; that is, it is the median of the data falling *below* the Q_2 position (and not including Q_2).
4. The third quartile Q_3 is the median of the upper half of the data; that is, it is the median of the data falling *above* the Q_2 position (and not including Q_2).

In short, all we do to find the quartiles is find three medians.

Interquartile range

The median, or second quartile, is a popular measure of the center utilizing relative position. A useful measure of data spread utilizing relative position is the *interquartile range (IQR)*. It is simply the difference between the third and first quartiles.

$$\textbf{Interquartile range} = Q_3 - Q_1$$

The interquartile range tells us the spread of the middle half of the data. Now let's look at an example to see how to compute all of these quantities.

EXAMPLE 9

QUARTILES

In a hurry? On the run? Hungry as well? How about an ice cream bar as a snack? Ice cream bars are popular among all age groups. *Consumer Reports* did a study of ice cream bars. Twenty-seven bars with taste ratings of at least "fair" were listed, and cost per bar was included in the report. Just how much does an ice cream bar cost? The inflation adjusted data, expressed in dollars, appear in Table 3-4. As you can see, the cost varies quite a bit, partly because the bars are not of uniform size.

TABLE 3-4 Cost of Ice Cream Bars (in dollars)

1.29	1.37	1.30	0.80	0.67	1.33	1.37	1.37
1.27	0.93	0.63	0.80	0.27	1.38	0.77	1.14
1.53	0.55	0.80	0.70	0.36	0.65	0.47	0.68
0.50	0.48	0.46					

TABLE 3-5 Ordered Cost of Ice Cream Bars (in dollars)

0.46	0.47	0.48	0.50	0.55	0.63	0.63	0.65
0.67	0.68	0.70	0.77	0.80	0.80	0.80	0.93
1.14	1.27	1.27	1.29	1.30	1.33	1.37	1.37
1.37	1.38	1.53					

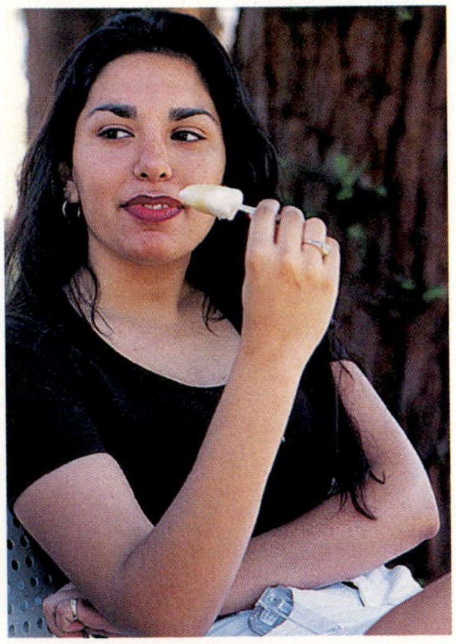

Gary Conner/PhotoEdit

(a) Find the quartiles.

SOLUTION: We first order the data from smallest to largest. Table 3-5 shows the data in order. Next, we find the median. Since the number of data values is 27, there are an odd number of data, and the median is simply the center or 14th value. The value is shown boxed in Table 3-5.

$$\text{Median} = Q_2 = 0.80$$

There are 13 values below the median position, and Q_1 is the median of these values. It is the middle or seventh value and is shaded in Table 3-5.

$$\text{First quartile} = Q_1 = 0.63$$

There are also 13 values above the median position. The median of these is the seventh value from the right end. This value is also shaded in Table 3-5.

$$\text{Third quartile} = Q_3 = 1.30$$

(b) Find the interquartile range.

SOLUTION:

$$\begin{aligned} IQR &= Q_3 - Q_1 \\ &= 1.30 - 0.63 \\ &= 0.67 \end{aligned}$$

This means that the middle half of the data has a cost spread of 67¢.

GUIDED EXERCISE 7 QUARTILES

Many people consider the number of calories in an ice cream bar as important as, if not more important than, the cost. The *Consumer Reports* article also included the calorie count of the rated ice cream bars (Table 3-6). There were 22 vanilla-flavored bars rated. Again, the bars varied in size, and some of the smaller bars had fewer calories. The calorie counts for the vanilla bars follow.

TABLE 3-6 **Calories in Vanilla-Flavored Ice Cream Bars**

342	377	319	353	295
234	294	286	377	182
310	439	111	201	182
197	209	147	190	151
131	151			

(a) Our first step is to order the data. Do so.

TABLE 3-7 **Ordered Data**

111	131	147	151	151	182
182	190	197	201	209	234
286	294	295	310	319	342
353	377	377	439		

(b) There are 22 data values. Find the median.

Average the 11th and 12th data values boxed together in Table 3-7.

$$\text{Median} = \frac{209 + 234}{2} = 221.5$$

(c) How many values are below the median position? Find Q_1.

Since the median lies halfway between the 11th and 12th values, there are 11 values below the median position. Q_1 is the median of these values.

$Q_1 = 182$

(d) There are the same number of data above as below the median. Use this fact to find Q_3.

Q_3 is the median of the upper half of the data. There are 11 values in the upper portion.

$Q_3 = 319$

(e) Find the interquartile range and comment on its meaning.

$$IQR = Q_3 - Q_1 = 319 - 182 = 137$$

The middle portion of the data has a spread of 137 calories.

BOX-AND-WHISKER PLOTS

Five-number summary

The quartiles together with the low and high data values give us a very useful *five-number summary* of the data and their spread.

FIVE-NUMBER SUMMARY

Lowest value, Q_1, median, Q_3, highest value

Box-and-whisker plot

We will use these five numbers to create a graphic sketch of the data called a *box-and-whisker plot.* Box-and-whisker plots provide another useful technique from exploratory data analysis (EDA) for describing data.

PROCEDURE

HOW TO MAKE A BOX-AND-WHISKER PLOT

1. Draw a vertical scale to include the lowest and highest data values.
2. To the right of the scale, draw a box from Q_1 to Q_3.
3. Include a solid line through the box at the median level.
4. Draw vertical lines, called *whiskers*, from Q_1 to the lowest value and from Q_3 to the highest value.

Whiskers

FIGURE 3-6

Box-and-Whisker Plot

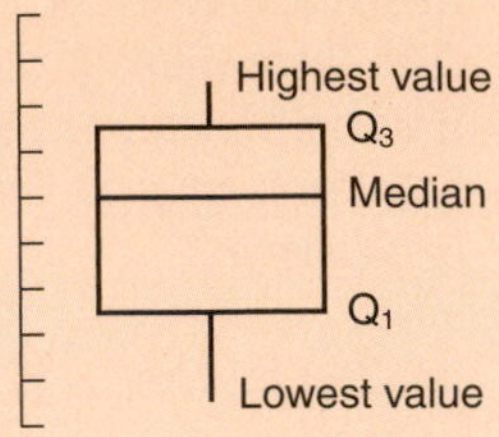

The next example demonstrates the process of making a box-and-whisker plot.

EXAMPLE 10

BOX-AND-WHISKER PLOT

Using the data from Guided Exercise 7, make a box-and-whisker plot showing the calories in vanilla-flavored ice cream bars. Use the plot to make observations about the distribution of calories.

(a) In Guided Exercise 7, we ordered the data (see Table 3-7) and found the values of the median, Q_1, and Q_3. From this previous work we have the following five-number summary:

$$\text{low value} = 111;\ Q_1 = 182;\ \text{median} = 221.5;\ Q_3 = 319;\ \text{high value} = 439$$

(b) We select an appropriate vertical scale and make the plot (Figure 3-7).

(c) ***Interpretation*** A quick glance at the box-and-whisker plot reveals the following:

(i) The box tells us where the middle half of the data lies, so we see that half of the ice cream bars have between 182 and 319 calories, with an interquartile range of 137 calories.

(ii) The median is slightly closer to the lower part of the box. This means that the lower calorie counts are more concentrated. The calorie counts above the median are more spread out, indicating that the distribution is slightly skewed toward the higher values.

(iii) The upper whisker is longer than the lower, which again emphasizes skewness toward the higher values.

FIGURE 3-7

Box-and-Whisker Plot for Calories in Vanilla-Flavored Ice Cream Bars

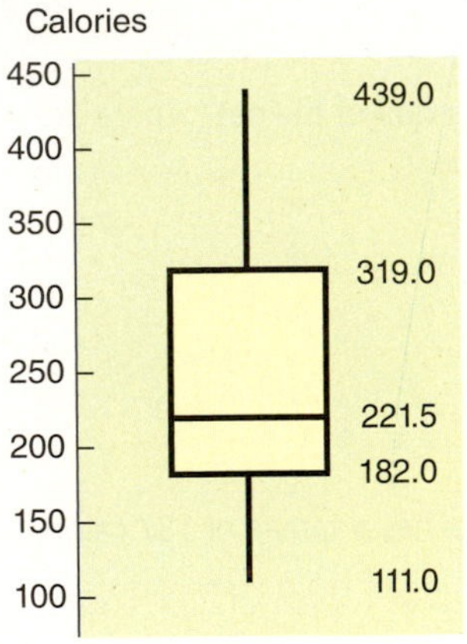

COMMENT In exploratory data analysis, *hinges* rather than quartiles are used to create the box. Hinges are computed in a manner similar to the method used to compute quartiles. However, in the case of an odd number of data values, include the median itself in both the lower and upper halves of the data (see *Applications, Basics, and Computing of Exploratory Data Analysis* by Paul Velleman and David Hoaglin, Duxbury Press). This has the effect of shrinking the box and moving the ends of the box slightly toward the median. For an even number of data, the quartiles as we computed them equal the hinges.

GUIDED EXERCISE 8 | Box-and-whisker plot

The Renata College Development Office sent salary surveys to alumni who graduated 2 and 5 years ago. The voluntary responses received are summarized in the box-and-whisker plots shown in Figure 3-8 on the next page.

(a) From Figure 3-8, estimate the median and extreme values of salaries of alumni graduating 2 years ago. In what range are the middle half of the salaries?

→ The median seems to be about $44,000. The extremes are about $33,000 and $54,000. The middle half of the salaries fall between $40,000 and $47,000.

FIGURE 3-8

Box-and-Whisker Plots for Alumni Salaries (in thousands of dollars)

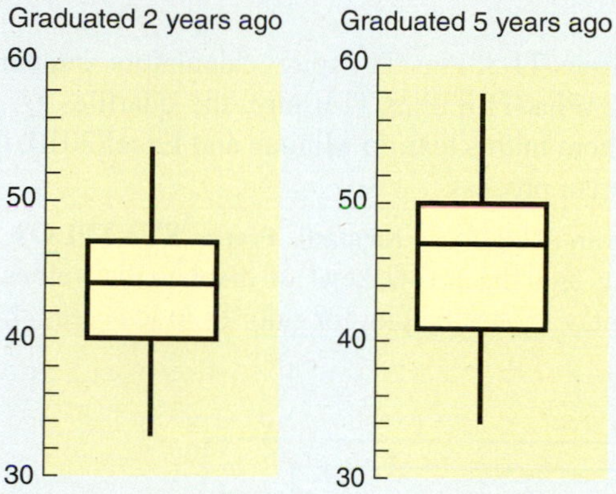

(b) From Figure 3-8, estimate the median and the extreme values of salaries of alumni graduating 5 years ago. What is the location of the middle half of the salaries?

→ The median seems to be $47,000. The extremes are $34,000 and $58,000. The middle half of the data is enclosed by the box with low side at $41,000 and high side at $50,000.

(c) *Interpretation* Compare the two box-and-whisker plots and make comments about the salaries of alumni graduating 2 and 5 years ago.

→ The salaries of the alumni graduating 5 years ago have a larger range. They begin slightly higher than and extend to levels about $4000 above the salaries of those graduating 2 years ago. The middle half of the data is also more spread out, with higher boundaries and a higher median.

CRITICAL THINKING

Problem 12 discusses how to identify possible outliers on a box-and-whisker plot.

Box-and-whisker plots provide a graphic display of the spread of data about the median. The box shows the location of the middle half of the data. One-quarter of the data is located along each whisker.

To the extent that the median is centered in the box and the whiskers are about the same length, the data distribution is symmetric around the median. If the median line is near one end of the box, the data are skewed toward the other end of the box.

We have developed the skeletal box-and-whisker display. Other variations include *fences*, which are marks placed on either side of the box to represent various portions of data. Values that lie beyond the fences are *outliers*. Problem 12 of this section discusses some criteria for locating fences and identifying outliers.

WHAT DOES A BOX-AND-WHISKER PLOT TELL US?

A box-and-whisker plot is a visual display of data spread around the *median*. It tells us

- the high value, low value, first quartile, median, and fourth quartile;
- how the data are spread around the median;
- the location of the middle half of the data;
- if there are outliers (*see* Problem 12 of this section).

TECH NOTES *Box-and-Whisker Plot*

Both Minitab and the TI-84Plus/TI-83Plus/TI-*n*spire calculators support box-and-whisker plots. On the TI-84Plus/TI-83Plus/TI-*n*spire, the quartiles Q_1 and Q_3 are calculated as we calculate them in this text. In Minitab and Excel 2010, they are calculated using a slightly different process.

TI-84Plus/TI-83Plus/TI-*n*spire (with TI-84Plus Keypad) Press **STATPLOT ➤On.** Highlight box plot. Use **Trace** and the arrow keys to display the values of the five-number summary. The display shows the plot for calories in ice cream bars.

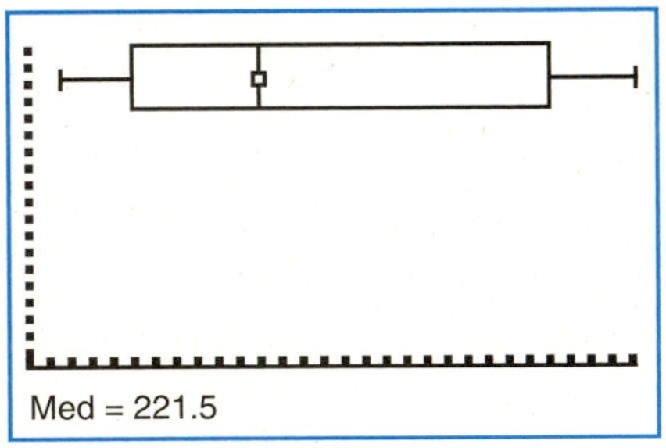

Excel 2010 Does not produce box-and-whisker plot. However, each value of the five-number summary can be found. On the **Home** ribbon, click the **Insert Function** f_x. In the dialogue box, select **Statistical** as the category and scroll to **Quartile**. In the dialogue box, enter the data location and then enter the number of the value you want. For instance, enter 1 in the quartile box for the first quartile.

Minitab Press **Graph ➤ Boxplot.** In the dialogue box, set Display to IQRange Box.

VIEWPOINT Is Shorter Higher?

Can you estimate a person's height *from the* pitch *of his or her voice? Is a soprano shorter than an alto? Is a bass taller than a tenor? A statistical study of singers in the New York Choral Society provided information. For more information, visit the Brase/Brase statistics site at* **http://www.cengage.com/statistics/brase11e** *and find the link to DASL, the Carnegie Mellon University Data and Story Library. From Data Subjects, select Music and then Singers. Methods of this chapter can be used with new methods we will learn in Chapters 8 and 9 to examine such questions from a statistical point of view.*

SECTION 3.3 PROBLEMS

1. *Statistical Literacy* Angela took a general aptitude test and scored in the 82nd percentile for aptitude in accounting. What percentage of the scores were at or below her score? What percentage were above?

2. *Statistical Literacy* One standard for admission to Redfield College is that the student rank in the upper quartile of his or her graduating high school class. What is the minimal percentile rank of a successful applicant?

3. *Critical Thinking* The town of Butler, Nebraska, decided to give a teacher-competency exam and defined the passing scores to be those in the 70th percentile or higher. The raw test scores ranged from 0 to 100. Was a raw score of 82 necessarily a passing score? Explain.

4. *Critical Thinking* Clayton and Timothy took different sections of Introduction to Economics. Each section had a different final exam. Timothy scored 83 out of 100 and had a percentile rank in his class of 72. Clayton scored 85 out of 100 but his percentile rank in his class was 70. Who performed better with respect to the rest of the students in the class, Clayton or Timothy? Explain your answer.

5. *Basic Computation: Five-Number Summary, Interquartile Range* Consider the following ordered data:

   2 5 5 6 7 7 8 9 10

   (a) Find the low, Q_1, median, Q_3, high.
   (b) Find the interquartile range.
   (c) Make a box-and-whisker plot.

6. *Basic Computation: Five-Number Summary, Interquartile Range* Consider the following ordered data:

   2 5 5 6 7 8 8 9 10 12

   (a) Find the low, Q_1, median, Q_3 high.
   (b) Find the interquartile range.
   (c) Make a box-and-whisker plot.

7. *Health Care: Nurses* At Center Hospital there is some concern about the high turnover of nurses. A survey was done to determine how long (in months) nurses had been in their current positions. The responses (in months) of 20 nurses were

23	2	5	14	25	36	27	42	12	8
7	23	29	26	28	11	20	31	8	36

   Make a box-and-whisker plot of the data. Find the interquartile range.

8. *Health Care: Staff* Another survey was done at Center Hospital to determine how long (in months) clerical staff had been in their current positions. The responses (in months) of 20 clerical staff members were

25	22	7	24	26	31	18	14	17	20
31	42	6	25	22	3	29	32	15	72

   (a) Make a box-and-whisker plot. Find the interquartile range.
   (b) Compare this plot with the one in Problem 7. Discuss the locations of the medians, the locations of the middle halves of the data banks, and the distances from Q_1 and Q_3 to the extreme values.

9. *Sociology: College Graduates* What percentage of the general U.S. population have bachelor's degrees? The *Statistical Abstract of the United States,* 120th Edition, gives the percentage of bachelor's degrees by state. For convenience, the data are sorted in increasing order.

17	18	18	18	19	20	20	20	21	21
21	21	22	22	22	22	22	22	23	23
24	24	24	24	24	24	24	24	25	26
26	26	26	26	26	27	27	27	27	27
28	28	29	31	31	32	32	34	35	38

(a) Make a box-and-whisker plot and find the interquartile range.
(b) Illinois has a bachelor's degree percentage rate of about 26%. Into what quartile does this rate fall?

10. *Sociology: High School Dropouts* What percentage of the general U.S. population are high school dropouts? The *Statistical Abstract of the United States,* 120th Edition, gives the percentage of high school dropouts by state. For convenience, the data are sorted in increasing order.

5	6	7	7	7	7	8	8	8	8
8	9	9	9	9	9	9	9	10	10
10	10	10	10	10	10	11	11	11	11
11	11	11	11	12	12	12	12	13	13
13	13	13	13	14	14	14	14	14	15

(a) Make a box-and-whisker plot and find the interquartile range.
(b) Wyoming has a dropout rate of about 7%. Into what quartile does this rate fall?

11. *Auto Insurance: Interpret Graphs* *Consumer Reports* rated automobile insurance companies and listed annual premiums for top-rated companies in several states. Figure 3-9 shows box-and-whisker plots for annual premiums for urban customers (married couple with one 17-year-old son) in three states. The box-and-whisker plots in Figure 3-9 were all drawn using the same scale on a TI-84Plus/TI-83Plus/TI-*n*spire calculator.

FIGURE 3-9

Insurance Premium (annual, urban)

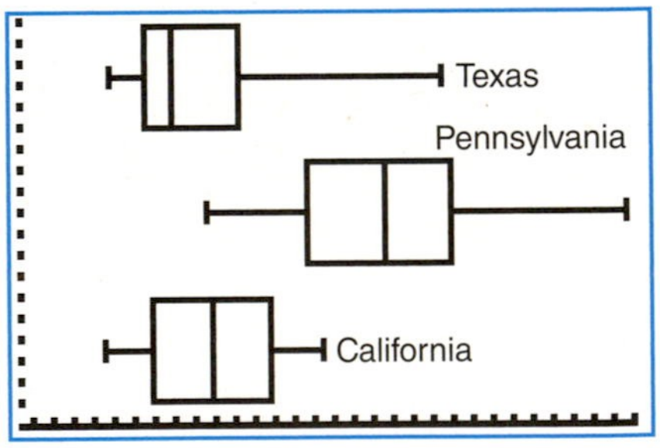

(a) Which state has the lowest premium? the highest?
(b) Which state has the highest median premium?
(c) Which state has the smallest range of premiums? the smallest interquartile range?
(d) Figure 3-10 gives the five-number summaries generated on the TI-84Plus/TI-83Plus/TI-*n*spire calculators for the box-and-whisker plots of Figure 3-9. Match the five-number summaries to the appropriate box-and-whisker plots.

FIGURE 3-10

Five-Number Summaries for Insurance Premiums

(a)

```
1-Var Stats
↑n=10
 minX=2382
 Q1=2758
 Med=2991
 Q3=3652
 maxX=5715
```

(b)

```
1-Var Stats
↑n=10
 minX=3314
 Q1=4326
 Med=5116.5
 Q3=5801
 maxX=7527
```

(c)

```
1-Var Stats
↑n=10
 minX=2323
 Q1=2801
 Med=3377.5
 Q3=3966
 maxX=4482
```

12. *Expand Your Knowledge: Outliers* Some data sets include values so high or so low that they seem to stand apart from the rest of the data. These data are called *outliers.* Outliers may represent data collection errors, data entry errors, or simply valid but unusual data values. It is important to identify outliers in the data set and examine the outliers carefully to determine if they are in error. One way to detect outliers is to use a box-and-whisker plot. Data values that fall beyond the limits,

Lower limit: $Q_1 - 1.5 \times (IQR)$

Upper limit: $Q_3 + 1.5 \times (IQR)$

where *IQR* is the interquartile range, are suspected outliers. In the computer software package Minitab, values beyond these limits are plotted with asterisks (*).

Students from a statistics class were asked to record their heights in inches. The heights (as recorded) were

65	72	68	64	60	55	73	71	52	63	61	74
69	67	74	50	4	75	67	62	66	80	64	65

(a) Make a box-and-whisker plot of the data.
(b) Find the value of the interquartile range (*IQR*).
(c) Multiply the *IQR* by 1.5 and find the lower and upper limits.
(d) Are there any data values below the lower limit? above the upper limit? List any suspected outliers. What might be some explanations for the outliers?

CHAPTER REVIEW

SUMMARY

To characterize numerical data, we use both measures of center and of spread.

- Commonly used measures of center are the arithmetic mean, the median, and the mode. The weighted average and trimmed mean are also used as appropriate.
- Commonly used measures of spread are the variance, the standard deviation, and the range. The variance and standard deviation are measures of spread about the mean.
- Chebyshev's theorem enables us to estimate the data spread about the mean.
- The coefficient of variation lets us compare the relative spreads of different data sets.
- Other measures of data spread include percentiles, which indicate the percentage of data falling at or below the specified percentile value.
- Box-and-whisker plots show how the data are distributed about the median and the location of the middle half of the data distribution.

In later work, the average we will use most often is the mean; the measure of variation we will use most often is the standard deviation.

IMPORTANT WORDS & SYMBOLS

Section 3.1

Average 90
Mode 90
Median 91
Mean 93
Summation symbol, Σ 93
Sample mean, $\bar{x}$ 93
Population mean, μ 93
Resistant measure 94
Trimmed mean 94
Weighted average 96
Harmonic mean 101
Geometric mean 102

Section 3.2

Range 102
Variance 103
Standard deviation 103
Sum of squares, $\Sigma(x - \bar{x})^2$ 103
Sample variance, s^2 103
Sample standard deviation, s 104
Population variance, σ^2 107
Population standard deviation, σ 107
Population size, N 107
Coefficient of variation, CV 109
Chebyshev's theorem 110
Outlier 112
Mean of grouped data 117
Standard deviation of grouped data 117
Moving average 118

Section 3.3

Percentile 122
Quartiles 123
Interquartile range, IQR 123
Five-number summary 125
Box-and-whisker plot 125
Whisker 126

VIEWPOINT The Fujita Scale

How do you measure a tornado? Professor Fujita and Allen Pearson (Director of the National Severe Storm Forecast Center) developed a measure based on wind speed and type of damage done by a tornado. The result is an excellent example of both descriptive and inferential statistical methods. For more information, visit the Brase/Brase statistics site at **http://www.cengage.com/statistics/brase11e** *and find the link to the tornado project. Then look up Fujita scale. If we group the data a little, the scale becomes*

FS	WS	%
F0 & F1	40–112	67
F2 & F3	113–206	29
F4 & F5	207–318	4

where FS represents Fujita scale; WS, wind speed in miles per hour; and %, percentage of all tornados. Out of 100 tornados, what would you estimate for the mean and standard deviation of wind speed?

CHAPTER REVIEW PROBLEMS

1. *Statistical Literacy*
   (a) What measures of variation indicate spread about the mean?
   (b) Which graphic display shows the median and data spread about the median?

2. *Critical Thinking* Look at the two histograms below. Each involves the same number of data. The data are all whole numbers, so the height of each bar represents the number of values equal to the corresponding midpoint shown on the horizontal axis. Notice that both distributions are symmetric.

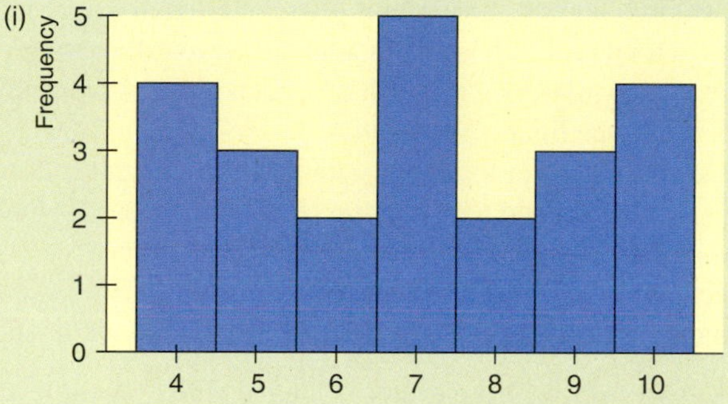

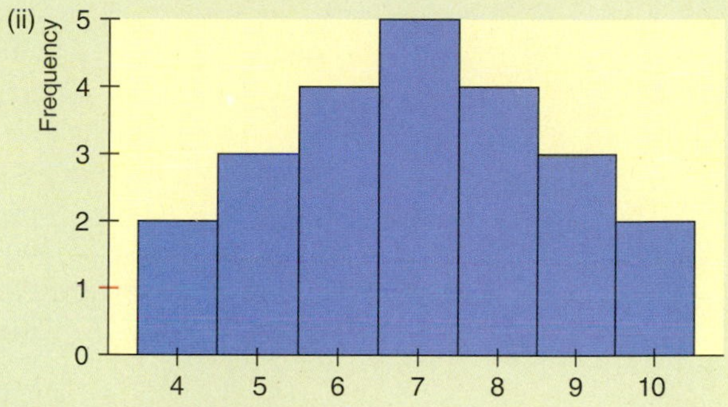

   (a) Estimate the mode, median, and mean for each histogram.
   (b) Which distribution has the larger standard deviation? Why?

3. *Critical Thinking* Consider the following Minitab display of two data sets.

```
Variable  N   Mean  SE Mean  StDev  Minimum     Q1  Median     Q3  Maximum
C1       20  20.00     1.62   7.26     7.00  15.00   20.00  25.00    31.00
C2       20  20.00     1.30   5.79     7.00  20.00   22.00  22.00    31.00
```

   (a) What are the respective means? the respective ranges?
   (b) Which data set seems more symmetric? Why?
   (c) Compare the interquartile ranges of the two sets. How do the middle halves of the data sets compare?

4. *Consumer: Radon Gas* "Radon: The Problem No One Wants to Face" is the title of an article appearing in *Consumer Reports*. Radon is a gas emitted from the ground that can collect in houses and buildings. At certain levels it can cause lung cancer. Radon concentrations are measured in picocuries per liter (pCi/L). A radon level of 4 pCi/L is considered "acceptable." Radon levels in a house vary from week to week. In one house, a sample of 8 weeks had the following readings for radon level (in pCi/L):

   1.9 2.8 5.7 4.2 1.9 8.6 3.9 7.2

   (a) Find the mean, median, and mode.
   (b) Find the sample standard deviation, coefficient of variation, and range.

5. *Political Science: Georgia Democrats* How Democratic is Georgia? County-by-county results are shown for a recent election. For your convenience, the data have been sorted in increasing order (Source: *County and City Data Book,* 12th Edition, U.S. Census Bureau).

Percentage of Democratic Vote by Counties in Georgia

31	33	34	34	35	35	35	36	38	38	38	39	40	40	40	40
41	41	41	41	41	41	41	42	42	43	44	44	44	45	45	46
46	46	46	47	48	49	49	49	49	50	51	52	52	53	53	53
53	53	55	56	56	57	57	59	62	66	66	68				

(a) Make a box-and-whisker plot of the data. Find the interquartile range.
(b) *Grouped Data* Make a frequency table using five classes. Then estimate the mean and sample standard deviation using the frequency table. Compute a 75% Chebyshev interval centered about the mean.
(c) If you have a statistical calculator or computer, use it to find the actual sample mean and sample standard deviation. Otherwise, use the values $\Sigma x = 2769$ and $\Sigma x^2 = 132{,}179$ to compute the sample mean and sample standard deviation.

6. *Grades: Weighted Average* Professor Cramer determines a final grade based on attendance, two papers, three major tests, and a final exam. Each of these activities has a total of 100 possible points. However, the activities carry different weights. Attendance is worth 5%, each paper is worth 8%, each test is worth 15%, and the final is worth 34%.
(a) What is the average for a student with 92 on attendance, 73 on the first paper, 81 on the second paper, 85 on test 1, 87 on test 2, 83 on test 3, and 90 on the final exam?
(b) Compute the average for a student with the above scores on the papers, tests, and final exam, but with a score of only 20 on attendance.

7. *General: Average Weight* An elevator is loaded with 16 people and is at its load limit of 2500 pounds. What is the mean weight of these people?

8. *Agriculture: Harvest Weight of Maize* The following data represent weights in kilograms of maize harvest from a random sample of 72 experimental plots on St. Vincent, an island in the Caribbean (Reference: B. G. F. Springer, *Proceedings, Caribbean Food Corps. Soc.,* Vol. 10, pp. 147–152). *Note:* These data are also available for download at the Online Study Center. For convenience, the data are presented in increasing order.

7.8	9.1	9.5	10.0	10.2	10.5	11.1	11.5	11.7	11.8
12.2	12.2	12.5	13.1	13.5	13.7	13.7	14.0	14.4	14.5
14.6	15.2	15.5	16.0	16.0	16.1	16.5	17.2	17.8	18.2
19.0	19.1	19.3	19.8	20.0	20.2	20.3	20.5	20.9	21.1
21.4	21.8	22.0	22.0	22.4	22.5	22.5	22.8	22.8	23.1
23.1	23.2	23.7	23.8	23.8	23.8	23.8	24.0	24.1	24.1
24.5	24.5	24.9	25.1	25.2	25.5	26.1	26.4	26.5	26.7
27.1	29.5								

(a) Compute the five-number summary.
(b) Compute the interquartile range.
(c) Make a box-and-whisker plot.
(d) *Interpretation* Discuss the distribution. Does the lower half of the distribution show more data spread than the upper half?

9. *Focus Problem: The Educational Advantage* Solve the focus problem at the beginning of this chapter.

10. *Agriculture: Bell Peppers* The pathogen *Phytophthora capsici* causes bell pepper plants to wilt and die. A research project was designed to study the effect of soil water content and the spread of the disease in fields of bell peppers (Source: *Journal of Agricultural, Biological, and Environmental Statistics*, Vol. 2, No. 2). It is thought that too much water helps spread the disease. The fields were divided into rows and quadrants. The soil water content (percent of water by volume of soil) was determined for each plot. An important first step in such a research project is to give a statistical description of the data.

Soil Water Content for Bell Pepper Study

15	14	14	14	13	12	11	11	11	11	10	11	13	16	10
9	15	12	9	10	7	14	13	14	8	9	8	11	13	13
15	12	9	10	9	9	16	16	12	10	11	11	12	15	6
10	10	10	11	9										

(a) Make a box-and-whisker plot of the data. Find the interquartile range.

(b) *Grouped Data* Make a frequency table using four classes. Then estimate the mean and sample standard deviation using the frequency table. Compute a 75% Chebyshev interval centered about the mean.

(c) If you have a statistical calculator or computer, use it to find the actual sample mean and sample standard deviation.

11. *Performance Rating: Weighted Average* A performance evaluation for new sales representatives at Office Automation Incorporated involves several ratings done on a scale of 1 to 10, with 10 the highest rating. The activities rated include new contacts, successful contacts, total contacts, dollar volume of sales, and reports. Then an overall rating is determined by using a weighted average. The weights are 2 for new contacts, 3 for successful contacts, 3 for total contacts, 5 for dollar value of sales, and 3 for reports. What would the overall rating be for a sales representative with ratings of 5 for new contacts, 8 for successful contacts, 7 for total contacts, 9 for dollar volume of sales, and 7 for reports?

DATA HIGHLIGHTS: GROUP PROJECTS

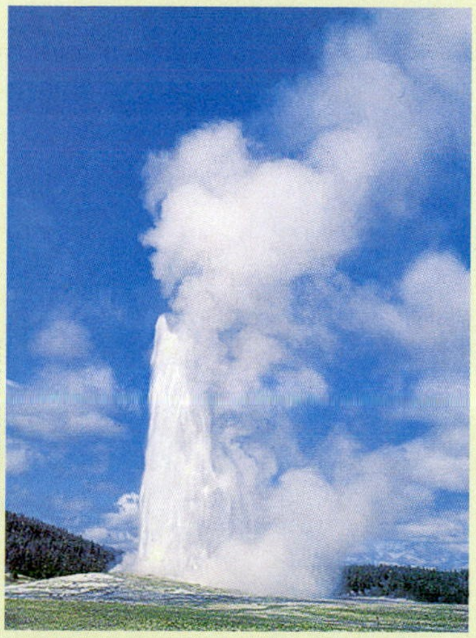

Old Faithful Geyser, Yellowstone National Park

Break into small groups and discuss the following topics. Organize a brief outline in which you summarize the main points of your group discussion.

1. *The Story of Old Faithful* is a short book written by George Marler and published by the Yellowstone Association. Chapter 7 of this interesting book talks about the effect of the 1959 earthquake on eruption intervals for Old Faithful Geyser. Dr. John Rinehart (a senior research scientist with the National Oceanic and Atmospheric Administration) has done extensive studies of the eruption intervals before and after the 1959 earthquake. Examine Figure 3-11. Notice the general shape. Is the graph more or less symmetrical? Does it have a single mode frequency? The mean interval between eruptions has remained steady at about 65 minutes for the past 100 years. Therefore, the 1959 earthquake did not significantly change the mean, but it did change the distribution of eruption intervals. Examine Figure 3-12. Would you say there are really two frequency modes, one shorter and the other longer? Explain. The overall mean is about the same for both graphs, but one graph has a much larger standard deviation (for eruption intervals) than the other. Do no calculations, just look at both graphs, and then explain which graph has the smaller and which has the larger standard deviation. Which distribution will have the larger coefficient of variation? In everyday terms, what would this mean if you were actually at Yellowstone waiting to see the next eruption of Old Faithful? Explain your answer.

FIGURE 3-11

Typical Behavior of Old Faithful Geyser Before 1959 Quake

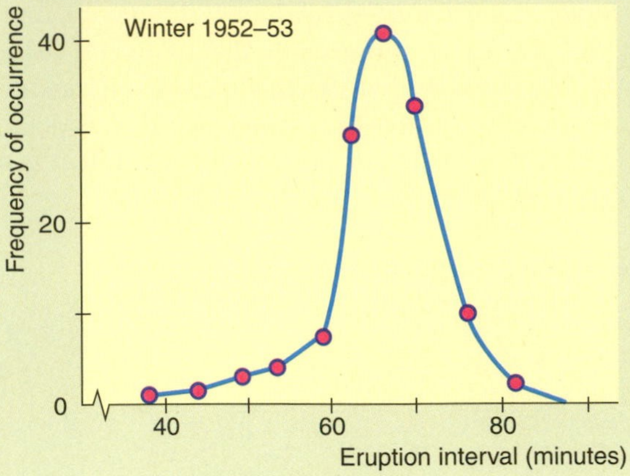

FIGURE 3-12

Typical Behavior of Old Faithful Geyser After 1959 Quake

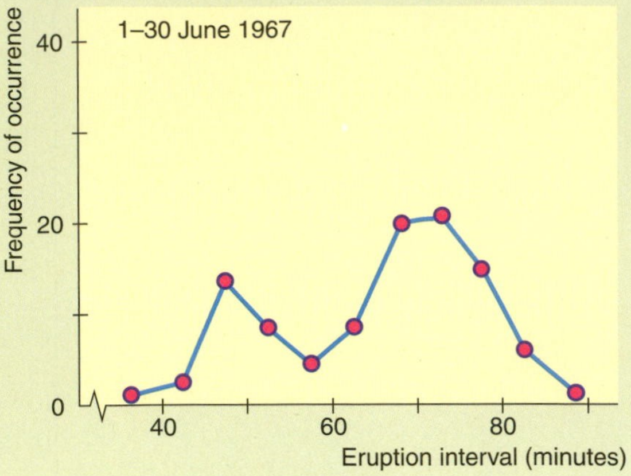

2. Most academic advisors tell students to major in a field they really love. After all, it is true that money cannot buy happiness! Nevertheless, it is interesting to at least look at some of the higher-paying fields of study. After all, a field like mathematics can be a lot of fun, once you get into it. We see that women's salaries tend to be less than men's salaries. However, women's salaries are rapidly catching up, and this benefits the entire workforce in different ways. Figure 3-13 shows the median incomes for college graduates with different majors. The employees in the sample are all at least 30 years old. Does it seem reasonable to assume that many of the employees are in jobs beyond the entry level? Explain. Compare the median incomes shown for all women aged 30 or older holding bachelor's degrees with the median incomes for men of similar age holding bachelor's degrees. Look at the particular majors listed. What percentage of men holding bachelor's degrees in mathematics make $52,316 or more? What percentage of women holding computer/information science degrees make $41,559 or more? How do median incomes for men and women holding engineering degrees compare? What about pharmacy degrees?

FIGURE 3-13

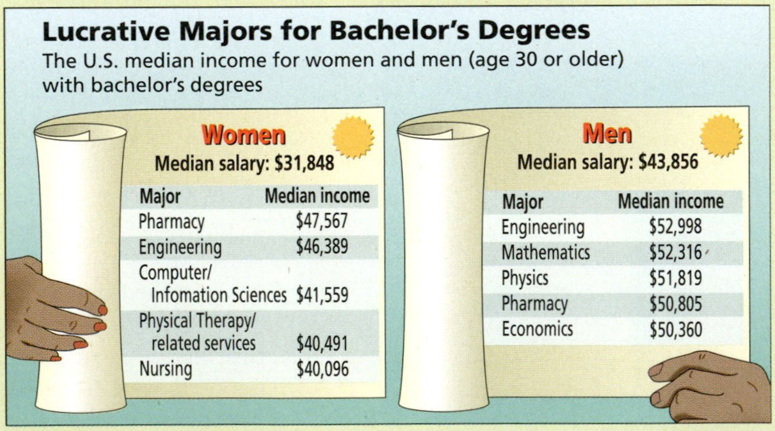
Lucrative Majors for Bachelor's Degrees
The U.S. median income for women and men (age 30 or older) with bachelor's degrees
Women
Median salary: $31,848
Major | Median income
Pharmacy $47,567
Engineering $46,389
Computer/ Infomation Sciences $41,559
Physical Therapy/ related services $40,491
Nursing $40,096
Men
Median salary: $43,856
Major | Median income
Engineering $52,998
Mathematics $52,316
Physics $51,819
Pharmacy $50,805
Economics $50,360

Source: Bureau of Labor Statistics

LINKING CONCEPTS: WRITING PROJECTS

Discuss each of the following topics in class or review the topics on your own. Then write a brief but complete essay in which you summarize the main points. Please include formulas and graphs as appropriate.

1. An average is an attempt to summarize a collection of data into just *one* number. Discuss how the mean, median, and mode all represent averages in this context. Also discuss the differences among these averages. Why is the mean a balance point? Why is the median a midway point? Why is the mode the most common data point? List three areas of daily life in which you think the mean, median, or mode would be the best choice to describe an "average."
2. Why do we need to study the variation of a collection of data? Why isn't the average by itself adequate? We have studied three ways to measure variation. The range, the standard deviation, and, to a large extent, a box-and-whisker plot all indicate the variation within a data collection. Discuss similarities and differences among these ways to measure data variation. Why would it seem reasonable to pair the median with a box-and-whisker plot and to pair the mean with the standard deviation? What are the advantages and disadvantages of each method of describing data spread? Comment on statements such as the following: (a) The range is easy to compute, but it doesn't give much information; (b) although the standard deviation is more complicated to compute, it has some significant applications; (c) the box-and-whisker plot is fairly easy to construct, and it gives a lot of information at a glance.
3. Why is the coefficient of variation important? What do we mean when we say that the coefficient of variation has no units? What advantage can there be in having no units? Why is *relative size* important?

   Consider robin eggs; the mean weight of a collection of robin eggs is 0.72 ounce and the standard deviation is 0.12 ounce. Now consider elephants; the mean weight of elephants in the zoo is 6.42 tons, with a standard deviation 1.07 tons. The units of measurement are different and there is a great deal of difference between the weight of an elephant and that of a robin's egg. Yet the coefficient of variation is about the same for both. Comment on this from the viewpoint of the size of the standard deviation relative to that of the mean.
4. What is Chebyshev's theorem? Suppose you have a friend who knows very little about statistics. Write a paragraph or two in which you describe Chebyshev's theorem for your friend. Keep the discussion as simple as possible, but be sure to get the main ideas across to your friend. Suppose he or she asks, "What is this stuff good for?" and suppose you respond (a little sarcastically) that Chebyshev's theorem applies to everything from butterflies to the orbits of the planets! Would you be correct? Explain.

USING TECHNOLOGY

Raw Data

Application

Using the software or calculator available to you, do the following.

1. Trade winds are one of the beautiful features of island life in Hawaii. The following data represent total air movement in miles per day over a weather station in Hawaii as determined by a continuous anemometer recorder. The period of observation is January 1 to February 15, 1971.

26	14	18	14
113	50	13	22
27	57	28	50
72	52	105	138
16	33	18	16
32	26	11	16
17	14	57	100
35	20	21	34
18	13	18	28
21	13	25	19
11	19	22	19
15	20		

Source: United States Department of Commerce, National Oceanic and Atmospheric Administration, Environmental Data Service. *Climatological Data, Annual Summary, Hawaii,* Vol. 67, No. 13. Asheville: National Climatic Center, 1971, pp. 11, 24.

(a) Use the computer to find the sample mean, median, and (if it exists) mode. Also, find the range, sample variance, and sample standard deviation.

(b) Use the five-number summary provided by the computer to make a box-and-whisker plot of total air movement over the weather station.

(c) Four data values are exceptionally high: 113, 105, 138, and 100. The strong winds of January 5 (113 reading) brought in a cold front that dropped snow on Haleakala National Park (at the 8000 ft elevation). The residents were so excited that they drove up to see the snow and caused such a massive traffic jam that the Park Service had to close the road. The winds of January 15, 16, and 28 (readings 105, 138, and 100) accompanied a storm with funnel clouds that did much damage. Eliminate these values (i.e., 100, 105, 113, and 138) from the data bank and redo parts (a) and (b). Compare your results with those previously obtained. Which average is most affected? What happens to the standard deviation? How do the two box-and-whisker plots compare?

Technology Hints: Raw Data

TI-84Plus/TI-83Plus/TI-*n*spire (with TI-84Plus keypad), Excel 2010, Minitab

The Tech Note of Section 3.2 gives brief instructions for finding summary statistics for raw data using the TI-84Plus/TI-83Plus/TI-*n*spire calculators, Excel 2010, and Minitab. The Tech Note of Section 3.3 gives brief instructions for constructing box-and-whisker plots using the TI-84Plus/TI-83Plus/TI-*n*spire calculators and Minitab.

SPSS

Many commands in SPSS provide an option to display various summary statistics. A direct way to display summary statistics is to use the menu choices **Analyze ➤ Descriptive Statistics ➤ Descriptives.** In the dialogue box, move the variable containing your data into the variables box. Click **Options...** and then check the summary statistics you wish to display. Click Continue and then OK. Notice that the median is not available. A more complete list of summary statistics is available with the menu choices **Analyze ➤ Descriptive Statistics ➤ Frequencies.** Click the **Statistics** button and check the summary statistics you wish to display.

For box-and-whisker plots, use the menu options **Graphs ➤ Legacy Dialogues ➤ Interactive ➤ Boxplot**. In the dialogue box, place the variable containing your data in the box along the vertical axis. After selecting the options you want, click OK.

SPSS Student Version 17 has other options for creating box-and-whisker plots. The Tech Guide for SPSS that accompanies this text has instructions for using the **Chart Builder** and **Graphboard Template Chooser** options found in the **Graphs** menu. Note that SPSS requires the appropriate level of measurement for processing data. In SPSS, the **scale level** corresponds to both interval and ratio level. The **nominal** and **ordinal** levels are as described in this text.

CUMULATIVE REVIEW PROBLEMS

CHAPTER 1-3

Critical Thinking and Literacy

1. Consider the following measures: mean, median, variance, standard deviation, percentile.
   (a) Which measures utilize relative position of the data values?
   (b) Which measures utilize actual data values regardless of relative position?
2. Describe how the presence of possible outliers might be identified on
   (a) histograms.
   (b) dotplots.
   (c) stem-and-leaf displays.
   (d) box-and-whisker plots.
3. Consider two data sets, A and B. The sets are identical except that the high value of data set B is three times greater than the high value of data set A.
   (a) How do the medians of the two data sets compare?
   (b) How do the means of the two data sets compare?
   (c) How do the standard deviations of the two data sets compare?
   (d) How do the box-and-whisker plots of the two data sets compare?
4. You are examining two data sets involving test scores, set A and set B. The score 86 appears in both data sets. In which of the following data sets does 86 represent a higher score? Explain.
   (a) The percentile rank of 86 is higher in set A than in set B.
   (b) The mean is 80 in both data sets, but set A has a higher standard deviation.

Applications

In west Texas, water is extremely important. The following data represent pH levels in ground water for a random sample of 102 west Texas wells. A pH less than 7 is acidic and a pH above 7 is alkaline. Scanning the data, you can see that water in this region tends to be hard (alkaline). Too high a pH means the water is unusable or needs expensive treatment to make it usable (Reference: C. E. Nichols and V. E. Kane, Union Carbide Technical Report K/UR-1).

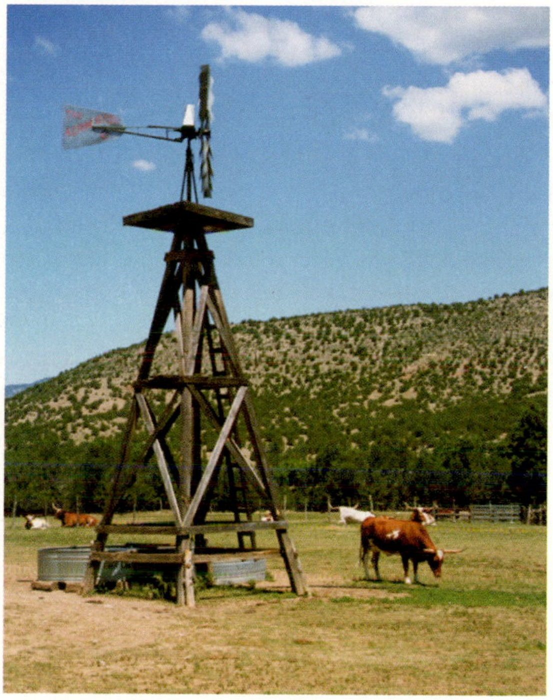

Dave G. Houser/Documentary Value/Corbis

These data are also available for download at the Online Study Center. For convenience, the data are presented in increasing order.

x: pH of Ground Water in 102 West Texas Wells

7.0	7.0	7.0	7.0	7.0	7.0	7.0	7.0	7.1	7.1	7.1	7.1
7.1	7.1	7.1	7.1	7.1	7.1	7.2	7.2	7.2	7.2	7.2	7.2
7.2	7.2	7.2	7.2	7.3	7.3	7.3	7.3	7.3	7.3	7.3	7.3
7.3	7.3	7.3	7.4	7.4	7.4	7.4	7.4	7.4	7.4	7.4	7.4
7.5	7.5	7.5	7.5	7.5	7.5	7.5	7.5	7.6	7.6	7.6	7.6
7.6	7.6	7.6	7.6	7.6	7.7	7.7	7.7	7.7	7.7	7.7	7.8
7.8	7.8	7.8	7.8	7.9	7.9	7.9	7.9	7.9	8.0	8.1	8.1
8.1	8.1	8.1	8.1	8.1	8.2	8.2	8.2	8.2	8.2	8.2	8.2
8.4	8.5	8.6	8.7	8.8	8.8						

5. Write a brief description in which you outline how you would obtain a random sample of 102 west Texas water wells. Explain how random numbers would be used in the selection process.
6. Is the given data nominal, ordinal, interval, or ratio? Explain.
7. Make a stem-and-leaf display. Use five lines per stem so that leaf values 0 and 1 are on one line, 2 and 3 are on the next line, 4 and 5 are on the next, 6 and 7 are on the next, and 8 and 9 are on the last line of the stem.
8. Make a frequency table, histogram, and relative-frequency histogram using five classes. Recall that for decimal data, we "clear the decimal" to determine classes for whole-number data and then reinsert the decimal to obtain the classes for the frequency table of the original data.
9. Make an ogive using five classes.
10. Compute the range, mean, median, and mode for the given data.
11. (a) Verify that $\Sigma x = 772.9$ and $\Sigma x^2 = 5876.6$.
    (b) Compute the sample variance, sample standard deviation, and coefficient of variation for the given data. Is the sample standard deviation small relative to the mean pH?
12. Compute a 75% Chebyshev interval centered on the mean.
13. Make a box-and-whisker plot. Find the interquartile range.

Interpretation

Wow! In Problems 5–13 you constructed a lot of information regarding the pH of west Texas ground water based on sample data. Let's continue the investigation.

14. Look at the histogram. Is the pH distribution for these wells symmetric or skewed? Are lower or higher values more common?
15. Look at the ogive. What percent of the wells have a pH less than 8.15? Suppose a certain crop can tolerate irrigation water with a pH between 7.35 and 8.55. What percent of the wells could be used for such a crop?
16. Look at the stem-and-leaf plot. Are there any unusually high or low pH levels in this sample of wells? How many wells are neutral (pH of 7)?

17. Use the box-and-whisker plot to describe how the data are spread about the median. Are the pH values above the median more spread out than those below? Is this observation consistent with the skew of the histogram?

18. Suppose you are working for the regional water commissioner. You have been asked to submit a brief report about the pH level in ground water in the west Texas region. Write such a report and include appropriate graphs.

4

Nathan Griffith/Corbis

North Wind/North Wind Picture Archive

We see that the theory of probabilities is at bottom only common sense reduced to calculation; it makes us appreciate with exactitude what reasonable minds feel by a sort of instinct, often without being able to account for it.

—Pierre-Simon Laplace

This quote explains how the great mathematician Pierre-Simon Laplace (1749–1827) described the theory of mathematical probability. The discovery of the mathematical theory of probability was shared by two Frenchmen: Blaise Pascal and Pierre Fermat. These 17th-century scholars were attracted to the subject by the inquiries of the Chevalier de Méré, a gentleman gambler.

Although the first applications of probability were to games of chance and gambling, today the subject seems to pervade almost every aspect of modern life. Everything from the orbits of spacecraft to the social behavior of woodchucks is described in terms of probabilities.

ELEMENTARY PROBABILITY THEORY

PREVIEW QUESTIONS

Why would anyone study probability? *Hint:* Most big issues in life involve uncertainty. (SECTION 4.1)

What are the basic definitions and rules of probability? (SECTION 4.2)

What are counting techniques, trees, permutations, and combinations? (SECTION 4.3)

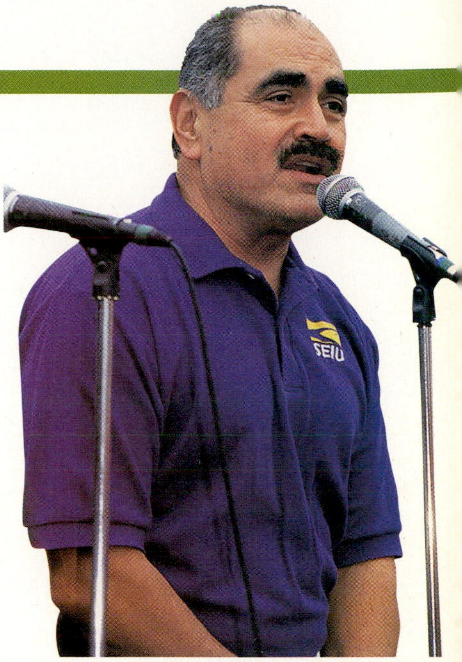

David Young-Wolff/PhotoEdit

FOCUS PROBLEM

How Often Do Lie Detectors Lie?

James Burke is an educator who is known for his interesting science-related radio and television shows aired by the British Broadcasting Corporation. His book *Chances: Risk and Odds in Everyday Life* (Virginia Books, London) contains a great wealth of fascinating information about probabilities. The following quote is from Professor Burke's book:

> *If I take a polygraph test and lie, what is the risk I will be detected?* According to some studies, there's about a 72 percent chance you will be caught by the machine.
>
> *What is the risk that if I take a polygraph test it will incorrectly say that I lied?* At least 1 in 15 will be thus falsely accused.

Both of these statements contain conditional probabilities, which we will study in Section 4.2. Information from that section will enable us to answer the following:

Suppose a person answers 10% of a long battery of questions with lies. Assume that the remaining 90% of the questions are answered truthfully.

1. Estimate the percentage of answers the polygraph will *wrongly* indicate as lies.
2. Estimate the percentage of answers the polygraph will *correctly* indicate as lies.

If the polygraph indicated that 30% of the questions were answered as lies, what would you estimate for the *actual* percentage of questions the person answered as lies? (See Problems 27 and 28 in Section 4.2.)

John Neubauer/PhotoEdit

For online student resources, visit the Brase/Brase, *Understandable Statistics*, 11th edition web site at **http://www.cengage.com/statistics/brase11e**

SECTION **4.1**

What Is Probability?

FOCUS POINTS

- Assign probabilities to events.
- Explain how the law of large numbers relates to relative frequencies.
- Apply basic rules of probability in everyday life.
- Explain the relationship between statistics and probability.

We encounter statements given in terms of probability all the time. An excited sports announcer claims that Sheila has a 90% chance of breaking the world record in the upcoming 100-yard dash. Henry figures that if he guesses on a true–false question, the probability of getting it right is 1/2. The Right to Health lobby claims that the probability is 0.40 of getting an erroneous report from a medical laboratory at a low-cost health center. It is consequently lobbying for a federal agency to license and monitor all medical laboratories.

When we use probability in a statement, we're using a *number between 0 and 1* to indicate the likelihood of an event.

Probability is a numerical measure between 0 and 1 that describes the likelihood that an event will occur. Probabilities closer to 1 indicate that the event is more likely to occur. Probabilities closer to 0 indicate that the event is less likely to occur.

Probability of an event *A*, *P*(*A*)

***P*(*A*)**, read "P of A," denotes the **probability of event *A*.**

If $P(A) = 1$, event *A* is certain to occur.

If $P(A) = 0$, event *A* is certain not to occur.

It is important to know what probability statements mean and how to determine probabilities of events, because probability is the language of inferential statistics.

PROBABILITY ASSIGNMENTS

Intuition

1. A probability assignment based on **intuition** incorporates past experience, judgment, or opinion to estimate the likelihood of an event.

Relative frequency

2. A probability assignment based on **relative frequency** uses the formula

$$\text{Probability of event} = \text{relative frequency} = \frac{f}{n} \quad (1)$$

where f is the frequency of the event occurrence in a sample of n observations.

Equally likely outcomes

3. A probability assignment based on **equally likely outcomes** uses the formula

$$\text{Probability of event} = \frac{\text{Number of outcomes favorable to event}}{\text{Total number of outcomes}} \quad (2)$$

EXAMPLE 1

PROBABILITY ASSIGNMENT

Consider each of the following events, and determine how the probability is assigned.

(a) A sports announcer claims that Sheila has a 90% chance of breaking the world record in the 100-yard dash.

SOLUTION: It is likely the sports announcer used intuition based on Sheila's past performance.

(b) Henry figures that if he guesses on a true–false question, the probability of getting it right is 0.50.

SOLUTION: In this case there are two possible outcomes: Henry's answer is either correct or incorrect. Since Henry is guessing, we assume the outcomes are equally likely. There are $n = 2$ equally likely outcomes, and only one is correct. By formula (2),

$$P(\text{correct answer}) = \frac{\text{Number of favorable outcomes}}{\text{Total number of outcomes}} = \frac{1}{2} = 0.50$$

(c) The Right to Health lobby claims that the probability of getting an erroneous medical laboratory report is 0.40, based on a random sample of 200 laboratory reports, of which 80 were erroneous.

SOLUTION: Formula (1) for relative frequency gives the probability, with sample size $n = 200$ and number of errors $f = 80$.

$$P(\text{error}) = \text{relative frequency} = \frac{f}{n} = \frac{80}{200} = 0.40$$

We've seen three ways to assign probabilities: intuition, relative frequency, and—when outcomes are equally likely—a formula. Which do we use? Most of the time it depends on the information that is at hand or that can be feasibly obtained. Our choice of methods also depends on the particular problem. In Guided Exercise 1, you will see three different situations, and you will decide the best way to assign the probabilities. *Remember, probabilities are numbers between 0 and 1, so don't assign probabilities outside this range.*

GUIDED EXERCISE 1

DETERMINE A PROBABILITY

Assign a probability to the indicated event on the basis of the information provided. Indicate the technique you used: intuition, relative frequency, or the formula for equally likely outcomes.

(a) A random sample of 500 students at Hudson College were surveyed and it was determined that 375 wear glasses or contact lenses. Estimate the probability that a Hudson College student selected at random wears corrective lenses.

➡ In this case we are given a sample size of 500, and we are told that 375 of these students wear corrective lenses. It is appropriate to use a relative frequency for the desired probability:

$$P(\text{student needs corrective lenses}) = \frac{f}{n} = \frac{375}{500} = 0.75$$

Continued

GUIDED EXERCISE 1 *continued*

(b) The Friends of the Library hosts a fundraising barbecue. George is on the cleanup committee. There are four members on this committee, and they draw lots to see who will clean the grills. Assuming that each member is equally likely to be drawn, what is the probability that George will be assigned the grill-cleaning job?

→ There are four people on the committee, and each is equally likely to be drawn. It is appropriate to use the formula for equally likely events. George can be drawn in only one way, so there is only one outcome favorable to the event.

$$P(\text{George}) = \frac{\text{No. of favorable outcomes}}{\text{Total no. of outcomes}} = \frac{1}{4} = 0.25$$

(c) Joanna photographs whales for Sea Life Adventure Films. On her next expedition, she is to film blue whales feeding. Based on her knowledge of the habits of blue whales, she is almost certain she will be successful. What specific number do you suppose she estimates for the probability of success?

→ Since Joanna is almost certain of success, she should make the probability close to 1. We could say $P(\text{success})$ is above 0.90 but less than 1. This probability assignment is based on intuition.

The technique of using the relative frequency of an event as the probability of that event is a common way of assigning probabilities and will be used a great deal in later chapters. The underlying assumption we make is that if events occurred a certain percentage of times in the past, they will occur about the same percentage of times in the future. In fact, this assumption can be strengthened to a very general statement called the *law of large numbers*.

Law of large numbers

LAW OF LARGE NUMBERS

In the long run, as the sample size increases and increases, the relative frequencies of outcomes get closer and closer to the theoretical (or actual) probability value.

The law of large numbers is the reason businesses such as health insurance, automobile insurance, and gambling casinos can exist and make a profit.

No matter how we compute probabilities, it is useful to know what outcomes are possible in a given setting. For instance, if you are going to decide the probability that Hardscrabble will win the Kentucky Derby, you need to know which other horses will be running.

To determine the possible outcomes for a given setting, we need to define a *statistical experiment*.

Statistical experiment

Event

Simple event

Sample space

A **statistical experiment** or **statistical observation** can be thought of as any random activity that results in a definite outcome.

An **event** is a collection of one or more outcomes of a statistical experiment or observation.

A **simple event** is one particular outcome of a statistical experiment.

The set of all simple events constitutes the **sample space** of an experiment.

EXAMPLE 2

USING A SAMPLE SPACE

Human eye color is controlled by a single pair of genes (one from the father and one from the mother) called a *genotype*. Brown eye color, B, is dominant over blue eye color, ℓ. Therefore, in the genotype Bℓ, consisting of one brown gene B and one blue gene ℓ, the brown gene dominates. A person with a Bℓ genotype has brown eyes.

If both parents have brown eyes and have genotype Bℓ, what is the probability that their child will have blue eyes? What is the probability the child will have brown eyes?

SOLUTION: To answer these questions, we need to look at the sample space of all possible eye-color genotypes for the child. They are given in Table 4-1.

TABLE 4-1 **Eye Color Genotypes for Child**

	Mother	
Father	**B**	**ℓ**
B	BB	$B\ell$
ℓ	ℓB	$\ell\ell$

According to genetics theory, the four possible genotypes for the child are equally likely. Therefore, we can use Formula (2) to compute probabilities. Blue eyes can occur only with the $\ell\ell$ genotype, so there is only one outcome favorable to blue eyes. By formula (2),

$$P(\text{blue eyes}) = \frac{\text{Number of favorable outcomes}}{\text{Total number of outcomes}} = \frac{1}{4}$$

Brown eyes occur with the three remaining genotypes: BB, Bℓ, and ℓB. By formula (2),

$$P(\text{brown eyes}) = \frac{\text{Number of favorable outcomes}}{\text{Total number of outcomes}} = \frac{3}{4}$$

GUIDED EXERCISE 2

USING A SAMPLE SPACE

Professor Gutierrez is making up a final exam for a course in literature of the southwest. He wants the last three questions to be of the true–false type. To guarantee that the answers do not follow his favorite pattern, he lists all possible true–false combinations for three questions on slips of paper and then picks one at random from a hat.

(a) Finish listing the outcomes in the given sample space.

TTT	FTT	TFT	____
TTF	FTF	TFF	____

→ The missing outcomes are FFT and FFF.

(b) What is the probability that all three items will be false? Use the formula

$$P(\text{all F}) = \frac{\text{No. of favorable outcomes}}{\text{Total no. of outcomes}}$$

→ There is only one outcome, FFF, favorable to all false, so

$$P(\text{all F}) = \frac{1}{8}$$

Continued

GUIDED EXERCISE 2 *continued*

(c) What is the probability that exactly two items will be true?

→ There are three outcomes that have exactly two true items: TTF, TFT, and FTT. Thus,

$$P(\text{two T}) = \frac{\text{No. of favorable outcomes}}{\text{Total no. of outcomes}} = \frac{3}{8}$$

There is another important point about probability assignments of simple events.

The **sum** of the probabilities of all simple events in a sample space must equal 1.

We can use this fact to determine the probability that an event will not occur. For instance, if you think the probability is 0.65 that you will win a tennis match, you assume the probability is 0.35 that your opponent will win.

The *complement* of an event A is the event that A *does not occur*. We use the notation A^c to designate the complement of event A. Figure 4-1 shows the event A and its complement A^c.

Notice that the two distinct events A and A^c make up the entire sample space. Therefore, the sum of their probabilities is 1.

Complement of an event A^c

The **complement of event** A is the event that A *does not occur*. A^c designates the complement of event A. Furthermore,

1. $P(A) + P(A^c) = 1$
2. $P(\text{event } A \text{ does } not \text{ occur}) = P(A^c) = 1 - P(A)$ (3)

EXAMPLE 3

COMPLEMENT OF AN EVENT

The probability that a college student who has not received a flu shot will get the flu is 0.45. What is the probability that a college student will *not* get the flu if the student has not had the flu shot?

SOLUTION: In this case, we have

$$P(\text{will get flu}) = 0.45$$

$$P(\text{will } not \text{ get flu}) = 1 - P(\text{will get flu}) = 1 - 0.45 = 0.55$$

FIGURE 4-1

The Event A and Its Complement A^c

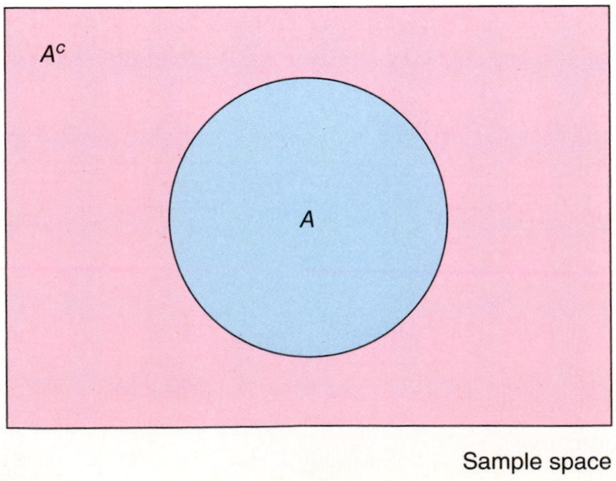

GUIDED EXERCISE 3 | COMPLEMENT OF AN EVENT

A veterinarian tells you that if you breed two cream-colored guinea pigs, the probability that an offspring will be pure white is 0.25. What is the probability that an offspring will not be pure white?

(a) $P(\text{pure white}) + P(\textit{not}\text{ pure white}) = $ ______ → 1

(b) $P(\textit{not}\text{ pure white}) = $ ______ → $1 - 0.25$, or 0.75

LOOKING FORWARD

The complement rule has special application in Chapter 5. There we study probabilities associated with binomial experiments. These are experiments that have only two outcomes, "success" and "failure." One feature of these experiments is that the outcomes can be categorized according to the number of "successes" out of n independent trials.

Simple Events of Sample Space of a Binomial Experiment
0 successes, 1 success, 2 successes, 3 successes, . . . , n successes

A convenient way to compute the probability of "at least one" success is to observe that the complement of "at least one success" is the event "0 successes." Then, by the complement rule

$$P(\text{at least one success}) = 1 - P(0 \text{ successes})$$

Problems 10, 11, and 12 of this section show a similar use of the complement rule.

Sometimes, instead of being told the probability of an event, we are given the odds for the event. Problems 21 and 22 discuss odds in favor and odds against an event.

SUMMARY: SOME IMPORTANT FACTS ABOUT PROBABILITY

1. A **statistical experiment** or **statistical observation** is any random activity that results in a definite outcome. A **simple event** consists of one and only one outcome of the experiment. The **sample space** is the set of all simple events. An **event *A*** is any subset of the sample space.
2. The probability of an event A is denoted by $P(A)$.
3. The probability of an event is a number between 0 and 1. The closer to 1 the probability is, the more likely it is the event will occur. The closer to 0 the probability is, the less likely it is the event will occur.
4. The sum of the probabilities of all simple events in a sample space is 1.
5. Probabilities can be assigned by using intuition, relative frequencies, or the formula for equally likely outcomes. Additional ways to assign probabilities will be introduced in later chapters.
6. The **complement** of an event A is denoted by A^c. So, A^c is the event that A does not occur.
7. $P(A) + P(A^c) = 1$

INTERPRETING PROBABILITIES

As stressed in item 3 in the summary about probabilities, the closer the probability is to 1, the more likely the event is to occur. The closer the probability is to 0, the less likely the event is to occur. However, just because the probability of an event is very high, it is not a certainty that the event will occur. Likewise, even though the probability of an event is very low, the event might still occur.

Events with low probability but big consequences are of special concern. Such events are sometimes referred to as "black swan events," as discussed in the very popular book *The Black Swan* by Nassim Nicholas Taleb. In his varied careers Taleb has been a stock trader, author, and professor at the University of Massachusetts. His special interests are mathematical problems of luck, uncertainty, probability, and general knowledge. In his book *The Black Swan*, Taleb discusses "the impact of the highly improbable." If an event has (a) little impact, bearing, or meaning on a person's life and (b) very low probability, then (c) it seems reasonable to ignore the event.

However, some of the really big mistakes in a person's life can result from misjudging either (a) the size of an event's impact or (b) the likelihood the event will occur. An event of great importance cannot be ignored even if it has a low probability of occurrence.

So what should a person do?

Taleb describes himself as a hyperskeptic when others are gullible, and gullible when others are skeptical. In particular, he recommends being skeptical about (mentally) confirming an event when the errors of confirmation are costly. Lots and lots of data should not be thought of as confirming an event when a simple instance can disconfirm an important and costly event. Taleb recommends that one be skeptical in the face of wild randomness (and big consequences) and gullible when the randomness (and consequences) are mild. One way to estimate the "randomness" of a situation is to estimate its probability. Estimating the "consequences" of an event in your own life must be a personal decision.

WHAT DOES THE PROBABILITY OF AN EVENT TELL US?

- The probability of an event A tells us the likelihood that event A will occur. If the probability is 1, the event A is certain to occur. If the probability is 0, the event A will not occur. Probabilities closer to 1 indicate the event A is more likely to occur, but it might not. Probabilities closer to 0 indicate event A is less likely to occur, but it might.
- The probability of event A applies only in the context of conditions surrounding the sample space containing event A. For example, consider the event A that a freshman student graduates with a bachelor's degree in 4 years or less. Do events in the sample space apply to entering freshmen in all colleges and universities in the United States or just a particular college? Are majors specified? Are minimal SAT or ACT scores specified? All these conditions of events in the sample space can affect the probability of the event. It would be inappropriate to apply the probability of event A to students outside those described in the sample space.
- If we know the probability of event A, then we can easily compute the probability of event *not* A in the context of the same sample space. $P(\text{not } A) = 1 - P(A)$.

PROBABILITY RELATED TO STATISTICS

We conclude this section with a few comments on the nature of statistics versus probability. Although statistics and probability are closely related fields of mathematics, they are nevertheless separate fields. It can be said that probability is the medium through which statistical work is done. In fact, if it were not for probability theory, inferential statistics would not be possible.

Put very briefly, probability is the field of study that makes statements about what will occur when samples are drawn from a *known population*. Statistics is the field of study that describes how samples are to be obtained and how inferences are to be made about *unknown populations*.

A simple but effective illustration of the difference between these two subjects can be made by considering how we treat the following examples.

EXAMPLE OF A PROBABILITY APPLICATION

Condition: We *know* the exact makeup of the *entire* population.

Example: Given 3 green marbles, 5 red marbles, and 4 white marbles in a bag, draw 6 marbles at random from the bag. What is the probability that none of the marbles is red?

EXAMPLE OF A STATISTICAL APPLICATION

Condition: We have only *samples* from an otherwise *unknown* population.

Example: Draw a random sample of 6 marbles from the (unknown) population of all marbles in a bag and observe the colors. Based on the sample results, make a conjecture about the colors and numbers of marbles in the entire population of all marbles in the bag.

In another sense, probability and statistics are like flip sides of the same coin. On the probability side, you know the overall description of the population. The central problem is to compute the likelihood that a specific outcome will happen. On the statistics side, you know only the results of a sample drawn from the population. The central problem is to describe the sample (descriptive statistic) and to draw conclusions about the population based on the sample results (inferential statistics).

In statistical work, the inferences we draw about an unknown population are not claimed to be absolutely correct. Since the population remains unknown (in a theoretical sense), we must accept a "best guess" for our conclusions and act using the most probable answer rather than absolute certainty.

Probability is the topic of this chapter. However, we will not study probability just for its own sake. Probability is a wonderful field of mathematics, but we will study mainly the ideas from probability that are needed for a proper understanding of statistics.

VIEWPOINT What Makes a Good Teacher?

A survey of 735 students at nine colleges in the United States was taken to determine instructor behaviors that help students succeed. Data from this survey can be found by visiting the Brase/Brase statistics site at **http://www.cengage.com/statistics/brase11e** *and finding the link to DASL, the Carnegie Mellon University Data and Story Library. Once at the DASL site, select Data Subjects, then Psychology, and then Instructor Behavior. You can estimate the probability of how a student would respond (very positive, neutral, very negative) to different instructor behaviors. For example, more than 90% of the students responded "very positive" to the instructor's use of real-world examples in the classroom.*

SECTION 4.1 PROBLEMS

1. *Statistical Literacy* List three methods of assigning probabilities.

2. *Statistical Literacy* Suppose the newspaper states that the probability of rain today is 30%. What is the complement of the event "rain today"? What is the probability of the complement?

3. *Statistical Literacy* What is the probability of
   (a) an event A that is certain to occur?
   (b) an event B that is impossible?

4. *Statistical Literacy* What is the law of large numbers? If you were using the relative frequency of an event to estimate the probability of the event, would it be better to use 100 trials or 500 trials? Explain.

5. *Interpretation* A Harris Poll indicated that of those adults who drive and have a cell phone, the probability that a driver between the ages of 18 and 24 sends or reads text messages is 0.51. Can this probability be applied to *all* drivers with cell phones? Explain.

6. *Interpretation* According to a recent Harris Poll of adults with pets, the probability that the pet owner cooks especially for the pet either frequently or occasionally is 0.24.
   (a) From this information, can we conclude that the probability a male owner cooks for the pet is the same as for a female owner? Explain.
   (b) According to the poll, the probability a male owner cooks for his pet is 0.27, whereas the probability a female owner does so is 0.22. Let's explore how such probabilities might occur. Suppose the pool of pet owners surveyed consisted of 200 pet owners, 100 of whom are male and 100 of whom are female. Of the pet owners, a total of 49 cook for their pets. Of the 49 who cook for their pets, 27 are male and 22 are female. Use relative frequencies to determine the probability a pet owner cooks for a pet, the probability a male owner cooks for his pet, and the probability a female owner cooks for her pet.

7. *Basic Computation: Probability as Relative Frequency* A recent Harris Poll survey of 1010 U.S. adults selected at random showed that 627 consider the occupation of firefighter to have very great prestige. Estimate the probability (to the nearest hundredth) that a U.S. adult selected at random thinks the occupation of firefighter has very great prestige.

8. *Basic Computation: Probability of Equally Likely Events* What is the probability that a day of the week selected at random will be a Wednesday?

9. *Interpretation* An investment opportunity boasts that the chance of doubling your money in 3 years is 95%. However, when you research the details of the investment, you estimate that there is a 3% chance that you could lose the entire investment. Based on this information, are you certain to make money on this investment? Are there risks in this investment opportunity?

10. *Interpretation* A sample space consists of 4 simple events: A, B, C, D. Which events comprise the complement of A? Can the sample space be viewed as having two events, A and A^c? Explain.

11. *Critical Thinking* Consider a family with 3 children. Assume the probability that one child is a boy is 0.5 and the probability that one child is a girl is also 0.5, and that the events "boy" and "girl" are independent.
    (a) List the equally likely events for the gender of the 3 children, from oldest to youngest.
    (b) What is the probability that all 3 children are male? Notice that the complement of the event "all three children are male" is "at least one of the children is female." Use this information to compute the probability that at least one child is female.

12. *Critical Thinking* Consider the experiment of tossing a fair coin 3 times. For each coin, the possible outcomes are heads or tails.
    (a) List the equally likely events of the sample space for the three tosses.
    (b) What is the probability that all three coins come up heads? Notice that the complement of the event "3 heads" is "at least one tail." Use this information to compute the probability that there will be at least one tail.

13. *Critical Thinking* On a single toss of a fair coin, the probability of heads is 0.5 and the probability of tails is 0.5. If you toss a coin twice and get heads on the first toss, are you guaranteed to get tails on the second toss? Explain.

14. *Critical Thinking*
    (a) Explain why -0.41 cannot be the probability of some event.
    (b) Explain why 1.21 cannot be the probability of some event.
    (c) Explain why 120% cannot be the probability of some event.
    (d) Can the number 0.56 be the probability of an event? Explain.

15. *Probability Estimate: Wiggle Your Ears* Can you wiggle your ears? Use the students in your statistics class (or a group of friends) to estimate the percentage of people who can wiggle their ears. How can your result be thought of as an estimate for the probability that a person chosen at random can wiggle his or her ears? *Comment*: National statistics indicate that about 13% of Americans can wiggle their ears (Source: Bernice Kanner, *Are You Normal?*, St. Martin's Press, New York).

16. *Probability Estimate: Raise One Eyebrow* Can you raise one eyebrow at a time? Use the students in your statistics class (or a group of friends) to estimate the percentage of people who can raise one eyebrow at a time. How can your result be thought of as an estimate for the probability that a person chosen at random can raise one eyebrow at a time? *Comment:* National statistics indicate that about 30% of Americans can raise one eyebrow at a time (see source in Problem 15).

17. *Myers–Briggs: Personality Types* Isabel Briggs Myers was a pioneer in the study of personality types. The personality types are broadly defined according to four main preferences. Do married couples choose similar or different personality types in their mates? The following data give an indication (Source: I. B. Myers and M. H. McCaulley, *A Guide to the Development and Use of the Myers–Briggs Type Indicators*).

© iStockphoto.com/Gunther Beck

Similarities and Differences in a Random Sample of 375 Married Couples

Number of Similar Preferences	Number of Married Couples
All four	34
Three	131
Two	124
One	71
None	15

Suppose that a married couple is selected at random.
    (a) Use the data to estimate the probability that they will have 0, 1, 2, 3, or 4 personality preferences in common.
    (b) Do the probabilities add up to 1? Why should they? What is the sample space in this problem?

18. *General: Roll a Die*
    (a) If you roll a single die and count the number of dots on top, what is the sample space of all possible outcomes? Are the outcomes equally likely?
    (b) Assign probabilities to the outcomes of the sample space of part (a). Do the probabilities add up to 1? Should they add up to 1? Explain.
    (c) What is the probability of getting a number less than 5 on a single throw?
    (d) What is the probability of getting 5 or 6 on a single throw?

19. *Psychology: Creativity* When do creative people get their *best* ideas? *USA Today* did a survey of 966 inventors (who hold U.S. patents) and obtained the following information:

Time of Day When Best Ideas Occur

Time	Number of Inventors
6 A.M.–12 noon	290
12 noon–6 P.M.	135
6 P.M.–12 midnight	319
12 midnight–6 A.M.	222

(a) Assuming that the time interval includes the left limit and all the times up to but not including the right limit, estimate the probability that an inventor has a best idea during each time interval: from 6 A.M. to 12 noon, from 12 noon to 6 P.M., from 6 P.M. to 12 midnight, from 12 midnight to 6 A.M.

(b) Do the probabilities of part (a) add up to 1? Why should they? What is the sample space in this problem?

20. *Agriculture: Cotton* A botanist has developed a new hybrid cotton plant that can withstand insects better than other cotton plants. However, there is some concern about the germination of seeds from the new plant. To estimate the probability that a seed from the new plant will germinate, a random sample of 3000 seeds was planted in warm, moist soil. Of these seeds, 2430 germinated.

(a) Use relative frequencies to estimate the probability that a seed will germinate. What is your estimate?

(b) Use relative frequencies to estimate the probability that a seed will *not* germinate. What is your estimate?

(c) Either a seed germinates or it does not. What is the sample space in this problem? Do the probabilities assigned to the sample space add up to 1? Should they add up to 1? Explain.

(d) Are the outcomes in the sample space of part (c) equally likely?

21. *Expand Your Knowledge: Odds in Favor* Sometimes probability statements are expressed in terms of odds.

The *odds in favor* of an event A are the ratio $\dfrac{P(A)}{P(notA)} = \dfrac{P(A)}{P(A^c)}$.

For instance, if $P(A) = 0.60$, then $P(A^c) = 0.40$ and the odds in favor of A are

$$\frac{0.60}{0.40} = \frac{6}{4} = \frac{3}{2}, \text{written as 3 to 2 or 3:2}$$

(a) Show that if we are given the *odds in favor* of event A as $n{:}m$, the probability of event A is given by $P(A) = \dfrac{n}{n+m}$. *Hint:* Solve the equation $\dfrac{n}{m} = \dfrac{P(A)}{1 - P(A)}$ for $P(A)$.

(b) A telemarketing supervisor tells a new worker that the odds of making a sale on a single call are 2 to 15. What is the probability of a successful call?

(c) A sports announcer says that the odds a basketball player will make a free throw shot are 3 to 5. What is the probability the player will make the shot?

22. *Expand Your Knowledge: Odds Against* Betting odds are usually stated against the event happening (against winning).

The *odds against* event W are the ratio $\dfrac{P(not\ W)}{P(W)} = \dfrac{P(W^c)}{P(W)}$.

Cynthia HUNTER/AFP PHOTO/Getty Images

In horse racing, the betting odds are based on the probability that the horse does *not* win.

(a) Show that if we are given the *odds against* an event W as $a{:}b$, the probability of *not* W is $P(W^c) = \frac{a}{a+b}$. *Hint:* Solve the equation $\frac{a}{b} = \frac{P(W^c)}{1 - P(W^c)}$ for $P(W^c)$.

(b) In a recent Kentucky Derby, the betting odds for the favorite horse, Point Given, were 9 to 5. Use these odds to compute the probability that Point Given would lose the race. What is the probability that Point Given would win the race?

(c) In the same race, the betting odds for the horse Monarchos were 6 to 1. Use these odds to estimate the probability that Monarchos would lose the race. What is the probability that Monarchos would win the race?

(d) Invisible Ink was a long shot, with betting odds of 30 to 1. Use these odds to estimate the probability that Invisible Ink would lose the race. What is the probability the horse would win the race? For further information on the Kentucky Derby, visit the Brase/Brase statistics site at **http://www.cengage.com/statistics/brase** and find the link to the Kentucky Derby.

23. *Business: Customers* John runs a computer software store. Yesterday he counted 127 people who walked by his store, 58 of whom came into the store. Of the 58, only 25 bought something in the store.

(a) Estimate the probability that a person who walks by the store will enter the store.

(b) Estimate the probability that a person who walks into the store will buy something.

(c) Estimate the probability that a person who walks by the store will come in *and* buy something.

(d) Estimate the probability that a person who comes into the store will buy nothing.

SECTION **4.2**

Some Probability Rules—Compound Events

FOCUS POINTS

- Compute probabilities of general compound events.
- Compute probabilities involving independent events or mutually exclusive events.
- Use survey results to compute conditional probabilities.

CONDITIONAL PROBABILITY AND MULTIPLICATION RULES

You roll two dice. What is the probability that you will get a 5 on each die? Consider a collection of 6 balls identical in every way except in color. There are 3 green balls, 2 blue balls, and 1 red ball. You draw two balls at random from the collection *without replacing* the first ball before drawing the second. What is the probability that both balls will be green?

It seems that these two problems are nearly alike. They are alike in the sense that in each case, you are to find the probability of two events occurring *together*.

In the first problem, you are to find

P(5 on 1st die *and* 5 on 2nd die)

In the second, you want

P(green ball on 1st draw *and* green ball on 2nd draw)

The two problems differ in one important aspect, however. In the dice problem, the outcome of a 5 on the first die does not have any effect on the probability of getting a 5 on the second die. Because of this, the events are *independent*.

Independent events

> Two events are **independent** if the occurrence or nonoccurrence of one event does *not* change the probability that the other event will occur.

In the problem concerning a collection of colored balls, the probability that the first ball is green is 3/6, since there are 6 balls in the collection and 3 of them are green. If you get a green ball on the first draw, the probability of getting a green ball on the second draw is changed to 2/5, because one green ball has already been drawn and only 5 balls remain. Therefore, the two events in the ball-drawing problem are *not* independent. They are, in fact, *dependent*, since the outcome of the first draw changes the probability of getting a green ball on the second draw.

Dependent events

Probability of A and B

Why does the *independence* or *dependence* of two events matter? The type of events determines the way we compute the probability of the two events happening together. If two events A and B are *independent*, then we use formula (4) to compute the probability of the event A and B:

MULTIPLICATION RULE FOR INDEPENDENT EVENTS

$$P(A \text{ and } B) = P(A) \cdot P(B) \quad (4)$$

If the events are *dependent*, then we must take into account the changes in the probability of one event caused by the occurrence of the other event. The notation $P(A, \text{ given } B)$ denotes the probability that event A will occur *given* that event B has occurred. This is called a *conditional probability*. We read $P(A, \text{ given } B)$ as "probability of A given B." If A and B are dependent events, then $P(A) \neq P(A, \text{ given } B)$ because the occurrence of event B has changed the probability that event A will occur. A standard notation for $P(A, \text{ given } B)$ is $P(A \mid B)$.

Conditional probability
$P(A \mid B)$

GENERAL MULTIPLICATION RULE FOR ANY EVENTS

$$P(A \text{ and } B) = P(A) \cdot P(B \mid A) \quad (5)$$

$$P(A \text{ and } B) = P(B) \cdot P(A \mid B) \quad (6)$$

We will use either formula (5) or formula (6) according to the information available.

Multiplication rules of probability

Formulas (4), (5), and (6) constitute the *multiplication rules of probability*. They help us compute the probability of events happening together when the sample space is too large for convenient reference or when it is not completely known.

Note: For conditional probability, observe that the multiplication rule

$$P(A \text{ and } B) = P(B) \cdot P(A \mid B)$$

can be solved for $P(A \mid B)$, leading to

CONDITIONAL PROBABILITY (WHEN $P(B) \neq 0$)

$$P(A \mid B) = \frac{P(A \text{ and } B)}{P(B)}$$

We will see some applications of this formula in later chapters.

Let's use the multiplication rules to complete the dice and ball-drawing problems just discussed. We'll compare the results with those obtained by using the sample space directly.

EXAMPLE 4

MULTIPLICATION RULE, INDEPENDENT EVENTS

Suppose you are going to throw two fair dice. What is the probability of getting a 5 on each die?

SOLUTION USING THE MULTIPLICATION RULE: The two events are independent, so we should use formula (4). $P(5 \text{ on 1st die } \textit{and} \text{ 5 on 2nd die}) = P(5 \text{ on 1st}) \cdot P(5 \text{ on 2nd})$. To finish the problem, we need to compute the probability of getting a 5 when we throw one die.

There are six faces on a die, and on a fair die each is equally likely to come up when you throw the die. Only one face has five dots, so by formula (2) for equally likely outcomes,

$$P(5 \text{ on die}) = \frac{1}{6}$$

Now we can complete the calculation.

$$P(5 \text{ on 1st die } \textit{and} \text{ 5 on 2nd die}) = P(5 \text{ on 1st}) \cdot P(5 \text{ on 2nd})$$
$$= \frac{1}{6} \cdot \frac{1}{6} = \frac{1}{36}$$

Philip Harvey/Spirit/Corbis

SOLUTION USING SAMPLE SPACE: The first task is to write down the sample space. Each die has six equally likely outcomes, and each outcome of the second die can be paired with each of the first. The sample space is shown in Figure 4-2. The total number of outcomes is 36, and only one is favorable to a 5 on the first die *and* a 5 on the second. The 36 outcomes are equally likely, so by formula (2) for equally likely outcomes,

$$P(5 \text{ on 1st } \textit{and} \text{ 5 on 2nd}) = \frac{1}{36}$$

The two methods yield the same result. The multiplication rule was easier to use because we did not need to look at all 36 outcomes in the sample space for tossing two dice.

FIGURE 4-2

Sample Space for Two Dice

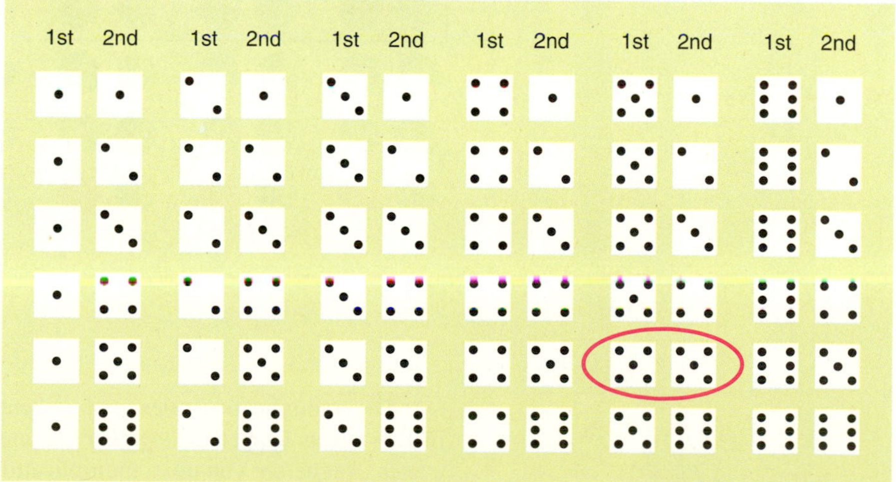

EXAMPLE 5

Multiplication rule, dependent events

Consider a collection of 6 balls that are identical except in color. There are 3 green balls, 2 blue balls, and 1 red ball. Compute the probability of drawing 2 green balls from the collection if the first ball is *not replaced* before the second ball is drawn.

MULTIPLICATION RULE METHOD: These events are *dependent*. The probability of a green ball on the first draw is 3/6, but on the second draw the probability of a green ball is only 2/5 if a green ball was removed on the first draw. By the multiplication rule for dependent events,

$$P\begin{pmatrix}\text{green ball on 1st draw } \textit{and} \\ \text{gree ball on 2nd draw}\end{pmatrix} = P(\text{green on 1st}) \cdot P(\text{green on 2nd} \mid \text{green on 1st})$$

$$= \frac{3}{6} \cdot \frac{2}{5} = \frac{1}{5} = 0.2$$

SAMPLE SPACE METHOD: Each of the 6 possible outcomes for the 1st draw must be paired with each of the 5 possible outcomes for the second draw. This means that there are a total of 30 possible pairs of balls. Figure 4-3 shows all the possible pairs of balls. In 6 of the pairs, both balls are green.

$$P(\text{green ball on 1st draw } \textit{and} \text{ green ball on 2nd draw}) = \frac{6}{30} = 0.2$$

Again, the two methods agree.

FIGURE 4-3

Sample Space of Drawing Two Balls Without Replacement from a Collection of 3 Green Balls, 2 Blue Balls, and 1 Red Ball

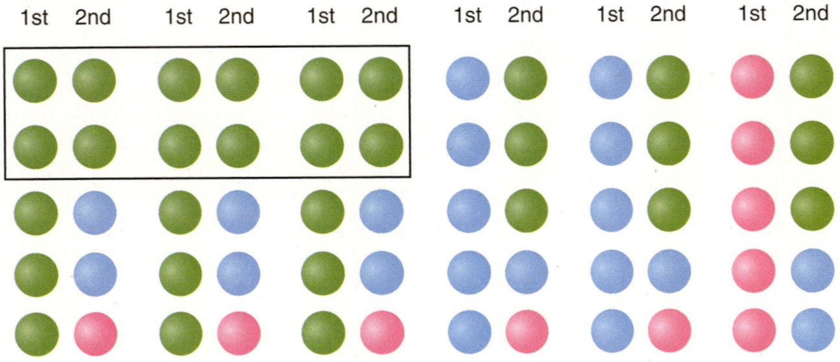

The multiplication rules apply whenever we wish to determine the probability of two events happening *together*. To indicate "together," we use *and* between the events. But before you use a multiplication rule to compute the probability of *A and B*, you must determine if *A* and *B* are independent or dependent events.

PROCEDURE

HOW TO USE THE MULTIPLICATION RULES

1. First determine whether A and B are independent events. If $P(A) = P(A \mid B)$, then the events are independent.
2. If A and B are independent events,

$$P(A \text{ and } B) = P(A) \cdot P(B) \tag{4}$$

3. If A and B are any events,

$$P(A \text{ and } B) = P(A) \cdot P(B \mid A) \tag{5}$$

$$\text{or} \quad P(A \text{ and } B) = P(B) \cdot P(A \mid B) \tag{6}$$

Let's practice using the multiplication rule.

GUIDED EXERCISE 4 MULTIPLICATION RULE

Andrew is 55, and the probability that he will be alive in 10 years is 0.72. Ellen is 35, and the probability that she will be alive in 10 years is 0.92. Assuming that the life span of one will have no effect on the life span of the other, what is the probability they will both be alive in 10 years?

(a) Are these events dependent or independent?

→ Since the life span of one does not affect the life span of the other, the events are independent.

(b) Use the appropriate multiplication rule to find *P*(Andrew alive in 10 years *and* Ellen alive in 10 years).

→ We use the rule for independent events:

$P(A \text{ and } B) = P(A) \cdot P(B)$

$P(\text{Andrew alive } \textit{and} \text{ Ellen alive})$

$= P(\text{Andrew alive}) \cdot P(\text{Ellen alive})$

$= (0.72)(0.92) \approx 0.66$

GUIDED EXERCISE 5 DEPENDENT EVENTS

A quality-control procedure for testing Ready-Flash digital cameras consists of drawing two cameras at random from each lot of 100 without replacing the first camera before drawing the second. If both are defective, the entire lot is rejected. Find the probability that both cameras are defective if the lot contains 10 defective cameras. Since we are drawing the cameras at random, assume that each camera in the lot has an equal chance of being drawn.

(a) What is the probability of getting a defective camera on the first draw?

→ The sample space consists of all 100 cameras. Since each is equally likely to be drawn and there are 10 defective ones,

$$P(\text{defective camera}) = \frac{10}{100} = \frac{1}{10}$$

Continued

GUIDED EXERCISE 5 *continued*

(b) The first camera drawn is not replaced, so there are only 99 cameras for the second draw. What is the probability of getting a defective camera on the second draw if the first camera was defective?

→ If the first camera was defective, then there are only 9 defective cameras left among the 99 remaining cameras in the lot.

$P(\text{defective on 2nd draw} \mid \text{defective on 1st})$

$$= \frac{9}{99} = \frac{1}{11}$$

(c) Are the probabilities computed in parts (a) and (b) different? Does drawing a defective camera on the first draw change the probability of getting a defective camera on the second draw? Are the events dependent?

→ The answer to all these questions is yes.

(d) Use the formula for dependent events,

$$P(A \text{ and } B) = P(A) \cdot P(B \mid A)$$

to compute $P(\text{1st camera defective } and \text{ 2nd camera defective})$.

→ $$P(\text{1st defective } and \text{ 2nd defective}) = \frac{1}{10} \cdot \frac{1}{11} = \frac{1}{110} \approx 0.009$$

WHAT DOES CONDITIONAL PROBABILITY TELL US?

Conditional probability of two events A and B tell us

- the probability that event A will happen under the assumption that event B has happened (or is guaranteed to happen in the future). This probability is designated $P(A \mid B)$ and is read "probability of event A *given* event B." Note that $P(A \mid B)$ might be larger or smaller than $P(A)$.
- the probability that event B will happen under the assumption that event A has happened. This probability is designated $P(B \mid A)$. Note that $P(A \mid B)$ and $P(B \mid A)$ are not necessarily equal.
- if $P(A \mid B) = P(A)$ or $P(B \mid A) = P(B)$, then events A and B are *independent* This means the occurrence of one of the events does not change the probability that the other event will occur.
- conditional probabilities enter into the calculations that two events A and B will both happen together.

  $$P(A \text{ and } B) = P(A) \cdot P(B \mid A)$$
  $$\text{also } P(A \text{ and } B) = P(B) \cdot P(A \mid B)$$

  In the case that events A and B are *independent*, then the formulas for $P(A \text{ and } B)$ simplify to

  $$P(A \text{ and } B) = P(A) \cdot P(B)$$

- if we know the values of $P(A \text{ and } B)$ and $P(B)$, then we can calculate the value of $P(A \mid B)$.

  $$P(A \mid B) = \frac{P(A \text{ and } B)}{P(B)} \quad \text{assuming } P(B) \neq 0$$

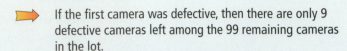

More than two independent events

The multiplication rule for independent events extends to *more than two independent events*. For instance, if the probability that a single seed will germinate is 0.85 and you plant 3 seeds, the probability that they will all germinate (assuming seed germinations are independent) is

$$P(\text{1st germinates } \textit{and} \text{ 2nd germinates } \textit{and} \text{ 3rd germinates}) = (0.85)(0.85)(0.85) \approx 0.614$$

ADDITION RULES

One of the multiplication rules can be used any time we are trying to find the probability of two events happening *together*. Pictorially, we are looking for the probability of the shaded region in Figure 4-4(a).

Probability of *A or B*

Another way to combine events is to consider the possibility of one event *or* another occurring. For instance, if a sports car saleswoman gets an extra bonus if she sells a convertible or a car with leather upholstery, she is interested in the probability that you will buy a car that is a convertible *or* that has leather upholstery. Of course, if you bought a convertible with leather upholstery, that would be fine, too. Pictorially, the shaded portion of Figure 4-4(b) represents the outcomes satisfying the *or* condition. Notice that the condition *A or B* is satisfied by any one of the following conditions:

1. Any outcome in *A* occurs.
2. Any outcome in *B* occurs.
3. Any outcome in both *A* and *B* occurs.

It is important to distinguish between the *or* combinations and the *and* combinations because we apply different rules to compute their probabilities.

FIGURE 4-4

(a) The Event *A* and *B*

(b) The Event *A* or *B*

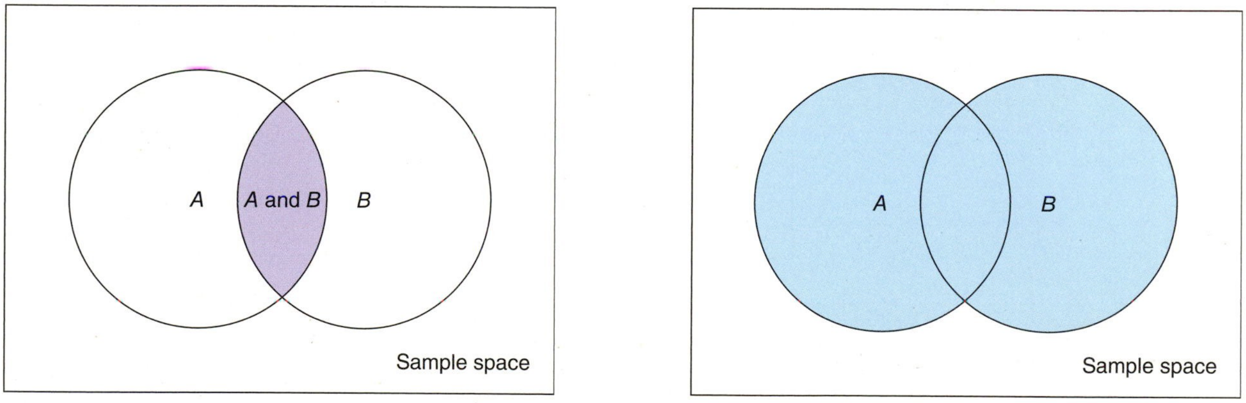

GUIDED EXERCISE 6 | COMBINING EVENTS

Indicate how each of the following pairs of events are combined. Use either the *and* combination or the *or* combination.

(a) Satisfying the humanities requirement by taking a course in the history of Japan or by taking a course in classical literature → Use the *or* combination.

Continued

GUIDED EXERCISE 6 *continued*

(b) Buying new tires and aligning the tires → Use the *and* combination.

(c) Getting an A not only in psychology but also in biology → Use the *and* combination.

(d) Having at least one of these pets: cat, dog, bird, rabbit → Use the *or* combination.

Once you decide that you are to find the probability of an *or* combination rather than an *and* combination, what formula do you use? It depends on the situation. In particular, it depends on whether or not the events being combined share any outcomes. Example 6 illustrates two situations.

EXAMPLE 6

Probability of events combined with *or*

Consider an introductory statistics class with 31 students. The students range from freshmen through seniors. Some students are male and some are female. Figure 4-5 shows the sample space of the class.

FIGURE 4-5

Sample Space for Statistics Class

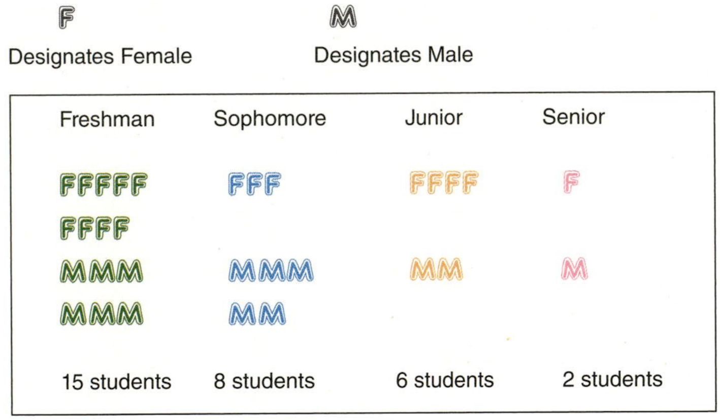

(a) Suppose we select one student at random from the class. Find the probability that the student is either a freshman or a sophomore.

Since there are 15 freshmen out of 31 students, $P(\text{freshmen}) = \frac{15}{31}$.

Since there are 8 sophomores out of 31 students, $P(\text{sophomore}) = \frac{8}{31}$.

$$P(\text{freshman } or \text{ sophomore}) = \frac{15}{31} + \frac{8}{31} = \frac{23}{31} \approx 0.742$$

Notice that we can simply add the probability of freshman to the probability of sophomore to find the probability that a student selected at random will be either a freshman or a sophomore. No student can be both a freshman and a sophomore at the same time.

(b) Select one student at random from the class. What is the probability that the student is either a male or a sophomore? Here we note that

$$P(\text{sophomore}) = \frac{8}{31} \quad P(\text{male}) = \frac{14}{31} \quad P(\text{sophomore } \textit{and} \text{ male}) = \frac{5}{31}$$

If we simply add P(sophomore) and P(male), we're including P(sophomore *and* male) twice in the sum. To compensate for this double summing, we simply subtract P(sophomore *and* male) from the sum. Therefore,

$$P(\text{sophomore } \textit{or} \text{ male}) = P(\text{sophomore}) + P(\text{male}) - P(\text{sophomore } \textit{and} \text{ male})$$

$$= \frac{8}{31} + \frac{14}{31} - \frac{5}{31} = \frac{17}{31} \approx 0.548$$

Mutually exclusive events

We say the events A and B are *mutually exclusive* or *disjoint* if they cannot occur together. This means that A and B have no outcomes in common or, put another way, that $P(A \textit{ and } B) = 0$.

Two events are **mutually exclusive** or **disjoint** if they cannot occur together. In particular, events A and B are mutually exclusive if $P(A \textit{ and } B) = 0$.

Formula (7) is the *addition rule for mutually exclusive events* A and B.

Addition rules

ADDITION RULE FOR *MUTUALLY EXCLUSIVE* EVENTS *A* AND *B*

$$P(A \text{ or } B) = P(A) + P(B) \qquad (7)$$

If the events are not mutually exclusive, we must use the more general formula (8), which is the *general addition rule for any events* A and B.

GENERAL ADDITION RULE FOR ANY EVENTS *A* AND *B*

$$P(A \text{ or } B) = P(A) + P(B) - P(A \text{ and } B) \qquad (8)$$

You may ask: Which formula should I use? The answer is: Use formula (7) only if you know that A and B are mutually exclusive (i.e., cannot occur together); if you do not know whether A and B are mutually exclusive, then use formula (8). Formula (8) is valid either way. Notice that when A and B are mutually exclusive, then $P(A \textit{ and } B) = 0$, so formula (8) reduces to formula (7).

PROCEDURE

HOW TO USE THE ADDITION RULES

1. First determine whether A and B are mutually exclusive events. If $P(A \textit{ and } B) = 0$, then the events are mutually exclusive.
2. If A and B are mutually exclusive events,
   $$P(A \textit{ or } B) = P(A) + P(B) \qquad (7)$$
3. If A and B are any events,
   $$P(A \textit{ or } B) = P(A) + P(B) - P(A \textit{ and } B) \qquad (8)$$

GUIDED EXERCISE 7 | Mutually exclusive events

The Cost Less Clothing Store carries remainder pairs of slacks. If you buy a pair of slacks in your regular waist size without trying them on, the probability that the waist will be too tight is 0.30 and the probability that it will be too loose is 0.10.

(a) Are the events "too tight" and "too loose" mutually exclusive?

➡ The waist cannot be both too tight and too loose at the same time, so the events are mutually exclusive.

(b) If you choose a pair of slacks at random in your regular waist size, what is the probability that the waist will be too tight or too loose?

➡ Since the events are mutually exclusive,

P(too tight *or* too loose)

= *P*(too tight) + *P*(too loose)

= 0.30 + 0.10

= 0.40

GUIDED EXERCISE 8 | General addition rule

Professor Jackson is in charge of a program to prepare people for a high school equivalency exam. Records show that 80% of the students need work in math, 70% need work in English, and 55% need work in both areas.

(a) Are the events "needs math" and "needs English" mutually exclusive?

➡ These events are not mutually exclusive, since some students need both. In fact,

P(needs math *and* needs English) = 0.55

(b) Use the appropriate formula to compute the probability that a student selected at random needs math *or* needs English.

➡ Since the events are not mutually exclusive, we use formula (8):

P(needs math *or* needs English)

= *P*(needs math) + *P*(needs English)

− *P*(needs math *and* English)

= 0.80 + 0.70 − 0.55

= 0.95

WHAT DOES THE FACT THAT TWO EVENTS ARE MUTUALLY EXCLUSIVE TELL US?

If two events *A* and *B* are *mutually exclusive,* then we know the occurrence of one of the events means the other event will not happen. In terms of calculations, this tell us

- $P(A \text{ and } B) = 0$ for mutually exclusive events.
- $P(A \text{ or } B) = P(A) + P(B)$ for mutually exclusive events.
- $P(A \mid B) = 0$ and $P(B \mid A) = 0$ for mutually exclusive events. That is, if event *B* occurs, then event *A* will not occur, and vice versa.

More than two mutually exclusive events

The addition rule for mutually exclusive events can be extended to apply to the situation in which we have more than two events, all of which are mutually exclusive to all the other events.

EXAMPLE 7

MUTUALLY EXCLUSIVE EVENTS

Laura is playing Monopoly. On her next move she needs to throw a sum bigger than 8 on the two dice in order to land on her own property and pass Go. What is the probability that Laura will roll a sum bigger than 8?

SOLUTION: When two dice are thrown, the largest sum that can come up is 12. Consequently, the only sums larger than 8 are 9, 10, 11, and 12. These outcomes are mutually exclusive, since only one of these sums can possibly occur on one throw of the dice. The probability of throwing more than 8 is the same as

$$P(9 \text{ or } 10 \text{ or } 11 \text{ or } 12)$$

Since the events are mutually exclusive,

$$\begin{aligned} P(9 \text{ or } 10 \text{ or } 11 \text{ or } 12) &= P(9) + P(10) + P(11) + P(12) \\ &= \frac{4}{36} + \frac{3}{36} + \frac{2}{36} + \frac{1}{36} \\ &= \frac{10}{36} = \frac{5}{18} \end{aligned}$$

To get the specific values of $P(9)$, $P(10)$, $P(11)$, and $P(12)$, we used the sample space for throwing two dice (see Figure 4-2 on page 157). There are 36 equally likely outcomes—for example, those favorable to 9 are 6, 3; 3, 6; 5, 4; and 4, 5. So $P(9) = 4/36$. The other values can be computed in a similar way.

LOOKING FORWARD

Chapters 5 and 6 involve several probability distributions. In these chapters, many of the events of interest are mutually exclusive or independent. This means we can compute probabilities by using the extended addition rule for mutually exclusive events and the extended multiplication rule for independent events.

FURTHER EXAMPLES USING CONTINGENCY TABLES

Most of us have been asked to participate in a survey. Schools, retail stores, news media, and government offices all conduct surveys. There are many types of surveys, and it is not our intention to give a general discussion of this topic. Let us study a very popular method called the *simple tally survey*. Such a survey consists of questions to which the responses can be recorded in the rows and columns of a table called a *contingency table*. These questions are appropriate to the information you want and are designed to cover the *entire* population of interest. In addition, the questions should be designed so that you can partition the sample space of responses into distinct (that is, mutually exclusive) sectors.

Contingency table

If the survey includes responses from a reasonably large random sample, then the results should be representative of your population. In this case, you can estimate simple probabilities, conditional probabilities, and the probabilities of some combinations of events directly from the results of the survey.

EXAMPLE 8

SURVEY

At Hopewell Electronics, all 140 employees were asked about their political affiliations. The employees were grouped by type of work, as executives or production workers. The results with row and column totals are shown in Table 4-2.

TABLE 4-2 **Employee Type and Political Affiliation**

Employee Type	Political Affiliation			
	Democrat (*D*)	Republican (*R*)	Independent (*I*)	Row Total
Executive (*E*)	5	34	9	48
Production worker (*PW*)	63	21	8	92
Column Total	68	55	17	140 Grand Total

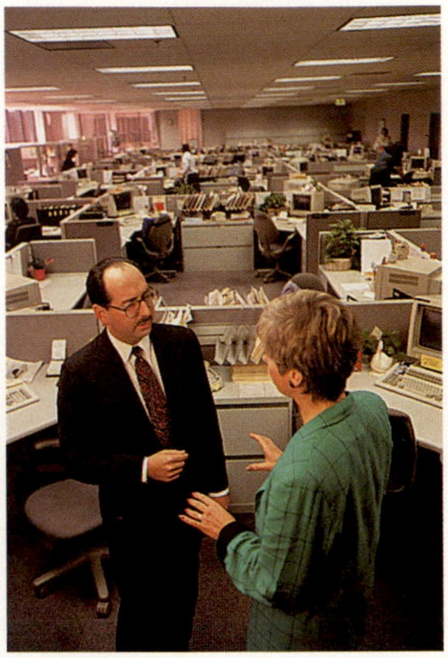
Mark Richards / PhotoEdit

Suppose an employee is selected at random from the 140 Hopewell employees. Let us use the following notation to represent different events of choosing: E = executive; PW = production worker; D = Democrat; R = Republican; I = Independent.

(a) Compute $P(D)$ and $P(E)$.

SOLUTION: To find these probabilities, we look at the *entire* sample space.

$$P(D) = \frac{\text{Number of Democrats}}{\text{Number of employees}} = \frac{68}{140} \approx 0.486$$

$$P(E) = \frac{\text{Number of executives}}{\text{Number of employees}} = \frac{48}{140} \approx 0.343$$

(b) Compute $P(D \mid E)$.

SOLUTION: For the conditional probability, we restrict our attention to the portion of the sample space satisfying the condition of being an executive.

$$P(D \mid E) = \frac{\text{Number of executives who are Democrats}}{\text{Number of executives}} = \frac{5}{48} \approx 0.104$$

(c) Are the events D and E independent?

SOLUTION: One way to determine if the events D and E are independent is to see if $P(D) = P(D \mid E)$ [or equivalently, if $P(E) = P(E \mid D)$]. Since $P(D) \approx 0.486$ and $P(D \mid E) \approx 0.104$, we see that $P(D) \neq P(D \mid E)$. This means that the events D and E are *not* independent. The probability of event D "depends on" whether or not event E has occurred.

(d) Compute $P(D \text{ and } E)$.

SOLUTION: This probability is not conditional, so we must look at the entire sample space.

$$P(D \textit{ and } E) = \frac{\text{Number of executives who are Democrats}}{\text{Total number of employees}} = \frac{5}{140} \approx 0.036$$

Let's recompute this probability using the rules of probability for dependent events.

$$P(D \textit{ and } E) = P(E) \cdot P(D \mid E) = \frac{48}{140} \cdot \frac{5}{48} = \frac{5}{140} \approx 0.036$$

The results using the rules are consistent with those using the sample space.

(e) Compute $P(D \text{ or } E)$.

SOLUTION: From part (d), we know that the events "Democrat" and "executive" are not mutually exclusive, because $P(D \text{ and } E) \neq 0$. Therefore,

$$P(D \textit{ or } E) = P(D) + P(E) - P(D \textit{ and } E)$$
$$= \frac{68}{140} + \frac{48}{140} - \frac{5}{140} = \frac{111}{140} \approx 0.793$$

GUIDED EXERCISE 9 | Survey

Using Table 4-2 on page 166, let's consider other probabilities regarding the types of employees at Hopewell and their political affiliations. This time let's consider the production worker and the affiliation of Independent. Suppose an employee is selected at random from the group of 140.

(a) Compute $P(I)$ and $P(PW)$.

$$P(I) = \frac{\text{No. of Independents}}{\text{Total no. of employees}} = \frac{17}{140} \approx 0.121$$

$$P(PW) = \frac{\text{No. of production workers}}{\text{Total no. of employees}} = \frac{92}{140} \approx 0.657$$

(b) Compute $P(I \mid PW)$. This is a conditional probability. Be sure to restrict your attention to production workers, since that is the condition given.

$$P(I \mid PW) = \frac{\text{No. of Independent production workers}}{\text{No. of production workers}} = \frac{8}{92} \approx 0.087$$

(c) Compute $P(I \text{ and } PW)$. In this case, look at the entire sample space and the number of employees who are both Independent and in production.

$$P(I \text{ and } PW) = \frac{\text{No. of Independent production workers}}{\text{Total no. employees}} = \frac{8}{140} \approx 0.057$$

(d) Use the multiplication rule for dependent events to calculate $P(I \text{ and } PW)$. Is the result the same as that of part (c)?

By the multiplication rule,

$$P(I \text{ and } PW) = P(PW) \cdot P(I \mid PW) = \frac{92}{140} \cdot \frac{8}{92} = \frac{8}{140} \approx 0.057$$

The results are the same.

(e) Compute $P(I \text{ or } PW)$. Are the events mutually exclusive?

Since the events are not mutully exclusive,

$$P(I \text{ or } PW) = P(I) + P(PW) - P(I \text{ and } PW) = \frac{17}{140} + \frac{92}{140} - \frac{8}{140} = \frac{101}{140} \approx 0.721$$

As you apply probability to various settings, keep the following rules in mind.

Basic probability rules

SUMMARY OF BASIC PROBABILITY RULES

A statistical experiment or statistical observation is any random activity that results in a recordable outcome. The sample space is the set of all simple events that are the outcomes of the statistical experiment and cannot be broken into other "simpler" events. A general event is any subset of the sample space. The notation $P(A)$ designates the probability of event A.

1. $P(\text{entire sample space}) = 1$
2. For any event A: $0 \le P(A) \le 1$
3. A^c designates the **complement** of A: $P(A^c) = 1 - P(A)$
4. Events A and B are **independent events** if $P(A) = P(A \mid B)$.
5. Multiplication Rules

   General: $P(A \text{ and } B) = P(A) \cdot P(B \mid A)$

   $P(A \text{ and } B) = P(B) \cdot P(A \mid B)$

   Independent events: $P(A \text{ and } B) = P(A) \cdot P(B)$
6. Conditional Probability: $P(A \mid B) = \dfrac{P(A \text{ and } B)}{P(B)}$
7. Events A and B are **mutually exclusive** if $P(A \text{ and } B) = 0$.
8. Addition Rules

   General: $P(A \text{ or } B) = P(A) + P(B) - P(A \text{ and } B)$

   Mutually exclusive events: $P(A \text{ or } B) = P(A) + P(B)$

CRITICAL THINKING

Translating events described by common English phrases into events described using *and, or, complement*, or *given* takes a bit of care. Table 4-3 shows some common phrases and their corresponding translations into symbols.

TABLE 4-3 English Phrases and Corresponding Symbolic Translations

Consider the following events for a person selected at random from the general population:

A = person is taking college classes

B = person is under 30 years old

Phrase	Symbolic Expression
1. The probability that a person is under 30 years old and is taking college classes is 40%.	$P(B \text{ and } A) = 0.40$ or $P(A \text{ and } B) = 0.40$
2. The probability that a person under 30 years old is taking college classes is 45%.	$P(A \mid B) = 0.45$
3. The probability is 45% that a person is taking college classes if the person is under 30.	$P(A \mid B) = 0.45$
4. The probability that a person taking college classes is under 30 is 0.60.	$P(B \mid A) = 0.60$
5. The probability that a person is not taking college classes or is under 30 years old is 0.75.	$P(A^c \text{ or } B) = 0.75$

In this section, we have studied some important rules that are valid in all probability spaces. The rules and definitions of probability are not only interesting but also have extensive *applications* in our everyday lives. If you are inclined to continue your study of probability a little further, we recommend reading about *Bayes's theorem* in Appendix I. The Reverend Thomas Bayes (1702–1761) was an English mathematician who discovered an important relation for conditional probabilities.

VIEWPOINT The Psychology of Odors

The Smell and Taste Treatment Research Foundation of Chicago collected data on the time required to complete a maze while subjects were smelling different scents. Data for this survey can be found by visiting the Brase/Brase statistics site at **http://www.cengage.com/statistics/brase11e** *and finding the link to DASL, the Carnegie Mellon University Data and Story Library. Once at the DASL site, select Data Subjects, then Psychology, and then Scents. You can estimate conditional probabilities regarding response times for smokers, nonsmokers, and types of scents.*

SECTION 4.2 PROBLEMS

1. *Statistical Literacy* If two events are mutually exclusive, can they occur concurrently? Explain.

2. *Statistical Literacy* If two events A and B are independent and you know that $P(A) = 0.3$, what is the value of $P(A \mid B)$?

3. *Basic Computation: Addition Rule* Given $P(A) = 0.3$ and $P(B) = 0.4$:
   (a) If A and B are mutually exclusive events, compute $P(A \text{ or } B)$.
   (b) If $P(A \text{ and } B) = 0.1$, compute $P(A \text{ or } B)$.

4. *Basic Computation: Addition Rule* Given $P(A) = 0.7$ and $P(B) = 0.4$:
   (a) Can events A and B be mutually exclusive? Explain.
   (b) If $P(A \text{ and } B) = 0.2$, compute $P(A \text{ or } B)$.

5. *Basic Computation: Multiplication Rule* Given $P(A) = 0.2$ and $P(B) = 0.4$:
   (a) If A and B are independent events, compute $P(A \text{ and } B)$.
   (b) If $P(A \mid B) = 0.1$, compute $P(A \text{ and } B)$.

6. *Basic Computation: Multiplication Rule* Given $P(A) = 0.7$ and $P(B) = 0.8$:
   (a) If A and B, are independent events, compute $P(A \text{ and } B)$.
   (b) If $P(B \mid A) = 0.9$, compute $P(A \text{ and } B)$.

7. *Basic Computations: Rules of Probability* Given $P(A) = 0.2$, $P(B) = 0.5$, $P(A \mid B) = 0.3$:
   (a) Compute $P(A \text{ and } B)$.
   (b) Compute $P(A \text{ or } B)$.

8. *Basic Computation: Rules of Probability* Given $P(A^c) = 0.8$, $P(B) = 0.3$, $P(B \mid A) = 0.2$:
   (a) Compute $P(A \text{ and } B)$.
   (b) Compute $P(A \text{ or } B)$.

9. *Critical Thinking* Lisa is making up questions for a small quiz on probability. She assigns these probabilities: $P(A) = 0.3$, $P(B) = 0.3$, $P(A \text{ and } B) = 0.4$. What is wrong with these probability assignments?

10. *Critical Thinking* Greg made up another question for a small quiz. He assigns the probabilities $P(A) = 0.6$, $P(B) = 0.7$, $P(A \mid B) = 0.1$ and asks for the probability $P(A \text{ or } B)$. What is wrong with the probability assignments?

11. *Critical Thinking* Suppose two events A and B are mutually exclusive, with $P(A) \neq 0$ and $P(B) \neq 0$. By working through the following steps, you'll see why two mutually exclusive events are not independent.
    (a) For mutually exclusive events, can event A occur if event B has occurred? What is the value of $P(A \mid B)$?
    (b) Using the information from part (a), can you conclude that events A and B are *not* independent if they are mutually exclusive? Explain.

12. *Critical Thinking* Suppose two events A and B are independent, with $P(A) \neq 0$ and $P(B) \neq 0$. By working through the following steps, you'll see why two independent events are not mutually exclusive.
    (a) What formula is used to compute $P(A \text{ and } B)$? Is $P(A \text{ and } B) \neq 0$? Explain.
    (b) Using the information from part (a), can you conclude that events A and B are *not* mutually exclusive?

13. *Critical Thinking* Consider the following events for a driver selected at random from the general population:

    A = driver is under 25 years old

    B = driver has received a speeding ticket

    Translate each of the following phrases into symbols.
    (a) The probability the driver has received a speeding ticket and is under 25 years old
    (b) The probability a driver who is under 25 years old has received a speeding ticket
    (c) The probability a driver who has received a speeding ticket is 25 years old or older
    (d) The probability the driver is under 25 years old or has received a speeding ticket
    (e) The probability the driver has not received a speeding ticket or is under 25 years old

14. *Critical Thinking* Consider the following events for a college student selected at random:

    A = student is female
    B = student is majoring in business

    Translate each of the following phrases into symbols.
    (a) The probability the student is male or is majoring in business
    (b) The probability a female student is majoring in business
    (c) The probability a business major is female
    (d) The probability the student is female and is not majoring in business
    (e) The probability the student is female and is majoring in business

15. *General: Candy Colors* M&M plain candies come in various colors. According to the M&M/Mars Department of Consumer Affairs (link to the Mars company web site from the Brase/Brase statistics site at **http://www.cengage.com/statistics /brase**), the distribution of colors for plain M&M candies is

Color	Purple	Yellow	Red	Orange	Green	Blue	Brown
Percentage	20%	20%	20%	10%	10%	10%	10%

    Suppose you have a large bag of plain M&M candies and you choose one candy at random. Find
    (a) P(green candy *or* blue candy). Are these outcomes mutually exclusive? Why?
    (b) P(yellow candy *or* red candy). Are these outcomes mutually exclusive? Why?
    (c) P(*not* purple candy)

16. *Environmental: Land Formations* Arches National Park is located in southern Utah. The park is famous for its beautiful desert landscape and its many natural sandstone arches. Park Ranger Edward McCarrick started an inventory (not yet complete) of natural arches within the park that have an opening of at least 3 feet. The following table is based on information taken from the book *Canyon Country Arches and Bridges* by F. A. Barnes. The height of the arch opening is rounded to the nearest foot.

Height of arch, feet	3–9	10–29	30–49	50–74	75 and higher
Number of arches in park	111	96	30	33	18

For an arch chosen at random in Arches National Park, use the preceding information to estimate the probability that the height of the arch opening is
(a) 3 to 9 feet tall
(b) 30 feet or taller
(c) 3 to 49 feet tall
(d) 10 to 74 feet tall
(e) 75 feet or taller

17. *General: Roll Two Dice* You roll two fair dice, a green one and a red one.
(a) Are the outcomes on the dice independent?
(b) Find P(5 on green die *and* 3 on red die).
(c) Find P(3 on green die *and* 5 on red die).
(d) Find P((5 on green die *and* 3 on red die) *or* (3 on green die *and* 5 on red die)).

18. *General: Roll Two Dice* You roll two fair dice, a green one and a red one.
(a) Are the outcomes on the dice independent?
(b) Find P(1 on green die *and* 2 on red die).
(c) Find P(2 on green die *and* 1 on red die).
(d) Find P((1 on green die *and* 2 on red die) *or* (2 on green die *and* 1 on red die)).

19. *General: Roll Two Dice* You roll two fair dice, a green one and a red one.
(a) What is the probability of getting a sum of 6?
(b) What is the probability of getting a sum of 4?
(c) What is the probability of getting a sum of 6 *or* 4? Are these outcomes mutually exclusive?

20. *General: Roll Two Dice* You roll two fair dice, a green one and a red one.
(a) What is the probability of getting a sum of 7?
(b) What is the probability of getting a sum of 11?
(c) What is the probability of getting a sum of 7 *or* 11? Are these outcomes mutually exclusive?

Problems 21–24 involve a standard deck of 52 playing cards. In such a deck of cards there are four suits of 13 cards each. The four suits are: hearts, diamonds, clubs, and spades. The 26 cards included in hearts and diamonds are red in color. The 26 cards included in clubs and spades are black in color. The 13 cards in each suit are: 2, 3, 4, 5, 6, 7, 8, 9, 10, Jack, Queen, King, and Ace. This means there are four Aces, four Kings, four Queens, four 10's, etc., down to four 2's in each deck.

21. *General: Deck of Cards* You draw two cards from a standard deck of 52 cards without replacing the first one before drawing the second.
(a) Are the outcomes on the two cards independent? Why?
(b) Find P(Ace on 1st card *and* King on 2nd).
(c) Find P(King on 1st card *and* Ace on 2nd).
(d) Find the probability of drawing an Ace *and* a King in either order.

22. *General: Deck of Cards* You draw two cards from a standard deck of 52 cards without replacing the first one before drawing the second.
(a) Are the outcomes on the two cards independent? Why?
(b) Find P(3 on 1st card *and* 10 on 2nd).
(c) Find P(10 on 1st card *and* 3 on 2nd).
(d) Find the probability of drawing a 10 *and* a 3 in either order.

23. *General: Deck of Cards* You draw two cards from a standard deck of 52 cards, but before you draw the second card, you put the first one back and reshuffle the deck.
(a) Are the outcomes on the two cards independent? Why?
(b) Find P(Ace on 1st card *and* King on 2nd).
(c) Find P(King on 1st card *and* Ace on 2nd).
(d) Find the probability of drawing an Ace *and* a King in either order.

24. *General: Deck of Cards* You draw two cards from a standard deck of 52 cards, but before you draw the second card, you put the first one back and reshuffle the deck.
(a) Are the outcomes on the two cards independent? Why?
(b) Find P(3 on 1st card *and* 10 on 2nd).
(c) Find P(10 on 1st card *and* 3 on 2nd).
(d) Find the probability of drawing a 10 *and* a 3 in either order.

25. *Marketing: Toys* *USA Today* gave the information shown in the table about ages of children receiving toys. The percentages represent all toys sold.

Age (years)	Percentage of Toys
2 and under	15%
3–5	22%
6–9	27%
10–12	14%
13 and over	22%

What is the probability that a toy is purchased for someone
(a) 6 years old or older?
(b) 12 years old or younger?
(c) between 6 and 12 years old?
(d) between 3 and 9 years old?

Interpretation A child between 10 and 12 years old looks at this probability distribution and asks, "Why are people more likely to buy toys for kids older than I am [13 and over] than for kids in my age group [10–12]?" How would you respond?

26. *Health Care: Flu* Based on data from the *Statistical Abstract of the United States*, 112th Edition, only about 14% of senior citizens (65 years old or older) get the flu each year. However, about 24% of the people under 65 years old get the flu each year. In the general population, there are 12.5% senior citizens (65 years old or older).
(a) What is the probability that a person selected at random from the general population is a senior citizen who will get the flu this year?
(b) What is the probability that a person selected at random from the general population is a person under age 65 who will get the flu this year?
(c) Answer parts (a) and (b) for a community that is 95% senior citizens.
(d) Answer parts (a) and (b) for a community that is 50% senior citizens.

27. *Focus Problem: Lie Detector Test* In this problem, you are asked to solve part of the Focus Problem at the beginning of this chapter. In his book *Chances: Risk and Odds in Everyday Life*, James Burke says that there is a 72% chance a polygraph test (lie detector test) will catch a person who is, in fact, lying. Furthermore, there is approximately a 7% chance that the polygraph will falsely accuse someone of lying.
(a) Suppose a person answers 90% of a long battery of questions truthfully. What percentage of the answers will the polygraph *wrongly* indicate are lies?

(b) Suppose a person answers 10% of a long battery of questions with lies. What percentage of the answers will the polygraph *correctly* indicate are lies?

(c) Repeat parts (a) and (b) if 50% of the questions are answered truthfully and 50% are answered with lies.

(d) Repeat parts (a) and (b) if 15% of the questions are answered truthfully and the rest are answered with lies.

28. *Focus Problem: Expand Your Knowledge* This problem continues the Focus Problem. The solution involves applying several basic probability rules and a little algebra to solve an equation.

(a) If the polygraph of Problem 27 indicated that 30% of the questions were answered with lies, what would you estimate for the actual percentage of lies in the answers? *Hint:* Let $B =$ event detector indicates a lie. We are given $P(B) = 0.30$. Let $A =$ event person is lying, so $A^c =$ event person is not lying. Then

$$P(B) = P(A \text{ and } B) + P(A^c \text{ and } B)$$

$$P(B) = P(A)P(B \mid A) + P(A^c)P(B \mid A^c)$$

Replacing $P(A^c)$ by $1 - P(A)$ gives

$$P(B) = P(A)\cdot P(B \mid A) + [1 - P(A)]\cdot P(B \mid A^c)$$

Substitute known values for $P(B)$, $P(B \mid A)$, and $P(B \mid A^c)$ into the last equation and solve for $P(A)$.

(b) If the polygraph indicated that 70% of the questions were answered with lies, what would you estimate for the actual percentage of lies?

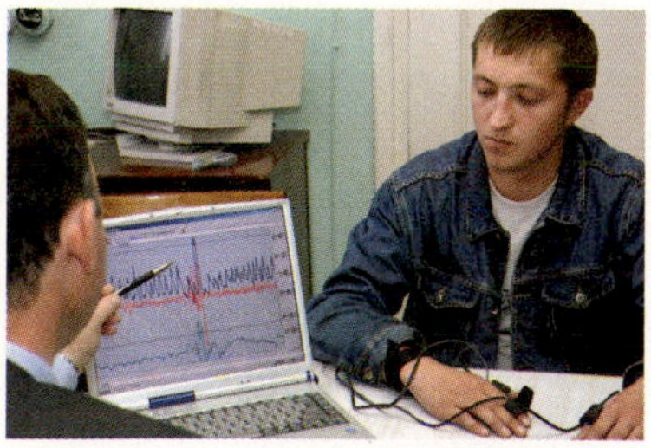

Fedoseyev Lev/ITAR-TASS/ Landov

29. *Survey: Sales Approach* In a sales effectiveness seminar, a group of sales representatives tried two approaches to selling a customer a new automobile: the aggressive approach and the passive approach. For 1160 customers, the following record was kept:

	Sale	No Sale	Row Total
Aggressive	270	310	580
Passive	416	164	580
Column Total	686	474	1160

Suppose a customer is selected at random from the 1160 participating customers. Let us use the following notation for events: $A =$ aggressive approach, $Pa =$ passive approach, $S =$ sale, $N =$ no sale. So, $P(A)$ is the probability that an aggressive approach was used, and so on.

(a) Compute $P(S)$, $P(S \mid A)$, and $P(S \mid Pa)$.

(b) Are the events $S =$ sale and $Pa =$ passive approach independent? Explain.

(c) Compute $P(A \text{ and } S)$ and $P(Pa \text{ and } S)$.

(d) Compute $P(N)$ and $P(N \mid A)$.

(e) Are the events $N =$ no sale and $A =$ aggressive approach independent? Explain.

(f) Compute $P(A \text{ or } S)$.

30. *Survey: Medical Tests* Diagnostic tests of medical conditions can have several types of results. The test result can be positive or negative, whether or not a patient has the condition. A positive test (+) indicates that the patient has the condition. A negative test (−) indicates that the patient does not have the condition. Remember, a positive test does not prove that the patient has the condition. Additional medical work may be required. Consider a random

sample of 200 patients, some of whom have a medical condition and some of whom do not. Results of a new diagnostic test for the condition are shown.

	Condition Present	Condition Absent	Row Total
Test Result +	110	20	130
Test Result −	20	50	70
Column Total	130	70	200

Assume the sample is representative of the entire population. For a person selected at random, compute the following probabilities:
(a) $P(+ \mid \text{condition present})$; this is known as the *sensitivity* of a test.
(b) $P(- \mid \text{condition present})$; this is known as the *false-negative rate*.
(c) $P(- \mid \text{condition absent})$; this is known as the *specificity* of a test.
(d) $P(+ \mid \text{condition absent})$; this is known as the *false-positive rate*.
(e) $P(\text{condition present } and\ +)$; this is the *predictive value* of the test.
(f) $P(\text{condition present } and\ -)$.

31. *Survey: Lung/Heart* In an article titled "Diagnostic accuracy of fever as a measure of postoperative pulmonary complications" (*Heart Lung*, Vol. 10, No. 1, p. 61), J. Roberts and colleagues discuss using a fever of 38°C or higher as a diagnostic indicator of postoperative atelectasis (collapse of the lung) as evidenced by x-ray observation. For fever ≥38°C as the diagnostic test, the results for postoperative patients are

	Condition Present	Condition Absent	Row Total
Test Result +	72	37	109
Test Result −	82	79	161
Column Total	154	116	270

For the meaning of + and −, see Problem 30.
Complete parts (a) through (f) from Problem 30.

32. *Survey: Customer Loyalty* Are customers more loyal in the east or in the west? The following table is based on information from *Trends in the United States*, published by the Food Marketing Institute, Washington, D.C. The columns represent length of customer loyalty (in years) at a primary supermarket. The rows represent regions of the United States.

	Less Than 1 Year	1–2 Years	3–4 Years	5–9 Years	10–14 Years	15 or More Years	Row Total
East	32	54	59	112	77	118	452
Midwest	31	68	68	120	63	173	523
South	53	92	93	158	106	158	660
West	41	56	67	78	45	86	373
Column Total	157	270	287	468	291	535	2008

What is the probability that a customer chosen at random
(a) has been loyal 10 to 14 years?
(b) has been loyal 10 to 14 years, given that he or she is from the east?
(c) has been loyal *at least* 10 years?
(d) has been loyal *at least* 10 years, given that he or she is from the west?
(e) is from the west, given that he or she has been loyal less than 1 year?
(f) is from the south, given that he or she has been loyal less than 1 year?
(g) has been loyal *1 or more years*, given that he or she is from the east?

(h) has been loyal *1 or more years*, given that he or she is from the west?
(i) Are the events "from the east" and "loyal 15 or more years" independent? Explain.

33. *Franchise Stores: Profits* Wing Foot is a shoe franchise commonly found in shopping centers across the United States. Wing Foot knows that its stores will not show a profit unless they gross over $940,000 per year. Let A be the event that a new Wing Foot store grosses over $940,000 its first year. Let B be the event that a store grosses over $940,000 its second year. Wing Foot has an administrative policy of closing a new store if it does not show a profit in *either* of the first 2 years. The accounting office at Wing Foot provided the following information: 65% of *all* Wing Foot stores show a profit the first year; 71% of *all* Wing Foot stores show a profit the second year (this includes stores that did not show a profit the first year); however, 87% of Wing Foot stores that showed a profit the first year also showed a profit the second year. Compute the following:
(a) $P(A)$
(b) $P(B)$
(c) $P(B \mid A)$
(d) $P(A \text{ and } B)$
(e) $P(A \text{ or } B)$
(f) What is the probability that a new Wing Foot store will not be closed after 2 years? What is the probability that a new Wing Foot store will be closed after 2 years?

34. *Education: College of Nursing* At Litchfield College of Nursing, 85% of incoming freshmen nursing students are female and 15% are male. Recent records indicate that 70% of the entering female students will graduate with a BSN degree, while 90% of the male students will obtain a BSN degree. If an incoming freshman nursing student is selected at random, find
(a) *P*(student will graduate | student is female).
(b) *P*(student will graduate *and* student is female).
(c) *P*(student will graduate | student is male).
(d) *P*(student will graduate *and* student is male).
(e) *P*(student will graduate). Note that those who will graduate are either males who will graduate or females who will graduate.
(f) The events described by the phrases "will graduate *and* is female" and "will graduate, *given* female" seem to be describing the same students. Why are the probabilities *P*(will graduate *and* is female) and *P*(will graduate | female) different?

35. *Medical: Tuberculosis* The state medical school has discovered a new test for tuberculosis. (If the test indicates a person has tuberculosis, the test is positive.) Experimentation has shown that the probability of a positive test is 0.82, given that a person has tuberculosis. The probability is 0.09 that the test registers positive, given that the person does not have tuberculosis. Assume that in the general population, the probability that a person has tuberculosis is 0.04. What is the probability that a person chosen at random will
(a) have tuberculosis and have a positive test?
(b) not have tuberculosis?
(c) not have tuberculosis and have a positive test?

36. *Therapy: Alcohol Recovery* The Eastmore Program is a special program to help alcoholics. In the Eastmore Program, an alcoholic lives at home but undergoes a two-phase treatment plan. Phase I is an intensive group-therapy program lasting 10 weeks. Phase II is a long-term counseling program lasting 1 year. Eastmore Programs are located in most major cities, and past data gave

the following information based on percentages of success and failure collected over a long period of time: The probability that a client will have a relapse in phase I is 0.27; the probability that a client will have a relapse in phase II is 0.23. However, if a client did not have a relapse in phase I, then the probability that this client will not have a relapse in phase II is 0.95. If a client did have a relapse in phase I, then the probability that this client will have a relapse in phase II is 0.70. Let A be the event that a client has a relapse in phase I and B be the event that a client has a relapse in phase II. Let C be the event that a client has no relapse in phase I and D be the event that a client has no relapse in phase II. Compute the following:

(a) $P(A)$, $P(B)$, $P(C)$, and $P(D)$
(b) $P(B \mid A)$ and $P(D \mid C)$
(c) $P(A \text{ and } B)$ and $P(C \text{ and } D)$
(d) $P(A \text{ or } B)$
(e) What is the probability that a client will go through both phase I and phase II without a relapse?
(f) What is the probability that a client will have a relapse in both phase I and phase II?
(g) What is the probability that a client will have a relapse in either phase I or phase II?

Brain Teasers Assume A and B are events such that $0 < P(A) < 1$ and $0 < P(B) < 1$. Answer questions 37–51 true or false and give a brief explanation for each answer. *Hint:* Review the summary of basic probability rules.

37. $P(A \text{ and } A^c) = 0$
38. $P(A \text{ or } A^c) = 0$
39. $P(A \mid A^c) = 1$
40. $P(A \text{ or } B) = P(A) + P(B)$
41. $P(A \mid B) \geq P(A \text{ and } B)$
42. $P(A \text{ or } B) \geq P(A)$ if A and B are independent events
43. $P(A \text{ and } B) \leq P(A)$
44. $P(A \mid B) > P(A)$ if A and B are independent events
45. $P(A^c \text{ and } B^c) \leq 1 - P(A)$
46. $P(A^c \text{ or } B^c) \leq 2 - P(A) - P(B)$
47. If A and B are independent events, they must also be mutually exclusive events.
48. If A and B are mutually exclusive, they must also be independent.
49. If A and B are both mutually exclusive and independent, then at least one of $P(A)$ or $P(B)$ must be zero.
50. If A and B are mutually exclusive, then $P(A \mid B) = 0$.
51. $P(A \mid B) + P(A^c \mid B) = 1$
52. *Brain Teaser* The Reverend Thomas Bayes (1702–1761) was an English mathematician who discovered an important rule of probability (*see* Bayes's theorem, Appendix I, part I). A key feature of Bayes's theorem is the formula

$$P(B) = P(B \mid A) \cdot P(A) + P(B \mid A^c) \cdot P(A^c)$$

Explain why this formula is valid. *Hint:* See Figure AI-1 in Appendix I.

SECTION **4.3**

Trees and Counting Techniques

FOCUS POINTS

- Organize outcomes in a sample space using tree diagrams.
- Compute number of ordered arrangements of outcomes using permutations.
- Compute number of (nonordered) groupings of outcomes using combinations.
- Explain how counting techniques relate to probability in everyday life.

When outcomes are equally likely, we compute the probability of an event by using the formula

$$P(A) = \frac{\text{Number of outcomes favorable to the event } A}{\text{Number of outcomes in the sample space}}$$

The probability formula requires that we be able to determine the number of outcomes in the sample space. In the problems we have done in previous sections, this task has not been difficult because the number of outcomes was small or the sample space consisted of fairly straightforward events. The tools we present in this section will help you count the number of possible outcomes in larger sample spaces or those formed by more complicated events.

When an outcome of an experiment is composed of a series of events, the multiplication rule gives us the *total number* of outcomes.

Multiplication rule of counting

MULTIPLICATION RULE OF COUNTING

Consider the series of events E_1 through E_m, where n_1 is the number of possible outcomes for event E_1, n_2 is the number of possible outcomes for event E_2, and n_m designates the number of possible outcomes for event E_m. Then the product

$$n_1 \times n_2 \times \cdots \times n_m$$

gives the total number of possible outcomes for the series of events E_1, followed by E_2, up through event E_m.

EXAMPLE 9

MULTIPLICATION RULE

Jacqueline is in a nursing program and is required to take a course in psychology and one in physiology (A and P) next semester. She also wants to take Spanish II. If there are two sections of psychology, two of A and P, and three of Spanish II, how many different class schedules can Jacqueline choose from? (Assume that the times of the sections do not conflict.)

SOLUTION: Creating a class schedule can be considered an experiment with a series of three events. There are two possible outcomes for the psychology section, two for the A and P section, and three for the Spanish II section. By the multiplication rule, the total number of class schedules possible is

$$2 \times 2 \times 3 = 12$$

Tree diagram

A *tree diagram* gives a visual display of the total number of outcomes of an experiment consisting of a series of events. From a tree diagram, we can determine not only the total number of outcomes, but also the individual outcomes.

EXAMPLE 10

TREE DIAGRAM

Using the information from Example 9, let's make a tree diagram that shows all the possible course schedules for Jacqueline.

SOLUTION: Figure 4-6 shows the tree diagram. Let's study the diagram. There are two branches from Start. These branches indicate the two possible choices for psychology sections. No matter which section of psychology Jacqueline chooses, she can choose from the two available A and P sections. Therefore, we have two branches leading from *each* psychology branch. Finally, after the psychology and A and P sections are selected, there are three choices for Spanish II. That is why there are three branches from *each* A and P section.

FIGURE 4-6

Tree Diagram for Selecting Class Schedules

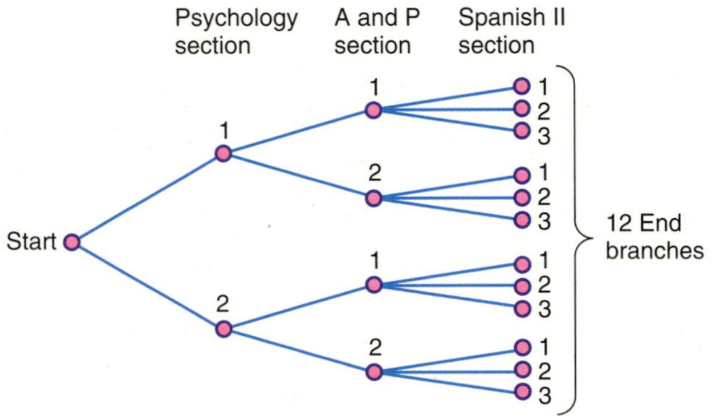

The tree ends with a total of 12 branches. The number of end branches tells us the number of possible schedules. The outcomes themselves can be listed from the tree by following each series of branches from Start to End. For instance, the top branch from Start generates the schedules shown in Table 4-4. The other six schedules can be listed in a similar manner, except they begin with the second section of psychology.

TABLE 4-4 **Schedules Utilizing Section 1 of Psychology**

Psychology Section	A and P Section	Spanish II Section
1	1	1
1	1	2
1	1	3
1	2	1
1	2	2
1	2	3

GUIDED EXERCISE 10 | TREE DIAGRAM AND MULTIPLICATION RULE

Louis plays three tennis matches. Use a tree diagram to list the possible win and loss sequences Louis can experience for the set of three matches.

(a) On the first match Louis can win or lose. From Start, indicate these two branches.

→ FIGURE 4-7 W = Win, L = Lose

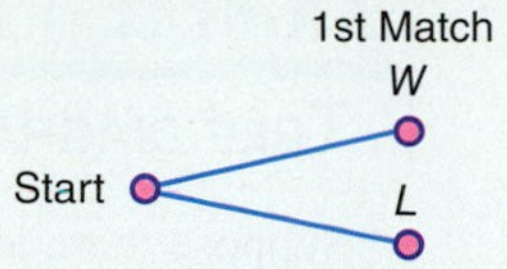

(b) Regardless of whether Louis wins or loses the first match, he plays the second and can again win or lose. Attach branches representing these two outcomes to *each* of the first match results.

→ FIGURE 4-8

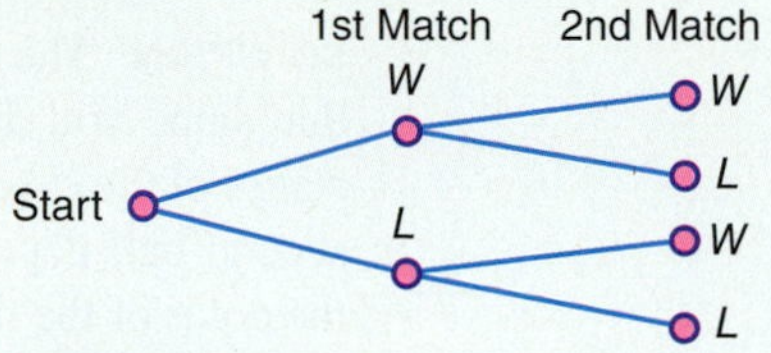

(c) Louis may win or lose the third match. Attach branches representing these two outcomes to *each* of the second match results.

→ FIGURE 4-9

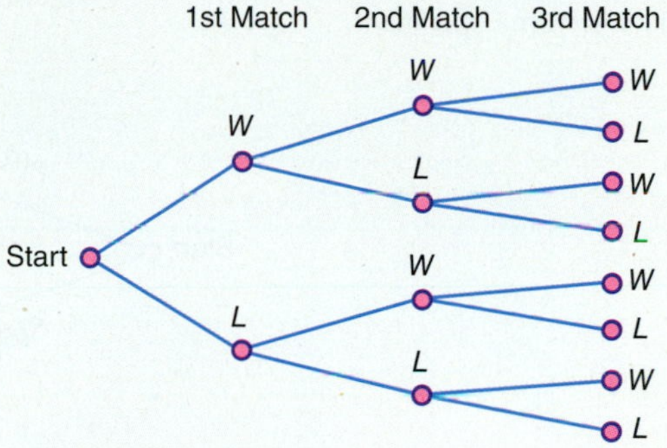

(d) How many possible win–lose sequences are there for the three matches?

→ Since there are eight branches at the end, there are eight sequences.

(e) Complete this list of win–lose sequences.

1st	2nd	3rd
W	W	W
W	W	L
W	L	W
W	L	L
____	____	____
____	____	____
____	____	____
____	____	____

→ The last four sequences all involve a loss on Match 1.

1st	2nd	3rd
L	W	W
L	W	L
L	L	W
L	L	L

(f) Use the multiplication rule to compute the total number of outcomes for the three matches.

→ The number of outcomes for a series of three events, each with two outcomes, is

$$2 \times 2 \times 2 = 8$$

Tree diagrams help us display the outcomes of an experiment involving several stages. If we label each branch of the tree with an appropriate probability, we can use the tree diagram to help us compute the probability of an outcome displayed on the tree. One of the easiest ways to illustrate this feature of tree diagrams is to use an experiment of drawing balls out of an urn. We do this in the next example.

EXAMPLE 11

TREE DIAGRAM AND PROBABILITY

Suppose there are five balls in an urn. They are identical except for color. Three of the balls are red and two are blue. You are instructed to draw out one ball, note its color, and set it aside. Then you are to draw out another ball and note its color. What are the outcomes of the experiment? What is the probability of each outcome?

SOLUTION: The tree diagram in Figure 4-10 will help us answer these questions. Notice that since you did not replace the first ball before drawing the second one, the two stages of the experiment are dependent. The probability associated with the color of the second ball depends on the color of the first ball. For instance, on the top branches, the color of the first ball drawn is red, so we compute the probabilities of the colors on the second ball accordingly. The tree diagram helps us organize the probabilities.

FIGURE 4-10

Tree Diagram for Urn Experiment

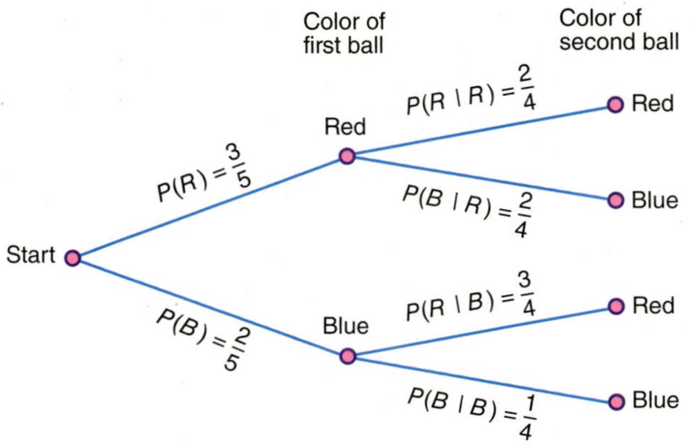

From the diagram, we see that there are four possible outcomes to the experiment. They are

RR = red on 1st *and* red on 2nd

RB = red on 1st *and* blue on 2nd

BR = blue on 1st *and* red on 2nd

BB = blue on 1st *and* blue on 2nd

To compute the probability of each outcome, we will use the multiplication rule for dependent events. As we follow the branches for each outcome, we will find the necessary probabilities.

$$P(R \text{ on 1st } \textit{and } R \text{ on 2nd}) = P(R)\cdot P(R \mid R) = \frac{3}{5}\cdot\frac{2}{4} = \frac{3}{10}$$

$$P(R \text{ on 1st } \textit{and } B \text{ and 2nd}) = P(R)\cdot P(B \mid R) = \frac{3}{5}\cdot\frac{2}{4} = \frac{3}{10}$$

$$P(B \text{ on 1st } \textit{and } R \text{ on 2nd}) = P(B)\cdot P(R \mid B) = \frac{2}{5}\cdot\frac{3}{4} = \frac{3}{10}$$

$$P(B \text{ on 1st } \textit{and } B \text{ and 2nd}) = P(B)\cdot P(B \mid B) = \frac{2}{5}\cdot\frac{1}{4} = \frac{1}{10}$$

Notice that the probabilities of the outcomes in the sample space add to 1, as they should.

Radius Images/Jupiter Images

Sometimes when we consider n items, we need to know the number of different *ordered arrangements* of the n items that are possible. The multiplication rules can help us find the number of possible ordered arrangements. Let's consider the classic example of determining the number of different ways in which eight people can be seated at a dinner table. For the first chair at the head of the table, there are eight choices. For the second chair, there are seven choices, since one person is already seated. For the third chair, there are six choices, since two people are already seated. By the time we get to the last chair, there is only one person left for that seat. We can view each arrangement as an outcome of a series of eight events. Event 1 is *fill the first chair*, event 2 is *fill the second chair*, and so forth. The multiplication rule will tell us the number of different outcomes.

Choices for	1st	2nd	3rd	4th	5th	6th	7th	8th	Chair position
	↓	↓	↓	↓	↓	↓	↓	↓	
	(8)	(7)	(6)	(5)	(4)	(3)	(2)	(1)	= 40,320

In all, there are 40,320 different seating arrangements for eight people. It is no wonder that it takes a little time to seat guests at a dinner table!

Factorial notation

The multiplication pattern shown above is not unusual. In fact, it is an example of the multiplication indicated by the *factorial notation* 8!.

! is read "factorial"

8! is read "8 factorial"

$8! = 8 \cdot 7 \cdot 6 \cdot 5 \cdot 4 \cdot 3 \cdot 2 \cdot 1$

In general, $n!$ indicates the product of n with each of the positive counting numbers less than n. *By special definition,* $0! = 1$.

FACTORIAL NOTATION

For a counting number n,

$0! = 1$

$1! = 1$

$n! = n(n-1)(n-2)\cdots 1$

GUIDED EXERCISE 11 | Factorial

(a) Evaluate 3!.

→ $3! = 3 \cdot 2 \cdot 1 = 6$

(b) In how many different ways can three objects be arranged in order? How many choices do you have for the first position? for the second position? for the third position?

→ You have three choices for the first position, two for the second position, and one for the third position. By the multiplication rule, you have

$(3)(2)(1) = 3! = 6$ arrangements

Permutations rule

We have considered the number of ordered arrangements of n objects taken as an entire group. But what if we don't arrange the entire group? Specifically, we considered a dinner party for eight and found the number of ordered seating arrangements for all eight people. However, suppose you have an open house and have only five chairs. How many ways can five of the eight people seat themselves in the chairs? The formula we use to compute this number is called the *permutation formula*. As we see in the next example, the *permutations rule* is really another version of the multiplication rule.

COUNTING RULE FOR PERMUTATIONS

The number of ways to *arrange in order n* distinct objects, taking them *r* at a time, is

$$P_{n,r} = \frac{n!}{(n-r)!} \tag{9}$$

where *n* and *r* are whole numbers and $n \geq r$. Another commonly used notation for permutations is nPr.

EXAMPLE 12

PERMUTATIONS RULE

Let's compute the number of possible ordered seating arrangements for eight people in five chairs.

SOLUTION: In this case, we are considering a total of $n = 8$ different people, and we wish to arrange $r = 5$ of these people. Substituting into formula (9), we have

$$P_{n,r} = \frac{n!}{(n-r)!}$$

$$P_{8,5} = \frac{8!}{(8-5)!} = \frac{8!}{3!} = \frac{40{,}320}{6} = 6720$$

Using the multiplication rule, we get the same results

Chair	1		2		3		4		5		
Choices for	8	×	7	×	6	×	5	×	4	=	6720

The permutations rule has the advantage of using factorials. Most scientific calculators have a factorial key (!) as well as a permutations key (nPr) (see Tech Notes).

TECH NOTES Most scientific calculators have a factorial key, often designated *x!* or *n!*. Many of these same calculators have the permutation function built in, often labeled nPr. They also have the combination function, which is discussed next. The combination function is often labeled nCr.

TI-84Plus/TI-83Plus/TI-*n*spire (with TI-84Plus keypad) The factorial, permutation, and combination functions are all under **MATH,** then **PRB.**

Excel 2010 Click on the **insert function** (f_x), then select **all** for the category. **Fact** gives factorials, **Permut** gives permutations, and **Combin** gives combinations.

In each of our previous counting formulas, we have taken the *order* of the objects or people into account. However, suppose that in your political science class you are given a list of 10 books. You are to select 4 to read during the semester. The order in which you read the books is not important. We are interested in the *different groupings* or *combinations* of 4 books from among the 10 on the list. The next formula tells us how to compute the number of different combinations.

COUNTING RULE FOR COMBINATIONS

The number of *combinations* of n objects taken r at a time is

$$C_{n,r} = \frac{n!}{r!(n-r)!} \tag{10}$$

where n and r are whole numbers and $n \geq r$. Other commonly used notations for combinations include nCr and $\binom{n}{r}$.

Notice the difference between the concepts of permutations and combinations. When we consider permutations, we are considering groupings *and order*. When we consider combinations, we are considering only the number of different groupings. For combinations, order within the groupings is not considered. As a result, the number of combinations of n objects taken r at a time is generally smaller than the number of permutations of the same n objects taken r at a time. In fact, the combinations formula is simply the permutations formula with the number of permutations of each distinct group divided out. In the formula for combinations, notice the factor of $r!$ in the denominator.

Combinations rule

Now let's look at an example in which we use the *combinations rule* to compute the number of *combinations* of 10 books taken 4 at a time.

EXAMPLE 13

COMBINATIONS

In your political science class, you are assigned to read any 4 books from a list of 10 books. How many different groups of 4 are available from the list of 10?

SOLUTION: In this case, we are interested in *combinations*, rather than permutations, of 10 books taken 4 at a time. Using $n = 10$ and $r = 4$, we have

$$C_{n,r} = \frac{n!}{r!(n-r)!} = \frac{10!}{4!(10-4)!} = 210$$

There are 210 different groups of 4 books that can be selected from the list of 10. An alternate solution method is to use the combinations key (often nCr or $C_{n,r}$) on a calculator.

LOOKING FORWARD

We will see the combinations rule again in Section 5.2 when we discuss the formula for the binomial probability distribution.

PROCEDURE

HOW TO DETERMINE THE NUMBER OF OUTCOMES OF AN EXPERIMENT

1. If the experiment consists of a series of stages with various outcomes, use the multiplication rule or a tree diagram.
2. If the outcomes consist of ordered subgroups of r items taken from a group of n items, use the permutations rule, $P_{n,r}$.

$$P_{n,r} = \frac{n!}{(n-r)!} \tag{9}$$

3. If the outcomes consist of nonordered subgroups of r items taken from a group of n items, use the combinations rule, $C_{n,r}$.

$$C_{n,r} = \frac{n!}{r!(n-r)!} \tag{10}$$

GUIDED EXERCISE 12 — Permutations and combinations

The board of directors at Belford Community Hospital has 12 members.

(i) Three officers—president, vice president, and treasurer—must be elected from among the members. How many different slates of officers are possible? We will view a slate of officers as a list of three people, with the president listed first, the vice president listed second, and the treasurer listed third. For instance, if Mr. Acosta, Ms. Hill, and Mr. Smith wish to be on a slate together, there are several different slates possible, depending on the person listed for each office. Not only are we asking for the number of different groups of three names for a slate, we are also concerned about order.

(a) Do we use the permutations rule or the combinations rule? What is the value of n? What is the value of r?

→ We use the permutations rule, since order is important. The size of the group from which the slates of officers are to be selected is n. The size of each slate is r.

$n = 12$ and $r = 3$

(b) Use the permutations rule with $n = 12$ and $r = 3$ to compute $P_{12,3}$.

→ $$P_{n,r} = \frac{n!}{(n-r)!} = \frac{12!}{(12-3!)} = 1320$$

An alternative is to use the permutations key on a calculator.

(ii) Three members from the group of 12 on the board of directors at Belford Community Hospital will be selected to go to a convention (all expenses paid) in Hawaii. How many different groups of 3 are there?

(c) Do we use the permutations rule or the combinations rule? What is the value of n? What is the value of r?

→ We use the combinations rule, because order is not important. The size of the board is $n = 12$ and the size of each group going to the convention is $r = 3$.

(d) Use the combinations rule with $n = 12$ and $r = 3$ to compute $C_{12,3}$.

→ $$C_{n,r} = \frac{n!}{r!(n-r)!} = \frac{12!}{3!(12-3)!} = 220$$

An alternative is to use the combinations key on a calculator.

WHAT DO COUNTING RULES TELL US?

Counting rules tell us the total number of outcomes created by combining a sequence of events in specified ways.

- The **multiplication rule** tells us the total number of possible outcomes for a sequence of events. **Tree diagrams** provide a visual display of all the resulting outcomes.
- The **permutation rule** tells us the total number of ways we can **arrange in order** n distinct objects into a group of size r.
- The **combination rule** tells us how many ways we can form n distinct objects into a group of size r. The order of the objects is irrelevant.

VIEWPOINT Powerball

Powerball is a multistate lottery game that consists of drawing five distinct whole numbers from the numbers 1 through 59 in any order. Then one more number from the numbers 1 through 35 is selected as the Powerball number (this number could be one of the original five). Powerball numbers are drawn every Wednesday and Saturday. If you match all six numbers, you win the jackpot, which is worth at least $40 million. Use methods of this section to show that there are 175,223,510 possible Powerball plays. For more information about the game of Powerball and the probability of winning different prizes, visit the Brase/Brase statistics site at **http://www.cengage.com/statistics/brase11e** *and find the link to the Multi-State Lottery Association. Then select Powerball.*

SECTION 4.3 PROBLEMS

1. *Statistical Literacy* What is the main difference between a situation in which the use of the permutations rule is appropriate and one in which the use of the combinations rule is appropriate?

2. *Statistical Literacy* Consider a series of events. How does a tree diagram help you list all the possible outcomes of a series of events? How can you use a tree diagram to determine the total number of outcomes of a series of events?

3. *Critical Thinking* For each of the following situations, explain why the combinations rule or the permutations rule should be used.
   (a) Determine the number of different groups of 5 items that can be selected from 12 distinct items.
   (b) Determine the number of different arrangements of 5 items that can be selected from 12 distinct items.

4. *Critical Thinking* You need to know the number of different arrangements possible for five distinct letters. You decide to use the permutations rule, but your friend tells you to use 5!. Who is correct? Explain.

5. *Tree Diagram*
   (a) Draw a tree diagram to display all the possible head–tail sequences that can occur when you flip a coin three times.
   (b) How many sequences contain exactly two heads?
   (c) *Probability Extension* Assuming the sequences are all equally likely, what is the probability that you will get exactly two heads when you toss a coin three times?

6. *Tree Diagram*
   (a) Draw a tree diagram to display all the possible outcomes that can occur when you flip a coin and then toss a die.
   (b) How many outcomes contain a head and a number greater than 4?
   (c) *Probability Extension* Assuming the outcomes displayed in the tree diagram are all equally likely, what is the probability that you will get a head *and* a number greater than 4 when you flip a coin and toss a die?

7. *Tree Diagram* There are six balls in an urn. They are identical except for color. Two are red, three are blue, and one is yellow. You are to draw a ball from the urn, note its color, and set it aside. Then you are to draw another ball from the urn and note its color.
   (a) Make a tree diagram to show all possible outcomes of the experiment. Label the probability associated with each stage of the experiment on the appropriate branch.
   (b) *Probability Extension* Compute the probability for each outcome of the experiment.

8. *Tree Diagram*
   (a) Make a tree diagram to show all the possible sequences of answers for three multiple-choice questions, each with four possible responses.
   (b) *Probability Extension* Assuming that you are guessing the answers so that all outcomes listed in the tree are equally likely, what is the probability that you will guess the one sequence that contains all three correct answers?

9. *Multiplication Rule* Four wires (red, green, blue, and yellow) need to be attached to a circuit board. A robotic device will attach the wires. The wires can be attached in any order, and the production manager wishes to determine which order would be fastest for the robot to use. Use the multiplication rule of counting to determine the number of possible sequences of assembly that must be tested. *Hint:* There are four choices for the first wire, three for the second, two for the third, and only one for the fourth.

10. *Multiplication Rule* A sales representative must visit four cities: Omaha, Dallas, Wichita, and Oklahoma City. There are direct air connections between each of the cities. Use the multiplication rule of counting to determine the number of different choices the sales representative has for the order in which to visit the cities. How is this problem similar to Problem 9?

11. *Counting: Agriculture* Barbara is a research biologist for Green Carpet Lawns. She is studying the effects of fertilizer type, temperature at time of application, and water treatment after application. She has four fertilizer types, three temperature zones, and three water treatments to test. Determine the number of different lawn plots she needs in order to test each fertilizer type, temperature range, and water treatment configuration.

12. *Counting: Outcomes* You toss a pair of dice.
    (a) Determine the number of possible pairs of outcomes. (Recall that there are six possible outcomes for each die.)
    (b) There are three even numbers on each die. How many outcomes are possible with even numbers appearing on each die?
    (c) *Probability extension:* What is the probability that both dice will show an even number?

13. Compute $P_{5,2}$.

14. Compute $P_{8,3}$.

15. Compute $P_{7,7}$.

16. Compute $P_{9,9}$.

17. Compute $C_{5,2}$.

18. Compute $C_{8,3}$.

19. Compute $C_{7,7}$.

20. Compute $C_{8,8}$.

21. *Counting: Hiring* There are three nursing positions to be filled at Lilly Hospital. Position 1 is the day nursing supervisor; position 2 is the night nursing supervisor; and position 3 is the nursing coordinator position. There are 15 candidates qualified for all three of the positions. Determine the number of different ways the positions can be filled by these applicants.

22. *Counting: Lottery* In the Cash Now lottery game there are 10 finalists who submitted entry tickets on time. From these 10 tickets, three grand prize winners will be drawn. The first prize is $1 million, the second prize is $100,000, and the third prize is $10,000. Determine the total number of different ways in which the winners can be drawn. (Assume that the tickets are not replaced after they are drawn.)

23. *Counting: Sports* The University of Montana ski team has five entrants in a men's downhill ski event. The coach would like the first, second, and third places to go to the team members. In how many ways can the five team entrants achieve first, second, and third places?

24. *Counting: Sales* During the Computer Daze special promotion, a customer purchasing a computer and printer is given a choice of 3 free software packages. There are 10 different software packages from which to select. How many different groups of software packages can be selected?

25. *Counting: Hiring* There are 15 qualified applicants for 5 trainee positions in a fast-food management program. How many different groups of trainees can be selected?

26. *Counting: Grading* One professor grades homework by randomly choosing 5 out of 12 homework problems to grade.
    (a) How many different groups of 5 problems can be chosen from the 12 problems?
    (b) *Probability Extension* Jerry did only 5 problems of one assignment. What is the probability that the problems he did comprised the group that was selected to be graded?
    (c) Silvia did 7 problems. How many different groups of 5 did she complete? What is the probability that one of the groups of 5 she completed comprised the group selected to be graded?

27. *Counting: Hiring* The qualified applicant pool for six management trainee positions consists of seven women and five men.
    (a) How many different groups of applicants can be selected for the positions?
    (b) How many different groups of trainees would consist entirely of women?
    (c) *Probability Extension* If the applicants are equally qualified and the trainee positions are selected by drawing the names at random so that all groups of six are equally likely, what is the probability that the trainee class will consist entirely of women?

28. *Counting: Powerball* The Viewpoint of this section, on page 185, describes how the lottery game of Powerball is played.
    (a) The first step is to select five distinct whole numbers between 1 and 59. Order is not important. Use the appropriate counting rule to determine the number of ways groups of five different numbers can be selected. *Note:* The winning group of five numbers is selected by random drawing of 5 white balls from a collection of 59 numbered balls.
    (b) The next step is to choose the Powerball number, which is any number between 1 and 35. The number need not be distinct from numbers chosen for the first five described in part (a). Use the appropriate counting rule to determine the number of possible distinct outcomes for the first five numbers, chosen as described in part (a) together with the Powerball number. *Note:* The Powerball number appears on a red ball that is drawn at random from a collection of 35 numbered balls.

CHAPTER REVIEW

SUMMARY

In this chapter we explored basic features of probability.

- The probability of an event A is a number between 0 and 1, inclusive. The more likely the event, the closer the probability of the event is to 1.
- Three main ways to determine the probability of an event are: the method of relative frequency, the method of equally likely outcomes, and intuition. Other important ways will be discussed later.
- The law of large numbers indicates that as the number of trials of a statistical experiment or observation increases, the relative frequency of a designated event becomes closer to the theoretical probability of that event.
- Events are mutually exclusive if they cannot occur together. Events are independent if the occurrence of one event does not change the probability of the occurrence of the other.
- Conditional probability is the probability that one event will occur, given that another event has occurred.
- The complement rule gives the probability that an event will not occur. The addition rule gives the probability that at least one of two specified events will occur. The multiplication rule gives the probability that two events will occur together.
- To determine the probability of equally likely events, we need to know how many outcomes are possible. Devices such as tree diagrams and counting rules—such as the multiplication rule of counting, the permutations rule, and the combinations rule—help us determine the total number of outcomes of a statistical experiment or observation.

In most of the statistical applications of later chapters, we will use the addition rule for mutually exclusive events and the multiplication rule for independent events.

IMPORTANT WORDS & SYMBOLS

Section 4.1

Probability of an event A, $P(A)$ 144
Intuition 144
Relative frequency 144
Equally likely outcomes 144
Law of large numbers 146
Statistical experiment 146
Event 146
Simple event 146
Sample space 146
Complement of event A^c 148

Section 4.2

Independent events 156
Dependent events 156
Probability of A *and* B 156
Event $A \mid B$
Conditional probability 156
$P(A \mid B)$ 156
Multiplication rules of probability (for independent and dependent events) 156
More than two independent events 161
Probability of A *or* B 161
Event A *and* B 161
Event A *or* B 161
Mutually exclusive events 163
Addition rules (for mutually exclusive and general events) 163
More than two mutually exclusive events 165
Basic probability rules 168

Section 4.3

Multiplication rule of counting 177
Tree diagram 178
Factorial notation 181
Permutations rule 181
Combinations rule 183

VIEWPOINT Deathday and Birthday

Can people really postpone death? If so, how much can the timing of death be influenced by psychological, social, or other influential factors? One special event is a birthday. Do famous people try to postpone their deaths until an important birthday? Both Thomas Jefferson and John Adams died on July 4, 1826, when the United States was celebrating its 50th birthday. Is this only a strange coincidence, or is there an unexpected connection between birthdays and deathdays? The probability associated with a death rate decline of famous people just before important birthdays has been studied by Professor D. P. Phillips of the State University of New York and is presented in the book Statistics, A Guide to the Unknown, *edited by J. M. Tanur.*

CHAPTER REVIEW PROBLEMS

1. *Statistical Literacy* Consider the following two events for an individual:

   A = owns a cell phone B = owns a laptop computer

   Translate each event into words.
   (a) A^c
   (b) A *and* B
   (c) A *or* B
   (d) $A \mid B$
   (e) $B \mid A$

2. *Statistical Literacy* If two events A and B are mutually exclusive, what is the value of $P(A \text{ and } B)$?

3. *Statistical Literacy* If two events A and B are independent, how do the probabilities $P(A)$ and $P(A \mid B)$ compare?

4. *Interpretation* You are considering two facial cosmetic surgeries. These are elective surgeries and their outcomes are independent. The probability of success for each surgery is 0.90. What is the probability of success for both surgeries? If the probability of success for both surgeries is less than 0.85, you will decide not to have the surgeries. Will you have the surgeries or not?

5. *Interpretation* You are applying for two jobs, and you estimate the probability of getting an offer for the first job is 0.70 while the probability of getting an offer for the second job is 0.80. Assume the job offers are independent.
   (a) Compute the probability of getting offers for both jobs. How does this probability compare to the probability of getting each individual job offer?
   (b) Compute the probability of getting an offer for either the first or the second job. How does this probability compare to the probability of getting each individual job offer? Does it seem worthwhile to apply for both jobs? Explain.

6. *Critical Thinking* You are given the information that $P(A) = 0.30$ and $P(B) = 0.40$.
   (a) Do you have enough information to compute $P(A \text{ or } B)$? Explain.
   (b) If you know that events A and B are mutually exclusive, do you have enough information to compute $P(A \text{ or } B)$? Explain.

7. *Critical Thinking* You are given the information that $P(A) = 0.30$ and $P(B) = 0.40$.
   (a) Do you have enough information to compute $P(A \text{ and } B)$? Explain.
   (b) If you know that events A and B are independent, do you have enough information to compute $P(A \text{ and } B)$? Explain.

8. *Critical Thinking* For a class activity, your group has been assigned the task of generating a quiz question that requires use of the formula for conditional probability to compute $P(B \mid A)$. Your group comes up with the following question: "If $P(A \text{ and } B) = 0.40$ and $P(A) = 0.20$, what is the value of $P(B \mid A)$?" What is wrong with this question? *Hint:* Consider the answer you get when using the correct formula, $P(B \mid A) = P(A \text{ and } B)/P(A)$.

9. *Salary Raise: Women* Does it pay to ask for a raise? A national survey of heads of households showed the percentage of those who asked for a raise and the percentage who got one (*USA Today*). According to the survey, of the women interviewed, 24% had asked for a raise, and of those women who had asked for a raise, 45% received the raise. If a woman is selected at random from the survey population of women, find the following probabilities: P(woman asked for a raise); P(woman received raise, *given* she asked for one); P(woman asked for raise *and* received raise).

10. *Salary Raise: Men* According to the same survey quoted in Problem 9, of the men interviewed, 20% had asked for a raise and 59% of the men who had asked for a raise received the raise. If a man is selected at random from the survey population of men, find the following probabilities: P(man asked for a raise); P(man received raise, *given* he asked for one); P(man asked for raise *and* received raise).

11. *General: Thumbtack* Drop a thumbtack and observe how it lands.
(a) Describe how you could use a relative frequency to estimate the probability that a thumbtack will land with its flat side down.
(b) What is the sample space of outcomes for the thumbtack?
(c) How would you make a probability assignment to this sample space if when you drop 500 tacks, 340 land flat side down?

12. *Survey: Reaction to Poison Ivy* Allergic reactions to poison ivy can be miserable. Plant oils cause the reaction. Researchers at Allergy Institute did a study to determine the effects of washing the oil off within 5 minutes of exposure. A random sample of 1000 people with known allergies to poison ivy participated in the study. Oil from the poison ivy plant was rubbed on a patch of skin. For 500 of the subjects, it was washed off *within* 5 minutes. For the other 500 subjects, the oil was washed off *after* 5 minutes. The results are summarized in Table 4-5.

TABLE 4-5 **Time Within Which Oil Was Washed Off**

Reaction	Within 5 Minutes	After 5 Minutes	Row Total
None	420	50	470
Mild	60	330	390
Strong	20	120	140
Column Total	500	500	1000

Let's use the following notation for the various events: W = washing oil off within 5 minute, A = washing oil off after 5 minutes, N = no reaction, M = mild reaction, S = strong reaction. Find the following probabilities for a person selected at random from this sample of 1000 subjects.
(a) $P(N)$, $P(M)$, $P(S)$
(b) $P(N \mid W)$, $P(S \mid W)$
(c) $P(N \mid A)$, $P(S \mid A)$
(d) $P(N \text{ and } W)$, $P(M \text{ and } W)$
(e) $P(N \text{ or } M)$. Are the events N = no reaction and M = mild reaction mutually exclusive? Explain.
(f) Are the events N = no reaction and W = washing oil off within 5 minutes independent? Explain.

13. *General: Two Dice* In a game of craps, you roll two fair dice. Whether you win or lose depends on the sum of the numbers appearing on the tops of the dice. Let x be the random variable that represents the sum of the numbers on the tops of the dice.
(a) What values can x take on?
(b) What is the probability distribution of these x values (that is, what is the probability that $x = 2, 3$, etc.)?

14. *Academic: Passing French* Class records at Rockwood College indicate that a student selected at random has probability 0.77 of passing French 101. For the student who passes French 101, the probability is 0.90 that he or she will pass French 102. What is the probability that a student selected at random will pass both French 101 and French 102?

15. *Combination: City Council* There is money to send two of eight city council members to a conference in Honolulu. All want to go, so they decide to choose the members to go to the conference by a random process. How many different combinations of two council members can be selected from the eight who want to go to the conference?

16. *Basic Computation* Compute. (a) $P_{7,2}$ (b) $C_{7,2}$ (c) $P_{3,3}$ (d) $C_{4,4}$

17. *Counting: Exam Answers* There are five multiple-choice questions on an exam, each with four possible answers. Determine the number of possible answer sequences for the five questions. Only one of the sets can contain all five correct answers. If you are guessing, so that you are as likely to choose one sequence of answers as another, what is the probability of getting all five answers correct?

18. *Scheduling: College Courses* A student must satisfy the literature, social science, and philosophy requirements this semester. There are four literature courses to select from, three social science courses, and two philosophy courses. Make a tree diagram showing all the possible sequences of literature, social science, and philosophy courses.

19. *General: Combination Lock* To open a combination lock, you turn the dial to the right and stop at a number; then you turn it to the left and stop at a second number. Finally, you turn the dial back to the right and stop at a third number. If you used the correct sequence of numbers, the lock opens. If the dial of the lock contains 10 numbers, 0 through 9, determine the number of different combinations possible for the lock. Note: The same number can be reused.

20. *General: Combination Lock* You have a combination lock. Again, to open it you turn the dial to the right and stop at a first number; then you turn it to the left and stop at a second number. Finally, you turn the dial to the right and stop at a third number. Suppose you remember that the three numbers for your lock are 2, 9, and 5, but you don't remember the order in which the numbers occur. How many sequences of these three numbers are possible?

DATA HIGHLIGHTS: GROUP PROJECTS

Break into small groups and discuss the following topics. Organize a brief outline in which you summarize the main points of your group discussion.

1. Look at Figure 4-11, "Who's Cracking the Books?"
   (a) Does the figure show the probability distribution of grade records for male students? for female students? Describe all the grade-record probability distributions shown in Figure 4-11. Find the probability that a male student selected at random has a grade record showing mostly A's.

FIGURE 4-11

Who's Cracking the Books?

Undergraduate Grade Record by Student Characteristic

Student characteristic		C's and D's or lower	B's and C's	Mostly B's	A's and B's	Mostly A's
Gender:	Men	38.8 %	16.6 %	22.6 %	9.6 %	12.4 %
	Women	29.4	16.2	26.2	12.0	16.2
Class level:	Graduating senior	15.8 %	21.9 %	34.5 %	14.7 %	13.0 %
	All other class levels	35.4	15.8	23.6	10.5	14.7
Age:	18 or younger	42.6 %	14.7 %	23.4 %	9.4 %	10.0 %
	19 to 23	38.1	19.0	25.1	9.4	8.3
	24 to 29	33.3	16.9	24.7	10.3	14.9
	30 to 39	23.1	13.2	25.9	14.8	23.0
	40 and older	20.1	10.1	22.0	14.8	33.0

Source: U.S. Department of Education

(b) Is the probability distribution shown for all students making mostly A's? Explain your answer. *Hint:* Do the percentages shown for mostly A's add up to 1? Can Figure 4-11 be used to determine the probability that a student selected at random has mostly A's? Can it be used to determine the probability that a female student selected at random has mostly A's? What is the probability?

(c) Can we use the information shown in the figure to determine the probability that a graduating senior has grades consisting of mostly B's or higher? What is the probability?

(d) Does Figure 4-11 give sufficient information to determine the probability that a student selected at random is in the age range 19 to 23 *and* has grades that are mostly B's? What is the probability that a student selected at random has grades that are mostly B's, *given* that he or she is in the age range 19 to 23?

(e) Suppose that 65% of the students at State University are between 19 and 23 years of age. What is the probability that a student selected at random is in this age range *and* has grades that are mostly B's?

2. Consider the information given in Figure 4-12, "Vulnerable Knees." What is the probability that an orthopedic case selected at random involves knee problems? Of those cases, estimate the probability that the case requires full knee replacement. Compute the probability that an orthopedic case selected at random involves a knee problem *and* requires a full knee replacement. Next, look at the probability distribution for ages of patients requiring full knee replacements. Medicare insurance coverage begins when a person reaches age 65. What is the probability that the age of a person receiving a knee replacement is 65 or older?

FIGURE 4-12

Vulnerable Knees

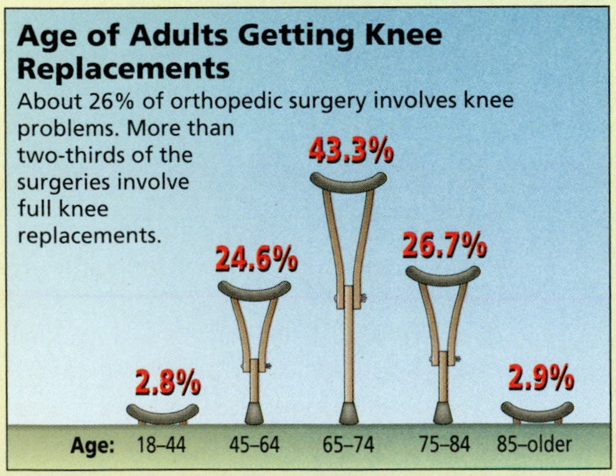

Source: American Academy of Orthopedic Surgeons

LINKING CONCEPTS: WRITING PROJECTS

Discuss each of the following topics in class or review the topics on your own. Then write a brief but complete essay in which you summarize the main points. Please include formulas as appropriate.

1. Discuss the following concepts and give examples from everyday life in which you might encounter each concept. *Hint:* For instance, consider the "experiment" of arriving for class. Some possible outcomes are not arriving (that is, missing class), arriving on time, and arriving late.
   (a) Sample space.
   (b) Probability assignment to a sample space. In your discussion, be sure to include answers to the following questions.
      (i) Is there more than one valid way to assign probabilities to a sample space? Explain and give an example.
      (ii) How can probabilities be estimated by relative frequencies? How can probabilities be computed if events are equally likely?
2. Discuss the concepts of mutually exclusive events and independent events. List several examples of each type of event from everyday life.
   (a) If A and B are mutually exclusive events, does it follow that A and B *cannot* be independent events? Give an example to demonstrate your answer. *Hint:* Discuss an election where only one person can win the election. Let A be the event that party A's candidate wins, and let B be the event that party B's candidate wins. Does the outcome of one event determine the outcome of the other event? Are A and B mutually exclusive events?
   (b) Discuss the conditions under which $P(A \text{ and } B) = P(A) \cdot P(B)$ is true. Under what conditions is this not true?
   (c) Discuss the conditions under which $P(A \text{ or } B) = P(A) + P(B)$ is true. Under what conditions is this not true?
3. Although we learn a good deal about probability in this course, the main emphasis is on statistics. Write a few paragraphs in which you talk about the distinction between probability and statistics. In what types of problems would probability be the main tool? In what types of problems would statistics be the main tool? Give some examples of both types of problems. What kinds of outcomes or conclusions do we expect from each type of problem?

USING TECHNOLOGY

Demonstration of the Law of Large Numbers

Computers can be used to simulate experiments. With packages such as Excel 2010, Minitab, and SPSS, programs using random-number generators can be designed (see the *Technology Guide*) to simulate activities such as tossing a die.

The following printouts show the results of the simulations for tossing a die 6, 500, 50,000, 500,000, and 1,000,000 times. Notice how the relative frequencies of the outcomes approach the theoretical probabilities of 1/6 or 0.16667 for each outcome. Do you expect the same results every time the simulation is done? Why or why not?

Results of tossing one die 6 times

Outcome	Number of Occurrences	Relative Frequency
⚀	0	0.00000
⚁	1	0.16667
⚂	2	0.33333
⚃	0	0.00000
⚄	1	0.16667
⚅	2	0.33333

Results of tossing one die 500 times

Outcome	Number of Occurrences	Relative Frequency
⚀	87	0.17400
⚁	83	0.16600
⚂	91	0.18200
⚃	69	0.13800
⚄	87	0.17400
⚅	83	0.16600

Results of tossing one die 50,000 times

Outcome	Number of Occurrences	Relative Frequency
⚀	8528	0.17056
⚁	8354	0.16708
⚂	8246	0.16492
⚃	8414	0.16828
⚄	8178	0.16356
⚅	8280	0.16560

Results of tossing one die 500,000 times

Outcome	Number of Occurrences	Relative Frequency
⚀	83644	0.16729
⚁	83368	0.16674
⚂	83398	0.16680
⚃	83095	0.16619
⚄	83268	0.16654
⚅	83227	0.16645

Results of tossing one die 1,000,000 times

Outcome	Number of Occurrences	Relative Frequency
⚀	166643	0.16664
⚁	166168	0.16617
⚂	167391	0.16739
⚃	165790	0.16579
⚄	167243	0.16724
⚅	166765	0.16677

5

National Portrait Gallery, Smithsonian Institution/Art Resource, NY

Education is the key to unlock the golden door of freedom.

—George Washington Carver

George Washington Carver (1859–1943) won international fame for agricultural research. After graduating from Iowa State College, he was appointed a faculty member in the Iowa State Botany Department. Carver took charge of the greenhouse and started a fungus collection that later included more than 20,000 species. This collection brought him professional acclaim in the field of botany.

At the invitation of his friend Booker T. Washington, Carver joined the faculty of the Tuskegee Institute, where he spent the rest of his long and distinguished career. Carver's creative genius accounted for more than 300 inventions from peanuts, 118 inventions from sweet potatoes, and 75 inventions from pecans.

Gathering and analyzing data were important components of Carver's work. Methods you will learn in this course are widely used in research in every field, including agriculture.

THE BINOMIAL PROBABILITY DISTRIBUTION AND RELATED TOPICS

Andresr/iStockphoto.com

PREVIEW QUESTIONS

What is a random variable? How do you compute μ and σ for a discrete random variable? How do you compute μ and σ for linear combinations of independent random variables? (SECTION 5.1)

Many of life's experiences consist of some successes together with some failures. Suppose you make n attempts to succeed at a certain project. How can you use the binomial probability distribution to compute the probability of r successes? (SECTION 5.2)

How do you compute μ and σ for the binomial distribution? (SECTION 5.3)

How is the binomial distribution related to other probability distributions, such as the geometric and Poisson? (SECTION 5.4)

FOCUS PROBLEM

Personality Preference Types: Introvert or Extrovert?

Mitch Wojnarowicz/The Image Works

Isabel Briggs Myers was a pioneer in the study of personality types. Her work has been used successfully in counseling, educational, and industrial settings. In the book *A Guide to the Development and Use of the Myers–Briggs Type Indicators,* by Myers and McCaully, it was reported that based on a very large sample (2282 professors), approximately 45% of all university professors are extroverted.

After completing this chapter, you will be able to answer the following questions. Suppose you have classes with six different professors.

(a) What is the probability that all six are extroverts?

(b) What is the probability that none of your professors is an extrovert?

(c) What is the probability that at least two of your professors are extroverts?

(d) In a group of six professors selected at random, what is the *expected number* of extroverts? What is the *standard deviation* of the distribution?

(e) Suppose you were assigned to write an article for the student newspaper and you were given a quota (by the editor) of interviewing at least three extroverted professors. How many professors selected at random would you need to interview to be at least 90% sure of filling the quota?

(See Problem 26 of Section 5.3.)

COMMENT Both extroverted and introverted professors can be excellent teachers.

For online student resources, visit the Brase/Brase, *Understandable Statistics*, 11th edition web site at http://www.cengage.com/statistics/brase11e

SECTION **5.1**

Introduction to Random Variables and Probability Distributions

FOCUS POINTS

- Distinguish between discrete and continuous random variables.
- Graph discrete probability distributions.
- Compute μ and σ for a discrete probability distribution.
- Compute μ and σ for a linear function of a random variable x.
- Compute μ and σ for a linear combination of two independent random variables.

RANDOM VARIABLES

For our purposes, we say that a *statistical experiment* or *observation* is any process by which measurements are obtained. For instance, you might count the number of eggs in a robin's nest or measure daily rainfall in inches. It is common practice to use the letter x to represent the quantitative result of an experiment or observation. As such, we call x a variable.

Random variable

Discrete random variable

Continuous random variable

A quantitative variable x is a **random variable** if the value that x takes on in a given experiment or observation is a chance or random outcome.

A **discrete random variable** can take on only a finite number of values or a countable number of values.

A **continuous random variable** can take on any of the countless number of values in a line interval.

The distinction between discrete and continuous random variables is important because of the different mathematical techniques associated with the two kinds of random variables.

In most of the cases we will consider, a *discrete random variable* will be the result of a count. The number of students in a statistics class is a discrete random variable. Values such as 15, 25, 50, and 250 are all possible. However, 25.5 students is not a possible value for the number of students.

Most of the *continuous random variables* we will see will occur as the result of a measurement on a continuous scale. For example, the air pressure in an automobile tire represents a continuous random variable. The air pressure could, in theory, take on any value from 0 lb/in^2 (psi) to the bursting pressure of the tire. Values such as 20.126 psi, 20.12678 psi, and so forth are possible.

GUIDED EXERCISE 1 — DISCRETE OR CONTINUOUS RANDOM VARIABLES

Which of the following random variables are discrete and which are continuous?

(a) *Measure* the time it takes a student selected at random to register for the fall term.

→ Time can take on any value, so this is a continuous random variable.

(b) *Count* the number of bad checks drawn on Upright Bank on a day selected at random.

→ The number of bad checks can be only a whole number such as 0, 1, 2, 3, etc. This is a discrete variable.

Continued

GUIDED EXERCISE 1 *continued*

(c) *Measure* the amount of gasoline needed to drive your car 200 miles. → We are measuring volume, which can assume any value, so this is a continuous random variable.

(d) Pick a random sample of 50 registered voters in a district and find the number who voted in the last county election. → This is a count, so the variable is discrete.

PROBABILITY DISTRIBUTION OF A DISCRETE RANDOM VARIABLE

A random variable has a probability distribution whether it is discrete or continuous.

Probability distribution

A **probability distribution** is an assignment of probabilities to each distinct value of a discrete random variable or to each interval of values of a continuous random variable.

FEATURES OF THE PROBABILITY DISTRIBUTION OF A DISCRETE RANDOM VARIABLE

1. The probability distribution has a probability assigned to *each* distinct value of the random variable.
2. The sum of all the assigned probabilities must be 1.

EXAMPLE 1

DISCRETE PROBABILITY DISTRIBUTION

Dr. Mendoza developed a test to measure boredom tolerance. He administered it to a group of 20,000 adults between the ages of 25 and 35. The possible scores were 0, 1, 2, 3, 4, 5, and 6, with 6 indicating the highest tolerance for boredom. The test results for this group are shown in Table 5-1.

(a) If a subject is chosen at random from this group, the probability that he or she will have a score of 3 is 6000/20,000, or 0.30. In a similar way, we can use relative frequencies to compute the probabilities for the other scores (Table 5-2).

TABLE 5-1 Boredom Tolerance Test Scores for 20,000 Subjects

Score	Number of Subjects
0	1400
1	2600
2	3600
3	6000
4	4400
5	1600
6	400

TABLE 5-2 Probability Distribution of Scores on Boredom Tolerance Test

Score x	Probability $P(x)$
0	0.07
1	0.13
2	0.18
3	0.30
4	0.22
5	0.08
6	0.02
	$\Sigma P(x) = 1$

FIGURE 5-1

Graph of the Probability Distribution of Test Scores

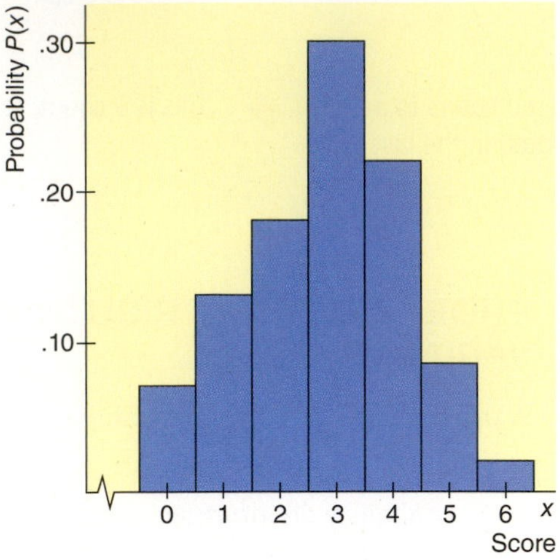

These probability assignments make up the probability distribution. Notice that the scores are mutually exclusive: No one subject has two scores. The sum of the probabilities of all the scores is 1.

(b) The graph of this distribution is simply a relative-frequency histogram (see Figure 5-1) in which the height of the bar over a score represents the probability of that score. Since each bar is one unit wide, the area of the bar over a score equals the height and thus represents the probability of that score. Since the sum of the probabilities is 1, the area under the graph is also 1.

(c) The Topnotch Clothing Company needs to hire someone with a score on the boredom tolerance test of 5 or 6 to operate the fabric press machine. Since the scores 5 and 6 are mutually exclusive, the probability that someone in the group who took the boredom tolerance test made either a 5 or a 6 is the sum

$$P(5 \text{ or } 6) = P(5) + P(6)$$
$$= 0.08 + 0.02 = 0.10$$

Notice that to find $P(5 \text{ or } 6)$, we could have simply added the *areas* of the bars over 5 and over 6. One out of 10 of the group who took the boredom tolerance test would qualify for the position at Topnotch Clothing.

GUIDED EXERCISE 2 | DISCRETE PROBABILITY DISTRIBUTION

A tool of cryptanalysis (science of code breaking) is to use relative frequencies of occurrence of letters to break codes. In addition to cryptanalysis, creation of word games also uses the technique. Oxford Dictionaries publishes dictionaries of English vocabulary. They did an analysis of letter frequencies in words listed in the main entries of the *Concise Oxford Dictionary* (11th edition). See **http://oxforddictionaries.com**. Suppose Langley took a random sample of 1000 words occurring in crossword puzzles. Table 5-3 shows the relative frequency of letters occurring in the sample.

(a) Use the relative frequencies to compute the omitted probabilities in Table 5-3.

→ (a) Table 5-4 shows the completion of Table 5-3.

Continued

GUIDED EXERCISE 2 *continued*

TABLE 5-3 **Frequencies of Letters in a 1000-Letter Sample**

Letter	Freq.	Prob.	Letter	Freq.	Prob.
A	85	____	N	66	0.066
B	21	0.021	O	72	____
C	45	0.045	P	32	0.032
D	34	0.034	Q	2	0.002
E	112	____	R	76	0.076
F	18	0.018	S	57	0.057
G	25	0.025	T	69	0.069
H	30	0.030	U	36	____
I	75	____	V	10	0.010
J	2	0.002	W	13	0.013
K	11	0.011	X	3	0.003
L	55	0.055	Y	18	0.018
M	30	0.030	Z	3	0.003

TABLE 5-4 **Entries for Table 5-3**

Letter	Relative Frequency	Probability
A	$\frac{85}{1000}$	0.085
E	$\frac{112}{1000}$	0.112
I	$\frac{75}{1000}$	0.075
O	$\frac{72}{1000}$	0.072
U	$\frac{36}{1000}$	0.036

(b) Do the probabilities of all the individual letters add up to 1?

→ (b) Yes

(c) If a letter is selected at random from a crossword puzzle, what is the probability The letter will be a vowel?

→ (c) If a letters is selected at random,

$$\begin{aligned} P(A, E, I, O, \text{or } U) &= P(A) + P(E) + P(I) + P(O) + P(U) \\ &= 0.085 + 0.112 + 0.075 + 0.072 + 0.036 \\ &= 0.380 \end{aligned}$$

Mean and standard deviation of a discrete probability distribution

A probability distribution can be thought of as a relative-frequency distribution based on a very large n. As such, it has a mean and standard deviation. If we are referring to the probability distribution of a *population*, then we use the Greek letters μ for the mean and σ for the standard deviation. When we see the Greek letters used, we know the information given is from the *entire population* rather than just a sample. If we have a sample probability distribution, we use $\bar{x}$ (x bar) and s, respectively, for the mean and standard deviation.

The **mean** and the **standard deviation of a discrete population probability distribution** are found by using these formulas:

$\mu = \Sigma xP(x)$; μ is called the expected value of x

$\sigma = \sqrt{\Sigma(x - \mu)^2 P(x)}$; σ is called the standard deviation of x

where x is the value of a random variable,
$P(x)$ is the probability of that variable, and
the sum Σ is taken for all the values of the random variable.

Note: μ is the *population mean* and σ is the underlying *population standard deviation* because the sum Σ is taken over *all* values of the random variable (i.e., the entire sample space).

Expected value

The mean of a probability distribution is often called the *expected value* of the distribution. This terminology reflects the idea that the mean represents a "central point" or "cluster point" for the entire distribution. Of course, the mean or expected value is an average value, and as such, it *need not be a point of the sample space.*

The standard deviation is often represented as a measure of *risk*. A larger standard deviation implies a greater likelihood that the random variable x is different from the expected value μ.

EXAMPLE 2

EXPECTED VALUE, STANDARD DEVIATION

Are we influenced to buy a product by an ad we saw on TV? National Infomercial Marketing Association determined the number of times *buyers* of a product had watched a TV infomercial *before* purchasing the product. The results are shown here:

Number of Times Buyers Saw Infomercial	1	2	3	4	5*
Percentage of Buyers	27%	31%	18%	9%	15%

*This category was 5 or more, but will be treated as 5 in this example.

Kelly-Mooney Photography/Encyclopedia/Corbis

We can treat the information shown as an estimate of the probability distribution because the events are mutually exclusive and the sum of the percentages is 100%. Compute the mean and standard deviation of the distribution.

SOLUTION: We put the data in the first two columns of a computation table and then fill in the other entries (see Table 5-5). The average number of times a buyer views the infomercial before purchase is

$$\mu = \Sigma xP(x) = 2.54 \text{ (sum of column 3)}$$

To find the standard deviation, we take the square root of the sum of column 6:

$$\sigma = \sqrt{\Sigma(x - \mu)^2 P(x)} \approx \sqrt{1.869} \approx 1.37$$

TABLE 5-5 Number of Times Buyers View Infomercial Before Making Purchase

x (number of viewings)	$P(x)$	$xP(x)$	$x - \mu$	$(x - \mu)^2$	$(x - \mu)^2 P(x)$
1	0.27	0.27	−1.54	2.372	0.640
2	0.31	0.62	−0.54	0.292	0.091
3	0.18	0.54	0.46	0.212	0.038
4	0.09	0.36	1.46	2.132	0.192
5	0.15	0.75	2.46	6.052	0.908
		$\mu = \Sigma xP(x) = 2.54$			$\Sigma(x - \mu)^2 P(x) = 1.869$

CALCULATOR NOTE Some calculators, including the TI-84Plus/TI-83Plus/TI-*n*spire (with TI-84Plus keypad) models, accept fractional frequencies. If yours does, you can get μ and σ directly by entering the outcomes in list L_1 with corresponding probabilities in list L_2. Then use **1-Var Stats L_1, L_2**.

GUIDED EXERCISE 3 EXPECTED VALUE

At a carnival, you pay \$2.00 to play a coin-flipping game with three fair coins. On each coin one side has the number 0 and the other side has the number 1. You flip the three coins at one time and you win \$1.00 for every 1 that appears on top. Are your expected earnings equal to the cost to play? We'll answer this question in several steps.

(a) In this game, the random variable of interest counts the number of 1's that show. What is the sample space for the values of this random variable?

→ The sample space is {0, 1, 2, 3}, since any of these numbers of 1's can appear.

(b) There are eight equally likely outcomes for throwing three coins. They are 000, 001, 010, 011, 100, 101, ______, and ______.

→ 110 and 111.

(c) Complete Table 5-6.

TABLE 5-6

Number of 1's, x	Frequency	$P(x)$	$xP(x)$
0	1	0.125	0
1	3	0.375	______
2	3	______	______
3	______	______	______

→ TABLE 5-7 **Completion of Table 5-6**

Number of 1's, x	Frequency	$P(x)$	$xP(x)$
0	1	0.125	0
1	3	0.375	0.375
2	3	0.375	0.750
3	1	0.125	0.375

(d) The expected value is the sum

$$\mu = \Sigma xP(x)$$

Sum the appropriate column of Table 5-6 to find this value. Are your expected earnings less than, equal to, or more than the cost of the game?

→ The expected value can be found by summing the last column of Table 5-7. The expected value is \$1.50. It cost \$2.00 to play the game; the expected value is less than the cost. The carnival is making money. In the long run, the carnival can expect to make an average of about 50 cents per player.

WHAT DOES A DISCRETE PROBABILITY DISTRIBUTION TELL US?

A discrete Probability distribution tell us

- the complete sample space on which the distribution is based.
- the corresponding probability of each event in the sample space.
- formulas tell us how to find the expected value μ and the standard deviation σ of the distribution

We have seen probability distributions of discrete variables and the formulas to compute the mean and standard deviation of a discrete population probability distribution. Probability distributions of continuous random variables are similar except that the probability assignments are made to intervals of values rather than to specific values of the random variable. We will see an important example of a discrete probability distribution, the binomial distribution, in the next section, and one of a continuous probability distribution in Chapter 6 when we study the normal distribution.

We conclude this section with some useful information about combining random variables.

LINEAR FUNCTIONS OF A RANDOM VARIABLE

Let a and b be any constants, and let x be a random variable. Then the new random variable $L = a + bx$ is called a *linear function of x*. Using some more advanced mathematics, the following can be proved.

Linear function of a random variable

Let x be a random variable with mean μ and standard deviation σ. Then the **linear function** $L = a + bx$ has mean, variance, and standard deviation as follows:

$$\mu_L = a + b\mu$$
$$\sigma_L^2 = b^2\sigma^2$$
$$\sigma_L = \sqrt{b^2\sigma^2} = |b|\sigma$$

LINEAR COMBINATIONS OF INDEPENDENT RANDOM VARIABLES

Linear combination of two independent random variables

Minimizing the variance of a linear combination of independent random variables is very useful. Problems 22 and 23 of this section show you how.

Suppose we have two random variables x_1 and x_2. These variables are *independent* if any event involving x_1 by itself is *independent* of any event involving x_2 by itself. Sometimes, we want to combine independent random variables and examine the mean and standard deviation of the resulting combination.

Let x_1 and x_2 be independent random variables, and let a and b be any constants. Then the new random variable $W = ax_1 + bx_2$ is called a *linear combination of x_1 and x_2*. Using some more advanced mathematics, the following can be proved.

Let x_1 and x_2 be independent random variables with respective means μ_1 and μ_2, and variances σ_1^2 and σ_2^2. For the **linear combination** $W = ax_1 + bx_2$, the mean, variance, and standard deviation are as follows:

$$\mu_W = a\mu_1 + b\mu_2$$
$$\sigma_W^2 = a^2\sigma_1^2 + b^2\sigma_2^2$$
$$\sigma_W = \sqrt{a^2\sigma_1^2 + b^2\sigma_2^2}$$

Note: The formula for the mean of a linear combination of random variables is valid regardless of whether the variables are independent. However, *the formulas for the variance and standard deviation are valid* only if x_1 and x_2 are *independent* random variables. In later work (Chapter 6 on), we will use independent random samples to ensure that the resulting variables (usually means, proportions, etc.) are statistically independent.

EXAMPLE 3

LINEAR COMBINATIONS OF INDEPENDENT RANDOM VARIABLES

Let x_1 and x_2 be independent random variables with respective means $\mu_1 = 75$ and $\mu_2 = 50$ and standard deviations $\sigma_1 = 16$ and $\sigma_2 = 9$.

(a) Let $L = 3 + 2x_1$. Compute the mean, variance, and standard deviation of L.

SOLUTION: L is a linear function of the random variable x_1. Using the formulas with $a = 3$ and $b = 2$, we have

$$\mu_L = 3 + 2\mu_1 = 3 + 2(75) = 153$$
$$\sigma_L^2 = 2^2\sigma_1^2 = 4(16)^2 = 1024$$
$$\sigma_L = |2|\sigma_1 = 2(16) = 32$$

Notice that the variance and standard deviation of the linear function are influenced only by the coefficient of x_1 in the linear function.

(b) Let $W = x_1 + x_2$. Find the mean, variance, and standard deviation of W.

SOLUTION: W is a linear combination of the independent random variables x_1 and x_2. Using the formulas with both a and b equal to 1, we have

$$\mu_W = \mu_1 + \mu_2 = 75 + 50 = 125$$
$$\sigma_W^2 = \sigma_1^2 + \sigma_2^2 = 16^2 + 9^2 = 337$$
$$\sigma_W = \sqrt{\sigma_1^2 + \sigma_2^2} = \sqrt{337} \approx 18.36$$

(c) Let $W = x_1 - x_2$. Find the mean, variance, and standard deviation of W.

SOLUTION: W is a linear combination of the independent random variables x_1 and x_2. Using the formulas with $a = 1$ and $b = -1$, we have

$$\mu_W = \mu_1 - \mu_2 = 75 - 50 = 25$$
$$\sigma_W^2 = 1^2\sigma_1^2 + (-1)^2\sigma_2^2 = 16^2 + 9^2 = 337$$
$$\sigma_W = \sqrt{\sigma_1^2 + \sigma_2^2} = \sqrt{337} \approx 18.36$$

(d) Let $W = 3x_1 - 2x_2$. Find the mean, variance, and standard deviation of W.

SOLUTION: W is a linear combination of the independent random variables x_1 and x_2. Using the formulas with $a = 3$ and $b = -2$, we have

$$\mu_W = 3\mu_1 - 2\mu_2 = 3(75) - 2(50) = 125$$
$$\sigma_W^2 = 3^2\sigma_1^2 + (-2)^2\sigma_2^2 = 9(16^2) + 4(9^2) = 2628$$
$$\sigma_W = \sqrt{2628} \approx 51.26$$

LOOKING FORWARD

Problem 24 of Section 9.1 shows how to find the mean, variance, and standard deviation of a linear combination of two *linearly dependent* random variables.

VIEWPOINT The Rosetta Project

Around 196 B.C., Egyptian priests inscribed a decree on a granite slab affirming the rule of 13-year-old Ptolemy V. The proclamation was in Egyptian hieroglyphics with a translation in a form of ancient Greek. By 1799, the meaning of Egyptian hieroglyphics had been lost for many centuries. However, Napoleon's troops discovered the granite slab (Rosetta Stone). Linguists then used the Rosetta Stone and their knowledge of ancient Greek to unlock the meaning of the Egyptian hieroglyphics.

Linguistic experts say that because of industrialization and globalization, by the year 2100 as many as 90% of the world's languages may be extinct. To help preserve some of these languages for future generations, 1000 translations of the first three chapters of Genesis have been inscribed in tiny text onto 3-inch nickel disks and encased in hardened glass balls that are expected to last at least 1000 years. Why Genesis? Because it is the most translated text in the world. The Rosetta Project is sending the disks to libraries and universities all over the world. It is very difficult to send information into the future. However, if in the year 2500 linguists are using the "Rosetta Disks" to unlock the meaning of a lost language, you may be sure they will use statistical methods of cryptanalysis (see Guided Exercise 2). To find out more about the Rosetta Project, visit the Brase/Brase statistics site at **http://www.cengage.com/statistics/brase11e** *and find the link to the Rosetta Project site.*

SECTION 5.1 PROBLEMS

1. *Statistical Literacy* Which of the following are continuous variables, and which are discrete?
(a) Number of traffic fatalities per year in the state of Florida
(b) Distance a golf ball travels after being hit with a driver
(c) Time required to drive from home to college on any given day
(d) Number of ships in Pearl Harbor on any given day
(e) Your weight before breakfast each morning

2. *Statistical Literacy* Which of the following are continuous variables, and which are discrete?
   (a) Speed of an airplane
   (b) Age of a college professor chosen at random
   (c) Number of books in the college bookstore
   (d) Weight of a football player chosen at random
   (e) Number of lightning strikes in Rocky Mountain National Park on a given day

3. *Statistical Literacy* Consider each distribution. Determine if it is a valid probability distribution or not, and explain your answer.

(a)

x	0	1	2
$P(x)$	0.25	0.60	0.15

(b)

x	0	1	2
$P(x)$	0.25	0.60	0.20

4. *Statistical Literacy* At state College all classes start on the hour, with the earliest start time at 7 A.M. and the latest at 8 P.M. A random sample of freshmen showed the percentages preferring the listed start times.

Start Time	7 or 8 A.M.	9,10,or 11A.M.	12 or 1 P.M.	1 P.M., or later	after 5 P.M.,
% preferring	10%	35%	28%	25%	15%

Can this information be used to make a discrete probability distribution? Explain.

5. *Statistical Literacy* Consider two discrete probability distribution with the same sample space and the same expected value. Are the standard deviations of the two distributions necessarily equal? Explain

6. *Statistical Literacy* Consider the probability distribution of a random variable x. Is the expected value of the distribution necessarily one of the possible values of x? Explain or give an example.

7. *Basic Computation: Expected Value and Standard Deviation* Consider the probability distribution shown in Problem 3(a). Compute the expected value and the standard deviation of the distribution.

8. *Basic Computation: Expected Value* For a fundraiser, 1000 raffle tickets are sold and the winner is chosen at random. There is only one prize, $500 in cash. You buy one ticket.
   (a) What is the probability you will win the prize of $500?
   (b) Your expected earnings can be found by multiplying the value of the prize by the probability you will win the prize. What are your expected earnings?
   (c) *Interpretation* If a ticket costs $2, what is the difference between your "costs" and "expected earnings"? How much are you effectively contributing to the fundraiser?

9. *Critical Thinking: Simulation* We can use the random-number table to simulate outcomes from a given discrete probability distribution. Jose plays basketball and has probability 0.7 of making a free-throw shot. Let x be the random variable that counts the number of successful shots out of 10 attempts. Consider the digits 0 through 9 of the random-number table. Since Jose has a 70% chance of making a shot, assign the digits 0 through 6 to "making a basket from the free throw line" and the digits 7 through 9 to "missing the shot."
   (a) Do 70% of the possible digits 0 through 9 represent "making a basket"?
   (b) Start at line 2, column 1 of the random-number table. Going across the row, determine the results of 10 "trials." How many free throw shots are successful in this simulation?

(c) Your friend decides to assign the digits 0 through 2 to "missing the shot" and the digits 3 through 9 to "making the basket." Is this assignment valid? Explain. Using this assignment, repeat part (b).

10. *Marketing: Age* What is the age distribution of promotion-sensitive shoppers? A *supermarket super shopper* is defined as a shopper for whom at least 70% of the items purchased were on sale or purchased with a coupon. The following table is based on information taken from *Trends in the United States* (Food Marketing Institute, Washington, D.C.).

Age range, years	18–28	29–39	40–50	51–61	62 and over
Midpoint x	23	34	45	56	67
Percent of super shoppers	7%	44%	24%	14%	11%

For the 62-and-over group, use the midpoint 67 years.

(a) Using the age midpoints x and the percentage of super shoppers, do we have a valid probability distribution? Explain.
(b) Use a histogram to graph the probability distribution of part (a).
(c) Compute the expected age μ of a super shopper.
(d) Compute the standard deviation σ for ages of super shoppers.

11. *Marketing: Income* What is the income distribution of super shoppers (see Problem 10). In the following table, income units are in thousands of dollars, and each interval goes up to but does not include the given high value. The midpoints are given to the nearest thousand dollars.

Income range	5–15	15–25	25–35	35–45	45–55	55 or more
Midpoint x	10	20	30	40	50	60
Percent of super shoppers	21%	14%	22%	15%	20%	8%

(a) Using the income midpoints x and the percent of super shoppers, do we have a valid probability distribution? Explain.
(b) Use a histogram to graph the probability distribution of part (a).
(c) Compute the expected income μ of a super shopper.
(d) Compute the standard deviation σ for the income of super shoppers.

12. *History: Florence Nightingale* What was the age distribution of nurses in Great Britain at the time of Florence Nightingale? Thanks to Florence Nightingale and the British census of 1851, we have the following information (based on data from the classic text *Notes on Nursing,* by Florence Nightingale). *Note:* In 1851 there were 25,466 nurses in Great Britain. Furthermore, Nightingale made a strict distinction between nurses and domestic servants.

Age range (yr)	20–29	30–39	40–49	50–59	60–69	70–79	80+
Midpoint x	24.5	34.5	44.5	54.5	64.5	74.5	84.5
Percent of nurses	5.7%	9.7%	19.5%	29.2%	25.0%	9.1%	1.8%

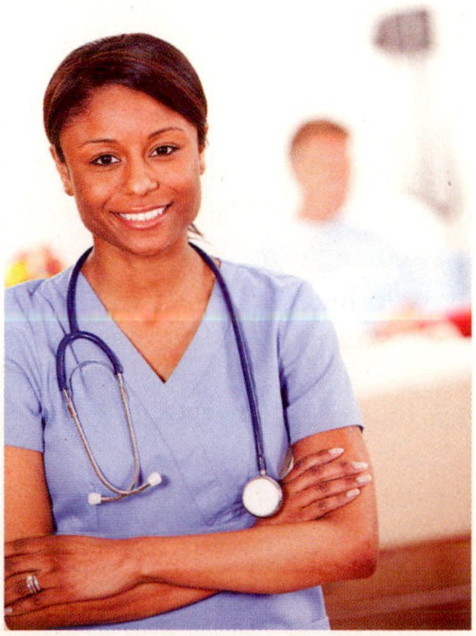
Sean Locke/iStockphoto.com

(a) Using the age midpoints x and the percent of nurses, do we have a valid probability distribution? Explain.
(b) Use a histogram to graph the probability distribution of part (a).
(c) Find the probability that a British nurse selected at random in 1851 was 60 years of age or older.
(d) Compute the expected age μ of a British nurse contemporary to Florence Nightingale.
(e) Compute the standard deviation σ for ages of nurses shown in the distribution.

13. *Fishing: Trout* The following data are based on information taken from *Daily Creel Summary,* published by the Paiute Indian Nation, Pyramid Lake, Nevada. Movie stars and U.S. presidents have fished Pyramid Lake. It is one of the best places in the lower 48 states to catch trophy cutthroat trout. In this table, x = number of fish caught in a 6-hour period. The percentage data are the percentages of fishermen who catch x fish in a 6-hour period while fishing from shore.

x	0	1	2	3	4 or more
%	44%	36%	15%	4%	1%

(a) Convert the percentages to probabilities and make a histogram of the probability distribution.
(b) Find the probability that a fisherman selected at random fishing from shore catches one or more fish in a 6-hour period.
(c) Find the probability that a fisherman selected at random fishing from shore catches two or more fish in a 6-hour period.
(d) Compute μ, the expected value of the number of fish caught per fisherman in a 6-hour period (round 4 or more to 4).
(e) Compute σ, the standard deviation of the number of fish caught per fisherman in a 6-hour period (round 4 or more to 4).

14. *Criminal Justice: Parole* *USA Today* reported that approximately 25% of all state prison inmates released on parole become repeat offenders while on parole. Suppose the parole board is examining five prisoners up for parole. Let x = number of prisoners out of five on parole who become repeat offenders. The methods of Section 5.2 can be used to compute the probability assignments for the x distribution:

x	0	1	2	3	4	5
$P(x)$	0.237	0.396	0.264	0.088	0.015	0.001

(a) Find the probability that one or more of the five parolees will be repeat offenders. How does this number relate to the probability that none of the parolees will be repeat offenders?
(b) Find the probability that two or more of the five parolees will be repeat offenders.
(c) Find the probability that four or more of the five parolees will be repeat offenders.
(d) Compute μ, the expected number of repeat offenders out of five.
(e) Compute σ, the standard deviation of the number of repeat offenders out of five.

15. *Fundraiser: Hiking Club* The college hiking club is having a fundraiser to buy new equipment for fall and winter outings. The club is selling Chinese fortune cookies at a price of $1 per cookie. Each cookie contains a piece of paper with a different number written on it. A random drawing will determine which number is the winner of a dinner for two at a local Chinese restaurant. The dinner is valued at $35. Since the fortune cookies were donated to the club, we can ignore the cost of the cookies. The club sold 719 cookies before the drawing.
(a) Lisa bought 15 cookies. What is the probability she will win the dinner for two? What is the probability she will not win?
(b) *Interpretation* Lisa's expected earnings can be found by multiplying the value of the dinner by the probability that she will win. What are Lisa's expected earnings? How much did she effectively contribute to the hiking club?

16. *Spring Break: Caribbean Cruise* The college student senate is sponsoring a spring break Caribbean cruise raffle. The proceeds are to be donated to the Samaritan Center for the Homeless. A local travel agency donated the cruise, valued at \$2000. The students sold 2852 raffle tickets at \$5 per ticket.
    (a) Kevin bought six tickets. What is the probability that Kevin will win the spring break cruise to the Caribbean? What is the probability that Kevin will not win the cruise?
    (b) *Interpretation* Expected earnings can be found by multiplying the value of the cruise by the probability that Kevin will win. What are Kevin's expected earnings? Is this more or less than the amount Kevin paid for the six tickets? How much did Kevin effectively contribute to the Samaritan Center for the Homeless?

17. *Expected Value: Life Insurance* Jim is a 60-year-old Anglo male in reasonably good health. He wants to take out a \$50,000 term (that is, straight death benefit) life insurance policy until he is 65. The policy will expire on his 65th birthday. The probability of death in a given year is provided by the Vital Statistics Section of the *Statistical Abstract of the United States* (116th Edition).

x = age	60	61	62	63	64
P(death at this age)	0.01191	0.01292	0.01396	0.01503	0.01613

Jim is applying to Big Rock Insurance Company for his term insurance policy.
    (a) What is the probability that Jim will die in his 60th year? Using this probability and the \$50,000 death benefit, what is the expected cost to Big Rock Insurance?
    (b) Repeat part (a) for years 61, 62, 63, and 64. What would be the total expected cost to Big Rock Insurance over the years 60 through 64?
    (c) *Interpretation* If Big Rock Insurance wants to make a profit of \$700 above the expected total cost paid out for Jim's death, how much should it charge for the policy?
    (d) *Interpretation* If Big Rock Insurance Company charges \$5000 for the policy, how much profit does the company expect to make?

18. *Expected Value: Life Insurance* Sara is a 60-year-old Anglo female in reasonably good health. She wants to take out a \$50,000 term (that is, straight death benefit) life insurance policy until she is 65. The policy will expire on her 65th birthday. The probability of death in a given year is provided by the Vital Statistics Section of the *Statistical Abstract of the United States* (116th Edition).

x = age	60	61	62	63	64
P(death at this age)	0.00756	0.00825	0.00896	0.00965	0.01035

Sara is applying to Big Rock Insurance Company for her term insurance policy.
    (a) What is the probability that Sara will die in her 60th year? Using this probability and the \$50,000 death benefit, what is the expected cost to Big Rock Insurance?
    (b) Repeat part (a) for years 61, 62, 63, and 64. What would be the total expected cost to Big Rock Insurance over the years 60 through 64?
    (c) *Interpretation* If Big Rock Insurance wants to make a profit of \$700 above the expected total cost paid out for Sara's death, how much should it charge for the policy?
    (d) *Interpretation* If Big Rock Insurance Company charges \$5000 for the policy, how much profit does the company expect to make?

19. *Combination of Random Variables: Golf* Norb and Gary are entered in a local golf tournament. Both have played the local course many times. Their scores are random variables with the following means and standard deviations.

$$\text{Norb}, x_1 : \mu_1 = 115; \sigma_1 = 12 \qquad \text{Gary}, x_2 : \mu_2 = 100; \sigma_2 = 8$$

Ranplett/iStockphoto.com

In the tournament, Norb and Gary are not playing together, and we will assume their scores vary independently of each other.

(a) The difference between their scores is $W = x_1 - x_2$. Compute the mean, variance, and standard deviation for the random variable W.
(b) The average of their scores is $W = 0.5x_1 + 0.5x_2$. Compute the mean, variance, and standard deviation for the random variable W.
(c) The tournament rules have a special handicap system for each player. For Norb, the handicap formula is $L = 0.8x_1 - 2$. Compute the mean, variance, and standard deviation for the random variable L.
(d) For Gary, the handicap formula is $L = 0.95x_2 - 5$. Compute the mean, variance, and standard deviation for the random variable L.

20. *Combination of Random Variables: Repair Service* A computer repair shop has two work centers. The first center examines the computer to see what is wrong, and the second center repairs the computer. Let x_1 and x_2 be random variables representing the lengths of time in minutes to examine a computer (x_1) and to repair a computer (x_2). Assume x_1 and x_2 are independent random variables. Long-term history has shown the following times:

$$\text{Examine computer}, x_1 : \mu_1 = 28.1 \text{ minutes}; \sigma_1 = 8.2 \text{ minutes}$$
$$\text{Repair computer}, x_2 : \mu_2 = 90.5 \text{ minutes}; \sigma_2 = 15.2 \text{ minutes}$$

(a) Let $W = x_1 + x_2$ be a random variable representing the total time to examine and repair the computer. Compute the mean, variance, and standard deviation of W.
(b) Suppose it costs \$1.50 per minute to examine the computer and \$2.75 per minute to repair the computer. Then $W = 1.50x_1 + 2.75x_2$ is a random variable representing the service charges (without parts). Compute the mean, variance, and standard deviation of W.
(c) The shop charges a flat rate of \$1.50 per minute to examine the computer, and if no repairs are ordered, there is also an additional \$50 service charge. Let $L = 1.5x_1 + 50$. Compute the mean, variance, and standard deviation of L.

21. *Combination of Random Variables: Insurance Risk* Insurance companies know the *risk* of insurance is greatly reduced if the company insures not just one person, but many people. How does this work? Let x be a random variable representing the expectation of life in years for a 25-year-old male (i.e., number of years until death). Then the mean and standard deviation of x are $\mu = 50.2$ years and $\sigma = 11.5$ years (Vital Statistics Section of the *Statistical Abstract of the United States,* 116th Edition).

Suppose Big Rock Insurance Company has sold life insurance policies to Joel and David. Both are 25 years old, unrelated, live in different states, and have about the same health record. Let x_1 and x_2 be random variables representing Joel's and David's life expectancies. It is reasonable to assume x_1 and x_2 are independent.

hfng/Shutterstock.com

$$\text{Joel}, x_1 : \mu_1 = 50.2; \sigma_1 = 11.5$$

$$\text{David}, x_2 : \mu_2 = 50.2; \sigma_2 = 11.5$$

If life expectancy can be predicted with more accuracy, Big Rock will have less risk in its insurance business. Risk in this case is measured by σ (larger σ means more risk).

Mark Aplet/Shutterstock.com

(a) The average life expectancy for Joel and David is $W = 0.5x_1 + 0.5x_2$. Compute the mean, variance, and standard deviation of W.

(b) Compare the mean life expectancy for a single policy (x_1) with that for two policies (W).

(c) Compare the standard deviation of the life expectancy for a single policy (x_1) with that for two policies (W).

(d) The mean life expectancy is the same for a single policy (x_1) as it is for two policies (W), but the standard deviation is smaller for two policies. What happens to the mean life expectancy and the standard deviation when we include more policies issued to people whose life expectancies have the same mean and standard deviation (i.e., 25-year-old males)? For instance, for three policies, $W = (\mu + \mu + \mu)/3 = \mu$ and $\sigma_W^2 = (1/3)^2\sigma^2 + (1/3)^2\sigma^2 + (1/3)^2\sigma^2 = (1/3)^2(3\sigma^2) = (1/3)\sigma^2$ and $\sigma_W = \frac{1}{\sqrt{3}}\sigma$. Likewise, for n such policies, $W = \mu$ and $\sigma_W^2 = (1/n)\sigma^2$ and $\sigma_W = \frac{1}{\sqrt{n}}\sigma$. Looking at the general result, is it appropriate to say that when we increase the number of policies to n, the risk decreases by a factor of $\sigma_W = \frac{1}{\sqrt{n}}$?

22. *Expand Your Knowledge: Minimal Variance; How to make a Smoother Blend—Superglue* This problem shows you how to make a better blend of almost anything.

Let $x_1, x_2, \ldots x_n$ be *independent* random variables with respective variances $\sigma_1^2, \sigma_2^2, \ldots, \sigma_n^2$.

Let $c_1, c_2, \ldots, c_n$ be constant weights such that $0 \le c_i \le 1$ and $c_1 + c_2 + \ldots + c_n = 1$. The linear combination $w = c_1x_1 + c_2x_2 + \ldots + c_nx_n$ is a random variable with variance

$$\sigma_w^2 = c_1^2\sigma_1^2 + c_2^2\sigma_2^2 + \cdots + c_n^2\sigma_n^2$$

(a) In your own words write a brief explanation regarding the following statement: The variance of w is a measure of the consistency or variability of performance or outcomes of the random variable w. To get a more consistent performance out of the blend w, choose weights c_i that make σ_w^2 as small as possible.

Now the question is how do we choose weights c_i to make σ_w^2 as small as possible? Glad you asked! A lot of mathematics can be used to show the following choice of weights will minimize σ_w^2.

$$c_i = \frac{\frac{1}{\sigma_i^2}}{\left[\frac{1}{\sigma_1^2} + \frac{1}{\sigma_2^2} + \cdots + \frac{1}{\sigma_n^2}\right]} \qquad \text{for } i = 1, 2, \cdots, n$$

Reference: *Introduction to Mathematical Statistics* 4th edition by Paul Hoel.

(b) Two types of epoxy resin are used to make a new blend of superglue. Both resins have about the same mean breaking strength and act independently. The question is how to blend the resins (with the hardener) to get the most consistent breaking strength. Why is this important, and why would this require minimal σ_w^2? *Hint:* We don't want some bonds to be really strong while others are very weak, resulting in inconsistent bonding.

Let x_1 and x_2 be random variables representing breaking strength (lb) of each resin under uniform testing conditions. If $\sigma_1 = 8$ lb $\sigma_2 = 12$ lb, show why a blend of about 69% resin 1 and 31% resin 2 will result in a superglue with smallest σ_w^2 and most consistent bond strength.

(c) Use $c_1 = 0.69$ and $c_2 = 0.31$ to compute σ_w and show that σ_w is less than both σ_1 and σ_2. The dictionary meaning of the word *synergetic* is "working together or cooperating for a better overall effect." Write a brief explanation of how the blend $w = c_1x_1 + c_2x_2$ has a synergetic effect for the purpose of reducing variance.

23. *Expand Your Knowledge: Minimal Variance; How to make a Smoother Blend—Jet Fuel* Three grades of jet fuel are blended to be used on long commercial flights. Each grade cost about the same and has about the same ignition or burn rate (gal/min) in a jet engine. The question is how to make the blend for the most consistent ignition rate. Assume the ignition rates of the three grades of fuel are statistically independent.
    (a) Write a brief explanation in which you discuss why consistent ignition rate is important. *Hint:* You sure do not want to run out of fuel up in the air! Also you don't want to land with a lot of extra fuel instead of payload cargo.
    (b) Let x_1, x_2, x_3 be random variables representing ignition rate (gal/min) for the three grades of fuel. After extensive testing (flying under different conditions) the standard deviations are found to be $\sigma_1 = 7$, $\sigma_2 = 12$ and $\sigma_3 = 15$. Explain why a blend of approximately 64% grade 1 fuel with 22% grade 2 fuel and 14% grade 3 fuel would give the most consistent ignition rate. *Hint:* See Problem 22.
    (c) Use $c_1 = 0.64$, $c_2 = 0.22$, and $c_3 = 0.14$ to compute σ_w and show that σ_w is less than each of σ_1, σ_2 and σ_3. Would you say the blended jet fuel has a more consistent ignition rate than each separate fuel? Explain.

SECTION **5.2**

Binomial Probabilities

FOCUS POINTS

- List the defining features of a binomial experiment.
- Compute binomial probabilities using the formula $P(r) = C_{n,r}p^rq^{n-r}$.
- Use the binomial table to find $P(r)$.
- Use the binomial probability distribution to solve real-world applications.

BINOMIAL EXPERIMENT

On a TV game show, each contestant has a try at the wheel of fortune. The wheel of fortune is a roulette wheel with 36 slots, one of which is gold. If the ball lands in the gold slot, the contestant wins $50,000. No other slot pays. What is the probability that the game show will have to pay the fortune to three contestants out of 100?

In this problem, the contestant and the game show sponsors are concerned about only two outcomes from the wheel of fortune: The ball lands on the gold, or the ball does not land on the gold. This problem is typical of an entire class of problems that are characterized by the feature that there are exactly two possible outcomes (for each trial) of interest. These problems are called *binomial experiments,* or *Bernoulli experiments,* after the Swiss mathematician Jacob Bernoulli, who studied them extensively in the late 1600s.

Binomial experiments

Features of a binomial experiment
Number of trials, n
Independent trials
Success, S
Failure, F

$P(S) = p$
$P(F) = q = 1 - p$

Number of successes, r

FEATURES OF A BINOMIAL EXPERIMENT

1. There is a *fixed number of trials.* We denote this number by the letter n.
2. The n trials are *independent* and repeated under identical conditions.
3. Each trial has only *two outcomes:* success, denoted by S, and failure, denoted by F.
4. For each individual trial, the *probability of success is the same.* We denote the probability of success by p and that of failure by q. Since each trial results in either success or failure, $p + q = 1$ and $q = 1 - p$.
5. The central problem of a binomial experiment is to find the *probability of r successes out of n trials.*

EXAMPLE 4

BINOMIAL EXPERIMENT

Let's see how the wheel of fortune problem meets the criteria of a binomial experiment. We'll take the criteria one at a time.

SOLUTION:

1. Each of the 100 contestants has a trial at the wheel, so there are $n = 100$ trials in this problem.
2. Assuming that the wheel is fair, the *trials are independent,* since the result of one spin of the wheel has no effect on the results of other spins.
3. We are interested in only two outcomes on each spin of the wheel: The ball either lands on the gold, or it does not. Let's call landing on the gold *success* (S) and not landing on the gold *failure* (F). In general, the assignment of the terms *success* and *failure* to outcomes does not imply good or bad results. These terms are assigned simply for the user's convenience.
4. On each trial the probability p of success (landing on the gold) is 1/36, since there are 36 slots and only one of them is gold. Consequently, the probability of failure is

$$q = 1 - p = 1 - \frac{1}{36} = \frac{35}{36}$$

on each trial.

5. We want to know the probability of 3 successes out of 100 trials, so $r = 3$ in this example. It turns out that the probability the quiz show will have to pay the fortune to 3 contestants out of 100 is about 0.23. Later in this section we'll see how this probability was computed.

Anytime we make selections from a population *without replacement, we do not have independent trials.* However, replacement is often not practical. If the number of trials is quite small with respect to the population, we *almost* have independent trials, and we can say the situation is *closely approximated* by a binomial experiment. For instance, suppose we select 20 tuition bills at random from a collection of 10,000 bills issued at one college and observe if each bill is in error or not. If 600 of the 10,000 bills are in error, then the probability that the first one selected is in error is 600/10,000, or 0.0600. If the first is in error, then the probability that the second is in error is 599/9999, or 0.0599. Even if the first 19 bills selected are in error, the probability that the 20th is also in error is 581/9981, or 0.0582. All these probabilities round to 0.06, and we can say that the independence condition is approximately satisfied.

GUIDED EXERCISE 4

BINOMIAL EXPERIMENT

Let's analyze the following binomial experiment to determine *p, q, n,* and *r*:

According to the *Textbook of Medical Physiology,* 5th Edition, by Arthur Guyton, 9% of the population has blood type B. Suppose we choose 18 people at random from the population and test the blood type of each. What is the probability that three of these people have blood type B? *Note:* Independence is approximated because 18 people is an extremely small sample with respect to the entire population.

(a) In this experiment, we are observing whether or not a person has type B blood. We will say we have a success if the person has type B blood. What is failure?

→ Failure occurs if a person does not have type B blood.

Continued

GUIDED EXERCISE 4 *continued*

(b) The probability of success is 0.09, since 9% of the population has type B blood. What is the probability of failure, q?	The probability of failure is $q = 1 - p$ $= 1 - 0.09 = 0.91$
(c) In this experiment, there are $n =$ ________ trials.	In this experiment, $n = 18$.
(d) We wish to compute the probability of 3 successes out of 18 trials. In this case, $r =$ ________.	In this case, $r = 3$.

Next, we will see how to compute the probability of r successes out of n trials when we have a binomial experiment.

COMPUTING PROBABILITIES FOR A BINOMIAL EXPERIMENT USING THE BINOMIAL DISTRIBUTION FORMULA

The central problem of a binomial experiment is finding the probability of r successes out of n trials. Now we'll see how to find these probabilities.

A model with three trials

Suppose you are taking a timed final exam. You have three multiple-choice questions left to do. Each question has four suggested answers, and only one of the answers is correct. You have only 5 seconds left to do these three questions, so you decide to mark answers on the answer sheet without even reading the questions. Assuming that your answers are randomly selected, what is the probability that you get zero, one, two, or all three questions correct?

This is a binomial experiment. Each question can be thought of as a trial, so there are $n = 3$ trials. The possible outcomes on each trial are success S, indicating a correct response, or failure F, meaning a wrong answer. The trials are independent—the outcome of any one trial does not affect the outcome of the others.

Probability of success $P(S) = p$

What is the *probability of success* on anyone question? Since you are guessing and there are four answers from which to select, the probability of a correct answer is 0.25. The probability q of a wrong answer is then 0.75. In short, we have a binomial experiment with $n = 3$, $p = 0.25$, and $q = 0.75$.

Now, what are the possible outcomes in terms of success or failure for these three trials? Let's use the notation SSF to mean success on the first question, success on the second, and failure on the third. There are eight possible combinations of S's and F's. They are

$$SSS \quad SSF \quad SFS \quad FSS \quad SFF \quad FSF \quad FFS \quad FFF$$

To compute the probability of each outcome, we can use the multiplication law because the trials are independent. For instance, the probability of success on the first two questions and failure on the last is

$$P(SSF) = P(S) \cdot P(S) \cdot P(F) = p \cdot p \cdot q = p^2q = (0.25)^2(0.75) \approx 0.047$$

In a similar fashion, we can compute the probability of each of the eight outcomes. These are shown in Table 5-8, along with the number of successes r associated with each outcome.

TABLE 5-8 **Outcomes for a Binomial Experiment with $n = 3$ Trials**

Outcome	Probability of Outcome	r (number of successes)
SSS	$P(SSS) = P(S)P(S)P(S) = p^3 = (0.25)^3 \approx 0.016$	3
SSF	$P(SSF) = P(S)P(S)P(F) = p^2q = (0.25)^2(0.75) \approx 0.047$	2
SFS	$P(SFS) = P(S)P(F)P(S) = p^2q = (0.25)^2(0.75) \approx 0.047$	2
FSS	$P(FSS) = P(F)P(S)P(S) = p^2q = (0.25)^2(0.75) \approx 0.047$	2
SFF	$P(SFF) = P(S)P(F)P(F) = pq^2 = (0.25)(0.75)^2 \approx 0.141$	1
FSF	$P(FSF) = P(F)P(S)P(F) = pq^2 = (0.25)(0.75)^2 \approx 0.141$	1
FFS	$P(FFS) = P(F)P(F)P(S) = pq^2 = (0.25)(0.75)^2 \approx 0.141$	1
FFF	$P(FFF) = P(F)P(F)P(F) = q^3 = (0.75)^3 \approx 0.422$	0

Now we can compute the probability of r successes out of three trials for $r = 0, 1, 2$, or 3. Let's compute $P(1)$. The notation $P(1)$ stands for the probability of one success. For three trials, there are three different outcomes that show exactly one success. They are the outcomes *SFF, FSF,* and *FFS*. Since the outcomes are mutually exclusive, we can add the probabilities. So,

$$\begin{aligned} P(1) = P(SFF \text{ or } FSF \text{ or } FFS) &= P(SFF) + P(FSF) + P(FFS) \\ &= pq^2 + pq^2 + pq^2 \\ &= 3pq^2 \\ &= 3(0.25)(0.75)^2 \\ &= 0.422 \end{aligned}$$

In the same way, we can find $P(0)$, $P(2)$, and $P(3)$. These values are shown in Table 5-9.

TABLE 5-9 **$P(r)$ for $n = 3$ Trials, $p = 0.25$**

r (number of successes)	$P(r)$ (probability of r successes in 3 trials)	$P(r)$ for $p = 0.25$
0	$P(0) = P(FFF) = q^3$	0.422
1	$P(1) = P(SFF) + P(FSF) + P(FFS) = 3pq^2$	0.422
2	$P(2) = P(SSF) + P(SFS) + P(FSS) = 3p^2q$	0.141
3	$P(3) = P(SSS) = p^3$	0.016

We have done quite a bit of work to determine your chances of $r = 0, 1, 2$, or 3 successes on three multiple-choice questions if you are just guessing. Now we see that there is only a small chance (about 0.016) that you will get them all correct.

Table 5-9 can be used as a model for computing the probability of r successes out of only *three* trials. How can we compute the probability of 7 successes out of 10 trials? We can develop a table for $n = 10$, but this would be a tremendous task because there are 1024 possible combinations of successes and failures on 10 trials. Fortunately, mathematicians have given us a direct formula to compute the probability of r successes for any number of trials.

General formula for binomial probability distribution

FORMULA FOR THE BINOMIAL PROBABILITY DISTRIBUTION

$$P(r) = \frac{n!}{r!(n-r)!}p^r q^{n-r} = C_{n,r}p^r q^{n-r}$$

where n = number of trials

p = probability of success on each trial

$q = 1 - p$ = probability of failure on each trial

r = random variable representing the number of successes out of n trials ($0 \le r \le n$)

$!$ = factorial notation. Recall from Section 4.3 that the factorial symbol $n!$ designates the product of all the integers between 1 and n. For instance, $4! = 4 \cdot 3 \cdot 2 \cdot 1 = 24$. Special cases are $1! = 1$ and $0! = 1$.

Table for $C_{n,r}$

$C_{n,r} = \frac{n!}{r!(n-r)!}$ is the binomial coefficient. Table 2 of Appendix II gives values of $C_{n,r}$ for select n and r. Many calculators have a key designated nCr that gives the value of $C_{n,r}$ directly.

Binomial coefficient $C_{n,r}$

Note: The binomial coefficient $C_{n,r}$ represents the number of combinations of n distinct objects (n = number of trials in this case) taken r at a time (r = number of successes). For more information about $C_{n,r}$, see Section 4.3.

Let's look more carefully at the formula for $P(r)$. There are two main parts. The expression $p^r q^{n-r}$ is the probability of getting one outcome with r successes and $n - r$ failures. The binomial coefficient $C_{n,r}$ counts the number of outcomes that have r successes and $n - r$ failures. For instance, in the case of $n = 3$ trials, we saw in Table 5-8 that the probability of getting an outcome with one success and two failures was pq^2. This is the value of $p^r q^{n-r}$ when $r = 1$ and $n = 3$. We also observed that there were three outcomes with one success and two failures, so $C_{3,1}$ is 3.

Now let's take a look at an application of the binomial distribution formula in Example 5.

EXAMPLE 5

COMPUTE $P(r)$ USING THE BINOMIAL DISTRIBUTION FORMULA

Privacy is a concern for many users of the Internet. One survey showed that 59% of Internet users are somewhat concerned about the confidentiality of their e-mail. Based on this information, what is the probability that for a random sample of 10 Internet users, 6 are concerned about the privacy of their e-mail?

SOLUTION:

(a) This is a binomial experiment with 10 trials. If we assign success to an Internet user being concerned about the privacy of e-mail, the probability of success is 59%. We are interested in the probability of 6 successes. We have

$$n = 10 \quad p = 0.59 \quad q = 0.41 \quad r = 6$$

By the formula,

$$P(6) = C_{10,6}(0.59)^6(0.41)^{10-6}$$

$$= 210(0.59)^6(0.41)^4 \qquad \text{Use Table 2 of Appendix II or a calculator.}$$

$$\approx 210(0.0422)(0.0283) \qquad \text{Use a calculator.}$$

$$\approx 0.25$$

FIGURE 5-2

TI-84Plus/TI-83Plus/TI-*n*spire (with TI-84Plus keypad) Display

```
10 nCr 6*.59^6*.
41^(10-6)
         .2503034245
```

There is a 25% chance that *exactly* 6 of the 10 Internet users are concerned about the privacy of e-mail.

(b) Many calculators have a built-in combinations function. On the TI-84Plus/TI-83Plus/TI-*n*spire (with TI-84Plus keypad) calculators, press the **MATH** key and select **PRB**. The combinations function is designated nCr. Figure 5-2 displays the process for computing $P(6)$ directly on these calculators.

USING A BINOMIAL DISTRIBUTION TABLE

In many cases we will be interested in the probability of a range of successes. In such cases, we need to use the addition rule for mutually exclusive events. For instance, for $n = 6$ and $p = 0.50$,

$$P(4 \textit{ or fewer } \text{successes}) = P(r \leq 4)$$
$$= P(r = 4 \textit{ or } 3 \textit{ or } 2 \textit{ or } 1 \textit{ or } 0)$$
$$= P(4) + P(3) + P(2) + P(1) + P(0)$$

It would be a bit of a chore to use the binomial distribution formula to compute all the required probabilities. Table 3 of Appendix II gives values of $P(r)$ for selected p values and values of n through 20. To use the table, find the appropriate section for n, and then use the entries in the columns headed by the p values and the rows headed by the r values.

Table 5-10 is an excerpt from Table 3 of Appendix II showing the section for $n = 6$. Notice that all possible r values between 0 and 6 are given as row headers. The value $p = 0.50$ is one of the column headers. For $n = 6$ and $p = 0.50$, you can find the value of $P(4)$ by looking at the entry in the row headed by 4 and the column headed by 0.50. Notice that $P(4) = 0.234$.

TABLE 5-10 Excerpt from Table 3 of Appendix II for $n = 6$

							p								
n	*r*	.01	.05	.10	...	.30	...	.50	...	.70	...	.85	.90	.95
⋮														
6	0	.941	.735	.531	...	.118	...	.016	...	.001	...	.000	.000	.000
	1	.057	.232	.354	...	.303	...	.094	...	.010	...	.000	.000	.000
	2	.001	.031	.098	...	.324	...	.234	...	.060	...	.006	.001	.000
	3	.000	.002	.015	...	.185	...	.312	...	.185	...	.042	.015	.002
	4	.000	.000	.001	...	.060	...	.234	...	.324		.176	.098	.031
	5	.000	.000	.000	...	.010	...	.094	...	.303	...	.399	.354	.232
	6	.000	.000	.000	...	.001	...	.016	...	.118	...	.377	.531	.735

Likewise, you can find other values of $P(r)$ from the table. In fact, for $n = 6$ and $p = 0.50$,

$$P(r \leq 4) = P(4) + P(3) + P(2) + P(1) + P(0)$$
$$= 0.234 + 0.312 + 0.234 + 0.094 + 0.016 = 0.890$$

Alternatively, to compute $P(r \leq 4)$ for $n = 6$, you can use the fact that the total of all $P(r)$ values for r between 0 and 6 is 1 and the complement rule. Since the complement of the event $r \leq 4$ is the event $r \geq 5$, we have

$$P(r \leq 4) = 1 - P(5) - P(6)$$
$$= 1 - 0.094 - 0.016 = 0.890$$

Note: In Table 3 of Appendix II, probability entries of 0.000 do not mean the probability is exactly zero. Rather, to three digits after the decimal, the probability rounds to 0.000.

CRITICAL THINKING

In Chapter 4, we saw the complement rule of probability. As we saw in the previous discussion, this rule provides a useful strategy to simplify binomial probability computations for a range of successes. For example, in a binomial experiment with $n = 7$ trials, the sample space for the number of successes r is

0 1 2 3 4 5 6 7

Notice that $r = 0$ successes is the complement of $r \geq 1$ successes. By the complement rule,

$$P(r \geq 1) = 1 - P(r = 0)$$

It is faster to compute or look up $P(r = 0)$ and subtract than it is to compute or look up all the probabilities $P(r = 1)$ through $P(r = 7)$.

Likewise, the outcome $r \leq 2$ is the complement of the outcome $r \geq 3$, so,

$$P(r \geq 3) = 1 - P(r \leq 2) = 1 - P(r = 2) - P(r = 1) - P(r = 0)$$

Before you use the complement rule for computing probabilities, be sure the outcomes used comprise complementary events. Complements can be expressed in several ways. For instance, the complement of the event $r \leq 2$ can be expressed as event $r > 2$ or event $r \geq 3$.

Because of rounding in the binomial probability table, probabilities computed by using the addition rule directly might differ slightly from corresponding probabilities computed by using the complement rule.

EXAMPLE 6

Using the Binomial Distribution Table to Find $P(r)$

A biologist is studying a new hybrid tomato. It is known that the seeds of this hybrid tomato have probability 0.70 of germinating. The biologist plants six seeds.

(a) What is the probability that *exactly* four seeds will germinate?

SOLUTION: This is a binomial experiment with $n = 6$ trials. Each seed planted represents an independent trial. We'll say germination is success, so the probability for success on each trial is 0.70.

$$n = 6 \quad p = 0.70 \quad q = 0.30 \quad r = 4$$

We wish to find P(4), the probability of exactly four successes.

In Table 3, Appendix II, find the section with $n = 6$ (excerpt is given in Table 5-10). Then find the entry in the column headed by $p = 0.70$ and the row headed by $r = 4$. This entry is 0.324.

$$P(4) = 0.324$$

Christina Richards/Shutterstock.com

(b) What is the probability that *at least* four seeds will germinate?

SOLUTION: In this case, we are interested in the probability of four or more seeds germinating. This means we are to compute $P(r \geq 4)$. Since the events are mutually exclusive, we can use the addition rule

$$P(r \geq 4) = P(r = 4 \text{ or } r = 5 \text{ or } r = 6) = P(4) + P(5) + P(6)$$

We already know the value of $P(4)$. We need to find $P(5)$ and $P(6)$.

Use the same part of the table but find the entries in the row headed by the r value 5 and then the r value 6. Be sure to use the column headed by the value of p, 0.70.

$$P(5) = 0.303 \text{ and } P(6) = 0.118$$

Now we have all the parts necessary to compute $P(r \geq 4)$.

$$\begin{aligned} P(r \geq 4) &= P(4) + P(5) + P(6) \\ &= 0.324 + 0.303 + 0.118 \\ &= 0.745 \end{aligned}$$

In Guided Exercise 5 you'll practice using the formula for $P(r)$ in one part, and then you'll use Table 3, Appendix II, for $P(r)$ values in the second part.

GUIDED EXERCISE 5 | FIND $P(r)$

A rarely performed and somewhat risky eye operation is known to be successful in restoring the eyesight of 30% of the patients who undergo the operation. A team of surgeons has developed a new technique for this operation that has been successful in four of six operations. Does it seem likely that the new technique is much better than the old? We'll use the binomial probability distribution to answer this question. We'll compute the probability of at least four successes in six trials for the old technique.

(a) Each operation is a binomial trial. In this case, $n =$ ____, $p =$ ____, $q =$ ____, $r =$ ____.

→ $n = 6$, $p = 0.30$, $q = 1 - 0.30 = 0.70$, $r = 4$

(b) Use your values of n, p, and q, as well as Table 2 of Appendix II (or your calculator), to compute $P(4)$ from the formula:
$P(r) = C_{n,r}p^r q^{n-r}$

→
$$\begin{aligned} P(4) &= C_{6,4}(0.30)^4(0.70)^2 \\ &= 15(0.0081)(0.490) \\ &\approx 0.060 \end{aligned}$$

(c) Compute the probability of *at least* four successes out of the six trials.

$$\begin{aligned} P(r \geq 4) &= P(r = 4 \text{ or } r = 5 \text{ or } r = 6) \\ &= P(4) + P(5) + P(6) \end{aligned}$$

Use Table 3 of Appendix II to find values of $P(4)$, $P(5)$, and $P(6)$. Then use these values to compute $P(r \geq 4)$.

→ To find $P(4)$, $P(5)$, and $P(6)$ in Table 3, we look in the section labeled $n = 6$. Then we find the column headed by $p = 0.30$. To find $P(4)$, we use the row labeled $r = 4$. For the values of $P(5)$ and $P(6)$, we look in the same column but in the rows headed by $r = 5$ and $r = 6$, respectively.

$$\begin{aligned} P(r \geq 4) &= P(4) + P(5) + P(6) \\ &= 0.060 + 0.010 + 0.001 = 0.071 \end{aligned}$$

(d) *Interpretation* Under the older operation technique, the probability that at least four patients out of six regain their eyesight is _______. Does it seem that the new technique is better than the old? Would you encourage the surgeon team to do more work on the new technique?

→ It seems the new technique is better than the old since, by pure chance, the probability of four or more successes out of six trials is only 0.071 for the old technique. This means one of the following two things may be happening:

(i) The new method is no better than the old method, and our surgeons have encountered a rare event (probability 0.071), or

(ii) The new method is in fact better. We think it is worth encouraging the surgeons to do more work on the new technique.

WHAT DOES A BINOMIAL PROBABILITY DISTRIBUTION TELL US?

When a binomial probability distribution is used, we know

- the sample space of events consists of a fixed number n of binomial trials. This means the trials are independent and performed under the same conditions. There are only two outcomes for each trial, success and failure. The probability of success on each trail is the same and is designated p The probability of failure on each trail is designated q and $q = 1 - p$.
- the binomial probability distribution gives the values of $P(r)$, the probability of r successes out of the n trials for each r between (and including) 0 and n
- the calculation of $P(r)$ depends on the probability p of success on each trial, the value of r, and the value of n.

USING TECHNOLOGY TO COMPUTE BINOMIAL PROBABILITIES

Cumulative probability

Some calculators and computer software packages support the binomial distribution. In general, these technologies will provide both the probability $P(r)$ for an exact number of successes r and the *cumulative probability* $P(r \leq k)$, where k is a specified value less than or equal to the number of trials n. Note that most of the technologies use the letter x instead of r for the random variable denoting the number of successes out of n trials.

TECH NOTES

The software packages Minitab and Excel 2010, as well as the TI-84Plus/TI-83Plus/TI-*n*spire calculators, include built-in binomial probability distribution options. These options give the probability $P(r)$ of a specific number of successes r, as well as the cumulative total probability for r or fewer successes.

TI-84Plus/TI-83Plus/TI-*n*spire (with TI-84Plus keypad) Press the **DISTR** key and scroll to **binompdf** (n, p, r). Enter the number of trials n, the probability of success on a single trial p, and the number of successes r. This gives $P(r)$. For the cumulative probability that there are r or fewer successes, use **binomcdf** (n, p, r).

	binompdf(6,.3,4)
$P(r = 4)$	.059535
	binomcdf(6,.3,4)
$P(r \leq 4)$	.989065

Excel 2010 Click the **Insert Function** (f_x). In the dialogue box, select **Statistical** for the category, and then select **Binom.dist**. In the next dialogue box, fill in the values r, n, and p. For $P(r)$, use false; for P(at most r successes), use true.

Minitab First, enter the r values 0, 1, 2, . . . , n in a column. Then use menu choice **Calc ➤ Probability Distribution ➤ Binomial**. In the dialogue box, select Probability for $P(r)$ or Cumulative for P(at most r successes). Enter the number of trials n, the probability of success p, and the column containing the r values. A sample printout is shown in Problem 25 at the end of this section.

LOOKING FORWARD

There are several ways to find the probability of r successes out of n binomial trials. In particular we have used the general formula for the binomial probability distribution, a table of binomial probabilities such as Table 3 of Appendix II, or results from calculator or computer software that is based on the formula. However, depending on the number of trials and the size of the probability of success, the calculations can become tedious, or rounding errors can become an issue (*see* Linking Concepts, Problem 2). In Section 6.6 we will see how to use the *normal* probability distribution to approximate the binomial distribution in the case of sufficiently large numbers of trials n. Section 5.4 shows how to use a *Poisson* probability distribution to approximate the binomial distribution when the probability of success on a single trial p is very small and the number of trials is 100 or more.

Common expressions and corresponding inequalities

Many times we are asked to compute the probability of a range of successes. For instance, in a binomial experiment with n trials, we may be asked to compute the probability of four or more successes. Table 5-11 shows how common English expressions such as "four or more successes" translate to inequalities involving r.

Problems 29 and 30 show how to compute conditional probabilities for binomial experiments. In these problems we see how to compute the probability of r successes out of n binomial trials, given that a certain number of successes will occur.

TABLE 5-11 Common English Expressions and Corresponding Inequalities (consider a binomial experiment with n trials and r successes)

Expression	Inequality
Four or more successes	$r \geq 4$
At least four successes	That is, $r = 4, 5, 6, \ldots, n$
No fewer than four successes	
Not less than four successes	
Four or fewer successes	$r \leq 4$
At most four successes	That is, $r = 0, 1, 2, 3,$ or 4
No more than four successes	
The number of successes does not exceed four	
More than four successes	$r > 4$
The number of successes exceeds four	That is, $r = 5, 6, 7, \ldots, n$
Fewer than four successes	$r < 4$
The number of successes is not as large as four	That is, $r = 0, 1, 2, 3$

Interpretation Often we are not interested in the probability of a *specific number* of successes out of n binomial trials. Rather, we are interested in a minimum number of successes, a maximum number of successes, or a range of a number of successes. In other words, we are interested in the probability of *at least* a certain number of successes or *no more than* a certain number of successes, or the probability that the number of successes is between two given values.

For instance, suppose engineers have determined that at least 3 of 5 rivets on a bridge connector need to hold. If the probability that a single rivet holds is 0.80, and the performances of the rivets are independent, then the engineers are interested in the probability that *at least* 3 of the rivets hold. Notice that $P(r \geq 3 \text{ rivets hold}) = 0.943$, while $P(r = 3 \text{ rivets hold}) = 0.205$. The probability of *at least* 3 successes is much higher than the probability of *exactly* 3 successes. Safety concerns require that 3 of the rivets hold. However, there is a greater margin of safety if more than 3 rivets hold.

What happens when we have independent trials but there are more then two possible outcomes for each trial and the probability of each outcomes is the same for each trial? In such case we use the multinomial distribution. Problems 31 and 32 of this section discuss the multinomial distribution.

As you consider binomial experiments, determine whether you are interested in a *specific number* of successes or a *range* of successes.

SAMPLING WITHOUT REPLACEMENT: USE OF THE HYPERGEOMETRIC PROBABILITY DISTRIBUTION

If the population is relatively small and we draw samples without replacement, the assumption of independent trials is not valid and we should not use the binomial distribution.

Hypergeometric probability distribution

The *hypergeometric distribution* is a probability distribution of a random variable that has two outcomes when sampling is done without replacement. This is the distribution that is appropriate when the sample size is so small that sampling without replacement results in trials that are not even approximately independent. A discussion of the hypergeometric distribution can be found in Appendix I.

LOOKING FORWARD

The binomial distribution requires that exactly one of two possible outcomes occur for each binomial experiment: success (*S*) failure (*F*). What if we consider more than just two possible outcomes? If this is the case, we would use the multinomial distribution. As we will soon see, there is a close relation between binomial and normal distributions. Later we will see there is also a close relation between multinomial and chi-square distributions. See Problems 31 and 32 at the end of this section and also the end of Section 10.1 for more details..

VIEWPOINT — Lies! Lies!! Lies!!! The Psychology of Deceit

This is the title of an intriguing book by C. V. Ford, professor of psychiatry. The book recounts the true story of Floyd "Buzz" Fay, who was falsely convicted of murder on the basis of a failed polygraph examination. During his $2\frac{1}{2}$ years of wrongful imprisonment, Buzz became a polygraph expert. He taught inmates, who freely confessed guilt, how to pass a polygraph examination. (For more information on this topic, see Problem 21.)

SECTION 5.2 PROBLEMS

1. *Statistical Literacy* What does the random variable for a binomial experiment of n trials measure?

2. *Statistical Literacy* What does it mean to say that the trials of an experiment are independent?

3. *Statistical Literacy* For a binomial experiment, how many outcomes are possible for each trial? What are the possible outcomes?

4. *Statistical Literacy* In a binomial experiment, is it possible for the probability of success to change from one trial to the next? Explain.

5. *Interpretation* Suppose you are a hospital manager and have been told that there is no need to worry that respirator monitoring equipment might fail because the probability any one monitor will fail is only 0.01. The hospital has 20 such monitors and they work independently. Should you be more concerned about the probability that *exactly one* of the 20 monitors fails, or that *at least one* fails? Explain.

6. *Interpretation* From long experience a landlord knows that the probability an apartment in a complex will not be rented is 0.10. There are 20 apartments in the complex, and the rental status of each apartment is independent of the status of the others. When a minimum of 16 apartment units are rented, the landlord can meet all monthly expenses. Which probability is more relevant to the landlord in terms of being able to meet expenses: the probability that there are *exactly four* unrented units or the probability that there are *four or fewer* unrented units? Explain.

7. *Critical Thinking* In an experiment, there are n independent trials. For each trial, there are three outcomes, A, B, and C. For each trial, the probability of outcome A is 0.40; the probability of outcome B is 0.50; and the probability of outcome C is 0.10. Suppose there are 10 trials.
   (a) Can we use the binomial experiment model to determine the probability of four outcomes of type A, five of type B, and one of type C? Explain.

(b) Can we use the binomial experiment model to determine the probability of four outcomes of type A and six outcomes that are not of type A? Explain. What is the probability of success on each trial?

8. *Critical Thinking* In a carnival game, there are six identical boxes, one of which contains a prize. A contestant wins the prize by selecting the box containing it. Before each game, the old prize is removed and another prize is placed at random in one of the six boxes. Is it appropriate to use the binomial probability distribution to find the probability that a contestant who plays the game five times wins exactly twice? Check each of the requirements of a binomial experiment and give the values of n, r, and p.

9. *Critical Thinking* According to the college registrar's office, 40% of students enrolled in an introductory statistics class this semester are freshmen, 25% are sophomores, 15% are juniors, and 20% are seniors. You want to determine the probability that in a random sample of five students enrolled in introductory statistics this semester, exactly two are freshmen.
   (a) Describe a trial. Can we model a trial as having only two outcomes? If so, what is success? What is failure? What is the probability of success?
   (b) We are sampling without replacement. If only 30 students are enrolled in introductory statistics this semester, is it appropriate to model 5 trials as independent, with the same probability of success on each trial? Explain. What other probability distribution would be more appropriate in this setting?

10. *Critical Thinking: Simulation* Central Eye Clinic advertises that 90% of its patients approved for LASIK surgery to correct vision problems have successful surgeries.
   (a) In the random-number table, assign the digits 0 through 8 to the event "successful surgery" and the digit 9 to the event "unsuccessful surgery." Does this assignment of digits simulate 90% successful outcomes?
   (b) Use the random-digit assignment model of part (a) to simulate the outcomes of 15 trials. Begin at column 1, line 2.
   (c) Your friend assigned the digits 1 through 9 to the event "successful surgery" and the digit 0 to the event "unsuccessful surgery." Does this assignment of digits simulate 90% successful outcomes? Using this digit assignment, repeat part (b).

In each of the following problems, the binomial distribution will be used. Answers may vary slightly depending on whether the binomial distribution formula, the binomial distribution table, or distribution results from a calculator or computer are used. Please answer the following questions and then complete the problem.

What makes up a trial? What is a success? What is a failure?

What are the values of n, p, and q?

11. *Basic Computation: Binomial Distribution* Consider a binomial experiment with $n = 7$ trials where the probability of success on a single trial is $p = 0.30$.
   (a) Find $P(r = 0)$.
   (b) Find $P(r \geq 1)$ by using the complement rule.

12. *Basic Computation: Binomial Distribution* Consider a binomial experiment with $n = 7$ trials where the probability of success on a single trial is $p = 0.60$.
   (a) Find $P(r = 7)$.
   (b) Find $P(r \leq 6)$ by using the complement rule.

13. *Basic Computation: Binomial Distribution* Consider a binomial experiment with $n = 6$ trials where the probability of success on a single trial is $p = 0.85$.
   (a) Find $P(r \leq 1)$.
   (b) *Interpretation* If you conducted the experiment and got fewer than 2 successes, would you be surprised? Why?

14. *Basic Computation: Binomial Distribution* Consider a binomial experiment with $n = 6$ trials where the probability of success on a single trial is $p = 0.20$.
(a) Find $P(0 < r \leq 2)$.
(b) *Interpretation* If you conducted the experiment and got 1 or 2 successes, would you be surprised? Why?

15. *Binomial Probabilities: Coin Flip* A fair quarter is flipped three times. For each of the following probabilities, use the formula for the binomial distribution and a calculator to compute the requested probability. Next, look up the probability in Table 3 of Appendix II and compare the table result with the computed result.
(a) Find the probability of getting exactly three heads.
(b) Find the probability of getting exactly two heads.
(c) Find the probability of getting two or more heads.
(d) Find the probability of getting exactly three tails.

16. *Binomial Probabilities: Multiple-Choice Quiz* Richard has just been given a 10-question multiple-choice quiz in his history class. Each question has five answers, of which only one is correct. Since Richard has not attended class recently, he doesn't know any of the answers. Assuming that Richard guesses on all 10 questions, find the indicated probabilities.
(a) What is the probability that he will answer all questions correctly?
(b) What is the probability that he will answer all questions incorrectly?
(c) What is the probability that he will answer at least one of the questions correctly? Compute this probability two ways. First, use the rule for mutually exclusive events and the probabilities shown in Table 3 of Appendix II. Then use the fact that $P(r \geq 1) = 1 - P(r = 0)$. Compare the two results. Should they be equal? Are they equal? If not, how do you account for the difference?
(d) What is the probability that Richard will answer at least half the questions correctly?

17. *Ecology: Wolves* The following is based on information taken from *The Wolf in the Southwest: The Making of an Endangered Species,* edited by David Brown (University of Arizona Press). Before 1918, approximately 55% of the wolves in the New Mexico and Arizona region were male, and 45% were female. However, cattle ranchers in this area have made a determined effort to exterminate wolves. From 1918 to the present, approximately 70% of wolves in the region are male, and 30% are female. Biologists suspect that male wolves are more likely than females to return to an area where the population has been greatly reduced.
(a) Before 1918, in a random sample of 12 wolves spotted in the region, what was the probability that 6 or more were male? What was the probability that 6 or more were female? What was the probability that fewer than 4 were female?
(b) Answer part (a) for the period from 1918 to the present.

18. *Sociology: Ethics* The one-time fling! Have you ever purchased an article of clothing (dress, sports jacket, etc.), worn the item *once* to a party, and then returned the purchase? This is called a *one-time fling.* About 10% of all adults deliberately do a one-time fling and feel no guilt about it (Source: *Are You Normal?,* by Bernice Kanner, St. Martin's Press). In a group of seven adult friends, what is the probability that
(a) no one has done a one-time fling?
(b) at least one person has done a one-time fling?
(c) no more than two people have done a one-time fling?

19. *Sociology: Mother-in-Law* Sociologists say that 90% of married women claim that their husband's mother is the biggest bone of contention in their marriages (sex and money are lower-rated areas of contention). (See the source in Problem 18.) Suppose that six married women are having coffee together one morning. What is the probability that
(a) all of them dislike their mother-in-law?
(b) none of them dislike their mother-in-law?
(c) at least four of them dislike their mother-in-law?
(d) no more than three of them dislike their mother-in-law?

20. *Sociology: Dress Habits* A research team at Cornell University conducted a study showing that approximately 10% of all businessmen who wear ties wear them so tightly that they actually reduce blood flow to the brain, diminishing cerebral functions (Source: *Chances: Risk and Odds in Everyday Life,* by James Burke). At a board meeting of 20 businessmen, all of whom wear ties, what is the probability that
(a) at least one tie is too tight?
(b) more than two ties are too tight?
(c) no tie is too tight?
(d) at least 18 ties are *not* too tight?

21. *Psychology: Deceit* Aldrich Ames is a convicted traitor who leaked American secrets to a foreign power. Yet Ames took routine lie detector tests and each time passed them. How can this be done? Recognizing control questions, employing unusual breathing patterns, biting one's tongue at the right time, pressing one's toes hard to the floor, and counting backward by 7 are countermeasures that are difficult to detect but can change the results of a polygraph examination (Source: *Lies! Lies!! Lies!!! The Psychology of Deceit,* by C. V. Ford, professor of psychiatry, University of Alabama). In fact, it is reported in Professor Ford's book that after only 20 minutes of instruction by "Buzz" Fay (a prison inmate), 85% of those trained were able to pass the polygraph examination even when guilty of a crime. Suppose that a random sample of nine students (in a psychology laboratory) are told a "secret" and then given instructions on how to pass the polygraph examination without revealing their knowledge of the secret. What is the probability that
(a) all the students are able to pass the polygraph examination?
(b) more than half the students are able to pass the polygraph examination?
(c) no more than four of the students are able to pass the polygraph examination?
(d) all the students fail the polygraph examination?

22. *Hardware Store: Income* Trevor is interested in purchasing the local hardware/sporting goods store in the small town of Dove Creek, Montana. After examining accounting records for the past several years, he found that the store has been grossing over $850 per day about 60% of the business days it is open. Estimate the probability that the store will gross over $850
(a) at least 3 out of 5 business days.
(b) at least 6 out of 10 business days.
(c) fewer than 5 out of 10 business days.
(d) fewer than 6 out of the next 20 business days. *Interpretation* If this actually happened, might it shake your confidence in the statement $p = 0.60$? Might it make you suspect that p is less than 0.60? Explain.
(e) more than 17 out of the next 20 business days. *Interpretation* If this actually happened, might you suspect that p is greater than 0.60? Explain.

23. *Psychology: Myers–Briggs* Approximately 75% of all marketing personnel are extroverts, whereas about 60% of all computer programmers are introverts (Source: *A Guide to the Development and Use of the Myers–Briggs Type Indicator,* by Myers and McCaulley).
(a) At a meeting of 15 marketing personnel, what is the probability that 10 or more are extroverts? What is the probability that 5 or more are extroverts? What is the probability that all are extroverts?
(b) In a group of 5 computer programmers, what is the probability that none are introverts? What is the probability that 3 or more are introverts? What is the probability that all are introverts?

24. *Business Ethics: Privacy* Are your finances, buying habits, medical records, and phone calls really private? A real concern for many adults is that computers and the Internet are reducing privacy. A survey conducted by Peter D. Hart Research Associates for the Shell Poll was reported in *USA Today.* According to the survey, 37% of adults are concerned that employers are monitoring phone calls. Use the binomial distribution formula to calculate the probability that
(a) out of five adults, none is concerned that employers are monitoring phone calls.
(b) out of five adults, all are concerned that employers are monitoring phone calls.
(c) out of five adults, exactly three are concerned that employers are monitoring phone calls.

25. *Business Ethics: Privacy* According to the same poll quoted in Problem 24, 53% of adults are concerned that Social Security numbers are used for general identification. For a group of eight adults selected at random, we used Minitab to generate the binomial probability distribution and the cumulative binomial probability distribution (menu selections ➤ **Calc** ➤ **Probability Distributions** ➤ **Binomial**).

Jeff Titcomb/Photographer's Choice/ Getty Images

Number	r	P(r)	P(<= r)
	0	0.002381	0.00238
	1	0.021481	0.02386
	2	0.084781	0.10864
	3	0.191208	0.29985
	4	0.269521	0.56937
	5	0.243143	0.81251
	6	0.137091	0.94960
	7	0.044169	0.99377
	8	0.006226	1.00000

Find the probability that out of eight adults selected at random,
(a) at most five are concerned about Social Security numbers being used for identification. Do the problem by adding the probabilities $P(r = 0)$ through $P(r = 5)$. Is this the same as the cumulative probability $P(r \le 5)$?
(b) more than five are concerned about Social Security numbers being used for identification. First, do the problem by adding the probabilities $P(r = 6)$ through $P(r = 8)$. Then do the problem by subtracting the cumulative probability $P(r \le 5)$ from 1. Do you get the same results?

26. *Health Care: Office Visits* What is the age distribution of patients who make office visits to a doctor or nurse? The following table is based on information taken from the Medical Practice Characteristics section of the *Statistical Abstract of the United States* (116th Edition).

Age group, years	Under 15	15–24	25–44	45–64	65 and older
Percent of office visitors	20%	10%	25%	20%	25%

Suppose you are a district manager of a health management organization (HMO) that is monitoring the office of a local doctor or nurse in general family practice. This morning the office you are monitoring has eight office visits on the schedule. What is the probability that
(a) at least half the patients are under 15 years old? First, explain how this can be modeled as a binomial distribution with 8 trials, where success is visitor age is under 15 years old and the probability of success is 20%.
(b) from 2 to 5 patients are 65 years old or older (include 2 and 5)?
(c) from 2 to 5 patients are 45 years old or older (include 2 and 5)? *Hint:* Success is 45 or older. Use the table to compute the probability of success on a single trial.
(d) all the patients are under 25 years of age?
(e) all the patients are 15 years old or older?

27. *Binomial Distribution Table: Symmetry* Study the binomial distribution table (Table 3, Appendix II). Notice that the probability of success on a single trial p ranges from 0.01 to 0.95. Some binomial distribution tables stop at 0.50 because of the symmetry in the table. Let's look for that symmetry. Consider the section of the table for which $n = 5$. Look at the numbers in the columns headed by $p = 0.30$ and $p = 0.70$. Do you detect any similarities? Consider the following probabilities for a binomial experiment with five trials.
(a) Compare P(3 successes), where $p = 0.30$, with P(2 successes), where $p = 0.70$.
(b) Compare P(3 or more successes), where $p = 0.30$, with P(2 or fewer successes), where $p = 0.70$.
(c) Find the value of P(4 successes), where $p = 0.30$. For what value of r is P(r successes) the same using $p = 0.70$?
(d) What column is symmetrical with the one headed by $p = 0.20$?

28. *Binomial Distribution: Control Charts* This problem will be referred to in the study of control charts (Section 6.1). In the binomial probability distribution, let the number of trials be $n = 3$, and let the probability of success be $p = 0.0228$. Use a calculator to compute
(a) the probability of two successes.
(b) the probability of three successes.
(c) the probability of two or three successes.

Tomas Bercic/iStockphoto.com

29. *Expand Your Knowledge: Conditional Probability* In the western United States, there are many dry land wheat farms that depend on winter snow and spring rain to produce good crops. About 65% of the years, there is enough moisture to produce a good wheat crop, depending on the region (Reference: *Agricultural Statistics*, U.S. Department of Agriculture).
(a) Let r be a random variable that represents the number of good wheat crops in $n = 8$ years. Suppose the Zimmer farm has reason to believe that at least 4 out of 8 years will be good. However, they need at least 6 good years out of 8 to survive financially. Compute the probability that the Zimmers will get at least 6 good years out of 8, given what they believe is true; that is, compute $P(6 \le r \mid 4 \le r)$. See part (d) for a hint.

(b) Let r be a random variable that represents the number of good wheat crops in $n = 10$ years. Suppose the Montoya farm has reason to believe that at least 6 out of 10 years will be good. However, they need at least 8 good years out of 10 to survive financially. Compute the probability that the Montoyas will get at least 8 good years out of 10, given what they believe is true; that is, compute $P(8 \le r \mid 6 \le r)$.

(c) List at least three other areas besides agriculture to which you think conditional binomial probabilities can be applied.

(d) *Hint for solution:* Review item 6, conditional probability, in the summary of basic probability rules at the end of Section 4.2. Note that

$$P(A \mid B) = \frac{P(A\,and\,B)}{P(B)}$$

and show that in part (a),

$$P(6 \le r \mid 4 \le r) = \frac{P((6 \le r)\,and\,(4 \le r))}{P(4 \le r)} = \frac{P(6 \le r)}{P(4 \le r)}$$

30. *Conditional Probability: Blood Supply* Only about 70% of all donated human blood can be used in hospitals. The remaining 30% cannot be used because of various infections in the blood. Suppose a blood bank has 10 newly donated pints of blood. Let r be a binomial random variable that represents the number of "good" pints that can be used.

(a) Based on questionnaires completed by the donors, it is believed that at least 6 of the 10 pints are usable. What is the probability that at least 8 of the pints are usable, given this belief is true? Compute $P(8 \le r \mid 6 \le r)$.

(b) Assuming the belief that at least 6 of the pints are usable is true, what is the probability that all 10 pints can be used? Compute $P(r = 10 \mid 6 \le r)$.

Hint: See Problem 29.

31. *Expand Your Knowledge: Multinomial Probability Distribution* Consider a *multinomial experiment*. This means that

1. The trials are independent and repeated under identical conditions.
2. The outcomes of each trial falls into exactly one of $k \ge 2$ categories.
3. The probability that the outcomes of a single trial will fall into ith category is p_i (where $i = 1, 2\ldots, k$) and remains the same for each trial. Furthermore, $p_1 + p_2 + \ldots + p_k = 1$.
4. Let r_i be a random variable that represents the number of trials in which the outcomes falls into category i. If you have n trials, then $r_1 + r_2 + \ldots + r_k = n$. The multinomial probability distribution is then

$$P(r_1, r_2, \cdots r_k) = \frac{n!}{r_1! r_2! \cdots r_k!} p_1^{r_1} p_2^{r_2} \cdots p_k^{r_2}$$

How are the multinomial distribution and the binomial distribution related? For the special case $k = 2$, we use the notation $r_1 = r$, $r_2 = n - r$, $p_1 = p$, and $p_2 = q$. In this special case, the multinomial distribution becomes the binomial distribution.

The city of Boulder, Colorado is having an election to determine the establishment of a new municipal electrical power plant. The new plant would emphasize renewable energy (e.g., wind, solar, geothermal). A recent large survey of Boulder voters showed 50% favor the new plant, 30% oppose it, and 20% are undecided. Let $p_1 = 0.5$, $p_2 = 0.3$, and $p_3 = 0.2$. Suppose a random sample of $n = 6$ Boulder voters is taken. What is the probability that

(a) $r_1 = 3$ favor, $r_2 = 2$ oppose, and $r_3 = 1$ are undecided regarding the new power plant?

(b) $r_1 = 4$ favor, $r_2 = 2$ oppose, and $r_3 = 0$ are undecided regarding the new power plant?

32. *Multinomial Distribution: Commercial Airliners* Aircraft inspectors (who specialize in mechanical engineering) report wing cracks in aircraft as non-existent, detectable (but still functional), or critical (needs immediate repair). For a particular model of commercial jet 10 years old, history indicates 75% of the planes had no wing cracks, 20% had detectable wing cracks, and 5% had critical wing cracks. Five planes that are 10 years old are randomly selected. What is the Probability that
(a) 4 have no cracks, 1 has detectable cracks, and 0 have no critical cracks?
(b) 3 have no cracks, 1 has detectable cracks, and 1 has critical cracks?

Hint: See Problem 31.

SECTION **5.3**

Additional Properties of the Binomial Distribution

FOCUS POINTS

- Make histograms for binomial distributions.
- Compute μ and σ for a binomial distribution.
- Compute the minimum number of trials n needed to achieve a given probability of success $P(r)$.

GRAPHING A BINOMIAL DISTRIBUTION

Any probability distribution may be represented in graphic form. How should we graph the binomial distribution? Remember, the binomial distribution tells us the probability of r successes out of n trials. Therefore, we'll place values of r along the horizontal axis and values of $P(r)$ on the vertical axis. The binomial distribution is a *discrete* probability distribution because r can assume only whole-number values such as 0, 1, 2, 3, . . . Therefore, a histogram is an appropriate graph of a binomial distribution.

PROCEDURE

HOW TO GRAPH A BINOMIAL DISTRIBUTION

1. Place r values on the horizontal axis.
2. Place $P(r)$ values on the vertical axis.
3. Construct a bar over each r value extending from $r - 0.5$ to $r + 0.5$. The height of the corresponding bar is $P(r)$.

Let's look at an example to see exactly how we'll make these histograms.

EXAMPLE 7

GRAPH OF A BINOMIAL DISTRIBUTION

A waiter at the Green Spot Restaurant has learned from long experience that the probability that a lone diner will leave a tip is only 0.7. During one lunch hour, the waiter serves six people who are dining by themselves. Make a graph of the binomial probability distribution that shows the probabilities that 0, 1, 2, 3, 4, 5, or all 6 lone diners leave tips.

SOLUTION: This is a binomial experiment with $n = 6$ trials. Success is achieved when the lone diner leaves a tip, so the probability of success is 0.7 and that of failure is 0.3:

$$n = 6 \quad p = 0.7 \quad q = 0.3$$

We want to make a histogram showing the probability of r successes when $r = 0, 1, 2, 3, 4, 5,$ or 6. It is easier to make the histogram if we first make a table of r values and the corresponding $P(r)$ values (Table 5-12). We'll use Table 3 of Appendix II to find the $P(r)$ values for $n = 6$ and $p = 0.70$.

FIGURE 5-3

Graph of the Binomial Distribution for $n = 6$ and $p = 0.7$

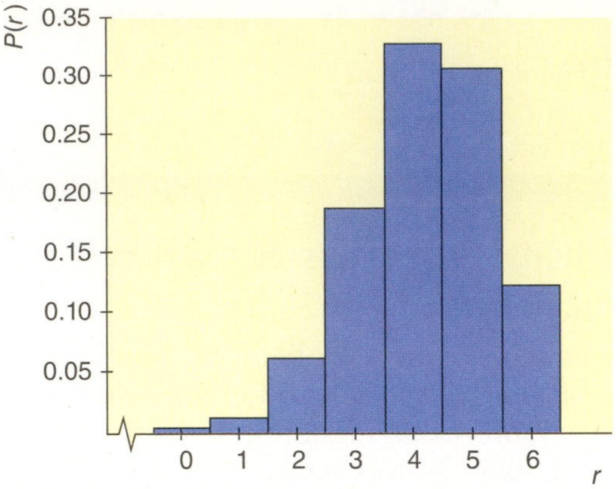

TABLE 5-12 **Binomial Distribution for $n = 6$ and $p = 0.70$**

r	$P(r)$
0	0.001
1	0.010
2	0.060
3	0.185
4	0.324
5	0.303
6	0.118

Mark Richards/PhotoEdit

To construct the histogram, we'll put r values on the horizontal axis and $P(r)$ values on the vertical axis. Our bars will be one unit wide and will be centered over the appropriate r value. The height of the bar over a particular r value tells the probability of that r (see Figure 5-3).

The probability of a particular value of r is given not only by the height of the bar over that r value but also by the *area* of the bar. Each bar is only one unit wide, so its area (area = height times width) equals its height. Since the area of each bar represents the probability of the r value under it, the sum of the areas of the bars must be 1. In this example, the sum turns out to be 1.001. It is not exactly equal to 1 because of rounding error.

Guided Exercise 6 illustrates another binomial distribution with $n = 6$ trials. The graph will be different from that of Figure 5-3 because the probability of success p is different.

GUIDED EXERCISE 6 | GRAPH OF A BINOMIAL DISTRIBUTION

Jim enjoys playing basketball. He figures that he makes about 50% of the field goals he attempts during a game. Make a histogram showing the probability that Jim will make 0, 1, 2, 3, 4, 5, or 6 shots out of six attempted field goals.

(a) This is a binomial experiment with $n =$ ______ trials. In this situation, we'll say success occurs when Jim makes an attempted field goal. What is the value of p?

→ In this example, $n = 6$ and $p = 0.5$.

Continued

GUIDED EXERCISE 6 *continued*

(b) Use Table 3 of Appendix II to complete Table 5-13 of $P(r)$ values for $n = 6$ and $p = 0.5$.

TABLE 5-13

r	$P(r)$
0	0.016
1	0.094
2	0.234
3	____
4	____
5	____
6	____

TABLE 5-14 **Completion of Table 5-13**

r	$P(r)$
.	.
.	.
.	.
3	0.312
4	0.234
5	0.094
6	0.016

(c) Use the values of $P(r)$ given in Table 5-14 to complete the histogram in Figure 5-4.

FIGURE 5-4 Beginning of Graph of Binomial Distribution for $n = 6$ and $p = 0.5$

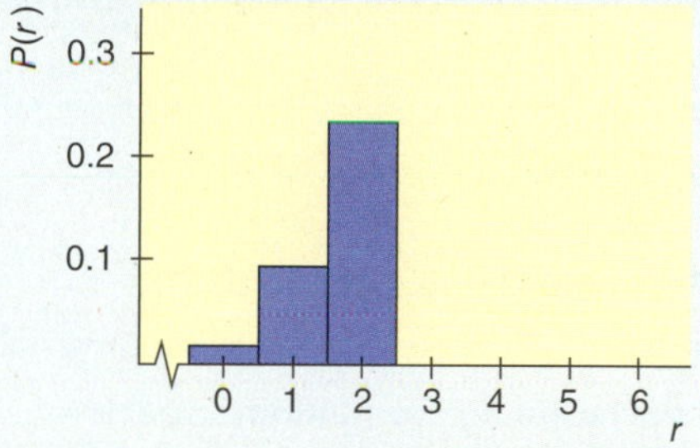

FIGURE 5-5 Completion of Figure 5-4

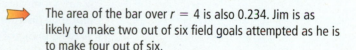

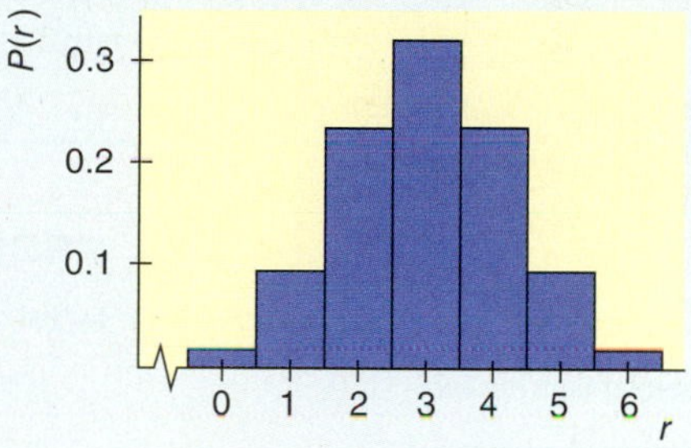

(d) The area of the bar over $r = 2$ is 0.234. What is the area of the bar over $r = 4$? How does the probability that Jim makes exactly two field goals out of six compare with the probability that he makes exactly four field goals out of six?

The area of the bar over $r = 4$ is also 0.234. Jim is as likely to make two out of six field goals attempted as he is to make four out of six.

In Example 7 and Guided Exercise 6, we see the graphs of two binomial distributions associated with $n = 6$ trials. The two graphs are different because the probability of success p is different in the two cases. In Example 7, $p = 0.7$ and the graph is skewed to the left—that is, the left tail is longer. In Guided Exercise 6, p is equal to 0.5 and the graph is symmetrical—that is, if we fold it in half, the two halves coincide exactly. Whenever *p equals 0.5, the graph of the binomial distribution will be symmetrical no matter how many trials we have.* In Chapter 6, we will see that if the number of trials n is quite large, the binomial distribution is almost symmetrical over the bars containing most of the area even when p is not close to 0.5.

WHAT DOES THE GRAPH OF A BINOMIAL PROBABILITY DISTRIBUTION TELL US?

Consider a binomial probability distribution with n trials and probability of success on a single trial p. The graph of the distribution tells us

- the probability of r successes for each r from 0 to n.
- if the graph is mound-shaped and symmetric, the probability of success p is close to 0.50.
- if the graph is skewed right, the probability of success p is less then 0.50.The more skewed right the graph, the closer p is to 0.
- if the graph is skewed left, the probability of success p is greater than 0.05. The more skewed left the graph, the closer p is to 1.

MEAN AND STANDARD DEVIATION OF A BINOMIAL DISTRIBUTION

Two other features that help describe the graph of any distribution are the balance point of the distribution and the spread of the distribution about that balance point. The *balance point* is the mean μ of the distribution, and the *measure of spread* that is most commonly used is the standard deviation σ. The mean μ is the *expected value* of the number of successes.

For the binomial distribution, we can use two special formulas to compute the mean μ and the standard deviation σ. These are easier to use than the general formulas in Section 5.1 for μ and σ of any discrete probability distribution.

PROCEDURE

HOW TO COMPUTE μ AND σ FOR A BINOMIAL DISTRIBUTION

Mean of a binomial distribution

Standard deviation of a binomial distribution

$\mu = np$ is the **expected number of successes** for the random variable r

$\sigma = \sqrt{npq}$ is the **standard deviation** for the random variable r

where

r is a random variable representing the number of successes in a binomial distribution,

n is the number of trials,

p is the probability of success on a single trial, and

$q = 1 - p$ is the probability of failure on a single trial.

EXAMPLE 8

COMPUTE μ AND σ

Let's compute the mean and standard deviation for the distribution of Example 7 that describes that probabilities of lone diners leaving tips at the Green Spot Restaurant.

SOLUTION: In Example 7,

$$n = 6 \quad p = 0.7 \quad q = 0.3$$

For the binomial distribution,

$$\mu = np = 6(0.7) = 4.2$$

The balance point of the distribution is at $\mu = 4.2$. The standard deviation is given by

$$\sigma = \sqrt{npq} = \sqrt{6(0.7)(0.3)} = \sqrt{1.26} \approx 1.12$$

The mean μ is not only the balance point of the distribution; it is also the *expected value* of r. Specifically, in Example 7, the waiter can expect 4.2 lone diners out of 6 to leave a tip. (The waiter would probably round the expected value to 4 tippers out of 6.)

GUIDED EXERCISE 7 | EXPECTED VALUE AND STANDARD DEVIATION

When Jim (of Guided Exercise 6) shoots field goals in basketball games, the probability that he makes a shot is only 0.5.

(a) The mean of the binomial distribution is the expected value of r successes out of n trials. Out of six throws, what is the expected number of goals Jim will make?

→ The expected value is the mean μ:

$\mu = np = 6(0.5) = 3$

Jim can expect to make three goals out of six tries.

(b) For six trials, what is the standard deviation of the binomial distribution of the number of successful field goals Jim makes?

→ $\sigma = \sqrt{npq} = \sqrt{6(0.5)(0.5)} = \sqrt{1.5} \approx 1.22$

CRITICAL THINKING

UNUSUAL VALUES

Chebyshev's theorem tells us that no matter what the data distribution looks like, at least 75% of the data will fall within 2 standard deviations of the mean. As we will see in Chapter 6, when the distribution is mound-shaped and symmetrical, about 95% of the data are within 2 standard deviations of the mean. Data values beyond 2 standard deviations from the mean are less common than those closer to the mean.

In fact, one indicator that a data value might be an outlier is that it is more than 2.5 standard deviations from the mean (Source: *Statistics*, by G. Upton and I. Cook, Oxford University Press).

UNUSUAL VALUES

For a binomial distribution, it is unusual for the number of successes r to be higher than $\mu + 2.5\sigma$ or lower than $\mu - 2.5\sigma$.

We can use this indicator to determine whether a specified number of successes out of n trials in a binomial experiment are unusual.

For instance, consider a binomial experiment with 20 trials for which probability of success on a single trial is $p = 0.70$. The expected number of successes is $\mu = 14$, with a standard deviation of $\sigma \approx 2$. A number of successes above 19 or below 9 would be considered unusual. However, such numbers of successes are possible.

QUOTA PROBLEMS: MINIMUM NUMBER OF TRIALS FOR A GIVEN PROBABILITY

In applications, you do not want to confuse the expected value of r with certain probabilities associated with r. Guided Exercise 8 illustrates this point.

GUIDED EXERCISE 8 | FIND THE MINIMUM VALUE OF N FOR A GIVEN $P(r)$

A satellite is powered by three solar cells. The probability that any one of these cells will fail is 0.15, and the cells operate or fail independently.

Part I: In this part, we want to find the least number of cells the satellite should have so that the *expected value* of the number of working cells is no smaller than 3. In this situation, n represents the number of cells, r is the number of successful or working cells, p is the probability that a cell will work, q is the probability that a cell will fail, and μ is the expected value, which should be no smaller than 3.

(a) What is the value of q? of p?

→ $q = 0.15$, as given in the problem. p must be 0.85, since $p = 1 - q$.

(b) The expected value μ for the number of working cells is given by $\mu = np$. The expected value of the number of working cells should be no smaller than 3, so

$$3 \le \mu = np$$

From part (a), we know the value of p. Solve the inequality $3 \le np$ for n.

→ $3 \le np$

$3 \le n(0.85)$

$\frac{3}{0.85} \le n$ Divide both sides by 0.85.

$3.53 \le n$

(c) Since n is between 3 and 4, should we round it to 3 or to 4 to make sure that μ is at least 3?

→ n should be at least 3.53. Since we can't have a fraction of a cell, we had best make $n = 4$. For $n = 4$, $\mu = 4(0.85) = 3.4$. This value satisfies the condition that μ be at least 3.

Part II: In this part, we want to find the smallest number of cells the satellite should have to be 97% sure that there will be adequate power—that is, that at least three cells work.

(a) The letter r has been used to denote the number of successes. In this case, r represents the number of working cells. We are trying to find the number n of cells necessary to ensure that (choose the correct statement)

(i) $P(r \ge 3) = 0.97$ or

(ii) $P(r \le 3) = 0.97$

→ $P(r \ge 3) = 0.97$

(b) We need to find a value for n such that

$P(r \ge 3) = 0.97$

Try $n = 4$. Then, $r \ge 3$ means $r = 3$ or 4, so,

→ $P(3) = 0.368$

$P(4) = 0.522$

$P(r \ge 3) = 0.368 + 0.522 = 0.890$

Continued

GUIDED EXERCISE 8 *continued*

$P(r \geq 3) = P(3) + P(4)$
Use Table 3 (Appendix II) with $n = 4$ and $p = 0.85$ to find values of $P(3)$ and $P(4)$. Then, compute $P(r \geq 3)$ for $n = 4$. Will $n = 4$ guarantee that $P(r \geq 3)$ is at least 0.97?

Thus, $n = 4$ is *not* sufficient to be 97% sure that at least three cells will work. For $n = 4$, the probability that at least three cells will work is only 0.890.

(c) Now try $n = 5$ cells. For $n = 5$,
$P(r \geq 3) = P(3) + P(4) + P(5)$
since r can be 3, 4, or 5. Are $n = 5$ cells adequate? [Be sure to find new values of $P(3)$ and $P(4)$, since we now have $n = 5$.]

$$P(r \geq 3) = P(3) + P(4) + P(5)$$
$$= 0.138 + 0.392 + 0.444$$
$$= 0.974$$

Thus, $n = 5$ cells are required if we want to be 97% sure that at least three cells will work.

In Part I and Part II, we got different values for *n*. Why? In Part I, we had $n = 4$ and $\mu = 3.4$. This means that if we put up lots of satellites with four cells, we can expect that an *average* of 3.4 cells will work per satellite. But for $n = 4$ cells, there is a probability of only 0.89 that at least three cells will work in any one satellite. In Part II, we are trying to find the number of cells necessary so that the probability is 0.97 that at least three cells will work in any *one* satellite. If we use $n = 5$ cells, then we can satisfy this requirement.

Quota problems

Quotas occur in many aspects of everyday life. The manager of a sales team gives every member of the team a weekly sales quota. In some districts, police have a monthly quota for the number of traffic tickets issued. Nonprofit organizations have recruitment quotas for donations or new volunteers. The basic ideas used to compute quotas also can be used in medical science (how frequently checkups should occur), quality control (how many production flaws should be expected), or risk management (how many bad loans a bank should expect in a certain investment group). In fact, Part II of Guided Exercise 8 is a *quota problem.* To have adequate power, a satellite must have a quota of three working solar cells. Such problems come from many different sources, but they all have one thing in common: They are solved using the binomial probability distribution.

To solve quota problems, it is often helpful to use equivalent formulas for expressing binomial probabilities. These formulas involve the complement rule and the fact that binomial events are independent. Equivalent probabilities will be used in Example 9.

PROCEDURE

HOW TO EXPRESS BINOMIAL PROBABILITIES USING EQUIVALENT FORMULAS

$P(\text{at least one success}) = P(r \geq 1) = 1 - P(0)$

$P(\text{at least two successes}) = P(r \geq 2) = 1 - P(0) - P(1)$

$P(\text{at least three successes}) = P(r \geq 3) = 1 - P(0) - P(1) - P(2)$

$P(\text{at least m successes}) = P(r \geq m) = 1 - P(0) - P(1) - \cdots - P(m - 1)$,
where $1 \leq m \leq$ number of trials

For a discussion of the mathematics behind these formulas, see Problem 28 at the end of this section.

Example 9 is a quota problem. Junk bonds are sometimes controversial. In some cases, junk bonds have been the salvation of a basically good company that has had a run of bad luck. From another point of view, junk bonds are not much more than a gambler's effort to make money by shady ethics.

The book *Liar's Poker,* by Michael Lewis, is an exciting and sometimes humorous description of his career as a Wall Street bond broker. Most bond brokers, including Mr. Lewis, are ethical people. However, the book does contain an interesting discussion of Michael Milken and shady ethics. In the book, Mr. Lewis says, "If it was a good deal, the brokers kept it for themselves; if it was a bad deal, they'd try to sell it to their customers." In Example 9, we use some binomial probabilities for a brief explanation of what Mr. Lewis's book is talking about.

EXAMPLE 9

QUOTA

Junk bonds can be profitable as well as risky. Why are investors willing to consider junk bonds? Suppose you can buy junk bonds at a tremendous discount. You try to choose "good" companies with a "good" product. The company should have done well but for some reason did not. Suppose you consider only companies with a 35% estimated risk of default, and your financial investment goal requires four bonds to be "good" bonds in the sense that they will not default before a certain date. Remember, junk bonds that do not default are usually very profitable because they carry a very high rate of return. The other bonds in your investment group can default (or not) without harming your investment plan. Suppose you want to be 95% certain of meeting your goal (quota) of at least four good bonds. How many junk bond issues should you buy to meet this goal?

SOLUTION: Since the probability of default is 35%, the probability of a "good" bond is 65%. Let success S be represented by a good bond. Let n be the number of bonds purchased, and let r be the number of good bonds in this group. We want

$$P(r \geq 4) \geq 0.95$$

This is equivalent to

$$1 - P(0) - P(1) - P(2) - P(3) \geq 0.95$$

Since the probability of success is $p = P(S) = 0.65$, we need to look in the binomial table under $p = 0.65$ and different values of n to find the *smallest value of n* that will satisfy the preceding relation. Table 3 of Appendix II shows that if $n = 10$ when $p = 0.65$, then,

$$1 - P(0) - P(1) - P(2) - P(3) = 1 - 0 - 0 - 0.004 - 0.021 = 0.975$$

The probability 0.975 satisfies the condition of being greater than or equal to 0.95. We see that 10 is the smallest value of n for which the condition

$$P(r \geq 4) \geq 0.95$$

is satisfied. Under the given conditions (a good discount on price, no more than 35% chance of default, and a fundamentally good company), you can be 95% sure of meeting your investment goal with $n = 10$ (carefully selected) junk bond issues.

In this example, we see that by carefully selecting junk bonds, there is a high probability of getting some good bonds that will produce a real profit. What do you do with the other bonds that aren't so good? Perhaps the quote from *Liar's Poker* will suggest what is sometimes attempted.

VIEWPOINT Kodiak Island, Alaska

Kodiak Island is famous for its giant brown bears. The sea surrounding the island is also famous for its king crab. The state of Alaska's Department of Fish and Game has collected a huge amount of data regarding ocean latitude, ocean longitude, and size of king crab. Of special interest to commercial fishing skippers is the size of crab. Those too small must be returned to the sea. To find locations and sizes of king crab catches near Kodiak Island, visit the Brase/Brase statistics site at **http://www.cengage.com/statistics/brase11e** *and find the link to the StatLib site hosted by the Department of Statistics at Carnegie Mellon University. Once at StatLib, go to crab data. From this information, it is possible to use methods of this chapter and Chapter 7 to estimate the proportion of legal crab in a sea skipper's catch.*

SECTION 5.3 PROBLEMS

1. *Statistical Literacy* What does the expected value of a binomial distribution with n trials tell you?

2. *Statistical Literacy* Consider two binomial distributions, with n trials each. The first distribution has a higher probability of success on each trial than the second. How does the expected value of the first distribution compare to that of the second?

3. *Basic Computation: Expected Value and Standard Deviation* Consider a binomial experiment with $n = 8$ trials and $p = 0.20$.
   (a) Find the expected value and the standard deviation of the distribution.
   (b) *Interpretation* Would it be unusual to obtain 5 or more successes? Explain. Confirm your answer by looking at the binomial probability distribution table.

4. *Basic Computation: Expected Value and Standard Deviation* Consider a binomial experiment with $n = 20$ trials and $p = 0.40$.
   (a) Find the expected value and the standard deviation of the distribution.
   (b) *Interpretation* Would it be unusual to obtain fewer than 3 successes? Explain. Confirm your answer by looking at the binomial probability distribution table.

5. *Critical Thinking* Consider a binomial distribution of 200 trials with expected value 80 and standard deviation of about 6.9. Use the criterion that it is unusual to have data values more than 2.5 standard deviations above the mean or 2.5 standard deviations below the mean to answer the following questions.
   (a) Would it be unusual to have more than 120 successes out of 200 trials? Explain.
   (b) Would it be unusual to have fewer than 40 successes out of 200 trials? Explain.
   (c) Would it be unusual to have from 70 to 90 successes out of 200 trials? Explain.

6. *Critical Thinking* Consider a binomial distribution with 10 trials. Look at Table 3 (Appendix II) showing binomial probabilities for various values of p, the probability of success on a single trial.
   (a) For what value of p is the distribution symmetric? What is the expected value of this distribution? Is the distribution centered over this value?
   (b) For small values of p, is the distribution skewed right or left?
   (c) For large values of p, is the distribution skewed right or left?

7. *Binomial Distribution: Histograms* Consider a binomial distribution with $n = 5$ trials. Use the probabilities given in Table 3 of Appendix II to make histograms showing the probabilities of $r = 0, 1, 2, 3, 4$, and 5 successes for each of the following. Comment on the skewness of each distribution.
   (a) The probability of success is $p = 0.50$.
   (b) The probability of success is $p = 0.25$.
   (c) The probability of success is $p = 0.75$.
   (d) What is the relationship between the distributions shown in parts (b) and (c)?
   (e) If the probability of success is $p = 0.73$, do you expect the distribution to be skewed to the right or to the left? Why?

8. *Binomial Distributions: Histograms* Figure 5-6 shows histograms of several binomial distributions with $n = 6$ trials. Match the given probability of success with the best graph.
   (a) $p = 0.30$ goes with graph ________.
   (b) $p = 0.50$ goes with graph ________.
   (c) $p = 0.65$ goes with graph ________.
   (d) $p = 0.90$ goes with graph ________.
   (e) In general, when the probability of success p is close to 0.5, would you say that the graph is more symmetrical or more skewed? In general, when the probability of success p is close to 1, would you say that the graph is skewed to the right or to the left? What about when p is close to 0?

FIGURE 5-6

Binomial Probability Distributions with $n = 6$ (generated on the TI-84Plus calculator)

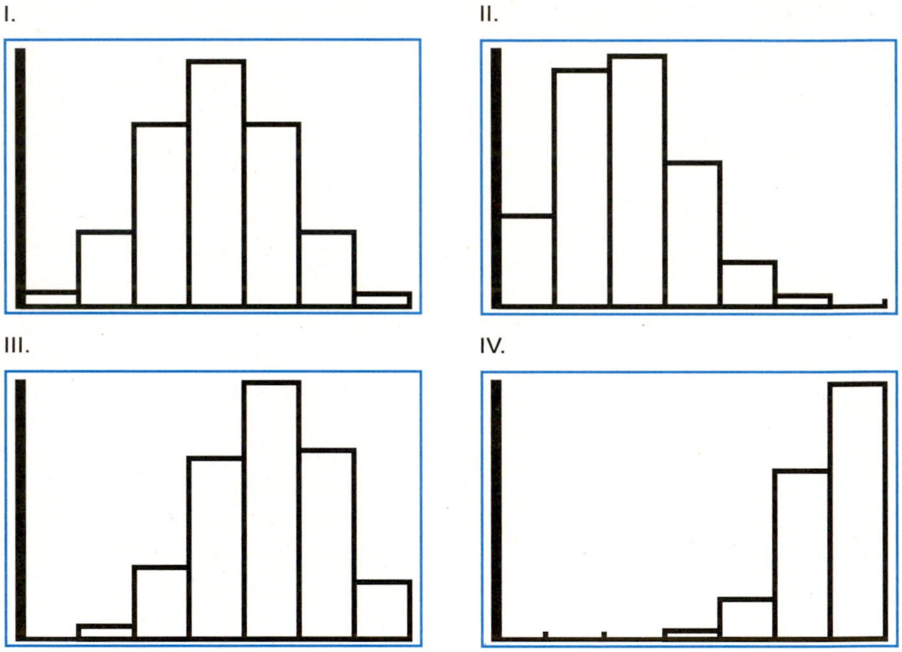

9. *Critical Thinking* Consider a binomial distribution with $n = 10$ trials and the probability of success on a single trial $p = 0.85$.
   (a) Is the distribution skewed left, skewed right, or symmetrical?
   (b) Compute the expected number of successes in 10 trials.
   (c) Given the high probability of success p on a single trial, would you expect $P(r \leq 3)$ to be very high or very low? Explain.
   (d) Given the high probability of success p on a single trial, would you expect $P(r \geq 8)$ to be very high or very low? Explain.

10. *Critical Thinking* Consider a binomial distribution with $n = 10$ trials and the probability of success on a single trial $p = 0.05$.
   (a) Is the distribution skewed left, skewed right, or symmetrical?
   (b) Compute the expected number of successes in 10 trials.

(c) Given the low probability of success p on a single trial, would you expect $P(r \leq 1)$ to be very high or very low? Explain.

(d) Given the low probability of success p on a single trial, would you expect $P(r \geq 8)$ to be very high or very low? Explain.

11. *Marketing: Photography* Does the *kid factor* make a difference? If you are talking photography, the answer may be yes! The following table is based on information from *American Demographics* (Vol. 19, No. 7).

Ages of children in household, years	Under 2	None under 21
Percent of U.S. households that buy photo gear	80%	50%

Let us say you are a market research person who interviews a random sample of 10 households.

(a) Suppose you interview 10 households with children under the age of 2 years. Let r represent the number of such households that buy photo gear. Make a histogram showing the probability distribution of r for $r = 0$ through $r = 10$. Find the mean and standard deviation of this probability distribution.

(b) Suppose that the 10 households are chosen to have no children under 21 years old. Let r represent the number of such households that buy photo gear. Make a histogram showing the probability distribution of r for $r = 0$ through $r = 10$. Find the mean and standard deviation of this probability distribution.

(c) *Interpretation* Compare the distributions in parts (a) and (b). You are designing TV ads to sell photo gear. Could you justify featuring ads of parents taking pictures of toddlers? Explain your answer.

12. *Quality Control: Syringes* The quality-control inspector of a production plant will reject a batch of syringes if two or more defective syringes are found in a random sample of eight syringes taken from the batch. Suppose the batch contains 1% defective syringes.

(a) Make a histogram showing the probabilities of $r = 0, 1, 2, 3, 4, 5, 6, 7$, and 8 defective syringes in a random sample of eight syringes.

(b) Find μ. What is the expected number of defective syringes the inspector will find?

(c) What is the probability that the batch will be accepted?

(d) Find σ.

13. *Private Investigation: Locating People* Old Friends Information Service is a California company that is in the business of finding addresses of long-lost friends. Old Friends claims to have a 70% success rate (Source: *The Wall Street Journal*). Suppose that you have the names of six friends for whom you have no addresses and decide to use Old Friends to track them.

(a) Make a histogram showing the probability of $r = 0$ to 6 friends for whom an address will be found.

(b) Find the mean and standard deviation of this probability distribution. What is the expected number of friends for whom addresses will be found?

(c) *Quota Problem* How many names would you have to submit to be 97% sure that at least two addresses will be found?

14. *Insurance: Auto* The Mountain States Office of State Farm Insurance Company reports that approximately 85% of all automobile damage liability claims are made by people under 25 years of age. A random sample of five automobile insurance liability claims is under study.

(a) Make a histogram showing the probability that $r = 0$ to 5 claims are made by people under 25 years of age.

(b) Find the mean and standard deviation of this probability distribution. For samples of size 5, what is the expected number of claims made by people under 25 years of age?

15. *Education: Illiteracy* *USA Today* reported that about 20% of all people in the United States are illiterate. Suppose you interview seven people at random off a city street.
   (a) Make a histogram showing the probability distribution of the number of illiterate people out of the seven people in the sample.
   (b) Find the mean and standard deviation of this probability distribution. Find the expected number of people in this sample who are illiterate.
   (c) *Quota Problem* How many people would you need to interview to be 98% sure that at least seven of these people can read and write (are not illiterate)?

16. *Rude Drivers: Tailgating* Do you tailgate the car in front of you? About 35% of all drivers will tailgate before passing, thinking they can make the car in front of them go faster (Source: Bernice Kanner, *Are You Normal?*, St. Martin's Press). Suppose that you are driving a considerable distance on a two-lane highway and are passed by 12 vehicles.
   (a) Let r be the number of vehicles that tailgate before passing. Make a histogram showing the probability distribution of r for $r = 0$ through $r = 12$.
   (b) Compute the expected number of vehicles out of 12 that will tailgate.
   (c) Compute the standard deviation of this distribution.

17. *Hype: Improved Products* *The Wall Street Journal* reported that approximately 25% of the people who are told a product is *improved* will believe that it is, in fact, improved. The remaining 75% believe that this is just hype (the same old thing with no real improvement). Suppose a marketing study consists of a random sample of eight people who are given a sales talk about a new, *improved* product.
   (a) Make a histogram showing the probability that $r = 0$ to 8 people believe the product is, in fact, improved.
   (b) Compute the mean and standard deviation of this probability distribution.
   (c) *Quota Problem* How many people are needed in the marketing study to be 99% sure that at least one person believes the product to be improved? *Hint:* Note that $P(r \geq 1) = 0.99$ is equivalent to $1 - P(0) = 0.99$, or $P(0) = 0.01$.

18. *Quota Problem: Archaeology* An archaeological excavation at Burnt Mesa Pueblo showed that about 10% of the flaked stone objects were finished arrow points (Source: *Bandelier Archaeological Excavation Project: Summer 1990 Excavations at Burnt Mesa Pueblo,* edited by Kohler, Washington State University). How many flaked stone objects need to be found to be 90% sure that at least one is a finished arrow point? *Hint:* Use a calculator and note that $P(r \geq 1) \geq 0.90$ is equivalent to $1 - P(0) \geq 0.90$, or $P(0) \leq 0.10$.

19. *Criminal Justice: Parole* *USA Today* reports that about 25% of all prison parolees become repeat offenders. Alice is a social worker whose job is to counsel people on parole. Let us say success means a person does not become a repeat offender. Alice has been given a group of four parolees.
   (a) Find the probability $P(r)$ of r successes ranging from 0 to 4.
   (b) Make a histogram for the probability distribution of part (a).
   (c) What is the expected number of parolees in Alice's group who will not be repeat offenders? What is the standard deviation?
   (d) *Quota Problem* How large a group should Alice counsel to be about 98% sure that three or more parolees will not become repeat offenders?

20. *Defense: Radar Stations* The probability that a single radar station will detect an enemy plane is 0.65.
   (a) *Quota Problem* How many such stations are required for 98% certainty that an enemy plane flying over will be detected by at least one station?

(b) If four stations are in use, what is the expected number of stations that will detect an enemy plane?

21. *Criminal Justice: Jury Duty* Have you ever tried to get out of jury duty? About 25% of those called will find an excuse (work, poor health, travel out of town, etc.) to avoid jury duty (Source: Bernice Kanner, *Are You Normal?*, St. Martin's Press, New York). If 12 people are called for jury duty,
(a) what is the probability that all 12 will be available to serve on the jury?
(b) what is the probability that 6 or more will *not* be available to serve on the jury?
(c) Find the expected number of those available to serve on the jury. What is the standard deviation?
(d) *Quota Problem* How many people n must the jury commissioner contact to be 95.9% sure of finding at least 12 people who are available to serve?

22. *Public Safety: 911 Calls* *The Denver Post* reported that a recent audit of Los Angeles 911 calls showed that 85% were not emergencies. Suppose the 911 operators in Los Angeles have just received four calls.
(a) What is the probability that all four calls are, in fact, emergencies?
(b) What is the probability that three or more calls are not emergencies?
(c) *Quota Problem* How many calls n would the 911 operators need to answer to be 96% (or more) sure that at least one call is, in fact, an emergency?

23. *Law Enforcement: Property Crime* Does crime pay? The *FBI Standard Survey of Crimes* shows that for about 80% of all property crimes (burglary, larceny, car theft, etc.), the criminals are never found and the case is never solved (Source: *True Odds,* by James Walsh, Merrit Publishing). Suppose a neighborhood district in a large city suffers repeated property crimes, not always perpetrated by the same criminals. The police are investigating six property crime cases in this district.
(a) What is the probability that none of the crimes will ever be solved?
(b) What is the probability that at least one crime will be solved?
(c) What is the expected number of crimes that will be solved? What is the standard deviation?
(d) *Quota Problem* How many property crimes n must the police investigate before they can be at least 90% sure of solving one or more cases?

24. *Security: Burglar Alarms* A large bank vault has several automatic burglar alarms. The probability is 0.55 that a single alarm will detect a burglar.
(a) *Quota Problem* How many such alarms should be used for 99% certainty that a burglar trying to enter will be detected by at least one alarm?
(b) Suppose the bank installs nine alarms. What is the expected number of alarms that will detect a burglar?

25. *Criminal Justice: Convictions* Innocent until proven guilty? In Japanese criminal trials, about 95% of the defendants are found guilty. In the United States, about 60% of the defendants are found guilty in criminal trials (Source: *The Book of Risks,* by Larry Laudan, John Wiley and Sons). Suppose you are a news reporter following seven criminal trials.
(a) If the trials were in Japan, what is the probability that all the defendants would be found guilty? What is this probability if the trials were in the United States?
(b) Of the seven trials, what is the expected number of guilty verdicts in Japan? What is the expected number in the United States? What is the standard deviation in each case?
(c) *Quota Problem* As a U.S. news reporter, how many trials n would you need to cover to be at least 99% sure of two or more convictions? How many trials n would you need if you covered trials in Japan?

26. *Focus Problem: Personality Types* We now have the tools to solve the Chapter Focus Problem. In the book *A Guide to the Development and Use of the Myers–Briggs Type Indicators* by Myers and McCaully, it was reported that approximately 45% of all university professors are extroverted. Suppose you have classes with six different professors.
   (a) What is the probability that all six are extroverts?
   (b) What is the probability that none of your professors is an extrovert?
   (c) What is the probability that at least two of your professors are extroverts?
   (d) In a group of six professors selected at random, what is the *expected number* of extroverts? What is the *standard deviation* of the distribution?
   (e) *Quota Problem* Suppose you were assigned to write an article for the student newspaper and you were given a quota (by the editor) of interviewing at least three extroverted professors. How many professors selected at random would you need to interview to be at least 90% sure of filling the quota?

27. *Quota Problem: Motel Rooms* The owners of a motel in Florida have noticed that in the long run, about 40% of the people who stop and inquire about a room for the night actually rent a room.
   (a) *Quota Problem* How many inquiries must the owner answer to be 99% sure of renting at least one room?
   (b) If 25 separate inquiries are made about rooms, what is the expected number of inquiries that will result in room rentals?

28. *Critical Thinking* Let r be a binomial random variable representing the number of successes out of n trials.
   (a) Explain why the sample space for r consists of the set $\{0, 1, 2, \ldots, n\}$ and why the sum of the probabilities of all the entries in the entire sample space must be 1.
   (b) Explain why $P(r \geq 1) = 1 - P(0)$.
   (c) Explain why $P(r \geq 2) = 1 - P(0) - P(1)$.
   (d) Explain why $P(r \geq m) = 1 - P(0) - P(1) - \cdots - P(m - 1)$ for $1 \leq m \leq n$.

SECTION **5.4**

The Geometric and Poisson Probability Distributions

FOCUS POINTS

- In many activities, the *first* to succeed wins everything! Use the geometric distribution to compute the probability that the *n*th trial is the first success.
- Use the Poisson distribution to compute the probability of the occurrence of events spread out over time or space.
- Use the Poisson distribution to approximate the binomial distribution when the number of trials is large and the probability of success is small.

In this chapter, we have studied binomial probabilities for the discrete random variable r, the number of successes in n binomial trials. Before we continue in Chapter 6 with continuous random variables, let us examine two other discrete probability distributions, the *geometric* and the *Poisson* probability distributions. These are both related to the binomial distribution.

GEOMETRIC DISTRIBUTION

Suppose we have an experiment in which we repeat binomial trials until we get our *first success,* and then we stop. Let n be the number of the trial on which we get our *first success.* In this context, n is not a fixed number. In fact, n could be any of the numbers 1, 2, 3, and so on. What is the probability that our first success comes on the nth trial? The answer is given by the *geometric probability distribution.*

Geometric probability distribution

GEOMETRIC PROBABILITY DISTRIBUTION

$$P(n) = p(1 - p)^{n-1}$$

where n is the number of the binomial trial on which the *first success* occurs ($n = 1, 2, 3, \ldots$) and p is the probability of success on each trial. *Note:* p must be the same for each trial.

Using some mathematics involving infinite series, it can be shown that the **population mean** and **standard deviation** of the geometric distribution are

$$\mu = \frac{1}{p} \text{ and } \sigma = \frac{\sqrt{1-p}}{p}$$

In many real-life situations, we keep on trying until we achieve success. This is true in areas as diverse as diplomacy, military science, real estate sales, general marketing strategies, medical science, engineering, and technology.

To Engineer Is Human: The Role of Failure in Successful Design is a fascinating book by Henry Petroski (a professor of engineering at Duke University). Reviewers for the *Los Angeles Times* describe this book as serious, amusing, probing, and sometimes frightening. The book examines topics such as the collapse of the Tacoma Narrows suspension bridge, the collapse of the Kansas City Hyatt Regency walkway, and the explosion of the space shuttle *Challenger*. Professor Petroski discusses such topics as "success in foreseeing failure" and the "limits of design." What is meant by expressions such as "foreseeing failure" and the "limits of design"? In the next example, we will see how the geometric probability distribution might help us "forecast" failure.

EXAMPLE 10

FIRST SUCCESS

An automobile assembly plant produces sheet-metal door panels. Each panel moves on an assembly line. As the panel passes a robot, a mechanical arm will perform spot welding at different locations. Each location has a magnetic dot painted where the weld is to be made. The robot is programmed to locate the magnetic dot and perform the weld. However, experience shows that on each trial the robot is only 85% successful at locating the dot. If it cannot locate the magnetic dot, it is programmed to *try again*. The robot will keep trying until it finds the dot (and does the weld) or the door panel passes out of the robot's reach.

(a) What is the probability that the robot's first success will be on attempts $n = 1, 2,$ or 3?

SOLUTION: Since the robot will keep trying until it is successful, the geometric distribution is appropriate. In this case, success S means that the robot finds the correct location. The probability of success is $p = P(S) = 0.85$. The probabilities are

n	$P(n) = p(1-p)^{n-1} = 0.85(0.15)^{n-1}$
1	$0.85(0.15)^0 = 0.85$
2	$0.85(0.15)^1 = 0.1275$
3	$0.85(0.15)^2 \approx 0.0191$

ricardo azoury/iStockphoto.com

(b) The assembly line moves so fast that the robot has a maximum of only three chances before the door panel is out of reach. What is the probability that the robot will be successful before the door panel is out of reach?

SOLUTION: Since $n = 1$ or 2 or 3 is mutually exclusive, then

$$\begin{aligned} P(n = 1 \text{ or } 2 \text{ or } 3) &= P(1) + P(2) + P(3) \\ &\approx 0.85 + 0.1275 + 0.0191 \\ &= 0.9966 \end{aligned}$$

This means that the weld should be correctly located about 99.7% of the time.

(c) ***Interpretation*** What is the probability that the robot will not be able to locate the correct spot within three tries? If 10,000 panels are made, what is the expected number of defectives? Comment on the meaning of this answer in the context of "forecasting failure" and the "limits of design."

SOLUTION: The probability that the robot will correctly locate the weld is 0.9966, from part (b). Therefore, the probability that it will not do so (after three unsuccessful tries) is $1 - 0.9966 = 0.0034$. If we made 10,000 panels, we would expect (forecast) $(10{,}000)(0.0034) = 34$ defectives. We could reduce this by inspecting every door, but such a solution is most likely too costly. If a defective weld of this type is not considered too dangerous, we can accept an expected 34 failures out of 10,000 panels due to the limits of our production design—that is, the speed of the assembly line and the accuracy of the robot. If this is not acceptable, a new (perhaps more costly) design is needed.

Problem 30 introduces the negative binomial distribution, with additional applications in Problems 31 and 32. Problem 33 explores the proof of the negative binomial distribution formula.

COMMENT The geometric distribution deals with binomial trials that are repeated until we have our *first success* on the *n*th trial. Suppose we repeat a binomial trial n times until we have k *successes* (not just one). In the literature, the probability distribution for the random variable n is called the *negative binomial distribution*. For more information on this topic, see Problems 30, 31, 32, and 33 at the end of this section.

POISSON PROBABILITY DISTRIBUTION

Poisson probability distribution

If we examine the binomial distribution as the number of trials n gets larger and larger while the probability of success p gets smaller and smaller, we obtain the *Poisson distribution.* Siméon Denis Poisson (1781–1840) was a French mathematician who studied probabilities of rare events that occur infrequently in space, time, volume, and so forth. The Poisson distribution applies to accident rates, arrival times, defect rates, the occurrence of bacteria in the air, and many other areas of everyday life.

As with the binomial distribution, we assume only two outcomes: A particular event occurs (success) or does not occur (failure) during the specified time period or space. The events need to be independent so that one success does not change the probability of another success during the specified interval. We are interested in computing the probability of r occurrences in the given time period, space, volume, or specified interval.

POISSON DISTRIBUTION

Let λ (Greek letter lambda) be the mean number of successes over time, volume, area, and so forth. Let r be the number of successes ($r = 0, 1, 2, 3, \ldots$) in a corresponding interval of time, volume, area, and so forth. Then the probability of r successes in the interval is

$$P(r) = \frac{e^{-\lambda}\lambda^r}{r!}$$

where e is approximately equal to 2.7183.

Using some mathematics involving infinite series, it can be shown that the **population mean** and **standard deviation** of the Poisson distribution are

$$\mu = \lambda \text{ and } \sigma = \sqrt{\lambda}$$

Note: e^x is a key found on most calculators. Simply use 1 as the exponent, and the calculator will display a decimal approximation for e.

There are many applications of the Poisson distribution. For example, if we take the point of view that waiting time can be subdivided into many small intervals, then the actual arrival (of whatever we are waiting for) during any one of the very short intervals could be thought of as an infrequent (or rare) event. This means that the Poisson distribution can be used as a mathematical model to describe the probabilities of arrivals such as cars to a gas station, planes to an airport, calls to a fire station, births of babies, and even fish on a fisherman's line.

EXAMPLE 11

POISSON DISTRIBUTION

Richard Rowan/Photo Researchers

Pyramid Lake, Nevada

Pyramid Lake is located in Nevada on the Paiute Indian Reservation. The lake is described as a lovely jewel in a beautiful desert setting. In addition to its natural beauty, the lake contains some of the world's largest cutthroat trout. Eight- to ten-pound trout are not uncommon, and 12- to 15-pound trophies are taken each season. The Paiute Nation uses highly trained fish biologists to study and maintain this famous fishery. In one of its publications, *Creel Chronicle* (Vol. 3, No. 2), the following information was given about the November catch for boat fishermen.

Total fish per hour $= 0.667$

Suppose you decide to fish Pyramid Lake for 7 hours during the month of November.

(a) Use the information provided by the fishery biologist in *Creel Chronicle* to find a probability distribution for r, the number of fish (of all sizes) you catch in a period of 7 hours.

SOLUTION: For fish of all sizes, the mean success rate per hour is 0.667.

$$\lambda = 0.667/1 \text{ hour}$$

Since we want to study a *7-hour interval,* we use a little arithmetic to adjust λ to 7 hours. That is, we adjust λ so that it represents the average number of fish expected in a 7-hour period.

$$\lambda = \frac{0.667}{1 \text{ hour}} \cdot \left(\frac{7}{7}\right) = \frac{4.669}{7 \text{ hours}}$$

For convenience, let us use the rounded value $\lambda = 4.7$ for a 7-hour period. Since r is the number of successes (fish caught) in the corresponding 7-hour period and $\lambda = 4.7$ for this period, we use the Poisson distribution to get

$$P(r) = \frac{e^{-\lambda}\lambda^r}{r!} = \frac{e^{-4.7}(4.7)^r}{r!}$$

Recall that $e \approx 2.7183\ldots$ Most calculators have e^x, y^x, and $n!$ keys (see your calculator manual), so the Poisson distribution is not hard to compute.

(b) What is the probability that in 7 hours you will get 0, 1, 2, or 3 fish of any size?

SOLUTION: Using the result of part (a), we get

$$P(0) = \frac{e^{-4.7}(4.7)^0}{0!} \approx 0.0091 \approx 0.01$$

$$P(1) = \frac{e^{-4.7}(4.7)^1}{1!} \approx 0.0427 \approx 0.04$$

$$P(2) = \frac{e^{-4.7}(4.7)^2}{2!} \approx 0.1005 \approx 0.10$$

$$P(3) = \frac{e^{-4.7}(4.7)^3}{3!} \approx 0.1574 \approx 0.16$$

The probabilities of getting 0, 1, 2, or 3 fish are about 1%, 4%, 10%, and 16%, respectively.

(c) What is the probability that you will get four or more fish in the 7-hour fishing period?

SOLUTION: The sample space of all r values is $r = 0, 1, 2, 3, 4, 5, \ldots$ The probability of the entire sample space is 1, and these events are mutually exclusive. Therefore,

$$1 = P(0) + P(1) + P(2) + P(3) + P(4) + P(5) + \cdots$$

So,

$$\begin{aligned} P(r \geq 4) &= P(4) + P(5) + \cdots = 1 - P(0) - P(1) - P(2) - P(3) \\ &\approx 1 - 0.01 - 0.04 - 0.10 - 0.16 \\ &= 0.69 \end{aligned}$$

There is about a 69% chance that you will catch four or more fish in a 7-hour period.

Use of tables

Table 4 of Appendix II is a table of the Poisson probability distribution for selected values of λ and the number of successes r. Table 5-15 is an excerpt from that table.

To find the value of $P(r = 2)$ when $\lambda = 0.3$, look in the column headed by 0.3 and the row headed by 2. From the table, we see that $P(2) = 0.0333$.

LOOKING FORWARD

In Example 11 we studied the probability distribution of r, the *number* of fish caught. Now, what about the probability distribution of x, the *waiting time* between catching one fish and the next? It turns out that the random variable x has a continuous distribution called an *exponential distribution*. Continuous distributions will be studied in Chapter 6. To learn more about the exponential distribution and waiting time between catching fish, please see Problem 20 at the end of Section 6.1.

TABLE 5-15 Excerpt from Appendix II, Table 4, "Poisson Probability Distribution"

			λ		
r	0.1	0.2	0.3	0.4	0.5 ....
0	.9048	.8187	.7408	.6703	.6065
1	.0905	.1637	.2222	.2681	.3033
2	.0045	.0164	.0333	.0536	.0758
3	.0002	.0011	.0033	.0072	.0126
4	.0000	.0001	.0003	.0007	.0016

TECH NOTES

The TI-84Plus/TI-83Plus/TI-*n*spire calculators have commands for the geometric and Poisson distributions. Excel 2010 and Minitab support the Poisson distribution. All the technologies have both the probability distribution and the cumulative distribution.

TI-84Plus/TI-83Plus/TI-*n*spire (with TI-84Plus keypad) Use the **DISTR** key and scroll to **geometpdf**(p,n) for the probability of first success on trial number n; scroll to **geomecdf**(p,n) for the probability of first success on trial number $\leq n$. Use **poissonpdf**(λ,r) for the probability of r successes. Use **poissoncdf**(λ,r) for the probability of at most r successes. For example, when $\lambda = 0.25$ and $r = 2$, we get the following results.

```
poissonpdf(.25,2)
                .0243375245        P(r = 2)
poissoncdf(.25,2)
                .9978385033        P(r ≤ 2)
```

Excel 2010 Click on the **Insert Function** f_x. In the dialogue box, select **Statistical** for the category, and then select **Poisson**. Enter the trial number *r*, and use λ for the mean. False gives the probability $P(r)$ and True gives the cumulative probability $P(\text{at most } r)$.

Minitab Put the *r* values in a column. Then use the menu choice **Calc ➤ Probability Distribution ➤ Poisson**. In the dialogue box, select probability or cumulative, enter the column number containing the *r* values, and use λ for the mean.

POISSON APPROXIMATION TO THE BINOMIAL PROBABILITY DISTRIBUTION

Poisson approximation to the binomial

In the preceding examples, we have seen how the Poisson distribution can be used over intervals of time, space, area, and so on. However, the Poisson distribution also can be used as a probability distribution for "rare" events. In the binomial distribution, if the number of trials n is large while the probability p of success is quite small, we call the event (success) a "rare" event. Put another way, it can be shown that for most practical purposes, the Poisson distribution will be a very good *approximation to the binomial distribution*, provided the number of trials n is larger than or equal to 100 and $\lambda = np$ is less than 10. As n gets larger and p gets smaller, the approximation becomes better and better.

Problems 28 and 29 show how to compute conditional probabilities for binomial experiments using the Poisson approximation for the binomial distribution. In these problems we see how to compute the probability of *r* successes out of *n* binomial trials, given that a certain number of successes occurs.

PROCEDURE

HOW TO APPROXIMATE BINOMIAL PROBABILITIES USING POISSON PROBABILITIES

Suppose you have a binomial distribution with

n = number of trials

r = number of successes

p = probability of success on each trial

If $n \geq 100$ and $np < 10$, then r has a binomial distribution that is approximated by a Poisson distribution with $\lambda = np$.

$$P(r) \approx \frac{e^{-\lambda}\lambda^r}{r!}$$

Note: $\lambda = np$ is the expected value of the binomial distribution.

EXAMPLE 12

POISSON APPROXIMATION TO THE BINOMIAL

Isabel Briggs Myers was a pioneer in the study of personality types. Today the Myers–Briggs Type Indicator is used in many career counseling programs as well as in many industrial settings where people must work closely together as a team. The 16 personality types are discussed in detail in the book *A Guide to the Development and Use of*

Ariel Skelley/Blend Images/Getty Images

the Myers–Briggs Type Indicators, by Myers and McCaulley. Each personality type has its own special contribution in any group activity. One of the more "rare" types is INFJ (introverted, intuitive, feeling, judgmental), which occurs in only about 2.1% of the population. Suppose a high school graduating class has 167 students, and suppose we call "success" the event that a student is of personality type INFJ.

(a) Let r be the number of successes (INFJ students) in the $n = 167$ trials (graduating class). If $p = P(S) = 0.021$, will the Poisson distribution be a good approximation to the binomial?

SOLUTION: Since $n = 167$ is greater than 100 and $\lambda = np = 167(0.021) \approx 3.5$ is less than 10, the Poisson distribution should be a good approximation to the binomial.

(b) Estimate the probability that this graduating class has 0, 1, 2, 3, or 4 people who have the INFJ personality type.

SOLUTION: Since Table 4 (Appendix II) for the Poisson distribution includes the values $\lambda = 3.5$ and $r = 0, 1, 2, 3$, or 4, we may simply look up the values for $P(r)$, $r = 0, 1, 2, 3, 4$:

$$P(r = 0) = 0.0302 \quad P(r = 1) = 0.1057$$
$$P(r = 2) = 0.1850 \quad P(r = 3) = 0.2158$$
$$P(r = 4) = 0.1888$$

Since the outcomes $r = 0, 1, 2, 3$, or 4 successes are mutually exclusive, we can compute the probability of 4 or fewer INFJ types by using the addition rule for mutually exclusive events:

$$\begin{aligned} P(r \le 4) &= P(0) + P(1) + P(2) + P(3) + P(4) \\ &= 0.0302 + 0.1057 + 0.1850 + 0.2158 + 0.1888 \\ &= 0.7255 \end{aligned}$$

The probability that the graduating class will have four or fewer INFJ personality types is about 0.73.

(c) Estimate the probability that this class has five or more INFJ personality types.

SOLUTION: Because the outcomes of a binomial experiment are all mutually exclusive, we have

$$P(r \le 4) + P(r \ge 5) = 1$$
$$\text{or } P(r \ge 5) = 1 - P(r \le 4) = 1 - 0.7255 = 0.2745$$

The probability is approximately 0.27 that there will be five or more INFJ personality types in the graduating class.

SUMMARY

In this section, we have studied two discrete probability distributions. The Poisson distribution gives us the probability of r successes in an interval of time or space. The Poisson distribution also can be used to approximate the binomial distribution when $n \ge 100$ and $np < 10$. The geometric distribution gives us the probability that our first success will occur on the nth trial. In the next guided exercise, we will see situations in which each of these distributions applies.

PROCEDURE

HOW TO IDENTIFY DISCRETE PROBABILITY DISTRIBUTIONS

Distribution	Conditions and Setting	Formulas
Binomial distribution	1. There are n independent trials, each repeated under identical conditions. 2. Each trial has two outcomes, S = success and F = failure. 3. $P(S) = p$ is the same for each trial, as is $P(F) = q = 1 - p$ 4. The random variable r represents the number of successes out of n trials. $0 \le r \le n$	The probability of exactly r successes out of n trials is $P(r) = \frac{n!}{r!(n-r)!}p^r q^{n-r} = C_{n,r}p^r q^{n-r}$ For r, $\mu = np$ and $\sigma = \sqrt{npq}$ Table 3 of Appendix II has $P(r)$ values for selected n and p.
Geometric distribution	1. There are n independent trials, each repeated under identical conditions. 2. Each trial has two outcomes, S = success and F = failure. 3. $P(S) = p$ is the same for each trial, as is $P(F) = q = 1 - p$ 4. The random variable n represents the number of the trial on which the *first success* occurs. $n = 1, 2, 3, \ldots$.	The probability that the first success occurs on the nth trial is $P(n) = pq^{n-1}$ For n, $\mu = \frac{1}{p}$ and $\sigma = \frac{\sqrt{q}}{p}$
Poisson distribution	1. Consider a random process that occurs over time, volume, area, or any other quantity that can (in theory) be subdivided into smaller and smaller intervals. 2. Identify success in the context of the interval (time, volume, area, . . .) you are studying. 3. Based on long-term experience, compute the mean or average number of successes that occur over the interval (time, volume, area, . . .) you are studying. λ = mean number of successes over designated interval 4. The random variable r represents the number of successes that occur over the interval on which you perform the random process. $r = 0, 1, 2, 3, \ldots$.	The probability of r successes in the interval is $P(r) = \frac{e^{-\lambda}\lambda^r}{r!}$ For r, $\mu = \lambda$ and $\sigma = \sqrt{\lambda}$ Table 4 of Appendix II gives $P(r)$ for selected values of λ and r.
Poisson approximation to the binomial distribution	1. There are n independent trials, each repeated under identical conditions. 2. Each trial has two outcomes, S = success and F = failure. 3. $P(S) = p$ is the same for each trial. 4. In addition, $n \ge 100$ and $np < 10$. 5. The random variable r represents the number of successes out of n trials in a binomial distribution.	$\lambda = np$ the expected value of r. The probability of r successes on n trials is $P(r) \approx \frac{e^{-\lambda}\lambda^r}{r!}$ Table 4 of Appendix II gives values of $P(r)$ for selected λ and r.

GUIDED EXERCISE 9 SELECT APPROPRIATE DISTRIBUTION

For each problem, first identify the type of probability distribution needed to solve the problem: binomial, geometric, Poisson, or Poisson approximation to the binomial. Then solve the problem.

(I) Denver, Colorado, is prone to severe hailstorms. Insurance agents claim that a homeowner in Denver can expect to replace his or her roof (due to hail damage) once every 10 years. What is the probability that in 12 years, a homeowner in Denver will need to replace the roof twice because of hail?

Continued

GUIDED EXERCISE 9 *continued*

(a) Consider the problem stated in part (I). What is success in this case? We are interested in the probability of two successes over a specified time interval of 12 years. Which distribution should we use?

Here we can say success is needing to replace a roof because of hail. Because we are interested in the probability of two successes over a time interval, we should use the Poisson distribution.

(b) In part (I), we are told that the average roof replacement is once every 10 years. What is the average number of times the roof needs to be replaced in 12 years? What is λ for the 12-year period?

We are given a value of $\lambda = 1$ for 10 years. To compute λ for 12 years, we convert the denominator to 12 years.

$$\lambda = \frac{0.1}{1\,\text{year}} \cdot \frac{12}{12} = \frac{1.2}{12\,\text{year}}$$

For 12 years, we have $\lambda = 1.2$.

(c) To finish part (I), use the Poisson distribution to find the probability of two successes in the 12-year period.

We may use Table 4 of Appendix II because $\lambda = 1.2$ and $r = 2$ are values in the table. The table gives $P(r = 2) = 0.2169$. Using the formula or a calculator gives the same result.

$$P(r) = \frac{e^{-\lambda}\lambda^r}{r!}$$

$$P(r = 2) = \frac{e^{-1.2}(1.2)^2}{2!} \approx 0.2169$$

There is about a 21.7% chance the roof will be damaged twice by hail during a 12-year period.

(II) A telephone network substation will keep trying to connect a long-distance call to a trunk line until the fourth attempt has been made. After the fourth unsuccessful attempt, the call number goes into a buffer memory bank, and the caller gets a recorded message to be patient. During peak calling periods, the probability of a call connecting into a trunk line is 65% on each try. What percentage of all calls made during peak times will wind up in the buffer memory bank?

(d) Consider part (II). What is success? We are interested in the probability that a call will wind up in the buffer memory. This will occur if success does not occur before which trial? Since we are looking at the probability of a first success by a specified trial number, which probability distribution do we use?

Success is connecting a long-distance call to a trunk line. The call will go into the buffer memory if success is not achieved during the first four attempts. In symbols, the call will go into the buffer if the trial number n of the first success is such that $n \geq 5$. We use the geometric distribution.

(e) What is the probability of success on a single trial? Use this information and the formula for the geometric distribution to compute the probability that the first success occurs on trial 1, 2, 3, or 4. Then use this information to compute $P(n \geq 5)$, where n is the trial number of the first success. What percentage of the calls go to the buffer?

Success means the call connects to the trunk line. According to the description in the problem,

$P(S) = 0.65 = p$

By the formula for the geometric probability distribution, where n represents the trial number of the first success,

$P(n) = p(1 - p)^{n-1}$

Therefore,

$$P(1) = (0.65)(0.35)^0 = 0.65$$

$$P(2) = (0.65)(0.35)^1 = 0.2275$$

Continued

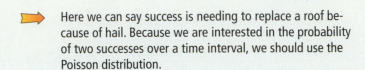

GUIDED EXERCISE 9 *continued*

$$P(3) = (0.65)(0.35)^2 \approx 0.0796$$

$$P(4) = (0.65)(0.35)^3 \approx 0.0279$$

$$P(n \geq 5) = 1 - P(1) - P(2) - P(3) - P(4)$$

$$\approx 1 - 0.65 - 0.2275 - 0.0796 - 0.0279$$

$$= 0.015$$

About 1.5% of the calls go to the buffer.

(III) The murder rate is 3.6 murders per 100,000 inhabitants (Reference: U.S. Department of Justice, Federal Bureau of Investigation). In a community of 1254 people, what is the probability that at least one person will be murdered?

(f) Consider part (III). What is success in this case? What is the value of n? Find p, the probability of success on a single trial, to six places after the decimal.

→ Success is a person being murdered.

$n = 1254$

$$p = \frac{3.6}{100{,}000} = 0.000036$$

(g) Compute np to three decimal places. Is it appropriate to use the Poisson approximation to the binomial? What is the value of λ to three decimal places?

→ $np \approx 0.045$

Yes.

$\lambda = np \approx 0.045$

(h) Estimate $P(r = 0)$ to three decimal places.

→ $$P(r) \approx \frac{e^{-\lambda}\lambda^r}{r!} \approx \frac{e^{-0.045}(0.045)^0}{0!} \approx 0.956$$

Recall that $0! = 1$.

(i) Use the relation $P(r \geq 1) = 1 - P(0)$ to estimate the probability that at least one person will be murdered.

→ $$P(r \geq 1) \approx 1 - 0.956 \approx 0.044$$

VIEWPOINT When Do Cracks Become Breakthroughs?

No one wants to learn by mistakes! However, learning by our successes will not take us beyond the state of the art! Each new idea, technology, social plan, or engineering structure can be considered a new trial. In the meantime, we the laypeople, whose spokesperson is often a poet or writer, will be threatened by both **failures** *and* **successes**. *This is the nature not only of science, technology, and engineering but also of all human endeavors. [For more discussion on this topic, see Problem 18, as well as* To Engineer Is Human: The Role of Failure in Successful Design *by Professor Petroski (Duke University Press).]*

SECTION 5.4 PROBLEMS

1. *Statistical Literacy* For a binomial experiment, what probability distribution is used to find the probability that the *first* success will occur on a specified trial?

2. *Statistical Literacy* When using the Poisson distribution, which parameter of the distribution is used in probability computations? What is the symbol used for this parameter?

3. *Critical Thinking* Suppose we have a binomial experiment with 50 trials, and the probability of success on a single trial is 0.02. Is it appropriate to use the Poisson distribution to approximate the probability of two successes? Explain.

4. *Critical Thinking* Suppose we have a binomial experiment, and the probability of success on a single trial is 0.02. If there are 150 trials, is it appropriate to use the Poisson distribution to approximate the probability of three successes? Explain.

5. *Basic Computation: Geometric Distribution* Given a binomial experiment with probability of success on a single trial $p = 0.40$, find the probability that the first success occurs on trial number $n = 3$.

6. *Basic Computation: Geometric Distribution* Given a binomial experiment with probability of success on a single trial $p = 0.30$, find the probability that the first success occurs on trial number $n = 2$.

7. *Basic Computation: Poisson Distribution* Given a binomial experiment with $n = 200$ trials and probability of success on a single trial $p = 0.04$, find the value of λ and then use the Poisson distribution to estimate the probability of $r = 8$ successes.

8. *Basic Computation: Poisson Distribution* Given a binomial experiment with $n = 150$ trials and probability of success on a single trial $p = 0.06$, find the value of λ and then use the Poisson distribution to estimate the probability of $r \leq 2$ successes.

9. *College: Core Requirement* Susan is taking Western Civilization this semester on a pass/fail basis. The department teaching the course has a history of passing 77% of the students in Western Civilization each term. Let $n = 1, 2, 3, \ldots$ represent the number of times a student takes western civilization until the *first* passing grade is received. (Assume the trials are independent.)
   (a) Write out a formula for the probability distribution of the random variable n.
   (b) What is the probability that Susan passes on the first try ($n = 1$)?
   (c) What is the probability that Susan first passes on the second try ($n = 2$)?
   (d) What is the probability that Susan needs three or more tries to pass western civilization?
   (e) What is the expected number of attempts at western civilization Susan must make to have her (first) pass? *Hint:* Use μ for the geometric distribution and round.

10. *Law: Bar Exam* Bob is a recent law school graduate who intends to take the state bar exam. According to the National Conference on Bar Examiners, about 57% of all people who take the state bar exam pass (Source: *The Book of Odds* by Shook and Shook, Signet). Let $n = 1, 2, 3, \ldots$ represent the number of times a person takes the bar exam until the *first* pass.
    (a) Write out a formula for the probability distribution of the random variable n.
    (b) What is the probability that Bob first passes the bar exam on the second try ($n = 2$)?
    (c) What is the probability that Bob needs three attempts to pass the bar exam?

(d) What is the probability that Bob needs more than three attempts to pass the bar exam?
(e) What is the expected number of attempts at the state bar exam Bob must make for his (first) pass? *Hint:* Use μ for the geometric distribution and round.

11. *Sociology: Hawaiians* On the leeward side of the island of Oahu, in the small village of Nanakuli, about 80% of the residents are of Hawaiian ancestry (Source: *The Honolulu Advertiser*). Let $n = 1, 2, 3, \ldots$ represent the number of people you must meet until you encounter the *first* person of Hawaiian ancestry in the village of Nanakuli.
(a) Write out a formula for the probability distribution of the random variable n.
(b) Compute the probabilities that $n = 1$, $n = 2$, and $n = 3$.
(c) Compute the probability that $n \geq 4$.
(d) In Waikiki, it is estimated that about 4% of the residents are of Hawaiian ancestry. Repeat parts (a), (b), and (c) for Waikiki.

12. *Agriculture: Apples* Approximately 3.6% of all (untreated) Jonathan apples had bitter pit in a study conducted by the botanists Ratkowsky and Martin (Source: *Australian Journal of Agricultural Research,* Vol. 25, pp. 783–790). (Bitter pit is a disease of apples resulting in a soggy core, which can be caused either by overwatering the apple tree or by a calcium deficiency in the soil.) Let n be a random variable that represents the first Jonathan apple chosen at random that has bitter pit.
(a) Write out a formula for the probability distribution of the random variable n.
(b) Find the probabilities that $n = 3$, $n = 5$, and $n = 12$.
(c) Find the probability that $n \geq 5$.
(d) What is the expected number of apples that must be examined to find the first one with bitter pit? *Hint:* Use μ for the geometric distribution and round.

13. *Fishing: Lake Trout* At Fontaine Lake Camp on Lake Athabasca in northern Canada, history shows that about 30% of the guests catch lake trout over 20 pounds on a 4-day fishing trip (Source: Athabasca Fishing Lodges, Saskatoon, Canada). Let n be a random variable that represents the *first* trip to Fontaine Lake Camp on which a guest catches a lake trout over 20 pounds.
(a) Write out a formula for the probability distribution of the random variable n.
(b) Find the probability that a guest catches a lake trout weighing at least 20 pounds for the *first* time on trip number 3.
(c) Find the probability that it takes more than three trips for a guest to catch a lake trout weighing at least 20 pounds.
(d) What is the expected number of fishing trips that must be taken to catch the first lake trout over 20 pounds? *Hint:* Use μ for the geometric distribution and round.

14. *Archaeology: Artifacts* At Burnt Mesa Pueblo, in one of the archaeological excavation sites, the artifact density (number of prehistoric artifacts per 10 liters of sediment) was 1.5 (Source: *Bandelier Archaeological Excavation Project: Summer 1990 Excavations at Burnt Mesa Pueblo and Casa del Rito,* edited by Kohler, Washington State University Department of Anthropology). Suppose you are going to dig up and examine 50 liters of sediment at this site. Let $r = 0, 1, 2, 3, \ldots$ be a random variable that represents the number of prehistoric artifacts found in your 50 liters of sediment.
(a) Explain why the Poisson distribution would be a good choice for the probability distribution of r. What is λ? Write out the formula for the probability distribution of the random variable r.
(b) Compute the probabilities that in your 50 liters of sediment you will find two prehistoric artifacts, three prehistoric artifacts, and four prehistoric artifacts.

(c) Find the probability that you will find three or more prehistoric artifacts in the 50 liters of sediment.

(d) Find the probability that you will find fewer than three prehistoric artifacts in the 50 liters of sediment.

15. *Ecology: River Otters* In his doctoral thesis, L. A. Beckel (University of Minnesota, 1982) studied the social behavior of river otters during the mating season. An important role in the bonding process of river otters is very short periods of social grooming. After extensive observations, Dr. Beckel found that one group of river otters under study had a frequency of initiating grooming of approximately 1.7 for every 10 minutes. Suppose that you are observing river otters for 30 minutes. Let $r = 0, 1, 2, \ldots$ be a random variable that represents the number of times (in a 30-minute interval) one otter initiates social grooming of another.

TRyburn/Shutterstock.com

(a) Explain why the Poisson distribution would be a good choice for the probability distribution of r. What is λ? Write out the formula for the probability distribution of the random variable r.

(b) Find the probabilities that in your 30 minutes of observation, one otter will initiate social grooming four times, five times, and six times.

(c) Find the probability that one otter will initiate social grooming four or more times during the 30-minute observation period.

(d) Find the probability that one otter will initiate social grooming fewer than four times during the 30-minute observation period.

16. *Law Enforcement: Shoplifting* *The Denver Post* reported that, on average, a large shopping center has had an incident of shoplifting caught by security once every three hours. The shopping center is open from 10 A.M. to 9 P.M. (11 hours). Let r be the number of shoplifting incidents caught by security in the 11-hour period during which the center is open.

(a) Explain why the Poisson probability distribution would be a good choice for the random variable r. What is λ?

(b) What is the probability that from 10 A.M. to 9 P.M. there will be at least one shoplifting incident caught by security?

(c) What is the probability that from 10 A.M. to 9 P.M. there will be at least three shoplifting incidents caught by security?

(d) What is the probability that from 10 A.M. to 9 P.M. there will be no shoplifting incidents caught by security?

17. *Vital Statistics: Birthrate* *USA Today* reported that the U.S. (annual) birthrate is about 16 per 1000 people, and the death rate is about 8 per 1000 people.

(a) Explain why the Poisson probability distribution would be a good choice for the random variable r = number of births (or deaths) for a community of a given population size.

(b) In a community of 1000 people, what is the (annual) probability of 10 births? What is the probability of 10 deaths? What is the probability of 16 births? 16 deaths?

(c) Repeat part (b) for a community of 1500 people. You will need to use a calculator to compute P(10 births) and P(16 births).

(d) Repeat part (b) for a community of 750 people.

18. *Engineering: Cracks* Henry Petroski is a professor of civil engineering at Duke University. In his book *To Engineer Is Human: The Role of Failure in Successful Design,* Professor Petroski says that up to 95% of all structural failures, including those of bridges, airplanes, and other commonplace products of technology, are believed to be the result of crack growth. In most cases, the cracks grow slowly. It is only when the cracks reach intolerable proportions and still go undetected that catastrophe can occur. In a cement retaining wall, occasional hairline cracks are normal and nothing to worry about. If these cracks are spread out and not too close together, the

Yuan Zhang/iStockphoto.com

wall is considered safe. However, if a number of cracks group together in a small region, there may be real trouble. Suppose a given cement retaining wall is considered safe if hairline cracks are evenly spread out and occur on the average of 4.2 cracks per 30-foot section of wall.

(a) Explain why a Poisson probability distribution would be a good choice for the random variable r = number of hairline cracks for a given length of retaining wall.

(b) In a 50-foot section of safe wall, what is the probability of three (evenly spread-out) hairline cracks? What is the probability of three *or more* (evenly spread-out) hairline cracks?

(c) Answer part (b) for a 20-foot section of wall.

(d) Answer part (b) for a 2-foot section of wall. Round λ to the nearest tenth.

(e) Consider your answers to parts (b), (c), and (d). If you had three hairline cracks evenly spread out over a 50-foot section of wall, should this be cause for concern? The probability is low. Could this mean that you are lucky to have so few cracks? On a 20-foot section of wall [part (c)], the probability of three cracks is higher. Does this mean that this distribution of cracks is closer to what we should expect? For part (d), the probability is very small. Could this mean you are not so lucky and have something to worry about? Explain your answers.

19. *Meteorology: Winter Conditions* Much of Trail Ridge Road in Rocky Mountain National Park is over 12,000 feet high. Although it is a beautiful drive in summer months, in winter the road is closed because of severe weather conditions. *Winter Wind Studies in Rocky Mountain National Park* by Glidden (published by Rocky Mountain Nature Association) states that sustained gale-force winds (over 32 miles per hour and often over 90 miles per hour) occur on the average of once every 60 hours at a Trail Ridge Road weather station.

(a) Let r = frequency with which gale-force winds occur in a given time interval. Explain why the Poisson probability distribution would be a good choice for the random variable r.

(b) For an interval of 108 hours, what are the probabilities that r = 2, 3, and 4? What is the probability that $r < 2$?

(c) For an interval of 180 hours, what are the probabilities that r = 3, 4, and 5? What is the probability that $r < 3$?

20. *Earthquakes: San Andreas Fault* *USA Today* reported that Parkfield, California, is dubbed the world's earthquake capital because it sits on top of the notorious San Andreas fault. Since 1857, Parkfield has had a major earthquake on the average of once every 22 years.

(a) Explain why a Poisson probability distribution would be a good choice for r = number of earthquakes in a given time interval.

(b) Compute the probability of at least one major earthquake in the next 22 years. Round λ to the nearest hundredth, and use a calculator.

(c) Compute the probability that there will be no major earthquake in the next 22 years. Round λ to the nearest hundredth, and use a calculator.

(d) Compute the probability of at least one major earthquake in the next 50 years. Round λ to the nearest hundredth, and use a calculator.

(e) Compute the probability of no major earthquakes in the next 50 years. Round λ to the nearest hundredth, and use a calculator.

21. *Real Estate: Sales* Jim is a real estate agent who sells large commercial buildings. Because his commission is so large on a single sale, he does not need to sell many buildings to make a good living. History shows that Jim has a record of selling an average of eight large commercial buildings every 275 days.

(a) Explain why a Poisson probability distribution would be a good choice for r = number of buildings sold in a given time interval.

(b) In a 60-day period, what is the probability that Jim will make no sales? one sale? two or more sales?
(c) In a 90-day period, what is the probability that Jim will make no sales? two sales? three or more sales?

22. *Law Enforcement: Burglaries* *The Honolulu Advertiser* stated that in Honolulu there was an average of 661 burglaries per 100,000 households in a given year. In the Kohola Drive neighborhood there are 316 homes. Let r = number of these homes that will be burglarized in a year.
(a) Explain why the Poisson approximation to the binomial would be a good choice for the random variable r. What is n? What is p? What is λ to the nearest tenth?
(b) What is the probability that there will be no burglaries this year in the Kohola Drive neighborhood?
(c) What is the probability that there will be no more than one burglary in the Kohola Drive neighborhood?
(d) What is the probability that there will be two or more burglaries in the Kohola Drive neighborhood?

23. *Criminal Justice: Drunk Drivers* *Harper's Index* reported that the number of Orange County, California convicted drunk drivers whose sentence included a tour of the morgue was 569, of which only 1 became a repeat offender.
(a) Suppose that of 1000 newly convicted drunk drivers, all were required to take a tour of the morgue. Let us assume that the probability of a repeat offender is still $p = 1/569$. Explain why the Poisson approximation to the binomial would be a good choice for r = number of repeat offenders out of 1000 convicted drunk drivers who toured the morgue. What is λ to the nearest tenth?
(b) What is the probability that $r = 0$?
(c) What is the probability that $r > 1$?
(d) What is the probability that $r > 2$?
(e) What is the probability that $r > 3$?

24. *Airlines: Lost Bags* *USA Today* reported that for all airlines, the number of lost bags was

May: 6.02 per 1000 passengers   December: 12.78 per 1000 passengers

Note: A passenger could lose more than one bag.
(a) Let r = number of bags lost per 1000 passengers in May. Explain why the Poisson distribution would be a good choice for the random variable r. What is λ to the nearest tenth?
(b) In the month of May, what is the probability that out of 1000 passengers, no bags are lost? that 3 or more bags are lost? that 6 or more bags are lost?
(c) In the month of December, what is the probability that out of 1000 passengers, no bags are lost? that 6 or more bags are lost? that 12 or more bags are lost? (Round λ to the nearest whole number.)

25. *Law Enforcement: Officers Killed* *Chances: Risk and Odds in Everyday Life*, by James Burke, reports that the probability a police officer will be killed in the line of duty is 0.5% (or less).
(a) In a police precinct with 175 officers, let r = number of police officers killed in the line of duty. Explain why the Poisson approximation to the binomial would be a good choice for the random variable r. What is n? What is p? What is λ to the nearest tenth?
(b) What is the probability that no officer in this precinct will be killed in the line of duty?
(c) What is the probability that one or more officers in this precinct will be killed in the line of duty?

(d) What is the probability that two or more officers in this precinct will be killed in the line of duty?

26. *Business Franchise: Shopping Center* *Chances: Risk and Odds in Everyday Life,* by James Burke, reports that only 2% of all local franchises are business failures. A Colorado Springs shopping complex has 137 franchises (restaurants, print shops, convenience stores, hair salons, etc.).
(a) Let r be the number of these franchises that are business failures. Explain why a Poisson approximation to the binomial would be appropriate for the random variable r. What is n? What is p? What is λ (rounded to the nearest tenth)?
(b) What is the probability that none of the franchises will be a business failure?
(c) What is the probability that two or more franchises will be business failures?
(d) What is the probability that four or more franchises will be business failures?

27. *Poisson Approximation to the Binomial: Comparisons*
(a) For $n = 100$, $p = 0.02$, and $r = 2$, compute $P(r)$ using the formula for the binomial distribution and your calculator:

$$P(r) = C_{n,r}p^r(1 - p)^{n-r}$$

(b) For $n = 100$, $p = 0.02$, and $r = 2$, estimate $P(r)$ using the Poisson approximation to the binomial.
(c) Compare the results of parts (a) and (b). Does it appear that the Poisson distribution with $\lambda = np$ provides a good approximation for $P(r = 2)$?
(d) Repeat parts (a) to (c) for $r = 3$.

28. *Expand Your Knowledge: Conditional Probability* Pyramid Lake is located in Nevada on the Paiute Indian Reservation. This lake is famous for large cutthroat trout. The mean number of trout (large and small) caught from a boat is 0.667 fish per hour (Reference: *Creel Chronicle,* Vol. 3, No. 2). Suppose you rent a boat and go fishing for 8 hours. Let r be a random variable that represents the number of fish you catch in the 8-hour period.
(a) Explain why a Poisson probability distribution is appropriate for r. What is λ for the 8-hour fishing trip? Round λ to the nearest tenth so that you can use Table 4 of Appendix II for Poisson probabilities.
(b) If you have already caught three trout, what is the probability you will catch a total of seven or more trout? Compute $P(r \geq 7 \mid r \geq 3)$. See Hint below.
(c) If you have already caught four trout, what is the probability you will catch a total of fewer than nine trout? Compute $P(r < 9 \mid r \geq 4)$. See Hint below.
(d) List at least three other areas besides fishing to which you think conditional Poisson probabilities can be applied.

Hint for solution: Review item 6, conditional probability, in the summary of basic probability rules at the end of Section 4.2. Note that

$$P(A \mid B) = \frac{P(A \text{ and } B)}{P(B)}$$

and show that in part (b),

$$P(r \geq 7 \mid r \geq 3) = \frac{P((r \geq 7) \text{ and } (r \geq 3))}{P(r \geq 3)} = \frac{P(r \geq 7)}{P(r \geq 3)}$$

29. *Conditional Probability: Hail Damage* In western Kansas, the summer density of hailstorms is estimated at about 2.1 storms per 5 square miles. In most cases, a hailstorm damages only a relatively small area in a square mile (Reference: *Agricultural Statistics,* U.S. Department of Agriculture). A crop insurance company has insured a tract of 8 square miles of Kansas wheat land

Hydromet/Shutterstock.com

against hail damage. Let r be a random variable that represents the number of hailstorms this summer in the 8-square-mile tract.

(a) Explain why a Poisson probability distribution is appropriate for r. What is λ for the 8-square-mile tract of land? Round λ to the nearest tenth so that you can use Table 4 of Appendix II for Poisson probabilities.

(b) If there already have been two hailstorms this summer, what is the probability that there will be a total of four or more hailstorms in this tract of land? Compute $P(r \geq 4 \mid r \geq 2)$.

(c) If there already have been three hailstorms this summer, what is the probability that there will be a total of fewer than six hailstorms? Compute $P(r < 6 \mid r \geq 3)$.

Hint: See Problem 28.

30. *Expand Your Knowledge: Negative Binomial Distribution* Suppose you have binomial trials for which the probability of success on each trial is p and the probability of failure is $q = 1 - p$. Let k be a fixed whole number greater than or equal to 1. Let n be the number of the trial on which the kth success occurs. This means that the first $k - 1$ successes occur within the first $n - 1$ trials, while the kth success actually occurs on the nth trial. Now, if we are going to have k successes, we must have at least k trials. So, $n = k, k + 1, k + 2, \ldots$ and n is a random variable. In the literature of mathematical statistics, the probability distribution for n is called the *negative binomial distribution.* The formula for the probability distribution of n is shown in the next display (see Problem 33 for a derivation).

Negative binomial distribution

NEGATIVE BINOMIAL DISTRIBUTION

Let $k \geq 1$ be a fixed whole number. The probability that the kth success occurs on trial number n is

$$P(n) = C_{n-1,\,k-1}\,p^k q^{n-k}$$

where $$C_{n-1,\,k-1} = \frac{(n-1)!}{(k-1)!(n-k)!}$$

$$n = k, k+1, k+2, \ldots$$

The expected value and standard deviation of this probability distribution are

$$\mu = \frac{k}{p} \text{ and } \sigma = \frac{\sqrt{kq}}{p}$$

Note: If $k = 1$, the negative binomial distribution is called the *geometric distribution.*

In eastern Colorado, there are many dry-land wheat farms. The success of a spring wheat crop is dependent on sufficient moisture in March and April. Assume that the probability of a successful wheat crop in this region is about 65%. So, the probability of success in a single year is $p = 0.65$, and the probability of failure is $q = 0.35$. The Wagner farm has taken out a loan and needs $k = 4$ successful crops to repay it. Let n be a random variable representing the year in which the fourth successful crop occurs (after the loan was made).

(a) Write out the formula for $P(n)$ in the context of this application.

(b) Compute $P(n = 4)$, $P(n = 5)$, $P(n = 6)$, and $P(n = 7)$.

(c) What is the probability that the Wagners can repay the loan within 4 to 7 years? *Hint:* Compute $P(4 \leq n \leq 7)$.

(d) What is the probability that the Wagners will need to farm for 8 or more years before they can repay the loan? *Hint:* Compute $P(n \geq 8)$.

(e) What are the expected value μ and standard deviation σ of the random variable n? Interpret these values in the context of this application.

31. *Negative Binomial Distribution: Marketing* Susan is a sales representative who has a history of making a successful sale from about 80% of her sales contacts. If she makes 12 successful sales this week, Susan will get a bonus. Let n be a random variable representing the number of contacts needed for Susan to get the 12th sale.

(a) Explain why a negative binomial distribution is appropriate for the random variable n. Write out the formula for $P(n)$ in the context of this application. *Hint:* See Problem 30.

(b) Compute $P(n = 12)$, $P(n = 13)$, and $P(n = 14)$.

(c) What is the probability that Susan will need from 12 to 14 contacts to get the bonus?

(d) What is the probability that Susan will need more than 14 contacts to get the bonus?

(e) What are the expected value μ and standard deviation σ of the random variable n? Interpret these values in the context of this application.

32. *Negative Binomial Distribution: Type A Blood Donors* Blood type A occurs in about 41% of the population (Reference: *Laboratory and Diagnostic Tests* by F. Fischbach). A clinic needs 3 pints of type A blood. A donor usually gives a pint of blood. Let n be a random variable representing the number of donors needed to provide 3 pints of type A blood.

(a) Explain why a negative binomial distribution is appropriate for the random variable n. Write out the formula for $P(n)$ in the context of this application. *Hint:* See Problem 30.

(b) Compute $P(n = 3)$, $P(n = 4)$, $P(n = 5)$, and $P(n = 6)$.

(c) What is the probability that the clinic will need from three to six donors to obtain the needed 3 pints of type A blood?

(d) What is the probability that the clinic will need more than six donors to obtain 3 pints of type A blood?

(e) What are the expected value μ and standard deviation σ of the random variable n? Interpret these values in the context of this application.

33. *Expand Your Knowledge: Brain Teaser* If you enjoy a little abstract thinking, you may want to derive the formula for the negative binomial probability distribution. Use the notation of Problem 30. Consider two events, A and B.

$$A = \{\text{event that the first } n = 1 \text{ trials contain } k - 1 \text{ successes}\}$$
$$B = \{\text{event that the } n\text{th trial is a success}\}$$

(a) Use the binomial probability distribution to show that the probability of A is $P(A) = C_{n-1,k-1}p^{k-1}q^{(n-1)-(k-1)}$.

(b) Show that the probability of B is that of a single trial in a binomial experiment, $P(B) = p$.

(c) Why is $P(A \text{ and } B) = P(A) \cdot P(B)$? *Hint:* Binomial trials are independent.

(d) Use parts (a), (b), and (c) to compute and simplify $P(A \text{ and } B)$.

(e) Compare $P(A \text{ and } B)$ with the negative binomial formula and comment on the meaning of your results.

CHAPTER REVIEW

SUMMARY

This chapter discusses random variables and important probability distributions associated with discrete random variables.

- The value of a *random variable* is determined by chance. Formulas for the mean, variance, and standard deviation of linear functions and linear combinations of independent random variables are given.
- Random variables are either *discrete* or *continuous*.
- A probability distribution of a discrete random variable x consists of all distinct values of x and the corresponding probabilities $P(x)$. For each x, $0 \le P(x) \le 1$ and $\Sigma P(x) = 1$.
- A discrete probability distribution can be displayed visually by a *probability histogram* in which the values of the random variable x are displayed on the horizontal axis, the height of each bar is $P(x)$, and each bar is 1 unit wide.
- For discrete probability distributions,

$$\mu = \Sigma x P(x) \quad \text{and} \quad \sigma = \sqrt{\Sigma(x - \mu)^2 P(x)}$$

- The mean μ is called the *expected value* of the probability distribution.
- A *binomial experiment* consists of a fixed number n of independent trials repeated under identical conditions. There are two outcomes for each trial, called *success* and *failure*. The probability p of success on each trial is the same.
- The number of successes r in a binomial experiment is the random variable for the binomial probability distribution. Probabilities can be computed using a formula or probability distribution outputs from a computer or calculator. Some probabilities can be found in Table 3 of Appendix II.
- For a binomial distribution,

$$\mu = np \quad \text{and} \quad \sigma = \sqrt{npq},$$

- where $q = 1 - p$.
- For a binomial experiment, the number of successes is usually within the interval from $\mu - 2.5\sigma$ to $\mu + 2.5\sigma$. A number of successes outside this range of values is unusual but can occur.
- The *geometric probability distribution* is used to find the probability that the first success of a binomial experiment occurs on trial number n.
- The *Poisson distribution* is used to compute the probability of r successes in an interval of time, volume, area, and so forth.
- The *Poisson distribution* can be used to approximate the binomial distribution when $n \ge 100$ and $np < 10$.
- The *hypergeometric distribution* (discussed in Appendix I) is a probability distribution of a random variable that has two outcomes when sampling is done without replacement.

It is important to check the conditions required for the use of each probability distribution.

IMPORTANT WORDS & SYMBOLS

Section 5.1

Random variable 198
  Discrete 198
  Continuous 198
Probability distribution 199
Mean μ of a probability distribution 201
Standard deviation σ of a probability distribution 201
Expected value μ 202
Linear function of a random variable 204
Linear combination of two independent random variables 204

Section 5.2

Binomial experiment 212
Number of trials, n 212
Independent trials 212
Successes and failures in a binomial experiment 212
Probability of success $P(S) = p$ 212
Probability of failure $P(F) = q = 1 - p$ 212
Number of successes, r 212
Binomial coefficient $C_{n,r}$ 216

Binomial probability distribution
$P(r) = C_{n,r}p^r q^{n-r}$ 219
Hypergeometric probability distribution (see Appendix I) 222
Multinomial probability distribution 228

Section 5.3

Mean for the binomial distribution $\mu = np$ 232
Standard deviation for the binomial distribution $\sigma = \sqrt{npq}$ 232
Quota problem 235

Section 5.4

Geometric probability distribution 242
Poisson probability distribution 244
Poisson approximation to the binomial 247
Negative binomial distribution 258

VIEWPOINT What's Your Type?

Are students and professors really compatible? One way of answering this question is to look at the Myers–Briggs Type Indicators for personality preferences. What is the probability that your professor is introverted and judgmental? What is the probability that you are extroverted and perceptive? Are most of the leaders in student government extroverted and judgmental? Is it true that members of Phi Beta Kappa have personality types more like those of the professors'? We will consider questions such as these in more detail in Chapter 7 (estimation) and Chapter 8 (hypothesis testing), where we will continue our work with binomial probabilities. In the meantime, you can find many answers regarding careers, probability, and personality types in Applications of the Myers–Briggs Type Indicator in Higher Education, *edited by J. Provost and S. Anchors.*

CHAPTER REVIEW PROBLEMS

1. *Statistical Literacy* What are the requirements for a probability distribution?

2. *Statistical Literacy* List the criteria for a binomial experiment. What does the random variable of a binomial experiment measure?

3. *Critical Thinking* For a binomial probability distribution, it is unusual for the number of successes to be less than $\mu - 2.5\sigma$ or greater than $\mu + 2.5\sigma$.
   (a) For a binomial experiment with 10 trials for which the probability of success on a single trial is 0.2, is it unusual to have more than five successes? Explain.
   (b) If you were simply guessing on a multiple-choice exam consisting of 10 questions with 5 possible responses for each question, would you be likely to get more than half of the questions correct? Explain.

4. *Critical Thinking* Consider a binomial experiment. If the number of trials is increased, what happens to the expected value? to the standard deviation? Explain.

5. *Probability Distribution: Auto Leases* Consumer Banker Association released a report showing the lengths of automobile leases for new automobiles. The results are as follows.

Lease Length in Months	Percent of Leases
13–24	12.7%
25–36	37.1%
37–48	28.5%
49–60	21.5%
More than 60	0.2%

   (a) Use the midpoint of each class, and call the midpoint of the last class 66.5 months, for purposes of computing the expected lease term. Also find the standard deviation of the distribution.
   (b) Sketch a graph of the probability distribution for the duration of new auto leases.

georgesanker.com/Alamy

6. *Ecology: Predator and Prey* Isle Royale, an island in Lake Superior, has provided an important study site of wolves and their prey. In the National Park Service Scientific Monograph Series 11, *Wolf Ecology and Prey Relationships on Isle Royale,* Peterson gives results of many wolf–moose studies. Of special interest is the study of the number of moose killed by wolves. In the period from 1958 to 1974, there were 296 moose deaths identified as wolf kills. The age distribution of the kills is as follows.

Age of Moose in Years	Number Killed by Wolves
Calf (0.5 yr)	112
1–5	53
6–10	73
11–15	56
16–20	2

(a) For each age group, compute the probability that a moose in that age group is killed by a wolf.
(b) Consider all ages in a class equal to the class midpoint. Find the expected age of a moose killed by a wolf and the standard deviation of the ages.

7. *Insurance: Auto* State Farm Insurance studies show that in Colorado, 55% of the auto insurance claims submitted for property damage are submitted by males under 25 years of age. Suppose 10 property damage claims involving automobiles are selected at random.
(a) Let r be the number of claims made by males under age 25. Make a histogram for the r-distribution probabilities.
(b) What is the probability that six or more claims are made by males under age 25?
(c) What is the expected number of claims made by males under age 25? What is the standard deviation of the r-probability distribution?

8. *Quality Control: Pens* A stationery store has decided to accept a large shipment of ball-point pens if an inspection of 20 randomly selected pens yields no more than two defective pens.
(a) Find the probability that this shipment is accepted if 5% of the total shipment is defective.
(b) Find the probability that this shipment is not accepted if 15% of the total shipment is defective.

9. *Criminal Justice: Inmates* According to *Harper's Index,* 50% of all federal inmates are serving time for drug dealing. A random sample of 16 federal inmates is selected.
(a) What is the probability that 12 or more are serving time for drug dealing?
(b) What is the probability that 7 or fewer are serving time for drug dealing?
(c) What is the expected number of inmates serving time for drug dealing?

10. *Airlines: On-Time Arrivals* *Consumer Reports* rated airlines and found that 80% of the flights involved in the study arrived on time (that is, within 15 minutes of scheduled arrival time). Assuming that the on-time arrival rate is representative of the entire commercial airline industry, consider a random sample of 200 flights. What is the expected number that will arrive on time? What is the standard deviation of this distribution?

11. *Agriculture: Grapefruit* It is estimated that 75% of a grapefruit crop is good; the other 25% have rotten centers that cannot be detected until the grapefruit are cut open. The grapefruit are sold in sacks of 10. Let r be the number of good grapefruit in a sack.
(a) Make a histogram of the probability distribution of r.
(b) What is the probability of getting no more than one bad grapefruit in a sack? What is the probability of getting at least one good grapefruit in a sack?

(c) What is the expected number of good grapefruit in a sack?
(d) What is the standard deviation of the r-probability distribution?

12. *Restaurants: Reservations* The Orchard Café has found that about 5% of the diners who make reservations don't show up. If 82 reservations have been made, how many diners can be expected to show up? Find the standard deviation of this distribution.

13. *College Life: Student Government* The student government claims that 85% of all students favor an increase in student fees to buy indoor potted plants for the classrooms. A random sample of 12 students produced 2 in favor of the project. What is the probability that 2 or fewer in the sample will favor the project, assuming the student government's claim is correct? ***Interpretation*** Do the data support the student government's claim, or does it seem that the percentage favoring the increase in fees is less than 85%?

14. *Quota Problem: Financial* Suppose you are a (junk) bond broker who buys only bonds that have a 50% chance of default. You want a portfolio with at least five bonds that do not default. You can dispose of the other bonds in the portfolio with no great loss. How many such bonds should you buy if you want to be 94.1% sure that five or more will not default?

15. *Theater: Coughs* A person with a cough is a *persona non grata* on airplanes, elevators, or at the theater. In theaters especially, the irritation level rises with each muffled explosion. According to Dr. Brian Carlin, a Pittsburgh pulmonologist, in any large audience you'll hear about 11 coughs per minute (Source: *USA Today*).
(a) Let r = number of coughs in a given time interval. Explain why the Poisson distribution would be a good choice for the probability distribution of r.
(b) Find the probability of three or fewer coughs (in a large auditorium) in a 1-minute period.
(c) Find the probability of at least three coughs (in a large auditorium) in a 30-second period.

16. *Accident Rate: Small Planes* Flying over the western states with mountainous terrain in a small aircraft is 40% riskier than flying over similar distances in flatter portions of the nation, according to a General Accounting Office study completed in response to a congressional request. The accident rate for small aircraft in the 11 mountainous western states is 2.4 accidents per 100,000 flight operations (Source: *The Denver Post*).
(a) Let r = number of accidents for a given number of operations. Explain why the Poisson distribution would be a good choice for the probability distribution of r.
(b) Find the probability of no accidents in 100,000 flight operations.
(c) Find the probability of at least 4 accidents in 200,000 flight operations.

17. *Banking: Loan Defaults* Records over the past year show that 1 out of 350 loans made by Mammon Bank have defaulted. Find the probability that 2 or more out of 300 loans will default. *Hint:* Is it appropriate to use the Poisson approximation to the binomial distribution?

18. *Car Theft: Hawaii* In Hawaii, the rate of motor vehicle theft is 551 thefts per 100,000 vehicles (Reference: U.S. Department of Justice, Federal Bureau of Investigation). A large parking structure in Honolulu has issued 482 parking permits.
(a) What is the probability that none of the vehicles with a permit will eventually be stolen?
(b) What is the probability that at least one of the vehicles with a permit will eventually be stolen?

(c) What is the probability that two or more of the vehicles with a permit will eventually be stolen?

Note: The vehicles may or may not be stolen from the parking structure. *Hint:* Is it appropriate to use the Poisson approximation to the binomial? Explain.

19. *General: Coin Flip* An experiment consists of tossing a coin a specified number of times and recording the outcomes.
(a) What is the probability that the *first* head will occur on the second trial? Does this probability change if we toss the coin three times? What if we toss the coin four times? What probability distribution model do we use to compute these probabilities?
(b) What is the probability that the *first* head will occur on the fourth trial? after the fourth trial?

20. *Testing: CPA Exam* Cathy is planning to take the Certified Public Accountant Examination (CPA exam). Records kept by the college of business from which she graduated indicate that 83% of the students who graduated pass the CPA exam. Assume that the exam is changed each time it is given. Let $n = 1, 2, 3, \ldots$ represent the number of times a person takes the CPA exam until the *first* pass. (Assume the trials are independent.)
(a) What is the probability that Cathy passes the CPA exam on the first try?
(b) What is the probability that Cathy passes the CPA exam on the second or third try?

DATA HIGHLIGHTS: GROUP PROJECTS

Break into small groups and discuss the following topics. Organize a brief outline in which you summarize the main points of your group discussion.

1. Powerball! Imagine, you could win a jackpot worth at least $40 million. Some jackpots have been worth more than $250 million! Powerball is a multistate lottery. To play Powerball, you purchase a $2 ticket. On the ticket you select five distinct white balls (numbered 1 through 59) and then one red Powerball (numbered 1 through 35). The red Powerball number may be any of the numbers 1 through 35, including any such numbers you selected for the white balls. Every Wednesday and Saturday there is a drawing. If your chosen numbers match those drawn, you win! Figure 5-7 shows all the prizes and the probability of winning each prize and specifies how many numbers on your ticket must match those drawn to win the prize. The Multi-State Lottery Association maintains a web site that displays the results of each drawing, as well as a history of the results of previous drawings. For updated Powerball data, visit the Brase/Brase statistics site at **http://www.cengage.com/statistics/brase11e** and find the link to the Multi-State Lottery Association.
   (a) Assume the jackpot is $40 million and there will be only one jackpot winner. Figure 5-7 lists the prizes and the probability of winning each prize. What is the probability of *not winning* any prize? Consider all the prizes and their respective probabilities, and the prize of $0 (no win) and its probability. Use all these values to estimate your expected winnings μ if you play one ticket. How much do you effectively contribute to the state in which you purchased the ticket (ignoring the overhead cost of operating Powerball)?
   (b) Suppose the jackpot increased to $100 million (and there was to be only one winner). Compute your expected winnings if you buy one ticket. Does the probability of winning the jackpot change because the jackpot is higher?
   (c) Pretend that you are going to buy 10 Powerball tickets when the jackpot is $40 million. Use the random-number table to select your numbers. Check the Multi-State Lottery Association web site (or any other Powerball site) for the most recent drawing results to see if you would have won a prize.
   (d) The probability of winning *any* prize is about 0.0314. Suppose you decide to buy five tickets. Use the binomial distribution to compute the probability

FIGURE 5-7

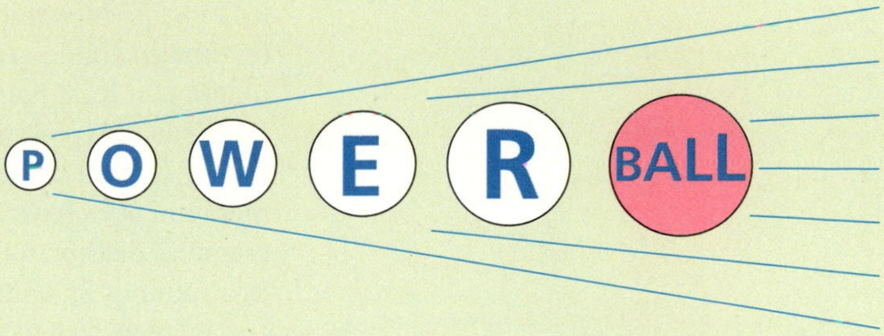

Match	Approximate Probability	Prize
5 white balls + Powerball	0.00000000571	Jackpot*
5 white balls	0.000000194	\$1,000,000
4 white balls + Powerball	0.00000154	\$10,000
4 white balls	0.0000524	\$100
3 white balls + Powerball	0.0000817	\$100
3 white balls	0.00278	\$7
2 white balls + Powerball	0.00142	\$7
1 white ball + Powerball	0.00902	\$4
0 white balls + Powerball	0.0180	\$4
Overall chance of winning	0.0314	

*The Jackpot will be divided equally (if necessary) among multiple winners and is paid in 30 annual installments or in a reduced lump sum.

of winning (any prize) at least once. *Note:* You will need to use the binomial formula. Carry at least three digits after the decimal.

(e) The probability of winning *any* prize is about 0.0314. Suppose you play Powerball 100 times. Explain why it is appropriate to use the Poisson approximation to the binomial to compute the probability of winning at least one prize. Compute $\lambda = np$. Use the Poisson table to estimate the probability of winning at least one prize.

2. Would you like to travel in space, if given a chance? According to Opinion Research for Space Day Partners, if your answer is yes, you are not alone. Forty-four percent of adults surveyed agreed that they would travel in space if given a chance. Look at Figure 5-8, and use the information presented to answer the following questions.

FIGURE 5-8

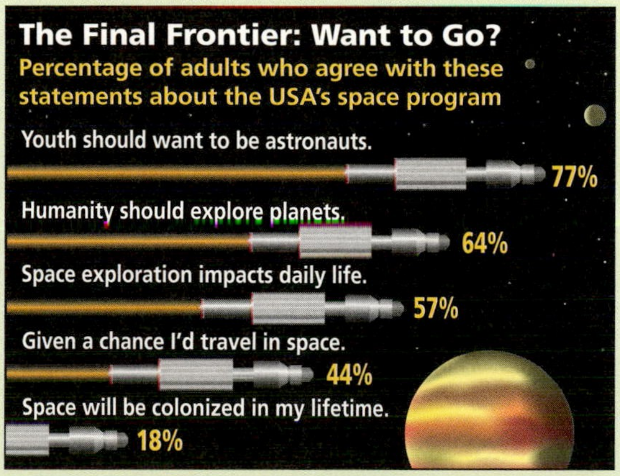

Source: Opinion Research for Space Day Partners

(a) According to Figure 5-8, the probability that an adult selected at random agrees with the statement that humanity should explore planets is 64%. Round this probability to 65%, and use this estimate with the binomial distribution table to determine the probability that of 10 adults selected at random, at least half would agree that humanity should explore planets.
(b) Does space exploration have an impact on daily life? Find the probability that of 10 adults selected at random, at least 9 would agree that space exploration does have an impact on daily life. *Hint:* Use the formula for the binomial distribution.
(c) In a room of 35 adults, what is the expected number who would travel in space, given a chance? What is the standard deviation?
(d) What is the probability that the first adult (selected at random) you asked would agree with the statement that space will be colonized in the person's lifetime? *Hint:* Use the geometric distribution.

LINKING CONCEPTS: WRITING PROJECTS

Discuss each of the following topics in class or review the topics on your own. Then write a brief but complete essay in which you summarize the main points. Please include formulas and graphs as appropriate.

1. Discuss what we mean by a binomial experiment. As you can see, a binomial process or binomial experiment involves a lot of assumptions! For example, all the trials are supposed to be independent and repeated under identical conditions. Is this always true? Can we always be completely certain that the probability of success does not change from one trial to the next? In the real world, there is almost nothing we can be absolutely sure about, so the *theoretical* assumptions of the binomial probability distribution often will not be completely satisfied. Does that mean we cannot use the binomial distribution to solve practical problems? Looking at this chapter, the answer seems to be that we can indeed use the binomial distribution even if not all the assumptions are *exactly* met. We find in practice that the conclusions are sufficiently accurate for our intended application. List three applications of the binomial distribution for which you think, although some of the assumptions are not exactly met, there is adequate reason to apply the binomial distribution anyhow.
2. Why do we need to learn the formula for the binomial probability distribution? Using the formula repeatedly can be very tedious. To cut down on tedious calculations, most people will use a binomial table such as the one found in Appendix II of this book.
   (a) However, there are many applications for which a table in the back of *any* book is not adequate. For instance, compute

   $$P(r = 3) \text{ where } n = 5 \text{ and } p = 0.735$$

   Can you find the result in the table? Do the calculation by using the formula. List some other situations in which a table might not be adequate to solve a particular binomial distribution problem.
   (b) The formula itself also has limitations. For instance, consider the difficulty of computing

   $$P(r \geq 285) \text{ where } n = 500 \text{ and } p = 0.6$$

   What are some of the difficulties you run into? Consider the calculation of $P(r = 285)$. You will be raising 0.6 and 0.4 to very high powers; this will give you very, very small numbers. Then you need to compute $C_{500,285}$, which is a very, very large number. When combining extremely large and extremely small numbers in the same calculation, most accuracy is lost unless you carry a huge number of significant digits. If this isn't tedious enough, consider the steps you need to compute

$$P(r \geq 285) = P(r = 285) + P(r = 286) + \cdots + P(r = 500)$$

Does it seem clear that we need a better way to estimate $P(r \geq 285)$? In Chapter 6, you will learn a much better way to estimate binomial probabilities when the number of trials is large.

3. In Chapter 3, we learned about means and standard deviations. In Section 5.1, we learned that probability distributions also can have a mean and standard deviation. Discuss what is meant by the expected value and standard deviation of a binomial distribution. How does this relate back to the material we learned in Chapter 3 and Section 5.1?
4. In Chapter 2, we looked at the shapes of distributions. Review the concepts of skewness and symmetry; then categorize the following distributions as to skewness or symmetry:
   (a) A binomial distribution with $n = 11$ trials and $p = 0.50$
   (b) A binomial distribution with $n = 11$ trials and $p = 0.10$
   (c) A binomial distribution with $n = 11$ trials and $p = 0.90$

   In general, does it seem true that binomial probability distributions in which the probability of success is close to 0 are skewed right, whereas those with probability of success close to 1 are skewed left?

USING TECHNOLOGY

Binomial Distributions

Although tables of binomial probabilities can be found in most libraries, such tables are often inadequate. Either the value of p (the probability of success on a trial) you are looking for is not in the table, or the value of n (the number of trials) you are looking for is too large for the table. In Chapter 6, we will study the normal approximation to the binomial. This approximation is a great help in many practical applications. Even so, we sometimes use the formula for the binomial probability distribution on a computer or graphing calculator to compute the probability we want.

Applications

The following percentages were obtained over many years of observation by the U.S. Weather Bureau. All data listed are for the month of December.

Location	Long-Term Mean % of Clear Days in Dec.
Juneau, Alaska	18%
Seattle, Washington	24%
Hilo, Hawaii	36%
Honolulu, Hawaii	60%
Las Vegas, Nevada	75%
Phoenix, Arizona	77%

Adapted from *Local Climatological Data*, U.S. Weather Bureau publication, "Normals, Means, and Extremes" Table.

In the locations listed, the month of December is a relatively stable month with respect to weather. Since weather patterns from one day to the next are more or less the same, it is reasonable to use a binomial probability model.

1. Let r be the number of clear days in December. Since December has 31 days, $0 \le r \le 31$. Using appropriate computer software or calculators available to you, find the probability $P(r)$ for each of the listed locations when $r = 0, 1, 2, \dots, 31$.
2. For each location, what is the expected value of the probability distribution? What is the standard deviation?

You may find that using cumulative probabilities and appropriate subtraction of probabilities, rather than addition of probabilities, will make finding the solutions to Applications 3 to 7 easier.

3. Estimate the probability that Juneau will have at most 7 clear days in December.
4. Estimate the probability that Seattle will have from 5 to 10 (including 5 and 10) clear days in December.
5. Estimate the probability that Hilo will have at least 12 clear days in December.
6. Estimate the probability that Phoenix will have 20 or more clear days in December.
7. Estimate the probability that Las Vegas will have from 20 to 25 (including 20 and 25) clear days in December.

Technology Hints

TI-84Plus/TI-83Plus/TI-*n*spire (with TI-84 Plus keypad), Excel 2010, Minitab

The Tech Note in Section 5.2 gives specific instructions for binomial distribution functions on the TI-84Plus/TI-83Plus/TI-*n*spire (with TI-84Plus keypad) calculators, Excel 2010, and Minitab.

SPSS

In SPSS, the function **PDF.BINOM(q,n,p)** gives the probability of q successes out of n trials, where p is the probability of success on a single trial. In the data editor, name a variable r and enter values 0 through n. Name another variable Prob_*r*. Then use the menu choices **Transform ➤ Compute**. In the dialogue box, use Prob_*r* for the target variable. In the function group, select **PDF and Noncentral PDF**. In the function box, select **PDF.BINOM(q,n,p)**. Use the variable r for q and appropriate values for n and p. Note that the function **CDF.BINOM(q,n,p)**, from the **CDF and Noncentral CDF** group, gives the cumulative probability of 0 through q successes.

6

WARNER BROS/THE KOBAL COLLECTION/Picture Desk

Time & Life Pictures/ Getty Images

One cannot escape the feeling that these mathematical formulas have an independent existence and an intelligence of their own, that they are wiser than we are, wiser even than their discoverers, that we get more out of them than was originally put into them.

—Heinrich Hertz

How can it be that mathematics, a product of human thought independent of experience, is so admirably adapted to the objects of reality?

—Albert Einstein

Heinrich Hertz (1857–1894) was a pioneer in the study of radio waves. His work and the later work of Maxwell and Marconi led the way to modern radio, television, and radar. Albert Einstein is world renowned for his great discoveries in relativity and quantum mechanics. Everyone who has worked in both mathematics and real-world applications cannot help but marvel at how the "pure thought" of the mathematical sciences can predict and explain events in other realms. In this chapter, we will study the most important type of probability distribution in all of mathematical statistics: the normal distribution. Why is the normal distribution so important? Two of the reasons are that it applies to a wide variety of situations and that other distributions tend to become normal under certain conditions.

NORMAL CURVES AND SAMPLING DISTRIBUTIONS

PREVIEW QUESTIONS

What are some characteristics of a normal distribution? What does the empirical rule tell you about data spread around the mean? How can this information be used in quality control? (SECTION 6.1)

Can you compare apples and oranges, or maybe elephants and butterflies? In most cases, the answer is no—unless you first standardize your measurements. What are a standard normal distribution and a standard *z* score? (SECTION 6.2)

How do you convert any normal distribution to a standard normal distribution? How do you find probabilities of "standardized events"? (SECTION 6.3)

As humans, our experiences are finite and limited. Consequently, most of the important decisions in our lives are based on sample (incomplete) information. What is a probability sampling distribution? How will sampling distributions help us make good decisions based on incomplete information? (SECTION 6.4)

There is an old saying: All roads lead to Rome. In statistics, we could recast this saying: All probability distributions average out to be normal distributions (as the sample size increases). How can we take advantage of this in our study of sampling distributions? (SECTION 6.5)

The binomial and normal distributions are two of the most important probability distributions in statistics. Under certain limiting conditions, the binomial can be thought to evolve (or envelope) into the normal distribution. How can you apply this concept in the real world? (SECTION 6.6)

Many issues in life come down to success or failure. In most cases, we will not be successful all the time, so proportions of successes are very important. What is the probability sampling distribution for proportions? (SECTION 6.6)

Dana White/PhotoEdit

FOCUS PROBLEM

Impulse Buying

Syracuse Newspapers/David Lassman/The Image Works Image

The Food Marketing Institute, Progressive Grocer, New Products News, and Point of Purchaser Advertising Institute are organizations that analyze supermarket sales. One of the interesting discoveries was that the average amount of impulse buying in a grocery store is very time-dependent. As reported in *The Denver Post,* "When you dilly dally in a store for 10 unplanned minutes, you can kiss nearly $20 good-bye." For this reason, it is in the best interest of the supermarket to keep you in the store longer. In the *Post* article, it was pointed out that long checkout lines (near end-aisle displays), "samplefest" events of free tasting, video kiosks, magazine and book sections, and so on, help keep customers in the store longer. On average, a single customer who strays from his or her

For online student resources, visit the Brase/Brase, *Understandable Statistics*, 11th edition web site at http://www.cengage.com/statistics/brase11e

grocery list can plan on impulse spending of \$20 for every 10 minutes spent wandering about in the supermarket.

Let x represent the dollar amount spent on supermarket impulse buying in a 10-minute (unplanned) shopping interval. Based on the *Post* article, the mean of the x distribution is about \$20 and the (estimated) standard deviation is about \$7.

(a) Consider a random sample of $n = 100$ customers, each of whom has 10 minutes of unplanned shopping time in a supermarket. From the central limit theorem, what can you say about the probability distribution of $\bar{x}$, the *average* amount spent by these customers due to impulse buying? Is the $\bar{x}$ distribution approximately normal? What are the mean and standard deviation of the $\bar{x}$ distribution? Is it necessary to make any assumption about the x distribution? Explain.
(b) What is the probability that $\bar{x}$ is between \$18 and \$22?
(c) Let us assume that x has a distribution that is approximately normal. What is the probability that x is between \$18 and \$22?
(d) In part (b), we used $\bar{x}$, the *average* amount spent, computed for 100 customers. In part (c), we used x, the amount spent by only *one* individual customer. The answers to parts (b) and (c) are very different. Why would this happen? In this example, $\bar{x}$ is a much more predictable or reliable statistic than x. Consider that almost all marketing strategies and sales pitches are designed for the *average* customer and *not* the *individual* customer. How does the central limit theorem tell us that the average customer is much more predictable than the individual customer? (See Problem 18 of Section 6.5.)

SECTION **6.1**

Graphs of Normal Probability Distributions

FOCUS POINTS

- Graph a normal curve and summarize its important properties.
- Apply the empirical rule to solve real-world problems.
- Use control limits to construct control charts. Examine the chart for three possible out-of-control signals.

Normal distribution

One of the most important examples of a continuous probability distribution is the *normal distribution*. This distribution was studied by the French mathematician Abraham de Moivre (1667–1754) and later by the German mathematician Carl Friedrich Gauss (1777–1855), whose work is so important that the normal distribution is sometimes called *Gaussian*. The work of these mathematicians provided a foundation on which much of the theory of statistical inference is based.

Applications of a normal probability distribution are so numerous that some mathematicians refer to it as "a veritable Boy Scout knife of statistics." However, before we can apply it, we must examine some of the properties of a normal distribution.

A rather complicated formula, presented later in this section, defines a normal distribution in terms of μ and σ, the mean and standard deviation of the population distribution. It is only through this formula that we can verify if a distribution is normal. However, we can look at the graph of a normal distribution and get a good pictorial idea of some of the essential features of any normal distribution.

Normal curve

The graph of a normal distribution is called a *normal curve*. It possesses a shape very much like the cross section of a pile of dry sand. Because of its shape, blacksmiths would sometimes use a pile of dry sand in the construction of a mold for a bell. Thus the normal curve is also called a *bell-shaped curve* (see Figure 6-1).

We see that a general normal curve is smooth and symmetrical about the vertical line extending upward from the mean μ. Notice that the highest point of the curve occurs over μ. If the distribution were graphed on a piece of sheet metal, cut out, and placed on a knife edge, the balance point would be at μ. We also see that the curve tends to level out and approach the horizontal (x axis) like a glider

FIGURE 6-1

A Normal Curve

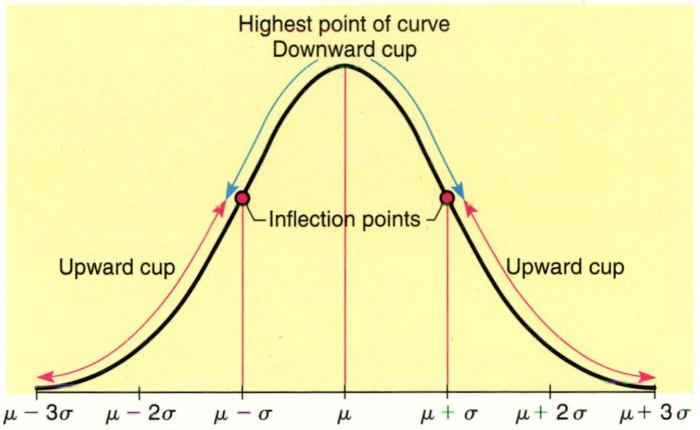

Two other continuous probability distributions are discussed in the section problems. Problems 16 and 17 discuss the *uniform probability distribution*. Problems 18 through 20 discuss the *exponential probability distribution*.

making a landing. However, in mathematical theory, such a glider would never quite finish its landing because a normal curve never touches the horizontal axis.

The parameter σ controls the spread of the curve. The curve is quite close to the horizontal axis at $\mu + 3\sigma$ and $\mu - 3\sigma$. Thus, if the standard deviation σ is large, the curve will be more spread out; if it is small, the curve will be more peaked. Figure 6-1 shows the normal curve cupped downward for an interval on either side of the mean μ. Then it begins to cup upward as we go to the lower part of the bell. The exact places where the *transition* between the upward and downward cupping occur are above the points $\mu + \sigma$ and $\mu - \sigma$. In the terminology of calculus, transition points such as these are called *inflection points*.

Downward cup
Upward cup

IMPORTANT PROPERTIES OF A NORMAL CURVE

Symmetry of normal curves

1. The curve is bell-shaped, with the highest point over the mean μ.
2. The curve is symmetrical about a vertical line through μ.
3. The curve approaches the horizontal axis but never touches or crosses it.
4. The inflection (transition) points between cupping upward and downward occur above $\mu + \sigma$ and $\mu - \sigma$.
5. The area under the entire curve is 1.

The parameters that control the shape of a normal curve are the mean μ and the standard deviation σ. When both μ and σ are specified, a specific normal curve is determined. In brief, μ locates the balance point and σ determines the extent of the spread.

GUIDED EXERCISE 1 | IDENTIFY μ AND σ ON A NORMAL CURVE

Look at the normal curves in Figure 6-2.

FIGURE 6-2

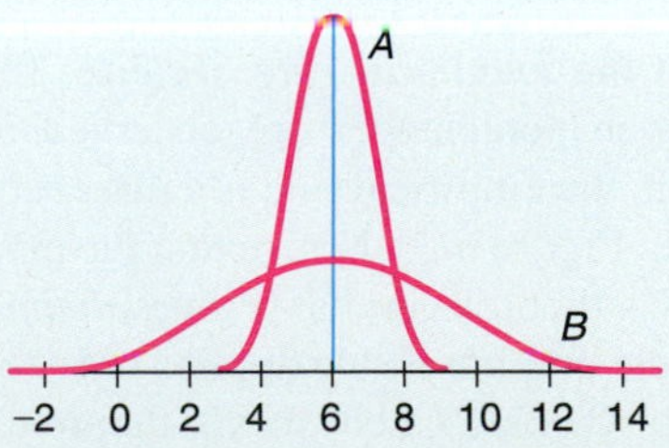

Continued

GUIDED EXERCISE 1 *continued*

(a) Do these distributions have the same mean? If so, what is it?	The means are the same, since both graphs have the high point over 6. $\mu = 6$.
(b) One of the curves corresponds to a normal distribution with $\sigma = 3$ and the other to one with $\sigma = 1$. Which curve has which σ?	Curve *A* has $\sigma = 1$ and curve *B* has $\sigma = 3$. (Since curve *B* is more spread out, it has the larger σ value.)

COMMENT The normal distribution curve is always above the horizontal axis. The area beneath the curve and above the axis is exactly 1. As such, the normal distribution curve is an example of a *density curve.* The formula used to generate the shape of the normal distribution curve is called the *normal density function.* If *x* is a normal random variable with mean μ and standard deviation σ, the formula for the normal density function is

$$f(x) = \frac{e^{(-1/2)((x-\mu)/\sigma)^2}}{\sigma\sqrt{2\pi}}$$

In this text, we will not use this formula explicitly. However, we will use tables of areas based on the normal density function.

The total area under any normal curve studied in this book will *always* be 1. The graph of the normal distribution is important because the portion of the *area* under the curve above a given interval represents the *probability* that a measurement will lie in that interval.

Empirical rule

In Section 3.2, we studied Chebyshev's theorem. This theorem gives us information about the *smallest* proportion of data that lies within 2, 3, or *k* standard deviations of the mean. This result applies to *any* distribution. However, for normal distributions, we can get a much more precise result, which is given by the *empirical rule.*

EMPIRICAL RULE

For a distribution that is symmetrical and bell-shaped (in particular, for a normal distribution):

Approximately 68% of the data values will lie within 1 standard deviation on each side of the mean.

Approximately 95% of the data values will lie within 2 standard deviations on each side of the mean.

Approximately 99.7% (or almost all) of the data values will lie within 3 standard deviations on each side of the mean.

The preceding statement is called the *empirical rule* because, for symmetrical, bell-shaped distributions, the given percentages are observed in practice. Furthermore, for the normal distribution, the empirical rule is a direct consequence of the very nature of the distribution (see Figure 6-3). Notice that the empirical rule is a stronger statement than Chebyshev's theorem in that it gives *definite percentages,* not just lower limits. Of course, the empirical rule applies only to normal or symmetrical, bell-shaped distributions, whereas Chebyshev's theorem applies to all distributions.

FIGURE 6-3

Area Under a Normal Curve

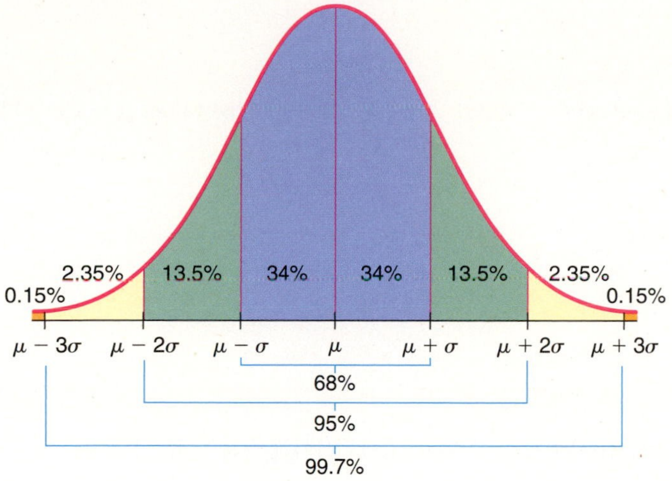

FIGURE 6-4

Distribution of Playing Times

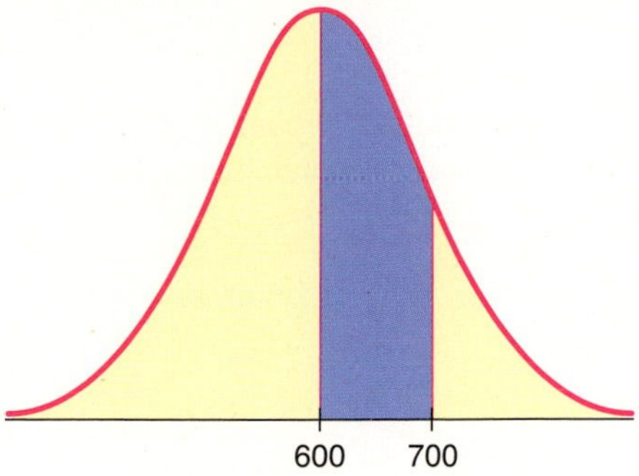

EXAMPLE 1

EMPIRICAL RULE

The playing life of a Sunshine radio is normally distributed with mean $\mu = 600$ hours and standard deviation $\sigma = 100$ hours. What is the probability that a radio selected at random will last from 600 to 700 hours?

SOLUTION: The probability that the playing life will be between 600 and 700 hours is equal to the percentage of the total area under the curve that is shaded in Figure 6-4. Since $\mu = 600$ and $\mu + \sigma = 600 + 100 = 700$, we see that the shaded area is simply the area between μ and $\mu + \sigma$. The area from μ to $\mu + \sigma$ is 34% of the total area. This tells us that the probability a Sunshine radio will last between 600 and 700 playing hours is about 0.34.

GUIDED EXERCISE 2

EMPIRICAL RULE

The yearly wheat yield per acre on a particular farm is normally distributed with mean $\mu = 35$ bushels and standard deviation $\sigma = 8$ bushels.

(a) Shade the area under the curve in Figure 6-5 that represents the probability that an acre will yield between 19 and 35 bushels.

→ See Figure 6-6.

FIGURE 6-5

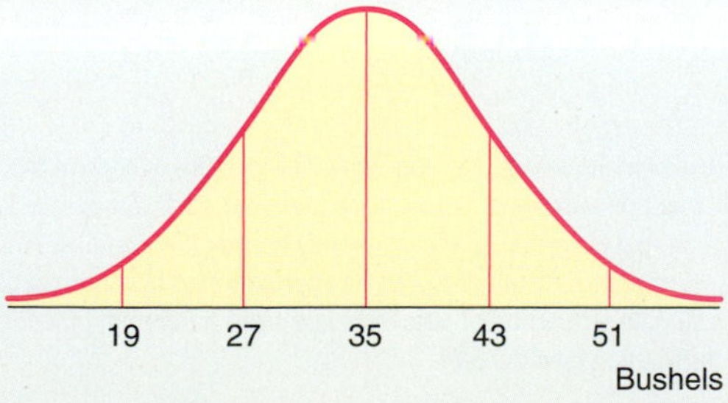

FIGURE 6-6 Completion of Figure 6-5

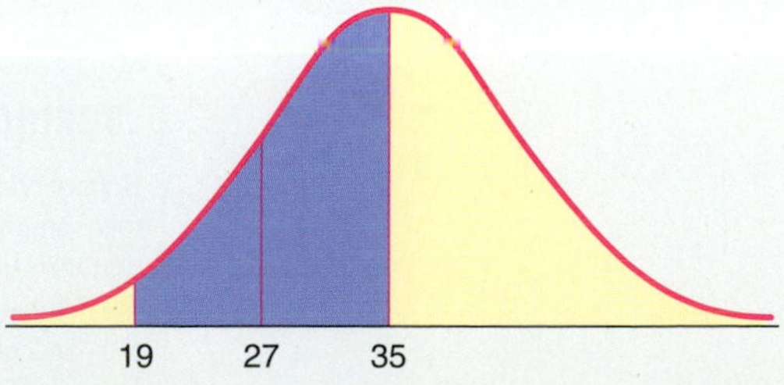

Continued

GUIDED EXERCISE 2 *continued*

(b) Is the area the same as the area between $\mu - 2\sigma$ and μ?

→ Yes, since $\mu = 35$ and $\mu - 2\sigma = 35 - 2(8) = 19$.

(c) Use Figure 6-3 to find the percentage of area over the interval between 19 and 35.

→ The area between the values $\mu - 2\sigma$ and μ is 47.5% of the total area.

(d) What is the probability that the yield will be between 19 and 35 bushels per acre?

→ It is 47.5% of the total area, which is 1. Therefore, the probability is 0.475 that the yield will be between 19 and 35 bushels.

WHAT DOES A NORMAL DISTRIBUTION TELL US?

If a continuous random variable has a normal distribution, then

- the area under the entire distribution is 1.
- the area over a specific interval of values from *a* to *b* is the probability that a randomly selected value falls between *a* and *b*.
- the distribution is symmetrical and mound-shaped and is centered over μ.
- most of the data (99.7%) range from $\mu - 3\sigma$ to $\mu + 3\sigma$.

TECH NOTES

We can graph normal distributions using the TI-84Plus/TI-83Plus/TI-*n*spire calculators, Excel 2010, and Minitab. In each technology, set the range of *x* values between $\mu - 3.5\sigma$ and $\mu + 3.5\sigma$. Then use the built-in normal density functions to generate the corresponding *y* values.

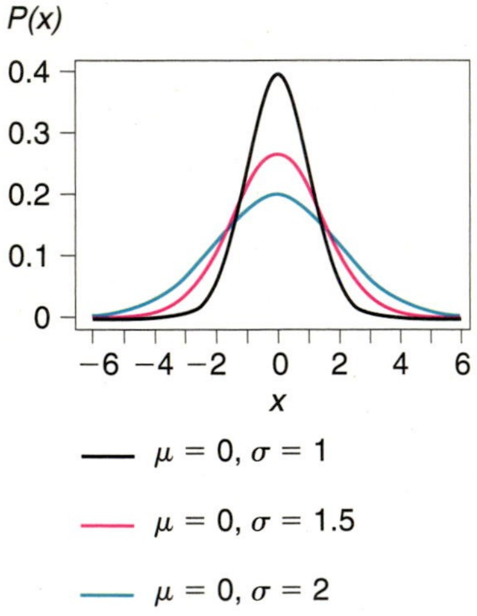

TI-84Plus/TI-83Plus/TI-*n*spire (with TI-84Plus keypad) Press the Y = key. Then, under **DISTR,** select **1:normalpdf (x,μ,σ)** and fill in desired μ and σ values. Press the **WINDOW** key. Set **Xmin** to $\mu - 3\sigma$ and **Xmax** to $\mu + 3\sigma$. Finally, press the **ZOOM** key and select option **0:ZoomFit.**

Excel 2010 In one column, enter *x* values from $\mu - 3.5\sigma$ to $\mu + 3.5\sigma$ in increments of 0.2σ. In the next column, generate *y* values by using the ribbon choices **Insert function** (f_x)

In the dialogue box, select **Statistical** for the Category and then for the Function, select **NORM.DIST** (x,μ,σ,false). Next click the **Insert** tab and in the **Charts** group, click **Scatter**. Select the scatter diagram with smooth lines (third grouping).

Minitab In one column, enter *x* values from -3.5σ to 3.5σ in increments of 0.2σ. In the next column, enter *y* values by using the menu choices **Calc ➤ Probability Distribution ➤ Normal.** Fill in the dialogue box. Next, use menu choices **Graph ➤ Plot.** Fill in the dialogue box. Under Display, select connect.

LOOKING FORWARD

Normal probability distributions will be used extensively in our later work. For instance, when we repeatedly take samples of the same size from a distribution and compute the same mean for each sample, we'll find that the sample means follow a distribution that is normal or approximately normal (Section 6.5). Also, when the number of trials is sufficiently large, the binomial distribution can be approximated by a normal distribution (Section 6.6). The distribution of the sample proportion of successes in a fixed number of binomial trials also can be approximated by a normal distribution (Section 6.6).

CONTROL CHARTS

Control charts

If we are examining data over a period of equally spaced time intervals or in some sequential order, then *control charts* are especially useful. Business managers and people in charge of production processes are aware that there exists an inherent amount of variability in any sequential set of data. The sugar content of bottled drinks taken sequentially off a production line, the extent of clerical errors in a bank from day to day, advertising expenses from month to month, or even the number of new customers from year to year are examples of sequential data. There is a certain amount of variability in each.

A random variable x is said to be in *statistical control* if it can be described by the *same* probability distribution when it is observed at successive points in time. Control charts combine graphic and numerical descriptions of data with probability distributions.

Control charts were invented in the 1920s by Walter Shewhart at Bell Telephone Laboratories. Since a control chart is a *warning device,* it is not absolutely necessary that our assumptions and probability calculations be precisely correct. For example, the x distributions need not follow a normal distribution exactly. Any mound-shaped and more or less symmetrical distribution will be good enough.

PROCEDURE

HOW TO MAKE A CONTROL CHART FOR THE RANDOM VARIABLE x

A control chart for a random variable x is a plot of observed x values in time sequence order.

1. Find the mean μ and standard deviation σ of the x distribution by
   (a) using past data from a period during which the process was "in control" or
   (b) using specified "target" values for μ and σ.
2. Create a graph in which the vertical axis represents x values and the horizontal axis represents time.
3. Draw a horizontal line at height μ and horizontal, dashed control-limit lines at $\mu \pm 2\sigma$ and $\mu \pm 3\sigma$.
4. Plot the variable x on the graph in time sequence order. Use line segments to connect the points in time sequence order.

How do we pick values for μ and σ? In most practical cases, values for μ (population mean) and σ (population standard deviation) are computed from past data for which the process we are studying was known to be *in control*. Methods for choosing the sample size to fit given error tolerances can be found in Chapter 7.

Sometimes values for μ and σ are chosen as *target values*. That is, μ and σ values are chosen as set goals or targets that reflect the production level or service level at which a company hopes to perform. To be realistic, such target assignments for μ and σ should be reasonably close to actual data taken when the process was operating at a satisfactory production level. In Example 2, we will make a control chart; then we will discuss ways to analyze it to see if a process or service is "in control."

EXAMPLE 2

CONTROL CHART

Susan Tamara is director of personnel at the Antlers Lodge in Denali National Park, Alaska. Every summer Ms. Tamara hires many part-time employees from all over the United States. Most are college students seeking summer employment. One of the biggest activities for the lodge staff is that of "making up" the rooms each day.

Although the rooms are supposed to be ready by 3:30 P.M., there are always some rooms not made up by this time because of high personnel turnover.

Every 15 days Ms. Tamara has a general staff meeting at which she shows a control chart of the number of rooms not made up by 3:30 P.M. each day. From extensive experience, Ms. Tamara is aware that the distribution of rooms not made up by 3:30 P.M. is approximately normal, with mean $\mu = 19.3$ rooms and standard deviation $\sigma = 4.7$ rooms. This distribution of x values is acceptable to the top administration of Antlers Lodge. For the past 15 days, the housekeeping unit has reported the number of rooms not ready by 3:30 P.M. (Table 6-1). Make a control chart for these data.

Mt. McKinley, Denali National Park

TABLE 6-1 **Number of Rooms x Not Made Up by 3:30 P.M.**

Day	1	2	3	4	5	6	7	8	9	10	11	12	13	14	15
x	11	20	25	23	16	19	8	25	17	20	23	29	18	14	10

SOLUTION: A control chart for a variable x is a plot of the observed x values (vertical scale) in time sequence order (the horizontal scale represents time). Place horizontal lines at

the mean $\mu = 19.3$
the control limits $\mu \pm 2\sigma = 19.3 \pm 2(4.7)$, or 9.90 and 28.70
the control limits $\mu \pm 3\sigma = 19.3 \pm 3(4.7)$, or 5.20 and 33.40
Then plot the data from Table 6-1. (See Figure 6-7.)

FIGURE 6-7

Number of Rooms Not Made Up by 3:30 P.M.

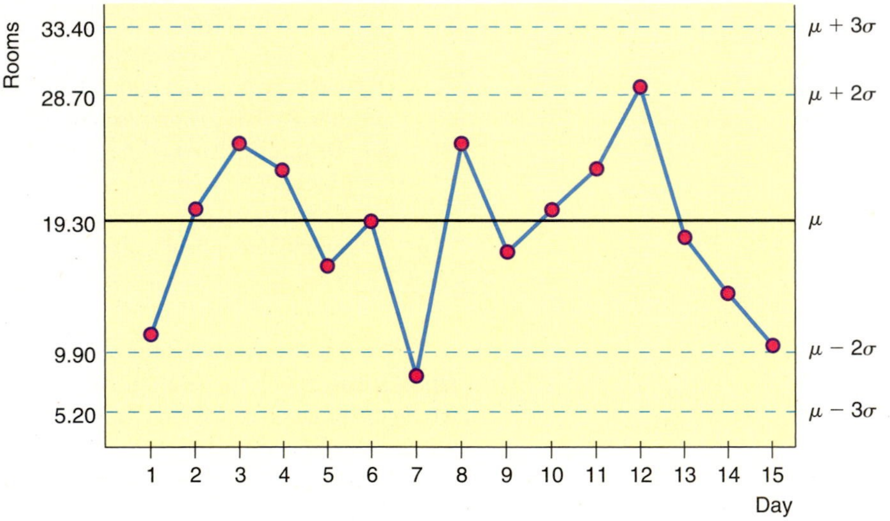

Once we have made a control chart, the main question is the following: As time goes on, is the x variable continuing in this same distribution, or is the distribution of x values changing? If the x distribution is continuing in more or less the same manner, we say it is *in statistical control*. If it is not, we say it is *out of control*.

Out-of-control warning signals

Many popular methods can set off a warning signal that a process is out of control. Remember, a random variable x is said to be *out of control* if successive time measurements of x indicate that it is no longer following the target probability distribution. We will assume that the target distribution is (approximately) normal and has (user-set) target values for μ and σ.

Three of the most popular warning signals are described next.

Out-of-Control Signal I

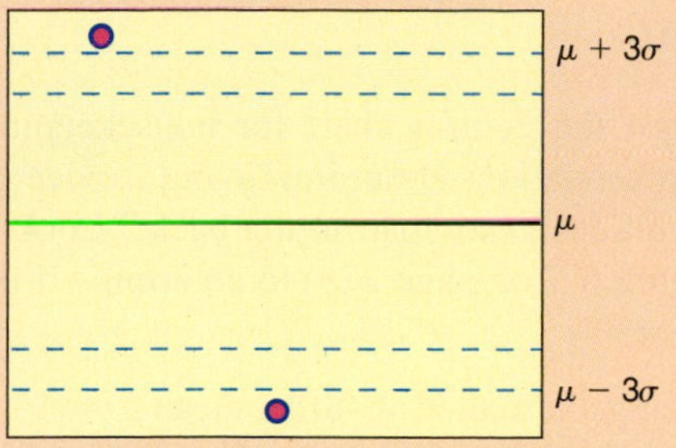

Out-of-Control Signal II

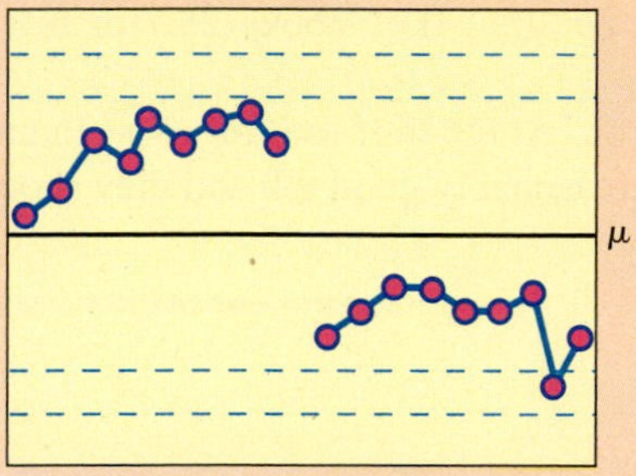

Out-of-Control Signal III

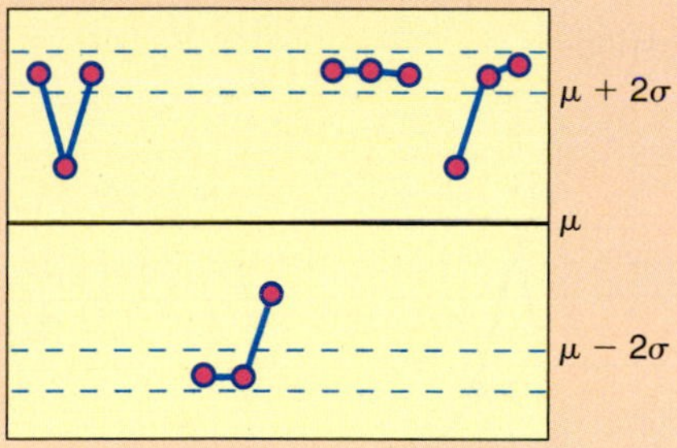

OUT-OF-CONTROL SIGNALS

1. **Out-of-Control Signal I: One point falls beyond the 3σ level**
   What is the probability that signal I will be a false alarm? By the empirical rule, the probability that a point lies within 3σ of the mean is 0.997. The probability that signal I will be a false alarm is $1 - 0.997 = 0.003$. Remember, a false alarm means that the *x* distribution is really on the target distribution, and we simply have a very rare (probability of 0.003) event.

2. **Out-of-Control Signal II: A run of nine consecutive points on one side of the center line (the line at target value μ)**
   To find the probability that signal II is a false alarm, we observe that if the *x* distribution and the target distribution are the same, then there is a 50% chance that the *x* values will lie above or below the center line at μ. Because the samples are (time) independent, the probability of a run of nine points on one side of the center line is $(0.5)^9 = 0.002$. If we consider both sides, this probability becomes 0.004. Therefore, the probability that signal II is a false alarm is approximately 0.004.

3. **Out-of-Control Signal III: At least two of three consecutive points lie beyond the 2σ level on the same side of the center line**
   To determine the probability that signal III will produce a false alarm, we use the empirical rule. By this rule, the probability that an *x* value will be above the 2σ level is about 0.023. Using the binomial probability distribution (with success being the point is above 2σ), the probability of two or more successes out of three trials is

$$\frac{3!}{2!1!}(0.023)^2(0.997) + \frac{3!}{3!0!}(0.023)^3 \approx 0.002$$

   Taking into account *both* above and below the center line, it follows that the probability that signal III is a false alarm is about 0.004.

Remember, a control chart is only a warning device, and it is possible to get a false alarm. A false alarm happens when one (or more) of the out-of-control signals occurs, but the *x* distribution is really on the target or assigned distribution. In this case, we simply have a rare event (probability of 0.003 or 0.004). In practice, whenever a control chart indicates that a process is out of control, it is usually a good precaution to examine what is going on. If the process is out of control, corrective steps can be taken before things get a lot worse. The rare false alarm is a small price to pay if we can avert what might become real trouble.

Type of Warning Signal	Probability of a False Alarm
Type I: Point beyond 3σ	0.003
Type II: Run of nine consecutive points, all below center line μ or all above center line μ	0.004
Type III: At least two out of three consecutive points beyond 2σ	0.004

From an intuitive point of view, signal I can be thought of as a blowup, something dramatically out of control. Signal II can be thought of as a slow drift out of control. Signal III is somewhere between a blowup and a slow drift.

EXAMPLE 3

INTERPRETING A CONTROL CHART

Ms. Tamara of the Antlers Lodge examines the control chart for housekeeping. During the staff meetings, she makes recommendations about improving service or, if all is going well, she gives her staff a well-deserved "pat on the back." Look at the control chart created in Example 2 (Figure 6-7 on page 278) to determine if the housekeeping process is out of control.

SOLUTION: The x values are more or less evenly distributed about the mean $\mu = 19.3$. None of the points are outside the $\mu \pm 3\sigma$ limit (i.e., above 33.40 or below 5.20 rooms). There is no run of nine consecutive points above or below μ. No two of three consecutive points are beyond the $\mu \pm 2\sigma$ limit (i.e., above 28.7 or below 9.90 rooms).

It appears that the x distribution is "in control." At the staff meeting, Ms. Tamara should tell her employees that they are doing a reasonably good job and they should keep up the fine work!

GUIDED EXERCISE 3

INTERPRETING A CONTROL CHART

Figures 6-8 and 6-9 show control charts of housekeeping reports for two other 15-day periods.

FIGURE 6-8 Report II

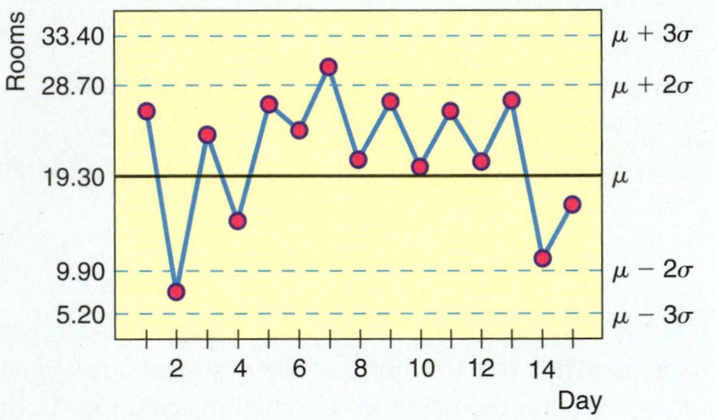

FIGURE 6-9 Report III

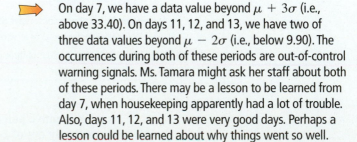

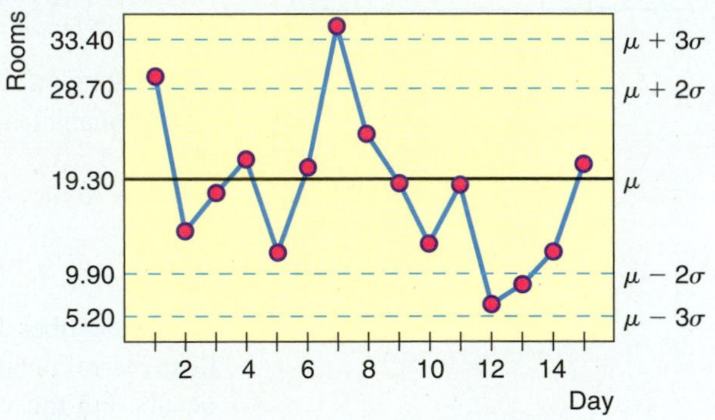

(a) ***Interpret*** the control chart in Figure 6-8.

Days 5 to 13 are above $\mu = 19.3$. We have nine consecutive days on one side of the mean. This is a warning signal! It would appear that the mean μ is slowly drifting up beyond the target value of 19.3. The chart indicates that housekeeping is "out of control." Ms. Tamara should take corrective measures at her staff meeting.

(b) ***Interpret*** the control chart in Figure 6-9.

On day 7, we have a data value beyond $\mu + 3\sigma$ (i.e., above 33.40). On days 11, 12, and 13, we have two of three data values beyond $\mu - 2\sigma$ (i.e., below 9.90). The occurrences during both of these periods are out-of-control warning signals. Ms. Tamara might ask her staff about both of these periods. There may be a lesson to be learned from day 7, when housekeeping apparently had a lot of trouble. Also, days 11, 12, and 13 were very good days. Perhaps a lesson could be learned about why things went so well.

COMMENT Uniform Distribution and Exponential Distribution

Normal distributions are the central topic of this chapter. Normal distributions are very important in general probability and statistics. However, there are other, more specialized distributions that also have many applications. Two such distributions are the *uniform distribution* and the *exponential distribution*. To learn more about these distributions, please see Problems 16 to 20 at the end of this section.

VIEWPOINT In Control? Out of Control?

If you care about quality, you also must care about control! Dr. Walter Shewhart invented control charts when he was working for Bell Laboratories. The great contribution of control charts is that they separate variation into two sources: (1) random or chance causes (in control) and (2) special or assignable causes (out of control). A process is said to be in statistical control *when it is no longer afflicted with special or assignable causes. The performance of a process that is in statistical control is predictable. Predictability and quality control tend to be closely associated.*

(Source: Adapted from the classic text *Statistical Methods from the Viewpoint of Quality Control*, by W. A. Shewhart, with foreword by W. E. Deming, Dover Publications.)

SECTION 6.1 PROBLEMS

1. *Statistical Literacy* Which, if any, of the curves in Figure 6-10 look(s) like a normal curve? If a curve is not a normal curve, tell why.

2. *Statistical Literacy* Look at the normal curve in Figure 6-11, and find μ, $\mu + \sigma$, and σ.

FIGURE 6-10

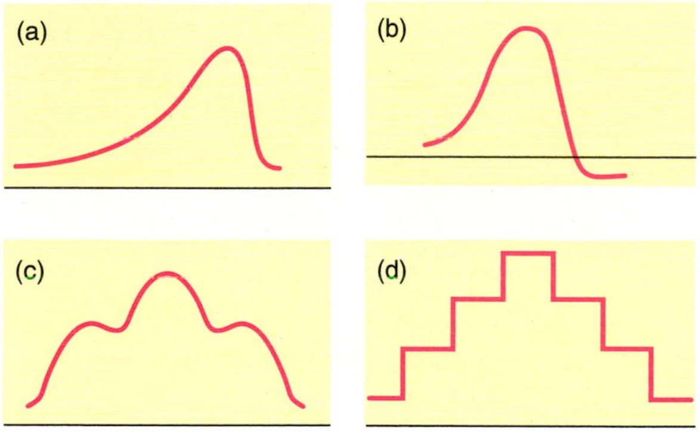

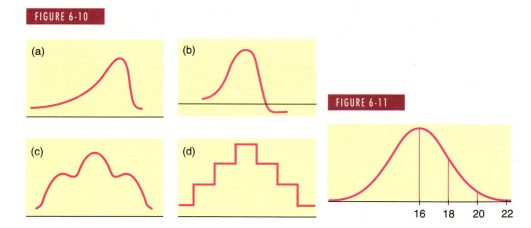

FIGURE 6-11

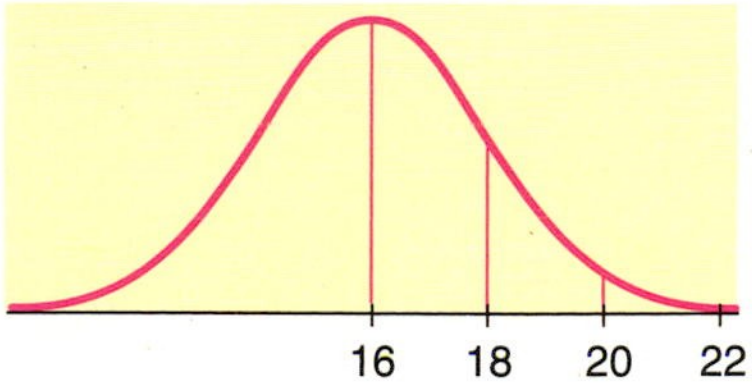

3. *Critical Thinking* Look at the two normal curves in Figures 6-12 and 6-13. Which has the larger standard deviation? What is the mean of the curve in Figure 6-12? What is the mean of the curve in Figure 6-13?

FIGURE 6-13

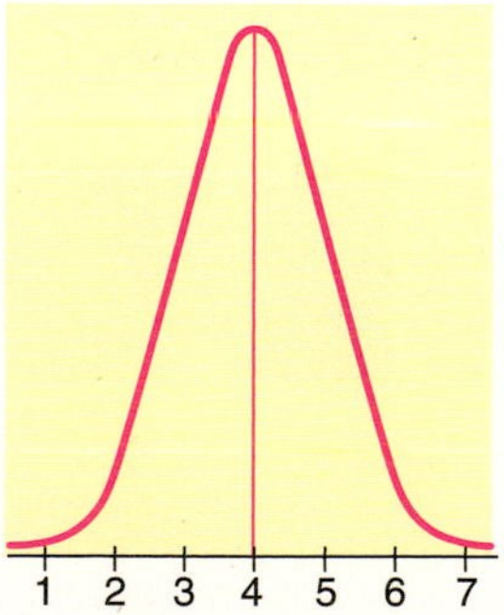

FIGURE 6-12

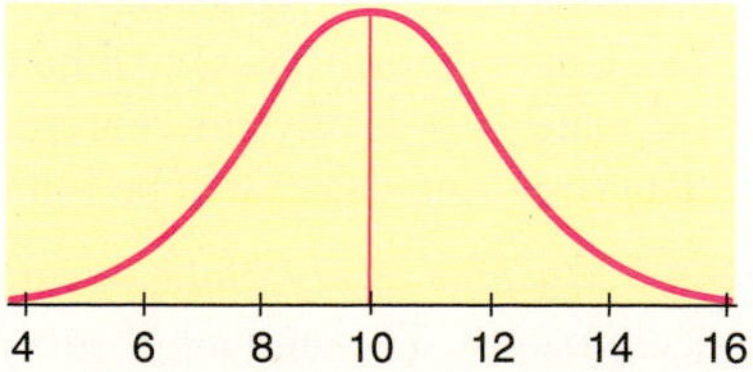

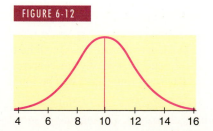

4. *Critical Thinking* Sketch a normal curve
(a) with mean 15 and standard deviation 2.
(b) with mean 15 and standard deviation 3.
(c) with mean 12 and standard deviation 2.
(d) with mean 12 and standard deviation 3.
(e) Consider two normal curves. If the first one has a larger mean than the second one, must it have a larger standard deviation as well? Explain your answer.

5. *Basic Computation: Empirical Rule* What percentage of the area under the normal curve lies
(a) to the left of μ?
(b) between $\mu - \sigma$ and $\mu + \sigma$?
(c) between $\mu - 3\sigma$ and $\mu + 3\sigma$?

6. *Basic Computation: Empirical Rule* What percentage of the area under the normal curve lies
(a) to the right of μ?
(b) between $\mu - 2\sigma$ and $\mu + 2\sigma$?
(c) to the right of $\mu + 3\sigma$?

7. *Distribution: Heights of Coeds* Assuming that the heights of college women are normally distributed with mean 65 inches and standard deviation 2.5 inches (based on information from *Statistical Abstract of the United States*, 112th Edition), answer the following questions. *Hint:* Use Problems 5 and 6 and Figure 6-3.
(a) What percentage of women are taller than 65 inches?
(b) What percentage of women are shorter than 65 inches?
(c) What percentage of women are between 62.5 inches and 67.5 inches?
(d) What percentage of women are between 60 inches and 70 inches?

8. *Distribution: Rhode Island Red Chicks* The incubation time for Rhode Island Red chicks is normally distributed with a mean of 21 days and standard deviation of approximately 1 day (based on information from *World Book Encyclopedia*). Look at Figure 6-3 and answer the following questions. If 1000 eggs are being incubated, how many chicks do we expect will hatch
(a) in 19 to 23 days?
(b) in 20 to 22 days?
(c) in 21 days or fewer?
(d) in 18 to 24 days? (Assume all eggs eventually hatch.)

Note: In this problem, let us agree to think of a single day or a succession of days as a continuous interval of time.

Raul Touzon/National Geographic Stock

9. *Archaeology: Tree Rings* At Burnt Mesa Pueblo, archaeological studies have used the method of tree-ring dating in an effort to determine when prehistoric people lived in the pueblo. Wood from several excavations gave a mean of (year) 1243 with a standard deviation of 36 years (*Bandelier Archaeological Excavation Project: Summer 1989 Excavations at Burnt Mesa Pueblo,* edited by Kohler, Washington State University Department of Anthropology). The distribution of dates was more or less mound-shaped and symmetrical about the mean. Use the empirical rule to
(a) estimate a range of years centered about the mean in which about 68% of the data (tree-ring dates) will be found.
(b) estimate a range of years centered about the mean in which about 95% of the data (tree-ring dates) will be found.
(c) estimate a range of years centered about the mean in which almost all the data (tree-ring dates) will be found.

10. *Vending Machine: Soft Drinks* A vending machine automatically pours soft drinks into cups. The amount of soft drink dispensed into a cup is normally

distributed with a mean of 7.6 ounces and standard deviation of 0.4 ounce. Examine Figure 6-3 and answer the following questions.

(a) Estimate the probability that the machine will overflow an 8-ounce cup.
(b) Estimate the probability that the machine will not overflow an 8-ounce cup.
(c) The machine has just been loaded with 850 cups. How many of these do you expect will overflow when served?

11. *Pain Management: Laser Therapy* "Effect of Helium-Neon Laser Auriculotherapy on Experimental Pain Threshold" is the title of an article in the journal *Physical Therapy* (Vol. 70, No. 1, pp. 24–30). In this article, laser therapy was discussed as a useful alternative to drugs in pain management of chronically ill patients. To measure pain threshold, a machine was used that delivered low-voltage direct current to different parts of the body (wrist, neck, and back). The machine measured current in milliamperes (mA). The pretreatment experimental group in the study had an average threshold of pain (pain was first detectable) at $\mu = 3.15$ mA with standard deviation $\sigma = 1.45$ mA. Assume that the distribution of threshold pain, measured in milliamperes, is symmetrical and more or less mound-shaped. Use the empirical rule to
(a) estimate a range of milliamperes centered about the mean in which about 68% of the experimental group had a threshold of pain.
(b) estimate a range of milliamperes centered about the mean in which about 95% of the experimental group had a threshold of pain.

12. *Control Charts: Yellowstone National Park* Yellowstone Park Medical Services (YPMS) provides emergency health care for park visitors. Such health care includes treatment for everything from indigestion and sunburn to more serious injuries. A recent issue of *Yellowstone Today* (National Park Service Publication) indicated that the average number of visitors treated each day by YPMS is 21.7. The estimated standard deviation is 4.2 (summer data). The distribution of numbers treated is approximately mound-shaped and symmetrical.
(a) For a 10-day summer period, the following data show the number of visitors treated each day by YPMS:

Day	1	2	3	4	5	6	7	8	9	10
Number treated	25	19	17	15	20	24	30	19	16	23

Make a control chart for the daily number of visitors treated by YPMS, and plot the data on the control chart. Do the data indicate that the number of visitors treated by YPMS is "in control"? Explain your answer.
(b) For another 10-day summer period, the following data were obtained:

Day	1	2	3	4	5	6	7	8	9	10
Number treated	20	15	12	21	24	28	32	36	35	37

Make a control chart, and plot the data on the chart. ***Interpretation*** Do the data indicate that the number of visitors treated by YPMS is "in control" or "out of control"? Explain your answer. Identify all out-of-control signals by type (I, II, or III). If you were the park superintendent, do you think YPMS might need some (temporary) extra help? Explain.

13. *Control Charts: Bank Loans* Tri-County Bank is a small independent bank in central Wyoming. This is a rural bank that makes loans on items as small as horses and pickup trucks to items as large as ranch land. Total monthly loan requests are used by bank officials as an indicator of economic business conditions in this rural community. The mean monthly loan request for the past several years has been 615.1 (in thousands of dollars) with a standard deviation of 11.2 (in thousands of dollars). The distribution of loan requests is approximately mound-shaped and symmetrical.

(a) For 12 months, the following monthly loan requests (in thousands of dollars) were made to Tri-County Bank:

Month	1	2	3	4	5	6
Loan request	619.3	625.1	610.2	614.2	630.4	615.9

Month	7	8	9	10	11	12
Loan request	617.2	610.1	592.7	596.4	585.1	588.2

Make a control chart for the total monthly loan requests, and plot the preceding data on the control chart. *Interpretation* From the control chart, would you say the local business economy is heating up or cooling down? Explain your answer by referring to any trend you may see on the control chart. Identify all out-of-control signals by type (I, II, or III).

(b) For another 12-month period, the following monthly loan requests (in thousands of dollars) were made to Tri-County Bank:

Month	1	2	3	4	5	6
Loan request	608.3	610.4	615.1	617.2	619.3	622.1

Month	7	8	9	10	11	12
Loan request	625.7	633.1	635.4	625.0	628.2	619.8

Make a control chart for the total monthly loan requests, and plot the preceding data on the control chart. *Interpretation* From the control chart, would you say the local business economy is heating up, cooling down, or about normal? Explain your answer by referring to the control chart. Identify all out-of-control signals by type (I, II, or III).

14. *Control Charts: Motel Rooms* The manager of Motel 11 has 316 rooms in Palo Alto, California. From observation over a long period of time, she knows that on an average night, 268 rooms will be rented. The long-term standard deviation is 12 rooms. This distribution is approximately mound-shaped and symmetrical.

(a) For 10 consecutive nights, the following numbers of rooms were rented each night:

Night	1	2	3	4	5	6
Number of rooms	234	258	265	271	283	267

Night	7	8	9	10
Number of rooms	290	286	263	240

Make a control chart for the number of rooms rented each night, and plot the preceding data on the control chart. *Interpretation* Looking at the control chart, would you say the number of rooms rented during this 10-night period has been unusually low? unusually high? about what you expected? Explain your answer. Identify all out-of-control signals by type (I, II, or III).

(b) For another 10 consecutive nights, the following numbers of rooms were rented each night:

Night	1	2	3	4	5	6
Number of rooms	238	245	261	269	273	250

Night	7	8	9	10
Number of rooms	241	230	215	217

Make a control chart for the number of rooms rented each night, and plot the preceding data on the control chart. *Interpretation* Would you say the room occupancy has been high? low? about what you expected? Explain your answer. Identify all out-of-control signals by type (I, II, or III).

15. *Control Chart: Air Pollution* The visibility standard index (VSI) is a measure of Denver air pollution that is reported each day in the *Rocky Mountain News*. The index ranges from 0 (excellent air quality) to 200 (very bad air quality). During winter months, when air pollution is higher, the index has a mean of about 90 (rated as fair) with a standard deviation of approximately 30. Suppose that for 15 days, the following VSI measures were reported each day:

Day	1	2	3	4	5	6	7	8	9	10	11	12	13	14	15
VSI	80	115	100	90	15	10	53	75	80	110	165	160	120	140	195

Make a control chart for the VSI, and plot the preceding data on the control chart. Identify all out-of-control signals (high or low) that you find in the control chart by type (I, II, or III).

16. *Expand Your Knowledge: Continuous Uniform Probability Distribution* Let α and β be any two constants such that $\alpha < \beta$. Suppose we choose a point x at random in the interval from α to β. In this context the phrase *at random* is taken to mean that the point x is as likely to be chosen from one particular part of the interval as any other part. Consider the rectangle.

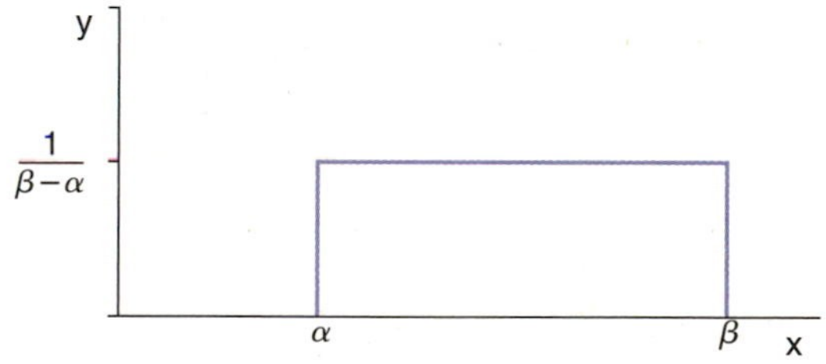

The base of the rectangle has length $\beta - \alpha$ and the height of the rectangle is $1/(\beta - \alpha)$, so the area of the rectangle is 1. As such, this rectangle's top can be thought of as part of a probability density curve. Since we specify that x must lie between α and β, the probability of a point occurring outside the interval $[\alpha, \beta]$ is, by definition, 0. From a geometric point of view, x chosen at random from α to β means we are equally likely to land anywhere in the interval from α to β. For this reason, the top of the (rectangle's) density curve is flat or uniform.

Now suppose that a and b are numbers such that $\alpha \le a < b \le \beta$. What is the probability that a number x chosen at random from α to β will fall in the interval $[a, b]$? Consider the graph

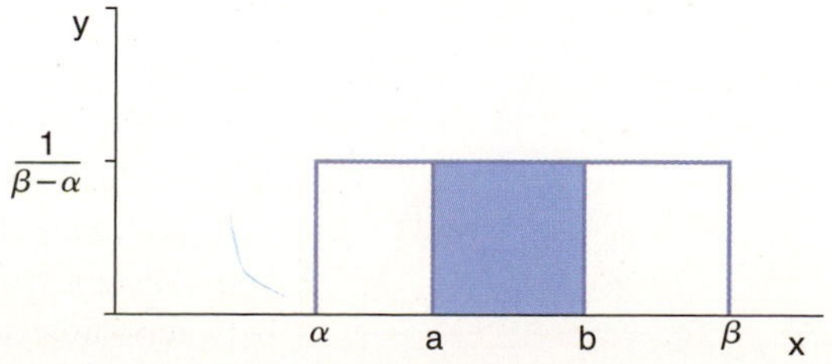

Because x is chosen at random from $[\alpha, \beta]$, the area of the rectangle that lies above $[a, b]$ is the probability that x lies in $[a, b]$. This area is

$$P(a < x < b) = \frac{b - a}{\beta - \alpha}$$

Continuous uniform distribution
Rectangular distribution

In this way we can assign a probability to any interval inside $|\alpha, \beta|$. This probability distribution is called the *continuous uniform distribution* (also called the rectangular distribution). Using some extra mathematics, it can be shown that if x is a random variable with this distribution, then the mean and standard deviation of x are

$$\mu = \frac{\alpha + \beta}{2} \quad \text{and} \quad \sigma = \frac{\beta - \alpha}{\sqrt{12}}$$

Sedimentation experiments are very important in the study of biology, medicine, hydrodynamics, petroleum engineering, civil engineering, and so on. The size (diameter) of approximately spherical particles is important since larger particles hinder and sometimes block the movement of smaller particles. Usually the size of sediment particles follows a uniform distribution (Reference: Y. Zimmels, "Theory of Kindred Sedimentation of Polydisperse Mixtures," *AIChE Journal*, Vol. 29, No. 4, pp. 669–676).

Suppose a veterinary science experiment injects very small, spherical pellets of low-level radiation directly into an animal's bloodstream. The purpose is to attempt to cure a form of recurring cancer. The pellets eventually dissolve and pass through the animal's system. Diameters of the pellets are uniformly distributed from 0.015 mm to 0.065 mm. If a pellet enters an artery, what is the probability that it will be the following sizes?

(a) 0.050 mm or larger. *Hint*: All particles are between 0.015 mm and 0.065 mm, so larger than 0.050 means $0.050 \le x \le 0.065$.
(b) 0.040 mm or smaller
(c) between 0.035 mm and 0.055 mm
(d) Compute the mean size of the particles.
(e) Compute the standard deviation of particle size.

17. *Uniform Distribution: Measurement Errors* Measurement errors from instruments are often modeled using the uniform distribution (see Problem 16). To determine the range of a large public address system, acoustical engineers use a method of triangulation to measure the shock waves sent out by the speakers. The time at which the waves arrive at the sensors must be measured accurately. In this context, a negative error means the signal arrived too early. A positive error means the signal arrived too late. Measurement errors in reading these times have a uniform distribution from -0.05 to $+0.05$ microseconds. (Reference: J. Perruzzi and E. Hilliard, "Modeling Time Delay Measurement Errors," *Journal of the Acoustical Society of America*, Vol. 75, No. 1, pp. 197–201.) What is the probability that such measurements will be in error by
(a) less than $+0.03$ microsecond (i.e., $-0.05 \le x < 0.03$)?
(b) more than -0.02 microsecond?
(c) between -0.04 and $+0.01$ microsecond?
(d) Find the mean and standard deviation of measurement errors. Measurements from an instrument are called *unbiased* if the mean of the measurement errors is zero. Would you say the measurements for these acoustical sensors are unbiased? Explain.

18. *Expand Your Knowledge: Exponential Distribution* The Poisson distribution (Section 5.4) gives the probability for the *number of occurrences* for a "rare" event. Now, let x be a random variable that represents the *waiting time* between rare events. Using some mathematics, it can be shown that x has an

Exponential distribution

exponential distribution. Let $x > 0$ be a random variable and let $\beta > 0$ be a constant. Then $y = \frac{1}{\beta}e^{-x/\beta}$ is a curve representing the exponential distribution. Areas under this curve give us exponential probabilities.

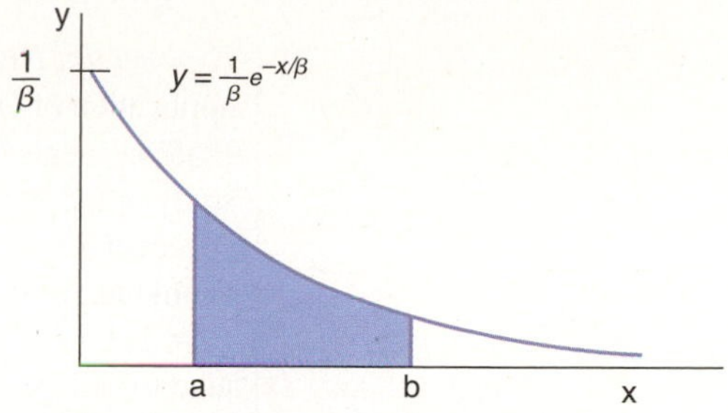

If a and b are any numbers such that $0 < a < b$, then using some extra mathematics, it can be shown that the area under the curve above the interval $[a, b]$ is

$$P(a < x < b) = e^{-a/\beta} - e^{-b/\beta}$$

Notice that by definition, x cannot be negative, so, $P(x < 0) = 0$. The random variable x is called an *exponential random variable.* Using some more mathematics, it can be shown that the mean and standard deviation of x are

$$\mu = \beta \text{ and } \sigma = \beta$$

Note: The number $e = 2.71828\ldots$ is used throughout probability, statistics and mathematics. The key e^x is conveniently located on most calculators.

Comment: The Poisson and exponential distributions have a special relationship. Specifically, it can be shown that the *waiting time* between successive Poisson arrivals (i.e., successes or rare events) has an exponential distribution with $\beta = 1/\lambda$, where λ is the average number of Poisson successes (rare events) per unit of time. For more on this topic, please see Problem 20.)

Fatal accidents on scheduled domestic passenger flights are rare events. In fact, airlines do all they possibly can to prevent such accidents. However, around the world such fatal accidents do occur. Let x be a random variable representing the waiting time between fatal airline accidents. Research has shown that x has an exponential distribution with a mean of approximately 44 days (Reference: R. Pyke, "Spacings," *Journal of the Royal Statistical Society* B, Vol. 27, No. 3, p. 426.)

We take the point of view that x (measured in days as units) is a continuous random variable. Suppose a fatal airline accident has just been reported on the news. What is the probability that the waiting time to the next reported fatal airline accident is

(a) less than 30 days (i.e., $0 \le x < 30$)?
(b) more than 50 days (i.e., $50 < x < \infty$)? *Hint:* $e^{-\infty} = 0$.
(c) between 20 and 60 days?
(d) What are the mean and the standard deviation of the waiting times x?

19. *Exponential Distribution: Supply and Demand* Another application for exponential distributions (see Problem 18) is supply/demand problems. The operator of a pumping station in a small Wyoming town has observed that demand for water on a typical summer afternoon is exponentially distributed with a mean of 75 cfs (cubic feet per second). Let x be a random variable that represents the town's demand for water (in cfs). What is the probability that on a typical summer afternoon, this town will have a water demand x

(a) more than 60 cfs (i.e., $60 < x < \infty$)? *Hint:* $e^{-\infty} = 0$.
(b) less than 140 cfs (i.e., $0 < x < 140$)?
(c) between 60 and 100 cfs?

(d) *Brain teaser* How much water c (in cfs) should the station pump to be 80% sure that the town demand x (in cfs) will not exceed the supply c? *Hint:* First explain why the equation $P(0 < x < c) = 0.80$ represents the problem as stated. Then solve for c.

20. *Exponential Distribution: Waiting Time Between Poisson Events* A common application of exponential distributions (see Problem 18) is waiting time between Poisson events (e.g., successes in the Poisson distribution; see Section 5.4). In our study of the Poisson distribution (Example 11, Section 5.4), we saw that the mean success rate per hour of catching a fish at Pyramid Lake is $\lambda = 0.667$ fish/hour. From this we see that the mean waiting time between fish can be thought of as $\beta = 1/\lambda = 1/0.667 \approx 1.5$ hours/fish. Remember, the fish at Pyramid Lake tend to be large. Suppose you have just caught a fish. Let x be a random variable representing the waiting time (in hours) to catch the next fish. Use the exponential distribution to determine the probability that the waiting time is
(a) less than half an hour (i.e., $0 < x < 0.5$).
(b) more than 3 hours (i.e., $3 < x < \infty$). *Hint:* $e^{-\infty} = 0$.
(c) between 1 and 3 hours (i.e., $1 < x < 3$).
(d) What are the mean and the standard deviation of the x distribution?

SECTION **6.2**

Standard Units and Areas Under the Standard Normal Distribution

FOCUS POINTS

- Given μ and σ, convert raw data to z scores.
- Given μ and σ, convert z scores to raw data.
- Graph the standard normal distribution, and find areas under the standard normal curve.

z SCORES AND RAW SCORES

Normal distributions vary from one another in two ways: The mean μ may be located anywhere on the x axis, and the bell shape may be more or less spread according to the size of the standard deviation σ. The differences among the normal distributions cause difficulties when we try to compute the area under the curve in a specified interval of x values and, hence, the probability that a measurement will fall into that interval.

It would be a futile task to try to set up a table of areas under the normal curve for each different μ and σ combination. We need a way to standardize the distributions so that we can use *one* table of areas for *all* normal distributions. We achieve this standardization by considering how many standard deviations a measurement lies from the mean. In this way, we can compare a value in one normal distribution with a value in another, different normal distribution. The next situation shows how this is done.

Suppose Tina and Jack are in two different sections of the same course. Each section is quite large, and the scores on the midterm exams of each section follow a normal distribution. In Tina's section, the average (mean) was 64 and her score was 74. In Jack's section, the mean was 72 and his score was 82. Both Tina and Jack were pleased that their scores were each 10 points above the average of each respective section. However, the fact that each was 10 points above average does not really tell us how each did *with respect to the other students in the section.* In Figure 6-14, we see the normal distribution of grades for each section.

Tina's 74 was higher than most of the other scores in her section, while Jack's 82 is only an upper-middle score in his section. Tina's score is far better with respect to her class than Jack's score is with respect to his class.

FIGURE 6-14

Distributions of Midterm Scores

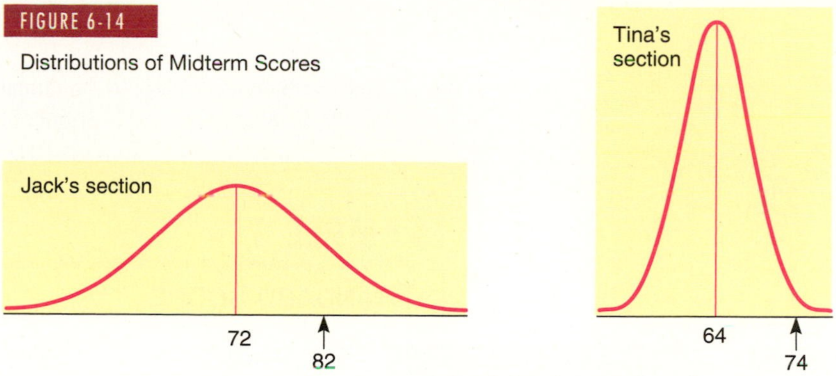

The preceding situation demonstrates that it is not sufficient to know the difference between a measurement (x value) and the mean of a distribution. We need also to consider the spread of the curve, or the standard deviation. What we really want to know is the number of standard deviations between a measurement and the mean. This "distance" takes both μ and σ into account.

We can use a simple formula to compute the number z of standard deviations between a measurement x and the mean μ of a normal distribution with standard deviation σ:

$$\left(\begin{array}{c}\text{Number of standard deviations}\\ \text{between the measurement and}\\ \text{the mean}\end{array}\right) = \left(\frac{\begin{array}{c}\text{Difference between the}\\ \text{measurement and the mean}\end{array}}{\text{Standard deviation}}\right)$$

z value, z score, or standard score

The **z value** or **z score** (also known as standard score) gives the number of standard deviations between the original measurement x and the mean μ of the x distribution.

$$z = \frac{x - \mu}{\sigma}$$

TABLE 6-2 **x Values and Corresponding z Values**

x Value in Original Distribution	Corresponding z Value or Standard Unit
$x = \mu$	$z = 0$
$x > \mu$	$z > 0$
$x < \mu$	$z < 0$

Standard units

The mean is a special value of a distribution. Let's see what happens when we convert $x = \mu$ to a z value:

$$z = \frac{x - \mu}{\sigma} = \frac{\mu - \mu}{\sigma} = 0$$

The mean of the original distribution is always zero, in *standard units*. This makes sense because the mean is zero standard variations from itself.

An x value in the original distribution that is *above* the mean μ has a corresponding z value that is *positive*. Again, this makes sense because a measurement above the mean would be a positive number of standard deviations from the mean. Likewise, an x value *below* the mean has a *negative* z value. (See Table 6-2.)

WHAT DOES A STANDARD SCORE TELL US?

A *standard score* or *z score* of a measurement tells us the number of standard deviations the measurement is from the mean.

- A standard score close to zero tells us the measurement is near the mean of the distribution.
- A positive standard score tells us the measurement is above the mean.
- A negative standard score tells us the measurement is below the mean.

NOTE

Unless otherwise stated, in the remainder of the book we will take the word *average* to be either the sample arithmetic mean $\bar{x}$ or the population mean μ.

EXAMPLE 4

STANDARD SCORE

Africa Studio/Shutterstock.com

A pizza parlor franchise specifies that the average (mean) amount of cheese on a large pizza should be 8 ounces and the standard deviation only 0.5 ounce. An inspector picks out a large pizza at random in one of the pizza parlors and finds that it is made with 6.9 ounces of cheese. Assume that the amount of cheese on a pizza follows a normal distribution. If the amount of cheese is below the mean by more than 3 standard deviations, the parlor will be in danger of losing its franchise.

How many standard deviations from the mean is 6.9? Is the pizza parlor in danger of losing its franchise?

SOLUTION: Since we want to know the number of standard deviations from the mean, we want to convert 6.9 to standard z units.

$$z = \frac{x - \mu}{\sigma} = \frac{6.9 - 8}{0.5} = -2.20$$

Interpretation The amount of cheese on the selected pizza is only 2.20 standard deviations below the mean. The fact that z is negative indicates that the amount of cheese is 2.20 standard deviations *below* the mean. The parlor will not lose its franchise based on this sample.

Raw score, *x*

We have seen how to convert from x measurements to standard units z. We can easily reverse the process to find the original *raw score* x if we know the mean and standard deviation of the original x distribution. Simply solve the z score formula for x.

Given an x distribution with mean μ and standard deviation σ, the **raw score x** corresponding to a z score is

$$x = z\sigma + \mu$$

GUIDED EXERCISE 4

STANDARD SCORE AND RAW SCORE

Rod figures that it takes an average (mean) of 17 minutes with a standard deviation of 3 minutes to drive from home, park the car, and walk to an early morning class.

(a) One day it took Rod 21 minutes to get to class. How many standard deviations from the average is that? Is the z value positive or negative? Explain why it should be either positive or negative.

→ The number of standard deviations from the mean is given by the z value:

$$z = \frac{x - \mu}{\sigma} = \frac{21 - 17}{3} \approx 1.33$$

The z value is positive. We should expect a positive z value, since 21 minutes is *more* than the mean of 17.

Continued

GUIDED EXERCISE 4 *continued*

(b) What commuting time corresponds to a standard score of $z = -2.5$? ***Interpretation*** Could Rod count on making it to class in this amount of time or less?

$$x = z\sigma + \mu$$
$$= (-2.5)(3) + 17$$
$$= 9.5 \text{ minutes}$$

No, commute times at or less than 2.5 standard deviations below the mean are rare.

LOOKING FORWARD

The basic structure of the formula for the standard score of a distribution is very general. When we verbalize the formula, we see it is

$$z = \frac{\text{measurement} - \text{mean of the distribution}}{\text{standard deviation of the distribution}}$$

We will see this general formula used again and again. In particular, when we look at sampling distributions for the mean (Section 6.4) and when we use the normal approximation of the binomial distribution (Section 6.6), we'll see this formula. We'll also see it when we discuss the sampling distribution for proportions (Section 6.6). Further uses occur in computations for confidence intervals (Chapter 7) and hypothesis testing (Chapter 8).

STANDARD NORMAL DISTRIBUTION

If the original distribution of x *values is normal,* then the corresponding z *values have a normal distribution as well.* The z distribution has a mean of 0 and a standard deviation of 1. The normal curve with these properties has a special name.

Standard normal distribution

The **standard normal distribution** is a normal distribution with mean $\mu = 0$ and standard deviation $\sigma = 1$ (Figure 6-15).

Any normal distribution of x values can be converted to the standard normal distribution by converting all x values to their corresponding z values. The resulting standard distribution will always have mean $\mu = 0$ and standard deviation $\sigma = 1$.

FIGURE 6-15

The Standard Normal Distribution ($\mu = 0, \sigma = 1$)

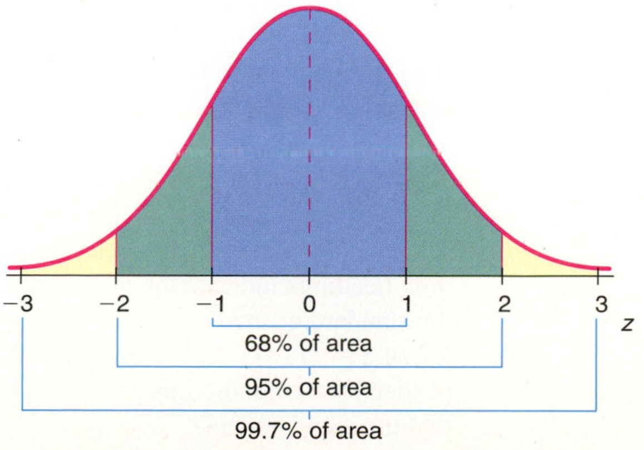

WHAT DOES THE STANDARD NORMAL DISTRIBUTION TELL US?

When we have the standard normal distribution, we know

- the mean is 0.
- the standard deviation is 1.
- any normal distribution can be converted to a standard normal distribution by converting all the measurements to standard z scores.

AREAS UNDER THE STANDARD NORMAL CURVE

Area under the standard normal curve

We have seen how to convert *any* normal distribution to the *standard* normal distribution. We can change any x value to a z value and back again. But what is the advantage of all this work? The advantage is that there are extensive tables that show the *area under the standard normal curve* for almost any interval along the z axis. The areas are important because each area is equal to the *probability* that the measurement of an item selected at random falls in this interval. Thus, the *standard* normal distribution can be a tremendously helpful tool.

USING A STANDARD NORMAL DISTRIBUTION TABLE

Using a table to find areas and probabilities associated with the standard normal distribution is a fairly straightforward activity. However, it is important to first observe the range of z values for which areas are given. This range is usually depicted in a picture that accompanies the table.

Left-tail style table

In this text, *we will use the left-tail style table*. This style table gives cumulative areas to the left of a specified z. Determining other areas under the curve utilizes the fact that the area under the entire curve is 1. Taking advantage of the symmetry of the normal distribution is also useful. The procedures you learn for using the left-tail style normal distribution table apply directly to cumulative normal distribution areas found on calculators and in computer software packages such as Excel 2010 and Minitab.

EXAMPLE 5

STANDARD NORMAL DISTRIBUTION TABLE

Use Table 5 of Appendix II to find the described areas under the standard normal curve.

(a) Find the area under the standard normal curve to the left of $z = -1.00$.

SOLUTION: First, shade the area to be found on the standard normal distribution curve, as shown in Figure 6-16. Notice that the z value we are using is negative. This means that we will look at the portion of Table 5 of Appendix II for which the z values are negative. In the upper-left corner of the table we see the letter z. The column under z gives us the units value and tenths value for z. The other column headings indicate the hundredths value of z. Table entries give areas under the standard normal curve to the left of the listed z values. To find the area to the left of $z = -1.00$, we use the row headed by -1.0 and then move to the column headed by the hundredths position, .00. This entry is shaded in Table 6-3. We see that the area is 0.1587.

FIGURE 6-16

Area to the Left of $z = -1.00$

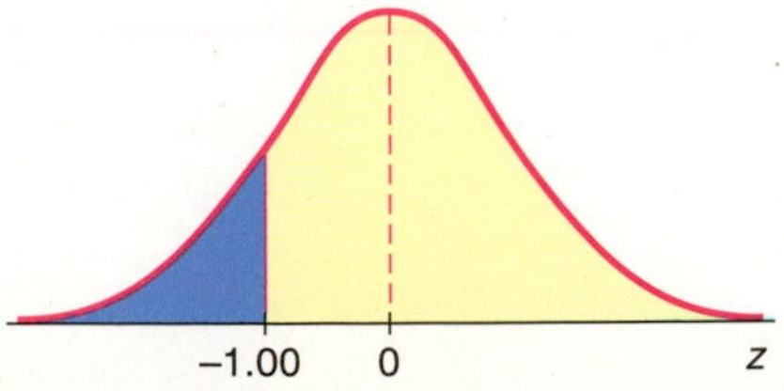

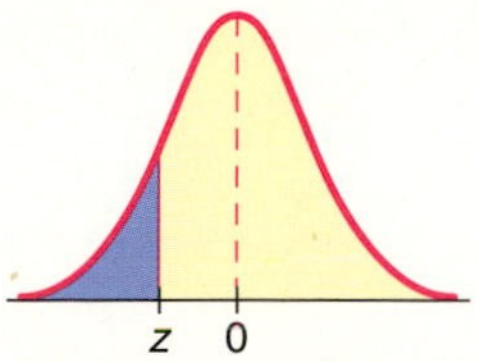

TABLE 6-3 Excerpt from Table 5 of Appendix II Showing Negative z Values

z	.00	.01	...	.07	.08	.09
−3.4	.0003	.0003	...	.0003	.0003	.0002
⋮						
−1.1	.1357	.1335	...	.1210	.1190	.1170
−1.0	.1587	.1562	...	.1423	.1401	.1379
−0.9	.1841	.1814	...	.1660	.1635	.1611
⋮						
−0.0	.5000	.4960	...	.4721	.4681	.4641

FIGURE 6-17

Area to the Left of $z = 1.18$

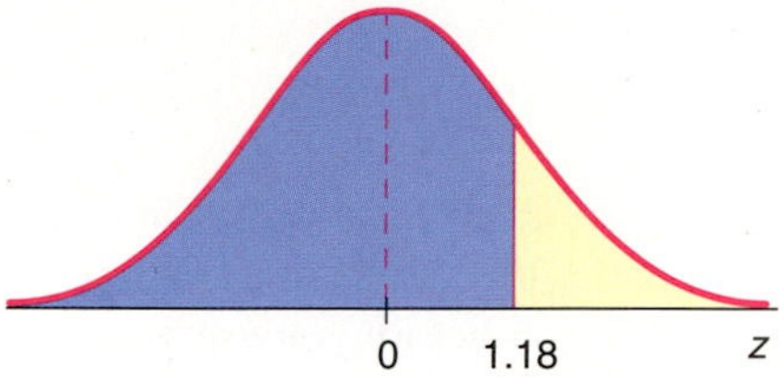

(b) Find the area to the left of $z = 1.18$, as illustrated in Figure 6-17.

SOLUTION: In this case, we are looking for an area to the left of a positive z value, so we look in the portion of Table 5 that shows positive z values. Again, we first sketch the area to be found on a standard normal curve, as shown in Figure 6-17. Look in the row headed by 1.1 and move to the column headed by .08. The desired area is shaded (see Table 6-4). We see that the area to the left of 1.18 is 0.8810.

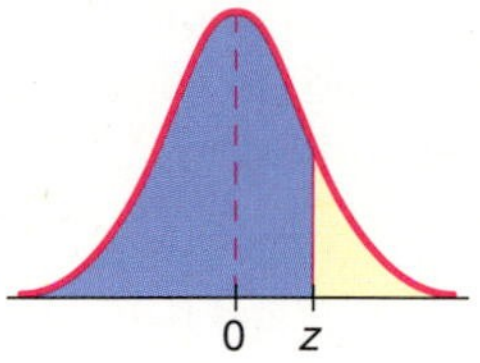

TABLE 6-4 Excerpt from Table 5 of Appendix II Showing Positive z Values

z	.00	.01	.02	...	.08	.09
0.0	.5000	.5040	.5080	...	.5319	.5359
⋮						
0.9	.8159	.8186	.8212	...	.8365	.8389
1.0	.8413	.8438	.8461	...	.8599	.8621
1.1	.8643	.8665	.8686	...	.8810	.8830
⋮						
3.4	.9997	.9997	.9997	...	.9997	.9998

GUIDED EXERCISE 5 USING THE STANDARD NORMAL DISTRIBUTION TABLE

Table 5, Areas of a Standard Normal Distribution, is located in Appendix II as well as in the endpapers of the text. Spend a little time studying the table, and then answer these questions.

(a) As z values increase, do the areas to the left of z increase?

➡ Yes. As z values increase, we move to the right on the normal curve, and the areas increase.

(b) If a z value is negative, is the area to the left of z less than 0.5000?

➡ Yes. Remember that a negative z value is on the left side of the standard normal distribution. The entire left half of the normal distribution has area 0.5, so any area to the left of $z = 0$ will be less than 0.5.

(c) If a z value is positive, is the area to the left of z greater than 0.5000?

➡ Yes. Positive z values are on the right side of the standard normal distribution, and any area to the left of a positive z value includes the entire left half of the normal distribution.

Using Table 5 to find other areas

Table 5 gives areas under the standard normal distribution that are to the *left of a z value.* How do we find other areas under the standard normal curve?

PROCEDURE

HOW TO USE A LEFT-TAIL STYLE STANDARD NORMAL DISTRIBUTION TABLE

1. For areas to the left of a specified z value, use the table entry directly.
2. For areas to the right of a specified z value, look up the table entry for z and subtract the area from 1.
   Note: Another way to find the same area is to use the symmetry of the normal curve and look up the table entry for $-z$.
3. For areas between two z values, z_1 and z_2 (where $z_2 > z_1$), *subtract* the table area for z_1 from the table area for z_2.

Figure 6-18 illustrates the procedure for using Table 5, Areas of a Standard Normal Distribution, to find any specified area under the standard normal distribution. Again, it is useful to sketch the area in question before you use Table 5.

COMMENT Notice that the z values shown in Table 5 of Appendix II are formatted to the hundredths position. It is convenient to *round or format z values to the hundredths position* before using the table. The areas are all given to four places after the decimal, so give your answers to four places after the decimal.

COMMENT The smallest z value shown in Table 5 is -3.49, while the largest value is 3.49. These values are, respectively, far to the left and far to the right on the standard normal distribution, with very little area beyond either value. We will follow the common convention of treating any area to the left of a z value smaller than -3.49 as 0.000. Similarly, we will consider any area to the right of a z value greater than 3.49 as 0.000. We understand that there is some area in these extreme tails. However, these areas are each less than 0.0002. Now let's get real about this! Some very specialized

FIGURE 6-18

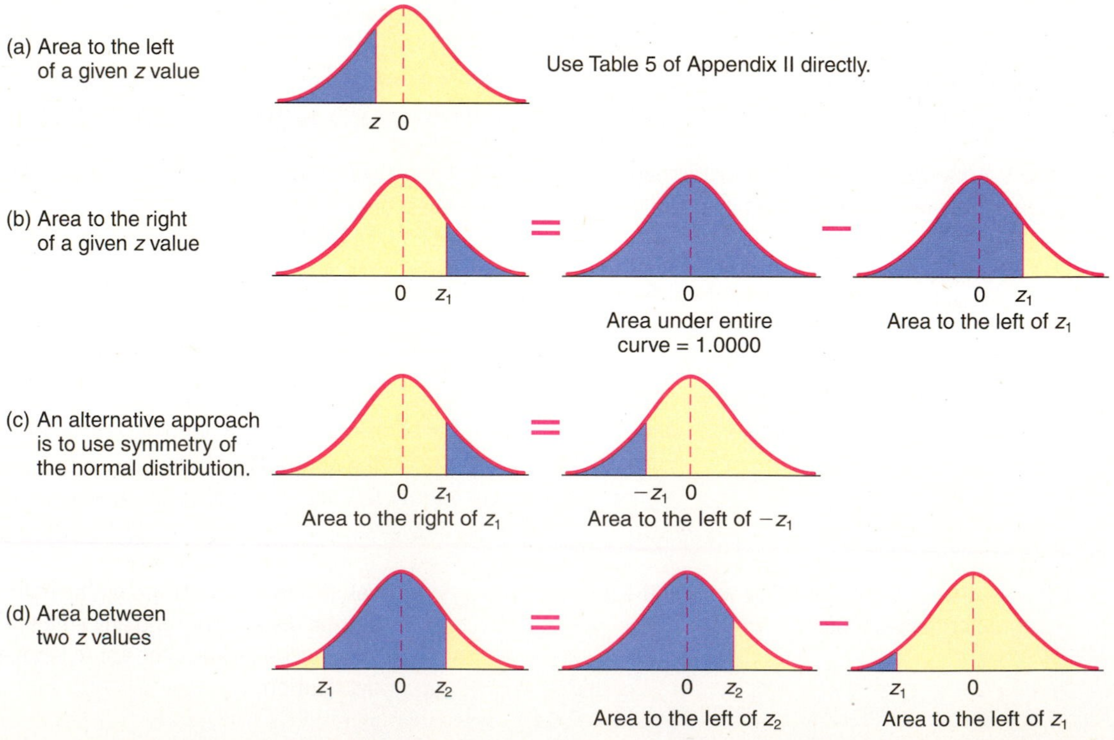

applications, beyond the scope of this book, do need to measure areas and corresponding probabilities in these extreme tails. But in most practical applications, *we follow the convention of treating the areas in the extreme tails as zero.*

CONVENTION FOR USING TABLE 5 OF APPENDIX II

1. Treat any area to the left of a z value smaller than -3.49 as 0.000.
2. Treat any area to the left of a z value greater than 3.49 as 1.000.

EXAMPLE 6

USING TABLE TO FIND AREAS

Use Table 5 of Appendix II to find the specified areas.

(a) Find the area between $z = 1.00$ and $z = 2.70$.

SOLUTION: First, sketch a diagram showing the area (see Figure 6-19). Because we are finding the area between two z values, we subtract corresponding table entries.

$$\begin{aligned}\text{(Area between 1.00 and 2.70)} &= \text{(Area left of 2.70)} - \text{(Area left of 1.00)}\\ &= 0.9965 - 08413\\ &= 0.1552\end{aligned}$$

(b) Find the area to the right of $z = 0.94$.

SOLUTION: First, sketch the area to be found (see Figure 6-20).

$$\begin{aligned}\text{(Area to right of 0.94)} &= \text{(Area under entire curve)} - \text{(Area to left of 0.94)}\\ &= 1.000 - 0.8264\\ &= 0.1736\end{aligned}$$

Alternatively,

$$\begin{aligned}\text{(Area to right of 0.94)} &= \text{(Area to left of } -0.94)\\ &= 0.1736\end{aligned}$$

FIGURE 6-19

Area from $z = 1.00$ to $z = 2.70$

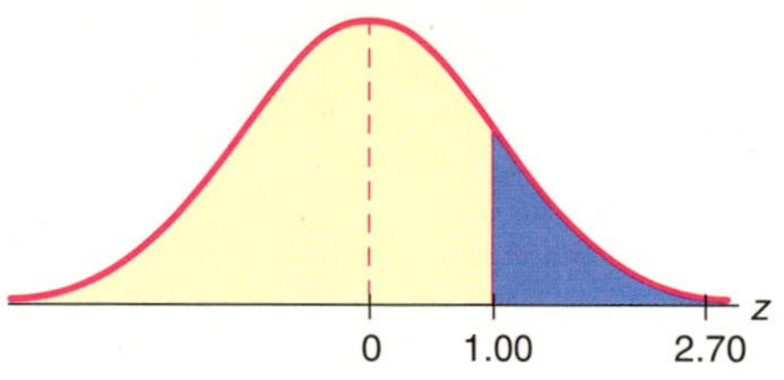

FIGURE 6-20

Area to the Right of $z = 0.94$

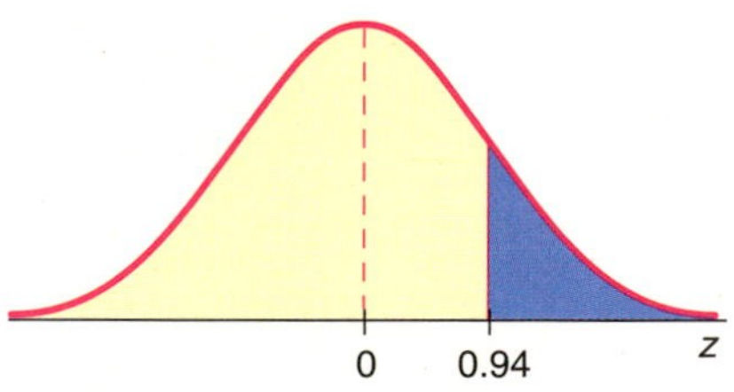

Probabilities associated with the standard normal distribution

We have practiced the skill of finding areas under the standard normal curve for various intervals along the z axis. This skill is important because *the probability* that z lies in an interval *is given by the area* under the standard normal curve above that interval.

Because the normal distribution is continuous, there is no area under the curve exactly over a specific z. Therefore, probabilities such as $P(z \geq z_1)$ are the same as $P(z > z_1)$. When dealing with probabilities or areas under a normal curve that are specified with inequalities, *strict inequality* symbols can be used *interchangeably* with *inequality or equal* symbols.

GUIDED EXERCISE 6 — PROBABILITIES ASSOCIATED WITH THE STANDARD NORMAL DISTRIBUTION

Let z be a random variable with a standard normal distribution.

(a) $P(z \geq 1.15)$ refers to the probability that z values lie to the right of 1.15. Shade the corresponding area under the standard normal curve (Figure 6-21) and find $P(z \geq 1.15)$.

FIGURE 6-21 Area to Be Found

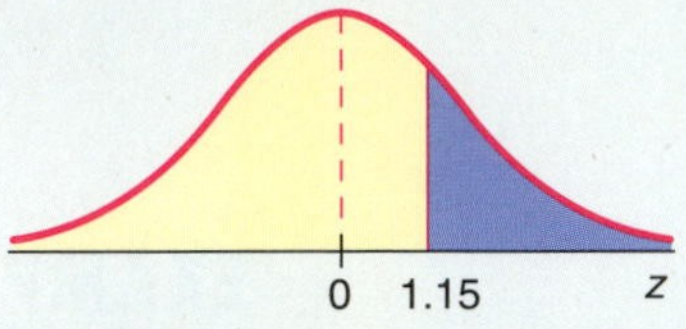

$P(z \geq 1.15) = 1.000 - P(z \leq 1.15) = 1.000 - 0.8749 = 0.1251$

Alternatively,

$P(z \geq 1.15) = P(z \leq -1.15) = 0.1251$

(b) Find $P(-1.78 \leq z \leq 0.35)$. First, sketch the area under the standard normal curve corresponding to the area (Figure 6-22).

FIGURE 6-22 Area to Be Found

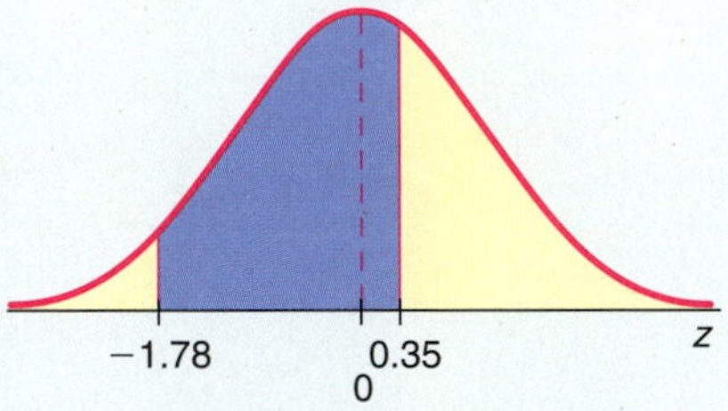

$$P(-1.78 \leq z \leq 0.35) = P(z \leq 0.35) - P(z \leq -1.78)$$
$$= 0.6368 - 0.0375 = 0.5993$$

TECH NOTES The TI-84Plus/TI-83Plus/TI-*n*spire calculators, Excel 2010, and Minitab all provide cumulative areas under any normal distribution, including the standard normal. The Tech Note of Section 6.3 shows examples.

VIEWPOINT — Mighty Oaks from Little Acorns Grow!

Just how big is that acorn? What if we compare it with other acorns? Is that oak tree taller than an average oak tree? How does it compare with other oak trees? What do you mean, this oak tree has a larger geographic range? Compared with what? Answers to questions such as these can be given only if we resort to standardized statistical units. Can you compare a single oak tree with an entire forest of oak trees? The answer is yes, if you use standardized z *scores. For more information about sizes of acorns, oak trees, and geographic locations, visit the Brase/Brase statistics site at* **http://www.cengage.com/statistics/brase11e** *and find the link to DASL, the Carnegie Mellon University Data and Story Library. From the DASL site, find Biology under Data Subjects, and select Acorns.*

SECTION 6.2 PROBLEMS

In these problems, assume that all distributions are *normal*. In all problems in Chapter 6, *average* is always taken to be the arithmetic mean $\bar{x}$ or μ.

1. *Statistical Literacy* What does a standard score measure?

2. *Statistical Literacy* Does a raw score less than the mean correspond to a positive or negative standard score? What about a raw score greater than the mean?

3. *Statistical Literacy* What is the value of the standard score for the mean of a distribution?

4. *Statistical Literacy* What are the values of the mean and standard deviation of a standard normal distribution?

5. *Basic Computation: z Score and Raw Score* A normal distribution has $\mu = 30$ and $\sigma = 5$.
   (a) Find the z score corresponding to $x = 25$.
   (b) Find the z score corresponding to $x = 42$.
   (c) Find the raw score corresponding to $z = -2$.
   (d) Find the raw score corresponding to $z = 1.3$.

6. *Basic Computation: z Score and Raw Score* A normal distribution has $\mu = 10$ and $\sigma = 2$.
   (a) Find the z score corresponding to $x = 12$.
   (b) Find the z score corresponding to $x = 4$.
   (c) Find the raw score corresponding to $z = 1.5$.
   (d) Find the raw score corresponding to $z = -1.2$.

7. *Critical Thinking* Consider the following scores:
   (i) Score of 40 from a distribution with mean 50 and standard deviation 10
   (ii) Score of 45 from a distribution with mean 50 and standard deviation 5

   How do the two scores compare relative to their respective distributions?

8. *Critical Thinking* Raul received a score of 80 on a history test for which the class mean was 70 with standard deviation 10. He received a score of 75 on a biology test for which the class mean was 70 with standard deviation 2.5. On which test did he do better relative to the rest of the class?

9. *z Scores: First Aid Course* The college physical education department offered an advanced first aid course last semester. The scores on the comprehensive final exam were normally distributed, and the z scores for some of the students are shown below:

Robert, 1.10	Juan, 1.70	Susan, −2.00
Joel, 0.00	Jan, −0.80	Linda, 1.60.

   (a) Which of these students scored above the mean?
   (b) Which of these students scored on the mean?
   (c) Which of these students scored below the mean?
   (d) If the mean score was $\mu = 150$ with standard deviation $\sigma = 20$, what was the final exam score for each student?

Comstock/Jupiter Images

10. *z Scores: Fawns* Fawns between 1 and 5 months old in Mesa Verde National Park have a body weight that is approximately normally distributed with mean $\mu = 27.2$ kilograms and standard deviation $\sigma = 4.3$ kilograms (based on information from *The Mule Deer of Mesa Verde National Park*, by G. W. Mierau and J. L. Schmidt, Mesa Verde Museum Association). Let x be the weight of a fawn in kilograms. Convert each of the following x intervals to z intervals.
    (a) $x < 30$ (b) $19 < x$ (c) $32 < x < 35$

Convert each of the following z intervals to x intervals.
(d) $-2.17 < z$ (e) $z < 1.28$ (f) $-1.99 < z < 1.44$
(g) *Interpretation* If a fawn weighs 14 kilograms, would you say it is an unusually small animal? Explain using z values and Figure 6-15.
(h) *Interpretation* If a fawn is unusually large, would you say that the z value for the weight of the fawn will be close to 0, -2, or 3? Explain.

11. *z Scores: Red Blood Cell Count* Let x = red blood cell (RBC) count in millions per cubic millimeter of whole blood. For healthy females, x has an approximately normal distribution with mean $\mu = 4.8$ and standard deviation $\sigma = 0.3$ (based on information from *Diagnostic Tests with Nursing Implications,* edited by S. Loeb, Springhouse Press). Convert each of the following x intervals to z intervals.
(a) $4.5 < x$ (b) $x < 4.2$ (c) $4.0 < x < 5.5$

Convert each of the following z intervals to x intervals.
(d) $z < -1.44$ (e) $1.28 < z$ (f) $-2.25 < z < -1.00$
(g) *Interpretation* If a female had an RBC count of 5.9 or higher, would that be considered unusually high? Explain using z values and Figure 6-15.

12. *Normal Curve: Tree Rings* Tree-ring dates were used extensively in archaeological studies at Burnt Mesa Pueblo (*Bandelier Archaeological Excavation Project: Summer 1989 Excavations at Burnt Mesa Pueblo,* edited by Kohler, Washington State University Department of Anthropology). At one site on the mesa, tree-ring dates (for many samples) gave a mean date of μ_1 = year 1272 with standard deviation σ_1 = 35 years. At a second, removed site, the tree-ring dates gave a mean of μ_2 = year 1122 with standard deviation σ_2 = 40 years. Assume that both sites had dates that were approximately normally distributed. In the first area, an object was found and dated as x_1 = year 1250. In the second area, another object was found and dated as x_2 = year 1234.
(a) Convert both x_1 and x_2 to z values, and locate both of these values under the standard normal curve of Figure 6-15.
(b) *Interpretation* Which of these two items is the more unusual as an archaeological find in its location?

Basic Computation: Finding Areas Under the Standard Normal Curve In Problems 13–30, sketch the areas under the standard normal curve over the indicated intervals and find the specified areas.

13. To the right of $z = 0$
14. To the left of $z = 0$
15. To the left of $z = -1.32$
16. To the left of $z = -0.47$
17. To the left of $z = 0.45$
18. To the left of $z = 0.72$
19. To the right of $z = 1.52$
20. To the right of $z = 0.15$
21. To the right of $z = -1.22$
22. To the right of $z = -2.17$
23. Between $z = 0$ and $z = 3.18$
24. Between $z = 0$ and $z = -1.93$
25. Between $z = -2.18$ and $z = 1.34$
26. Between $z = -1.40$ and $z = 2.03$
27. Between $z = 0.32$ and $z = 1.92$
28. Between $z = 1.42$ and $z = 2.17$
29. Between $z = -2.42$ and $z = -1.77$
30. Between $z = -1.98$ and $z = -0.03$

Basic Computation: Finding Probabilities In Problems 31–50, let z be a random variable with a standard normal distribution. Find the indicated probability, and shade the corresponding area under the standard normal curve.

31. $P(z \le 0)$
32. $P(z \ge 0)$
33. $P(z \le -0.13)$
34. $P(z \le -2.15)$
35. $P(z \le 1.20)$
36. $P(z \le 3.20)$

37. $P(z \geq 1.35)$
38. $P(z \geq 2.17)$
39. $P(z \geq -1.20)$
40. $P(z \geq -1.50)$
41. $P(-1.20 \leq z \leq 2.64)$
42. $P(-2.20 \leq z \leq 1.40)$
43. $P(-2.18 \leq z \leq -0.42)$
44. $P(-1.78 \leq z \leq -1.23)$
45. $P(0 \leq z \leq 1.62)$
46. $P(0 \leq z \leq 0.54)$
47. $P(-0.82 \leq z \leq 0)$
48. $P(-2.37 \leq z \leq 0)$
49. $P(-0.45 \leq z \leq 2.73)$
50. $P(-0.73 \leq z \leq 3.12)$

SECTION **6.3**

Areas Under Any Normal Curve

FOCUS POINTS

- Compute the probability of "standardized events."
- Find a z score from a given normal probability (inverse normal).
- Use the inverse normal to solve guarantee problems.

NORMAL DISTRIBUTION AREAS

In many applied situations, the original normal curve is not the standard normal curve. Generally, there will not be a table of areas available for the original normal curve. This does not mean that we cannot find the probability that a measurement x will fall into an interval from a to b. What we must do is *convert* the original measurements x, a, and b to z values.

To compute a z score, we need to know the standard deviation of the distribution. But what if we don't know σ? If we have a range of data values, we can estimate σ. Problems 31 to 34 show how this is done.

PROCEDURE

HOW TO WORK WITH NORMAL DISTRIBUTIONS

To find areas and probabilities for a random variable x that follows a normal distribution with mean μ and standard deviation σ, convert x values to z values using the formula

$$z = \frac{x - \mu}{\sigma}$$

Then use Table 5 of Appendix II to find corresponding areas and probabilities.

EXAMPLE 7

NORMAL DISTRIBUTION PROBABILITY

Let x have a normal distribution with $\mu = 10$ and $\sigma = 2$. Find the probability that an x value selected at random from this distribution is between 11 and 14. In symbols, find $P(11 \leq x \leq 14)$.

SOLUTION: Since probabilities correspond to areas under the distribution curve, we want to find the area under the x curve above the interval from $x = 11$ to $x = 14$. To do so, we will convert the x values to standard z values and then use Table 5 of Appendix II to find the corresponding area under the standard curve.

We use the formula

$$z = \frac{x - \mu}{\sigma}$$

to convert the given x interval to a z interval.

$$z_1 = \frac{11 - 10}{2} = 0.50 \qquad \text{(Use } x = 11, \mu = 10, \sigma = 2.)$$

$$z_2 = \frac{14 - 10}{2} = 2.00 \qquad \text{(Use } x = 14, \mu = 10, \sigma = 2.)$$

The corresponding areas under the x and z curves are shown in Figure 6-23. From Figure 6-23 we see that

$$\begin{aligned} P(11 \le x \le 14) &= P(0.50 \le z \le 2.00) \\ &= P(z \le 2.00) - P(z \le 0.50) \\ &= 0.9772 - 0.6915 \text{ (From Table 5, Appendix II)} \\ &= 0.2857 \end{aligned}$$

Interpretation The probability is 0.2857 that an x value selected at random from a normal distribution with mean 10 and standard deviation 2 lies between 11 and 14.

FIGURE 6-23

Corresponding Areas Under the x Curve and z Curve

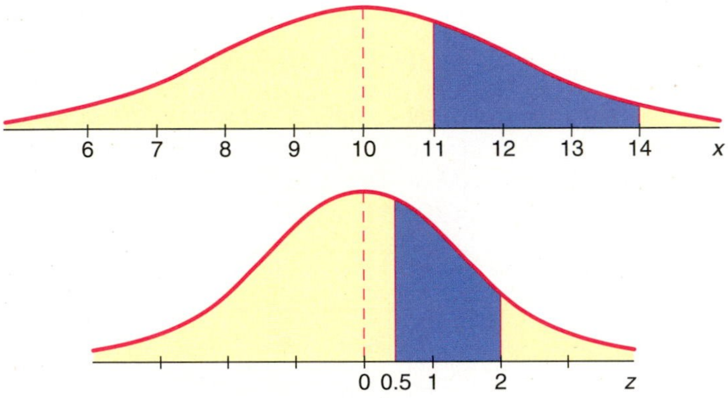

GUIDED EXERCISE 7 | NORMAL DISTRIBUTION PROBABILITY

The life span of a rechargeable battery is the time before the battery must be replaced because it no longer holds a charge. One tablet computer model has a battery with a life span that is normally distributed with a mean of 2.3 years and a standard deviation of 0.4 year. What is the probability that the battery will have to be replaced during the guarantee period of 2 years?

(a) Let x represent the battery life span. The statement that the battery needs to be replaced during the 2-year guarantee period means the life span is less than 2 years, or $x \le 2$. Convert this statement to a statement about z.

➡ $z = \dfrac{x - \mu}{\sigma} = \dfrac{2 - 2.3}{0.4} = -0.75$

So, $x \le 2$ means $z \le -0.75$.

(b) Indicate the area to be found in Figure 6-24. Does this area correspond to the probability that $z \le -0.75$?

➡ See Figure 6-25.
Yes, the shaded area does correspond to the probability that $z \le -0.75$.

Continued

GUIDED EXERCISE 7 *continued*

FIGURE 6-24

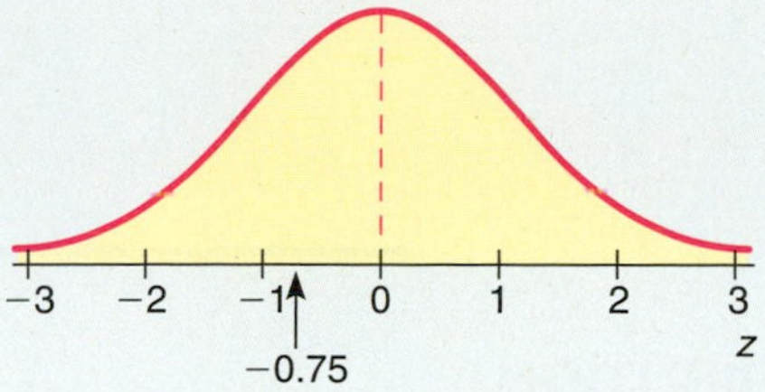

FIGURE 6-25 $z \leq -0.75$

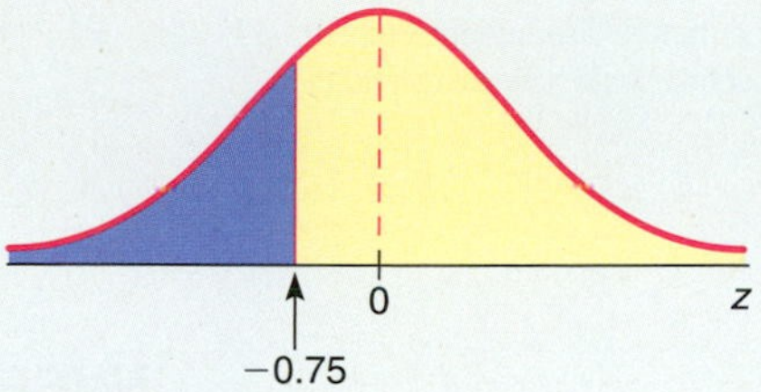

(c) Use Table 5 of Appendix II to find $P(z \leq -0.75)$. → 0.2266

(d) ***Interpretation*** What is the probability that the battery will fail before the end of the guarantee period? [*Hint:* $P(x \leq 2) = P(z \leq -0.75)$.] → The probability is

$$P(x \leq 2) = P(z \leq -0.75) = 0.2266$$

This means that the company will repair or replace about 23% of the batteries.

TECH NOTES

The TI-84Plus/TI-83Plus/TI-*n*spire calculators, Excel 2010, and Minitab all provide areas under any normal distribution. Excel 2010 and Minitab give the left-tail area to the left of a specified x value. The TI-84Plus/TI-83Plus/TI-*n*spire has you specify an interval from a lower bound to an upper bound and provides the area under the normal curve for that interval. For example, to solve Guided Exercise 7 regarding the probability a battery will fail during the guarantee period, we find $P(x \leq 2)$ for a normal distribution with $\mu = 2.3$ and $\sigma = 0.4$.

TI-84Plus/TI-83Plus/TI-*n*spire (with TI-84Plus keypad) Press the **DISTR** key, select **2:normalcdf (lower bound, upper bound, μ, σ)** and press Enter. Type in the specified values. For a left-tail area, use a lower bound setting at about 4 standard deviations below the mean. Likewise, for a right-tail area, use an upper bound setting about 4 standard deviations above the mean. For our example, use a lower bound of $\mu - 4\sigma = 2.3 - 4(0.4) = 0.7$.

```
normalcdf(.7,2,2.3,
.4)
              .2265955934
```

Excel 2010 Select **Insert Function** (fx) In the dialogue box, select **Statistical** for the Category and then for the Function, select **NORM.DIST**. Fill in the dialogue box using True for cumulative.

fx =NORM.DIST(2,2.3,0.4, TRUE)

C	D	E
0.226627		

Minitab Use the menu selection **Calc ➤ Probability Distribution ➤ Normal.** Fill in the dialogue box, marking cumulative.

What if we need to compute conditional probabilities based on the normal distribution? Problems 39 and 40 of this section show how we can do this.

```
Cumulative Distribution Function
Normal with mean = 2.3 and
standard deviation = 0.4
     x    P(X< = x)
   2.0    0.2266
```

INVERSE NORMAL DISTRIBUTION

Finding *z* or *x*, given a probability

Inverse normal probability distribution

Sometimes we need to find z or x values that correspond to a given area under the normal curve. This situation arises when we want to specify a guarantee period such that a given percentage of the total products produced by a company last at least as long as the duration of the guarantee period. In such cases, we use the standard normal distribution table "in reverse." When we look up an area and find the corresponding z value, we are using the *inverse normal probability distribution*.

EXAMPLE 8

FIND *x*, GIVEN PROBABILITY

Magic Video Games, Inc., sells an expensive video games package. Because the package is so expensive, the company wants to advertise an impressive guarantee for the life expectancy of its computer control system. The guarantee policy will refund the full purchase price if the computer fails during the guarantee period. The research department has done tests that show that the mean life for the computer is 30 months, with standard deviation of 4 months. The computer life is normally distributed. How long can the guarantee period be if management does not want to refund the purchase price on more than 7% of the Magic Video packages?

SOLUTION: Let us look at the distribution of lifetimes for the computer control system, and shade the portion of the distribution in which the computer lasts fewer months than the guarantee period. (See Figure 6-26.)

If a computer system lasts fewer months than the guarantee period, a full-price refund will have to be made. The lifetimes requiring a refund are in the shaded region in Figure 6-26. This region represents 7% of the total area under the curve.

We can use Table 5 of Appendix II to find the z value such that 7% of the total area under the *standard* normal curve lies to the left of the z value. Then we convert the z value to its corresponding x value to find the guarantee period.

We want to find the z value with 7% of the area under the standard normal curve to the left of z. Since we are given the area in a left tail, we can use Table 5 of Appendix II directly to find z. The area value is 0.0700. However, this area is not in our table, so we use the closest area, which is 0.0694, and the corresponding z value of $z = -1.48$ (see Table 6-5).

Pierre Arsenault/Alamy

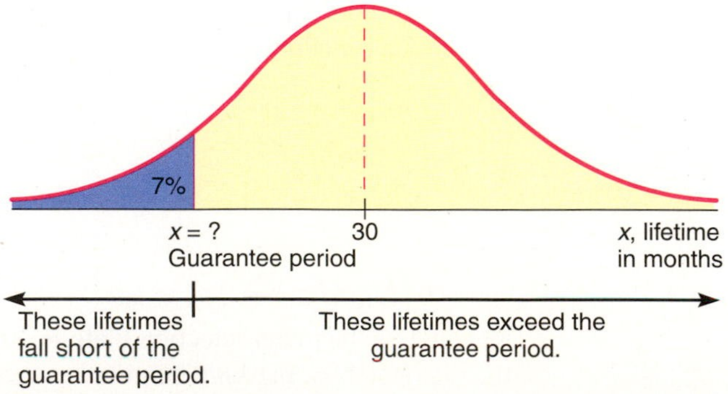

FIGURE 6-26

7% of the Computers Have a Lifetime Less Than the Guarantee Period

TABLE 6-5 Excerpt from Table 5 of Appendix II

z	.00	...	.07		.08	.09
⋮						
−1.4	.0808		.0708	↑	.0694	.0681
				0.0700		

To translate this value back to an x value (in months), we use the formula

$$\begin{aligned} x &= z\sigma + \mu \\ &= -1.48(4) + 30 \quad \text{(Use } \sigma = 4 \text{ months and } \mu = 30 \text{ months.)} \\ &= 24.08 \text{ months} \end{aligned}$$

Interpretation The company can guarantee the Magic Video Games package for $x = 24$ months. For this guarantee period, they expect to refund the purchase price of no more than 7% of the video games packages.

Example 8 had us find a z value corresponding to a given area to the left of z. What if the specified area is to the right of z or between $-z$ and z? Figure 6-27 shows us how to proceed.

COMMENT When we use Table 5 of Appendix II to find a z value corresponding to a given area, we usually use the nearest area value rather than interpolating between values. However, when the area value given is exactly halfway between two area values of the table, we use the z value halfway between the z values of the corresponding table areas. Example 9 demonstrates this procedure. However, this interpolation convention is not always used, especially if the area is changing slowly, as it does in the tail ends of the distribution. *When the z value corresponding to an area is smaller than −2, the standard convention is to use the z value*

FIGURE 6-27

Inverse Normal: Use Table 5 of Appendix II to Find z Corresponding to a Given Area A ($0 < A < 1$)

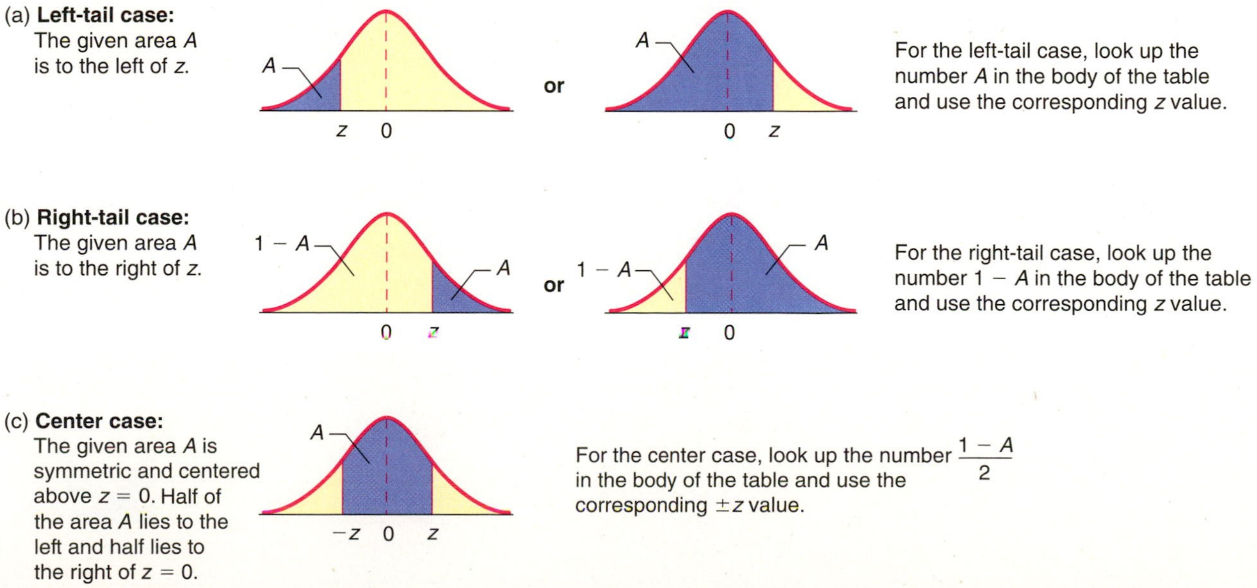
(a) Left-tail case:
The given area A is to the left of z.
A
z 0
or
A
0 z
For the left-tail case, look up the number A in the body of the table and use the corresponding z value.
(b) Right-tail case:
The given area A is to the right of z.
1 − A
A
0 z
or
1 − A
A
z 0
For the right-tail case, look up the number 1 − A in the body of the table and use the corresponding z value.
(c) Center case:
The given area A is symmetric and centered above z = 0. Half of the area A lies to the left and half lies to the right of z = 0.
A
−z 0 z
For the center case, look up the number (1 − A)/2 in the body of the table and use the corresponding ±z value.

corresponding to the smaller area. Likewise, when the z value is larger than 2, the standard convention is to use the z value corresponding to the larger area. We will see an example of this special case in Example 1 of Chapter 7.

EXAMPLE 9

FIND Z

Find the z value such that 90% of the area under the standard normal curve lies between $-z$ and z.

SOLUTION: Sketch a picture showing the described area (see Figure 6-28).

FIGURE 6-28

Area Between $-z$ and z Is 90%

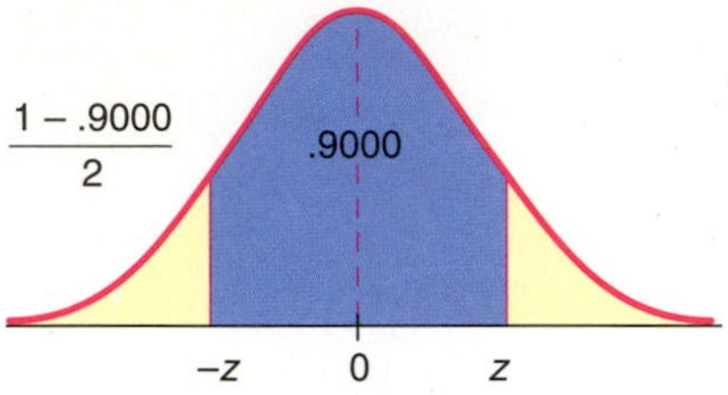

We find the corresponding area in the left tail.

$$(\text{Area left of } -z) = \frac{1 - 0.9000}{2}$$
$$= 0.0500$$

Looking in Table 6-6, we see that 0.0500 lies exactly between areas 0.0495 and 0.0505. The halfway value between $z = -1.65$ and $z = -1.64$ is $z = -1.645$. Therefore, we conclude that 90% of the area under the standard normal curve lies between the z values -1.645 and 1.645.

TABLE 6-6 **Excerpt from Table 5 of Appendix II**

z	...	.04		.05
⋮				
−1.6		.0505		.0495
			↑	
			0.0500	

GUIDED EXERCISE 8

FIND Z

Find the z value such that 3% of the area under the standard normal curve lies to the right of z.

(a) Draw a sketch of the standard normal distribution showing the described area (Figure 6-29).

FIGURE 6-29 3% of Total Area Lies to the Right of z

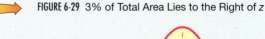

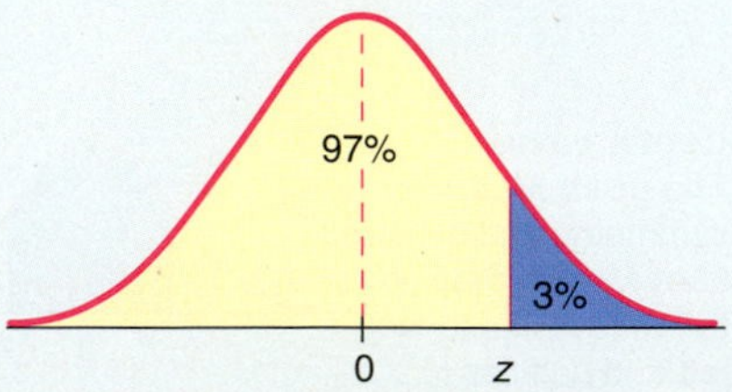

Continued

GUIDED EXERCISE 8 *continued*

(b) Find the area to the left of z.

Area to the left of $z = 1 - 0.0300 = 0.9700$.

(c) Look up the area in Table 6-7 and find the corresponding z.

The closest area is 0.9699. This area is to the left of $z = 1.88$.

TABLE 6-7 **Excerpt from Table 5 of Appendix II**

z	.00	.01	.02	.03	.04	.05	.06	.07	.08	.09
1.8	.9641	.9649	.9656	.9664	.9671	.9678	.9686	.9693	.9699	.9706
1.9	.9713	.9719	.9726	.9732	.9738	.9744	.9750	.9756	.9761	.9767

(d) Suppose the time to complete a test is normally distributed with $\mu = 40$ minutes and $\sigma = 5$ minutes. After how many minutes can we expect all but about 3% of the tests to be completed?

We are looking for an x value such that 3% of the normal distribution lies to the right of x. In part (c), we found that 3% of the standard normal curve lies to the right of $z = 1.88$. We convert $z = 1.88$ to an x value.

$$x = z\sigma + \mu$$
$$= 1.88(5) + 40 = 49.4 \text{ minutes}$$

All but about 3% of the tests will be complete after 50 minutes.

(e) Use Table 6-8 to find a z value such that 3% of the area under the standard normal curve lies to the left of z.

The closest area is 0.0301. This is the area to the left of $z = -1.88$.

TABLE 6-8 **Excerpt from Table 5 of Appendix II**

z	.00	.01	.02	.03	.04	.05	.06	.07	.08	.09
−1.9	.0287	.0281	.0274	.0268	.0262	.0256	.0250	.0244	.0239	.0233
−1.8	.0359	.0351	.0344	.0336	.0329	.0322	.0314	.0307	.0301	.0294

(f) Compare the z value of part (c) with the z value of part (e). Is there any relationship between the z values?

One z value is the negative of the other. This result is expected because of the symmetry of the normal distribution.

TECH NOTES When we are given a z value and we find an area to the left of z, we are using a normal distribution function. When we are given an area to the left of z and we find the corresponding z, we are using an inverse normal distribution function. The TI-84Plus/TI-83Plus/TI-*n*spire calculators, Excel 2010, and Minitab all have inverse normal distribution functions for any normal distribution. For instance, to find an x value from a normal distribution with mean 40 and standard deviation 5 such that 97% of the area lies to the left of x, use the described instructions.

TI-84Plus/TI-83Plus/TI-*n*spire (with TI-84Plus keypad) Press the **DISTR** key and select **3:invNorm(area,μ,σ)**.

```
invNorm(.97,40,5)
         49.40396805
```

Excel 2010 Select **Insert Function** (f_x). In the dialogue box, select **Statistical** for the Category and then for the Function, select **NORM.INV**. Fill in the dialogue box.

f_x	=NORM.INV(0.97,40,5)
C	D
49.40395	

LOOKING FORWARD

In our work with confidence intervals (Chapter 7), we will use inverse probability distributions for the normal distribution and for the Student's *t* distribution (a similar distribution introduced in Chapter 7). Just as in Example 9, we'll use inverse probability distributions to identify values such that 90%, 95%, or 99% of the area under the distribution graph centered over the mean falls between the values.

Minitab Use the menu selection **Calc ➤ Probability Distribution ➤ Normal.** Fill in the dialogue box, marking Inverse Cumulative.

```
Inverse Cumulative Distribution Function
Normal with mean = 40.000 and
  standard deviation = 5.00000
        P(X< = x)       x
            0.9700   49.4040
```

CRITICAL THINKING

CHECKING FOR NORMALITY

How can we tell if data follow a normal distribution? There are several checks we can make. The following procedure lists some guidelines.

PROCEDURE

HOW TO DETERMINE WHETHER DATA HAVE A NORMAL DISTRIBUTION

The following guidelines represent some useful devices for determining whether or not data follow a normal distribution.

1. **Histogram:** Make a histogram. For a normal distribution, the histogram should be roughly bell-shaped.
2. **Outliers:** For a normal distribution, there should not be more than one outlier. One way to check for outliers is to use a box-and-whisker plot. Recall that outliers are those data values that are

   above Q_3 by an amount greater than $1.5 \times$ interquartile range
   below Q_1 by an amount greater than $1.5 \times$ interquartile range

3. **Skewness:** Normal distributions are symmetric. One measure of skewness for sample data is given by Pearson's index:

$$\text{Pearson's index} = \frac{3(\bar{x} - \text{median})}{s}$$

   An index value greater than 1 or less than -1 indicates skewness. Skewed distributions are not normal.

Continued

4. **Normal quantile plot (or normal probability plot):** This plot is provided through statistical software on a computer or graphing calculator. The Using Technology feature, Application 1, at the end of this chapter gives a brief description of how such plots are constructed. The section also gives commands for producing such plots on the TI-84Plus/TI-83Plus/TI-*n*spire calculators, Minitab, or SPSS.
   Examine a normal quantile plot of the data.

   If the points lie close to a straight line, the data come from a distribution that is approximately normal.

   If the points do not lie close to a straight line or if they show a pattern that is not a straight line, the data are likely to come from a distribution that is not normal.

EXAMPLE 10

ASSESSING NORMALITY

Consider the following data, which are rounded to the nearest integer.

19	19	19	16	21	14	23	17	19	20	18	24	20	13	16
17	19	18	19	17	21	24	18	23	19	21	22	20	20	20
24	17	20	22	19	22	21	18	20	22	16	15	21	23	21
18	18	20	15	25										

(a) Look at the histogram and box-and-whisker plot generated by Minitab in Figure 6-30 and comment about normality of the data from these indicators.

FIGURE 6-30
Histogram and Box-and-Whisker Plot

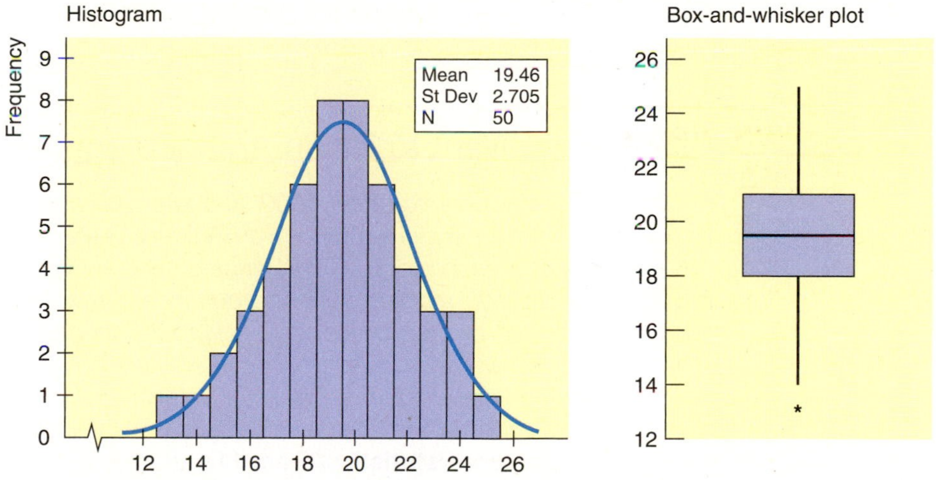

SOLUTION: Note that the histogram is approximately normal. The box-and-whisker plot shows just one outlier. Both of these graphs indicate normality.

(b) Use Pearson's index to check for skewness.

SOLUTION: Summary statistics from Minitab:

```
Variable   N  N*    Mean  Se Mean  StDev  Minimum      Q1  Median      Q3
C2        50   0  19.460    0.382  2.705   13.000  18.000  19.500  21.000
Variable    Maximum
C2           25.000
```

We see that $\bar{x} = 19.46$, median $= 19.5$, and $s = 2.705$.

$$\text{Pearson's index} = \frac{3(19.46 - 19.5)}{2.705} \approx -0.04$$

Since the index is between -1 and 1, we detect no skewness. The data appear to be symmetric.

(c) Look at the normal quantile plot in Figure 6-31 and comment on normality.

FIGURE 6-31

Normal Quantile Plot

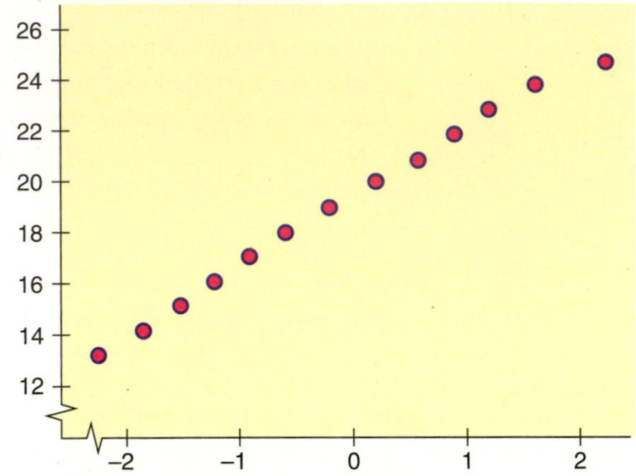

SOLUTION: The data fall close to a straight line, so the data appear to come from a normal distribution.

(d) ***Interpretation*** Interpret the results.

SOLUTION: The histogram is roughly bell-shaped, there is only one outlier, Pearson's index does not indicate skewness, and the points on the normal quantile plot lie fairly close to a straight line. It appears that the data are likely from a distribution that is approximately normal.

VIEWPOINT Want to Be an Archaeologist?

Each year about 4500 students work with professional archaeologists in scientific research at the Crow Canyon Archaeological Center, Cortez, Colorado. In fact, Crow Canyon was included in The Princeton Review Guide to America's Top 100 Internships. *The nonprofit, multidisciplinary program at Crow Canyon enables students and laypeople with little or no archaeology background to get started in archaeological research. The only requirement is that you be interested in Native American culture and history. By the way, a knowledge of introductory statistics could come in handy for this internship. For more information about the program, visit the Brase/Brase statistics site at* **http://www.cengage.com/statistics/brase11e** *and find the link to Crow Canyon.*

SECTION 6.3 PROBLEMS

1. *Statistical Literacy* Consider a normal distribution with mean 30 and standard deviation 2. What is the probability a value selected at random from this distribution is greater than 30?

2. *Statistical Literacy* Suppose 5% of the area under the standard normal curve lies to the right of z. Is z positive or negative?

3. *Statistical Literacy* Suppose 5% of the area under the standard normal curve lies to the left of z. Is z positive or negative?

4. *Critical Thinking: Normality* Consider the following data. The summary statistics, histogram, and normal quantile plot were generated by Minitab.

27	27	27	28	28	28	28	28	28	29	29	29	29	29	29
29	29	29	29	30	30	30	30	30	30	30	30	30	30	30
30	31	31	31	31	31	31	31	31	32	32	32	32	33	33
33	33	33	34	34										

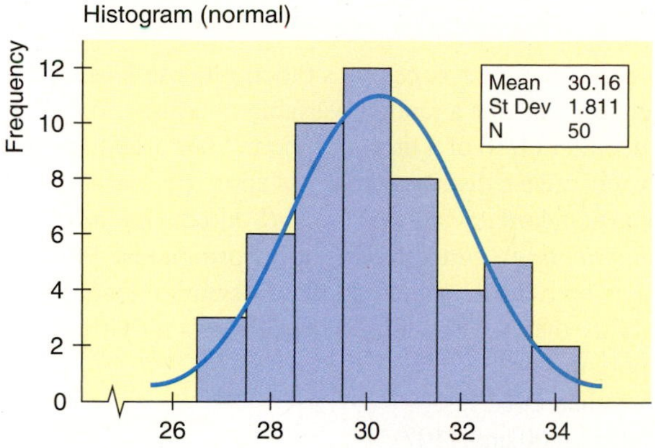

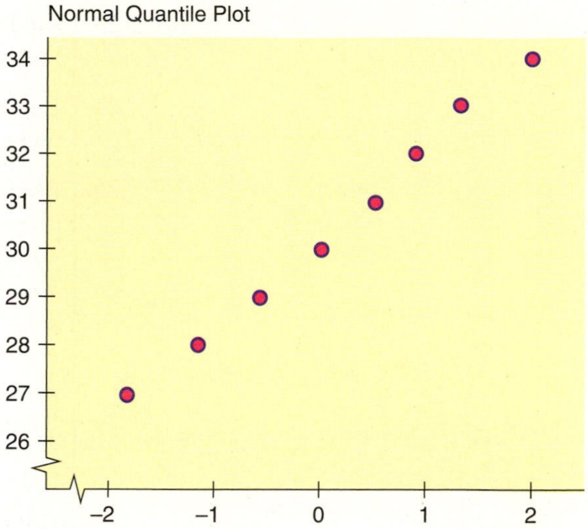


```
Variable  N  N*   Mean  Se Mean  StDev  Minimum      Q1  Median      Q3
Data     50   0  30.160   0.256  1.811   27.000  29.000  30.000  31.000
Variable   Maximum
Data        24.000
```

(a) Does the histogram indicate normality for the data distribution? Explain.
(b) Does the normal quantile plot indicate normality for the data distribution? Explain.
(c) Compute the interquartile range and check for outliers.
(d) Compute Pearson's index. Does the index value indicate skewness?
(e) Using parts (a) through (d), would you say the data are from a normal distribution?

Basic Computation: Find Probabilities In Problems 5–14, assume that x has a normal distribution with the specified mean and standard deviation. Find the indicated probabilities.

5. $P(3 \le x \le 6)$; $\mu = 4$; $\sigma = 2$
6. $P(10 \le x \le 26)$; $\mu = 15$; $\sigma = 4$
7. $P(50 \le x \le 70)$; $\mu = 40$; $\sigma = 15$
8. $P(7 \le x \le 9)$; $\mu = 5$; $\sigma = 1.2$
9. $P(8 \le x \le 12)$; $\mu = 15$; $\sigma = 3.2$
10. $P(40 \le x \le 47)$; $\mu = 50$; $\sigma = 15$
11. $P(x \ge 30)$; $\mu = 20$; $\sigma = 3.4$
12. $P(x \ge 120)$; $\mu = 100$; $\sigma = 15$
13. $P(x \ge 90)$; $\mu = 100$; $\sigma = 15$
14. $P(x \ge 2)$; $\mu = 3$; $\sigma = 0.25$

Basic Computation: Find z Values In Problems 15–24, find the z value described and sketch the area described.

15. Find z such that 6% of the standard normal curve lies to the left of z.
16. Find z such that 5.2% of the standard normal curve lies to the left of z.
17. Find z such that 55% of the standard normal curve lies to the left of z.

18. Find z such that 97.5% of the standard normal curve lies to the left of z.

19. Find z such that 8% of the standard normal curve lies to the right of z.

20. Find z such that 5% of the standard normal curve lies to the right of z.

21. Find z such that 82% of the standard normal curve lies to the right of z.

22. Find z such that 95% of the standard normal curve lies to the right of z.

23. Find the z value such that 98% of the standard normal curve lies between $-z$ and z.

24. Find the z value such that 95% of the standard normal curve lies between $-z$ and z.

25. *Medical: Blood Glucose* A person's blood glucose level and diabetes are closely related. Let x be a random variable measured in milligrams of glucose per deciliter (1/10 of a liter) of blood. After a 12-hour fast, the random variable x will have a distribution that is approximately normal with mean $\mu = 85$ and standard deviation $\sigma = 25$ (Source: *Diagnostic Tests with Nursing Implications,* edited by S. Loeb, Springhouse Press). *Note:* After 50 years of age, both the mean and standard deviation tend to increase. What is the probability that, for an adult (under 50 years old) after a 12-hour fast,
(a) x is more than 60?
(b) x is less than 110?
(c) x is between 60 and 110?
(d) x is greater than 140 (borderline diabetes starts at 140)?

26. *Medical: Blood Protoplasm* Porphyrin is a pigment in blood protoplasm and other body fluids that is significant in body energy and storage. Let x be a random variable that represents the number of milligrams of porphyrin per deciliter of blood. In healthy adults, x is approximately normally distributed with mean $\mu = 38$ and standard deviation $\sigma = 12$ (see reference in Problem 25). What is the probability that
(a) x is less than 60?
(b) x is greater than 16?
(c) x is between 16 and 60?
(d) x is more than 60? (This may indicate an infection, anemia, or another type of illness.)

27. *Archaeology: Hopi Village* Thickness measurements of ancient prehistoric Native American pot shards discovered in a Hopi village are approximately normally distributed, with a mean of 5.1 millimeters (mm) and a standard deviation of 0.9 mm (Source: *Homol'ovi II: Archaeology of an Ancestral Hopi Village, Arizona,* edited by E. C. Adams and K. A. Hays, University of Arizona Press). For a randomly found shard, what is the probability that the thickness is
(a) less than 3.0 mm?
(b) more than 7.0 mm?
(c) between 3.0 mm and 7.0 mm?

28. *Law Enforcement: Police Response Time* Police response time to an emergency call is the difference between the time the call is first received by the dispatcher and the time a patrol car radios that it has arrived at the scene (based on information from *The Denver Post*). Over a long period of time, it has been determined that the police response time has a normal distribution with a mean of 8.4 minutes and a standard deviation of 1.7 minutes. For a randomly received emergency call, what is the probability that the response time will be
(a) between 5 and 10 minutes?
(b) less than 5 minutes?
(c) more than 10 minutes?

29. *Guarantee: Batteries* Quick Start Company makes 12-volt car batteries. After many years of product testing, the company knows that the average life of a Quick Start battery is normally distributed, with a mean of 45 months and a standard deviation of 8 months.
    (a) If Quick Start guarantees a full refund on any battery that fails within the 36-month period after purchase, what percentage of its batteries will the company expect to replace?
    (b) *Inverse Normal Distribution* If Quick Start does not want to make refunds for more than 10% of its batteries under the full-refund guarantee policy, for how long should the company guarantee the batteries (to the nearest month)?

30. *Guarantee: Watches* Accrotime is a manufacturer of quartz crystal watches. Accrotime researchers have shown that the watches have an average life of 28 months before certain electronic components deteriorate, causing the watch to become unreliable. The standard deviation of watch lifetimes is 5 months, and the distribution of lifetimes is normal.
    (a) If Accrotime guarantees a full refund on any defective watch for 2 years after purchase, what percentage of total production should the company expect to replace?
    (b) *Inverse Normal Distribution* If Accrotime does not want to make refunds on more than 12% of the watches it makes, how long should the guarantee period be (to the nearest month)?

31. *Expand Your Knowledge: Estimating the Standard Deviation Consumer Reports* gave information about the ages at which various household products are replaced. For example, color TVs are replaced at an average age of $\mu = 8$ years after purchase, and the (95% of data) range was from 5 to 11 years. Thus, the range was $11 - 5 = 6$ years. Let x be the age (in years) at which a color TV is replaced. Assume that x has a distribution that is approximately normal.
    (a) The empirical rule (Section 6.1) indicates that for a symmetrical and bell-shaped distribution, approximately 95% of the data lies within two standard deviations of the mean. Therefore, a 95% range of data values extending from $\mu - 2\sigma$ to $\mu + 2\sigma$ is often used for "commonly occurring" data values. Note that the interval from $\mu - 2\sigma$ to $\mu + 2\sigma$ is 4σ in length. This leads to a "rule of thumb" for estimating the standard deviation from a 95% range of data values.

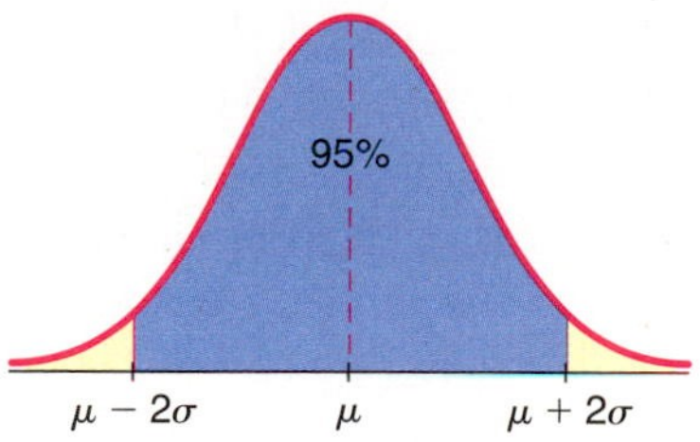

ESTIMATING THE STANDARD DEVIATION

For a symmetric, bell-shaped distribution,

$$\textbf{standard deviation} \approx \frac{\text{range}}{4} \approx \frac{\text{high value} - \text{low value}}{4}$$

where it is estimated that about 95% of the commonly occurring data values fall into this range.

Use this "rule of thumb" to approximate the standard deviation of x values, where x is the age (in years) at which a color TV is replaced.
    (b) What is the probability that someone will keep a color TV more than 5 years before replacement?
    (c) What is the probability that someone will keep a color TV fewer than 10 years before replacement?
    (d) *Inverse Normal Distribution* Assume that the average life of a color TV is 8 years with a standard deviation of 1.5 years before it breaks. Suppose that a company guarantees color TVs and will replace a TV that breaks while under guarantee with a new one. However, the

company does not want to replace more than 10% of the TVs under guarantee. For how long should the guarantee be made (rounded to the nearest tenth of a year)?

32. *Estimating the Standard Deviation: Refrigerator Replacement* *Consumer Reports* indicated that the average life of a refrigerator before replacement is $\mu = 14$ years with a (95% of data) range from 9 to 19 years. Let x = age at which a refrigerator is replaced. Assume that x has a distribution that is approximately normal.
   (a) Find a good approximation for the standard deviation of x values. *Hint:* See Problem 31.
   (b) What is the probability that someone will keep a refrigerator fewer than 11 years before replacement?
   (c) What is the probability that someone will keep a refrigerator more than 18 years before replacement?
   (d) *Inverse Normal Distribution* Assume that the average life of a refrigerator is 14 years, with the standard deviation given in part (a) before it breaks. Suppose that a company guarantees refrigerators and will replace a refrigerator that breaks while under guarantee with a new one. However, the company does not want to replace more than 5% of the refrigerators under guarantee. For how long should the guarantee be made (rounded to the nearest tenth of a year)?

33. *Estimating the Standard Deviation: Veterinary Science* The resting heart rate for an adult horse should average about $\mu = 46$ beats per minute with a (95% of data) range from 22 to 70 beats per minute, based on information from the *Merck Veterinary Manual* (a classic reference used in most veterinary colleges). Let x be a random variable that represents the resting heart rate for an adult horse. Assume that x has a distribution that is approximately normal.
   (a) Estimate the standard deviation of the x distribution. *Hint:* See Problem 31.
   (b) What is the probability that the heart rate is fewer than 25 beats per minute?
   (c) What is the probability that the heart rate is greater than 60 beats per minute?
   (d) What is the probability that the heart rate is between 25 and 60 beats per minute?
   (e) *Inverse Normal Distribution* A horse whose resting heart rate is in the upper 10% of the probability distribution of heart rates may have a secondary infection or illness that needs to be treated. What is the heart rate corresponding to the upper 10% cutoff point of the probability distribution?

34. *Estimating the Standard Deviation: Veterinary Science* How much should a healthy kitten weigh? A healthy 10-week-old (domestic) kitten should weigh an average of $\mu = 24.5$ ounces with a (95% of data) range from 14 to 35 ounces. (See reference in Problem 33.) Let x be a random variable that represents the weight (in ounces) of a healthy 10-week-old kitten. Assume that x has a distribution that is approximately normal.
   (a) Estimate the standard deviation of the x distribution. *Hint:* See Problem 31.
   (b) What is the probability that a healthy 10-week-old kitten will weigh less than 14 ounces?
   (c) What is the probability that a healthy 10-week-old kitten will weigh more than 33 ounces?
   (d) What is the probability that a healthy 10-week-old kitten will weigh between 14 and 33 ounces?
   (e) *Inverse Normal Distribution* A kitten whose weight is in the bottom 10% of the probability distribution of weights is called *undernourished.* What is the cutoff point for the weight of an undernourished kitten?

Max Dannenbaum/Stone/Getty Images

35. *Insurance: Satellites* A relay microchip in a telecommunications satellite has a life expectancy that follows a normal distribution with a mean of 90 months and a standard deviation of 3.7 months. When this computer-relay microchip malfunctions, the entire satellite is useless. A large London insurance company is going to insure the satellite for \$50 million. Assume that the only part of the satellite in question is the microchip. All other components will work indefinitely.
    (a) *Inverse Normal Distribution* For how many months should the satellite be insured to be 99% confident that it will last beyond the insurance date?
    (b) If the satellite is insured for 84 months, what is the probability that it will malfunction before the insurance coverage ends?
    (c) If the satellite is insured for 84 months, what is the expected loss to the insurance company?
    (d) If the insurance company charges \$3 million for 84 months of insurance, how much profit does the company expect to make?

36. *Convertion Center: Exhibition Show Attendance* Attendance at large exhibition shows in Denver averages about 8000 people per day, with standard deviation of about 500. Assume that the daily attendance figures follow a normal distribution.
    (a) What is the probability that the daily attendance will be fewer than 7200 people?
    (b) What is the probability that the daily attendance will be more than 8900 people?
    (c) What is the probability that the daily attendance will be between 7200 and 8900 people?

37. *Exhibition Shows: Inverse Normal Distribution* Most exhibition shows open in the morning and close in the late evening. A study of Saturday arrival times showed that the average arrival time was 3 hours and 48 minutes after the doors opened, and the standard deviation was estimated at about 52 minutes. Assume that the arrival times follow a normal distribution.
    (a) At what time after the doors open will 90% of the people who are coming to the Saturday show have arrived?
    (b) At what time after the doors open will only 15% of the people who are coming to the Saturday show have arrived?
    (c) Do you think the probability distribution of arrival times for Friday might be different from the distribution of arrival times for Saturday? Explain.

38. *Budget: Maintenance* The amount of money spent weekly on cleaning, maintenance, and repairs at a large restaurant was observed over a long period of time to be approximately normally distributed, with mean $\mu = \$615$ and standard deviation $\sigma = \$42$.
    (a) If \$646 is budgeted for next week, what is the probability that the actual costs will exceed the budgeted amount?
    (b) *Inverse Normal Distribution* How much should be budgeted for weekly repairs, cleaning, and maintenance so that the probability that the budgeted amount will be exceeded in a given week is only 0.10?

39. *Expand Your Knowledge: Conditional Probability* Suppose you want to eat lunch at a popular restaurant. The restaurant does not take reservations, so there is usually a waiting time before you can be seated. Let x represent the length of time waiting to be seated. From past experience, you know that the mean waiting time is $\mu = 18$ minutes with $\sigma = 4$ minutes. You assume that the x distribution is approximately normal.

(a) What is the probability that the waiting time will *exceed* 20 minutes, given that it has exceeded 15 minutes? *Hint:* Compute $P(x > 20 \mid x > 15)$.

(b) What is the probability that the waiting time will exceed 25 minutes, given that it has exceeded 18 minutes? *Hint:* Compute $P(x > 25 \mid x > 18)$.

(c) *Hint for solution:* Review item 6, conditional probability, in the summary of basic probability rules at the end of Section 4.2. Note that

$$P(A \mid B) = \frac{P(A \text{ and } B)}{P(B)}$$

and show that in part (a),

$$P(x > 20 \mid x > 15) = \frac{P((x > 20) \text{ and } (x > 15))}{P(x > 15)} = \frac{P(x > 20)}{P(x > 15)}$$

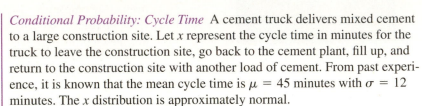

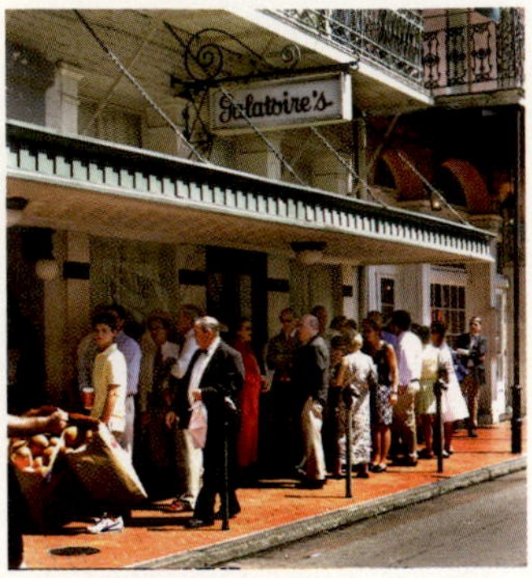

AP Photo/Judi Bottoni

40. *Conditional Probability: Cycle Time* A cement truck delivers mixed cement to a large construction site. Let x represent the cycle time in minutes for the truck to leave the construction site, go back to the cement plant, fill up, and return to the construction site with another load of cement. From past experience, it is known that the mean cycle time is $\mu = 45$ minutes with $\sigma = 12$ minutes. The x distribution is approximately normal.

(a) What is the probability that the cycle time will *exceed* 60 minutes, given that it has exceeded 50 minutes? *Hint:* See Problem 39, part (c).

(b) What is the probability that the cycle time will exceed 55 minutes, given that it has exceeded 40 minutes?

SECTION **6.4**

Sampling Distributions

FOCUS POINTS

- Review such commonly used terms as *random sample, relative frequency, parameter, statistic,* and *sampling distribution*.
- From raw data, construct a relative frequency distribution for $\bar{x}$ values and compare the result to a theoretical sampling distribution.

Let us begin with some common statistical terms. Most of these have been discussed before, but this is a good time to review them.

From a statistical point of view, a *population* can be thought of as a complete set of measurements (or counts), either existing or conceptual. We discussed populations at some length in Chapter 1. A *sample* is a subset of measurements from the population. For our purposes, the most important samples are *random samples,* which were discussed in Section 1.2.

When we compute a descriptive measure such as an average, it makes a difference whether it was computed from a population or from a sample.

Statistic
Parameter

A **statistic** is a numerical descriptive measure of a *sample*.
A **parameter** is a numerical descriptive measure of a *population*.

It is important to notice that for a given population, a specified parameter is a fixed quantity. On the other hand, the value of a statistic might vary depending on which sample has been selected.

SOME COMMONLY USED STATISTICS AND CORRESPONDING PARAMETERS

Measure	*Statistic*	*Parameter*
Mean	$\bar{x}$ (x bar)	μ (mu)
Variance	s^2	σ^2 (sigma squared)
Standard deviation	s	σ (sigma)
Proportion	$\hat{p}$ (p hat)	p

Population parameter

Often we do not have access to all the measurements of an entire population because of constraints on time, money, or effort. So, we must use measurements from a sample instead. In such cases, we will use a statistic (such as $\bar{x}$, s, or $\hat{p}$) to make *inferences* about a corresponding *population parameter* (e.g., μ, σ, or p). The principal types of inferences we will make are the following.

TYPES OF INFERENCES

1. **Estimation:** In this type of inference, we estimate the *value* of a population parameter.
2. **Testing:** In this type of inference, we formulate a *decision* about the value of a population parameter.
3. **Regression:** In this type of inference, we make *predictions* or *forecasts* about the value of a statistical variable (discussed in Chapter 9).

Sampling distribution

To evaluate the reliability of our inferences, we will need to know the probability distribution for the statistic we are using. Such a probability distribution is called a *sampling distribution.* Perhaps Example 11 below will help clarify this discussion.

A **sampling distribution** is a probability distribution of a sample statistic based on all possible simple random samples of the *same* size from the same population.

EXAMPLE 11

SAMPLING DISTRIBUTION FOR $\bar{x}$

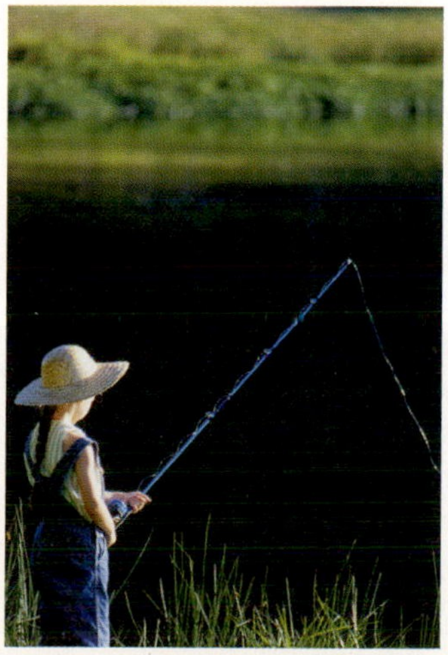
Philip James Corwin/Documentary/Corbis

Pinedale, Wisconsin, is a rural community with a children's fishing pond. Posted rules state that all fish under 6 inches must be returned to the pond, only children under 12 years old may fish, and a limit of five fish may be kept per day. Susan is a college student who was hired by the community last summer to make sure the rules were obeyed and to see that the children were safe from accidents. The pond contains only rainbow trout and has been well stocked for many years. Each child has no difficulty catching his or her limit of five trout.

As a project for her biometrics class, Susan kept a record of the lengths (to the nearest inch) of all trout caught last summer. Hundreds of children visited the pond and caught their limit of five trout, so Susan has a lot of data. To make Table 6-9, Susan selected 100 children at random and listed the lengths of each of the five trout caught by a child in the sample. Then, for each child, she listed the mean length of the five trout that child caught.

Now let us turn our attention to the following question: What is the average (mean) length of a trout taken from the Pinedale children's pond last summer?

TABLE 6-9 Length Measurements of Trout Caught by a Random Sample of 100 Children at the Pinedale Children's Pond

Sample	Length (to nearest inch)					$\bar{x}$ = Sample Mean	Sample	Length (to nearest inch)					$\bar{x}$ = Sample Mean
1	11	10	10	12	11	10.8	51	9	10	12	10	9	10.0
2	11	11	9	9	9	9.8	52	7	11	10	11	10	9.8
3	12	9	10	11	10	10.4	53	9	11	9	11	12	10.4
4	11	10	13	11	8	10.6	54	12	9	8	10	11	10.0
5	10	10	13	11	12	11.2	55	8	11	10	9	10	9.6
6	12	7	10	9	11	9.8	56	10	10	9	9	13	10.2
7	7	10	13	10	10	10.0	57	9	8	10	10	12	9.8
8	10	9	9	9	10	9.4	58	10	11	9	8	9	9.4
9	10	10	11	12	8	10.2	59	10	8	9	10	12	9.8
10	10	11	10	7	9	9.4	60	11	9	9	11	11	10.2
11	12	11	11	11	13	11.6	61	11	10	11	10	11	10.6
12	10	11	10	12	13	11.2	62	12	10	10	9	11	10.4
13	11	10	10	9	11	10.2	63	10	10	9	11	7	9.4
14	10	10	13	8	11	10.4	64	11	11	12	10	11	11.0
15	9	11	9	10	10	9.8	65	10	10	11	10	9	10.0
16	13	9	11	12	10	11.0	66	8	9	10	11	11	9.8
17	8	9	7	10	11	9.0	67	9	11	11	9	8	9.6
18	12	12	8	12	12	11.2	68	10	9	10	9	11	9.8
19	10	8	9	10	10	9.4	69	9	9	11	11	11	10.2
20	10	11	10	10	10	10.2	70	13	11	11	9	11	11.0
21	11	10	11	9	12	10.6	71	12	10	8	8	9	9.4
22	9	12	9	10	9	9.8	72	13	7	12	9	10	10.2
23	8	11	10	11	10	10.0	73	9	10	9	8	9	9.0
24	9	12	10	9	11	10.2	74	11	11	10	9	10	10.2
25	9	9	8	9	10	9.0	75	9	11	14	9	11	10.8
26	11	11	12	11	11	11.2	76	14	10	11	12	12	11.8
27	10	10	10	11	13	10.8	77	8	12	10	10	9	9.8
28	8	7	9	10	8	8.4	78	8	10	13	9	8	9.6
29	11	11	8	10	11	10.2	79	11	11	11	13	10	11.2
30	8	11	11	9	12	10.2	80	12	10	11	12	9	10.8
31	11	9	12	10	10	10.4	81	10	9	10	10	13	10.4
32	10	11	10	11	12	10.8	82	11	10	9	9	12	10.2
33	12	11	8	8	11	10.0	83	11	11	10	10	10	10.4
34	8	10	10	9	10	9.4	84	11	10	11	9	9	10.0
35	10	10	10	10	11	10.2	85	10	11	10	9	7	9.4
36	10	8	10	11	13	10.4	86	7	11	10	9	11	9.6
37	11	10	11	11	10	10.6	87	10	11	10	10	10	10.2
38	7	13	9	12	11	10.4	88	9	8	11	10	12	10.0
39	11	11	8	11	11	10.4	89	14	9	12	10	9	10.8
40	11	10	11	12	9	10.6	90	9	12	9	10	10	10.0
41	11	10	9	11	12	10.6	91	10	10	8	6	11	9.0
42	11	13	10	12	9	11.0	92	8	9	11	9	10	9.4
43	10	9	11	10	11	10.2	93	8	10	9	9	11	9.4
44	10	9	11	10	9	9.8	94	12	11	12	13	10	11.6
45	12	11	9	11	12	11.0	95	11	11	9	9	9	9.8
46	13	9	11	8	8	9.8	96	8	12	8	11	10	9.8
47	10	11	11	11	10	10.6	97	13	11	11	12	8	11.0
48	9	9	10	11	11	10.0	98	10	11	8	10	11	10.0
49	10	9	9	10	10	9.6	99	13	10	7	11	9	10.0
50	10	10	6	9	10	9.0	100	9	9	10	12	12	10.4

TABLE 6-10 **Frequency Table for 100 Values of $\bar{x}$**

Class	Class Limits: Lower	Class Limits: Upper	f = Frequency	$f/100$ = Relative Frequency
1	8.39	8.76	1	0.01
2	8.77	9.14	5	0.05
3	9.15	9.52	10	0.10
4	9.53	9.90	19	0.19
5	9.91	10.28	27	0.27
6	10.29	10.66	18	0.18
7	10.67	11.04	12	0.12
8	11.05	11.42	5	0.05
9	11.43	11.80	3	0.03

SOLUTION: We can get an idea of the average length by looking at the far-right column of Table 6-9. But just looking at 100 of the $\bar{x}$ values doesn't tell us much. Let's organize our $\bar{x}$ values into a frequency table. We used a class width of 0.38 to make Table 6-10.

Note: Techniques of Section 2.1 dictate a class width of 0.4. However, this choice results in the tenth class being beyond the data. Consequently, we shortened the class width slightly and also started the first class with a value slightly smaller than the smallest data value.

The far-right column of Table 6-10 contains relative frequencies $f/100$. Recall that relative frequencies may be thought of as probabilities, so we effectively have a probability distribution. Because $\bar{x}$ represents the mean length of a trout (based on samples of five trout caught by each child), we estimate the probability of $\bar{x}$ falling into each class by using the relative frequencies. Figure 6-32 is a relative-frequency or probability distribution of the $\bar{x}$ values.

FIGURE 6-32

Estimates of Probabilities of $\bar{x}$ Values

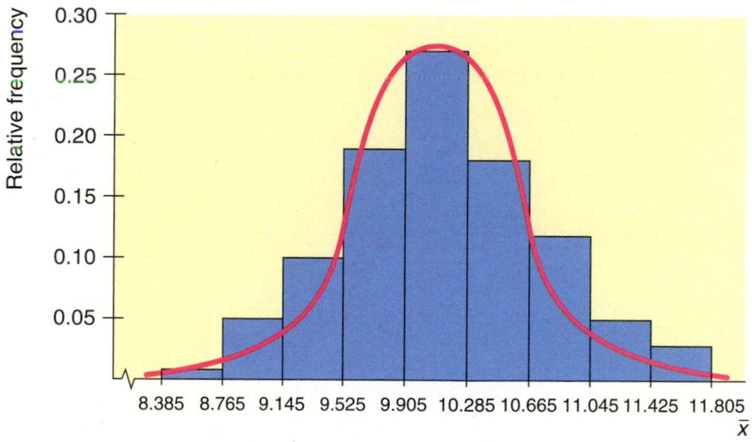

The bars of Figure 6-32 represent our estimated probabilities of $\bar{x}$ values based on the data of Table 6-9. The bell-shaped curve represents the theoretical probability distribution that would be obtained if the number of children (i.e., number of $\bar{x}$ values) were much larger.

Figure 6-32 represents a *probability sampling distribution* for the sample mean $\bar{x}$ of trout lengths based on random samples of size 5. We see that the distribution is mound-shaped and even somewhat bell-shaped. Irregularities are due to the small number of samples used (only 100 sample means) and the rather small sample size (five trout per child). These irregularities would become less obvious and even disappear if the sample of children became much larger, if we used a larger number of classes in Figure 6-32, and if the number of trout in each sample became larger. In fact, the curve would eventually become a perfect bell-shaped curve. We will discuss this property at some length in the next section, which introduces the *central limit theorem.*

There are other sampling distributions besides the $\bar{x}$ distribution. Section 6.6 shows the sampling distribution for $\hat{p}$. In the chapters ahead, we will see that other statistics have different sampling distributions. However, the $\bar{x}$ sampling distribution is very important. It will serve us well in our inferential work in Chapters 7 and 8 on estimation and testing.

WHAT DOES A SAMPLING DISTRIBUTION TELL US?

A sampling distribution gives us information regarding sample statistics such as $\bar{x}$ or $\hat{p}$.

- The sampling distribution is based on values of the sample statistics from *all* samples of a specified size *n*.
- Sampling distributions are used to obtain information about corresponding population parameters. In the next sections and chapters we will see how to obtain information about the value of the parameter μ from sampling distributions of $\bar{x}$. We will also study sampling distributions of $\hat{p}$, the sample probability of success in a binomial trial.

Let us summarize the information about sampling distributions in the following exercise.

GUIDED EXERCISE 9 | TERMINOLOGY

(a) What is a population parameter? Give an example.

→ A population parameter is a numerical descriptive measure of a population. Examples are μ, σ, and *p*. (There are many others.)

(b) What is a sample statistic? Give an example.

→ A sample statistic or statistic is a numerical descriptive measure of a sample. Examples are $\bar{x}$, *s*, and $\hat{p}$.

(c) What is a sampling distribution?

→ A sampling distribution is a probability distribution for the sample statistic we are using based on all possible samples of the same size.

(d) In Table 6-9, what makes up the members of the sample? What is the sample statistic corresponding to each sample? What is the sampling distribution? To which population parameter does this sampling distribution correspond?

→ There are 100 samples, each of which comprises five trout lengths. In the first sample, the five trout have lengths 11, 10, 10, 12, and 11. The sample statistic is the sample mean $\bar{x} = 10.8$. The sampling distribution is shown in Figure 6-32. This sampling distribution relates to the population mean μ of all lengths of trout taken from the Pinedale children's pond (i.e., trout over 6 inches long).

Continued

GUIDED EXERCISE 9 *continued*

(e) Where will sampling distributions be used in our study of statistics?

→ Sampling distributions will be used for statistical inference. (Chapter 7 will concentrate on a method of inference called *estimation.* Chapter 8 will concentrate on a method of inference called *testing.*)

VIEWPOINT "Chance Favors the Prepared Mind"

—Louis Pasteur

It also has been said that a discovery is nothing more than an accident that meets a prepared mind. Sampling can be one of the best forms of preparation. In fact, sampling may be the primary way we humans venture into the unknown. Probability sampling distributions can provide new information for the sociologist, scientist, or economist. In addition, ordinary human sampling of life can help writers and artists develop preferences, styles, and insights. Ansel Adams became famous for photographing lyrical, unforgettable land scapes such as "Moonrise, Hernandez, New Mexico." Adams claimed that he was a strong believer in the quote by Pasteur. In fact, he claimed that the Hernandez photograph was just such a favored chance happening that his prepared mind readily grasped. During his lifetime, Adams made over $25 million from sales and royalties on the Hernandez photograph.

SECTION 6.4 PROBLEMS

This is a good time to review several important concepts, some of which we have studied earlier. Please write out a careful but brief answer to each of the following questions.

1. *Statistical Literacy* What is a population? Give three examples.

2. *Statistical Literacy* What is a random sample from a population? *Hint:* See Section 1.2.

3. *Statistical Literacy* What is a population parameter? Give three examples.

4. *Statistical Literacy* What is a sample statistic? Give three examples.

5. *Statistical Literacy* What is the meaning of the term *statistical inference*? What types of inferences will we make about population parameters?

6. *Statistical Literacy* What is a sampling distribution?

7. *Critical Thinking* How do frequency tables, relative frequencies, and histograms showing relative frequencies help us understand sampling distributions?

8. *Critical Thinking* How can relative frequencies be used to help us estimate probabilities occurring in sampling distributions?

9. *Critical Thinking* Give an example of a specific sampling distribution we studied in this section. Outline other possible examples of sampling distributions from areas such as business administration, economics, finance, psychology, political science, sociology, biology, medical science, sports, engineering, chemistry, linguistics, and so on.

SECTION **6.5**

The Central Limit Theorem

FOCUS POINTS

- For a normal distribution, use μ and σ to construct the theoretical sampling distribution for the statistic $\bar{x}$.
- For large samples, use sample estimates to construct a good approximate sampling distribution for the statistic $\bar{x}$.
- Learn the statement and underlying meaning of the central limit theorem well enough to explain it to a friend who is intelligent, but (unfortunately) doesn't know much about statistics.

THE $\bar{x}$ DISTRIBUTION, GIVEN x IS NORMAL

In Section 6.4, we began a study of the distribution of $\bar{x}$ values, where $\bar{x}$ was the (sample) mean length of five trout caught by children at the Pinedale children's fishing pond. Let's consider this example again in the light of a very important theorem of mathematical statistics.

THEOREM 6.1 **For a Normal Probability Distribution** Let x be a random variable with a *normal distribution* whose mean is μ and whose standard deviation is σ. Let $\bar{x}$ be the sample mean corresponding to random samples of size n taken from the x distribution. Then the following are true:

(a) The $\bar{x}$ distribution is a *normal distribution.*

(b) The mean of the $\bar{x}$ distribution is μ.

(c) The standard deviation of the $\bar{x}$ distribution is $\sigma/\sqrt{n}$.

We conclude from Theorem 6.1 that when x has a normal distribution, the $\bar{x}$ distribution will be normal *for any sample size n.* Furthermore, we can convert the $\bar{x}$ distribution to the standard normal z distribution using the following formulas.

$$\mu_{\bar{x}} = \mu$$

$$\sigma_{\bar{x}} = \frac{\sigma}{\sqrt{n}}$$

$$z = \frac{\bar{x} - \mu_{\bar{x}}}{\sigma_{\bar{x}}} = \frac{\bar{x} - \mu}{\sigma/\sqrt{n}}$$

where n is the sample size,
μ is the mean of the x distribution, and
σ is the standard deviation of the x distribution.

Theorem 6.1 is a wonderful theorem! It states that the $\bar{x}$ distribution will be normal provided the x distribution is normal. The sample size n could be 2, 3, 4, or any other (fixed) sample size we wish. Furthermore, the mean of the $\bar{x}$ distribution is μ (same as for the x distribution), but the standard deviation is $\sigma/\sqrt{n}$ (which is, of course, smaller than σ). The next example illustrates Theorem 6.1.

EXAMPLE 12

PROBABILITY REGARDING x AND $\bar{x}$

Suppose a team of biologists has been studying the Pinedale children's fishing pond. Let x represent the length of a single trout taken at random from the pond. This group of biologists has determined that x has a normal distribution with mean $\mu = 10.2$ inches and standard deviation $\sigma = 1.4$ inches.

(a) What is the probability that a *single trout* taken at random from the pond is between 8 and 12 inches long?

SOLUTION: We use the methods of Section 6.3, with $\mu = 10.2$ and $\sigma = 1.4$, to get

$$z = \frac{x - \mu}{\sigma} = \frac{x - 10.2}{1.4}$$

Therefore,

$$\begin{aligned} P(8 < x < 12) &= P\left(\frac{8 - 10.2}{1.4} < z < \frac{12 - 10.2}{1.4}\right) \\ &= P(-1.57 < z < 1.29) \\ &= 0.9015 - 0.0582 = 0.8433 \end{aligned}$$

Therefore, the probability is about 0.8433 that a single trout taken at random is between 8 and 12 inches long.

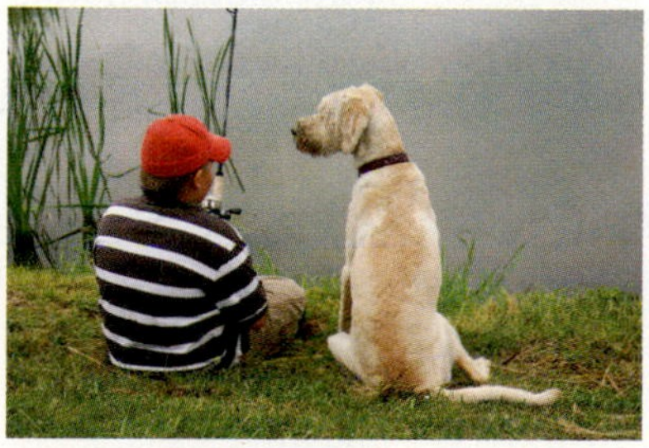

appletat/iStockphoto.com

(b) What is the probability that the *mean length* $\bar{x}$ of five trout taken at random is between 8 and 12 inches?

SOLUTION: If we let $\mu_{\bar{x}}$ represent the mean of the distribution, then Theorem 6.1, part (b), tells us that

$$\mu_{\bar{x}} = \mu = 10.2$$

If $\sigma_{\bar{x}}$ represents the standard deviation of the $\sigma_{\bar{x}}$ distribution, then Theorem 6.1, part (c), tells us that

$$\sigma_{\bar{x}} = \sigma/\sqrt{n} = 1.4/\sqrt{5} \approx 0.63$$

To create a standard z variable from $\bar{x}$, we subtract $\mu_{\bar{x}}$ and divide by $\sigma_{\bar{x}}$:

$$z = \frac{\bar{x} - \mu_{\bar{x}}}{\sigma_{\bar{x}}} = \frac{\bar{x} - \mu}{\sigma/\sqrt{n}} \approx \frac{\bar{x} - 10.2}{0.63}$$

To standardize the interval $8 < \bar{x} < 12$, we use 8 and then 12 in place of $\bar{x}$ in the preceding formula for z.

$$\begin{gathered} 8 < \bar{x} < 12 \\ \frac{8 - 10.2}{0.63} < z < \frac{12 - 10.2}{0.63} \\ -3.49 < z < 2.86 \end{gathered}$$

Theorem 6.1, part (a), tells us that $\bar{x}$ has a normal distribution. Therefore,

$$P(8 < \bar{x} < 12) = P(-3.49 < z < 2.86) = 0.9979 - 0.0002 = 0.9977$$

The probability is about 0.9977 that the mean length based on a sample size of 5 is between 8 and 12 inches.

(c) Looking at the results of parts (a) and (b), we see that the probabilities (0.8433 and 0.9977) are quite different. Why is this the case?

SOLUTION: According to Theorem 6.1, both x and $\bar{x}$ have a normal distribution, and both have the same mean of 10.2 inches. The difference is in the standard deviations for x and $\bar{x}$. The standard deviation of the x distribution is $\sigma = 1.4$. The standard deviation of the $\bar{x}$ distribution is

$$\sigma_{\bar{x}} = \sigma/\sqrt{n} = 1.4/\sqrt{5} \approx 0.63$$

The standard deviation of the $\bar{x}$ distribution is less than half the standard deviation of the x distribution. Figure 6-33 shows the distributions of x and $\bar{x}$.

FIGURE 6-33

General Shapes of the x and $\bar{x}$ Distributions

(a) The x Distribution with $\mu = 10.2$ and $\sigma = 1.4$

(b) The $\bar{x}$ Distribution with $\mu_{\bar{x}} = 10.2$ and $\sigma_{\bar{x}} = 0.63$ for Samples of Size $n = 5$

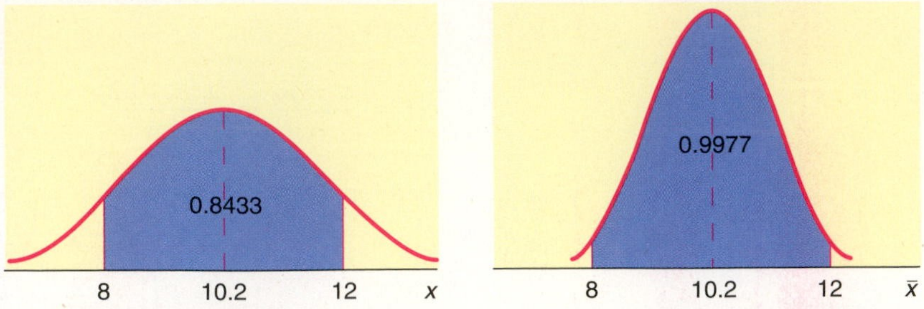

Looking at Figure 6-33(a) and (b), we see that both curves use the same scale on the horizontal axis. The means are the same, and the shaded area is above the interval from 8 to 12 on each graph. It becomes clear that the smaller standard deviation of the $\bar{x}$ distribution has the effect of gathering together much more of the total probability into the region over its mean. Therefore, the region from 8 to 12 has a much higher probability for the $\bar{x}$ distribution.

Standard error of the mean

Theorem 6.1 describes the distribution of a particular statistic: namely, the distribution of sample mean $\bar{x}$. The standard deviation of a statistic is referred to as the *standard error* of that statistic.

The **standard error** is the standard deviation of a sampling distribution. For the $\bar{x}$ sampling distribution,

$$\textbf{standard error} = \sigma_{\bar{x}} = \sigma/\sqrt{n}$$

Statistical software

The expression *standard error* appears commonly on printouts and refers to the standard deviation of the sampling distribution being used. (In Minitab, the expression SE MEAN refers to the standard error of the mean.)

THE $\bar{x}$ DISTRIBUTION, GIVEN x FOLLOWS ANY DISTRIBUTION

Central limit theorem

Theorem 6.1 gives complete information about the $\bar{x}$ distribution, provided the original x distribution is known to be normal. What happens if we don't have information about the shape of the original x distribution? The *central limit theorem* tells us what to expect.

THEOREM 6.2 The Central Limit Theorem for Any Probability Distribution If x possesses *any* distribution with mean μ and standard deviation σ, then the sample mean $\bar{x}$ based on a random sample of size n will have a distribution that approaches the distribution of a normal random variable with mean μ and standard deviation $\sigma/\sqrt{n}$ as n increases without limit.

The central limit theorem is indeed surprising! It says that x can have *any* distribution whatsoever, but that as the sample size gets larger and larger, the distribution of $\bar{x}$ will approach a *normal* distribution. From this relation, we begin to appreciate the scope and significance of the normal distribution.

Large sample

In the central limit theorem, the degree to which the distribution of $\bar{x}$ values fits a normal distribution depends on both the selected value of n and the original distribution of x values. A natural question is: How large should the sample size be if we want to apply the central limit theorem? After a great deal of theoretical as well as empirical study, statisticians agree that if n is 30 or larger, the $\bar{x}$ distribution will appear to be normal and the central limit theorem will apply. However, this rule should not be applied blindly. If the x distribution is definitely not symmetrical about its mean, then the $\bar{x}$ distribution also will display a lack of symmetry. In such a case, a sample size larger than 30 may be required to get a reasonable approximation to the normal.

In practice, it is a good idea, when possible, to make a histogram of sample x values. If the histogram is approximately mound-shaped, and if it is more or less symmetrical, then we may be assured that, for all practical purposes, the $\bar{x}$ distribution will be well approximated by a normal distribution and the central limit theorem will apply when the sample size is 30 or larger. The main thing to remember is that in almost all practical applications, a sample size of 30 or more is adequate for the central limit theorem to hold. However, in a few rare applications, you may need a sample size larger than 30 to get reliable results.

WHAT DOES THE CENTRAL LIMIT THEOREM TELL US?

The central limit theorem gives us information about the characteristics of the $\bar{x}$ sampling distribution based on all samples of size n. When n is sufficiently large ($n \geq 30$ in most cases), the central limit tells us that

- the $\bar{x}$ distribution is approximately normal.
- the mean of the $\bar{x}$ distribution is μ, the mean of the original x distribution.
- the standard deviation (also known as the standard error) of the $\bar{x}$ distribution is $\sigma/\sqrt{n}$, where σ is the standard deviation of the x distribution.

Let's summarize this information for convenient reference: For almost all x distributions, if we use a random sample of size 30 or larger, the $\bar{x}$ distribution will be approximately normal. The larger the sample size becomes, the closer the $\bar{x}$ distribution gets to the normal. Furthermore, we may convert the $\bar{x}$ distribution to a standard normal distribution using the following formulas.

USING THE CENTRAL LIMIT THEOREM TO CONVERT THE $\bar{x}$ DISTRIBUTION TO THE STANDARD NORMAL DISTRIBUTION

$$\mu_{\bar{x}} = \mu$$

$$\sigma_{\bar{x}} = \frac{\sigma}{\sqrt{n}}$$

$$z = \frac{\bar{x} - \mu_{\bar{x}}}{\sigma_{\bar{x}}} = \frac{\bar{x} - \mu}{\sigma/\sqrt{n}}$$

where n is the sample size ($n \geq 30$),
μ is the mean of the x distribution, and
σ is the standard deviation of the x distribution.

Guided Exercise 10 shows how to standardize when appropriate. Then, Example 13 demonstrates the use of the central limit theorem in a decision-making process.

GUIDED EXERCISE 10 CENTRAL LIMIT THEOREM

(a) Suppose x has a *normal* distribution with mean $\mu = 18$ and standard deviation $\sigma = 13$. If you draw random samples of size 5 from the x distribution and $\bar{x}$ represents the sample mean, what can you say about the $\bar{x}$ distribution? How could you standardize the $\bar{x}$ distribution?

Since the x distribution is given to be *normal,* the $\bar{x}$ distribution also will be normal even though the sample size is much less than 30. The mean is $\mu_{\bar{x}} = \mu = 18$. The standard deviation is

$$\sigma_{\bar{x}} = \sigma/\sqrt{n} = 13/\sqrt{5} \approx 1.3$$

We could standardize $\bar{x}$ as follows:

$$z = \frac{\bar{x} - \mu}{\sigma/\sqrt{n}} \approx \frac{\bar{x} - 18}{1.3}$$

(b) Suppose you know that the x distribution has mean $\mu = 75$ and standard deviation $\sigma = 12$, but you have no information as to whether or not the x distribution is normal. If you draw samples of size 30 from the x distribution and $\bar{x}$ represents the sample mean, what can you say about the $\bar{x}$ distribution? How could you standardize the $\bar{x}$ distribution?

Since the sample size is large enough, the $\bar{x}$ distribution will be an approximately normal distribution. The mean of the $\bar{x}$ distribution is

$$\mu_{\bar{x}} = \mu = 75$$

The standard deviation of the $\bar{x}$ distribution is

$$\sigma_{\bar{x}} = \sigma/\sqrt{n} = 12/\sqrt{30} \approx 2.2$$

We could standardize $\bar{x}$ as follows:

$$z = \frac{\bar{x} - \mu}{\sigma/\sqrt{n}} \approx \frac{\bar{x} - 75}{2.2}$$

(c) Suppose you did not know that x had a normal distribution. Would you be justified in saying that the $\bar{x}$ distribution is approximately normal if the sample size were $n = 8$?

No, the sample size should be 30 or larger if we don't know that x has a normal distribution.

EXAMPLE 13

CENTRAL LIMIT THEOREM

A certain strain of bacteria occurs in all raw milk. Let x be the bacteria count per milliliter of milk. The health department has found that if the milk is not contaminated, then x has a distribution that is more or less mound-shaped and symmetrical. The mean of the x distribution is $\mu = 2500$, and the standard deviation is $\sigma = 300$. In a large commercial dairy, the health inspector takes 42 random samples of the milk produced each day. At the end of the day, the bacteria count in each of the 42 samples is averaged to obtain the sample mean bacteria count $\bar{x}$.

(a) Assuming the milk is not contaminated, what is the distribution of $\bar{x}$?

SOLUTION: The sample size is $n = 42$. Since this value exceeds 30, the central limit theorem applies, and we know that $\bar{x}$ will be approximately normal, with mean and standard deviation

$$\mu_{\bar{x}} = \mu = 2500$$

$$\sigma_{\bar{x}} = \sigma/\sqrt{n} = 300/\sqrt{42} \approx 46.3$$

(b) Assuming the milk is not contaminated, what is the probability that the average bacteria count $\bar{x}$ for one day is between 2350 and 2650 bacteria per milliliter?

SOLUTION: We convert the interval

$$2350 \le \bar{x} \le 2650$$

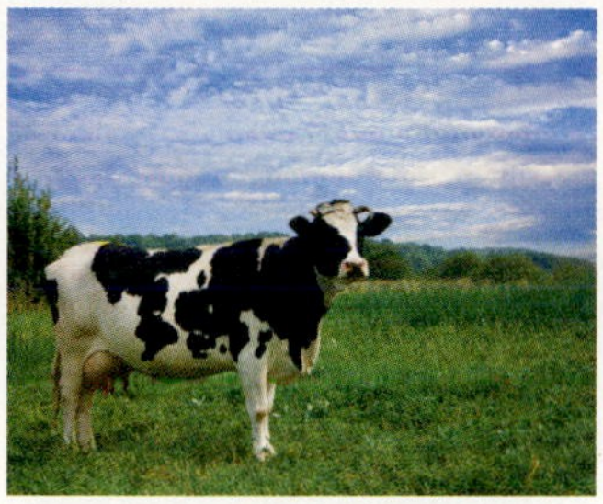

Aleksas Kvedoras/iStockphoto.com

to a corresponding interval on the standard z axis.

$$z = \frac{\bar{x} - \mu}{\sigma/\sqrt{n}} \approx \frac{\bar{x} - 2500}{46.3}$$

$$\bar{x} = 2350 \text{ converts to } z = \frac{2350 - 2500}{46.3} \approx -3.24$$

$$\bar{x} = 2650 \text{ converts to } z = \frac{2650 - 2500}{46.3} \approx 3.24$$

Therefore,

$$\begin{aligned} P(2350 \le \bar{x} \le 2650) &= P(-3.24 \le z \le 3.24) \\ &= 0.9994 - 0.0006 \\ &= 0.9988 \end{aligned}$$

The probability is 0.9988 that $\bar{x}$ is between 2350 and 2650.

(c) ***Interpretation*** At the end of each day, the inspector must decide to accept or reject the accumulated milk that has been held in cold storage awaiting shipment. Suppose the 42 samples taken by the inspector have a mean bacteria count $\bar{x}$ that is *not* between 2350 and 2650. If you were the inspector, what would be your comment on this situation?

SOLUTION: The probability that $\bar{x}$ is between 2350 and 2650 for milk that is not contaminated is very high. If the inspector finds that the average bacteria count for the 42 samples is not between 2350 and 2650, then it is reasonable to conclude that there is something wrong with the milk. If $\bar{x}$ is less than 2350, you might suspect someone added chemicals to the milk to artificially reduce the bacteria count. If $\bar{x}$ is above 2650, you might suspect some other kind of biologic contamination.

In Problems 21, 22, and 23, we'll apply the central limit theorem to solve problems involving a *sum* of random variables.

PROCEDURE

HOW TO FIND PROBABILITIES REGARDING $\bar{x}$

Given a probability distribution of x values where

n = sample size

μ = mean of the x distribution

σ = standard deviation of the x distribution

1. If the x distribution is *normal*, then the $\bar{x}$ distribution is *normal*.
2. Even if the x distribution is *not* normal, if the *sample* size $n \ge 30$, then, by the central limit theorem, the $\bar{x}$ distribution is *approximately normal*.
3. Convert $\bar{x}$ to z using the formula

$$z = \frac{\bar{x} - \mu_{\bar{x}}}{\sigma_{\bar{x}}} = \frac{\bar{x} - \mu}{\sigma/\sqrt{n}}$$

4. Use the standard normal distribution to find the corresponding probabilities of events regarding $\bar{x}$.

GUIDED EXERCISE 11 Probability regarding $\bar{x}$

In mountain country, major highways sometimes use tunnels instead of long, winding roads over high passes. However, too many vehicles in a tunnel at the same time can cause a hazardous situation. Traffic engineers are studying a long tunnel in Colorado. If x represents the time for a vehicle to go through the tunnel, it is known that the x distribution has mean $\mu = 12.1$ minutes and standard deviation $\sigma = 3.8$ minutes under ordinary traffic conditions. From a histogram of x values, it was found that the x distribution is mound-shaped with some symmetry about the mean.

Engineers have calculated that, *on average*, vehicles should spend from 11 to 13 minutes in the tunnel. If the time is less than 11 minutes, traffic is moving too fast for safe travel in the tunnel. If the time is more than 13 minutes, there is a problem of bad air quality (too much carbon monoxide and other pollutants).

Under ordinary conditions, there are about 50 vehicles in the tunnel at one time. What is the probability that the mean time for 50 vehicles in the tunnel will be from 11 to 13 minutes?
We will answer this question in steps.

(a) Let $\bar{x}$ represent the sample mean based on samples of size 50. Describe the $\bar{x}$ distribution.

From the central limit theorem, we expect the $\bar{x}$ distribution to be approximately normal, with mean and standard deviation

$$\mu_{\bar{x}} = \mu = 12.1 \quad \sigma_{\bar{x}} = \frac{\sigma}{\sqrt{n}} = \frac{3.8}{\sqrt{50}} \approx 0.54$$

(b) Find $P(11 < \bar{x} < 13)$.

Jupiter Images

We convert the interval

$$11 < \bar{x} < 13$$

to a standard z interval and use the standard normal probability table to find our answer. Since

$$z = \frac{\bar{x} - \mu}{\sigma/\sqrt{n}} \approx \frac{\bar{x} - 12.1}{0.54}$$

$\bar{x} = 11$ converts to $z \approx \dfrac{11 - 12.1}{0.54} = -2.04$

and $\bar{x} = 13$ converts to $z \approx \dfrac{13 - 12.1}{0.54} = 1.67$

Therefore,

$$\begin{aligned} P(11 < \bar{x} < 13) &= P(-2.04 < z < 1.67) \\ &= 0.9525 - 0.0207 \\ &= 0.9318 \end{aligned}$$

(c) ***Interpret*** your answer to part (b).

It seems that about 93% of the time, there should be no safety hazard for average traffic flow.

CRITICAL THINKING

BIAS AND VARIABILITY

Whenever we use a sample statistic as an estimate of a population parameter, we need to consider both *bias* and *variability* of the statistic.

A sample statistic is **unbiased** if the mean of its sampling distribution equals the value of the parameter being estimated.

Continued

The spread of the sampling distribution indicates the **variability of the statistic.** The spread is affected by the sampling method and the sample size. Statistics from larger random samples have spreads that are smaller.

LOOKING FORWARD

Sampling distributions for the mean $\bar{x}$ and for proportions $\hat{p}$ (Section 6.6) will form the basis of our work with estimation (Chapter 7) and hypothesis testing (Chapter 8). With these inferential statistics methods, we will be able to use information from a sample to make statements regarding the population. These statements will be made in terms of probabilities derived from the underlying sampling distributions.

We see from the central limit theorem that the sample mean $\bar{x}$ is an unbiased estimator of the mean μ when $n \geq 30$. The variability of $\bar{x}$ decreases as the sample size increases.

In Section 6.6, we will see that the sample proportion $\hat{p}$ is an unbiased estimator of the population proportion of successes p in binomial experiments with sufficiently large numbers of trials n. Again, we will see that the variability of $\hat{p}$ decreases with increasing numbers of trials.

The sample variance s^2 is an unbiased estimator for the population variance σ^2.

VIEWPOINT Chaos!

Is there a different side to random sampling? Can sampling be used as a weapon? According to The Wall Street Journal, *the answer could be yes! The acronym for Create Havoc Around Our System is CHAOS. The Association of Flight Attendants (AFA) is a union that successfully used CHAOS against Alaska Airlines in 1994 as a negotiation tool. CHAOS involves a small sample of random strikes—a few flights at a time—instead of a mass walkout. The president of the AFA claims that by striking randomly, "We take control of the schedule." The entire schedule becomes unreliable, and that is some thing management cannot tolerate. In 1986, TWA flight attendants struck in a mass walkout, and all were permanently replaced! Using CHAOS, only a few jobs are put at risk, and these are usually not lost. It appears that random sampling can be used as a weapon.*

SECTION 6.5 PROBLEMS

In these problems, the word *average* refers to the arithmetic mean $\bar{x}$ or μ, as appropriate.

1. *Statistical Literacy* What is the standard error of a sampling distribution?

2. *Statistical Literacy* What is the standard deviation of a sampling distribution called?

3. *Statistical Literacy* List two unbiased estimators and their corresponding parameters.

4. *Statistical Literacy* Describe how the variability of the $\bar{x}$ distribution changes as the sample size increases.

5. *Basic Computation: Central Limit Theorem* Suppose x has a distribution with a mean of 8 and a standard deviation of 16. Random samples of size $n = 64$ are drawn.
   (a) Describe the $\bar{x}$ distribution and compute the mean and standard deviation of the distribution.
   (b) Find the z value corresponding to $\bar{x} = 9$.
   (c) Find $P(\bar{x} > 9)$.
   (d) *Interpretation* Would it be unusual for a random sample of size 64 from the x distribution to have a sample mean greater than 9? Explain.

6. *Basic Computation: Central Limit Theorem* Suppose x has a distribution with a mean of 20 and a standard deviation of 3. Random samples of size $n = 36$ are drawn.
   (a) Describe the $\bar{x}$ distribution and compute the mean and standard deviation of the distribution.
   (b) Find the z value corresponding to $\bar{x} = 19$.
   (c) Find $P(\bar{x} < 19)$.
   (d) *Interpretation* Would it be unusual for a random sample of size 36 from the x distribution to have a sample mean less than 19? Explain.

7. *Statistical Literacy*
   (a) If we have a distribution of x values that is more or less mound-shaped and somewhat symmetrical, what is the sample size needed to claim that the distribution of sample means $\bar{x}$ from random samples of that size is approximately normal?
   (b) If the original distribution of x values is known to be normal, do we need to make any restriction about sample size in order to claim that the distribution of sample means $\bar{x}$ taken from random samples of a given size is normal?

8. *Critical Thinking* Suppose x has a distribution with $\mu = 72$ and $\sigma = 8$.
   (a) If random samples of size $n = 16$ are selected, can we say anything about the $\bar{x}$ distribution of sample means?
   (b) If the original x distribution is *normal*, can we say anything about the $\bar{x}$ distribution of random samples of size 16? Find $P(68 \le \bar{x} \le 73)$.

9. *Critical Thinking* Consider two $\bar{x}$ distributions corresponding to the same x distribution. The first $\bar{x}$ distribution is based on samples of size $n = 100$ and the second is based on samples of size $n = 225$. Which $\bar{x}$ distribution has the smaller standard error? Explain.

10. *Critical Thinking* Consider an x distribution with standard deviation $\sigma = 12$.
   (a) If specifications for a research project require the standard error of the corresponding $\bar{x}$ distribution to be 2, how large does the sample size need to be?
   (b) If specifications for a research project require the standard error of the corresponding $\bar{x}$ distribution to be 1, how large does the sample size need to be?

11. *Critical Thinking* Suppose x has a distribution with $\mu = 15$ and $\sigma = 14$.
   (a) If a random sample of size $n = 49$ is drawn, find $\mu_{\bar{x}}$, $\sigma_{\bar{x}}$, and $P(15 \le \bar{x} \le 17)$.
   (b) If a random sample of size $n = 64$ is drawn, find $\mu_{\bar{x}}$, $\sigma_{\bar{x}}$, and $P(15 \le \bar{x} \le 17)$.
   (c) Why should you expect the probability of part (b) to be higher than that of part (a)? *Hint:* Consider the standard deviations in parts (a) and (b).

12. *Critical Thinking* Suppose an x distribution has mean $\mu = 5$. Consider two corresponding $\bar{x}$ distributions, the first based on samples of size $n = 49$ and the second based on samples of size $n = 81$.
   (a) What is the value of the mean of each of the two $\bar{x}$ distributions?
   (b) For which $\bar{x}$ distribution is $P(\bar{x} > 6)$ smaller? Explain.
   (c) For which $\bar{x}$ distribution is $P(4 < \bar{x} < 6)$ greater? Explain.

13. *Coal: Automatic Loader* Coal is carried from a mine in West Virginia to a power plant in New York in hopper cars on a long train. The automatic hopper car loader is set to put 75 tons of coal into each car. The actual weights of coal loaded into each car are *normally distributed,* with mean $\mu = 75$ tons and standard deviation $\sigma = 0.8$ ton.
   (a) What is the probability that one car chosen at random will have less than 74.5 tons of coal?

(b) What is the probability that 20 cars chosen at random will have a mean load weight $\bar{x}$ of less than 74.5 tons of coal?
(c) *Interpretation* Suppose the weight of coal in one car was less than 74.5 tons. Would that fact make you suspect that the loader had slipped out of adjustment? Suppose the weight of coal in 20 cars selected at random had an average $\bar{x}$ of less than 74.5 tons. Would that fact make you suspect that the loader had slipped out of adjustment? Why?

14. *Vital Statistics: Heights of Men* The heights of 18-year-old men are approximately *normally distributed,* with mean 68 inches and standard deviation 3 inches (based on information from *Statistical Abstract of the United States,* 112th Edition).
(a) What is the probability that an 18-year-old man selected at random is between 67 and 69 inches tall?
(b) If a random sample of nine 18-year-old men is selected, what is the probability that the mean height $\bar{x}$ is between 67 and 69 inches?
(c) *Interpretation* Compare your answers to parts (a) and (b). Is the probability in part (b) much higher? Why would you expect this?

15. *Medical: Blood Glucose* Let x be a random variable that represents the level of glucose in the blood (milligrams per deciliter of blood) after a 12-hour fast. Assume that for people under 50 years old, x has a distribution that is approximately normal, with mean $\mu = 85$ and estimated standard deviation $\sigma = 25$ (based on information from *Diagnostic Tests with Nursing Applications,* edited by S. Loeb, Springhouse). A test result $x < 40$ is an indication of severe excess insulin, and medication is usually prescribed.
(a) What is the probability that, on a single test, $x < 40$?
(b) Suppose a doctor uses the average $\bar{x}$ for two tests taken about a week apart. What can we say about the probability distribution of $\bar{x}$? *Hint:* See Theorem 6.1. What is the probability that $\bar{x} < 40$?
(c) Repeat part (b) for $n = 3$ tests taken a week apart.
(d) Repeat part (b) for $n = 5$ tests taken a week apart.
(e) *Interpretation* Compare your answers to parts (a), (b), (c), and (d). Did the probabilities decrease as n increased? Explain what this might imply if you were a doctor or a nurse. If a patient had a test result of $\bar{x} < 40$ based on five tests, explain why either you are looking at an extremely rare event or (more likely) the person has a case of excess insulin.

16. *Medical: White Blood Cells* Let x be a random variable that represents white blood cell count per cubic milliliter of whole blood. Assume that x has a distribution that is approximately normal, with mean $\mu = 7500$ and estimated standard deviation $\sigma = 1750$ (see reference in Problem 15). A test result of $x < 3500$ is an indication of leukopenia. This indicates bone marrow depression that may be the result of a viral infection.
(a) What is the probability that, on a single test, x is less than 3500?
(b) Suppose a doctor uses the average $\bar{x}$ for two tests taken about a week apart. What can we say about the probability distribution of $\bar{x}$? What is the probability of $\bar{x} < 3500$?
(c) Repeat part (b) for $n = 3$ tests taken a week apart.
(d) *Interpretation* Compare your answers to parts (a), (b), and (c). How did the probabilities change as n increased? If a person had $\bar{x} < 3500$ based on three tests, what conclusion would you draw as a doctor or a nurse?

17. *Wildlife: Deer* Let x be a random variable that represents the weights in kilograms (kg) of healthy adult female deer (does) in December in Mesa Verde National Park. Then x has a distribution that is approximately normal, with mean $\mu = 63.0$ kg and standard deviation $\sigma = 7.1$ kg (Source: *The Mule Deer of Mesa Verde National Park,* by G. W. Mierau and J. L. Schmidt, Mesa

Steve Krull/iStockphoto.com

Verde Museum Association). Suppose a doe that weighs less than 54 kg is considered undernourished.

(a) What is the probability that a single doe captured (weighed and released) at random in December is undernourished?

(b) If the park has about 2200 does, what number do you expect to be undernourished in December?

(c) *Interpretation* To estimate the health of the December doe population, park rangers use the rule that the average weight of $n = 50$ does should be more than 60 kg. If the average weight is less than 60 kg, it is thought that the entire population of does might be undernourished. What is the probability that the average weight $\bar{x}$ for a random sample of 50 does is less than 60 kg (assume a healthy population)?

(d) *Interpretation* Compute the probability that $\bar{x} < 64.2$ kg for 50 does (assume a healthy population). Suppose park rangers captured, weighed, and released 50 does in December, and the average weight was $\bar{x} = 64.2$ kg. Do you think the doe population is undernourished or not? Explain.

18. *Focus Problem: Impulse Buying* Let x represent the dollar amount spent on supermarket impulse buying in a 10-minute (unplanned) shopping interval. Based on a *Denver Post* article, the mean of the x distribution is about \$20 and the estimated standard deviation is about \$7.

(a) Consider a random sample of $n = 100$ customers, each of whom has 10 minutes of unplanned shopping time in a supermarket. From the central limit theorem, what can you say about the probability distribution of $\bar{x}$, the average amount spent by these customers due to impulse buying? What are the mean and standard deviation of the $\bar{x}$ distribution? Is it necessary to make any assumption about the x distribution? Explain.

(b) What is the probability that $\bar{x}$ is between \$18 and \$22?

(c) Let us assume that x has a distribution that is approximately normal. What is the probability that x is between \$18 and \$22?

(d) *Interpretation:* In part (b), we used $\bar{x}$, the *average* amount spent, computed for 100 customers. In part (c), we used x, the amount spent by only *one* customer. The answers to parts (b) and (c) are very different. Why would this happen? In this example, $\bar{x}$ is a much more predictable or reliable statistic than x. Consider that almost all marketing strategies and sales pitches are designed for the *average* customer and *not the individual* customer. How does the central limit theorem tell us that the average customer is much more predictable than the individual customer?

19. *Finance: Templeton Funds* Templeton World is a mutual fund that invests in both U.S. and foreign markets. Let x be a random variable that represents the monthly percentage return for the Templeton World fund. Based on information from the *Morningstar Guide to Mutual Funds* (available in most libraries), x has mean $\mu = 1.6\%$ and standard deviation $\sigma = 0.9\%$.

(a) Templeton World fund has over 250 stocks that combine together to give the overall monthly percentage return x. We can consider the monthly return of the stocks in the fund to be a sample from the population of monthly returns of all world stocks. Then we see that the overall monthly return x for Templeton World fund is itself an average return computed using all 250 stocks in the fund. Why would this indicate that x has an approximately normal distribution? Explain. *Hint:* See the discussion after Theorem 6.2.

(b) After 6 months, what is the probability that the *average* monthly percentage return $\bar{x}$ will be between 1% and 2%? *Hint:* See Theorem 6.1, and assume that x has a normal distribution as based on part (a).

(c) After 2 years, what is the probability that $\bar{x}$ will be between 1% and 2%?

(d) Compare your answers to parts (b) and (c). Did the probability increase as n (number of months) increased? Why would this happen?

(e) *Interpretation:* If after 2 years the average monthly percentage return $\bar{x}$ was less than 1%, would that tend to shake your confidence in the statement that $\mu = 1.6\%$? Might you suspect that μ has slipped below 1.6%? Explain.

20. *Finance: European Growth Fund* A European growth mutual fund specializes in stocks from the British Isles, continental Europe, and Scandinavia. The fund has over 100 stocks. Let x be a random variable that represents the monthly percentage return for this fund. Based on information from *Morningstar* (see Problem 19), x has mean $\mu = 1.4\%$ and standard deviation $\sigma = 0.8\%$.

(a) Let's consider the monthly return of the stocks in the European growth fund to be a sample from the population of monthly returns of all European stocks. Is it reasonable to assume that x (the average monthly return on the 100 stocks in the European growth fund) has a distribution that is approximately normal? Explain. *Hint:* See Problem 19, part (a).

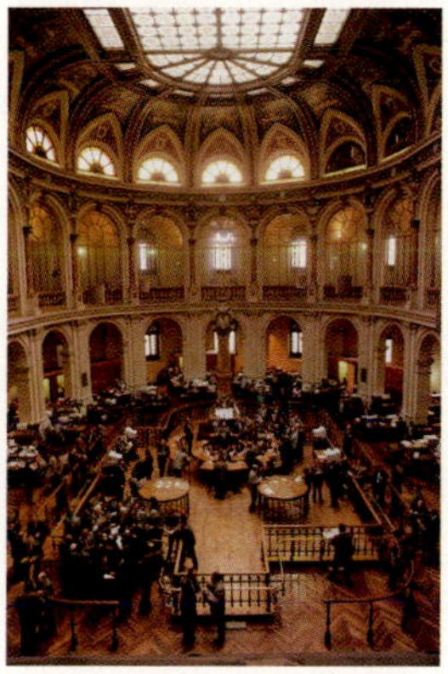

Mark L. Stephenson/CORBIS

(b) After 9 months, what is the probability that the *average* monthly percentage return $\bar{x}$ will be between 1% and 2%? *Hint:* See Theorem 6.1 and the results of part (a).

(c) After 18 months, what is the probability that the *average* monthly percentage return $\bar{x}$ will be between 1% and 2%?

(d) Compare your answers to parts (b) and (c). Did the probability increase as n (number of months) increased? Why would this happen?

(e) *Interpretation:* If after 18 months the average monthly percentage return $\bar{x}$ is more than 2%, would that tend to shake your confidence in the statement that $\mu = 1.4\%$? If this happened, do you think the European stock market might be heating up? Explain.

21. *Expand Your Knowledge: Totals Instead of Averages* Let x be a random variable that represents checkout time (time spent in the actual checkout process) in minutes in the express lane of a large grocery. Based on a consumer survey, the mean of the x distribution is about $\mu = 2.7$ minutes, with standard deviation $\sigma = 0.6$ minute. Assume that the express lane always has customers waiting to be checked out and that the distribution of x values is more or less symmetrical and mound-shaped. What is the probability that the *total* checkout time for the next 30 customers is less than 90 minutes? Let us solve this problem in steps.

(a) Let x_i (for $i = 1, 2, 3, \ldots, 30$) represent the checkout time for each customer. For example, x_1 is the checkout time for the first customer, x_2 is the checkout time for the second customer, and so forth. Each x_i has mean $\mu = 2.7$ minutes and standard deviation $\sigma = 0.6$ minute. Let $w = x_1 + x_2 + \cdots + x_{30}$. Explain why the problem is asking us to compute the probability that w is less than 90.

(b) Use a little algebra and explain why $w < 90$ is mathematically equivalent to $w/30 < 3$. Since w is the total of the 30 x values, then $w/30 = \bar{x}$. Therefore, the statement $\bar{x} < 3$ is equivalent to the statement $w < 90$. From this we conclude that the probabilities $P(\bar{x} < 3)$ and $P(w < 90)$ are equal.

(c) What does the central limit theorem say about the probability distribution of $\bar{x}$? Is it approximately normal? What are the mean and standard deviation of the $\bar{x}$ distribution?

(d) Use the result of part (c) to compute $P(\bar{x} < 3)$. What does this result tell you about $P(w < 90)$?

22. *Totals Instead of Averages: Airplane Takeoff Time* The taxi and takeoff time for commercial jets is a random variable x with a mean of 8.5 minutes and a standard deviation of 2.5 minutes. Assume that the distribution of taxi and takeoff times is approximately normal. You may assume that the jets are lined

up on a runway so that one taxies and takes off immediately after another, and that they take off one at a time on a given runway. What is the probability that for 36 jets on a given runway, total taxi and takeoff time will be
(a) less than 320 minutes?
(b) more than 275 minutes?
(c) between 275 and 320 minutes?
Hint: See Problem 21.

23. *Totals Instead of Averages: Escape Dunes* It's true—sand dunes in Colorado rival sand dunes of the Great Sahara Desert! The highest dunes at Great Sand Dunes National Monument can exceed the highest dunes in the Great Sahara, extending over 700 feet in height. However, like all sand dunes, they tend to move around in the wind. This can cause a bit of trouble for temporary structures located near the "escaping" dunes. Roads, parking lots, campgrounds, small buildings, trees, and other vegetation are destroyed when a sand dune moves in and takes over. Such dunes are called "escape dunes" in the sense that they move out of the main body of sand dunes and, by the force of nature (prevailing winds), take over whatever space they choose to occupy. In most cases, dune movement does not occur quickly. An escape dune can take years to relocate itself. Just how fast does an escape dune move? Let x be a random variable representing movement (in feet per year) of such sand dunes (measured from the crest of the dune). Let us assume that x has a normal distribution with $\mu = 17$ feet per year and $\sigma = 3.3$ feet per year. (For more information, see *Hydrologic, Geologic, and Biologic Research at Great Sand Dunes National Monument and Vicinity, Colorado,* proceedings of the National Park Service Research Symposium.)

Under the influence of prevailing wind patterns, what is the probability that
(a) an escape dune will move a total distance of more than 90 feet in 5 years?
(b) an escape dune will move a total distance of less than 80 feet in 5 years?
(c) an escape dune will move a total distance of between 80 and 90 feet in 5 years?
Hint: See Problem 21 and Theorem 6.1.

SECTION **6.6**

Normal Approximation to Binomial Distribution and to $\hat{p}$ Distribution

FOCUS POINTS

- State the assumptions needed to use the normal approximation to the binomial distribution.
- Compute μ and σ for the normal approximation.
- Use the continuity correction to convert a range of r values to a corresponding range of normal x values.
- Convert the x values to a range of standardized z scores and find desired probabilities.
- Describe the sampling distribution for proportions $\hat{p}$.

The probability that a new vaccine will protect adults from cholera is known to be 0.85. The vaccine is administered to 300 adults who must enter an area where the disease is prevalent. What is the probability that more than 280 of these adults will be protected from cholera by the vaccine?

This question falls into the category of a binomial experiment, with the number of trials n equal to 300, the probability of success p equal to 0.85, and the number of successes r greater than 280. It is possible to use the formula for the binomial distribution to compute the probability that r is greater than 280. However, this approach would involve a number of tedious and long calculations. There is an easier way to do this problem, for under the conditions stated below, the normal distribution can be used to approximate the binomial distribution.

NORMAL APPROXIMATION TO THE BINOMIAL DISTRIBUTION

Consider a binomial distribution where

n = number of trials
r = number of successes
p = probability of success on a single trial
$q = 1 - p$ = probability of failure on a single trial

Criteria $np > 5$ and $nq > 5$

If $np > 5$ and $nq > 5$, then r has a binomial distribution that is approximated by a normal distribution with

$$\mu = np \quad \text{and} \quad \sigma = \sqrt{npq}$$

Note: As n increases, the approximation becomes better.

Example 14 demonstrates that as n increases, the normal approximation to the binomial distribution improves.

EXAMPLE 14

BINOMIAL DISTRIBUTION GRAPHS

Graph the binomial distributions for which $p = 0.25$, $q = 0.75$, and the number of trials is first $n = 3$, then $n = 10$, then $n = 25$, and finally $n = 50$.

SOLUTION: The authors used a computer program to obtain the binomial distributions for the given values of p, q, and n. The results have been organized and graphed in Figures 6-34, 6-35, 6-36, and 6-37.

When $n = 3$, the outline of the histogram does not even begin to take the shape of a normal curve. But when $n = 10$, 25, or 50, it does begin to take a normal shape, indicated by the red curve. From a theoretical point of view, the histograms in

FIGURE 6-34

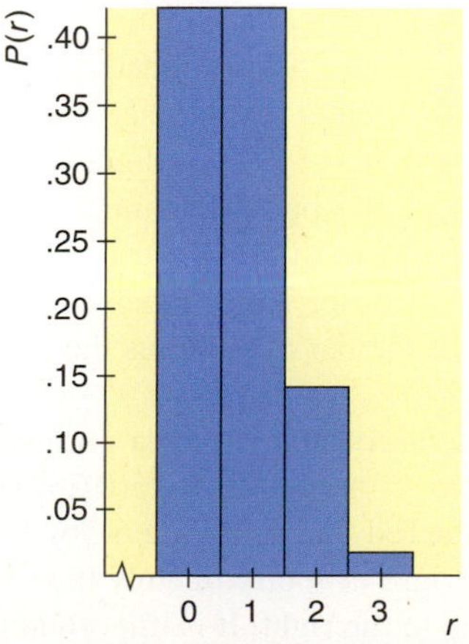

$n = 3$
$p = 0.25$
$q = 0.75$
$\mu = np = 0.75$
$s = \sqrt{npq} = 0.75$

FIGURE 6-35

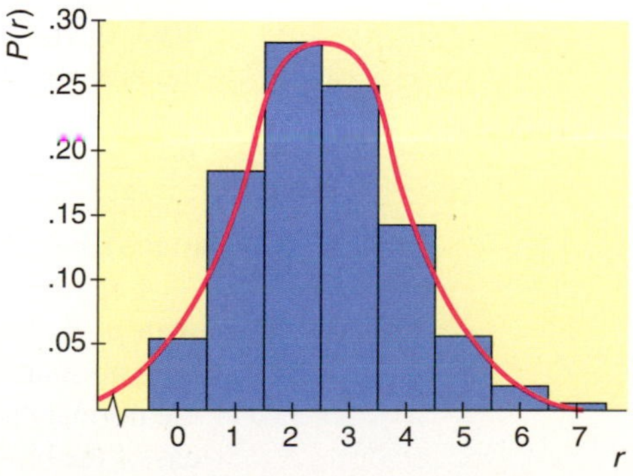

$n = 10$
$p = 0.25$
$q = 0.75$
$\mu = np = 2.5$
$\sigma = \sqrt{npq} = 1.37$

FIGURE 6-36

Good Normal Approximation; $np > 5$ and $nq > 5$

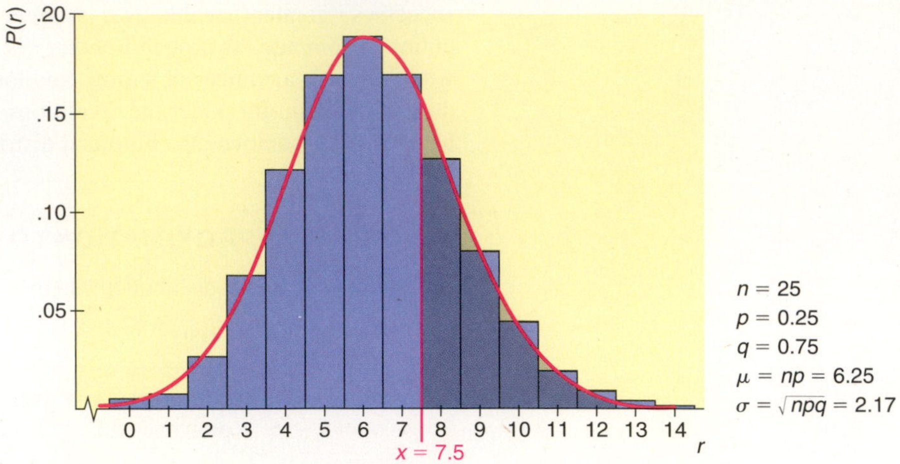

FIGURE 6-37

Good Normal Approximation; $np > 5$ and $nq > 5$

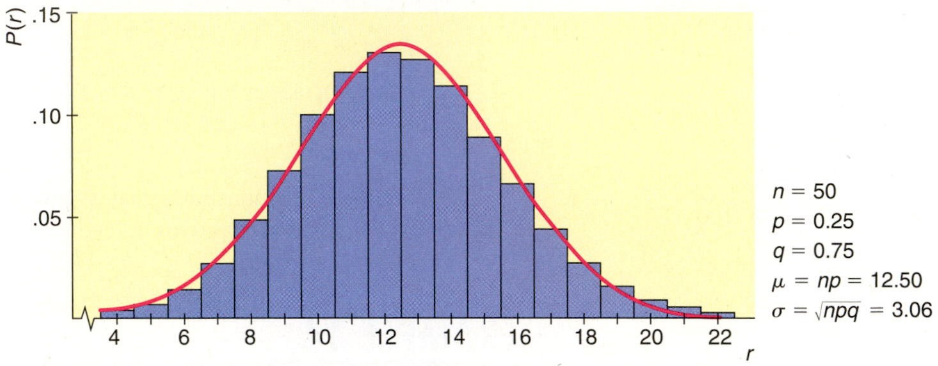

Figures 6-35, 6-36, and 6-37 would have bars for all values of r from $r = 0$ to $r = n$. However, in the construction of these histograms, the bars of height less than 0.001 unit have been omitted—that is, in this example, probabilities less than 0.001 have been rounded to 0.

EXAMPLE 15

NORMAL APPROXIMATION

The owner of a new apartment building must install 25 water heaters. From past experience in other apartment buildings, she knows that Quick Hot is a good brand. A Quick Hot heater is guaranteed for 5 years only, but from the owner's past experience, she knows that the probability it will last 10 years is 0.25.

(a) What is the probability that 8 or more of the 25 water heaters will last at least 10 years? Define success to mean a water heater that lasts at least 10 years.

SOLUTION: In this example, $n = 25$ and $p = 0.25$, so Figure 6-36 represents the probability distribution we will use. Let r be the binomial random variable corresponding to the number of successes out of $n = 25$ trials. We want to find $P(r \geq 8)$ by using the normal approximation. This probability is represented graphically (Figure 6-36) by the area of the bar over 8 plus the areas of all bars to the right of the bar over 8.

Let x be a normal random variable corresponding to a normal distribution, with $\mu = np = 25(0.25) = 6.25$ and $\sigma = \sqrt{npq} = \sqrt{25(0.25)(0.75)} \approx 2.17$. This normal curve is represented by the red line in Figure 6-36. The area under the normal curve from $x = 7.5$ to the right is approximately the same as the areas of the bars from the bar over $r = 8$ to the right. It is important to notice that

Marcus Lindstrøm/ iStockphoto.com

we start with $x = 7.5$ because the bar over $r = 8$ really starts at $x = 7.5$.

The areas of the bars and the area under the corresponding red (normal) curve are approximately equal, so we conclude that $P(r \geq 8)$ is approximately equal to $P(x \geq 7.5)$.

When we convert $x = 7.5$ to standard units, we get

$$z = \frac{x - \mu}{\sigma} = \frac{7.5 - 6.25}{2.17} \approx 0.58 \quad \text{(Use } \mu = 6.25 \text{ and } \sigma = 2.17.)$$

The probability we want is

$$P(x \geq 7.5) = P(z \geq 0.58) = 1 - P(z \leq 0.58) = 1 - 0.7190 = 0.2810$$

(b) How does this result compare with the result we can obtain by using the formula for the binomial probability distribution with $n = 25$ and $p = 0.25$?

SOLUTION: Using the binomial distribution function on the TI-84Plus/TI-83Plus/ TI-*n*spire model calculators, the authors computed that $P(r \geq 8) \approx 0.2735$. This means that the probability is approximately 0.27 that 8 or more water heaters will last at least 10 years.

(c) How do the results of parts (a) and (b) compare?

SOLUTION: The error of approximation is the difference between the approximate normal value (0.2810) and the binomial value (0.2735). The error is only $0.2810 - 0.2735 = 0.0075$, which is negligible for most practical purposes.

We knew in advance that the normal approximation to the binomial probability would be good, since $np = 25(0.25) = 6.25$ and $nq = 25(0.75) = 18.75$ are both greater than 5. These are the conditions that assure us that the normal approximation will be sufficiently close to the binomial probability for most practical purposes.

Continuity correction: Converting *r* values to *x* values

Remember that when we use the normal distribution to approximate the binomial, we are computing the areas under bars. The bar over the discrete variable r extends from $r - 0.5$ to $r + 0.5$. This means that the corresponding continuous normal variable x extends from $r - 0.5$ to $r + 0.5$. Adjusting the values of discrete random variables to obtain a corresponding range for a continuous random variable is called making a *continuity correction*.

PROCEDURE

HOW TO MAKE THE CONTINUITY CORRECTION

Convert the discrete random variable r (number of successes) to the continuous normal random variable x by doing the following:

1. If r is a **left point** of an interval, subtract 0.5 to obtain the corresponding normal variable x; that is, $x = r - 0.5$.
2. If r is a **right point** of an interval, add 0.5 to obtain the corresponding normal variable x; that is, $x = r + 0.5$.

For instance, $P(6 \leq r \leq 10)$, where r is a binomial random variable, is approximated by $P(5.5 \leq x \leq 10.5)$, where x is the corresponding normal random variable (see Figure 6-38 on the following page).

COMMENT Both the binomial and Poisson distributions are for *discrete* random variables. Therefore, adding or subtracting 0.5 to r was not necessary when we approximated the binomial distribution by the Poisson distribution (Section 5.4). However, the normal distribution is for a *continuous* random variable. In this case, adding or subtracting 0.5 to or from (as appropriate) r will improve the approximation of the normal to the binomial distribution.

FIGURE 6-38

$P(6 \leq r \leq 10)$ Is Approximately Equal to $P(5.5 \leq x \leq 10.5)$

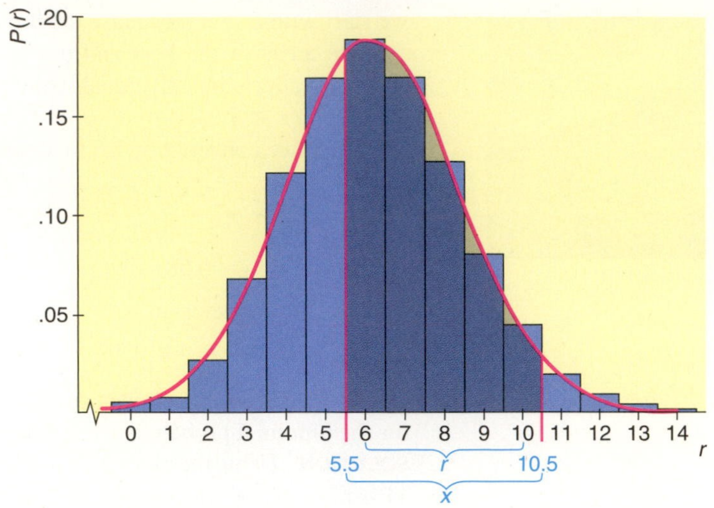

WHAT DOES A NORMAL APPROXIMATION TO THE BINOMIAL TELL US?

A normal distribution can be used to approximate the probability of r successes out of n binomial trials. In order to use the normal approximation,

- the products np and nq must both exceed 5, where n is the number of binomial trials, p is the probability of success on a single trial, and $q = 1 - p$.
- unless n is fairly large, a continuity correction may be necessary in order to improve the approximation.

GUIDED EXERCISE 12 CONTINUITY CORRECTION

From many years of observation, a biologist knows that the probability is only 0.65 that any given Arctic tern will survive the migration from its summer nesting area to its winter feeding grounds. A random sample of 500 Arctic terns were banded at their summer nesting area. Use the normal approximation to the binomial and the following steps to find the probability that between 310 and 340 of the banded Arctic terns will survive the migration. Let r be the number of surviving terns.

William Mullins/Photo Researchers

Arctic tern

(a) To approximate $P(310 \leq r \leq 340)$, we use the normal curve with $\mu =$ ________ and $\sigma =$ ________.

→ We use the normal curve with

$\mu = np = 500(0.65) = 325$ and

$\sigma = \sqrt{npq} = \sqrt{500(0.65)(0.35)} \approx 10.67$

(b) $P(310 \leq r \leq 340)$ is approximately equal to $P($________ $\leq x \leq$ ________$)$, where x is a variable from the normal distribution described in part (a).

→ Since 310 is the left endpoint, we subtract 0.5, and since 340 is the right endpoint, we add 0.5. Consequently,

$P(310 \leq r \leq 340) \approx P(309.5 \leq x \leq 340.5)$

Continued

GUIDED EXERCISE 12 *continued*

(c) Convert the condition $309.5 \le x \le 340.5$ to a condition in standard units.

➡ Since $\mu = 325$ and $\sigma \approx 10.67$, the condition $309.5 \le x \le 340.5$ becomes

$$\frac{309.5 - 325}{10.67} \le z \le \frac{340.5 - 325}{10.67}$$

or

$$-1.45 \le z \le 1.45$$

(d) $P(310 \le r \le 340) = P(309.5 \le x \le 340.5)$
$= P(-1.45 \le z \le 1.45)$
$=$ ________

➡ $P(-1.45 \le z \le 1.45) = P(z \le 1.45) - P(z \le -1.45)$
$= 0.9265 - 0.0735$
$= 0.8530$

(e) Will the normal distribution make a good approximation to the binomial for this problem? Explain your answer.

➡ Since

$np = 500(0.65) = 35$

and

$nq = 500(0.35) = 175$

are both greater than 5, the normal distribution will be a good approximation to the binomial.

SAMPLING DISTRIBUTIONS FOR PROPORTIONS

In Sections 6.4 and 6.5, we studied the sampling distribution for the mean. Now we have the tools to look at sampling distributions for proportions. Suppose we repeat a binomial experiment with n trials again and again and, for each n trials, record the sample proportion of successes $\hat{p} = r/n$. The $\hat{p}$ values form a sampling distribution for proportions.

Sampling distribution for $\hat{p}$

SAMPLING DISTRIBUTION FOR THE PROPORTION $\hat{p} = \frac{r}{n}$

Given $n =$ number of binomial trials (fixed constant)
$r =$ number of successes
$p =$ probability of success on each trial
$q = 1 - p =$ probability of failure on each trial

If $np > 5$ and $nq > 5$, then the random variable $\hat{p} = r/n$ can be approximated by a normal random variable (x) with mean and standard deviation

$$\mu_{\hat{p}} = p \text{ and } \sigma_{\hat{p}} = \sqrt{\frac{pq}{n}}$$

Standard error of a proportion

TERMINOLOGY The *standard error* for the $\hat{p}$ distribution is the standard deviation $\sigma_{\hat{p}}$ of the $\hat{p}$ sampling distribution.

COMMENT To obtain the information regarding the sampling distribution for the proportion $\hat{p} = r/n$, we consider the sampling distribution for r, the number of successes out of n binomial trials. Earlier we saw that when $np > 5$ and $nq > 5$, the r distribution is approximately normal, with mean $\mu_r = np$ and standard deviation $\sigma_r = \sqrt{npq}$. Notice that $\hat{p} = r/n$ is a linear function of r. This means that the

$\hat{p}$ distribution is also approximately normal when np and nq are both greater than 5. In addition, from our work in Section 5.1 with linear functions of random variables, we know that $\mu_{\hat{p}} = \mu_{r}/n = np/n = p$ and $\sigma_{\hat{p}} = \sigma_r/n = \sqrt{npq}/n = \sqrt{pq/n}$. Although the values r/n are discrete for a fixed n, we do not use a continuity correction for the $\hat{p}$ distribution. This is the accepted standard practice for applications in inferential statistics, especially considering the requirements that $np > 5$ and $nq >$ are met.

We see from the sampling distribution for proportions that the mean of the $\hat{p}$ distribution is p, the population proportion of successes. This means that $\hat{p}$ is an *unbiased* estimator for p.

WHAT DOES A $\hat{p}$ SAMPLING DISTRIBUTION TELL US?

Consider n binomial trials. We use $\hat{p} = r/n$ to estimate p, the population probability of success on a single trial where r is the number of successes out of the n binomial trials. To create a $\hat{p}$ distribution, repeated sets of n binomial trials must be conducted to determine individual sample statistics $\hat{p}$. Fortunately, mathematics can be used to obtain the following information.

- For $np > 5$ and $nq > 5$, the $\hat{p}$ distribution is approximately normal.
- The mean of the resulting $\hat{p}$ distribution is p.
- The standard deviation (also known as standard error) is $\sqrt{pq/n}$, where $q = 1 - p$.

LOOKING FORWARD

We will use the sampling distribution for proportions in our work with estimation (Chapter 7) and hypothesis testing (Chapter 8).

EXAMPLE 16

SAMPLING DISTRIBUTION OF $\hat{p}$

The annual crime rate in the Capital Hill neighborhood of Denver is 111 victims per 1000 residents. This means that 111 out of 1000 residents have been the victim of at least one crime (Source: *Neighborhood Facts,* Piton Foundation). For more information, visit the Brase/Brase statistics site at **http://www.cengage.com/statistics/brase11e** and find the link to the Piton Foundation. These crimes range from relatively minor crimes (stolen hubcaps or purse snatching) to major crimes (murder). The Arms is an apartment building in this neighborhood that has 50 year-round residents. Suppose we view each of the $n = 50$ residents as a binomial trial. The random variable r (which takes on values 0, 1, 2, . . . , 50) represents the number of victims of at least one crime in the next year.

(a) What is the population probability p that a resident in the Capital Hill neighborhood will be the victim of a crime next year? What is the probability q that a resident will not be a victim?

SOLUTION: Using the Piton Foundation report, we take

$$p = 111/1000 = 0.111 \text{ and } q = 1 - p = 0.889$$

Beyza Sultan Durna/iStockphoto.com

(b) Consider the random variable

$$\hat{p} = \frac{r}{n} = \frac{r}{50}$$

Can we approximate the $\hat{p}$ distribution with a normal distribution? Explain.

SOLUTION:

$$np = 50(0.111) = 5.55$$

$$nq = 50(0.889) = 44.45$$

Since both np and nq are greater than 5, we can approximate the $\hat{p}$ distribution with a normal distribution.

(c) What are the mean and standard deviation for the $\hat{p}$ distribution?

SOLUTION: $\mu_{\hat{p}} = p = 0.111$

$$\sigma_{\hat{p}} = \sqrt{\frac{pq}{n}}$$

$$= \sqrt{\frac{(0.111)(0.889)}{50}} \approx 0.044$$

VIEWPOINT Sunspots, Tree Rings, and Statistics

Ancient Chinese astronomers recorded extreme sunspot activity, with a peak around 1200 A.D. Mesa Verde tree rings in the period between 1276 and 1299 were unusually narrow, indicating a drought and/or a severe cold spell in the region at that time. A cooling trend could have narrowed the window of frost-free days below the approximately 80 days needed for cultivation of aboriginal corn and beans. Is this the reason the ancient Anasazi dwellings in Mesa Verde were abandoned? Is there a connection to the extreme sunspot activity? Much research and statistical work continue to be done on this topic.

Reference: *Prehistoric Astronomy in the Southwest,* by J. McKim Malville and C. Putnam, Department of Astronomy, University of Colorado.

SECTION 6.6 PROBLEMS

Note: When we say *between a and b,* we mean every value from *a* to *b, including a* and *b.* Due to rounding, your answers might vary slightly from answers given in the text.

1. *Statistical Literacy* Binomial probability distributions depend on the number of trials n of a binomial experiment and the probability of success p on each trial. Under what conditions is it appropriate to use a normal approximation to the binomial?

2. *Statistical Literacy* When we use a normal distribution to approximate a binomial distribution, why do we make a continuity correction?

3. *Basic Computation: Normal Approximation to a Binomial Distribution* Suppose we have a binomial experiment with $n = 40$ trials and a probability of success $p = 0.50$.
   (a) Is it appropriate to use a normal approximation to this binomial distribution? Why?
   (b) Compute μ and σ of the approximating normal distribution.
   (c) Use a continuity correction factor to convert the statement $r \geq 23$ successes to a statement about the corresponding normal variable x.
   (d) Estimate $P(r \geq 23)$.
   (e) *Interpretation* Is it unusual for a binomial experiment with 40 trials and probability of success 0.50 to have 23 or more successes? Explain.

4. *Basic Computation: Normal Approximation to a Binomial Distribution* Suppose we have a binomial experiment with $n = 40$ trials and probability of success $p = 0.85$.
   (a) Is it appropriate to use a normal approximation to this binomial distribution? Why?
   (b) Compute μ and σ of the approximating normal distribution.

(c) Use a continuity correction factor to convert the statement $r < 30$ successes to a statement about the corresponding normal variable x.
(d) Estimate $P(r < 30)$.
(e) *Interpretation* Is it unusual for a binomial experiment with 40 trials and probability of success 0.85 to have fewer than 30 successes? Explain.

5. *Critical Thinking* You need to compute the probability of 5 or fewer successes for a binomial experiment with 10 trials. The probability of success on a single trial is 0.43. Since this probability of success is not in the table, you decide to use the normal approximation to the binomial. Is this an appropriate strategy? Explain.

6. *Critical Thinking* Consider a binomial experiment with 20 trials and probability 0.45 of success on a single trial.
(a) Use the binomial distribution to find the probability of exactly 10 successes.
(b) Use the normal distribution to approximate the probability of exactly 10 successes.
(c) Compare the results of parts (a) and (b).

In the following problems, check that it is appropriate to use the normal approximation to the binomial. Then use the normal distribution to estimate the requested probabilities.

7. *Health: Lead Contamination* More than a decade ago, high levels of lead in the blood put 88% of children at risk. A concerted effort was made to remove lead from the environment. Now, according to the *Third National Health and Nutrition Examination Survey* (*NHANES III*) conducted by the Centers for Disease Control and Prevention, only 9% of children in the United States are at risk of high blood-lead levels.
(a) In a random sample of 200 children taken more than a decade ago, what is the probability that 50 or more had high blood-lead levels?
(b) In a random sample of 200 children taken now, what is the probability that 50 or more have high blood-lead levels?

8. *Insurance: Claims* Do you try to *pad* an insurance claim to cover your deductible? About 40% of all U.S. adults will try to pad their insurance claims! (Source: *Are You Normal?*, by Bernice Kanner, St. Martin's Press.) Suppose that you are the director of an insurance adjustment office. Your office has just received 128 insurance claims to be processed in the next few days. What is the probability that
(a) half or more of the claims have been padded?
(b) fewer than 45 of the claims have been padded?
(c) from 40 to 64 of the claims have been padded?
(d) more than 80 of the claims have *not* been padded?

9. *Longevity: 90th Birthday* It is estimated that 3.5% of the general population will live past their 90th birthday (*Statistical Abstract of the United States*, 112th Edition). In a graduating class of 753 high school seniors, what is the probability that
(a) 15 or more will live beyond their 90th birthday?
(b) 30 or more will live beyond their 90th birthday?
(c) between 25 and 35 will live beyond their 90th birthday?
(d) more than 40 will live beyond their 90th birthday?

10. *Fishing: Billfish* Ocean fishing for billfish is very popular in the Cozumel region of Mexico. In *World Record Game Fishes* (published by the International Game Fish Association), it was stated that in the Cozumel region, about 44% of strikes (while trolling) result in a catch. Suppose that on a given day a fleet of fishing boats got a total of 24 strikes. What is the probability that the number of fish caught was

(a) 12 or fewer?
(b) 5 or more?
(c) between 5 and 12?

11. *Grocery Stores: New Products* *The Denver Post* stated that 80% of all new products introduced in grocery stores fail (are taken off the market) within 2 years. If a grocery store chain introduces 66 new products, what is the probability that within 2 years
(a) 47 or more fail?
(b) 58 or fewer fail?
(c) 15 or more succeed?
(d) fewer than 10 succeed?

12. *Crime: Murder* What are the chances that a person who is murdered actually knew the murderer? The answer to this question explains why a lot of police detective work begins with relatives and friends of the victim! About 64% of people who are murdered actually knew the person who committed the murder (*Chances: Risk and Odds in Everyday Life,* by James Burke). Suppose that a detective file in New Orleans has 63 current unsolved murders. What is the probability that
(a) at least 35 of the victims knew their murderers?
(b) at most 48 of the victims knew their murderers?
(c) fewer than 30 victims did *not* know their murderers?
(d) more than 20 victims did *not* know their murderers?

13. *Supermarkets: Free Samples* Do you take the free samples offered in supermarkets? About 60% of all customers will take free samples. Furthermore, of those who take the free samples, about 37% will buy what they have sampled. (See reference in Problem 8.) Suppose you set up a counter in a supermarket offering free samples of a new product. The day you are offering free samples, 317 customers pass by your counter.
(a) What is the probability that more than 180 take your free sample?
(b) What is the probability that fewer than 200 take your free sample?
(c) What is the probability that a customer takes a free sample *and* buys the product? *Hint:* Use the multiplication rule for *dependent* events. Notice that we are given the conditional probability $P(\text{buy} \mid \text{sample}) = 0.37$, while $P(\text{sample}) = 0.60$.
(d) What is the probability that between 60 and 80 customers will take the free sample *and* buy the product? *Hint:* Use the probability of success calculated in part (c).

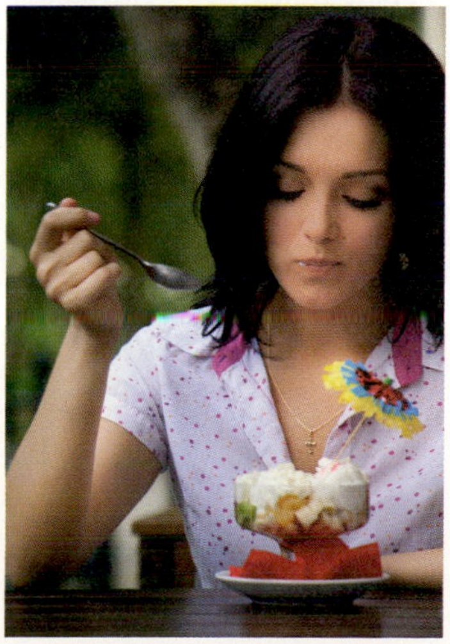
Kirill P/Shutterstock.com

14. *Ice Cream: Flavors* What's your favorite ice cream flavor? For people who buy ice cream, the all-time favorite is still vanilla. About 25% of ice cream sales are vanilla. Chocolate accounts for only 9% of ice cream sales. (See reference in Problem 8.) Suppose that 175 customers go to a grocery store in Cheyenne, Wyoming today to buy ice cream.
(a) What is the probability that 50 or more will buy vanilla?
(b) What is the probability that 12 or more will buy chocolate?
(c) A customer who buys ice cream is not limited to one container or one flavor. What is the probability that someone who is buying ice cream will buy chocolate or vanilla? *Hint:* Chocolate flavor and vanilla flavor are not mutually exclusive events. Assume that the choice to buy one flavor is independent of the choice to buy another flavor. Then use the multiplication rule for independent events, together with the addition rule for events that are not mutually exclusive, to compute the requested probability. (See Section 4.2.)
(d) What is the probability that between 50 and 60 customers will buy chocolate or vanilla ice cream? *Hint:* Use the probability of success computed in part (c).

15. *Airline Flights: No-Shows* Based on long experience, an airline has found that about 6% of the people making reservations on a flight from Miami to Denver do not show up for the flight. Suppose the airline overbooks this flight by selling 267 ticket reservations for an airplane with only 255 seats.
(a) What is the probability that a person holding a reservation will show up for the flight?
(b) Let $n = 267$ represent the number of ticket reservations. Let r represent the number of people with reservations who show up for the flight. Which expression represents the probability that a seat will be available for everyone who shows up holding a reservation?

$$P(255 \leq r);\ P(r \leq 255);\ P(r \leq 267);\ P(r = 255)$$

(c) Use the normal approximation to the binomial distribution and part (b) to answer the following question: What is the probability that a seat will be available for every person who shows up holding a reservation?

16. *General: Approximations* We have studied *two* approximations to the binomial, the normal approximation and the Poisson approximation (Section 5.4). Write a brief but complete essay in which you discuss and summarize the *conditions* under which each approximation would be used, the *formulas* involved, and the *assumptions* made for each approximation. Give details and examples in your essay. How could you apply these statistical methods in your everyday life?

17. *Statistical Literacy* Under what conditions is it appropriate to use a normal distribution to approximate the $\hat{p}$ distribution?

18. *Statistical Literacy* What is the formula for the standard error of the normal approximation to the $\hat{p}$ distribution? What is the mean of the $\hat{p}$ distribution?

19. *Statistical Literacy* Is $\hat{p}$ an unbiased estimator for p when $np > 5$ and $nq > 5$? Recall that a statistic is an unbiased estimator of the corresponding parameter if the mean of the sampling distribution equals the parameter in question.

20. *Basic Computation: $\hat{p}$ Distribution* Suppose we have a binomial experiment in which success is defined to be a particular quality or attribute that interests us.
(a) Suppose $n = 33$ and $p = 0.21$. Can we approximate the $\hat{p}$ distribution by a normal distribution? Why? What are the values of $\mu_{\hat{p}}$ and $\sigma_{\hat{p}}$?
(b) Suppose $n = 25$ and $p = 0.15$. Can we safely approximate the $\hat{p}$ distribution by a normal distribution? Why or why not?
(c) Suppose $n = 48$ and $p = 0.15$. Can we approximate the $\hat{p}$ distribution by a normal distribution? Why? What are the values of $\mu_{\hat{p}}$ and $\sigma_{\hat{p}}$?

21. *Basic Computation: $\hat{p}$ Distribution* Suppose we have a binomial experiment in which success is defined to be a particular quality or attribute that interests us.
(a) Suppose $n = 100$ and $p = 0.23$. Can we safely approximate the $\hat{p}$ distribution by a normal distribution? Why? Compute $\mu_{\hat{p}}$ and $\sigma_{\hat{p}}$.
(b) Suppose $n = 20$ and $p = 0.23$. Can we safely approximate the $\hat{p}$ distribution by a normal distribution? Why or why not?

CHAPTER REVIEW

SUMMARY

In this chapter, we examined properties and applications of the normal probability distribution.

- A normal probability distribution is a distribution of a continuous random variable. Normal distributions are bell-shaped and symmetric around the mean. The high point occurs over the mean, and most of the area occurs within 3 standard deviations of the mean. The mean and median are equal.
- The empirical rule for normal distributions gives areas within 1, 2, and 3 standard deviations of the mean. Approximately
  - 68% of the data lie within the interval $\mu \pm \sigma$
  - 95% of the data lie within the interval $\mu \pm 2\sigma$
  - 99.7% of the data lie within the interval $\mu \pm 3\sigma$
- For symmetric, bell-shaped distributions,

$$\text{standard deviation} \approx \frac{\text{range of data}}{4}$$

- A z score measures the number of standard deviations a raw score x lies from the mean.

$$z = \frac{x - \mu}{\sigma} \quad \text{and} \quad x = z\sigma + \mu$$

- For the standard normal distribution, $\mu = 0$ and $\sigma = 1$.
- Table 5 of Appendix II gives areas under a standard normal distribution that are to the left of a specified value of z.
- After raw scores x have been converted to z scores, the standard normal distribution table can be used to find probabilities associated with intervals of x values from any normal distribution.
- The inverse normal distribution is used to find z values associated with areas to the left of z. Table 5 of Appendix II can be used to find approximate z values associated with specific probabilities.
- Tools for assessing the normality of a data distribution include:
  - Histogram of the data. A roughly bell-shaped histogram indicates normality.
  - Presence of outliers. A limited number indicates normality.
  - Skewness. For normality, Pearson's index is between -1 and 1.
  - Normal quantile plot. For normality, points lie close to a straight line.
- Control charts are an important application of normal distributions.
- Sampling distributions give us the basis for inferential statistics. By studying the distribution of a sample statistic, we can learn about the corresponding population parameter.
- For random samples of size n, the $\bar{x}$ distribution is the sampling distribution for the sample mean of an x distribution with population mean μ and population standard deviation σ. If the x distribution is normal, then the corresponding $\bar{x}$ distribution is normal.

  By the central limit theorem, when n is sufficiently large ($n \geq 30$), the $\bar{x}$ distribution is approximately normal even if the original x distribution is not normal.

  In both cases,

$$\mu_{\bar{x}} = \mu$$

$$\sigma_{\bar{x}} = \frac{\sigma}{\sqrt{n}}$$

- For n binomial trials with probability of success p on each trial, the $\hat{p}$ distribution is the sampling distribution of the sample proportion of successes. When $np > 5$ and $nq > 5$, the $\hat{p}$ distribution is approximately normal with

$$\mu_{\hat{p}} = p$$

$$\sigma_{\hat{p}} = \sqrt{\frac{pq}{n}}$$

- The binomial distribution can be approximated by a normal distribution with $\mu = np$ and $\sigma = \sqrt{npq}$ provided

$$np > 5 \text{ and } nq > 5, \text{ with } q = 1 - p$$

and a continuity correction is made.

Data from many applications follow distributions that are approximately normal. We will see normal distributions used extensively in later chapters.

IMPORTANT WORDS & SYMBOLS

Section 6.1

Normal distributions 272
Normal curves 272
Downward cup and upward cup on normal curves 273
Symmetry of normal curves 273
Normal density function 274
Empirical rule 274
Control chart 277
Out-of-control signals 278
Uniform distribution 281
Exponential distribution 281

Section 6.2

z value, z score or standard score 289
Standard units 289
Raw score, x 290
Standard normal distribution ($\mu = 0$ and $\sigma = 1$) 291
Area under the standard normal curve 292

Section 6.3

Areas under any normal curve 299
Inverse normal distribution 302
Normality indicators 306

Section 6.4

Sample statistic 314
Population parameter 315
Sampling distribution 315

Section 6.5

$\mu_{\bar{x}}$ 320
$\sigma_{\bar{x}}$ 320
Standard error of the mean 322
Central limit theorem 322
Large sample 322

Section 6.6

Normal approximation to the binomial distribution 333
Continuity correction 335
Sampling distribution for $\hat{p}$ 337
$\mu_{\hat{p}}$ 337
$\sigma_{\hat{p}}$ 337
Standard error of a proportion 337

VIEWPOINT Nenana Ice Classic

The Nenana Ice Classic is a betting pool offering a large cash prize to the lucky winner who can guess the time, to the nearest minute, of the ice breakup on the Tanana River in the town of Nenana, Alaska. Official breakup time is defined as the time when the surging river dislodges a tripod on the ice. This breaks an attached line and stops a clock set to Yukon Standard Time. The event is so popular that the first state legislature of Alaska (1959) made the Nenana Ice Classic an official statewide lottery. Since 1918, the earliest breakup has been April 20, 1940, at 3:27 P.M., and the latest recorded breakup was May 20, 1964, at 11:41 A.M. Want to make a statistical guess predicting when the ice will break up? Breakup times from the years 1918 to 1996 are recorded in The Alaska Almanac, *published by Alaska Northwest Books, Anchorage.*

CHAPTER REVIEW PROBLEMS

1. *Statistical Literacy* Describe a normal probability distribution.
2. *Statistical Literacy* According to the empirical rule, approximately what percentage of the area under a normal distribution lies within 1 standard deviation of the mean? within 2 standard deviations? within 3 standard deviations?
3. *Statistical Literacy* Is a process in control if the corresponding control chart for data having a normal distribution shows a value beyond 3 standard deviations of the mean?
4. *Statistical Literacy* Can a normal distribution always be used to approximate a binomial distribution? Explain.
5. *Statistical Literacy* What characteristic of a normal quantile plot indicates that the data follow a distribution that is approximately normal?

6. *Statistical Literacy* For a normal distribution, is it likely that a data value selected at random is more than 2 standard deviations above the mean?

7. *Statistical Literacy* Give the formula for the *standard error* of the sample mean $\bar{x}$ distribution, based on samples of size n from a distribution with standard deviation σ.

8. *Statistical Literacy* Give the formula for the *standard error* of the sample proportion $\hat{p}$ distribution, based on n binomial trials with probability of success p on each trial.

9. *Critical Thinking* Let x be a random variable representing the amount of sleep each adult in New York City got last night. Consider a sampling distribution of sample means $\bar{x}$.
   (a) As the sample size becomes increasingly large, what distribution does the $\bar{x}$ distribution approach?
   (b) As the sample size becomes increasingly large, what value will the mean $\mu_{\bar{x}}$ of the $\bar{x}$ distribution approach?
   (c) What value will the standard deviation $\sigma_{\bar{x}}$ of the sampling distribution approach?
   (d) How do the two $\bar{x}$ distributions for sample size $n = 50$ and $n = 100$ compare?

10. *Critical Thinking* If x has a normal distribution with mean $\mu = 15$ and standard deviation $\sigma = 3$, describe the distribution of $\bar{x}$ values for sample size n, where $n = 4$, $n = 16$ and $n = 100$. How do the $\bar{x}$ distributions compare for the various sample sizes?

11. *Basic Computation: Probability* Given that x is a normal variable with mean $\mu = 47$ and standard deviation $\sigma = 6.2$, find
    (a) $P(x \leq 60)$ (b) $P(x \geq 50)$ (c) $P(50 \leq x \leq 60)$

12. *Basic Computation: Probability* Given that x is a normal variable with mean $\mu = 110$ and standard deviation $\sigma = 12$, find
    (a) $P(x \leq 120)$ (b) $P(x \geq 80)$ (c) $P(108 \leq x \leq 117)$

13. *Basic Computation: Inverse Normal* Find z such that 5% of the area under the standard normal curve lies to the right of z.

14. *Basic Computation: Inverse Normal* Find z such that 99% of the area under the standard normal curve lies between $-z$ and z.

15. *Medical: Blood Type* Blood type AB is found in only 3% of the population (*Textbook of Medical Physiology,* by A. Guyton, M.D.). If 250 people are chosen at random, what is the probability that
    (a) 5 or more will have this blood type?
    (b) between 5 and 10 will have this blood type?

16. *Customer Complaints: Time* The Customer Service Center in a large New York department store has determined that the amount of time spent with a customer about a complaint is normally distributed, with a mean of 9.3 minutes and a standard deviation of 2.5 minutes. What is the probability that for a randomly chosen customer with a complaint, the amount of time spent resolving the complaint will be
    (a) less than 10 minutes?
    (b) longer than 5 minutes?
    (c) between 8 and 15 minutes?

17. *Recycling: Aluminum Cans* One environmental group did a study of recycling habits in a California community. It found that 70% of the aluminum cans sold in the area were recycled.
    (a) If 400 cans are sold today, what is the probability that 300 or more will be recycled?

(b) Of the 400 cans sold, what is the probability that between 260 and 300 will be recycled?

18. *Guarantee: Disc Players* Future Electronics makes compact disc players. Its research department found that the life of the laser beam device is normally distributed, with mean 5000 hours and standard deviation 450 hours.
(a) Find the probability that the laser beam device will wear out in 5000 hours or less.
(b) *Inverse Normal Distribution* Future Electronics wants to place a guarantee on the players so that no more than 5% fail during the guarantee period. Because the laser pickup is the part most likely to wear out first, the guarantee period will be based on the life of the laser beam device. How many playing hours should the guarantee cover? (Round to the next playing hour.)

19. *Guarantee: Package Delivery* Express Courier Service has found that the delivery time for packages is normally distributed, with mean 14 hours and standard deviation 2 hours.
(a) For a package selected at random, what is the probability that it will be delivered in 18 hours or less?
(b) *Inverse Normal Distribution* What should be the guaranteed delivery time on all packages in order to be 95% sure that the package will be delivered before this time? *Hint:* Note that 5% of the packages will be delivered at a time beyond the guaranteed time period.

20. *Control Chart: Landing Gear* Hydraulic pressure in the main cylinder of the landing gear of a commercial jet is very important for a safe landing. If the pressure is not high enough, the landing gear may not lower properly. If it is too high, the connectors in the hydraulic line may spring a leak.

In-flight landing tests show that the actual pressure in the main cylinders is a variable with mean 819 pounds per square inch and standard deviation 23 pounds per square inch. Assume that these values for the mean and standard deviation are considered safe values by engineers.

Jeff Greenberg/PhotoEdit

(a) For nine consecutive test landings, the pressure in the main cylinder is recorded as follows:

Landing number	1	2	3	4	5	6	7	8	9
Pressure	870	855	830	815	847	836	825	810	792

Make a control chart for the pressure in the main cylinder of the hydraulic landing gear, and plot the data on the control chart. Looking at the control chart, would you say the pressure is "in control" or "out of control"? Explain your answer. Identify any out-of-control signals by type (I, II, or III).

(b) For 10 consecutive test landings, the pressure was recorded on another plane as follows:

Landing number	1	2	3	4	5	6	7	8	9	10
Pressure	865	850	841	820	815	789	801	765	730	725

Make a control chart and plot the data on the chart. Would you say the pressure is "in control" or not? Explain your answer. Identify any out-of-control signals by type (I, II, or III).

21. *Job Interview: Length* The personnel office at a large electronics firm regularly schedules job interviews and maintains records of the interviews. From the past records, they have found that the length of a first interview is normally distributed, with mean $\mu = 35$ minutes and standard deviation $\sigma = 7$ minutes.
(a) What is the probability that a first interview will last 40 minutes or longer?
(b) Nine first interviews are usually scheduled per day. What is the probability that the average length of time for the nine interviews will be 40 minutes or longer?

22. *Drugs: Effects* A new muscle relaxant is available. Researchers from the firm developing the relaxant have done studies that indicate that the time lapse between administration of the drug and beginning effects of the drug is normally distributed, with mean $\mu = 38$ minutes and standard deviation $\sigma = 5$ minutes.
    (a) The drug is administered to one patient selected at random. What is the probability that the time it takes to go into effect is 35 minutes or less?
    (b) The drug is administered to a random sample of 10 patients. What is the probability that the average time before it is effective for all 10 patients is 35 minutes or less?
    (c) Comment on the differences of the results in parts (a) and (b).

23. *Psychology: IQ Scores* Assume that IQ scores are normally distributed, with a standard deviation of 15 points and a mean of 100 points. If 100 people are chosen at random, what is the probability that the sample mean of IQ scores will not differ from the population mean by more than 2 points?

24. *Hatchery Fish: Length* A large tank of fish from a hatchery is being delivered to a lake. The hatchery claims that the mean length of fish in the tank is 15 inches, and the standard deviation is 2 inches. A random sample of 36 fish is taken from the tank. Let $\bar{x}$ be the mean sample length of these fish. What is the probability that $\bar{x}$ is within 0.5 inch of the claimed population mean?

25. *Basic Computation: $\hat{p}$ Distribution* Suppose we have a binomial distribution with $n = 24$ trials and probability of success $p = 0.4$ on each trial. The sample proportion of successes is $\hat{p} = r/n$.
    (a) Is it appropriate to approximate the $\hat{p}$ distribution with a normal distribution? Explain.
    (b) What are the values of $\mu_{\hat{p}}$ and $\sigma_{\hat{p}}$?

26. *Green Behavior: Purchasing Habits* A recent Harris Poll on green behavior showed that 25% of adults often purchase used items instead of new ones. Consider a random sample of 75 adults. Let $\hat{p}$ be the sample proportion of adults who often purchase used instead of new items.
    (a) Is it appropriate to approximate the $\hat{p}$ distribution with a normal distribution? Explain.
    (b) What are the values of $\mu_{\hat{p}}$ and $\sigma_{\hat{p}}$?

DATA HIGHLIGHTS: GROUP PROJECTS

Neil Burton/Shutterstock.com

Wild iris

Break into small groups and discuss the following topics. Organize a brief outline in which you summarize the main points of your group discussion.

Iris setosa is a beautiful wildflower that is found in such diverse places as Alaska, the Gulf of St. Lawrence, much of North America, and even in English meadows and parks. R. A. Fisher, with his colleague Dr. Edgar Anderson, studied these flowers extensively. Dr. Anderson described how he collected information on irises:

> I have studied such irises as I could get to see, in as great detail as possible, measuring iris standard after iris standard and iris fall after iris fall, sitting squat-legged with record book and ruler in mountain meadows, in cypress swamps, on lake beaches, and in English parks. [E. Anderson, "The Irises of the Gaspé Peninsula," *Bulletin, American Iris Society*, Vol. 59 pp. 2–5, 1935.]

The data in Table 6-11 were collected by Dr. Anderson and were published by his friend and colleague R. A. Fisher in a paper titled "The Use of Multiple Measurements in Taxonomic Problems" (*Annals of Eugenics,* part II, pp. 179–188, 1936). To find these data, visit the Brase/Brase statistics site at **http://www.cengage.com/statistics/brase11e** and find the link to DASL, the Carnegie Mellon University Data and Story Library. From the DASL site, look under famous data sets.

Let x be a random variable representing petal length. Using a TI-84Plus/TI-83Plus/TI-*n*spire calculator, it was found that the sample mean is $\bar{x} = 1.46$ centimeters (cm) and the sample standard deviation is $s = 0.17$ cm. Figure 6-39 shows a histogram for the given data generated on a TI-84Plus/TI-83Plus/TI-*n*spire calculator.

(a) Examine the histogram for petal lengths. Would you say that the distribution is approximately mound-shaped and symmetrical? Our sample has only 50 irises; if many thousands of irises had been used, do you think the distribution would look even more like a normal curve? Let x be the petal length of *Iris setosa*. Research has shown that x has an approximately normal distribution, with mean $\mu = 1.5$ cm and standard deviation $\sigma = 0.2$ cm.

(b) Use the empirical rule with $\mu = 1.5$ and $\sigma = 0.2$ to get an interval into which approximately 68% of the petal lengths will fall. Repeat this for 95% and 99.7%. Examine the raw data and compute the percentage of the raw data that actually fall into each of these intervals (the 68% interval, the 95% interval, and the 99.7% interval). Compare your computed percentages with those given by the empirical rule.

(c) Compute the probability that a petal length is between 1.3 and 1.6 cm. Compute the probability that a petal length is greater than 1.6 cm.

(d) Suppose that a random sample of 30 irises is obtained. Compute the probability that the average petal length for this sample is between 1.3 and 1.6 cm. Compute the probability that the average petal length is greater than 1.6 cm.

(e) Compare your answers to parts (c) and (d). Do you notice any differences? Why would these differences occur?

TABLE 6-11 **Petal Length in Centimeters for *Iris setosa***

1.4	1.4	1.3	1.5	1.4
1.7	1.4	1.5	1.4	1.5
1.5	1.6	1.4	1.1	1.2
1.5	1.3	1.4	1.7	1.5
1.7	1.5	1	1.7	1.9
1.6	1.6	1.5	1.4	1.6
1.6	1.5	1.5	1.4	1.5
1.2	1.3	1.4	1.3	1.5
1.3	1.3	1.3	1.6	1.9
1.4	1.6	1.4	1.5	1.4

FIGURE 6-39

Petal Length (cm) for *Iris setosa* (TI-84Plus/ TI-83Plus/TI-*n*spire)

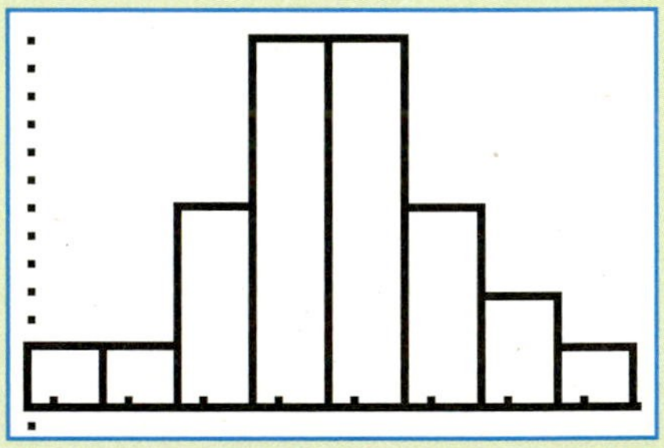

LINKING CONCEPTS: WRITING PROJECTS

Discuss each of the following topics in class or review the topics on your own. Then write a brief but complete essay in which you summarize the main points. Please include formulas and graphs as appropriate.

1. If you look up the word *empirical* in a dictionary, you will find that it means "relying on experiment and observation rather than on theory." Discuss the empirical rule in this context. The empirical rule certainly applies to the normal distribution, but does it also apply to a wide variety of other distributions that are not *exactly* (theoretically) normal? Discuss the terms *mound-shaped* and *symmetrical*. Draw several sketches of distributions that are mound-shaped *and* symmetrical. Draw sketches of distributions that are not mound-shaped or symmetrical. To which distributions will the empirical rule apply?
2. Why are standard z values so important? Is it true that z values have no units of measurement? Why would this be desirable for comparing data sets with

different units of measurement? How can we assess differences in quality or performance by simply comparing z values under a standard normal curve? Examine the formula for computing standard z values. Notice that it involves *both* the mean and the standard deviation. Recall that in Chapter 3 we commented that the mean of a data collection is not entirely adequate to describe the data; you need the standard deviation as well. Discuss this topic again in light of what you now know about normal distributions and standard z values.

3. Most companies that manufacture a product have a division responsible for quality control or quality assurance. The purpose of the quality-control division is to make reasonably certain that the products manufactured are up to company standards. Write a brief essay in which you describe how the statistics you have learned so far could be applied to an industrial application (such as control charts and the Antlers Lodge example).

4. Most people would agree that increased information should give better predictions. Discuss how sampling distributions actually enable better predictions by providing more information. Examine Theorem 6.1 again. Suppose that x is a random variable with a *normal* distribution. Then $\bar{x}$, the sample mean based on random samples of size n, also will have a normal distribution for *any* value of $n = 1, 2, 3, \ldots$.

   What happens to the standard deviation of the $\bar{x}$ distribution as n (the sample size) increases? Consider the following table for different values of n.

n	1	2	3	4	10	50	100
$\sigma/\sqrt{n}$,	1σ	0.71σ	0.58σ	0.50σ	0.32σ	0.14σ	0.10σ

   In this case, "increased information" means a larger sample size n. Give a brief explanation as to why a *large* standard deviation will usually result in poor statistical predictions, whereas a *small* standard deviation usually results in much better predictions. Since the standard deviation of the sampling distribution $\bar{x}$ is $\sigma/\sqrt{n}$, we can decrease the standard deviation by increasing n. In fact, if we look at the preceding table, we see that if we use a sample size of only $n = 4$, we cut the standard deviation of $\bar{x}$ by 50% of the standard deviation σ of x. If we were to use a sample size of $n = 100$, we would cut the standard deviation of $\bar{x}$ to 10% of the standard deviation σ of x.

   Give the preceding discussion some thought and explain why you should get much better predictions for μ by using $\bar{x}$ from a sample of size n rather than by just using x. Write a brief essay in which you explain why sampling distributions are an important tool in statistics.

5. In a way, the central limit theorem can be thought of as a kind of "grand central station." It is a connecting hub or center for a great deal of statistical work. We will use it extensively in Chapters 7, 8, and 9. Put in a very elementary way, the central limit theorem states that as the sample size n increases, the distribution of the sample mean $\bar{x}$ will always approach a normal distribution, no matter where the original x variable came from. For most people, it is the complete generality of the central limit theorem that is so awe inspiring: It applies to practically everything. List and discuss at least three variables from everyday life for which you expect the variable x itself *not* to follow a normal or bell-shaped distribution. Then discuss what would happen to the sampling distribution $\bar{x}$ if the sample size were increased. Sketch diagrams of the $\bar{x}$ distributions as the sample size n increases.

USING TECHNOLOGY

Application 1

How can we determine if data originated from a normal distribution? We can look at a stem-and-leaf plot or histogram of the data to check for general symmetry, skewness, clusters of data, or outliers. However, a more sensitive way to check that a distribution is normal is to look at a special graph called a *normal quantile plot* (or a variation of this plot called a *normal probability plot* in some software packages). It really is not feasible to make a normal quantile plot by hand, but statistical software packages provide such plots. A simple version of the basic idea behind normal quantile plots involves the following process:

(a) Arrange the observed data values in order from smallest to largest, and determine the percentile occupied by each value. For instance, if there are 20 data values, the smallest datum is at the 5% point, the next smallest is at the 10% point, and so on.

(b) Find the z values that correspond to the percentile points. For instance, the z value that corresponds to the percentile 5% (i.e., percent in the left tail of the distribution) is $z = -1.645$.

(c) Plot each data value x against the corresponding percentile z score. If the data are close to a normal distribution, the plotted points will lie close to a straight line. (If the data are close to a standard normal distribution, the points will lie close to the line $x = z$.)

The actual process that statistical software packages use to produce the z scores for the data is more complicated.

INTERPRETING NORMAL QUANTILE PLOTS

If the points of a normal quantile plot lie close to a straight line, the plot indicates that the data follow a normal distribution. Systematic deviations from a straight line or bulges in the plot indicate that the data distribution is not normal. Individual points off the line may be outliers.

Consider Figure 6-40. This figure shows Minitab-generated quantile plots for two data sets. The black dots show the normal quantile plot for monthly salary data. The red dots show the normal quantile plot for a random sample of 42 data values drawn from a theoretical normal distribution with the same mean and standard deviation as the salary data ($\mu \approx 3421$, $\sigma \approx 709$).

FIGURE 6-40 Normal Quantile Plots

- Salary data for (city) government employees
- A random sample of 42 values from a theoretical normal distribution with the same mean and standard deviation as the salary data

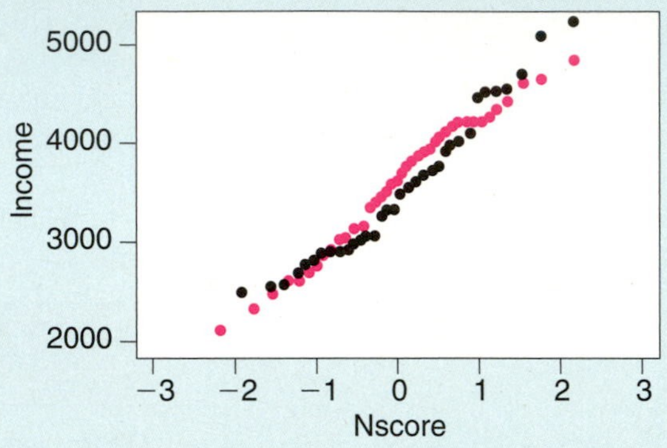

(a) Do the black dots lie close to a straight line? Do the salaries appear to follow a normal distribution? Are there any outliers on the low or high side? Would you say that any of the salaries are "out of line" for a normal distribution?

(b) Do the red dots lie close to a straight line? We know the red dots represent a sample drawn from a normal distribution. Is the normal quantile plot for the red dots consistent with this fact? Are there any outliers shown?

Technology Hints

TI-84Plus/TI-83Plus/TI-*n*spire

Enter the data in list **L1** and corresponding z scores in list **L2**. Press **STATPLOT** and select one of the plots. Highlight **ON**. Then highlight the sixth plot option. To get a plot similar to that of Figure 6-40, choose Y as the data axis.

Minitab

Minitab has several types of normal quantile plots that use different types of scales. To create a normal quantile plot similar to that of Figure 6-40, enter the data in column C1. Then use the menu choices **Calc ➤ Calculator.** In the dialogue box listing the functions, scroll to **Normal Scores.** Use **NSCOR(C1)** and store the results in column C2. Finally, use the menu choices **Graph ➤ Plot.** In the dialogue box, use C1 for variable y and C2 for variable x.

SPSS

Enter the data. Use the menu choices **Analyze ➤ Descriptive Statistics ➤ Explore.** In the dialogue box, move your data variable to the dependent list. Click **Plots . . . .** Check "Normality plots with tests." The graph appears in the output window.

Application 2

As we have seen in this chapter, the value of a sample statistic such as $\bar{x}$ varies from one sample to another. The central limit theorem describes the distribution of the sample statistic $\bar{x}$ when samples are sufficiently large.

We can use technology tools to generate samples of the same size from the same population. Then we can look at the statistic $\bar{x}$ for each sample, and the resulting $\bar{x}$ distribution.

Project Illustrating the Central Limit Theorem

Step 1: Generate random samples of specified size n from a population.

The random-number table enables us to sample from the uniform distribution of digits 0 through 9. Use either the random-number table or a random-number generator to generate 30 samples of size 10.

Step 2: Compute the sample mean $\bar{x}$ of the digits in each sample.

Step 3: Compute the sample mean of the means (i.e., $\bar{x}_{\bar{x}}$) as well as the standard deviation $s_{\bar{x}}$ of the sample means.

The population mean of the uniform distribution of digits from 0 through 9 is 4.5. How does $\bar{x}_{\bar{x}}$ compare to this value?

Step 4: Compare the sample distribution of $\bar{x}$ values to a normal distribution having the mean and standard deviation computed in Step 3.

(a) Use the values of $\bar{x}_{\bar{x}}$ and $s_{\bar{x}}$ computed in Step 3 to create the intervals shown in column 1 of Table 6-12.

(b) Tally the sample means computed in Step 2 to determine how many fall into each interval of column 2. Then compute the percent of data in each interval and record the results in column 3.

(c) The percentages listed in column 4 are those from a normal distribution (see Figure 6-3 showing the empirical rule). Compare the percentages in column 3 to those in column 4. How do the sample percentages compare with the hypothetical normal distribution?

Step 5: Create a histogram showing the sample means computed in Step 2.

Look at the histogram and compare it to a normal distribution with the mean and standard deviation of the $\bar{x}$s (as computed in Step 3).

Step 6: Compare the results of this project to the central limit theorem.

Increase the sample size of Step 1 to 20, 30, and 40 and repeat Steps 1 to 5.

TABLE 6-12 Frequency Table of Sample Means

1. Interval	2. Frequency	3. Percent	4. Hypothetical Normal Distribution
$\bar{x} - 3s$ to $\bar{x} - 2s$	Tally the sample	Compute	2 or 3%
$\bar{x} - 2s$ to $\bar{x} - s$	means computed	percents	13 or 14%
$\bar{x} - s$ to $\bar{x}$	in Step 2 and	from	About 34%
$\bar{x}$ to $\bar{x} + s$	place here.	column 2	About 34%
$\bar{x} + s$ to $\bar{x} + 2s$		and place	13 or 14%
$\bar{x} + 2s$ to $\bar{x} + 3s$		here.	2 or 3%

FIGURE 6-41
SPSS-Generated Histograms for Samples of Size 30 and Size 100

(a) $n = 30$

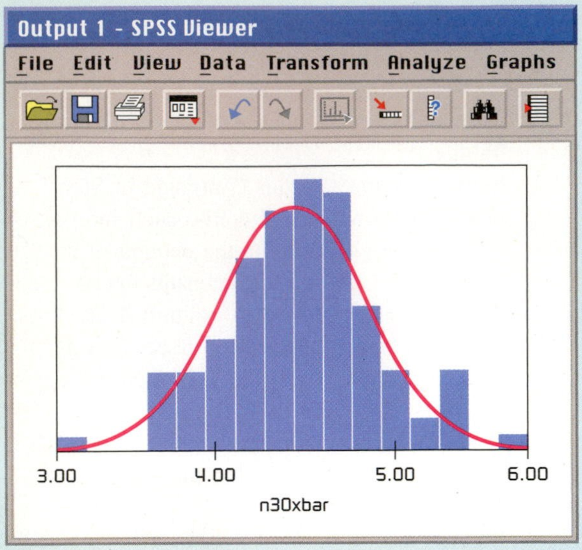

(b) $n = 100$

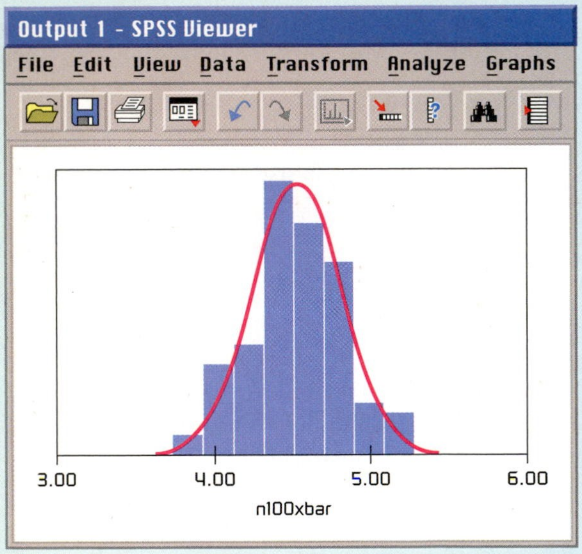

Technology Hints

The TI-84Plus/TI-83Plus/TI-*n*spire calculators, Excel 2010, Minitab, and SPSS all support the process of drawing random samples from a variety of distributions. Macros can be written in Excel 2010, Minitab, and the professional version of SPSS to repeat the six steps of the project. Figure 6-41 shows histograms generated by SPSS for random samples of size 30 and size 100. The samples are taken from a uniform probability distribution.

TI-84Plus/TI-83Plus/TI-*n*spire

You can generate random samples from uniform, normal, and binomial distributions. Press **MATH** and select **PRB.** Selection **5:randInt(lower, upper, sample size m)** generates m random integers from the specified interval. Selection **6:randNorm(**μ, σ, sample size m) generates m random numbers from a normal distribution with mean μ and standard deviation σ. Selection **7:randBin(number of trials n, p, sample size m)** generates m random values (number of successes out of n trials) for a binomial distribution with probability of success p on each trial. You can put these values in lists by using **Edit** under **Stat.** Highlight the list header, press Enter, and then select one of the options discussed.

Excel 2010

On the Home screen, click on the **Data** tab. In the **Analysis** group, select **Data Analysis**. In the dialogue box, select **Random Number Generator.** The next dialogue box provides choices for the population distribution, including uniform, binomial, and normal distributions. Fill in the required parameters and designate the location for the output.

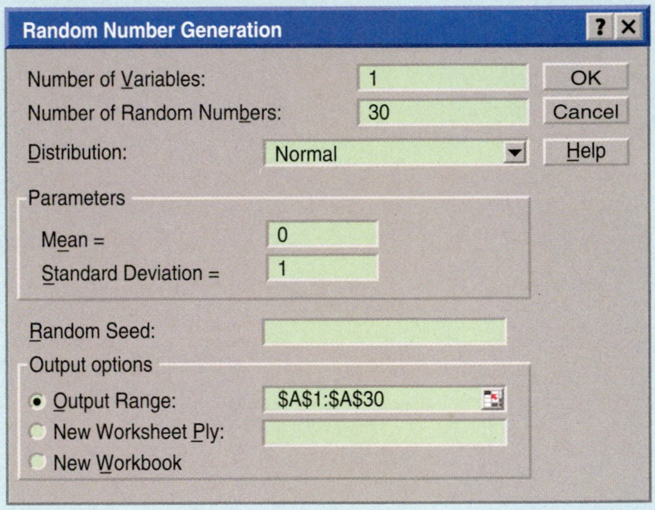

Minitab

Use the menu selections **Calc ➤ Random Data.** Then select the population distribution. The choices include uniform, binomial, and normal distributions. Fill in the dialogue box, where the number of rows indicates the number of data in the sample.

SPSS

SPSS supports random samples from a variety of distributions, including binomial, normal, and uniform. In data view, generate a column of consecutive integers from 1 to *n,* where *n* is the sample size. In variable view, name the variables sample1, sample2, and so on, through sample30. These variables head the columns containing each of the 30 samples of size *n.* Then use the menu choices **Transform ➤ Compute.** In the dialogue box, use sample1 as the target variable for the first sample, and so forth.

In the function group, select **Random Numbers** and in the functions and special variables group, select **Rv.Uniform** for samples from a uniform distribution. Functions **Rv.Normal** and **Rv.Binom** provide random samples from normal and binomial distributions, respectively. For each function, the necessary parameters are described.

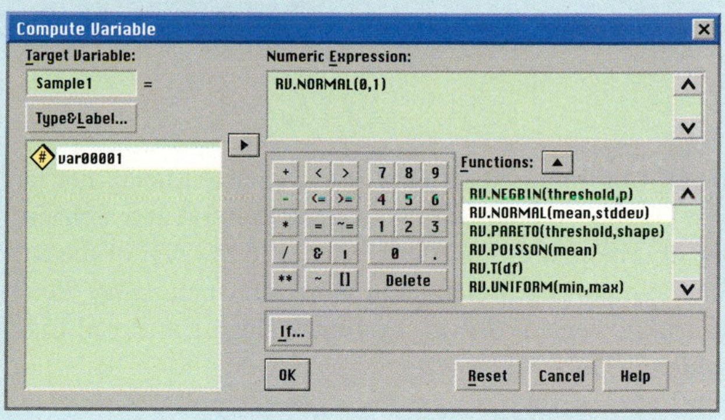

Hubert Stadler/Corbis

CUMULATIVE REVIEW PROBLEMS

CHAPTER 4-6

The Hill of Tara is located in south-central Meath, not far from Dublin, Ireland. Tara is of great cultural and archaeological importance, since it is by legend the seat of the ancient high kings of Ireland. For more information, see *Tara: An Archaeological Survey,* by Conor Newman, Royal Irish Academy, Dublin.

Hubert Stadler/Corbis

Magnetic surveying is one technique used by archaeologists to determine anomalies arising from variations in magnetic susceptibility. Unusual changes in magnetic susceptibility might (or might not) indicate an important archaeological discovery. Let x be a random variable that represents a magnetic susceptibility (MS) reading for a randomly chosen site on the Hill of Tara. A random sample of 120 sites gave the readings shown in Table A below.

TABLE A **Magnetic Susceptibility Readings, centimeter-gram-second $\times 10^{-6}$ (cmg $\times 10^{-6}$)**

Comment	Magnetic Susceptibility	Number of Readings	Estimated Probability
"cool"	$0 \le x < 10$	30	$30/120 = 0.25$
"neutral"	$10 \le x < 20$	54	$54/120 = 0.45$
"warm"	$20 \le x < 30$	18	$18/20 = 0.15$
"very interesting"	$30 \le x < 40$	12	$12/120 = 0.10$
"hot spot"	$40 \le x$	6	$6/120 = 0.05$

Answers may vary slightly due to rounding.

1. *Statistical Literacy: Sample Space* What is a statistical experiment? How could the magnetic susceptibility intervals $0 \le x < 10$, $10 \le x < 20$, and so on, be considered events in the sample space of all possible readings?

2. *Statistical Literacy: Probability* What is probability? What do we mean by relative frequency as a probability estimate for events? What is the law of large numbers? How would the law of large numbers apply in this context?

3. *Statistical Literacy: Probability Distribution* Do the probabilities shown in Table A add up to 1? Why should they total to 1?

4. *Probability Rules* For a site chosen at random, estimate the following probabilities.
   (a) $P(0 \le x < 30)$ (b) $P(10 \le x < 40)$
   (c) $P(x < 20)$ (d) $P(x \ge 20)$
   (e) $P(30 \le x)$ (f) P(x not less than 10)
   (g) $P(0 \le x < 10 \text{ or } 40 \le x)$
   (h) $P(40 \le x \text{ and } 20 \le x)$

5. *Conditional Probability* Suppose you are working in a "warm" region in which all MS readings are 20 or higher. In this same region, what is the probability that you will find a "hot spot" in which the readings are 40 or higher? Use conditional probability to estimate $P(40 \le x \mid 20 \le x)$. *Hint*: See Problem 39 of Section 6.3.

6. *Discrete Probability Distribution* Consider the midpoint of each interval. Assign the value 45 as the midpoint for the interval $40 \le x$. The midpoints constitute the sample space for a discrete random variable. Using Table A, compute the expected value μ and the standard deviation σ.

Midpoint x	5	15	25	35	45
$P(x)$					

7. *Binomial Distribution* Suppose a reading between 30 and 40 is called "very interesting" from an archaeological point of view. Let us say you take readings at $n = 12$ sites chosen at random. Let r be a binomial random variable that represents the number of "very interesting" readings from these 12 sites.
   (a) Let us call "very interesting" a binomial success. Use Table A to find p, the probability of success on a single trial, where $p = P(\text{success}) = P(30 \leq x < 40)$.
   (b) What is the expected value μ and standard deviation σ for the random variable r?
   (c) What is the probability that you will find *at least* one "very interesting" reading in the 12 sites?
   (d) What is the probability that you will find *fewer than* three "very interesting" readings in the 12 sites?

8. *Geometric Distribution* Suppose a "hot spot" is a site with a reading of 40 or higher.
   (a) In a binomial setting, let us call success a "hot spot." Use Table A to find $p = P(\text{success}) = P(40 \leq x)$ for a single trial.
   (b) Suppose you decide to take readings at random until you get your *first* "hot spot." Let n be a random variable representing the trial on which you get your first "hot spot." Use the geometric probability distribution to write out a formula for $P(n)$.
   (c) What is the probability that you will need more than four readings to find the first "hot spot"? Compute $P(n > 4)$.

9. *Poisson Approximation to the Binomial* Suppose an archaeologist is looking for geomagnetic "hot spots" in an unexplored region of Tara. As in Problem 8, we have a binomial setting where success is a "hot spot." In this case, the probability of success is $p = P(40 \leq x)$. The archaeologist takes $n = 100$ magnetic susceptibility readings in the new, unexplored area. Let r be a binomial random variable representing the number of "hot spots" in the 100 readings.
   (a) We want to approximate the binomial random variable r by a Poisson distribution. Is this appropriate? What requirements must be satisfied before we can do this? Do you think these requirements are satisfied in this case? Explain. What is the value of λ?
   (b) What is the probability that the archaeologists will find six or fewer "hot spots?" *Hint*: Use Table 4 of Appendix II.
   (c) What is the probability that the archaeologists will find more than eight "hot spots"?

10. *Normal Approximation to the Binomial* Consider a binomial setting in which "neutral" is defined to be a success. So, $p = P(\text{success}) = P(10 \leq x < 20)$. Suppose $n = 65$ geomagnetic readings are taken. Let r be a binomial random variable that represents the number of "neutral" geomagnetic readings.
   (a) We want to approximate the binomial random variable r by a normal variable x. Is this appropriate? What requirements must be satisfied before we can do this? Do you think these requirements are satisfied in this case? Explain.
   (b) What is the probability that there will be at least 20 "neutral" readings out of these 65 trials?
   (c) Why would the Poisson approximation to the binomial *not* be appropriate in this case? Explain.

11. *Normal Distribution* *Oxygen demand* is a term biologists use to describe the oxygen needed by fish and other aquatic organisms for survival. The Environmental Protection Agency conducted a study of a wetland area in Marin County, California. In this wetland environment, the mean oxygen demand was $\mu = 9.9$ mg/L with 95% of the data ranging from 6.5 mg/L to 13.3 mg/L (Reference: EPA Report 832-R-93-005). Let x be a random variable that represents oxygen demand in this wetland environment. Assume x has a probability distribution that is approximately normal.
   (a) Use the 95% data range to estimate the standard deviation for oxygen demand. *Hint:* See Problem 31 of Section 6.3.
   (b) An oxygen demand below 8 indicates that some organisms in the wetland environment may be dying. What is the probability that the oxygen demand will fall below 8 mg/L?
   (c) A high oxygen demand can also indicate trouble. An oxygen demand above 12 may indicate an overabundance of organisms that endanger some types of plant life. What is the probability that the oxygen demand will exceed 12 mg/L?

12. *Statistical Literacy* Please give a careful but brief answer to each of the following questions.
   (a) What is a population? How do you get a simple random sample? Give examples.
   (b) What is a sample statistic? What is a sampling distribution? Give examples.
   (c) Give a careful and complete statement of the central limit theorem.
   (d) List at least three areas of everyday life to which the above concepts can be applied. Be specific.

13. *Sampling Distribution* $\bar{x}$ Workers at a large toxic cleanup project are concerned that their white blood cell counts may have been reduced. Let x be a random variable that represents white blood cell count per cubic millimeter of whole blood in a healthy adult. Then $\mu = 7500$ and $\sigma \approx 1750$ (Reference: *Diagnostic Tests with Nursing Applications*, S. Loeb). A random sample of $n = 50$ workers from the toxic cleanup site were given a blood test that showed $\bar{x} = 6820$. What is the probability that, for healthy adults, $\bar{x}$ will be this low or lower?
    (a) How does the central limit theorem apply? Explain.
    (b) Compute $P(\bar{x} \leq 6820)$.
    (c) *Interpretation* Based on your answer to part (b), would you recommend that additional facts be obtained, or would you recommend that the workers' concerns be dismissed? Explain.

14. *Sampling Distribution* $\hat{p}$ Do you have a great deal of confidence in the advice given to you by your medical doctor? About 45% of all adult Americans claim they do have a great deal of confidence in their M.D.s (Reference: *National Opinion Research Center*, University of Chicago). Suppose a random sample of $n = 32$ adults in a health insurance program are asked about their confidence in the medical advice their doctors give.
    (a) Is the normal approximation to the proportion $\hat{p} = r/n$ valid?
    (b) Find the values of $\mu_{\hat{p}}$ and $\sigma_{\hat{p}}$.

15. *Summary* Write a brief but complete essay in which you describe the probability distributions you have studied so far. Which apply to discrete random variables? Which apply to continuous random variables? Under what conditions can the binomial distribution be approximated by the normal? by the Poisson?

7

We dance round in a ring and suppose,
But the Secret sits in the middle and knows.

—Robert Frost, "The Secret Sits"*

In Chapter 1, we said that statistics is the study of how to collect, organize, analyze, and interpret numerical data. That part of statistics concerned with analysis, interpretation, and forming conclusions about the source of the data is called *statistical inference*. Problems of statistical inference require us to draw a *random sample* of observations from a larger *population*. A sample usually contains incomplete information, so in a sense we must "dance round in a ring and suppose," to quote the words of the celebrated American poet Robert Lee Frost (1874–1963).

Nevertheless, conclusions about the population can be obtained from sample data by the use of statistical estimates. This chapter introduces you to several widely used methods of estimation.

*"The Secret Sits," from *The Poetry of Robert Frost*, edited by Edward Connery Lathem. Copyright 1942 by Robert Frost, © 1970 by Lesley Frost Ballantine, © 1969 by Henry Holt and Company, Inc. Reprinted by permission of Henry Holt and Company, Inc.

ESTIMATION

Bobby Deal/RealDealPhoto/
Shutterstock.com

PREVIEW QUESTIONS

How do you estimate the expected value of a random variable? What assumptions are needed? How much confidence should be placed in such estimates? (SECTION 7.1)

At the beginning design stage of a statistical project, how large a sample size should you plan to get? (SECTION 7.1)

What famous statistician worked for Guinness brewing company in Ireland? What has this got to do with constructing estimates from sample data? (SECTION 7.2)

How do you estimate the proportion p of successes in a binomial experiment? How does the normal approximation fit into this process? (SECTION 7.3)

Sometimes even small differences can be extremely important. How do you estimate differences? (SECTION 7.4)

FOCUS PROBLEM

The Trouble with Wood Ducks

Photo Researchers

The National Wildlife Federation published an article titled "The Trouble with Wood Ducks" (*National Wildlife*, Vol. 31, No. 5). In this article, wood ducks are described as beautiful birds living in forested areas such as the Pacific Northwest and southeast United States. Because of overhunting and habitat destruction, these birds were in danger of extinction. A federal ban on hunting wood ducks in 1918 helped save the species from extinction. Wood ducks like to nest in tree cavities. However, many such trees were disappearing due to heavy timber cutting. For a period of time it seemed that nesting boxes were the solution to disappearing trees. At first, the wood duck population grew, but after a few seasons, the population declined sharply. Good biology research combined with good statistics provided an answer to this disturbing phenomenon.

Cornell University professors of ecology Paul Sherman and Brad Semel found that the nesting boxes were placed too close to each other. Female wood ducks prefer a secluded nest that is a considerable distance from the next wood duck nest. In fact, female wood duck behavior changed when the nests were too close to each other. Some females would lay their eggs in another female's nest. The result was too many eggs in one nest. The biologists found that if there were too many eggs in a nest, the proportion of eggs that hatched was considerably reduced. In the long run, this meant a decline in the population of wood ducks.

For online student resources, visit the Brase/Brase, *Understandable Statistics*, 11th edition web site at **http://www.cengage.com/statistics/brase11e**

In their study, Sherman and Semel used two placements of nesting boxes. Group I boxes were well separated from each other and well hidden by available brush. Group II boxes were highly visible and grouped closely together.

In group I boxes, there were a total of 474 eggs, of which a field count showed that about 270 hatched. In group II boxes, there were a total of 805 eggs, of which a field count showed that, again, about 270 hatched.

The material in Chapter 7 will enable us to answer many questions about the hatch ratios of eggs from nests in the two groups.

(a) Find a point estimate $\hat{p}_1$ for p_1, the proportion of eggs that hatch in group I nest box placements. Find a 95% confidence interval for p_1.
(b) Find a point estimate $\hat{p}_2$ for p_2, the proportion of eggs that hatch in group II nest box placements. Find a 95% confidence interval for p_2.
(c) Find a 95% confidence interval for $p_1 - p_2$. Does the interval indicate that the proportion of eggs hatched from group I nest box placements is higher than, lower than, or equal to the proportion of eggs hatched from group II nest boxes?
(d) What conclusions about placement of nest boxes can be drawn? In the article, additional concerns are raised about the higher cost of placing and maintaining group I nest boxes. Also at issue is the cost efficiency per successful wood duck hatch. Data in the article do not include information that would help us answer questions of cost efficiency. However, the data presented do help us answer questions about the proportions of successful hatches in the two nest box configurations. (See Problem 26 of Section 7.4.)

SECTION **7.1**

Estimating μ When σ Is Known

FOCUS POINTS

- Explain the meanings of confidence level, error of estimate, and critical value.
- Find the critical value corresponding to a given confidence level.
- Compute confidence intervals for μ when σ is known. Interpret the results.
- Compute the sample size to be used for estimating a mean μ.

Because of time and money constraints, difficulty in finding population members, and so forth, we usually do not have access to *all* measurements of an *entire* population. Instead we rely on information from a sample.

In this section, we develop techniques for estimating the population mean μ using sample data. We assume the population standard deviation σ is known.

Let's begin by listing some basic assumptions used in the development of our formulas for estimating μ when σ is known.

ASSUMPTIONS ABOUT THE RANDOM VARIABLE x

1. We have a *simple random sample* of size n drawn from a population of x values.
2. The value of σ, the population standard deviation of x, *is known*.
3. If the *x distribution is normal*, then our methods work for *any sample size n*.
4. If x has an unknown distribution, then we require a *sample size $n \geq 30$*. However, if the x distribution is distinctly skewed and definitely not mound-shaped, a sample of size 50 or even 100 or higher may be necessary.

Point estimate

An estimate of a population parameter given by a single number is called a *point estimate* for that parameter. It will come as no great surprise that we use $\bar{x}$ (the sample mean) as the point estimate for μ (the population mean).

A **point estimate** of a population parameter is an estimate of the parameter using a single number.

$\bar{x}$ is the **point estimate** for μ.

Margin of error

Even with a large random sample, the value of $\bar{x}$ usually is not *exactly* equal to the population mean μ. The *margin of error* is the magnitude of the difference between the sample point estimate and the true population parameter value.

When using $\bar{x}$ as a point estimate for μ, the **margin of error** is the magnitude of $\bar{x} - \mu$ or $|\bar{x} - \mu|$.

Confidence level, c

FIGURE 7-1

Confidence Level c and Corresponding Critical Value z_c Shown on the Standard Normal Curve

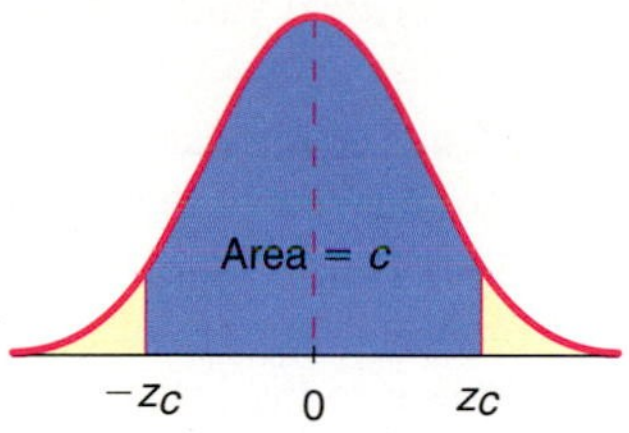

We cannot say exactly how close $\bar{x}$ is to μ when μ is unknown. Therefore, the exact margin of error is unknown when the population parameter is unknown. Of course, μ is usually not known, or there would be no need to estimate it. In this section, we will use the language of probability to give us an idea of the size of the margin of error when we use $\bar{x}$ as a point estimate for μ.

First, we need to learn about *confidence levels*. The reliability of an estimate will be measured by the confidence level.

Suppose we want a confidence level of c (see Figure 7-1). Theoretically, we can choose c to be any value between 0 and 1, but usually c is equal to a number such as 0.90, 0.95, or 0.99. In each case, the value z_c is the number such that the area under the standard normal curve falling between $-z_c$ and z_c is equal to c. The value z_c is called the *critical value* for a confidence level of c.

Critical value, z_c

For a confidence level c, the **critical value z_c** is the number such that the area under the standard normal curve between $-z_c$ and z_c equals c.

The area under the normal curve from $-z_c$ to z_c is the probability that the standardized normal variable z lies in that interval. This means that

$$P(-z_c < z < z_c) = c$$

EXAMPLE 1

FIND A CRITICAL VALUE

Let us use Table 5 of Appendix II to find a number $z_{0.99}$ such that 99% of the area under the standard normal curve lies between $-z_{0.99}$ and $z_{0.99}$. That is, we will find $z_{0.99}$ such that

$$P(-z_{0.99} < z < z_{0.99}) = 0.99$$

FIGURE 7-2

Area Between $-z$ and z Is 0.99

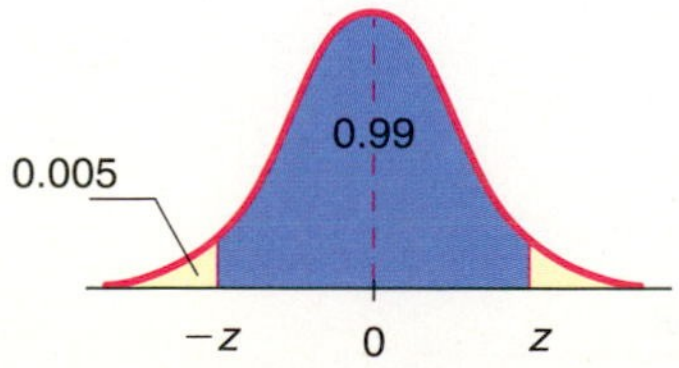

SOLUTION: In Section 6.3, we saw how to find the z value when we were given an area between $-z$ and z. The first thing we did was to find the corresponding area to the left of $-z$. If A is the area between $-z$ and z, then $(1 - A)/2$ is the area to the left of z. In our case, the area between $-z$ and z is 0.99. The corresponding area in the left tail is $(1 - 0.99)/2 = 0.005$ (see Figure 7-2).

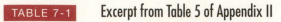

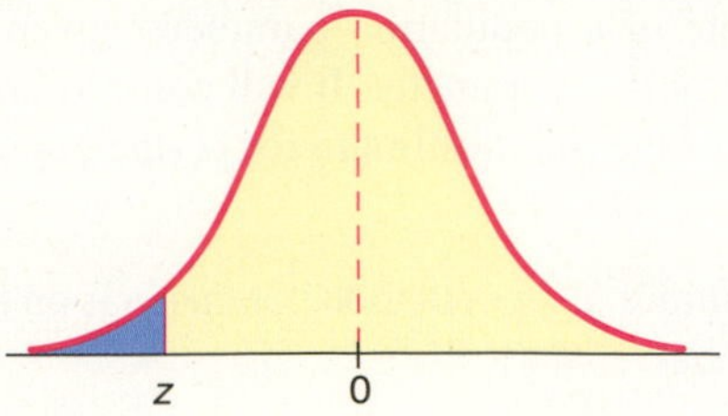

TABLE 7-1 **Excerpt from Table 5 of Appendix II**

z	.00	...	.07	.08	.09
−3.4	.0003		0003	.0003	.0002
⋮					
−2.5	.0062		.0051	.0049	.0048

↑
.0050

Next, we use Table 5 of Appendix II to find the z value corresponding to a left-tail area of 0.0050. Table 7-1 shows an excerpt from Table 5 of Appendix II.

From Table 7-1, we see that the desired area, 0.0050, is exactly halfway between the areas corresponding to $z = -2.58$ and $z = -2.57$. Because the two area values are so close together, we use the more conservative z value -2.58 rather than interpolate. In fact, $z_{0.99} \approx 2.576$. However, to two decimal places, we use $z_{0.99} = 2.58$ as the critical value for a confidence level of $c = 0.99$. We have

$$P(-2.58 < z < 2.58) \approx 0.99$$

The results of Example 1 will be used a great deal in our later work. For convenience, Table 7-2 gives some levels of confidence and corresponding critical values z_c. The same information is provided in Table 5(b) of Appendix II.

An estimate is not very valuable unless we have some kind of measure of how "good" it is. The language of probability can give us an idea of the size of the margin of error caused by using the sample mean $\bar{x}$ as an estimate for the population mean.

Remember that $\bar{x}$ is a random variable. Each time we draw a sample of size n from a population, we can get a different value for $\bar{x}$. According to the central limit theorem, if the sample size is large, then $\bar{x}$ has a distribution that is approximately normal with mean $\mu_{\bar{x}} = \mu$, the population mean we are trying to estimate. The standard deviation is $\sigma_{\bar{x}} = \sigma/\sqrt{n}$. If x has a normal distribution, these results are true *for any sample size*. (See Theorem 6.1.)

This information, together with our work on confidence levels, leads us (as shown in the optional derivation that follows) to the probability statement

$$P\left(-z_c\frac{\sigma}{\sqrt{n}} < \bar{x} - \mu < z_c\frac{\sigma}{\sqrt{n}}\right) = c \qquad (1)$$

TABLE 7-2 **Some Levels of Confidence and Their Corresponding Critical Values**

Level of Confidence c	Critical Value z_c
0.70, or 70%	1.04
0.75, or 75%	1.15
0.80, or 80%	1.28
0.85, or 85%	1.44
0.90, or 90%	1.645
0.95, or 95%	1.96
0.98, or 98%	2.33
0.99, or 99%	2.58

FIGURE 7-3

Distribution of Sample Means $\bar{x}$

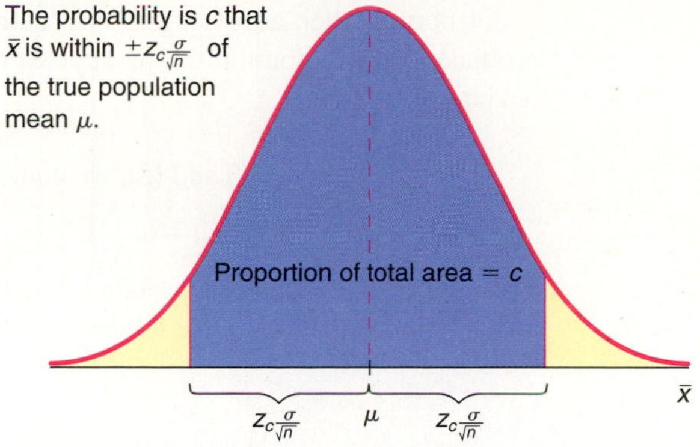
The probability is c that $\bar{x}$ is within $\pm z_c\frac{\sigma}{\sqrt{n}}$ of the true population mean μ.
Proportion of total area = c
$z_c\frac{\sigma}{\sqrt{n}}$
μ
$z_c\frac{\sigma}{\sqrt{n}}$
$\bar{x}$

Equation (1) uses the language of probability to give us an idea of the size of the margin of error for the corresponding confidence level c. In words, Equation (1) states that the probability is c that our point estimate $\bar{x}$ is within a distance $\pm z_c(\sigma/\sqrt{n})$ of the population mean μ. This relationship is shown in Figure 7-3.

In the following optional discussion, we derive Equation (1). If you prefer, you may jump ahead to the discussion about the margin of error.

OPTIONAL DERIVATION OF EQUATION (1)

For a c confidence level, we know that

$$P(-z_c < z < z_c) = c \qquad (2)$$

This statement gives us information about the size of z, but we want information about the size of $\bar{x} - \mu$. Is there a relationship between z and $\bar{x} - \mu$? The answer is yes since, by the central limit theorem, $\bar{x}$ has a distribution that is approximately normal, with mean μ and standard deviation $\sigma/\sqrt{n}$. We can convert $\bar{x}$ to a standard z score by using the formula

$$z = \frac{\bar{x} - \mu}{\sigma/\sqrt{n}} \qquad (3)$$

Substituting this expression for z into Equation (2) gives

$$P\left(-z_c < \frac{\bar{x} - \mu}{\sigma/\sqrt{n}} < z_c\right) = c \qquad (4)$$

Multiplying all parts of the inequality in (4) by $\sigma/\sqrt{n}$ gives us

$$P\left(-z_c\frac{\sigma}{\sqrt{n}} < \bar{x} - \mu < z_c\frac{\sigma}{\sqrt{n}}\right) = c \qquad (1)$$

Equation (1) is precisely the equation we set out to derive.

The *margin of error* (or absolute error) using $\bar{x}$ as a point estimate for μ is $|\bar{x} - \mu|$. In most practical problems, μ is unknown, so the margin of error is also unknown. However, Equation (1) allows us to compute an *error tolerance* E that serves as a bound on the margin of error. Using a c% level of confidence, we can say that the point estimate $\bar{x}$ differs from the population mean μ by a *maximal margin of error*

Maximal margin of error, E

$$E = z_c\frac{\sigma}{\sqrt{n}} \qquad (5)$$

Note: Formula (5) for E is based on the fact that the sampling distribution for $\bar{x}$ is exactly normal, with mean μ and standard deviation $\sigma/\sqrt{n}$. This occurs whenever the x distribution is normal with mean μ and standard deviation σ. If the x distribution is not normal, then according to the central limit theorem, large samples ($n \geq 30$) produce an $\bar{x}$ distribution that is approximately normal, with mean μ and standard deviation $\sigma/\sqrt{n}$.

Using Equations (1) and (5), we conclude that

$$P(-E < \bar{x} - \mu < E) = c \qquad (6)$$

Equation (6) states that the probability is c that the difference between $\bar{x}$ and μ is no more than the maximal error tolerance E. If we use a little algebra on the inequality

$$-E < \bar{x} - \mu < E \qquad (7)$$

for μ, we can rewrite it in the following mathematically equivalent way:

$$\bar{x} - E < \mu < \bar{x} + E \qquad (8)$$

Since formulas (7) and (8) are mathematically equivalent, their probabilities are the same. Therefore, from (6), (7), and (8), we obtain

Confidence interval for μ with σ known

$$P(\bar{x} - E < \mu < \bar{x} + E) = c \qquad (9)$$

Equation (9) states that there is a chance c that the interval from $\bar{x} - E$ to $\bar{x} + E$ contains the population mean μ. We call this interval a *c confidence interval for* μ.

A ***c* confidence interval for μ** is an interval computed from sample data in such a way that *c* is the probability of generating an interval containing the actual value of μ. In other words, *c* is the proportion of confidence intervals, based on random samples of size *n*, that actually contain μ.

We may get a different confidence interval for each different sample that is taken. Some intervals will contain the population mean μ and others will not. However, in the long run, the proportion of confidence intervals that contain μ is c.

PROCEDURE

HOW TO FIND A CONFIDENCE INTERVAL FOR μ WHEN σ IS KNOWN

Requirements

Let x be a random variable appropriate to your application. Obtain a simple random sample (of size n) of x values from which you compute the sample mean $\bar{x}$. The value of σ is already known (perhaps from a previous study).

If you can assume that x has a normal distribution, then any sample size n will work. If you cannot assume this, then use a sample size of $n \geq 30$.

Confidence interval for μ when σ is known

$$\bar{x} - E < \mu < \bar{x} + E \qquad (10)$$

where $\bar{x}$ = sample mean of a simple random sample

$E = z_c \dfrac{\sigma}{\sqrt{n}}$

c = confidence level ($0 < c < 1$)

z_c = critical value for confidence level c based on the standard normal distribution (see Table 5(b) of Appendix II for frequently used values).

EXAMPLE 2

CONFIDENCE INTERVAL FOR μ WITH σ KNOWN

Julia enjoys jogging. She has been jogging over a period of several years, during which time her physical condition has remained constantly good. Usually, she jogs 2 miles per day. The standard deviation of her times is $\sigma = 1.80$ minutes. During the past year, Julia has recorded her times to run 2 miles. She has a random sample of 90 of these times. For these 90 times, the mean was $\bar{x} = 15.60$ minutes. Let μ be the mean jogging time for the entire distribution of Julia's 2-mile running times (taken over the past year). Find a 0.95 confidence interval for μ.

Myrleen Ferguson Cate/PhotoEdit

SOLUTION: *Check Requirements* We have a simple random sample of running times, and the sample size $n = 90$ is large enough for the $\bar{x}$ distribution to be approximately normal. We also know σ. It is appropriate to use the normal distribution to compute a confidence interval for μ.

To compute E for the 95% confidence interval $\bar{x} - E$ to $\bar{x} + E$, we use the fact that $z_c = 1.96$ (see Table 7-2), together with the values $n = 90$ and $\sigma = 1.80$. Therefore,

$$E = z_c \frac{\sigma}{\sqrt{n}}$$

$$E = 1.96\left(\frac{1.80}{\sqrt{90}}\right)$$

$$E \approx 0.37$$

Using Equation (10), the given value of $\bar{x}$, and our computed value for E, we get the 95% confidence interval for μ.

$$\bar{x} - E < \mu < \bar{x} + E$$

$$15.60 - 0.37 < \mu < 15.60 + 0.37$$

$$15.23 < \mu < 15.97$$

Interpretation We conclude with 95% confidence that the interval from 15.23 minutes to 15.97 minutes is one that contains the population mean μ of jogging times for Julia.

CRITICAL THINKING

INTERPRETING CONFIDENCE INTERVALS

A few comments are in order about the general meaning of the term *confidence interval*.

- Since $\bar{x}$ is a random variable, the endpoints $\bar{x} \pm E$ are also random variables. Equation (9) states that we have a chance c of obtaining a sample such that the interval, once it is computed, will contain the parameter μ.
- After the confidence interval is numerically fixed for a specific sample, it either does or does not contain μ. So, the probability is 1 or 0 that the interval, when it is fixed, will contain μ.

A nontrivial probability statement can be made only about variables, not constants.

- Equation (9), $P(\bar{x} - E < \mu < \bar{x} + E) = c$, really states that if we draw many random samples of size n and get lots of confidence intervals, then the proportion of all intervals that will turn out to contain the mean μ is c.

Continued

CRITICAL THINKING *continued*

For example, in Figure 7-4 the horizontal lines represent 0.90 confidence intervals for various samples of the same size from an x distribution. Some of these intervals contain μ and others do not. Since the intervals are 0.90 confidence intervals, about 90% of all such intervals should contain μ. For each sample, the interval goes from $\bar{x} - E$ to $\bar{x} + E$.

- Once we have a *specific* confidence interval for μ, such as $3 < \mu < 5$, all we can say is that we are c% confident that we have one of the intervals that actually contains μ. Another appropriate statement is that at the c confidence level, our interval contains μ.

FIGURE 7-4

0.90 Confidence Intervals for Samples of the Same Size

For each sample, the interval goes from $\bar{x} - E$ to $\bar{x} + E$

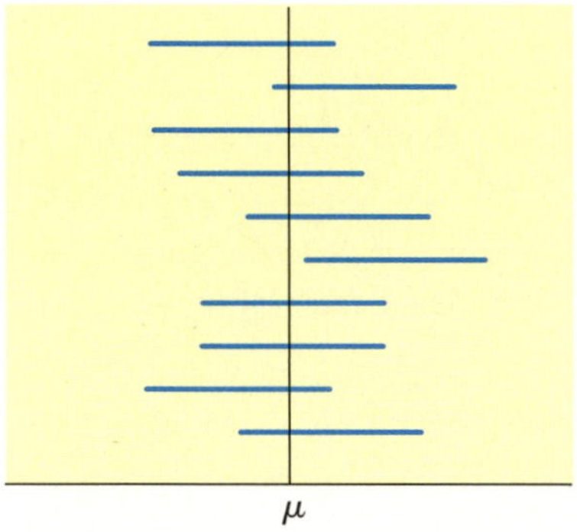

COMMENT Please see Using Technology at the end of this chapter for a computer demonstration of this discussion about confidence intervals.

WHAT DOES A CONFIDENCE INTERVAL TELL US?

A confidence interval gives us a range of values for a parameter.

- A confidence interval for a parameter is based on a corresponding sample statistic from a random sample of a specified size n.
- A confidence interval depends on a confidence level c. Common values for c are 0.90 and 0.95.
- If we take all possible samples of size n from a population and compute a c confidence interval from each sample, then c% of all such confidence intervals will actually contain the population parameter in question. Similarly $(1 - c)$% of the confidence intervals will not contain the parameter.

GUIDED EXERCISE 1 CONFIDENCE INTERVAL FOR μ WITH σ KNOWN

Walter usually meets Julia at the track. He prefers to jog 3 miles. From long experience, he knows that $\sigma = 2.40$ minutes for his jogging times. For a random sample of 90 jogging sessions, the mean time was $\bar{x} = 22.50$ minutes. Let μ be the mean jogging time for the entire distribution of Walter's 3-mile running times over the past several years. Find a 0.99 confidence interval for μ.

Continued

GUIDED EXERCISE 1 *continued*

(a) ***Check Requirements*** Is the $\bar{x}$ distribution approximately normal? Do we know σ?

→ Yes; we know this from the central limit theorem. Yes, $\sigma = 2.40$ minutes.

(b) What is the value of $z_{0.99}$? (See Table 7-2.)

→ $z_{0.99} = 2.58$

(c) What is the value of E?

→ $E = z_c \dfrac{\sigma}{\sqrt{n}} = 2.58\left(\dfrac{2.40}{\sqrt{90}}\right) \approx 0.65$

(d) What are the endpoints for a 0.99 confidence interval for μ?

→ The endpoints are given by

$$\bar{x} - E \approx 22.50 - 0.65 = 21.85$$
$$\bar{x} + E \approx 22.50 + 0.65 = 23.15$$

(e) ***Interpretation*** Explain what the confidence interval tells us.

→ We are 99% certain that the interval from 21.85 to 23.15 is an interval that contains the population mean time μ.

When we use samples to estimate the mean of a population, we generate a small error. However, samples are useful even when it is possible to survey the entire population, because the use of a sample may yield savings of time or effort in collecting data.

LOOKING FORWARD

The basic structure for most confidence intervals for population parameters is

$$\text{sample statistic} - E < \text{population parameter} < \text{sample statistic} + E$$

where E is the maximal margin of error based on the sample statistic distribution and the level of confidence c. We will see this same format used for confidence intervals of the mean when σ is unknown (Section 7.2), for proportions (Section 7.3), for differences of means from independent samples (Section 7.4), for differences of proportions (Section 7.4), and for parameters of linear regression (Chapter 9). This structure for confidence intervals is so basic that some software packages, such as Excel, simply give the value of E for a confidence interval and expect the user to finish the computation.

TECH NOTES The TI-84Plus/TI-83Plus/TI-*n*spire calculators, Excel 2010, and Minitab all support confidence intervals for μ from large samples. The level of support varies according to the technology. When a confidence interval is given, the standard mathematical notation (lower value, upper value) is used. For instance, the notation (15.23, 15.97) means the interval from 15.23 to 15.97.

TI-84Plus/TI-83Plus/TI-*n*spire (with TI-84Plus keypad) These calculators give the most extensive support. The user can opt to enter raw data or just summary statistics. In each case, the value of σ must be specified. Press the **STAT** key, then select **TESTS**, and use **7:ZInterval**. The TI-84Plus/TI-83Plus/TI-*n*spire output shows the results for Example 2.

```
ZInterval
 Inpt:Data Stats
 σ:1.8
 x̄:15.6
 n:90
 C-Level:95
 Calculate
```

```
ZInterval
 (15.228, 15.972)
 x̄=15.6
 n=90
```

Excel 2010 Excel gives only the value of the maximal error of estimate E. On the **Home** screen click the **Insert Function** (f_x). In the dialogue box, select **Statistical** for the category, and then select **Confidence.Norm** In the resulting dialogue box, the value of **alpha** is 1 − confidence level. For example, alpha is 0.05 for a 95% confidence interval. The values of σ and n are also required. The Excel output shows the value of E for Example 2.

f_x =CONFIDENCE.NORM(0.05,1.8,90)

C	D	E
0.371876		

An alternate approach incorporating raw data (using the Student's t distribution presented in the next section) uses a selection from the Data Analysis package. Click the **Data** tab on the home ribbon. From the Analysis group, select **Data Analysis.** In the dialogue box, select **Descriptive Statistics.** Check the box by **Confidence Level for Mean,** and enter the confidence level. Again, the value of E for the interval is given.

Minitab Raw data are required. Use the menu choices **Stat ➤ Basic Statistics ➤ 1-SampleZ.**

SAMPLE SIZE FOR ESTIMATING THE MEAN μ

In the design stages of statistical research projects, it is a good idea to decide in advance on the confidence level you wish to use and to select the *maximal* margin of error E you want for your project. How you choose to make these decisions depends on the requirements of the project and the practical nature of the problem.

Whatever specifications you make, the next step is to determine the sample size. Solving the formula that gives the maximal margin of error E for n enables us to determine the minimal sample size.

PROCEDURE

HOW TO FIND THE SAMPLE SIZE n FOR ESTIMATING μ WHEN σ IS KNOWN

Requirements

The distribution of sample means $\bar{x}$ is approximately normal.

Formula for sample size

$$n = \left(\frac{z_c \sigma}{E}\right)^2 \tag{11}$$

where E = specified maximal margin of error

σ = population standard deviation

z_c = critical value from the normal distribution for the desired confidence level c. Commonly used values of z_c can be found in Table 5(b) of Appendix II.

If n is not a whole number, increase n to the next higher whole number. Note that n is the minimal sample size for a specified confidence level and maximal error of estimate E.

COMMENT If you have a *preliminary study* involving a sample size of 30 or larger, then for most practical purposes it is safe to approximate σ with the sample standard deviation s in the formula for sample size.

EXAMPLE 3

SAMPLE SIZE FOR ESTIMATING μ

A wildlife study is designed to find the mean weight of salmon caught by an Alaskan fishing company. A preliminary study of a random sample of 50 salmon showed $s \approx 2.15$ pounds. How large a sample should be taken to be 99% confident that the sample mean $\bar{x}$ is within 0.20 pound of the true mean weight μ?

Photo Researchers

Salmon moving upstream

SOLUTION: In this problem, $z_{0.99} = 2.58$ (see Table 7-2) and $E = 0.20$. The preliminary study of 50 fish is large enough to permit a good approximation of σ by $s = 2.15$. Therefore, Equation (11) becomes

$$n = \left(\frac{z_c \sigma}{E}\right)^2 \approx \left(\frac{(2.58)(2.15)}{0.20}\right)^2 = 769.2$$

Note: In determining sample size, any fractional value of n is always rounded to the *next higher whole number*. We conclude that a sample size of 770 will be large enough to satisfy the specifications. Of course, a sample size larger than 770 also works.

VIEWPOINT Music and Techno Theft

Performing rights organizations ASCAP (American Society of Composers, Authors, and Publishers) and BMI (Broadcast Music, Inc.) collect royalties for songwriters and music publishers. Radio, television, cable, nightclubs, restaurants, elevators, and even beauty salons play music that is copyrighted by a composer or publisher. The royalty payment for this music turns out to be more than a billion dollars a year (Source: The Wall Street Journal*). How do ASCAP and BMI know who is playing what music? The answer is, they don't! Instead of tracking exactly what gets played, they use random sampling and* confidence intervals*. For example, each radio station (there are more than 10,000 in the United States) has randomly chosen days of programming analyzed every year. The results are used to assess royalty fees. In fact, Deloitte & Touche (a financial services company) administers the sampling process.*

Although the system is not perfect, it helps bring order into an otherwise chaotic accounting system. Such methods of "copyright policing" help prevent techno theft, ensuring that many songwriters and recording artists get a reasonable return for their creative work.

SECTION 7.1 PROBLEMS

In Problems 1–8, answer true or false. Explain your answer.

1. *Statistical Literacy* The value z_c is a value from the standard normal distribution such that $P(-z_c < x < z_c) = c$.

2. *Statistical Literacy* The point estimate for the population mean μ of an x distribution is $\bar{x}$, computed from a random sample of the x distribution.

3. *Statistical Literacy* Consider a random sample of size n from an x distribution. For such a sample, the margin of error for estimating μ is the magnitude of the difference between $\bar{x}$ and μ.

4. *Statistical Literacy* Every random sample of the same size from a given population will produce exactly the same confidence interval for μ.

5. *Statistical Literacy* A larger sample size produces a longer confidence interval for μ.

6. *Statistical Literacy* If the original x distribution has a relatively small standard deviation, the confidence interval for μ will be relatively short.

7. *Statistical Literacy* If the sample mean $\bar{x}$ of a random sample from an x distribution is relatively small, then the confidence interval for μ will be relatively short.

8. *Statistical Literacy* For the same random sample, when the confidence level c is reduced, the confidence interval for μ becomes shorter.

9. *Critical Thinking* Sam computed a 95% confidence interval for μ from a specific random sample. His confidence interval was $10.1 < \mu < 12.2$. He claims that the probability that μ is in this interval is 0.95. What is wrong with his claim?

10. *Critical Thinking* Sam computed a 90% confidence interval for μ from a specific random sample of size n. He claims that at the 90% confidence level, his confidence interval contains μ. Is his claim correct? Explain.

Answers may vary slightly due to rounding.

11. *Basic Computation: Confidence Interval* Suppose x has a normal distribution with $\sigma = 6$. A random sample of size 16 has sample mean 50.
    (a) *Check Requirements* Is it appropriate to use a normal distribution to compute a confidence interval for the population mean μ? Explain.
    (b) Find a 90% confidence interval for μ.
    (c) *Interpretation* Explain the meaning of the confidence interval you computed.

12. *Basic Computation: Confidence Interval* Suppose x has a mound-shaped distribution with $\sigma = 9$. A random sample of size 36 has sample mean 20.
    (a) *Check Requirements* Is it appropriate to use a normal distribution to compute a confidence interval for the population mean μ? Explain.
    (b) Find a 95% confidence interval for μ.
    (c) *Interpretation* Explain the meaning of the confidence interval you computed.

13. *Basic Computation: Sample Size* Suppose x has a mound-shaped distribution with $\sigma = 3$.
    (a) Find the minimal sample size required so that for a 95% confidence interval, the maximal margin of error is $E = 0.4$.
    (b) *Check Requirements* Based on this sample size, can we assume that the $\bar{x}$ distribution is approximately normal? Explain.

14. *Basic Computation: Sample Size* Suppose x has a normal distribution with $\sigma = 1.2$.
    (a) Find the minimal sample size required so that for a 90% confidence interval, the maximal margin of error is $E = 0.5$.
    (b) *Check Requirements* Based on this sample size and the x distribution, can we assume that the $\bar{x}$ distribution is approximately normal? Explain.

15. *Zoology: Hummingbirds* Allen's hummingbird (*Selasphorus sasin*) has been studied by zoologist Bill Alther (Reference: *Hummingbirds* by K. Long and W. Alther). A small group of 15 Allen's hummingbirds has been under study in Arizona. The average weight for these birds is $\bar{x} = 3.15$ grams. Based on previous studies, we can assume that the weights of Allen's hummingbirds have a normal distribution, with $\sigma = 0.33$ gram.
    (a) Find an 80% confidence interval for the average weights of Allen's hummingbirds in the study region. What is the margin of error?
    (b) What conditions are necessary for your calculations?
    (c) *Interpret* your results in the context of this problem.

(d) *Sample Size* Find the sample size necessary for an 80% confidence level with a maximal margin of error $E = 0.08$ for the mean weights of the hummingbirds.

16. *Diagnostic Tests: Uric Acid* Overproduction of uric acid in the body can be an indication of cell breakdown. This may be an advance indication of illness such as gout, leukemia, or lymphoma (Reference: *Manual of Laboratory and Diagnostic Tests* by F. Fischbach). Over a period of months, an adult male patient has taken eight blood tests for uric acid. The mean concentration was $\bar{x} = 5.35$ mg/dl. The distribution of uric acid in healthy adult males can be assumed to be normal, with $\sigma = 1.85$ mg/dl.
(a) Find a 95% confidence interval for the population mean concentration of uric acid in this patient's blood. What is the margin of error?
(b) What conditions are necessary for your calculations?
(c) *Interpret* your results in the context of this problem.
(d) *Sample Size* Find the sample size necessary for a 95% confidence level with maximal margin of error $E = 1.10$ for the mean concentration of uric acid in this patient's blood.

17. *Diagnostic Tests: Plasma Volume* Total plasma volume is important in determining the required plasma component in blood replacement therapy for a person undergoing surgery. Plasma volume is influenced by the overall health and physical activity of an individual. (Reference: See Problem 16.) Suppose that a random sample of 45 male firefighters are tested and that they have a plasma volume sample mean of $\bar{x} = 37.5$ ml/kg (milliliters plasma per kilogram body weight). Assume that $\sigma = 7.50$ ml/kg for the distribution of blood plasma.
(a) Find a 99% confidence interval for the population mean blood plasma volume in male firefighters. What is the margin of error?
(b) What conditions are necessary for your calculations?
(c) *Interpret* your results in the context of this problem.
(d) *Sample Size* Find the sample size necessary for a 99% confidence level with maximal margin of error $E = 2.50$ for the mean plasma volume in male firefighters.

18. *Agriculture: Watermelon* What price do farmers get for their watermelon crops? In the third week of July, a random sample of 40 farming regions gave a sample mean of $\bar{x} = \$6.88$ per 100 pounds of watermelon. Assume that σ is known to be \$1.92 per 100 pounds (Reference: *Agricultural Statistics*, U.S. Department of Agriculture).
(a) Find a 90% confidence interval for the population mean price (per 100 pounds) that farmers in this region get for their watermelon crop. What is the margin of error?
(b) *Sample Size* Find the sample size necessary for a 90% confidence level with maximal margin of error $E = 0.3$ for the mean price per 100 pounds of watermelon.
(c) A farm brings 15 tons of watermelon to market. Find a 90% confidence interval for the population mean cash value of this crop. What is the margin of error? *Hint:* 1 ton is 2000 pounds.

19. *FBI Report: Larceny* Thirty small communities in Connecticut (population near 10,000 each) gave an average of $\bar{x} = 138.5$ reported cases of larceny per year. Assume that σ is known to be 42.6 cases per year (Reference: *Crime in the United States*, Federal Bureau of Investigation).
(a) Find a 90% confidence interval for the population mean annual number of reported larceny cases in such communities. What is the margin of error?
(b) Find a 95% confidence interval for the population mean annual number of reported larceny cases in such communities. What is the margin of error?

(c) Find a 99% confidence interval for the population mean annual number of reported larceny cases in such communities. What is the margin of error?
(d) Compare the margins of error for parts (a) through (c). As the confidence levels increase, do the margins of error increase?
(e) *Critical Thinking* Compare the lengths of the confidence intervals for parts (a) through (c). As the confidence levels increase, do the confidence intervals increase in length?

20. *Confidence Intervals: Values of σ* A random sample of size 36 is drawn from an x distribution. The sample mean is 100.
(a) Suppose the x distribution has $\sigma = 30$. Compute a 90% confidence interval for μ. What is the value of the margin of error?
(b) Suppose the x distribution has $\sigma = 20$. Compute a 90% confidence interval for μ. What is the value of the margin of error?
(c) Suppose the x distribution has $\sigma = 10$. Compute a 90% confidence interval for μ. What is the value of the margin of error?
(d) Compare the margins of error for parts (a) through (c). As the standard deviation decreases, does the margin of error decrease?
(e) *Critical Thinking* Compare the lengths of the confidence intervals for parts (a) through (c). As the standard deviation decreases, does the length of a 90% confidence interval decrease?

21. *Confidence Intervals: Sample Size* A random sample is drawn from a population with $\sigma = 12$. The sample mean is 30.
(a) Compute a 95% confidence interval for μ based on a sample of size 49. What is the value of the margin of error?
(b) Compute a 95% confidence interval for μ based on a sample of size 100. What is the value of the margin of error?
(c) Compute a 95% confidence interval for μ based on a sample of size 225. What is the value of the margin of error?
(d) Compare the margins of error for parts (a) through (c). As the sample size increases, does the margin of error decrease?
(e) *Critical Thinking* Compare the lengths of the confidence intervals for parts (a) through (c). As the sample size increases, does the length of a 90% confidence interval decrease?

22. *Ecology: Sand Dunes* At wind speeds above 1000 centimeters per second (cm/sec), significant sand-moving events begin to occur. Wind speeds below 1000 cm/sec deposit sand, and wind speeds above 1000 cm/sec move sand to new locations. The cyclic nature of wind and moving sand determines the shape and location of large dunes (Reference: *Hydraulic, Geologic, and Biologic Research at Great Sand Dunes National Monument and Vicinity, Colorado*, Proceedings of the National Park Service Research Symposium). At a test site, the prevailing direction of the wind did not change noticeably. However, the velocity did change. Sixty wind speed readings gave an average velocity of $\bar{x} = 1075$ cm/sec. Based on long-term experience, σ can be assumed to be 265 cm/sec.
(a) Find a 95% confidence interval for the population mean wind speed at this site.
(b) *Interpretation* Does the confidence interval indicate that the population mean wind speed is such that the sand is always moving at this site? Explain.

23. *Profits: Banks* Jobs and productivity! How do banks rate? One way to answer this question is to examine annual profits per employee. *Forbes Top Companies*, edited by J. T. Davis (John Wiley & Sons), gave the following data about annual profits per employee (in units of one thousand dollars per

employee) for representative companies in financial services. Companies such as Wells Fargo, First Bank System, and Key Banks were included. Assume $\sigma \approx 10.2$ thousand dollars.

42.9	43.8	48.2	60.6	54.9	55.1	52.9	54.9	42.5	33.0	33.6
36.9	27.0	47.1	33.8	28.1	28.5	29.1	36.5	36.1	26.9	27.8
28.8	29.3	31.5	31.7	31.1	38.0	32.0	31.7	32.9	23.1	54.9
43.8	36.9	31.9	25.5	23.2	29.8	22.3	26.5	26.7		

(a) Use a calculator or appropriate computer software to verify that, for the preceding data, $\bar{x} \approx 36.0$.

(b) Let us say that the preceding data are representative of the entire sector of (successful) financial services corporations. Find a 75% confidence interval for μ, the average annual profit per employee for all successful banks.

(c) *Interpretation* Let us say that you are the manager of a local bank with a large number of employees. Suppose the annual profits per employee are less than 30 thousand dollars per employee. Do you think this might be somewhat low compared with other successful financial institutions? Explain by referring to the confidence interval you computed in part (b).

(d) *Interpretation* Suppose the annual profits are more than 40 thousand dollars per employee. As manager of the bank, would you feel somewhat better? Explain by referring to the confidence interval you computed in part (b).

(e) Repeat parts (b), (c), and (d) for a 90% confidence level.

24. *Profits: Retail* Jobs and productivity! How do retail stores rate? One way to answer this question is to examine annual profits per employee. The following data give annual profits per employee (in units of one thousand dollars per employee) for companies in retail sales. (See reference in Problem 23.) Companies such as Gap, Nordstrom, Dillards, JCPenney, Sears, Wal-Mart, Office Depot, and Toys "Я" Us are included. Assume $\sigma \approx 3.8$ thousand dollars.

4.4	6.5	4.2	8.9	8.7	8.1	6.1	6.0	2.6	2.9	8.1	−1.9
11.9	8.2	6.4	4.7	5.5	4.8	3.0	4.3	−6.0	1.5	2.9	4.8
−1.7	9.4	5.5	5.8	4.7	6.2	15.0	4.1	3.7	5.1	4.2	

(a) Use a calculator or appropriate computer software to verify that, for the preceding data, $\bar{x} \approx 5.1$.

(b) Let us say that the preceding data are representative of the entire sector of retail sales companies. Find an 80% confidence interval for μ, the average annual profit per employee for retail sales.

(c) *Interpretation* Let us say that you are the manager of a retail store with a large number of employees. Suppose the annual profits per employee are less than 3 thousand dollars per employee. Do you think this might be low compared with other retail stores? Explain by referring to the confidence interval you computed in part (b).

(d) *Interpretation* Suppose the annual profits are more than 6.5 thousand dollars per employee. As store manager, would you feel somewhat better? Explain by referring to the confidence interval you computed in part (b).

(e) Repeat parts (b), (c), and (d) for a 95% confidence interval.

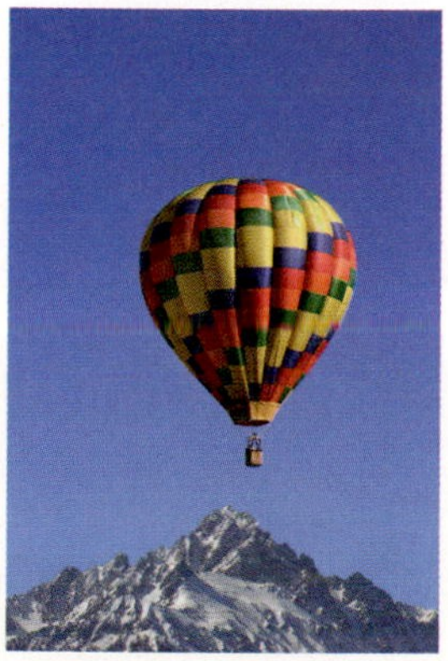

Images Etc Ltd /Getty images

25. *Ballooning: Air Temperature* How hot is the air in the top (crown) of a hot air balloon? Information from *Ballooning: The Complete Guide to Riding the Winds* by Wirth and Young (Random House) claims that the air in the crown should be an average of 100°C for a balloon to be in a state of equilibrium. However, the temperature does not need to be exactly 100°C. What is a reasonable and safe range of temperatures? This range may vary with the size

and (decorative) shape of the balloon. All balloons have a temperature gauge in the crown. Suppose that 56 readings (for a balloon in equilibrium) gave a mean temperature of $\bar{x} = 97°C$. For this balloon, $\sigma \approx 17°C$.

(a) Compute a 95% confidence interval for the average temperature at which this balloon will be in a steady-state equilibrium.

(b) *Interpretation* If the average temperature in the crown of the balloon goes above the high end of your confidence interval, do you expect that the balloon will go up or down? Explain.

SECTION **7.2**

Estimating μ When σ Is Unknown

FOCUS POINTS

- Learn about degrees of freedom and Student's *t* distributions.
- Find critical values using degrees of freedom and confidence levels.
- Compute confidence intervals for μ when σ is unknown. What does this information tell you?

In order to use the normal distribution to find confidence intervals for a population mean μ, we need to know the value of σ, the population standard deviation. However, much of the time, when μ is unknown, σ is unknown as well. In such cases, we use the sample standard deviation s to approximate σ. When we use s to approximate σ, the sampling distribution for $\bar{x}$ follows a new distribution called a *Student's t distribution*.

Student's *t* distribution

STUDENT'S *t* DISTRIBUTIONS

Student's *t* distributions were discovered in 1908 by W. S. Gosset. He was employed as a statistician by Guinness brewing company, a company that discouraged publication of research by its employees. As a result, Gosset published his research under the pseudonym *Student*. Gosset was the first to recognize the importance of developing statistical methods for obtaining reliable information from samples of populations with unknown σ. Gosset used the variable t when he introduced the distribution in 1908. To this day and in his honor, it is still called a Student's t distribution. It might be more fitting to call this distribution *Gosset's t distribution*; however, in the literature of mathematical statistics, it is known as a *Student's t distribution*.

The variable t is defined as follows. A Student's t distribution depends on sample size n.

Assume that x has a normal distribution with mean μ. For samples of size n with sample mean $\bar{x}$ and sample standard deviation s, the ***t* variable**

$$t = \frac{\bar{x} - \mu}{\frac{s}{\sqrt{n}}} \qquad (12)$$

has a **Student's *t* distribution** with **degrees of freedom** $d.f. = n - 1$.

Degrees of freedom, *d.f.*

If many random samples of size n are drawn, then we get many t values from Equation (12). These t values can be organized into a frequency table, and a histogram can be drawn, thereby giving us an idea of the shape of the t distribution (for a given n).

Fortunately, all this work is not necessary because mathematical theorems can be used to obtain a formula for the t distribution. However, it is important to observe that these theorems say that the shape of the t distribution depends only on n, provided the basic variable x has a normal distribution. So, *when we use a t distribution, we will assume that the x distribution is normal.*

Table 6 of Appendix II gives values of the variable t corresponding to what we call the number of *degrees of freedom*, abbreviated *d.f.* For the methods used in this section, the number of degrees of freedom is given by the formula

$$d.f. = n - 1 \tag{13}$$

where *d.f.* stands for the degrees of freedom and n is the sample size. Each choice for *d.f.* gives a different t distribution.

The graph of a t distribution is always symmetrical about its mean, which (as for the z distribution) is 0. The main observable difference between a t distribution and the standard normal z distribution is that a t distribution has somewhat thicker tails.

Figure 7-5 shows a standard normal z distribution and Student's t distribution with $d.f. = 3$ and $d.f. = 5$.

FIGURE 7-5

A Standard Normal Distribution and Student's t Distribution with $d.f. = 3$ and $d.f. = 5$

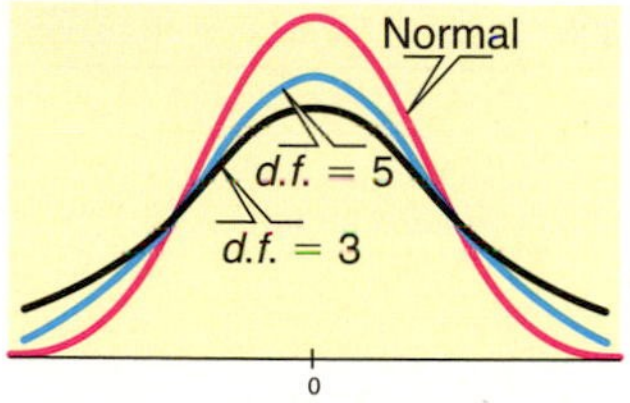

PROPERTIES OF A STUDENT'S *t* DISTRIBUTION

1. The distribution is *symmetric* about the mean 0.
2. The distribution depends on the *degrees of freedom, d.f.* ($d.f. = n - 1$ for μ confidence intervals).
3. The distribution is *bell-shaped*, but has thicker tails than the standard normal distribution.
4. As the degrees of freedom increase, the t distribution *approaches* the standard normal distribution.
5. The area under the entire curve is 1.

USING TABLE 6 TO FIND CRITICAL VALUES FOR CONFIDENCE INTERVALS

Critical values t_c

Table 6 of Appendix II gives various t values for different degrees of freedom *d.f.* We will use this table to find *critical values* t_c for a c confidence level. In other words, we want to find t_c such that an area equal to c under the t distribution for a given number of degrees of freedom falls between $-t_c$ and t_c. In the language of probability, we want to find t_c such that

$$P(-t_c < t < t_c) = c$$

This probability corresponds to the shaded area in Figure 7-6.

FIGURE 7-6

Area Under the t Curve Between $-t_c$ and t_c

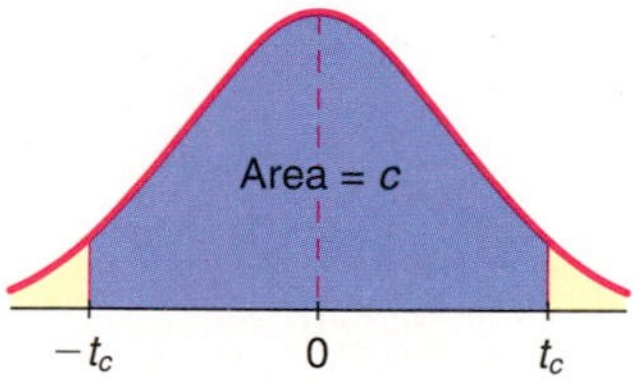

Table 6 of Appendix II has been arranged so that c is one of the column headings, and the degrees of freedom *d.f.* are the row headings. To find t_c for any specific c, we find the column headed by that c value and read down until we reach the row headed by the appropriate number of degrees of freedom *d.f.* (You will notice two other column headings: one-tail area and two-tail area. We will use these later, but for the time being, just ignore them.)

CONVENTION FOR USING A STUDENT'S *t* DISTRIBUTION TABLE

If the degrees of freedom *d.f.* you need are not in the table, use the closest *d.f.* in the table that is *smaller*. This procedure results in a critical value t_c that is more conservative, in the sense that it is larger. The resulting confidence interval will be longer and have a probability that is slightly higher than c.

EXAMPLE 4

STUDENT'S *t* DISTRIBUTION

Use Table 7-3 (an excerpt from Table 6 of Appendix II) to find the critical value t_c for a 0.99 confidence level for a t distribution with sample size $n = 5$.

SOLUTION:

(a) First, we find the column with c heading 0.990.

(b) Next, we compute the number of degrees of freedom: $d.f. = n - 1 = 5 - 1 = 4$.

(c) We read down the column under the heading $c = 0.99$ until we reach the row headed by 4 (under $d.f.$). The entry is 4.604. Therefore, $t_{0.99} = 4.604$.

TABLE 7-3 **Student's *t* Distribution Critical Values (Excerpt from Table 6, Appendix II)**

one-tail area	—	—	—	—
two-tail area	—	—	—	—
d.f. \ *c*	. . . 0.900	0.950	0.980	0.990 . . .
⋮				
3	. . . 2.353	3.182	4.541	5.841 . . .
4	. . . 2.132	2.776	3.747	4.604 . . .
⋮				
7	. . . 1.895	2.365	2.998	3.449 . . .
8	. . . 1.860	2.306	2.896	3.355 . . .

GUIDED EXERCISE 2

STUDENT'S *t* DISTRIBUTION TABLE

Use Table 6 of Appendix II (or Table 7-3, showing an excerpt from the table) to find t_c for a 0.90 confidence level for a t distribution with sample size $n = 9$.

(a) We find the column headed by $c =$ __________. → $c = 0.900$.

(b) The degrees of freedom are given by $d.f. = n - 1 =$ __________. → $d.f. = n - 1 = 9 - 1 = 8$.

(c) Read down the column found in part (a) until you reach the entry in the row headed by $d.f. = 8$. The value of $t_{0.90}$ is __________ for a sample of size 9. → $t_{0.90} = 1.860$ for a sample of size $n = 9$.

(d) Find t_c for a 0.95 confidence level for a t distribution with sample size $n = 9$. → $t_{0.95} = 2.306$ for a sample of size $n = 9$.

TECH NOTES The TI-84Plus/TI-83Plus calculators with operating system v2.55MP, (available January 2011), Excel 2010, and Minitab all support the inverse t distribution. With the inverse t distribution function, you enter the area (cumulative probability) to the left of the unknown t value and the degrees of freedom. The output is the t value corresponding to the designated cumulative probability. For instance, to find a t value from a Student's t distribution with degrees of freedom 25 such that 95% of the area lies to the left of t, use the described instructions.

Using technology to find the critical value t_c for a c confidence interval is not difficult. However, we need to find the total area to the left of t_c. This area includes the area above the confidence interval as well as the area to the left of $-t_c$.

$$\text{Area to the left of } t_c = c + \frac{1-c}{2}$$

So for a 0.90 confidence interval, the area to the left of $t_{0.90}$ is 0.90 + 0.10/2, or 0.95. The following instructions show how to find $t_{0.90}$ for a distribution with 25 degrees of freedom. Again, 95% of the area lies to the left of $t_{0.90}$.

TI-84Plus/TI-83Plus/TI-*n*spire (with TI-84Plus keypad) The new operating system v2.55MP for the TI-84Plus/TI-83Plus series calculator provides the InvT function. Press the **DISTR** and select **4:invT(area,df)**. Enter the area and the degrees of freedom. Note that with the v2.55MP operating system, the calculator prompts you to enter the area and the *d.f.* The resulting t value is shown.

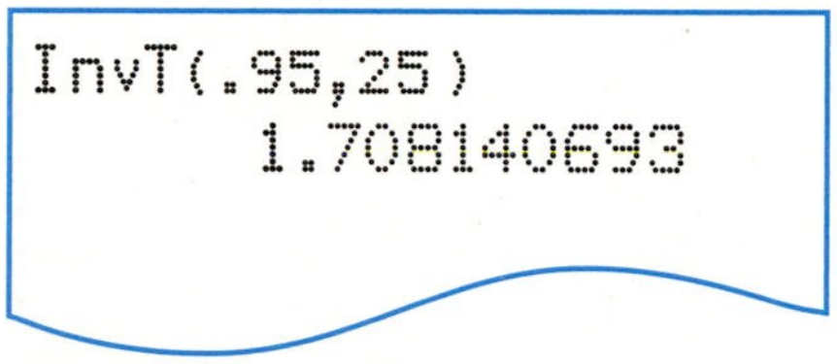

Excel 2010 Select the **Insert Function** (f_x). In the dialogue box select **Statistical** for the Category, and then for the Function select **T.INV.** Fill in the dialogue box.

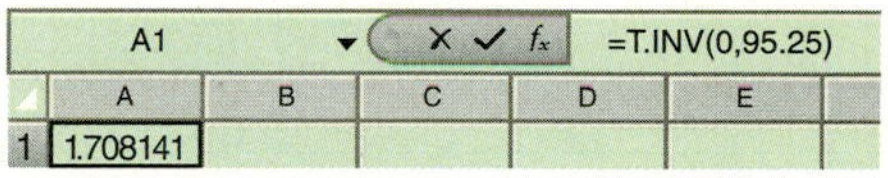

Minitab Use the menu selection **Calc ➤ Probability Distribution ➤ t...** Fill in the dialogue box, selecting inverse cumulative probability. Enter 0 for the noncentrality parameter. Enter the degrees of freedom. Select input constant and enter the area (cumulative probability) to the left of the desired t.

```
Inverse Cumulative Distribution Function
Student's t distribution with 25 DF
P(X <= x)          x
   0.95      1.70814
```

LOOKING FORWARD

Student's t distributions will be used again in Chapter 8 when testing μ and when testing differences of means. The distributions are also used for confidence intervals and testing of parameters of linear regression (Sections 9.3 and 9.4).

CONFIDENCE INTERVALS FOR μ WHEN σ IS UNKNOWN

Maximal margin of error, *E*

In Section 7.1, we found bounds $\pm E$ on the margin of error for a c confidence level. Using the same basic approach, we arrive at the conclusion that

$$E = t_c \frac{s}{\sqrt{n}}$$

is the maximal margin of error for a c confidence level when σ is unknown (i.e., $|\bar{x} - \mu| < E$ with probability c). The analogue of Equation (1) in Section 7.1 is

$$P\left(-t_c\frac{s}{\sqrt{n}} < \bar{x} - \mu < t_c\frac{s}{\sqrt{n}}\right) = c \qquad (14)$$

COMMENT Comparing Equation (14) with Equation (1) in Section 7.1, it becomes evident that we are using the same basic method on the t distribution that we used on the z distribution.

Likewise, for samples from normal populations with unknown σ, Equation (9) of Section 7.1 becomes

$$P(\bar{x} - E < \mu < \bar{x} + E) = c \quad (15)$$

where $E = t_c(s/\sqrt{n})$. Let us organize what we have been doing in a convenient summary.

Confidence interval for μ with σ unknown

PROCEDURE

HOW TO FIND A CONFIDENCE INTERVAL FOR μ WHEN σ IS UNKNOWN

Requirements

Let x be a random variable appropriate to your application. Obtain a simple random sample (of size n) of x values from which you compute the sample mean $\bar{x}$ and the sample standard deviation s.

If you can assume that x has a normal distribution or simply a mound-shaped, symmetric distribution, then any sample size n will work. If you cannot assume this, then use a sample size of $n \geq 30$.

Confidence interval for μ when σ is unknown

$$\bar{x} - E < \mu < \bar{x} + E \quad (16)$$

where $\bar{x}$ = sample mean of a simple random sample

$E = t_c \dfrac{s}{\sqrt{n}}$

c = confidence level $(0 < c < 1)$

t_c = critical value for confidence level c and degrees of freedom $d.f. = n - 1$ (See Table 6 of Appendix II.)

Problem 23 presents an alternate method for computing confidence intervals when σ is unknown but $n \geq 30$. The alternate method approximates σ by the sample standard deviation s and utilizes the normal distribution.

COMMENT In our applications of Student's t distributions, we have made the basic assumption that x has a normal distribution. However, the same methods apply even if x is only approximately normal. In fact, the main requirement for using a Student's t distribution is that the distribution of x values be reasonably symmetrical and mound-shaped. If this is the case, then the methods we employ with the t distribution can be considered valid for most practical applications.

EXAMPLE 5

CONFIDENCE INTERVAL FOR μ, σ UNKNOWN

Suppose an archaeologist discovers seven fossil skeletons from a previously unknown species of miniature horse. Reconstructions of the skeletons of these seven miniature horses show the shoulder heights (in centimeters) to be

45.3	47.1	44.2	46.8	46.5	45.5	47.6

For these sample data, the mean is $\bar{x} \approx 46.14$ and the sample standard deviation is $s \approx 1.19$. Let μ be the mean shoulder height (in centimeters) for this entire species of miniature horse, and assume that the population of shoulder heights is approximately normal.

Find a 99% confidence interval for μ, the mean shoulder height of the entire population of such horses.

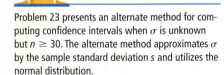

SOLUTION: *Check Requirements* We assume that the shoulder heights of the reconstructed skeletons form a random sample of shoulder heights for all the miniature horses of the unknown species. The x distribution is assumed to be approximately normal. Since σ is unknown, it is appropriate to use a Student's t distribution and sample information to compute a confidence interval for μ.

In this case, $n = 7$, so $d.f. = n - 1 = 7 - 1 = 6$. For $c = 0.990$, Table 6 of Appendix II gives $t_{0.99} = 3.707$ (for $d.f. = 6$). The sample standard deviation is $s \approx 1.19$.

$$E = t_c \frac{s}{\sqrt{n}} = (3.707)\frac{1.19}{\sqrt{7}} \approx 1.67$$

The 99% confidence interval is

$$\bar{x} - E < \mu < \bar{x} + E$$
$$46.14 - 1.67 < \mu < 46.14 + 1.67$$
$$44.5 < \mu < 47.8$$

Interpretation The archaeologist can be 99% confident that the interval from 44.5 cm to 47.8 cm is an interval that contains the population mean μ for shoulder height of this species of miniature horse.

GUIDED EXERCISE 3 CONFIDENCE INTERVAL FOR μ, σ UNKNOWN

A company has a new process for manufacturing large artificial sapphires. In a trial run, 37 sapphires are produced. The distribution of weights is mound-shaped and symmetric. The mean weight for these 37 gems is $\bar{x} = 6.75$ carats, and the sample standard deviation is $s = 0.33$ carat. Let μ be the mean weight for the distribution of all sapphires produced by the new process.

(a) *Check Requirements* Is it appropriate to use a Student's t distribution to compute a confidence interval for μ?

→ Yes, we assume that the 37 sapphires constitute a simple random sample of all sapphires produced under the new process. The requirement that the x distribution be approximately normal can be dropped since the sample size is large enough to ensure that the $\bar{x}$ distribution is approximately normal.

(b) What is $d.f.$ for this setting?

→ $d.f. = n - 1$, where n is the sample size. Since $n = 37$, $d.f. = 37 - 1 = 36$.

(c) Use Table 6 of Appendix II to find $t_{0.95}$. Note that $d.f. = 36$ is not in the table. Use the $d.f.$ closest to 36 that is *smaller* than 36.

→ $d.f. = 35$ is the closest $d.f.$ in the table that is *smaller* than 36. Using $d.f. = 35$ and $c = 0.95$, we find $t_{0.95} = 2.030$.

(d) Find E.

→ $$E = t_{0.95}\frac{s}{\sqrt{n}}$$
$$\approx 2.030\frac{0.33}{\sqrt{37}} \approx 0.11 \text{ carat}$$

(e) Find a 95% confidence interval for μ.

→ $$\bar{x} - E < \mu < \bar{x} + E$$
$$6.75 - 0.11 < \mu < 6.75 + 0.11$$
$$6.64 \text{ carats} < \mu < 6.86 \text{ carats}$$

(f) *Interpretation* What does the confidence interval tell us in the context of the problem?

→ The company can be 95% confident that the interval from 6.64 to 6.86 is an interval that contains the population mean weight of sapphires produced by the new process.

We have several formulas for confidence intervals for the population mean μ. How do we choose an appropriate one? We need to look at the sample size, the distribution of the original population, and whether or not the population standard deviation σ is known.

SUMMARY: CONFIDENCE INTERVALS FOR THE MEAN

Assume that you have a random sample of size n from an x distribution and that you have computed $\bar{x}$ and s. A confidence interval for μ is

$$\bar{x} - E < \mu < \bar{x} + E$$

where E is the margin of error. How do you find E? It depends on how much you know about the x distribution.

Situation I (most common)

You don't know the population standard deviation σ. In this situation, you use the t distribution with margin of error

$$E = t_c \frac{s}{\sqrt{n}}$$

where degrees of freedom

$$d.f. = n - 1$$

Although a t distribution can be used in many situations, you need to observe some guidelines. If n is less than 30, x should have a distribution that is mound-shaped and approximately symmetric. It's even better if the x distribution is normal. If n is 30 or more, the central limit theorem (Chapter 6) implies that these restrictions can be relaxed.

Situation II (almost never happens!)

You actually know the population value of σ. In addition, you know that x has a normal distribution. If you don't know that the x distribution is normal, then your sample size n must be 30 or larger. In this situation, you use the standard normal z distribution with margin of error

$$E = z_c \frac{\sigma}{\sqrt{n}}$$

Which distribution should you use for $\bar{x}$?

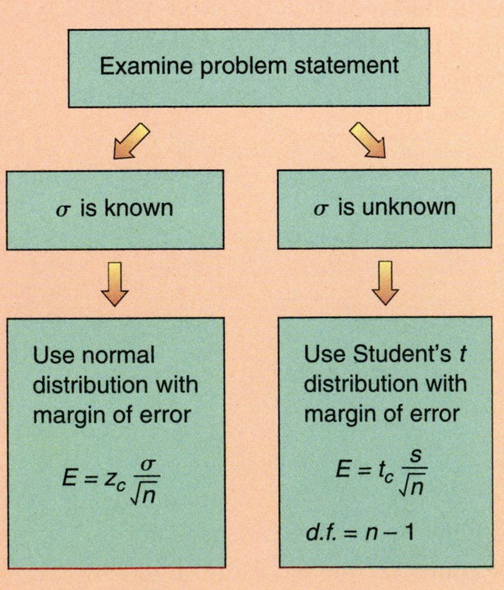

COMMENT To find confidence intervals for μ based on small samples, we need to know that the population distribution is approximately normal. What if this is not the case? A procedure called *bootstrap* utilizes computer power to generate an approximation for the $\bar{x}$ sampling distribution. Essentially, the bootstrap method treats the sample as if it were the population. Then, using repetition, it takes many samples (often thousands) from the original sample. This process is called *resampling*. The sample mean $\bar{x}$ is computed for each resample and a distribution of sample means is created. For example, a 95% confidence interval reflects the range for the middle 95% of the bootstrap $\bar{x}$ distribution. If you read Using Technology at the end of this chapter, you will find one (of many) bootstrap methods (Reference: *An Introduction to the Bootstrap* by B. Efron and R. Tibshirani).

TECH NOTES

The TI-84Plus/TI-83Plus/TI-*n*spire calculators, Excel 2010, and Minitab support confidence intervals using the Student's t distribution.

TI-84Plus/TI-83Plus/TI-*n*spire (with TI-84Plus keypad) Press the **STAT** key, select **TESTS**, and choose the option **8:TInterval**. You may use either raw data in a list or summary statistics.

Excel 2010 Excel gives only the value of the maximal margin of error E. You can easily construct the confidence interval by computing $\bar{x} - E$ and $\bar{x} + E$. On the Home screen, click the **Data** tab. In the Analysis group, click **Data Analysis**. In the dialogue box, select **Descriptive Statistics**. In the dialogue box, check summary statistics and check confidence level for mean. Then set the desired confidence level. Under these choices, Excel uses the Student's t distribution.

An alternate approach utilizing only the sample standard deviation and sample size is to use the **Insert Function** (f_x) button on the Home screen. In the dialogue box, select **Statistical** for the category and then **CONFIDENCE.T** for the function. In the next dialogue box, the value of **alpha** is (1 − the confidence level). For example, alpha is 0.05 for a 95% confidence interval. Enter the standard deviation and the sample size. The output is the value of the maximal margin of error E.

Minitab Use the menu choices **Stat ➤ Basic Statistics ➤ 1-Sample t**. In the dialogue box, indicate the column that contains the raw data. The Minitab output shows the confidence interval for Example 5.

T Confidence Intervals

Variable	N	Mean	StDev	SE Mean	99.0 % CI
C1	7	46.143	1.190	0.450	(44.475, 47.810)

VIEWPOINT Earthquakes!

California, Washington, Nevada, and even Yellowstone National Park all have earthquakes. Some earthquakes are severe! Earthquakes often bring fear and anxiety to people living near the quake. Is San Francisco due for a really big quake like the 1906 major earthquake? How big are the sizes of recent earthquakes compared with really big earthquakes? What is the duration of an earthquake? How long is the time span between major earthquakes? One way to answer questions such as these is to use existing data to estimate confidence intervals for the average size, duration, and time interval between quakes. Recent data sets for computing such confidence intervals can be found at the National Earthquake Information Service of the U.S. Geological Survey web site. To access the site, visit the Brase/Brase statistics site at **http://www.cengage.com/statistics/brase11e** *and find the link to National Earthquake Information Service.*

SECTION 7.2 PROBLEMS

1. Use Table 6 of Appendix II to find t_c for a 0.95 confidence level when the sample size is 18.

2. Use Table 6 of Appendix II to find t_c for a 0.99 confidence level when the sample size is 4.

3. Use Table 6 of Appendix II to find t_c for a 0.90 confidence level when the sample size is 22.

4. Use Table 6 of Appendix II to find t_c for a 0.95 confidence level when the sample size is 12.

5. *Statistical Literacy* Student's t distributions are symmetric about a value of t. What is that t value?

6. *Statistical Literacy* As the degrees of freedom increase, what distribution does the Student's t distribution become more like?

7. *Critical Thinking* Consider a 90% confidence interval for μ. Assume σ is not known. For which sample size, $n = 10$ or $n = 20$, is the critical value t_c larger?

8. *Critical Thinking* Consider a 90% confidence interval for μ. Assume σ is not known. For which sample size, $n = 10$ or $n = 20$, is the confidence interval longer?

9. *Critical Thinking* Lorraine computed a confidence interval for μ based on a sample of size 41. Since she did not know σ, she used s in her calculations. Lorraine used the normal distribution for the confidence interval instead of a Student's t distribution. Was her interval longer or shorter than one obtained by using an appropriate Student's t distribution? Explain.

10. *Critical Thinking* Lorraine was in a hurry when she computed a confidence interval for μ. Because σ was not known, she used a Student's t distribution. However, she accidentally used degrees of freedom n instead of $n - 1$. Was her confidence interval longer or shorter than one found using the correct degrees of freedom $n - 1$? Explain.

Answers may vary slightly due to rounding.

11. *Basic Computation: Confidence Interval* Suppose x has a mound-shaped distribution. A random sample of size 16 has sample mean 10 and sample standard deviation 2.
    (a) *Check Requirements* Is it appropriate to use a Student's t distribution to compute a confidence interval for the population mean μ? Explain.
    (b) Find a 90% confidence interval for μ.
    (c) *Interpretation* Explain the meaning of the confidence interval you computed.

12. *Basic Computation: Confidence Interval* A random sample of size 81 has sample mean 20 and sample standard deviation 3.
    (a) *Check Requirements* Is it appropriate to use a Student's t distribution to compute a confidence interval for the population mean μ? Explain.
    (b) Find a 95% confidence interval for μ.
    (c) *Interpretation* Explain the meaning of the confidence interval you computed.

In Problems 13–19, assume that the population of x values has an approximately normal distribution.

13. *Archaeology: Tree Rings* At Burnt Mesa Pueblo, the method of tree-ring dating gave the following years A.D. for an archaeological excavation site (*Bandelier Archaeological Excavation Project: Summer 1990 Excavations at Burnt Mesa Pueblo*, edited by Kohler, Washington State University):

1189	1271	1267	1272	1268	1316	1275	1317	1275

(a) Use a calculator with mean and standard deviation keys to verify that the sample mean year is $\bar{x} \approx 1272$, with sample standard deviation $s \approx 37$ years.
(b) Find a 90% confidence interval for the mean of all tree-ring dates from this archaeological site.
(c) *Interpretation* What does the confidence interval mean in the context of this problem?

14. *Camping: Cost of a Sleeping Bag* How much does a sleeping bag cost? Let's say you want a sleeping bag that should keep you warm in temperatures from 20°F to 45°F. A random sample of prices ($) for sleeping bags in this temperature range was taken from *Backpacker Magazine: Gear Guide* (Vol. 25, Issue 157, No. 2). Brand names include American Camper, Cabela's, Camp 7, Caribou, Cascade, and Coleman.

80	90	100	120	75	37	30	23	100	110
105	95	105	60	110	120	95	90	60	70

(a) Use a calculator with mean and sample standard deviation keys to verify that $\bar{x} \approx \$83.75$ and $s \approx \$28.97$.
(b) Using the given data as representative of the population of prices of all summer sleeping bags, find a 90% confidence interval for the mean price μ of all summer sleeping bags.
(c) *Interpretation* What does the confidence interval mean in the context of this problem?

15. *Wildlife: Mountain Lions* How much do wild mountain lions weigh? *The 77th Annual Report of the New Mexico Department of Game and Fish*, edited by Bill Montoya, gave the following information. Adult wild mountain lions (18 months or older) captured and released for the first time in the San Andres Mountains gave the following weights (pounds):

68	104	128	122	60	64

(a) Use a calculator with mean and sample standard deviation keys to verify that $\bar{x} = 91.0$ pounds and $s \approx 30.7$ pounds.
(b) Find a 75% confidence interval for the population average weight μ of all adult mountain lions in the specified region.
(c) *Interpretation* What does the confidence interval mean in the context of this problem?

16. *Franchise: Candy Store* Do you want to own your own candy store? With some interest in running your own business and a decent credit rating, you can probably get a bank loan on startup costs for franchises such as Candy Express, The Fudge Company, Karmel Corn, and Rocky Mountain Chocolate Factory. Startup costs (in thousands of dollars) for a random sample of candy stores are given below (Source: *Entrepreneur Magazine*, Vol. 23, No. 10).

95	173	129	95	75	94	116	100	85

(a) Use a calculator with mean and sample standard deviation keys to verify that $\bar{x} \approx 106.9$ thousand dollars and $s \approx 29.4$ thousand dollars.
(b) Find a 90% confidence interval for the population average startup costs μ for candy store franchises.
(c) *Interpretation* What does the confidence interval mean in the context of this problem?

17. *Diagnostic Tests: Total Calcium* Over the past several months, an adult patient has been treated for tetany (severe muscle spasms). This condition is associated with an average total calcium level below 6 mg/dl (Reference: *Manual of Laboratory and Diagnostic Tests* by F. Fischbach). Recently, the patient's total calcium tests gave the following readings (in mg/dl).

9.3	8.8	10.1	8.9	9.4	9.8	10.0
9.9	11.2	12.1				

(a) Use a calculator to verify that $\bar{x} = 9.95$ and $s \approx 1.02$.
(b) Find a 99.9% confidence interval for the population mean of total calcium in this patient's blood.
(c) *Interpretation* Based on your results in part (b), does it seem that this patient still has a calcium deficiency? Explain.

18. *Hospitals: Charity Care* What percentage of hospitals provide at least some charity care? The following problem is based on information taken from *State Health Care Data: Utilization, Spending, and Characteristics* (American Medical Association). Based on a random sample of hospital reports from eastern states, the following information was obtained (units in percentage of hospitals providing at least some charity care):

57.1 56.2 53.0 66.1 59.0 64.7 70.1 64.7 53.5 78.2

(a) Use a calculator with mean and sample standard deviation keys to verify that $\bar{x} \approx 62.3\%$ and $s \approx 8.0\%$.
(b) Find a 90% confidence interval for the population average μ of the percentage of hospitals providing at least some charity care.
(c) *Interpretation* What does the confidence interval mean in the context of this problem?

19. *Critical Thinking: Boxplots and Confidence Intervals* The distribution of heights of 18-year-old men in the United States is approximately normal, with mean 68 inches and standard deviation 3 inches (U.S. Census Bureau). In Minitab, we can simulate the drawing of random samples of size 20 from this population (**➤ Calc ➤ Random Data ➤ Normal**, with 20 rows from a distribution with mean 68 and standard deviation 3). Then we can have Minitab compute a 95% confidence interval and draw a boxplot of the data (**➤ Stat ➤ Basic Statistics ➤ 1—Sample t**, with boxplot selected in the graphs). The boxplots and confidence intervals for four different samples are shown in the accompanying figures. The four confidence intervals are

VARIABLE	N	MEAN	STDEV	SEMEAN	95.0 % CI
Sample 1	20	68.050	2.901	0.649	(66.692 , 69.407)
Sample 2	20	67.958	3.137	0.702	(66.490 , 69.426)
Sample 3	20	67.976	2.639	0.590	(66.741 , 69.211)
Sample 4	20	66.908	2.440	0.546	(65.766 , 68.050)

(a) Examine the figure [parts (a) to (d)]. How do the boxplots for the four samples differ? Why should you expect the boxplots to differ?

95% Confidence Intervals for Mean Height of 18-Year-Old Men (Sample size 20)

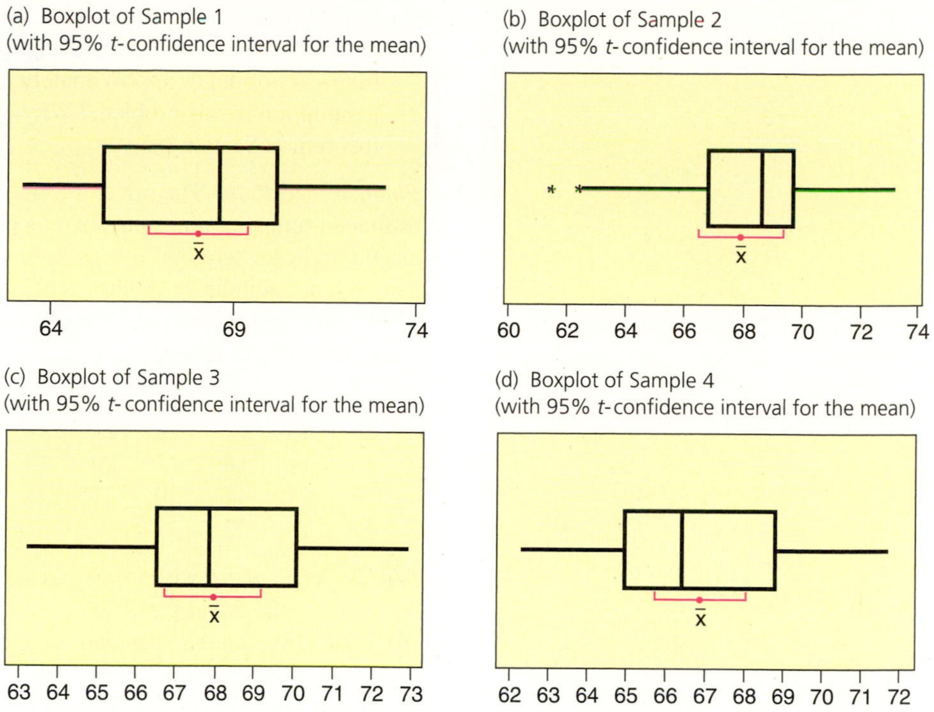

John Lund/Sam Diephuis/Blend Images/Getty Images

(b) Examine the 95% confidence intervals for the four samples shown in the printout. Do the intervals differ in length? Do the intervals all contain the expected population mean of 68 inches? If we draw more samples, do you expect all of the resulting 95% confidence intervals to contain $\mu = 68$? Why or why not?

20. *Crime Rate: Denver* The following data represent crime rates per 1000 population for a random sample of 46 Denver neighborhoods (Reference: *The Piton Foundation*, Denver, Colorado).

63.2	36.3	26.2	53.2	65.3	32.0	65.0
66.3	68.9	35.2	25.1	32.5	54.0	42.4
77.5	123.2	66.3	92.7	56.9	77.1	27.5
69.2	73.8	71.5	58.5	67.2	78.6	33.2
74.9	45.1	132.1	104.7	63.2	59.6	75.7
39.2	69.9	87.5	56.0	154.2	85.5	77.5
84.7	24.2	37.5	41.1			

(a) Use a calculator with mean and sample standard deviation keys to verify that $\bar{x} \approx 64.2$ and $s \approx 27.9$ crimes per 1000 population.

(b) Let us say that the preceding data are representative of the population crime rates in Denver neighborhoods. Compute an 80% confidence interval for μ, the population mean crime rate for all Denver neighborhoods.

(c) *Interpretation* Suppose you are advising the police department about police patrol assignments. One neighborhood has a crime rate of 57 crimes per 1000 population. Do you think that this rate is below the average population crime rate and that fewer patrols could safely be assigned to this neighborhood? Use the confidence interval to justify your answer.

(d) *Interpretation* Another neighborhood has a crime rate of 75 crimes per 1000 population. Does this crime rate seem to be higher than the population average? Would you recommend assigning more patrols to this neighborhood? Use the confidence interval to justify your answer.

(e) Repeat parts (b), (c), and (d) for a 95% confidence interval.

(f) *Check Requirement* In previous problems, we assumed the x distribution was normal or approximately normal. Do we need to make such an assumption in this problem? Why or why not? *Hint:* See the central limit theorem in Section 6.5.

21. *Finance: P/E Ratio* The price of a share of stock divided by a company's estimated future earnings per share is called the P/E ratio. High P/E ratios usually indicate "growth" stocks, or maybe stocks that are simply overpriced. Low P/E ratios indicate "value" stocks or bargain stocks. A random sample of 51 of the largest companies in the United States gave the following P/E ratios (Reference: *Forbes*).

11	35	19	13	15	21	40	18	60	72	9	20
29	53	16	26	21	14	21	27	10	12	47	14
33	14	18	17	20	19	13	25	23	27	5	16
8	49	44	20	27	8	19	12	31	67	51	26
19	18	32									

(a) Use a calculator with mean and sample standard deviation keys to verify that $\bar{x} \approx 25.2$ and $s \approx 15.5$.

(b) Find a 90% confidence interval for the P/E population mean μ of all large U.S. companies.

(c) Find a 99% confidence interval for the P/E population mean μ of all large U.S. companies.

(d) *Interpretation* Bank One (now merged with J.P. Morgan) had a P/E of 12, AT&T Wireless had a P/E of 72, and Disney had a P/E of 24. Examine the confidence intervals in parts (b) and (c). How would you describe these stocks at the time the sample was taken?

(e) *Check Requirements* In previous problems, we assumed the x distribution was normal or approximately normal. Do we need to make such an assumption in this problem? Why or why not? *Hint:* See the central limit theorem in Section 6.5.

22. *Baseball: Home Run Percentage* The home run percentage is the number of home runs per 100 times at bat. A random sample of 43 professional baseball players gave the following data for home run percentages (Reference: *The Baseball Encyclopedia*, Macmillan).

1.6	2.4	1.2	6.6	2.3	0.0	1.8	2.5	6.5	1.8
2.7	2.0	1.9	1.3	2.7	1.7	1.3	2.1	2.8	1.4
3.8	2.1	3.4	1.3	1.5	2.9	2.6	0.0	4.1	2.9
1.9	2.4	0.0	1.8	3.1	3.8	3.2	1.6	4.2	0.0
1.2	1.8	2.4							

(a) Use a calculator with mean and standard deviation keys to verify that $\bar{x} \approx 2.29$ and $s \approx 1.40$.

(b) Compute a 90% confidence interval for the population mean μ of home run percentages for all professional baseball players. *Hint:* If you use Table 6 of Appendix II, be sure to use the closest *d.f.* that is *smaller*.

(c) Compute a 99% confidence interval for the population mean μ of home run percentages for all professional baseball players.

(d) *Interpretation* The home run percentages for three professional players are

Tim Huelett, 2.5 Herb Hunter, 2.0 Jackie Jensen, 3.8

Examine your confidence intervals and describe how the home run percentages for these players compare to the population average.

(e) *Check Requirements* In previous problems, we assumed the x distribution was normal or approximately normal. Do we need to make such an assumption in this problem? Why or why not? *Hint:* See the central limit theorem in Section 6.5.

23. *Expand Your Knowledge: Alternate Method for Confidence Intervals* When σ is unknown and the sample is of size $n \geq 30$, there are two methods for computing confidence intervals for μ.

Method 1: Use the Student's t distribution with $d.f. = n - 1$.
This is the method used in the text. It is widely employed in statistical studies. Also, most statistical software packages use this method.

Method 2: When $n \geq 30$, use the sample standard deviation s as an estimate for σ, and then use the standard normal distribution.
This method is based on the fact that for large samples, s is a fairly good approximation for σ. Also, for large n, the critical values for the Student's t distribution approach those of the standard normal distribution.

Consider a random sample of size $n = 31$, with sample mean $\bar{x} = 45.2$ and sample standard deviation $s = 5.3$.

(a) Compute 90%, 95%, and 99% confidence intervals for μ using Method 1 with a Student's t distribution. Round endpoints to two digits after the decimal.

(b) Compute 90%, 95%, and 99% confidence intervals for μ using Method 2 with the standard normal distribution. Use s as an estimate for σ. Round endpoints to two digits after the decimal.

(c) Compare intervals for the two methods. Would you say that confidence intervals using a Student's t distribution are more conservative in the sense that they tend to be longer than intervals based on the standard normal distribution?

(d) Repeat parts (a) through (c) for a sample of size $n = 81$. With increased sample size, do the two methods give respective confidence intervals that are more similar?

SECTION **7.3**

Estimating *p* in the Binomial Distribution

FOCUS POINTS

- Compute the maximal margin of error for proportions using a given level of confidence.
- Compute confidence intervals for p and interpret the results.
- Interpret poll results.
- Compute the sample size to be used for estimating a proportion p when we have an estimate for p.
- Compute the sample size to be used for estimating a proportion p when we have no estimate for p.

The binomial distribution is completely determined by the number of trials n and the probability p of success on a single trial. For most experiments, the number of trials is chosen in advance. Then the distribution is completely determined by p. In this section, we will consider the problem of estimating p under the assumption that n has already been selected.

We are employing what are called *large-sample methods*. We will assume that the normal curve is a good approximation to the binomial distribution, and when necessary, we will use sample estimates for the standard deviation. Empirical studies have shown that these methods are quite good, provided *both*

Basic criteria

$$np > 5 \text{ and } nq > 5, \text{ where } q = 1 - p$$

Point estimate for p

Let r be the number of successes out of n trials in a binomial experiment. We will take the sample proportion of successes $\hat{p}$ (read "p hat") $= r/n$ as our *point estimate for p*, the population proportion of successes.

> The **point estimates for p and q** are
>
> $$\hat{p} = \frac{r}{n}$$
>
> $$\hat{q} = 1 - \hat{p}$$
>
> where n = number of trials and r = number of successes.

For example, suppose that 800 students are selected at random from a student body of 20,000 and that they are each given a shot to prevent a certain type of flu. These 800 students are then exposed to the flu, and 600 of them do not get the flu. What is the probability p that the shot will be successful for any single student selected at random from the entire population of 20,000 students? We estimate p for the entire student body by computing r/n from the sample of 800 students. The value $\hat{p} = r/n$ is 600/800, or 0.75. The value $\hat{p} = 0.75$ is then the *point estimate* for p.

Margin of error

The difference between the actual value of p and the estimate $\hat{p}$ is the size of our error caused by using $\hat{p}$ as a point estimate for p. The magnitude of $\hat{p} - p$ is called the *margin of error* for using $\hat{p} = r/n$ as a point estimate for p. In absolute value notation, the margin of error is $|\hat{p} - p|$.

To compute the bounds for the margin of error, we need some information about the distribution of $\hat{p} = r/n$ values for different samples of the same size n. It turns out that, for large samples, the distribution of $\hat{p}$ values is well approximated by a *normal curve* with

$$\text{mean} = \mu = p \qquad \text{and} \qquad \text{standard error } \sigma = \sqrt{pq/n}$$

Since the distribution of $\hat{p} = r/n$ is approximately normal, we use features of the standard normal distribution to find the bounds for the difference $\hat{p} - p$. Recall that z_c is the number such that an area equal to c under the standard normal curve falls between $-z_c$ and z_c. Then, in terms of the language of probability,

$$P\left(-z_c\sqrt{\frac{pq}{n}} < \hat{p} - p < z_c\sqrt{\frac{pq}{n}}\right) = c \tag{17}$$

Equation (17) says that the chance is c that the numerical difference between $\hat{p}$ and p is between $-z_c\sqrt{pq/n}$ and $z_c\sqrt{pq/n}$. With the c confidence level, our estimate $\hat{p}$ differs from p by no more than

Maximal margin of error, E

$$E = z_c\sqrt{pq/n}$$

As in Section 7.1, we call E the *maximal margin of error*.

OPTIONAL DERIVATION OF EQUATION (17)

First, we need to show that $\hat{p} = r/n$ has a distribution that is approximately normal, with $\mu = p$ and $\sigma = \sqrt{pq/n}$. From Section 6.6, we know that for sufficiently large n, the binomial distribution can be approximated by a normal distribution with

Continued

mean $\mu = np$ and standard deviation $\sigma = \sqrt{npq}$. If r is the number of successes out of n trials of a binomial experiment, then r is a binomial random variable with a binomial distribution. When we convert r to standard z units, we obtain

$$z = \frac{r - \mu}{\sigma} = \frac{r - np}{\sqrt{npq}}$$

For sufficiently large n, r will be approximately normally distributed, so z will be too.

If we divide both numerator and denominator of the last expression by n, the value of z will not change.

$$z = \frac{\dfrac{r - np}{n}}{\dfrac{\sqrt{npq}}{n}} \quad \text{Simplified, we find } z = \frac{\dfrac{r}{n} - p}{\sqrt{\dfrac{pq}{n}}}. \tag{18}$$

Equation (18) tells us that the $\hat{p} = r/n$ distribution is approximated by a normal curve with $\mu = p$ and $\sigma = \sqrt{pq/n}$.

The probability is c that z lies in the interval between $-z_c$ and z_c because an area equal to c under the standard normal curve lies between $-z_c$ and z_c. Using the language of probability, we write

$$P(-z_c < z < z_c) = c$$

From Equation (18), we know that

$$z = \frac{\hat{p} - p}{\sqrt{\dfrac{pq}{n}}}$$

If we put this expression for z into the preceding equation, we obtain

$$P\left(-z_c < \frac{\hat{p} - p}{\sqrt{\dfrac{pq}{n}}} < z_c\right) = c$$

If we multiply all parts of the inequality by $\sqrt{pq/n}$, we obtain the equivalent statement

$$P\left(-z_c\sqrt{\frac{pq}{n}} < \hat{p} - p < z_c\sqrt{\frac{pq}{n}}\right) = c \tag{17}$$

Confidence interval for *p*

To find a c confidence interval for p, we will use E in place of the expression $z_c\sqrt{pq/n}$ in Equation (17). Then we get

$$P(-E < \hat{p} - p < E) = c \tag{19}$$

Some algebraic manipulation produces the mathematically equivalent statement

$$P(\hat{p} - E < p < \hat{p} + E) = c \tag{20}$$

Equation (20) states that the probability is c that p lies in the interval from $\hat{p} - E$ to $\hat{p} + E$. Therefore, the interval from $\hat{p} - E$ to $\hat{p} + E$ is the c confidence interval for p that we wanted to find.

There is one technical difficulty in computing the c confidence interval for p. The expression $E = z_c\sqrt{pq/n}$ requires that we know the values of p and q. In most situations, we will not know the actual values of p or q, so we will use our point estimates

$$p \approx \hat{p} \quad \text{and} \quad q = 1 - p \approx 1 - \hat{p}$$

to estimate E. These estimates are reliable for most practical purposes, since we are dealing with large-sample theory ($np > 5$ and $nq > 5$).

For convenient reference, we'll summarize the information about c confidence intervals for p, the probability of success in a binomial distribution.

PROCEDURE

HOW TO FIND A CONFIDENCE INTERVAL FOR A PROPORTION p

Requirements

Consider a binomial experiment with n trials, where p represents the population probability of success on a single trial and $q = 1 - p$ represents the population probability of failure. Let r be a random variable that represents the number of successes out of the n binomial trials.

The point estimates for p and q are

$$\hat{p} = \frac{r}{n} \quad \text{and} \quad \hat{q} = 1 - \hat{p}$$

The number of trials n should be sufficiently large so that both $n\hat{p} > 5$ and $n\hat{q} > 5$.

Confidence interval for p

$$\hat{p} - E < p < \hat{p} + E$$

where $E \approx z_c\sqrt{\frac{\hat{p}\hat{q}}{n}} = z_c\sqrt{\frac{\hat{p}(1-\hat{p})}{n}}$

c = confidence level ($0 < c < 1$)

z_c = critical value for confidence level c based on the standard normal distribution (See Table 5(b) of Appendix II for frequently used values.)

Problem 28 of this section shows the method for computing a "plus four confidence interval for p." This is an alternate method that generally results in a slightly smaller confidence interval than the ones computed with standard methods presented in this section and used in most statistical technology packages.

COMMENT Problem 6 asks you to show that the two conditions $n\hat{p} > 5$ and $n\hat{q} > 5$ are equivalent to the two conditions that the number of successes $r > 5$ and the number of failures $n - r > 5$.

EXAMPLE 6

CONFIDENCE INTERVAL FOR p

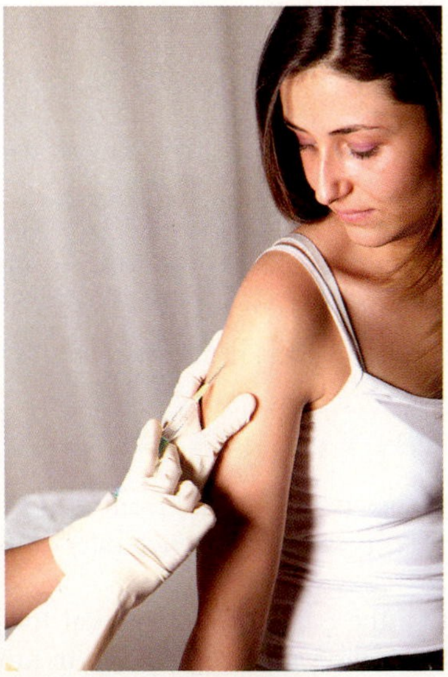

Christine Langer-Pueschel/Shutterstock

Let's return to our flu shot experiment described at the beginning of this section. Suppose that 800 students were selected at random from a student body of 20,000 and given shots to prevent a certain type of flu. All 800 students were exposed to the flu, and 600 of them did not get the flu. Let p represent the probability that the shot will be successful for any single student selected at random from the entire population of 20,000. Let q be the probability that the shot is not successful.

(a) What is the number of trials n? What is the value of r?

SOLUTION: Since each of the 800 students receiving the shot may be thought of as a trial, then $n = 800$, and $r = 600$ is the number of successful trials.

(b) What are the point estimates for p and q?

SOLUTION: We estimate p by the sample point estimate

$$\hat{p} = \frac{r}{n} = \frac{600}{800} = 0.75$$

We estimate q by

$$\hat{q} = 1 - \hat{p} = 1 - 0.75 = 0.25$$

(c) ***Check Requirements*** Would it seem that the number of trials is large enough to justify a normal approximation to the binomial?

SOLUTION: Since $n = 800$, $p \approx 0.75$, and $q \approx 0.25$, then

$$np \approx (800)(0.75) = 600 > 5 \quad \text{and} \quad np \approx (800)(0.25) = 200 > 5$$

A normal approximation is certainly justified.

(d) Find a 99% confidence interval for p.

SOLUTION:

$$z_{0.99} = 2.58 \text{ (see Table 7-2 or Table 5(b) of Appendix II)}$$

$$E \approx z_{0.99}\sqrt{\frac{\hat{p}(1-\hat{p})}{n}} \approx 2.58\sqrt{\frac{(0.75)(0.25)}{800}} \approx 0.0395$$

The 99% confidence interval is then

$$\hat{p} - E < p < \hat{p} + E$$
$$0.75 - 0.0395 < p < 0.75 + 0.0395$$
$$0.71 < p < 0.79$$

Interpretation We are 99% confident that the probability a flu shot will be effective for a student selected at random is between 0.71 and 0.79.

GUIDED EXERCISE 4 | CONFIDENCE INTERVAL FOR p

A random sample of 188 books purchased at a local bookstore showed that 66 of the books were murder mysteries. Let p represent the proportion of books sold by this store that are murder mysteries.

(a) What is a point estimate for p?

→ $\hat{p} = \frac{r}{n} = \frac{66}{188} \approx 0.35$

(b) Find a 90% confidence interval for p.

→ $E = z_c\sqrt{\frac{\hat{p}(1-\hat{p})}{n}}$

$$\approx 1.645\sqrt{\frac{(0.35)(1-0.35)}{188}} \approx 0.0572$$

The confidence interval is

$$\hat{p} - E < p < \hat{p} + E$$
$$0.35 - 0.0572 < p < 0.35 + 0.0572$$
$$0.29 < p < 0.41$$

(c) ***Interpretation*** What does the confidence interval you just computed mean in the context of this application?

→ If we had computed the interval for many different sets of 188 books, we would have found that about 90% of the intervals actually contained p, the population proportion of mysteries. Consequently, we can be 90% confident that our interval is one of the intervals that contain the unknown value p.

Continued

GUIDED EXERCISE 4 *continued*

(d) ***Check Requirements*** To compute the confidence interval, we used a normal approximation. Does this seem justified?	$n = 188$; $p \approx 0.35$; $q \approx 0.65$ Since $np \approx 65.8 > 5$ and $np \approx 122.2 > 5$, the approximation is justified.

It is interesting to note that our sample point estimate $\hat{p} = r/n$ and the confidence interval for the population proportion p do not depend on the size of the population. In our bookstore example, it made no difference how many books the store sold. On the other hand, the size of the sample does affect the accuracy of a statistical estimate. At the end of this section, we will study the effect of sample size on the reliability of our estimate.

TECH NOTES The TI-84Plus/TI-83Plus/TI-*n*spire calculators and Minitab provide confidence intervals for proportions.

TI-84Plus/TI-83Plus/TI-*n*spire (with TI-84Plus keypad) Press the **STAT** key, select **TESTS**, and choose option **A:1-PropZInt**. The letter x represents the number of successes r. The TI-84Plus/TI-83Plus/TI-*n*spire output shows the results for Guided Exercise 4.

```
1-PropZInt
 (.29381,.40832)
 p̂=.3510638298
 n=188
```

Minitab Use the menu selections **Stat ➤ Basic Statistics ➤ 1 Proportion**. In the dialogue box, select Summarized Data and fill in the number of trials and the number of successes. Under Options, select a confidence interval. Minitab uses the binomial distribution directly unless Normal is checked. The Minitab output shows the results for Guided Exercise 4. Information from Chapter 8 material is also shown.

```
Test and Confidence Interval for One Proportion (Using Binomial)
Test of p = 0.5 vs p not = 0.5
                                                            Exact
Sample    X      N    Sample p          90.0 % CI          P-Value
  1      66    188    0.351064     (0.293222, 0.412466)     0.000
Test and Confidence Interval for One Proportion (Using Normal)
Test of p = 0.5 vs p not = 0.5
Sample   X     N   Sample p         90.0 % CI          Z-Value  P-Value
1       66   188   0.351064  (0.293805, 0.408323)   -4.08    0.000
```

INTERPRETING RESULTS FROM A POLL

Newspapers frequently report the results of an opinion poll. In articles that give more information, a statement about the margin of error accompanies the poll results. In most polls, the margin of error is given for a *95% confidence interval*.

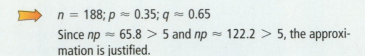

Margin of error for polls

GENERAL INTERPRETATION OF POLL RESULTS

1. When a poll states the results of a survey, the proportion reported to respond in the designated manner is $\hat{p}$, the sample estimate of the population proportion.
2. The *margin of error* is the maximal error E of a 95% confidence interval for p.
3. A 95% confidence interval for the population proportion p is
poll report $\hat{p}$ − margin of error $E < p <$ poll report $\hat{p}$ + margin of error E

COMMENT Leslie Kish, a statistician at the University of Michigan, was the first to apply the term *margin of error*. He was a pioneer in the study of population sampling techniques. His book *Survey Sampling* is still widely used all around the world.

Some articles clarify the meaning of the margin of error further by saying that it is an error due to sampling. For instance, the following comments accompany results of a political poll reported in an issue of *The Wall Street Journal*.

How Poll Was Conducted

The Wall Street Journal/NBC News poll was based on nationwide telephone interviews of 1508 adults conducted last Friday through Tuesday by the polling organizations of Peter Hart and Robert Teeter.

The sample was drawn from 315 randomly selected geographic points in the continental United States. Each region was represented in proportion to its population. Households were selected by a method that gave all telephone numbers . . . an equal chance of being included.

One adult, 18 years or older, was selected from each household by a procedure to provide the correct number of male and female respondents.

Chances are 19 of 20 that if all adults with telephones in the United States had been surveyed, the findings would differ from these poll results by no more than 2.6 percentage points in either direction.

GUIDED EXERCISE 5 | READING A POLL

Read the last paragraph of the article excerpt "How Poll Was Conducted."

(a) What confidence level corresponds to the phrase "chances are 19 of 20 that if . . ."?

$$\frac{19}{20} = 0.95$$

A 95% confidence interval is being discussed.

(b) The complete article indicates that everyone in the sample was asked the question "Which party, the Democratic Party or the Republican Party, do you think would do a better job handling . . . education?" Possible responses were "Democrats," "neither," "both," or "Republicans." The poll reported that 32% of the respondents said, "Democrats." Does 32% represent the sample statistic $\hat{p}$ or the population parameter p for the proportion of adults responding, "Democrats"?

32% represents a sample statistic $\hat{p}$ because 32% represents the percentage of the adults in the *sample* who responded, "Democrats."

Continued

GUIDED EXERCISE 5 *continued*

(c) Continue reading the last paragraph of the article. It goes on to state, "... if all adults with telephones in the U.S. had been surveyed, the findings would differ from these poll results by no more than 2.6 percentage points in either direction." Use this information, together with parts (a) and (b), to find a 95% confidence interval for the proportion p of the specified population who responded, "Democrats" to the question.

The value 2.6 percentage points represents the margin of error. Since the margin of error is for a 95% confidence interval, the confidence interval is

$$32\% - 2.6\% < p < 32\% + 2.6\%$$
$$29.4\% < p < 34.6\%$$

The poll indicates that at the time of the poll, between 29.4% and 34.6% of the specified population thought Democrats would do a better job handling education.

SAMPLE SIZE FOR ESTIMATING p

Suppose you want to specify the maximal margin of error in advance for a confidence interval for p at a given confidence level c. What sample size do you need? The answer depends on whether or not you have a preliminary estimate for the population probability of success p in a binomial distribution.

PROCEDURE

HOW TO FIND THE SAMPLE SIZE n FOR ESTIMATING A PROPORTION p

$$n = p(1-p)\left(\frac{z_c}{E}\right)^2 \text{ if you have a preliminary estimate for } p \quad (21)$$

$$n = \frac{1}{4}\left(\frac{z_c}{E}\right)^2 \text{ if you do } \textit{not} \text{ have a preliminary estimate for } p \quad (22)$$

where E = specified maximal error of estimate

z_c = critical value from the normal distribution for the desired confidence level c. Commonly used value of z_c can be found in Table 5(b) of Appendix II.

If n is not a whole number, increase n to the next higher whole number. Also, if necessary, increase the sample size n to ensure that both $np > 5$ and $nq > 5$. Note that n is the minimal sample size for a specified confidence level and maximal error of estimate.

COMMENT To obtain Equation (21), simply solve the formula that gives the maximal error of estimate E of p for the sample size n. When you don't have an estimate for p, a little algebra can be used to show that the maximum value of $p(1-p)$ is 1/4. See Problem 27.

EXAMPLE 7

SAMPLE SIZE FOR ESTIMATING p

A company is in the business of selling wholesale popcorn to grocery stores. The company buys directly from farmers. A buyer for the company is examining a large amount of corn from a certain farmer. Before the purchase is made, the buyer wants to estimate p, the probability that a kernel will pop.

Suppose a random sample of n kernels is taken and r of these kernels pop. The buyer wants to be 95% sure that the point estimate $\hat{p} = r/n$ for p will be in error either way by less than 0.01.

(a) If no preliminary study is made to estimate p, how large a sample should the buyer use?

SOLUTION: In this case, we use Equation (22) with $z_{0.95} = 1.96$ (see Table 7-2) and $E = 0.01$.

$$n = \frac{1}{4}\left(\frac{z_c}{E}\right)^2 = \frac{1}{4}\left(\frac{1.96}{0.01}\right)^2 = 0.25(38{,}416) = 9604$$

The buyer would need a sample of $n = 9604$ kernels.

(b) A preliminary study showed that p was approximately 0.86. If the buyer uses the results of the preliminary study, how large a sample should he use?

SOLUTION: In this case, we use Equation (21) with $p \approx 0.86$. Again, from Table 7-2, $z_{0.95} = 1.96$, and from the problem, $E = 0.01$.

$$n = p(1-p)\left(\frac{z_c}{E}\right)^2 = (0.86)(0.14)\left(\frac{1.96}{0.01}\right)^2 = 4625.29$$

The sample size should be at least $n = 4626$. This sample is less than half the sample size necessary without the preliminary study.

VIEWPOINT "Band-Aid Surgery"

Faster recovery time and less pain. Sounds great! An alternate surgical technique called laparoscopic ("Band-Aid") surgery *involves small incisions through which tiny video cameras and long surgical instruments are maneuvered. Instead of a 10-inch incision, surgeons might use four little stabs of about $\frac{1}{2}$-inch in length. However, not every such surgery is successful. An article in the Health section of* The Wall Street Journal *recommends using a surgeon who has done at least 50 such surgeries. Then the prospective patient should ask about the* rate of conversion—*that is, the proportion* p *of times the surgeon has been forced by complications to switch in midoperation to conventional surgery. A confidence interval for the proportion* p *would be useful patient information!*

SECTION 7.3 PROBLEMS

For all these problems, carry at least four digits after the decimal in your calculations. Answers may vary slightly due to rounding.

1. *Statistical Literacy* For a binomial experiment with r successes out of n trials, what value do we use as a point estimate for the probability of success p on a single trial?

2. *Statistical Literacy* In order to use a normal distribution to compute confidence intervals for p, what conditions on np and nq need to be satisfied?

3. *Critical Thinking* Results of a poll of a random sample of 3003 American adults showed that 20% did not know that caffeine contributes to dehydration. The poll was conducted for the Nutrition Information Center and had a margin of error of ±1.4%.
   (a) Does the margin of error take into account any problems with the wording of the survey question, interviewer errors, bias from sequence of questions, and so forth?
   (b) What does the margin of error reflect?

4. *Critical Thinking* You want to conduct a survey to determine the proportion of people who favor a proposed tax policy. How does increasing the sample size affect the size of the margin of error?

5. *Critical Thinking* Jerry tested 30 laptop computers owned by classmates enrolled in a large computer science class and discovered that 22 were infected with keystroke-tracking spyware. Is it appropriate for Jerry to use his data to estimate the proportion of all laptops infected with such spyware? Explain.

6. *Critical Thinking: Brain Teaser* A requirement for using the normal distribution to approximate the $\hat{p}$ distribution is that both $np > 5$ and $nq > 5$. Since we usually do not know p, we estimate p by $\hat{p}$ and q by $\hat{q} = 1 - \hat{p}$. Then we require that $n\hat{p} > 5$ and $n\hat{q} > 5$. Show that the conditions $n\hat{p} > 5$ and $n\hat{q} > 5$ are equivalent to the condition that out of n binomial trials, both the number of successes r and the number of failures $n - r$ must exceed 5. *Hint:* In the inequality $n\hat{p} > 5$, replace $\hat{p}$ by r/n and solve for r. In the inequality $n\hat{q} > 5$, replace $\hat{q}$ by $(n - r)/n$ and solve for $n - r$.

7. *Basic Computation: Confidence Interval for p* Consider $n = 100$ binomial trials with $r = 30$ successes.
   (a) *Check Requirements* Is it appropriate to use a normal distribution to approximate the $\hat{p}$ distribution?
   (b) Find a 90% confidence interval for the population proportion of successes p.
   (c) *Interpretation* Explain the meaning of the confidence interval you computed.

8. *Basic Computation: Confidence Interval for p* Consider $n = 200$ binomial trials with $r = 80$ successes.
   (a) *Check Requirements* Is it appropriate to use a normal distribution to approximate the $\hat{p}$ distribution?
   (b) Find a 95% confidence interval for the population proportion of successes p.
   (c) *Interpretation* Explain the meaning of the confidence interval you computed.

9. *Basic Computation: Sample Size* What is the minimal sample size needed for a 95% confidence interval to have a maximal margin of error of 0.1
   (a) if a preliminary estimate for p is 0.25?
   (b) if there is no preliminary estimate for p?

10. *Basic Computation: Sample Size* What is the minimal sample size needed for a 99% confidence interval to have a maximal margin of error of 0.06
    (a) if a preliminary estimate for p is 0.8?
    (b) if there is no preliminary estimate for p?

11. *Myers–Briggs: Actors* Isabel Myers was a pioneer in the study of personality types. The following information is taken from *A Guide to the Development and Use of the Myers–Briggs Type Indicator* by Myers and McCaulley (Consulting Psychologists Press). In a random sample of 62 professional actors, it was found that 39 were extroverts.
    (a) Let p represent the proportion of all actors who are extroverts. Find a point estimate for p.
    (b) Find a 95% confidence interval for p. Give a brief interpretation of the meaning of the confidence interval you have found.
    (c) *Check Requirements* Do you think the conditions $np > 5$ and $nq > 5$ are satisfied in this problem? Explain why this would be an important consideration.

12. *Myers–Briggs: Judges* In a random sample of 519 judges, it was found that 285 were introverts. (See reference in Problem 11.)
    (a) Let p represent the proportion of all judges who are introverts. Find a point estimate for p.
    (b) Find a 99% confidence interval for p. Give a brief interpretation of the meaning of the confidence interval you have found.
    (c) *Check Requirements* Do you think the conditions $np > 5$ and $nq > 5$ are satisfied in this problem? Explain why this would be an important consideration.

13. *Navajo Lifestyle: Traditional Hogans* A random sample of 5222 permanent dwellings on the entire Navajo Indian Reservation showed that 1619 were traditional Navajo hogans (*Navajo Architecture: Forms, History, Distributions* by Jett and Spencer, University of Arizona Press).
    (a) Let p be the proportion of all permanent dwellings on the entire Navajo Reservation that are traditional hogans. Find a point estimate for p.
    (b) Find a 99% confidence interval for p. Give a brief interpretation of the confidence interval.
    (c) *Check Requirements* Do you think that $np > 5$ and $nq > 5$ are satisfied for this problem? Explain why this would be an important consideration.

14. *Archaeology: Pottery* Santa Fe black-on-white is a type of pottery commonly found at archaeological excavations in Bandelier National Monument. At one excavation site a sample of 592 potsherds was found, of which 360 were identified as Santa Fe black-on-white (*Bandelier Archaeological Excavation Project: Summer 1990 Excavations at Burnt Mesa Pueblo and Casa del Rito*, edited by Kohler and Root, Washington State University).
    (a) Let p represent the population proportion of Santa Fe black-on-white potsherds at the excavation site. Find a point estimate for p.
    (b) Find a 95% confidence interval for p. Give a brief statement of the meaning of the confidence interval.
    (c) *Check Requirements* Do you think the conditions $np > 5$ and $nq > 5$ are satisfied in this problem? Why would this be important?

15. *Health Care: Colorado Physicians* A random sample of 5792 physicians in Colorado showed that 3139 provide at least some charity care (i.e., treat poor people at no cost). These data are based on information from *State Health Care Data: Utilization, Spending, and Characteristics* (American Medical Association).
    (a) Let p represent the proportion of all Colorado physicians who provide some charity care. Find a point estimate for p.
    (b) Find a 99% confidence interval for p. Give a brief explanation of the meaning of your answer in the context of this problem.
    (c) *Check Requirements* Is the normal approximation to the binomial justified in this problem? Explain.

16. *Law Enforcement: Escaped Convicts* Case studies showed that out of 10,351 convicts who escaped from U.S. prisons, only 7867 were recaptured (*The Book of Odds* by Shook and Shook, Signet).
    (a) Let p represent the proportion of all escaped convicts who will eventually be recaptured. Find a point estimate for p.
    (b) Find a 99% confidence interval for p. Give a brief statement of the meaning of the confidence interval.
    (c) *Check Requirements* Is use of the normal approximation to the binomial justified in this problem? Explain.

17. *Fishing: Barbless Hooks* In a combined study of northern pike, cutthroat trout, rainbow trout, and lake trout, it was found that 26 out of 855 fish died when caught and released using barbless hooks on flies or lures. All hooks were removed from the fish (Source: *A National Symposium on Catch and Release Fishing*, Humboldt State University Press).
    (a) Let p represent the proportion of all pike and trout that die (i.e., p is the mortality rate) when caught and released using barbless hooks. Find a point estimate for p.
    (b) Find a 99% confidence interval for p, and give a brief explanation of the meaning of the interval.
    (c) *Check Requirements* Is the normal approximation to the binomial justified in this problem? Explain.

18. *Physicians: Solo Practice* A random sample of 328 medical doctors showed that 171 have a solo practice (Source: *Practice Patterns of General Internal Medicine*, American Medical Association).
(a) Let p represent the proportion of all medical doctors who have a solo practice. Find a point estimate for p.
(b) Find a 95% confidence interval for p. Give a brief explanation of the meaning of the interval.
(c) *Interpretation* As a news writer, how would you report the survey results regarding the percentage of medical doctors in solo practice? What is the margin of error based on a 95% confidence interval?

19. *Marketing: Customer Loyalty* In a marketing survey, a random sample of 730 women shoppers revealed that 628 remained loyal to their favorite supermarket during the past year (i.e., did not switch stores) (Source: *Trends in the United States: Consumer Attitudes and the Supermarket*, The Research Department, Food Marketing Institute).
(a) Let p represent the proportion of all women shoppers who remain loyal to their favorite supermarket. Find a point estimate for p.
(b) Find a 95% confidence interval for p. Give a brief explanation of the meaning of the interval.
(c) *Interpretation* As a news writer, how would you report the survey results regarding the percentage of women supermarket shoppers who remained loyal to their favorite supermarket during the past year? What is the margin of error based on a 95% confidence interval?

20. *Marketing: Bargain Hunters* In a marketing survey, a random sample of 1001 supermarket shoppers revealed that 273 always stock up on an item when they find that item at a real bargain price. (See reference in Problem 19.)
(a) Let p represent the proportion of all supermarket shoppers who always stock up on an item when they find a real bargain. Find a point estimate for p.
(b) Find a 95% confidence interval for p. Give a brief explanation of the meaning of the interval.
(c) *Interpretation* As a news writer, how would you report the survey results on the percentage of supermarket shoppers who stock up on real-bargain items? What is the margin of error based on a 95% confidence interval?

21. *Lifestyle: Smoking* In a survey of 1000 large corporations, 250 said that, given a choice between a job candidate who smokes and an equally qualified nonsmoker, the nonsmoker would get the job (*USA Today*).
(a) Let p represent the proportion of all corporations preferring a nonsmoking candidate. Find a point estimate for p.
(b) Find a 0.95 confidence interval for p.
(c) *Interpretation* As a news writer, how would you report the survey results regarding the proportion of corporations that hire the equally qualified nonsmoker? What is the margin of error based on a 95% confidence interval?

22. *Opinion Poll: Crime and Violence* A *New York Times*/CBS poll asked the question, "What do you think is the most important problem facing this country today?" Nineteen percent of the respondents answered, "Crime and violence." The margin of sampling error was plus or minus 3 percentage points. Following the convention that the margin of error is based on a 95% confidence interval, find a 95% confidence interval for the percentage of the population that would respond, "Crime and violence" to the question asked by the pollsters.

23. *Medical: Blood Type* A random sample of medical files is used to estimate the proportion p of all people who have blood type B.
(a) If you have no preliminary estimate for p, how many medical files should you include in a random sample in order to be 85% sure that the point estimate $\hat{p}$ will be within a distance of 0.05 from p?

(b) Answer part (a) if you use the preliminary estimate that about 8 out of 90 people have blood type B (Reference: *Manual of Laboratory and Diagnostic Tests* by F. Fischbach).

24. *Business: Phone Contact* How hard is it to reach a businessperson by phone? Let p be the proportion of calls to businesspeople for which the caller reaches the person being called on the *first* try.
(a) If you have no preliminary estimate for p, how many business phone calls should you include in a random sample to be 80% sure that the point estimate $\hat{p}$ will be within a distance of 0.03 from p?
(b) *The Book of Odds* by Shook and Shook (Signet) reports that businesspeople can be reached by a single phone call approximately 17% of the time. Using this (national) estimate for p, answer part (a).

25. *Campus Life: Coeds* What percentage of your campus student body is female? Let p be the proportion of women students on your campus.
(a) If no preliminary study is made to estimate p, how large a sample is needed to be 99% sure that a point estimate $\hat{p}$ will be within a distance of 0.05 from p?
(b) The *Statistical Abstract of the United States*, 112th Edition, indicates that approximately 54% of college students are female. Answer part (a) using this estimate for p.

26. *Small Business: Bankruptcy* The National Council of Small Businesses is interested in the proportion of small businesses that declared Chapter 11 bankruptcy last year. Since there are so many small businesses, the National Council intends to estimate the proportion from a random sample. Let p be the proportion of small businesses that declared Chapter 11 bankruptcy last year.
(a) If no preliminary sample is taken to estimate p, how large a sample is necessary to be 95% sure that a point estimate $\hat{p}$ will be within a distance of 0.10 from p?
(b) In a preliminary random sample of 38 small businesses, it was found that six had declared Chapter 11 bankruptcy. How many *more* small businesses should be included in the sample to be 95% sure that a point estimate $\hat{p}$ will be within a distance of 0.10 from p?

27. *Brain Teaser: Algebra* Why do we use 1/4 in place of $p(1 - p)$ in formula (22) for sample size when the probability of success p is unknown?
(a) Show that $p(1 - p) = 1/4 - (p - 1/2)^2$.
(b) Why is $p(1 - p)$ never greater than 1/4?

28. *Expand Your Knowledge: Plus Four Confidence Interval for a Single Proportion* One of the technical difficulties that arises in the computation of confidence intervals for a single proportion is that the exact formula for the maximal margin of error requires knowledge of the population proportion of success p. Since p is usually not known, we use the sample estimate $\hat{p} = r/n$ in place of p. As discussed in the article "How Much Confidence Should You Have in Binomial Confidence Intervals?" appearing in issue no. 45 of the magazine *STATS* (a publication of the American Statistical Association), use of $\hat{p}$ as an estimate for p means that the actual confidence level for the intervals may in fact be smaller than the specified level c. This problem arises even when n is large, especially if p is not near 1/2.

A simple adjustment to the formula for the confidence intervals is the *plus four estimate*, first suggested by Edwin Bidwell Wilson in 1927. It is also called the Agresti–Coull confidence interval. This adjustment works best for 95% confidence intervals.

The plus four adjustment has us add two successes and two failures to the sample data. This means that r, the number of successes, is increased by 2, and n, the sample size, is increased by 4. We use the symbol $\tilde{p}$, read "p tilde," for the resulting sample estimate of p. So, $\tilde{p} = (r + 2)/(n + 4)$.

PROCEDURE

HOW TO COMPUTE A PLUS FOUR CONFIDENCE INTERVAL FOR *p*

Requirements

Consider a binomial experiment with n trials, where p represents the population probability of success and $q = 1 - p$ represents the population probability of failure. Let r be a random variable that represents the number of successes out of the n binomial trials.

The plus four point estimates for p and q are

$$\tilde{p} = \frac{r + 2}{n + 4} \quad \text{and} \quad \tilde{q} = 1 - \tilde{p}$$

The number of trials n should be at least 10.

Approximate confidence interval for p

$$\tilde{p} - E < p < \tilde{p} + E$$

$$\text{where} \quad E \approx z_c\sqrt{\frac{\tilde{p}\tilde{q}}{n + 4}} = z_c\sqrt{\frac{\tilde{p}(1 - \tilde{p})}{n + 4}}$$

c = confidence level ($0 < c < 1$)

z_c = critical value for confidence level c based on the standard normal distribution

(a) Consider a random sample of 50 trials with 20 successes. Compute a 95% confidence interval for p using the plus four method.

(b) Compute a traditional 95% confidence interval for p using a random sample of 50 trials with 20 successes.

(c) Compare the lengths of the intervals obtained using the two methods. Is the point estimate closer to 1/2 when using the plus four method? Is the margin of error smaller when using the plus four method?

SECTION **7.4**

Estimating $\mu_1 - \mu_2$ and $p_1 - p_2$

FOCUS POINTS

- Distinguish between independent and dependent samples.
- Compute confidence intervals for $\mu_1 - \mu_2$ when σ_1 and σ_2 are known.
- Compute confidence intervals for $\mu_1 - \mu_2$ when σ_1 and σ_2 are unknown.
- Compute confidence intervals for $p_1 - p_2$ using the normal approximation.
- Interpret the meaning and implications of an all-positive, all-negative, or mixed confidence interval.

INDEPENDENT SAMPLES AND DEPENDENT SAMPLES

How can we tell if two populations are different? One way is to compare the difference in population means or the difference in population proportions. In this section, we will use samples from two populations to create confidence intervals for the difference between population parameters.

To make a statistical estimate about the difference between two population parameters, we need to have a sample from each population. Samples may be *independent* or *dependent* according to how they are selected.

Independent samples

Two samples are **independent** if sample data drawn from one population are completely unrelated to the selection of sample data from the other population.

Dependent samples

Two samples are **dependent** if each data value in one sample can be paired with a corresponding data value in the other sample.

Dependent samples and data pairs occur very naturally in "before and after" situations in which the *same object* or item is *measured twice*. We will devote an entire section (8.4) to the study of dependent samples and paired data. However, in this section, we will confine our interest to independent samples.

Independent samples occur very naturally when we draw *two random samples*, one from the first population and one from the second population. Because *both* samples are random samples, there is no pairing of measurements between the two populations. All the examples of this section will involve independent random samples.

GUIDED EXERCISE 6 Distinguishing between independent and dependent samples

For each experiment, categorize the sampling as independent or dependent, and explain your choice.

(a) In many medical experiments, a sample of subjects is randomly divided into two groups. One group is given a specific treatment, and the other group is given a placebo. After a certain period of time, both groups are measured for the same condition. Do the measurements from these two groups constitute independent or dependent samples?

→ Since the subjects are *randomly assigned* to the two treatment groups (one receives a treatment, the other a placebo), the resulting measurements form independent samples.

(b) In an accountability study, a group of students in an English composition course is given a pretest. After the course, the same students are given a posttest covering similar material. Are the two groups of scores independent or dependent?

→ Since the pretest scores and the posttest scores are from the same students, the samples are dependent. Each student has both a pretest score and a posttest score, so there is a natural pairing of data values.

CONFIDENCE INTERVALS FOR $\mu_1 - \mu_2$ (σ_1 AND σ_2 KNOWN)

The $\bar{x}_1 - \bar{x}_2$ sampling distribution

In this section, we will use probability distributions that arise from a difference of means (or proportions). How do we obtain such distributions? Suppose we have two statistical variables x_1 and x_2, each with its own distribution. We take *independent* random samples of size n_1 from the x_1 distribution and of size n_2 from the x_2 distribution. Then we compute the respective means $\bar{x}_1$ and $\bar{x}_2$. Now consider the difference $\bar{x}_1 - \bar{x}_2$. This expression represents a difference of means. If we repeat this sampling process over and over, we will create lots of $\bar{x}_1 - \bar{x}_2$ values. Figure 7-7 illustrates the sampling distribution of $\bar{x}_1 - \bar{x}_2$.

The values of $\bar{x}_1 - \bar{x}_2$ that come from repeated (independent) sampling of populations 1 and 2 can be arranged in a relative-frequency table and a relative-frequency

FIGURE 7-7

Sampling Distribution of $\bar{x}_1 - \bar{x}_2$

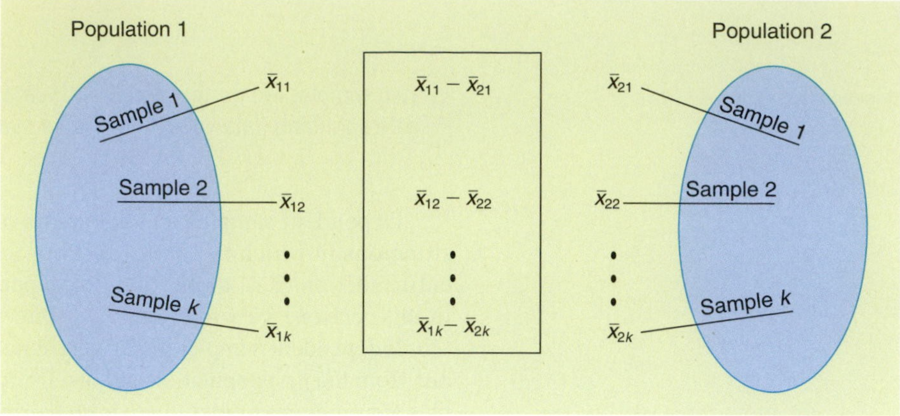

histogram (see Section 2.1). This would give us an experimental idea of the theoretical probability distribution of $\bar{x}_1 - \bar{x}_2$.

Fortunately, it is not necessary to carry out this lengthy process for each example. The results have been worked out mathematically. The next theorem presents the main results.

THEOREM 7.1 Let x_1 and x_2 have normal distributions with means μ_1 and μ_2 and standard deviations σ_1 and σ_2, respectively. If we take independent random samples of size n_1 from the x_1 distribution and of size n_2 from the x_2 distribution, then the variable $\bar{x}_1 - \bar{x}_2$ has

1. a normal distribution
2. mean $\mu_1 - \mu_2$
3. standard deviation $\sqrt{\dfrac{\sigma_1^2}{n_1} + \dfrac{\sigma_2^2}{n_2}}$

COMMENT The theorem requires that x_1 and x_2 have *normal* distributions. However, if *both* n_1 and n_2 are 30 or larger, then the central limit theorem (Section 6.5) assures us that $\bar{x}_1$ and $\bar{x}_2$ are approximately normally distributed. In this case, the conclusions of the theorem are again valid even if the original x_1 and x_2 distributions are not exactly normal.

Confidence intervals for $\mu_1 - \mu_2$ (σ_1 and σ_2 known)

If we use Theorem 7.1, then a discussion similar to that of Section 7.1 gives the following information.

PROCEDURE

HOW TO FIND A CONFIDENCE INTERVAL FOR $\mu_1 - \mu_2$ WHEN BOTH σ_1 AND σ_2 ARE KNOWN

Requirements

Let σ_1 and σ_2 be the population standard deviations of populations 1 and 2. Obtain two independent random samples from populations 1 and 2, where

$\bar{x}_1$ and $\bar{x}_2$ are sample means from populations 1 and 2

n_1 and n_2 are sample sizes from populations 1 and 2

If you can assume that both population distributions 1 and 2 are normal, any sample sizes n_1 and n_2 will work. If you cannot assume this, then use sample sizes $n_1 \geq 30$ and $n_2 \geq 30$.

Continued

Problem 28 of this section shows how to determine minimal sample sizes for a specified maximal error of estimate E.

Confidence interval for $\mu_1 - \mu_2$

$$(\bar{x}_1 - \bar{x}_2) - E < \mu_1 - \mu_2 < (\bar{x}_1 - \bar{x}_2) + E$$

where $E = z_c\sqrt{\dfrac{\sigma_1^2}{n_1} + \dfrac{\sigma_2^2}{n_2}}$

c = confidence level $(0 < c < 1)$

z_c = critical value for confidence level c based on the standard normal distribution. (See Table 5(b) of Appendix II for commonly used values.)

EXAMPLE 8

CONFIDENCE INTERVAL FOR $\mu_1 - \mu_2$, σ_1 AND σ_2 KNOWN

In the summer of 1988, Yellowstone National Park had some major fires that destroyed large tracts of old timber near many famous trout streams. Fishermen were concerned about the long-term effects of the fires on these streams. However, biologists claimed that the new meadows that would spring up under dead trees would produce a lot more insects, which would in turn mean better fishing in the years ahead. Guide services registered with the park provided data about the daily catch for fishermen over many years. Ranger checks on the streams also provided data about the daily number of fish caught by fishermen. *Yellowstone Today* (a national park publication) indicates that the biologists' claim is basically correct and that Yellowstone anglers are delighted by their average increased catch.

Mark R/Shutterstock.Com

Yellowstone National Park

Suppose you are a biologist studying fishing data from Yellowstone streams before and after the fire. Fishing reports include the number of trout caught per day per fisherman. A random sample of $n_1 = 167$ reports from the period before the fire showed that the average catch was $\bar{x}_1 = 5.2$ trout per day. Assume that the standard deviation of daily catch per fisherman during this period was $\sigma_1 = 1.9$. Another random sample of $n_2 = 125$ fishing reports 5 years after the fire showed that the average catch per day was $\bar{x}_2 = 6.8$ trout. Assume that the standard deviation during this period was $\sigma_2 = 2.3$.

(a) ***Check Requirements*** For each sample, what is the population? Are the samples dependent or independent? Explain. Is it approriate to use a normal distribution for the $\bar{x}_1 - \bar{x}_2$ distribution? Explain.

SOLUTION: The population for the first sample is the number of trout caught per day by fishermen before the fire. The population for the second sample is the number of trout caught per day after the fire. Both samples were random samples taken in their respective time periods. There was no effort to pair individual data values. Therefore, the samples can be thought of as independent samples.

A normal distribution is appropriate for the $\bar{x}_1 - \bar{x}_2$ distribution because sample sizes are sufficiently large and we know both σ_1 and σ_2.

(b) Compute a 95% confidence interval for $\mu_1 - \mu_2$, the difference of population means.

SOLUTION: Since $n_1 = 167$, $\bar{x}_1 = 5.2$, $\sigma_1 = 1.9$, $n_2 = 125$, $\bar{x}_2 = 6.8$, $\sigma_2 = 2.3$, and $z_{0.95} = 1.96$ (see Table 7-2), then

$$E = z_c\sqrt{\frac{\sigma_1^2}{n_1} + \frac{\sigma_2^2}{n_2}}$$

$$= 1.96\sqrt{\frac{(1.9)^2}{167} + \frac{(2.3)^2}{125}} \approx 1.96\sqrt{0.0639} \approx 0.4955 \approx 0.50$$

The 95% confidence interval is

$$(\bar{x}_1 - \bar{x}_2) - E < \mu_1 - \mu_2 < (\bar{x}_1 - \bar{x}_2) + E$$
$$(5.2 - 6.8) - 0.50 < \mu_1 - \mu_2 < (5.2 - 6.8) + 0.50$$
$$-2.10 < \mu_1 - \mu_2 < -1.10$$

(c) ***Interpretation*** What is the meaning of the confidence interval computed in part (b)?
SOLUTION: We are 95% confident that the interval -2.10 to -1.10 fish per day is one of the intervals containing the population difference $\mu_1 - \mu_2$, where μ_1 represents the population average daily catch before the fire and μ_2 represents the population average daily catch after the fire. Put another way, since the confidence interval contains only *negative values*, we can be 95% sure that $\mu_1 - \mu_2 < 0$. This means we are 95% sure that $\mu_1 < \mu_2$. In words, we are 95% sure that the average catch before the fire was less than the average catch after the fire.

COMMENT In the case of large samples ($n_1 \geq 30$ and $n_2 \geq 30$), it is not unusual to see σ_1 and σ_2 approximated by s_1 and s_2. Then Theorem 7.1 is used as a basis for approximating confidence intervals for $\mu_1 - \mu_2$. In other words, when samples are large, sample estimates for σ_1 and σ_2 can be used together with the standard normal distribution to find confidence intervals for $\mu_1 - \mu_2$. However, in this text, we follow the more common convention of using a Student's t distribution whenever σ_1 and σ_2 are unknown.

CONFIDENCE INTERVALS FOR $\mu_1 - \mu_2$ WHEN σ_1 AND σ_2 ARE UNKNOWN

When σ_1 and σ_2 are unknown, we turn to a Student's t distribution. As before, when we use a Student's t distribution, we require that our populations be normal or approximately normal (mound-shaped and symmetric) when the sample sizes n_1 and n_2 are less than 30. We also replace σ_1 by s_1 and σ_2 by s_2. Then we consider the approximate t value attributed to Welch (*Biometrika*, Vol. 29, pp. 350–362).

$$t \approx \frac{(\bar{x}_1 - \bar{x}_2) - (\mu_1 - \mu_2)}{\sqrt{\dfrac{s_1^2}{n_1} + \dfrac{s_2^2}{n_2}}}$$

Unfortunately, this approximation is *not exactly* a Student's t distribution. However, it will be a good approximation provided we adjust the degrees of freedom by one of the following methods.

Problem 30 of this section shows the formula for Satterthwaite's approximation for degrees of freedom.

1. The adjustment for the degrees of freedom is calculated from sample data. The formula, called *Satterthwaite's approximation*, is rather complicated. Satterthwaite's approximation is used in statistical software packages such as Minitab and in the TI-84Plus/TI-83Plus/TI-*n*spire calculators. See Problem 30 for the formula.
2. An alternative method, which is much simpler, is to approximate the degrees of freedom using the *smaller* of $n_1 - 1$ and $n_2 - 1$.

For confidence intervals, we take the degrees of freedom *d.f.* to be the smaller of $n_1 - 1$ and $n_2 - 1$. This commonly used choice for the degrees of freedom is more conservative than Satterthwaite's approximation in the sense that the former produces a slightly larger margin of error. The resulting confidence interval will be *at least* at the c level, or a little higher.

Confidence intervals for $\mu_1 - \mu_2$ (σ_1 and σ_2 unknown)

Applying methods similar to those used to find confidence intervals for μ when σ is unknown, and using the Welch approximation for t, we obtain the following results.

PROCEDURE

HOW TO FIND A CONFIDENCE INTERVAL FOR $\mu_1 - \mu_2$ WHEN σ_1 AND σ_2 ARE UNKNOWN

Requirements

Obtain two independent random samples from populations 1 and 2, where

$\bar{x}_1$ and $\bar{x}_2$ are sample means from populations 1 and 2
s_1 and s_2 are sample standard deviations from populations 1 and 2
n_1 and n_2 are sample sizes from populations 1 and 2

If you can assume that both population distributions 1 and 2 are normal or at least mound-shaped and symmetric, then any sample sizes n_1 and n_2 will work. If you cannot assume this, then use sample sizes $n_1 \geq 30$ and $n_2 \geq 30$.

Confidence interval for $\mu_1 - \mu_2$

$$(\bar{x}_1 - \bar{x}_2) - E < \mu_1 - \mu_2 < (\bar{x}_1 - \bar{x}_2) + E$$

where $E \approx t_c\sqrt{\frac{s_1^2}{n_1} + \frac{s_2^2}{n_2}}$

c = confidence level $(0 < c < 1)$

t_c = critical value for confidence level c

$d.f.$ = *smaller* of $n_1 - 1$ and $n_2 - 1$. Note that statistical software gives a more accurate and larger $d.f.$ based on Satterthwaite's approximation.

EXAMPLE 9

CONFIDENCE INTERVAL FOR $\mu_1 - \mu_2$, σ_1 AND σ_2 UNKNOWN

Alexander Borbely is a professor at the Medical School of the University of Zurich, where he is director of the Sleep Laboratory. Dr. Borbely and his colleagues are experts on sleep, dreams, and sleep disorders. In his book *Secrets of Sleep*, Dr. Borbely discusses brain waves, which are measured in hertz, the number of oscillations per second. Rapid brain waves (wakefulness) are in the range of 16 to 25 hertz. Slow brain waves (sleep) are in the range of 4 to 8 hertz. During normal sleep, a person goes through several cycles (each cycle is about 90 minutes) of brain waves, from rapid to slow and back to rapid. During deep sleep, brain waves are at their slowest.

In his book, Professor Borbely comments that alcohol is a *poor* sleep aid. In one study, a number of subjects were given 1/2 liter of red wine before they went to sleep. The subjects fell asleep quickly but did not remain asleep the entire night. Toward morning, between 4 and 6 A.M., they tended to wake up and have trouble going back to sleep.

Suppose that a random sample of 29 college students was randomly divided into two groups. The first group of $n_1 = 15$ people was given 1/2 liter of red wine before going to sleep. The second group of $n_2 = 14$ people was given no alcohol before going to sleep. Everyone in both groups went to sleep at 11 P.M. The average brain wave activity (4 to 6 A.M.) was determined for each individual in the groups. Assume the average brain wave distribution in each group is mound-shaped and symmetric. The results follow:

Group 1 (x_1 values): $n_1 = 15$ (with alcohol)
Average brain wave activity in the hours 4 to 6 A.M.

16.0	19.6	19.9	20.9	20.3	20.1	16.4	20.6
20.1	22.3	18.8	19.1	17.4	21.1	22.1	

For group 1, we have the sample mean and standard deviation of

$\bar{x}_1 \approx 19.65$ and $s_1 \approx 1.86$

Group 2 (x_2 values): $n_2 = 14$ (no alcohol)
Average brain wave activity in the hours 4 to 6 A.M.

8.2	5.4	6.8	6.5	4.7	5.9	2.9
7.6	10.2	6.4	8.8	5.4	8.3	5.1

For group 2, we have the sample mean and standard deviation of

$\bar{x}_2 \approx 6.59$ and $s_2 \approx 1.91$

(a) ***Check Requirements*** Are the samples independent or dependent? Explain. Is it appropriate to use a Student's t distribution to approximate the $\bar{x}_1 - \bar{x}_2$ distribution? Explain.

SOLUTION: Since the original random sample of 29 students was randomly divided into two groups, it is reasonable to say that the samples are independent. A Student's t distribution is appropriate for the $\bar{x}_1 - \bar{x}_2$ distribution because both original distributions are mound-shaped and symmetric. We don't know population standard deviations, but we can compute s_1 and s_2.

(b) Compute a 90% confidence interval for $\mu_1 - \mu_2$, the difference of population means.

SOLUTION: First we find $t_{0.90}$. We approximate the degrees of freedom *d.f.* by using the smaller of $n_1 - 1$ and $n_2 - 1$. Since n_2 is smaller, $d.f. = n_2 - 1 = 14 - 1 = 13$. This gives us $t_{0.90} \approx 1.771$. The margin of error is then

$$E \approx t_c\sqrt{\frac{s_1^2}{n_1} + \frac{s_2^2}{n_2}} = 1.771\sqrt{\frac{1.86^2}{15} + \frac{1.91^2}{14}} \approx 1.24$$

The c confidence interval is

$$(\bar{x}_1 - \bar{x}_2) - E < \mu_1 - \mu_2 < (\bar{x}_1 - \bar{x}_2) + E$$
$$(19.65 - 6.59) - 1.24 < \mu_1 - \mu_2 < (19.65 - 6.59) + 1.24$$
$$11.82 < \mu_1 - \mu_2 < 14.30$$

After further rounding we have

11.8 hertz $< \mu_1 - \mu_2 <$ 14.3 hertz

(c) ***Interpretation*** What is the meaning of the confidence interval you computed in part (b)?

SOLUTION: μ_1 represents the population average brain wave activity for people who drank 1/2 liter of wine before sleeping. μ_2 represents the population average brain wave activity for people who took no alcohol before sleeping. Both periods of measurement are from 4 to 6 A.M. We are 90% confident that the interval between 11.8 and 14.3 hertz is one that contains the difference $\mu_1 - \mu_2$. It would seem reasonable to conclude that people who drink before sleeping might wake up in the early morning and have trouble going back to sleep. Since the confidence interval from 11.8 to 14.3 contains only *positive values*, we could express this by saying that we are 90% confident that $\mu_1 - \mu_2$ is *positive*. This means that $\mu_1 - \mu_2 > 0$. Thus, we are 90% confident that $\mu_1 > \mu_2$ (that is, average brain wave activity from 4 to 6 A.M. for the group drinking wine was more than average brain wave activity for the group not drinking).

There is another method of constructing confidence intervals for $\mu_1 - \mu_2$ when σ_1 and σ_2 are unknown. Suppose the sample values s_1 and s_2 are sufficiently close and there is reason to believe that $\sigma_1 = \sigma_2$. Methods shown in Section 10.4 use sample standard deviations s_1 and s_2 to determine if $\sigma_1 = \sigma_2$. When you can assume that $\sigma_1 = \sigma_2$, it is best to use a *pooled standard deviation* to compute the margin of error. The $\bar{x}_1 - \bar{x}_2$ distribution has an *exact* Student's t distribution with $d.f. = n_1 + n_2 - 2$. Problem 31 of this section gives the details.

The formulas and requirements for using a pooled standard deviation to compute a confidence interval for $\mu_1 - \mu_2$ can be found in Problem 31 at the end of this section.

Sometimes we want several confidence intervals to hold concurrently. Problem 19 in the Chapter Review Problems discusses this situation.

SUMMARY

What should a person do? You have independent random samples from two populations. You can compute $\bar{x}_1$, $\bar{x}_2$, s_1, and s_2 and you have the sample sizes n_1 and n_2. In any case, a confidence interval for the difference $\mu_1 - \mu_2$ of population means is

$$(\bar{x}_1 - \bar{x}_2) - E < \mu_1 - \mu_2 < (\bar{x}_1 - \bar{x}_2) + E$$

where E is the margin of error. How do you compute E? The answer depends on how much you know about the x_1 and x_2 distributions.

Situation I (the usual case)

You simply don't know the population values of σ_1 and σ_2. In this situation, you use a t distribution with margin of error

$$E = t_c\sqrt{\frac{s_1^2}{n_1} + \frac{s_2^2}{n_2}}$$

where a conservative estimate for the degrees of freedom is

$$d.f. = \text{minimum of } n_1 - 1 \text{ and } n_2 - 1$$

Like a good friend, the t distribution has a reputation for being robust and forgiving. Nevertheless, some guidelines should be observed. If n_1 and n_2 are both less than 30, then x_1 and x_2 should have distributions that are mound-shaped and approximately symmetrical (or, even better, normal). If both n_1 and n_2 are 30 or more, the central limit theorem (Chapter 6) implies that these restrictions can be relaxed.

Situation II (almost never happens)

You actually know the population values of σ_1 and σ_2. In addition, you know that x_1 and x_2 have normal distributions. If you know σ_1 and σ_2 but are not sure about the x_1 and x_2 distributions, then you must have $n_1 \geq 30$ and $n_2 \geq 30$. In this situation, you use a z distribution with margin of error

$$E = z_c\sqrt{\frac{\sigma_1^2}{n_1} + \frac{\sigma_2^2}{n_2}}$$

Situation III (yes, this does sometimes occur)

You don't know σ_1 and σ_2, but the sample values s_1 and s_2 are close to each other and there is reason to believe that $\sigma_1 = \sigma_2$. This can happen when you make a slight change or alteration to a known process or method of production. The

Continued

standard deviation may not change much, but the outputs or means could be very different. In this situation, you are advised to use a t distribution with a pooled standard deviation. See Problem 31 at the end of this section.

Which distribution should you use for $\bar{x}_1 - \bar{x}_2$

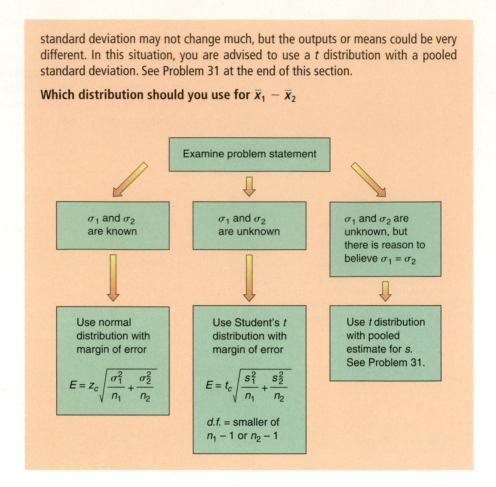

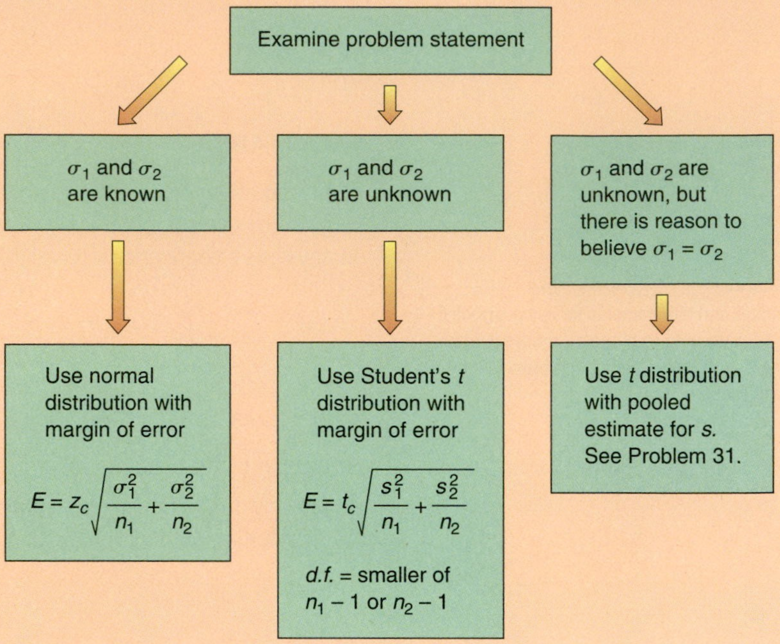

ESTIMATING THE DIFFERENCE OF PROPORTIONS $p_1 - p_2$

We conclude this section with a discussion of confidence intervals for $p_1 - p_2$, the difference of two proportions from binomial probability distributions. The main result on this topic is the following theorem.

THEOREM 7.2 Consider two binomial probability distributions

Distribution 1	*Distribution 2*
n_1 = number of trials	n_2 = number of trials
r_1 = number of successes out of n_1 trials	r_2 = number of successes out of n_2 trials
p_1 = probability of success on each trial	p_2 = probability of success on each trial
$q_1 = 1 - p_1$ = probability of failure on each trial	$q_2 = 1 - p_2$ = probability of failure on each trial
$\hat{p}_1 = \frac{r_1}{n_1}$ = point estimate for p_1	$\hat{p}_2 = \frac{r_2}{n_2}$ = point estimate for p_2
$\hat{q}_1 = 1 - \frac{r_1}{n_1}$ = point estimate for q_1	$\hat{q}_2 = 1 - \frac{r_2}{n_2}$ = point estimate for q_2

For most practical applications, if the four quantities

$$n_1\hat{p}_1 \quad n_1\hat{q}_1 \quad n_2\hat{p}_2 \quad n_2\hat{q}_2$$

are all larger than 5 (see Section 6.6), then the following statements are true about the random variable $\frac{r_1}{n_1} - \frac{r_2}{n_2}$:

1. $\frac{r_1}{n_1} - \frac{r_2}{n_2}$ has an approximately normal distribution.
2. The mean is $p_1 - p_2$.
3. The standard deviation is approximately

$$\hat{\sigma} = \sqrt{\frac{\hat{p}_1\hat{q}_1}{n_1} + \frac{\hat{p}_2\hat{q}_2}{n_2}}$$

Confidence intervals for $p_1 - p_2$

Based on the preceding theorem, we can find confidence intervals for $p_1 - p_2$ in the following way:

PROCEDURE

HOW TO FIND A CONFIDENCE INTERVAL FOR $p_1 - p_2$

Requirements

Consider two independent binomial experiments.

Binomial Experiment 1

n_1 = number of trials

r_1 = number of successes out of n_1 trials

$\hat{p}_1 = \frac{r_1}{n_1}$; $\hat{q}_1 = 1 - \hat{p}_1$

p_1 = population probability of success

Binomial Experiment 2

n_2 = number of trials

r_2 = number of successes out of n_2 trials

$\hat{p}_2 = \frac{r_2}{n_2}$; $\hat{q}_2 = 1 - \hat{p}_2$

p_2 = population probability of success

The number of trials should be sufficiently large so that all four of the following inequalities are true:

$$n_1\hat{p}_1 > 5; \quad n_1\hat{q}_1 > 5; \quad n_2\hat{p}_2 > 5; \quad n_2\hat{q}_2 > 5$$

Confidence interval for $p_1 - p_2$

$$(\hat{p}_1 - \hat{p}_2) - E \le p_1 - p_2 \le (\hat{p}_1 - \hat{p}_2) + E$$

where

$$E = z_c\hat{\sigma} = z_c\sqrt{\frac{\hat{p}_1\hat{q}_1}{n_1} + \frac{\hat{p}_2\hat{q}_2}{n_2}}$$

c = confidence level, $0 < c < 1$

z_c = critical value for confidence level c based on the standard normal distribution (See Table 5(b) of Appendix II for commonly used values.)

Problem 29 shows how to compute minimal sample sizes for a specified margin of error E.

EXAMPLE 10

CONFIDENCE INTERVAL FOR $p_1 - p_2$

In his book *Secrets of Sleep*, Professor Borbely describes research on dreams in the Sleep Laboratory at the University of Zurich Medical School. During normal sleep, there is a phase known as *REM* (rapid eye movement). For most people, REM sleep occurs about every 90 minutes or so, and it is thought that dreams occur just before or during the REM phase. Using electronic equipment in the Sleep Laboratory, it is possible to detect the REM phase in a sleeping person. If a person is wakened immediately after the REM phase, he or she usually can describe a dream that has just taken place. Based on a study of over 650 people in the Zurich Sleep Laboratory, it was found that about one-third of all dream reports contain feelings of fear, anxiety, or aggression. There is a conjecture that if a person is in a good mood when going to sleep, the proportion of "bad" dreams (fear, anxiety, aggression) might be reduced.

Suppose that two groups of subjects were randomly chosen for a sleep study. In group I, before going to sleep, the subjects spent 1 hour watching a comedy movie. In this group, there were a total of $n_1 = 175$ dreams recorded, of which $r_1 = 49$ were dreams with feelings of anxiety, fear, or aggression. In group II, the subjects did not watch a movie but simply went to sleep. In this group, there were a total of $n_2 = 180$ dreams recorded, of which $r_2 = 63$ were dreams with feelings of anxiety, fear, or aggression.

(a) ***Check Requirements*** Why could groups I and II be considered independent binomial distributions? Why do we have a "large-sample" situation?

SOLUTION: Since the two groups were chosen randomly, it is reasonable to assume that neither group's responses would be related to the other's. In both groups, each recorded dream could be thought of as a trial, with success being a dream with feelings of fear, anxiety, or aggression.

$$\hat{p}_1 = \frac{r_1}{n_1} = \frac{49}{175} = 0.28 \quad \text{and} \quad \hat{q}_1 = 1 - \hat{p}_1 = 0.72$$

$$\hat{p}_2 = \frac{r_2}{n_2} = \frac{63}{180} = 0.35 \quad \text{and} \quad \hat{q}_2 = 1 - \hat{p}_2 = 0.65$$

Since

$$n_1\hat{p}_1 = 49 > 5 \quad n_1\hat{q}_1 = 126 > 5$$
$$n_2\hat{p}_2 = 63 > 5 \quad n_2\hat{q}_2 = 117 > 5$$

large-sample theory is appropriate.

(b) What is $p_1 - p_2$? Compute a 95% confidence interval for $p_1 - p_2$.

SOLUTION: p_1 is the population proportion of successes (bad dreams) for all people who watched comedy movies before bed. Thus, p_1 can be thought of as the percentage of bad dreams for all people who were in a "good mood" when they went to bed. Likewise, p_2 is the percentage of bad dreams for the population of all people who just went to bed (no movie). The difference $p_1 - p_2$ is the population difference.

To find a confidence interval for $p_1 - p_2$, we need the values of z_c, $\hat{\sigma}$, and then E. From Table 7-2, we see that $z_{0.95} = 1.96$, so

$$\hat{\sigma} = \sqrt{\frac{\hat{p}_1\hat{q}_1}{n_1} + \frac{\hat{p}_2\hat{q}_2}{n_2}} = \sqrt{\frac{(0.28)(0.72)}{175} + \frac{(0.35)(0.65)}{180}}$$
$$\approx \sqrt{0.0024} \approx 0.0492$$
$$E = z_c\hat{\sigma} = 1.96(0.0492) \approx 0.096$$
$$(\hat{p}_1 - \hat{p}_2) - E < p_1 - p_2 < (\hat{p}_1 - \hat{p}_2) + E$$
$$(0.28 - 0.35) - 0.096 < p_1 - p_2 < (0.28 - 0.35) + 0.096$$
$$-0.166 < p_1 - p_2 < 0.026$$

(c) ***Interpretation*** What is the meaning of the confidence interval constructed in part (b)?

SOLUTION: We are 95% sure that the interval between −16.6% and 2.6% is one that contains the percentage difference of "bad" dreams for group I and group II. Since the interval −0.166 to 0.026 is not all negative (or all positive), we cannot say that $p_1 - p_2 < 0$ (or $p_1 - p_2 > 0$). Thus, at the 95% confidence level, we *cannot* conclude that $p_1 < p_2$ or $p_1 > p_2$. The comedy movies before bed helped some people reduce the percentage of "bad" dreams, but at the 95% confidence level, we cannot say that the *population difference* is reduced.

CRITICAL THINKING

INTERPRETING CONFIDENCE INTERVALS FOR DIFFERENCES

As we have seen in the preceding examples, at the c confidence level we can determine how two means or proportions from independent random samples are related. The next procedure summarizes the results.

PROCEDURE

HOW TO INTERPRET CONFIDENCE INTERVALS FOR DIFFERENCES

Suppose we construct a c% confidence interval for $\mu_1 - \mu_2$ (or $p_1 - p_2$). Then three cases arise:

1. The c% confidence interval contains only *negative values* (see Example 8). In this case, we conclude that $\mu_1 - \mu_2 < 0$ (or $p_1 - p_2 < 0$), and we are therefore c% confident that $\mu_1 < \mu_2$ (or $p_1 < p_2$).
2. The c% confidence interval contains only *positive values* (see Example 9). In this case, we conclude that $\mu_1 - \mu_2 > 0$ (or $p_1 - p_2 > 0$), and we can be c% confident that $\mu_1 > \mu_2$ (or $p_1 > p_2$).
3. The c% confidence interval contains *both positive and negative values* (see Example 10). In this case, we cannot at the c% confidence level conclude that either μ_1 or μ_2 (or p_1 or p_2) is larger. However, if we *reduce* the confidence level c to a *smaller value*, then the confidence interval will, in general, be shorter (explain why). Another approach (when possible) is to increase the sample sizes n_1 and n_2. This would also tend to make the confidence interval shorter (explain why). A shorter confidence interval *might* put us back into case 1 or case 2 above (again, explain why).

In Section 8.5, we will see another method to determine if two means or proportions from independent random samples are equal.

GUIDED EXERCISE 7 **INTERPRETING A CONFIDENCE INTERVAL**

(a) A study reported a 90% confidence interval for the difference of means to be

$10 < \mu_1 - \mu_2 < 20$

For this interval, what can you conclude about the respective values of μ_1 and μ_2?

→ At a 90% level of confidence, we can say that the difference $\mu_1 - \mu_2$ is positive, so $\mu_1 - \mu_2 > 0$ and $\mu_1 > \mu_2$.

(b) A study reported a 95% confidence interval for the difference of proportions to be

$-0.32 < p_1 - p_2 < 0.16$

For this interval, what can you conclude about the respective values of p_1 and p_2?

→ At the 95% confidence level, we see that the difference of proportions ranges from negative to positive values. We cannot tell from this interval if p_1 is greater than p_2 or p_1 is less than p_2.

TECH NOTES The TI-84Plus/TI-83Plus/TI-*n*spire calculators and Minitab supply confidence intervals for the difference of means and for the difference of proportions.

TI-84Plus/TI-83Plus/TI-*n*spire (with TI-84Plus keypad) Use the **STAT** key and highlight **TESTS**. The choice **9:2-SampZInt** finds confidence intervals for differences of means when σ_1 and σ_2 are known. Choice **0:2-SampTInt** finds confidence intervals for differences of means when σ_1 and σ_2 are unknown. In general, use No for Pooled. However, if $\sigma_1 \approx \sigma_2$, use Yes for Pooled. Choice **B:2-PropZInt** provides confidence intervals for proportions.

Minitab Use the menu choice **STAT ➤ Basic Statistics ➤ 2 sample t** or **2 proportions**. Minitab always uses the Student's *t* distribution for $\mu_1 - \mu_2$ confidence intervals. If the variances are equal, check "assume equal variances."

VIEWPOINT What's the Difference?

Will two 15-minute piano lessons a week significantly improve a child's analytical reasoning skills? Why piano? Why not computer keyboard instruction or maybe voice lessons? Professor Frances Rauscher, University of Wisconsin, and Professor Gordon Shaw, University of California at Irvine, claim there is a difference! How could this be measured? A large number of piano students were given complicated tests of mental ability. Independent control groups of other students were given the same tests. Techniques involving the study of differences of means were used to draw the conclusion that students taking piano lessons did better on tests measuring analytical reasoning skills. (See Music Training Causes Long-Term Enhancement of Preschool Chrildren's Spatial-Temporal Reasoning by Rausher, Shaw, and others, *Neurological Research Volume 19.)*

SECTION 7.4 PROBLEMS

Answers may vary slightly due to rounding.

1. *Statistical Literacy* When are two random samples independent?

2. *Statistical Literacy* When are two random samples dependent?

3. *Critical Thinking* Josh and Kendra each calculated a 90% confidence interval for the difference of means using a Student's *t* distribution for random samples of size $n_1 = 20$ and $n_2 = 31$. Kendra followed the convention of using the smaller sample size to compute $d.f. = 19$. Josh used his calculator and Satterthwaite's approximation and obtained $d.f. \approx 36.3$. Which confidence interval is shorter? Which confidence interval is more conservative in the sense that the margin of error is larger?

4. *Critical Thinking* If a 90% confidence interval for the difference of means $\mu_1 - \mu_2$ contains all positive values, what can we conclude about the relationship between μ_1 and μ_2 at the 90% confidence level?

5. *Critical Thinking* If a 90% confidence interval for the difference of means $\mu_1 - \mu_2$ contains all negative values, what can we conclude about the relationship between μ_1 and μ_2 at the 90% confidence level?

6. *Critical Thinking* If a 90% confidence interval for the difference of proportions contains some positive and some negative values, what can we conclude about the relationship between p_1 and p_2 at the 90% confidence level?

7. *Basic Computation: Confidence Interval for $\mu_1 - \mu_2$* Consider two independent normal distributions. A random sample of size $n_1 = 20$ from the first distribution showed $\bar{x}_1 = 12$ and a random sample of size $n_2 = 25$ from the second distribution showed $\bar{x}_2 = 14$.

(a) *Check Requirements* If σ_1 and σ_2 are known, what distribution does $\bar{x}_1 - \bar{x}_2$ follow? Explain.
(b) Given $\sigma_1 = 3$ and $\sigma_2 = 4$, find a 90% confidence interval for $\mu_1 - \mu_2$
(c) *Check Requirements* Suppose σ_1 and σ_2 are both unknown, but from the random samples, you know $s_1 = 3$ and $s_2 = 4$. What distribution approximates the $\bar{x}_1 - \bar{x}_2$ distribution? What are the degrees of freedom? Explain.
(d) With $s_1 = 3$ and $s_2 = 4$, find a 90% confidence interval for $\mu_1 - \mu_2$.
(e) If you have an appropriate calculator or computer software, find a 90% confidence interval for $\mu_1 - \mu_2$ using degrees of freedom based on Satterthwaite's approximation.
(f) *Interpretation* Based on the confidence intervals you computed, can you be 90% confident that μ_1 is smaller than μ_2? Explain.

8. *Basic Computation: Confidence Interval for $\mu_1 - \mu_2$* Consider two independent distributions that are mound-shaped. A random sample of size $n_1 = 36$ from the first distribution showed $\bar{x}_1 = 15$, and a random sample of size $n_2 = 40$ from the second distribution showed $\bar{x}_2 = 14$.
(a) *Check Requirements* If σ_1 and σ_2 are known, what distribution does $\bar{x}_1 - \bar{x}_2$ follow? Explain.
(b) Given $\sigma_1 = 3$ and $\sigma_2 = 4$, find a 95% confidence interval for $\mu_1 - \mu_2$.
(c) *Check Requirements* Suppose σ_1 and σ_2 are both unknown, but from the random samples, you know $s_1 = 3$ and $s_2 = 4$. What distribution approximates the $\bar{x}_1 - \bar{x}_2$ distribution? What are the degrees of freedom? Explain.
(d) With $s_1 = 3$ and $s_2 = 4$, find a 95% confidence interval for $\mu_1 - \mu_2$.
(e) If you have an appropriate calculator or computer software, find a 95% confidence interval for $\mu_1 - \mu_2$ using degrees of freedom based on Satterthwaite's approximation.
(f) *Interpretation* Based on the confidence intervals you computed, can you be 95% confident that μ_1 is larger than μ_2? Explain.

9. *Basic Computation: Confidence Interval for $p_1 - p_2$* Consider two independent binomial experiments. In the first one, 40 trials had 10 successes. In the second one, 50 trials had 15 successes.
(a) *Check Requirements* Is it appropriate to use a normal distribution to approximate the $\hat{p}_1 - \hat{p}_2$ distribution? Explain.
(b) Find a 90% confidence interval for $p_1 - p_2$.
(c) *Interpretation* Based on the confidence interval you computed, can you be 90% confident that p_1 is less than p_2? Explain.

10. *Basic Computation: Confidence Interval for $p_1 - p_2$* Consider two independent binomial experiments. In the first one, 40 trials had 15 successes. In the second one, 60 trials had 6 successes.
(a) *Check Requirements* Is it appropriate to use a normal distribution to approximate the $\hat{p}_1 - \hat{p}_2$ distribution? Explain.
(b) Find a 95% confidence interval for $p_1 - p_2$.
(c) *Interpretation* Based on the confidence interval you computed, can you be 95% confident that p_1 is more than p_2? Explain.

11. *Archaeology: Ireland* Inorganic phosphorous is a naturally occurring element in all plants and animals, with concentrations increasing progressively up the food chain (fruit $<$ vegetables $<$ cereals $<$ nuts $<$ corpse). Geochemical surveys take soil samples to determine phosphorous content (in ppm, parts per million). A high phosphorous content may or may not indicate an ancient burial site, food storage site, or even a garbage dump. The Hill of Tara is a very important archaeological site in Ireland. It is by legend the seat of Ireland's ancient high kings (Reference: *Tara, An Archaeological Survey* by Conor Newman, Royal Irish Academy, Dublin). Independent random samples from two regions in Tara gave the following phosphorous measurements (in ppm). Assume the population distributions of phosphorous are mound-shaped and symmetric for these two regions.

Region I: x_1; $n_1 = 12$

540	810	790	790	340	800
890	860	820	640	970	720

Region II: x_2; $n_2 = 16$

750	870	700	810	965	350	895	850
635	955	710	890	520	650	280	993

(a) Use a calculator with mean and standard deviation keys to verify that $\bar{x}_1 \approx 747.5$, $s_1 \approx 170.4$, $\bar{x}_2 \approx 738.9$, and $s_2 \approx 212.1$.
(b) Let μ_1 be the population mean for x_1 and let μ_2 be the population mean for x_2. Find a 90% confidence interval for $\mu_1 - \mu_2$.
(c) *Interpretation* Explain what the confidence interval means in the context of this problem. Does the interval consist of numbers that are all positive? all negative? of different signs? At the 90% level of confidence, is one region more interesting than the other from a geochemical perspective?
(d) *Check Requirements* Which distribution (standard normal or Student's t) did you use? Why?

12. *Archaeology: Ireland* Please see the setting and reference in Problem 11. Independent random samples from two regions (not those cited in Problem 11) gave the following phosphorous measurements (in ppm). Assume the distribution of phosphorous is mound-shaped and symmetric for these two regions.

Region I: x_1; $n_1 = 15$

855	1550	1230	875	1080	2330	1850	1860
2340	1080	910	1130	1450	1260	1010	

Region II: x_2; $n_2 = 14$

540	810	790	1230	1770	960	1650	860
890	640	1180	1160	1050	1020		

(a) Use a calculator with mean and standard deviation keys to verify that $\bar{x}_1 \approx 1387.3$, $s_1 \approx 498.3$, $\bar{x}_2 \approx 1039.3$, and $s_2 \approx 346.7$.
(b) Let μ_1 be the population mean for x_1 and let μ_2 be the population mean for x_2. Find an 80% confidence interval for $\mu_1 - \mu_2$.
(c) *Interpretation* Explain what the confidence interval means in the context of this problem. Does the interval consist of numbers that are all positive? all negative? of different signs? At the 80% level of confidence, is one region more interesting than the other from a geochemical perspective?
(d) *Check Requirements* Which distribution (standard normal or Student's t) did you use? Why?

13. *Large U.S. Companies: Foreign Revenue* For large U.S. companies, what percentage of their total income comes from foreign sales? A random sample of technology companies (IBM, Hewlett-Packard, Intel, and others) gave the following information.

Technology companies, % foreign revenue: x_1; $n_1 = 16$

62.8	55.7	47.0	59.6	55.3	41.0	65.1	51.1
53.4	50.8	48.5	44.6	49.4	61.2	39.3	41.8

Another independent random sample of basic consumer product companies (Goodyear, Sarah Lee, H.J. Heinz, Toys "Я" Us) gave the following information.

Basic consumer product companies, % foreign revenue: x_2; $n_2 = 17$

28.0	30.5	34.2	50.3	11.1	28.8	40.0	44.9
40.7	60.1	23.1	21.3	42.8	18.0	36.9	28.0
32.5							

(Reference: *Forbes Top Companies*.) Assume that the distributions of percentage foreign revenue are mound-shaped and symmetric for these two company types.

(a) Use a calculator with mean and standard deviation keys to verify that $\bar{x}_1 \approx 51.66$, $s_1 \approx 7.93$, $\bar{x}_2 \approx 33.60$, and $s_2 \approx 12.26$.

(b) Let μ_1 be the population mean for x_1 and let μ_2 be the population mean for x_2. Find an 85% confidence interval for $\mu_1 - \mu_2$.

(c) *Interpretation* Examine the confidence interval and explain what it means in the context of this problem. Does the interval consist of numbers that are all positive? all negative? of different signs? At the 85% level of confidence, do technology companies have a greater percentage foreign revenue than basic consumer product companies?

(d) *Check Requirements* Which distribution (standard normal or Student's t) did you use? Why?

14. *Pro Football and Basketball: Weights of Players* Independent random samples of professional football and basketball players gave the following information (References: *Sports Encyclopedia of Pro Football* and *Official NBA Basketball Encyclopedia*). *Note:* These data are also available for download at the Online Study Center. Assume that the weight distributions are mound-shaped and symmetric.

Weights (in lb) of pro football players: x_1; $n_1 = 21$

245	262	255	251	244	276	240	265	257	252	282
256	250	264	270	275	245	275	253	265	270	

Weights (in lb) of pro basketball players: x_2; $n_2 = 19$

205	200	220	210	191	215	221	216	228	207
225	208	195	191	207	196	181	193	201	

(a) Use a calculator with mean and standard deviation keys to verify that $\bar{x}_1 \approx 259.6$, $s_1 \approx 12.1$, $\bar{x}_2 \approx 205.8$, and $s_2 \approx 12.9$.

(b) Let μ_1 be the population mean for x_1 and let μ_2 be the population mean for x_2. Find a 99% confidence interval for $\mu_1 - \mu_2$.

(c) *Interpretation* Examine the confidence interval and explain what it means in the context of this problem. Does the interval consist of numbers that are all positive? all negative? of different signs? At the 99% level of confidence, do professional football players tend to have a higher population mean weight than professional basketball players?

(d) Which distribution (standard normal or Student's t) did you use? Why?

15. *Pro Football and Basketball: Heights of Players* Independent random samples of professional football and basketball players gave the following information (References: *Sports Encyclopedia of Pro Football* and *Official NBA Basketball Encyclopedia*). *Note:* These data are also available for download at the Online Study Center.

Heights (in ft) of pro football players: x_1; $n_1 = 45$

6.33	6.50	6.50	6.25	6.50	6.33	6.25	6.17	6.42	6.33
6.42	6.58	6.08	6.58	6.50	6.42	6.25	6.67	5.91	6.00
5.83	6.00	5.83	5.08	6.75	5.83	6.17	5.75	6.00	5.75
6.50	5.83	5.91	5.67	6.00	6.08	6.17	6.58	6.50	6.25
6.33	5.25	6.67	6.50	5.83					

Heights (in ft) of pro basketball players: x_2; $n_2 = 40$

6.08	6.58	6.25	6.58	6.25	5.92	7.00	6.41	6.75	6.25
6.00	6.92	6.83	6.58	6.41	6.67	6.67	5.75	6.25	6.25
6.50	6.00	6.92	6.25	6.42	6.58	6.58	6.08	6.75	6.50
6.83	6.08	6.92	6.00	6.33	6.50	6.58	6.83	6.50	6.58

(a) Use a calculator with mean and standard deviation keys to verify that $\bar{x}_1 \approx 6.179$, $s_1 \approx 0.366$, $\bar{x}_2 \approx 6.453$, and $s_2 \approx 0.314$.
(b) Let μ_1 be the population mean for x_1 and let μ_2 be the population mean for x_2. Find a 90% confidence interval for $\mu_1 - \mu_2$.
(c) *Interpretation* Examine the confidence interval and explain what it means in the context of this problem. Does the interval consist of numbers that are all positive? all negative? of different signs? At the 90% level of confidence, do professional football players tend to have a higher population mean height than professional basketball players?
(d) *Check Requirements* Which distribution (standard normal or Student's t) did you use? Why? Do you need information about the height distributions? Explain.

16. *Botany: Iris* The following data represent petal lengths (in cm) for independent random samples of two species of iris (Reference: E. Anderson, *Bulletin American Iris Society*). *Note:* These data are also available for download at the Online Study Center.

Petal length (in cm) of *Iris virginica*: x_1; $n_1 = 35$

5.1 5.8 6.3 6.1 5.1 5.5 5.3 5.5 6.9 5.0 4.9 6.0 4.8 6.1 5.6 5.1
5.6 4.8 5.4 5.1 5.1 5.9 5.2 5.7 5.4 4.5 6.1 5.3 5.5 6.7 5.7 4.9
4.8 5.8 5.1

Petal length (in cm) of *Iris setosa*: x_2; $n_2 = 38$

1.5 1.7 1.4 1.5 1.5 1.6 1.4 1.1 1.2 1.4 1.7 1.0 1.7 1.9 1.6 1.4
1.5 1.4 1.2 1.3 1.5 1.3 1.6 1.9 1.4 1.6 1.5 1.4 1.6 1.2 1.9 1.5
1.6 1.4 1.3 1.7 1.5 1.7

(a) Use a calculator with mean and standard deviation keys to verify that $\bar{x}_1 \approx 5.48$, $s_1 \approx 0.55$, $\bar{x}_2 \approx 1.49$, and $s_2 \approx 0.21$.
(b) Let μ_1 be the population mean for x_1 and let μ_2 be the population mean for x_2. Find a 99% confidence interval for $\mu_1 - \mu_2$.
(c) *Interpretation* Explain what the confidence interval means in the context of this problem. Does the interval consist of numbers that are all positive? all negative? of different signs? At the 99% level of confidence, is the population mean petal length of *Iris virginica* longer than that of *Iris setosa*?
(d) *Check Requirements* Which distribution (standard normal or Student's t) did you use? Why? Do you need information about the petal length distributions? Explain.

17. *Myers–Briggs: Marriage Counseling* Isabel Myers was a pioneer in the study of personality types. She identified four basic personality preferences, which are described at length in the book *A Guide to the Development and Use of the Myers–Briggs Type Indicator* by Myers and McCaulley (Consulting Psychologists Press). Marriage counselors know that couples who have none of the four preferences in common may have a stormy marriage. Myers took a random sample of 375 married couples and found that 289 had two or more personality preferences in common. In another random sample of 571 married couples, it was found that only 23 had no preferences in common. Let p_1 be the population proportion of all married couples who have two or more personality preferences in common. Let p_2 be the population proportion of all married couples who have no personality preferences in common.
(a) *Check Requirements* Can a normal distribution be used to approximate the $\hat{p}_1 - \hat{p}_2$ distribution? Explain.
(b) Find a 99% confidence interval for $p_1 - p_2$.
(c) *Interpretation* Explain the meaning of the confidence interval in part (a) in the context of this problem. Does the confidence interval contain all positive, all negative, or both positive and negative numbers? What does

this tell you (at the 99% confidence level) about the proportion of married couples with two or more personality preferences in common compared with the proportion of married couples sharing no personality preferences in common?

18. *Myers–Briggs: Marriage Counseling* Most married couples have two or three personality preferences in common (see reference in Problem 17). Myers used a random sample of 375 married couples and found that 132 had three preferences in common. Another random sample of 571 couples showed that 217 had two personality preferences in common. Let p_1 be the population proportion of all married couples who have three personality preferences in common. Let p_2 be the population proportion of all married couples who have two personality preferences in common.
(a) *Check Requirements* Can a normal distribution be used to approximate the $\hat{p}_1 - \hat{p}_2$ distribution? Explain.
(b) Find a 90% confidence interval for $p_1 - p_2$.
(c) *Interpretation* Examine the confidence interval in part (a) and explain what it means in the context of this problem. Does the confidence interval contain all positive, all negative, or both positive and negative numbers? What does this tell you about the proportion of married couples with three personality preferences in common compared with the proportion of couples with two preferences in common (at the 90% confidence level)?

19. *Yellowstone National Park: Old Faithful Geyser* The U.S. Geological Survey compiled historical data about Old Faithful Geyser (Yellowstone National Park) from 1870 to 1987. Some of these data are published in the book *The Story of Old Faithful*, by G. D. Marler (Yellowstone Association Press). Let x_1 be a random variable that represents the time interval (in minutes) between Old Faithful's eruptions for the years 1948 to 1952. Based on 9340 observations, the sample mean interval was $\bar{x}_1 = 63.3$ minutes. Let x_2 be a random variable that represents the time interval in minutes between Old Faithful's eruptions for the years 1983 to 1987. Based on 25,111 observations, the sample mean time interval was $\bar{x}_2 = 72.1$ minutes. Historical data suggest that $\sigma_1 = 9.17$ minutes and $\sigma_2 = 12.67$ minutes. Let μ_1 be the population mean of x_1 and let μ_2 be the population mean of x_2.
(a) *Check Requirements* Which distribution, normal or Student's t, do we use to approximate the $\bar{x}_1 - \bar{x}_2$ distribution? Explain.
(b) Compute a 99% confidence interval for $\mu_1 - \mu_2$.
(c) *Interpretation* Comment on the meaning of the confidence interval in the context of this problem. Does the interval consist of positive numbers only? negative numbers only? a mix of positive and negative numbers? Does it appear (at the 99% confidence level) that a change in the interval length between eruptions has occurred? Many geologic experts believe that the distribution of eruption times of Old Faithful changed after the major earthquake that occurred in 1959.

20. *Psychology: Parental Sensitivity* "Parental Sensitivity to Infant Cues: Similarities and Differences Between Mothers and Fathers" by M. V. Graham (*Journal of Pediatric Nursing*, Vol. 8, No. 6) reports a study of parental empathy for sensitivity cues and baby temperament (higher scores mean more empathy). Let x_1 be a random variable that represents the score of a mother on an empathy test (as regards her baby). Let x_2 be the empathy score of a father. A random sample of 32 mothers gave a sample mean of $\bar{x}_1 = 69.44$. Another random sample of 32 fathers gave $\bar{x}_2 = 59$. Assume that $\sigma_1 = 11.69$ and $\sigma_2 = 11.60$.
(a) *Check Requirements* Which distribution, normal or Student's t, do we use to approximate the $\bar{x}_1 - \bar{x}_2$ distribution? Explain.
(b) Let μ_1 be the population mean of x_1 and let μ_2 be the population mean of x_2. Find a 99% confidence interval for $\mu_1 - \mu_2$.

(c) *Interpretation* Examine the confidence interval and explain what it means in the context of this problem. Does the confidence interval contain all positive, all negative, or both positive and negative numbers? What does this tell you about the relationship between average empathy scores for mothers compared with those for fathers at the 99% confidence level?

21. *Navajo Culture: Traditional Hogans* S. C. Jett is a professor of geography at the University of California, Davis. He and a colleague, V. E. Spencer, are experts on modern Navajo culture and geography. The following information is taken from their book *Navajo Architecture: Forms, History, Distributions* (University of Arizona Press). On the Navajo Reservation, a random sample of 210 permanent dwellings in the Fort Defiance region showed that 65 were traditional Navajo hogans. In the Indian Wells region, a random sample of 152 permanent dwellings showed that 18 were traditional hogans. Let p_1 be the population proportion of all traditional hogans in the Fort Defiance region, and let p_2 be the population proportion of all traditional hogans in the Indian Wells region.
(a) *Check Requirements* Can a normal distribution be used to approximate the $\hat{p}_1 - \hat{p}_2$ distribution? Explain.
(b) Find a 99% confidence interval for $p_1 - p_2$.
(c) *Interpretation* Examine the confidence interval and comment on its meaning. Does it include numbers that are all positive? all negative? mixed? What if it is hypothesized that Navajo who follow the traditional culture of their people tend to occupy hogans? Comment on the confidence interval for $p_1 - p_2$ in this context.

22. *Archaeology: Cultural Affiliation* "Unknown cultural affiliations and loss of identity at high elevations." These words are used to propose the hypothesis that archaeological sites tend to lose their identity as altitude extremes are reached. This idea is based on the notion that prehistoric people tended *not* to take trade wares to temporary settings and/or isolated areas (Source: *Prehistoric New Mexico: Background for Survey*, by D. E. Stuart and R. P. Gauthier, University of New Mexico Press). As elevation zones of prehistoric people (in what is now the state of New Mexico) increased, there seemed to be a loss of artifact identification. Consider the following information.

Elevation Zone	Number of Artifacts	Number Unidentified
7000–7500 ft	112	69
5000–5500 ft	140	26

Let p_1 be the population proportion of unidentified archaeological artifacts at the elevation zone 7000–7500 feet in the given archaeological area. Let p_2 be the population proportion of unidentified archaeological artifacts at the elevation zone 5000–5500 feet in the given archaeological area.
(a) *Check Requirements* Can a normal distribution be used to approximate the $\hat{p}_1 - \hat{p}_2$ distribution? Explain.
(b) Find a 99% confidence interval for $p_1 - p_2$.
(c) *Interpretation* Explain the meaning of the confidence interval in the context of this problem. Does the confidence interval contain all positive numbers? all negative numbers? both positive and negative numbers? What does this tell you (at the 99% confidence level) about the comparison of the population proportion of unidentified artifacts at high elevations (7000–7500 feet) with the population proportion of unidentified artifacts at lower elevations (5000–5500 feet)? How does this relate to the stated hypothesis?

23. *Wildlife: Wolves* David E. Brown is an expert in wildlife conservation. In his book *The Wolf in the Southwest: The Making of an Endangered Species* (University of Arizona Press), he lists the following weights of adult gray wolves from two regions in Old Mexico.

Chihuahua region: x_1 variable in pounds

86	75	91	70	79
80	68	71	74	64

Durango region: x_2 variable in pounds

68	72	79	68	77	89	62	55	68
68	59	63	66	58	54	71	59	67

(a) Use a calculator with mean and standard deviation keys to verify that $\bar{x}_1 = 75.80$ pounds, $s_1 = 8.32$ pounds, $\bar{x}_2 = 66.83$ pounds, and $s_2 = 8.87$ pounds.
(b) *Check Requirements* Assuming that the original distribution of the weights of wolves are mound-shaped and symmetric, what distribution can be used to approximate the $\bar{x}_1 - \bar{x}_2$ distribution? Explain.
(c) Let μ_1 be the mean weight of the population of all gray wolves in the Chihuahua region. Let μ_2 be the mean weight of the population of all gray wolves in the Durango region. Find an 85% confidence interval for $\mu_1 - \mu_2$.
(d) *Interpretation* Examine the confidence interval and explain what it means in the context of this problem. Does the interval consist of numbers that are all positive? all negative? of different signs? At the 85% level of confidence, what can you say about the comparison of the average weight of gray wolves in the Chihuahua region with the average weight of gray wolves in the Durango region?

24. *Medical: Plasma Compress* At Community Hospital, the burn center is experimenting with a new plasma compress treatment. A random sample of $n_1 = 316$ patients with minor burns received the plasma compress treatment. Of these patients, it was found that 259 had no visible scars after treatment. Another random sample of $n_2 = 419$ patients with minor burns received no plasma compress treatment. For this group, it was found that 94 had no visible scars after treatment. Let p_1 be the population proportion of all patients with minor burns receiving the plasma compress treatment who have no visible scars. Let p_2 be the population proportion of all patients with minor burns not receiving the plasma compress treatment who have no visible scars.
(a) *Check Requirements* Can a normal distribution be used to approximate the $\hat{p}_1 - \hat{p}_2$ distribution? Explain.
(b) Find a 95% confidence interval for $p_1 - p_2$.
(c) *Interpretation* Explain the meaning of the confidence interval found in part (b) in the context of the problem. Does the interval contain numbers that are all positive? all negative? both positive and negative? At the 95% level of confidence, does treatment with plasma compresses seem to make a difference in the proportion of patients with visible scars from minor burns?

25. *Psychology: Self-Esteem* Female undergraduates in randomized groups of 15 took part in a self-esteem study ("There's More to Self-Esteem than Whether It Is High or Low: The Importance of Stability of Self-Esteem," by M. H. Kernis et al., *Journal of Personality and Social Psychology*, Vol. 65, No. 6). The study measured an index of self-esteem from the points of view of competence, social acceptance, and physical attractiveness. Let x_1, x_2, and x_3 be random variables representing the measure of self-esteem through x_1 (competence), x_2 (social acceptance), and x_3 (attractiveness). Higher index values mean a more positive influence on self-esteem.

Variable	Sample Size	Mean $\bar{x}$	Standard Deviation s	Population Mean
x_1	15	19.84	3.07	μ_1
x_2	15	19.32	3.62	μ_2
x_3	15	17.88	3.74	μ_3

(a) Find an 85% confidence interval for $\mu_1 - \mu_2$.
(b) Find an 85% confidence interval for $\mu_1 - \mu_3$.
(c) Find an 85% confidence interval for $\mu_2 - \mu_3$.
(d) *Interpretation* Comment on the meaning of each of the confidence intervals found in parts (a), (b), and (c). At the 85% confidence level, what can you say about the average differences in influence on self-esteem between competence and social acceptance? between competence and attractiveness? between social acceptance and attractiveness?

26. *Focus Problem: Wood Duck Nests* In the Focus Problem at the beginning of this chapter, a study was described comparing the hatch ratios of wood duck nesting boxes. Group I nesting boxes were well separated from each other and well hidden by available brush. There were a total of 474 eggs in group I boxes, of which a field count showed about 270 had hatched. Group II nesting boxes were placed in highly visible locations and grouped closely together. There were a total of 805 eggs in group II boxes, of which a field count showed about 270 had hatched.
(a) Find a point estimate $\hat{p}_1$ for p_1, the proportion of eggs that hatched in group I nest box placements. Find a 95% confidence interval for p_1.
(b) Find a point estimate $\hat{p}_2$ for p_2, the proportion of eggs that hatched in group II nest box placements. Find a 95% confidence interval for p_2.
(c) Find a 95% confidence interval for $p_1 - p_2$. Does the interval indicate that the proportion of eggs hatched from group I nest boxes is higher than, lower than, or equal to the proportion of eggs hatched from group II nest boxes?
(d) *Interpretation* What conclusions about placement of nest boxes can be drawn? In the article discussed in the Focus Problem, additional concerns are raised about the higher cost of placing and maintaining group I nest box placements. Also at issue is the cost efficiency per successful wood duck hatch.

27. *Critical Thinking: Different Confidence Levels*
(a) Suppose a 95% confidence interval for the difference of means contains both positive and negative numbers. Will a 99% confidence interval based on the same data necessarily contain both positive and negative numbers? Explain. What about a 90% confidence interval? Explain.
(b) Suppose a 95% confidence interval for the difference of proportions contains all positive numbers. Will a 99% confidence interval based on the same data necessarily contain all positive numbers as well? Explain. What about a 90% confidence interval? Explain.

28. *Expand Your Knowledge: Sample Size, Difference of Means* What about sample size? If we want a confidence interval with maximal margin of error E and level of confidence c, then Section 7.1 shows us which formulas to apply for a *single* mean μ and Section 7.3 shows us formulas for a *single* proportion p.
(a) How about a *difference of means*? When σ_1 and σ_2 are known, the margin of error E for a c% confidence interval is

$$E = z_c\sqrt{\frac{\sigma_1^2}{n_1} + \frac{\sigma_2^2}{n_2}}$$

Let us make the simplifying assumption that we have *equal sample sizes* n so that $n = n_1 = n_2$. We also assume that $n \geq 30$. In this context, we get

$$E = z_c\sqrt{\frac{\sigma_1^2}{n} + \frac{\sigma_2^2}{n}} = \frac{z_c}{\sqrt{n}}\sqrt{\sigma_1^2 + \sigma_2^2}$$

Solve this equation for n and show that

$$n = \left(\frac{z_c}{E}\right)^2(\sigma_1^2 + \sigma_2^2)$$

(b) In Problem 15 (football and basketball player heights), suppose we want to be 95% sure that our estimate $\bar{x}_1 - \bar{x}_2$ for the difference $\mu_1 - \mu_2$ has a margin of error $E = 0.05$ foot. How large should the sample size be (assuming equal sample size—i.e., $n = n_1 = n_2$)? Since we do not know σ_1 or σ_2 and $n \geq 30$, use s_1 and s_2, respectively, from the preliminary sample of Problem 15.

(c) In Problem 16 (petal lengths of two iris species), suppose we want to be 90% sure that our estimate $\bar{x}_1 - \bar{x}_2$ for the difference $\mu_1 - \mu_2$ has a margin of error $E = 0.1$ cm. How large should the sample size be (assuming equal sample size—i.e., $n = n_1 = n_2$)? Since we do not know σ_1 or σ_2 and $n \geq 30$, use s_1 and s_2, respectively, from the preliminary sample of Problem 16.

29. *Expand Your Knowledge: Sample Size, Difference of Proportions* What about the sample size n for confidence intervals for the difference of proportions $p_1 - p_2$? Let us make the following assumptions: *equal sample sizes* $n = n_1 = n_2$ and *all four quantities* $n_1\hat{p}_1$, $n_1\hat{q}_1$, $n_2\hat{p}_2$, and $n_2\hat{q}_2$ *are greater than 5*. Those readers familiar with algebra can use the procedure outlined in Problem 28 to show that if we have preliminary estimates $\hat{p}_1$ and $\hat{p}_2$ and a given maximal margin of error E for a specified confidence level c, then the sample size n should be at least

$$n = \left(\frac{z_c}{E}\right)^2 (\hat{p}_1\hat{q}_1 + \hat{p}_2\hat{q}_2)$$

However, if we have no preliminary estimates for $\hat{p}_1$ and $\hat{p}_2$, then the theory similar to that used in this section tells us that the sample size n should be at least

$$n = \frac{1}{2}\left(\frac{z_c}{E}\right)^2$$

(a) In Problem 17 (Myers–Briggs personality type indicators in common for married couples), suppose we want to be 99% confident that our estimate $\hat{p}_1 - \hat{p}_2$ for the difference $p_1 - p_2$ has a maximal margin of error $E = 0.04$. Use the preliminary estimates $\hat{p}_1 = 289/375$ for the proportion of couples sharing two personality traits and $\hat{p}_2 = 23/571$ for the proportion having no traits in common. How large should the sample size be (assuming equal sample size—i.e., $n = n_1 = n_2$)?

(b) Suppose that in Problem 17 we have no preliminary estimates for $\hat{p}_1$ and $\hat{p}_2$ and we want to be 95% confident that our estimate $\hat{p}_1 - \hat{p}_2$ for the difference $p_1 - p_2$ has a maximal margin of error $E = 0.05$. How large should the sample size be (assuming equal sample size—i.e., $n = n_1 = n_2$)?

30. *Expand Your Knowledge: Software Approximation for Degrees of Freedom* Given x_1 and x_2 distributions that are normal or approximately normal with unknown σ_1 and σ_2, the value of t corresponding to $\bar{x}_1 - \bar{x}_2$ has a distribution that is approximated by a Student's t distribution. We use the convention that the degrees of freedom are approximately the smaller of $n_1 - 1$ and $n_2 - 1$. However, a more accurate estimate for the appropriate degrees of freedom is given by *Satterthwaite's formula*

$$d.f. \approx \frac{\left(\frac{s_1^2}{n_1} + \frac{s_2^2}{n_2}\right)^2}{\frac{1}{n_1 - 1}\left(\frac{s_1^2}{n_1}\right)^2 + \frac{1}{n_2 - 1}\left(\frac{s_2^2}{n_2}\right)^2}$$

where s_1, s_2, n_1, and n_2 are the respective sample standard deviations and sample sizes of independent random samples from the x_1 and x_2 distributions. This is the approximation used by most statistical software. When both n_1 and n_2 are 5 or larger, it is quite accurate. The degrees of freedom computed from this formula are either truncated or not rounded.

(a) Use the data of Problem 14 (weights of pro football and pro basketball players) to compute *d.f.* using the formula. Compare the result to 36, the value generated by Minitab. Did Minitab truncate?
(b) Compute a 99% confidence interval using $d.f. \approx 36$. (Using Table 6 requires using $d.f. = 35$.) Compare this confidence interval to the one you computed in Problem 14. Which *d.f.* gives the longer interval?

31. *Expand Your Knowledge: Pooled Two-Sample Procedures* Under the condition that both populations have equal standard deviations ($\sigma_1 = \sigma_2$), we can pool the standard deviations and use a Student's *t* distribution with degrees of freedom $d.f. = n_1 + n_2 - 2$ to find the margin of error of a *c* confidence interval for $\mu_1 - \mu_2$. This technique demonstrates another commonly used method of computing confidence intervals for $\mu_1 - \mu_2$.

PROCEDURE

HOW TO FIND A CONFIDENCE INTERVAL FOR $\mu_1 - \mu_2$ WHEN $\sigma_1 = \sigma_2$

Requirements

Consider two *independent* random samples, where

$\bar{x}_1$ and $\bar{x}_2$ are sample means from populations 1 and 2
s_1 and s_2 are sample standard deviations from populations 1 and 2
n_1 and n_2 are sample sizes from populations 1 and 2

If you can assume that both population distributions 1 and 2 are normal or at least mound-shaped and symmetric, then any sample sizes n_1 and n_2 will work. If you cannot assume this, then use sample sizes $n_1 \geq 30$ and $n_2 \geq 30$.

Confidence interval for $\mu_1 - \mu_2$ when $\sigma_1 = \sigma_2$

$$(\hat{x}_1 - \hat{x}_2) - E < \mu_1 - \mu_2 < (\bar{x}_1 - \bar{x}_2) + E$$

where

$$E = t_c s\sqrt{\frac{1}{n_1} + \frac{1}{n_2}}$$

Pooled standard deviation

$$s = \sqrt{\frac{(n_1 - 1)s_1^2 + (n_2 - 1)s_2^2}{n_1 + n_2 - 2}} \quad \text{(pooled standard deviation)}$$

c = confidence level ($0 < c < 1$)
t_c = critical value for confidence level *c* and degrees of freedom $d.f. = n_1 + n_2 - 2$ (See Table 6 of Appendix II.)

Note: With statistical software, select pooled variance or equal variance options.

(a) There are many situations in which we want to compare means from populations having standard deviations that are equal. The pooled standard deviation method applies even if the standard deviations are known to be only approximately equal. (See Section 10.4 for methods to test that $\sigma_1 = \sigma_2$.) Consider Problem 23 regarding weights of gray wolves in two regions. Notice that $s_1 = 8.32$ pounds and $s_2 = 8.87$ pounds are fairly close. Use the method of pooled standard deviation to find an 85% confidence interval for the difference in population mean weights of gray wolves in the Chihuahua region compared with those in the Durango region.
(b) Compare the confidence interval computed in part (a) with that computed in Problem 23. Which method has the larger degrees of freedom? Which method has the longer confidence interval?

CHAPTER REVIEW

SUMMARY

How do you get information about a population by looking at a random sample? One way is to use point estimates and confidence intervals.

- Point estimates and their corresponding parameters are

  $\bar{x}$ for μ $\quad$ $\bar{x}_1 - \bar{x}_2$ for $\mu_1 - \mu_1$

  $\hat{p}$ for p $\quad$ $\hat{p}_1 - \hat{p}_2$ for $p_1 - p_2$

- Confidence intervals are of the form

  point estimate $- E <$ parameter $<$ point estimate $+ E$

- E is the maximal margin of error. Specific values of E depend on the parameter, level of confidence, whether population standard deviations are known, sample size, and the shapes of the original population distributions.

  For μ: $E = z_c \dfrac{\sigma}{\sqrt{n}}$ when σ is known

  $E = t_c \dfrac{s}{\sqrt{n}}$ with $d.f. = n - 1$ when σ is unknown

  For p: $E = z_c \sqrt{\dfrac{\hat{p}(1-\hat{p})}{n}}$ when $n\hat{p} > 5$ and $n\hat{q} > 5$

  For $\mu_1 - \mu_2$: $E = z_c \sqrt{\dfrac{\sigma_1^2}{n_1} + \dfrac{\sigma_2^2}{n_2}}$: when σ_1 and σ_2 are known

  $E = t_c \sqrt{\dfrac{s_1^2}{n_1} + \dfrac{s_2^2}{n_2}}$ when σ_1 or σ_2 is unknown with $d.f.$ = smaller of $n_1 - 1$ or $n_2 - 1$

  Software uses Satterthwaite's approximation for $d.f.$

  For $p_1 - p_2$: $E = z_c \sqrt{\dfrac{\hat{p}_1\hat{q}_1}{n_1} + \dfrac{\hat{p}_2\hat{q}_2}{n_2}}$ for sufficiently large n

- Confidence intervals have an associated probability c called the confidence level. For a given sample size, the proportion of all corresponding confidence intervals that contain the parameter in question is c.

IMPORTANT WORDS & SYMBOLS

Section 7.1

Point estimate for μ 361
Confidence level c 361
Critical values z_c 361
Maximal margin of error E 363
c confidence interval
Sample size for estimating μ 364

Section 7.2

Student's t distribution 374
Degrees of freedom $d.f.$ 374
Critical values t_c 375

Section 7.3

Point estimate for p, $\hat{p}$ 388
Confidence interval for p 389
Margin of error for polls 393
Sample size for estimating p 394

Section 7.4

Independent samples 401
Dependent samples 401
$\bar{x}_1 - \bar{x}_2$ sampling distribution 401
Confidence interval for $\mu_1 - \mu_2$ (σ_1 and σ_2 known) 402
Confidence interval for $\mu_1 - \mu_2$ (σ_1 and σ_2 unknown) 404
Confidence interval for $p_1 - p_2$ 409
Satterthwaite's Formula for $d.f.$ 421
Pooled standard deviation 422

VIEWPOINT

All Systems Go?

On January 28, 1986, the space shuttle Challenger *caught fire and blew up only seconds after launch. A great deal of good engineering had gone into the design of the* Challenger. *However, when a system has several confidence levels operating at once, it can happen, in rare cases, that risks will increase rather than cancel out. (See*

Chapter Review Problem 19.) Diane Vaughn is a professor of sociology at Boston College and author of the book The Challenger Launch Decision *(University of Chicago Press). Her book contains an excellent discussion of risks, the normalization of deviants, and cost/safety tradeoffs. Vaughn's book is described as "a remarkable and important analysis of how social structures can induce consequential errors in a decision process" (Robert K. Merton, Columbia University).*

CHAPTER REVIEW PROBLEMS

1. *Statistical Literacy* In your own words, carefully explain the meanings of the following terms: *point estimate*, *critical value*, *maximal margin of error*, *confidence level*, and *confidence interval*.

2. *Critical Thinking* Suppose you are told that a 95% confidence interval for the average price of a gallon of regular gasoline in your state is from \$3.15 to \$3.45. Use the fact that the confidence interval for the mean has the form $\bar{x} - E$ to $\bar{x} + E$ to compute the sample mean and the maximal margin of error E.

3. *Critical Thinking* If you have a 99% confidence interval for μ based on a simple random sample,
   (a) is it correct to say that the *probability* that μ is in the specified interval is 99%? Explain.
   (b) is it correct to say that in the long run, if you computed many, many confidence intervals using the prescribed method, about 99% of such intervals would contain μ? Explain.

For Problems 4–19, categorize each problem according to the parameter being estimated: proportion p, mean μ, difference of means $\mu_1 - \mu_2$, or difference of proportions $p_1 - p_2$. Then solve the problem.

4. *Auto Insurance: Claims* Anystate Auto Insurance Company took a random sample of 370 insurance claims paid out during a 1-year period. The average claim paid was \$1570. Assume $\sigma =$ \$250. Find 0.90 and 0.99 confidence intervals for the mean claim payment.

5. *Psychology: Closure* Three experiments investigating the relationship between need for cognitive closure and persuasion were reported in "Motivated Resistance and Openness to Persuasion in the Presence or Absence of Prior Information" by A. W. Kruglanski (*Journal of Personality and Social Psychology*, Vol. 65, No. 5, pp. 861–874). Part of the study involved administering a "need for closure scale" to a group of students enrolled in an introductory psychology course. The "need for closure scale" has scores ranging from 101 to 201. For the 73 students in the highest quartile of the distribution, the mean score was $\bar{x} = 178.70$. Assume a population standard deviation of $\sigma = 7.81$. These students were all classified as high on their need for closure. Assume that the 73 students represent a random sample of all students who are classified as high on their need for closure. Find a 95% confidence interval for the population mean score μ on the "need for closure scale" for all students with a high need for closure.

6. *Psychology: Closure* How large a sample is needed in Problem 5 if we wish to be 99% confident that the sample mean score is within 2 points of the population mean score for students who are high on the need for closure?

7. *Archaeology: Excavations* The Wind Mountain archaeological site is located in southwestern New Mexico. Wind Mountain was home to an ancient culture of prehistoric Native Americans called Anasazi. A random sample of excavations at Wind Mountain gave the following depths (in centimeters) from present-day surface grade to the location of significant archaeological artifacts (Source: *Mimbres Mogollon Archaeology*, by A. Woosley and A. McIntyre, University of New Mexico Press).

85	45	120	80	75	55	65	60
65	95	90	70	75	65	68	

(a) Use a calculator with mean and sample standard deviation keys to verify that $\bar{x} \approx 74.2$ cm and $s \approx 18.3$ cm.
(b) Compute a 95% confidence interval for the mean depth μ at which archaeological artifacts from the Wind Mountain excavation site can be found.

8. *Archaeology: Pottery* Shards of clay vessels were put together to reconstruct rim diameters of the original ceramic vessels found at the Wind Mountain archaeological site (see source in Problem 7). A random sample of ceramic vessels gave the following rim diameters (in centimeters):

15.9	13.4	22.1	12.7	13.1	19.6	11.7	13.5	17.7	18.1

(a) Use a calculator with mean and sample standard deviation keys to verify that $\bar{x} \approx 15.8$ cm and $s \approx 3.5$ cm.
(b) Compute an 80% confidence interval for the population mean μ of rim diameters for such ceramic vessels found at the Wind Mountain archaeological site.

9. *Telephone Interviews: Survey* The National Study of the Changing Work Force conducted an extensive survey of 2958 wage and salaried workers on issues ranging from relationships with their bosses to household chores. The data were gathered through hour-long telephone interviews with a nationally representative sample (*The Wall Street Journal*). In response to the question "What does success mean to you?" 1538 responded, "Personal satisfaction from doing a good job." Let p be the population proportion of all wage and salaried workers who would respond the same way to the stated question. Find a 90% confidence interval for p.

10. *Telephone Interviews: Survey* How large a sample is needed in Problem 9 if we wish to be 95% confident that the sample percentage of those equating success with personal satisfaction is within 1% of the population percentage? *Hint:* Use $p \approx 0.52$ as a preliminary estimate.

11. *Archaeology: Pottery* Three-circle, red-on-white is one distinctive pattern painted on ceramic vessels of the Anasazi period found at the Wind Mountain archaeological site (see source for Problem 7). At one excavation, a sample of 167 potsherds indicated that 68 were of the three-circle, red-on-white pattern.
(a) Find a point estimate $\hat{p}$ for the proportion of all ceramic potsherds at this site that are of the three-circle, red-on-white pattern.
(b) Compute a 95% confidence interval for the population proportion p of all ceramic potsherds with this distinctive pattern found at the site.

12. *Archaeology: Pottery* Consider the three-circle, red-on-white pattern discussed in Problem 11. How many ceramic potsherds must be found and identified if we are to be 95% confident that the sample proportion $\hat{p}$ of such potsherds is within 6% of the population proportion of three-circle, red-on-white patterns found at this excavation site? *Hint:* Use the results of Problem 11 as a preliminary estimate.

13. *Agriculture: Bell Peppers* The following data represent soil water content (percent water by volume) for independent random samples of soil taken from two experimental fields growing bell peppers (Reference: *Journal of Agricultural, Biological, and Environmental Statistics*). *Note:* These data are also available for download at the Online Study Center.

Soil water content from field I: x_1; $n_1 = 72$

15.1	11.2	10.3	10.8	16.6	8.3	9.1	12.3	9.1	14.3
10.7	16.1	10.2	15.2	8.9	9.5	9.6	11.3	14.0	11.3
15.6	11.2	13.8	9.0	8.4	8.2	12.0	13.9	11.6	16.0
9.6	11.4	8.4	8.0	14.1	10.9	13.2	13.8	14.6	10.2
11.5	13.1	14.7	12.5	10.2	11.8	11.0	12.7	10.3	10.8
11.0	12.6	10.8	9.6	11.5	10.6	11.7	10.1	9.7	9.7
11.2	9.8	10.3	11.9	9.7	11.3	10.4	12.0	11.0	10.7
8.8	11.1								

Soil water content from field II: x_2; $n_2 = 80$

12.1	10.2	13.6	8.1	13.5	7.8	11.8	7.7	8.1	9.2
14.1	8.9	13.9	7.5	12.6	7.3	14.9	12.2	7.6	8.9
13.9	8.4	13.4	7.1	12.4	7.6	9.9	26.0	7.3	7.4
14.3	8.4	13.2	7.3	11.3	7.5	9.7	12.3	6.9	7.6
13.8	7.5	13.3	8.0	11.3	6.8	7.4	11.7	11.8	7.7
12.6	7.7	13.2	13.9	10.4	12.8	7.6	10.7	10.7	10.9
12.5	11.3	10.7	13.2	8.9	12.9	7.7	9.7	9.7	11.4
11.9	13.4	9.2	13.4	8.8	11.9	7.1	8.5	14.0	14.2

(a) Use a calculator with mean and standard deviation keys to verify that $\bar{x}_1 \approx 11.42$, $s_1 \approx 2.08$, $\bar{x}_2 \approx 10.65$, and $s_2 \approx 3.03$.
(b) Let μ_1 be the population mean for x_1 and let μ_2 be the population mean for x_2. Find a 95% confidence interval for $\mu_1 - \mu_2$.
(c) *Interpretation* Explain what the confidence interval means in the context of this problem. Does the interval consist of numbers that are all positive? all negative? of different signs? At the 95% level of confidence, is the population mean soil water content of the first field higher than that of the second field?
(d) Which distribution (standard normal or Student's t) did you use? Why? Do you need information about the soil water content distributions?

14. *Stocks: Retail and Utility* How profitable are different sectors of the stock market? One way to answer such a question is to examine profit as a percentage of stockholder equity. A random sample of 32 retail stocks such as Toys "Я" Us, Best Buy, and Gap was studied for x_1, profit as a percentage of stockholder equity. The result was $\bar{x}_1 = 13.7$. A random sample of 34 utility (gas and electric) stocks such as Boston Edison, Wisconsin Energy, and Texas Utilities was studied for x_2, profit as a percentage of stockholder equity. The result was $\bar{x}_2 = 10.1$ (Source: *Fortune 500*, Vol. 135, No. 8). Assume that $\sigma_1 = 4.1$ and $\sigma_2 = 2.7$.
(a) Let μ_1 represent the population mean profit as a percentage of stockholder equity for retail stocks, and let μ_2 represent the population mean profit as a percentage of stockholder equity for utility stocks. Find a 95% confidence interval for $\mu_1 - \mu_2$.
(b) *Interpretation* Examine the confidence interval and explain what it means in the context of this problem. Does the interval consist of numbers that are all positive? all negative? of different signs? At the 95% level of confidence, does it appear that the profit as a percentage of stockholder equity for retail stocks is higher than that for utility stocks?

15. *Wildlife: Wolves* A random sample of 18 adult male wolves from the Canadian Northwest Territories gave an average weight $\bar{x}_1 = 98$ pounds, with estimated sample standard deviation $s_1 = 6.5$ pounds. Another sample of 24 adult male wolves from Alaska gave an average weight $\bar{x}_2 = 90$ pounds, with estimated sample standard deviation $s_2 = 7.3$ pounds (Source: *The Wolf* by L. D. Mech, University of Minnesota Press).
(a) Let μ_1 represent the population mean weight of adult male wolves from the Northwest Territories, and let μ_2 represent the population mean weight of adult male wolves from Alaska. Find a 75% confidence interval for $\mu_1 - \mu_2$.
(b) *Interpretation* Examine the confidence interval and explain what it means in the context of this problem. Does the interval consist of numbers that are all positive? all negative? of different signs? At the 75% level of confidence, does it appear that the average weight of adult male wolves from the Northwest Territories is greater than that of the Alaska wolves?

16. *Wildlife: Wolves* A random sample of 17 wolf litters in Ontario, Canada, gave an average of $\bar{x}_1 = 4.9$ wolf pups per litter, with estimated sample standard deviation $s_1 = 1.0$. Another random sample of 6 wolf litters in Finland gave an average of $\bar{x}_2 = 2.8$ wolf pups per litter, with sample standard deviation $s_2 = 1.2$ (see source for Problem 15).

(a) Find an 85% confidence interval for $\mu_1 - \mu_2$, the difference in population mean litter size between Ontario and Finland.

(b) *Interpretation* Examine the confidence interval and explain what it means in the context of this problem. Does the interval consist of numbers that are all positive? all negative? of different signs? At the 85% level of confidence, does it appear that the average litter size of wolf pups in Ontario is greater than the average litter size in Finland?

17. *Survey Response: Validity* The book *Survey Responses: An Evaluation of Their Validity* by E. J. Wentland and K. Smith (Academic Press), includes studies reporting accuracy of answers to questions from surveys. A study by Locander et al. considered the question "Are you a registered voter?" Accuracy of response was confirmed by a check of city voting records. Two methods of survey were used: a face-to-face interview and a telephone interview. A random sample of 93 people were asked the voter registration question face-to-face. Seventy-nine respondents gave accurate answers (as verified by city records). Another random sample of 83 people were asked the same question during a telephone interview. Seventy-four respondents gave accurate answers. Assume the samples are representative of the general population.

(a) Let p_1 be the population proportion of all people who answer the voter registration question accurately during a face-to-face interview. Let p_2 be the population proportion of all people who answer the question accurately during a telephone interview. Find a 95% confidence interval for $p_1 - p_2$.

(b) *Interpretation* Does the interval contain numbers that are all positive? all negative? mixed? Comment on the meaning of the confidence interval in the context of this problem. At the 95% level, do you detect any difference in the proportion of accurate responses from face-to-face interviews compared with the proportion of accurate responses from telephone interviews?

18. *Survey Response: Validity* Locander et al. (see reference in Problem 17) also studied the accuracy of responses on questions involving more sensitive material than voter registration. From public records, individuals were identified as having been charged with drunken driving not less than 6 months or more than 12 months from the starting date of the study. Two random samples from this group were studied. In the first sample of 30 individuals, the respondents were asked in a face-to-face interview if they had been charged with drunken driving in the last 12 months. Of these 30 people interviewed face-to-face, 16 answered the question accurately. The second random sample consisted of 46 people who had been charged with drunken driving. During a telephone interview, 25 of these responded accurately to the question asking if they had been charged with drunken driving during the past 12 months. Assume the samples are representative of all people recently charged with drunken driving.

(a) Let p_1 represent the population proportion of all people with recent charges of drunken driving who respond accurately to a face-to-face interview asking if they have been charged with drunken driving during the past 12 months. Let p_2 represent the population proportion of people who respond accurately to the question when it is asked in a telephone interview. Find a 90% confidence interval for $p_1 - p_2$.

(b) *Interpretation* Does the interval found in part (a) contain numbers that are all positive? all negative? mixed? Comment on the meaning of the confidence interval in the context of this problem. At the 90% level, do you detect any differences in the proportion of accurate responses to the question from face-to-face interviews as compared with the proportion of accurate responses from telephone interviews?

19. *Expand Your Knowledge: Two Confidence Intervals* What happens if we want several confidence intervals to hold at the same time (concurrently)? Do we still have the same level of confidence we had for *each* individual interval?

(a) Suppose we have two independent random variables x_1 and x_2 with respective population means μ_1 and μ_2. Let us say that we use sample data to construct two 80% confidence intervals.

Confidence Interval	Confidence Level
$A_1 < \mu_1 < B_1$	0.80
$A_2 < \mu_2 < B_2$	0.80

Now, what is the probability that *both* intervals hold at the same time? Use methods of Section 4.2 to show that

$$P(A_1 < \mu_1 < B_1 \textit{ and } A_2 < \mu_2 < B_2) = 0.64$$

Hint: You are combining independent events. If the confidence is 64% that both intervals hold concurrently, explain why the risk that at least one interval does not hold (i.e., fails) must be 36%.

(b) Suppose we want *both* intervals to hold with 90% confidence (i.e., only 10% risk level). How much confidence c should each interval have to achieve this combined level of confidence? (Assume that each interval has the same confidence level c.)

Hint: $P(A_1 < \mu_1 < B_1 \textit{ and } A_2 < \mu_2 < B_2) = 0.90$

$P(A_1 < \mu_1 < B_1) \times P(A_2 < \mu_2 < B_2) = 0.90$

$c \times c = 0.90$

Now solve for c.

(c) If we want *both* intervals to hold at the 90% level of confidence, then the individual intervals must hold at a *higher* level of confidence. Write a brief but detailed explanation of how this could be of importance in a large, complex engineering design such as a rocket booster or a spacecraft.

DATA HIGHLIGHTS: GROUP PROJECTS

Judy Griesedieck/Encyclopedia/Corbis

Digging clams

Break into small groups and discuss the following topics. Organize a brief outline in which you summarize the main points of your group discussion.

1. Garrison Bay is a small bay in Washington state. A popular recreational activity in the bay is clam digging. For several years, this harvest has been monitored and the size distribution of clams recorded. Data for lengths and widths of little neck clams (*Protothaca staminea*) were recorded by a method of systematic sampling in a study done by S. Scherba and V. F. Gallucci ("The Application of Systematic Sampling to a Study of Infaunal Variation in a Soft Substrate Intertidal Environment," *Fishery Bulletin*, Vol. 74, pp. 937–948). The data in Tables 7-4 and 7-5 give lengths and widths for 35 little neck clams.
   (a) Use a calculator to compute the sample mean and sample standard deviation for the lengths and widths. Compute the coefficient of variation for each.
   (b) Compute a 95% confidence interval for the population mean length of all Garrison Bay little neck clams.
   (c) How many more little neck clams would be needed in a sample if you wanted to be 95% sure that the sample mean length is within a maximal margin of error of 10 mm of the population mean length?

TABLE 7-4 **Lengths of Little Neck Clams (mm)**

530	517	505	512	487	481	485	479	452	468
459	449	472	471	455	394	475	335	508	486
474	465	420	402	410	393	389	330	305	169
91	537	519	509	511					

TABLE 7-5 **Widths of Little Neck Clams (mm)**

494	477	471	413	407	427	408	430	395	417
394	397	402	401	385	338	422	288	464	436
414	402	383	340	349	333	356	268	264	141
77	498	456	433	447					

(d) Compute a 95% confidence interval for the population mean width of all Garrison Bay little neck clams.

(e) How many more little neck clams would be needed in a sample if you wanted to be 95% sure that the sample mean width is within a maximal margin of error of 10 mm of the population mean width?

(f) The *same* 35 clams were used for measures of length and width. Are the sample measurements length and width independent or dependent? Why?

2. Examine Figure 7-8, "Fall Back."

(a) Of the 1024 adults surveyed, 66% were reported to favor daylight saving time. How many people in the sample preferred daylight saving time? Using the statistic $\hat{p} = 0.66$ and sample size $n = 1024$, find a 95% confidence interval for the proportion of people p who favor daylight saving time. How could you report this information in terms of a margin of error?

(b) Look at Figure 7-8 to find the sample statistic $\hat{p}$ for the proportion of people preferring standard time. Find a 95% confidence interval for the population proportion p of people who favor standard time. Report the same information in terms of a margin of error.

3. Examine Figure 7-9,"Coupons: Limited Use."

(a) Use Figure 7-9 to estimate the percentage of merchandise coupons that were redeemed. Also estimate the percentage dollar value of the coupons that were redeemed. Are these numbers approximately equal?

(b) Suppose you are a marketing executive working for a national chain of toy stores. You wish to estimate the percentage of coupons that will be redeemed for the toy stores. How many coupons should you check to be 95% sure that the percentage of coupons redeemed is within 1% of the population proportion of all coupons redeemed for the toy store?

(c) Use the results of part (a) as a preliminary estimate for p, the percentage of coupons that are redeemed, and redo part (b).

(d) Suppose you sent out 937 coupons and found that 27 were redeemed. Explain why you could be 95% confident that the proportion of such coupons redeemed in the future would be between 1.9% and 3.9%.

(e) Suppose the dollar value of a collection of coupons was \$10,000. Use the data in Figure 7-9 to find the expected value and standard deviation of the dollar value of the redeemed coupons. What is the probability that between \$225 and \$275 (out of the \$10,000) is redeemed?

FIGURE 7-8

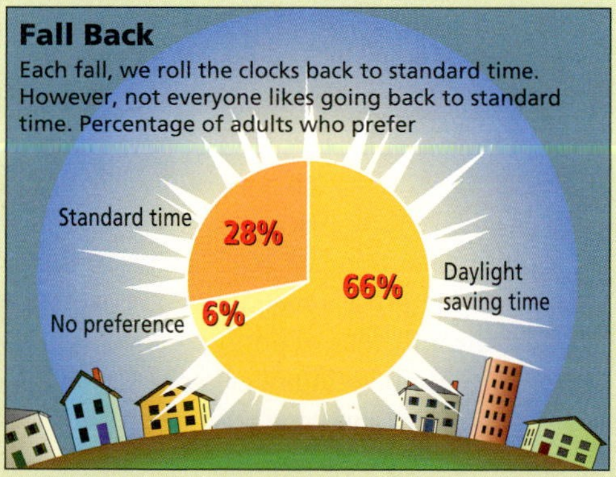

Source: Hilton Time Survey of 1024 adults

FIGURE 7-9 Coupons: Limited Use

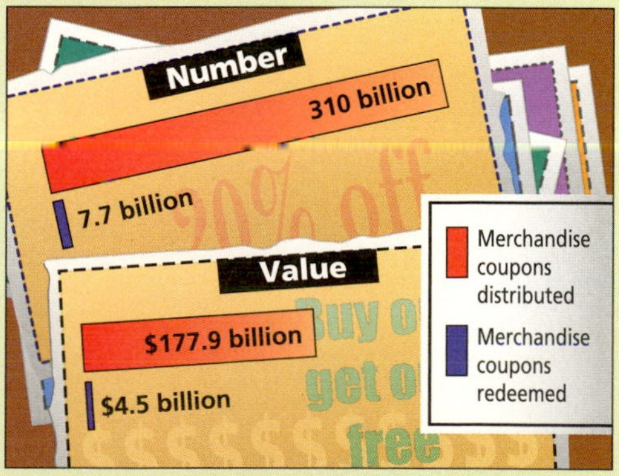

Source: NCH Promotional Services

LINKING CONCEPTS: WRITING PROJECTS

Discuss each of the following topics in class or review the topics on your own. Then write a brief but complete essay in which you summarize the main points. Please include formulas and graphs as appropriate.

1. In this chapter, we have studied confidence intervals. Carefully read the following statements about confidence intervals:
   (a) Once the endpoints of the confidence interval are numerically fixed, the parameter in question (either μ or p) does or does not fall inside the "fixed" interval.
   (b) A given fixed interval either does or does not contain the parameter μ or p; therefore, the probability is 1 or 0 that the parameter is in the interval.

   Next, read the following statements. Then discuss all four statements in the context of what we actually mean by a confidence interval.
   (c) Nontrivial probability statements can be made only about variables, not constants.
   (d) The confidence level c represents the proportion of all (fixed) intervals that would contain the parameter if we repeated the process many, many times.
2. Throughout Chapter 7, we have used the normal distribution, the central limit theorem, and the Student's t distribution.
   (a) Give a brief outline describing how confidence intervals for means use the normal distribution or Student's t distribution in their basic construction.
   (b) Give a brief outline describing how the normal approximation to the binomial distribution is used in the construction of confidence intervals for a proportion p.
   (c) Give a brief outline describing how the sample size for a predetermined error tolerance and level of confidence is determined from the normal distribution.
3. When the results of a survey or a poll are published, the sample size is usually given, as well as the margin of error. For example, suppose the *Honolulu Star Bulletin* reported that it surveyed 385 Honolulu residents and 78% said they favor mandatory jail sentences for people convicted of driving under the influence of drugs or alcohol (with margin of error of 3 percentage points in either direction). Usually the confidence level of the interval is not given, but it is standard practice to use the margin of error for a 95% confidence interval when no other confidence level is given.
   (a) The paper reported a point estimate of 78%, with margin of error of $\pm 3\%$. Write this information in the form of a confidence interval for p, the population proportion of residents favoring mandatory jail sentences for people convicted of driving under the influence. What is the assumed confidence level?
   (b) The margin of error is simply the error due to using a sample instead of the entire population. It does not take into account the bias that might be introduced by the wording of the question, by the truthfulness of the respondents, or by other factors. Suppose the question was asked in this fashion: "Considering the devastating injuries suffered by innocent victims in auto accidents caused by drunken or drugged drivers, do you favor a mandatory jail sentence for those convicted of driving under the influence of drugs or alcohol?" Do you think the wording of the question would influence the respondents? Do you think the population proportion of those favoring mandatory jail sentences would be accurately represented by a confidence interval based on responses to such a question? Explain your answer.

       If the question had been "Considering existing overcrowding of our prisons, do you favor a mandatory jail sentence for people convicted of driving under the influence of drugs or alcohol?" do you think the population proportion of those favoring mandatory sentences would be accurately represented by a confidence interval based on responses to such a question? Explain.

USING TECHNOLOGY

Application 1

Finding a Confidence Interval for a Population Mean μ

Cryptanalysis, the science of breaking codes, makes extensive use of language patterns. The frequency of various letter combinations is an important part of the study. A letter combination consisting of a single letter is a monograph, while combinations consisting of two letters are called digraphs, and those with three letters are called trigraphs. In the English language, the most frequent digraph is the letter combination TH.

The *characteristic rate* of a letter combination is a measurement of its rate of occurrence. To compute the characteristic rate, count the number of occurrences of a given letter combination and divide by the number of letters in the text. For instance, to estimate the characteristic rate of the digraph TH, you could select a newspaper text and pick a random starting place. From that place, mark off 2000 letters and count the number of times that TH occurs. Then divide the number of occurrences by 2000.

The characteristic rate of a digraph can vary slightly depending on the style of the author, so to estimate an overall characteristic frequency, you want to consider several samples of newspaper text by different authors. Suppose you did this with a random sample of 15 articles and found the characteristic rates of the digraph TH in the articles. The results follow.

0.0275	0.0230	0.0300	0.0255
0.0280	0.0295	0.0265	0.0265
0.0240	0.0315	0.0250	0.0265
0.0290	0.0295	0.0275	

(a) Find a 95% confidence interval for the mean characteristic rate of the digraph TH.
(b) Repeat part (a) for a 90% confidence interval.
(c) Repeat part (a) for an 80% confidence interval.
(d) Repeat part (a) for a 70% confidence interval.
(e) Repeat part (a) for a 60% confidence interval.
(f) For each confidence interval in parts (a)–(e), compute the length of the given interval. Do you notice a relation between the confidence level and the length of the interval?

A good reference for cryptanalysis is a book by Sinkov:

Sinkov, Abraham. *Elementary Cryptanalysis*. New York: Random House.

In the book, other common digraphs and trigraphs are given.

Application 2

Confidence Interval Demonstration

When we generate different random samples of the same size from a population, we discover that $\bar{x}$ varies from sample to sample. Likewise, different samples produce different confidence intervals for μ. The endpoints $\bar{x} \pm E$ of a confidence interval are statistical variables. A 90% confidence interval tells us that if we obtain lots of confidence intervals (for the same sample size), then the proportion of all intervals that will turn out to contain μ is 90%.

(a) Use the technology of your choice to generate 10 large random samples from a population with a known mean μ.
(b) Construct a 90% confidence interval for the mean for each sample.
(c) Examine the confidence intervals and note the percentage of the intervals that contain the population mean μ. We have 10 confidence intervals. Will exactly 90% of 10 intervals always contain μ? Explain. What if we have 1000 intervals?

Technology Hints for Confidence Interval Demonstration

TI-84Plus/TI-83Plus/TI-*n*spire

The TI-84Plus/TI-83Plus/TI-*n*spire (with TI-84Plus keypad) generates random samples from uniform, normal, and binomial distributions. Press the **MATH** key and select **PRB**. Choice **5:randInt(lower, upper, sample size *n*)** generates random samples of size n from the integers between the specified lower and upper values. Choice **6:randNorm(μ, σ, sample size *n*)** generates random samples of size n from a normal distribution with specified mean and standard deviation. Choice **7:randBin(number of trials, *p*, sample size)** generates samples of the specified size from the designated binomial distribution. Under **STAT**, select **EDIT** and highlight the list name, such as L1. At the = sign, use the **MATH** key to access the desired population distribution. Finally, use **Zinterval** under the **TESTS** option of the **STAT** key to generate 90% confidence intervals.

Excel 2010

On the **Home** screen, click the **Data** tab. Then in the Analysis Group, click **Data Analysis**. In the resulting dialogue box, select **Random Number Generator**. In that dialogue box, the number of variables refers to the number of samples. The number of random numbers refers to the number of data values in each sample. Select the population distribution (uniform, normal, binomial, etc.). When you click **OK** the data appear in columns on a spreadsheet, with each sample appearing in a separate column. Click on the **Insert Function** (fx). In the dialogue box, select **Statistical** for the category and then select **Confidence**. In the dialogue box for Confidence, **alpha** = $1 - c$, so for a 90% confidence interval, enter 0.10 for alpha. Then enter the population standard deviation $\boldsymbol{\sigma}$, and the **sample size**. The resulting output gives the value of the maximal margin of error E for the confidence interval for the mean μ. Note that if you use the population standard deviation σ in the function, the value of E will be the same for all samples of the same size. Next, find the sample mean $\bar{x}$ for each sample (use **Insert function** (fx) with **Statistical** for category in the dialogue box and select **Average**). Finally, construct the endpoints $\bar{x} \pm E$ of the confidence interval for each sample.

Minitab

Minitab provides options for sampling from a variety of distributions. To generate random samples from a specific distribution, use the menu selection **Calc ➤ Random Data ➤** and then select the population distribution. In the dialogue box, the *number of rows of data* represents the *sample size*. The *number of samples* corresponds to the number of columns selected for data storage. For example, C1–C10 in data storage produces 10 different random samples of the specified size. Use the menu selection **Stat ➤ Basic Statistics ➤ 1 sample** z to generate confidence intervals for the mean μ from each sample. In the variables box, list all the columns containing your samples. For instance, using C1–C10 in the variables list will produce confidence intervals for each of the 10 samples stored in columns C1 through C10.

The Minitab display shows 90% confidence intervals for 10 different random samples of size 50 taken from a normal distribution with $\mu = 30$ and $\sigma = 4$. Notice that, as expected, 9 out of 10 of the intervals contain $\mu = 30$.

Minitab Display

Z Confidence Intervals (Samples from a Normal Population with $\mu = 30$ and $\sigma = 4$)

The assumed sigma = 4.00

Variable	N	Mean	StDev	SE Mean	90.0 % CI
C1	50	30.265	4.300	0.566	(29.334, 31.195)
C2	50	31.040	3.957	0.566	(30.109, 31.971)
C3	50	29.940	4.195	0.566	(29.010, 30.871)
C4	50	30.753	3.842	0.566	(29.823, 31.684)
C5	50	30.047	4.174	0.566	(29.116, 30.977)
C6	50	29.254	4.423	0.566	(28.324, 30.185)
C7	50	29.062	4.532	0.566	(28.131, 29.992)
C8	50	29.344	4.487	0.566	(28.414, 30.275)
C9	50	30.062	4.199	0.566	(29.131, 30.992)
C10	50	29.989	3.451	0.566	(29.058, 30.919)

SPSS

SPSS uses a Student's t distribution to generate confidence intervals for the mean and difference of means. Use the menu choices **Analyze ➤ Compare Means** and then **One-Sample T Test** or **Independent-Sample T Tests** for confidence intervals for a single mean or difference of means, respectively. In the dialogue box, use 0 for the test value. Click **Options . . .** to provide the confidence level.

To generate 10 random samples of size $n = 30$ from a normal distribution with $\mu = 30$ and $\sigma = 4$, first enter consecutive integers from 1 to 30 in a column of the data editor. Then, under variable view, enter the variable names Sample1 through Sample10. Use the menu choices **Transform ➤ Compute Variable**. In the dialogue box, use Sample1 for the target variable. In the function group select **Random Numbers**. Then select the function **Rv.Normal**. Use 30 for the mean and 4 for the standard deviation. Continue until you have 10 samples. To sample from other distributions, use appropriate functions in the Compute dialogue box.

The SPSS display shows 90% confidence intervals for 10 different random samples of size 30 taken from a normal distribution with $\mu = 30$ and $\sigma = 4$. Notice that, as expected, 9 of the 10 intervals contain the population mean $\mu = 30$.

SPSS Display

90% t-confidence intervals for random samples of size $n = 30$ from a normal distribution with $\mu = 30$ and $\sigma = 4$.

	t	df	Sig(2-tail)	Mean	Lower	Upper
SAMPLE1	42.304	29	.000	29.7149	28.5214	30.9084
SAMPLE2	43.374	29	.000	30.1552	28.9739	31.3365
SAMPLE3	53.606	29	.000	31.2743	30.2830	32.2656
SAMPLE4	35.648	29	.000	30.1490	28.7120	31.5860
SAMPLE5	47.964	29	.000	31.0161	29.9173	32.1148
SAMPLE6	34.718	29	.000	30.3519	28.8665	31.8374
SAMPLE7	34.698	29	.000	30.7665	29.2599	32.2731
SAMPLE8	39.731	29	.000	30.2388	28.9456	31.5320
SAMPLE9	44.206	29	.000	29.7256	28.5831	30.8681
SAMPLE10	49.981	29	.000	29.7273	28.7167	30.7379

Application 3

Resampling (Bootstrap) Demonstration

Resampling (also called bootstrap) can be used to construct confidence intervals for μ when traditional methods cannot be used. For example, if the sample size is small and the sample shows extreme outliers or extreme lack of symmetry, use of the Student's t distribution is inappropriate. Bootstrap makes no assumptions about the population.

Consider the following random sample of size 20:

12	15	21	2	6	3	15	51	22	18
37	12	25	19	33	15	14	17	12	27

A stem-and-leaf display shows that the data are skewed with one outlier.

0	2 represents 2
0	236
1	2224555789
2	1257
3	37
4	
5	1

We can use Minitab to model the bootstrap method for constructing confidence intervals for μ. (The Professional edition of Minitab is required because of spreadsheet size and other limitations of the Student edition.) This demonstration uses only 1000 samples. Bootstrap uses many thousands.

Step 1: Create 1000 new samples, each of size 20, by sampling *with replacement* from the original data. To do this in Minitab, we enter the original 20 data values in column C1. Then, in column C2, we place equal probabilities of 0.05 beside each of the original data values. Use the menu choices **Calc ➤ Random Data ➤ Discrete**. In the dialogue box, fill in 1000 as the number of rows, store the data in columns C11–C30, and use column C1 for values and column C2 for probabilities.

Step 2: Find the sample mean of each of the 1000 samples. To do this in Minitab, use the menu choices **Calc ➤ Row Statistics**. In the dialogue box, select **mean**. Use columns C11–C30 as the input variables and store the results in column C31.

Step 3: Order the 1000 means from smallest to largest. In Minitab, use the menu choices **Manip ➤ Sort**. In the dialogue box, indicate C31 as the column to be sorted. Store the results in column C32. Sort by values in column C31.

Step 4: Create a 95% confidence interval by finding the boundaries for the middle 95% of the data. In other words, you need to find the values of the 2.5 percentile ($P_{2.5}$) and the 97.5 percentile ($P_{97.5}$). Since there are 1000 data values, the 2.5 percentile is the data value in position 25, while the 97.5 percentile is the data value in position 975. The confidence interval is $P_{2.5} < \mu < P_{97.5}$.

Demonstration Results

Figure 7-10 shows a histogram of the 1000 $\bar{x}$ values from one bootstrap simulation. Three bootstrap simulations produced the following 95% confidence intervals.

13.90 to 23.90
14.00 to 24.15
14.05 to 23.8

Using the *t* distribution on the sample data, Minitab produced the interval 13.33 to 24.27. The results of the bootstrap simulations and the *t* distribution method are quite close.

FIGURE 7-10 Bootstrap Simulation, $\bar{x}$ Distribution

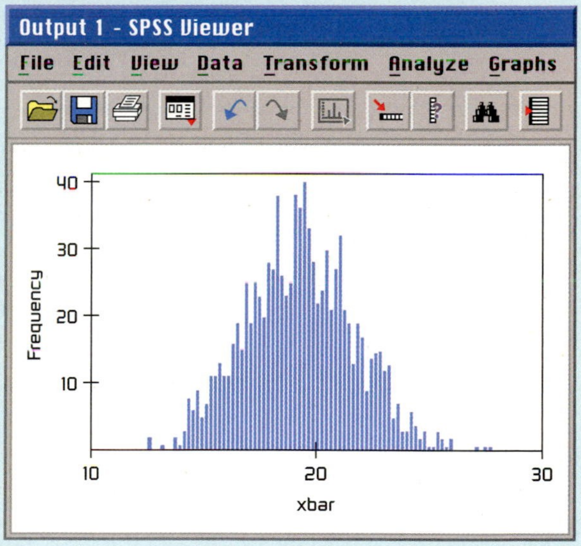

8

"Would you tell me, please, which way I ought to go from here?"

"That depends a good deal on where you want to get to," said the Cat.

"I don't much care where—" said Alice.

"Then it doesn't matter which way you go," said the Cat.

—Lewis Carroll
Alice's Adventures in Wonderland

Charles Lutwidge Dodgson (1832–1898) was an English mathematician who loved to write children's stories in his free time. The dialogue between Alice and the Cheshire Cat occurs in the masterpiece *Alice's Adventures in Wonderland*, written by Dodgson under the pen name Lewis Carroll. These lines relate to our study of hypothesis testing. Statistical tests cannot answer all of life's questions. They cannot always tell us "where to go," but after this decision is made on other grounds, they can help us find the best way to get there.

HYPOTHESIS TESTING

Comstock Images/Jupiter Images

PREVIEW QUESTIONS

Many of life's questions require a yes or no answer. When you must act on incomplete (sample) information, how do you decide whether to accept or reject a proposal? (SECTION 8.1)

What is the p-value of a statistical test? What does this measurement have to do with performance reliability? (SECTION 8.1)

How do you construct statistical tests for μ? Does it make a difference whether σ is known or unknown? (SECTION 8.2)

How do you construct statistical tests for the proportion p of successes in a binomial experiment? (SECTION 8.3)

What are the advantages of pairing data values? How do you construct statistical tests for paired differences? (SECTION 8.4)

How do you construct statistical tests for differences of independent random variables? (SECTION 8.5)

FOCUS PROBLEM

Benford's Law: The Importance of Being Number 1

Benford's Law states that in a wide variety of circumstances, numbers have "1" as their first nonzero digit disproportionately often. Benford's Law applies to such diverse topics as the drainage areas of rivers; properties of chemicals; populations of towns; figures in newspapers, magazines, and government reports; and the half-lives of radioactive atoms!

Specifically, such diverse measurements begin with "1" about 30% of the time, with "2" about 18% of time, and with "3" about 12.5% of the time. Larger digits occur less often. For example, less than 5% of the numbers in circumstances such as these begin with the digit 9. This is in dramatic contrast to a random sampling situation, in which each of the digits 1 through 9 has an equal chance of appearing.

Corbis

The first nonzero digits of numbers taken from large bodies of numerical records such as tax returns, population studies, government records, and so forth, show the probabilities of occurrence as displayed in the table on the next page.

For online student resources, visit the Brase/Brase, *Understandable Statistics*, 11th edition web site at http://www.cengage.com/statistics/brase11e

First nonzero digit	1	2	3	4	5	6	7	8	9
Probability	0.301	0.176	0.125	0.097	0.079	0.067	0.058	0.051	0.046

More than 100 years ago, the astronomer Simon Newcomb noticed that books of logarithm tables were much dirtier near the fronts of the tables. It seemed that people were more frequently looking up numbers with a low first digit. This was regarded as an odd phenomenon and a strange curiosity. The phenomenon was rediscovered in 1938 by physicist Frank Benford (hence the name *Benford's Law*).

More recently, Ted Hill, a mathematician at the Georgia Institute of Technology, studied situations that might demonstrate Benford's Law. Professor Hill showed that such probability distributions are likely to occur when we have a "distribution of distributions." Put another way, large random collections of random samples tend to follow Benford's Law. This seems to be especially true for samples taken from large government data banks, accounting reports for large corporations, large collections of astronomical observations, and so forth. For more information, see *American Scientist,* Vol. 86, pp. 358–363, and *Chance,* American Statistical Association, Vol. 12, No. 3, pp. 27–31.

Can Benford's Law be applied to help solve a real-world problem? Well, one application might be accounting fraud! Suppose the first nonzero digits of the entries in the accounting records of a large corporation (such as Enron or WorldCom) do not follow Benford's Law. Should this set off an accounting alarm for the FBI or the stockholders? How "significant" would this be? Such questions are the subject of statistics.

In Section 8.3, you will see how to use sample data to test whether the proportion of first nonzero digits of the entries in a large accounting report follows Benford's Law. Problems 7 and 8 of Section 8.3 relate to Benford's Law and accounting discrepancies. In one problem, you are asked to use sample data to determine if accounting books have been "cooked" by "pumping numbers up" to make the company look more attractive or perhaps to provide a cover for money laundering. In the other problem, you are asked to determine if accounting books have been "cooked" by artificially lowered numbers, perhaps to hide profits from the Internal Revenue Service or to divert company profits to unscrupulous employees. (See Problems 7 and 8 of Section 8.3.)

SECTION 8.1 Introduction to Statistical Tests

FOCUS POINTS

- Understand the rationale for statistical tests.
- Identify the null and alternate hypotheses in a statistical test.
- Identify right-tailed, left-tailed, and two-tailed tests.
- Use a test statistic to compute a *P*-value.
- Recognize types of errors, level of significance, and power of a test.
- Understand the meaning and risks of rejecting or not rejecting the null hypothesis.

In Chapter 1, we emphasized the fact that one of a statistician's most important jobs is to draw inferences about populations based on samples taken from the populations. Most statistical inference centers around the parameters of a population (often the mean or probability of success in a binomial trial). Methods for drawing inferences about parameters are of two types: Either we make decisions concerning the value of the parameter, or we actually estimate the value of the parameter. When we estimate the value (or location) of a parameter, we are using methods of estimation such as those studied in Chapter 7. Decisions concerning the value of a parameter are obtained by *hypothesis testing*, the topic we shall study in this chapter.

Hypothesis testing

Hypothesis

Students often ask which method should be used on a particular problem—that is, should the parameter be estimated, or should we test a *hypothesis* involving the parameter? The answer lies in the practical nature of the problem and the questions posed about it. Some people prefer to test theories concerning the parameters. Others prefer to express their inferences as estimates. Both estimation and hypothesis testing are found extensively in the literature of statistical applications.

STATING HYPOTHESES

Null hypothesis H_0

Our first step is to establish a working hypothesis about the population parameter in question. This hypothesis is called the *null hypothesis*, denoted by the symbol H_0. The value specified in the null hypothesis is often a historical value, a claim, or a production specification. For instance, if the average height of a professional male basketball player was 6.5 feet 10 years ago, we might use a null hypothesis H_0: $\mu = 6.5$ feet for a study involving the average height of this year's professional male basketball players. If television networks claim that the average length of time devoted to commercials in a 60-minute program is 12 minutes, we would use H_0: $\mu = 12$ minutes as our null hypothesis in a study regarding the average length of time devoted to commercials. Finally, if a repair shop claims that it should take an average of 25 minutes to install a new muffler on a passenger automobile, we would use H_0: $\mu = 25$ minutes as the null hypothesis for a study of how well the repair shop is conforming to specified average times for a muffler installation.

Alternate hypothesis H_1

Any hypothesis that differs from the null hypothesis is called an *alternate hypothesis*. An alternate hypothesis is constructed in such a way that it is the hypothesis to be accepted when the null hypothesis must be rejected. The alternate hypothesis is denoted by the symbol H_1. For instance, if we believe the average height of professional male basketball players is taller than it was 10 years ago, we would use an alternate hypothesis H_1: $\mu > 6.5$ feet with the null hypothesis H_0: $\mu = 6.5$ feet.

> **Null hypothesis H_0:** This is the statement that is under investigation or being tested. Usually the null hypothesis represents a statement of "no effect," "no difference," or, put another way, "things haven't changed."
> **Alternate hypothesis H_1:** This is the statement you will adopt in the situation in which the evidence (data) is so strong that you reject H_0. A statistical test is designed to assess the strength of the evidence (data) against the null hypothesis.

EXAMPLE 1

NULL AND ALTERNATE HYPOTHESES

A car manufacturer advertises that its new subcompact models get 47 miles per gallon (mpg). Let μ be the mean of the mileage distribution for these cars. You assume that the manufacturer will not underrate the car, but you suspect that the mileage might be overrated.

(a) What shall we use for H_0?

SOLUTION: We want to see if the manufacturer's claim that $\mu = 47$ mpg can be rejected. Therefore, our null hypothesis is simply that $\mu = 47$ mpg. We denote the null hypothesis as

$$H_0: \mu = 47 \text{ mpg}$$

(b) What shall we use for H_1?

SOLUTION: From experience with this manufacturer, we have every reason to believe that the advertised mileage is too high. If μ is not 47 mpg, we are sure it is less than 47 mpg. Therefore, the alternate hypothesis is

$$H_1\text{: } \mu < 47 \text{ mpg}$$

GUIDED EXERCISE 1 NULL AND ALTERNATE HYPOTHESES

A company manufactures ball bearings for precision machines. The average diameter of a certain type of ball bearing should be 6.0 mm. To check that the average diameter is correct, the company formulates a statistical test.

(a) What should be used for H_0? (*Hint:* What is the company trying to test?)

→ If μ is the mean diameter of the ball bearings, the company wants to test whether $\mu = 6.0$ mm. Therefore, H_0: $\mu = 6.0$ mm.

(b) What should be used for H_1? (*Hint:* An error either way, too small or too large, would be serious.)

→ An error either way could occur, and it would be serious. Therefore, H_1: $\mu \neq 6.0$ mm (μ is either smaller than or larger than 6.0 mm).

COMMENT: NOTATION REGARDING THE NULL HYPOTHESIS In statistical testing, the null hypothesis H_0 always contains the equals symbol. However, in the null hypothesis, some statistical software packages and texts also include the inequality symbol that is opposite that shown in the alternate hypothesis. For instance, if the alternate hypothesis is "μ is less than 3" ($\mu < 3$), then the corresponding null hypothesis is sometimes written as "μ is greater than or equal to 3" ($\mu \geq 3$). The mathematical construction of a statistical test uses the null hypothesis to assign a specific number (rather than a range of numbers) to the parameter μ in question. The null hypothesis establishes a single fixed value for μ, so we are working with a single distribution having a specific mean. In this case, H_0 assigns $\mu = 3$. So, when H_1: $\mu < 3$ is the alternate hypothesis, we follow the commonly used convention of writing the null hypothesis simply as H_0: $\mu = 3$.

TYPES OF TESTS

The null hypothesis H_0 always states that the parameter of interest *equals* a specified value. The alternate hypothesis H_1 states that the parameter is *less than, greater than,* or simply *not equal to* the same value. We categorize a statistical test as *left-tailed, right-tailed,* or *two-tailed* according to the alternate hypothesis.

Left-tailed test
Right-tailed test
Two-tailed test

TYPES OF STATISTICAL TESTS

A statistical test is:

left-tailed if H_1 states that the parameter is less than the value claimed in H_0

right-tailed if H_1 states that the parameter is greater than the value claimed in H_0

two-tailed if H_1 states that the parameter is different from (or not equal to) the value claimed in H_0

TABLE 8-1 **The Null and Alternate Hypotheses for Tests of the Mean μ**

Null Hypothesis	Alternate Hypotheses and Type of Test		
Claim about μ or historical value of μ	You believe that μ is less than value stated in H_0.	You believe that μ is more than value stated in H_0.	You believe that μ is different from value stated in H_0.
H_0: $\mu = k$	H_1: $\mu < k$	H_1: $\mu > k$	H_1: $\mu \neq k$
	Left-tailed test	Right-tailed test	Two-tailed test

In this introduction to statistical tests, we discuss tests involving a population mean μ. However, you should keep an open mind and be aware that the methods outlined apply to testing other parameters as well (e.g., p, σ, $\mu_1 - \mu_2$, $p_1 - p_2$, and so on). Table 8-1 shows how tests of the mean μ are categorized.

HYPOTHESIS TESTS OF μ, GIVEN x IS NORMAL AND σ IS KNOWN

Once you have selected the null and alternate hypotheses, how do you decide which hypothesis is likely to be valid? Data from a simple random sample and the sample test statistic, together with the corresponding sampling distribution of the test statistic, will help you decide. Example 2 leads you through the decision process.

First, a quick review of Section 6.4 is in order. Recall that a population *parameter* is a numerical descriptive measurement of the entire population. Examples of population parameters are μ, p, and σ. It is important to remember that for a given population, the parameters are *fixed* values. They do not vary! The null hypothesis H_0 makes a statement about a population parameter.

A *statistic* is a numerical descriptive measurement of a sample. Examples of statistics are $\bar{x}$, $\hat{p}$, and s. Statistics usually *vary* from one sample to the next. The probability distribution of the statistic we are using is called a *sampling distribution.*

Sample test statistic for μ, given x normal and σ known

For hypothesis testing, we take a *simple* random sample and compute a *sample test statistic* corresponding to the parameter in H_0. Based on the sampling distribution of the statistic, we can assess how compatible the sample test statistic is with H_0.

In this section, we use hypothesis tests about the mean to introduce the concepts and vocabulary of hypothesis testing. In particular, let's suppose that x has *a normal distribution* with mean μ and standard deviation σ. Then, Theorem 6.1 tells us that $\bar{x}$ has a *normal distribution* with mean μ and standard deviation $\sigma/\sqrt{n}$.

PROCEDURE

Requirements The x distribution is *normal* with known standard deviation σ. Then $\bar{x}$ has a normal distribution. The standardized test statistic is

$$\textbf{test statistic} = z = \frac{\bar{x} - \mu}{\sigma/\sqrt{n}}$$

where $\bar{x}$ = mean of a simple random sample
μ = value stated in H_0
n = sample size

EXAMPLE 2

STATISTICAL TESTING PREVIEW

Rosie is an aging sheep dog in Montana who gets regular checkups from her owner, the local veterinarian. Let x be a random variable that represents Rosie's resting heart rate (in beats per minute). From past experience, the vet knows that x has a normal

PictureQuest/Jupiter Images

distribution with $\sigma = 12$. The vet checked the *Merck Veterinary Manual* and found that for dogs of this breed, $\mu = 115$ beats per minute.

Over the past 6 weeks, Rosie's heart rate (beats/min) measured

93	109	110	89	112	117

The sample mean is $\bar{x} = 105.0$. The vet is concerned that Rosie's heart rate may be slowing. Do the data indicate that this is the case?

SOLUTION:

(a) Establish the null and alternate hypotheses.
If "nothing has changed" from Rosie's earlier life, then her heart rate should be nearly average. This point of view is represented by the null hypothesis

$$H_0: \mu = 115$$

However, the vet is concerned about Rosie's heart rate slowing. This point of view is represented by the alternate hypothesis

$$H_1: \mu < 115$$

(b) Are the observed sample data compatible with the null hypothesis?
Are the six observations of Rosie's heart rate compatible with the null hypothesis H_0: $\mu = 115$? To answer this question, we need to know the *probability* of obtaining a sample mean of 105.0 or less from a population with true mean $\mu = 115$. If this probability is small, we conclude that H_0: $\mu = 115$ is not the case. Rather, H_1: $\mu < 115$ and Rosie's heart rate is slowing.

(c) How do we compute the probability in part (b)?
Well, you probably guessed it! We use the sampling distribution for $\bar{x}$ and compute $P(\bar{x} < 105.0)$. Figure 8-1 shows the $\bar{x}$ distribution and the corresponding standard normal distribution with the desired probability shaded.

Check Requirements Since x has a normal distribution, $\bar{x}$ will also have a normal distribution for any sample size n and given σ (see Theorem 6.1).

Note that using $\mu = 115$ from H_0, $\sigma = 12$, and $n = 6$ the sample $\bar{x} = 105.0$ converts to

$$\text{test statistic} = z = \frac{\bar{x} - \mu}{\sigma/\sqrt{n}} = \frac{105.0 - 115}{12/\sqrt{6}} \approx -2.04$$

Using the standard normal distribution table, we find that

$$P(\bar{x} < 105.0) = P(z < -2.04) = 0.0207$$

The area in the left tail that is more extreme than $\bar{x} = 105.0$ is called the *P-value* of the test. In this example, P-value $= 0.0207$. We will learn more about P-values later.

FIGURE 8-1

Sampling Distribution for $\bar{x}$ and Corresponding z Distribution

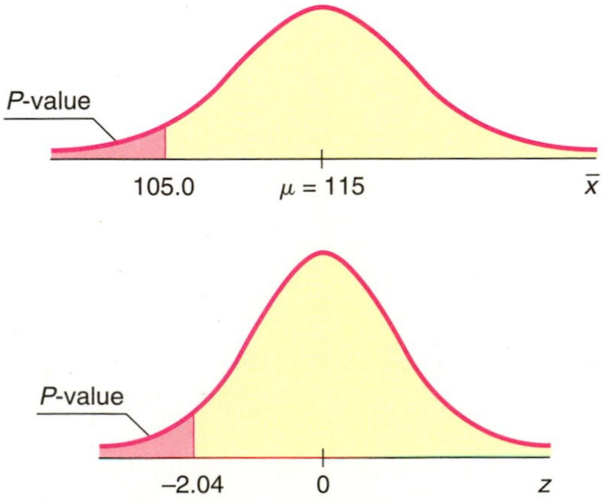

(d) ***Interpretation*** What conclusion can be drawn about Rosie's average heart rate?

If H_0: $\mu = 115$ is in fact true, the probability of getting a sample mean of $\bar{x} \leq 105.0$ is only about 2%. Because this probability is small, we reject H_0: $\mu = 115$ and conclude that H_1: $\mu < 115$ Rosie's average heart rate seems to be slowing.

(e) Have we proved H_0: $\mu = 115$ to be false and H_1: $\mu < 115$ to be true?

No! The sample data do not *prove* H_0 to be false and H_1 to be true! We do say that H_0 has been "discredited" by a small P-value of 0.0207. Therefore, we abandon the claim H_0: $\mu = 115$ and adopt the claim H_1: $\mu < 115$.

THE *P*-VALUE OF A STATISTICAL TEST

Rosie the sheep dog has helped us to "sniff out" an important statistical concept.

P-VALUE

Assuming H_0 is true, the *probability* that the test statistic will take on values as extreme as or more extreme than the observed test statistic (computed from sample data) is called the ***P*-value** of the test. The smaller the P-value computed from sample data, the stronger the evidence against H_0.

The P-value, sometimes called the *probability of chance*, can be thought of as the probability that the results of a statistical experiment are due only to chance. The lower the P-value, the greater the likelihood of obtaining the same (or very similar) results in a repetition of the statistical experiment. Thus, a low P-value is a good indication that your results are not due to random chance alone.

The P-value associated with the observed test statistic takes on different values depending on the alternate hypothesis and the type of test. Let's look at P-values and types of tests when the test involves the mean and standard normal distribution. Notice that in Example 2, part (c), we computed a P-value for a left-tailed test. Guided Exercise 3 asks you to compute a P-value for a two-tailed test.

P-VALUES AND TYPES OF TESTS

Let $z_{\bar{x}}$ represent the standardized sample test statistic for testing a mean μ using the standard normal distribution. That is, $z_{\bar{x}} = (\bar{x} - \mu)/(\sigma/\sqrt{n})$.

I. Left-tailed Test

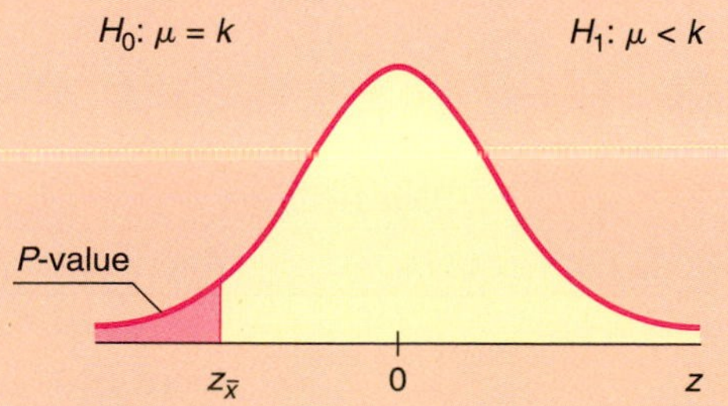

P-value $= P(z < z_{\bar{x}})$

This is the probability of getting a test statistic as low as or lower than $z_{\bar{x}}$

Continued

II. Right-tailed Test

H_0: $\mu = k$ H_1: $\mu > k$

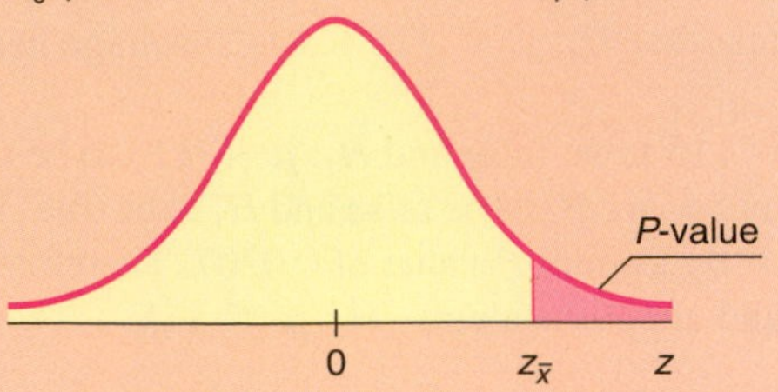

$P\text{-value} = P(z > z_{\bar{x}})$

This is the probability of getting a test statistic as high as or higher than $z_{\bar{x}}$

III. Two-tailed Test

H_0: $\mu = k$ H_1: $\mu \neq k$

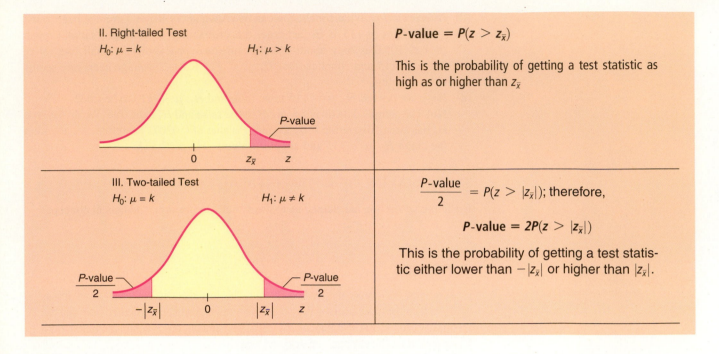

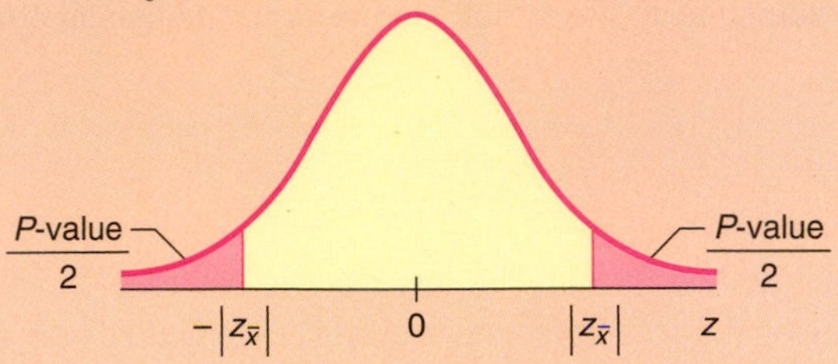

$$\frac{P\text{-value}}{2} = P(z > |z_{\bar{x}}|)\text{; therefore,}$$

$$P\text{-value} = 2P(z > |z_{\bar{x}}|)$$

This is the probability of getting a test statistic either lower than $-|z_{\bar{x}}|$ or higher than $|z_{\bar{x}}|$.

TYPES OF ERRORS

Type I error

Type II error

If we *reject the null hypothesis when it is,* in fact, *true,* we have made an error that is called a *type I error.* On the other hand, if we *accept the null hypothesis when it is,* in fact, *false,* we have made an error that is called a *type II error.* Table 8-2 indicates how these errors occur.

For tests of hypotheses to be well constructed, they must be designed to minimize possible errors of decision. (Usually, we do not know if an error has been made, and therefore, we can talk only about the probability of making an error.) Usually, for a given sample size, an attempt to reduce the probability of one type of error results in an increase in the probability of the other type of error. In practical applications, one type of error may be more serious than another. In such a case, careful attention is given to the more serious error. If we increase the sample size, it is possible to reduce both types of errors, but increasing the sample size may not be possible.

Level of significance α

Good statistical practice requires that we announce in advance how much evidence against H_0 will be required to reject H_0. The probability with which we are willing to risk a type I error is called the *level of significance* of a test. The level of significance is denoted by the Greek letter α (pronounced "alpha").

The **level of significance α** is the probability of rejecting H_0 when it is true. This is the probability of a type I error.

TABLE 8-2 **Type I and Type II Errors**

	Our Decision	
Truth of H_0	**And if we do not reject H_0**	**And if we reject H_0**
If H_0 is true	Correct decision; no error	Type I error
If H_0 is false	Type II error	Correct decision; no error

TABLE 8-3 **Probabilities Associated with a Statistical Test**

	Our Decision	
Truth of H_0	**And if we accept H_0 as true**	**And if we reject H_0 as false**
If H_0 is true	Correct decision, with corresponding probability $1 - \alpha$	Type I error, with corresponding probability α, called the *level of significance of the test*
If H_0 is false	Type II error, with corresponding probability β	Correct decision; with corresponding probability $1 - \beta$, called the *power of the test*

Probability of a type II error β

The *probability of making a type II error* is denoted by the Greek letter $\boldsymbol{\beta}$ (pronounced "beta").

Methods of hypothesis testing require us to choose α and β values to be as small as possible. In elementary statistical applications, we usually choose α first.

Power of a test $(1 - \beta)$

The quantity $\mathbf{1 - \boldsymbol{\beta}}$ is called the *power of a test* and represents the probability of rejecting H_0 when it is, in fact, false.

For a given level of significance, how much power can we expect from a test? The actual value of the power is usually difficult (and sometimes impossible) to obtain, since it requires us to know the H_1 distribution. However, we can make the following general comments:

1. The power of a statistical test increases as the level of significance α increases. A test performed at the $\alpha = 0.05$ level has more power than one performed at $\alpha = 0.01$. This means that the less stringent we make our significance level α, the more likely we will be to reject the null hypothesis when it is false.
2. Using a larger value of α will increase the power, but it also will increase the probability of a type I error. Despite this fact, most business executives, administrators, social scientists, and scientists use *small* α values. This choice reflects the conservative nature of administrators and scientists, who are usually more willing to make an error by failing to reject a claim (i.e., H_0) than to make an error by accepting another claim (i.e., H_1) that is false. Table 8-3 summarizes the probabilities of errors associated with a statistical test.

COMMENT Since the calculation of the probability of a type II error is treated in advanced statistics courses, we will restrict our attention to the probability of a type I error.

GUIDED EXERCISE 2 | Types of Errors

Let's reconsider Guided Exercise 1, in which we were considering the manufacturing specifications for the diameter of ball bearings. The hypotheses were

H_0: $\mu = 6.0$ mm (manufacturer's specification) H_1: $\mu \neq 6.0$ mm (cause for adjusting process)

(a) Suppose the manufacturer requires a 1% level of significance. Describe a type I error, its consequence, and its probability.

→ A type I error is caused when sample evidence indicates that we should reject H_0 when, in fact, the average diameter of the ball bearings being produced is 6.0 mm. A type I error will cause a needless adjustment and delay of the manufacturing process. The probability of such an error is 1% because $\alpha = 0.01$.

Continued

GUIDED EXERCISE 2 *continued*

(b) Discuss a type II error and its consequences.

A type II error occurs if the sample evidence tells us not to reject the null hypothesis H_0: $\mu = 6.0$ mm when, in fact, the average diameter of the ball bearing is either too large or too small to meet specifications. Such an error would mean that the production process would not be adjusted even though it really needed to be adjusted. This could possibly result in a large production of ball bearings that do not meet specifications.

CONCLUDING A STATISTICAL TEST

Usually, α is specified in advance before any samples are drawn so that results will not influence the choice for the level of significance. To conclude a statistical test, we compare our α value with the P-value computed using sample data and the sampling distribution.

PROCEDURE

HOW TO CONCLUDE A TEST USING THE *P*-VALUE AND LEVEL OF SIGNIFICANCE α

Statistical significance

If P-value $\leq \alpha$, we reject the null hypothesis and say the data are **statistically significant** at the level α.

If P-value $> \alpha$, we do not reject the null hypothesis.

In what sense are we using the word *significant*? *Webster's Dictionary* gives two interpretations of *significance*: (1) having or signifying *meaning*: or (2) being important or momentous.

In statistical work, significance does not necessarily imply momentous importance. For us, "significant" at the α level has a special meaning. It says that at the α level of risk, the evidence (sample data) against the null hypothesis H_0 is sufficient to discredit H_0, so we adopt the alternate hypothesis H_1.

In any case, we do not claim that we have "proved" or "disproved" the null hypothesis H_0. We can say that the probability of a type I error (rejecting H_0 when it is, in fact, true) is α.

BASIC COMPONENTS OF A STATISTICAL TEST

A statistical test can be thought of as a package of five basic ingredients.

1. **Null hypothesis H_0, alternate hypothesis H_1, and preset level of significance α**
   If the evidence (sample data) against H_0 is strong enough, we reject H_0 and adopt H_1. The level of significance α is the probability of rejecting H_0 when it is, in fact, true.
2. **Test statistic and sampling distribution**
   These are mathematical tools used to measure compatibility of sample data and the null hypothesis.
3. ***P*-value**
   This is the probability of obtaining a test statistic from the sampling distribution that is as extreme as, or more extreme (as specified by H_1) than, the sample test statistic computed from the data under the assumption that H_0 is true.
4. **Test conclusion**
   If P-value $\leq \alpha$, we reject H_0 and say that the data are significant at level α. If P-value $> \alpha$, we do not reject H_0.
5. **Interpretation of the test results**
   Give a simple explanation of your conclusions in the context of the application.

GUIDED EXERCISE 3 CONSTRUCTING A STATISTICAL TEST FOR μ (NORMAL DISTRIBUTION)

The Environmental Protection Agency has been studying Miller Creek regarding ammonia nitrogen concentration. For many years, the concentration has been 2.3 mg/L. However, a new golf course and new housing developments are raising concern that the concentration may have changed because of lawn fertilizer. Any change (either an increase or a decrease) in the ammonia nitrogen concentration can affect plant and animal life in and around the creek (Reference: *EPA Report* 832-R-93-005). Let x be a random variable representing ammonia nitrogen concentration (in mg/L). Based on recent studies of Miller Creek, we may assume that x has a normal distribution with $\sigma = 0.30$ Recently, a random sample of eight water tests from the creek gave the following x values.

2.1 2.5 2.2 2.8 3.0 2.2 2.4 2.9

The sample mean is $\bar{x} \approx 2.51$.

Let us construct a statistical test to examine the claim that the concentration of ammonia nitrogen has changed from 2.3 mg/L. Use level of significance $\alpha = 0.01$.

(a) What is the null hypothesis? What is the alternate hypothesis? What is the level of significance α?

→ $H_0: \mu = 2.3$

$H_1: \mu \neq 2.3$

$\alpha = 0.01$

(b) Is this a right-tailed, left-tailed, or two-tailed test?

→ Since $H_1: \mu \neq 2.3$, this is a two-tailed test.

(c) ***Check Requirements*** What sampling distribution shall we use? Note that the value of m is given in the null hypothesis, H_0.

→ Since the x distribution is normal and σ is known, we use the standard normal distribution with

$$z = \frac{\bar{x} - \mu}{\frac{\sigma}{\sqrt{n}}} = \frac{\bar{x} - 2.3}{\frac{0.3}{\sqrt{8}}}$$

(d) What is the sample test statistic? Convert the sample mean $\bar{x}$ to a standard z value.

→ The sample of eight measurements has mean $\bar{x} = 2.51$. Converting this measurement to z, we have

$$\text{test statistic} = z = \frac{2.51 - 2.3}{\frac{0.3}{\sqrt{8}}} \approx 1.98$$

(e) Draw a sketch showing the P-value area on the standard normal distribution. Find the P-value.

→ $P\text{-value} = 2P(z > 1.98) = 2(0.0239) = 0.0478$

FIGURE 8-2

P-value

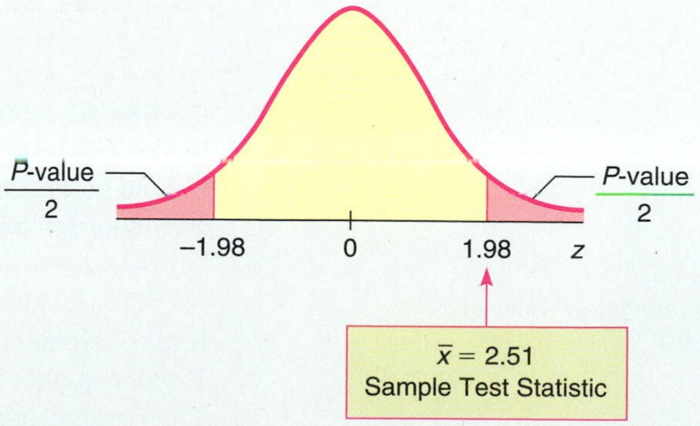

Continued

GUIDED EXERCISE 3 *continued*

(f) Compare the level of significance α and the P-value. What is your conclusion?

Since P-value $0.0478 \geq 0.01$, we see that P-value $> \alpha$. We fail to reject H_0.

(g) *Interpret* your results in the context of this problem.

The sample data are not significant at the $\alpha = 1\%$ level. At this point in time, there is not enough evidence to conclude that the ammonia nitrogen concentration has changed in Miller Creek.

Interpretation of level of significance

In most statistical applications, the level of significance is specified to be $\alpha = 0.05$ or $\alpha = 0.01$, although other values can be used. If $\alpha = 0.05$, then we say we are using a 5% level of significance. This means that in 100 similar situations, H_0 will be rejected 5 times, on average, when it should not have been rejected.

Meaning of accepting H_0

Using Technology at the end of this chapter shows a simulation of this phenomenon.

When we accept (or fail to reject) the null hypothesis, we should understand that we are *not proving the null hypothesis.* We are saying only that the sample evidence (data) is not strong enough to justify rejection of the null hypothesis. The word *accept* sometimes has a stronger meaning in common English usage than we are willing to give it in our application of statistics. Therefore, we often use the expression *fail to reject* H_0 instead of *accept* H_0. "*Fail to reject* the null hypothesis" simply means that the evidence in favor of rejection was not strong enough (see Table 8-4). Often, in the case that H_0 cannot be rejected, a confidence interval is used to estimate the parameter in question. The confidence interval gives the statistician a range of possible values for the parameter.

Fail to reject H_0

Reject H_0

TABLE 8-4 **Meaning of the Terms *Fail to Reject* H_0 and *Reject* H_0**

Term	Meaning
Fail to reject H_0	There is not enough evidence in the data (and the test being used) to justify a rejection of H_0. This means that we retain H_0 with the understanding that we have not proved it to be true beyond all doubt.
Reject H_0	There is enough evidence in the data (and the test employed) to justify rejection of H_0. This means that we choose the alternate hypothesis H_1 with the understanding that we have not proved H_1 to be true beyond all doubt.

COMMENT Some comments about P-values and level of significance α should be made. The level of significance α should be a fixed, prespecified value. Usually, α is chosen before any samples are drawn. The level of significance α is the probability of a type I error. So, α is the probability of rejecting H_0 when, in fact, H_0 is true.

Interpreting the P-value of a test statistic

The P-value should *not* be interpreted as the probability of a type I error. The level of significance (in theory) is set in advance before any samples are drawn. The P-value cannot be set in advance, since it is determined from the random sample. The P-value, together with α, should be regarded as tools used to conclude the test. If P-value $\leq \alpha$, then reject H_0, and if P-value $> \alpha$, then do not reject H_0.

In most computer applications and journal articles, only the *P*-value is given. It is understood that the person using this information will supply an appropriate level of significance α. From an historical point of view, the English statistician F. Y. Edgeworth (1845–1926) was one of the first to use the term *significant* to imply that the sample data indicate a "meaningful" difference from a previously held view.

In this book, we are using the most popular method of testing, which is called the *P-value method.* At the end of the next section, you will learn about another (equivalent) method of testing called the *critical region method.* An extensive discussion regarding the *P*-value method of testing versus the critical region method can be found in *The American Statistician,* Vol. 57, No. 3, pp. 171–178, American Statistical Association.

WHAT DO COMPONENTS OF A HYPOTHESIS TEST TELL US?

- The *null hypothesis* H_0 specifies the value of a parameter. We assume this specified value is correct unless sample evidence is strong enough to discredit the claim.
- The *alternate hypothesis* H_1 tells us about the value of the parameter in question if sample evidence is sufficient to discredit the null hypothesis.
- The *level of significance* α is the probability of rejecting H_0 when it is in fact true. The value of α is preset and is often 0.01 or 0.05.
- *Sample evidence* to test the null hypothesis is based on a random sample and the computed value of the sample statistic corresponding to the specified parameter in H_0.
- The *P-value* of the sample statistic provides the basis for deciding whether or not to reject H_0. To compute the *P*-value of the sample test statistic, we assume that H_0 is true. Then based on the sample size, the alternate hypothesis H_1, the value of the sample statistic, and the sampling distribution, we compute the *P*-value. The *P*-value tells us how likely we are to get another sample statistic that is as extreme or more extreme than the test statistic computed from the random sample we already have. If the *P*-value is less than or equal to the level of significance α we reject H_0 and say the results are *significant*. Otherwise, we retain H_0 and say the evidence was not sufficient to reject H_0.
- If the sample evidence tells us not to reject H_0, we have not proved H_0 to be true beyond all doubt. We simply do not have enough evidence to reject H_0 at the specified level of significance α and the computed *P*-value.

VIEWPOINT Lovers, Take Heed!!!

If you are going to whisper sweet nothings to your sweetheart, be sure to whisper them in the left ear. Professor Sim of Sam Houston State University (Huntsville, Texas) found that emotionally loaded words have a higher recall rate when spoken into a person's left ear, not the right. Professor Sim presented his findings at the British Psychology Society European Congress. He told the Congress that his findings are consistent with the hypothesis that the brain's right hemisphere has more influence in the processing of emotional stimuli. (The left ear is controlled by the right side of the brain.) Sim's research involved statistical tests like the ones you will study in this chapter.

SECTION 8.1 PROBLEMS

1. *Statistical Literacy* Discuss each of the following topics in class or review the topics on your own. Then write a brief but complete essay in which you answer the following questions.
   (a) What is a null hypothesis H_0?
   (b) What is an alternate hypothesis H_1?
   (c) What is a type I error? a type II error?
   (d) What is the level of significance of a test? What is the probability of a type II error?

2. *Statistical Literacy* In a statistical test, we have a choice of a left-tailed test, a right-tailed test, or a two-tailed test. Is it the null hypothesis or the alternate hypothesis that determines which type of test is used? Explain your answer.

3. *Statistical Literacy* If we fail to reject (i.e., "accept") the null hypothesis, does this mean that we have *proved* it to be true beyond *all* doubt? Explain your answer.

4. *Statistical Literacy* If we reject the null hypothesis, does this mean that we have *proved* it to be false beyond *all* doubt? Explain your answer.

5. *Statistical Literacy* What terminology do we use for the probability of rejecting the null hypothesis when it is true? What symbol do we use for this probability? Is this the probability of a type I or a type II error?

6. *Statistical Literacy* What terminology do we use for the probability of rejecting the null hypothesis when it is, in fact, false?

7. *Statistical Literacy* If the P-value in a statistical test is greater than the level of significance for the test, do we reject or fail to reject H_0?

8. *Statistical Literacy* If the P-value in a statistical test is less than or equal to the level of significance for the test, do we reject or fail to reject H_0?

9. *Statistical Literacy* Suppose the P-value in a right-tailed test is 0.0092. Based on the same population, sample, and null hypothesis, what is the P-value for a corresponding two-tailed test?

10. *Statistical Literacy* Suppose the P-value in a two-tailed test is 0.0134. Based on the same population, sample, and null hypothesis, and assuming the test statistic z is negative, what is the P-value for a corresponding left-tailed test?

11. *Basic Computation: Setting Hypotheses* Suppose you want to test the claim that a population mean equals 40.
    (a) State the null hypothesis.
    (b) State the alternate hypothesis if you have no information regarding how the population mean might differ from 40.
    (c) State the alternate hypothesis if you believe (based on experience or past studies) that the population mean may exceed 40.
    (d) State the alternate hypothesis if you believe (based on experience or past studies) that the population mean may be less than 40.

12. *Basic Computation: Setting Hypotheses* Suppose you want to test the claim that a population mean equals 30.
    (a) State the null hypothesis.
    (b) State the alternate hypothesis if you have no information regarding how the population mean might differ from 30.
    (c) State the alternate hypothesis if you believe (based on experience or past studies) that the population mean may be greater than 30.
    (d) State the alternate hypothesis if you believe (based on experience or past studies) that the population mean may not be as large as 30.

13. *Basic Computation: Find Test Statistic, Corresponding P-value, and Conclude Test* A random sample of size 20 from a normal distribution with $\sigma = 4$ produced a sample mean of 8.
(a) *Check Requirements* Is the $\bar{x}$ distribution normal? Explain.
(b) Compute the sample test statistic z under the null hypothesis H_0: $\mu = 7$.
(c) For H_1: $\mu \neq 7$, estimate the P-value of the test statistic.
(d) For a level of significance of 0.05 and the hypotheses of parts (b) and (c), do you reject or fail to reject the null hypothesis? Explain.

14. *Basic Computation: Find the Test Statistic, Corresponding P-value, and Conclude Test* A random sample of size 16 from a normal distribution with $\sigma = 3$ produced a sample mean of 4.5.
(a) *Check Requirements* Is the $\bar{x}$ distribution normal? Explain.
(b) Compute the sample test statistic z under the null hypothesis H_0: $\mu = 6.3$.
(c) For H_1: $\mu < 6.3$, estimate the P-value of the test statistic.
(d) For a level of significance of 0.01 and the hypotheses of parts (b) and (c), do you reject or fail to reject the null hypothesis? Explain.

15. *Veterinary Science: Colts* The body weight of a healthy 3-month-old colt should be about $\mu = 60$ kg (Source: *The Merck Veterinary Manual,* a standard reference manual used in most veterinary colleges).
(a) If you want to set up a statistical test to challenge the claim that $\mu = 60$ kg, what would you use for the null hypothesis H_0?
(b) In Nevada, there are many herds of wild horses. Suppose you want to test the claim that the average weight of a wild Nevada colt (3 months old) is less than 60 kg. What would you use for the alternate hypothesis H_1?
(c) Suppose you want to test the claim that the average weight of such a wild colt is greater than 60 kg. What would you use for the alternate hypothesis?
(d) Suppose you want to test the claim that the average weight of such a wild colt is *different* from 60 kg. What would you use for the alternate hypothesis?
(e) For each of the tests in parts (b), (c), and (d), would the area corresponding to the P-value be on the left, on the right, or on both sides of the mean? Explain your answer in each case.

16. *Marketing: Shopping Time* How much customers buy is a direct result of how much time they spend in a store. A study of average shopping times in a large national housewares store gave the following information (Source: *Why We Buy: The Science of Shopping* by P. Underhill):

Women with female companion: 8.3 min.

Women with male companion: 4.5 min.

Suppose you want to set up a statistical test to challenge the claim that a woman with a female friend spends an average of 8.3 minutes shopping in such a store.
(a) What would you use for the null and alternate hypotheses if you believe the average shopping time is less than 8.3 minutes? Is this a right-tailed, left-tailed, or two-tailed test?
(b) What would you use for the null and alternate hypotheses if you believe the average shopping time is different from 8.3 minutes? Is this a right-tailed, left-tailed, or two-tailed test?

Stores that sell mainly to women should figure out a way to engage the interest of men—perhaps comfortable seats and a big TV with sports programs! Suppose such an entertainment center was installed and you now wish to challenge the claim that a woman with a male friend spends only 4.5 minutes shopping in a housewares store.

(c) What would you use for the null and alternate hypotheses if you believe the average shopping time is more than 4.5 minutes? Is this a right-tailed, left-tailed, or two-tailed test?
(d) What would you use for the null and alternate hypotheses if you believe the average shopping time is different from 4.5 minutes? Is this a right-tailed, left-tailed, or two-tailed test?

17. *Meteorology: Storms* *Weatherwise* magazine is published in association with the American Meteorological Society. Volume 46, Number 6 has a rating system to classify Nor'easter storms that frequently hit New England states and can cause much damage near the ocean coast. A *severe* storm has an average peak wave height of 16.4 feet for waves hitting the shore. Suppose that a Nor'easter is in progress at the severe storm class rating.
(a) Let us say that we want to set up a statistical test to see if the wave action (i.e., height) is dying down or getting worse. What would be the null hypothesis regarding average wave height?
(b) If you wanted to test the hypothesis that the storm is getting worse, what would you use for the alternate hypothesis?
(c) If you wanted to test the hypothesis that the waves are dying down, what would you use for the alternate hypothesis?
(d) Suppose you do not know whether the storm is getting worse or dying out. You just want to test the hypothesis that the average wave height is *different* (either higher or lower) from the severe storm class rating. What would you use for the alternate hypothesis?
(e) For each of the tests in parts (b), (c), and (d), would the area corresponding to the *P*-value be on the left, on the right, or on both sides of the mean? Explain your answer in each case.

18. *Chrysler Concorde: Acceleration* *Consumer Reports* stated that the mean time for a Chrysler Concorde to go from 0 to 60 miles per hour is 8.7 seconds.
(a) If you want to set up a statistical test to challenge the claim of 8.7 seconds, what would you use for the null hypothesis?
(b) The town of Leadville, Colorado, has an elevation over 10,000 feet. Suppose you wanted to test the claim that the average time to accelerate from 0 to 60 miles per hour is longer in Leadville (because of less oxygen). What would you use for the alternate hypothesis?
(c) Suppose you made an engine modification and you think the average time to accelerate from 0 to 60 miles per hour is reduced. What would you use for the alternate hypothesis?
(d) For each of the tests in parts (b) and (c), would the *P*-value area be on the left, on the right, or on both sides of the mean? Explain your answer in each case.

For Problems 19–24, please provide the following information.
(a) What is the level of significance? State the null and alternate hypotheses. Will you use a left-tailed, right-tailed, or two-tailed test?
(b) *Check Requirements* What sampling distribution will you use? Explain the rationale for your choice of sampling distribution. Compute the value of the sample test statistic.
(c) Find (or estimate) the *P*-value. Sketch the sampling distribution and show the area corresponding to the *P*-value.
(d) Based on your answers in parts (a) to (c), will you reject or fail to reject the null hypothesis? Are the data statistically significant at level α?
(e) *Interpret* your conclusion in the context of the application.

19. *Dividend Yield: Australian Bank Stocks* Let x be a random variable representing dividend yield of Australian bank stocks. We may assume that x has a normal distribution with $\sigma = 2.4\%$. A random sample of 10 Australian bank stocks gave the following yields.

5.7	4.8	6.0	4.9	4.0	3.4	6.5	7.1	5.3	6.1

The sample mean is $\bar{x} = 5.38\%$. For the entire Australian stock market, the mean dividend yield is $\mu = 4.7\%$ (Reference: *Forbes*). Do these data indicate that the dividend yield of all Australian bank stocks is higher than 4.7%? Use $\alpha = 0.01$.

20. *Glucose Level: Horses* Gentle Ben is a Morgan horse at a Colorado dude ranch. Over the past 8 weeks, a veterinarian took the following glucose readings from this horse (in mg/100 ml).

93	88	82	105	99	110	84	89

The sample mean is $\bar{x} \approx 93.8$. Let x be a random variable representing glucose readings taken from Gentle Ben. We may assume that x has a normal distribution, and we know from past experience that $\sigma = 12.5$. The mean glucose level for horses should be $\mu = 85$ mg/100 ml (Reference: *Merck Veterinary Manual*). Do these data indicate that Gentle Ben has an overall average glucose level higher than 85? Use $\alpha = 0.05$.

21. *Ecology: Hummingbirds* Bill Alther is a zoologist who studies Anna's hummingbird (*Calypte anna*) (Reference: *Hummingbirds* by K. Long and W. Alther). Suppose that in a remote part of the Grand Canyon, a random sample of six of these birds was caught, weighed, and released. The weights (in grams) were

3.7	2.9	3.8	4.2	4.8	3.1

The sample mean is $\bar{x} = 3.75$ grams. Let x be a random variable representing weights of Anna's hummingbirds in this part of the Grand Canyon. We assume that x has a normal distribution and $\sigma = 0.70$ gram. It is known that for the population of all Anna's hummingbirds, the mean weight is $\mu = 4.55$ grams. Do the data indicate that the mean weight of these birds in this part of the Grand Canyon is less than 4.55 grams? Use $\alpha = 0.01$.

22. *Finance: P/E of Stocks* The price-to-earnings (P/E) ratio is an important tool in financial work. A random sample of 14 large U.S. banks (J.P. Morgan, Bank of America, and others) gave the following P/E ratios (Reference: *Forbes*).

24	16	22	14	12	13	17
22	15	19	23	13	11	18

The sample mean is $\bar{x} \approx 17.1$. Generally speaking, a low P/E ratio indicates a "value" or bargain stock. A recent copy of *The Wall Street Journal* indicated that the P/E ratio of the entire S&P 500 stock index is $\mu = 19$. Let x be a random variable representing the P/E ratio of all large U.S. bank stocks. We assume that x has a normal distribution and $\sigma = 4.5$. Do these data indicate that the P/E ratio of all U.S. bank stocks is less than 19? Use $\alpha = 0.05$.

23. *Insurance: Hail Damage* Nationally, about 11% of the total U.S. wheat crop is destroyed each year by hail (Reference: *Agricultural Statistics,* U.S. Department of Agriculture). An insurance company is studying wheat hail damage claims in Weld County, Colorado. A random sample of 16 claims in Weld County gave the following data (% wheat crop lost to hail).

15	8	9	11	12	20	14	11
7	10	24	20	13	9	12	5

The sample mean is $\bar{x} = 12.5\%$. Let x be a random variable that represents the percentage of wheat crop in Weld County lost to hail. Assume that x has a normal distribution and $\sigma = 5.0\%$. Do these data indicate that the percentage of wheat crop lost to hail in Weld County is different (either way) from the national mean of 11%? Use $\alpha = 0.01$.

24. *Medical: Red Blood Cell Volume* Total blood volume (in ml) per body weight (in kg) is important in medical research. For healthy adults, the red blood cell volume mean is about $\mu = 28$ ml/kg (Reference: *Laboratory and Diagnostic Tests* by F. Fischbach). Red blood cell volume that is too low or too high can indicate a medical problem (see reference). Suppose that Roger has had seven blood tests, and the red blood cell volumes were

    32 25 41 35 30 37 29

    The sample mean is $\bar{x} \approx 32.7$ ml/kg. Let x be a random variable that represents Roger's red blood cell volume. Assume that x has a normal distribution and $\sigma = 4.75$. Do the data indicate that Roger's red blood cell volume is different (either way) from $\mu = 28$ ml/kg? Use a 0.01 level of significance.

SECTION **8.2**

Testing the Mean μ

FOCUS POINTS

- Review the general procedure for testing using *P*-values.
- Test μ when σ is known using the normal distribution.
- Test μ when σ is unknown using a Student's *t* distribution.
- Understand the "traditional" method of testing that uses critical regions and critical values instead of *P*-values.

In this section, we continue our study of testing the mean μ. The method we are using is called the *P*-value method. It was used extensively by the famous statistician R. A. Fisher and is the most popular method of testing in use today. At the end of this section, we present another method of testing called the *critical region method* (or *traditional method*). The critical region method was used extensively by the statisticians J. Neyman and E. Pearson. In recent years, the use of this method has been declining. It is important to realize that for a fixed, preset level of significance α, both methods are logically equivalent.

In Section 8.1, we discussed the vocabulary and method of hypothesis testing using *P*-values. Let's quickly review the basic process.

1. We first state a proposed value for a population parameter in the null hypothesis H_0. The alternate hypothesis H_1 states alternative values of the parameter, either $<$, $>$, or $\neq$ the value proposed in H_0. We also set the level of significance α. This is the risk we are willing to take of committing a type I error. That is, α is the probability of rejecting H_0 when it is, in fact, true.
2. We use a corresponding sample statistic from a simple random sample to challenge the statement made in H_0. We convert the sample statistic to a test statistic, which is the corresponding value of the appropriate sampling distribution.
3. We use the sampling distribution of the test statistic and the type of test to compute the *P*-value of this statistic. Under the assumption that the null hypothesis is true, the *P*-value is the probability of getting a sample statistic as extreme as or more extreme than the observed statistic from our random sample.
4. Next, we conclude the test. If the *P*-value is very small, we have evidence to reject H_0 and adopt H_1. What do we mean by "very small"? We compare the

P-value to the preset level of significance α. If the P-value $\leq \alpha$, then we say that we have evidence to reject H_0 and adopt H_1. Otherwise, we say that the sample evidence is insufficient to reject H_0.

5. Finally, we *interpret* the results in the context of the application.

Knowing the sampling distribution of the sample test statistic is an essential part of the hypothesis testing process. For tests of μ, we use one of two sampling distributions for $\bar{x}$: the standard normal distribution or a Student's t distribution. As discussed in Chapters 6 and 7, the appropriate distribution depends upon our knowledge of the population standard deviation σ, the nature of the x distribution, and the sample size.

PART I: TESTING μ WHEN σ IS KNOWN

In most real-world situations, σ is simply not known. However, in some cases a preliminary study or other information can be used to get a realistic and accurate value for σ.

PROCEDURE

HOW TO TEST μ WHEN σ IS KNOWN

Requirements

Let x be a random variable appropriate to your application. Obtain a simple random sample (of size n) of x values from which you compute the sample mean $\bar{x}$. The value of σ is already known (perhaps from a previous study). If you can assume that x has a normal distribution, then any sample size n will work. If you cannot assume this, then use a sample size $n \geq 30$.

Procedure

1. In the context of the application, state the *null and alternate hypotheses* and set the *level of significance* α.
2. Use the known σ, the sample size n, the value of $\bar{x}$ from the sample, and μ from the null hypothesis to compute the standardized *sample test statistic*.

$$z = \frac{\bar{x} - \mu}{\dfrac{\sigma}{\sqrt{n}}}$$

3. Use the standard normal distribution and the type of test, one-tailed or two-tailed, to find the P-value corresponding to the test statistic.
4. *Conclude* the test. If P-value $\leq \alpha$, then reject H_0. If P-value $> \alpha$, then do not reject H_0.
5. *Interpret your conclusion* in the context of the application.

In Section 8.1, we examined P-value tests for normal distributions with relatively small sample sizes ($n < 30$). The next example does not assume a normal distribution, but has a large sample size ($n \geq 30$).

EXAMPLE 3

TESTING μ, σ KNOWN

Sunspots have been observed for many centuries. Records of sunspots from ancient Persian and Chinese astronomers go back thousands of years. Some archaeologists think sunspot activity may somehow be related to prolonged periods of drought in the southwestern United States. Let x be a random variable representing the average number of sunspots observed in a four-week period. A random sample of

FIGURE 8-3

P-value Area

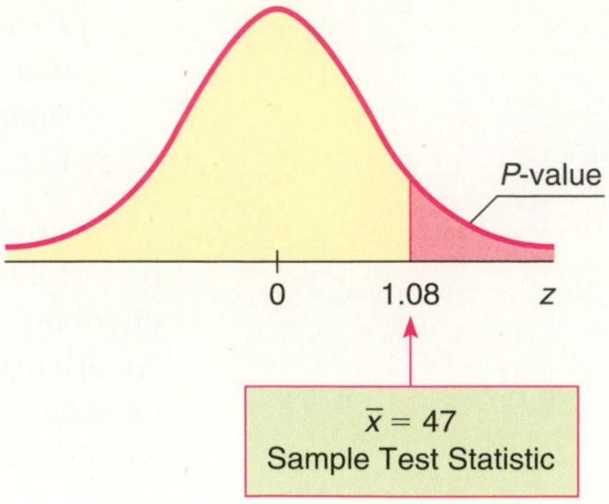

Comstock Images/Jupiter Images

40 such periods from Spanish colonial times gave the following data (Reference: M. Waldmeir, *Sun Spot Activity*, International Astronomical Union Bulletin).

12.5	14.1	37.6	48.3	67.3	70.0	43.8	56.5	59.7	24.0
12.0	27.4	53.5	73.9	104.0	54.6	4.4	177.3	70.1	54.0
28.0	13.0	6.5	134.7	114.0	72.7	81.2	24.1	20.4	13.3
9.4	25.7	47.8	50.0	45.3	61.0	39.0	12.0	7.2	11.3

The sample mean is $\bar{x} \approx 47.0$. Previous studies of sunspot activity during this period indicate that $\sigma = 35$. It is thought that for thousands of years, the mean number of sunspots per 4-week period was about $\mu = 41$. Sunspot activity above this level may (or may not) be linked to gradual climate change. Do the data indicate that the mean sunspot activity during the Spanish colonial period was higher than 41? Use $\alpha = 0.05$.

SOLUTION:

(a) Establish the null and alternate hypotheses.
Since we want to know whether the average sunspot activity during the Spanish colonial period was higher than the long-term average of $\mu = 41$,

$$H_0\text{: } \mu = 41 \quad \text{and} \quad H_1\text{: } \mu > 41$$

(b) ***Check Requirements*** What distribution do we use for the sample test statistic? Compute the test statistic from the sample data. Since $n \geq 30$ and we know σ, we use the standard normal distribution. Using $\bar{x} = 47$ from the sample, $\sigma = 35$, $\mu = 41$ from H_0, and $n = 40$,

$$z = \frac{\bar{x} - \mu}{\sigma/\sqrt{n}} \approx \frac{47 - 41}{35/\sqrt{40}} \approx 1.08$$

(c) Find the *P*-value of the test statistic.
Figure 8-3 shows the *P*-value. Since we have a right-tailed test, the *P*-value is the area to the right of $z = 1.08$ shown in Figure 8-3. Using Table 5 of Appendix II, we find that

$$P\text{-value} = P(z > 1.08) \approx 0.1401$$

(d) Conclude the test.
Since the *P*-value of $0.1401 > 0.05$ for α, we do not reject H_0.

(e) ***Interpretation*** Interpret the results in the context of the problem.
At the 5% level of significance, the evidence is not sufficient to reject H_0. Based on the sample data, we do not think the average sunspot activity during the Spanish colonial period was higher than the long-term mean.

PART II: TESTING μ WHEN σ IS UNKNOWN

In many real-world situations, you have only a random sample of data values. In addition, you may have some limited information about the probability distribution of your data values. Can you still test μ under these circumstances? In most cases, the answer is yes!

PROCEDURE

HOW TO TEST μ WHEN σ IS UNKNOWN

Requirements

Let x be a random variable appropriate to your application. Obtain a simple random sample (of size n) of x values from which you compute the sample mean $\bar{x}$ and the sample standard deviation s. If you can assume that x has a normal distribution or simply a mound-shaped and symmetric distribution, then any sample size n will work. If you cannot assume this, use a sample size $n \geq 30$.

Procedure

1. In the context of the application, state the *null and alternate hypotheses* and set the *level of significance* α.
2. Use $\bar{x}$, s, and n from the sample, with μ from H_0, to compute the *sample test statistic.*

d.f. for testing μ when σ unknown

$$t = \frac{\bar{x} - \mu}{\frac{s}{\sqrt{n}}} \quad \text{with degrees of freedom } d.f. = n - 1$$

3. Use the Student's t distribution and the type of test, one-tailed or two-tailed, to find (or estimate) the *P-value* corresponding to the test statistic.
4. *Conclude* the test. If P-value $\leq \alpha$, then reject H_0. If P-value $> \alpha$, then do not reject H_0.
5. *Interpret your conclusion* in the context of the application.

Using the Student's t table to estimate P-values

In Sections 7.2 and 7.4, we used Table 6 of Appendix II, Student's t Distribution, to find critical values t_c for confidence intervals. The critical values are in the body of the table. We find P-values in the *rows* headed by "one-tail area" and "two-tail area," depending on whether we have a one-tailed or two-tailed test. If the test statistic t for the sample statistic $\bar{x}$ is negative, look up the P-value for the corresponding *positive* value of t (i.e., look up the P-value for $|t|$).

Note: In Table 6, areas are given in *one tail* beyond positive t on the right or negative t on the left, and in *two tails* beyond $\pm t$. Notice that in each column, two-tail area $= 2$(one-tail area). Consequently, we use *one-tail areas* as endpoints of the interval containing the P-value for *one-tailed tests.* We use *two-tail areas* as endpoints of the interval containing the P-value for *two-tailed tests.* (See Figure 8-4.)

Example 4 and Guided Exercise 4 show how to use Table 6 of Appendix II to find an interval containing the P-value corresponding to a test statistic t.

FIGURE 8-4

P-value for One-Tailed Tests and for Two-Tailed Tests

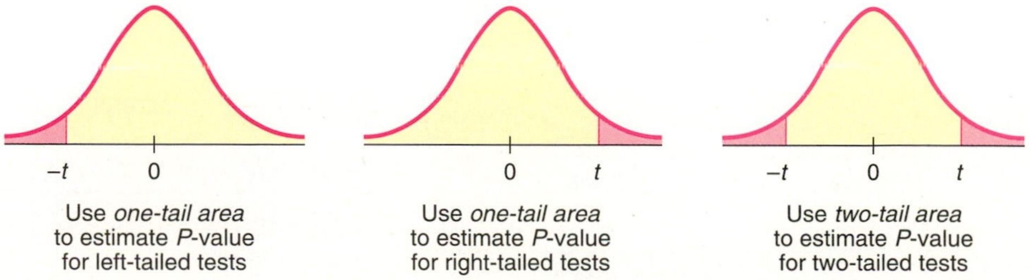

EXAMPLE 4

TESTING μ, σ UNKNOWN

The drug 6-mP (6-mercaptopurine) is used to treat leukemia. The following data represent the remission times (in weeks) for a random sample of 21 patients using 6-mP (Reference: E. A. Gehan, University of Texas Cancer Center).

10	7	32	23	22	6	16	34	32	25	11
20	19	6	17	35	6	13	9	6	10	

The sample mean is $\bar{x} \approx 17.1$ weeks, with sample standard deviation $s \approx 10.0$. Let x be a random variable representing the remission time (in weeks) for all patients using 6-mP. Assume the x distribution is mound-shaped and symmetric. A previously used drug treatment had a mean remission time of $\mu = 12.5$ weeks. Do the data indicate that the mean remission time using the drug 6-mP is different (either way) from 12.5 weeks? Use $\alpha = 0.01$.

SOLUTION:

(a) Establish the null and alternate hypotheses.
Since we want to determine if the drug 6-mP provides a mean remission time that is different from that provided by a previously used drug having $\mu = 12.5$ weeks,

$$H_0: \mu = 12.5 \text{ weeks and } H_1: \mu \neq 12.5 \text{ weeks}$$

(b) ***Check Requirements*** What distribution do we use for the sample test statistic t? Compute the sample test statistic from the sample data.
The x distribution is assumed to be mound-shaped and symmetric. Because we don't know σ, we use a Student's t distribution with $d.f. = 20$. Using $\bar{x} \approx 17.1$ and $s \approx 10.0$ from the sample data, $\mu = 12.5$ from H_0, and $n = 21$,

$$t \approx \frac{\bar{x} - \mu}{s/\sqrt{n}} \approx \frac{17.1 - 12.5}{10.0/\sqrt{21}} \approx 2.108$$

(c) Find the P-value or the interval containing the P-value.
Figure 8-5 shows the P-value. Using Table 6 of Appendix II, we find an interval containing the P-value. Since this is a two-tailed test, we use entries from the row headed by *two-tail area*. Look up the t value in the row headed by $d.f. = n - 1 = 21 - 1 = 20$. The sample statistic $t = 2.108$ falls between 2.086 and 2.528. The P-value for the sample t falls between the corresponding two-tail areas 0.050 and 0.020. (See Table 8-5.)

$$0.020 < P\text{-value} < 0.050$$

(d) Conclude the test.
The diagram on the next page shows the interval that contains the single P-value corresponding to the test statistic. Note that there is just one P-value corresponding to the test statistic. Table 6 of Appendix II does not give that specific value, but it does give a range that contains that specific P-value. As the diagram shows,

FIGURE 8-5

P-value

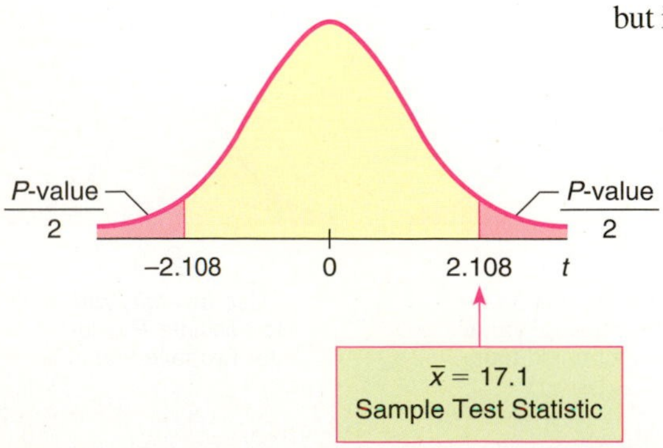

TABLE 8-5 **Excerpt from Student's *t* Distribution (Table 6, Appendix II)**

one-tail area	...	...
✓ two-tail area	0.050	0.020
$d.f. = 20$	2.086	2.528
	↑ Sample $t = 2.108$	

the entire range is greater than α. This means the specific P-value is greater than α, so we cannot reject H_0.

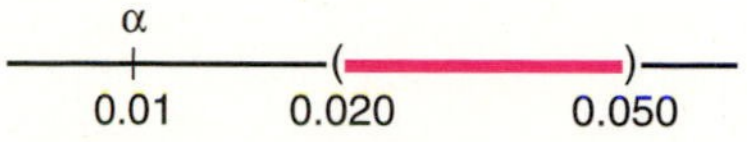

Note: Using the raw data, computer software gives P-value ≈ 0.048. This value is in the interval we estimated. It is larger than the α value of 0.01, so we do not reject H_0.

(e) ***Interpretation*** Interpret the results in the context of the problem.
At the 1% level of significance, the evidence is not sufficient to reject H_0. Based on the sample data, we cannot say that the drug 6-mP provides a different average remission time than the previous drug.

GUIDED EXERCISE 4 — TESTING μ, σ UNKNOWN

Archaeologists become excited when they find an anomaly in discovered artifacts. The anomaly may (or may not) indicate a new trading region or a new method of craftsmanship. Suppose the lengths of projectile points (arrowheads) at a certain archaeological site have mean length $\mu = 2.6$ cm. A random sample of 61 recently discovered projectile points in an adjacent cliff dwelling gave the following lengths (in cm) (Reference: A. Woosley and A. McIntyre, *Mimbres Mogollon Archaeology,* University of New Mexico Press).

3.1	4.1	1.8	2.1	2.2	1.3	1.7	3.0	3.7	2.3	2.6	2.2	2.8	3.0
3.2	3.3	2.4	2.8	2.8	2.9	2.9	2.2	2.4	2.1	3.4	3.1	1.6	3.1
3.5	2.3	3.1	2.7	2.1	2.0	4.8	1.9	3.9	2.0	5.2	2.2	2.6	1.9
4.0	3.0	3.4	4.2	2.4	3.5	3.1	3.7	3.7	2.9	2.6	3.6	3.9	3.5
1.9	4.0	4.0	4.6	1.9									

The sample mean is $\bar{x} \approx 2.92$ cm and the sample standard deviation is $s \approx 0.85$, where x is a random variable that represents the lengths (in cm) of all projectile points found at the adjacent cliff dwelling site. Do these data indicate that the mean length of projectile points in the adjacent cliff dwelling is longer than 2.6 cm? Use a 1% level of significance.

(a) State H_0, H_1, and α.

→ H_0: $\mu = 2.6$ cm; H_1: $\mu > 2.6$ cm; $\alpha = 0.01$

(b) *Check Requirements* What sampling distribution should you use? What is the value of the sample test statistic t?

→ Because $n \geq 30$ and σ is unknown, use the Student's t distribution with $d.f. = n - 1 = 61 - 1 = 60$. Using $\bar{x} \approx 2.92$, $s \approx 0.85$, $\mu = 2.6$ from H_0 and $n = 61$,

$$t = \frac{\bar{x} - \mu}{\sigma/\sqrt{n}} \approx \frac{2.92 - 2.6}{0.85/\sqrt{61}} \approx 2.940$$

(c) When you use Table 6, Appendix II, to find an interval containing the P-value, do you use one-tail or two-tail areas? Why? Sketch a figure showing the P-value. Find an interval containing the P-value.

→ This is a right-tailed test, so use a one-tail area.

FIGURE 8-6 P-value

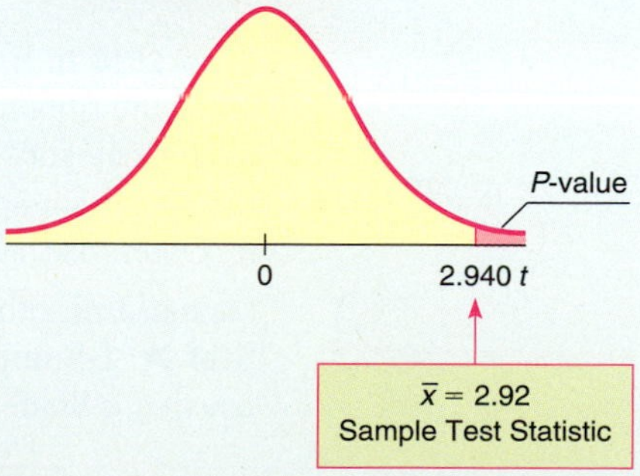

TABLE 8-6 **Excerpt from Student's t Table**

✓ **one-tail area**	**. . . 0.050**	**0.0005**
two-tail area	**. . . 0.010**	**0.0010**
$d.f. = 60$	. . . 2.660	3.460

↑ Sample $t = 2.940$

Continued

GUIDED EXERCISE 4 *continued*

Using *d.f.* $= 60$, we find that the sample $t = 2.940$ is between the critical values 2.660 and 3.460. The sample *P*-value is then between the one-tail areas 0.005 and 0.0005.

$0.0005 < P\text{-value} < 0.005$

(d) Do we reject or fail to reject H_0?

Since the interval containing the *P*-value lies to the left of $\alpha = 0.01$, we reject H_0.

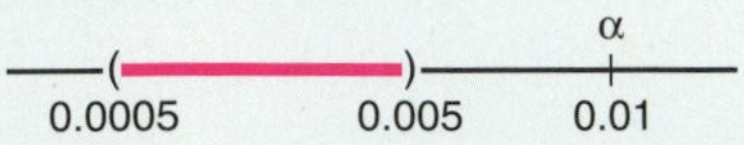

Note: Using the raw data, computer software gives *P*-value ≈ 0.0022. This value is in our estimated range and is less than $\alpha = 0.01$, so we reject H_0.

(e) *Interpretation* Interpret your results in the context of the application.

At the 1% level of significance, sample evidence is sufficiently strong to reject H_0 and conclude that the average projectile point length at the adjacent cliff dwelling site is longer than 2.6 cm.

TECH NOTES

The TI-84Plus/TI-83Plus/TI-*n*spire calculators, Excel 2010, and Minitab all support testing of μ using the standard normal distribution. The TI-84Plus/TI-83Plus/TI-*n*spire and Minitab support testing of μ using a Student's *t* distribution. All the technologies return a *P*-value for the test.

TI-84Plus/TI-83Plus/TI-*n*spire (with TI-84Plus keypad) You can select to enter raw data **(Data)** or summary statistics **(Stats).** Enter the value of $\boldsymbol{\mu_0}$ used in the null hypothesis H_0: $\boldsymbol{\mu} = \boldsymbol{\mu_0}$. Select the symbol used in the alternate hypothesis ($\neq\boldsymbol{\mu_0}$, $<\boldsymbol{\mu_0}$, $>\boldsymbol{\mu_0}$). To test μ using the standard normal distribution, press **Stat,** select **Tests,** and use option **1:Z-Test.** The value for σ is required. To test μ using a Student's *t* distribution, use option **2:T-Test.** Using data from Example 4 regarding remission times, we have the following displays. The *P*-value is given as p.

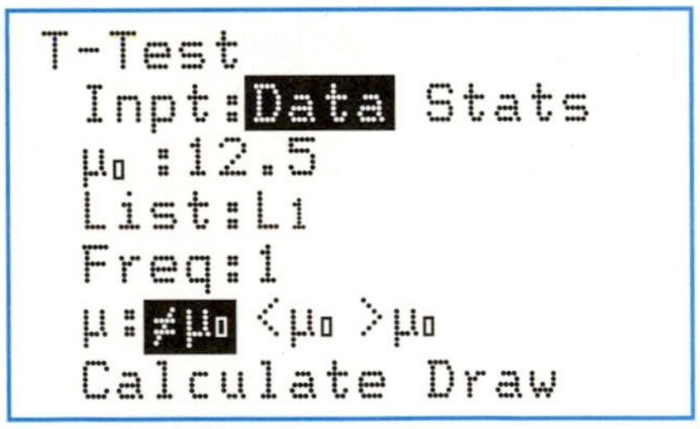

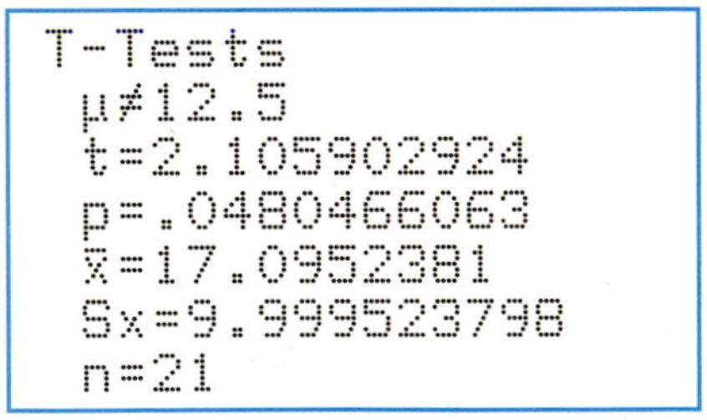

Excel 2010 In Excel, the **Z.TEST** function finds the *P*-values for a right-tailed test. Click the ribbon choice **Insert Function** (f_x). In the dialogue box, select **Statistical** for the category and **Z.TEST** for the function. In the next dialogue box, give the cell range containing your data for the array. Use the value of μ stated in H_0 for *x*. Provide σ. Otherwise, Excel uses the sample standard deviation computed from the data.

Minitab Enter the raw data from a sample. Use the menu selections **Stat ➤ Basic Stat ➤ 1-Sample z** for tests using the standard normal distribution. For tests of μ using a Student's *t* distribution, select **1-Sample t.**

PART III: TESTING μ USING CRITICAL REGIONS (TRADITIONAL METHOD)

Critical region method

The most popular method of statistical testing is the *P*-value method. For that reason, the *P*-value method is emphasized in this book. Another method of testing is called the *critical region method* or *traditional method.*

For a fixed, preset value of the level of significance α, both methods are logically equivalent. Because of this, we treat the traditional method as an "optional" topic and consider only the case of testing μ when σ is known.

Consider the null hypothesis H_0: $\mu = k$. We use information from a random sample, together with the sampling distribution for $\bar{x}$ and the level of significance α, to determine whether or not we should reject the null hypothesis. The essential question is, "How much can $\bar{x}$ vary from $\mu = k$ before we suspect that H_0: $\mu = k$ is false and reject it?"

The answer to the question regarding the relative sizes of $\bar{x}$ and μ, as stated in the null hypothesis, depends on the sampling distribution of $\bar{x}$, the alternate hypothesis H_1, and the level of significance α. If the sample test statistic $\bar{x}$ is sufficiently different from the claim about μ made in the null hypothesis, we reject the null hypothesis.

Another method for concluding two-tailed tests involves the use of confidence intervals. Problems 25 and 26 at the end of this section discuss the confidence interval method.

The values of $\bar{x}$ for which we reject H_0 are called the *critical region* of the $\bar{x}$ distribution. Depending on the alternate hypothesis, the critical region is located on the left side, the right side, or both sides of the $\bar{x}$, distribution. Figure 8-7 shows the relationship of the critical region to the alternate hypothesis and the level of significance α.

Notice that the total area in the critical region is preset to be the level of significance α. This is *not* the *P*-value discussed earlier! In fact, you cannot set the *P*-value in advance because it is determined from a random sample. Recall that the level of significance α should (in theory) be a fixed, preset number assigned before drawing any samples.

Critical values

The most commonly used levels of significance are $\alpha = 0.05$ and $\alpha = 0.01$. Critical regions of a standard normal distribution are shown for these levels of significance in Figure 8-8. *Critical values* are the boundaries of the critical region. Critical values designated as z_0 for the standard normal distribution are shown in Figure 8-8 on the next page. For easy reference, they are also included in Table 5 of Appendix II, Areas of a Standard Normal Distribution.

The procedure for hypothesis testing using critical regions follows the same first two steps as the procedure using *P*-values. However, instead of finding a *P*-value for the sample test statistic, we check if the sample test statistic falls in the critical region. If it does, we reject H_0. Otherwise, we do not reject H_0.

FIGURE 8-7

Critical Regions for H_0: $\mu = k$

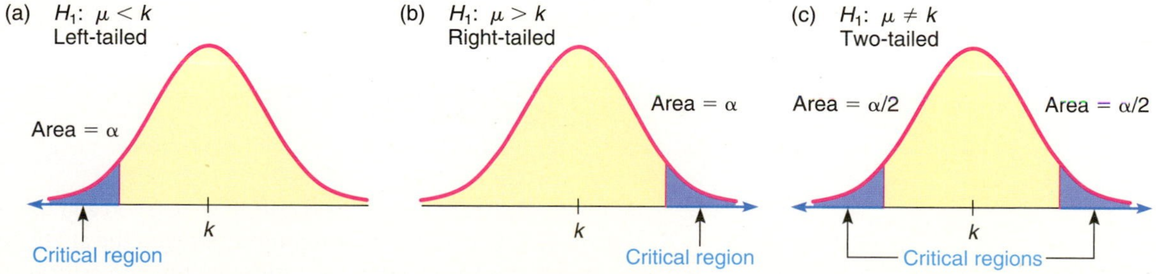

FIGURE 8-8

Critical Values z_0 for Tests Involving a Mean (Large Samples)

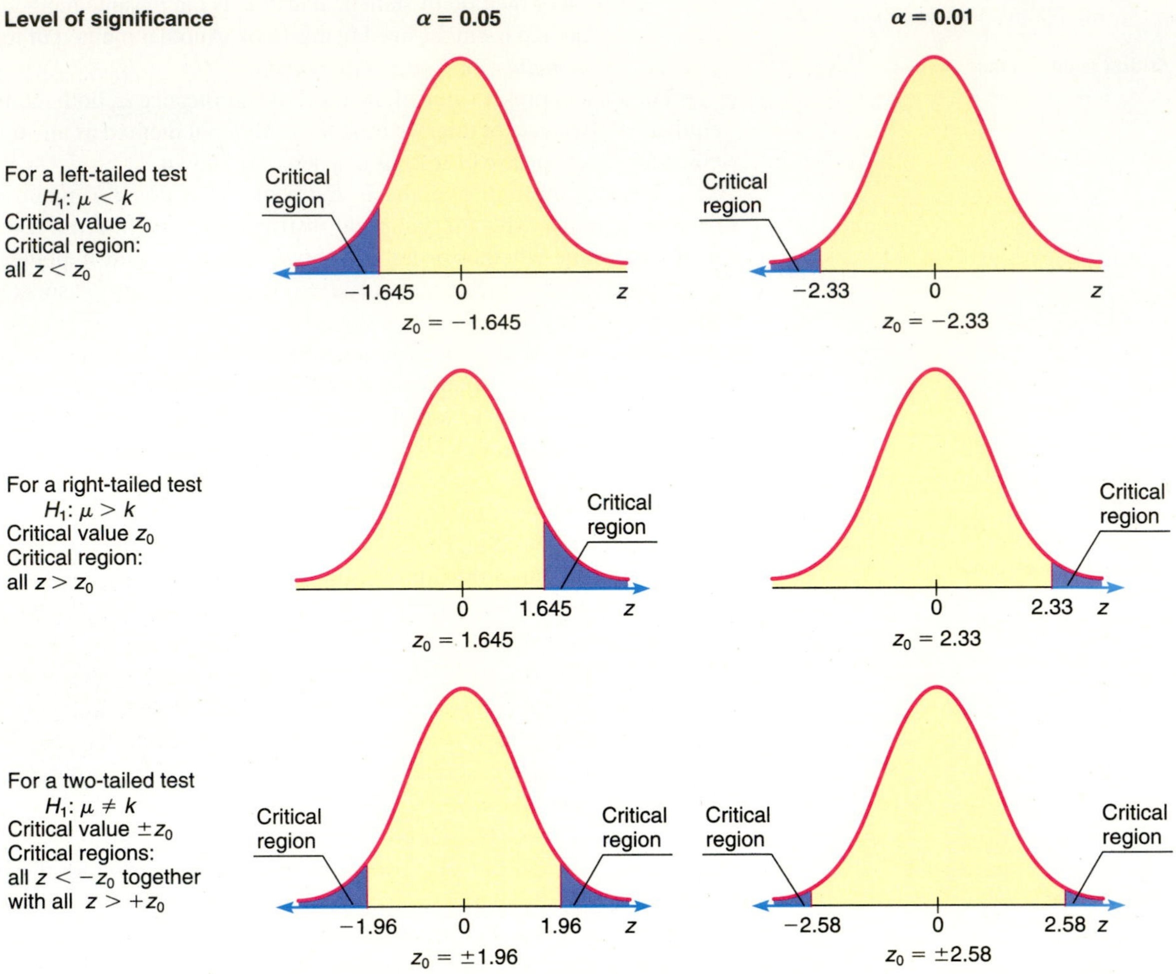

PROCEDURE

HOW TO TEST μ WHEN σ IS KNOWN (CRITICAL REGION METHOD)

Let x be a random variable appropriate to your application. Obtain a simple random sample (of size n) of x values from which you compute the sample mean $\bar{x}$. The value of σ is already known (perhaps from a previous study). If you can assume that x has a normal distribution, then any sample size n will work. If you cannot assume this, use a sample size $n \geq 30$. Then $\bar{x}$ follows a distribution that is normal or approximately normal.

1. In the context of the application, state the *null and alternate hypotheses* and set the *level of significance* α. We use the most popular choices, $\alpha = 0.05$ or $\alpha = 0.01$.
2. Use the known σ, the sample size n, the value of $\bar{x}$ from the sample, and μ from the null hypothesis to compute the standardized sample *test statistic.*

$$z = \frac{\bar{x} - \mu}{\dfrac{\sigma}{\sqrt{n}}}$$

3. Show the *critical region* and *critical value(s)* on a graph of the sampling distribution. The level of significance α and the alternate hypothesis determine the locations of critical regions and critical values.

Continued

4. *Conclude* the test. If the test statistic z computed in Step 2 is in the critical region, then reject H_0. If the test statistic z is not in the critical region, then do not reject H_0.
5. *Interpret your conclusion* in the context of the application.

EXAMPLE 5

CRITICAL REGION METHOD OF TESTING μ

Consider Example 3 regarding sunspots. Let x be a random variable representing the number of sunspots observed in a four-week period. A random sample of 40 such periods from Spanish colonial times gave the number of sunspots per period. The raw data are given in Example 3. The sample mean is $\bar{x} \approx 47.0$. Previous studies indicate that for this period, $\sigma = 35$. It is thought that for thousands of years, the mean number of sunspots per 4-week period was about $\mu = 41$. Do the data indicate that the mean sunspot activity during the Spanish colonial period was higher than 41? Use $\alpha = 0.05$.

SOLUTION:

(a) Set the null and alternate hypotheses.
As in Example 3, we use H_0: $\mu = 41$ and H_1: $\mu > 41$.

(b) Compute the sample test statistic.
As in Example 3, we use the standard normal distribution, with $\bar{x} = 47$, $\sigma = 35$, $\mu = 41$ from H_0, and $n = 40$.

$$z = \frac{\bar{x} - \mu}{\sigma/\sqrt{n}} \approx \frac{47 - 41}{35/\sqrt{40}} \approx 1.08$$

(c) Determine the critical region and critical value based on H_1 and $\alpha = 0.05$.
Since we have a right-tailed test, the critical region is the rightmost 5% of the standard normal distribution. According to Figure 8-8, the critical value is $z_0 = 1.645$.

(d) Conclude the test.
We conclude the test by showing the critical region, critical value, and sample test statistic $z = 1.08$ on the standard normal curve. For a right-tailed test with $\alpha = 0.05$ the critical value is $z_0 = 1.645$. Figure 8-9 shows the critical region. As we can see, the sample test statistic does not fall in the critical region. Therefore, we fail to reject H_0.

FIGURE 8-9

Critical Region, $\alpha = 0.05$

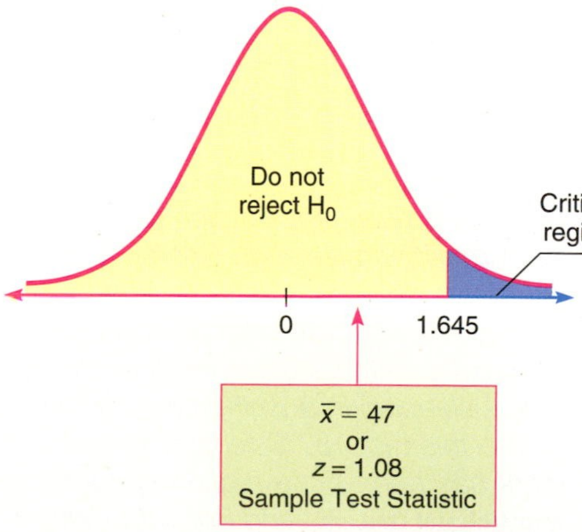

Continued

(e) ***Interpretation*** Interpret the results in the context of the application.
At the 5% level of significance, the sample evidence is insufficient to justify rejecting H_0. It seems that the average sunspot activity during the Spanish colonial period was the same as the historical average.

(f) How do results of the critical region method compare to the results of the P-value method for a 5% level of significance?
The results, as expected, are the same. In both cases, we fail to reject H_0.

The critical region method of testing as outlined applies to tests of other parameters. As with the P-value method, you need to know the sampling distribution of the sample test statistic. Critical values for distributions are usually found in tables rather than in computer software outputs. For example, Table 6 of Appendix II provides critical values for Student's t distributions.

The critical region method of hypothesis testing is very general. The following procedure box outlines the process of concluding a hypothesis test using the critical region method.

PROCEDURE

HOW TO CONCLUDE TESTS USING THE CRITICAL REGION METHOD

1. Compute the sample test statistic using an appropriate sampling distribution.
2. Using the same sampling distribution, find the critical value(s) as determined by the level of significance α and the nature of the test: right-tailed, left-tailed, or two-tailed.
3. Compare the sample test statistic to the critical value(s).
   (a) For a right-tailed test,
      i. if sample test statistic $\geq$ critical value, reject H_0.
      ii. if sample test statistic $<$ critical value, fail to reject H_0.
   (b) For a left-tailed test,
      i. if sample test statistic $\leq$ critical value, reject H_0.
      ii. if sample test statistic $>$ critical value, fail to reject H_0.
   (c) For a two-tailed test,
      i. if sample test statistic lies at or beyond critical values, reject H_0.
      ii. if sample test statistic lies between critical values, fail to reject H_0.

VIEWPOINT Predator or Prey?

Consider animals such as the arctic fox, gray wolf, desert lion, and South American jaguar. Each animal is a predator. What are the total sleep time (hours per day), maximum life span (years), and overall danger index from other animals? Now consider prey such as rabbits, deer, wild horses, and the Brazilian tapir (a wild pig). Is there a statistically significant difference in average sleep time, life span, and danger index? What about other variables such as the ratio of brain weight to body weight or the sleep exposure index (sleeping in a well-protected den or out in the open)? How did prehistoric humans fit into this picture? Scientists have collected a lot of data, and a great deal of statistical work has been done regarding such questions. For more information, see the web site **http://lib.stat.cmu.edu/** *and follow the links to Datasets and then Sleep.*

SECTION 8.2 PROBLEMS

1. *Statistical Literacy* For the same sample data and null hypothesis, how does the P-value for a two-tailed test of μ compare to that for a one-tailed test?

2. *Statistical Literacy* To test μ for an x distribution that is mound-shaped using sample size $n \geq 30$, how do you decide whether to use the normal or the Student's t distribution?

3. *Statistical Literacy* When using the Student's t distribution to test μ, what value do you use for the degrees of freedom?

4. *Critical Thinking* Consider a test for μ. If the P-value is such that you can reject H_0 at the 5% level of significance, can you always reject H_0 at the 1% level of significance? Explain.

5. *Critical Thinking* Consider a test for μ. If the P-value is such that you can reject H_0 for $\alpha = 0.01$, can you always reject H_0 for $\alpha = 0.05$? Explain.

6. *Critical Thinking* If sample data is such that for a one-tailed test of μ you can reject H_0 at the 1% level of significance, can you always reject H_0 for a two-tailed test at the same level of significance? Explain.

7. *Basic Computation: P-value Corresponding to t Value* For a Student's t distribution with $d.f. = 10$ and $t = 2.930$,
   (a) find an interval containing the corresponding P-value for a two-tailed test.
   (b) find an interval containing the corresponding P-value for a right-tailed test.

8. *Basic Computation: P-value Corresponding to t Value* For a Student's t distribution with $d.f. = 16$ and $t = -1.830$,
   (a) find an interval containing the corresponding P-value for a two-tailed test.
   (b) find an interval containing the corresponding P-value for a left-tailed test.

9. *Basic Computation: Testing μ, σ Unknown* A random sample of 25 values is drawn from a mound-shaped and symmetrical distribution. The sample mean is 10 and the sample standard deviation is 2. Use a level of significance of 0.05 to conduct a two-tailed test of the claim that the population mean is 9.5.
   (a) *Check Requirements* Is it appropriate to use a Student's t distribution? Explain. How many degrees of freedom do we use?
   (b) What are the hypotheses?
   (c) Compute the sample test statistic t.
   (d) Estimate the P-value for the test.
   (e) Do we reject or fail to reject H_0?
   (f) ***Interpret*** the results.

10. *Basic Computation: Testing μ, σ Unknown* A random sample has 49 values. The sample mean is 8.5 and the sample standard deviation is 1.5. Use a level of significance of 0.01 to conduct a left-tailed test of the claim that the population mean is 9.2.
   (a) *Check Requirements* Is it appropriate to use a Student's t distribution? Explain. How many degrees of freedom do we use?
   (b) What are the hypotheses?
   (c) Compute the sample test statistic t.
   (d) Estimate the P-value for the test.
   (e) Do we reject or fail to reject H_0?
   (f) ***Interpret*** the results.

Please provide the following information for Problems 11–22.
(a) What is the level of significance? State the null and alternate hypotheses.
(b) *Check Requirements* What sampling distribution will you use? Explain the rationale for your choice of sampling distribution. Compute the value of the sample test statistic.

(c) Find (or estimate) the P-value. Sketch the sampling distribution and show the area corresponding to the P-value.
(d) Based on your answers in parts (a) to (c), will you reject or fail to reject the null hypothesis? Are the data statistically significant at level α?
(e) ***Interpret*** your conclusion in the context of the application.
Note: For degrees of freedom *d.f.* not given in the Student's t table, use the closest *d.f.* that is *smaller.* In some situations, this choice of *d.f.* may increase the P-value by a small amount and therefore produce a slightly more "conservative" answer.

11. *Meteorology: Storms* *Weatherwise* is a magazine published by the American Meteorological Society. One issue gives a rating system used to classify Nor'easter storms that frequently hit New England and can cause much damage near the ocean. A severe storm has an average peak wave height of $\mu = 16.4$ feet for waves hitting the shore. Suppose that a Nor'easter is in progress at the severe storm class rating. Peak wave heights are usually measured from land (using binoculars) off fixed cement piers. Suppose that a reading of 36 waves showed an average wave height of $\bar{x} = 17.3$ feet. Previous studies of severe storms indicate that $\sigma = 3.5$ feet. Does this information suggest that the storm is (perhaps temporarily) increasing above the severe rating? Use $\alpha = 0.01$.

12. *Medical: Blood Plasma* Let x be a random variable that represents the pH of arterial plasma (i.e., acidity of the blood). For healthy adults, the mean of the x distribution is $\mu = 7.4$ (Reference: *The Merck Manual,* a commonly used reference in medical schools and nursing programs). A new drug for arthritis has been developed. However, it is thought that this drug may change blood pH. A random sample of 31 patients with arthritis took the drug for 3 months. Blood tests showed that $\bar{x} = 8.1$ with sample standard deviation $s = 1.9$. Use a 5% level of significance to test the claim that the drug has changed (either way) the mean pH level of the blood.

13. *Wildlife: Coyotes* A random sample of 46 adult coyotes in a region of northern Minnesota showed the average age to be $\bar{x} = 2.05$ years, with sample standard deviation $s = 0.82$ years (based on information from the book *Coyotes: Biology, Behavior and Management* by M. Bekoff, Academic Press). However, it is thought that the overall population mean age of coyotes is $\mu = 1.75$. Do the sample data indicate that coyotes in this region of northern Minnesota tend to live longer than the average of 1.75 years? Use $\alpha = 0.01$.

14. *Fishing: Trout* Pyramid Lake is on the Paiute Indian Reservation in Nevada. The lake is famous for cutthroat trout. Suppose a friend tells you that the average length of trout caught in Pyramid Lake is $\mu = 19$ inches. However, the *Creel Survey* (published by the Pyramid Lake Paiute Tribe Fisheries Association) reported that of a random sample of 51 fish caught, the mean length was $\bar{x} = 18.5$ inches, with estimated standard deviation $s = 3.2$ inches. Do these data indicate that the average length of a trout caught in Pyramid Lake is less than $\mu = 19$ inches? Use $\alpha = 0.05$.

15. *Investing: Stocks* Socially conscious investors screen out stocks of alcohol and tobacco makers, firms with poor environmental records, and companies with poor labor practices. Some examples of "good," socially conscious companies are Johnson and Johnson, Dell Computers, Bank of America, and Home Depot. The question is, are such stocks overpriced? One measure of value is the P/E, or price-to-earnings, ratio. High P/E ratios may indicate a stock is overpriced. For the S&P stock index of all major stocks, the mean P/E ratio is $\mu = 19.4$. A random sample of 36 "socially conscious" stocks gave a P/E ratio sample mean of $\bar{x} = 17.9$, with sample standard deviation

$s = 5.2$ (Reference: *Morningstar,* a financial analysis company in Chicago). Does this indicate that the mean P/E ratio of all socially conscious stocks is different (either way) from the mean P/E ratio of the S&P stock index? Use $\alpha = 0.05$.

16. *Agriculture: Ground Water* Unfortunately, arsenic occurs naturally in some ground water (Reference: *Union Carbide Technical Report K/UR-1*). A mean arsenic level of $\mu = 8.0$ parts per billion (ppb) is considered safe for agricultural use. A well in Texas is used to water cotton crops. This well is tested on a regular basis for arsenic. A random sample of 37 tests gave a sample mean of $\bar{x} = 7.2$ ppb arsenic, with $s = 1.9$ ppb. Does this information indicate that the mean level of arsenic in this well is less than 8 ppb? Use $\alpha = 0.01$.

17. *Medical: Red Blood Cell Count* Let x be a random variable that represents red blood cell (RBC) count in millions of cells per cubic millimeter of whole blood. Then x has a distribution that is approximately normal. For the population of healthy female adults, the mean of the x distribution is about 4.8 (based on information from *Diagnostic Tests with Nursing Implications,* Springhouse Corporation). Suppose that a female patient has taken six laboratory blood tests over the past several months and that the RBC count data sent to the patient's doctor are

4.9	4.2	4.5	4.1	4.4	4.3

   i. Use a calculator with sample mean and sample standard deviation keys to verify that $\bar{x} = 4.40$ and $s \approx 0.28$.
   ii. Do the given data indicate that the population mean RBC count for this patient is lower than 4.8? Use $\alpha = 0.05$.

18. *Medical: Hemoglobin Count* Let x be a random variable that represents hemoglobin count (HC) in grams per 100 milliliters of whole blood. Then x has a distribution that is approximately normal, with population mean of about 14 for healthy adult women (see reference in Problem 17). Suppose that a female patient has taken 10 laboratory blood tests during the past year. The HC data sent to the patient's doctor are

15	18	16	19	14	12	14	17	15	11

   i. Use a calculator with sample mean and sample standard deviation keys to verify that $\bar{x} = 15.1$ and $s \approx 2.51$.
   ii. Does this information indicate that the population average HC for this patient is higher than 14? Use $\alpha = 0.01$.

19. *Ski Patrol: Avalanches* Snow avalanches can be a real problem for travelers in the western United States and Canada. A very common type of avalanche is called the slab avalanche. These have been studied extensively by David McClung, a professor of civil engineering at the University of British Columbia. Slab avalanches studied in Canada have an average thickness of $\mu = 67$ (Source: *Avalanche Handbook* by D. McClung and P. Schaerer). The ski patrol at Vail, Colorado, is studying slab avalanches in its region. A random sample of avalanches in spring gave the following thicknesses (in cm):

59	51	76	38	65	54	49	62
68	55	64	67	63	74	65	79

   i. Use a calculator with mean and standard deviation keys to verify that $\bar{x} \approx 61.8$ and $s \approx 10.6$ cm.
   ii. Assume the slab thickness has an approximately normal distribution. Use a 1% level of significance to test the claim that the mean slab thickness in the Vail region is different from that in Canada.

20. *Longevity: Honolulu* *USA Today* reported that the state with the longest mean life span is Hawaii, where the population mean life span is 77 years. A random sample of 20 obituary notices in the *Honolulu Advertizer* gave the following information about life span (in years) of Honolulu residents:

72	68	81	93	56	19	78	94	83	84
77	69	85	97	75	71	86	47	66	27

i. Use a calculator with mean and standard deviation keys to verify that $\bar{x} = 71.4$ years and $s \approx 20.65$ years.
ii. Assuming that life span in Honolulu is approximately normally distributed, does this information indicate that the population mean life span for Honolulu residents is less than 77 years? Use a 5% level of significance.

21. *Fishing: Atlantic Salmon* Homser Lake, Oregon, has an Atlantic salmon catch and release program that has been very successful. The average fisherman's catch has been $\mu = 8.8$ Atlantic salmon per day (Source: *National Symposium on Catch and Release Fishing,* Humboldt State University). Suppose that a new quota system restricting the number of fishermen has been put into effect this season. A random sample of fishermen gave the following catches per day:

12	6	11	12	5	0	2
7	8	7	6	3	12	12

i. Use a calculator with mean and sample standard deviation keys to verify that $\bar{x} = 7.36$ and $s \approx 4.03$.
ii. Assuming the catch per day has an approximately normal distribution, use a 5% level of significance to test the claim that the population average catch per day is now different from 8.8.

22. *Archaeology: Tree Rings* Tree-ring dating from archaeological excavation sites is used in conjunction with other chronologic evidence to estimate occupation dates of prehistoric Indian ruins in the southwestern United States. It is thought that Burnt Mesa Pueblo was occupied around 1300 A.D. (based on evidence from potsherds and stone tools). The following data give tree-ring dates (A.D.) from adjacent archaeological sites (*Bandelier Archaeological Excavation Project: Summer 1990 Excavations at Burnt Mesa Pueblo,* edited by T. Kohler, Washington State University Department of Anthropology, 1992):

1189	1267	1268	1275	1275
1271	1272	1316	1317	1230

i. Use a calculator with mean and standard deviation keys to verify that $\bar{x} = 1268$ and $s \approx 37.29$ years.
ii. Assuming the tree-ring dates in this excavation area follow a distribution that is approximately normal, does this information indicate that the population mean of tree-ring dates in the area is different from (either higher or lower than) that in 1300 A.D.? Use a 1% level of significance.

23. *Critical Thinking: One-Tailed versus Two-Tailed Tests*
(a) For the same data and null hypothesis, is the P-value of a one-tailed test (right or left) larger or smaller than that of a two-tailed test? Explain.
(b) For the same data, null hypothesis, and level of significance, is it possible that a one-tailed test results in the conclusion to reject H_0 while a two-tailed test results in the conclusion to fail to reject H_0? Explain.
(c) For the same data, null hypothesis, and level of significance, if the conclusion is to reject H_0 based on a two-tailed test, do you also reject H_0 based on a one-tailed test? Explain.

(d) If a report states that certain data were used to reject a given hypothesis, would it be a good idea to know what type of test (one-tailed or two-tailed) was used? Explain.

24. *Critical Thinking: Comparing Hypothesis Tests with U.S. Courtroom System* Compare statistical testing with legal methods used in a U.S. court setting. Then discuss the following topics in class or consider the topics on your own. Please write a brief but complete essay in which you answer the following questions.

(a) In a court setting, the person charged with a crime is initially considered to be innocent. The claim of innocence is maintained until the jury returns with a decision. Explain how the claim of innocence could be taken to be the null hypothesis. Do we assume that the null hypothesis is true throughout the testing procedure? What would the alternate hypothesis be in a court setting?

(b) The court claims that a person is innocent if the evidence against the person is not adequate to find him or her guilty. This does not mean, however, that the court has necessarily *proved* the person to be innocent. It simply means that the evidence against the person was not adequate for the jury to find him or her guilty. How does this situation compare with a statistical test for which the conclusion is "do not reject" (i.e., accept) the null hypothesis? What would be a type II error in this context?

(c) If the evidence against a person is adequate for the jury to find him or her guilty, then the court claims that the person is guilty. Remember, this does not mean that the court has necessarily *proved* the person to be guilty. It simply means that the evidence against the person was strong enough to find him or her guilty. How does this situation compare with a statistical test for which the conclusion is to "reject" the null hypothesis? What would be a type I error in this context?

(d) In a court setting, the final decision as to whether the person charged is innocent or guilty is made at the end of the trial, usually by a jury of impartial people. In hypothesis testing, the final decision to reject or not reject the null hypothesis is made at the end of the test by using information or data from an (impartial) random sample. Discuss these similarities between statistical hypothesis testing and a court setting.

(e) We hope that you are able to use this discussion to increase your understanding of statistical testing by comparing it with something that is a well-known part of our American way of life. However, all analogies have weak points, and it is important not to take the analogy between statistical hypothesis testing and legal court methods too far. For instance, the judge does not set a level of significance and tell the jury to determine a verdict that is wrong only 5% or 1% of the time. Discuss some of these weak points in the analogy between the court setting and hypothesis testing.

25. *Expand Your Knowledge: Confidence Intervals and Two-Tailed Hypothesis Tests* Is there a relationship between confidence intervals and two-tailed hypothesis tests? Let c be the level of confidence used to construct a confidence interval from sample data. Let α be the level of significance for a two-tailed hypothesis test. The following statement applies to hypothesis tests of the mean.

For a two-tailed hypothesis test with level of significance α and null hypothesis H_0: $\mu = k$, we *reject* H_0 whenever k falls *outside* the $c = 1 - \alpha$ confidence interval for μ based on the sample data. When k falls within the $c = 1 - \alpha$ confidence interval, we do not reject H_0.

(A corresponding relationship between confidence intervals and two-tailed hypothesis tests also is valid for other parameters, such as p, $\mu_1 - \mu_2$, and $p_1 - p_2$, which we will study in Sections 8.3 and 8.5.) Whenever the value of k given in the null hypothesis falls *outside* the $c = 1 - \alpha$ confidence interval for the parameter, we *reject* H_0. For example, consider a two-tailed hypothesis test with $\alpha = 0.01$ and

$$H_0: \mu = 20 \qquad H_1: \mu \neq 20$$

A random sample of size 36 has a sample mean $\bar{x} = 22$ from a population with standard deviation $\sigma = 4$.

(a) What is the value of $c = 1 - \alpha$? Using the methods of Chapter 7, construct a $1 - \alpha$ confidence interval for μ from the sample data. What is the value of μ given in the null hypothesis (i.e., what is k)? Is this value in the confidence interval? Do we reject or fail to reject H_0 based on this information?

(b) Using methods of this chapter, find the P-value for the hypothesis test. Do we reject or fail to reject H_0? Compare your result to that of part (a).

26. *Confidence Intervals and Two-Tailed Hypothesis Tests* Change the null hypotheses of Problem 25 to H_0: $\mu = 21$ and H_1: $\mu \neq 21$. Repeat parts (a) and (b).

27. *Critical Region Method: Standard Normal* Solve Problem 11 using the critical region method of testing (i.e., traditional method). Compare your conclusion with the conclusion obtained by using the P-value method. Are they the same?

28. *Critical Region Method: Student's t* Table 6 of Appendix II gives critical values for the Student's t distribution. Use an appropriate *d.f.* as the row header. For a *right-tailed* test, the column header is the value of α found in the *one-tail area* row. For a *left-tailed* test, the column header is the value of α found in the *one-tail area* row, but you must change the sign of the critical value t to $-t$. For a *two-tailed* test, the column header is the value of α from the *two-tail area* row. The critical values are the $\pm t$ values shown. Solve Problem 12 using the critical region method of testing. Compare your conclusion with the conclusion obtained by using the P-value method. Are they the same?

29. *Critical Region Method: Student's t* Solve Problem 13 using the critical region method of testing. *Hint:* See Problem 28. Compare your conclusion with the conclusion obtained by using the P-value method. Are they the same?

30. *Critical Region Method: Student's t* Solve Problem 14 using the critical region method of testing. *Hint:* See Problem 28. Compare your conclusion with the conclusion obtained by using the P-value method. Are they the same?

SECTION **8.3**

Testing a Proportion p

FOCUS POINTS

- Identify the components needed for testing a proportion.
- Compute the sample test statistic.
- Find the *P*-value and conclude the test.

Many situations arise that call for tests of proportions or percentages rather than means. For instance, a college registrar may want to determine if the proportion of students wanting 3-week intensive courses has increased.

How can we make such a test? In this section, we will study tests involving proportions (i.e., percentages or proportions). Such tests are similar to those in

Sections 8.1 and 8.2. The main difference is that we are working with a distribution of proportions.

Tests for a single proportion

Throughout this section, we will assume that the situations we are dealing with satisfy the conditions underlying the binomial distribution. In particular, we will let r be a binomial random variable. This means that r is the number of successes out of n independent binomial trials (for the definition of a binomial trial, see Section 5.2). We will use $\hat{p} = r/n$ as our estimate for p, the population probability of success on each trial. The letter q again represents the population probability of failure on each trial, and so $q = 1 - p$. We also assume that the samples are large (i.e., $np > 5$ and $nq > 5$).

Criteria for using normal approximation to binomial $np > 5$ and $nq > 5$

For large samples, $np > 5$ and $nq > 5$, the distribution of $\hat{p} = r/n$ values is well approximated by a *normal curve* with mean μ and standard deviation σ as follows:

$$\mu = p$$

$$\sigma = \sqrt{\frac{pq}{n}}$$

Hypothesis for testing p

The null and alternate hypotheses for tests of proportions are

Left-Tailed Test	**Right-Tailed Test**	**Two-Tailed Test**
$H_0: p = k$	$H_0: p = k$	$H_0: p = k$
$H_1: p < k$	$H_1: p > k$	$H_1: p \neq k$

depending on what is asked for in the problem. Notice that since p is a probability, the value k must be between 0 and 1.

Sample test statistic $\hat{p}$

For tests of proportions, we need to convert the sample test statistic $\hat{p}$ to a z value. Then we can find a P-value appropriate for the test. The $\hat{p}$ distribution is approximately normal, with mean p and standard deviation $\sqrt{pq/n}$. Therefore, the conversion of $\hat{p}$ to z follows the formula

$$z = \frac{\hat{p} - p}{\sqrt{\dfrac{pq}{n}}}$$

where $\hat{p} = r/n$ is the sample test statistic and r is the number of successes
n = number of trials
p = proportion specified in H_0
$q = 1 - p$

Using this mathematical information about the sampling distribution for $\hat{p}$, the basic procedure is similar to tests you have conducted before.

PROCEDURE

HOW TO TEST A PROPORTION *P*

Requirements

Consider a binomial experiment with n trials, where p represents the population probability of success and $q = 1 - p$ represents the population probability of failure. Let r be a random variable that represents the number of successes out of the n binomial trials. The number of trials n should be sufficiently large so that both $np > 5$ and $nq > 5$ (use p from the null hypothesis). In this case, the sample test statistic $\hat{p} = r/n$ can be approximated by the normal distribution.

Procedure

1. In the context of the application, state the *null and alternate hypotheses* and set the *level of significance* α.

Continued

2. Compute the standardized *sample test statistic*

$$z = \frac{\hat{p} - p}{\sqrt{\frac{pq}{n}}}$$

where p is the value specified in H_0 and $q = 1 - p$.
3. Use the standard normal distribution and the type of test, one-tailed or two-tailed, to find the *P-value* corresponding to the test statistic.
4. *Conclude* the test. If P-value $\leq \alpha$, then reject H_0. If P-value $> \alpha$, then do not reject H_0.
5. *Interpret your conclusion* in the context of the application.

EXAMPLE 6

TESTING P

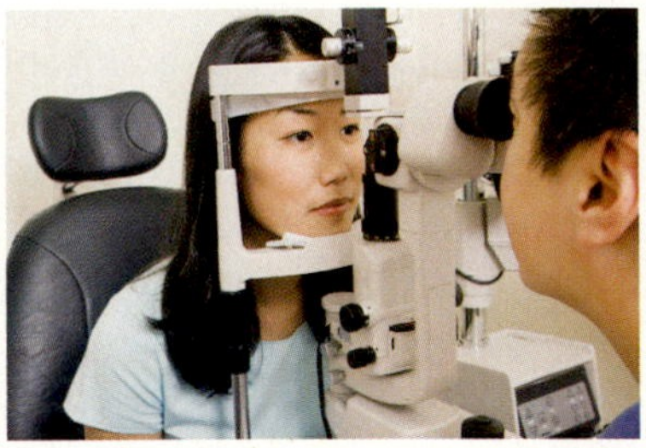
Ned Frisk/Blend Images/ Jupiter images

A team of eye surgeons has developed a new technique for a risky eye operation to restore the sight of people blinded from a certain disease. Under the old method, it is known that only 30% of the patients who undergo this operation recover their eyesight.

Suppose that surgeons in various hospitals have performed a total of 225 operations using the new method and that 88 have been successful (i.e., the patients fully recovered their sight). Can we justify the claim that the new method is better than the old one? (Use a 1% level of significance.)

SOLUTION:

(a) Establish H_0 and H_1 and note the level of significance.
The level of significance is $\alpha = 0.01$. Let p be the probability that a patient fully recovers his or her eyesight. The null hypothesis is that p is still 0.30, even for the new method. The alternate hypothesis is that the new method has improved the chances of a patient recovering his or her eyesight. Therefore,

$$H_0: p = 0.30 \quad \text{and} \quad H_1: p > 0.30$$

(b) ***Check Requirements*** Is the sample sufficiently large to justify use of the normal distribution for $\hat{p}$? Find the sample test statistic $\hat{p}$ and convert it to a z value, if appropriate.
Using p from H_0 we note that $np = 225(0.3) = 67.5$ is greater than 5 and that $nq = 225(0.7) = 157.5$ is also greater than 5, so we can use the normal distribution for the sample statistic $\hat{p}$.

$$\hat{p} = \frac{r}{n} = \frac{88}{225} \approx 0.39$$

The z value corresponding to $\hat{p}$ is

$$z = \frac{\hat{p} - p}{\sqrt{\frac{pq}{n}}} \approx \frac{0.39 - 0.30}{\sqrt{\frac{0.30(0.70)}{225}}} \approx 2.95$$

In the formula, the value for p is from the null hypothesis. H_0 specifies that $p = 0.30$, so $q = 1 - 0.30 = 0.70$.

(c) Find the P-value of the test statistic.
Figure 8-10 shows the P-value. Since we have a right-tailed test, the P-value is the area to the right of $z = 2.95$. Using the normal distribution (Table 5 of Appendix II), we find that P-value $= P(z > 2.95) \approx 0.0016$.

FIGURE 8-10

P-value Area

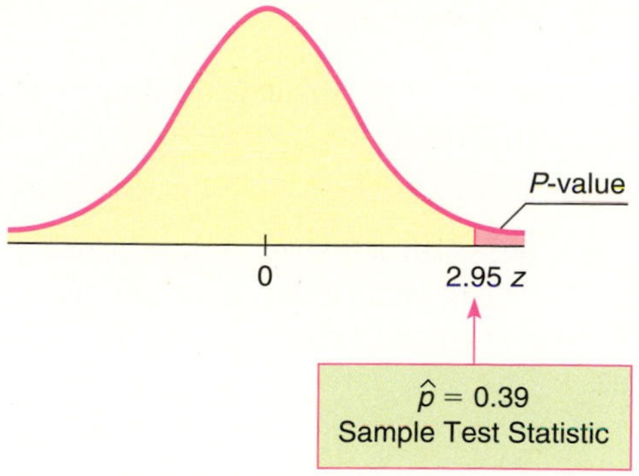

(d) Conclude the test.
Since the P-value of $0.0016 \le 0.01$ for α, we reject H_0.

(e) ***Interpretation*** Interpret the results in the context of the problem.
At the 1% level of significance, the evidence shows that the population probability of success for the new surgery technique is higher than that of the old technique.

GUIDED EXERCISE 5 Testing p

A botanist has produced a new variety of hybrid wheat that is better able to withstand drought than other varieties. The botanist knows that for the parent plants, the proportion of seeds germinating is 80%. The proportion of seeds germinating for the hybrid variety is unknown, but the botanist claims that it is 80%. To test this claim, 400 seeds from the hybrid plant are tested, and it is found that 312 germinate. Use a 5% level of significance to test the claim that the proportion germinating for the hybrid is 80%.

(a) Let p be the proportion of hybrid seeds that will germinate. Notice that we have no prior knowledge about the germination proportion for the hybrid plant. State H_0 and H_1. What is the required level of significance?

→ H_0: $p = 0.80$; H_1: $p \neq 0.80$; $\alpha = 0.05$

(b) ***Check Requirements*** Using the value of p in H_0, are both $np > 5$ and $nq > 5$? Can we use the normal distribution for $\hat{p}$?

→ From H_0, $p = 0.80$ and $q = 1 - p = 0.20$
$np = 400(0.8) = 320 > 5$
$nq = 400(0.2) = 80 > 5$
So, we can use the normal distribution for $\hat{p}$.

Calculate the sample test statistic $\hat{p}$.

→ The number of trials is $n = 400$, and the number of successes is $r = 312$. Thus,

$$\hat{p} = \frac{r}{n} = \frac{312}{400} = 0.78$$

(c) Next, we convert the sample test statistic $\hat{p} = 0.78$ to a z value. Based on our choice for H_0, what value should we use for p in our formula? Since $q = 1 - p$, what value should we use for q? Using these values for p and q, convert $\hat{p}$ to a z value.

→ According to H_0, $p = 0.80$. Then $q = 1 - p = 0.20$. Using these values in the following formula gives

$$z = \frac{\hat{p} - p}{\sqrt{\frac{pq}{n}}} = \frac{0.78 - 0.80}{\sqrt{\frac{0.80(0.20)}{400}}} = -1.00$$

Continued

GUIDED EXERCISE 5 *continued*

CALCULATOR NOTE If you evaluate the denominator separately, be sure to carry at least four digits after the decimal.

(d) Is the test right-tailed, left-tailed, or two-tailed? Find the *P*-value of the sample test statistic and sketch a standard normal curve showing the *P*-value.

→ For a two-tailed test, using the normal distribution (Table 5 of Appendix II), we find that

$P\text{-value} = 2P(z < -1.00) = 2(0.1587) = 0.3174$

FIGURE 8-11 *P*-value

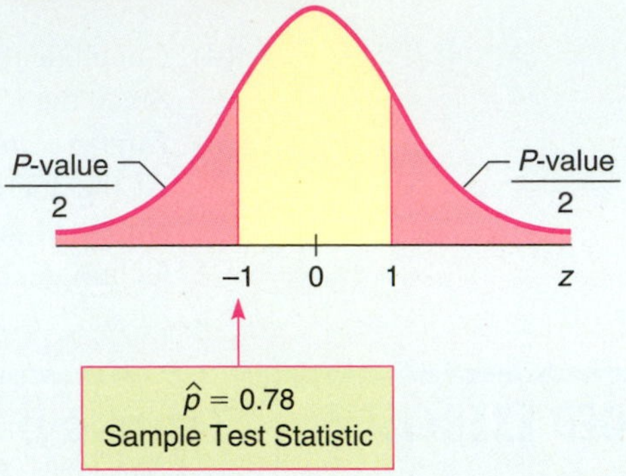

(e) Do we reject or fail to reject H_0?

→ We fail to reject H_0 because *P*-value of $0.3171 > 0.05$ for α.

(f) ***Interpretation*** Interpret your conclusion in the context of the application.

→ At the 5% level of significance, there is insufficient evidence to conclude that the germination rate is not 80%.

Critical region method

Since the $\hat{p}$ sampling distribution is approximately normal, we use Table 5, "Areas of a Standard Normal Distribution," in Appendix II to find critical values.

EXAMPLE 7

CRITICAL REGION METHOD FOR TESTING *p*

Let's solve Guided Exercise 5 using the critical region approach. In that problem, 312 of 400 seeds from a hybrid wheat variety germinated. For the parent plants, the proportion of germinating seeds is 80%. Use a 5% level of significance to test the claim that the population proportion of germinating seeds from the hybrid wheat is different from that of the parent plants.

Zeljko Radojko/Shutterstock.com

SOLUTION:

(a) As in Guided Exercise 5, we have $\alpha = 0.05$, H_0: $p = 0.80$, and H_1: $p \neq 0.80$. The next step is to find $\hat{p}$ and the corresponding sample test statistic z. This was done in Guided Exercise 5, where we found that $\hat{p} = 0.78$, with corresponding $z = -1.00$.

(b) Now we find the critical value z_0 for a two-tailed test using $\alpha = 0.05$. This means that we want the total area 0.05 divided between two tails, one to the right of z_0 and one to the left of $-z_0$. As shown in Figure 8-8 of Section 8.2, the critical value(s) are ± 1.96. (See also Table 5, part (c), of Appendix II for critical values of the z distribution.)

(c) Figure 8-12 shows the critical regions and the location of the sample test statistic.

FIGURE 8-12

Critical Regions, $\alpha = 0.05$

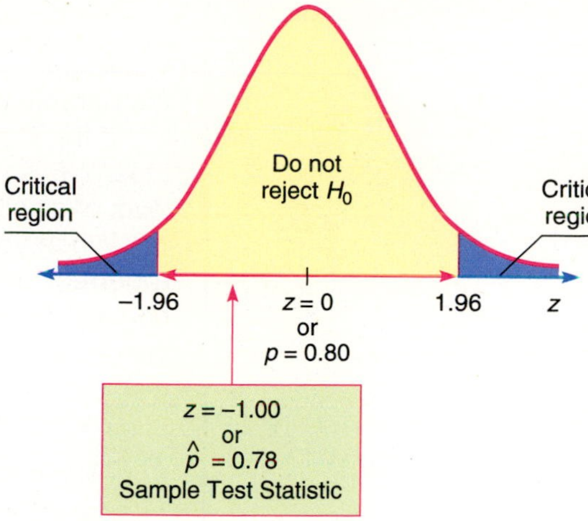

(d) Finally, we conclude the test and compare the results to Guided Exercise 5. Since the sample test statistic does not fall in the critical region, we fail to reject H_0 and conclude that, at the 5% level of significance, the evidence is not strong enough to reject the botanist's claim. This result, as expected, is consistent with the conclusion obtained by using the P-value method.

TECH NOTES

The TI-84Plus/TI-83Plus/TI-*n*spire calculators and Minitab support tests of proportions. The output for both technologies includes the sample proportion $\hat{p}$ and the P-value of $\hat{p}$. Minitab also includes the z value corresponding to $\hat{p}$.

TI-84Plus/TI-83Plus/TI-*n*spire (with TI-84Plus keypad) Press **STAT,** select **TESTS,** and use option **5:1-PropZTest.** The value of p_0 is from the null hypothesis H_0: $p = p_0$. The number of successes is the value for x.

Minitab Menu selections: **Stat ➤ Basic Statistics ➤ 1 Proportion.** Under options, set the test proportion as the value in H_0. Choose to use the normal distribution.

CRITICAL THINKING

ISSUES RELATED TO HYPOTHESIS TESTING

Through our work with hypothesis tests of μ and p, we've gained experience in setting up, performing, and interpreting results of such tests.

We know that different random samples from the same population are very likely to have sample statistics $\bar{x}$ or $\hat{p}$ that differ from their corresponding parameters μ or p. Some values of a statistic from a random sample will be close to the corresponding parameter. Others may be farther away simply because we happened to draw a random sample of more extreme data values.

The central question in hypothesis testing is whether or not you think the value of the sample test statistic is too far away from the value of the population parameter proposed in H_0 to occur by chance alone.

This is where the P-value of the sample test statistic comes into play. The P-value of the sample test statistic tells you the probability that you

Continued

CRITICAL THINKING *continued*

would get a sample statistic as far away as, or farther from, the value of the parameter as stated in the null hypothesis H_0.

If the *P*-value is very small, you reject H_0. But what does "very small" mean? It is customary to define "very small" as smaller than the preset level of significance α.

When you reject H_0, are you absolutely certain that you are making a correct decision? The answer is no! You are simply willing to take a chance that you are making a mistake (a type I error). The level of significance α describes the chance of making a mistake if you reject H_0 when it is, in fact, true.

Several issues come to mind:

1. What if the *P*-value is so close to α that we "barely" reject or fail to reject H_0? In such cases, researchers might attempt to clarify the results by
   - increasing the sample size.
   - controlling the experiment to reduce the standard deviation.

   Both actions tend to increase the magnitude of the *z* or *t* value of the sample test statistic, resulting in a smaller corresponding *P*-value.

2. How reliable is the study and the measurements in the sample?
   - When reading results of a statistical study, be aware of the source of the data and the reliability of the organization doing the study.
   - Is the study sponsored by an organization that might profit or benefit from the stated conclusions? If so, look at the study carefully to ensure that the measurements, sampling technique, and handling of data are proper and meet professional standards.

VIEWPOINT Who Did What?

Art, music, literature, and science share a common need to classify things: Who painted that picture? Who composed that music? Who wrote that document? Who should get that patent? In statistics, such questions are called classification problems. *For example, the Federalist Papers were published anonymously in 1787–1788 by Alexander Hamilton, John Jay, and James Madison. But who wrote what? That question is addressed by F. Mosteller (Harvard University) and D. Wallace (University of Chicago) in the book* Statistics: A Guide to the Unknown, *edited by J. M. Tanur. Other scholars have studied authorship regarding Plato's* Republic *and Plato's* Dialogues, *including the* Symposium. *For more information on this topic, see the source in Problems 15 and 16 of this exercise set.*

SECTION 8.3 PROBLEMS

1. *Statistical Literacy* To use the normal distribution to test a proportion *p*, the conditions $np > 5$ and $nq > 5$ must be satisfied. Does the value of *p* come from H_0, or is it estimated by using $\hat{p}$ from the sample?

2. *Statistical Literacy* Consider a binomial experiment with *n* trials and *r* successes. For a test for a proportion *p*, what is the formula for the sample test statistic? Describe each symbol used in the formula.

3. *Critical Thinking* In general, if sample data are such that the null hypothesis is rejected at the $\alpha = 1\%$ level of significance based on a two-tailed test, is H_0 also rejected at the $\alpha = 1\%$ level of significance for a corresponding one-tailed test? Explain.

4. *Critical Thinking* An article in a newspaper states that the proportion of traffic accidents involving road rage is higher this year than it was last year, when it was 15%. Reconstruct the information of the study in terms of a hypothesis test. Discuss possible hypotheses, possible issues about the sample, possible levels of significance, and the "absolute truth" of the conclusion.

5. *Basic Computation: Testing p* A random sample of 30 binomials trials resulted in 12 successes. Test the claim that the population proportion of successes does not equal 0.50. Use a level of significance of 0.05.
   (a) *Check Requirements* Can a normal distribution be used for the $\hat{p}$ distribution? Explain.
   (b) State the hypotheses.
   (c) Compute $\hat{p}$ and the corresponding standardized sample test statistic.
   (d) Find the *P*-value of the test statistic.
   (e) Do you reject or fail to reject H_0? Explain.
   (f) *Interpretation* What do the results tell you?

6. *Basic Computation: Testing p* A random sample of 60 binomials trials resulted in 18 successes. Test the claim that the population proportion of successes exceeds 18%. Use a level of significance of 0.01.
   (a) *Check Requirements* Can a normal distribution be used for the $\hat{p}$ distribution? Explain.
   (b) State the hypotheses.
   (c) Compute $\hat{p}$ and the corresponding standardized sample test statistic.
   (d) Find the *P*-value of the test statistic.
   (e) Do you reject or fail to reject H_0? Explain.
   (f) *Interpretation* What do the results tell you?

For Problems 7–21, please provide the following information.
(a) What is the level of significance? State the null and alternate hypotheses.
(b) *Check Requirements* What sampling distribution will you use? Do you think the sample size is sufficiently large? Explain. Compute the value of the sample test statistic.
(c) Find the *P*-value of the test statistic. Sketch the sampling distribution and show the area corresponding to the *P*-value.
(d) Based on your answers in parts (a) to (c), will you reject or fail to reject the null hypothesis? Are the data statistically significant at level α?
(e) ***Interpret*** your conclusion in the context of the application.

7. *Focus Problem: Benford's Law* Please read the Focus Problem at the beginning of this chapter. Recall that Benford's Law claims that numbers chosen from very large data files tend to have "1" as the first nonzero digit disproportionately often. In fact, research has shown that if you randomly draw a number from a very large data file, the probability of getting a number with "1" as the leading digit is about 0.301 (see the reference in this chapter's Focus Problem).

   Now suppose you are an auditor for a very large corporation. The revenue report involves millions of numbers in a large computer file. Let us say you took a random sample of $n = 215$ numerical entries from the file and $r = 46$ of the entries had a first nonzero digit of 1. Let p represent the population proportion of all numbers in the corporate file that have a first nonzero digit of 1.
   i. Test the claim that p is less than 0.301. Use $\alpha = 0.01$.
   ii. If p is in fact less than 0.301, would it make you suspect that there are not enough numbers in the data file with leading 1's? Could this indicate that the books have been "cooked" by "pumping up" or inflating the numbers? Comment from the viewpoint of a stockholder. Comment from the perspective of the Federal Bureau of Investigation as it looks for money laundering in the form of false profits.

iii. Comment on the following statement: "If we reject the null hypothesis at level of significance α, we have not *proved* H_0 to be false. We can say that the probability is α that we made a mistake in rejecting H_0." Based on the outcome of the test, would you recommend further investigation before accusing the company of fraud?

8. *Focus Problem: Benford's Law* Again, suppose you are the auditor for a very large corporation. The revenue file contains millions of numbers in a large computer data bank (see Problem 7). You draw a random sample of $n = 228$ numbers from this file and $r = 92$ have a first nonzero digit of 1. Let p represent the population proportion of all numbers in the computer file that have a leading digit of 1.
   i. Test the claim that p is more than 0.301. Use $\alpha = 0.01$.
   ii. If p is in fact larger than 0.301, it would seem there are too many numbers in the file with leading 1's. Could this indicate that the books have been "cooked" by artificially lowering numbers in the file? Comment from the point of view of the Internal Revenue Service. Comment from the perspective of the Federal Bureau of Investigation as it looks for "profit skimming" by unscrupulous employees.
   iii. Comment on the following statement: "If we reject the null hypothesis at level of significance α, we have not *proved* H_0 to be false. We can say that the probability is α that we made a mistake in rejecting H_0." Based on the outcome of the test, would you recommend further investigation before accusing the company of fraud?

9. *Sociology: Crime Rate* Is the national crime rate really going down? Some sociologists say yes! They say that the reason for the decline in crime rates in the 1980s and 1990s is demographics. It seems that the population is aging, and older people commit fewer crimes. According to the FBI and the Justice Department, 70% of all arrests are of males aged 15 to 34 years (Source: *True Odds* by J. Walsh, Merritt Publishing). Suppose you are a sociologist in Rock Springs, Wyoming, and a random sample of police files showed that of 32 arrests last month, 24 were of males aged 15 to 34 years. Use a 1% level of significance to test the claim that the population proportion of such arrests in Rock Springs is different from 70%.

10. *College Athletics: Graduation Rate* Women athletes at the University of Colorado, Boulder, have a long-term graduation rate of 67% (Source: *Chronicle of Higher Education*). Over the past several years, a random sample of 38 women athletes at the school showed that 21 eventually graduated. Does this indicate that the population proportion of women athletes who graduate from the University of Colorado, Boulder, is now less than 67%? Use a 5% level of significance.

11. *Highway Accidents: DUI* The U.S. Department of Transportation, National Highway Traffic Safety Administration, reported that 77% of all fatally injured automobile drivers were intoxicated. A random sample of 27 records of automobile driver fatalities in Kit Carson County, Colorado, showed that 15 involved an intoxicated driver. Do these data indicate that the population proportion of driver fatalities related to alcohol is less than 77% in Kit Carson County? Use $\alpha = 0.01$.

12. *Preference: Color* What is your favorite color? A large survey of countries, including the United States, China, Russia, France, Turkey, Kenya, and others, indicated that most people prefer the color blue. In fact, about 24% of the population claim blue as their favorite color (Reference: Study by J. Bunge and A. Freeman-Gallant, Statistics Center, Cornell University). Suppose a random sample of $n = 56$ college students were surveyed and $r = 12$ of

them said that blue is their favorite color. Does this information imply that the color preference of all college students is different (either way) from that of the general population? Use $\alpha = 0.05$.

13. *Wildlife: Wolves* The following is based on information from *The Wolf in the Southwest: The Making of an Endangered Species* by David E. Brown (University of Arizona Press). Before 1918, the proportion of female wolves in the general population of all southwestern wolves was about 50%. However, after 1918, southwestern cattle ranchers began a widespread effort to destroy wolves. In a recent sample of 34 wolves, there were only 10 females. One theory is that male wolves tend to return sooner than females to their old territories where their predecessors were exterminated. Do these data indicate that the population proportion of female wolves is now less than 50% in the region? Use $\alpha = 0.01$.

14. *Fishing: Northern Pike* Athabasca Fishing Lodge is located on Lake Athabasca in northern Canada. In one of its recent brochures, the lodge advertises that 75% of its guests catch northern pike over 20 pounds. Suppose that last summer 64 out of a random sample of 83 guests did, in fact, catch northern pike weighing over 20 pounds. Does this indicate that the population proportion of guests who catch pike over 20 pounds is different from 75% (either higher or lower)? Use $\alpha = 0.05$.

15. *Plato's* **Republic:** *Syllable Patterns* Prose rhythm is characterized by the occurrence of five-syllable sequences in long passages of text. This characterization may be used to assess the similarity among passages of text and sometimes the identity of authors. The following information is based on an article by D. Wishart and S. V. Leach appearing in *Computer Studies of the Humanities and Verbal Behavior* (Vol. 3, pp. 90–99). Syllables were categorized as long or short. On analyzing Plato's *Republic,* Wishart and Leach found that about 26.1% of the five-syllable sequences are of the type in which two are short and three are long. Suppose that Greek archaeologists have found an ancient manuscript dating back to Plato's time (about 427–347 B.C.). A random sample of 317 five-syllable sequences from the newly discovered manuscript showed that 61 are of the type two short and three long. Do the data indicate that the population proportion of this type of five-syllable sequence is different (either way) from the text of Plato's *Republic*? Use $\alpha = 0.01$.

16. *Plato's* **Dialogues:** *Prose Rhythm* *Symposium* is part of a larger work referred to as Plato's *Dialogues*. Wishart and Leach (see source in Problem 15) found that about 21.4% of five-syllable sequences in *Symposium* are of the type in which four are short and one is long. Suppose an antiquities store in Athens has a very old manuscript that the owner claims is part of Plato's *Dialogues*. A random sample of 493 five-syllable sequences from this manuscript showed that 136 were of the type four short and one long. Do the data indicate that the population proportion of this type of five-syllable sequence is higher than that found in Plato's *Symposium*? Use $\alpha = 0.01$.

17. *Consumers: Product Loyalty* *USA Today* reported that about 47% of the general consumer population in the United States is loyal to the automobile manufacturer of their choice. Suppose Chevrolet did a study of a random sample of 1006 Chevrolet owners and found that 490 said they would buy another Chevrolet. Does this indicate that the population proportion of consumers loyal to Chevrolet is more than 47%? Use $\alpha = 0.01$.

18. *Supermarket: Prices* *Harper's Index* reported that 80% of all supermarket prices end in the digit 9 or 5. Suppose you check a random sample of 115 items in a supermarket and find that 88 have prices that end in 9 or 5. Does this indicate that less than 80% of the prices in the store end in the digits 9 or 5? Use $\alpha = 0.05$.

19. *Medical: Hypertension* This problem is based on information taken from *The Merck Manual* (a reference manual used in most medical and nursing schools). Hypertension is defined as a blood pressure reading over 140 mm Hg systolic and/or over 90 mm Hg diastolic. Hypertension, if not corrected, can cause long-term health problems. In the college-age population (18–24 years), about 9.2% have hypertension. Suppose that a blood donor program is taking place in a college dormitory this week (final exams week). Before each student gives blood, the nurse takes a blood pressure reading. Of 196 donors, it is found that 29 have hypertension. Do these data indicate that the population proportion of students with hypertension during final exams week is higher than 9.2%? Use a 5% level of significance.

20. *Medical: Hypertension* Diltiazem is a commonly prescribed drug for hypertension (see source in Problem 19). However, diltiazem causes headaches in about 12% of patients using the drug. It is hypothesized that regular exercise might help reduce the headaches. If a random sample of 209 patients using diltiazem exercised regularly and only 16 had headaches, would this indicate a reduction in the population proportion of patients having headaches? Use a 1% level of significance.

21. *Myers–Briggs: Extroverts* Are most student government leaders extroverts? According to Myers–Briggs estimates, about 82% of college student government leaders are extroverts (Source: *Myers–Briggs Type Indicator Atlas of Type Tables*). Suppose that a Myers–Briggs personality preference test was given to a random sample of 73 student government leaders attending a large national leadership conference and that 56 were found to be extroverts. Does this indicate that the population proportion of extroverts among college student government leaders is different (either way) from 82%? Use $\alpha = 0.01$.

22. *Critical Region Method: Testing Proportions* Solve Problem 9 using the critical region method of testing. Since the sampling distribution of $\hat{p}$ is the normal distribution, you can use critical values from the standard normal distribution as shown in Figure 8-8 or part (c) of Table 5, Appendix II. Compare your conclusions with the conclusions obtained by using the P-value method. Are they the same?

23. *Critical Region Method: Testing Proportions* Solve Problem 11 using the critical region method of testing. *Hint:* See Problem 22. Compare your conclusions with the conclusions obtained by using the *P*-value method. Are they the same?

24. *Critical Region Method: Testing Proportions* Solve Problem 17 using the critical region method of testing. *Hint:* See Problem 22. Compare your conclusions with the conclusions obtained by using the *P*-value method. Are they the same?

SECTION **8.4**

Tests Involving Paired Differences (Dependent Samples)

FOCUS POINTS

- Identify paired data and dependent samples.
- Explain the advantages of paired data tests.
- Compute differences and the sample test statistic.
- Estimate the *P*-value and conclude the test.

PAIRED DATA

Paired data

Creating data pairs

Many statistical applications use *paired data* samples to draw conclusions about the difference between two population means. *Data pairs* occur very naturally in "before and after" situations, where the *same* object or item is measured both before and after a treatment. Applied problems in social science, natural science, and business administration frequently involve a study of matching pairs. Psychological studies of identical twins; biological studies of plant growth on plots of land matched for soil type, moisture, and sun exposure; and business studies on sales of matched inventories are examples of paired data studies.

When working with paired data, it is very important to have a definite and uniform method of creating data pairs that clearly utilizes a natural matching of characteristics. The next example and Guided Exercise demonstrate this feature.

EXAMPLE 8

PAIRED DATA

A shoe manufacturer claims that among the general population of adults in the United States, the average length of the left foot is longer than that of the right. To compare the average length of the left foot with that of the right, we can take a random sample of 15 U.S. adults and measure the length of the left foot and then the length of the right foot for each person in the sample. Is there a natural way of pairing the measurements? How many pairs will we have?

SOLUTION: In this case, we can pair each left-foot measurement with the same person's right-foot measurement. The person serves as the "matching link" between the two distributions. We will have 15 pairs of measurements.

GUIDED EXERCISE 6

PAIRED DATA

A psychologist has developed a series of exercises called the Instrumental Enrichment (IE) Program, which he claims is useful in overcoming cognitive deficiencies in mentally retarded children. To test the program, extensive statistical tests are being conducted. In one experiment, a random sample of 10-year-old students with IQ scores

Continued

GUIDED EXERCISE 6 *continued*

below 80 was selected. An IQ test was given to these students before they spent 2 years in an IE Program, and an IQ test was given to the same students after the program.

(a) On what basis can you pair the IQ scores?

→ Take the "before and after" IQ scores of each individual student.

(b) If there were 20 students in the sample, how many data pairs would you have?

→ Twenty data pairs. Note that there would be 40 IQ scores, but only 20 pairs.

COMMENT To compare two populations, we cannot always employ paired data tests, but when we can, what are the advantages? Using matched or paired data often can reduce the danger of introducing extraneous or uncontrollable factors into our sample measurements because the matched or paired data have essentially the *same* characteristics except for the *one* characteristic that is being measured. Furthermore, it can be shown that pairing data has the theoretical effect of reducing measurement variability (i.e., variance), which increases the accuracy of statistical conclusions.

When we wish to compare the means of two samples, the first item to be determined is whether or not there is a natural pairing between the data in the two samples. Again, data pairs are created from "before and after" situations, or from matching data by using studies of the same object, or by a process of taking measurements of closely matched items.

Testing the differences d

When testing *paired* data, we take the difference d of the data pairs *first* and look at the mean difference $\bar{d}$. Then we use a test on $\bar{d}$. Theorem 8.1 provides the basis for our work with paired data.

THEOREM 8.1 Consider a random sample of n data pairs. Suppose the differences d between the first and second members of each data pair are (approximately) normally distributed, with population mean μ_d. Then the t values

$$t = \frac{\bar{d} - \mu_d}{s_d/\sqrt{n}}$$

where $\bar{d}$ is the sample mean of the d values, n is the number of data pairs, and

$$s_d = \sqrt{\frac{\Sigma(d - \bar{d})^2}{n - 1}}$$

is the sample standard deviation of the d values, follow a Student's t distribution with degrees of freedom $d.f. = n - 1$.

Hypotheses for testing the mean of paired differences

When testing the mean of the differences of paired data values, the null hypothesis is that there is no difference among the pairs. That is, the mean of the differences μ_d is zero.

$$H_0: \mu_d = 0$$

The alternate hypothesis depends on the problem and can be

$H_1: \mu_d < 0$	$H_1: \mu_d > 0$	$H_1: \mu_d \neq 0$
(left-tailed)	(right-tailed)	(two-tailed)

Sample test statistic

For paired difference tests, we make our decision regarding H_0 according to the evidence of the sample mean $\bar{d}$ of the differences of measurements. By Theorem 8.1, we convert the sample test statistic $\bar{d}$ to a t value using the formula

$$t = \frac{\bar{d} - \mu_d}{(s_d/\sqrt{n})} \text{ with } d.f. = n - 1$$

where s_d = sample standard deviation of the differences d
n = number of data pairs
$\mu_d = 0$, as specified in H_0

P-values from Table 6 of Appendix II

To find the P-value (or an interval containing the P-value) corresponding to the test statistic t computed from $\bar{d}$, we use the Student's t distribution table (Table 6, Appendix II). Recall from Section 8.2 that we find the test statistic t (or, if t is negative, $|t|$) in the row headed by $d.f. = n - 1$, where n is the number of data pairs. The P-value for the test statistic is the column entry in the *one-tail area* row for one-tailed tests (right or left). For two-tailed tests, the P-value is the column entry in the *two-tail area* row. Usually the exact test statistic t is not in the table, so we obtain an interval that contains the P-value by using adjacent entries in the table. Table 8-7 gives the basic structure for using the Student's t distribution table to find the P-value or an interval containing the P-value.

TABLE 8-7 Using Student's t Distribution Table for P-values

For one-tailed tests:	one-tail area	P-value	P-value
For two-tailed tests:	two-tail area	P-value	P-value
Use row header	$d.f. = n - 1$	↑ Find t value	

With the preceding information, you are now ready to test paired differences. First let's summarize the procedure.

PROCEDURE

HOW TO TEST PAIRED DIFFERENCES USING THE STUDENT'S t DISTRIBUTION

Requirements

Obtain a simple random sample of n matched data pairs A, B. Let d be a random variable representing the difference between the values in a matched data pair. Compute the sample mean $\bar{d}$ and sample standard deviation s_d. If you can assume that d has a normal distribution or simply has a mound-shaped, symmetric distribution, then any sample size n will work. If you cannot assume this, then use a sample size $n \geq 30$.

Procedure

1. Use the *null hypothesis* of *no difference*, H_0: $\mu_d = 0$. In the context of the application, choose the *alternate hypothesis* to be H_1: $\mu_d > 0$, H_1: $\mu_d < 0$, or H_1: $\mu_d \neq 0$. Set the *level of significance* α.
2. Use $\bar{d}$, s_d, the sample size n, and $\mu_d = 0$ from the null hypothesis to compute the *sample test statistic*

$$t = \frac{\bar{d} - 0}{\frac{s_d}{\sqrt{n}}} = \frac{\bar{d}\sqrt{n}}{s_d}$$

with degrees of freedom $d.f. = n - 1$.

Continued

Problem 22 at the end of this section shows how to find confidence intervals for μ_d.

3. Use the Student's t distribution and the type of test, one-tailed or two-tailed, to find (or estimate) the *P-value* corresponding to the test statistic.
4. *Conclude* the test. If P-value $\leq \alpha$, then reject H_0. If P-value $> \alpha$, then do not reject H_0.
5. *Interpret your conclusion* in the context of the application.

EXAMPLE 9

PAIRED DIFFERENCE TEST

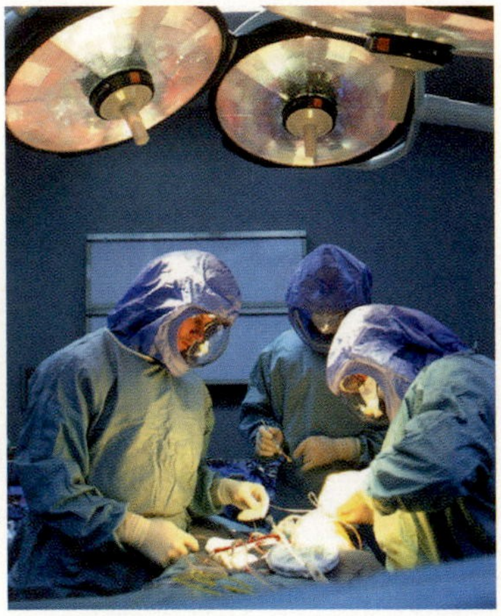

The Image Works

A team of heart surgeons at Saint Ann's Hospital knows that many patients who undergo corrective heart surgery have a dangerous buildup of anxiety before their scheduled operations. The staff psychiatrist at the hospital has started a new counseling program intended to reduce this anxiety. A test of anxiety is given to patients who know they must undergo heart surgery. Then each patient participates in a series of counseling sessions with the staff psychiatrist. At the end of the counseling sessions, each patient is retested to determine anxiety level. Table 8-8 on the next page, indicates the results for a random sample of nine patients. Higher scores mean higher levels of anxiety. Assume the distribution of differences is mound-shaped and symmetric.

From the given data, can we conclude that the counseling sessions reduce anxiety? Use a 0.01 level of significance.

SOLUTION: Before we answer this question, let us notice two important points: (1) We have a *random sample* of nine patients, and (2) we have a *pair* of measurements taken on the same patient before and after counseling sessions. In our problem, the sample size is $n = 9$ pairs (i.e., patients), and the d values are found in the fourth column of Table 8-8.

(a) Note the level of significance and set the hypotheses.
In the problem statement, $\alpha = 0.01$. We want to test the claim that the counseling sessions reduce anxiety. This means that the anxiety level before counseling is expected to be higher than the anxiety level after counseling. In symbols, $d = B - A$ should tend to be positive, and the population mean of differences μ_d also should be positive. Therefore, we have

$$H_0: \mu_d = 0 \text{ and } H_1: \mu_d > 0$$

(b) ***Check Requirements*** Is it appropriate to use a Student's t distribution for the sample test statistic? Explain. What degrees of freedom are used?
We have a random sample of paired differences d. Under the assumption that the d distribution is mound-shaped and symmetric, we use a Student's t distribution with $d.f. = n - 1 = 9 - 1 = 8$.

(c) Find the sample test statistic $\bar{d}$ and convert it to a corresponding test statistic t.
First we need to compute $\bar{d}$ and s_d. Using formulas or a calculator and the d values shown in Table 8-8, we find that

$$\bar{d} \approx 33.33 \text{ and } s_d \approx 22.92$$

Using these values together with $n = 9$ and $\mu_d = 0$, we have

$$t = \frac{\bar{d} - 0}{(s_d/\sqrt{n})} \approx \frac{33.33}{22.92/\sqrt{9}} \approx 4.363$$

TABLE 8-8

Patient	*B* Score Before Counseling	*A* Score After Counseling	$d = B - A$ Difference
Jan	121	76	45
Tom	93	93	0
Diane	105	64	41
Barbara	115	117	−2
Mike	130	82	48
Bill	98	80	18
Frank	142	79	63
Carol	118	67	51
Alice	125	89	36

(d) Find the *P*-value for the test statistic and sketch the *P*-value on the *t* distribution. Since we have a right-tailed test, the *P*-value is the area to the right of $t = 4.363$, as shown in Figure 8-13. In Table 6 of Appendix II, we find an interval containing the *P*-value. Use entries from the row headed by $d.f. = n - 1 = 9 - 1 = 8$. The test statistic $t = 4.363$ falls between 3.355 and 5.041. The *P*-value for the sample *t* falls between the corresponding one-tail areas 0.005 and 0.0005. (See Table 8-9, excerpt from Table 6, Appendix II.)

$$0.0005 < P\text{-value} < 0.005$$

TABLE 8-9 **Excerpt from Student's *t* Distribution Table(Table 6, Appendix II)**

✓ one-tail area	0.005		0.0005
two-tail area	0.010		0.0010
d.f. = 8	3.355	↑ Sample $t = 4.363$	5.041

FIGURE 8-13

P-value

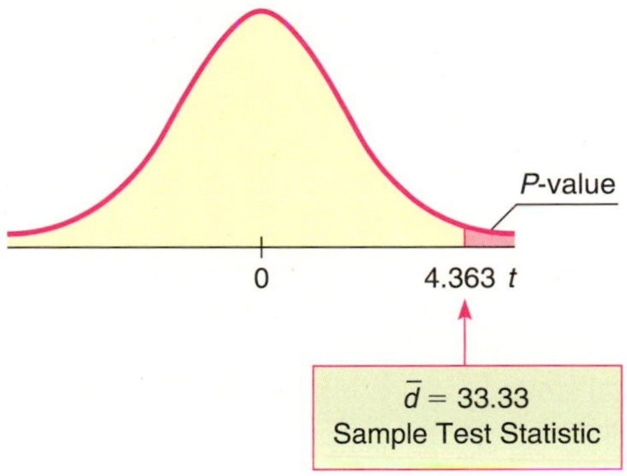

(e) Conclude the test.

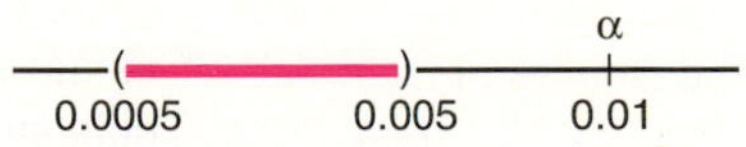

Since the interval containing the P-value lies to the left of $\alpha = 0.01$, we reject H_0. *Note:* Using the raw data and software, P-value ≈ 0.0012.

(f) ***Interpretation*** Interpret the results in the context of the application.
At the 1% level of significance, we conclude that the counseling sessions reduce the average anxiety level of patients about to undergo corrective heart surgery.

The problem we have just solved is a paired difference problem of the "before and after" type. The next guided exercise demonstrates a paired difference problem of the "matched pair" type.

GUIDED EXERCISE 7 PAIRED DIFFERENCE TEST

Do educational toys make a difference in the age at which a child learns to read? To study this question, researchers designed an experiment in which one group of preschool children spent 2 hours each day (for 6 months) in a room well supplied with "educational" toys such as alphabet blocks, puzzles, ABC readers, coloring books featuring letters, and so forth. A control group of children spent 2 hours a day for 6 months in a "noneducational" toy room. It was anticipated that IQ differences and home environment might be uncontrollable factors unless identical twins could be used. Therefore, six pairs of identical twins of preschool age were randomly selected. From each pair, one member was randomly selected to participate in the experimental (i.e., educational toy room) group and the other in the control (i.e., noneducational toy room) group. For each twin, the data item recorded is the age in months at which the child began reading at the primary level (Table 8-10). Assume the distribution of differences is mound-shaped and symmetric.

TABLE 8-10 **Reading Ages for Identical Twins (in Months)**

Twin Pair	Experimental Group B = Reading Age	Control Group A = Reading Age	Difference $d = B - A$
1	58	60	
2	61	64	
3	53	52	
4	60	65	
5	71	75	
6	62	63	

(a) Compute the entries in the $d = B - A$ column of Table 8-10. Using formulas for the mean and sample standard deviation or a calculator with mean and sample standard deviation keys, compute $\bar{d}$ and s_d.

Pair	$d = B - A$
1	−2
2	−3
3	1
4	−5
5	−4
6	−1

$\bar{d} \approx -2.33$
$s_d \approx 2.16$

(b) What is the null hypothesis?

H_0: $\mu_d = 0$

(c) To test the claim that the experimental group learned to read at a *different age* (either younger or older), what should the alternate hypothesis be?

H_1: $\mu_d \neq 0$

(d) ***Check Requirements*** What distribution does the sample test statistic t follow? Find the degrees of freedom.

Because the d distribution is mound-shaped and symmetric and we have a random sample of $n = 6$ paired differences, the sample test statistic follows a Student's t distribution with $d.f. = n - 1 = 6 - 1 = 5$.

Continued

GUIDED EXERCISE 7 *continued*

(e) Convert the sample test statistic $\bar{d}$ to a t value.

Using $\mu_d = 0$ from H_0, $\bar{d} = -2.33$, $n = 6$ and $s_d = 2.16$, we get

$$t = \frac{\bar{d} - \mu_d}{(s_d/\sqrt{n})} \approx \frac{-2.33 - 0}{(2.16/\sqrt{6})} \approx -2.642$$

(f) When we use Table 6 of Appendix II to find an interval containing the P-value, do we use one-tail or two-tail areas? Why? Sketch a figure showing the P-value. Find an interval containing the P-value.

This is a two-tailed test, so we use two-tail areas.

FIGURE 8-14 *P*-value

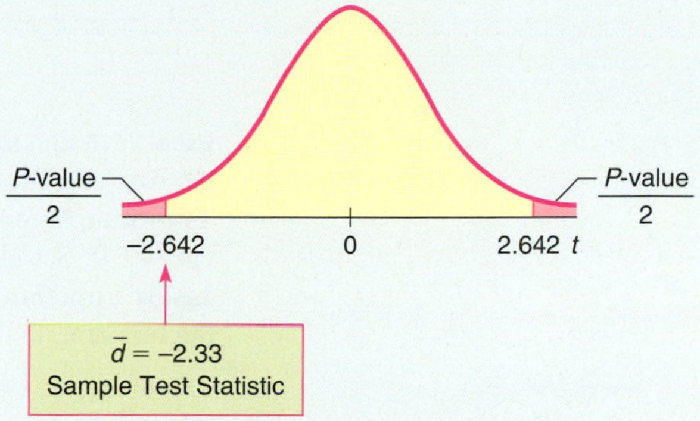

TABLE 8-11 **Excerpt from Student's *t* Table**

one-tail area	**0.025**	**0.010**
✓ **two-tail area**	**0.050**	**0.020**
d.f. = 5	2.571	3.365
	↑ Sample t = 2.642	

The sample t is between 2.571 and 3.365.
$0.020 < P\text{-value} < 0.050$
Note: Using the raw data and software, P-value ≈ 0.0457.

(g) Using $\alpha = 0.05$, do we reject or fail to reject H_0?

Since the interval containing the P-value has values that are all smaller than or equal to 0.05, we reject H_0.

(h) ***Interpretation*** What do the results mean in the context of this application?

At the 5% level of significance, the experiment indicates that educational toys make a difference in the age at which a child learns to read.

LOOKING FORWARD

The test for paired differences used in this section is called a *parametric test*. Such tests usually require certain assumptions, such as a normal distribution or a large sample size. In Section 11.1, you will find the sign test for matched pairs. This is a *nonparametric test*. Such tests are useful when you cannot make assumptions about the population distribution. The disadvantage of nonparametric tests is that they tend to accept the null hypothesis more often than they should. That is, they are less sensitive tests. Which type of test should you use? If it is reasonable to assume that the underlying population is normal (or at least mound-shaped and symmetric), or if you have a large sample size, then use the more powerful parametric test described in this section.

TECH NOTES

Both Excel 2010 and Minitab support paired difference tests directly. On the TI-84Plus/TI-83Plus/TI-*n*spire calculators, construct a column of differences and then do a *t* test on the data in that column. For each technology, be sure to relate the alternate hypothesis to the "before and after" assignments. All the displays show the results for the data of Guided Exercise 7.

TI-84Plus/TI-83Plus/TI-*n*spire (with TI-84Plus keypad) Enter the "before" data in column L1 and the "after" data in column L2. Highlight L3, type L1 − L2, and press Enter. The column L3 now contains the $B - A$ differences. To conduct the test, press **STAT,** select **TESTS,** and use option **2:T-Test.** Note that the letter x is used in place of d.

L1	L2	L3 3
58	60	-2
61	64	-3
53	52	1
60	65	-5
71	75	-4
62	63	-1
L3(7) =		

```
T-Test
 μ≠0
 t=-2.645751311
 p=.0456591238
 x̄=-2.333333333
 Sx=2.160246899
 n=6
```

Excel 2010 Enter the data in two columns. On the home ribbon, click the **Data** tab. In the Analysis Group, select **Data Analysis**. In the dialogue box, select **t-Test: Paired Two-Sample for Means.** Fill in the dialogue box with the hypothesized mean difference of 0. Set alpha. Another way to conduct a paired-difference t test is to press **Insert Function** (f_x) and select **Statistical** for the category and then **T. TEST** for the function. In the dialogue box, use **1** (paired) for the type of test.

	B	C	D
	t-Test: Paired Two Sample for Means		
		Variable 1	*Variable 2*
	Mean	60.83333	63.16666667
	Variance	34.96667	55.76666667
	Observations	6	6
	Pearson Correlation	0.974519	
	Hypothesized Mean Difference	0	
	df	5	
	t Stat	−2.64575	
	P(T<=t) one-tail	0.02283	
	t Critical one-tail	2.015049	
P-value →	P(T<=t) two-tail	0.045659	
	t Critical two-tail	2.570578	

Minitab Enter the data in two columns. Use the menu selection **Stat ➤ Basic Statistics ➤ Paired *t*.** Under Options, set the null and alternate hypotheses.

```
Paired T-Test and Confidence Interval
Paired T for B − A
                N       Mean          StDevSE        Mean
B               6       60.83         5.91           2.41
A               6       63.17         7.47           3.05
Difference      6       −2.333        2.160
0.882
95% CI for mean difference:(−4.601, −0.066)
T-Test of mean difference = 0 (vs not = 0):
      T-Value = −2.65 P-Value = 0.046
```

EXAMPLE 10

CRITICAL REGION METHOD

Let's revisit Guided Exercise 7 regarding educational toys and reading age and conclude the test using the critical region method. Recall that there were six pairs of twins. One twin of each set was given educational toys and the other was not. The difference d in reading ages for each pair of twins was measured, and $\alpha = 0.05$.

SOLUTION: From Guided Exercise 7, we have

$$H_0: \mu_d = 0 \text{ and } H_1: \mu_d \neq 0$$

We computed $\bar{d} \approx -2.33$ with corresponding sample test statistic $t \approx -2.642$.

(a) Find the critical values for $\alpha = 0.05$.
Since the number of pairs is $n = 6$, $d.f. = n - 1 = 5$. In the Student's t distribution table (Table 6, Appendix II), look in the row headed by 5. To find the column, locate $\alpha = 0.05$ in the *two-tail area* row, since we have a two-tailed test. The critical values are $\pm t_0 = \pm 2.571$.

(b) Sketch the critical regions and place the t value of the sample test statistic $\bar{d}$ on the sketch. Conclude the test. Compare the result to the result given by the P-value method of Guided Exercise 7.
Since the sample test statistic falls in the critical region (see Figure 8-15), we reject H_0 at the 5% level of significance. At this level, educational toys seem to make a difference in reading age. Notice that this conclusion is consistent with the conclusion obtained using the P-value.

FIGURE 8-15

Critical Region, with $\alpha = 0.05$, $d.f. = 5$

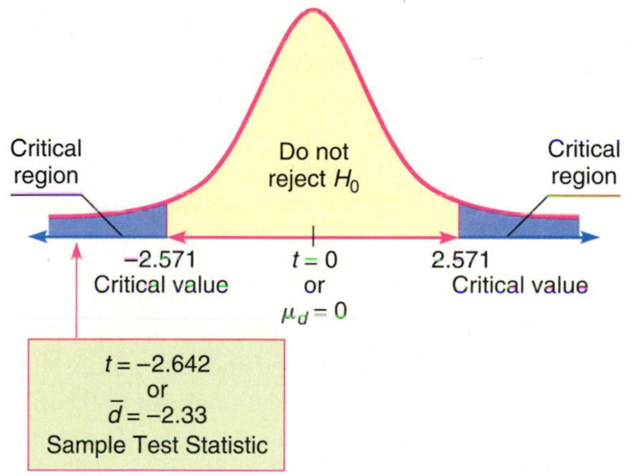

VIEWPOINT | DUI

DUI usually means "driving under the influence" of alcohol, *but driving under the influence of* sleep loss *can be just as dangerous. Researchers in Australia have found that after staying awake for 24 hours straight, a person will be about as impaired as if he or she had had enough alcohol to be legally drunk in most U.S. states (Source:* Rocky Mountain News*). Using driver-simulation exams and statistical tests (paired difference tests) found in this section, it is possible to show that the null hypothesis $H_0: \mu_d = 0$ cannot be rejected. Or, put another way, the average level of impairment for a given individual from alcohol (at the DUI level) is about the same as the average level of impairment from sleep loss (24 hours without sleep).*

SECTION 8.4 PROBLEMS

1. *Statistical Literacy* Are data that can be paired independent or dependent?

2. *Statistical Literacy* Consider a set of data pairs. What is the first step in processing the data for a paired differences test? What is the formula for the sample test statistic t? Describe each symbol used in the formula.

3. *Statistical Literacy* When testing the difference of means for paired data, what is the null hypothesis?

4. *Statistical Literacy* When conducting a paired differences test, what is the value of n?

5. *Statistical Literacy* When using a Student's t distribution for a paired differences test with n data pairs, what value do you use for the degrees of freedom?

6. *Critical Thinking* Alisha is conducting a paired differences test for a "before (B score) and after (A score)" situation. She is interested in testing whether the average of the "before" scores is higher than that of the "after" scores.
   (a) To use a right-tailed test, how should Alisha construct the differences between the "before" and "after" scores?
   (b) To use a left-tailed test, how should she construct the differences between the "before" and "after" scores?

7. *Basic Computation: Paired Differences Test* For a random sample of 36 data pairs, the sample mean of the differences was 0.8. The sample standard deviation of the differences was 2. At the 5% level of significance, test the claim that the population mean of the differences is different from 0.
   (a) *Check Requirements* Is it appropriate to use a Student's t distribution for the sample test statistic? Explain. What degrees of freedom are used?
   (b) State the hypotheses.
   (c) Compute the sample test statistic.
   (d) Estimate the P-value of the sample test statistic.
   (e) Do we reject or fail to reject the null hypothesis? Explain.
   (f) *Interpretation* What do your results tell you?

8. *Basic Computation: Paired Differences Test* For a random sample of 20 data pairs, the sample mean of the differences was 2. The sample standard deviation of the differences was 5. Assume that the distribution of the differences is mound-shaped and symmetric. At the 1% level of significance, test the claim that the population mean of the differences is positive.
   (a) *Check Requirements* Is it appropriate to use a Student's t distribution for the sample test statistic? Explain. What degrees of freedom are used?
   (b) State the hypotheses.
   (c) Compute the sample test statistic.
   (d) Estimate the P-value of the sample test statistic.
   (e) Do we reject or fail to reject the null hypothesis? Explain.
   (f) *Interpretation* What do your results tell you?

For Problems 9–21 assume that the distribution of differences d is mound-shaped and symmetric.

Please provide the following information for Problems 9–21.
(a) What is the level of significance? State the null and alternate hypotheses. Will you use a left-tailed, right-tailed, or two-tailed test?
(b) *Check Requirements* What sampling distribution will you use? What assumptions are you making? Compute the value of the sample test statistic.
(c) Find (or estimate) the P-value. Sketch the sampling distribution and show the area corresponding to the P-value.

(d) Based on your answers in parts (a) to (c), will you reject or fail to reject the null hypothesis? Are the data statistically significant at level α?
(e) *Interpret* your conclusion in the context of the application.

In these problems, assume that the distribution of differences is approximately normal.

Note: For degrees of freedom *d.f.* not in the Student's t table, use the closest *d.f.* that is *smaller.* In some situations, this choice of *d.f.* may increase the P-value by a small amount and therefore produce a slightly more "conservative" answer.

9. *Business: CEO Raises* Are America's top chief executive officers (CEOs) really worth all that money? One way to answer this question is to look at row *B,* the annual company percentage increase in revenue, versus row *A,* the CEO's annual percentage salary increase in that same company (Source: *Forbes,* Vol. 159, No. 10). A random sample of companies such as John Deere & Co., General Electric, Union Carbide, and Dow Chemical yielded the following data:

B: Percent increase for company	24	23	25	18	6	4	21	37
A: Percent increase for CEO	21	25	20	14	−4	19	15	30

Do these data indicate that the population mean percentage increase in corporate revenue (row *B*) is different from the population mean percentage increase in CEO salary? Use a 5% level of significance.

10. *Fishing: Shore or Boat?* Is fishing better from a boat or from the shore? Pyramid Lake is located on the Paiute Indian Reservation in Nevada. Presidents, movie stars, and people who just want to catch fish go to Pyramid Lake for really large cutthroat trout. Let row *B* represent hours per fish caught fishing from the shore, and let row *A* represent hours per fish caught using a boat. The following data are paired by month from October through April (Source: *Pyramid Lake Fisheries,* Paiute Reservation, Nevada).

	Oct.	Nov.	Dec.	Jan.	Feb.	March	April
B: Shore	1.6	1.8	2.0	3.2	3.9	3.6	3.3
A: Boat	1.5	1.4	1.6	2.2	3.3	3.0	3.8

Use a 1% level of significance to test if there is a difference in the population mean hours per fish caught using a boat compared with fishing from the shore.

11. *Ecology: Rocky Mountain National Park* The following is based on information taken from *Winter Wind Studies in Rocky Mountain National Park* by D. E. Glidden (Rocky Mountain Nature Association). At five weather stations on Trail Ridge Road in Rocky Mountain National Park, the peak wind gusts (in miles per hour) for January and April are recorded below.

Weather Station	1	2	3	4	5
January	139	122	126	64	78
April	104	113	100	88	61

Does this information indicate that the peak wind gusts are higher in January than in April? Use $\alpha = 0.01$.

12. *Wildlife: Highways* The western United States has a number of four-lane interstate highways that cut through long tracts of wilderness. To prevent car accidents with wild animals, the highways are bordered on both sides with 12-foot-high woven wire fences. Although the fences prevent accidents, they also disturb the winter migration pattern of many animals. To compensate for this disturbance, the highways have frequent wilderness underpasses designed for exclusive use by deer, elk, and other animals.

In Colorado, there is a large group of deer that spend their summer months in a region on one side of a highway and survive the winter months in a lower region on the other side. To determine if the highway has disturbed deer migration to the winter feeding area, the following data were gathered on a random sample of 10 wilderness districts in the winter feeding area. Row B represents the average January deer count for a 5-year period before the highway was built, and row A represents the average January deer count for a 5-year period after the highway was built. The highway department claims that the January population has not changed. Test this claim against the claim that the January population has dropped. Use a 5% level of significance. Units used in the table are hundreds of deer.

Wilderness District	1	2	3	4	5	6	7	8	9	10
B: Before highway	10.3	7.2	12.9	5.8	17.4	9.9	20.5	16.2	18.9	11.6
A: After highway	9.1	8.4	10.0	4.1	4.0	7.1	15.2	8.3	12.2	7.3

13. *Wildlife: Wolves* In environmental studies, sex ratios are of great importance. Wolf society, packs, and ecology have been studied extensively at different locations in the United States and foreign countries. Sex ratios for eight study sites in northern Europe are shown in the following table. (based on *The Wolf* by L. D. Mech, University of Minnesota Press).

Gender Study of Large Wolf Packs

Location of Wolf Pack	% Males (Winter)	% Males (Summer)
Finland	72	53
Finland	47	51
Finland	89	72
Lapland	55	48
Lapland	64	55
Russia	50	50
Russia	41	50
Russia	55	45

It is hypothesized that in winter, "loner" males (not present in summer packs) join the pack to increase survival rate. Use a 5% level of significance to test the claim that the average percentage of males in a wolf pack is higher in winter.

14. *Demographics: Birthrate and Death Rate* In the following data pairs, A represents birthrate and B represents death rate per 1000 resident population. The data are paired by counties in the Midwest. A random sample of 16 counties gave the following information (Reference: *County and City Data Book*, U.S. Department of Commerce).

A:	12.7	13.4	12.8	12.1	11.6	11.1	14.2	15.1
B:	9.8	14.5	10.7	14.2	13.0	12.9	10.9	10.0

A:	12.5	12.3	13.1	15.8	10.3	12.7	11.1	15.7
B:	14.1	13.6	9.1	10.2	17.9	11.8	7.0	9.2

Do the data indicate a difference (either way) between population average birthrate and death rate in this region? Use $\alpha = 0.01$.

15. *Navajo Reservation: Hogans* The following data are based on information taken from the book *Navajo Architecture: Forms, History, Distributions* by S. C. Jett and V. E. Spencer (University of Arizona Press). A survey of houses and traditional hogans was made in a number of different regions of the modern Navajo Indian Reservation. The following table is the result of a random sample of eight regions on the Navajo Reservation.

Area on Navajo Reservation	Number of Inhabited Houses	Number of Inhabited Hogan
Bitter Springs	18	13
Rainbow Lodge	16	14
Kayenta	68	46
Red Mesa	9	32
Black Mesa	11	15
Canyon de Chelly	28	47
Cedar Point	50	17
Burnt Water	50	18

Does this information indicate that the population mean number of inhabited houses is greater than that of hogans on the Navajo Reservation? Use a 5% level of significance.

16. *Archaeology: Stone Tools* The following is based on information taken from *Bandelier Archaeological Excavation Project: Summer 1990 Excavations at Burnt Mesa Pueblo and Casa del Rito,* edited by T. A. Kohler (Washington State University, Department of Anthropology). The artifact frequency for an excavation of a kiva in Bandelier National Monument gave the following information.

Stratum	Flaked Stone Tools	Nonflaked Stone Tools
1	7	3
2	3	2
3	10	1
4	1	3
5	4	7
6	38	32
7	51	30
8	25	12

Does this information indicate that there tend to be more flaked stone tools than nonflaked stone tools at this excavation site? Use a 5% level of significance.

17. *Economics: Cost of Living Index* In the following data pairs, A represents the cost of living index for housing and B represents the cost of living index for groceries. The data are paired by metropolitan areas in the United States. A random sample of 36 metropolitan areas gave the following information (Reference: *Statistical Abstract of the United States,* 121st Edition).

A:	132	109	128	122	100	96	100	131	97
B:	125	118	139	104	103	107	109	117	105

A:	120	115	98	111	93	97	111	110	92
B:	110	109	105	109	104	102	100	106	103

A:	85	109	123	115	107	96	108	104	128
B:	98	102	100	95	93	98	93	90	108

A:	121	85	91	115	114	86	115	90	113
B:	102	96	92	108	117	109	107	100	95

i. Let d be the random variable $d = A - B$. Use a calculator to verify that $\bar{d} \approx 2.472$ and $s_d \approx 12.124$.

ii. Do the data indicate that the U.S. population mean cost of living index for housing is higher than that for groceries in these areas? Use $\alpha = 0.05$.

18. *Economics: Cost of Living Index* In the following data pairs, A represents the cost of living index for utilities and B represents the cost of living index for transportation. The data are paired by metropolitan areas in the United States. A random sample of 46 metropolitan areas gave the following information (Reference: *Statistical Abstract of the United States,* 121st Edition).

A:	90	84	85	106	83	101	89	125	105
B:	100	91	103	103	109	109	94	114	113

A:	118	133	104	84	80	77	90	92	90
B:	120	130	117	109	107	104	104	113	101

A:	106	95	110	112	105	93	119	99	109
B:	96	109	103	107	103	102	101	86	94

A:	109	113	90	121	120	85	91	91	97
B:	88	100	104	119	116	104	121	108	86

A:	95	115	99	86	88	106	80	108	90	87
B:	100	83	88	103	94	125	115	100	96	127

i. Let d be the random variable $d = A - B$. Use a calculator to verify that $\bar{d} \approx -5.739$ and $s_d \approx 15.910$.

ii. Do the data indicate that the U.S. population mean cost of living index for utilities is less than that for transportation in these areas? Use $\alpha = 0.05$.

19. *Golf: Tournaments* Do professional golfers play better in their first round? Let row B represent the score in the fourth (and final) round, and let row A represent the score in the first round of a professional golf tournament. A

random sample of finalists in the British Open gave the following data for their first and last rounds in the tournament (Source: *Golf Almanac*).

B: Last	73	68	73	71	71	72	68	68	74
A: First	66	70	64	71	65	71	71	71	71

Do the data indicate that the population mean score on the last round is higher than that on the first? Use a 5% level of significance.

20. *Psychology: Training Rats* The following data are based on information from the Regis University Psychology Department. In an effort to determine if rats perform certain tasks more quickly if offered larger rewards, the following experiment was performed. On day 1, a group of three rats was given a reward of one food pellet each time they ran a maze. A second group of three rats was given a reward of five food pellets each time they ran the maze. On day 2, the groups were reversed, so the first group now got five food pellets for running the maze and the second group got only one pellet for running the same maze. The average times in seconds for each rat to run the maze 30 times are shown in the following table.

Rat	*A*	*B*	*C*	*D*	*E*	*F*
Time with one food pellet	3.6	4.2	2.9	3.1	3.5	3.9
Time with five food pellets	3.0	3.7	3.0	3.3	2.8	3.0

Do these data indicate that rats receiving larger rewards tend to run the maze in less time? Use a 5% level of significance.

21. *Psychology: Training Rats* The same experimental design discussed in Problem 20 was used to test rats trained to climb a sequence of short ladders. Times in seconds for eight rats to perform this task are shown in the following table.

Rat	*A*	*B*	*C*	*D*	*E*	*F*	*G*	*H*
Time 1 pellet	12.5	13.7	11.4	12.1	11.0	10.4	14.6	12.3
Time 5 pellets	11.1	12.0	12.2	10.6	11.5	10.5	12.9	11.0

Do these data indicate that rats receiving larger rewards tend to climb the ladder in less time? Use a 5% level of significance.

22. *Expand Your Knowledge: Confidence Intervals for μ_d* Using techniques from Section 7.2, we can find a confidence interval for μ_d. Consider a random sample of n matched data pairs A, B. Let $d = B - A$ be a random variable representing the difference between the values in a matched data pair. Compute the sample mean $\bar{d}$ of the differences and the sample standard deviation s_d. If d has a normal distribution or is mound-shaped, or if $n \geq 30$, then a confidence interval for μ_d is

$$\bar{d} - E < \mu_d < \bar{d} + E$$

where $E = t_c \dfrac{s_d}{\sqrt{n}}$

c = confidence level ($0 < c < 1$)
t_c = critical value for confidence level c and $d.f. = n - 1$

(a) Using the data of Problem 9, find a 95% confidence interval for the mean difference between percentage increase in company revenue and percentage increase in CEO salary.

(b) Use the confidence interval method of hypothesis testing outlined in Problem 25 of Section 8.2 to test the hypothesis that population mean percentage increase in company revenue is different from that of CEO salary. Use a 5% level of significance.

23. *Critical Region Method: Student's t* Solve Problem 9 using the critical region method of testing. Compare your conclusions with the conclusion obtained by using the *P*-value method. Are they the same?

24. *Critical Region Method: Student's t* Solve Problem 11 using the critical region method of testing. Compare your conclusions with the conclusion obtained by using the *P*-value method. Are they the same?

SECTION **8.5**

Testing $\mu_1 - \mu_2$ and $p_1 - p_2$ (Independent Samples)

FOCUS POINTS

- Identify independent samples and sampling distributions.
- Compute the sample test statistic and *P*-value for $\mu_1 - \mu_2$ and conclude the test.
- Compute the sample test statistic and *P*-value for $p_1 - p_2$ and conclude the test.

INDEPENDENT SAMPLES

Many practical applications of statistics involve a comparison of two population means or two population proportions. In Section 8.4, we considered tests of differences of means for *dependent samples*. With dependent samples, we could pair the data and then consider the difference of data measurements *d*. In this section, we will turn our attention to tests of differences of means from *independent samples*. We will see new techniques for testing the difference of means from independent samples.

Independent Samples

First, let's consider independent samples. We say that two sampling distributions are *independent* if there is no relationship whatsoever between specific values of the two distributions.

EXAMPLE 11

INDEPENDENT SAMPLE

A teacher wishes to compare the effectiveness of two teaching methods. Students are randomly divided into two groups: The first group is taught by method 1; the second group, by method 2. At the end of the course, a comprehensive exam is given to all students, and the mean score $\bar{x}_1$ for group 1 is compared with the mean score $\bar{x}_2$ for group 2. Are the samples independent or dependent?

SOLUTION: Because the students were *randomly* divided into two groups, it is reasonable to say that the $\bar{x}_1$ distribution is independent of the $\bar{x}_2$ distribution.

EXAMPLE 12

DEPENDENT SAMPLE

In Section 8.4, we considered a situation in which a shoe manufacturer claimed that for the general population of adult U.S. citizens, the average length of the left foot is longer than the average length of the right foot. To study this claim, the manufacturer gathers data in this fashion: Sixty adult U.S. citizens are drawn at random, and for these 60 people, both their left and right feet are measured. Let $\bar{x}_1$ be the mean length of the left feet and $\bar{x}_2$ be the mean length of the right feet. Are the $\bar{x}_1$ and $\bar{x}_2$ distributions independent for this method of collecting data?

SOLUTION: In this method, there is only *one* random sample of people drawn, and both the left and right feet are measured from this sample. The length of a person's left foot is usually related to the length of the person's right foot, so in this case the $\bar{x}_1$ and $\bar{x}_2$ distributions are *not* independent. In fact, we could pair the data and consider the distribution of the differences, left foot length minus right foot length. Then we would use the techniques of paired difference tests as found in Section 8.4.

GUIDED EXERCISE 8

INDEPENDENT SAMPLE

Suppose the shoe manufacturer of Example 12 gathers data in the following way: Sixty adult U.S. citizens are drawn at random and their left feet are measured; then another 60 adult U.S. citizens are drawn at random and their right feet are measured. Again, $\bar{x}_1$ is the mean of the left foot measurements and $\bar{x}_2$ is the mean of the right foot measurements.

Are the $\bar{x}_1$ and $\bar{x}_2$ distributions independent for this method of collecting data?

→ For this method of gathering data, two random samples are drawn: one for the left-foot measurements and one for the right-foot measurements. The first sample is not related to the second sample. The $\bar{x}_1$ and $\bar{x}_2$ distributions are independent.

PART I: TESTING $\mu_1 - \mu_2$ WHEN σ_1 AND σ_2 ARE KNOWN

Properties of $\bar{x}_1 - \bar{x}_2$ distribution, σ_1 and σ_2 known

In this part, we will use distributions that arise from a difference of means from independent samples. How do we obtain such distributions? If we have two statistical variables x_1 and x_2, each with its own distribution, we take independent random samples of size n_1 from the x_1 distribution and of size n_2 from the x_2 distribution. Then we can compute the respective means $\bar{x}_1$ and $\bar{x}_2$. Consider the difference $\bar{x}_1 - \bar{x}_2$. This represents a difference of means. If we repeat the sampling process over and over, we will come up with lots of $\bar{x}_1 - \bar{x}_2$ values. These values can be arranged in a frequency table, and we can make a histogram for the distribution of $\bar{x}_1 - \bar{x}_2$ values. This will give us an experimental idea of the theoretical distribution of $\bar{x}_1 - \bar{x}_2$.

Fortunately, it is not necessary to carry out this lengthy process for each example. The results have already been worked out mathematically. The next theorem presents the main results.

THEOREM 8.2 Let x_1 have a normal distribution with mean μ_1 and standard deviation σ_1. Let x_2 have a normal distribution with mean μ_2 and standard deviation

σ_2. If we take independent random samples of size n_1 from the x_1 distribution and of size n_2 from the x_2 distribution, then the variable $\bar{x}_1 - \bar{x}_2$ has

1. A normal distribution
2. Mean $\mu_1 - \mu_2$
3. Standard deviation

$$\sqrt{\frac{\sigma_1^2}{n_1} + \frac{\sigma_2^2}{n_2}}$$

COMMENT Theorem 8.2 requires that x_1 and x_2 have normal distributions. However, if both n_1 and n_2 are 30 or larger, then for most practical applications, the central limit theorem assures us that $\bar{x}_1$ and $\bar{x}_2$ are approximately normally distributed. In this case, the conclusions of the theorem are again valid, even if the original x_1 and x_2 distributions are not normal.

Hypotheses for testing difference of means

When testing the difference of means, it is customary to use the null hypothesis:

$$H_0: \mu_1 - \mu_2 = 0 \text{ or, equivalently, } H_0: \mu_1 = \mu_2$$

As mentioned in Section 8.1, the null hypothesis is set up to see if it can be rejected. When testing the difference of means, we first set up the hypothesis H_0 that there is no difference. The alternate hypothesis could then be any of the ones listed in Table 8-12. The alternate hypothesis and consequent type of test used depend on the particular problem. Note that μ_1 is always listed first.

TABLE 8-12 Alternate Hypotheses and Type of Test: Difference of Two Means

	H_1		Type of Test
$H_1: \mu_1 - \mu_2 < 0$	or equivalently	$H_1: \mu_1 < \mu_2$	Left-tailed test
$H_1: \mu_1 - \mu_2 > 0$	or equivalently	$H_1: \mu_1 > \mu_2$	Right-tailed test
$H_1: \mu_1 - \mu_2 \neq 0$	or equivalently	$H_1: \mu_1 \neq \mu_2$	Two-tailed test

Using Theorem 8.2 and the central limit theorem (Section 6.5), we can summarize the procedure for testing $\mu_1 - \mu_2$ when both σ_1 and σ_2 are known.

PROCEDURE

HOW TO TEST $\mu_1 - \mu_2$ WHEN BOTH σ_1 AND σ_2 ARE KNOWN

Requirements

Let σ_1 and σ_2 be the known population standard deviations of populations 1 and 2. Obtain two independent random samples from populations 1 and 2, where

$\bar{x}_1$ and $\bar{x}_2$ are sample means from populations 1 and 2
n_1 and n_2 are sample sizes from populations 1 and 2

If you can assume that both population distributions 1 and 2 are normal, any sample sizes n_1 and n_2 will work. If you cannot assume this, then use sample sizes $n_1 \geq 30$ and $n_2 \geq 30$.

Procedure

1. In the context of the application, state the *null and alternate hypotheses* and set the *level of significance* α. It is customary to use $H_0: \mu_1 - \mu_2 = 0$.

Continued

2. Use $\mu_1 - \mu_2 = 0$ from the null hypothesis together with $\bar{x}_1, \bar{x}_2, \sigma_1, \sigma_2, n_1$, and n_2 to compute the *sample test statistic.*

$$z = \frac{(\bar{x}_1 - \bar{x}_2) - (\mu_1 - \mu_2)}{\sqrt{\frac{\sigma_1^2}{n_1} + \frac{\sigma_2^2}{n_2}}} = \frac{\bar{x}_1 - \bar{x}_2}{\sqrt{\frac{\sigma_1^2}{n_1} + \frac{\sigma_2^2}{n_2}}}$$

3. Use the standard normal distribution and the type of test, one-tailed or two-tailed, to find the *P-value* corresponding to the sample test statistic.
4. *Conclude the test.* If P-value $\le \alpha$, then reject H_0. If P-value $> \alpha$, then do not reject H_0.
5. *Interpret your conclusion* in the context of the application.

EXAMPLE 13

TESTING THE DIFFERENCE OF MEANS (σ_1 AND σ_2 KNOWN)

Bob Winsett/CORBIS

A consumer group is testing camp stoves. To test the heating capacity of a stove, it measures the time required to bring 2 quarts of water from 50°F to boiling (at sea level). Two competing models are under consideration. Ten stoves of the first model and 12 stoves of the second model are tested. The following results are obtained.

Model 1: Mean time $\bar{x}_1 = 11.4$ min; $\sigma_1 = 2.5$ min; $n_1 = 10$
Model 2: Mean time $\bar{x}_2 = 9.9$ min; $\sigma_2 = 3.0$ min; $n_2 = 12$

Assume that the time required to bring water to a boil is normally distributed for each stove. Is there any difference (either way) between the performances of these two models? Use a 5% level of significance.

SOLUTION:

(a) State the null and alternate hypotheses and note the value of α.
Let μ_1 and μ_2 be the means of the distributions of times for models 1 and 2, respectively. We set up the null hypothesis to state that there is no difference:

$$H_0: \mu_1 = \mu_2 \text{ or } H_0: \mu_1 - \mu_2 = 0$$

The alternate hypothesis states that there is a difference:

$$H_1: \mu_1 \ne \mu_2 \text{ or } H_1: \mu_1 - \mu_2 \ne 0$$

The level of significance is $\alpha = 0.05$.

(b) ***Check Requirements*** What distribution does the sample test statistic follow?
The sample test statistic z follows a standard normal distribution because the original distributions from which the samples are drawn are normal. In addition, the population standard deviations of the original distributions are known and the samples are independent.

(c) Compute $\bar{x}_1 - \bar{x}_2$ and then convert it to the sample test statistic z. We are given the values $\bar{x}_1 = 11.4$ and $\bar{x}_2 = 9.9$. Therefore, $\bar{x}_1 - \bar{x}_2 = 11.4 - 9.9 = 1.5$. To convert this to a z value, we use the values $\sigma_1 = 2.5$, $\sigma_2 = 3.0$, $n_1 = 10$, and $n_2 = 12$. From the null hypothesis, $\mu_1 - \mu_2 = 0$

$$z = \frac{(\bar{x}_1 - \bar{x}_2) - (\mu_1 - \mu_2)}{\sqrt{\frac{\sigma_1^2}{n_1} + \frac{\sigma_2^2}{n_2}}} = \frac{1.5}{\sqrt{\frac{2.5^2}{10} + \frac{3.0^2}{12}}} \approx 1.28$$

(d) Find the P-value and sketch the area on the standard normal curve. Figure 8-16 on the next page, shows the P-value. Use the standard normal distribution (Table 5 of Appendix II) and the fact that we have a two-tailed test.

$$P\text{-value} \approx 2(0.1003) = 0.2006$$

FIGURE 8-16

P-value

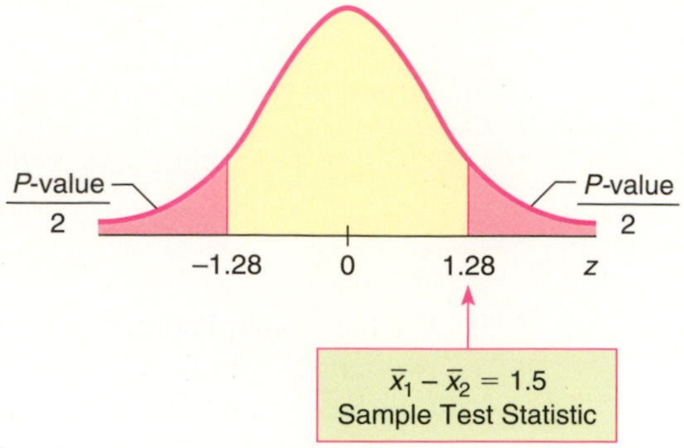

(e) Conclude the test.
The *P*-value is 0.2006 and $\alpha = 0.05$. Since *P*-value $> \alpha$, do not reject H_0.

(f) ***Interpretation*** Interpret the results.
At the 5% level of significance, the sample data do not indicate any difference in the population mean times for boiling water for the two stove models.

GUIDED EXERCISE 9 TESTING THE DIFFERENCE OF MEANS (σ_1 AND σ_2 KNOWN)

Let us return to Example 11 at the beginning of this section. A teacher wishes to compare the effectiveness of two teaching methods for her students. Students are randomly divided into two groups. The first group is taught by method 1; the second group, by method 2. At the end of the course, a comprehensive exam is given to all students.

The first group consists of $n_1 = 49$ students, with a mean score of $\bar{x}_1 = 74.8$ points. The second group has $n_2 = 50$ students, with a mean score of $\bar{x}_2 = 81.3$ points. The teacher claims that the second method will increase the mean score on the comprehensive exam. Is this claim justified at the 5% level of significance? Earlier research for the two methods indicates that $\sigma_1 = 14$ points and $\sigma_2 = 15$ points.

Let μ_1 and μ_2 be the mean scores of the distribution of all scores using method 1 and method 2, respectively.

(a) What is the null hypothesis?

→ H_0: $\mu_1 = \mu_2$ or H_0: $\mu_1 - \mu_2 = 0$

(b) ***Check Requirements*** What distribution does the sample test statistic follow? Explain.

→ The sample test statistic follows a standard normal distribution because the samples are large, both σ_1 and σ_2 are known, and the samples are independent.

(c) To examine the validity of the teacher's claim, what should the alternate hypothesis be? What is α?

→ H_1: $\mu_1 < \mu_2$ (the second method gives a higher average score) or H_1: $\mu_1 - \mu_2 < 0$.

$\alpha = 0.05$

(d) Compute $\bar{x}_1 - \bar{x}_2$.

→ $\bar{x}_1 - \bar{x}_2 = 74.8 - 81.3 = -6.5$

(e) Convert $\bar{x}_1 - \bar{x}_2 = -6.5$ to the sample test statistic z.

→ Using $\sigma_1 = 14$, $\sigma_2 = 15$, $n_1 = 49$, $n_2 = 50$, and $\mu_1 - \mu_2 = 0$ from H_0, we have

$$z = \frac{(\bar{x}_1 - \bar{x}_2) - (\mu_1 - \mu_2)}{\sqrt{\frac{\sigma_1^2}{n_1} + \frac{\sigma_2^2}{n_2}}} = \frac{-6.5 - 0}{\sqrt{\frac{14^2}{49} + \frac{15^2}{50}}} \approx -2.23$$

Continued

GUIDED EXERCISE 9 *continued*

(f) Find the *P*-value and sketch the area on the standard normal curve.

→ Figure 8-17 shows the *P*-value. It is the area to the left of $z = -2.23$. Using Table 5 of Appendix II, we find $P\text{-value} = P(z < -2.23) \approx 0.0129$.

FIGURE 8-17 *P*-value

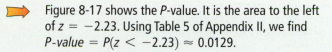

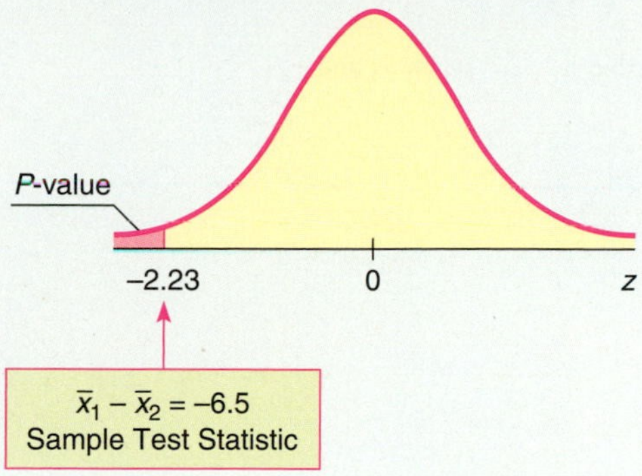

(g) Conclude the test.

→ Since *P*-value of $0.0129 \le 0.05$ for α, reject H_0.

(h) ***Interpret*** the results in the context of the application.

→ At the 5% level of significance, there is sufficient evidence to show that the second teaching method increased the population mean score on the exam.

PART II: TESTING $\mu_1 - \mu_2$ WHEN σ_1 AND σ_2 ARE UNKNOWN

To test $\mu_1 - \mu_2$ when σ_1 and σ_2 are unknown, we use distribution methods similar to those in Chapter 7 for estimating $\mu_1 - \mu_2$. In particular, if the two distributions are normal or approximately mound-shaped, or if both sample sizes are large (≥ 30), we use a Student's t distribution. Let's summarize the method of testing $\mu_1 - \mu_2$.

PROCEDURE

HOW TO TEST $\mu_1 - \mu_2$ WHEN σ_1 AND σ_2 ARE UNKNOWN

Requirements

Obtain two independent random samples from populations 1 and 2, where

$\bar{x}_1$ and $\bar{x}_2$ are sample means from populations 1 and 2
s_1 and s_2 are sample standard deviations from populations 1 and 2
n_1 and n_2 are sample sizes from populations 1 and 2

If you can assume that both population distributions 1 and 2 are normal or at least mound-shaped and symmetric, then any sample sizes n_1 and n_2 will work. If you cannot assume this, then use sample sizes $n_1 \ge 30$ and $n_2 \ge 30$.

Procedure

1. In the context of the application, state the *null and alternate hypotheses* and set the *level of significance* α. It is customary to use H_0: $\mu_1 - \mu_2 = 0$.

Continued

Problem 27 at the end of this section gives the formula for Satterthwaite's approximation for degrees of freedom.

2. Use $\mu_1 - \mu_2 = 0$ from the null hypothesis together with $\bar{x}_1, \bar{x}_2, s_1, s_2, n_1$, and n_2 to compute the *sample test statistic.*

$$t = \frac{(\bar{x}_1 - \bar{x}_2) - (\mu_1 - \mu_2)}{\sqrt{\dfrac{s_1^2}{n_1} + \dfrac{s_2^2}{n_2}}} = \frac{\bar{x}_1 - \bar{x}_2}{\sqrt{\dfrac{s_1^2}{n_1} + \dfrac{s_2^2}{n_2}}}$$

The sample test statistic distribution is approximately that of a Student's t with *degrees of freedom d.f.* = *smaller* of $n_1 - 1$ and $n_2 - 1$.

d.f. for testing $\mu_1 - \mu_2$ when σ_1 and σ_2 unknown

Note that statistical software gives a more accurate and larger *d.f.* based on Satterthwaite's approximation (see Problem 27).

3. Use a Student's t distribution and the type of test, one-tailed or two-tailed, to find the *P-value* corresponding to the sample test statistic.
4. *Conclude the test.* If P-value $\le \alpha$, then reject H_0. If P-value $> \alpha$, then do not reject H_0.
5. *Interpret your conclusion* in the context of the application.

EXAMPLE 14

TESTING THE DIFFERENCE OF MEANS (σ_1 AND σ_2 UNKNOWN)

Two competing headache remedies claim to give fast-acting relief. An experiment was performed to compare the mean lengths of time required for bodily absorption of brand A and brand B headache remedies.

Twelve people were randomly selected and given an oral dosage of brand A. Another 12 were randomly selected and given an equal dosage of brand B. The lengths of time in minutes for the drugs to reach a specified level in the blood were recorded. The means, standard deviations, and sizes of the two samples follow.

Brand A: $\bar{x}_1 = 21.8$ min;	$s_1 = 8.7$ min;	$n_1 = 12$
Brand B: $\bar{x}_2 = 18.9$ min;	$s_2 = 7.5$ min;	$n_2 = 12$

Past experience with the drug composition of the two remedies permits researchers to assume that both distributions are approximately normal. Let us use a 5% level of significance to test the claim that there is no difference between the two brands in the mean time required for bodily absorption. Also, find or estimate the P-value of the sample test statistic.

SOLUTION:

(a) $\alpha = 0.05$. The null hypothesis is

$$H_0: \mu_1 = \mu_2 \quad \text{or} \quad H_0: \mu_1 - \mu_2 = 0$$

Since we have no prior knowledge about which brand is faster, the alternate hypothesis is

$$H_1: \mu_1 \ne \mu_2 \quad \text{or} \quad H_1: \mu_1 - \mu_2 \ne 0$$

(b) ***Check Requirements*** What distribution does the sample test statistic follow? We use a Student's t distribution with $d.f.$ = smaller sample size $- 1 = 12 - 1 = 11$. Note that both samples are of size 12. Degrees of freedom can be computed also by Satterthwaite's approximation. A Student's t distribution is appropriate because the original populations are approximately normal, the population standard deviations are not known, and the samples are independent.

(c) Compute the sample test statistic.

We're given $\bar{x}_1 = 21.8$ and $\bar{x}_2 = 18.9$, so the sample difference is $\bar{x}_1 - \bar{x}_2 = 21.8 - 18.9 = 2.9$. Using $s_1 = 8.7$, $s_2 = 7.5$, $n_1 = 12$, $n_2 = 12$, and $\mu_1 - \mu_2 = 0$ from H_0, we compute the sample test statistic.

$$t = \frac{(\bar{x}_1 - \bar{x}_2) - (\mu_1 - \mu_2)}{\sqrt{\dfrac{s_1^2}{n_1} + \dfrac{s_2^2}{n_2}}} = \frac{2.9}{\sqrt{\dfrac{8.7^2}{12} + \dfrac{7.5^2}{12}}} \approx 0.875$$

(d) Estimate the *P*-value and sketch the area on a *t* graph.
Figure 8-18 shows the *P*-value. The degrees of freedom are $d.f. = 11$ (since both samples are of size 12). Because the test is a two-tailed test, the *P*-value is the area to the right of 0.875 together with the area to the left of -0.875. In the Student's *t* distribution table (Table 6 of Appendix II), we find an interval containing the *P*-value. Find 0.875 in the row headed by $d.f. = 11$. The test statistic 0.875 falls between the entries 0.697 and 1.214. Because this is a two-tailed test, we use the corresponding *P*-values 0.500 and 0.250 from the *two-tail* area row (see Table 8-13, Excerpt from Table 6). The *P*-value for the sample *t* is in the interval

$$0.250 < P\text{-value} < 0.500$$

(e) Conclude the test.

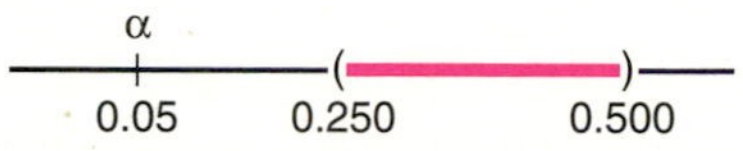

Since the interval containing the *P*-value lies to the right of $\alpha = 0.05$, we fail to reject H_0.
Note: Using the raw data and a calculator with Satterthwaite's approximation for the degrees of freedom $d.f. \approx 21.53$, the *P*-value ≈ 0.3915. This value is in the interval we computed.

FIGURE 8-18

P-value

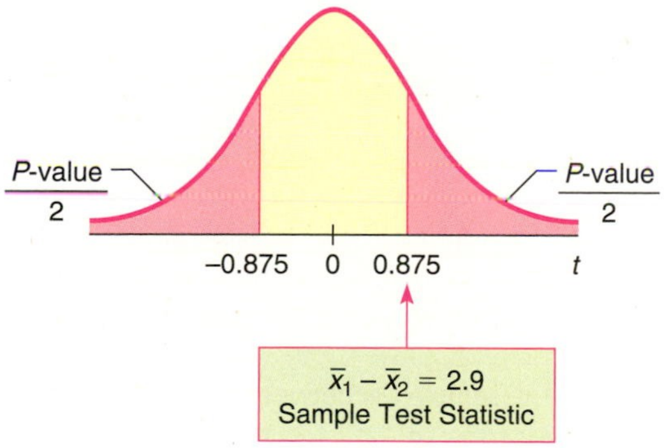

TABLE 8-13 Excerpt from Table 6, Appendix II

one-tail area	0.250	0.125
✓two-tail area	0.500	0.250
$d.f. = 11$	0.697 ↑ Sample $t = 0.875$	1.214

(f) ***Interpretation*** Interpret the results.
At the 5% level of significance, there is insufficient evidence to conclude that there is a difference in mean times for the remedies to reach the specified level in the bloodstream.

GUIDED EXERCISE 10 TESTING THE DIFFERENCE OF MEANS (σ_1 AND σ_2 UNKNOWN)

Suppose the experiment to measure the times in minutes for the headache remedies to enter the bloodstream (Example 14) yielded sample means, sample standard deviations, and sample sizes as follows:

Brand A: $\bar{x}_1 = 20.1$ min; $s_1 = 8.7$ min; $n_1 = 12$
Brand B: $\bar{x}_2 = 11.2$ min; $s_2 = 7.5$ min; $n_2 = 8$

Continued

GUIDED EXERCISE 10 *continued*

Brand B claims to be faster. Is this claim justified at the 5% level of significance? (Use the following steps to obtain the answer.)

(a) What is α? State H_0 and H_1.

$\alpha = 0.05$

H_0: $\mu_1 = \mu_2$ *or* H_0: $\mu_1 - \mu_2 = 0$

H_1: $\mu_1 > \mu_2$ (*or* H_1: $\mu_1 - \mu_2 > 0$). This states that the mean time for brand B is less than the mean time for brand A.

(b) ***Check Requirements*** What distribution does the sample test statistic follow? Explain.

Student's t distribution with $d.f. = n_2 - 1 = 8 - 1 = 7$ since $n_2 < n_1$. From Example 14, we know that the original distributions are approximately normal. Both σ_1 and σ_2 are unknown and the samples are independent.

(c) Compute $\bar{x}_1 - \bar{x}_2$ and the corresponding sample test statistic t.

$\bar{x}_1 - \bar{x}_2 = 20.1 - 11.2 = 8.9$. Using $s_1 = 8.7$, $s_2 = 7.5$, $n_1 = 12$, $n_2 = 8$, and $\mu_1 - \mu_2 = 0$ from H_0, we have

$$t = \frac{\bar{x}_1 - \bar{x}_2}{\sqrt{\dfrac{s_1^2}{n_1} + \dfrac{s_2^2}{n_2}}} = \frac{8.9}{\sqrt{\dfrac{8.7^2}{12} + \dfrac{7.5^2}{8}}} \approx 2.437$$

(d) What degrees of freedom do you use? To find an interval containing the P-value, do you use one-tail or two-tail areas of Table 6, Appendix II? Sketch a figure showing the P-value. Find an interval for the P-value.

Since $n_2 < n_1$, $d.f. = n_2 - 1 = 8 - 1 = 7$. Use *one-tail area* of Table 6.

FIGURE 8-19 *P*-value

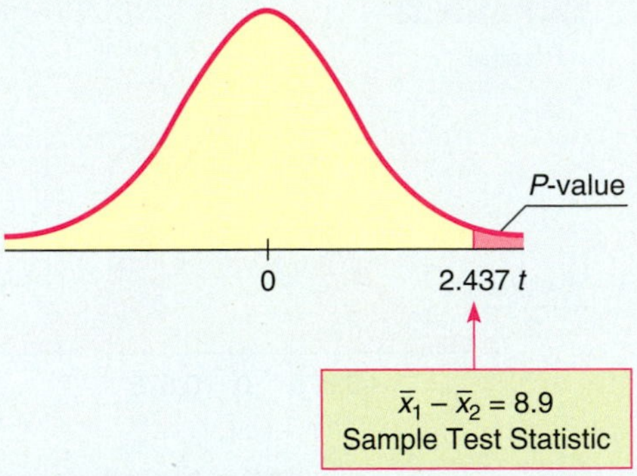

TABLE 8-14 **Excerpt from Student's *t* Table**

✓ **one-tail area**	**0.025**	**0.010**
two-tail area	**0.050**	**0.020**
d.f. = 7	2.365	2.998
	↑ Sample t = 2.437	

The sample t is between 2.365 and 2.998.

$0.010 < P\text{-value} < 0.025$

(e) Do we reject or fail to reject H_0?

Since the interval containing the P-value has values that are all less than 0.05, we reject H_0.

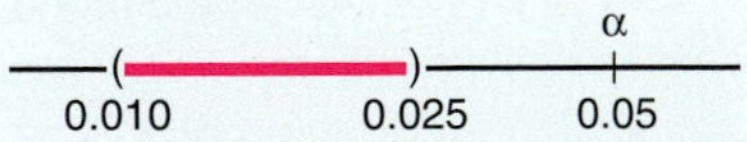

Note: On the calculator with Satterthwaite's approximation for *d.f.*, we have $d.f. = 16.66$ and P-value ≈ 0.013.

(f) ***Interpret*** the results in the context of the application.

At the 5% level of significance, there is sufficient evidence to conclude that the mean time for brand B to enter the bloodstream is less than that for brand A.

Alternate method using pooled standard deviation

Problem 28 at the end of this section discusses the pooled standard deviation method for testing the difference of means when population standard deviations are equal. Section 10.4 has techniques for testing whether two standard deviations or two variances are equal.

Pooled standard deviation

There is another method of testing $\mu_1 - \mu_2$ when σ_1 and σ_2 are unknown. Suppose that the sample values s_1 and s_2 are sufficiently close and that there is reason to believe $\sigma_1 = \sigma_2$ (or the standard deviations are approximately equal). This situation can happen when you make a slight change or alteration to a known process or method of production. The standard deviation may not change much, but the outputs or means could be very different. When there is reason to believe that $\sigma_1 = \sigma_2$, it is best to use a *pooled standard deviation.* The sample test statistic $\bar{x}_1 - \bar{x}_2$ has a corresponding t variable with an *exact* Student's t distribution and degrees of freedom $d.f. = n_1 + n_2 - 2$. Problem 28 at the end of the section provides the details.

Under the null hypothesis H_0: $\mu_1 = \mu_2$, which distribution should you use for the sample test statistic $\bar{x}_1 - \bar{x}_2$?

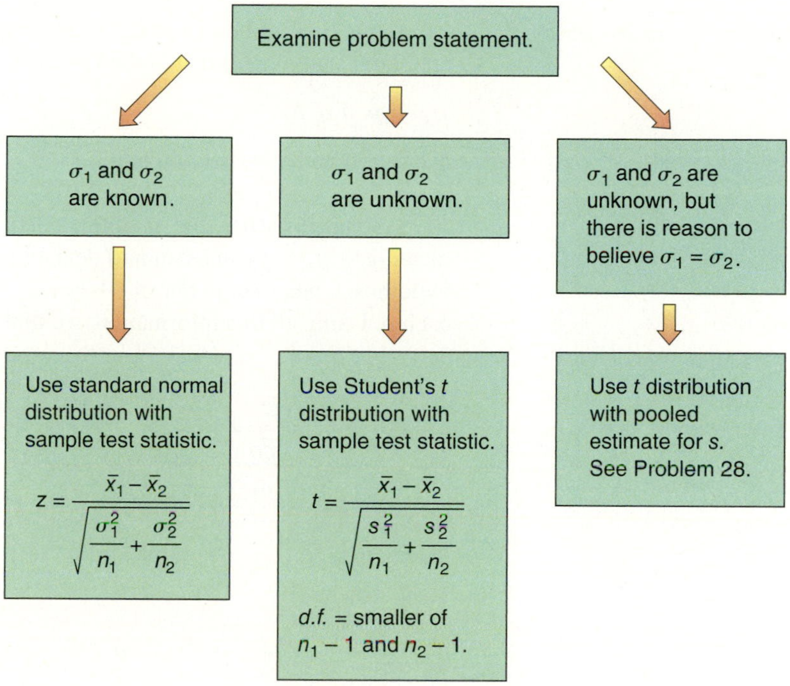

PART III: TESTING A DIFFERENCE OF PROPORTIONS $p_1 - p_2$

Suppose we have two independent binomial experiments. That is, outcomes from one binomial experiment are in no way paired with outcomes from the other. We use the notation

Binomial Experiment 1	*Binomial Experiment 2*
n_1 = number of trials	n_2 = number of trials
r_1 = number of successes	r_2 = number of successes
p_1 = population probability of success on a single trial	p_2 = population probability of success on a single trial

For *large* values of n_1 and n_2, the distribution of sample differences

$$\hat{p}_1 - \hat{p}_2 = \frac{r_1}{n_1} - \frac{r_2}{n_2}$$

is closely approximated by a *normal distribution* with mean μ and standard deviation σ as shown:

$$\mu = p_1 - p_2 \qquad \sigma = \sqrt{\frac{p_1q_1}{n_1} + \frac{p_2q_2}{n_2}}$$

where $q_1 = 1 - p_1$ and $q_2 = 1 - p_2$.

For most practical problems involving a comparison of two binomial populations, the experimenters will want to test the null hypothesis H_0: $p_1 = p_2$. Consequently, this is the type of test we shall consider. Since the values of p_1 and p_2 are unknown, and since specific values are not assumed under the null hypothesis H_0: $p_1 = p_2$, the best estimate for the common value is the total number of successes $(r_1 + r_2)$ divided by the total number of trials $(n_1 + n_2)$. If we denote this *pooled estimate of proportion* by $\bar{p}$ (read "p bar"), then

Pooled estimate of proportion $\bar{p}$

$$\bar{p} = \frac{r_1 + r_2}{n_1 + n_2}$$

This formula gives the best sample estimate $\bar{p}$ for p_1 and p_2 *under the assumption that* $p_1 = p_2$. Also, $\bar{q} = 1 - \bar{p}$.

Criteria for using the normal approximation to the binomial

COMMENT For most practical applications, the sample sizes n_1 and n_2 are considered large samples if each of the four quantities

$$n_1\bar{p} \quad n_1\bar{q} \quad n_2\bar{p} \quad n_2\bar{q}$$

is larger than 5 (see Section 6.6).

As stated earlier, the sample statistic $\hat{p}_1 - \hat{p}_2$ has a normal distribution with mean $\mu = p_1 - p_2$ and standard deviation $\sigma = \sqrt{p_1q_1/n_1 + p_2q_2/n_2}$. Under the null hypothesis, we assume that $p_1 = p_2$ and then use the pooled estimate $\bar{p}$ in place of each p. Using all this information, we find that the *sample test statistic* is

Sample test statistic

$$z = \frac{\hat{p}_1 - \hat{p}_2}{\sqrt{\dfrac{\bar{p}\bar{q}}{n_1} + \dfrac{\bar{p}\bar{q}}{n_2}}}$$

where $\bar{p} = \dfrac{r_1 + r_2}{n_1 + n_2}$ and $\bar{q} = 1 - \bar{p}$

$\hat{p}_1 = \dfrac{r_1}{n_1}$ and $\hat{p}_2 = \dfrac{r_2}{n_2}$

Using this information, we summarize the procedure for testing $p_1 - p_2$.

PROCEDURE

HOW TO TEST A DIFFERENCE OF PROPORTIONS $P_1 - P_2$

Procedure

Consider two independent binomial experiments.

Binomial Experiment 1

n_1 = number of trials

r_1 = number of successes out of n_1 trials

$\hat{p}_1 = \dfrac{r_1}{n_1}$

p_1 = population probability of success on a single trial

Binomial Experiment 2

n_2 = number of trials

r_2 = number of successes out of n_2 trials

$\hat{p}_2 = \dfrac{r_2}{n_2}$

p_2 = population probability of success on a single trial

1. Use the *null hypothesis* of no difference, H_0: $p_1 - p_2 = 0$. In the context of the application, choose the *alternate hypothesis*. Set the *level of significance* α.

Continued

2. The null hypothesis claims that $p_1 = p_2$; therefore, *pooled best estimates* for the population probabilities of success and failure are

$$\bar{p} = \frac{r_1 + r_2}{n_1 + n_2} \quad \text{and} \quad \bar{q} = 1 - \bar{p}$$

Requirements

The number of trials should be sufficiently large so that each of the four quantities $n_1\bar{p}$, $n_1\bar{q}$, $n_2\bar{p}$, and $n_2\bar{q}$ is larger than 5.

In this case, you compute the *sample test statistic*

$$z = \frac{\hat{p}_1 - \hat{p}_2}{\sqrt{\frac{\bar{p}\bar{q}}{n_1} + \frac{\bar{p}\bar{q}}{n_2}}}$$

3. Use the standard normal distribution and a type of test, one-tailed or two-tailed, to find the *P-value* corresponding to the sample test statistic.
4. *Conclude the test.* If P-value $\leq \alpha$, then reject H_0. If P-value $> \alpha$, then do not reject H_0.
5. *Interpret your conclusion* in the context of the application.

EXAMPLE 15

TESTING THE DIFFERENCE OF PROPORTIONS

The Macek County Clerk wishes to improve voter registration. One method under consideration is to send reminders in the mail to all citizens in the county who are eligible to register. As part of a pilot study to determine if this method will actually improve voter registration, a random sample of 1250 potential voters was taken. Then this sample was randomly divided into two groups.

Group 1: There were 625 people in this group. No reminders to register were sent to them. The number of potential voters from this group who registered was 295.

Group 2: This group also contained 625 people. Reminders were sent in the mail to each member in the group, and the number who registered to vote was 350.

The county clerk claims that the proportion of people who registered was significantly greater in group 2. On the basis of this claim, the clerk recommends that the project be funded for the entire population of Macek County. Use a 5% level of significance to test the claim that the proportion of potential voters who registered was greater in group 2, the group that received reminders.

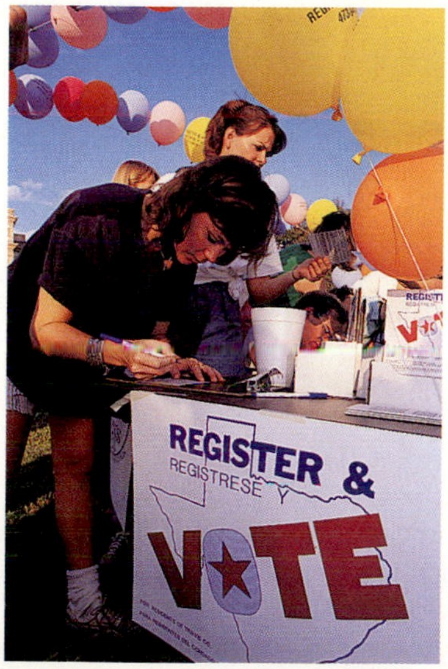

The Image Works

SOLUTION:

(a) Note that $\alpha = 0.05$. Let p_1 be the proportion of voters from group 1 who registered, and let p_2 be the proportion from group 2 who registered. The null hypothesis is that there is no difference in proportions, so,

$$H_0\text{: } p_1 = p_2 \quad \text{or} \quad H_0\text{: } p_1 - p_2 = 0$$

The alternate hypothesis is that the proportion of voters who registered was greater in the group that received reminders.

$$H_1\text{: } p_1 < p_2 \quad \text{or} \quad H_1\text{: } p_1 - p_2 < 0$$

(b) Calculate the *pooled estimates* $\bar{p}$ and $\bar{q}$.
Under the null hypothesis that $p_1 = p_2$ we find

$$\bar{p} = \frac{r_1 + r_2}{n_1 + n_2} = \frac{295 + 350}{625 + 625} \approx 0.516 \text{ and } \bar{q} = 1 - \bar{p} \approx 0.484$$

(c) ***Check Requirements*** What distribution does the sample test statistic follow?
The sample test statistic follows a standard normal distribution because the number of binomial trials is large enough that each of the products $n_1\bar{p}$, $n_1\bar{q}$, $n_2\bar{p}$, $n_2\bar{q}$ exceeds 5.

(d) Compute $\hat{p}_1 - \hat{p}_2$ and convert it to the sample statistic z.
CALCULATOR NOTE Carry the values for $\hat{p}_1$, $\hat{p}_2$, and the pooled estimates $\bar{p}$ and $\bar{q}$ out to at least three places after the decimal. Then round the z value of the corresponding test statistic to two places after the decimal.

For the first group, the number of successes is $r_1 = 295$ out of $n_1 = 625$ trials. For the second group, there are $r_2 = 350$ successes out of $n_2 = 625$ trials. Since

$$\hat{p}_1 = \frac{r_1}{n_1} = \frac{295}{625} = 0.472 \quad \text{and} \quad \hat{p}_2 = \frac{r_2}{n_2} = \frac{350}{625} = 0.560$$

then

$$\hat{p}_1 - \hat{p}_2 = 0.472 - 0.560 = -0.088$$

We computed the pooled estimate for $\bar{p}$ and $\bar{q}$ in part (b).

$$\bar{p} = 0.516 \text{ and } \bar{q} = 1 - \bar{p} = 0.484$$

Using these values, we find the sample test statistic

$$z = \frac{\hat{p}_1 - \hat{p}_2}{\sqrt{\frac{\bar{p}\bar{q}}{n_1} + \frac{\bar{p}\bar{q}}{n_2}}} = \frac{-0.088}{\sqrt{\frac{(0.516)(0.484)}{625} + \frac{(0.516)(0.484)}{625}}} \approx -3.11$$

(e) Find the P-value and sketch the area on the standard normal curve.
Figure 8-20 shows the P-value. This is a left-tailed test, so the P-value is the area to the left of -3.11. Using the standard normal distribution (Table 5 of Appendix II), we find P-value $= P(z < -3.11) \approx 0.0009$.

FIGURE 8-20

P-value

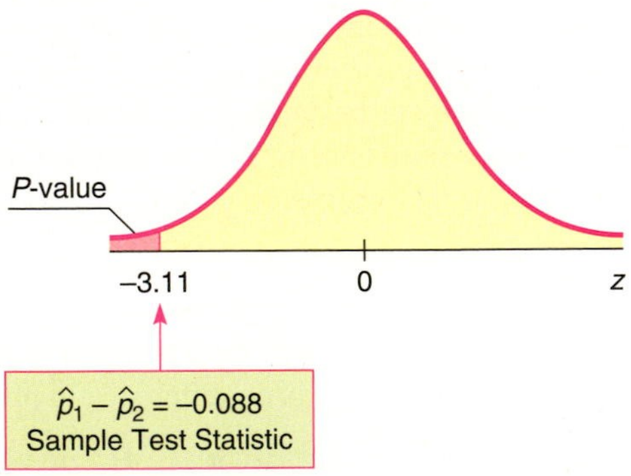

(f) Conclude the test.
Since P-value of $0.0009 \le 0.05$ for α, we reject H_0.

(g) ***Interpretation*** Interpret the results in the context of the application.
At the 5% level of significance, the data indicate that the population proportion of potential voters who registered was greater in group 2, the group that received reminders.

GUIDED EXERCISE 11 TESTING THE DIFFERENCE OF PROPORTIONS

In Example 15 about voter registration, suppose that a random sample of 1100 potential voters was randomly divided into two groups.

Group 1: 500 potential voters; no registration reminders sent; 248 registered to vote
Group 2: 600 potential voters; registration reminders sent; 332 registered to vote

Do these data support the claim that the proportion of voters who registered was greater in the group that received reminders than in the group that did not? Use a 1% level of significance.

(a) What is α? State H_0 and H_1.

→ $\alpha = 0.01$. As before, H_0: $p_1 = p_2$ and H_1: $p_1 < p_2$.

(b) Under the null hypothesis $p_1 = p_2$, calculate the *pooled estimates* $\bar{p}$ and $\bar{q}$

→ $n_1 = 500, r_1 = 248; n_2 = 600, r_2 = 332$

$$\bar{p} = \frac{r_1 + r_2}{n_1 + n_2} = \frac{248 + 332}{500 + 600} \approx 0.527$$

$$\bar{q} = 1 - \bar{p} \approx 1 - 0.527 \approx 0.473$$

(c) ***Check Requirements*** What distribution does the sample test statistic follow?

→ The sample test statistic z follows a standard normal distribution because the number of binomial trials is large enough that each of the products $n_1\bar{p}, n_1\bar{q}, n_2\bar{p}, n_2\bar{q}$ exceeds 5.

(d) Compute $\hat{p}_1 - \hat{p}_2$.

→ $\hat{p}_1 = \dfrac{r_1}{n_1} = \dfrac{248}{500} = 0.496$ $\hat{p}_2 = \dfrac{r_2}{n_2} = \dfrac{332}{600} \approx 0.553$

$\hat{p}_1 - \hat{p}_2 \approx -0.057$

(e) Find the sample test statistic.

→
$$z = \frac{\hat{p}_1 - \hat{p}_2}{\sqrt{\dfrac{\bar{p}\,\bar{q}}{n_1} + \dfrac{\bar{p}\,\bar{q}}{n_2}}} = \frac{-0.057}{\sqrt{\dfrac{(0.527)(0.473)}{500} + \dfrac{(0.527)(0.473)}{600}}}$$

$$\approx -1.89$$

(f) Find the P-value and sketch the area on the standard normal curve.

→ Figure 8-21 shows the P-value. It is the area to the left of $z = -1.89$. Using Table 5 of Appendix II, we find P-value $= P(z < -1.89) = 0.0294$.

FIGURE 8-21 *P*-value

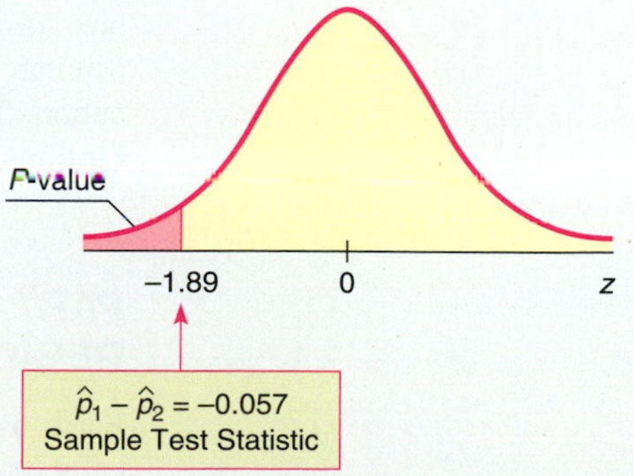

Continued

GUIDED EXERCISE 11 *continued*

(g) Conclude the test. ➡ Since *P*-value of 0.0294 $>$ 0.01 for α, we cannot reject H_0.

(h) ***Interpret*** the results in the context of the application. ➡ At the 1% level of significance, the data do not support the claim that the reminders increased the proportion of registered voters.

TECH NOTES The TI-84Plus/TI-83Plus/TI-*n*spire calculators, Excel 2010, and Minitab all support testing the difference of means for independent samples. The TI-84Plus/TI-83Plus/TI-*n*spire calculators and Minitab also support testing the difference of proportions. When testing the difference of means using the normal distribution, the technologies require the population standard deviations for the distributions of the two samples. When testing the difference of means using a Student's t distribution, the technologies give the option of using the pooled standard deviation. As discussed in Problem 28 of this section, the pooled standard deviation is appropriate when the standard deviations of the two populations are approximately equal. If the pooled standard deviation option is not selected, the technologies compute the sample test statistic for $\bar{x}_1 - \bar{x}_2$ using the procedures described in this section. However, Satterthwaite's approximation for the degrees of freedom is used.

TI-84Plus/TI-83Plus/TI-*n*spire (with TI-84Plus keypad) Enter the data. Press **STAT** and select **TESTS.** Options **3: 2-SampZTest, 4: 2-SampTTest, 6: 2-PropZTest** perform tests for the difference of means using the normal distribution, difference of means using a Student's t distribution, and difference of proportions, respectively. The calculator uses the symbol $\hat{p}$ to designate the pooled estimate for p.

Excel 2010 Enter the data in columns. On the home ribbon, click the **Data** tab. Then in the Analysis Group, select **Data Analysis**. The choice **z-Test Two Sample Means** conducts a test for the difference of means using the normal distribution. The choice **t-Test: Two-Sample Assuming Unequal Variances** conducts a test using a Student's t distribution with Satterthwaite's approximation for the degrees of freedom. The choice **t-Test: Two-Sample Assuming Equal Variances** conducts a test using the pooled standard deviation. Another way to conduct a paired-difference-of-means test is to press **Insert Function** (f_x) and select **Statistical** for the category and then **T.TEST** for the function. In the dialogue box for type of test, use **2** for the t test with Satterthwaite's approximation for degrees of freedom or **3** for tests with pooled standard deviation.

Minitab Enter the data in two columns. Use the menu choices **Stat ➤ Basic Statistics.** The choice **2 sample t** tests the difference of means using a Student's t distribution. In the dialogue box, leaving the box **Assume equal variances** unchecked produces a test using Satterthwaite's approximation for the degrees of freedom. Checking the box **Assume equal variances** produces a test using the pooled standard deviation. Minitab does not support testing the difference of means using the normal distribution. The menu item **2 Proportions** tests the difference of proportions. Under **Options**, select the null and alternate hypotheses.

PART IV: TESTING $\mu_1 - \mu_2$ AND $p_1 - p_2$ USING CRITICAL REGIONS

For a fixed, preset level of significance α, the *P*-value method of testing is logically equivalent to the critical region method of testing. This book emphasizes the *P*-value method because of its great popularity and because it is readily compatible with most computer software. However, for completeness, we provide an optional example utilizing the critical region method.

Recall that the critical region method and the P-value method of testing share a number of steps. Both methods use the same *null and alternate hypotheses,* the same *sample test statistic,* and the same *sampling distribution* for the test statistic. However, instead of computing the P-value and comparing it to the level of significance α, the critical region method compares the sample test statistic to a *critical value* from the sampling distribution that is based on α and the alternate hypothesis (left-tailed, right-tailed, or two-tailed).

Test conclusions based on critical values

For a *right-tailed test*, if the *sample test statistic* $\geq$ *critical value*, reject H_0.

For a *left-tailed test*, if the *sample test statistic* $\leq$ *critical value*, reject H_0.

For a *two-tailed test*, if the *sample test statistic lies at or beyond the critical values* (that is, $\leq$ negative critical value or $\geq$ positive critical value), reject H_0.

Otherwise, in each case, do not reject H_0.

Critical values z_0 for tests using the standard normal distribution can be found in Table 5(c) of Appendix II. Critical values t_0 for tests using a Student's t distribution are found in Table 6 of Appendix II. Use the row headed by the appropriate degrees of freedom and the column that includes the value of α (level of significance) in the *one-tail area* row for one-tailed tests or the *two-tail area* row for two-tailed tests.

EXAMPLE 16

CRITICAL REGION METHOD

Use the critical region method to solve the application in Example 14 (testing $\mu_1 - \mu_2$ when σ_1 and σ_2 are unknown).

SOLUTION: Example 14 involves testing the difference in average time for two headache remedies to reach the bloodstream. For brand A, $\bar{x}_1 = 21.8$ min, $s_1 = 8.7$ min, and $n_1 = 12$; for brand B, $\bar{x}_2 = 18.9$ min, $s_2 = 7.5$ min, and $n_2 = 12$. The level of significance α is 0.05. The Student's t distribution is appropriate because both populations are approximately normal.

(a) To use the critical region method to test for a difference in average time, we use the same hypotheses and the same sample test statistic as in Example 14.

$$H_0\text{: } \mu_1 = \mu_2;\ H_1\text{: } \mu_1 \neq \mu_2;$$

$$\bar{x}_1 - \bar{x}_2 = 2.9 \text{ min}; \text{ sample test statistic } t = 0.875$$

(b) Instead of finding the P-value of the sample test statistic, we use α and H_1 to find the critical values in Table 6 of Appendix II. We have $d.f. = 11$ (since both samples are of size 12). We find $\alpha = 0.05$ in the *two-tail area* row, since we have a two-tailed test. The critical values are $\pm t_0 = \pm 2.201$.

Next compare the sample test statistic $t = 0.875$ to the critical values. Figure 8-22 shows the critical regions and the sample test statistic. We see that the sample test statistic falls in the "do not reject H_0" region. At the 5% level of significance, sample evidence does not show a difference in the time for the drugs to reach the bloodstream. The result is consistent with the result obtained by the P-value method of Example 14.

FIGURE 8-22

Critical Regions $\alpha = 0.05$; $d.f. = 11$

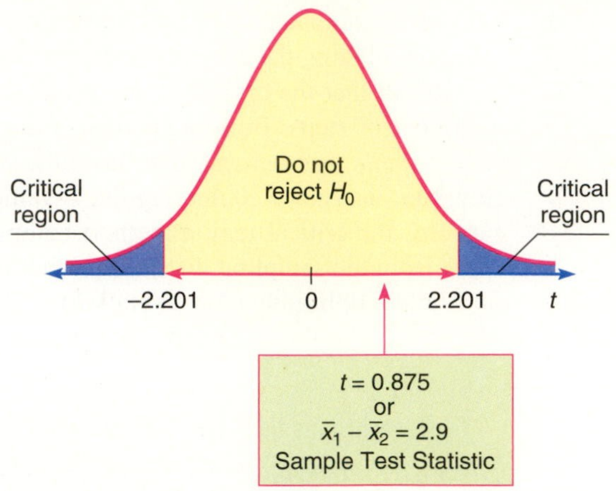

VIEWPOINT Temper! Temper!

In her book Red Ink Behaviors, *Jean Hollands discusses inappropriate, problem behaviors of professional employees in the corporate business world. Temper tantrums, flaming e-mails, omitting essential information, sabotaging fellow workers, and the arrogant opinion that others are "dumb and dispensable" create personnel problems that cost companies a lot in the form of wasted time, reduced productivity, and lost revenues. A study of major industries in the Silicon Valley area gave Hollands data for estimating just how much time and money are wasted by such "red ink behaviors." For more information, see Problems 25 and 26 in this section.*

SECTION 8.5 PROBLEMS

1. *Statistical Literacy* Consider a hypothesis test of difference of means for two independent populations x_1 and x_2.
   (a) What does the null hypothesis say about the relationship between the two population means?
   (b) In the case that the sample test statistic is z, give the formula for z.
   (c) In the case that the sample test statistic is t, give the formula for t.

2. *Statistical Literacy* Consider a hypothesis test of difference of means for two independent populations x_1 and x_2. Suppose that both sample sizes are greater than 30 and that you know σ_1 but not σ_2. Is it standard practice to use the normal distribution or a Student's t distribution?

3. *Statistical Literacy* Consider a hypothesis test of difference of means for two independent populations x_1 and x_2. What are two ways of expressing the null hypothesis?

4. *Statistical Literacy* Consider a hypothesis test of difference of proportions for two independent populations. Suppose random samples produce r_1 successes out of n_1 trials for the first population and r_2 successes out of n_2 trials for the second population.
   (a) What does the null hypothesis claim about the relationship between the proportions of successes in the two populations?
   (b) What is the formula for the sample test statistic?

5. *Statistical Literacy* Consider a hypothesis test of difference of proportions for two independent populations. Suppose random samples produce r_1 successes out of n_1 trials for the first population and r_2 successes out of n_2 trials for the second population. What is the best pooled estimate $\bar{p}$ for the population probability of success using H_0: $p_1 = p_2$?

6. *Critical Thinking* Consider use of a Student's t distribution to test the difference of means for independent populations using random samples of sizes n_1 and n_2.
   (a) Which process gives the larger degrees of freedom, Satterthwaite's approximation or using the smaller of $n_1 - 1$ and $n_2 - 1$? Which method is more conservative? What do we mean by "conservative"? Note that computer programs and other technologies commonly use Satterthwaite's approximation.
   (b) Using the same hypotheses and sample data, is the P-value smaller for larger degrees of freedom? How might a larger P-value impact the significance of a test?

7. *Critical Thinking* When conducting a test for the difference of means for two independent populations x_1 and x_2, what alternate hypothesis would indicate that the mean of the x_2 population is smaller than that of the x_1 population? Express the alternate hypothesis in two ways.

8. *Critical Thinking* When conducting a test for the difference of means for two independent populations x_1 and x_2, what alternate hypothesis would indicate that the mean of the x_2 population is larger than that of the x_1 population? Express the alternate hypothesis in two ways.

9. *Basic Computation: Testing $\mu_1 - \mu_2$* A random sample of 49 measurements from one population had a sample mean of 10, with sample standard deviation 3. An independent random sample of 64 measurements from a second population had a sample mean of 12, with sample standard deviation 4. Test the claim that the population means are different. Use level of significance 0.01.
   (a) *Check Requirements* What distribution does the sample test statistic follow? Explain.
   (b) State the hypotheses.
   (c) Compute $\bar{x}_1 - \bar{x}_2$ and the corresponding sample test statistic.
   (d) Estimate the P-value of the sample test statistic.
   (e) Conclude the test.
   (f) ***Interpret*** the results.

10. *Basic Computation: Testing $\mu_1 - \mu_2$* Two populations have mound-shaped, symmetric distributions. A random sample of 16 measurements from the first population had a sample mean of 20, with sample standard deviation 2. An independent random sample of 9 measurements from the second population had a sample mean of 19, with sample standard deviation 3. Test the claim that the population mean of the first population exceeds that of the second. Use a 5% level of significance.
    (a) *Check Requirements* What distribution does the sample test statistic follow? Explain.
    (b) State the hypotheses.
    (c) Compute $\bar{x}_1 - \bar{x}_2$ and the corresponding sample test statistic.
    (d) Estimate the P-value of the sample test statistic.
    (e) Conclude the test.
    (f) ***Interpret*** the results.

11. *Basic Computation: Testing $\mu_1 - \mu_2$* A random sample of 49 measurements from a population with population standard deviation 3 had a sample mean of 10. An independent random sample of 64 measurements from a second population with population standard deviation 4 had a sample mean of 12. Test the claim that the population means are different. Use level of significance 0.01.
    (a) *Check Requirements* What distribution does the sample test statistic follow? Explain.
    (b) State the hypotheses.
    (c) Compute $\bar{x}_1 - \bar{x}_2$ and the corresponding sample test statistic.
    (d) Find the P-value of the sample test statistic.

(e) Conclude the test.
(f) ***Interpret*** the results.

12. *Basic Computation: Testing* $\mu_1 - \mu_2$ Two populations have normal distributions. The first has population standard deviation 2 and the second has population standard deviation 3. A random sample of 16 measurements from the first population had a sample mean of 20. An independent random sample of 9 measurements from the second population had a sample mean of 19. Test the claim that the population mean of the first population exceeds that of the second. Use a 5% level of significance.
(a) *Check Requirements* What distribution does the sample test statistic follow? Explain.
(b) State the hypotheses.
(c) Compute $\bar{x}_1 - \bar{x}_2$ and the corresponding sample test statistic.
(d) Find the *P*-value of the sample test statistic.
(e) Conclude the test
(f) ***Interpret*** the results.

13. *Basic Computation: Test* $p_1 - p_2$ For one binomial experiment, 75 binomial trials produced 45 successes. For a second independent binomial experiment, 100 binomial trials produced 70 successes. At the 5% level of significance, test the claim that the probabilities of success for the two binomial experiments differ.
(a) Compute the pooled probability of success for the two experiments.
(b) *Check Requirements* What distribution does the sample test statistic follow? Explain.
(c) State the hypotheses.
(d) Compute $\hat{p}_1 - \hat{p}_2$ and the corresponding sample test statistic.
(e) Find the *P*-value of the sample test statistic.
(f) Conclude the test.
(g) ***Interpret*** the results.

14. *Basic Computation: Test* $p_1 - p_2$ For one binomial experiment, 200 binomial trials produced 60 successes. For a second independent binomial experiment, 400 binomial trials produced 156 successes. At the 5% level of significance, test the claim that the probability of success for the second binomial experiment is greater than that for the first.
(a) Compute the pooled probability of success for the two experiments.
(b) *Check Requirements* What distribution does the sample test statistic follow? Explain.
(c) State the hypotheses.
(d) Compute $\hat{p}_1 - \hat{p}_2$ and the corresponding sample test statistic.
(e) Find the *P*-value of the sample test statistic.
(f) Conclude the test.
(g) ***Interpret*** the results.

Please provide the following information for Problems 15–26 and 29–35.
(a) What is the level of significance? State the null and alternate hypotheses.
(b) *Check Requirements* What sampling distribution will you use? What assumptions are you making? Compute the sample test statistic.
(c) Find (or estimate) the *P*-value. Sketch the sampling distribution and show the area corresponding to the *P*-value.
(d) Based on your answers in parts (a) to (c), will you reject or fail to reject the null hypothesis? Are the data statistically significant at level α?
(e) ***Interpret*** your conclusion in the context of the application.
Note: For degrees of freedom *d.f.* not in the Student's *t* table, use the closest *d.f.* that is *smaller*. In some situations, this choice of *d.f.* may increase the *P*-value a small amount and therefore produce a slightly more "conservative" answer.

Answers may vary due to rounding.

15. *Medical: REM Sleep* REM (rapid eye movement) sleep is sleep during which most dreams occur. Each night a person has both REM and non-REM sleep. However, it is thought that children have more REM sleep than adults (Reference: *Secrets of Sleep* by Dr. A. Borbely). Assume that REM sleep time is normally distributed for both children and adults. A random sample of $n_1 = 10$ children (9 years old) showed that they had an average REM sleep time of $\bar{x}_1 = 2.8$ hours per night. From previous studies, it is known that $\sigma_1 = 0.5$ hour. Another random sample of $n_2 = 10$ adults showed that they had an average REM sleep time of $\bar{x}_2 = 2.1$ hours per night. Previous studies show that $\sigma_2 = 0.7$ hour. Do these data indicate that, on average, children tend to have more REM sleep than adults? Use a 1% level of significance.

16. *Environment: Pollution Index* Based on information from *The Denver Post*, a random sample of $n_1 = 12$ winter days in Denver gave a sample mean pollution index of $\bar{x}_1 = 43$. Previous studies show that $\sigma_1 = 21$. For Englewood (a suburb of Denver), a random sample of $n_2 = 14$ winter days gave a sample mean pollution index of $\bar{x}_2 = 36$. Previous studies show that $\sigma_2 = 15$. Assume the pollution index is normally distributed in both Englewood and Denver. Do these data indicate that the mean population pollution index of Englewood is different (either way) from that of Denver in the winter? Use a 1% level of significance.

17. *Survey: Outdoor Activities* A Michigan study concerning preference for outdoor activities used a questionnaire with a 6-point Likert-type response in which 1 designated "not important" and 6 designated "extremely important." A random sample of $n_1 = 46$ adults were asked about fishing as an outdoor activity. The mean response was $\bar{x}_1 = 4.9$. Another random sample of $n_2 = 51$ adults were asked about camping as an outdoor activity. For this group, the mean response was $\bar{x}_2 = 4.3$. From previous studies, it is known that $\sigma_1 = 1.5$ and $\sigma_2 = 1.2$. Does this indicate a difference (either way) regarding preference for camping versus preference for fishing as an outdoor activity? Use a 5% level of significance. *Note:* A *Likert scale* usually has to do with approval of or agreement with a statement in a questionnaire. For example, respondents are asked to indicate whether they "strongly agree," "agree," "disagree," or "strongly disagree" with the statement.

18. *Generation Gap: Education* Education influences attitude and lifestyle. Differences in education are a big factor in the "generation gap." Is the younger generation really better educated? Large surveys of people age 65 and older were taken in $n_1 = 32$ U.S. cities. The sample mean for these cities showed that $\bar{x}_1 = 15.2\%$ of the older adults had attended college. Large surveys of young adults (ages 25–34) were taken in $n_2 = 35$ U.S. cities. The sample mean for these cities showed that $\bar{x}_2 = 19.7\%$ of the young adults had attended college. From previous studies, it is known that $\sigma_1 = 7.2\%$ and $\sigma_2 = 5.2\%$ (Reference: *American Generations* by S. Mitchell). Does this information indicate that the population mean percentage of young adults who attended college is higher? Use $\alpha = 0.05$.

19. *Crime Rate: FBI* A random sample of $n_1 = 10$ regions in New England gave the following violent crime rates (per million population).

x_1: New England Crime Rate

3.5	3.7	4.0	3.9	3.3	4.1	1.8	4.8	2.9	3.1

Another random sample of $n_2 = 12$ regions in the Rocky Mountain states gave the following violent crime rates (per million population).

x_2: Rocky Mountain States

3.7	4.3	4.5	5.3	3.3	4.8	3.5	2.4	3.1	3.5	5.2	2.8

(Reference: *Crime in the United States*, Federal Bureau of Investigation.) Assume that the crime rate distribution is approximately normal in both regions.

i. Use a calculator to verify that $\bar{x}_1 \approx 3.51$, $s_1 \approx 0.81$, $\bar{x}_2 \approx 3.87$, and $s_2 \approx 0.94$.

ii. Do the data indicate that the violent crime rate in the Rocky Mountain region is higher than that in New England? Use $\alpha = 0.01$.

20. *Medical: Hay Fever* A random sample of $n_1 = 16$ communities in western Kansas gave the following information for people under 25 years of age.

x_1: Rate of hay fever per 1000 population for people under 25

98	90	120	128	92	123	112	93
125	95	125	117	97	122	127	88

A random sample of $n_2 = 14$ regions in western Kansas gave the following information for people over 50 years old.

x_2: Rate of hay fever per 1000 population for people over 50

95	110	101	97	112	88	110
79	115	100	89	114	85	96

(Reference: National Center for Health Statistics.)

i. Use a calculator to verify that $\bar{x}_1 \approx 109.50$, $s_1 \approx 15.41$, $\bar{x}_2 \approx 99.36$, and $s_2 \approx 11.57$.

ii. Assume that the hay fever rate in each age group has an approximately normal distribution. Do the data indicate that the age group over 50 has a lower rate of hay fever? Use $\alpha = 0.05$.

21. *Education: Tutoring* In the journal *Mental Retardation*, an article reported the results of a peer tutoring program to help mildly mentally retarded children learn to read. In the experiment, the mildly retarded children were randomly divided into two groups: the experimental group received peer tutoring along with regular instruction, and the control group received regular instruction with no peer tutoring. There were $n_1 = n_2 = 30$ children in each group. The Gates–MacGintie Reading Test was given to both groups before instruction began. For the experimental group, the mean score on the vocabulary portion of the test was $\bar{x}_1 = 344.5$, with sample standard deviation $s_1 = 49.1$. For the control group, the mean score on the same test was $\bar{x}_2 = 354.2$, with sample standard deviation $s_2 = 50.9$. Use a 5% level of significance to test the hypothesis that there was no difference in the vocabulary scores of the two groups before the instruction began.

22. *Education: Tutoring* In the article cited in Problem 21, the results of the following experiment were reported. Form 2 of the Gates–MacGintie Reading Test was administered to both an experimental group and a control group after 6 weeks of instruction, during which the experimental group received peer tutoring and the control group did not. For the experimental group $n_1 = 30$ children, the mean score on the vocabulary portion of the test was $\bar{x}_1 = 368.4$, with sample standard deviation $s_1 = 39.5$. The average score on the vocabulary portion of the test for the $n_2 = 30$ subjects in the control group was $\bar{x}_2 = 349.2$, with sample standard deviation $s_2 = 56.6$. Use a 1% level of significance to test the claim that the experimental group performed better than the control group.

23. *Wildlife: Fox Rabies* A study of fox rabies in southern Germany gave the following information about different regions and the occurrence of rabies in each region (Reference: B. Sayers et al., "A Pattern Analysis Study of a

Wildlife Rabies Epizootic," *Medical Informatics,* Vol. 2, pp. 11–34). Based on information from this article, a random sample of $n_1 = 16$ locations in region I gave the following information about the number of cases of fox rabies near that location.

x_1: Region I data	1	8	8	8	7	8	8	1
	3	3	3	2	5	1	4	6

A second random sample of $n_2 = 15$ locations in region II gave the following information about the number of cases of fox rabies near that location.

x_2: Region II data	1	1	3	1	4	8	5	4
	4	4	2	2	5	6	9	

i. Use a calculator with sample mean and sample standard deviation keys to verify that $\bar{x}_1 = 4.75$ with $s_1 \approx 2.82$ in region I and $\bar{x}_2 \approx 3.93$ with $s_2 \approx 2.43$ in region II.
ii. Does this information indicate that there is a difference (either way) in the mean number of cases of fox rabies between the two regions? Use a 5% level of significance. (Assume the distribution of rabies cases in both regions is mound-shaped and approximately normal.)

24. *Agriculture: Bell Peppers* The pathogen *Phytophthora capsici* causes bell peppers to wilt and die. Because bell peppers are an important commercial crop, this disease has undergone a great deal of agricultural research. It is thought that too much water aids the spread of the pathogen. Two fields are under study. The first step in the research project is to compare the mean soil water content for the two fields (Source: *Journal of Agricultural, Biological, and Environmental Statistics*, Vol. 2, No. 2). Units are percent water by volume of soil.

Field A samples, x_1:

10.2	10.7	15.5	10.4	9.9	10.0	16.6
15.1	15.2	13.8	14.1	11.4	11.5	11.0

Field B samples, x_2:

8.1	8.5	8.4	7.3	8.0	7.1	13.9	12.2
13.4	11.3	12.6	12.6	12.7	12.4	11.3	12.5

i. Use a calculator with mean and standard deviation keys to verify that $\bar{x}_1 \approx 12.53$, $s_1 \approx 2.39$, $\bar{x}_2 \approx 10.77$, and $s_2 = 2.40$.
ii. Assuming the distribution of soil water content in each field is mound-shaped and symmetric, use a 5% level of significance to test the claim that field A has, on average, a higher soil water content than field B.

25. *Management: Lost Time* In her book *Red Ink Behaviors*, Jean Hollands reports on the assessment of leading Silicon Valley companies regarding a manager's lost time due to inappropriate behavior of employees. Consider the following independent random variables. The first variable x_1 measures a manager's hours per week lost due to hot tempers, flaming e-mails, and general unproductive tensions:

x_1:	1	5	8	4	2	4	10

The variable x_2 measures a manager's hours per week lost due to disputes regarding technical workers' superior attitudes that their colleagues are "dumb and dispensable":

x_2:	10	5	4	7	9	4	10	3

i. Use a calculator with sample mean and standard deviation keys to verify that $\bar{x}_1 \approx 4.86$, $s_1 \approx 3.18$, $\bar{x}_2 = 6.5$, and $s_2 \approx 2.88$.
ii. Does the information indicate that the population mean time lost due to hot tempers is different (either way) from population mean time lost due to disputes arising from technical workers' superior attitudes? Use $\alpha = 0.05$. Assume that the two lost-time population distributions are mound-shaped and symmetric.

26. *Management: Intimidators and Stressors* This problem is based on information regarding productivity in leading Silicon Valley companies (see reference in Problem 25). In large corporations, an "intimidator" is an employee who tries to stop communication, sometimes sabotages others, and, above all, likes to listen to him- or herself talk. Let x_1 be a random variable representing productive hours per week lost by peer employees of an intimidator.

x_1:	8	3	6	2	2	5	2

A "stressor" is an employee with a hot temper that leads to unproductive tantrums in corporate society. Let x_2 be a random variable representing productive hours per week lost by peer employees of a stressor.

x_2:	3	3	10	7	6	2	5	8

i. Use a calculator with mean and standard deviation keys to verify that $\bar{x}_1 = 4.00$, $s_1 \approx 2.38$, $\bar{x}_2 = 5.5$, and $s_2 \approx 2.78$.
ii. Assuming the variables x_1 and x_2 are independent, do the data indicate that the population mean time lost due to stressors is greater than the population mean time lost due to intimidators? Use a 5% level of significance. (Assume the population distributions of time lost due to intimidators and time lost due to stressors are each mound-shaped and symmetric.)

27. *Expand Your Knowledge: Software Approximation for Degrees of Freedom* Given x_1 and x_2 distributions that are normal or approximately normal with unknown σ_1 and σ_2, the value of t corresponding to $\bar{x}_1 - \bar{x}_2$ has a distribution that is approximated by a Student's t distribution. We use the convention that the degrees of freedom are approximately the smaller of $n_1 - 1$ and $n_2 - 1$. However, a more accurate estimate for the appropriate degrees of freedom is given by Satterthwaite's formula:

$$d.f. \approx \frac{\left(\dfrac{s_1^2}{n_1} + \dfrac{s_2^2}{n_2}\right)^2}{\dfrac{1}{n_1 - 1}\left(\dfrac{s_1^2}{n_1}\right)^2 + \dfrac{1}{n_2 - 1}\left(\dfrac{s_2^2}{n_2}\right)^2}$$

where s_1, s_2, n_1, and n_2 are the respective sample standard deviations and sample sizes of independent random samples from the x_1 and x_2 distributions. This is the approximation used by most statistical software. When both n_1 and n_2 are 5 or larger, it is quite accurate. The degrees of freedom computed from this formula are either truncated or not rounded.

(a) In Problem 19, we tested whether the population average crime rate μ_2 in the Rocky Mountain region is higher than that in New England, μ_1. The data were $n_1 = 10$, $\bar{x}_1 \approx 3.51$, $s_1 \approx 0.81$, $n_2 = 12$, $\bar{x}_2 \approx 3.87$ and $s_2 \approx 0.94$. Use Satterthwaite's formula to compute the degrees of freedom for the Student's t distribution.

(b) When you did Problem 19, you followed the convention that degrees of freedom $d.f.$ = *smaller* of $n_1 = 1$ and $n_2 = 1$. Compare this value of $d.f.$ with that found with Satterthwaite's formula.

28. *Expand Your Knowledge: Pooled Two-Sample Procedure* Consider independent random samples from two populations that are normal or approximately normal, or the case in which both sample sizes are at least 30. Then, if σ_1 and σ_2 are unknown but we have reason to believe that $\sigma_1 = \sigma_2$, we can pool the standard deviations. Using sample sizes n_1 and n_2, the sample test statistic $\bar{x}_1 - \bar{x}_2$ has a Student's t distribution, where

$$t = \frac{\bar{x}_1 - \bar{x}_2}{s\sqrt{\frac{1}{n_1} + \frac{1}{n_2}}} \text{ with degrees of freedom } d.f. = n_1 + n_2 - 2$$

and where the **pooled standard deviation s** is

$$s = \sqrt{\frac{(n_1 - 1)s_1^2 + (n_2 - 1)s_2^2}{n_1 + n_2 - 2}}$$

Note: With statistical software, select the pooled variance or equal variance options.

(a) There are many situations in which we want to compare means from populations having standard deviations that are equal. This method applies even if the standard deviations are known to be only approximately equal (see Section 10.4 for methods to test that $\sigma_1 = \sigma_2$). Consider Problem 23 regarding average incidence of fox rabies in two regions. For region I, $n_1 = 16$, $\bar{x}_1 = 4.75$, and $s_1 \approx 2.82$ and for region II, $n_2 = 15$, $\bar{x}_2 \approx 3.93$, and $s_2 \approx 2.43$. The two sample standard deviations are sufficiently close that we can assume $\sigma_1 = \sigma_2$. Use the method of pooled standard deviation to redo Problem 23, where we tested if there was a difference in population mean average incidence of rabies at the 5% level of significance.

(b) Compare the t value calculated in part (a) using the pooled standard deviation with the t value calculated in Problem 23 using the unpooled standard deviation. Compare the degrees of freedom for the sample test statistic. Compare the conclusions.

29. *Federal Tax Money: Art Funding* Would you favor spending more federal tax money on the arts? This question was asked by a research group on behalf of *The National Institute* (Reference: *Painting by Numbers*, J. Wypijewski, University of California Press). Of a random sample of $n_1 = 220$ women, $r_1 = 59$ responded yes. Another random sample of $n_2 = 175$ men showed that $r_2 = 56$ responded yes. Does this information indicate a difference (either way) between the population proportion of women and the population proportion of men who favor spending more federal tax dollars on the arts? Use $\alpha = 0.05$.

30. *Art Funding: Politics* Would you favor spending more federal tax money on the arts? This question was asked by a research group on behalf of *The National Institute* (Reference: *Painting by Numbers*, J. Wypijewski, University of California Press). Of a random sample of $n_1 = 93$ politically conservative voters, $r_1 = 21$ responded yes. Another random sample of $n_2 = 83$ politically moderate voters showed that $r_2 = 22$ responded yes. Does this information indicate that the population proportion of conservative voters inclined to spend more federal tax money on funding the arts is less than the proportion of moderate voters so inclined? Use $\alpha = 0.05$.

31. *Sociology: High School Dropouts* This problem is based on information taken from *Life in America's Fifty States* by G. S. Thomas. A random sample of $n_1 = 153$ people ages 16 to 19 was taken from the island of Oahu, Hawaii, and 12 were found to be high school dropouts. Another random sample of $n_2 = 128$ people ages 16 to 19 was taken from Sweetwater County, Wyoming, and 7 were found to be high school dropouts. Do these data indicate that the population proportion of high school dropouts on Oahu is different (either way) from that of Sweetwater County? Use a 1% level of significance.

32. *Political Science: Voters* A random sample of $n_1 = 288$ voters registered in the state of California showed that 141 voted in the last general election. A random sample of $n_2 = 216$ registered voters in the state of Colorado showed that 125 voted in the most recent general election. (See reference in Problem 31.) Do these data indicate that the population proportion of voter turnout in Colorado is higher than that in California? Use a 5% level of significance.

33. *Extraterrestrials: Believe It?* Based on information from *Harper's Index*, $r_1 = 37$ people out of a random sample of $n_1 = 100$ adult Americans who did not attend college believe in extraterrestrials. However, out of a random sample of $n_2 = 100$ adult Americans who did attend college, $r_2 = 47$ claim that they believe in extraterrestrials. Does this indicate that the proportion of people who attended college and who believe in extraterrestrials is higher than the proportion who did not attend college but believe in extraterrestrials? Use $\alpha = 0.01$.

34. *Art: Politics* Do you prefer paintings in which the people are fully clothed? This question was asked by a professional survey group on behalf of the National Arts Society (see reference in Problem 30). A random sample of $n_1 = 59$ people who are conservative voters showed that $r_1 = 45$ said yes. Another random sample of $n_2 = 62$ people who are liberal voters showed that $r_2 = 36$ said yes. Does this indicate that the population proportion of conservative voters who prefer art with fully clothed people is higher than that of liberal voters? Use $\alpha = 0.05$.

35. *Sociology: Trusting People* Generally speaking, would you say that most people can be trusted? A random sample of $n_1 = 250$ people in Chicago ages 18–25 showed that $r_1 = 45$ said yes. Another random sample of $n_2 = 280$ people in Chicago ages 35–45 showed that $r_2 = 71$ said yes (based on information from the *National Opinion Research Center,* University of Chicago). Does this indicate that the population proportion of trusting people in Chicago is higher for the older group? Use $\alpha = 0.05$.

36. *Critical Region Method: Testing $\mu_1 - \mu_2$ σ_1, σ_2 Known* Redo Problem 15 using the critical region method and compare your results to those obtained using the P-value method.

37. *Critical Region Method: Testing $\mu_1 - \mu_2$ σ_1, σ_2 Unknown* Redo Problem 19 using the critical region method and compare your results to those obtained using the P-value method.

38. *Critical Region Method: Testing $p_1 - p_2$* Redo Problem 29 using the critical region method and compare your results to those obtained using the P-value method.

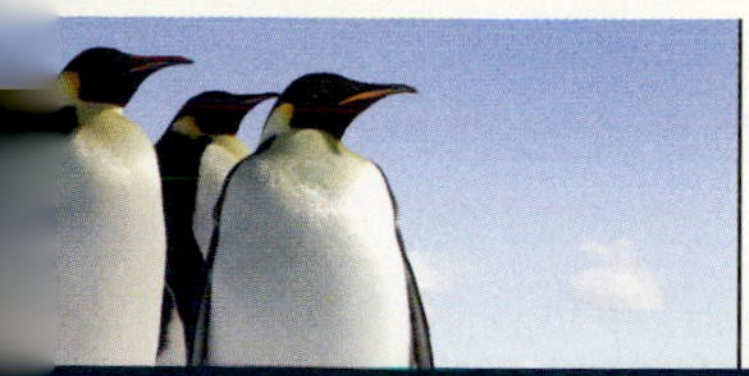

CHAPTER REVIEW

SUMMARY

Hypothesis testing is a major component of inferential statistics. In hypothesis testing, we propose a specific value for the population parameter in question. Then we use sample data from a random sample and probability to determine whether or not to reject this specific value for the parameter.

Basic components of a hypothesis test are:

- The *null hypothesis* H_0 states that a parameter equals a specific value.
- The *alternate hypothesis* H_1 states that the parameter is greater than, less than, or simply not equal to the value specified in H_0.
- The *level of significance* α of the test is the probability of rejecting H_0 when it is true.
- The *sample test statistic* corresponding to the parameter in H_0 is computed from a random sample and appropriate sampling distribution.
- Assuming H_0 is true, the probability that a sample test statistic will take on a value as extreme as, or more extreme than, the observed sample test statistic is the *P-value* of the test. The P-value is computed by using the sample test statistic, the corresponding sampling distribution, H_0, and H_1.
- If P-value $\leq \alpha$, we reject H_0. If P-value $> \alpha$, we fail to reject H_0.
- We say that sample data are *significant* if we can reject H_0.

An alternative way to conclude a test of hypotheses is to use critical regions based on the alternate hypothesis and α. Critical values z_0 are found in Table 5(c) of Appendix II. Critical values t_0 are found in Table 6 of Appendix II. If the sample test statistic falls beyond the critical values—that is, in the critical region—we reject H_0.

The methods of hypothesis testing are very general, and we will see them used again in later chapters. In this chapter, we looked at tests involving

- Parameter μ. Use standard normal or Student's t distribution. See procedure displays in Section 8.2.
- Parameter p. Use standard normal distribution. See procedure displays in Section 8.3.
- Paired difference test for difference of means from dependent populations. Use Student's t distribution. See procedure displays in Section 8.4.
- Parameter $\mu_1 - \mu_2$ from independent populations. Use standard normal or Student's t distribution. See procedure displays in Section 8.5.
- Parameter $p_1 - p_2$ from independent populations. Use standard normal distribution. See procedure displays in Section 8.5.

FINDING THE *P*-VALUE CORRESPONDING TO A SAMPLE TEST STATISTIC

Use the appropriate sampling distribution as described in procedure displays for each of the various tests.

Left-Tailed Test

P-value = area to the left of the sample test statistic.

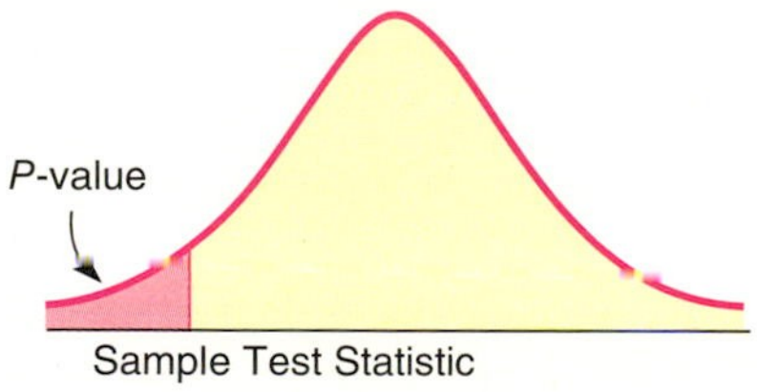

Right-Tailed Test

P-value = area to the right of the sample test statistic.

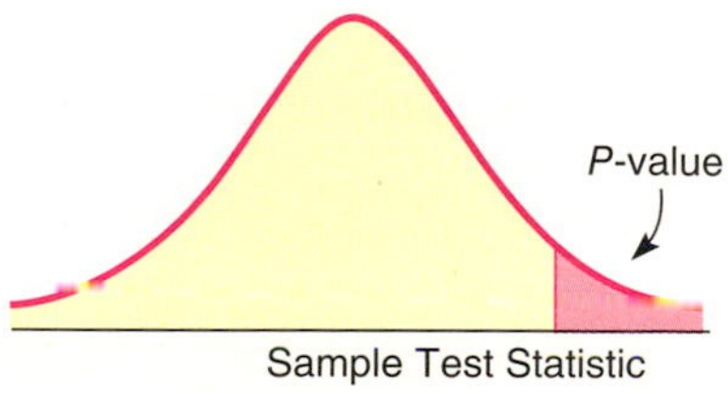

Two-Tailed Test

Sample test statistic lies to *left* of center P-value = twice the area to the left of sample test statistic

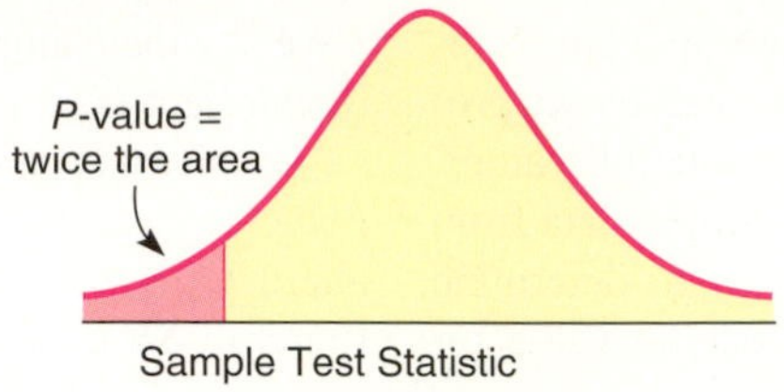

Sample test statistic lies to *right* of center P-value = twice area to the right of sample test statistic

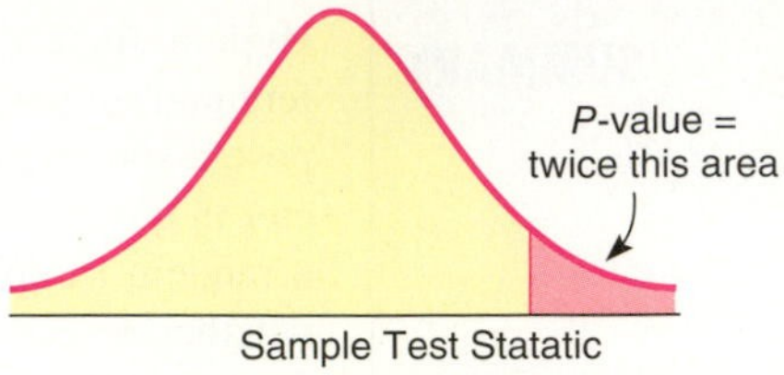

Sampling Distributions for Inferences Regarding μ or p

Parameter	Condition	Sampling Distribution
μ	σ is known and x has a normal distribution or $n \geq 30$	Normal Distribution
μ	σ is not known and x has a normal or mound-shaped symmetric distribution or $n \geq 30$	Student's t Distribution with $d.f. = n - 1$
p	$np > 5$ and $n(1 - p) > 5$	Normal Distribution

IMPORTANT WORDS & SYMBOLS

Section 8.1

Hypothesis testing 437
Hypothesis 437
Null hypothesis H_0 437
Alternate hypothesis H_1 437
Right-tailed test 438
Two-tailed test 438
Left-tailed test 438
Sample test statistic 439
P-value 441
Type I error 442
Type II error 442
Level of significance of a test and the probability of a type I error, α 442
Probability of a type II error, β 443
Power of a test $(1 - \beta)$ 443
Statistical significance 444

Section 8.2

d.f. for testing μ when σ is unknown 455
Critical region method 459
Critical value 459

Section 8.3

Criteria for using normal approximation to binomial, $np > 5$ and $nq > 5$ 469

Section 8.4

Paired data 479
Dependent samples 479

Section 8.5

Independent samples 494
d.f. for testing $\mu_1 - \mu_2$ when σ_1 and σ_2 are unknown 500
Pooled standard deviation 503
Pooled estimate of proportion $\bar{p}$ 504

VIEWPOINT Will It Rain?

Do cloud-seed experiments ever work? If you seed the clouds, will it rain? If it does rain, who will benefit? Who will be displeased by the rain? If you seed the clouds and nothing happens, will taxpayers (who support the effort) complain or rejoice? Maybe this should be studied over a remote island—such as Tasmania (near Australia). Using what you already know about statistical testing, you can conduct your own tests, given the appropriate data. Remember, there are sociological questions (pleased/displeased with result) as well as technical questions (number of inches of rain produced). For data regarding cloud-seeding experiments over Tasmania, visit the Brase/Brase statistics site at **http://www.cengage.com/statistics/brase11e** *and find the link to DASL, the Carnegie Mellon University Data and Story Library. From the DASL site, look under Datasets for Cloud.*

CHAPTER REVIEW PROBLEMS

1. *Statistical Literacy* When testing μ or the difference of means $\mu_1 - \mu_2$ from independent populations, how do we decide whether to use the standard normal distribution or a Student's t distribution?

2. *Statistical Literacy* What do we mean when we say a test is *significant*? Does this necessarily mean the results are important?

3. *Critical Thinking* All other conditions being equal, does a larger sample size increase or decrease the corresponding magnitude of the z or t value of the sample test statistic?

4. *Critical Thinking* All other conditions being equal, does a z or t value with larger magnitude have a larger or smaller corresponding P-value?

Before you solve each problem below, first categorize it by answering the following question: Are we testing a single mean, a difference of means, a paired difference, a single proportion, or a difference of proportions? Assume underlying population distributions are mound-shaped and symmetric for problems with small samples that involve testing a mean or difference of means. Then provide the following information for Problems 5–18.

(a) What is the level of significance? State the null and alternate hypotheses.
(b) *Check Requirements* What sampling distribution will you use? What assumptions are you making? What is the value of the sample test statistic?
(c) Find (or estimate) the P-value. Sketch the sampling distribution and show the area corresponding to the P-value.
(d) Based on your answers in parts (a) to (c), will you reject or fail to reject the null hypothesis? Are the data statistically significant at level α?
(e) ***Interpret*** your conclusion in the context of the application.
Note: For degrees of freedom *d.f.* not in the Student's t table, use the closest *d.f.* that is *smaller*. In some situations, this choice of *d.f.* may increase the P-value by a small amount and therefore produce a slightly more "conservative" answer. Answers may vary due to rounding.

5. *Vehicles: Mileage* Based on information in *Statistical Abstract of the United States* (116th Edition), the average annual miles driven per vehicle in the United States is 11.1 thousand miles, with $\sigma \approx 600$ miles. Suppose that a random sample of 36 vehicles owned by residents of Chicago showed that the average mileage driven last year was 10.8 thousand miles. Does this indicate that the average miles driven per vehicle in Chicago is different from (higher or lower than) the national average? Use a 0.05 level of significance.

6. *Student Life: Employment* Professor Jennings claims that only 35% of the students at Flora College work while attending school. Dean Renata thinks that the professor has underestimated the number of students with part-time or full-time jobs. A random sample of 81 students shows that 39 have jobs. Do the data indicate that more than 35% of the students have jobs? (Use a 5% level of significance.)

7. *Toys: Electric Trains* The Toylot Company makes an electric train with a motor that it claims will draw an average of only 0.8 ampere (A) under a normal load. A sample of nine motors was tested, and it was found that the mean current was $\bar{x} = 1.4$ A, with a sample standard deviation of $s = 0.41$ A. Do the data indicate that the Toylot claim of 0.8 A is too low? (Use a 1% level of significance.)

8. *Highways: Reflective Paint* The highway department is testing two types of reflecting paint for concrete bridge end pillars. The two kinds of paint are alike in every respect except that one is orange and the other is yellow. The orange paint is applied to 12 bridges, and the yellow paint is applied to 12 bridges. After a period of 1 year, reflectometer readings were made on all these bridge end pillars. (A higher reading means better visibility.) For the orange paint, the

mean reflectometer reading was $\bar{x}_1 = 9.4$, with standard deviation $s_1 = 2.1$. For the yellow paint, the mean was $\bar{x}_2 = 6.9$, with standard deviation $s_2 = 2.0$. Based on these data, can we conclude that the yellow paint has less visibility after 1 year? (Use a 1% level of significance.)

9. *Medical: Plasma Compress* A hospital reported that the normal death rate for patients with extensive burns (more than 40% of skin area) has been significantly reduced by the use of new fluid plasma compresses. Before the new treatment, the mortality rate for extensively burned patients was about 60%. Using the new compresses, the hospital found that only 40 of 90 patients with extensive burns died. Use a 1% level of significance to test the claim that the mortality rate has dropped.

10. *Bus Lines: Schedules* A comparison is made between two bus lines to determine if arrival times of their regular buses from Denver to Durango are off schedule by the same amount of time. For 51 randomly selected runs, bus line A was observed to be off schedule an average time of 53 minutes, with standard deviation 19 minutes. For 60 randomly selected runs, bus line B was observed to be off schedule an average of 62 minutes, with standard deviation 15 minutes. Do the data indicate a significant difference in average off-schedule times? Use a 5% level of significance.

11. *Matches: Number per Box* The Nero Match Company sells matchboxes that are supposed to have an average of 40 matches per box, with $\sigma = 9$. A random sample of 94 Nero matchboxes shows the average number of matches per box to be 43.1. Using a 1% level of significance, can you say that the average number of matches per box is more than 40?

12. *Magazines: Subscriptions* A study is made of residents in Phoenix and its suburbs concerning the proportion of residents who subscribe to *Sporting News*. A random sample of 88 urban residents showed that 12 subscribed, and a random sample of 97 suburban residents showed that 18 subscribed. Does this indicate that a higher proportion of suburban residents subscribe to *Sporting News*? (Use a 5% level of significance.)

13. *Archaeology: Arrowheads* The Wind Mountain archaeological site is in southwest New Mexico. Prehistoric Native Americans called Anasazi once lived and hunted small game in this region. A stemmed projectile point is an arrowhead that has a notch on each side of the base. Both stemmed and stemless projectile points were found at the Wind Mountain site. A random sample of $n_1 = 55$ stemmed projectile points showed the mean length to be $\bar{x}_1 = 3.0$ cm, with sample standard deviation $s_1 = 0.8$ cm. Another random sample of $n_2 = 51$ stemless projectile points showed the mean length to be $\bar{x}_2 = 2.7$ cm, with $s_2 = 0.9$ cm (Source: *Mimbres Mogollon Archaeology*, by A. I. Woosley and A. J. McIntyre, University of New Mexico Press). Do these data indicate a difference (either way) in the population mean length of the two types of projectile points? Use a 5% level of significance.

14. *Civil Service: College Degrees* The Congressional Budget Office reports that 36% of federal civilian employees have a bachelor's degree or higher (*The Wall Street Journal*). A random sample of 120 employees in the private sector showed that 33 have a bachelor's degree or higher. Does this indicate that the percentage of employees holding bachelor's degrees or higher in the private sector is less than that in the federal civilian sector? Use $\alpha = 0.05$.

15. *Vending Machines: Coffee* A machine in the student lounge dispenses coffee. The average cup of coffee is supposed to contain 7.0 ounces. Eight cups of coffee from this machine show the average content to be 7.3 ounces with a standard deviation of 0.5 ounce. Do you think that the machine has slipped out of adjustment and that the average amount of coffee per cup is different from 7 ounces? Use a 5% level of significance.

16. *Psychology: Creative Thinking* Six sets of identical twins were randomly selected from a population of identical twins. One child was taken at random from each pair to form an experimental group. These children participated in a program designed to promote creative thinking. The other child from each pair was part of the control group that did not participate in the program to promote creative thinking. At the end of the program, a creative problem-solving test was given, with the results shown in the following table:

Twin pair	A	B	C	D	E	F
Experimental group	53	35	12	25	33	47
Control group	39	21	5	18	21	42

Higher scores indicate better performance in creative problem solving. Do the data support the claim that the program of the experimental group did promote creative problem solving? (Use $\alpha = 0.01$.)

17. *Marketing: Sporting Goods* A marketing consultant was hired to visit a random sample of five sporting goods stores across the state of California. Each store was part of a large franchise of sporting goods stores. The consultant taught the managers of each store better ways to advertise and display their goods. The net sales for 1 month before and 1 month after the consultant's visit were recorded as follows for each store (in thousands of dollars):

Store	1	2	3	4	5
Before visit	57.1	94.6	49.2	77.4	43.2
After visit	63.5	101.8	57.8	81.2	41.9

Do the data indicate that the average net sales improved? (Use $\alpha = 0.05$.)

18. *Sports Car: Fuel Injection* The manufacturer of a sports car claims that the fuel injection system lasts 48 months before it needs to be replaced. A consumer group tests this claim by surveying a random sample of 10 owners who had the fuel injection system replaced. The ages of the cars at the time of replacement were (in months):

29 42 49 48 53 46 30 51 42 52

i. Use your calculator to verify that the mean age of a car when the fuel injection system fails is $\bar{x} = 44.2$ months, with standard deviation $s \approx 8.61$ months.

ii. Test the claim that the fuel injection system lasts less than an average of 48 months before needing replacement. Use a 5% level of significance.

DATA HIGHLIGHTS: GROUP PROJECTS

Break into small groups and discuss the following topics. Organize a brief outline in which you summarize the main points of your group discussion.

1. "With Sampling, There Is Too a Free Lunch"—This is a headline that appeared in *The Wall Street Journal.* The article is about food product samples available at grocery stores. Giving out food samples is expensive and labor-intensive. It clogs supermarket aisles. It is risky. What if a customer tries an item and spits it out on the floor or says the product is awful? It creates litter. Some customers drop toothpicks or small paper cups on the floor or spill the product. However, the budget that companies are willing to spend to have their products sampled is growing. The director of communications for Bigg's "hypermarket" (a combination grocery and general-merchandise store) says that more than 60% of customers sample products and about 37% of those who sample buy the product.
   (a) Let's test the hypothesis that 60% of customers sample a particular product. What is the null hypothesis? Do you believe that the percentage of customers who sample products is less than, more than, or just different from 60%? What will you use for the alternate hypothesis?
   (b) Choose a level of significance α.
   (c) Go to a grocery store when special products are being sampled (not just the usual in-house store samples often available at the deli or bakery). Count the number of customers going by the display when a sample is available and the number of customers who try the sample. Be sure the number of customers n is large enough to use the normal distribution to approximate the binomial.
   (d) Using your sample data, conclude the hypothesis test. What is your conclusion?
   (e) Do you think different food products might have a higher or lower percentage of customers trying them? For instance, does a higher percentage of customers try samples of pizza than samples of yogurt? How could you use statistics to justify your answer?
   (f) Do you want to include young children in your sample? Do they pick up items to include in the customer's basket, or do they just munch the samples?
2. "Sweets May Not Be Culprit in Hyper Kids"—This is a *USA Today* headliner reporting results of a study that appeared in the *New England Journal of Medicine.* In this study, the subjects were 25 normal preschoolers, aged 3 to 5, and 23 kids, aged 6 to 10, who had been described as "sensitive to sugar." The kids and their families were put on three different diets for 3 weeks each. One diet was high in sugar, one was low in sugar and contained aspartame, and one was low in sugar and contained saccharin. The diets were all free of additives, artificial food coloring, preservatives, and chocolate. All food in the household was removed, and the meals were delivered to the families. Researchers gathered information about the kids' behavior from parents, babysitters, and teachers. In addition, researchers tested the kids for memory, concentration, reading, and math skills. The result: "We couldn't find any difference in terms of their behavior or their learning on any of the three diets," says Mark Wolraich, professor of pediatrics at Vanderbilt University Medical Center who oversaw the project. In another interview, Dr. Wolraich is quoted as saying, "Our study would say there is no evidence sugar has an adverse effect on children's behavior."
   (a) This research involved comparing several means, not just two. (An introduction to such methods, called *analysis of variance,* is found in Chapter 10.) However, let us take a simplified view of the problem and consider the difference of behavior when children consumed the diet with sugar compared with their behavior when they consumed the diet with aspartame and low sugar. List some variables that might be measured to reflect the behavior of the children.

(b) Let's assume that the general null hypothesis was that there is no difference in children's behavior when they have a diet high in sugar. Was the evidence sufficient to allow the researchers to reject the null hypothesis and conclude that there are differences in children's behavior when they have a diet high in sugar? When we cannot reject H_0, have we *proved* that H_0 is true? In your own words, paraphrase the comments made by Dr. Wolraich.

LINKING CONCEPTS: WRITING PROJECTS

Discuss each of the following topics in class or review the topics on your own. Then write a brief but complete essay in which you summarize the main points. Please include formulas and graphs as appropriate.

The most important questions in life usually cannot be answered with absolute certainty. Many important questions are answered by giving an estimate and a measure of confidence in the estimate. This was the focus of Chapter 7. However, sometimes important questions must be answered in a more straightforward manner by a simple yes or no. Hypothesis testing is the statistical process of answering questions with a straightforward yes or no *and* providing an estimate of the risk in accepting the answer.

1. Review and discuss type I and type II errors associated with hypothesis testing.
2. Review and discuss the level of significance and power of a statistical test.
3. The following statements are very important. Give them some careful thought and discuss them.
   (a) When we fail to reject the null hypothesis, we do not claim that it is absolutely true. We simply claim that at the given level of significance, the data were not sufficient to reject the null hypothesis.
   (b) When we accept the alternate hypothesis, we do not claim that the null hypothesis is absolutely false. We do claim that at the given level of significance, the data presented enough evidence to reject the null hypothesis.
4. In the text, it is said that a statistical test is a package of five basic ingredients. List these ingredients, discuss them in class, and write a short description of how these ingredients relate to the above discussion questions.
5. As access to computers becomes more and more prevalent, we see the P-value reported in hypothesis testing more frequently. Review the use of the P-value in hypothesis testing. What is the difference between the level of significance of a test and the P-value? Considering both the P-value and level of significance, under what conditions do we reject or fail to reject the null hypothesis?

USING TECHNOLOGY

Simulation

Recall that the level of significance α is the probability of mistakenly rejecting a true null hypothesis. If $\alpha = 0.05$, then we expect to mistakenly reject a true null hypothesis about 5% of the time. The following simulation conducted with Minitab demonstrates this phenomenon.

We draw 40 random samples of size 50 from a population that is normally distributed with mean $\mu = 30$ and standard deviation $\sigma = 2.5$. The display shows the results of a hypothesis test with

$$H_0: \mu = 30 \quad H_1: \mu > 30$$

for each of the 40 samples labeled C1 through C40. Because each of the 40 samples is drawn from a population with mean $\mu = 30$, the null hypothesis H_0: $\mu = 30$ is true for the test based on each sample. However, as the display shows, for some samples we reject the true null hypothesis.

(a) How many of the 40 samples have a sample mean $\bar{x}$ above $\mu = 30$? below $\mu = 30$?

(b) Look at the P-value of the sample statistic $\bar{x}$ in each of the 40 samples. How many P-values are less than or equal to $\alpha = 0.05$? What percent of the P-values are less than or equal to α? What percent of the samples have us reject H_0 when, in fact, each of the samples was drawn from a normal distribution with $\mu = 30$, as hypothesized in the null hypothesis?

(c) If you have access to computer or calculator technology that creates random samples from a normal distribution with a specified mean and standard deviation, repeat this simulation. Do you expect to get the same results? Why or why not?

Minitab Display: Random samples of size 50 from a normal population with $\mu = 30$ and $\sigma = 2.5$

Z-Test

Test of mu = 30.000 vs. mu > 30.000

The assumed sigma = 2.50

Variable	N	Mean	StDev	SE Mean	Z	P
C1	50	30.002	2.776	0.354	0.01	0.50
C2	50	30.120	2.511	0.354	0.34	0.37
C3	50	30.032	2.721	0.354	0.09	0.46
C4	50	30.504	2.138	0.354	1.43	0.077
C5	50	29.901	2.496	0.354	−0.28	0.61
C6	50	30.059	2.836	0.354	0.17	0.43
C7	50	30.443	2.519	0.354	1.25	0.11
C8	50	29.775	2.530	0.354	−0.64	0.74
C9	50	30.188	2.204	0.354	0.53	0.30
C10	50	29.907	2.302	0.354	−0.26	0.60
C11	50	30.036	2.762	0.354	0.10	0.46
C12	50	30.656	2.399	0.354	1.86	0.032
C13	50	30.158	2.884	0.354	0.45	0.33
C14	50	29.830	3.129	0.354	−0.48	0.68
C15	50	30.308	2.241	0.354	0.87	0.19
C16	50	29.751	2.165	0.354	−0.70	0.76
C17	50	29.833	2.358	0.354	−0.47	0.68
C18	50	29.741	2.836	0.354	−0.73	0.77
C19	50	30.441	2.194	0.354	1.25	0.11
C20	50	29.820	2.156	0.354	−0.51	0.69
C21	50	29.611	2.360	0.354	−1.10	0.86
C22	50	30.569	2.659	0.354	1.61	0.054
C23	50	30.294	2.302	0.354	0.83	0.20
C24	50	29.978	2.298	0.354	−0.06	0.53
C25	50	29.836	2.438	0.354	−0.46	0.68
C26	50	30.102	2.322	0.354	0.29	0.39
C27	50	30.066	2.266	0.354	0.19	0.43
C28	50	29.071	2.219	0.354	−2.63	1.00
C29	50	30.597	2.426	0.354	1.69	0.046
C30	50	30.092	2.296	0.354	0.26	0.40
C31	50	29.803	2.495	0.354	−0.56	0.71
C32	50	29.546	2.335	0.354	−1.28	0.90
C33	50	29.702	1.902	0.354	−0.84	0.80
C34	50	29.233	2.657	0.354	−2.17	0.98
C35	50	30.097	2.472	0.354	0.28	0.39
C36	50	29.733	2.588	0.354	−0.76	0.78
C37	50	30.379	2.976	0.354	1.07	0.14
C38	50	29.424	2.827	0.354	−1.63	0.95
C39	50	30.288	2.396	0.354	0.81	0.21
C40	50	30.195	3.051	0.354	0.55	0.29

Technology Hints

TI-84Plus/TI-83Plus/TI-*n*spire (with TI-84Plus keypad)

Press **STAT** and select **EDIT.** Highlight the list name, such as L1. Then press **MATH,** select **PRB,** and highlight **6:randNorm(μ, σ, sample size**). Press enter. Fill in the values of $\mu = 30$, $\sigma = 2.5$, and sample size = 50. Press enter. Now list L1 contains a random sample from the normal distribution specified.

To test the hypothesis H_0: $\mu = 30$ against H_1: $\mu > 30$, press **STAT,** select **TESTS,** and use option **1:Z-Test.** Fill in the value 30 for μ_0, 2.5 for σ, and $> \mu_0$. The output provides the value of the sample statistic $\bar{x}$, its corresponding z value, and the P-value of the sample statistic.

Excel 2010

To draw random samples from a normal distribution with $\mu = 30$ and $\sigma = 2.5$, click the **Data** tab on the home ribbon and select **Data Analysis** in the Analysis Group. In the dialogue box, select **Random Number Generator.** In the first dialogue box, select **Normal** for the Distribution. In the resulting dialogue box shown below, the number of variables is the number of samples. Fill in the rest of the dialogue box as shown.

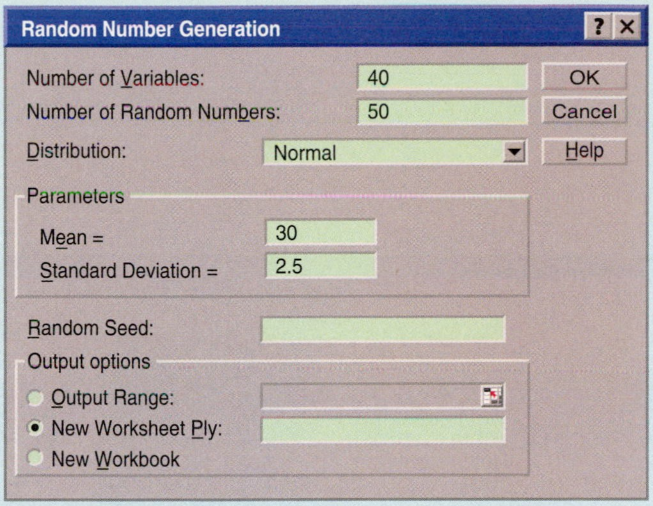

To conduct a hypothesis test of H_0: $\mu = 30$ against H_1: $\mu > 30$, click the **Insert Function** (f_x) on the home screen. For the category, select **Statistical** and then select the function **Z.TEST**. Fill in the dialogue box with the array containing the random numbers, 30 for X, the value in H_0, and 2.5 for sigma.

Minitab

To generate random samples from a normal distribution, use the menu choices **Calc ➤ Random Data ➤ Normal.** In the dialogue box, the number of rows refers to the sample size. Use 50 rows. Then designate the columns for the samples. Using C1–C40 will generate 40 random samples and put the samples in columns C1 through C40.

To test the hypothesis H_0: $\mu = 30$ against H_1: $\mu > 30$, use the menu choices **Stat ➤ Basic Statistics ➤ 1-SampleZ.** Use columns C1–C40 as the variables. Fill in 30 for the test mean, use "greater than" for the alternate hypothesis, and use 2.5 for sigma.

SPSS

SPSS uses a Student's t distribution to test the mean and difference of means. SPSS uses the sample standard deviation s even if the population σ is known. Use the menu choices **Analyze ➤ Compare Means** and then **One- Sample T Test** or **Independent-Sample T Tests** for tests of a single mean or a difference of means, respectively. In the dialogue box, fill in the test value of the null hypothesis.

To generate 40 random samples of size $n = 50$ from a normal distribution with $\mu = 30$ and $\sigma = 2.5$, first enter consecutive integers from 1 to 50 in a column of the data editor. Then, under variable view, enter the variable names Sample1 through Sample40. Use the menu choices **Transform ➤ Compute Variable**. In the dialogue box, use Sample1 for the target variable, select **Random Numbers** for the Function Group and then select the function **Rv.Normal(mean, stddev).** Use 30 for the mean and 2.5 for the standard deviation. Continue until you have 40 samples. To sample from other distributions, use appropriate functions in the Compute dialogue box.

The SPSS display shows the test results (H_0: $\mu = 30$; H_1: $\mu \neq 30$) for a sample of size $n = 50$ drawn from a normal distribution with $\mu = 30$ and $\sigma = 2.5$. The P-value is given as the significance for a two-tailed test. For a one-tailed test, divide the significance by 2. In the display, the significance is 0.360 for a two-tailed test. So, for a one-tailed test, the P-value is 0.360/2, or 0.180.

SPSS Display

T-Test

One-Sample Statistics

	N	Mean	Std. Deviation	Std. Error Mean
SAMPLE1	50	30.3228	2.46936	.34922

One-Sample Test

	Test Value = 30					
					90% Confidence Interval of the Difference	
	t	df	Sig. (2-tailed)	Mean Difference	Lower	Upper
SAMPLE1	.924	49	.360	.3228	−.2627	.9083

9

Mangostock/Shutterstock.com

Portrait of Rene Descartes c. 1649 (after) Frans Hals/Louvre/Giraudon/The Bridgeman Art Library

When it is not in our power to determine what is true, we ought to follow what is most probable.

—René Descartes

It is important to realize that statistics and probability do not deal in the realm of certainty. If there is any realm of human knowledge where genuine certainty exists, you may be sure that our statistical methods are not needed there. In most human endeavors, and in almost all of the natural world around us, the element of chance happenings cannot be avoided. When we cannot expect something with true certainty, we must rely on probability to be our guide. In this chapter, we will study regression, correlation, and forecasting. One of the tools we use is a scatter plot. René Decartes (1596–1650) was the first mathematician to systematically use rectangular coordinate plots. For this reason, such a coordinate axis is called a Cartesian axis.

CORRELATION AND REGRESSION

David Young-Wolff/PhotoEdit

PREVIEW QUESTIONS

How can you use a scatter diagram to visually estimate the degree of linear correlation of two random variables? (SECTION 9.1)

How do you compute the correlation coefficient and what does it tell you about the strength of the linear relationship between two random variables? (SECTION 9.1)

What is the least-squares criterion? How do you find the equation of the least-squares line? (SECTION 9.2)

What is the coefficient of determination and what does it tell you about explained variation of y in a random sample of data pairs (x, y)? (SECTION 9.2)

How do you determine if the sample correlation coefficient is statistically significant? (SECTION 9.3)

How do you find a confidence interval for predictions based on the least-squares model? (SECTION 9.3)

How do you test the slope β of the population least-squares line? How do you construct a confidence interval for β? (SECTION 9.3)

What if you have more than two random variables? How do you construct a linear regression model for three, four, or more random variables? (SECTION 9.4)

FOCUS PROBLEM

Changing Populations and Crime Rate

lofoto/Shutterstock.com

Is the crime rate higher in neighborhoods where people might not know each other very well? Is there a relationship between crime rate and population change? If so, can we make predictions based on such a relationship? Is the relationship statistically significant? Is it possible to predict crime rates from population changes?

Denver is a city that has had a lot of growth and consequently a lot of population change in recent years. Sociologists studying population changes and crime rates could find a wealth of information in Denver statistics. Let x be a random variable representing percentage change in neighborhood population in the past few years, and let y be a random variable representing crime rate (crimes per 1000 population). A random sample of six Denver neighborhoods gave the following information (Source: *Neighborhood Facts*, The Piton Foundation). To find out more about the Piton Foundation, visit the Brase/Brase statistics site at **http://www.cengage.com/statistics/brase11e** and find the link to the Piton Foundation.

For online student resources, visit the Brase/Brase, *Understandable Statistics*, 11th edition web site at http://www.cengage.com/statistics/brase11e

x	29	2	11	17	7	6
y	173	35	132	127	69	53

Using information presented in this chapter, you will be able to analyze the relationship between the variables x and y using the following tools.

- Scatter diagram
- Sample correlation coefficient and coefficient of determination
- Least-squares line equation
- Predictions for y using the least-squares line
- Tests of population correlation coefficient and of slope of least-squares line
- Confidence intervals for slope and for predictions

(See Problem 10 in the Chapter Review Problems.)

SECTION **9.1**

Scatter Diagrams and Linear Correlation

FOCUS POINTS

- Make a scatter diagram.
- Visually estimate the location of the "best-fitting" line for a scatter diagram.
- Use sample data to compute the sample correlation coefficient r.
- Investigate the meaning of the correlation coefficient r.

Paired data values

Studies of correlation and regression of two variables usually begin with a graph of *paired data values* (x, y). We call such a graph a *scatter diagram*.

Scatter diagram
Explanatory variable
Response variable

A **scatter diagram** is a graph in which data pairs (x, y) are plotted as individual points on a grid with horizontal axis x and vertical axis y. We call x the **explanatory variable** and y the **response variable**.

By looking at a scatter diagram of data pairs, you can observe whether there seems to be a linear relationship between the x and y values.

EXAMPLE 1

SCATTER DIAGRAM

Phosphorous is a chemical used in many household and industrial cleaning compounds. Unfortunately, phosphorous tends to find its way into surface water, where it can kill fish, plants, and other wetland creatures. Phosphorous-reduction programs are required by law and are monitored by the Environmental Protection Agency (EPA) (Reference: *EPA Case Study 832-R-93-005*).

A random sample of eight sites in a California wetlands study gave the following information about phosphorous reduction in drainage water. In this study, x is a random variable that represents phosphorous concentration (in 100 mg/L) at the inlet of a passive biotreatment facility, and y is a random variable that represents total phosphorous concentration (in 100 mg/L) at the outlet of the passive biotreatment facility.

x	5.2	7.3	6.7	5.9	6.1	8.3	5.5	7.0
y	3.3	5.9	4.8	4.5	4.0	7.1	3.6	6.1

FIGURE 9-1

Phosphorous Reduction (100 mg/L)

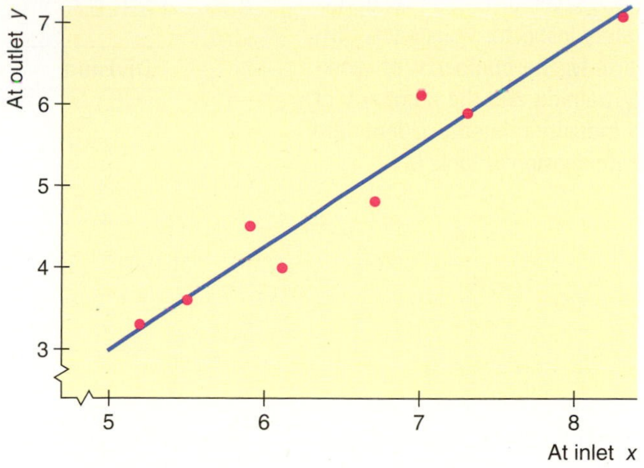

Cindy Kassab/Corbis Edge/Corbis

(a) Make a scatter diagram for these data.

SOLUTION: Figure 9-1 shows points corresponding to the given data pairs. These plotted points constitute the scatter diagram. To make the diagram, first scan the data and decide on an appropriate scale for each axis. Figure 9-1 shows the scatter diagram (points) along with a line segment showing the basic trend. Notice a "jump scale" on both axes.

(b) ***Interpretation*** Comment on the relationship between x and y shown in Figure 9-1.

SOLUTION: By inspecting the figure, we see that smaller values of x are associated with smaller values of y and larger values of x tend to be associated with larger values of y. Roughly speaking, the general trend seems to be reasonably well represented by an upward-sloping line segment, as shown in the diagram.

Of course, it is possible to draw many curves close to the points in Figure 9-1, but a straight line is the simplest and most widely used in elementary studies of paired data. We can draw many lines in Figure 9-1, but in some sense, the "best" line should be the one that comes closest to each of the points of the scatter diagram. To single out one line as the "best-fitting line," we must find a mathematical criterion for this line and a formula representing the line. This will be done in Section 9.2 using the *method of least squares*.

Introduction to linear correlation

Another problem precedes that of finding the "best-fitting line." That is the problem of determining how well the points of the scatter diagram are suited for fitting *any* line. Certainly, if the points are a very poor fit to any line, there is little use in trying to find the "best" line.

No linear correlation

If the points of a scatter diagram are located so that *no* line is realistically a "good" fit, we then say that the points possess *no linear correlation*. We see some examples of scatter diagrams for which there is no linear correlation in Figure 9-2.

FIGURE 9-2

Scatter Diagrams with No Linear Correlation

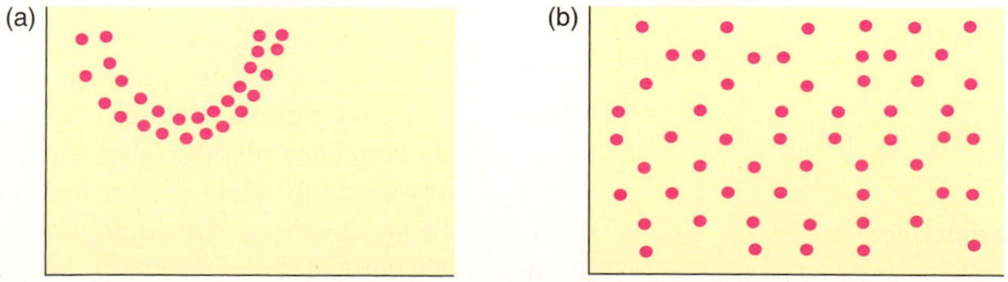

GUIDED EXERCISE 1 | Scatter Diagram

A large industrial plant has seven divisions that do the same type of work. A safety inspector visits each division of 20 workers quarterly. The number x of work-hours devoted to safety training and the number y of work-hours lost due to industry-related accidents are recorded for each separate division in Table 9-1.

TABLE 9-1 **Safety Report**

Division	x	y
1	10.0	80
2	19.5	65
3	30.0	68
4	45.0	55
5	50.0	35
6	65.0	10
7	80.0	12

(a) Make a scatter diagram for these pairs. Place the x values on the horizontal axis and the y values on the vertical axis.

FIGURE 9-3 Scatter Diagram for Safety Report

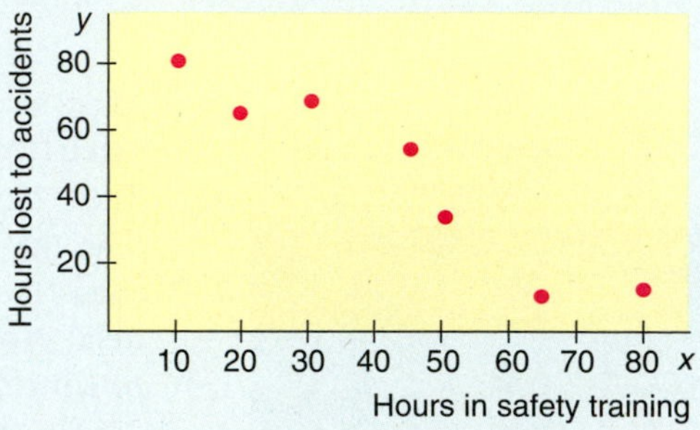

(b) As the number of hours spent on safety training increases, what happens to the number of hours lost due to industry-related accidents?

In general, as the number of hours in safety training goes up, the number of hours lost due to accidents goes down.

(c) ***Interpretation*** Does a line fit the data reasonably well?

A line fits reasonably well.

(d) Draw a line that you think "fits best."

Use a downward-sloping line that lies close to the points. Later, you will find the equation of the line that is a "best fit."

Perfect linear correlation

If the points seem close to a straight line, we say the linear correlation is moderate to high, depending on how close the points lie to the line. If all the points do, in fact, lie on a line, then we have *perfect linear correlation*. In Figure 9-4, we see some diagrams with perfect linear correlation. In statistical applications, perfect linear correlation almost never occurs.

Positive correlation

Negative correlation

The variables x and y are said to have *positive correlation* if low values of x are associated with low values of y and high values of x are associated with high values of y. Figure 9-4 parts (a) and (c) show scatter diagrams in which the variables are positively correlated. On the other hand, if low values of x are associated with high values of y and high values of x are associated with low values of y, the variables are said to be *negatively correlated*. Figure 9-4 parts (b) and (d) show variables that are negatively correlated.

FIGURE 9-4

Scatter Diagrams with Moderate and Perfect Linear Correlation

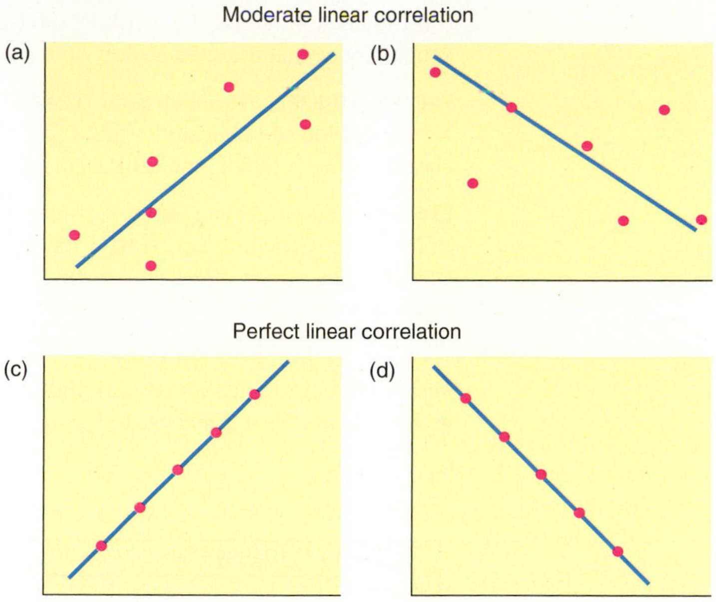

Scatter diagrams for the same data look different from one another when they are graphed using different scales. Problem 19 at the end of this section explores how changing scales affect the look of a scatter diagram.

GUIDED EXERCISE 2 | SCATTER DIAGRAM AND LINEAR CORRELATION

Examine the scatter diagrams in Figure 9-5 and then answer the following questions.

FIGURE 9-5 Scatter Diagrams

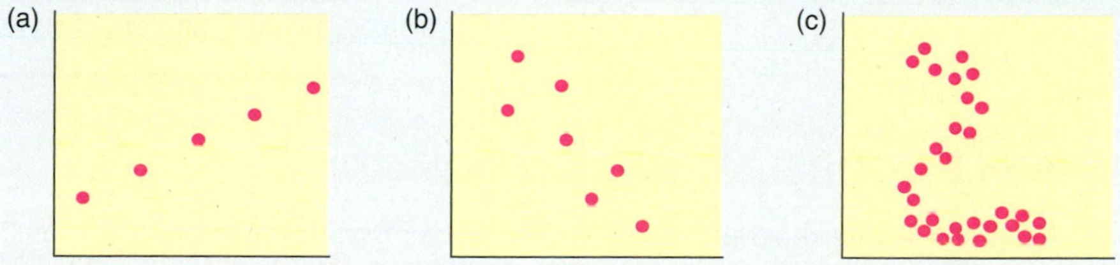

(a) Which diagram has no linear correlation? → Figure 9-5(c) has no linear correlation. No straight-line fit should be attempted.

(b) Which has perfect linear correlation? → Figure 9-5(a) has perfect linear correlation and can be fitted exactly by a straight line.

(c) Which can be reasonably fitted by a straight line? → Figure 9-5(b) can be reasonably fitted by a straight line.

WHAT DOES A SCATTER DIAGRAM TELL US?

A scatter diagram shows the relationship between two paired variables x and y. For each pair (x, y) of the data set, we plot the point on a grid with horizontal axis x and vertical axis y.

- To the extent the data points on the graph fall closer to a straight line, we can say that the relationship between x and y is more linear.
- To the extent the data pairs are scattered over the graph, we can say no linear relationship between x and y is apparent.

TECH NOTES The TI-84Plus/TI-83Plus/TI-*n*spire calculators, Excel 2010, and Minitab all produce scatter plots. For each technology, enter the x values in one column and the corresponding y values in another column. The displays on the next page show the data from Guided Exercise 1 regarding safety training and hours lost because of accidents. Notice that the scatter plots do not necessarily show the origin.

TI-84Plus/TI-83Plus/TI-*n*spire (with TI-84Plus keypad) Enter the data into two columns. Use **Stat Plot** and choose the first type. Use option **9: ZoomStat** under **Zoom**. To check the scale, look at the settings displayed under **Window**.

Excel 2010 Enter the data into two columns. On the home screen, click the **Insert** tab. In the Chart Group, select **Scatter** and choose the first type. In the next ribbon, the Chart Layout Group offers options for including titles and axes labels. Right clicking on data points provides other options such as data labels. Changing the size of the diagram box changes the scale on the axes.

Minitab Enter the data into two columns. Use the menu selections **Stat ➤ Regression ➤ Fitted Line Plot**. The best-fit line is automatically plotted on the scatter diagram.

TI-84Plus/TI-83Plus/TI-*n*spire Display

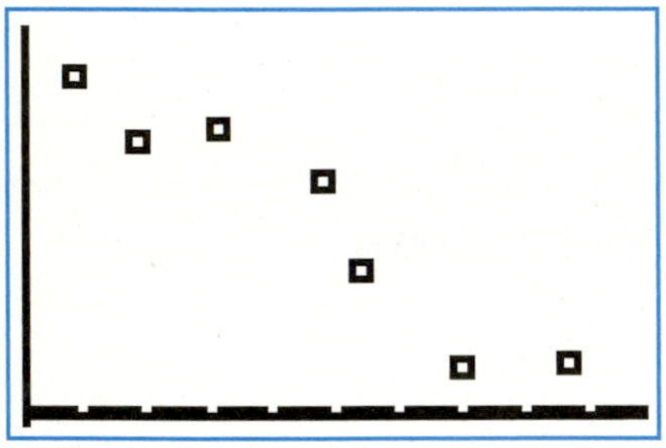

Excel Display

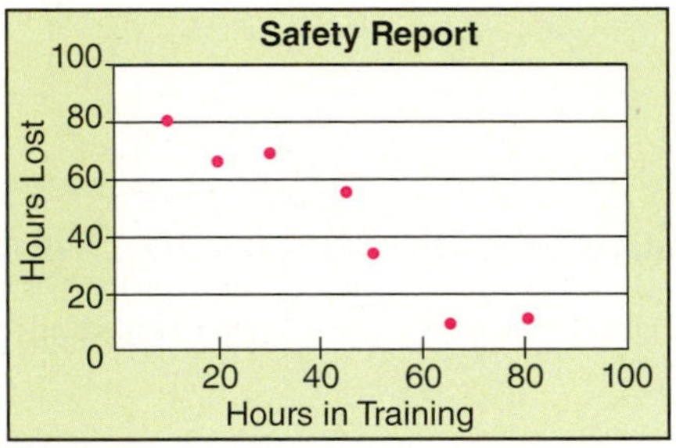

Minitab Display

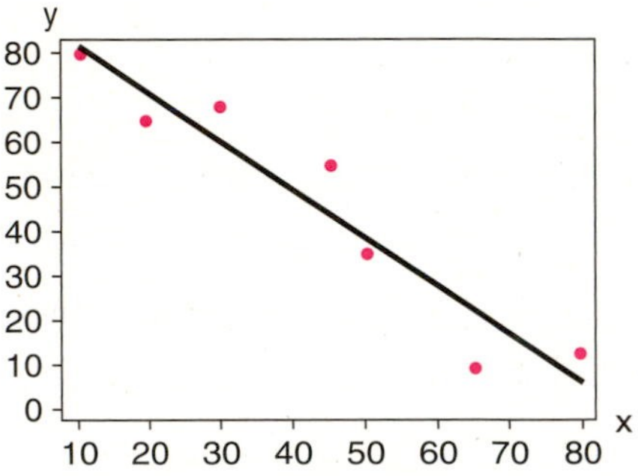

SAMPLE CORRELATION COEFFICIENT r

Looking at a scatter diagram to see whether a line best describes the relationship between the values of data pairs is useful. In fact, whenever you are looking for a relationship between two variables, making a scatter diagram is a good first step.

Sample correlation coefficient r

There is a mathematical measurement that describes the strength of the linear association between two variables. This measure is the *sample correlation coefficient* r. The full name for r is the *Pearson product-moment correlation coefficient*, named in honor of the English statistician Karl Pearson (1857–1936), who is credited with formulating r.

The **sample correlation coefficient r** is a numerical measurement that assesses the strength of a *linear* relationship between two variables x and y.

1. r is a unitless measurement between -1 and 1. In symbols, $-1 \leq r \leq 1$. If $r = 1$, there is perfect positive linear correlation. If $r = -1$, there is perfect negative linear correlation. If $r = 0$, there is no linear correlation. The closer r is to 1 or -1, the better a line describes the relationship between the two variables x and y.
2. Positive values of r imply that as x increases, y tends to increase. Negative values of r imply that as x increases, y tends to decrease.
3. The value of r is the same regardless of which variable is the explanatory variable and which is the response variable. In other words, the value of r is the same for the pairs (x, y) and the corresponding pairs (y, x).
4. The value of r does not change when either variable is converted to different units.

We'll develop the defining formula for r and then give a more convenient computation formula.

DEVELOPMENT OF FORMULA FOR *r*

If there is a *positive* linear relation between variables x and y, then high values of x are paired with high values of y, and low values of x are paired with low values of y. [See Figure 9-6(a).] In the case of *negative* linear correlation, high values of x are paired with low values of y, and low values of x are paired with high values of y. This relation is pictured in Figure 9-6(b). If there is *little or no linear correlation* between x and y, however, then we will find both high and low x values sometimes paired with high y values and sometimes paired with low y values. This relation is shown in Figure 9-6(c).

These observations lead us to the development of the formula for the sample correlation coefficient r. Taking *high* to mean "above the mean," we can express the relationships pictured in Figure 9-6 by considering the products

$$(x - \bar{x})(y - \bar{y})$$

If both x and y are high, both factors will be positive, and the product will be positive as well. The sign of this product will depend on the relative values of x and y compared with their respective means.

$$(x - \bar{x})(y - \bar{y}) \begin{cases} \text{is positive if } x \text{ and } y \text{ are both "high"} \\ \text{is positive if } x \text{ and } y \text{ are both "low"} \\ \\ \text{is negative if } x \text{ is "low," but } y \text{ is "high"} \\ \text{is negative if } x \text{ is "high," but } y \text{ is "low"} \end{cases}$$

In the case of positive linear correlation, most of the products $(x - \bar{x})(y - \bar{y})$ will be positive, and so will the sum over all the data pairs

$$\Sigma(x - \bar{x})(y - \bar{y})$$

For negative linear correlation, the products will tend to be negative, so the sum also will be negative. On the other hand, in the case of little, if any, linear correlation, the sum will tend to be zero.

One trouble with the preceding sum is that it increases or decreases, depending on the units of x and y. Because we want r to be unitless, we standardize both x and y of a data pair by dividing each factor $(x - \bar{x})$ by the sample standard deviation s_x and each factor $(y - \bar{y})$ by s_y. Finally, we take an average of all the products. For

FIGURE 9-6

Patterns for Linear Correlation

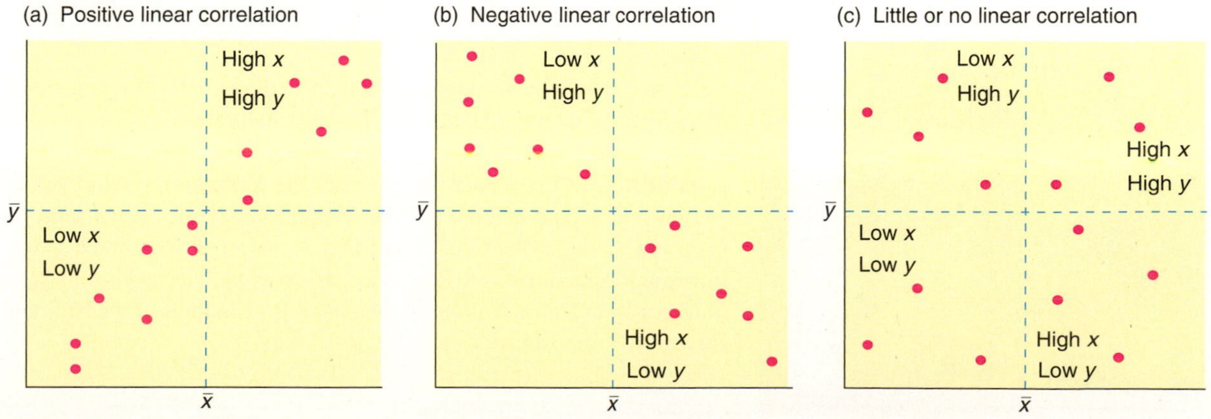

technical reasons, we take the average by dividing by $n - 1$ instead of by n. This process leads us to the desired measurement, r.

$$r = \frac{1}{n-1} \Sigma \frac{(x - \bar{x})}{s_x} \cdot \frac{(y - \bar{y})}{s_y} \quad (1)$$

COMPUTATION FORMULA FOR *r*

The defining formula for r shows how the mean and standard deviation of each variable in the data pair enter into the formulation of r. However, the defining formula is technically difficult to work with because of all the subtractions and products. A computation formula for r uses the raw data values of x and y directly.

PROCEDURE

HOW TO COMPUTE THE SAMPLE CORRELATION COEFFICIENT *r*

Requirements

Obtain a random sample of n data pairs (x, y). The data pairs should have a *bivariate normal distribution*. This means that for a fixed value of x, the y values should have a normal distribution (or at least a mound-shaped and symmetric distribution), and for a fixed y, the x values should have their own (approximately) normal distribution.

Procedure

1. Using the data pairs, compute Σx, Σy, Σx^2, Σy^2, and Σxy.
2. With n = sample size, Σx, Σy, Σx^2, Σy^2, and Σxy, you are ready to compute the sample correlation coefficient r using the computation formula

$$r = \frac{n\Sigma xy - (\Sigma x)(\Sigma y)}{\sqrt{n\Sigma x^2 - (\Sigma x)^2}\sqrt{n\Sigma y^2 - (\Sigma y)^2}} \quad (2)$$

Be careful! The notation Σx^2 means first square x and then calculate the sum, whereas $(\Sigma x)^2$ means first sum the x values and then square the result.

When x and y values of the data pairs are exchanged, the sample correlation coefficient r remains the same. Problem 20 explores this result.

Interpretation It can be shown mathematically that r is always a number between $+1$ and -1 $(-1 \le r \le +1)$. Table 9-2 gives a quick summary of some basic facts about r.

For most applications, you will use a calculator or computer software to compute r directly. However, to build some familiarity with the structure of the sample correlation coefficient, it is useful to do some calculations for yourself. Example 2 and Guided Exercise 3 show how to use the computation formula to compute r.

EXAMPLE 2

COMPUTING *r*

David Muench/Terra/CORBIS

Sand driven by wind creates large, beautiful dunes at the Great Sand Dunes National Monument, Colorado. Of course, the same natural forces also create large dunes in the Great Sahara and Arabia. Is there a linear correlation between wind velocity and sand drift rate? Let x be a random variable representing wind velocity (in 10 cm/sec) and let y be a random variable representing drift rate of sand (in 100 gm/cm/sec). A test site at the Great Sand Dunes National Monument gave the following information about x and y (Reference: *Hydrologic, Geologic, and Biologic Research at Great Sand Dunes National Monument*, Proceedings of the National Park Service Research Symposium).

TABLE 9-2 **Some Facts about the Correlation Coefficient**

If r Is	Then	The Scatter Diagram Might Look Something Like
0	There is no linear relation among the points of the scatter diagram.	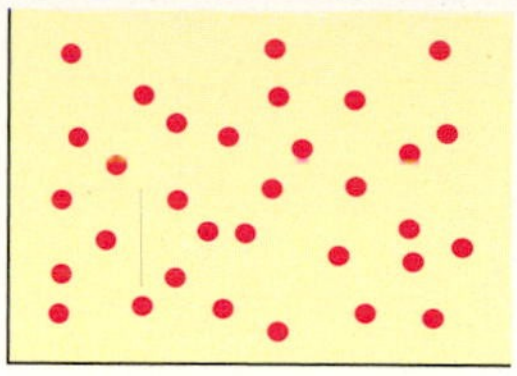
1 or -1	There is a perfect linear relation between x and y values; all points lie on the least-squares line.	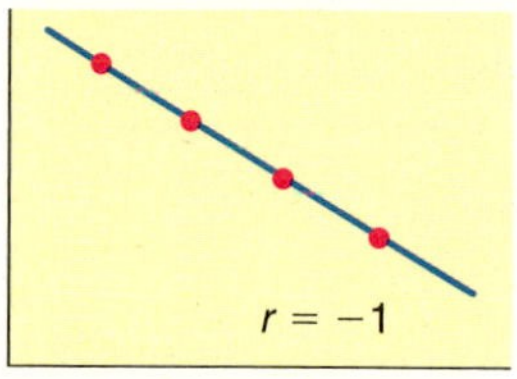 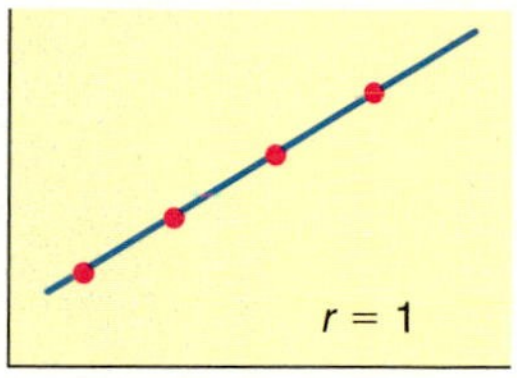
Between 0 and 1 ($0 < r < 1$)	The x and y values have a *positive correlation*. By this, we mean that *large* x values are associated with *large* y values, and *small* x values are associated with *small* y values.	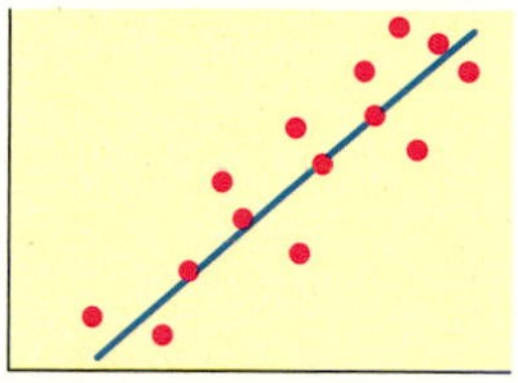 As we go from left to right, the least-squares line goes *up*.
Between -1 and 0 ($-1 < r < 0$)	The x and y values have a *negative correlation*. By this, we mean that *large* x values are associated with *small* y values, and *small* x values are associated with *large* y values.	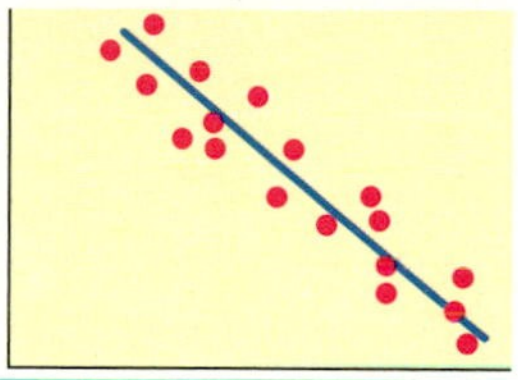 As we go from left to right, the least-squares line goes *down*.

x	70	115	105	82	93	125	88
y	3	45	21	7	16	62	12

(a) Construct a scatter diagram. Do you expect r to be positive?

SOLUTION: Figure 9-7 displays the scatter diagram. From the scatter diagram, it appears that as x values increase, y values also tend to increase. Therefore, r should be positive.

FIGURE 9-7

Wind Velocity (10 cm/sec) and Drift Rate of Sand (100 gm/cm/sec)

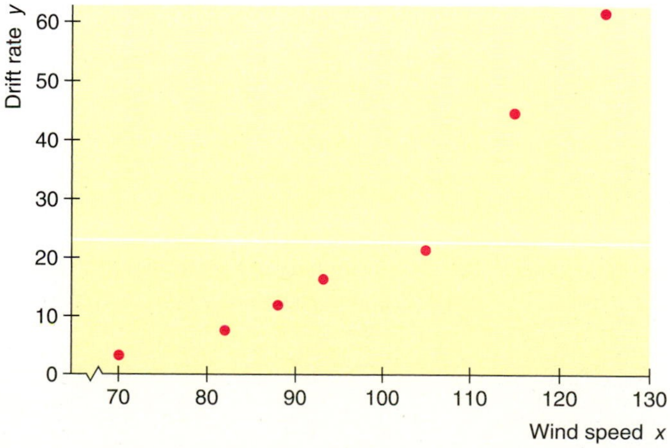

TABLE 9-3 Computation Table

x	y	x^2	y^2	xy
70	3	4900	9	210
115	45	13,225	2025	5175
105	21	11,025	441	2205
82	7	6724	49	574
93	16	8649	256	1488
125	62	15,625	3844	7750
88	12	7744	144	1056
$\Sigma x = 678$	$\Sigma y = 166$	$\Sigma x^2 = 67{,}892$	$\Sigma y^2 = 6768$	$\Sigma xy = 18{,}458$

(b) Compute r using the computation formula (formula 2).

SOLUTION: To find r, we need to compute Σx, Σx^2, Σy, Σy^2, and Σxy. It is convenient to organize the data in a table of five columns (Table 9-3) and then sum the entries in each column. Of course, many calculators give these sums directly. Using the computation formula for r, the sums from Table 9-3, and $n = 7$, we have

$$r = \frac{n\Sigma xy - (\Sigma x)(\Sigma y)}{\sqrt{n\Sigma x^2 - (\Sigma x)^2}\sqrt{n\Sigma y^2 - (\Sigma y)^2}} \qquad (2)$$

$$= \frac{7(18{,}458) - (678)(166)}{\sqrt{7(67{,}892) - (678)^2}\sqrt{7(6768) - (166)^2}} \approx \frac{16{,}658}{(124.74)(140.78)} \approx 0.949$$

Note: Using a calculator to compute r directly gives 0.949 to three places after the decimal.

(c) ***Interpretation*** What does the value of r tell you?

SOLUTION: Since r is very close to 1, we have an indication of a strong positive linear correlation between wind velocity and drift rate of sand. In other words, we expect that higher wind speeds tend to mean greater drift rates. Because r is so close to 1, the association between the variables appears to be linear.

Because it is quite a task to compute r for even seven data pairs, the use of columns as in Example 2 is extremely helpful. Your value for r should always be between -1 and 1, inclusive. Use a scatter diagram to get a rough idea of the value of r. If your computed value of r is outside the allowable range, or if it disagrees quite a bit with the scatter diagram, recheck your calculations. Be sure you distinguish between expressions such as (Σx^2) and $(\Sigma x)^2$. Negligible rounding errors may occur, depending on how you (or your calculator) round.

GUIDED EXERCISE 3 Computing *r*

In one of the Boston city parks, there has been a problem with muggings in the summer months. A police cadet took a random sample of 10 days (out of the 90-day summer) and compiled the following data. For each day, x represents the number of police officers on duty in the park and y represents the number of reported muggings on that day.

x	10	15	16	1	4	6	18	12	14	7
y	5	2	1	9	7	8	1	5	3	6

Continued

GUIDED EXERCISE 3 *continued*

(a) Construct a scatter diagram of x and y values.

Figure 9-8 shows the scatter diagram.

FIGURE 9-8 Scatter Diagram for Number of Police Officers versus Number of Muggings

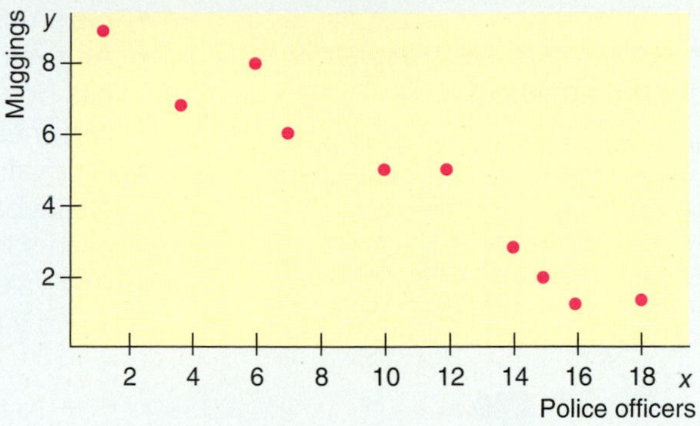

(b) From the scatter diagram, do you think the computed value of r will be positive, negative, or zero? Explain.

r will be negative. The general trend is that large x values are associated with small y values, and vice versa. From left to right, the least-squares line goes down.

(c) Complete Table 9-4.

TABLE 9-4

x	y	x^2	y^2	xy
10	5	100	25	50
15	2	225	4	30
16	1	256	1	16
1	9	1	81	9
4	7	16	49	28
6	8	___	___	___
18	1	___	___	___
12	5	___	___	___
14	3	___	___	___
7	6	49	36	42
$\Sigma x = 103$	$\Sigma y = 47$	$\Sigma x^2 =$ ___	$\Sigma y^2 =$ ___	$\Sigma xy =$ ___
$(\Sigma x)^2 =$ ___	$(\Sigma y)^2 =$ ___			

TABLE 9-5 **Completion of Table 9-4**

x	y	x^2	y^2	xy
6	8	36	64	48
18	1	324	1	18
12	5	144	25	60
14	3	196	9	42
		$\Sigma x^2 = 1347$	$\Sigma y^2 = 295$	$\Sigma xy = 343$
$(\Sigma x)^2 = 10{,}609$		$(\Sigma y)^2 = 2209$		

(d) Compute r. Alternatively, find the value of r directly by using a calculator or computer software.

$$r = \frac{n\Sigma xy - (\Sigma x)(\Sigma y)}{\sqrt{n\Sigma x^2 - (\Sigma x)^2}\sqrt{n\Sigma y^2 - (\Sigma y)^2}}$$

$$= \frac{10(343) - (103)(47)}{\sqrt{10(1347) - (103)^2}\sqrt{10(295) - (47)^2}}$$

$$\approx \frac{-1411}{(53.49)(27.22)} \approx -0.969$$

(e) ***Interpretation*** What does the value of r tell you about the relationship between the number of police officers and the number of muggings in the park?

There is a strong negative linear relationship between the number of police officers and the number of muggings. It seems that the more officers there are in the park, the fewer the number of muggings.

WHAT DOES THE CORRELATION COEFFICIENT *r* TELL US?

The correlation coefficient r is a sample statistic from a data set of ordered pairs (x, y). It is a measurement indicating the strength of a *linear* relationship between x and y.

- r is a unitless measurement ranging from -1 to 1.
- An r value close to 1 indicates that a positive linear relationship exists between x and y. This means that as x increases, y increases in a linear fashion.
- An r value close to -1 indicates that a negative linear relationship exists between x and y. This means that as x increases, y decreases in a linear fashion.
- An r value close to 0 indicates that the relationship (if any) between x and y is *not* linear.

LOOKING FORWARD

If the scatter diagram and the value of the sample correlation coefficient r indicate a linear relationship between the data pairs, how do we find a suitable linear equation for the data? This process, called linear regression, is presented in the next section, Section 9.2.

TECH NOTES Most calculators that support two-variable statistics provide the value of the sample correlation coefficient r directly. Statistical software provides r, r^2, or both.

TI-84Plus/TI-83Plus/TI-*n*spire (with TI-84Plus keypad) First use **CATALOG,** find **DiagnosticOn,** and press **Enter** twice. Then, when you use **STAT, CALC,** option **8:LinReg(a + bx),** the value of r will be given (data from Example 2). In the next section, we will discuss the line $y = a + bx$ and the meaning of r^2.

Excel 2010 Excel gives the value of the sample correlation coefficient r in several outputs. One way to find the value of r is to click the **Insert** function f_x. Then in the dialogue box, select **Statistical** for the category and **Correl** for the function.

Minitab Use the menu selection **Stat ➤ Basic Statistics ➤ Correlation**.

```
LinReg
 y=a+bx
 a=-79.97763496
 b=1.070565553
 r²=.8997719968
 r=.9485631222
```

LOOKING FORWARD

When the data are ranks (without ties) instead of measurements, the Pearson product-moment correlation coefficient can be reduced to a simpler equation called the Spearman rank correlation coefficient. This coefficient is used with nonparametric methods and is discussed in Section 11.3.

CRITICAL THINKING

Sample correlation compared to population correlation

CAUTIONS ABOUT CORRELATION

The correlation coefficient can be thought of as a measure of how well a linear model fits the data points on a scatter diagram. The closer r is to $+1$ or -1, the better a line "fits" the data. Values of r close to 0 indicate a poor fit to any line.

Usually a scatter diagram does not contain *all* possible data points that could be gathered. Most scatter diagrams represent only a *random sample* of data pairs taken from a very large population of all possible pairs. Because r is computed on the basis of a random sample of (x, y) pairs, we expect the values of r to vary from one sample to the next (much as the sample mean $\bar{x}$ varies from sample to sample). This brings up the question of the *significance* of r. Or, put another way, what are the chances that our random sample of data pairs indicates a high correlation when, in fact, the population's x and y values are not so strongly correlated? Right now, let's just say that the significance of r is a separate issue that will be treated in Section 9.3, where we test the *population correlation coefficient* ρ (Greek letter *rho*, pronounced "row").

Continued

Problem 21 demonstrates an informal process for determining whether or not r is significant. Problem 22 explores the effect of sample size on the significance of r. Problem 24 uses the value of ρ between two dependent variables to find the mean and standard deviation of a linear combination of the two variables.

r = **sample** correlation coefficient computed from a random sample of (x, y) data pairs.

ρ = **population** correlation coefficient computed from all population data pairs (x, y).

There is a less formal way to address the significance of r using a table of "critical values" or "cut-off values" based on the r distribution and the number of data pairs. Problem 21 at the end of this section discusses this method.

Extrapolation

The value of the sample correlation coefficient r and the strength of the linear relationship between variables is computed based on the sample data. The situation may change for measurements larger than or smaller than the data values included in the sample. For instance, for infants, there may be a high positive correlation between age in months and weight. However, that correlation might not apply for people ages 20 to 30 years.

Causation

The correlation coefficient is a mathematical tool for measuring the strength of a linear relationship between two variables. As such, it makes no implication about cause or effect. The fact that two variables tend to increase or decrease together does not mean that a change in one is *causing* a change in the other. A strong correlation between x and y is sometimes due to other (either known or unknown) variables. Such variables are called *lurking variables*.

Lurking variables

In ordered pairs (x, y), x is called the **explanatory** variable and y is called the **response** variable. When r indicates a linear correlation between x and y, changes in values of y tend to respond to changes in values of x according to a linear model. A **lurking variable** is a variable that is neither an explanatory nor a response variable. Yet, a lurking variable may be responsible for changes in both x and y.

EXAMPLE 3

CAUSATION AND LURKING VARIABLES

Over a period of years, the population of a certain town increased. It was observed that during this period the correlation between x, the number of people attending church, and y, the number of people in the city jail, was $r = 0.90$. Does going to church *cause* people to go to jail? Is there a *lurking variable* that might cause both variables x and y to increase?

SOLUTION: We hope church attendance does not cause people to go to jail! During this period, there was an increase in population. Therefore, it is not too surprising that both the number of people attending church and the number of people in jail increased. The high correlation between x and y is likely due to the lurking variable of population increase.

Correlation between averages

The correlation between two variables consisting of averages is usually higher than the correlation between two variables representing corresponding raw data. One reason is that the use of averages reduces the variation that exists between individual

Problem 23 at the end of this section explores the correlation of averages.

measurements (see Section 6.5 and the central limit theorem). A high correlation based on two variables consisting of averages does not necessarily imply a high correlation between two variables consisting of individual measurements. See Problem 23 at the end of this section.

VIEWPOINT Low on Credit, High on Cost!!!

How do you measure automobile insurance risk? One way is to use a little statistics and customer credit ratings. Insurers say statistics show that drivers who have a history of bad credit are more likely to be in serious car accidents. According to a high-level executive at Allstate Insurance Company, financial instability is an extremely powerful predictor of future insurance losses. In short, there seems to be a strong correlation between bad credit ratings and auto insurance claims. Consequently, insurance companies want to charge higher premiums to customers with bad credit ratings. Consumer advocates object strongly because they say bad credit does not cause automobile accidents, and more than 20 states prohibit or restrict the use of credit ratings to determine auto insurance premiums. Insurance companies respond by saying that your best defense is to pay your bills on time!

SECTION 9.1 PROBLEMS

Note: Answers may vary due to rounding.

1. *Statistical Literacy* When drawing a scatter diagram, along which axis is the explanatory variable placed? Along which axis is the response variable placed?

2. *Statistical Literacy* Suppose two variables are positively correlated. Does the response variable increase or decrease as the explanatory variable increases?

3. *Statistical Literacy* Suppose two variables are negatively correlated. Does the response variable increase or decrease as the explanatory variable increases?

4. *Statistical Literacy* Describe the relationship between two variables when the correlation coefficient r is
   (a) near -1.
   (b) near 0.
   (c) near 1.

5. *Critical Thinking: Linear Correlation* Look at the following diagrams. Does each diagram show high linear correlation, moderate or low linear correlation, or no linear correlation?

(a)
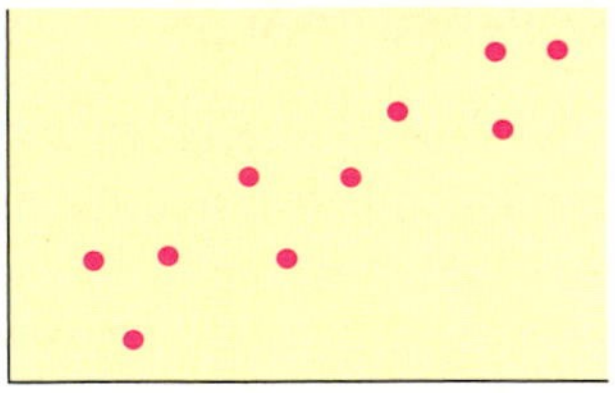
(b)
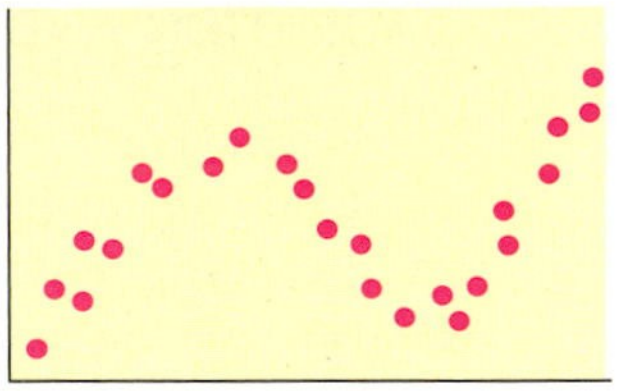
(c)
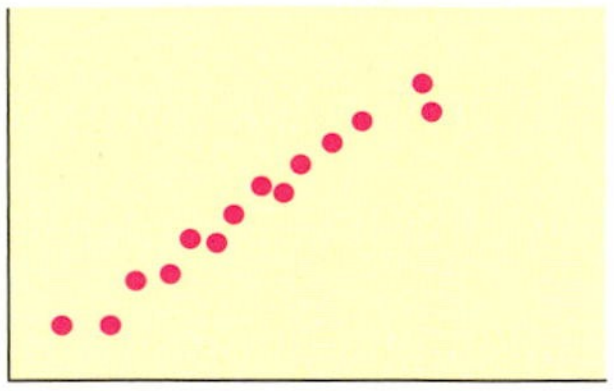

6. *Critical Thinking: Linear Correlation* Look at the following diagrams. Does each diagram show high linear correlation, moderate or low linear correlation, or no linear correlation?

(a)
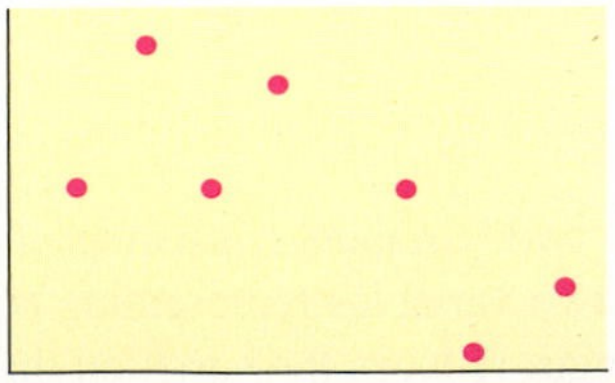
(b)
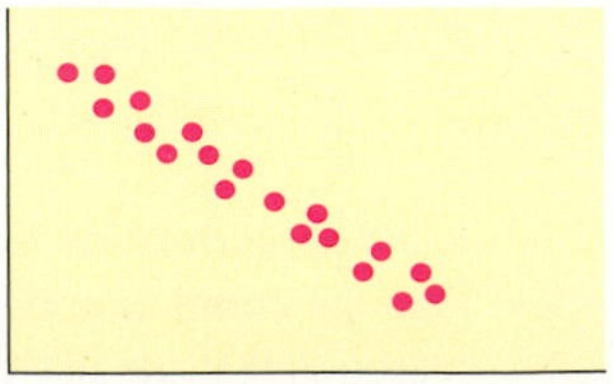
(c)
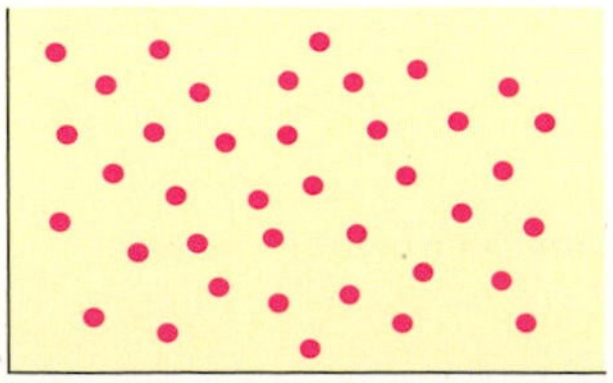

7. *Critical Thinking: Lurking Variables* Over the past few years, there has been a strong positive correlation between the annual consumption of diet soda drinks and the number of traffic accidents.
   (a) Do you think increasing consumption of diet soda drinks causes traffic accidents? Explain.
   (b) What lurking variables might be causing the increase in one or both of the variables? Explain.

8. *Critical Thinking: Lurking Variables* Over the past decade, there has been a strong positive correlation between teacher salaries and prescription drug costs.
   (a) Do you think paying teachers more causes prescription drugs to cost more? Explain.
   (b) What lurking variables might be causing the increase in one or both of the variables? Explain.

9. *Critical Thinking: Lurking Variables* Over the past 50 years, there has been a strong negative correlation between average annual income and the record time to run 1 mile. In other words, average annual incomes have been rising while the record time to run 1 mile has been decreasing.
   (a) Do you think increasing incomes cause decreasing times to run the mile? Explain.
   (b) What lurking variables might be causing the change in one or both of the variables? Explain.

10. *Critical Thinking: Lurking Variables* Over the past 30 years in the United States, there has been a strong negative correlation between the number of infant deaths at birth and the number of people over age 65.
    (a) Is the fact that people are living longer causing a decrease in infant mortalities at birth?
    (b) What lurking variables might be causing the increase in one or both of the variables? Explain.

11. *Interpretation* Trevor conducted a study and found that the correlation between the price of a gallon of gasoline and gasoline consumption has a linear correlation coefficient of -0.7. What does this result say about the relationship between price of gasoline and consumption? The study included gasoline prices ranging from $2.70 to $5.30 per gallon. Is it reliable to apply the results of this study to prices of gasoline higher than $5.30 per gallon? Explain.

12. *Interpretation* Do people who spend more time on social networking sites spend more time using Twitter? Megan conducted a study and found that the correlation between the times spent on the two activities was 0.8. What does this result say about the relationship between times spent on the two activities? If someone spends more time than average on a social networking site, can you automatically conclude that he or she spends more time than average using Twitter? Explain.

13. *Veterinary Science: Shetland Ponies* How much should a healthy Shetland pony weigh? Let x be the age of the pony (in months), and let y be the average weight of the pony (in kilograms). The following information is based on data taken from *The Merck Veterinary Manual* (a reference used in most veterinary colleges).

x	3	6	12	18	24
y	60	95	140	170	185

   (a) Make a scatter diagram and draw the line you think best fits the data.
   (b) Would you say the correlation is low, moderate, or strong? positive or negative?

(c) Use a calculator to verify that $\Sigma x = 63$, $\Sigma x^2 = 1089$, $\Sigma y = 650$, $\Sigma y^2 = 95{,}350$, and $\Sigma xy = 9930$. Compute r. As x increases from 3 to 24 months, does the value of r imply that y should tend to increase or decrease? Explain.

14. *Health Insurance: Administrative Cost* The following data are based on information from *Domestic Affairs*. Let x be the average number of employees in a group health insurance plan, and let y be the average administrative cost as a percentage of claims.

x	3	7	15	35	75
y	40	35	30	25	18

(a) Make a scatter diagram and draw the line you think best fits the data.
(b) Would you say the correlation is low, moderate, or strong? positive or negative?
(c) Use a calculator to verify that $\Sigma x = 135$, $\Sigma x^2 = 7133$, $\Sigma y = 148$, $\Sigma y^2 = 4674$, and $\Sigma xy = 3040$. Compute r. As x increases from 3 to 75, does the value of r imply that y should tend to increase or decrease? Explain.

15. *Meteorology: Cyclones* Can a low barometer reading be used to predict maximum wind speed of an approaching tropical cyclone? Data for this problem are based on information taken from *Weatherwise* (Vol. 46, No. 1), a publication of the American Meteorological Society. For a random sample of tropical cyclones, let x be the lowest pressure (in millibars) as a cyclone approaches, and let y be the maximum wind speed (in miles per hour) of the cyclone.

x	1004	975	992	935	985	932
y	40	100	65	145	80	150

(a) Make a scatter diagram and draw the line you think best fits the data.
(b) Would you say the correlation is low, moderate, or strong? positive or negative?
(c) Use a calculator to verify that $\Sigma x = 5823$, $\Sigma x^2 = 5{,}655{,}779$, $\Sigma y = 580$, $\Sigma y^2 = 65{,}750$, and $\Sigma xy = 556{,}315$. Compute r. As x increases, does the value of r imply that y should tend to increase or decrease? Explain.

16. *Geology: Earthquakes* Is the magnitude of an earthquake related to the depth below the surface at which the quake occurs? Let x be the magnitude of an earthquake (on the Richter scale), and let y be the depth (in kilometers) of the quake below the surface at the epicenter. The following is based on information taken from the National Earthquake Information Service of the U.S. Geological Survey. Additional data may be found by visiting the Brase/Brase statistics site at **http://www.cengage.com/statistics/brase11e** and finding the link to Earthquakes.

x	2.9	4.2	3.3	4.5	2.6	3.2	3.4
y	5.0	10.0	11.2	10.0	7.9	3.9	5.5

(a) Make a scatter diagram and draw the line you think best fits the data.
(b) Would you say the correlation is low, moderate, or strong? positive or negative?
(c) Use a calculator to verify that $\Sigma x = 24.1$, $\Sigma x^2 = 85.75$, $\Sigma y = 53.5$, $\Sigma y^2 = 458.31$, and $\Sigma xy = 190.18$. Compute r. As x increases, does the value of r imply that y should tend to increase or decrease? Explain.

17. *Baseball: Batting Averages and Home Runs* In baseball, is there a linear correlation between batting average and home run percentage? Let x represent the batting average of a professional baseball player, and let y represent the player's home run percentage (number of home runs per 100 times at bat). A random

sample of $n = 7$ professional baseball players gave the following information (Reference: *The Baseball Encyclopedia*, Macmillan Publishing Company).

x	0.243	0.259	0.286	0.263	0.268	0.339	0.299
y	1.4	3.6	5.5	3.8	3.5	7.3	5.0

(a) Make a scatter diagram and draw the line you think best fits the data.
(b) Would you say the correlation is low, moderate, or high? positive or negative?
(c) Use a calculator to verify that $\Sigma x = 1.957$, $\Sigma x^2 \approx 0.553$, $\Sigma y = 30.1$, $\Sigma y^2 = 150.15$, and $\Sigma xy \approx 8.753$. Compute r. As x increases, does the value of r imply that y should tend to increase or decrease? Explain.

18. *University Crime: FBI Report* Do larger universities tend to have more property crime? University crime statistics are affected by a variety of factors. The surrounding community, accessibility given to outside visitors, and many other factors influence crime rates. Let x be a variable that represents student enrollment (in thousands) on a university campus, and let y be a variable that represents the number of burglaries in a year on the university campus. A random sample of $n = 8$ universities in California gave the following information about enrollments and annual burglary incidents (Reference: *Crime in the United States*, Federal Bureau of Investigation).

x	12.5	30.0	24.5	14.3	7.5	27.7	16.2	20.1
y	26	73	39	23	15	30	15	25

(a) Make a scatter diagram and draw the line you think best fits the data.
(b) Would you say the correlation is low, moderate, or high? positive or negative?
(c) Using a calculator, verify that $\Sigma x = 152.8$, $\Sigma x^2 = 3350.98$, $\Sigma y = 246$, $\Sigma y^2 = 10{,}030$, and $\Sigma xy = 5488.4$. Compute r. As x increases, does the value of r imply that y should tend to increase or decrease? Explain.

19. *Expand Your Knowledge: Effect of Scale on Scatter Diagram* The initial visual impact of a scatter diagram depends on the scales used on the x and y axes. Consider the following data:

x	1	2	3	4	5	6
y	1	4	6	3	6	7

(a) Make a scatter diagram using the same scale on both the x and y axes (i.e., make sure the unit lengths on the two axes are equal).
(b) Make a scatter diagram using a scale on the y axis that is twice as long as that on the x axis.
(c) Make a scatter diagram using a scale on the y axis that is half as long as that on the x axis.
(d) On each of the three graphs, draw the straight line that you think best fits the data points. How do the slopes (or directions) of the three lines appear to change? *Note:* The actual slopes will be the same; they just appear different because of the choice of scale factors.

20. *Expand Your Knowledge: Effect on r of Exchanging x and y Values* Examine the computation formula for r, the sample correlation coefficient [formulas (1) and (2) of this section].
(a) In the formula for r, if we exchange the symbols x and y, do we get a different result or do we get the same (equivalent) result? Explain.
(b) If we have a set of x and y data values and we exchange corresponding x and y values to get a new data set, should the sample correlation coefficient be the same for both sets of data? Explain.

(c) Compute the sample correlation coefficient r for each of the following data sets and show that r is the same for both.

x	1	3	4
y	2	1	6

x	2	1	6
y	1	3	4

21. *Expand Your Knowledge: Using a Table to Test ρ* The correlation coefficient r is a *sample* statistic. What does it tell us about the value of the population correlation coefficient ρ (Greek letter *rho*)? We will build the formal structure of hypothesis tests of ρ in Section 9.3. However, there is a quick way to determine if the sample evidence based on r is strong enough to conclude that there is some population correlation between the variables. In other words, we can use the value of r to determine if $\rho \neq 0$. We do this by comparing the value $|r|$ to an entry in Table 9-6. The value of α in the table gives us the probability of concluding that $\rho \neq 0$ when, in fact, $\rho = 0$ and there is no population correlation. We have two choices for α: $\alpha = 0.05$ or $\alpha = 0.01$.

PROCEDURE

HOW TO USE TABLE 9-6 TO TEST ρ

1. First compute r from a random sample of n data pairs (x, y).
2. Find the table entry in the row headed by n and the column headed by your choice of α. Your choice of α is the risk you are willing to take of mistakenly concluding that $\rho \neq 0$ when, in fact, $\rho = 0$.
3. Compare $|r|$ to the table entry.
   (a) If $|r| \geq$ table entry, then there is sufficient evidence to conclude that $\rho \neq 0$, and we say that r is **significant**. In other words, we conclude that there is some population correlation between the two variables x and y.
   (b) If $|r| <$ table entry, then the evidence is insufficient to conclude that $\rho \neq 0$, and we say that r is **not significant**. We do not have enough evidence to conclude that there is any correlation between the two variables x and y.

TABLE 9-6 **Critical Values for Correlation Coefficient r**

n	$\alpha = 0.05$	$\alpha = 0.01$	n	$\alpha = 0.05$	$\alpha = 0.01$	n	$\alpha = 0.05$	$\alpha = 0.01$
3	1.00	1.00	13	0.53	0.68	23	0.41	0.53
4	0.95	0.99	14	0.53	0.66	24	0.40	0.52
5	0.88	0.96	15	0.51	0.64	25	0.40	0.51
6	0.81	0.92	16	0.50	0.61	26	0.39	0.50
7	0.75	0.87	17	0.48	0.61	27	0.38	0.49
8	0.71	0.83	18	0.47	0.59	28	0.37	0.48
9	0.67	0.80	19	0.46	0.58	29	0.37	0.47
10	0.63	0.76	20	0.44	0.56	30	0.36	0.46
11	0.60	0.73	21	0.43	0.55			
12	0.58	0.71	22	0.42	0.54			

(a) Look at Problem 13 regarding the variables x = age of a Shetland pony and y = weight of that pony. Is the value of $|r|$ large enough to conclude that weight and age of Shetland ponies are correlated? Use $\alpha = 0.05$.

(b) Look at Problem 15 regarding the variables x = lowest barometric pressure as a cyclone approaches and y = maximum wind speed of the cyclone. Is the value of $|r|$ large enough to conclude that lowest barometric pressure and wind speed of a cyclone are correlated? Use $\alpha = 0.01$.

22. *Expand Your Knowledge: Sample Size and Significance of Correlation* In this problem, we use Table 9-6 to explore the significance of r based on different sample sizes. See Problem 21.
(a) Is a sample correlation coefficient $r = 0.820$ significant at the $\alpha = 0.01$ level based on a sample size of $n = 7$ data pairs? What about $n = 9$ data pairs?
(b) Is a sample correlation coefficient $r = 0.40$ significant at the $\alpha = 0.05$ level based on a sample size of $n = 20$ data pairs? What about $n = 27$ data pairs?
(c) Is it true that in order to be significant, an r value must be larger than 0.90? larger than 0.70? larger than 0.50? What does sample size have to do with the significance of r? Explain.

23. *Expand Your Knowledge: Correlation of Averages* Fuming because you are stuck in traffic? Roadway congestion is a costly item, in both time wasted and fuel wasted. Let x represent the *average* annual hours per person spent in traffic delays and let y represent the *average* annual gallons of fuel wasted per person in traffic delays. A random sample of eight cities showed the following data (Reference: *Statistical Abstract of the United States*, 122nd Edition).

x (hr)	28	5	20	35	20	23	18	5
y (gal)	48	3	34	55	34	38	28	9

(a) Draw a scatter diagram for the data. Verify that $\Sigma x = 154$, $\Sigma x^2 = 3712$, $\Sigma y = 249$, $\Sigma y^2 = 9959$, and $\Sigma xy = 6067$. Compute r.

The data in part (a) represent *average* annual hours lost per person and *average* annual gallons of fuel wasted per person in traffic delays. Suppose that instead of using average data for different cities, you selected one person at random from each city and measured the annual number of hours lost x for that person and the annual gallons of fuel wasted y for the same person.

x (hr)	20	4	18	42	15	25	2	35
y (gal)	60	8	12	50	21	30	4	70

(b) Compute $\bar{x}$ and $\bar{y}$ for both sets of data pairs and compare the averages. Compute the sample standard deviations s_x and s_y for both sets of data pairs and compare the standard deviations. In which set are the standard deviations for x and y larger? Look at the defining formula for r, Equation 1. Why do smaller standard deviations s_x and s_y tend to increase the value of r?
(c) Make a scatter diagram for the second set of data pairs. Verify that $\Sigma x = 161$, $\Sigma x^2 = 4583$, $\Sigma y = 255$, $\Sigma y^2 = 12{,}565$, and $\Sigma xy = 7071$. Compute r.
(d) Compare r from part (a) with r from part (c). Do the data for averages have a higher correlation coefficient than the data for individual measurements? List some reasons why you think hours lost per individual and fuel wasted per individual might vary more than the same quantities averaged over all the people in a city.

24. *Expand Your Knowledge: Dependent Variables* In Section 5.1, we studied linear combinations of *independent* random variables. What happens if the variables are not independent? A lot of mathematics can be used to prove the following:

Let x and y be random variables with means μ_x and μ_y, variances σ_x^2 and σ_y^2, and population correlation coefficient ρ (the Greek letter *rho*). Let a and b be any constants and let $w = ax + by$. Then,

$$\mu_w = a\mu_x + b\mu_y$$
$$\sigma_w^2 = a^2\sigma_x^2 + b^2\sigma_y^2 + 2ab\sigma_x\sigma_y\rho$$

Covariance

In this formula, ρ is the population correlation coefficient, theoretically computed using the population of all (x, y) data pairs. The expression $\sigma_x\sigma_y\rho$ is called the *covariance* of x and y. If x and y are independent, then $\rho = 0$ and the formula for σ_w^2 reduces to the appropriate formula for independent variables (see Section 5.1). In most real-world applications, the population parameters are not known, so we use sample estimates with the understanding that our conclusions are also estimates.

Do you have to be rich to invest in bonds and real estate? No, mutual fund shares are available to you even if you aren't rich. Let x represent annual percentage return (after expenses) on the Vanguard Total Bond Index Fund, and let y represent annual percentage return on the Fidelity Real Estate Investment Fund. Over a long period of time, we have the following population estimates (based on *Morningstar Mutual Fund Report*).

$$\mu_x \approx 7.32 \quad \sigma_x \approx 6.59 \quad \mu_y \approx 13.19 \quad \sigma_y \approx 18.56 \quad \rho \approx 0.424$$

(a) Do you think the variables x and y are independent? Explain.
(b) Suppose you decide to put 60% of your investment in bonds and 40% in real estate. This means you will use a weighted average $w = 0.6x + 0.4y$. Estimate your expected percentage return μ_w and risk σ_w.
(c) Repeat part (b) if $w = 0.4x + 0.6y$.
(d) Compare your results in parts (b) and (c). Which investment has the higher expected return? Which has the greater risk as measured by σ_w?

SECTION 9.2 Linear Regression and the Coefficient of Determination

FOCUS POINTS

- State the least-squares criterion.
- Use sample data to find the equation of the least-squares line. Graph the least-squares line.
- Use the least-squares line to predict a value of the response variable y for a specified value of the explanatory variable x.
- Explain the difference between interpolation and extrapolation.
- Explain why extrapolation beyond the sample data range might give results that are misleading or meaningless.
- Use r^2 to determine *explained* and *unexplained* variation of the response variable y.

In Denali National Park, Alaska, the wolf population is dependent on a large, strong caribou population. In this wild setting, caribou are found in very large herds. The well-being of an entire caribou herd is not threatened by wolves. In fact, it is thought that wolves keep caribou herds strong by helping prevent overpopulation. Can the caribou population be used to predict the size of the wolf population?

Let x be a random variable that represents the fall caribou population (in hundreds) in Denali National Park, and let y be a random variable that represents the late-winter wolf population in the park. A random sample of recent years gave the following information (Reference: U.S. Department of the Interior, National Biological Service).

x	30	34	27	25	17	23	20
y	66	79	70	60	48	55	60

Looking at the scatter diagram in Figure 9-9, we can ask some questions.

1. Do the data indicate a linear relationship between x and y?
2. Can you find an equation for the best-fitting line relating x and y? Can you use this relationship to predict the size of the wolf population when you know the size of the caribou population?
3. What fractional part of the variability in y can be associated with the variability in x? What fractional part of the variability in y is not associated with a corresponding variability in x?

FIGURE 9-9

Caribou and Wolf Populations

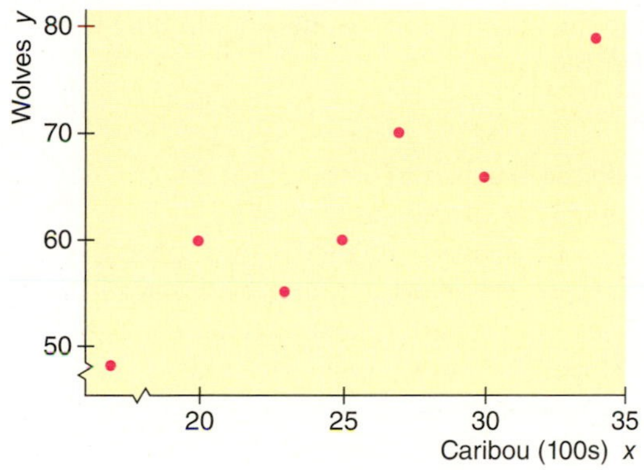

FIGURE 9-10

Least-Squares Criterion

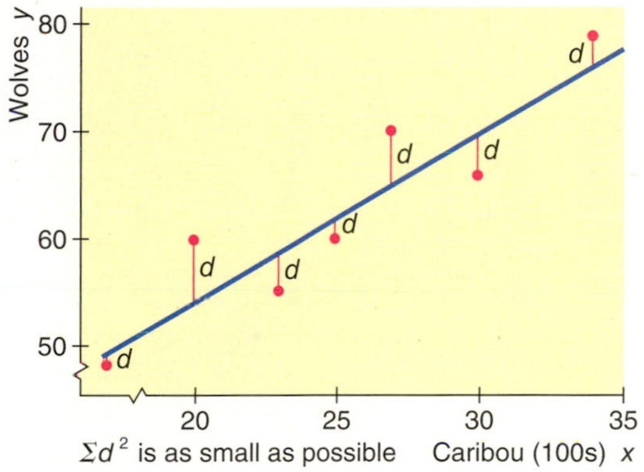

The first step in answering these questions is to try to express the relationship as a mathematical equation. There are many possible equations, but the simplest and most widely used is the linear equation, or the equation of a straight line. Because we will be using this line to predict the y values from the x values, we call x the *explanatory variable* and y the *response variable*.

Explanatory variable
Response variable
Least-squares criterion

Our job is to find the linear equation that "best" represents the points of the scatter diagram. For our criterion of best-fitting line, we use the *least-squares criterion*, which states that the line we fit to the data points must be such that *the sum of the squares of the vertical distances from the points to the line be made as small as possible*. The least-squares criterion is illustrated in Figure 9-10.

LEAST-SQUARES CRITERION

The sum of the squares of the vertical distances from the data points (x, y) to the line is made as small as possible.

In Figure 9-10, d represents the difference between the y coordinate of the data point and the corresponding y coordinate on the line. Thus, if the data point lies above the line, d is positive, but if the data point is below the line, d is negative. As a result, the sum of the d values can be small even if the points are widely spread in the scatter diagram. However, the squares d^2 cannot be negative. By minimizing the sum of the squares, we are, in effect, not allowing positive and negative d values to "cancel out" one another in the sum. It is in this way that we can meet the least-squares criterion of minimizing the sum of the squares of the vertical distances between the points and the line over *all* points in the scatter diagram.

Least-squares line

We use the notation $\hat{y} = a + bx$ for the *least-squares line*. A little algebra tells us that b is the slope and a is the intercept of the line. In this context, $\hat{y}$ (read "y hat") represents the value of the response variable y estimated using the least-squares line and a given value of the explanatory variable x.

Techniques of calculus can be applied to show that a and b may be computed using the following procedure.

Problem 21 demonstrates that for the same data set, the least-squares lines for predicting y or for predicting x are essentially different.

PROCEDURE

HOW TO FIND THE EQUATION OF THE LEAST-SQUARES LINE $\hat{y} = a + bx$

Requirements for Statistical Inference

Obtain a random sample of n data pairs (x, y), where x is the *explanatory variable* and y is the *response variable*. The data pairs should have a *bivariate normal distribution*. This means that for a fixed value of x, the y values should have a normal distribution (or at least a mound-shaped and symmetric distribution), and for a fixed y, the x values should have their own (approximately) normal distribution.

Procedure

1. Using the data pairs, compute Σx, Σy, Σx^2, Σy^2, and Σxy. Then compute the sample means $\bar{x}$ and $\bar{y}$.
2. With n = sample size, Σx, Σy, Σx^2, Σy^2, Σxy, $\bar{x}$, and $\bar{y}$, you are ready to compute the slope b and intercept a using the computation formulas

Slope b

$$\text{Slope: } b = \frac{n\Sigma xy - (\Sigma x)(\Sigma y)}{n\Sigma x^2 - (\Sigma x)^2} \tag{3}$$

Intercept a

$$\text{Intercept: } a = \bar{y} - b\bar{x} \tag{4}$$

   Be careful! The notation Σx^2 means first square x and then calculate the sum, whereas (Σx^2) means first sum the x values and then square the result.

3. The equation of the least-squares line computed from your sample data is

$$\hat{y} = a + bx \tag{5}$$

For data following exponential growth or power law models, logarithmic transformations can be used to transform the data into linear models. Then linear regression can be used on the transformed data. Problems 22–25 show these methods.

COMMENT The computation formulas for the slope of the least-squares line, the sample correlation coefficient r, and the standard deviations s_x and s_y use many of the same sums. There is, in fact, a relationship between the sample correlation coefficient r and the slope of the least-squares line b. In instances where we know r, s_x, and s_y, we can use the following formula to compute b.

$$b = r\left(\frac{s_y}{s_x}\right) \tag{6}$$

COMMENT In other mathematics courses, the slope-intercept form of the equation of a line is usually given as $y = mx + b$, where m refers to the slope of the line and b to the y coordinate of the y intercept. In statistics, when there is only one explanatory variable, it is common practice to use the letter b to designate the slope of the least-squares line and the letter a to designate the y coordinate of the intercept. For example, these are the symbols used on the TI-84Plus/TI-83Plus/TI-*n*spire calculators as well as on many other calculators.

Using the formulas to find the values of *a* and *b*

For most applications, you can use a calculator or computer software to compute a and b directly. However, to build some familiarity with the structure of the computation formulas, it is useful to do some calculations yourself. Example 4 shows how to use the computation formulas to find the values of a and b and the equation of the least-squares line $\hat{y} = a + bx$.

Note: If you are using your calculator to find the values of a and b directly, you may omit the discussion regarding use of the formulas. Go to the margin header "Using the values of a and b to construct the equation of the least-squares line."

EXAMPLE 4

LEAST-SQUARES LINE

Joe McDonald/Encyclopedia/Corbis

Let's find the least-squares equation relating the variables x = size of caribou population (in hundreds) and y = size of wolf population in Denali National Park. Use x as the explanatory variable and y as the response variable.

(a) Use the computation formulas to find the slope b of the least-squares line and the y intercept a.

SOLUTION: Table 9-7 gives the data values x and y along with the values x^2, y^2, and xy. First compute the sample means.

$$\bar{x} = \frac{\Sigma x}{n} = \frac{176}{7} \approx 25.14 \text{ and } \bar{y} = \frac{\Sigma y}{n} = \frac{438}{7} \approx 62.57$$

Next compute the slope b.

$$b = \frac{n\Sigma xy - (\Sigma x)(\Sigma y)}{n\Sigma x^2 - (\Sigma x)^2} = \frac{7(11{,}337) - (176)(438)}{7(4628) - (176)^2} = \frac{2271}{1420} \approx 1.60$$

Use the values of b, $\bar{x}$ and $\bar{y}$ to compute the y intercept a.

$$a = \bar{y} - b\bar{x} \approx 62.57 - 1.60(25.14) \approx 22.35$$

Note that calculators give the values $b \approx 1.599$ and $a \approx 22.36$. These values differ slightly from those you computed using the formulas because of rounding.

Using the values of *a* and *b* to construct the equation of the least-squares line

(b) Use the values of a and b (either computed or obtained from a calculator) to find the equation of the least-squares line.

SOLUTION:

$$\hat{y} = a + bx$$

$$\hat{y} \approx 22.35 + 1.60x \text{ since } a \approx 22.35 \text{ and } b \approx 1.60$$

TABLE 9-7 Sums for Computing b, $\bar{x}$ and $\bar{y}$

x	y	x^2	y^2	xy
30	66	900	4356	1980
34	79	1156	6241	2686
27	70	729	4900	1890
25	60	625	3600	1500
17	48	289	2304	816
23	55	529	3025	1265
20	60	400	3600	1200
$\Sigma x = 176$	$\Sigma y = 438$	$\Sigma x^2 = 4628$	$\Sigma y^2 = 28{,}026$	$\Sigma xy = 11{,}337$

Graphing the least-squares line

(c) Graph the equation of the least-squares line on a scatter diagram.

SOLUTION: To graph the least-squares line, we have several options available. The slope-intercept method of algebra is probably the quickest, but may not always be convenient if the intercept is not within the range of the sample data values. It is just as easy to select two x values in the range of the x data values and then use the least-squares line to compute two corresponding $\hat{y}$ values.

In fact, we already have the coordinates of one point on the least-squares line. By the formula for the intercept [Equation (4)], the point $(\bar{x}, \bar{y})$ is always on the least-squares line. For our example, $(\bar{x}, \bar{y}) = (25.14, 62.57)$.

The point $(\bar{x}, \bar{y})$ is always on the least-squares line.

Another x value within the data range is $x = 34$. Using the least-squares line to compute the corresponding $\hat{y}$ value gives

$$\hat{y} \approx 22.35 + 1.60(34) \approx 76.75$$

We place the two points (25.14, 62.57) and (34, 76.75) on the scatter diagram (using a different symbol than that used for the sample data points) and connect the points with a line segment (Figure 9-11).

Meaning of slope

In the equation $\hat{y} = a + bx$, the slope b tells us how many units $\hat{y}$ changes for each unit change in x. In Example 4 regarding size of wolf and caribou populations,

$$\hat{y} \approx 22.35 + 1.60x$$

The slope 1.60 tells us that if the number of caribou (in hundreds) changes by 1 (hundred), then we expect the sustainable wolf population to change by 1.60. In other words, our model says that an increase of 100 caribou will increase the predicted wolf population by 1.60. If the caribou population decreases by 400, we predict the sustainable wolf population to decrease by 6.4.

FIGURE 9-11

Caribou and Wolf Populations

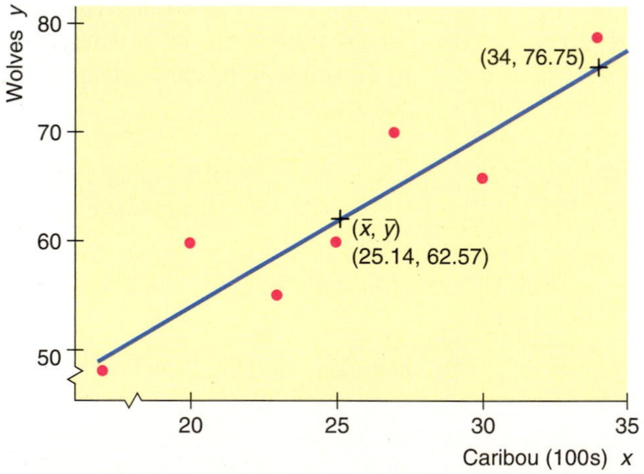

Marginal change

The slope of the least-squares line tells us how many units the response variable is expected to change for each unit change in the explanatory variable. The number of units change in the response variable for each unit change in the explanatory variable is called the **marginal change** of the response variable.

Some points in the data set have a strong influence on the equation of the least-squares line.

Influential points

A data pair is **influential** if removing it would substantially change the equation of the least-squares line or other calculations associated with linear regression. An influential point often has an x-value near the extreme high or low value of the data set.

Figure 9-12 shows two scatter diagrams produced in Excel. Figure 9-12(a) has an influential point. Figure 9-12(b) shows the scatter diagram with the influential point removed. Notice that the equations for the least-squares lines are quite different.

FIGURE 9-12

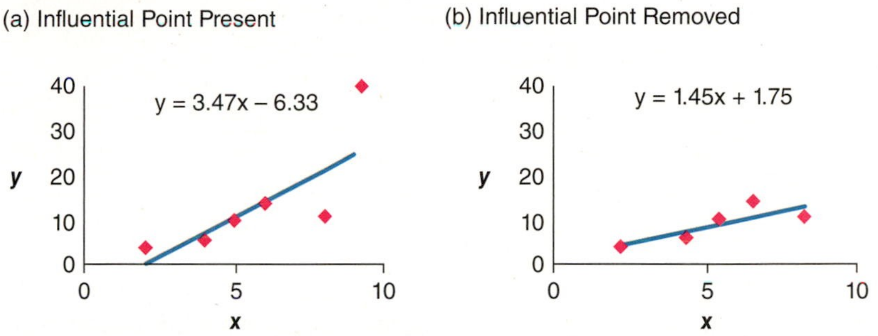

If a data set has an influential point, look at the influential point carefully to make sure it is not the result of a data collection error or a data-recording error. A valid influential point affects the equation of the least-squares line. The group project in Data Highlights at the end of this chapter further explores influential points.

CRITICAL THINKING

Predicting y for a specified x

USING THE LEAST-SQUARES LINE FOR PREDICTION

Making predictions is one of the main applications of linear regression. In other words, you use the equation of the least-squares line to predict the $\hat{y}$ value for a specified x value. The accuracy of the prediction depends on several components.

How well does the least-squares line fit the original data points?

The accuracy of the prediction depends on how well the least-squares line fits the original raw data points. Here are some tools to assess the fit of the line.

- Look at the scatter diagram, taking into account the scale of each axis.
- See if there are any influential points.
- Consider the value of the sample correlation coefficient r. The closer r is to 1 or -1, the better the least-squares line fits the data.
- Consider the value of the coefficient of determination r^2 (as discussed later in this section).
- Look at the residuals and a residual plot (see Problem 19 for a discussion of residual plots).

Residual

The **residual** is the difference between the y value in a specified data pair (x, y) and the value $\hat{y} = a + bx$ predicted by the least-squares line for the same x.

$y - \hat{y}$ is the **residual**.

Problems 19 and 20 at the end of this section show how to make a residual plot.

Continued

CRITICAL THINKING *continued*

If the residuals seem random about 0, the least-squares line provides a reasonable model for the data. Later in this section you will see the residual used in the development of the *coefficient of determination*, another important measurement associated with linear regression.

Does the prediction involve interpolation or extrapolation?

Interpolation
Extrapolation

Another issue that affects the validity of predictions is whether you are *interpolating* or *extrapolating*.

Predicting $\hat{y}$ values for x values that are **between** observed x values in the data set is called **interpolation**.

Predicting $\hat{y}$ values for x values that are **beyond** observed x values in the data set is called **extrapolation.** Extrapolation may produce unrealistic forecasts.

The least-squares line may not reflect the relationship between x and y for values of x outside the data range. For example, there is a fairly high correlation between height and age for boys ages 1 year to 10 years. In general, the older the boy, the taller the boy. A least-squares line based on such data would give good predictions of height for boys of ages between 1 and 10. However, it would be fairly meaningless to use the same linear regression line to predict the height of a 20-year-old or 50-year-old man.

The data are sample data.

Another consideration when working with predictions is the fact that the least-squares line is based on sample data. Each different sample will produce a slightly different equation for the least-squares line. Just as there are confidence intervals for parameters such as population means, there are confidence intervals for the prediction of y for a given x. We will examine confidence intervals for predictions in Section 9.3.

The least-squares line uses x as the explanatory variable and y as the response variables.

One more important fact about predictions: The least-squares line is developed with x as the explanatory variable and y as the response variable. This model can be used only to predict y values from specified x values. If you wish to begin with y values and predict corresponding x values, you must start all over and compute a new equation. Such an equation would be developed using a model with x as the response variable and y as the explanatory variable. See Problem 21 at the end of this section. Note that the equation for predicting x values *cannot* be derived from the least-squares line predicting y simply by solving the equation for x.

The least-squares line developed with x as the explanatory variable and y as the response variable can be used only to predict y values from specified x values.

The next example shows how to use the least-squares line for predictions.

EXAMPLE 5

PREDICTIONS

Joe McDonald/Corbis

We continue with Example 4 regarding size of the wolf population as it relates to size of the caribou population. Suppose you want to predict the size of the wolf population when the size of the caribou population is 21 (hundred).

(a) In the least-squares model developed in Example 4, which is the explanatory variable and which is the response variable? Can you use the equation to predict the size of the wolf population for a specified size of caribou population?

SOLUTION: The least-squares line $\hat{y} \approx 22.35 + 1.60x$ was developed using x = size of caribou population (in hundreds) as the explanatory variable and y = size of wolf population as the response variable. We can use the equation to predict the y value for a specified x value.

(b) The sample data pairs have x values ranging from 17 (hundred) to 34 (hundred) for the size of the caribou population. To predict the size of the wolf population when the size of the caribou population is 21 (hundred), will you be interpolating or extrapolating?

SOLUTION: Interpolating, since 21 (hundred) falls within the range of sample x values.

(c) Predict the size of the wolf population when the caribou population is 21 (hundred).

SOLUTION: Using the least-squares line from Example 4 and the value 21 in place of x gives

$$\hat{y} \approx 22.35 + 1.60x \approx 22.35 + 1.60(21) \approx 55.95$$

Rounding up to a whole number gives a prediction of 56 for the size of the wolf population.

GUIDED EXERCISE 4

LEAST-SQUARES LINE

The Quick Sell car dealership has been using 1-minute spot ads on a local TV station. The ads always occur during the evening hours and advertise the different models and price ranges of cars on the lot that week. During a 10-week period, a Quick Sell dealer kept a weekly record of the number x of TV ads versus the number y of cars sold. The results are given in Table 9-8.

The manager decided that Quick Sell can afford only 12 ads per week. At that level of advertisement, how many cars can Quick Sell expect to sell each week? We'll answer this question in several steps.

TABLE 9-8

x	y
6	15
20	31
0	10
14	16
25	28
16	20
28	40
18	25
10	12
8	15

(a) Draw a scatter diagram for the data.

→ The scatter diagram is shown in Figure 9-13 on the next page. The plain red dots in Figure 9-13 are the points of the scatter diagram. Notice that the least-squares line is also shown with two extra points used to position that line.

Continued

GUIDED EXERCISE 4 *continued*

FIGURE 9-13

Scatter Diagram and Least-Squares Line for Table 9-8

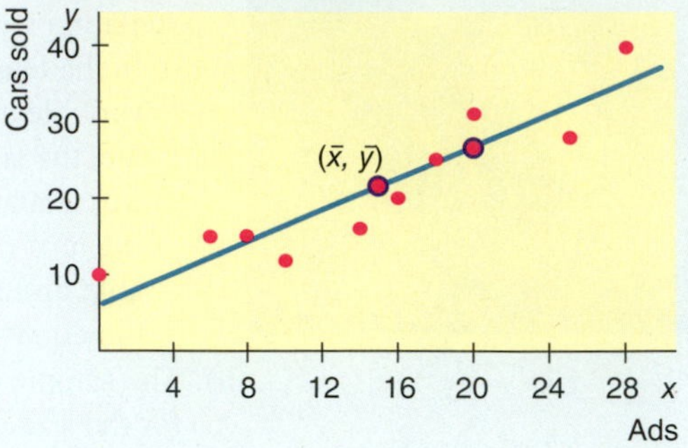

(b) Look at Equations (3) to (5) pertaining to the least-squares line (page 550). Two of the quantities that we need to find b are (Σx) and (Σxy). List the others.

→ We also need n, (Σy), (Σx^2), and $(\Sigma x)^2$.

(c) Complete Table 9-9(a).

TABLE 9-9(a)

x	y	x^2	xy
6	15	36	90
20	31	400	620
0	10	0	0
14	16	196	224
25	28	625	700
16	20	256	320
28	40	___	___
18	25	___	___
10	12	___	___
8	15	64	120
$\Sigma x = 145$	$\Sigma y = 212$	$\Sigma x^2 =$ ___	$\Sigma xy =$ ___

→ The missing table entries are shown in Table 9-9(b).

TABLE 9-9(b)

x^2	xy
$(28)^2 = 784$	$28(40) = 1120$
$(18)^2 = 324$	$18(25) = 450$
$(10)^2 = 100$	$10(12) = 120$
$\Sigma x^2 = 2785$	$\Sigma xy = 3764$

(d) Compute the sample means $\bar{x}$ and $\bar{y}$

→

$$\bar{x} = \frac{\Sigma x}{n} = \frac{145}{10} = 14.5$$

$$\bar{y} = \frac{\Sigma y}{n} = \frac{212}{10} = 21.2$$

(e) Compute a and b for the equation $\hat{y} = a + bx$ of the least-squares line.

→

$$b = \frac{n\Sigma xy - (\Sigma x)(\Sigma y)}{n\Sigma x^2 - (\Sigma x)^2}$$

$$= \frac{10(3764) - (145)(212)}{10(2785) - (145)^2} = \frac{6900}{6825} \approx 1.01$$

$$a = \bar{y} - b\bar{x}$$

$$\approx 21.2 - 1.01(14.5) \approx 6.56$$

Continued

GUIDED EXERCISE 4 *continued*

(f) What is the equation of the least-squares line $\hat{y} = a + bx$?

→ Using the values of a and b computed in part (e) or values of a and b obtained directly from a calculator,

$\hat{y} \approx 6.56 + 1.01x$

(g) Plot the least-squares line on your scatter diagram.

→ The least-squares line goes through the point $(\bar{x}, \bar{y}) = (14.5, 21.2)$. To get another point on the line, select a value for x and compute the corresponding $\hat{y}$ value using the equation $\hat{y} = 6.56 + 1.01x$. For $x = 20$, we get $\hat{y} = 6.56 + 1.01(20) = 26.8$, so the point (20, 26.8) is also on the line. The least-squares line is shown in Figure 9-13.

(h) Read the $\hat{y}$ value for $x = 12$ from your graph. Then use the equation of the least-squares line to calculate $\hat{y}$ when $x = 12$. How many cars can the manager expect to sell if 12 ads per week are aired on TV?

→ The graph gives $\hat{y} \approx 19$. From the equation, we get

$$\begin{aligned}\hat{y} &\approx 6.56 + 1.01x \\ &\approx 6.56 + 1.01(12) \quad \text{using 12 in place of } x \\ &\approx 18.68\end{aligned}$$

To the nearest whole number, the manager can expect to sell 19 cars when 12 ads are aired on TV each week.

(i) ***Interpretation*** How reliable do you think the prediction is? Explain. (Guided Exercise 5 will show that $r \approx 0.919$.)

→ The prediction should be fairly reliable. The prediction involves interpolation, and the scatter diagram shows that the data points are clustered around the least-squares line. From the next Guided Exercise we have the information that r is close to one. Of course, other variables might affect the value of y for $x = 12$.

WHAT DOES A LEAST-SQUARES LINE TELL US?

The equation of a least-squares line is based on a data set of ordered pairs (x, y) and the least-squares criterion. The least-squares criterion minimizes the sum of the square vertical distances between the data points and the line. The least-squares line expresses a linear relationship between x and y.

- The least-squares equation can be used to compute predicted $\hat{y}$ values corresponding to a given x value within the range of x values in the data pairs. The accuracy of such predictions depends on how well the least-squares line fits the data points.
- Under many circumstances it is appropriate to use the least-squares equation to compute predicted $\hat{y}$ values for x values outside but close to the range of x values in the data pairs.
- The slope of the least-squares line tells us how many units $\hat{y}$ changes for each unit change in x.

TECH NOTES When we have more data pairs, it is convenient to use a technology tool such as the TI-84Plus/TI-83Plus/TI-*n*spire calculators, Excel 2010, or Minitab to find the equation of the least-squares line. The displays show results for the data of Guided Exercise 4 regarding car sales and ads.

TI-84Plus/TI-83Plus/TI-*n*spire (with TI-84Plus keypad) Press **STAT,** choose **Calculate,** and use option **8:LinReg (a + bx).** For a graph showing the scatter plot and the least-squares line, press the **STAT PLOT** key, turn on a plot, and highlight the first type. Then press the **Y =** key. To enter the equation of the least-squares line, press **VARS,** select **5:Statistics,** highlight **EQ,** and then highlight **1:RegEQ.** Press **ENTER.** Finally, press **ZOOM** and choose **9:ZoomStat.**

Excel 2010 There are several ways to find the equation of the least-squares line in Excel. One way is to make a scatter plot. On the home screen, click the **Insert** tab. In the Chart Group, select **Scatter** and choose the first type. In the next ribbon, the Chart Layout Group offers options for including titles and axes labels. Right click on any data point and select **Add Trendline**. In the dialogue box, select **linear** and check **Display Equation on Chart**. To display the value of the coefficient of determination, check **Display R-squared Value on Chart**.

TI-84Plus/TI-83/TI-*n*spire Display

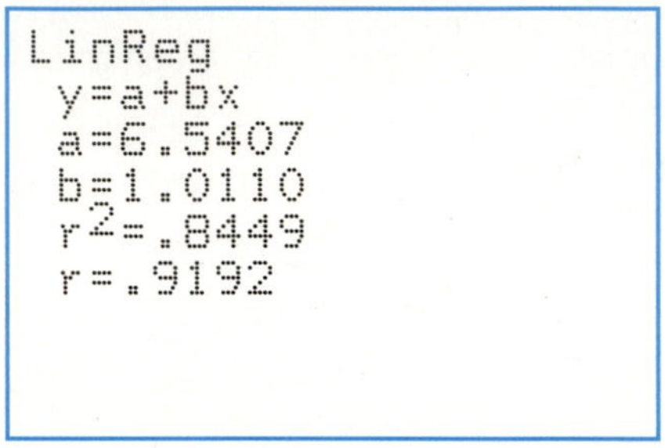

Excel Display

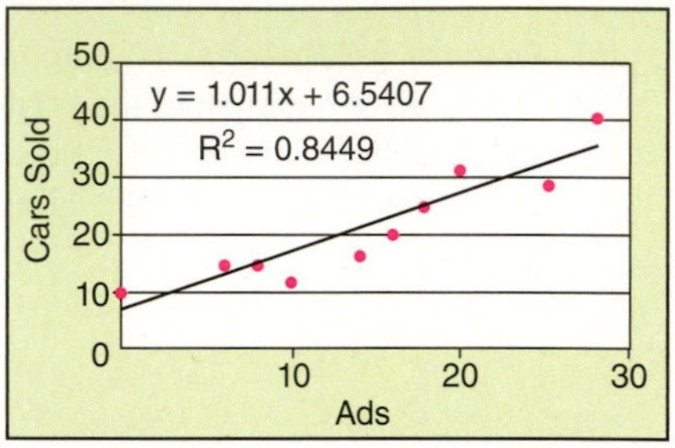

Minitab There are a number of ways to generate the least-squares line. One way is to use the menu selection **Stat ➤ Regression ➤ Fitted Line Plot.** The least-squares equation is shown with the diagram.

COEFFICIENT OF DETERMINATION

Coefficient of determination r^2

There is another way to answer the question "How good is the least-squares line as an instrument of regression?" The *coefficient of determination* r^2 is the square of the sample correlation coefficient r.

Suppose we have a scatter diagram and corresponding least-squares line, as shown in Figure 9-14.

FIGURE 9-14

Explained and Unexplained Deviations

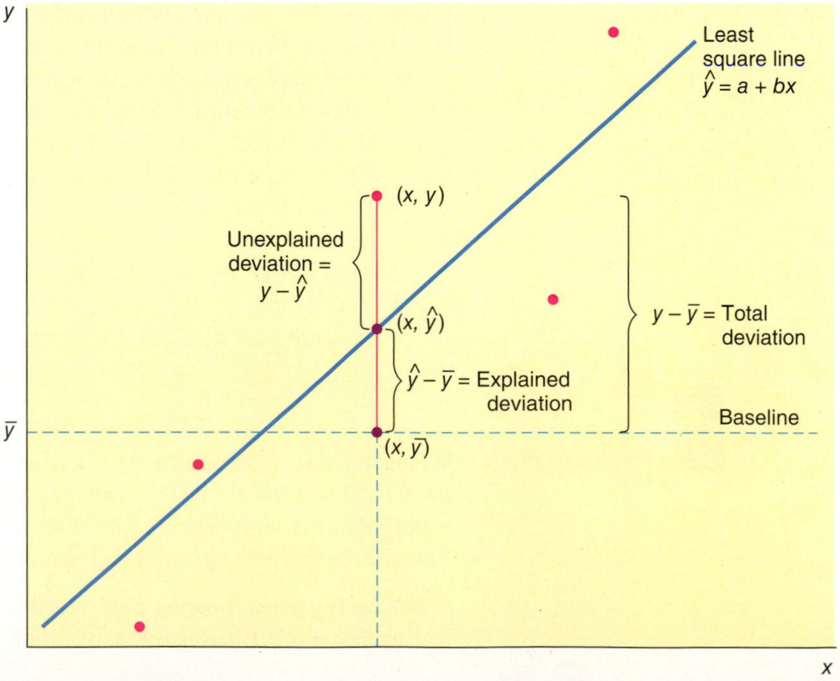

Let us take the point of view that $\bar{y}$ is a kind of baseline for the y values. If we were given an x value, and if we were completely ignorant of regression and correlation but we wanted to predict a value of y corresponding to the given x, a reasonable guess for y would be the mean $\bar{y}$. However, since we do know how to construct the least-squares regression line, we can calculate $\hat{y} = a + bx$, the predicted value corresponding to x. In most cases, the predicted value $\hat{y}$ on the least-squares line will not be the same as the actual data value y. We will measure deviations (or differences) from the baseline $\bar{y}$. (See Figure 9-14.)

$$\text{Total deviation} = y - \bar{y}$$

$$\text{Explained deviation} = \hat{y} - \bar{y}$$

$$\text{Unexplained deviation} = y - \hat{y} \qquad \text{(also known as the } residual\text{)}$$

The total deviation $y - \bar{y}$ is a measure of how far y is from the baseline $\bar{y}$. This can be broken into two parts. The explained deviation $\hat{y} - \bar{y}$ tells us how far the estimated y value "should" be from the baseline $\bar{y}$. (The "explanation" of this part of the deviation is the least-squares line, so to speak.) The unexplained deviation $y - \hat{y}$ tells us how far our data value y is "off." This amount is called *unexplained* because it is due to random chance and other factors that the least-squares line cannot account for.

$$(y - \bar{y}) = (\hat{y} - \bar{y}) + (y - \hat{y})$$

$$\begin{pmatrix}\text{Total}\\\text{deviation}\end{pmatrix} = \begin{pmatrix}\text{Explained}\\\text{deviation}\end{pmatrix} + \begin{pmatrix}\text{Unexplained}\\\text{deviation}\end{pmatrix}$$

At this point, we wish to include all the data pairs and we wish to deal only with nonnegative values (so that positive and negative deviations won't cancel out). Therefore, we construct the following equation for the sum of squares. This equation can be derived using some lengthy algebraic manipulations, which we omit.

Explained variation

Unexplained variation

$$\Sigma(y - \bar{y})^2 = \Sigma(\hat{y} - \bar{y})^2 + \Sigma(y - \hat{y})^2$$

$$\begin{pmatrix}\text{Total}\\\text{variation}\end{pmatrix} = \begin{pmatrix}\text{Explained}\\\text{variation}\end{pmatrix} + \begin{pmatrix}\text{Unexplained}\\\text{variation}\end{pmatrix}$$

Note that the sum of *squares* is taken over all data points and is then referred to as *variation* (not deviation).

The preceding concepts are connected together in the following important statement (whose proof we omit):

If r is the sample correlation coefficient [see Equation (2)], then it can be shown that

$$r^2 = \frac{\Sigma(\hat{y} - \bar{y})^2}{\Sigma(y - \bar{y})^2} = \frac{\text{Explained variation}}{\text{Total variation}}$$

r^2 is called the *coefficient of determination*.

Let us summarize our discussion.

COEFFICIENT OF DETERMINATION r^2

1. Compute the sample correlation coefficient r using the procedure of Section 9.1. Then simply compute r^2, the sample coefficient of determination.
2. ***Interpretation*** The value r^2 is the ratio of explained variation over total variation. That is, r^2 is the fractional amount of total variation in y that can be explained by using the linear model $\hat{y} = a + bx$.
3. ***Interpretation*** Furthermore, $1 - r^2$ is the fractional amount of total variation in y that is due to random chance or to the possibility of lurking variables that influence y.

In other words, the coefficient of determination r^2 is a measure of the proportion of variation in y that is explained by the regression line, using x as the explanatory variable. If $r = 0.90$, then $r^2 = 0.81$ is the coefficient of determination. We can say that about 81% of the (variation) behavior of the y variable can be explained by the corresponding (variation) behavior of the x variable if we use the equation of the least-squares line. The remaining 19% of the (variation) behavior of the y variable is due to random chance or to the possibility of lurking variables that influence y.

GUIDED EXERCISE 5 — Coefficient of Determination r^2

In Guided Exercise 4, we looked at the relationship between x = number of 1-minute spot ads on TV advertising different models of cars and y = number of cars sold each week by the sponsoring car dealership.

(a) Using the sums found in Guided Exercise 4, compute the sample correlation coefficient r. $n = 10$, $\Sigma x = 145$, $\Sigma y = 212$, $\Sigma x^2 = 2785$, and $\Sigma xy = 3764$. You also need $\Sigma y^2 = 5320$.

$$r = \frac{n\Sigma xy - (\Sigma x)(\Sigma y)}{\sqrt{n\Sigma x^2 - (\Sigma x)^2}\sqrt{n\Sigma y^2 - (\Sigma y)^2}}$$

→ $$r = \frac{10(3764) - (145)(212)}{\sqrt{10(2785) - (145)^2}\sqrt{10(5320) - (212)^2}}$$

$$\approx \frac{6900}{(82.61)(90.86)}$$

$$\approx 0.919$$

(b) Compute the coefficient of determination r^2.

→ $r^2 \approx 0.845$

(c) ***Interpretation*** What percentage of the variation in the number of car sales can be explained by the ads and the least-squares line?

→ 84.5%

(d) ***Interpretation*** What percentage of the variation in the number of car sales is not explained by the ads and the least-squares line?

→ 100% − 84.5%, or 15.5%

VIEWPOINT — It's Freezing!

Can you use average temperatures in January to predict how bad the rest of the winter will be? Can you predict the number of days with freezing temperatures for the entire calendar year using conditions in January? How good would such a forecast be for predicting growing season or number of frost-free days? Methods of this section can help you answer such questions. For more information, visit the Brase/Brase statistics site at **http://www.cengage.com/statistics/brase11e** *and find the link to Temperatures.*

SECTION 9.2 PROBLEMS

1. *Statistical Literacy* In the least-squares line $\hat{y} = 5 - 2x$, what is the value of the slope? When x changes by 1 unit, by how much does $\hat{y}$ change?

2. *Statistical Literacy* In the least squares line $\hat{y} = 5 + 3x$, what is the marginal change in $\hat{y}$ for each unit change in x?

3. *Critical Thinking* When we use a least-squares line to predict y values for x values beyond the range of x values found in the data, are we extrapolating or interpolating? Are there any concerns about such predictions?

4. *Critical Thinking* If two variables have a negative linear correlation, is the slope of the least-squares line positive or negative?

5. *Critical Thinking: Interpreting Computer Printouts* We use the form $\hat{y} = a + bx$ for the least-squares line. In some computer printouts, the least-squares equation is not given directly. Instead, the value of the constant a is given, and the coefficient b of the explanatory or predictor variable is displayed. Sometimes a is referred to as the constant, and sometimes as the intercept. Data from *Climatology Report No. 77-3* of the Department of Atmospheric Science, Colorado State University, showed the following relationship between elevation (in thousands of feet) and average number of frost-free days per year in Colorado locations.

A Minitab printout provides

Predictor	Coef	SE Coef	T	P
Constant	318.16	28.31	11.24	0.002
Elevation	−30.878	3.511	−8.79	0.003

```
s = 11.8603    R-Sq = 96.3%
```

Notice that "Elevation" is listed under "Predictor." This means that elevation is the explanatory variable x. Its coefficient is the slope b. "Constant" refers to a in the equation $\hat{y} = a + bx$.

(a) Use the printout to write the least-squares equation.
(b) For each 1000-foot increase in elevation, how many fewer frost-free days are predicted?
(c) The printout gives the value of the coefficient of determination r^2. What is the value of r? Be sure to give the correct sign for r based on the sign of b.
(d) ***Interpretation*** What percentage of the variation in y can be *explained* by the corresponding variation in x and the least-squares line? What percentage is *unexplained*?

6. *Critical Thinking: Interpreting Computer Printouts* Refer to the description of a computer display for regression described in Problem 5. The following Minitab display gives information regarding the relationship between the body weight of a child (in kilograms) and the metabolic rate of the child (in 100 kcal/24 hr). The data is based on information from *The Merck Manual* (a commonly used reference in medical schools and nursing programs).

Predictor	Coef	SE Coef	T	P
Constant	0.8565	0.4148	2.06	0.084
Weight	0.40248	0.02978	13.52	0.000

```
s = 0.517508    R-Sq = 96.8%
```

(a) Write out the least-squares equation.
(b) For each 1-kilogram increase in weight, how much does the metabolic rate of a child increase?
(c) What is the value of the sample correlation coefficient r?
(d) ***Interpretation*** What percentage of the variation in y can be *explained* by the corresponding variation in x and the least-squares line? What percentage is *unexplained*?

For Problems 7–18, please do the following.
(a) Draw a scatter diagram displaying the data.
(b) Verify the given sums Σx, Σy, Σx^2, Σy^2, and Σxy and the value of the sample correlation coefficient r.

(c) Find $\bar{x}$, $\bar{y}$, a, and b. Then find the equation of the least-squares line $\hat{y} = a + bx$.
(d) Graph the least-squares line on your scatter diagram. Be sure to use the point $(\bar{x}, \bar{y})$ as one of the points on the line.
(e) ***Interpretation*** Find the value of the coefficient of determination r^2. What percentage of the variation in y can be *explained* by the corresponding variation in x and the least-squares line? What percentage is *unexplained*?

Answers may vary slightly due to rounding.

7. *Economics: Entry-Level Jobs* An economist is studying the job market in Denver-area neighborhoods. Let x represent the total number of jobs in a given neighborhood, and let y represent the number of entry-level jobs in the same neighborhood. A sample of six Denver neighborhoods gave the following information (units in hundreds of jobs).

x	16	33	50	28	50	25
y	2	3	6	5	9	3

Source: Neighborhood Facts, The Piton Foundation. To find out more, visit the Brase/Brase statistics site at http://www.cengage.com/statistics/brase11e *and find the link to the Piton Foundation.*

Complete parts (a) through (e), given $\Sigma x = 202$, $\Sigma y = 28$, $\Sigma x^2 = 7754$, $\Sigma y^2 = 164$, $\Sigma xy = 1096$, and $r \approx 0.860$.
(f) For a neighborhood with $x = 40$ jobs, how many are predicted to be entry-level jobs?

8. *Ranching: Cattle* You are the foreman of the Bar-S cattle ranch in Colorado. A neighboring ranch has calves for sale, and you are going to buy some to add to the Bar-S herd. How much should a healthy calf weigh? Let x be the age of the calf (in weeks), and let y be the weight of the calf (in kilograms). The following information is based on data taken from *The Merck Veterinary Manual* (a reference used by many ranchers).

x	1	3	10	16	26	36
y	42	50	75	100	150	200

Complete parts (a) through (e), given $\Sigma x = 92$, $\Sigma y = 617$, $\Sigma x^2 = 2338$, $\Sigma y^2 = 82{,}389$, $\Sigma xy = 13{,}642$, and $r \approx 0.998$.
(f) The calves you want to buy are 12 weeks old. What does the least-squares line predict for a healthy weight?

9. *Weight of Car: Miles per Gallon* Do heavier cars really use more gasoline? Suppose a car is chosen at random. Let x be the weight of the car (in hundreds of pounds), and let y be the miles per gallon (mpg). The following information is based on data taken from *Consumer Reports* (Vol. 62, No. 4).

x	27	44	32	47	23	40	34	52
y	30	19	24	13	29	17	21	14

Complete parts (a) through (e), given $\Sigma x = 299$, $\Sigma y = 167$, $\Sigma x^2 = 11{,}887$, $\Sigma y^2 = 3773$, $\Sigma xy = 5814$, and $r \approx -0.946$.
(f) Suppose a car weighs $x = 38$ (hundred pounds). What does the least-squares line forecast for y = miles per gallon?

10. *Basketball: Fouls* Data for this problem are based on information from *STATS Basketball Scoreboard*. It is thought that basketball teams that make too many fouls in a game tend to lose the game even if they otherwise play well. Let x be the number of fouls that were more than (i.e., over and above) the number of fouls made the opposing team made. Let y be the percentage of times the team with the larger number of fouls won the game.

x	0	2	5	6
y	50	45	33	26

Complete parts (a) through (e), given $\Sigma x = 13$, $\Sigma y = 154$, $\Sigma x^2 = 65$, $\Sigma y^2 = 6290$, $\Sigma xy = 411$, and $r \approx -0.988$.

(f) If a team had $x = 4$ fouls over and above the opposing team, what does the least-squares equation forecast for y?

11. *Auto Accidents: Age* Data for this problem are based on information taken from *The Wall Street Journal*. Let x be the age in years of a licensed automobile driver. Let y be the percentage of all fatal accidents (for a given age) due to speeding. For example, the first data pair indicates that 36% of all fatal accidents involving 17-year-olds are due to speeding.

x	17	27	37	47	57	67	77
y	36	25	20	12	10	7	5

Complete parts (a) through (e), given $\Sigma x = 329$, $\Sigma y = 115$, $\Sigma x^2 = 18{,}263$, $\Sigma y^2 = 2639$, $\Sigma xy = 4015$, and $r \approx -0.959$.

(f) Predict the percentage of all fatal accidents due to speeding for 25-year-olds.

12. *Auto Accidents: Age* Let x be the age of a licensed driver in years. Let y be the percentage of all fatal accidents (for a given age) due to failure to yield the right-of-way. For example, the first data pair states that 5% of all fatal accidents of 37-year-olds are due to failure to yield the right-of-way. *The Wall Street Journal* article referenced in Problem 11 reported the following data:

x	37	47	57	67	77	87
y	5	8	10	16	30	43

Complete parts (a) through (e), given $\Sigma x = 372$, $\Sigma y = 112$, $\Sigma x^2 = 24{,}814$, $\Sigma y^2 = 3194$, $\Sigma xy = 8254$, and $r \approx -0.943$.

(f) Predict the percentage of all fatal accidents due to failing to yield the right-of-way for 70-year-olds.

13. *Income: Medical Care* Let x be per capita income in thousands of dollars. Let y be the number of medical doctors per 10,000 residents. Six small cities in Oregon gave the following information about x and y (based on information from *Life in America's Small Cities* by G. S. Thomas, Prometheus Books).

x	8.6	9.3	10.1	8.0	8.3	8.7
y	9.6	18.5	20.9	10.2	11.4	13.1

Complete parts (a) through (e), given $\Sigma x = 53$, $\Sigma y = 83.7$, $\Sigma x^2 = 471.04$, $\Sigma y^2 = 1276.83$, $\Sigma xy = 755.89$, and $r \approx 0.934$.

(f) Suppose a small city in Oregon has a per capita income of 10 thousand dollars. What is the predicted number of M.D.s per 10,000 residents?

14. *Violent Crimes: Prisons* Does prison really deter violent crime? Let x represent percent change in the rate of violent crime and y represent percent change in the rate of imprisonment in the general U.S. population. For 7 recent years, the following data have been obtained (Source: *The Crime Drop in America*, edited by Blumstein and Wallman, Cambridge University Press).

x	6.1	5.7	3.9	5.2	6.2	6.5	11.1
y	−1.4	−4.1	−7.0	−4.0	3.6	−0.1	−4.4

Complete parts (a) through (e), given $\Sigma x = 44.7$, $\Sigma y = -17.4$, $\Sigma x^2 = 315.85$, $\Sigma y^2 = 116.1$, $\Sigma xy = -107.18$, and $r \approx 0.084$.

(f) *Critical Thinking* Considering the values of r and r^2, does it make sense to use the least-squares line for prediction? Explain.

15. *Education: Violent Crime* The following data are based on information from the book *Life in America's Small Cities* (by G. S. Thomas, Prometheus Books). Let x be the percentage of 16- to 19-year-olds not in school and not high school graduates. Let y be the reported violent crimes per 1000 residents. Six small cities in Arkansas (Blytheville, El Dorado, Hot Springs, Jonesboro, Rogers, and Russellville) reported the following information about x and y:

x	24.2	19.0	18.2	14.9	19.0	17.5
y	13.0	4.4	9.3	1.3	0.8	3.6

Complete parts (a) through (e), given $\Sigma x = 112.8$, $\Sigma y = 32.4$, $\Sigma x^2 = 2167.14$, $\Sigma y^2 = 290.14$, $\Sigma xy = 665.03$, and $r \approx 0.764$.

(f) If the percentage of 16- to 19-year-olds not in school and not graduates reaches 24% in a similar city, what is the predicted rate of violent crimes per 1000 residents?

16. *Research: Patents* The following data are based on information from the *Harvard Business Review* (Vol. 72, No. 1). Let x be the number of different research programs, and let y be the mean number of patents per program. As in any business, a company can spread itself too thin. For example, too many research programs might lead to a decline in overall research productivity. The following data are for a collection of pharmaceutical companies and their research programs:

x	10	12	14	16	18	20
y	1.8	1.7	1.5	1.4	1.0	0.7

Complete parts (a) through (e), given $\Sigma x = 90$, $\Sigma y = 8.1$, $\Sigma x^2 = 1420$, $\Sigma y^2 = 11.83$, $\Sigma xy = 113.8$, and $r \approx -0.973$.

(f) Suppose a pharmaceutical company has 15 different research programs. What does the least-squares equation forecast for y = mean number of patents per program?

17. *Archaeology: Artifacts* Data for this problem are based on information taken from *Prehistoric New Mexico: Background for Survey* (by D. E. Stuart and R. P. Gauthier, University of New Mexico Press). It is thought that prehistoric Indians did not take their best tools, pottery, and household items when they visited higher elevations for their summer camps. It is hypothesized that archaeological sites tend to lose their cultural identity and specific cultural affiliation as the elevation of the site increases. Let x be the elevation (in thousands of feet) of an archaeological site in the southwestern United States. Let y be the percentage of unidentified artifacts (no specific cultural affiliation) at a given elevation. The following data were obtained for a collection of archaeological sites in New Mexico:

x	5.25	5.75	6.25	6.75	7.25
y	19	13	33	37	62

Complete parts (a) through (e), given $\Sigma x = 31.25$, $\Sigma y = 164$, $\Sigma x^2 = 197.813$, $\Sigma y^2 = 6832$, $\Sigma xy = 1080$, and $r \approx 0.913$.

(f) At an archaeological site with elevation 6.5 (thousand feet), what does the least-squares equation forecast for y = percentage of culturally unidentified artifacts?

18. *Cricket Chirps: Temperature* Anyone who has been outdoors on a summer evening has probably heard crickets. Did you know that it is possible to use the cricket as a thermometer? Crickets tend to chirp more frequently as temperatures increase. This phenomenon was studied in detail by George W. Pierce, a physics professor at Harvard. In the following data, x is a random variable representing chirps per second and y is a random variable representing temperature (°F). These data are also available for download at the Online Study Center.

Dorling Kindersley/Getty Images

x	20.0	16.0	19.8	18.4	17.1	15.5	14.7	17.1
y	88.6	71.6	93.3	84.3	80.6	75.2	69.7	82.0

x	15.4	16.2	15.0	17.2	16.0	17.0	14.4
y	69.4	83.3	79.6	82.6	80.6	83.5	76.3

Source: Reprinted by permission of the publisher from *The Songs of Insects* by George W. Pierce, Cambridge, Mass.: Harvard University Press, Copyright © 1948 by the President and Fellows of Harvard College.

Complete parts (a) through (e), given $\Sigma x = 249.8$, $\Sigma y = 1200.6$, $\Sigma x^2 = 4200.56$, $\Sigma y^2 = 96{,}725.86$, $\Sigma xy = 20{,}127.47$, and $r \approx 0.835$.
(f) What is the predicted temperature when $x = 19$ chirps per second?

19. *Expand Your Knowledge: Residual Plot* The least-squares line usually does not go through all the sample data points (x, y). In fact, for a specified x value from a data pair (x, y), there is usually a difference between the predicted value and the y value paired with x. This difference is called the *residual*.

The **residual** is the difference between the y value in a specified data pair (x, y) and the value $\hat{y} = a + bx$ predicted by the least-squares line for the same x.

$y - \hat{y}$ is the **residual**.

Residual plot

One way to assess how well a least-squares line serves as a model for the data is a **residual plot**. To make a residual plot, we put the x values in order on the horizontal axis and plot the corresponding residuals $y - \hat{y}$ in the vertical direction. Because the mean of the residuals is always zero for a least-squares model, we place a horizontal line at zero. The accompanying figure shows a residual plot for the data of Guided Exercise 4, in which the relationship between the number of ads run per week and the number of cars sold that week was explored. To make the residual plot, first compute all the residuals. Remember that x and y are the given data values, and $\hat{y}$ is computed from the least-squares line $\hat{y} \approx 6.56 + 1.01x$.

Residual

x	y	$\hat{y}$	$y - \hat{y}$
6	15	12.6	2.4
20	31	26.8	4.2
0	10	6.6	3.4
14	16	20.7	−4.7
25	28	31.8	−3.8

Residual

x	y	$\hat{y}$	$y - \hat{y}$
16	20	22.7	−2.7
28	40	34.8	5.2
18	25	24.7	0.3
10	12	16.7	−4.7
8	15	14.6	0.4

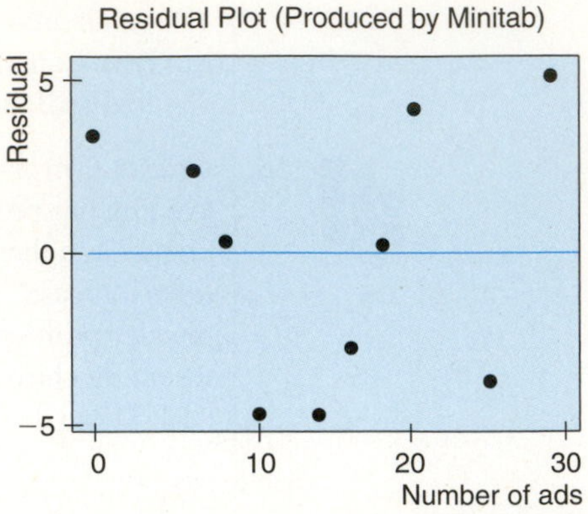

(a) If the least-squares line provides a reasonable model for the data, the pattern of points in the plot will seem random and unstructured about the horizontal line at 0. Is this the case for the residual plot?

(b) If a point on the residual plot seems far outside the pattern of other points, it might reflect an unusual data point (x, y), called an *outlier*. Such points may have quite an influence on the least-squares model. Do there appear to be any outliers in the data for the residual plot?

20. *Residual Plot: Miles per Gallon* Consider the data of Problem 9.

(a) Make a residual plot for the least-squares model.

(b) Use the residual plot to comment about the appropriateness of the least-squares model for these data. See Problem 19.

21. *Critical Thinking: Exchange x and y in Least-Squares Equation*

(a) Suppose you are given the following (x, y) data pairs:

x	1	3	4
y	2	1	6

Show that the least-squares equation for these data is $y = 0.066 + 1.393x$ (rounded to three digits after the decimal).

(b) Now suppose you are given these (x, y) data pairs:

x	2	1	6
y	1	3	4

Show that the least-squares equation for these data is $y = 0.971 + 0.494x$ (rounded to three digits after the decimal).

(c) In the data for parts (a) and (b), did we simply exchange the x and y values of each data pair?

(d) Solve $y = 0.066 + 1.393x$ for x. Do you get the least-squares equation of part (b) with the symbols x and y exchanged?

(e) In general, suppose we have the least-squares equation $y = a + bx$ for a set of data pairs (x, y). If we solve this equation for x, will we *necessarily* get the least-squares equation for the set of data pairs (y, x) (with x and y exchanged)? Explain using parts (a) through (d).

22. *Expand Your Knowledge: Logarithmic Transformations, Exponential Growth Model* There are several extensions of linear regression that apply to exponential growth and power law models. Problems 22–25 will outline some of these extensions. First of all, recall that a variable grows *linearly* over time if it *adds* a fixed increment during each equal time period. *Exponential* growth occurs when a variable is *multiplied* by a fixed number during each time period. This means that exponential growth increases by a fixed multiple or percentage of the previous amount. College algebra can be used to show

that if a variable grows exponentially, then its logarithm grows linearly. The exponential growth model is $y = \alpha\beta^x$, where α and β are fixed constants to be estimated from data.

How do we know when we are dealing with exponential growth, and how can we estimate α and β? Please read on. Populations of living things such as bacteria, locusts, fish, panda bears, and so on, tend to grow (or decline) exponentially. However, these populations can be restricted by outside limitations such as food, space, pollution, disease, hunting, and so on. Suppose we have data pairs (x, y) for which there is reason to believe the scatter plot is not linear, but rather exponential, as described above. This means the increase in y values begins rather slowly but then seems to explode. *Note:* For exponential growth models, we assume all $y > 0$.

x	1	2	3	4	5
y	3	12	22	51	145

Consider the following data, where x = time in hours and y = number of bacteria in a laboratory culture at the end of x hours.

(a) Look at the Excel graph of the scatter diagram of the (x, y) data pairs. Do you think a straight line will be a good fit to these data? Do the y values seem almost to explode as time goes on?

(b) Now consider a transformation $y' = \log y$. We are using common logarithms of base 10 (however, natural logarithms of base e would work just as well).

x	1	2	3	4	5
$y' = \log y$	0.477	1.079	1.342	1.748	2.161

Look at the Excel graph of the scatter diagram of the (x, y') data pairs and compare this diagram with the diagram in part (a). Which graph appears to better fit a straight line?

Excel Graphs

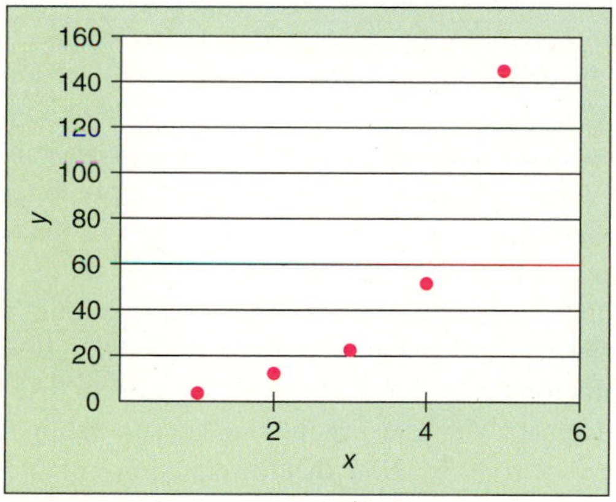

Part (a) Model with (x, y) Data Pairs

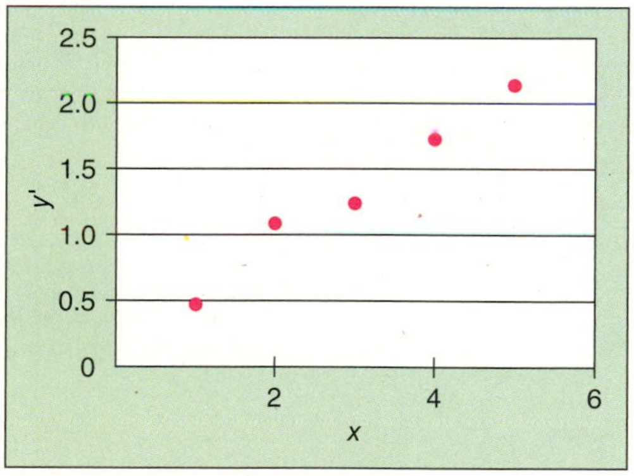

Part (b) Model with (x, y') Data Pairs

(c) Use a calculator with regression keys to verify the linear regression equation for the (x, y) data pairs, $\hat{y} = -50.3 + 32.3x$, with sample correlation coefficient $r = 0.882$.

(d) Use a calculator with regression keys to verify the linear regression equation for the (x, y') data pairs, $y' = 0.150 + 0.404x$, with sample correlation coefficient $r = 0.994$. The sample correlation coefficient $r = 0.882$ for the (x, y) pairs is not bad. But the sample correlation coefficient $r = 0.994$ for the (x, y') pairs is a lot better!

(e) The exponential growth model is $y = \alpha\beta^x$. Let us use the results of part (d) to estimate α and β for this strain of laboratory bacteria. The equation

$y' = a + bx$ is the same as $\log y = a + bx$. If we raise both sides of this equation to the power 10 and use some college algebra, we get $y = 10^a(10^b)^x$. Thus, $\alpha \approx 10^a$ and $\beta \approx 10^b$. Use these results to approximate α and β and write the exponential growth equation for our strain of bacteria.

Note: The TI-84Plus/TI-83Plus/TI-*n*spire calculators fully support the exponential growth model. Place the original x data in list L1 and the corresponding y data in list L2. Then press **STAT,** followed by **CALC,** and scroll down to option **0: ExpReg.** The output gives values for α, β, and the sample correlation coefficient r.

23. *Expand Your Knowledge: Logarithmic Transformations, Exponential Growth Model* Let x = day of observation and y = number of locusts per square meter during a locust infestation in a region of North Africa.

x	2	3	5	8	10
y	2	3	12	125	630

(a) Draw a scatter diagram of the (x, y) data pairs. Do you think a straight line will be a good fit to these data? Do the y values almost seem to explode as time goes on?
(b) Now consider a transformation $y' = \log y$. We are using common logarithms of base 10. Draw a scatter diagram of the (x, y') data pairs and compare this diagram with the diagram of part (a). Which graph appears to better fit a straight line?
(c) Use a calculator with regression keys to find the linear regression equation for the data pairs (x, y'). What is the correlation coefficient?
(d) The exponential growth model is $y = \alpha\beta^x$. Estimate α and β and write the exponential growth equation. *Hint:* See Problem 22.

24. *Expand Your Knowledge: Logarithmic Transformations, Power Law Model* When we take measurements of the same general type, a power law of the form $y = \alpha x^\beta$ often gives an excellent fit to the data. A lot of research has been conducted as to why power laws work so well in business, economics, biology, ecology, medicine, engineering, social science, and so on. Let us just say that if you do not have a good straight-line fit to data pairs (x, y), and the scatter plot does not rise dramatically (as in exponential growth), then a power law is often a good choice. College algebra can be used to show that power law models become linear when we apply logarithmic transformations to both variables. To see how this is done, please read on. *Note:* For power law models, we assume all $x > 0$ and all $y > 0$.

Suppose we have data pairs (x, y) and we want to find constants α and β such that $y = \alpha x^\beta$ is a good fit to the data. First, make the logarithmic transformations $x' = \log x$ and $y' = \log y$. Next, use the (x', y') data pairs and a calculator with linear regression keys to obtain the least-squares equation $y' = a + bx'$. Note that the equation $y' = a + bx'$ is the same as $\log y = a + b(\log x)$. If we raise both sides of this equation to the power 10 and use some college algebra, we get $y = 10^a(x)^b$. In other words, for the power law model, we have $\alpha \approx 10^a$ and $\beta \approx b$.

In the electronic design of a cell phone circuit, the buildup of electric current (Amps) is an important function of time (microseconds). Let x = time in microseconds and let y = Amps built up in the circuit at time x.

x	2	4	6	8	10
y	1.81	2.90	3.20	3.68	4.11

(a) Make the logarithmic transformations $x' = \log x$ and $y' = \log y$. Then make a scatter plot of the (x', y') values. Does a linear equation seem to be a good fit to this plot?

(b) Use the (x', y') data points and a calculator with regression keys to find the least-squares equation $y' = a + bx'$. What is the sample correlation coefficient?

(c) Use the results of part (b) to find estimates for α and β in the power law $y = \alpha x^{\beta}$. Write the power law giving the relationship between time and Amp buildup.

Note: The TI-84Plus/TI-83Plus/TI-*n*spire calculators fully support the power law model. Place the original x data in list L1 and the corresponding y data in list L2. Then press **STAT,** followed by **CALC,** and scroll down to option **A: PwrReg.** The output gives values for α, β, and the sample correlation coefficient r.

25. *Expand Your Knowledge: Logarithmic Transformations, Power Law Model* Let x = boiler steam pressure in 100 lb/in.^{2} and let y = critical sheer strength of boiler plate steel joints in tons/in.^{2}. We have the following data for a series of factory boilers.

x	4	5	6	8	10
y	3.4	4.2	6.3	10.9	13.3

(a) Make the logarithmic transformations $x' = \log x$ and $y' = \log y$. Then make a scatter plot of the (x', y') values. Does a linear equation seem to be a good fit to this plot?

(b) Use the (x', y') data points and a calculator with regression keys to find the least-squares equation $y' = a + bx'$. What is the sample correlation coefficient?

(c) Use the results of part (b) to find estimates for α and β in the power law $y = \alpha x^{\beta}$. Write the power equation for the relationship between steam pressure and sheer strength of boiler plate steel. *Hint:* See Problem 24.

SECTION **9.3**

Inferences for Correlation and Regression

FOCUS POINTS

- Test the correlation coefficient ρ.
- Use sample data to compute the standard error of estimate S_e.
- Find a confidence interval for the value of y predicted for a specified value of x.
- Test the slope β of the least-squares line.
- Find a confidence interval for the slope β of the least-squares line and interpret its meaning.

Learn more, earn more! We have probably all heard this platitude. The question is whether or not there is some truth in this statement. Do college graduates have an improved chance at a better income? Is there a trend in the general population to support the "learn more, earn more" statement?

Consider the following variables: x = percentage of the population 25 or older with at least four years of college and y = percentage *growth* in per capita income over the past 7 years. A random sample of six communities in Ohio gave the information (based on *Life in America's Small Cities* by G. S. Thomas) shown in Table 9-10 on the next page.

If we use what we learned in Sections 9.1 and 9.2, we can compute the sample correlation coefficient r and the least-squares line $\hat{y} = a + bx$ using the data of Table 9-10. However, r is only a *sample* correlation coefficient, and $\hat{y} = a + bx$ is only a "*sample-based*" least-squares line. What if we used *all* possible data pairs (x, y) from *all* U.S. cities, not just six towns in Ohio? If we accomplished this seemingly impossible task, we would have the *population* of all (x, y) pairs.

Population correlation coefficient ρ

From this population of (x, y) pairs, we could (in theory) compute the *population correlation coefficient*, which we call ρ (Greek letter *rho*, pronounced like "row"). We could also compute the least-squares line for the entire population, which we denote as $y = \alpha + \beta x$ using more Greek letters, α (*alpha*) and β (*beta*).

Sample Statistic		Population Parameter
r	$\rightarrow$	ρ
a	$\rightarrow$	α
b	$\rightarrow$	β
$\hat{y} = a + bx$	$\rightarrow$	$y = \alpha + \beta x$

Requirements for inferences concerning linear regression

Requirements for statistical inference

To make inferences regarding correlation and linear regression, we need to be sure that

(a) The set (x, y) of ordered pairs is a *random sample* from the population of all possible such (x, y) pairs.

(b) For each fixed value of x, the y values have a normal distribution. All of the y distributions have the same variance, and, for a given x value, the distribution of y values has a mean that lies on the least-squares line. We also assume that for a fixed y, each x has its own normal distribution. In most cases the results are still accurate if the distributions are simply mound-shaped and symmetric and the y variances are approximately equal.

We assume these conditions are met for all inferences presented in this section.

TESTING THE CORRELATION COEFFICIENT

The first topic we want to study is the statistical significance of the sample correlation coefficient r. To do this, we construct a statistical test of ρ, the population correlation coefficient. The test will be based on the following theorem.

THEOREM 9.1 Let r be the sample correlation coefficient computed using data pairs (x, y). We use the null hypothesis

H_0: x and y have no linear correlation, so $\rho = 0$

The alternate hypothesis may be

$H_1: \rho > 0$ or $H_1: \rho < 0$ or $H_1: \rho \neq 0$

The conversion of r to a Student's t distribution is

$$t = \frac{r\sqrt{n-2}}{\sqrt{1-r^2}} \quad \text{with } d.f. = n - 2$$

where n is the number of sample data pairs (x, y) $(n \geq 3)$.

Problem 13 at the end of this section discusses how sample size might affect the significance of r.

PROCEDURE

HOW TO TEST THE POPULATION CORRELATION COEFFICIENT ρ

1. Use the *null hypothesis* H_0: $\rho = 0$. In the context of the application, state the *alternate hypothesis* ($\rho > 0$ or $\rho < 0$ or $\rho \neq 0$) and set the *level of significance* α.

Continued

2. Obtain a random sample of $n \geq 3$ data pairs (x, y) and compute the *sample test statistic*

$$t = \frac{r\sqrt{n-2}}{\sqrt{1-r^2}} \text{ with degrees of freedom } d.f. = n - 2$$

where r is the sample correlation coefficient
n is the sample size

3. Use a Student's t distribution and the type of test, one-tailed or two-tailed, to find (or estimate) the *P-value* corresponding to the test statistic.
4. *Conclude* the test. If P-value $\leq \alpha$, then reject H_0. If P-value $> \alpha$, then do not reject H_0.
5. *Interpret your conclusion* in the context of the application.

Serial correlation (also known as *autocorrelation*) describes the extent to which the result of one time period of a time series is related to the result in the next period. Problems 15-17 at the end of this section discuss serial correlation.

EXAMPLE 6

TESTING ρ

Let's return to our data from Ohio regarding the percentage of the population with at least four years of college and the percentage of growth in per capita income (Table 9-10). We'll develop a test for the population correlation coefficient ρ.

TABLE 9-10 **Education and Income Growth Percentages**

x	9.9	11.4	8.1	14.7	8.5	12.6
y	37.1	43.0	33.4	47.1	26.5	40.2

SOLUTION: First, we compute the sample correlation coefficient r. Using a calculator, statistical software, or a "by-hand" calculation from Section 9.1, we find

$$r \approx 0.887$$

Now we test the population correlation coefficient ρ. Remember that x represents percentage college graduates and y represents percentage salary increases in the general population. We suspect the population correlation is positive, $\rho > 0$. Let's use a 1% level of significance:

H_0: $\rho = 0$ (no linear correlation)
H_1: $\rho > 0$ (positive linear correlation)

Convert the sample test statistic $r = 0.887$ to t using $n = 6$.

$$t = \frac{r\sqrt{n-2}}{\sqrt{1-r^2}} = \frac{0.887\sqrt{6-2}}{\sqrt{1-0.887^2}} \approx 3.84 \text{ with } d.f. = n - 2 = 6 - 2 = 4$$

The P-value for the sample test statistic $t = 3.84$ is shown in Figure 9-15. Since we have a right-tailed test, we use the one-tail area in the Student's t distribution (Table 6 of Appendix II).

From Table 9-11, we see that

$$0.005 < P\text{-value} < 0.010$$

Since the interval containing the P-value is less than or equal the level of significance $\alpha = 0.01$, we reject H_0 and conclude that the population correlation coefficient between x and y is positive. Technology gives P-value ≈ 0.0092.

TABLE 9-11 Excerpt from Student's t Distribution

✓ one-tail area	0.010	0.005
two-tail area	0.020	0.010
$d.f. = 4$	3.747	4.604
	↑ Sample $t = 3.84$	

FIGURE 9-15

P-value

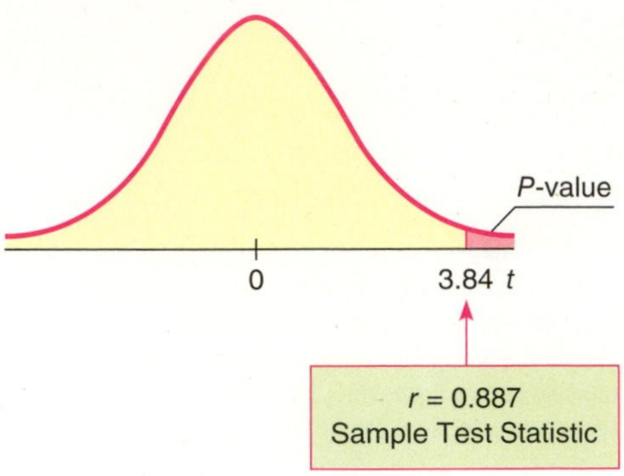

Caution: Although we have shown that x and y are positively correlated, we have not shown that an increase in education *causes* an increase in earnings.

GUIDED EXERCISE 6 Testing ρ

A medical research team is studying the effect of a new drug on red blood cells. Let x be a random variable representing milligrams of the drug given to a patient. Let y be a random variable representing red blood cells per cubic milliliter of whole blood. A random sample of $n = 7$ volunteer patients gave the following results.

x	9.2	10.1	9.0	12.5	8.8	9.1	9.5
y	5.0	4.8	4.5	5.7	5.1	4.6	4.2

Use a calculator to verify that $r \approx 0.689$. Then use a 1% level of significance to test the claim that $\rho \neq 0$.

(a) State the null and alternate hypotheses. What is the level of significance α?

➡ $H_0: \rho = 0$; $H_1: \rho \neq 0$; $\alpha = 0.01$

(b) Compute the sample test statistic.

➡ $$t = \frac{r\sqrt{n-2}}{\sqrt{1-r^2}} \approx \frac{0.689\sqrt{7-2}}{\sqrt{1-0.689^2}} \approx \frac{1.5406}{0.7248} \approx 2.126$$

(c) Use the Student's t distribution, Table 6 of Appendix II, to estimate the P-value.

➡ $d.f. = n - 2 = 7 - 2 = 5$; two-tailed test

✓ two-tail area	0.100	0.050
$d.f. = 5$	2.015	2.571
	↑ Sample $t = 2.126$	

$0.050 < P\text{-value} < 0.100$

(d) Do we reject or fail to reject H_0?

➡ Since the interval containing the P-value lies to the right of $\alpha = 0.01$, we do not reject H_0. Technology gives P-value ≈ 0.0866.

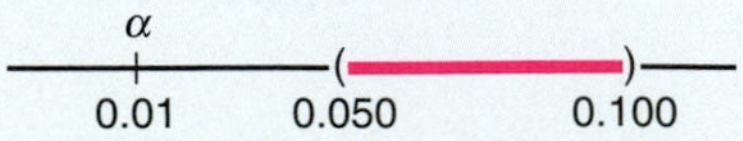

Continued

GUIDED EXERCISE 6 *continued*

(e) ***Interpret*** the conclusion in the context of the application.	At the 1% level of significance, the evidence is not strong enough to indicate any correlation between the amount of drug administered and the red blood cell count.

STANDARD ERROR OF ESTIMATE

Sometimes a scatter diagram clearly indicates the existence of a linear relationship between x and y, but it can happen that the points are widely scattered about the least-squares line. We need a method (besides just looking) for measuring the spread of a set of points about the least-squares line. There are three common methods of measuring the spread. One method uses the *standard error of estimate*. The others use the *coefficient of correlation* and the *coefficient of determination*.

For the standard error of estimate, we use a measure of spread that is in some ways like the standard deviation of measurements of a single variable. Let

$$\hat{y} = a + bx$$

FIGURE 9-16

The Distance Between Points (x, y) and $(x, \hat{y})$

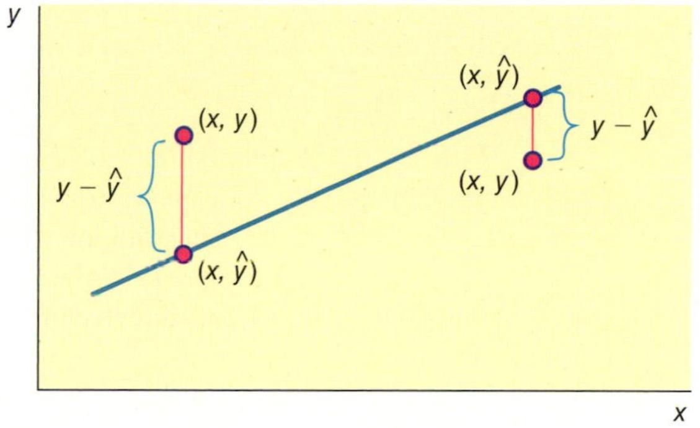

be the predicted value of y from the least-squares line. Then $y - \hat{y}$ is the difference between the y value of the *data point* (x, y) shown on the scatter diagram (Figure 9-16) and the $\hat{y}$ value of the point on the *least-squares line* with the same x value. The quantity $y - \hat{y}$ is known as the *residual*. To avoid the difficulty of having some positive and some negative values, we square the quantity $(y - \hat{y})$. Then we sum the squares and, for technical reasons, divide this sum by $n - 2$. Finally, we take the square root to obtain the *standard error of estimate*, denoted by S_e.

Residual

Standard error of estimate S_e

$$\textbf{Standard error of estimate} = S_e = \sqrt{\frac{\Sigma(y - \hat{y})^2}{n - 2}} \qquad (7)$$

where $\hat{y} = a + bx$ and $n \geq 3$.

Note: To compute the standard error of estimate, we require that there be at least three points on the scatter diagram. If we had only two points, the line would be a perfect fit, since two points determine a line. In such a case, there would be no need to compute S_e.

The nearer the scatter points lie to the least-squares line, the smaller S_e will be. In fact, if $S_e = 0$, it follows that each $y - \hat{y}$ is also zero. This means that all the scatter points lie *on* the least-squares line if $S_e = 0$. The larger S_e becomes, the more scattered the points are.

The formula for the standard error of estimate is reminiscent of the formula for the standard deviation, which is also a measure of dispersion. However, the standard deviation involves differences of data values from a mean, whereas the standard error of estimate involves the differences between experimental and predicted y values for a given x (i.e., $y - \hat{y}$).

The actual computation of S_e using Equation (7) is quite long because the formula requires us to use the least-squares line equation to compute a predicted value $\hat{y}$ for *each* x value in the data pairs. There is a computational formula that we strongly recommend you use. However, as with all the computation formulas, be careful about rounding. This formula is sensitive to rounding, and you should carry as many digits as seem reasonable for your problem. Answers will vary, depending on the rounding used. We give the formula here and follow it with an example of its use.

PROCEDURE

HOW TO FIND THE STANDARD ERROR OF ESTIMATE S_e

1. Obtain a random sample of $n \geq 3$ data pairs (x, y).
2. Use the procedures of Section 9.2 to find a and b from the sample least-squares line $\hat{y} = a + bx$.
3. The standard error of estimate is

$$S_e = \sqrt{\frac{\Sigma y^2 - a\Sigma y - b\Sigma xy}{n - 2}} \tag{8}$$

Computation formula for S_e

With a considerable amount of algebra, Equations (7) and (8) can be shown to be mathematically equivalent. Equation (7) shows the strong similarity between the standard error of estimate and the standard deviation. Equation (8) is a shortcut calculation formula because it involves few subtractions. The sums Σx, Σy, Σx^2, Σy^2, and Σxy are provided directly on most calculators that support two-variable statistics.

In the next example, we show you how to compute the standard error of estimate using the computation formula.

EXAMPLE 7

LEAST-SQUARES LINE AND S_e

Photo Researchers

June and Jim are partners in the chemistry lab. Their assignment is to determine how much copper sulfate ($CuSO_4$) will dissolve in water at 10, 20, 30, 40, 50, 60, and 70°C. Their lab results are shown in Table 9-12, where y is the weight in grams of copper sulfate that will dissolve in 100 grams of water at x°C.

Sketch a scatter diagram, find the equation of the least-squares line, and compute S_e.

SOLUTION: Figure 9-17 includes a scatter diagram for the data of Table 9-12. To find the equation of the least-squares line and the value of S_e, we set up a computational table (Table 9-13).

$$\bar{x} = \frac{\Sigma x}{n} = \frac{280}{7} = 40 \text{ and } \bar{y} = \frac{\Sigma y}{n} = \frac{213}{7} \approx 30.429$$

$$b = \frac{n\Sigma xy - (\Sigma x)(\Sigma y)}{n\Sigma x^2 - (\Sigma x)^2} = \frac{7(9940) - (280)(213)}{7(14{,}000) - (280)^2} = \frac{9940}{19{,}600} \approx 0.50714$$

$$a = \bar{y} - b\bar{x} \approx 30.429 - 0.507(40) \approx 10.149$$

The equation of the least-squares line is

$$\hat{y} = a + bx$$

$$\hat{y} \approx 10.14 + 0.51x$$

TABLE 9-12 Lab Results (x = °C, y = amount of $CuSo_4$)

x	y
10	17
20	21
30	25
40	28
50	33
60	40
70	49

FIGURE 9-17

Scatter Diagram and Least-Squares Line for Chemistry Experiment

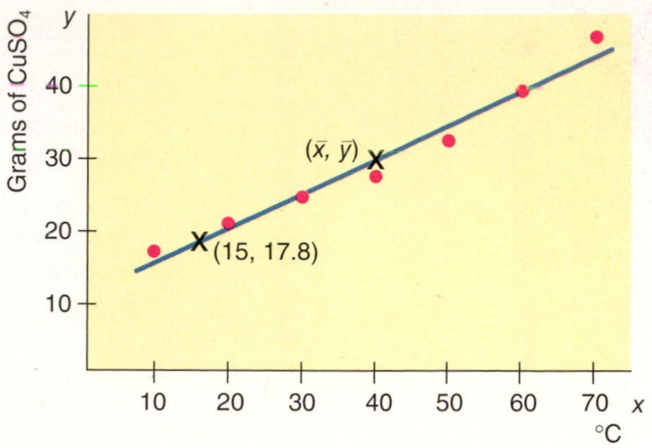

TABLE 9-13 Computational Table

x	y	x^2	y^2	xy
10	17	100	289	170
20	21	400	441	420
30	25	900	625	750
40	28	1600	784	1120
50	33	2500	1089	1650
60	40	3600	1600	2400
70	49	4900	2401	3430
$\Sigma x = 280$	$\Sigma y = 213$	$\Sigma x^2 = 14{,}000$	$\Sigma y^2 = 7229$	$\Sigma xy = 9940$

The graph of the least-squares line is shown in Figure 9-17. Notice that it passes through the point $(\bar{x}, \bar{y}) = (40, 30.4)$. Another point on the line can be found by using $x = 15$ in the equation of the line $\hat{y} = 10.14 + 0.51x$. When we use 15 in place of x, we obtain $\hat{y} = 10.14 + 0.51(15) = 17.8$. The point (15, 17.8) is the other point we used to graph the least-squares line in Figure 9-17.

The standard error of estimate is computed using the computational formula

$$S_e = \sqrt{\frac{\Sigma y^2 - a\Sigma y - b\Sigma xy}{n - 2}}$$

$$\approx \sqrt{\frac{7229 - 10.149(213) - 0.507(9940)}{7 - 2}} \approx \sqrt{\frac{27.683}{5}} \approx 2.35$$

Note: This formula is very sensitive to rounded values of a and b.

TECH NOTES Although many calculators that support two-variable statistics and linear regression do not provide the value of the standard error of estimate S_e directly, they do provide the sums required for the calculation of S_e. The TI-84Plus/TI-83Plus/TI-*n*spire, Excel 2010, and Minitab all provide the value of S_e.

TI-84Plus/TI-83Plus/TI-*n* spire (with TI-84Plus keypad) The value for S_e is given as s under **STAT, TEST,** option **E: LinRegTTest.**

Excel 2010 Click the **Insert Function** f_x. In the dialogue box, use **Statistical** for the category, and select the function **STEYX.**

Minitab Use the menu choices **Stat ➤ Regression ➤ Regression**. The value for S_e is given as *s* in the display.

CONFIDENCE INTERVALS FOR *y*

The least-squares line gives us a predicted value $\hat{y}$ for a specified x value. However, we used sample data to get the equation of the line. The line derived from the population of all data pairs is likely to have a slightly different slope, which we designate by the symbol β for population slope, and a slightly different y intercept, which we designate by the symbol α for population intercept. In addition, there is some random error ε, so the true y value is

Population slope β

$$y = \alpha + \beta x + \varepsilon$$

Because of the random variable ε, for each x value there is a corresponding distribution of y values. The methods of linear regression were developed so that the distribution of y values for a given x is centered on the population regression line. Furthermore, the distributions of y values corresponding to each x value all have the same standard deviation, estimated by the standard error of estimate S_e.

Confidence interval for predicted *y*

Using all this background, the theory tells us that for a specific x, a *c confidence interval for y* is given by the next procedure.

PROCEDURE

HOW TO FIND A CONFIDENCE INTERVAL FOR A PREDICTED *y* FROM THE LEAST-SQUARES LINE

1. Obtain a random sample of $n \geq 3$ data pairs (x, y).
2. Use the procedure of Section 9.2 to find the least-squares line $\hat{y} = a + bx$. You also need to find $\bar{x}$ from the sample data and the standard error of estimate S_e using Equation (8) of this section.
3. The ***c* confidence interval for *y*** for a **specified value of *x*** is

$$\hat{y} - E < y < \hat{y} + E$$

where

$$E = t_c S_e \sqrt{1 + \frac{1}{n} + \frac{n(x - \bar{x})^2}{n\Sigma x^2 - (\Sigma x)^2}}$$

$\hat{y} = a + bx$ is the predicted value of y from the least-squares line for a *specified* x value

c = confidence level $(0 < c < 1)$

n = number of data pairs $(n \geq 3)$

t_c = critical value from Student's t distribution for c confidence level using $d.f. = n - 2$

S_e = standard error of estimate

The formulas involved in the computation of a c confidence interval look complicated. However, they involve quantities we have already computed or values we can easily look up in tables. The next example illustrates this point.

EXAMPLE 8

CONFIDENCE INTERVAL FOR PREDICTION

Using the data of Table 9-13 of Example 7, find a 95% confidence interval for the amount of copper sulfate that will dissolve in 100 grams of water at 45°C.

SOLUTION: First, we need to find $\hat{y}$ for $x = 45°C$. We use the equation of the least-squares line that we found in Example 7.

$$\hat{y} \approx 10.14 + 0.51x \quad \text{from Example 7}$$
$$\hat{y} \approx 10.14 + 0.51(45) \quad \text{use 45 in place of } x$$
$$\hat{y} \approx 33$$

A 95% confidence interval for y is then

$$\hat{y} - E < y < \hat{y} + E$$
$$33 - E < y < 33 + E$$

where $E = t_c S_e \sqrt{1 + \frac{1}{n} + \frac{n(x - \bar{x})^2}{n\Sigma x^2 - (\Sigma x)^2}}$.

From Example 7, we have $n = 7$, $\Sigma x = 280$, $\Sigma x^2 = 14{,}000$, $\bar{x} = 40$, and $S_e \approx 2.35$. Using $n - 2 = 7 - 2 = 5$ degrees of freedom, we find from Table 6 of Appendix II that $t_{0.95} = 2.571$.

$$E \approx (2.571)(2.35)\sqrt{1 + \frac{1}{7} + \frac{7(45 - 40)^2}{7(14{,}000) - (280)^2}}$$
$$\approx (2.571)(2.35)\sqrt{1.15179} \approx 6.5$$

A 95% confidence interval for y is

$$33 - 6.5 \le y \le 33 + 6.5$$
$$26.5 \le y \le 39.5$$

This means we are 95% sure that the interval between 26.5 grams and 39.5 grams is one that contains the predicted amount of copper sulfate that will dissolve in 100 grams of water at 45°C. The interval is fairly wide but would decrease with more sample data.

GUIDED EXERCISE 7 | CONFIDENCE INTERVAL FOR PREDICTION

Let's use the data of Example 7 to compute a 95% confidence interval for y = amount of copper sulfate that will dissolve at $x = 15°C$.

(a) From Example 7, we have
$\hat{y} \approx 10.14 + 0.51x$
Evaluate $\hat{y}$ for $x = 15$.

→ $\hat{y} \approx 10.14 + 0.51x$
$\approx 10.14 + 0.51(15)$
≈ 17.8

(b) The bound E on the error of estimate is

$$E = t_c S_e \sqrt{1 + \frac{1}{n} + \frac{n(x - \bar{x})^2}{n\Sigma x^2 - (\Sigma x)^2}}$$

From Example 7, we know that $S_e \approx 2.35$, $\Sigma x = 280$, $\Sigma x^2 = 14{,}000$, $\bar{x} = 40$, and $n = 7$. Find $t_{0.95}$ and compute E.

→ $t_{0.95} = 2.571$ *for d.f.* $= n - 2 = 5$

$$E \approx (2.571)(2.35)\sqrt{1 + \frac{1}{7} + \frac{7(15 - 40)^2}{7(14{,}000) - (280)^2}}$$
$$\approx (2.571)(2.35)\sqrt{1.366071} \approx 7.1$$

(c) Find a 95% confidence interval for y.
$\hat{y} - E \le y \le \hat{y} + E$

→ The confidence interval is
$$17.8 - 7.1 \le y \le 17.8 + 7.1$$
$$10.7 \le y \le 24.9$$

As we compare the results of Guided Exercise 7 and Example 8, we notice that the 95% confidence interval of y values for $x = 15°C$ is 7.1 units above and below the least-squares line, while the 95% confidence interval of y values for $x = 45°C$ is only 6.5 units above and below the least-squares line. This comparison reflects the general property that confidence intervals for y are narrower the nearer we are to the mean $\bar{x}$ of the x values. As we move near the extremes of the x distribution, the confidence intervals for y become wider. This is another reason that we should not try to use the least-squares line to predict y values for x values beyond the data extremes of the sample x distribution.

Confidence prediction band

If we were to compute a 95% confidence interval for all x values in the range of the sample x values, the *confidence interval band* would curve away from the least-squares line, as shown in Figure 9-18.

FIGURE 9-18

95% Confidence Band for Predicted Values $\hat{y}$

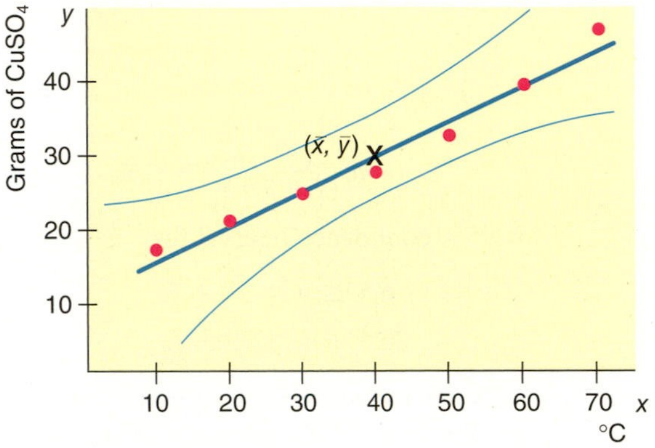

TECH NOTES Minitab provides confidence intervals for predictions. Use the menu selection **Stat ➤ Regression ➤ Regression**. Under **Options**, enter the observed x value and set the confidence level. In the output, the confidence interval for predictions is designated by %PI.

INFERENCES ABOUT THE SLOPE β

Recall that $\hat{y} = a + bx$ is the sample-based least-squares line and that $y = \alpha + \beta x$ is the population-based least-squares line computed (in theory) from the population of all (x, y) data pairs. In many real-world applications, the slope β is very important because β measures the rate at which y changes per unit change in x. Our next topic is to develop statistical tests and confidence intervals for β. Our work is based on the following theorem.

THEOREM 9.2 Let b be the slope of the sample least-squares line $\hat{y} = a + bx$ computed from a random sample of $n \geq 3$ data pairs (x, y). Let β be the slope of the population least-squares line $y = \alpha + \beta x$, which is in theory computed from the population of all (x, y) data pairs. Let S_e be the standard error of estimate computed from the sample. Then

$$t = \frac{b - \beta}{S_e \Big/ \sqrt{\Sigma x^2 - \frac{1}{n}(\Sigma x)^2}}$$

has a Student's t distribution with degrees of freedom $d.f. = n - 2$.

COMMENT The expression $S_e \Big/ \sqrt{\Sigma x^2 - \frac{1}{n}(\Sigma x)^2}$ is called the *standard error* for b.

Using this theorem, we can construct procedures for statistical tests and confidence intervals for β.

PROCEDURE

HOW TO TEST β AND FIND A CONFIDENCE INTERVAL FOR β

Requirements

Obtain a random sample of $n \geq 3$ data pairs (x, y). Use the procedure of Section 9.2 to find b, the slope of the sample least-squares line. Use Equation (8) of this section to find S_e, the standard error of estimate.

Procedure

Hypothesis test for β

For a statistical test of β

1. Use the *null hypothesis* H_0: $\beta = 0$. Use an *alternate hypothesis* H_1 appropriate to your application ($\beta > 0$ or $\beta < 0$ or $\beta \neq 0$). Set the level of significance α.
2. Use the null hypothesis H_0: $\beta = 0$ and the values of S_e, n, Σx, Σx^2, and b to compute the *sample test statistic.*

$$t = \frac{b}{S_e}\sqrt{\Sigma x^2 - \frac{1}{n}(\Sigma x)^2} \qquad \text{with } d.f. = n - 2$$

3. Use a Student's t distribution and the type of test, one-tailed or two-tailed, to find (or estimate) the *P-value* corresponding to the test statistic.
4. *Conclude* the test. If P-value $\leq \alpha$, then reject H_0. If P-value $> \alpha$, then do not reject H_0.
5. *Interpret your conclusion* in the context of the application.

Confidence interval for β

To find a confidence interval for β

$$b - E < \beta < b + E$$

where $E = \dfrac{t_c S_e}{\sqrt{\Sigma x^2 - \frac{1}{n}(\Sigma x)^2}}$

c = confidence level ($0 < c < 1$)

n = number of data pairs (x, y), $n \geq 3$

t_c = Student's t distribution critical value for confidence level c and $d.f. = n - 2$

S_e = standard error of estimate

EXAMPLE 9

TESTING β AND FINDING A CONFIDENCE INTERVAL FOR β

Plate tectonics and the spread of the ocean floor are very important topics in modern studies of earthquakes and earth science in general. A random sample of islands in the Indian Ocean gave the following information.

Photo Researchers

x = age of volcanic island in the Indian Ocean (units in 10^6 years)
y = distance of the island from the center of the midoceanic ridge (units in 100 kilometers)

x	120	83	60	50	35	30	20	17
y	30	16	15.5	14.5	22	18	12	0

Source: From King, Cuchaine A. M. *Physical Geography.* Oxford: Basil Blackwell, 1980, pp. 77–86 and 196–206. Reprinted by permission of the publisher.

(a) Starting from raw data values (x, y), the first step is simple but tedious. In short, you may verify (if you wish) that

$$\Sigma x = 415,\ \Sigma y = 128,\ \Sigma x^2 = 30{,}203,\ \Sigma y^2 = 2558.5,$$
$$\Sigma xy = 8133,\ \bar{x} = 51.875,\ \text{and}\ \bar{y} = 16$$

(b) The next step is to compute b, a, and S_e. Using a calculator, statistical software, or the formulas, we get

$$b \approx 0.1721 \text{ and } a \approx 7.072$$

Since $n = 8$, we get

$$S_e = \sqrt{\frac{\Sigma y^2 - a\Sigma y - b\Sigma xy}{n-2}}$$
$$\approx \sqrt{\frac{2558.5 - 7.072(128) - 0.1721(8133)}{8-2}} \approx 6.50$$

(c) Use an $\alpha = 5\%$ level of significance to test the claim that β is positive.
SOLUTION: $\alpha = 0.05$; $H_0: \beta = 0$; $H_1: \beta > 0$. The sample test statistic is

$$t = \frac{b}{S_e}\sqrt{\Sigma x^2 - \frac{1}{n}(\Sigma x)^2} \approx \frac{0.1721}{6.50}\sqrt{30{,}203 - \frac{(415)^2}{8}} \approx 2.466$$

with $d.f. = n - 2 = 8 - 2 = 6$.

We use the Student's t distribution (Table 6 of Appendix II) to find an interval containing the P-value. The test is a one-tailed test. Technology gives P-value ≈ 0.0244.

TABLE 9-14 Excerpt from Table 6, Appendix II

✓ one-tail area	0.025	0.010
d.f. = 6	2.447	3.143
	↑ Sample $t = 2.466$	

$$0.010 < P\text{-value} < 0.025$$

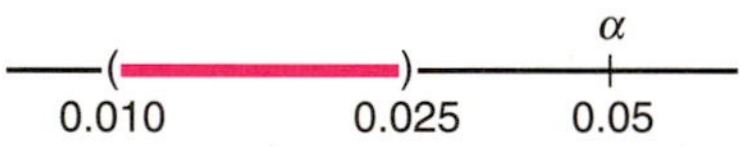

Since the interval containing the P-value is less than $\alpha = 0.05$, we reject H_0 and conclude that, at the 5% level of significance, the slope is positive.

(d) Find a 75% confidence interval for β.
SOLUTION: For $c = 0.75$ and $d.f. = n - 2 = 8 - 2 = 6$, the critical value $t_c = 1.273$. The margin of error E for the confidence interval is

$$E = \frac{t_c S_e}{\sqrt{\Sigma x^2 - \frac{1}{n}(\Sigma x)^2}} \approx \frac{1.273(6.50)}{\sqrt{30{,}203 - \frac{(415)^2}{8}}} \approx 0.0888$$

Using $b \approx 0.17$, a 75% confidence interval for β is

$$b - E < \beta < b + E$$
$$0.17 - 0.09 < \beta < 0.17 + 0.09$$
$$0.08 < \beta < 0.26$$

(e) *Interpretation* What does the confidence interval mean?
Recall the units involved (x in 10^6 years and y in 100 kilometers). It appears that, in this part of the world, we can be 75% confident that we have an interval showing that the ocean floor is moving at a rate of between 8 mm and 26 mm per year.

GUIDED EXERCISE 8 | INFERENCE FOR β

How fast do puppies grow? That depends on the puppy. How about male wolf pups in the Helsinki Zoo (Finland)? Let x = age in weeks and y = weight in kilograms for a random sample of male wolf pups. The following data are based on the article "Studies of the Wolf in Finland *Canis lupus L*" (*Ann. Zool. Fenn.*, Vol. 2, pp. 215–259) by E. Pulliainen, University of Helsinki.

outdoorsman/Shutterstock.com

x	8	10	14	20	28	40	45
y	7	13	17	23	30	34	35

$\Sigma x = 165$, $\Sigma y = 159$, $\Sigma y^2 = 5169$, $\Sigma y^2 = 4317$, $\Sigma xy = 4659$

(a) Verify the following values.
$\bar{x} \approx 23.571$ $\quad \bar{y} \approx 22.714$
$b \approx 0.7120$ $\quad a \approx 5.932$
$S_e \approx 3.368$

Use the formulas for $\bar{x}, \bar{y}, b, a$ and S_e or find the results directly using your calculator or computer software.

(b) Use a 1% level of significance to test the claim that $\beta \neq 0$, and ***interpret*** the results in the context of this application.

$\alpha = 0.01$; $H_0: \beta = 0$; $H_1: \beta \neq 0$
Convert $b \approx 0.7120$ to a t value.

$$t = \frac{b}{S_e}\sqrt{\Sigma x^2 - \frac{1}{n}(\Sigma x)^2}$$

$$\approx \frac{0.7120}{3.368}\sqrt{5169 - \frac{(165)^2}{7}} \approx 7.563$$

From Table 6, Appendix II, for a two-tailed test with $d.f. = n - 2 = 7 - 2 = 5$,

✓ two-tail area	0.001
$d.f. = 5$	6.869
	↑ Sample $t = 7.563$

Noting that areas decrease as t values increase, we have $0.001 > P$-value. Technology gives P-value ≈ 0.0006.

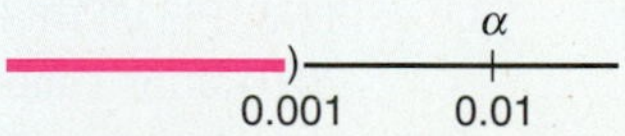

Since the P-value is less than $\alpha = 0.01$, we reject H_0 and conclude that the population slope β is not zero.

Continued

GUIDED EXERCISE 8 *continued*

(c) Compute an 80% confidence interval for β and ***interpret*** the results in the context of this application.

$d.f. = 5$. For an 80% confidence interval, the critical value $t_c = 1.476$. The confidence interval is

$$b - E < \beta < b + E$$

where $b = 0.712$ and

$$E = \frac{t_c S_e}{\sqrt{\Sigma x^2 - \frac{(\Sigma x)^2}{n}}} \approx \frac{1.476(3.368)}{\sqrt{5169 - \frac{(165)^2}{7}}}$$

$$\approx 0.139$$

The interval is from 0.57 kg to 0.85 kg. We can be 80% confident that the interval computed is one that contains β. For each week's change in age, the weight change is between 0.57 kg and 0.85 kg.

COMPUTATION HINTS FOR SAMPLE TEST STATISTIC t USED IN TESTING ρ AND TESTING β

Problem 14 discusses the fact that for the same data, the values of the sample test statistics for r and b are equal.

In Section 9.2 we saw that the values for the sample correlation coefficient r and slope b of the least-squares line are related by the formula

$$b = r\left(\frac{s_y}{s_x}\right)$$

Using this relationship and some algebra, it can be shown that

For the same sample, the sample correlation coefficient r and the slope b of the least-squares line have the same sample test statistic t, with $d.f. = n - 2$, where n is the number of data pairs.

Consequently, when doing calculations or using results from technology, we can use the following strategies.

(a) ***Calculations "by hand":*** Find the sample test statistic t corresponding to r. The sample test statistic t corresponding to b is the same.

(b) ***Using computer results:*** Most computer-based statistical packages provide the sample test statistic t corresponding to b. The sample test statistic t corresponding to r is the same.

TECH NOTES The sample test statistic t corresponding to the sample correlation coefficient r is the same as the t value corresponding to b, the slope of the least-squares line (see Problem 14 at the end of this section). Consequently, the two tests H_0: $\rho = 0$ and H_0: $\beta = 0$ (with similar corresponding alternate hypotheses) have the same conclusions. The TI-84Plus/TI-83Plus/TI-*n*spire calculators use this fact explicitly. Minitab and Excel 2010 show t and the two-tailed P-value for the slope b of the least-squares line. Excel also shows confidence intervals for β. The displays show data from Guided Exercise 8 regarding the age and weight of wolf pups.

TI-84Plus/TI-83Plus/TI-*n*spire (with TI-84Plus keypad) Under **STAT,** select **TEST** and use option **E:LinRegTTest.**

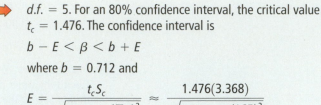

```
LinRegTTest
 y=a+bx
 β≠0 and ρ≠0
↑b=.7120
 s=3.3676
 r²=.9196
 r=.9590
```

```
LinRegTTest
 y=a+bx
 β≠0 and ρ≠0
 t=7.5632
 p=6.4075E-4
 df=5.0000
↓a=5.9317
```

Note that the value of S_e is given as s.

On TI calculators with operating system v2.55MP, the linear regression t test is located at option F instead of E under Statistical Tests. In addition, option **G:LinRegTInt** provides confidence intervals for the slope b of the least-squares regression line.

Excel 2010 On the home screen, click the **Data** tab. Select **Data Analysis** in the Analysis group. In the dialogue box, select **Regression**. Widen columns of the output as necessary to see all the results.

Regression Statistics	
Multiple R	0.958966516
R Square	0.919616778
Adjusted R Square	0.903540133
Standard Error	3.367628886
Observations	7

← Value of S_e (Standard Error)

	Coefficients	*Standard Error*	*t Stat*	*P-value*	*Lower 95%*	*Upper 95%*
Intercept	5.931681179	2.558126184	2.318760198	0.068158803	-0.644180779	12.50754314
X Variable 1	0.711989283	0.094138596	7.563202697	0.000640746	0.469998714	0.953979853

↑ b (Coefficients) ↑ for two-tailed test (P-value)

Minitab Use the menu selection **Stat ➤ Regression ➤ Regression**. The value of S_e is S; P is the P-value of a two-tailed test. For a one-tailed test, divide the P-value by 2.

```
Regression Analysis
The regression equation is
y = 5.93 + 0.712 x
Predictor        Coef        StDev       T        P
Constant        5.932        2.558     2.32    0.068
x             0.71199      0.09414     7.56    0.001
S = 3.368      R-Sq = 92.0%       R-Sq(adj) = 90.4%
```

VIEWPOINT

Hawaiian Island Hopping!

Suppose you want to go camping in Hawaii. Yes! Hawaii has both state and federal parks where you can enjoy camping on the beach or in the mountains. However, you will probably need to rent a car to get to the different campgrounds. How much will the car rental cost? That depends on the islands you visit. For car rental data and regression statistics you can compute regarding costs on different Hawaiian Islands, visit the Brase/Brase statistics site at **http://www.cengage.com/statistics/brase11e** *and find the link to Hawaiian Islands.*

SECTION 9.3 PROBLEMS

1. *Statistical Literacy* What is the symbol used for the population correlation coefficient?

2. *Statistical Literacy* What is the symbol used for the slope of the population least-squares line?

3. *Statistical Literacy* For a fixed confidence level, how does the length of the confidence interval for predicted values of y change as the corresponding x values become further away from $\bar{x}$?

4. *Statistical Literacy* How does the t value for the sample correlation coefficient r compare to the t value for the corresponding slope b of the sample least-squares line?

Using Computer Printouts Problems 5 and 6 use the following information. Prehistoric pottery vessels are usually found as sherds (broken pieces) and are carefully reconstructed if enough sherds can be found. Information taken from *Mimbres Mogollon Archaeology* by A. I. Woosley and A. J. McIntyre (University of New Mexico Press) provides data relating x = body diameter in centimeters and y = height in centimeters of prehistoric vessels reconstructed from sherds found at a prehistoric site. The following Minitab printout provides an analysis of the data.

Predictor	Coef	SE Coef	T	P
Constant	−0.223	2.429	−0.09	0.929
Diameter	0.7848	0.1471	5.33	0.001

S = 4.07980   R-Sq = 80.3%

5. *Critical Thinking: Using Information from a Computer Display to Test for Significance* Refer to the Minitab printout regarding prehistoric pottery.
   (a) Minitab calls the explanatory variable the predictor variable. Which is the predictor variable, the diameter of the pot or the height?
   (b) For the least-squares line $\hat{y} = a + bx$, what is the value of the constant a? What is the value of the slope b? (*Note*: The slope is the coefficient of the predictor variable.) Write the equation of the least-squares line.
   (c) The P-value for a two-tailed test corresponding to each coefficient is listed under "P." The t value corresponding to the coefficient is listed under "T." What is the P-value of the slope? What are the hypotheses for a two-tailed test of $\beta = 0$? Based on the P-value in the printout, do we reject or fail to reject the null hypothesis for $\alpha = 0.01$?
   (d) Recall that the t value and resulting P-value of the slope b equal the t value and resulting P-value of the corresponding sample correlation coefficient r. To find the value of the sample correlation coefficient r, take the square root of the R-Sq value shown in the display. What is the value of r? Consider a two-tailed test for ρ. Based on the P-value shown in the Minitab display, is the correlation coefficient significant at the 1% level of significance?

6. *Critical Thinking: Using Information from a Computer Display to Find a Confidence Interval* Refer to the Minitab printout regarding prehistoric pottery.
   (a) The standard error S_e of the linear regression model is given in the printout as "S." What is the value of S_e?
   (b) The standard error of the coefficient of the predictor variable is found under "SE Coef." Recall that the standard error for b is $S_e \Big/ \sqrt{\Sigma x^2 - \frac{1}{n}(\Sigma x)^2}$.
   From the Minitab display, what is the value of the standard error for the slope b?

(c) The formula for the margin of error E for a $c\%$ confidence interval for the slope β can be written as $E = t_c(\text{SE Coef})$. The Minitab display is based on $n = 9$ data pairs. Find the critical value t_c for a 95% confidence interval in Table 6 of Appendix II. Then find a 95% confidence interval for the population slope β.

In Problems 7–12, parts (a) and (b) relate to testing ρ. Part (c) requests the value of S_e. Parts (d) and (e) relate to confidence intervals for prediction. Parts (f) and (g) relate to testing β and finding confidence intervals for β.

Answers may vary due to rounding.

7. *Basketball: Free Throws and Field Goals* Let x be a random variable that represents the percentage of successful free throws a professional basketball player makes in a season. Let y be a random variable that represents the percentage of successful field goals a professional basketball player makes in a season. A random sample of $n = 6$ professional basketball players gave the following information (Reference: *The Official NBA Basketball Encyclopedia*, Villard Books).

x	67	65	75	86	73	73
y	44	42	48	51	44	51

(a) Verify that $\Sigma x = 439$, $\Sigma y = 280$, $\Sigma x^2 = 32{,}393$, $\Sigma y^2 = 13{,}142$, $\Sigma xy = 20{,}599$, and $r \approx 0.784$.
(b) Use a 5% level of significance to test the claim that $\rho > 0$.
(c) Verify that $S_e \approx 2.6964$, $a \approx 16.542$, $b \approx 0.4117$, and $\bar{x} \approx 73.167$.
(d) Find the predicted percentage $\hat{y}$ of successful field goals for a player with $x = 70\%$ successful free throws.
(e) Find a 90% confidence interval for y when $x = 70$.
(f) Use a 5% level of significance to test the claim that $\beta > 0$.
(g) Find a 90% confidence interval for β and *interpret* its meaning.

8. *Baseball: Batting Average and Strikeouts* Let x be a random variable that represents the batting average of a professional baseball player. Let y be a random variable that represents the percentage of strikeouts of a professional baseball player. A random sample of $n = 6$ professional baseball players gave the following information (Reference: *The Baseball Encyclopedia*, Macmillan).

x	0.328	0.290	0.340	0.248	0.367	0.269
y	3.2	7.6	4.0	8.6	3.1	11.1

(a) Verify that $\Sigma x = 1.842$, $\Sigma y = 37.6$, $\Sigma x^2 = 0.575838$, $\Sigma y^2 = 290.78$, $\Sigma xy = 10.87$, and $r \approx -0.891$.
(b) Use a 5% level of significance to test the claim that $\rho \neq 0$.
(c) Verify that $S_e \approx 1.6838$, $a \approx 26.247$, and $b \approx -65.081$.
(d) Find the predicted percentage of strikeouts for a player with an $x = 0.300$ batting average.
(e) Find an 80% confidence interval for y when $x = 0.300$.
(f) Use a 5% level of significance to test the claim that $\beta \neq 0$.
(g) Find a 90% confidence interval for β and *interpret* its meaning.

9. *Scuba Diving: Depth* What is the optimal amount of time for a scuba diver to be on the bottom of the ocean? That depends on the depth of the dive. The U.S. Navy has done a lot of research on this topic. The Navy defines the "optimal time" to be the time at each depth for the best balance between length of work period and decompression time after surfacing. Let x = depth of dive in meters, and let y = optimal time in hours. A random sample of divers gave the

following data (based on information taken from *Medical Physiology* by A. C. Guyton, M.D.).

x	14.1	24.3	30.2	38.3	51.3	20.5	22.7
y	2.58	2.08	1.58	1.03	0.75	2.38	2.20

(a) Verify that $\Sigma x = 201.4$, $\Sigma y = 12.6$, $\Sigma x^2 = 6735.46$, $\Sigma y^2 = 25.607$, $\Sigma xy = 311.292$, and $r \approx -0.976$.
(b) Use a 1% level of significance to test the claim that $\rho < 0$.
(c) Verify that $S_e \approx 0.1660$, $a \approx 3.366$, and $b \approx -0.0544$.
(d) Find the predicted optimal time in hours for a dive depth of $x = 18$ meters.
(e) Find an 80% confidence interval for y when $x = 18$ meters.
(f) Use a 1% level of significance to test the claim that $\beta < 0$.
(g) Find a 90% confidence interval for β and *interpret* its meaning.

10. *Physiology: Oxygen* Aviation and high-altitude physiology is a specialty in the study of medicine. Let x = partial pressure of oxygen in the alveoli (air cells in the lungs) when breathing naturally available air. Let y = partial pressure when breathing pure oxygen. The (x, y) data pairs correspond to elevations from 10,000 feet to 30,000 feet in 5000-foot intervals for a random sample of volunteers. Although the medical data were collected using airplanes, they apply equally well to Mt. Everest climbers (summit 29,028 feet).

x	6.7	5.1	4.2	3.3	2.1 (units: mm Hg/10)
y	43.6	32.9	26.2	16.2	13.9 (units: mm Hg/10)

(Based on information taken from *Medical Physiology* by A. C. Guyton, M.D.)
(a) Verify that $\Sigma x = 21.4$, $\Sigma y = 132.8$, $\Sigma x^2 = 103.84$, $\Sigma y^2 = 4125.46$, $\Sigma xy = 652.6$, and $r \approx 0.984$.
(b) Use a 1% level of significance to test the claim that $\rho > 0$.
(c) Verify that $S_e \approx 2.5319$, $a \approx -2.869$, and $b \approx 6.876$.
(d) Find the predicted pressure when breathing pure oxygen if the pressure from breathing available air is $x = 4.0$.
(e) Find a 90% confidence interval for y when $x = 4.0$.
(f) Use a 1% level of significance to test the claim that $\beta > 0$.
(g) Find a 95% confidence interval for β and *interpret* its meaning.

11. *New Car: Negotiating Price* Suppose you are interested in buying a new Toyota Corolla. You are standing on the sales lot looking at a model with different options. The list price is on the vehicle. As a salesperson approaches, you wonder what the dealer invoice price is for this model with its options. The following data are based on information taken from *Consumer Guide* (Vol. 677). Let x be the list price (in thousands of dollars) for a random selection of Toyota Corollas of different models and options. Let y be the dealer invoice (in thousands of dollars) for the given vehicle.

x	12.6	13.0	12.8	13.6	13.4	14.2
y	11.6	12.0	11.5	12.2	12.0	12.8

(a) Verify that $\Sigma x = 79.6$, $\Sigma y = 72.1$, $\Sigma x^2 = 1057.76$, $\Sigma y^2 = 867.49$, $\Sigma xy = 957.84$, and $r \approx 0.956$.
(b) Use a 1% level of significance to test the claim that $\rho > 0$.
(c) Verify that $S_e \approx 0.1527$, $a \approx 1.965$, and $b \approx 0.758$.
(d) Find the predicted dealer invoice when the list price is $x = 14$ (thousand dollars).

(e) Find an 85% confidence interval for y when $x = 14$ (thousand dollars).
(f) Use a 1% level of significance to test the claim that $\beta > 0$.
(g) Find a 95% confidence interval for β and *interpret* its meaning.

12. *New Car: Negotiating Price* Suppose you are interested in buying a new Lincoln Navigator or Town Car. You are standing on the sales lot looking at a model with different options. The list price is on the vehicle. As a salesperson approaches, you wonder what the dealer invoice price is for this model with its options. The following data are based on information taken from *Consumer Guide* (Vol. 677). Let x be the list price (in thousands of dollars) for a random selection of these cars of different models and options. Let y be the dealer invoice (in thousands of dollars) for the given vehicle.

x	32.1	33.5	36.1	44.0	47.8
y	29.8	31.1	32.0	42.1	42.2

(a) Verify that $\Sigma x = 193.5$, $\Sigma y = 177.2$, $\Sigma x^2 = 7676.71$, $\Sigma y^2 = 6432.5$, $\Sigma xy = 7023.19$, and $r \approx 0.977$.
(b) Use a 1% level of significance to test the claim that $\rho > 0$.
(c) Verify that $S_e \approx 1.5223$, $a \approx 1.4084$, and $b \approx 0.8794$.
(d) Find the predicted dealer invoice when the list price is $x = 40$ (thousand dollars).
(e) Find a 95% confidence interval for y when $x = 40$ (thousand dollars).
(f) Use a 1% level of significance to test the claim that $\beta > 0$.
(g) Find a 90% confidence interval for β and *interpret* its meaning.

13. *Expand Your Knowledge: Sample Size and Significance of r*
(a) Suppose $n = 6$ and the sample correlation coefficient is $r = 0.90$. Is r significant at the 1% level of significance (based on a two-tailed test)?
(b) Suppose $n = 10$ and the sample correlation coefficient is $r = 0.90$. Is r significant at the 1% level of significance (based on a two-tailed test)?
(c) Explain why the test results of parts (a) and (b) are different even though the sample correlation coefficient $r = 0.90$ is the same in both parts. Does it appear that sample size plays an important role in determining the significance of a correlation coefficient? Explain.

14. *Expand Your Knowledge: Student's t Value for Sample r and for Sample b* It is not obvious from the formulas, but the values of the sample test statistic t for the correlation coefficient and for the slope of the least-squares line are equal for the same data set. This fact is based on the relation

$$b = r\frac{s_y}{s_x}$$

where s_y and s_x are the sample standard deviations of the x and y values, respectively.
(a) Many computer software packages give the t value and corresponding P-value for b. If β is significant, is ρ significant?
(b) When doing statistical tests "by hand," it is easier to compute the sample test statistic t for the sample correlation coefficient r than it is to compute the sample test statistic t for the slope b of the sample least-squares line. Compare the results of parts (b) and (f) for Problems 7–12 of this problem set. Is the sample test statistic t for r the same as the corresponding test statistic for b? If you conclude that ρ is positive, can you conclude that β is positive at the same level of significance? If you conclude that ρ is not significant, is β also not significant at the same level of significance?

15. *Expand Your Knowledge: Time Series and Serial Correlation* *Serial correlation,* also known as *autocorrelation*, describes the extent to which the result in one period of a time series is related to the result in the next period. A time series with high serial correlation is said to be very predictable from one period to the next. If the serial correlation is low (or near zero), the time series is considered to be much less predictable. For more information about serial correlation, see the book *ibbotson SBBI* published by Morningstar.

A research veterinarian at a major university has developed a new vaccine to protect horses from West Nile virus. An important question is: How predictable is the buildup of antibodies in the horse's blood after the vaccination is given? A large random sample of horses from Wyoming were given the vaccination. The average antibody buildup factor (as determined from blood samples) was measured each week after the vaccination for 8 weeks. Results are shown in the following time series:

Original Time Series

Week	1	2	3	4	5	6	7	8
Buildup Factor	2.4	4.7	6.2	7.5	8.0	9.1	10.7	12.3

To construct a serial correlation, we simply use data pairs (x, y) where x = original buildup factor data and y = original data shifted ahead by 1 week. This gives us the following data set. Since we are shifting 1 week ahead, we now have 7 data pairs (not 8).

Data for Serial Correlation

x	2.4	4.7	6.2	7.5	8.0	9.1	10.7
y	4.7	6.2	7.5	8.0	9.1	10.7	12.3

For convenience, we are given the following sums:

$\Sigma x = 48.6 \quad \Sigma y = 58.5 \quad \Sigma x^2 = 383.84 \quad \Sigma y^2 = 529.37 \quad \Sigma xy = 448.7$

(a) Use the sums provided (or a calculator with least-squares regression) to compute the equation of the sample least-squares line, $\hat{y} = a + bx$. If the buildup factor was $x = 5.8$ one week, what would you predict the buildup factor to be the next week?

(b) Compute the sample correlation coefficient r and the coefficient of determination r^2. Test $\rho > 0$ at the 1% level of significance. Would you say the time series of antibody buildup factor is relatively predictable from one week to the next? Explain.

16. *Expand Your Knowledge: Time Series and Serial Correlation* An interner advertising agency is studying the number of "hits" on a certain website during an advertising campaign. It is hoped that as the campaign progresses, the number of hits on the website will also increase in a predictable way from one day to the next. For 10 days of the campaign, the number of hits $\times 10^5$ is shown:

Original Time Series

Day	1	2	3	4	5	6	7	8	9	10
Hits $\times 10^5$	1.2	3.5	4.4	7.2	6.9	8.3	9.0	11.2	13.1	14.6

(a) To construct a serial correlation, we use data pairs (x, y) where x = original data and y = original data shifted ahead by one time period. Verify that the data set (x, y) for serial correlation is shown here. (For discussion of serial correlation, see Problem 15.)

x	1.2	3.5	4.4	7.2	6.9	8.3	9.0	11.2	13.1
y	3.5	4.4	7.2	6.9	8.3	9.0	11.2	13.1	14.6

(b) For the (x, y) data set of part (a), compute the equation of the sample least-squares line $\hat{y} = a + bx$. If the number of hits was 9.3 ($\times 10^5$) one day, what do you predict for the number of hits the next day?

(c) Compute the sample correlation coefficient r and the coefficient of determination r^2. Test $\rho > 0$ at the 1% level of significance. Would you say the time series of website hits is relatively predictable from one day to the next? Explain.

17. *Expand Your Knowledge: Time Series and Serial Correlation* A company that produces and markets video games want to estimate the predictability of per capita consumer spending on video games in the United States. For the most recent 7 years, the amount of annual spending per person per year is shown here (Reference: Statistical *Abstract of the United States*, 128th Edition):

Year	1	2	3	4	5	6	7
$ per capita	32.23	34.03	37.84	43.34	44.64	49.61	51.89

(a) To construct a serial correlation, we use data pairs (x, y) where x = original data and y = original data shifted ahead by one time period. Verify that the data set (x, y) for serial correlation is shown here. (For discussion of serial correlation, see Problem 15.)

x	32.23	34.03	37.84	43.34	44.64	49.61
y	34.03	37.84	43.34	44.64	49.61	51.89

(b) For the (x, y) data set of part (a) compute the equation of the sample least-squares line $\hat{y} = a + bx$. If the per capita spending was x = \$42 one year, what do you predict for the spending the next year?

(c) Compute the sample correlation coefficient r and the coefficient of determination r^2. Test $\rho > 0$ at the 1% level of significance. Would you say the time series of per capita spending on video games is relatively predictable from one year to the next? Explain.

SECTION **9.4**

Multiple Regression

FOCUS POINTS

- Learn about the advantages of multiple regression.
- Learn the basic ingredients that go into a multiple regression model.
- Discuss standard error for computed coefficients and the coefficient of multiple determination.
- Test coefficients in the model for statistical significance.
- Compute confidence intervals for predictions.

ADVANTAGES OF MULTIPLE REGRESSION

There are many examples in statistics in which one variable can be predicted very accurately in terms of another *single* variable. However, predictions usually improve if we consider additional relevant information. For example, the sugar content y of golden delicious apples taken from an apple orchard in Colorado could be predicted from x_1 = number of days in growing season. If we also included information regarding x_2 = soil quality rating and x_3 = amount of available water, then we would expect our prediction of y = sugar content to be more accurate.

Likewise, the annual net income y of a new franchise auto parts store could be predicted using only x_1 = population size of sales district. However, we would probably get a better prediction of y values if we included the explanatory variables x_2 = size of store inventory, x_3 = dollar amount spent on advertising in local newspapers, and x_4 = number of competing stores in the sales district.

Multiple regression

For most statistical applications, we gain a definite advantage in the reliability of our predictions if we include more *relevant* data and corresponding (relevant) random variables in the computation of our predictions. In this section, we will give you an idea of how this can be done by methods of *multiple regression*. You should be aware that an in-depth study of multiple regression requires the use of advanced mathematics. However, if you are willing to let the computer be a "friend who gives you useful information," then you will learn a great deal about multiple regression in this section. We will let the computer do most of the calculating work while we interpret the results.

BASIC TERMINOLOGY AND NOTATION

In statistics, the most commonly used mathematical formulas for expressing linear relationships among more than two variables are *equations* of the form

$$y = b_0 + b_1x_1 + b_2x_2 + \cdots + b_kx_k \qquad (9)$$

Here, y is the variable that we want to predict or forecast. We will employ the usual terminology and call y the *response variable*. The k variables $x_1, x_2, \ldots, x_k$ are specified variables on which the predictions are going to be based. Once again, we will employ the popular terminology and call $x_1, x_2, \ldots, x_k$ the *explanatory variables*. This terminology is easy to remember if you just think of the explanatory variables $x_1, x_2, \ldots, x_k$ as "explaining" the response y.

In Equation (9), $b_0, b_1, b_2, \ldots, b_k$ are numerical constants (called *coefficients*) that must be mathematically determined from given data. The numerical values of these coefficients are obtained from the *least-squares criterion*, which we will discuss after the following exercise.

GUIDED EXERCISE 9 | COMPONENTS OF MULTIPLE REGRESSION EQUATION

An industrial psychologist working for a hospital-supply company is studying the following variables for a random sample of company employees:

x_1 = number of years the employee has been with the company
x_2 = job-training level (0 = lowest level and 5 = highest level)
x_3 = interpersonal skills (0 = lowest level and 10 = highest level)
y = job-performance rating from supervisor (1 = lowest rating, 20 = highest rating)

The psychologist wants to predict y using x_1, x_2, and x_3 together in a least-squares equation.

(a) Identify the response variable and the explanatory variables.

→ The response variable is what we want to predict. This is y, job performance. The explanatory variables are years of experience x_1, training level x_2, and interpersonal skills x_3. In a sense, these variables "explain" the response variable.

(b) After collecting data, the psychologist used a computer with appropriate software to obtain the least-squares linear equation

$$y = 1 + 0.2x_1 + 2.3x_2 + 0.7x_3$$

Identify the constant term and each of the coefficients with its corresponding variable.

→ The constant term is 1.

Explanatory Variable	Coefficient
x_1	0.2
x_2	2.3
x_3	0.7

Continued

GUIDED EXERCISE 9 *continued*

(c) Use the equation to predict the job-performance rating of an employee with 3 years of experience, a training level of 4, and an interpersonal skill rating of 2.

Substituting $x_1 = 3$, $x_2 = 4$, and $x_3 = 2$ into the least-squares equation and multiplying by the respective coefficients, we obtain the predicted job performance rating of

$$y = 1 + 0.2(3) + 2.3(4) + 0.7(2) = 12.2$$

Of course, the *predicted* value for job performance might differ from the actual rating given by the supervisor.

THEORY FOR THE LEAST-SQUARES CRITERION (OPTIONAL)

This material is a little sophisticated, so you may wish to skip ahead to the discussion of regression models and computers and omit the following explanation of basic theory.

In multiple regression, the least-squares criterion states that the following sum (over all data points),

$$\Sigma[y_i - (b_0 + b_1x_{1i} + b_2x_{2i} + \cdots + b_kx_{ki})]^2 \quad (10)$$

must be made as small as possible. In this formula,

$$\begin{aligned} y_i &= i\text{th data value for } y \\ x_{1i} &= i\text{th data value for } x_1 \\ x_{2i} &= i\text{th data value for } x_2 \\ &\vdots \\ x_{ki} &= i\text{th data value for } x_k \end{aligned}$$

Recall that Equation (9) gives the predicted y value; therefore,

$$y_i - (b_0 + b_1x_{1i} + b_2x_{2i} + \cdots + b_kx_{ki}) \quad (11)$$

represents the *difference* between the *observed* y value (that is, y_i) and the *predicted* y value based on the data values $x_{1i}, x_{2i}, \ldots, x_{ki}$. When we square this difference, total the result over all data points, and choose the values of $b_0, b_1, b_2, \ldots, b_k$ to minimize the sum [i.e., minimize Equation (10)], then we are satisfying the least-squares criterion.

Residual

COMMENT The algebraic expression in Equation (11) is very important. In fact, it has a special name in the theory of regression. It is called a *residual*. The residual is simply the difference between the actual data value and the predicted value of the response variable based on given data values for the explanatory variables. Advanced topics in the theory of regression study residuals in great detail. Such a detailed treatment is beyond the scope of this text. However, from the discussion presented so far, we see that the method of least squares chooses the values of the coefficients b_i to make the sum of the squares of the residuals as small as possible.

After a good deal of mathematics has been done (involving a considerable amount of calculus), the least-squares criterion can be reduced to solving a system of linear equations. These are usually called *normal equations* (not to be confused with the normal distribution).

In the simplest case, where there are only *two* explanatory variables x_1 and x_2 and we want to fit the equation

$$y = b_0 + b_1x_1 + b_2x_2$$

Continued

to given data, there are three normal equations that must be solved for b_0, b_1, and b_2. These normal equations are

$$\left.\begin{aligned} \Sigma y_i &= nb_0 + b_1(\Sigma x_{1i}) + b_2(\Sigma x_{2i}) \\ \Sigma x_{1i}y_i &= b_0(\Sigma x_{1i}) + b_1(\Sigma x_{1i}{}^2) + b_2(\Sigma x_{1i}x_{2i}) \\ \Sigma x_{2i}y_i &= b_0(\Sigma x_{2i}) + b_1(\Sigma x_{1i}x_{2i}) + b_2(\Sigma x_{2i}{}^2) \end{aligned}\right\} \quad (12)$$

In the system of Equations (12), n represents the number of data points and x_{1i}, x_{2i}, and y_i all represent given data values.

Therefore, the only unknowns are the coefficients b_0, b_1, and b_2; we can use the system of Equations (12) to solve for these unknowns. This is the procedure that lets us obtain the least-squares regression equation in Equation (9) when we have only *two* explanatory variables.

As you can see, this is all rather complicated, and the more explanatory variables $x_1, x_2, \ldots, x_k$ we have, the more involved the calculations become. In the general case, if you have k explanatory variables, there will be $k + 1$ normal equations that must be solved for the coefficients $b_0, b_1, b_2, \ldots, b_k$.

REGRESSION MODELS AND COMPUTERS

As you can see from the preceding optional discussion, the work required to find an equation satisfying the least-squares criterion is tremendous and can be very complex. Today, such work is conveniently left to computers. In this text, we use two computer software packages that specialize in statistical applications.

Minitab is a widely used statistical software package. It fully supports multiple regression. Excel 2010 has a multiple regression component that performs much of the multiple regression analysis. We will use Minitab in our example. Many other software packages, including SPSS, support multiple regression and have outputs similar to those of Minitab.

Ingredients of the regression model

In this section, we will often refer to a *regression model*. What do we mean by this? We mean a mathematical package that consists of the following ingredients:

1. The model will have a collection of random variables, *one* of which has been identified as the response variable, with *any or all* of the remaining variables being identified as explanatory variables.
2. Associated with a given application will be a collection of numerical data values for each of the variables of part 1.
3. Using the numerical data values, the least-squares criterion, and the declared response and explanatory variables, a *least-squares equation* (also called a *regression equation*) will be constructed. In Section 9.2, we were able to construct the least-squares equation using only a hand calculator. However, in multiple regression, we will use a computer to construct the least-squares equation.
4. The model usually includes additional information about the variables used, the coefficients and regression equation, and a measure of "goodness of fit" of the regression equation to the data values. In modern practice, this information usually comes to you in the form of computer displays.
5. Finally, the regression model enables you to supply given values of the explanatory variables for the purpose of predicting or forecasting the corresponding value of the response variable. You also should be able to construct a $c\%$ confidence interval for your least-squares prediction. In multiple regression, this will be done by the computer at your request.

Problem 7 at the end of this section discusses *curvilinear regression* (also known as *polynomial regression*).

The next example demonstrates computer applications of a typical multiple regression problem. In the context of the example, we will introduce some of the basic techniques of multiple regression.

EXAMPLE UTILIZING MINITAB

EXAMPLE 10

MULTIPLE REGRESSION

Antelope are beautiful and graceful animals that live on the high plains of the western United States. Thunder Basin National Grasslands in Wyoming is home to hundreds of antelope. The Bureau of Land Management (BLM) has been studying the Thunder Basin antelope population for the past 8 years. The variables used are

x_1 = spring fawn count (in hundreds of fawns)

x_2 = size of adult antelope population (in hundreds)

x_3 = annual precipitation (in inches)

x_4 = winter severity index (1 = mild and 5 = extremely severe) (This is an index based on temperature and wind chill factors.)

The data obtained in the study over the 8-year period are shown in Table 9-15.

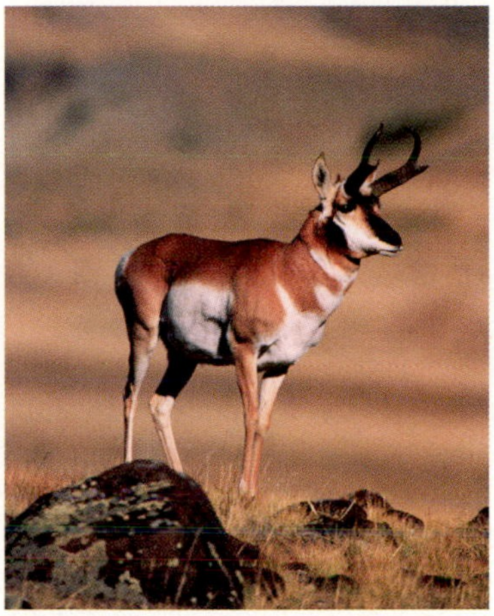

D. Robert & Lorri Franz/CORBIS

TABLE 9-15 **Data for Thunder Basin Antelope Study**

Year	x_1	x_2	x_3	x_4
1	2.9	9.2	13.2	2
2	2.4	8.7	11.5	3
3	2.0	7.2	10.8	4
4	2.3	8.5	12.3	2
5	3.2	9.6	12.6	3
6	1.9	6.8	10.6	5
7	3.4	9.7	14.1	1
8	2.1	7.9	11.2	3

Summary Statistics for Each Variable

It is a good idea to first look at the summary statistics for each variable. Figure 9-19 shows the Minitab display of the summary statistics.

Menu selection: **Stat ➤ Basic Statistic ➤ Display Descriptive Statistics**

FIGURE 9-19

Minitab Display of Summary Statistics for Each Variable

```
Descriptive Statistics
Variable    N       Mean      Median     TrMean     StDev     SE Mean
x1          8      2.525       2.350      2.525     0.570       0.202
x2          8      8.450       8.600      8.450     1.076       0.380
x3          8     12.037      11.900     12.037     1.229       0.435
x4          8      2.875       3.000      2.875     1.246       0.441
Variable    Minimum     Maximum        Q1        Q3
x1            1.900       3.400     2.025     3.125
x2            6.800       9.700     7.375     9.500
x3           10.600      14.100    10.900    13.050
x4            1.000       5.000     2.000     3.750
```

This type of information can be very useful because it tells you basic information about the variables you are studying. Sample means and sample standard deviations with a Student's t distribution are essential ingredients for estimating or testing population means (Chapters 7 and 8).

For example, if μ_2 represents the *population mean* of x_2 (adult antelope population), then by using the methods of Section 7.2 we can quickly estimate a 90% confidence interval for μ_2:

$$7.729 < \mu_2 < 9.171$$

Since our units are in hundreds, this means we can be 90% sure that the *population mean* μ_2 of adult antelope in the Thunder Basin National Grasslands is between 773 and 917.

Correlation Between Variables

It is also useful to examine how the variables relate to each other. Figure 9-20 shows the sample correlation coefficients r between each of the two variables. A natural question arises: Which of the variables are closely related to each other, and which are not as closely related? Recall (from Section 9.1) that if the correlation coefficient is near 1 or -1, then the corresponding variables have a lot in common. If the correlation coefficient is near zero, the variables have much less influence on each other.

Menu selection: **Stat ➤ Basic Statistics ➤ Correlation**

FIGURE 9-20

Minitab Display of Correlation Coefficients Between Variables

Correlations (Pearson)

	x1	x2	x3
x2	0.939		
x3	0.924	0.903	
x4	−0.739	−0.836	−0.901

Look at Figure 9-20. Which of the variables has the greatest influence on x_1? The sample correlation coefficient between x_1 and x_2 is $r = 0.939$, with a corresponding coefficient of determination of $r^2 \approx 0.88$. This means that if we consider only x_1 and x_2 (and none of the other variables), then about 88% of the variation in x_1 can be explained by the corresponding variation in x_2 (by itself). Similarly, if we consider only x_1 and x_3, we see the sample correlation coefficient $r = 0.924$, with a corresponding coefficient of determination of $r^2 \approx 0.85$. About 85% of the variation in x_1 can be explained by the corresponding variation in x_3. The variable x_4 has much less influence on x_1 because the sample correlation coefficient between these two variables is $r = -0.739$, with corresponding coefficient of determination $r^2 \approx 0.55$, or only 55%.

These relationships are very reasonable in the context of our problem. It is common sense that the number of spring fawns x_1 is strongly related to x_2, the size of the adult antelope population. Furthermore, the spring fawn count x_1 is very much influenced by available food for the fawn (and its mother). Thunder Basin National Grasslands is a semiarid region, and available food (grass) is almost completely determined by annual precipitation x_3. Antelope are naturally strong and hardy animals. Therefore, the temperature and wind chill index x_4 will have much less effect on the adult does and corresponding number of spring fawns provided there is plenty of available food.

Least-Squares Equation

Figure 9-21 shows a display that gives an expression for the actual least-squares equation and a lot of information about the equation. To get this display or a similar display, the user needs to declare which variable is the response variable and which are the explanatory variables. For Figure 9-21, we designated x_1 as the response variable. This means that x_1 is the variable we choose to predict. We also designated variables x_2, x_3, and x_4 as explanatory variables. This means that x_2, x_3, and x_4 will be used *together* to predict x_1. There is a lot of flexibility here. We could have designated any *one* of the variables x_1, x_2, x_3, x_4 as the response variable and *any or all* of the remaining variables as explanatory variables. So there are several possible regression models the computer can construct for you, depending on the type of information you want. In this example, we want to predict x_1 (spring fawn count) by using x_2 (adult population), x_3 (annual precipitation), and x_4 (winter index) *together*.

Menu selection: **Stat ➤ Regression ➤ Regression.** In the dialogue box, select x_1 as the response and x_2, x_3, x_4 as the predictors.

FIGURE 9-21

Minitab Display of Regression Analysis

```
Regression Analysis
The regression equation is
x1 = -5.92 + 0.338 x2 + 0.402 x3 + 0.263 x4
Predictor        Coef       StDev        T        P
Constant       -5.922       1.256    -4.72    0.009
x2            0.33822     0.09947     3.40    0.027
x3             0.4015      0.1099     3.65    0.022
x4            0.26295     0.08514     3.09    0.037
S = 0.1209    R-Sq = 97.4%     R-Sq(adj) = 95.5%
```

The least-squares regression equation is given near the top of the display. Then more information is given about the constant and coefficients. The parts of the equation are

$$x_1 = -5.92 + 0.338x_2 + 0.402x_3 + 0.263x_4 \tag{13}$$

response variable (x_1); constant (-5.92); coefficient of associated explanatory variable (0.338, 0.402, 0.263)

COMMENT In the case of a simple regression model in which we have only one explanatory variable, the coefficient of that variable is the *slope* of the least-squares line. This slope (or coefficient) represents the change in the response variable per unit change in the explanatory variable. In a multiple regression model such as Equation (13), the coefficients also can be thought of as a slope, *provided* we hold the other variables as arbitrary and fixed constants. For example, the coefficient of x_2 in Equation (13) is $b_2 = 0.338$. This means that if x_3 (precipitation) and x_4 (winter index) are taken into account but held constant, then $b_2 = 0.338$ represents the change in x_1 (spring fawn count) per unit change in x_2 (adult antelope count). Since our units are in hundreds, this indicates that if x_3 and x_4 are taken into account as arbitrary but fixed values, then an increase of 100 adult antelope would give an expected increase of 33.8, or 34, spring fawns.

A natural question arises: How good a fit is the least-squares regression Equation (13) for our given data?

Coefficient of multiple determination

One way to answer this question is to examine the *coefficient of multiple determination*. The coefficient of multiple determination is a direct generalization of the concept of coefficient of determination (between *two* variables) as discussed in Section 9.2, and it has essentially the same meaning. The coefficient of multiple determination is given in the display of Figure 9-21 as a percent. We see R-sq = 97.4%. This means that about 97.4% of the variation in the response variable x_1 can be explained from the least-squares Equation (13) and the corresponding *joint* variation of the variables x_2, x_3, and x_4 taken together. The remaining 100% − 97.4% = 2.6% of the variation in x_1 is due to random chance or possibly the presence of other variables not included in this regression equation. (We will discuss the *standard error* associated with each coefficient later in this section.)

Predictions

Let's use the current regression model to predict the response variable x_1. Recall that in Section 9.2 we first made predictions from the least-squares line and then constructed a confidence interval for our predictions. Although the exact details are beyond the scope of this text, this process can be generalized to multiple regression. The calculations are very tedious, but that's why we use a computer!

Suppose we ask the following question: In a year when $x_2 = 8.2$ (hundreds of adult antelope), $x_3 = 11.7$ (inches of precipitation), and $x_4 = 3$ (winter index), what do we predict for x_1 (spring fawn count)? Furthermore, let's suppose we want an 85% confidence interval for our prediction.

To answer this question, we look at Figure 9-22, which shows the Minitab prediction result for x_1 from the specified values of x_2, x_3, and x_4.

FIGURE 9-22

Minitab Display Showing the Predicted Value of x_1

```
Predicted Values
Fit        StDev Fit        85.0% CI              85.0% PI
2.3378        0.0472     (2.2539, 2.4217)      (2.1069, 2.5687)
```

Menu selection: **Stat ➤ Regression ➤ Regression.** In the dialogue box, select Options. List the new observations for x_2, x_3, and x_4 in order, separated by spaces. Specify the confidence level. Be sure that Fit Intercept is checked.

The value for Fit is 2.3378. This is the predicted value for x_1. The 85% confidence interval for the prediction is designated as 85% PI. We see that the interval for x_1 (rounded to two digits after the decimal) is $2.11 \le x_1 \le 2.57$. This means we are 85% confident that the number of spring fawns will be in the range of 211 to 257.

Please note that this is *not* a confidence interval for the population mean of x_1. Rather, we have constructed a confidence interval for the *actual value* of x_1 under the conditions $x_2 = 8.2$, $x_3 = 11.7$, and $x_4 = 3$.

COMMENT Extrapolation much beyond the data range for any of the variables in a multiple regression model can produce results that might be meaningless and unrealistic. Many computer software packages warn about computing a confidence interval for a prediction when some of the values of the explanatory variables are beyond the data range in either direction.

TESTING A COEFFICIENT FOR SIGNIFICANCE

In applications of multiple regression, it is possible to have many different variables. Occasionally, you might suspect that one of the explanatory variables x_i is not very useful as a tool for predicting the response variable. It simply may not influence the response variable much at all. To decide whether or not this is the case, we construct a test for the significance of the coefficient of x_i in the least-squares equation.

Recall that the general least-squares equation is

$$y = b_0 + b_1x_1 + b_2x_2 + \cdots + b_kx_k \tag{14}$$

where y = response variable
x_i = explanatory variable for $i = 1, 2, \ldots, k$
b_i = numerical coefficient for $i = 0, 1, 2, \ldots, k$

Equation (14) was constructed from given data. Usually, the data are only a small subset of all possible data that could have been collected.

Let us suppose (in theory) that we used *all possible data* that could ever be obtained for our regression problem and that we constructed the regression equation using the entire population of all possible data. Then we would get the *theoretical* regression equation

$$y = \beta_0 + \beta_1x_1 + \beta_2x_2 + \cdots + \beta_kx_k \tag{15}$$

where y and x_i are as in Equation (14), but β_i is the *theoretical* coefficient of x_i.

Now look back at the regression analysis in Figure 9-21. Beside the constant and each coefficient is a number in the StDev column. This is the *standard error* corresponding to that coefficient. The standard error can be thought of as similar to a standard deviation that corresponds to the coefficient. The calculation of the number is beyond the scope of this text, but it is available on computer printouts, and we will use it to construct our test.

Let us call S_i the standard error for coefficient x_i (S_0 is the standard error for the constant). Under very basic and general assumptions, it can be proved that

$$t = \frac{b_i - \beta_i}{S_i} \tag{16}$$

has a Student's t distribution with degrees of freedom $d.f. = n - k - 1$, where n = number of data points and k = number of explanatory variables in the least-squares equation.

Now let us return to the question: Is x_i useful as an explanatory variable in the least-squares equation?

The answer is that it is *not* useful if $\beta_i = 0$. In that case, the (theoretical) coefficient of x_i would be zero and x_i would contribute nothing to the least-squares equation. However, if $\beta_i \neq 0$, then the explanatory variable x_i does contribute information in the least-squares equation.

Consider the following hypotheses,

$$H_0\text{: } \beta_i = 0 \quad \text{and} \quad H_1\text{: } \beta_i \neq 0$$

If we accept H_0, we conclude that $\beta_i = 0$ and x_i probably should be dropped as an explanatory variable in the least-squares equation. If we accept H_1, we conclude that $\beta_i \neq 0$ and x_i should be included as an explanatory variable in the least-squares equation.

EXAMPLE 11

TESTING A COEFFICIENT

We'll use the data and printouts of Example 10 and test the significance of x_3 as an explanatory variable using level of significance $\alpha = 0.05$.

$$H_0\text{: } \beta_3 = 0 \quad \text{and} \quad H_1\text{: } \beta_3 \neq 0$$

To find the t value corresponding to b_3, we use Equation (16) and the null hypothesis H_0: $\beta_3 = 0$. This gives us the equation

$$t = \frac{b_3}{S_3} \tag{17}$$

In the regression analysis shown in Figure 9-21, we see a t value for the constant and each coefficient. This t value is exactly the value of $t = b_i/S_i$. This is the t value corresponding to the sample test statistic. For the coefficient of x_3, we see

$$t \text{ value} \approx 3.65$$

Notice in Figure 9-21 that we are also given the P-value based on a two-tailed test of the sample test statistic for each coefficient. This is the value in the column headed "P." For the sample test statistic $t \approx 3.65$, the corresponding P-value is 0.022. Since the P-value is less than the level of significance $\alpha = 0.05$, we reject H_0. In other words, at the 5% level of significance, we can say that the population correlation coefficient β_3 of x_3 is not 0.

We conclude at the 5% level of significance that x_3 (annual precipitation) should be included as an explanatory variable in the least-squares equation. Notice that Figure 9-21 also gives the P-value for each ratio, so we can conclude the test using P-values directly. Using the P-values, we see that x_2 and x_4 are also significant at the 5% level.

CONFIDENCE INTERVALS FOR COEFFICIENTS

Equation (16) also gives us the basis for finding *confidence intervals* for β_i. A c% confidence interval for β_i will be

$$b_i - tS_i < \beta_i < b_i + tS_i$$

where $d.f. = n - k - 1$, t is selected according to the specified confidence level, b_i is the numerical value of the coefficient from Figure 9-21, S_i is the numerical value of the standard error from Figure 9-21, n is the number of data points, and k is the number of explanatory variables in the least-squares equation.

EXAMPLE 12

CONFIDENCE INTERVAL FOR A COEFFICIENT

Suppose we want to compute a 90% confidence interval for β_2, the coefficient of x_2. From Figure 9-21, we have (rounding to three digits after the decimal)

$$b_2 = 0.338, \quad S_2 = 0.099, \quad \text{and } d.f. = 4$$

From the t table (Table 6, Appendix II), we find $t = 2.132$, so,

$$b_2 - tS_2 < \beta_2 < b_2 + tS_2$$
$$0.338 - 2.132(0.099) < \beta_2 < 0.338 + 2.132(0.099)$$
$$0.127 < \beta_2 < 0.549$$

EXCEL 2010 DISPLAYS

Although Excel gives information very similar to that supplied by Minitab, the least-squares equation is not explicitly displayed. However, the intercept (constant) and coefficients of the variables are shown with the corresponding standard errors and t values with P-values (two-tailed test). Excel shows the confidence interval for each coefficient. However, there is no built-in function to provide predicted values or confidence intervals for predicted values. Note that as in the Minitab regression analysis, we will not make use of the ANOVA information in the Excel display.

On the home screen, click the **Data** tab. Select **Data Analysis** from the Analysis group. In the dialogue box, select **Regression**. Note that when you enter data into the worksheet, all the explanatory variables must be together in a block. Figure 9-23 shows the Excel display for Examples 10 and 11.

FIGURE 9-23

Excel Display of Regression Analysis

Regression Statistics						
Multiple R	0.987060478					
R Square	0.974288388					
Adjusted R Square	0.955004679					
Standard Error	0.120927579					
Observations	8					
ANOVA						
	df	*SS*	*MS*	*F*	*Significance F*	
Regression	3	2.216506083	0.738835361	50.5239104	0.001228863	
Residual	4	0.058493917	0.014623479			
Total	7	2.275				
	Coefficients	*Standard Error*	*t Stat*	*P-value*	*Lower 95%*	*Upper 95%*
Intercept	-5.922011616	1.255623292	-4.716391972	0.009196085	-9.40818798	-2.435835251
x2	0.338217487	0.099470083	3.400193085	0.027272474	0.062043691	0.614391283
x3	0.401503945	0.109900277	3.653347874	0.021707246	0.096371226	0.706636664
x4	0.262946128	0.085136028	3.088541172	0.036626194	0.02657013	0.499322125

VIEWPOINT Synoptic Climatology

Synoptic means "giving a summary from the same basic point of view." In this case, the point of view is Niwot Ridge, high above the timberline in the Rocky Mountains. Vegetation, water, temperature, and wind all affect the delicate balance of this alpine environment. How do these elements of nature interact to sustain life in such a harsh land? One answer can be found by collecting data at the location and using multiple regression to study the interaction of variables. For more information, visit the Brase/Brase statistics site at **http://www.cengage.com/statistics/brase11e** *and find the link to Niwot Ridge Climate Study.*

SECTION 9.4 PROBLEMS

1. *Statistical Literacy* Given the linear regression equation

$$x_1 = 1.6 + 3.5x_2 - 7.9x_3 + 2.0x_4$$

(a) Which variable is the response variable? Which variables are the explanatory variables?
(b) Which number is the constant term? List the coefficients with their corresponding explanatory variables.
(c) If $x_2 = 2$, $x_3 = 1$, and $x_4 = 5$, what is the predicted value for x_1?
(d) Explain how each coefficient can be thought of as a "slope" under certain conditions. Suppose x_3 and x_4 were held at fixed but arbitrary values and x_2 was increased by 1 unit. What would be the corresponding change in x_1? Suppose x_2 increased by 2 units. What would be the expected change in x_1? Suppose x_2 decreased by 4 units. What would be the expected change in x_1?
(e) Suppose that $n = 12$ data points were used to construct the given regression equation and that the standard error for the coefficient of x_2 is 0.419. Construct a 90% confidence interval for the coefficient of x_2.
(f) Using the information of part (e) and level of significance 5%, test the claim that the coefficient of x_2 is different from zero. Explain how the conclusion of this test would affect the regression equation.

2. *Statistical Literacy* Given the linear regression equation

$$x_3 = -16.5 + 4.0x_1 + 9.2x_4 - 1.1x_7$$

(a) Which variable is the response variable? Which variables are the explanatory variables?
(b) Which number is the constant term? List the coefficients with their corresponding explanatory variables.
(c) If $x_1 = 10$, $x_4 = -1$, and $x_7 = 2$, what is the predicted value for x_3?
(d) Explain how each coefficient can be thought of as a "slope." Suppose x_1 and x_7 were held as fixed but arbitrary values. If x_4 increased by 1 unit, what would we expect the corresponding change in x_3 to be? If x_4 increased by 3 units, what would be the corresponding expected change in x_3? If x_4 decreased by 2 units, what would we expect for the corresponding change in x_3?
(e) Suppose that $n = 15$ data points were used to construct the given regression equation and that the standard error for the coefficient of x_4 is 0.921. Construct a 90% confidence interval for the coefficient of x_4.
(f) Using the information of part (e) and level of significance 1%, test the claim that the coefficient of x_4 is different from zero. Explain how the conclusion has a bearing on the regression equation.

For Problems 3–6, use appropriate multiple regression software of your choice and enter the data. Note that the data are also available for download at the Online Study Center in formats for Excel, Minitab portable files, SPSS files, and ASCII files.

3. *Medical: Blood Pressure* The systolic blood pressure of individuals is thought to be related to both age and weight. For a random sample of 11 men, the following data were obtained:

Systolic Blood Pressure x_1	Age (years) x_2	Weight (pounds) x_3	Systolic Blood Pressure x_1	Age (years) x_2	Weight (pounds) x_3
132	52	173	137	54	188
143	59	184	149	61	188
153	67	194	159	65	207
162	73	211	128	46	167
154	64	196	166	72	217
168	74	220			

(a) Generate summary statistics, including the mean and standard deviation of each variable. Compute the coefficient of variation (see Section 3.2) for each variable. Relative to its mean, which variable has the greatest spread of data values? Which variable has the smallest spread of data values relative to its mean?

(b) For each pair of variables, generate the sample correlation coefficient r. Compute the corresponding coefficient of determination r^2. Which variable (other than x_1) has the greatest influence (by itself) on x_1? Would you say that both variables x_2 and x_3 show a strong influence on x_1? Explain your answer. What percent of the variation in x_1 can be explained by the corresponding variation in x_2? Answer the same question for x_3.

(c) Perform a regression analysis with x_1 as the response variable. Use x_2 and x_3 as explanatory variables. Look at the coefficient of multiple determination. What percentage of the variation in x_1 can be explained by the corresponding variations in x_2 and x_3 *taken together*?

(d) Look at the coefficients of the regression equation. Write out the regression equation. Explain how each coefficient can be thought of as a slope. If age were held fixed, but a person put on 10 pounds, what would you expect for the corresponding change in systolic blood pressure? If a person kept the same weight but got 10 years older, what would you expect for the corresponding change in systolic blood pressure?

(e) Test each coefficient to determine if it is zero or not zero. Use level of significance 5%. Why would the outcome of each test help us determine whether or not a given variable should be used in the regression model?

(f) Find a 90% confidence interval for each coefficient.

(g) Suppose Michael is 68 years old and weighs 192 pounds. Predict his systolic blood pressure, and find a 90% confidence range for your prediction (if your software produces prediction intervals).

4. *Education: Exam Scores* Professor Gill has taught general psychology for many years. During the semester, she gives three multiple-choice exams, each worth 100 points. At the end of the course, Dr. Gill gives a comprehensive final worth 200 points. Let x_1, x_2, and x_3 represent a student's scores on exams 1, 2, and 3, respectively. Let x_4 represent the student's score on the final exam. Last semester Dr. Gill had 25 students in her class. The student exam scores are shown on the next page.

x_1	x_2	x_3	x_4	x_1	x_2	x_3	x_4	x_1	x_2	x_3	x_4
73	80	75	152	79	70	88	164	81	90	93	183
93	88	93	185	69	70	73	141	88	92	86	177
89	91	90	180	70	65	74	141	78	83	77	159
96	98	100	196	93	95	91	184	82	86	90	177
73	66	70	142	79	80	73	152	86	82	89	175
53	46	55	101	70	73	78	148	78	83	85	175
69	74	77	149	93	89	96	192	76	83	71	149
47	56	60	115	78	75	68	147	96	93	95	192
87	79	90	175								

Since Professor Gill has not changed the course much from last semester to the present semester, the preceding data should be useful for constructing a regression model that describes this semester as well.

(a) Generate summary statistics, including the mean and standard deviation of each variable. Compute the coefficient of variation (see Section 3.2) for each variable. Relative to its mean, would you say that each exam had about the same spread of scores? Most professors do not wish to give an exam that is extremely easy or extremely hard. Would you say that all of the exams were about the same level of difficulty? (Consider both means and spread of test scores.)

(b) For each pair of variables, generate the sample correlation coefficient r. Compute the corresponding coefficient of determination r^2. Of the three exams 1, 2, and 3, which do you think had the most influence on the final exam 4? Although one exam had more influence on the final exam, did the other two exams still have a lot of influence on the final? Explain each answer.

(c) Perform a regression analysis with x_4 as the response variable. Use x_1, x_2, and x_3 as explanatory variables. Look at the coefficient of multiple determination. What percentage of the variation in x_4 can be explained by the corresponding variations in x_1, x_2, and x_3 taken together?

(d) Write out the regression equation. Explain how each coefficient can be thought of as a slope. If a student were to study "extra hard" for exam 3 and increase his or her score on that exam by 10 points, what corresponding change would you expect on the final exam? (Assume that exams 1 and 2 remain "fixed" in their scores.)

(e) Test each coefficient in the regression equation to determine if it is zero or not zero. Use level of significance 5%. Why would the outcome of each hypothesis test help us decide whether or not a given variable should be used in the regression equation?

(f) Find a 90% confidence interval for each coefficient.

(g) This semester Susan has scores of 68, 72, and 75 on exams 1, 2, and 3, respectively. Make a prediction for Susan's score on the final exam and find a 90% confidence interval for your prediction (if your software supports prediction intervals).

5. *Entertainment: Movies* A motion picture industry analyst is studying movies based on epic novels. The following data were obtained for 10 Hollywood movies made in the past 5 years. Each movie was based on an epic novel. For these data, x_1 = first-year box office receipts of the movie, x_2 = total production costs of the movie, x_3 = total promotional costs of the movie, and x_4 = total book sales prior to movie release. All units are in millions of dollars.

x_1	x_2	x_3	x_4	x_1	x_2	x_3	x_4
85.1	8.5	5.1	4.7	30.3	3.5	1.2	3.5
106.3	12.9	5.8	8.8	79.4	9.2	3.7	9.7
50.2	5.2	2.1	15.1	91.0	9.0	7.6	5.9
130.6	10.7	8.4	12.2	135.4	15.1	7.7	20.8
54.8	3.1	2.9	10.6	89.3	10.2	4.5	7.9

(a) Generate summary statistics, including the mean and standard deviation of each variable. Compute the coefficient of variation (see Section 3.2) for each variable. Relative to its mean, which variable has the largest spread of data values? Why would a variable with a large coefficient of variation be expected to change a lot relative to its average value? Although x_1 has the largest standard deviation, it has the smallest coefficient of variation. How does the mean of x_1 help explain this?

(b) For each pair of variables, generate the sample correlation coefficient r. Compute the corresponding coefficient of determination r^2. Which of the three variables x_2, x_3, and x_4 has the *least* influence on box office receipts? What percent of the variation in box office receipts can be attributed to the corresponding variation in production costs?

(c) Perform a regression analysis with x_1 as the response variable. Use x_2, x_3, and x_4 as explanatory variables. Look at the coefficient of multiple determination. What percentage of the variation in x_1 can be explained by the corresponding variations in x_2, x_3, and x_4 taken together?

(d) Write out the regression equation. Explain how each coefficient can be thought of as a slope. If x_2 (production costs) and x_4 (book sales) were held fixed but x_3 (promotional costs) was increased by \$1 million, what would you expect for the corresponding change in x_1 (box office receipts)?

(e) Test each coefficient in the regression equation to determine if it is zero or not zero. Use level of significance 5%. Explain why book sales x_4 probably are not contributing much information in the regression model to forecast box office receipts x_1.

(f) Find a 90% confidence interval for each coefficient.

(g) Suppose a new movie (based on an epic novel) has just been released. Production costs were $x_2 = 11.4$ million; promotion costs were $x_3 = 4.7$ million; book sales were $x_4 = 8.1$ million. Make a prediction for $x_1 =$ first-year box office receipts and find an 85% confidence interval for your prediction (if your software supports prediction intervals).

(h) Construct a new regression model with x_3 as the response variable and x_1, x_2, and x_4 as explanatory variables. Suppose Hollywood is planning a new epic movie with projected box office sales $x_1 = 100$ million and production costs $x_2 = 12$ million. The book on which the movie is based had sales of $x_4 = 9.2$ million. Forecast the dollar amount (in millions) that should be budgeted for promotion costs x_3 and find an 80% confidence interval for your prediction.

6. *Franchise Business: Market Analysis* All Greens is a franchise store that sells house plants and lawn and garden supplies. Although All Greens is a franchise, each store is owned and managed by private individuals. Some friends have asked you to go into business with them to open a new All Greens store in the suburbs of San Diego. The national franchise headquarters sent you the following information at your request. These data are about 27 All Greens stores in California. Each of the 27 stores has been doing very well, and you would like to use the information to help set up your own new store. The variables for which we have data are

x_1 = annual net sales, in thousands of dollars
x_2 = number of square feet of floor display in store, in thousands of square feet

x_3 = value of store inventory, in thousands of dollars
x_4 = amount spent on local advertising, in thousands of dollars
x_5 = size of sales district, in thousands of families
x_6 = number of competing or similar stores in sales district

A sales district was defined to be the region within a 5-mile radius of an All Greens store.

x_1	x_2	x_3	x_4	x_5	x_6	x_1	x_2	x_3	x_4	x_5	x_6
231	3	294	8.2	8.2	11	65	1.2	168	4.7	3.3	11
156	2.2	232	6.9	4.1	12	98	1.6	151	4.6	2.7	10
10	0.5	149	3	4.3	15	398	4.3	342	5.5	16.0	4
519	5.5	600	12	16.1	1	161	2.6	196	7.2	6.3	13
437	4.4	567	10.6	14.1	5	397	3.8	453	10.4	13.9	7
487	4.8	571	11.8	12.7	4	497	5.3	518	11.5	16.3	1
299	3.1	512	8.1	10.1	10	528	5.6	615	12.3	16.0	0
195	2.5	347	7.7	8.4	12	99	0.8	278	2.8	6.5	14
20	1.2	212	3.3	2.1	15	0.5	1.1	142	3.1	1.6	12
68	0.6	102	4.9	4.7	8	347	3.6	461	9.6	11.3	6
570	5.4	788	17.4	12.3	1	341	3.5	382	9.8	11.5	5
428	4.2	577	10.5	14.0	7	507	5.1	590	12.0	15.7	0
464	4.7	535	11.3	15.0	3	400	8.6	517	7.0	12.0	8
15	0.6	163	2.5	2.5	14						

(a) Generate summary statistics, including the mean and standard deviation of each variable. Compute the coefficient of variation (see Section 3.2) for each variable. Relative to its mean, which variable has the largest spread of data values? Which variable has the least spread of data values relative to its mean?

(b) For each pair of variables, generate the sample correlation coefficient r. For all pairs involving x_1, compute the corresponding coefficient of determination r^2. Which variable has the greatest influence on annual net sales? Which variable has the least influence on annual net sales?

(c) Perform a regression analysis with x_1 as the response variable. Use x_2, x_3, x_4, x_5, and x_6 as explanatory variables. Look at the coefficient of multiple determination. What percentage of the variation in x_1 can be explained by the corresponding variations in x_2, x_3, x_4, x_5, and x_6 taken together?

(d) Write out the regression equation. If two new competing stores moved into the sales district but the other explanatory variables did not change, what would you expect for the corresponding change in annual net sales? Explain your answer. If you increased the local advertising by a thousand dollars but the other explanatory variables did not change, what would you expect for the corresponding change in annual net sales? Explain.

(e) Test each coefficient to determine if it is or is not zero. Use level of significance 5%.

(f) Suppose you and your business associates rent a store, get a bank loan to start up your business, and do a little research on the size of your sales district and the number of competing stores in the district. If $x_2 = 2.8$, $x_3 = 250$, $x_4 = 3.1$, $x_5 = 7.3$, and $x_6 = 2$, use a computer to forecast x_1 = annual net sales and find an 80% confidence interval for your forecast (if your software produces prediction intervals).

(g) Construct a new regression model with x_4 as the response variable and x_1, x_2, x_3, x_5, and x_6 as explanatory variables. Suppose an All Greens store in Sonoma, California, wants to estimate a range of advertising costs appropriate to its store. If it spends too little on advertising, it will not reach enough customers. However, it does not want to overspend on advertising for this type and size of store. At this store, $x_1 = 163$, $x_2 = 2.4$, $x_3 = 188$, $x_5 = 6.6$, and $x_6 = 10$. Use these data to predict x_4 (advertising costs) and find an 80% confidence interval for your prediction. At the 80% confidence level, what range of advertising costs do you think is appropriate for this store?

7. *Expand Your Knowledge: Curvilinear Polynomial Regression* In this section we studied multiple linear regression. Our basic linear model has been

$$y = b_0 + b_1x_1 + b_2x_2 + \cdots + b_kx_k$$

Since all the variables $x_1, x_2, \ldots, x_k$ are of first degree, this is an example of linear regression. However, the same basic methods of linear regression can be used for *curvilinear regression* (also known as *polynomial regression*). The interested reader can find a great deal of information on this topic in the book *Applied Numerical Methods* by Carnahan, Luther, and Wilkes from page 573 on.

Assume we have at least $k + 2$ data pairs (x, y) and we want to approximate y using a polynomial of degree k. To do this, we make the following identification.

$$x_1 = x; x_2 = x^2; x_3 = x^3; \ldots ; x_k = x^k$$

Then we use our known methods of multiple regression to obtain coefficients $b_0, b_1, b_2, b_3, \ldots, b_k$ and the equation $y = b_0 + b_1x_1 + b_2x^2 + b_3x^3 + \cdots + b_kx^k$. This is called the *least-squares curvilinear regression model.*

Marketing studies show that price increases often have a point of diminishing returns. For a popular product, the price can often increase with sales. However, when the price becomes too high, sales start to drop off. In the following graph, P = point of diminishing returns.

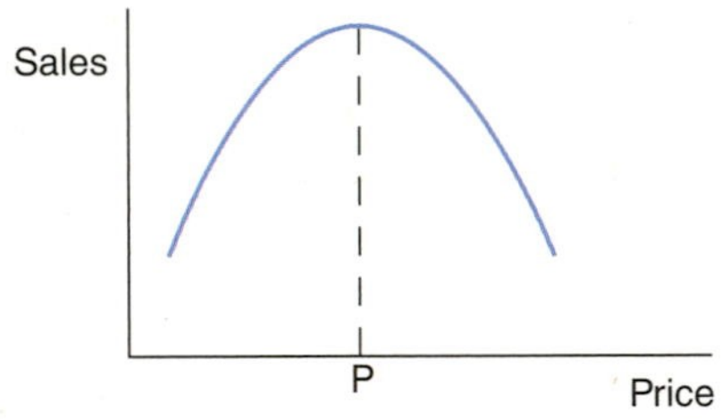

To estimate the point of diminishing returns, we use a quadratic polynomial, $y = b_0 + b_1x + b_2x^2$. A very popular women's knit T-shirt was tested for sales appeal and price in six large department stores. In each city, the T-shirts were advertised extensively in the local media, so price and sales initially went up. However, as price increased, sales eventually dropped off. Let x = price per T-shirt in dollars and y = number of T-shirts sold in a day at that price. We have the following data.

City	A	B	C	D	E	F
x	12.97	13.88	15.95	18.50	19.99	22.50
y	23	31	33	29	25	17

To construct our quadratic polynomial, we use multilinear regression with the following table of data values.

$x_1 = x$	12.97	13.88	15.95	18.50	19.99	22.50
$x_2 = x^2$	168.22	192.65	254.40	342.25	399.60	506.25
y	23	31	33	29	25	17

Computer software gives us coefficients for the model $y = b_0 + b_1x_1 + b_2x_2 = b_0 + b_1x + b_2x^2$, which becomes $y = -93.80 + 15.10x - 0.45x^2$. The coefficient of determination is $r^2 = 0.88$ (not too bad!). The curvilinear regression equation $y = -93.80 + 15.10x - 0.45x^2$ is a quadratic curve that opens downward. A little extra mathematics shows that the top point on the curve (point of diminishing returns) occurs when the cost per shirt of $x = \$16.78$ with $y = 32.87$ shirts sold per day. This suggests the knit T-shirts should be priced at \$16.78 and that about 33 of them will sell per day in a large department store.

Use the Internet, school library, popular magazines, or any other source to collect (x, y) data pairs regarding variables of interest to you. Construct a curvilinear regression model from your data and interpret the results.

TECH NOTES In the Using Technology section at the end of this chapter, you will find a "mini case study" of seven important variables from the economy of the United States for the years 1976 to 1987. Readers interested in applications of multiple regression and the U.S. economy are referred to this material.

CHAPTER REVIEW

SUMMARY

This chapter discusses linear regression models and inferences related to these models.

- A scatter diagram of data pairs (x, y) gives a graphical display of the relationship (if any) between x and y data. We are looking for a linear relationship.
- For data pairs (x, y), x is called the *explanatory variable* and is plotted along the horizontal axis. The *response variable* y is plotted along the vertical axis.
- The Pearson product-moment *correlation coefficient* r gives a numerical measurement assessing the strength of a linear relationship between x and y. It is based on a random sample of (x, y) data pairs.
- The value of r ranges from -1 to 1, with 1 indicating perfect positive linear correlation, -1 indicating perfect negative linear correlation, and 0 indicating no linear correlation.
- If the scatter diagram and sample correlation coefficient r indicate a linear relationship between x and y values of the data pairs, we use the least-squares criteria to develop the equation of the least-squares line

  $$\hat{y} = a + bx$$

  where $\hat{y}$ is the value of y predicted by the least-squares line for a given x value, a is the y intercept, and b is the slope.
- Methods of testing the population correlation coefficient ρ show whether or not the sample statistic r is significant. We test the null hypothesis H_0: $\rho = 0$ against a suitable alternate hypothesis ($\rho > 0$, $\rho < 0$, or $\rho \neq 0$).
- Methods of testing the population slope β show whether or not the sample slope b is significant. We test the null hypothesis H_0: $\beta = 0$ against a suitable alternate hypothesis ($\beta > 0$, $\beta < 0$, or $\beta \neq 0$).
- Confidence intervals for β give us a range of values for β based on the sample statistic b and specified confidence level c.
- Confidence intervals for the predicted value of y give us a range of values for y for a specific x value. The interval is based on the sample prediction $\hat{y}$ and confidence level c.
- The *coefficient of determination* r^2 is a value that measures the proportion of variation in y explained by the least-squares line, the linear regression model, and the variation in the explanatory variable x.
- The difference $y - \hat{y}$ between the y value in the data pair (x, y) and the corresponding predicted value $\hat{y}$ for the same x is called the *residual*.
- The *standard error of estimate* S_e is a measure of data spread about the least-squares line. It is based on the residuals.
- Techniques of multiple regression (with computer assistance) help us analyze a linear relation involving several variables.

IMPORTANT WORDS & SYMBOLS

Section 9.1

Paired data values 530
Scatter diagram 530
Explanatory variable 530
Response variable 530
No linear correlation 531
Perfect linear correlation 532
Positive correlation 532
Negative correlation 532
Sample correlation coefficient r 534
Extrapolation 541
Causation 541
Lurking variable 541

Section 9.2

Least-squares criterion 549
Least-squares line $\hat{y} = a + bx$ 550
Slope b 550
Intercept a 550
Meaning of slope 552
Marginal change 552
Influential point 553
Residual 553
Interpolation 554
Extrapolation 554
Coefficient of determination r^2 558
Explained variation 559
Unexplained variation 559
Residual plot 565

Section 9.3

Population correlation coefficient ρ 570
Standard error of estimate S_e 573
Population slope β 576
Confidence prediction band 578
Serial correlation (autocorrelation) 588

Section 9.4

Multiple regression 590
Coefficient of multiple determination 595
Curvilinear (polynomial) regression 604

VIEWPOINT Living Arrangements

Male, female, married, single, living alone, living with friends or relatives—all these categories are of interest to the U.S. Census Bureau. In addition to these categories, there are others, such as age, income, and health needs. How strongly correlated are these variables? Can we use one or more of these variables to predict the others? How good is such a prediction? Methods of this chapter can help you answer such questions. For more information regarding such data, visit the Brase/Brase statistics site at **http://www.cengage.com/statistics/brase11e** *and find the link to Census Bureau.*

CHAPTER REVIEW PROBLEMS

1. *Statistical Literacy* Suppose the scatter diagram of a random sample of data pairs (x, y) shows no linear relationship between x and y. Do you expect the value of the sample correlation coefficient r to be close to 1, -1, or 0?

2. *Statistical Literacy* What does it mean to say that the sample correlation coefficient r is significant?

3. *Statistical Literacy* When using the least-squares line for prediction, are results usually more reliable for extrapolation or interpolation?

4. *Statistical Literacy* Suppose that for $x = 3$, the predicted value is $\hat{y} = 6$. The data pair (3, 8) is part of the sample data. What is the value of the residual for $x = 3$?

In Problems 5–10, parts (a)–(e) involve scatter diagrams, least-squares lines, correlation coefficients with coefficients of determination, tests of ρ, and predictions. Parts (f)–(i) involve standard error of estimate, confidence intervals for predictions, tests of β, and confidence intervals for β.

When solving problems involving the standard error of estimate, testing of the correlation coefficient, or testing of β or confidence intervals for β, make the assumption that x and y are normally distributed random variables. Answers may vary slightly due to rounding.

5. *Desert Ecology: Wildlife* Bighorn sheep are beautiful wild animals found throughout the western United States. Data for this problem are based on information taken from *The Desert Bighorn*, edited by Monson and Sumner (University of Arizona Press). Let x be the age of a bighorn sheep (in years), and let y be the mortality rate (percent that die) for this age group. For example, $x = 1$, $y = 14$ means that 14% of the bighorn sheep between 1 and 2 years old die. A random sample of Arizona bighorn sheep gave the following information:

x	1	2	3	4	5
y	14	18.9	14.4	19.6	20.0

$\Sigma x = 15$; $\Sigma y = 86.9$; $\Sigma x^2 = 55$; $\Sigma y^2 = 1544.73$; $\Sigma xy = 273.4$

(a) Draw a scatter diagram.
(b) Find the equation of the least-squares line.
(c) Find r. Find the coefficient of determination r^2. Explain what these measures mean in the context of the problem.
(d) Test the claim that the population correlation coefficient is positive at the 1% level of significance.
(e) Given the lack of significance of r, is it practical to find estimates of y for a given x value based on the least-squares line model? Explain.

6. *Sociology: Job Changes* A sociologist is interested in the relation between x = number of job changes and y = annual salary (in thousands of dollars) for people living in the Nashville area. A random sample of 10 people employed in Nashville provided the following information:

x (Number of job changes)	4	7	5	6	1	5	9	10	10	3
y (Salary in \$1000)	33	37	34	32	32	38	43	37	40	33

$\Sigma x = 60$; $\Sigma y = 359$; $\Sigma x^2 = 442$; $\Sigma y^2 = 13{,}013$; $\Sigma xy = 2231$

(a) Draw a scatter diagram for the data.
(b) Find $\bar{x}$, $\bar{y}$, b, and the equation of the least-squares line. Plot the line on the scatter diagram of part (a).
(c) Find the sample correlation coefficient r and the coefficient of determination. What percentage of variation in y is explained by the least-squares model?
(d) Test the claim that the population correlation coefficient ρ is positive at the 5% level of significance.
(e) If someone had $x = 2$ job changes, what does the least-squares line predict for y, the annual salary?
(f) Verify that $S_e \approx 2.56$.
(g) Find a 90% confidence interval for the annual salary of an individual with $x = 2$ job changes.
(h) Test the claim that the slope β of the population least-squares line is positive at the 5% level of significance.
(i) Find a 90% confidence interval for β and interpret its meaning.

7. *Medical: Fat Babies* Modern medical practice tells us not to encourage babies to become too fat. Is there a positive correlation between the weight x of a 1-year-old baby and the weight y of the mature adult (30 years old)? A random sample of medical files produced the following information for 14 females:

x (lb)	21	25	23	24	20	15	25	21	17	24	26	22	18	19
y (lb)	125	125	120	125	130	120	145	130	130	130	130	140	110	115

$\Sigma x = 300$; $\Sigma y = 1775$; $\Sigma x^2 = 6572$; $\Sigma y^2 = 226{,}125$; $\Sigma xy = 38{,}220$

(a) Draw a scatter diagram for the data.
(b) Find $\bar{x}$, $\bar{y}$, b, and the equation of the least-squares line. Plot the line on the scatter diagram of part (a).
(c) Find the sample correlation coefficient r and the coefficient of determination. What percentage of the variation in y is explained by the least-squares model?
(d) Test the claim that the population correlation coefficient ρ is positive at the 1% level of significance.
(e) If a female baby weighs 20 pounds at 1 year, what do you predict she will weigh at 30 years of age?
(f) Verify that $S_e \approx 8.38$.
(g) Find a 95% confidence interval for weight at age 30 of a female who weighed 20 pounds at 1 year of age.
(h) Test the claim that the slope β of the population least-squares line is positive at the 1% level of significance.
(i) Find an 80% confidence interval for β and interpret its meaning.

8. *Sales: Insurance* Dorothy Kelly sells life insurance for the Prudence Insurance Company. She sells insurance by making visits to her clients' homes. Dorothy believes that the number of sales should depend, to some degree, on the number of visits made. For the past several years, she has kept careful records of the number of visits (x) she makes each week and the number of people (y) who buy insurance that week. For a random sample of 15 such weeks, the x and y values follow:

x	11	19	16	13	28	5	20	14	22	7	15	29	8	25	16
y	3	11	8	5	8	2	5	6	8	3	5	10	6	10	7

$\Sigma x = 248$; $\Sigma y = 97$; $\Sigma x^2 = 4856$; $\Sigma y^2 = 731$; $\Sigma xy = 1825$

(a) Draw a scatter diagram for the data.
(b) Find $\bar{x}$, $\bar{y}$, b, and the equation of the least-squares line. Plot the line on the scatter diagram of part (a).
(c) Find the sample correlation coefficient r and the coefficient of determination. What percentage of the variation in y is explained by the least-squares model?
(d) Test the claim that the population correlation coefficient ρ is positive at the 1% level of significance.
(e) In a week during which Dorothy makes 18 visits, how many people do you predict will buy insurance from her?
(f) Verify that $S_e \approx 1.731$.
(g) Find a 95% confidence interval for the number of sales Dorothy would make in a week during which she made 18 visits.
(h) Test the claim that the slope β of the population least-squares line is positive at the 1% level of significance.
(i) Find an 80% confidence interval for β and interpret its meaning.

9. *Marketing: Coupons* Each box of Healthy Crunch breakfast cereal contains a coupon entitling you to a free package of garden seeds. At the Healthy Crunch home office, they use the weight of incoming mail to determine how many of their employees are to be assigned to collecting coupons and mailing out seed packages on a given day. (Healthy Crunch has a policy of answering all its mail on the day it is received.)

Let x = weight of incoming mail and y = number of employees required to process the mail in one working day. A random sample of 8 days gave the following data:

x (lb)	11	20	16	6	12	18	23	25
y (Number of employees)	6	10	9	5	8	14	13	16

$\Sigma x = 131$; $\Sigma y = 81$; $\Sigma x^2 = 2435$; $\Sigma y^2 = 927$; $\Sigma xy = 1487$

(a) Draw a scatter diagram for the data.
(b) Find $\bar{x}$, $\bar{y}$, b, and the equation of the least-squares line. Plot the line on the scatter diagram of part (a).
(c) Find the sample correlation coefficient r and the coefficient of determination. What percentage of the variation in y is explained by the least-squares model?
(d) Test the claim that the population correlation coefficient ρ is positive at the 1% level of significance.
(e) If Healthy Crunch receives 15 pounds of mail, how many employees should be assigned mail duty that day?
(f) Verify that $S_e \approx 1.726$.
(g) Find a 95% confidence interval for the number of employees required to process mail for 15 pounds of mail.
(h) Test the claim that the slope β of the population least-squares line is positive at the 1% level of significance.
(i) Find an 80% confidence interval for β and interpret its meaning.

10. *Focus Problem: Changing Population and Crime Rate* Let x be a random variable representing percentage change in neighborhood population in the past few years, and let y be a random variable representing crime rate (crimes per 1000 population). A random sample of six Denver neighborhoods gave the following information (Source: *Neighborhood Facts*, The Piton Foundation).

x	29	2	11	17	7	6
y	173	35	132	127	69	53

$\Sigma x = 72$; $\Sigma y = 589$; $\Sigma x^2 = 1340$; $\Sigma y^2 = 72{,}277$; $\Sigma xy = 9499$

(a) Draw a scatter diagram for the data.
(b) Find $\bar{x}$, $\bar{y}$, b, and the equation of the least-squares line. Plot the line on the scatter diagram of part (a).
(c) Find the sample correlation coefficient r and the coefficient of determination. What percentage of the variation in y is explained by the least-squares model?
(d) Test the claim that the population correlation coefficient ρ is not zero at the 1% level of significance.
(e) For a neighborhood with $x = 12\%$ change in population in the past few years, predict the change in the crime rate (per 1000 residents).
(f) Verify that $S_e \approx 22.5908$.
(g) Find an 80% confidence interval for the change in crime rate when the percentage change in population is $x = 12\%$.
(h) Test the claim that the slope β of the population least-squares line is not zero at the 1% level of significance.
(i) Find an 80% confidence interval for β and interpret its meaning.

DATA HIGHLIGHTS: GROUP PROJECTS

Break into small groups and discuss the following topics. Organize a brief outline in which you summarize the main points of your group discussion.

Scatter diagrams! Are they really useful? Scatter diagrams give a first impression of a data relationship and help us assess whether a linear relation provides a reasonable model for the data. In addition, we can spot *influential points*. A data point with an extreme x value can heavily influence the position of the least-squares line. In this project, we look at data sets with an influential point.

x	1	4	5	9	10	15
y	3	7	6	10	12	4

(a) Compute r and b, the slope of the least-squares line. Find the equation of the least-squares line, and sketch the line on the scatter diagram.

(b) Notice the point boxed in blue in Figure 9-24. Does it seem to lie away from the linear pattern determined by the other points? The coordinates of that point are (15, 4). Is it an influential point? Remove that point from the model and recompute r, b, and the equation of the least-squares line. Sketch this least-squares line on the diagram. How does the removal of the influential point affect the values of r and b and the position of the least-squares line?

(c) Consider the scatter diagram of Figure 9-25. Is there an influential point? If you remove the influential point, will the slope of the new least-squares line be larger or smaller than the slope of the line from the original data? Will the correlation coefficient be larger or smaller?

FIGURE 9-24

Scatter Diagram

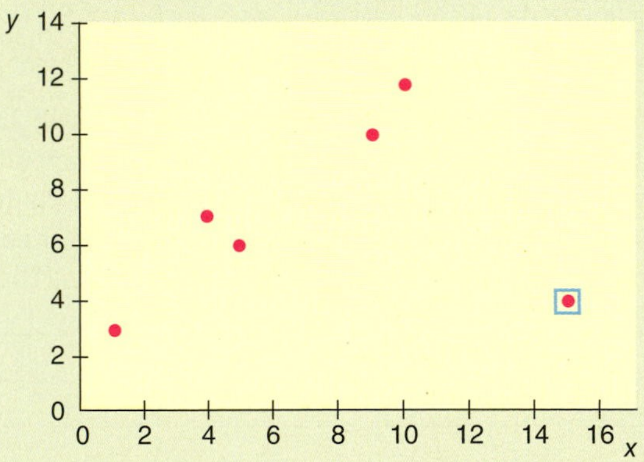

FIGURE 9-25

Scatter Diagram

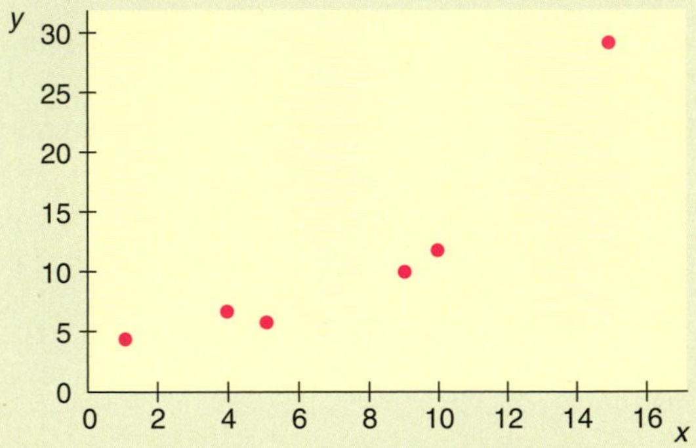

LINKING CONCEPTS: WRITING PROJECTS

Discuss each of the following topics in class or review the topics on your own. Then write a brief but complete essay in which you summarize the main points. Please include formulas and graphs as appropriate.

1. What do we mean when we say that two variables have a strong positive (or negative) linear correlation? What would a scatter diagram for these variables look like? Is it possible that two variables could be strongly related somehow but have a low *linear* correlation? Explain and draw a scatter diagram to demonstrate your point.
2. What do we mean by the least-squares criterion? Give a very general description of how the least-squares criterion is involved in the construction of the least-squares line. Why do we say the least-squares line is the "best-fitting" line for the data set?
3. In this chapter, we discussed three measures for "goodness of fit" of the least-squares line for given data. These measures were standard error of estimate, correlation coefficient, and coefficient of determination. Discuss the ways in which

these measurements are different and the ways in which they are similar to each other. Be sure to include a discussion of explained variation, unexplained variation, and total variation in your answer. Draw a sketch and include appropriate formulas.

4. Look at the formula for confidence bounds for least-squares predictions. Which of the following conditions do you think will result in a *shorter* confidence interval for a prediction?
   (a) Larger or smaller values for the standard error of estimate
   (b) Larger or smaller number of data pairs
   (c) A value of x near $\bar{x}$ or a value of x far away from $\bar{x}$

   Why would a shorter confidence interval for a prediction be more desirable than a longer interval?
5. If you did not cover Section 9.4, Multiple Regression, omit this problem.
   For many applications in statistics, more data lead to more accurate results. In multiple regression, we have more variables (and data) than we have in most simple regression problems. Why will this usually lead to more accurate predictions? Will additional variables *always* lead to more accurate predictions? Explain your answer. Discuss the coefficient of multiple determination and its meaning in the context of multiple regression. How do we know if an explanatory variable has a statistically significant influence on the response variable? What do we mean by a regression model?
6. Use the Internet or go to the library and find a magazine or journal article in your field of major interest to which the content of this chapter could be applied. List the variables used, method of data collection, and general type of information and conclusions drawn.

USING TECHNOLOGY

Simple Linear Regression (One Explanatory Variable)

Application 1

The data in this section are taken from this source:

Based on King, Cuchlaine A. M. *Physical Geography*. Oxford: Basil Blackwell, 1980, pp. 77–86, 196–206.

Throughout the world, natural ocean beaches are beautiful sights to see. If you have visited natural beaches, you may have noticed that when the gradient or dropoff is steep, the grains of sand tend to be larger. In fact, a man-made beach with the "wrong" size granules of sand tends to be washed away and eventually replaced when the proper size grain is selected by the action of the ocean and the gradient of the bottom. Since man-made beaches are expensive, grain size is an important consideration.

In the data that follow, x = median diameter (in millimeters) of granules of sand, and y = gradient of beach slope in degrees on natural ocean beaches.

x	y
0.17	0.63
0.19	0.70
0.22	0.82
0.235	0.88
0.235	1.15
0.30	1.50
0.35	4.40
0.42	7.30
0.85	11.30

1. Find the sample mean and standard deviation for x and y.
2. Make a scatter plot. Would you expect a moderately high correlation and a good fit for the least-squares line?
3. Find the equation of the least-squares line, and graph the line on the scatter plot.
4. Find the sample correlation coefficient r and the coefficient of determination r^2. Is r significant at the 1% level of significance (two-tailed test)?
5. Test that $\beta > 0$ at the 1% level of significance. Find the standard error of estimate S_e and form an 80% confidence interval for β. As the diameter of granules of sand changes by 0.10 mm, by how much does the gradient of beach slope change?
6. Suppose you have a truckload of sifted sand in which the median size of granules is 0.38 mm. If you want to put this sand on a beach and you don't want the sand to wash away, then what does the least-squares line predict for the angle of the beach? *Note:* Heavy storms that produce abnormal waves may also wash out the sand. However, in the long run, the size of sand granules that remain on the beach or that are brought back to the beach by long-term wave action are determined to a large extent by the angle at which the beach drops off. What range of angles should the beach have if we want to be 90% confident that we are matching the size of our sand granules (0.38 mm) to the proper angle of the beach?
7. Suppose we now have a truckload of sifted sand in which the median size of the granules is 0.45 mm. Repeat Problem 6.

Technology Hints (Simple Regression)

TI-84Plus/TI-83Plus/TI-*n*spire (with TI-84Plus keypad)

Be sure to set **DiagnosticOn** (under **Catalog**).

(a) Scatter diagram: Use **STAT PLOT,** select the first type, use **ZOOM** option **9:ZoomStat.**

(b) Least-squares line and r: Use **STAT, CALC,** option **8:LinReg**(a + bx).

(c) Graph least-squares line and predict: Press Y=. Then, under **VARS**, select **5:Statistics,** then select **EQ,** and finally select item **1:RegEQ.** Press enter. This sequence of steps will automatically set $Y_1 =$ your regression equation. Press **GRAPH.** To find a predicted value, when the graph is showing press the **CALC** key and select item **1:Value**. Enter the x value, and the corresponding y value will appear.

(d) Testing ρ and β, value for S_e: Use **STAT, TEST,** option **E:LinRegTTest**. The value of S_e is in the display as s.

(e) Confidence intervals for β or predictions: Use formulas from Section 9.3.

Excel 2010

(a) Scatter plot, least-squares line, r^2: On home screen, click **Insert** tab. In the Charts group, select **Scatter** and choose the first type. Once plot is displayed, *right* click on any data point. Select **trend line.** Under options, check display line and display r^2.

(b) Prediction: Use **insert function** (fx) ➤ **Statistical** ➤ **Forecast**.

(c) Coefficient r: Use (fx) ➤ **Statistical** ➤ **Correl.**

(d) Testing β and confidence intervals for b: Use menu selection **Tools** ➤ **Data Analysis** ➤ **Regression.**

(e) Confidence interval for prediction: Use formulas from Section 9.3.

Minitab

(a) Scatter plot, least-squares line, r^2, S_e: Use menu selection **Stat** ➤ **Regression** ➤ **Fitted line plot.** The value of S_e is displayed as the value of s.

(b) Coefficient r: Use menu selection **Stat** ➤ **Basic Statistics** ➤ **Correlation.**

(c) Testing β, predictions, confidence interval for predictions: Use menu selection **Stat** ➤ **Regression** ➤ **Regression.**

(d) Confidence interval for β: Use formulas from Section 9.3.

SPSS

SPSS offers several options for finding the correlation coefficient r and the equation of the least-squares line. First enter the data in the data editor and label the variables appropriately in the variable view window. Use the menu choices **Analyze** ➤ **Regression** ➤ **Linear** and select dependent and independent variables. The output includes the correlation coefficient, the standard error of estimate, the constant, and the coefficient of the dependent variable with corresponding t values and P-values for two-tailed tests. The display shows the results for the data in this chapter's Focus Problem regarding crime rate and percentage change in population.

SPSS Display

Model Summary

Model	R	R Square	Adjusted R Square	Std. Error of the Estimate
1	.927[a]	.859	.823	22.59076

a. Predictors: (Constant), % change in population

Coefficients[a]

Model		Unstandardized Coefficients B	Unstandardized Coefficients Std. Error	Standardized Coefficients Beta	t	Sig.
1	(Constant)	36.881	15.474		2.383	.076
	% change in population	5.107	1.035	.927	4.932	.008

a. Dependent Variable: Crime rate per 1,000

With the menu choices **Graph ➤ Legacy Dialogues ➤ Interactive ➤ Scatterplot,** SPSS produces a scatter diagram with the least-squares line, least-squares equation, coefficient of determination r^2, and optional prediction bands. In the dialogue box, move the dependent variable to the box along the vertical axis and the independent variable to the box along the horizontal axis. Click the "fit" tab, highlight Regression as method, and check the box to include the constant in the equation. For optional prediction band, check individual, enter the confidence level, and check total. The following display shows a scatter diagram for the data in this chapter's Focus Problem regarding crime rate and percentage change in population.

SPSS Display for Focus Problem

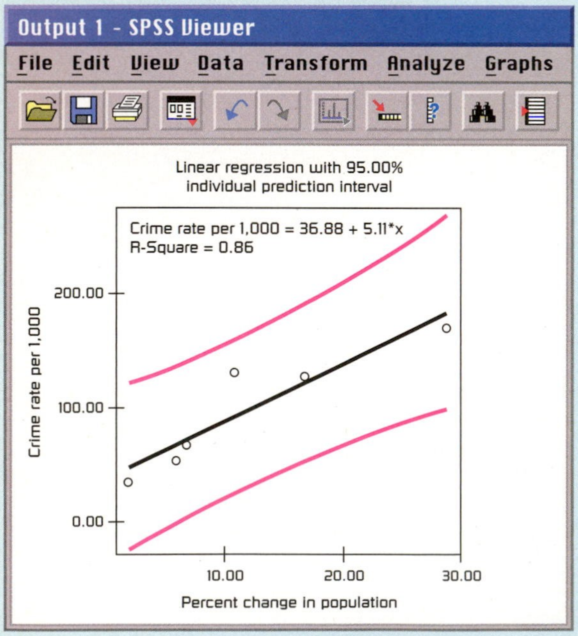

Multiple Regression

Application 2

Data values in the following study are taken from *Statistical Abstract of the United States*, U.S. Department of Commerce, 103rd and 109th Editions (see Table 9-16). All data values represent annual averages as determined by the U.S. Department of Commerce.

1. Construct a regression model with

   Response variable: x_3 (foreign investments)

   Explanatory variables: x_5 (GNP), x_6 (U.S. dollar), and x_7 (consumer credit)

What is the coefficient of multiple determination?

(a) Use a 1% level of significance and test each coefficient for significance (two-tailed test).

(b) Examine the coefficients of the regression equation. Then explain why you think the following statement is true or false: "If the purchasing power of the U.S. dollar did not change and the GNP did not change, then an increase in consumer credit would likely be accompanied by a reduction in foreign investments."

(c) Suppose $x_5 = 3500$, $x_6 = 0.975$, and $x_7 = 450$. Predict the level of foreign investment. Find a 90% confidence interval for your prediction.

We will use the following notation:

x_1 = price of a barrel of crude oil, in dollars per barrel
x_2 = percent interest on 10-year U.S. Treasury notes
x_3 = total foreign investments in U.S., in billions of dollars
x_4 = Dow Jones Industrial Average (DJIA)
x_5 = Gross National Product, GNP, in billions of dollars
x_6 = purchasing power of U.S. dollar with base 1983 corresponding to $1.000
x_7 = consumer credit (i.e., consumer debt), in billions of dollars

TABLE 9-16 **Economic Data, 1976–1987 (on the data disk)**

Year	x_1	x_2	x_3	x_4	x_5	x_6	x_7
1976	10.9	7.61	31	974.9	1718	1.757	234.4
1977	12.0	7.42	35	894.6	1918	1.649	263.8
1978	12.5	8.41	42	820.2	2164	1.532	308.3
1979	17.7	9.44	54	844.4	2418	1.380	347.5
1980	28.1	11.46	83	891.4	2732	1.215	349.4
1981	35.6	13.91	109	932.9	3053	1.098	366.6
1982	31.8	13.00	125	884.4	3166	1.035	381.1
1983	29.0	11.11	137	1190.3	3406	1.000	430.4
1984	28.6	12.44	165	1178.5	3772	0.961	511.8
1985	26.8	10.62	185	1328.2	4015	0.928	592.4
1986	14.6	7.68	209	1792.8	4240	0.913	646.1
1987	17.9	8.38	244	2276.0	4527	0.880	685.5

2. Construct a new regression model with

   Response variable: x_4 (DJIA)

   Explanatory variables: x_3 (foreign investments), x_5 (GNP), and x_7 (consumer credit)

What is the coefficient of multiple determination?

(a) Use a 5% level of significance and test each coefficient for significance (two-tailed test).

(b) Examine the coefficients of the regression equation; then explain why you think the following statement is true or false: "If the GNP and consumer credit didn't change but foreign investments increased, the DJIA would likely show a strong increase."

(c) Suppose $x_3 = 210$, $x_5 = 4260$, and $x_7 = 650$. Predict the DJIA and find an 85% confidence interval for your prediction.

3. Construct a new regression model with

   Response variable: x_7 (consumer credit)

   Explanatory variables: x_3 (foreign investments), x_5 (GNP), and x_6 (U.S. dollar)

What is the coefficient of multiple determination?

(a) Use a 1% level of significance and test each coefficient for significance (two-tailed test).

(b) Examine the coefficients of the regression equation; then explain why you think each of the following statements is true or false: "If both GNP and purchasing power of the U.S. dollar didn't change, then an increase in foreign investments would likely be accompanied by a reduction in consumer credit." "If both foreign investments and purchasing power of the U.S. dollar remained fixed, then an increase in GNP would likely be accompanied by an increase in consumer credit."

(c) Suppose $x_3 = 88$, $x_5 = 2750$, and $x_6 = 1.250$. Predict consumer credit and find an 80% confidence interval for your prediction.

Technology Hints (Multiple Regression)

TI-84Plus/TI-83Plus/TI-*n*spire

Does not support multiple regression.

Excel 2010

On the home screen, click the **Data** tab. Select **Data Analysis** from the Analysis group. In the dialogue box, select **Regression**. On the spreadsheet, the columns containing the explanatory variables need to be adjacent.

Minitab

Use the menu selection **Stat ➤ Regression ➤ Regression.**

SPSS

Use the menu selection **Analyze ➤ Regression ➤ Linear** and select dependent and independent variables.

CUMULATIVE REVIEW PROBLEMS

CHAPTER 7-9

In Problems 1–6, please use the following steps (i) through (v) for all hypothesis tests.

(i) What is the level of significance? State the null and alternate hypotheses.

(ii) *Check Requirements* What sampling distribution will you use? What assumptions are you making? What is the value of the sample test statistic?

(iii) Find (or estimate) the *P*-value. Sketch the sampling distribution and show the area corresponding to the *P*-value.

(iv) Based on your answers in parts (i) to (iii), will you reject or fail to reject the null hypothesis? Are the data statistically significant at level α?

(v) *Interpret* your conclusion in the context of the application.

Note: For degrees of freedom *d.f.* not in the Student's *t* table, use the closest *d.f.* that is *smaller*. In some situations, this choice of *d.f.* may increase the *P*-value a small amount and thereby produce a slightly more "conservative" answer.

1. *Testing and Estimating μ, σ Known* Let x be a random variable that represents micrograms of lead per liter of water (μg/L). An industrial plant discharges water into a creek. The Environmental Protection Agency has studied the discharged water and found x to have a normal distribution, with $\sigma = 0.7$ μg/L (Reference: *EPA Wetlands Case Studies*).

   (a) The industrial plant says that the population mean value of x is $\mu = 2.0$ μg/L. However, a random sample of $n = 10$ water samples showed that $\bar{x} = 2.56$ μg/L. Does this indicate that the lead concentration population mean is higher than the industrial plant claims? Use $\alpha = 1\%$.

   (b) Find a 95% confidence interval for μ using the sample data and the EPA value for σ.

   (c) How large a sample should be taken to be 95% confident that the sample mean $\bar{x}$ is within a margin of error $E = 0.2$ μg/L of the population mean?

2. *Testing and Estimating μ, σ Unknown* Carboxyhemoglobin is formed when hemoglobin is exposed to carbon monoxide. Heavy smokers tend to have a high percentage of carboxyhemoglobin in their blood (Reference: *Laboratory and Diagnostic Tests* by F. Fishbach). Let x be a random variable representing percentage of carboxyhemoglobin in the blood. For a person who is a regular heavy smoker, x has a distribution that is approximately normal. A random sample of $n = 12$ blood tests given to a heavy smoker gave the following results (percent carboxyhemoglobin in the blood).

9.1	9.5	10.2	9.8	11.3	12.2
11.6	10.3	8.9	9.7	13.4	9.9

   (a) Use a calculator to verify that $\bar{x} \approx 10.49$ and $s \approx 1.36$.

   (b) A long-term population mean $\mu = 10\%$ is considered a health risk. However, a long-term population mean above 10% is considered a clinical alert that the person may be asymptomatic. Do the data indicate that the population mean percentage is higher than 10% for this patient? Use $\alpha = 0.05$.

   (c) Use the given data to find a 99% confidence interval for μ for this patient.

3. *Testing and Estimating a Proportion p* Although older Americans are most afraid of crime, it is young people who are more likely to be the actual victims of crime. It seems that older people are more cautious about the people with whom they associate. A national

survey showed that 10% of all people ages 16–19 have been victims of crime (Reference: *Bureau of Justice Statistics*). At Jefferson High School, a random sample of $n = 68$ students (ages 16–19) showed that $r = 10$ had been victims of a crime.

(a) Do these data indicate that the population proportion of students in this school (ages 16–19) who have been victims of a crime is different (either way) from the national rate for this age group? Use $\alpha = 0.05$. Do you think the conditions $np > 5$ and $nq > 5$ are satisfied in this setting? Why is this important?

(b) Find a 95% confidence interval for the proportion of students in this school (ages 16–19) who have been victims of a crime.

(c) How large a sample size should be used to be 95% sure that the sample proportion $\hat{p}$ is within a margin of error $E = 0.05$ of the population proportion of all students in this school (ages 16–19) who have been victims of a crime? *Hint:* Use sample data $\hat{p}$ as a preliminary estimate for p.

4. *Testing Paired Differences* Phosphorous is a chemical that is found in many household cleaning products. Unfortunately, phosphorous also finds its way into surface water, where it can harm fish, plants, and other wildlife. Two methods of phosphorous reduction are being studied. At a random sample of 7 locations, both methods were used and the total phosphorous reduction (mg/L) was recorded (Reference: *Environmental Protection Agency Case Study 832-R-93-005).*

Site	1	2	3	4	5	6	7
Method I:	0.013	0.030	0.015	0.055	0.007	0.002	0.010
Method II:	0.014	0.058	0.017	0.039	0.017	0.001	0.013

Do these data indicate a difference (either way) in the average reduction of phosphorous between the two methods? Use $\alpha = 0.05$.

5. *Testing and Estimating $\mu_1 - \mu_2$, σ_1 and σ_2 Unknown* In the airline business, "on-time" flight arrival is important for connecting flights and general customer satisfaction. Is there a difference between summer and winter average on-time flight arrivals? Let x_1 be a random variable that represents percentage of on-time arrivals at major airports in the summer. Let x_2 be a random variable that represents percentage of on-time arrivals at major airports in the winter. A random sample of $n_1 = 16$ major airports showed that $\bar{x}_1 = 74.8\%$, with $s_1 = 5.2\%$. A random sample of $n_2 = 18$ major airports showed that $\bar{x}_2 = 70.1\%$, with $s_2 = 8.6\%$ (Reference: *Statistical Abstract of the United States*).

(a) Does this information indicate a difference (either way) in the population mean percentage of on-time arrivals for summer compared to winter? Use $\alpha = 0.05$.

(b) Find a 95% confidence interval for $\mu_1 - \mu_2$.

(c) What assumptions about the original populations have you made for the methods used?

6. *Testing and Estimating a Difference of Proportions $p_1 - p_2$* How often do you go out dancing? This question was asked by a professional survey group on behalf of the National Arts Survey. A random sample of $n_1 = 95$ single men showed that $r_1 = 23$ went out dancing occasionally. Another random sample of $n_2 = 92$ single women showed that $r_2 = 19$ went out dancing occasionally.

Corbis

(a) Do these data indicate that the proportion of single men who go out dancing occasionally is higher than the proportion of single women who do so? Use a 5% level of significance. List the assumptions you made in solving this problem. Do you think these assumptions are realistic?
(b) Compute a 90% confidence interval for the population difference of proportions $p_1 - p_2$ of single men and single women who occasionally go out dancing.

7. *Essay and Project* In Chapters 7 and 8 you studied, estimation and hypothesis testing.
(a) Write a brief essay in which you discuss using information from samples to infer information about populations. Be sure to include methods of estimation and hypothesis testing in your discussion. What two sampling distributions are used in estimation and hypothesis testing of population means, proportions, paired differences, differences of means, and differences of proportions? What are the criteria for determining the appropriate sampling distribution? What is the level of significance of a test? What is the *P*-value? How is the *P*-value related to the alternate hypothesis? How is the null hypothesis related to the sample test statistic? Explain.
(b) Suppose you want to study the length of time devoted to commercial breaks for two different types of television programs. Identify the types of programs you want to study (e.g., sitcoms, sports events, movies, news, children's programs, etc.). Write a brief outline for your study. Consider whether you will use paired data (such as same time slot on two different channels) or independent samples. Discuss how to obtain random samples. How large should the sample be for a specified margin of error? Describe the protocol you will follow to measure the times of the commercial breaks. Determine whether you are going to compare the average time devoted to commercials or the proportion of time devoted to commercials. What assumptions will you make regarding population distributions? What graphics might be appropriate? What methods of estimation will you use? What methods of testing will you use?

8. *Critical Thinking* Explain hypothesis testing to a friend, using the following scenario as a model. Describe the hypotheses, the sample statistic, the *P*-value, the meanings of type I and type II errors, and the level of significance. Discuss the significance of the results. Formulas are not required.

A team of research doctors designed a new knee surgery technique utilizing much smaller incisions than the standard method. They believe recovery times are shorter when the new method is used. Under the old method, the average recovery time for full use of the knee is 4.5 months. A random sample of 38 surgeries using the new method showed the average recovery time to be 3.6 months, with sample standard deviation of 1.7 months. The *P*-value for the test is 0.0011. The research team states that the results are statistically significant at the 1% level of significance.

9. *Linear Regression: Blood Glucose* Let x be a random variable that represents blood glucose level after a 12-hour fast. Let y be a random variable representing blood glucose level 1 hour after drinking sugar water (after the 12-hour fast). Units are in mg/10 ml. A random sample of eight adults gave the following information (Reference: *American Journal of Clinical Nutrition*, Vol. 19, pp. 345–351).

JGI/Daniel Grill/Blend Images/Corbis

$\Sigma x = 63.8$; $\Sigma x^2 = 521.56$; $\Sigma y = 90.7$;
$\Sigma y^2 = 1070.87$; $\Sigma xy = 739.65$

x	6.2	8.4	7.0	7.5	8.1	6.9	10.0	9.7
y	9.8	10.7	10.3	11.9	14.2	7.0	14.6	12.2

(a) Draw a scatter diagram for the data.
(b) Find the equation of the least-squares line and graph it on the scatter diagram.
(c) Find the sample correlation coefficient r and the sample coefficient of determination r^2. Explain the meaning of r^2 in the context of the application.
(d) If $x = 9.0$, use the least-squares line to predict y. Find an 80% confidence interval for your prediction.
(e) Use level of significance 1% and test the claim that the population correlation coefficient ρ is not zero. Interpret the results.
(f) Find an 85% confidence interval for the slope β of the population-based least-squares line. Explain its meaning in the context of the application.

10

Norman Rockwell

John Bryson/Time Life Pictures/Getty Images

"So what!"

—Anonymous

We have all heard the exclamation, "So what!" Philologists (people who study cultural linguistics) tell us that this expression is a shortened version of "So what is the difference!" They also tell us that there are similar popular or slang expressions about differences in all languages and cultures. Norman Rockwell (1894–1978) painted everyday people and situations. In this cover, "Girl with Black Eye," for the *Saturday Evening Post* (May 23, 1953), a young lady is about to have a conference with her school principal. So what! It is human nature to challenge the claim that something is better, worse, or just simply different. In this chapter, we will focus on this very human theme by studying a variety of topics regarding questions of whether or not differences exist between two population variances or among several population means.

CHI-SQUARE AND *F* DISTRIBUTIONS

How do you decide if random variables are dependent or independent? (SECTION 10.1)

How do you decide if different populations share the same proportions of specified characteristics? (SECTION 10.1)

How do you decide if two distributions are not only dependent, but actually the same distribution? (SECTION 10.2)

How do you compute confidence intervals and tests for σ? (SECTION 10.3)

How do you test two variances σ_1^2 and σ_2^2 ? (SECTION 10.4)

What is one-way ANOVA? Where is it used? (SECTION 10.5)

What about two-way ANOVA? Where is it used? (SECTION 10.6)

The Image Works

Mesa Verde National Park

FOCUS PROBLEM

Archaeology in Bandelier National Monument

Archaeologists at Washington State University did an extensive summer excavation at Burnt Mesa Pueblo in Bandelier National Monument. Their work is published in the book *Bandelier Archaeological Excavation Project: Summer 1990 Excavations at Burnt Mesa Pueblo and Casa del Rito,* edited by T. A. Kohler.

One question the archaeologists asked was: Is raw material used by prehistoric Indians for stone tool manufacture independent of the archaeological excavation site? Two different excavation sites at Burnt Mesa Pueblo gave the information in the table below.

Use a chi-square test with 5% level of significance to test the claim that raw material used for construction of stone tools and excavation site are independent. (See Problem 17 of Section 10.1.)

Stone Tool Construction Material, Burnt Mesa Pueblo

Material	Site A	Site B	Row Total
Basalt	731	584	1315
Obsidian	102	93	195
Pedernal chert	510	525	1035
Other	85	94	179
Column Total	1428	1296	2724

Robert Brenner/PhotoEdit

Archaeological excavation site

For online student resources, visit the Brase/Brase, *Understandable Statistics*, 11th edition web site at http://www.cengage.com/statistics/brase11e

PART I: INFERENCES USING THE CHI-SQUARE DISTRIBUTION

OVERVIEW OF THE CHI-SQUARE DISTRIBUTION

So far, we have used several probability distributions for hypothesis testing and confidence intervals, with the most frequently used being the normal distribution and the Student's t distribution. In this chapter, we will use two other probability distributions, namely, the chi-square distribution (where *chi* is pronounced like the first two letters in the word *kite*) and the F distribution. In Part I, we will see applications of the chi-square distribution, whereas in Part II, we will see some important applications of the F distribution.

Chi is a Greek letter denoted by the symbol χ, so chi-square is denoted by the symbol χ^2. Because the distribution is of chi-*square* values, the χ^2 values begin at 0 and then are all positive. The graph of the χ^2 distribution is not symmetrical, and like the Student's t distribution, it depends on the number of degrees of freedom. Figure 10-1 shows χ^2 distributions for several degrees of freedom (*d.f.*).

As the degrees of freedom increase, the graph of the chi-square distribution becomes more bell-like and begins to look more and more symmetric.

> The **mode (high point)** of a chi-square distribution with n degrees of freedom occurs over $n - 2$ (for $n \geq 3$).

Table 7 of Appendix II shows critical values of chi-square distributions for which a designated area falls to the *right* of the critical value. Table 10-1 gives an excerpt from Table 7. Notice that the row headers are degrees of freedom, and the column headers are areas in the *right* tail of the distribution. For instance, according to the table, for a χ^2 distribution with 3 degrees of freedom, the area occurring to the *right* of $\chi^2 = 0.072$ is 0.995. For a χ^2 distribution with 4 degrees of freedom, the area falling to the *right* of $\chi^2 = 13.28$ is 0.010.

In the next three sections, we will see how to apply the chi-square distribution to different applications.

FIGURE 10-1

The χ^2 Distribution

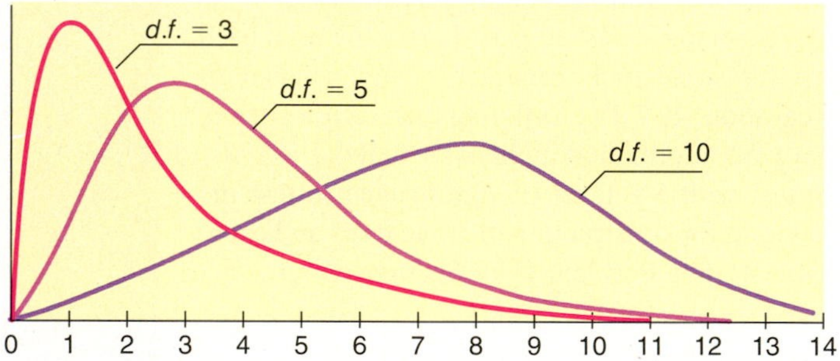

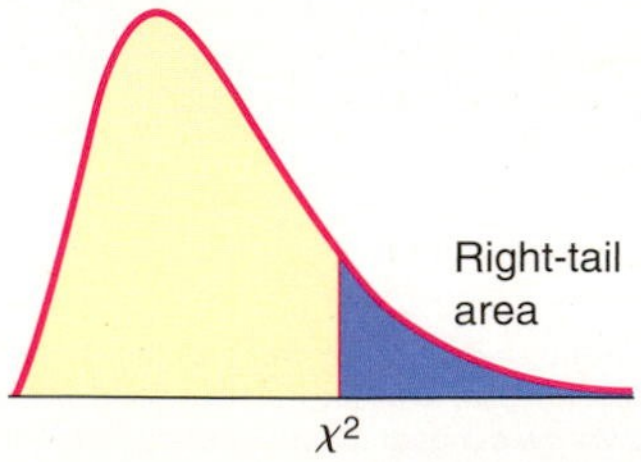

TABLE 10-1 Excerpt from Table 7 (Appendix II): The χ^2 Distribution

	Area of the Right Tail					
d.f.	0.995	0.990	0.975	. . .	0.010	0.005
⋮	⋮	⋮	⋮		⋮	⋮
3	0.072	0.115	0.216		11.34	12.84
4	0.207	0.297	0.484		13.28	14.86

SECTION **10.1**

Chi-Square: Tests of Independence and of Homogeneity

FOCUS POINTS

- Set up a test to investigate independence of random variables.
- Use contingency tables to compute the sample χ^2 statistic.
- Find or estimate the *P*-value of the sample χ^2 statistic and complete the test.
- Conduct a test of homogeneity of populations.

Innovative Machines Incorporated has developed two new letter arrangements for computer keyboards. The company wishes to see if there is any relationship between the arrangement of letters on the keyboard and the number of hours it takes a new typing student to learn to type at 20 words per minute. Or, from another point of view, is the time it takes a student to learn to type *independent* of the arrangement of the letters on a keyboard?

To answer questions of this type, we test the hypotheses

Hypotheses

H_0: Keyboard arrangement and learning times *are independent*.
H_1: Keyboard arrangement and learning times *are not independent*.

Test of independence
Chi-square distribution

In problems of this sort, we are testing the *independence* of two factors. The probability distribution we use to make the decision is the *chi-square distribution*. Recall from the overview of the chi-square distribution that *chi* is pronounced like the first two letters of the word *kite* and is a Greek letter denoted by the symbol χ. Thus, chi-square is denoted by χ^2.

Innovative Machines' first task is to gather data. Suppose the company took a random sample of 300 beginning typing students and randomly assigned them to learn to type on one of three keyboards. The learning times for this sample are shown in Table 10-2. These learning times are the observed frequencies O.

Observed frequency, O
Contingency table

Table 10-2 is called a *contingency table*. The *shaded boxes* that contain observed frequencies are called *cells*. The row and column totals are not considered to be cells. This contingency table is of size 3 × 3 (read "three-by-three") because there are three rows of cells and three columns. When giving the size of a contingency table, we always list the number of *rows first*.

To determine the **size** of a contingency table, count the number of rows containing data and the number of columns containing data. The size is

Number of rows × Number of columns

where the symbol "×" is read "by." The number of rows is always given first.

TABLE 10-2 **Keyboard versus Time to Learn to Type at 20 wpm**

Keyboard	21–40 h	41–60 h	61–80 h	Row Total
A	#1 25	#2 30	#3 25	80
B	#4 30	#5 71	#6 19	120
Standard	#7 35	#8 49	#9 16	100
Column Total	90	150	60	300 Sample size

GUIDED EXERCISE 1 | SIZE OF CONTINGENCY TABLE

Give the sizes of the contingency tables in Figures 10-2(a) and (b). Also, count the number of cells in each table. (Remember, each pink shaded box is a cell.)

(a) **FIGURE 10-2(a)** Contingency Table

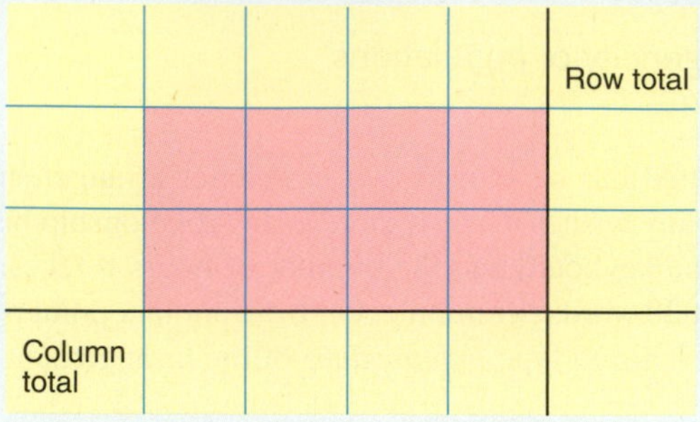

(b) **FIGURE 10-2(b)** Contingency Table

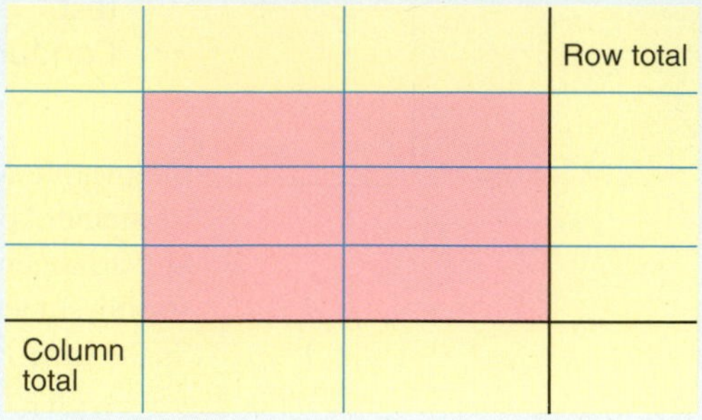

(a) ➡ There are two rows and four columns, so this is a 2×4 table. There are eight cells.

(b) ➡ Here we have three rows and two columns, so this is a 3×2 table with six cells.

Expected frequency, E

We are testing the null hypothesis that the keyboard arrangement and the time it takes a student to learn to type are *independent*. We use this hypothesis to determine the *expected frequency* of each cell.

For instance, to compute the expected frequency of cell 1 in Table 10-2, we observe that cell 1 consists of all the students in the sample who learned to type on keyboard A and who mastered the skill at the 20-words-per-minute level in 21 to 40 hours. By the assumption (null hypothesis) that the two events are independent, we use the multiplication law to obtain the probability that a student is in cell 1.

$$\begin{aligned} P(\text{cell } 1) &= P(\text{keyboard A } \textit{and} \text{ skill in } 21-40 \text{ h}) \\ &= P(\text{keyboard A}) \cdot P(\text{skill in } 21-40 \text{ h}) \end{aligned}$$

Because there are 300 students in the sample and 80 used keyboard A,

$$P(\text{keyboard A}) = \frac{80}{300}$$

Also, 90 of the 300 students learned to type in 21–40 hours, so

$$P(\text{skill in } 21-40 \text{ h}) = \frac{90}{300}$$

Using these two probabilities and the assumption of independence,

$$P(\text{keyboard A } \textit{and} \text{ skill in } 21-40 \text{ h}) = \frac{80}{300} \cdot \frac{90}{300}$$

Finally, because there are 300 students in the sample, we have the *expected frequency* E for cell 1.

$$\begin{aligned} E &= P(\text{student in cell } 1) \cdot (\text{no. of students in sample}) \\ &= \frac{80}{300} \cdot \frac{90}{300} \cdot 300 = \frac{80 \cdot 90}{300} = 24 \end{aligned}$$

We can repeat this process for each cell. However, the last step yields an easier formula for the expected frequency E.

Stephen Coburn/Shutterstock.com

Row total
Column total

Formula for expected frequency E

$$E = \frac{\text{(Row total)(Column total)}}{\text{Sample size}}$$

Note: If the expected value is not a whole number, do *not* round it to the nearest whole number.

Let's use this formula in Example 1 to find the expected frequency for cell 2.

EXAMPLE 1

EXPECTED FREQUENCY

Find the expected frequency for cell 2 of contingency Table 10-2.

SOLUTION: Cell 2 is in row 1 and column 2. The *row total* is 80, and the *column total* is 150. The size of the sample is still 300.

$$E = \frac{\text{(Row total)(Column total)}}{\text{Sample size}}$$

$$= \frac{(80)(150)}{300} = 40$$

GUIDED EXERCISE 2

EXPECTED FREQUENCY

Table 10-3 contains the *observed frequencies O* and *expected frequencies E* for the contingency table giving keyboard arrangement and number of hours it takes a student to learn to type at 20 words per minute. Fill in the missing expected frequencies.

TABLE 10-3 Complete Contingency Table of Keyboard Arrangement and Time to Learn to Type

Keyboard	21–40 h	41–60 h	61–80 h	Row Total
A	#1 $O = 25$ $E = 24$	#2 $O = 30$ $E = 40$	#3 $O = 25$ $E =$ ___	80
B	#4 $O = 30$ $E = 36$	#5 $O = 71$ $E =$ ___	#6 $O = 19$ $E =$ ___	120
Standard	#7 $O = 35$ $E =$ ___	#8 $O = 49$ $E = 50$	#9 $O = 16$ $E = 20$	100
Column Total	90	150	60	300 Sample Size

For cell 3, we have

$$E = \frac{(80)(60)}{300} = 16$$

For cell 5, we have

$$E = \frac{(120)(150)}{300} = 60$$

For cell 6, we have

$$E = \frac{(120)(60)}{300} = 24$$

For cell 7, we have

$$E = \frac{(100)(90)}{300} = 30$$

Computing the sample test statistic χ^2

Now we are ready to compute the sample statistic χ^2 for the typing students. The χ^2 value is a measure of the sum of the differences between *observed frequency O* and *expected frequency E* in each cell. These differences are listed in Table 10-4.

As you can see, if we sum the differences between the observed frequencies and the expected frequencies of the cells, we get the value zero. This total certainly does not

TABLE 10-4 **Differences Between Observed and Expected Frequencies**

Cell	Observed O	Expected E	Difference $(O - E)$
1	25	24	1
2	30	40	−10
3	25	16	9
4	30	36	−6
5	71	60	11
6	19	24	−5
7	35	30	5
8	49	50	−1
9	16	20	−4
			$\Sigma(O - E) = 0$

reflect the fact that there were differences between the observed and expected frequencies. To obtain a measure whose sum does reflect the magnitude of the differences, we square the differences and work with the quantities $(O - E)^2$. But instead of using the terms $(O - E)^2$, we use the values $(O - E)^2/E$.

We use this expression because a small difference between the observed and expected frequencies is not nearly as important when the expected frequency is large as it is when the expected frequency is small. For instance, for both cells 1 and 8, the squared difference $(O - E)^2$ is 1. However, this difference is more meaningful in cell 1, where the expected frequency is 24, than it is in cell 8, where the expected frequency is 50. When we divide the quantity $(O - E)^2$ by E, we take the size of the difference with respect to the size of the expected value. We use the sum of these values to form the sample statistic χ^2:

Sample test statistic χ^2

$$\chi^2 = \Sigma \frac{(O - E)^2}{E}$$

where the sum is over all cells in the contingency table.

COMMENT If you look up the word *irony* in a dictionary, you will find one of its meanings to be "the difference between actual (or observed) results and expected results." Because irony is so prevalent in much of our human experience, it is not surprising that statisticians have incorporated a related chi-square distribution into their work.

GUIDED EXERCISE 3 SAMPLE χ^2

(a) Complete Table 10-5.

TABLE 10-5 **Data of Table 10-4**

Cell	O	E	$O - E$	$(O - E)^2$	$(O - E)^2/E$
1	25	24	1	1	0.04
2	30	40	−10	100	2.50
3	25	16	9	81	5.06
4	30	36	−6	36	1.00
5	71	60	11	121	2.02
6	19	24	−5	25	1.04
7	35	30	5	25	0.83
8	49	50	____	____	____
9	16	20	____	____	____

$$\Sigma \frac{(O - E)^2}{E} = \_\_\_\_$$

The last two rows of Table 10-5 are

Cell	O	E	$O - E$	$(O - E)^2$	$(O - E)^2/E$
8	49	50	−1	1	0.02
9	16	20	−4	16	0.80

$$\Sigma \frac{(O - E)^2}{E} = \text{total of last column} = 13.31$$

Continued

GUIDED EXERCISE 3 *continued*

(b) Compute the statistic χ^2 for this sample. ➡ Since $\chi^2 = \Sigma\frac{(O-E)^2}{E}$, then $\chi^2 = 13.31$.

Notice that when the observed frequency and the expected frequency are very close, the quantity $(O - E)^2$ is close to zero, and so the statistic χ^2 is near zero. As the difference increases, the statistic χ^2 also increases. To determine how large the sample statistic can be before we must reject the null hypothesis of independence, we find the *P*-value of the statistic in the chi-square distribution, Table 7 of Appendix II, and compare it to the specified level of significance α. The *P*-value depends on the number of degrees of freedom. To test independence, the degrees of freedom *d.f.* are determined by the following formula.

Degrees of freedom for test of independence

DEGREES OF FREEDOM FOR TEST OF INDEPENDENCE

Degrees of freedom = (Number of rows − 1) · (Number of columns − 1)

or $d.f. = (R - 1)(C - 1)$

where R = number of cell rows

C = number of cell columns

GUIDED EXERCISE 4 Degrees of Freedom

Determine the number of degrees of freedom in the example of keyboard arrangements (see Table 10-2). Recall that the contingency table had three rows and three columns.

➡ $d.f. = (R - 1)(C - 1)$
$= (3 - 1)(3 - 1) = (2)(2) = 4$

Finding the *P*-value for tests of independence

To test the hypothesis that the letter arrangement on a keyboard and the time it takes to learn to type at 20 words per minute are independent at the $\alpha = 0.05$ level of significance, we estimate the *P*-value shown in Figure 10-3 below for the sample test statistic $\chi^2 = 13.31$ (calculated in Guided Exercise 3). We then compare the *P*-value to the specified level of significance α.

Type of test: right-tailed

For tests of independence, we always use a *right-tailed* test on the chi-square distribution. This is because we are testing to see if the χ^2 measure of the difference between the observed and expected frequencies is too large to be due to chance alone.

In Guided Exercise 4, we found that the degrees of freedom for the example of keyboard arrangements is 4. From Table 7 of Appendix II, in the row headed by $d.f. = 4$, we see that the sample $\chi^2 = 13.31$ falls between the entries 13.28 and 14.86.

FIGURE 10-3

P-value

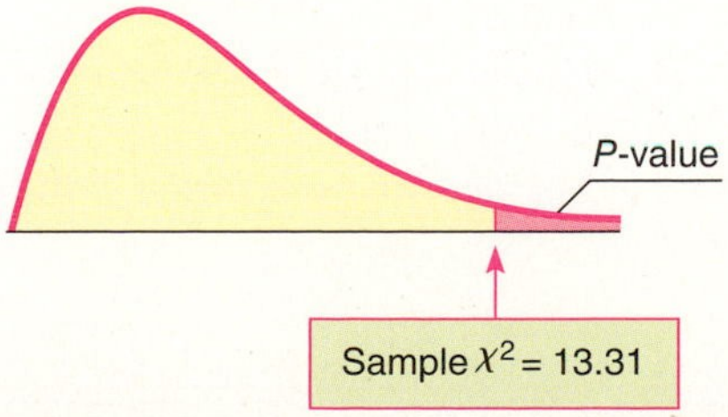

Right-tail Area	0.010	0.005
$d.f. = 4$	13.28	14.86
	↑ Sample $\chi^2 = 13.31$	

The corresponding P-value falls between 0.005 and 0.010. From technology, we get P-value $\approx$ 0.0098.

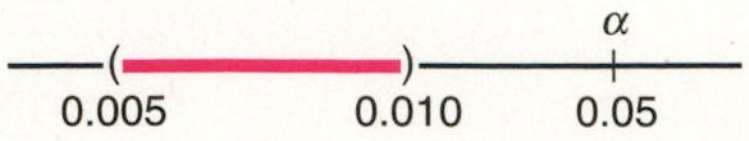

Since the P-value is less than the level of significance $\alpha = 0.05$, we reject the null hypothesis of independence and conclude that keyboard arrangement and learning time are *not* independent.

Tests of independence for two statistical variables involve a number of steps. A summary of the procedure follows.

PROCEDURE

HOW TO TEST FOR INDEPENDENCE OF TWO STATISTICAL VARIABLES

Setup

Construct a contingency table in which the rows represent one statistical variable and the columns represent the other. Obtain a random sample of observations, which are assigned to the cells described by the rows and columns. These assignments are called the **observed values *O*** from the sample.

Procedure

1. Set the level of significance α and use the hypotheses

   H_0: The variables are independent.
   H_1: The variables are not independent.

2. For each cell, compute the **expected frequency *E*** (do not round but give as a decimal number).

$$E = \frac{\text{(Row total)(Column total)}}{\text{Sample size}}$$

Requirement

You need a sample size large enough so that, for each cell, $E \geq 5$.

Now each cell has two numbers, the observed frequency O from the sample and the expected frequency E.

Next, compute the sample *chi-square test statistic*

$$\chi^2 = \Sigma \frac{(O - E)^2}{E} \quad \text{with degrees of freedom } d.f. = (R - 1)(C - 1)$$

where the sum is over all cells in the contingency table and

R = number of rows in contingency table

C = number of columns in contingency table

3. Use the chi-square distribution (Table 7 of Appendix II) and a *right-tailed test* to find (or estimate) the P-value corresponding to the test statistic.
4. *Conclude* the test. If P-value $\leq \alpha$, then reject H_0. If P-value $> \alpha$, then do not reject H_0.
5. *Interpret your conclusion* in the context of the application.

GUIDED EXERCISE 5 | TESTING INDEPENDENCE

Super Vending Machines Company is to install soda pop machines in elementary schools and high schools. The market analysts wish to know if flavor preference and school level are independent. A random sample of 200 students was taken. Their school level and soda pop preferences are given in Table 10-6. Is independence indicated at the $\alpha = 0.01$ level of significance?

STEP 1: State the null and alternate hypotheses.

→ H_0: School level and soda pop preference are independent.

H_1: School level and soda pop preference are not independent.

STEP 2:

(a) Complete the contingency Table 10-6 by filling in the required expected frequencies.

TABLE 10-6 **School Level and Soda Pop Preference**

Soda Pop	High School	Elementary School	Row Total
Kula Kola	$O = 33$ #1 $E = 36$	$O = 57$ #2 $E = 54$	90
Mountain Mist	$O = 30$ #3 $E = 20$	$O = 20$ #4 $E = 30$	50
Jungle Grape	$O = 5$ #5 $E =$ ____	$O = 35$ #6 $E =$ ____	40
Diet Pop	$O = 12$ #7 $E =$ ____	$O = 8$ #8 $E =$ ____	20
Column Total	80	120	200 Sample Size

→ The expected frequency

for cell 5 is $\dfrac{(40)(80)}{200} = 16$

for cell 6 is $\dfrac{(40)(120)}{200} = 24$

for cell 7 is $\dfrac{(20)(80)}{200} = 8$

for cell 8 is $\dfrac{(20)(120)}{200} = 12$

Note: In this example, the expected frequencies are all whole numbers. If the expected frequency has a decimal part, such as 8.45, do *not* round the value to the nearest whole number; rather, give the expected frequency as the decimal number.

(b) Fill in Table 10-7 and use the table to find the sample statistic χ^2.

TABLE 10-7 **Computational Table for χ^2**

Cell	O	E	$O - E$	$(O - E)^2$	$(O - E)^2/E$
1	33	36	−3	9	0.25
2	57	54	3	9	0.17
3	30	20	10	100	5.00
4	20	30	−10	100	3.33
5	5	16	−11	121	7.56
6	35	24	11	____	____
7	12	8	____	____	____
8	8	12	____	____	____

→ The last three rows of Table 10-7 should read as follows:

Cell	O	E	$O - E$	$(O - E)^2$	$(O - E)^2/E$
6	35	24	11	121	5.04
7	12	8	4	16	2.00
8	8	12	−4	16	1.33

$\chi^2 =$ total of last column

$$= \Sigma \frac{(O - E)^2}{E} = 24.68$$

(c) What is the size of the contingency table? Use the number of rows and the number of columns to determine the degrees of freedom.

→ The contingency table is of size 4×2. Since there are four rows and two columns,
$d.f. = (4 - 1)(2 - 1) = 3$

Continued

GUIDED EXERCISE 5 *continued*

STEP 3: Use Table 7 of Appendix II to estimate the *P*-value of the sample statistic $\chi^2 = 24.68$ with $d.f. = 3$.

Right-tail Area	**0.005**
$d.f. = 3$	12.84
	↑ Sample $\chi^2 = 24.68$

As the χ^2 values increase, the area to the right decreases, so *P*-value < 0.005

STEP 4: Conclude the test by comparing the *P*-value of the sample statistic to the level of significance $\alpha = 0.01$.

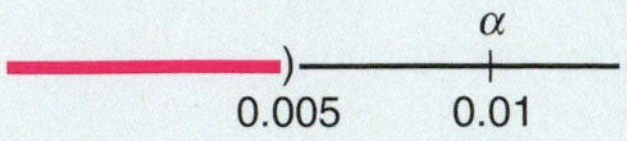

Since the *P*-value is less than α, we reject the null hypothesis of independence. Technology gives *P*-value ≈ 0.00002.

STEP 5: ***Interpret*** the test result in the context of the application.

At the 1% level of significance, we conclude that school level and soda pop preference are dependent.

TECH NOTES The TI-84Plus/TI-83Plus/TI-*n*spire calculators, Excel 2010, and Minitab all support chi-square tests of independence. In each case, the observed data are entered in the format of the contingency table.

TI-84Plus/TI-83Plus/TI-*n*spire (with TI-84Plus keypad) Enter the observed data into a matrix. Set the dimension of matrix [**B**] to match that of the matrix of observed values. Expected values will be placed in matrix [**B**]. Press **STAT, TESTS,** and select option **C:**χ^2**-Test.** The output gives the sample χ^2 with the *P*-value.

Excel 2010 Enter the table of observed values. Use the formulas of this section to compute the expected values. Enter the corresponding table of expected values. Finally, click **insert function** (f_x). Select **Statistical** for the category and **CHI-SQ.TEST** for the function. Excel returns the *P*-value of the sample χ^2 value.

Minitab Enter the contingency table of observed values. Use the menu selection **Stat ➤ Tables ➤ Chi-Square Test.** The output shows the contingency table with expected values and the sample χ^2 with *P*-value.

TESTS OF HOMOGENEITY

We've seen how to use contingency tables and the chi-square distribution to test for independence of two random variables. The same process enables us to determine whether several populations share the same proportions of distinct categories. Such a test is called a *test of homogeneity*.

According to the dictionary, among the definitions of the word *homogeneous* are "of the same structure" and "composed of similar parts." In statistical jargon, this translates as a test of homogeneity to see if two or more populations share specified characteristics in the same proportions.

Test of homogeneity

A **test of homogeneity** tests the claim that *different populations* share the *same proportions* of specified characteristics.

The computational processes for conducting tests of independence and tests of homogeneity are the same. However, there are two main differences in the initial setup of the two types of tests, namely, the sampling method and the hypotheses.

Tests of independence compared to tests of homogeneity

1. **Sampling method**
   For tests of independence, we use one random sample and observe how the sample members are distributed among distinct categories.
   For tests of homogeneity, we take random samples from each different population and see how members of each population are distributed over distinct categories.
2. **Hypotheses**
   For tests of independence,

   H_0: The variables are independent.
   H_1: The variables are not independent.

   For tests of homogeneity,

   H_0: Each population shares respective characteristics in the same proportion.
   H_1: Some populations have different proportions of respective characteristics.

EXAMPLE 2

TEST OF HOMOGENEITY

Pets—who can resist a cute kitten or puppy? Tim is doing a research project involving pet preferences among students at his college. He took random samples of 300 female and 250 male students. Each sample member responded to the survey question "If you could own only one pet, what kind would you choose?" The possible responses were: "dog," "cat," "other pet," "no pet." The results of the study follow.

Image Source/Getty Images

Pet Preference

Gender	Dog	Cat	Other Pet	No Pet
Female	120	132	18	30
Male	135	70	20	25

Does the same proportion of males as females prefer each type of pet? Use a 1% level of significance.

We'll answer this question in several steps.

(a) First make a cluster bar graph showing the percentages of females and the percentages of males favoring each category of pet. From the graph, does it appear that the proportions are the same for males and females?

SOLUTION: The cluster graph shown in Figure 10-4 was created using Minitab. Looking at the graph, it appears that there are differences in the proportions

FIGURE 10-4
Pet Preference by Gender

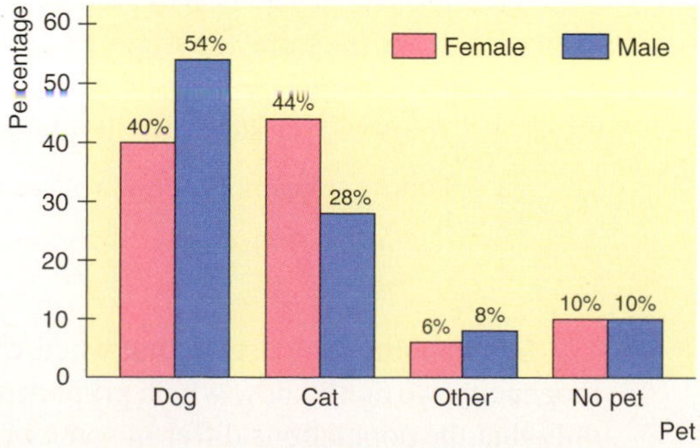

of females and males preferring each type of pet. However, let's conduct a statistical test to verify our visual impression.

(b) Is it appropriate to use a test of homogeneity?

SOLUTION: Yes, since there are separate random samples for each designated population, male and female. We also are interested in whether each population shares the same proportion of members favoring each category of pet.

(c) State the hypotheses and conclude the test by using the Minitab printout.

H_0: The proportions of females and males naming each pet preference are the same.

H_1: The proportions of females and males naming each pet preference are not the same.

```
Chi-Square Test: Dog, Cat, Other, No Pet
Expected counts are printed below observed counts
Chi-Square contributions are printed below expected counts
              Dog           Cat         Other        No Pet       Total
1             120           132            18            30         300
           139.09        110.18         20.73         30.00
            2.620         4.320         0.359         0.000

2             135            70            20            25         250
           115.91         91.82         17.27         25.00
            3.144         5.185         0.431         0.000

Total         255           202            38            55         550
Chi-Sq = 16.059, DF = 3, P-Value = 0.001
```

Since the *P*-value is less than α, we reject H_0 at the 1% level of significance.

(d) *Interpret* the results.

SOLUTION: It appears from the sample data that male and female students at Tim's college have different preferences when it comes to selecting a pet.

PROCEDURE

HOW TO TEST FOR HOMOGENEITY OF POPULATIONS

Setup

Obtain random samples from each of the populations. For each population, determine the number of members that share a distinct specified characteristic. Make a contingency table with the different populations as the rows (or columns) and the characteristics as the columns (or rows). The values recorded in the cells of the table are the **observed values *O*** taken from the samples.

Procedure

1. Set the level of significance and use the hypotheses

   H_0: The proportion of each population sharing specified characteristics is the same for all populations.

   H_1: The proportion of each population sharing specified characteristics is not the same for all populations.

2. Follow steps 2–5 of the procedure used to test for independence.

It is important to observe that when we reject the null hypothesis in a test of homogeneity, we don't know which proportions differ among the populations. We know only that the populations differ in some of the proportions sharing a characteristic.

MULTINOMIAL EXPERIMENTS (OPTIONAL READING)

Here are some observations that may be considered "brain teasers." In Chapters 6, 7, and 8, you studied normal approximations to binomial experiments. This concept resulted in some important statistical applications. Is it possible to extend this idea and obtain even more applications? Well, read on!

Consider a *binomial experiment* with n trials. The probability of success on each trial is p, and the probability of failure is $q = 1 - p$. If r is the number of successes out of n trials, then, from Chapter 5, you know that

$$P(r) = \frac{n!}{r!(n-r)!}p^r q^{n-r}$$

The binomial setting has just two outcomes: success or failure. What if you want to consider more than just two outcomes on each trial (for instance, the outcomes shown in a contingency table)? Well, you need a new statistical tool.

Multinomial experiments

Consider a *multinomial experiment.* This means that

1. The trials are independent and repeated under identical conditions.
2. The outcome on each trial falls into exactly one of $k \geq 2$ categories or cells.
3. The probability that the outcome of a single trial will fall into the ith category or cell is p_i (where $i = 1, 2, \ldots, k$) and remains the same for each trial. Furthermore, $p_1 + p_2 + \cdots + p_k = 1$.
4. Let r_i be a random variable that represents the number of trials in which the outcome falls into category or cell i. If you have n trials, then $r_1 + r_2 + \cdots + r_k = n$. The multinomial probability distribution is then

$$P(r_1, r_2, \ldots r_k) = \frac{n!}{r_1! r_2! \ldots r_k!} p_1^{r_1} p_2^{r_2} \cdots p_k^{r_k}$$

How are the multinomial distribution and the binomial distribution related? For the special case $k = 2$, we use the notation $r_1 = r$, $r_2 = n - r$, $p_1 = p$, and $p_2 = q$. In this special case, the multinomial distribution becomes the binomial distribution.

There are two important tests regarding the cell probabilities of a multinomial distribution.

I. **Test of Independence** (Section 10.1)
In this test, the null hypothesis of independence claims that each cell probability p_i will equal the product of its respective row and column probabilities. The alternate hypothesis claims that this is not so.

II. **Goodness-of-Fit Test** (Section 10.2)
In this test, the null hypothesis claims that each category or cell probability p_i will equal a prespecified value. The alternate hypothesis claims that this is not so.

So why don't we use the multinomial probability distribution in Sections 10.1 and 10.2? The reason is that the exact calculation of probabilities associated with type I errors using the multinomial distribution is very tedious and cumbersome. Fortunately, the British statistician Karl Pearson discovered that the chi-square distribution can be used for this purpose, provided the expected value of each cell or category is at least 5.

It is Pearson's chi-square methods that are presented in Sections 10.1 and 10.2. In a sense, you have seen a similar application to statistical tests in Section 8.3, where you used the normal approximation to the binomial when np, the expected number of successes, and nq, the expected number of failures, were both at least 5.

VIEWPOINT Loyalty! Going, Going, Gone!

Was there a time in the past when people worked for the same company all their lives, regularly purchased the same brand names, always voted for candidates from the same political party, and loyally cheered for the same sports team? One way to look at this question is to consider tests of statistical independence. Is customer loyalty independent of company profits? Can a company maintain its productivity independent of loyal workers? Can politicians do whatever they please independent of the voters back home? Americans may be ready to act on a pent-up desire to restore a sense of loyalty in their lives. For more information, see American Demographics, *Vol. 19, No. 9.*

SECTION 10.1 PROBLEMS

1. *Statistical Literacy* In general, are chi-square distributions symmetric or skewed? If skewed, are they skewed right or left?

2. *Statistical Literacy* For chi-square distributions, as the number of degrees of freedom increases, does any skewness increase or decrease? Do chi-square distributions become more symmetric (and normal) as the number of degrees of freedom becomes larger and larger?

3. *Statistical Literacy* For chi-square tests of independence and of homogeneity, do we use a right-tailed, left-tailed, or two-tailed test?

4. *Critical Thinking* In general, how do the hypotheses for chi-square tests of independence differ from those for chi-square tests of homogeneity? Explain.

5. *Critical Thinking* Zane is interested in the proportion of people who recycle each of three distinct products: paper, plastic, electronics. He wants to test the hypothesis that the proportion of people recycling each type of product differs by age group: 12–18 years old, 19–30 years old, 31–40 years old, over 40 years old. Describe the sampling method appropriate for a test of homogeneity regarding recycled products and age.

6. *Critical Thinking* Charlotte is doing a study on fraud and identity theft based both on source (checks, credit cards, debit cards, online banking/finance sites, other) and on gender of the victim. Describe the sampling method appropriate for a test of independence regarding source of fraud and gender.

7. *Interpretation: Test of Homogeneity* Consider Zane's study regarding products recycled and age group (see Problem 5). Suppose he found a sample $\chi^2 = 16.83$.
   (a) How many degrees of freedom are used? Recall that there were 4 age groups and 3 products specified. Approximate the *P*-value and conclude the test at the 1% level of significance. Does it appear that the proportion of people who recycle each of the specified products differ by age group? Explain.
   (b) From this study, can Zane identify how the different age groups differ regarding the proportion of those recycling the specified product? Explain.

8. *Interpretation: Test of Independence* Consider Charlotte's study of source of fraud/identity theft and gender (see Problem 6). She computed sample $\chi^2 = 10.2$.
   (a) How many degrees of freedom are used? Recall that there were 5 sources of fraud/identity theft and, of course, 2 genders. Approximate the *P*-value and conclude the test at the 5% level of significance. Would it seem that gender and source of fraud/identity theft are independent?
   (b) From this study, can Charlotte identify which source of fraud/identity theft is dependent with respect to gender? Explain.

For Problems 9–19, please provide the following information.

(a) What is the level of significance? State the null and alternate hypotheses.

(b) Find the value of the chi-square statistic for the sample. Are all the expected frequencies greater than 5? What sampling distribution will you use? What are the degrees of freedom?

(c) Find or estimate the P-value of the sample test statistic.

(d) Based on your answers in parts (a) to (c), will you reject or fail to reject the null hypothesis of independence?

(e) *Interpret* your conclusion in the context of the application.

Use the expected values E to the hundredths place.

9. *Psychology: Myers–Briggs* The following table shows the Myers–Briggs personality preferences for a random sample of 406 people in the listed professions (*Atlas of Type Tables* by Macdaid, McCaulley, and Kainz). E refers to extroverted and I refers to introverted.

	Personality Preference Type		
Occupation	**E**	**I**	**Row Total**
Clergy (all denominations)	62	45	107
M.D.	68	94	162
Lawyer	56	81	137
Column Total	186	220	406

Use the chi-square test to determine if the listed occupations and personality preferences are independent at the 0.05 level of significance.

10. *Psychology: Myers–Briggs* The following table shows the Myers–Briggs personality preferences for a random sample of 519 people in the listed professions (*Atlas of Type Tables* by Macdaid, McCaulley, and Kainz). T refers to thinking and F refers to feeling.

	Personality Preference Type		
Occupation	**T**	**F**	**Row Total**
Clergy (all denominations)	57	91	148
M.D.	77	82	159
Lawyer	118	94	212
Column Total	252	267	519

Use the chi-square test to determine if the listed occupations and personality preferences are independent at the 0.01 level of significance.

11. *Archaeology: Pottery* The following table shows site type and type of pottery for a random sample of 628 sherds at a location in Sand Canyon Archaeological Project, Colorado (*The Sand Canyon Archaeological Project,* edited by Lipe).

	Pottery Type			
Site Type	**Mesa Verde Black-on-White**	**McElmo Black-on-White**	**Mancos Black-on-White**	**Row Total Black-on-White**
Mesa Top	75	61	53	189
Cliff-Talus	81	70	62	213
Canyon Bench	92	68	66	226
Column Total	248	199	181	628

Use a chi-square test to determine if site type and pottery type are independent at the 0.01 level of significance.

12. *Archaeology: Pottery* The following table shows ceremonial ranking and type of pottery sherd for a random sample of 434 sherds at a location in the Sand Canyon Archaeological Project, Colorado (*The Architecture of Social Integration in Prehistoric Pueblos,* edited by Lipe and Hegmon).

Ceremonial Ranking	Cooking Jar Sherds	Decorated Jar Sherds (Noncooking)	Row Total
A	86	49	135
B	92	53	145
C	79	75	154
Column Total	257	177	434

Use a chi-square test to determine if ceremonial ranking and pottery type are independent at the 0.05 level of significance.

13. *Ecology: Buffalo* The following table shows age distribution and location of a random sample of 166 buffalo in Yellowstone National Park (based on information from *The Bison of Yellowstone National Park,* National Park Service Scientific Monograph Series).

Age	Lamar District	Nez Perce District	Firehole District	Row Total
Calf	13	13	15	41
Yearling	10	11	12	33
Adult	34	28	30	92
Column Total	57	52	57	166

Use a chi-square test to determine if age distribution and location are independent at the 0.05 level of significance.

14. *Psychology: Myers–Briggs* The following table shows the Myers–Briggs personality preference and area of study for a random sample of 519 college students (*Applications of the Myers–Briggs Type Indicator in Higher Education,* edited by Provost and Anchors). In the table, IN refers to introvert, intuitive; EN refers to extrovert, intuitive; IS refers to introvert, sensing; and ES refers to extrovert, sensing.

Myers–Briggs Preference	Arts & Science	Business	Allied Health	Row Total
IN	64	15	17	96
EN	82	42	30	154
IS	68	35	12	115
ES	75	42	37	154
Column Total	289	134	96	519

Use a chi-square test to determine if Myers–Briggs preference type is independent of area of study at the 0.05 level of significance.

15. *Sociology: Movie Preference* Mr. Acosta, a sociologist, is doing a study to see if there is a relationship between the age of a young adult (18 to 35 years old) and the type of movie preferred. A random sample of 93 young adults

revealed the following data. Test whether age and type of movie preferred are independent at the 0.05 level.

Movie	Person's Age 18–23 yr	24–29 yr	30–35 yr	Row Total
Drama	8	15	11	34
Science fiction	12	10	8	30
Comedy	9	8	12	29
Column Total	29	33	31	93

16. *Sociology: Ethnic Groups* After a large fund drive to help the Boston City Library, the following information was obtained from a random sample of contributors to the library fund. Using a 1% level of significance, test the claim that the amount contributed to the library fund is independent of ethnic group.

Ethnic Group	Number of People Making Contribution					Row Total
	$1–50	$51–100	$101–150	$151–200	Over $200	
A	83	62	53	35	18	251
B	94	77	48	25	20	264
C	78	65	51	40	32	266
D	105	89	63	54	29	340
Column Total	360	293	215	154	99	1121

17. *Focus Problem: Archaeology* The Focus Problem at the beginning of the chapter refers to excavations at Burnt Mesa Pueblo in Bandelier National Monument. One question the archaeologists asked was: Is raw material used by prehistoric Indians for stone tool manufacture independent of the archaeological excavation site? Two different excavation sites at Burnt Mesa Pueblo gave the information in the following table. Use a chi-square test with 5% level of significance to test the claim that raw material used for construction of stone tools and excavation site are independent.

Material	Stone Tool Construction Material, Burnt Mesa Pueblo Site A	Site B	Row Total
Basalt	731	584	1315
Obsidian	102	93	195
Pedernal chert	510	525	1035
Other	85	94	179
Column Total	1428	1296	2724

18. *Political Affiliation: Spending* Two random samples were drawn from members of the U.S. Congress. One sample was taken from members who are Democrats and the other from members who are Republicans. For each sample, the number of dollars spent on federal projects in each congressperson's home district was recorded.
    (i) Make a cluster bar graph showing the percentages of Congress members from each party who spent each designated amount in their respective home districts.
    (ii) Use a 1% level of significance to test whether congressional members of each political party spent designated amounts in the same proportions.

Party	Dollars Spent on Federal Projects in Home Districts: Less than 5 Billion	5 to 10 Billion	More than 10 Billion	Row Total
Democratic	8	15	22	45
Republican	12	19	16	47
Column Total	20	34	38	92

19. *Sociology: Methods of Communication* Random samples of people ages 15–24 and 25–34 were asked about their preferred method of (remote) communication with friends. The respondents were asked to select one of the methods from the following list: cell phone, instant message, e-mail, other.
 (i) Make a cluster bar graph showing the percentages in each age group who selected each method.
 (ii) Test whether the two populations share the same proportions of preferences for each type of communication method. Use $\alpha = 0.05$.

Age	Preferred Communication Method: Cell Phone	Instant Message	E-mail	Other	Row Total
15–24	48	40	5	7	100
25–34	41	30	15	14	100
Column Total	89	70	20	21	200

SECTION **10.2**

Chi-Square: Goodness of Fit

FOCUS POINTS

- Set up a test to investigate how well a sample distribution fits a given distribution.
- Use observed and expected frequencies to compute the sample χ^2 statistic.
- Find or estimate the *P*-value and complete the test.

Last year, the labor union bargaining agents listed five categories and asked each employee to mark the *one* most important to her or him. The categories and corresponding percentages of favorable responses are shown in Table 10-8. The bargaining agents need to determine if the *current* distribution of responses "fits" last year's distribution or if it is different.

In questions of this type, we are asking whether a population follows a specified distribution. In other words, we are testing the hypotheses

TABLE 10-8 **Bargaining Categories (last year)**

Category	Percentage of Favorable Responses
Vacation time	4%
Salary	65%
Safety regulations	13%
Health and retirement benefits	12%
Overtime policy and pay	6%

Hypotheses

H_0: The population fits the given distribution.

H_1: The population has a different distribution.

Computing sample χ^2

We use the chi-square distribution to test "goodness-of-fit" hypotheses. Just as with tests of independence, we compute the sample statistic:

$$\chi^2 = \Sigma \frac{(O - E)^2}{E} \text{ with degrees of freedom} = k - 1$$

where $E =$ expected frequency

$O =$ observed frequency

$\frac{(O - E)^2}{E}$ is summed for each category in the distribution

$k =$ number of categories in the distribution

Next we use the chi-square distribution table (Table 7, Appendix II) to estimate the P-value of the sample χ^2 statistic. Finally, we compare the P-value to the level of significance α and conclude the test.

Goodness-of-fit test

In the case of a *goodness-of-fit test,* we use the null hypothesis to compute the expected values for the categories. Let's look at the bargaining category problem to see how this is done.

In the bargaining category problem, the two hypotheses are

H_0: The present distribution of responses is the same as last year's.

H_1: The present distribution of responses is different.

The null hypothesis tells us that the *expected frequencies* of the present response distribution should follow the percentages indicated in last year's survey. To test this hypothesis, a random sample of 500 employees was taken. If the null hypothesis is true, then there should be 4%, or 20 responses, out of the 500 rating vacation time as the most important bargaining issue. Table 10-9 gives the other expected values and all the information necessary to compute the sample statistic χ^2. We see that the sample statistic is

$$\chi^2 = \Sigma \frac{(O - E)^2}{E} = 14.15$$

Steve Skjold/PhotoEdit

Larger values of the sample statistic χ^2 indicate greater differences between the proposed distribution and the distribution followed by the sample. The larger the χ^2 statistic, the stronger the evidence to reject the null hypothesis that the population distribution fits the given distribution. Consequently, goodness-of-fit tests are always *right-tailed* tests.

TABLE 10-9 Observed and Expected Frequencies for Bargaining Categories

Category	O	E	$(O - E)^2$	$(O - E)^2/E$
Vacation time	30	4% of 500 = 20	100	5.00
Salary	290	65% of 500 = 325	1225	3.77
Safety	70	13% of 500 = 65	25	0.38
Health and retirement	70	12% of 500 = 60	100	1.67
Overtime	40	6% of 500 = 30	100	3.33
	$\Sigma O = 500$	$\Sigma E = 500$	$\Sigma \frac{(O - E)^2}{E} = 14.15$	

Type of test: right-tailed

For *goodness-of-fit tests*, we use a *right-tailed* test on the chi-square distribution. This is because we are testing to see if the χ^2 measure of the difference between the observed and expected frequencies is too large to be due to chance alone.

Degrees of freedom

To test the hypothesis that the present distribution of responses to bargaining categories is the same as last year's, we use the chi-square distribution (Table 7 of Appendix II) to estimate the *P*-value of the sample statistic $\chi^2 = 14.15$. To estimate the *P*-value, we need to know the number of *degrees of freedom*. In the case of a goodness-of-fit test, the degrees of freedom are found by the following formula.

Degrees of freedom for goodness-of-fit test

$$d.f. = k - 1$$

where k = number of categories

Notice that when we compute the expected values E, we must use the null hypothesis to compute all but the last one. To compute the last one, we can subtract the previous expected values from the sample size. For instance, for the bargaining issues, we could have found the number of responses for overtime policy by adding the other expected values and subtracting that sum from the sample size 500. We would again get an expected value of 30 responses. The degrees of freedom, then, is the number of E values that *must* be computed by using the null hypothesis.

For the bargaining issues, we have

$$d.f. = 5 - 1 = 4$$

where $k = 5$ is the number of categories.

***P*-value**

We now have the tools necessary to use Table 7 of Appendix II to estimate the *P*-value of $\chi^2 = 14.15$. Figure 10-5 shows the *P*-value. In Table 7, we use the row headed by $d.f. = 4$. We see that $\chi^2 = 14.15$ falls between the entries 13.28 and 14.86. Therefore, the *P*-value falls between the corresponding right-tail areas 0.005 and 0.010. Technology gives the *P*-value ≈ 0.0068.

To test the hypothesis that the distribution of responses to bargaining issues is the same as last year's at the 1% level of significance, we compare the *P*-value of the statistic to $\alpha = 0.01$.

We see that the *P*-value is less than α, so we reject the null hypothesis that the distribution of responses to bargaining issues is the same as last year's.

Interpretation At the 1% level of significance, we can say that the evidence supports the conclusion that this year's responses to the issues are different from last year's.

Goodness-of-fit tests involve several steps that can be summarized as follows.

FIGURE 10-5

P-value

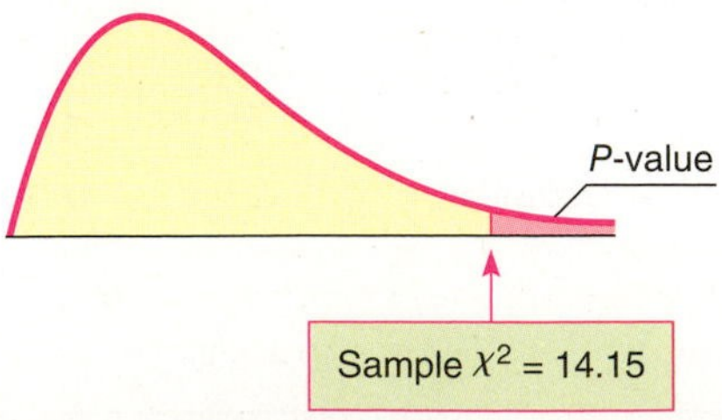

Right-tail Area	0.010	0.005
$d.f. = 4$	13.28	14.86
	↑ Sample $\chi^2 = 14.15$	

PROCEDURE

HOW TO TEST FOR GOODNESS OF FIT

Setup

First, each member of a population needs to be classified into exactly one of several different categories. Next, you need a specific (theoretical) distribution that assigns a fixed probability (or percentage) that a member of the population will fall into one of the categories. You then need a random sample size n from the population. Let O represent the *observed number* of data from the sample that fall into each category. Let E represent the *expected number* of data from the sample that, in theory, would fall into each category.

$O =$ observed frequency count of a category using sample data

$E =$ expected frequency of a category

$=$ (sample size n)(probability assigned to category)

Requirement

The sample size n should be large enough that $E \geq 5$ in each category.

Procedure

1. Set the *level of significance* α and use the *hypotheses*

   H_0: The population fits the specified distribution of categories.

   H_1: The population has a different distribution.

2. For each category, compute $(O - E)^2/E$, and then compute the *sample test statistic*

   $$\chi^2 = \Sigma \frac{(O - E)^2}{E} \text{ with } d.f. = k - 1$$

   where the sum is taken over all categories and $k =$ number of categories.
3. Use the chi-square distribution (Table 7 of Appendix II) and a *right-tailed test* to find (or estimate) the *P-value* corresponding to the sample test statistic.
4. *Conclude* the test. If P-value $\leq \alpha$, then reject H_0. If P-value $> \alpha$, then do not reject H_0.
5. *Interpret your conclusion* in the context of the application.

One important application of goodness-of-fit tests is to genetics theory. Such an application is shown in Guided Exercise 6.

GUIDED EXERCISE 6 | GOODNESS-OF-FIT TEST

According to genetics theory, red-green colorblindness in humans is a recessive sex-linked characteristic. In this case, the gene is carried on the X chromosome only. We will denote an X chromosome with the colorblindness gene by X_c and one without the gene by X_n. Women have two X chromosomes, and they will be red-green colorblind only if both chromosomes have the gene, designated X_cX_c. A woman can have normal vision but still carry the colorblind gene if only one of the chromosomes has the gene, designated X_cX_n. A man carries an X and a Y chromosome; if the X chromosome carries the colorblind gene (X_cY), the man is colorblind.

According to genetics theory, if a man with normal vision (X_nY) and a woman carrier (X_cX_n) have a child, the probabilities that the child will have red-green colorblindness, will have normal vision and not carry the gene, or will have normal vision and carry the gene are given by the *equally likely* events in Table 10-10.

Continued

GUIDED EXERCISE 6 *continued*

$P(\text{child has normal vision and is not a carrier}) = P(X_nY) + P(X_nX_n) = \frac{1}{2}$

$P(\text{child has normal vision and is a carrier}) = P(X_cX_n) = \frac{1}{4}$

$P(\text{child is red-green colorblind}) = P(X_cY) = \frac{1}{4}$

TABLE 10-10 **Red-Green Colorblindness**

	Father	
Mother	**Xn**	**Y**
Xc	XcXn	XcY
Xn	XnXn	XnY

To test this genetics theory, Genetics Labs took a random sample of 200 children whose mothers were carriers of the colorblind gene and whose fathers had normal vision. The results are shown in Table 10-11. We wish to test the hypothesis that the population follows the distribution predicted by the genetics theory (see Table 10-10). Use a 1% level of significance.

(a) State the null and alternate hypotheses. What is α?

→ H_0: The population fits the distribution predicted by genetics theory.
H_1: The population does not fit the distribution predicted by genetics theory.
$\alpha = 0.01$

(b) Fill in the rest of Table 10-11 and use the table to compute the sample statistic χ^2.

TABLE 10-11 **Colorblindness Sample**

Event	O	E	$(O - E)^2$	$(O - E)^2/E$
Red-green colorblind	35	50	225	4.50
Normal vision, noncarrier	105	____	____	____
Normal vision, carrier	60	____	____	____

→ TABLE 10-12 **Completion of Table 10-11**

Event	O	E	$(O - E)^2$	$(O - E)^2/E$
Red-green colorblind	35	50	225	4.50
Normal vision, noncarrier	105	100	25	0.25
Normal vision, carrier	60	50	100	2.00

The sample statistic is $\chi^2 = \Sigma \frac{(O - E)^2}{E} = 6.75$

(c) There are $k = 3$ categories listed in Table 10-11. Use this information to compute the degrees of freedom.

→ $d.f. = k - 1$
$= 3 - 1 = 2$

(d) Find the P-value for $\chi^2 = 6.75$.

→ Using Table 7 of Appendix II and the fact that goodness-of-fit tests are right-tailed tests, we see that

Right-tail area	**0.050**	**0.025**
$d.f. = 2$	5.99	7.38
	↑ Sample $\chi^2 = 6.75$	

$0.025 < P\text{-value} < 0.050$
Technology gives P-value ≈ 0.0342.

Continued

GUIDED EXERCISE 6 *continued*

(e) Conclude the test for $\alpha = 0.01$.

For $\alpha = 0.01$, we have

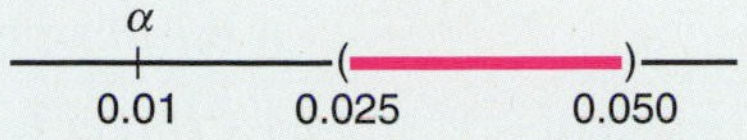

Since P-value $> \alpha$, do not reject H_0.

(f) ***Interpret*** the conclusion in the context of the application.

At the 1% level of significance, there is insufficient evidence to conclude that the population follows a distribution different from that predicted by genetics theory.

TECH NOTES The TI-84Plus/TI-83Plus/TI-*n*spire calculators with operating system v2.55MP support goodness of fit tests.

TI-84Plus/TI-83Plus/TI-*n*spire (with TI-84Plus keypad) Enter observed values in one list and corresponding expected values in another list. Press **STAT, TESTS,** and select option **D:χ^2 GOF-Test.** Enter the list containing the observed values, the list containing corresponding expected values, and the degrees of freedom. The output provides the sample x^2 value, the P-value of the sample x^2 value, and the values of each $(O - E)^2/E$ component.

VIEWPOINT Run! Run! Run!

What description would you use for marathon runners? How about age distribution? Body weight? Length of stride? Heart rate? Blood pressure? What countries do these runners come from? What are their best running times? Make your own estimated distribution for these variables, and then consider a goodness-of-fit test for your distribution compared with available data. For more information on marathon runners, visit the Brase/Brase statistics site at **http://www.cengage.com/statistics/brase11e** *and find links to the Honolulu marathon site and to the* Runners World *site.*

SECTION 10.2 PROBLEMS

1. *Statistical Literacy* For a chi-square goodness-of-fit test, how are the degrees of freedom computed?

2. *Statistical Literacy* How are expected frequencies computed for goodness-of-fit tests?

3. *Statistical Literacy* Explain why goodness-of-fit tests are always right-tailed tests.

4. *Critical Thinking* When the sample evidence is sufficient to justify rejecting the null hypothesis in a goodness-of-fit test, can we tell exactly how the distribution of observed values over the specified categories differs from the expected distribution? Explain.

For Problems 5–16, please provide the following information.

(a) What is the level of significance? State the null and alternate hypotheses.

(b) Find the value of the chi-square statistic for the sample. Are all the expected frequencies greater than 5? What sampling distribution will you use? What are the degrees of freedom?

(c) Find or estimate the P-value of the sample test statistic.

(d) Based on your answers in parts (a) to (c), will you reject or fail to reject the null hypothesis that the population fits the specified distribution of categories?
(e) *Interpret* your conclusion in the context of the application.

5. *Census: Age* The age distribution of the Canadian population and the age distribution of a random sample of 455 residents in the Indian community of Red Lake Village (Northwest Territories) are shown below (based on *U.S. Bureau of the Census, International Data Base*).

Age (years)	Percent of Canadian Population	Observed Number in Red Lake Village
Under 5	7.2%	47
5 to 14	13.6%	75
15 to 64	67.1%	288
65 and older	12.1%	45

Use a 5% level of significance to test the claim that the age distribution of the general Canadian population fits the age distribution of the residents of Red Lake Village.

6. *Census: Type of Household* The type of household for the U.S. population and for a random sample of 411 households from the community of Dove Creek, Montana, are shown (based on *Statistical Abstract of the United States*).

Type of Household	Percent of U.S. Households	Observed Number of Households in Dove Creek
Married with children	26%	102
Married, no children	29%	112
Single parent	9%	33
One person	25%	96
Other (e.g., roommates, siblings)	11%	68

Use a 5% level of significance to test the claim that the distribution of U.S. households fits the Dove Creek distribution.

7. *Archaeology: Stone Tools* The types of raw materials used to construct stone tools found at the archaeological site Casa del Rito are shown below (*Bandelier Archaeological Excavation Project,* edited by Kohler and Root). A random sample of 1486 stone tools was obtained from a current excavation site.

Raw Material	Regional Percent of Stone Tools	Observed Number of Tools at Current Excavation Site
Basalt	61.3%	906
Obsidian	10.6%	162
Welded tuff	11.4%	168
Pedernal chert	13.1%	197
Other	3.6%	53

Use a 1% level of significance to test the claim that the regional distribution of raw materials fits the distribution at the current excavation site.

8. *Ecology: Deer* The types of browse favored by deer are shown in the following table (*The Mule Deer of Mesa Verde National Park,* edited by Mierau and Schmidt). Using binoculars, volunteers observed the feeding habits of a random sample of 320 deer.

Type of Browse	Plant Composition in Study Area	Observed Number of Deer Feeding on This Plant
Sage brush	32%	102
Rabbit brush	38.7%	125
Salt brush	12%	43
Service berry	9.3%	27
Other	8%	23

Use a 5% level of significance to test the claim that the natural distribution of browse fits the deer feeding pattern.

9. *Meteorology: Normal Distribution* The following problem is based on information from the *National Oceanic and Atmospheric Administration (NOAA) Environmental Data Service.* Let x be a random variable that represents the average daily temperature (in degrees Fahrenheit) in July in the town of Kit Carson, Colorado. The x distribution has a mean μ of approximately 75°F and standard deviation σ of approximately 8°F. A 20-year study (620 July days) gave the entries in the rightmost column of the following table.

I	II	III	IV
Region under Normal Curve	x°F	Expected % from Normal Curve	Observed Number of Days in 20 Years
$\mu - 3\sigma \le x < \mu - 2\sigma$	$51 \le x < 59$	2.35%	16
$\mu - 2\sigma \le x < \mu - \sigma$	$59 \le x < 67$	13.5%	78
$\mu - \sigma \le x < \mu$	$67 \le x < 75$	34%	212
$\mu \le x < \mu + \sigma$	$75 \le x < 83$	34%	221
$\mu + \sigma \le x < \mu + 2\sigma$	$83 \le x < 91$	13.5%	81
$\mu + 2\sigma \le x < \mu + 3\sigma$	$91 \le x < 99$	2.35%	12

(i) Remember that $\mu = 75$ and $\sigma = 8$. Examine Figure 6-5 in Chapter 6. Write a brief explanation for Columns I, II, and III in the context of this problem.
(ii) Use a 1% level of significance to test the claim that the average daily July temperature follows a normal distribution with $\mu = 75$ and $\sigma = 8$.

10. *Meteorology: Normal Distribution* Let x be a random variable that represents the average daily temperature (in degrees Fahrenheit) in January for the town of Hana, Maui. The x variable has a mean μ of approximately 68°F and standard deviation σ of approximately 4°F (see reference in Problem 9). A 20-year study (620 January days) gave the entries in the rightmost column of the following table.

I	II	III	IV
Region Under Normal Curve	x°F	Expected % from Normal Curve	Observed Number of Days in 20 Years
$\mu - 3\sigma \le x < \mu - 2\sigma$	$56 \le x < 60$	2.35%	14
$\mu - 2\sigma \le x < \mu - \sigma$	$60 \le x < 64$	13.5%	86
$\mu - \sigma \le x < \mu$	$64 \le x < 68$	34%	207
$\mu \le x < \mu + \sigma$	$68 \le x < 76$	34%	215
$\mu + \sigma \le x < \mu + 2\sigma$	$72 \le x < 76$	13.5%	83
$\mu + 2\sigma \le x < \mu + 3\sigma$	$76 \le x < 80$	2.35%	15

(i) Remember that $\mu = 68$ and $\sigma = 4$. Examine Figure 6-5 in Chapter 6. Write a brief explanation for Columns I, II, and III in the context of this problem.
(ii) Use a 1% level of significance to test the claim that the average daily January temperature follows a normal distribution with $\mu = 68$ and $\sigma = 4$.

11. *Ecology: Fish* The Fish and Game Department stocked Lake Lulu with fish in the following proportions: 30% catfish, 15% bass, 40% bluegill, and 15% pike. Five years later it sampled the lake to see if the distribution of fish had changed. It found that the 500 fish in the sample were distributed as follows.

Catfish	Bass	Bluegill	Pike
120	85	220	75

In the 5-year interval, did the distribution of fish change at the 0.05 level?

12. *Library: Book Circulation* The director of library services at Fairmont College did a survey of types of books (by subject) in the circulation library. Then she used library records to take a random sample of 888 books checked out last term and classified the books in the sample by subject. The results are shown below.

Subject Area	Percent of Books in Circulation Library on This Subject	Number of Books in Sample on This Subject
Business	32%	268
Humanities	25%	214
Natural science	20%	215
Social science	15%	115
All other subjects	8%	76

Using a 5% level of significance, test the claim that the subject distribution of books in the library fits the distribution of books checked out by students.

13. *Census: California* The accuracy of a census report on a city in southern California was questioned by some government officials. A random sample of 1215 people living in the city was used to check the report, and the results are shown here:

Ethnic Origin	Census Percent	Sample Result
Black	10%	127
Asian	3%	40
Anglo	38%	480
Latino/Latina	41%	502
Native American	6%	56
All others	2%	10

Using a 1% level of significance, test the claim that the census distribution and the sample distribution agree.

14. *Marketing: Compact Discs* Snoop Incorporated is a firm that does market surveys. The Rollum Sound Company hired Snoop to study the age distribution of people who buy compact discs. To check the Snoop report, Rollum used a random sample of 519 customers and obtained the following data:

Customer Age (years)	Percent of Customers from Snoop Report	Number of Customers from Sample
Younger than 14	12%	88
14–18	29%	135
19–23	11%	52
24–28	10%	40
29–33	14%	76
Older than 33	24%	128

Using a 1% level of significance, test the claim that the distribution of customer ages in the Snoop report agrees with that of the sample report.

15. *Accounting Records: Benford's Law* Benford's law states that the first nonzero digits of numbers drawn at random from a large complex data file have the following probability distribution (Reference: American Statistical Association, *Chance,* Vol. 12, No. 3, pp. 27–31; see also the Focus Problem of Chapter 9).

First nonzero digit	1	2	3	4	5	6	7	8	9
Probability	0.301	0.176	0.125	0.097	0.079	0.067	0.058	0.051	0.046

Suppose that $n = 275$ numerical entries were drawn at random from a large accounting file of a major corporation. The first nonzero digits were recorded for the sample.

First nonzero digit	1	2	3	4	5	6	7	8	9
Sample frequency	83	49	32	22	25	18	13	17	16

Use a 1% level of significance to test the claim that the distribution of first nonzero digits in this accounting file follows Benford's law.

16. *Fair Dice: Uniform Distribution* A gambler complained about the dice. They seemed to be loaded! The dice were taken off the table and tested one at a time. One die was rolled 300 times and the following frequencies were recorded.

Outcome	1	2	3	4	5	6
Observed frequency O	62	45	63	32	47	51

Do these data indicate that the die is unbalanced? Use a 1% level of significance.
Hint: If the die is balanced, all outcomes should have the same expected frequency.

17. *Highway Accidents: Poisson Distribution* A civil engineer has been studying the frequency of vehicle accidents on a certain stretch of interstate highway. Long-term history indicates that there has been an average of 1.72 accidents per day on this section of the interstate. Let r be a random variable that represents number of accidents per day. Let O represent the number of observed accidents per day based on local highway patrol reports. A random sample of 90 days gave the following information.

r	0	1	2	3	4 or more
O	22	21	15	17	15

(a) The civil engineer wants to use a Poisson distribution to represent the probability of *r*, the number of accidents per day. The Poisson distribution is

$$P(r) = \frac{e^{-\lambda}\lambda^r}{r!}$$

where $\lambda = 1.72$ is the average number of accidents per day. Compute $P(r)$ for $r = 0, 1, 2, 3$, and 4 or more.

(b) Compute the expected number of accidents $E = 90P(r)$ for $r = 0, 1, 2, 3$, and 4 or more.

(c) Compute the sample statistic $\chi^2 = \Sigma \frac{(O - E)^2}{E}$ and the degrees of freedom.

(d) Test the statement that the Poisson distribution fits the sample data. Use a 1% level of significance.

18. *Bacteria Colonies: Poisson Distribution* A pathologist has been studying the frequency of bacterial colonies within the field of a microscope using samples of throat cultures from healthy adults. Long-term history indicates that there is an average of 2.80 bacteria colonies per field. Let *r* be a random variable that represents the number of bacteria colonies per field. Let *O* represent the number of observed bacteria colonies per field for throat cultures from healthy adults. A random sample of 100 healthy adults gave the following information.

r	0	1	2	3	4	5 or more
O	12	15	29	18	19	7

(a) The pathologist wants to use a Poisson distribution to represent the probability of *r*, the number of bacteria colonies per field. The Poisson distribution is

$$P(r) = \frac{e^{-\lambda}\lambda^r}{r!}$$

where $\lambda = 2.80$ is the average number of bacteria colonies per field. Compute $P(r)$ for $r = 0, 1, 2, 3, 4$, and 5 or more.

(b) Compute the expected number of colonies $E = 100P(r)$ for $r = 0, 1, 2, 3, 4$, and 5 or more.

(c) Compute the sample statistic $\chi^2 = \Sigma \frac{(O - E)^2}{E}$ and the degrees of freedom.

(d) Test the statement that the Poisson distribution fits the sample data. Use a 5% level of significance.

SECTION 10.3 Testing and Estimating a Single Variance or Standard Deviation

FOCUS POINTS

- Set up a test for a single variance σ^2.
- Compute the sample χ^2 statistic.
- Use the χ^2 distribution to estimate a *P*-value and conclude the test.
- Compute confidence intervals for σ^2 or σ.

TESTING σ^2

Many problems arise that require us to make decisions about variability. In this section, we will study two kinds of problems: (1) we will test hypotheses about the variance (or standard deviation) of a population, and (2) we will find confidence intervals for the variance (or standard deviation) of a population. It is customary to talk about variance instead of standard deviation because our techniques employ the sample variance rather than the standard deviation. Of course, the standard deviation is just the square root of the variance, so any discussion about variance is easily converted to a similar discussion about standard deviation.

Test of variance

Let us consider a specific example in which we might wish to test a hypothesis about the variance. Almost everyone has had to wait in line. In a grocery store, bank, post office, or registration center, there are usually several checkout or service areas. Frequently, each service area has its own independent line. However, many businesses and government offices are adopting a "single-line" procedure.

In a single-line procedure, there is only one waiting line for everyone. As any service area becomes available, the next person in line gets served. The old, independent-lines procedure has a line at each service center. An incoming customer simply picks the shortest line and hopes it will move quickly. In either procedure, the number of clerks and the rate at which they work is the same, so the average waiting time is the *same*. What is the advantage of the single-line procedure? The difference is in the *attitudes* of people who wait in the lines. A lengthy waiting line will be more acceptable, even though the average waiting time is the same, if the variability of waiting times is smaller. When the variability is small, the inconvenience of waiting (although it might not be reduced) does become more predictable. This means impatience is reduced and people are happier.

To test the hypothesis that variability is less in a single-line process, we use the chi-square distribution. The next theorem tells us how to use the sample and population variance to compute values of χ^2.

THEOREM 10.1 If we have a *normal* population with variance σ^2 and a random sample of n measurements is taken from this population with sample variance s^2, then

$$\chi^2 = \frac{(n-1)s^2}{\sigma^2}$$

is a value of a random variable having a chi-square distribution with degrees of freedom $d.f. = n - 1$.

Recall that the chi-square distribution is *not* symmetrical and that there are different chi-square distributions for different degrees of freedom. Table 7 of Appendix II gives chi-square values for which the area α is to the *right* of the given chi-square value.

EXAMPLE 3

χ^2 DISTRIBUTION

(a) Find the χ^2 value such that the area to the right of χ^2 is 0.05 when $d.f. = 10$.
SOLUTION: Since the area to the right of χ^2 is to be 0.05, we look in the right-tail area = 0.050 column and the row with $d.f. = 10$. $\chi^2 = 18.31$ (see Figure 10-6a).

(b) Find the χ^2 value such that the area to the *left* of χ^2 is 0.05 when $d.f. = 10$.
SOLUTION: When the area to the left of χ^2 is 0.05, the corresponding area to the *right* is $1 - 0.05 = 0.95$, so we look in the right-tail area = 0.950 column and the row with $d.f. = 10$. We find $\chi^2 = 3.94$ (see Figure 10-6b).

FIGURE 10-6

χ^2 Distribution with $d.f. = 10$.

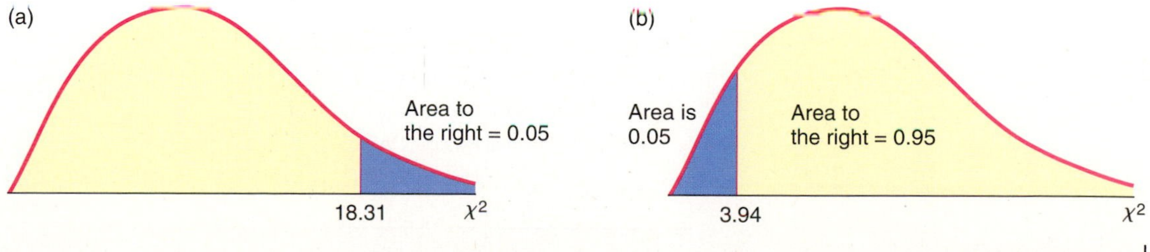

GUIDED EXERCISE 7 χ^2 DISTRIBUTION

(a) Find the area to the *right* of $\chi^2 = 37.57$ when $d.f. = 20$.

→ Use Table 7 of Appendix II. In the row headed by $d.f. = 20$, find the column with $\chi^2 = 37.57$. The column header 0.010 is the area to the right (see Figure 10-7).

FIGURE 10-7 χ^2 Distribution with $d.f. = 20$

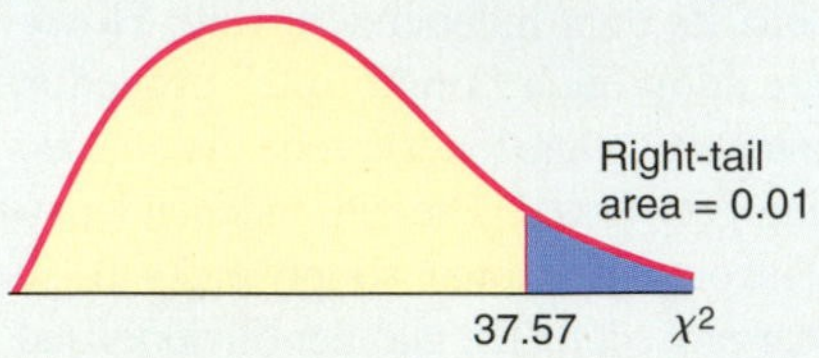

(b) Find the area to the *left* of $\chi^2 = 8.26$ when $d.f. = 20$.

(i) First use Table 7 of Appendix II to find the area to the right of $\chi^2 = 8.26$.

(ii) To get the area to the *left* of $\chi^2 = 8.26$, subtract the area to the right from 1.

→ Find the column with $\chi^2 = 8.26$ in the row headed by $d.f. = 20$. The column header 0.990 gives the area to the *right* of $\chi^2 = 8.26$.

Next, subtract the area to the right from 1.
Area to *left* = 1 − Area to *right* = 1 − 0.990 = 0.010 (see Figure 10-8).

FIGURE 10-8 χ^2 Distribution with $d.f. = 20$

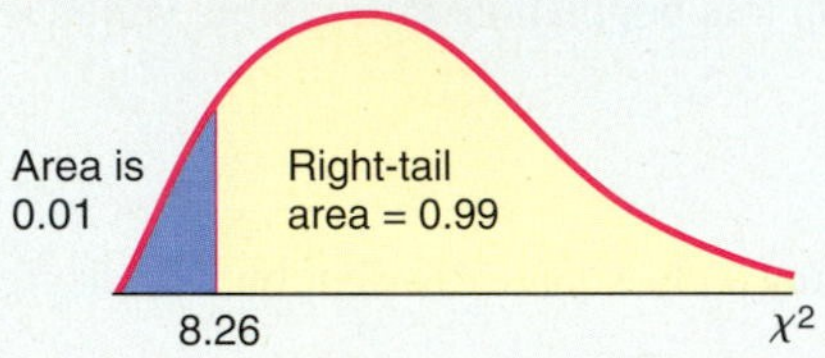

Table 10-13 summarizes the techniques for using the chi-square distribution (Table 7 of Appendix II) to find *P*-values for a right-tailed test, a left-tailed test, and a two-tailed test. Example 4 demonstrates the technique of finding *P*-values for a left-tailed test. Example 5 demonstrates the technique for a two-tailed test, and Guided Exercise 8 uses a right-tailed test.

TABLE 10-13 ***P*-values for Chi-Square Distribution Table (Table 7, Appendix II)**

(a) Two-tailed test

Remember that the *P*-value is the probability of getting a test statistic as extreme as, or more extreme than, the test statistic computed from the sample. For a two-tailed test, we need to account for corresponding equal areas in both the upper and lower tails. This means that in each tail, we have an area of *P*-value/2. The total *P*-value is then

$$P\text{-value} = 2\left(\frac{P\text{-value}}{2}\right)$$

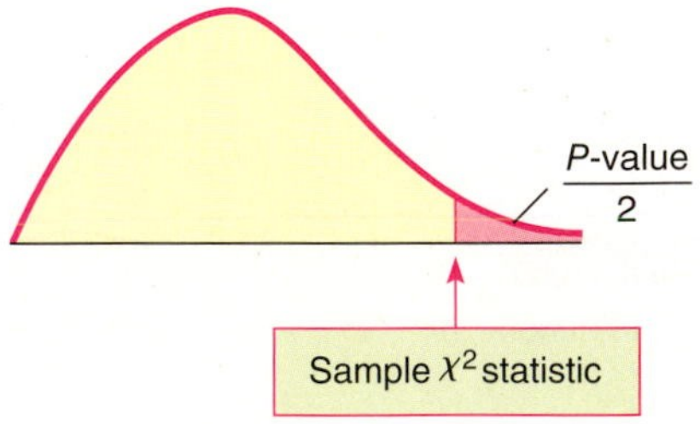

TABLE 10-13 ***P*-values for Chi-Square Distribution Table (Table 7, Appendix II)** *(continued)*

<table>
<tr><td></td><td>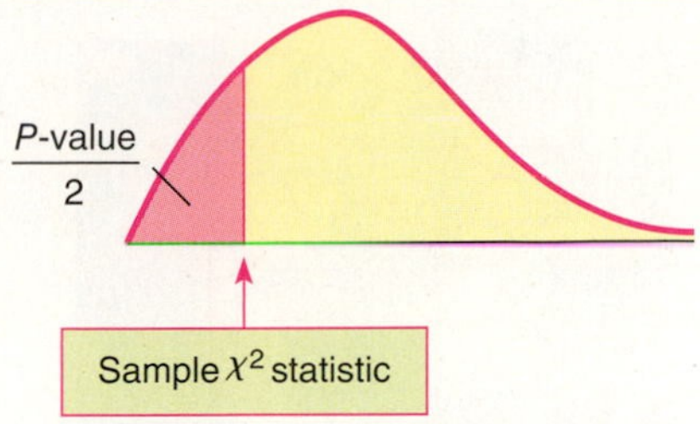

Be sure to choose the area in the appropriate tail (left or right) so that $\frac{P\text{-value}}{2} \le 0.5$.</td></tr>
<tr><td>(b) Right-tailed test
Since the chi-square table gives right-tail probabilities, you can use the table directly to find or estimate the P-value.</td><td>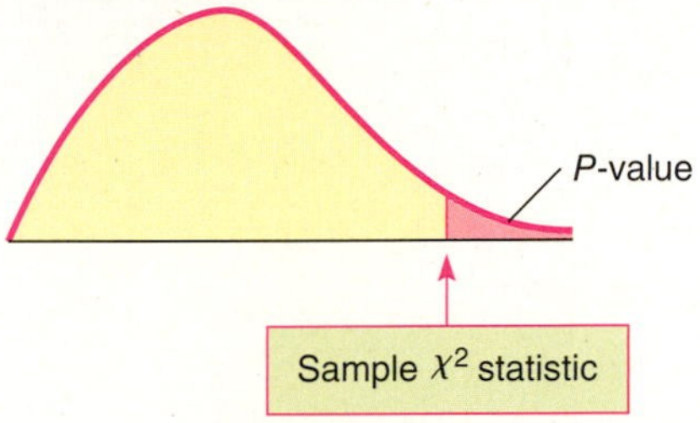
</td></tr>
<tr><td>(c) Left-tailed test
Since the chi-square table gives right-tail probabilities, you first find or estimate the quantity 1 − (P-value) right tail. Then subtract from 1 to get the P-value of the left-tail.</td><td>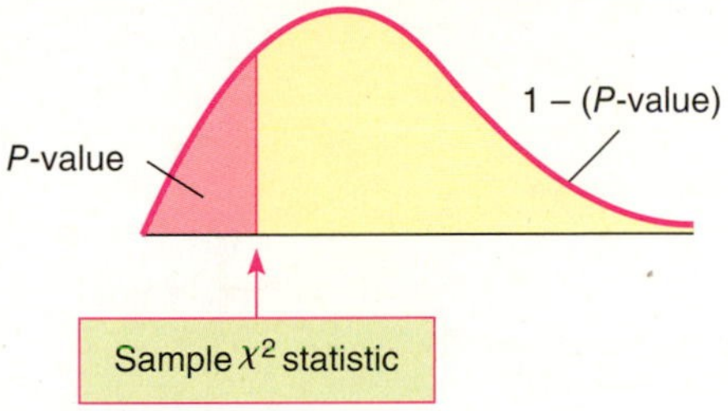
</td></tr>
</table>

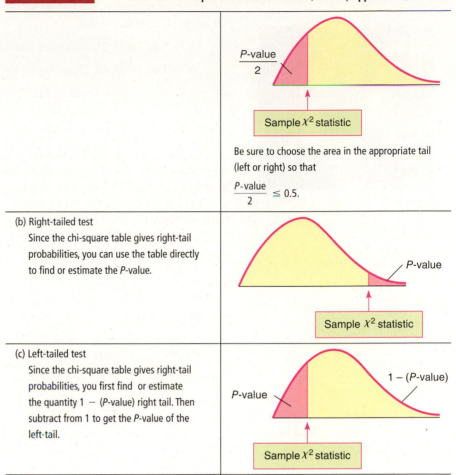

Now let's use Theorem 10.1 and our knowledge of the chi-square distribution to determine if a single-line procedure has less variance of waiting times than independent lines.

EXAMPLE 4

TESTING THE VARIANCE (LEFT-TAILED TEST)

For years, a large discount store has used independent lines to check out customers. Historically, the standard deviation of waiting times is 7 minutes. The manager tried a new, single-line procedure. A random sample of 25 customers using the single-line procedure was monitored, and it was found that the standard deviation for waiting times was only $s = 5$ minutes. Use $\alpha = 0.05$ to test the claim that the variance in waiting times is reduced for the single-line method. Assume the waiting times are normally distributed.

Establish H_0 and H_1

SOLUTION: As a null hypothesis, we assume that the variance of waiting times is the same as that of the former independent-lines procedure. The alternate hypothesis is that the variance for the single-line procedure is less than that for the independent-lines procedure. If we let σ be the standard deviation of waiting times for the single-line procedure, then σ^2 is the variance, and we have

$$H_0: \sigma^2 = 49 \qquad H_1: \sigma^2 < 49 \qquad (\text{use } 7^2 = 49)$$

Sample χ^2 value and degrees of freedom

Mark Richards/PhotoEdit

Checkout lines

We use the chi-square distribution to test the hypotheses. Assuming that the waiting times are normally distributed, we compute our observed value of χ^2 by using Theorem 10.1, with $n = 25$.

$$s = 5 \text{ so } s^2 = 25 \text{ (observed from sample)}$$

$$\sigma = 7 \text{ so } \sigma^2 = 49 \ (\textit{from } H_0\text{: } \sigma^2 = 49)$$

$$\chi^2 = \frac{(n-1)s^2}{\sigma^2} = \frac{(25-1)25}{49} \approx 12.24$$

$$d.f. = n - 1 = 25 - 1 = 24$$

P-value

Next we estimate the P-value for $\chi^2 = 12.24$. Since we have a left-tailed test, the P-value is the area of the chi-square distribution that lies to the *left* of $\chi^2 = 12.24$, as shown in Figure 10-9 on the next page.

To estimate the P-value on the left, we consider the fact that the area of the right tail is between 0.975 and 0.990. To find an estimate for the area of the left tail, we *subtract* each right-tail endpoint from 1. The P-value (area of the left tail) is in the interval

$$1 - 0.990 < P\text{-value of left tail} < 1 - 0.975$$
$$0.010 < P\text{-value} < 0.025$$

Test conclusion

To conclude the test, we compare the P-value to the level of significance $\alpha = 0.05$.

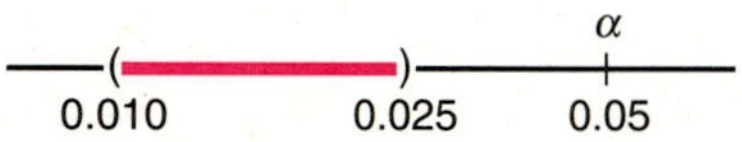

Since the P-value is less than α, we reject H_0.

Interpretation

Interpretation At the 5% level of significance, we conclude that the variance of waiting times for a single line is less than the variance of waiting times for multiple lines.

FIGURE 10-9

P-value

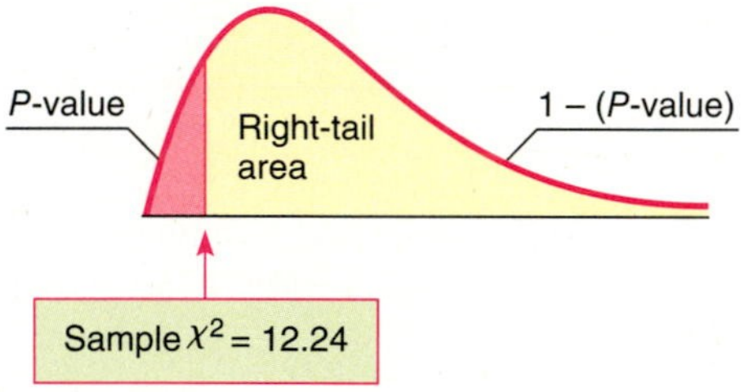

Right-tail Area	0.990	0.975
$d.f. = 4$	10.86	12.40
	↑ Sample $\chi^2 = 12.24$	

The steps used in Example 4 for testing the variance σ^2 are summarized as follows.

PROCEDURE

HOW TO TEST σ^2

Requirements

You first need to know that a random variable x has a normal distribution. In testing σ^2, the normal assumption must be strictly observed (whereas in testing means, we can say "normal" or "approximately normal"). Next you need a random sample (size $n \geq 2$) of values from the x distribution for which you compute the sample variance s^2.

Procedure

1. In the context of the problem, state the *null hypothesis* H_0 and the *alternate hypothesis* H_1, and set the *level of significance* α.

Continued

2. Use the value of σ^2 given in the null hypothesis H_0, the sample variance s^2, and the sample size n to compute the *sample test statistic*

$$\chi^2 = \frac{(n-1)s^2}{\sigma^2} \quad \text{with degrees of freedom } d.f. = n - 1$$

3. Use a chi-square distribution and the type of test to find or estimate the *P-value*. Use the procedures shown in Table 10-13 and Table 7 of Appendix II.
4. *Conclude* the test. If P-value $\leq \alpha$, then reject H_0. If P-value $> \alpha$, then do not reject H_0.
5. *Interpret your conclusion* in the context of the application.

EXAMPLE 5

TESTING THE VARIANCE (TWO-TAILED TEST)

Let x be a random variable that represents weight loss (in pounds) after following a certain diet for 6 months. After extensive study, it is found that x has a normal distribution with $\sigma = 5.7$ pounds. A new modification of the diet has been implemented. A random sample of $n = 21$ people use the modified diet for 6 months. For these people, the sample standard deviation of weight loss is $s = 4.1$ pounds. Does this result indicate that the variance of weight loss for the modified diet is different (either way) from the variance of weight loss for the original diet? Use $\alpha = 0.01$. Assume weight loss for each diet follows a normal distribution.

(a) What is the level of significance? State the null and alternate hypotheses.

SOLUTION: We are using $\alpha = 0.01$. The standard deviation of weight loss for the original diet is $\sigma = 5.7$ pounds, so the variance is $\sigma^2 = 32.49$. The null hypothesis is that the weight loss variance for the modified diet is the same as that for the original diet. The alternate hypothesis is that the variance is different.

$$H_0: \sigma^2 = 32.49 \quad H_1: \sigma^2 \neq 32.49$$

(b) Compute the sample test statistic χ^2 and the degrees of freedom.

SOLUTION: Using sample size $n = 21$, sample standard deviation $s = 4.1$ pounds, and $\sigma^2 = 32.49$ from the null hypothesis, we have

$$\chi^2 = \frac{(n-1)s^2}{\sigma^2} = \frac{(21-1)4.1^2}{32.49} \approx 10.35$$

with degrees of freedom $d.f. = n - 1 = 21 - 1 = 20$.

(c) Use the chi-square distribution (Table 7 of Appendix II) to estimate the P-value.

SOLUTION: For a *two-tailed* test, the area beyond the sample χ^2 represents *half* the total P-value or (P-value)/2. Figure 10-10 shows this region, which is to the left of $\chi^2 \approx 10.35$. However, Table 7 of Appendix II gives the areas in the *right*

FIGURE 10-10

P-value

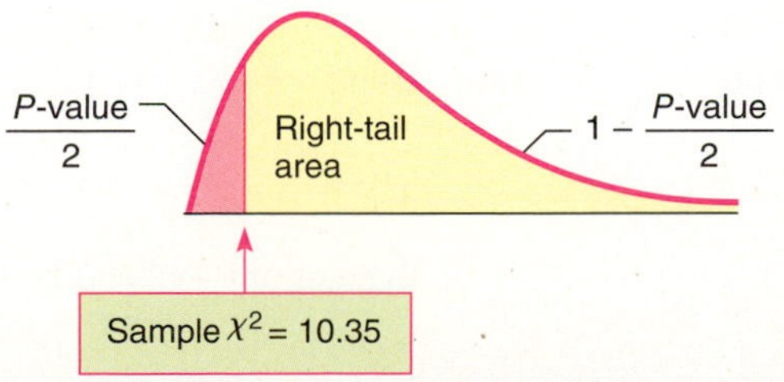

Right-tail Area	0.975	0.950
d.f. = 20	8.59	10.85
	↑ Sample $\chi^2 = 10.35$	

tail. We use Table 7 to find the area in the right tail and then subtract from 1 to find the corresponding area in the left tail.

From the table, we see that the right-tail area falls in the interval between 0.950 and 0.975. Subtracting each endpoint of the interval from 1 gives us an interval containing (P-value)/2. Multiplying by 2 gives an interval for the P-value.

$$1 - 0.975 < \frac{P\text{-value}}{2} < 1 - 0.950 \quad \text{Subtract right-tail-area endpoints from 1.}$$

$$0.025 < \frac{P\text{-value}}{2} < 0.050$$

$$0.05 < P\text{-value} < 0.10 \quad \text{Multiply each part by 2.}$$

(d) *Conclude* the test.

SOLUTION: The P-value is greater than $\alpha = 0.01$, so we do not reject H_0.

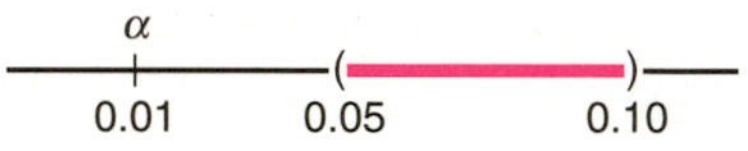

(e) *Interpret* the conclusion in the context of the application.

SOLUTION: At the 1% level of significance, there is insufficient evidence to conclude that the variance of weight loss using the modified diet is different from the variance of weight loss using the original diet.

GUIDED EXERCISE 8 TESTING THE VARIANCE (RIGHT-TAILED TEST)

Certain industrial machines require overhaul when wear on their parts introduces too much variability to pass inspection. A government official is visiting a dentist's office to inspect the operation of an x-ray machine. If the machine emits too little radiation, clear photographs cannot be obtained. However, too much radiation can be harmful to the patient. Government regulations specify an average emission of 60 millirads with standard deviation σ of 12 millirads, and the machine has been set for these readings. After examining the machine, the inspector is satisfied that the average emission is still 60 millirads. However, there is wear on certain mechanical parts. To test variability, the inspector takes a random sample of 30 x-ray emissions and finds the sample standard deviation to be $s = 15$ millirads. Does this support the claim that the variance is too high (i.e., the machine should be overhauled)? Use a 1% level of significance. Assume the emissions follow a normal distribution.

Let σ be the (population) standard deviation of emissions (in millirads) of the machine in its present condition.

(a) What is α? State H_0 and H_1.

→ $\alpha = 0.01$. Government regulations specify that $\sigma = 12$. This means that the variance $\sigma^2 = 144$. We are to test the claim that the variance is higher than government specifications allow.

H_0: $\sigma^2 = 144$ and H_1: $\sigma^2 > 144$

(b) Compute the sample test statistic χ^2 and corresponding degrees of freedom.

→ Using $n = 30$, $s = 15$, and $\sigma^2 = 144$ from H_0,

$$\chi^2 = \frac{(n-1)s^2}{\sigma^2} = \frac{(30-1)15^2}{144} \approx 45.3$$

Degrees of freedom $d.f. = n - 1 = 30 - 1 = 29$

Continued

GUIDED EXERCISE 8 *continued*

(c) Estimate the *P*-value for the sample $\chi^2 = 45.3$ with $d.f. = 29$.

FIGURE 10-11

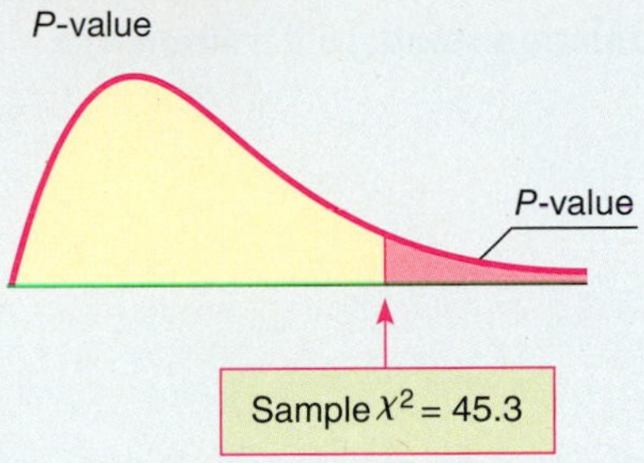

➡ Since this is a *right-tailed* test, we look up *P*-values directly in the chi-square table (Table 7 of Appendix II).

Right-tail Area	0.050	0.025
$d.f. = 29$	42.56	45.72
	↑ Sample $\chi^2 = 45.3$	

$0.025 < P\text{-value} < 0.050$

(d) *Conclude* the test.

➡ The *P*-value for $\chi^2 = 45.3$ is greater than $\alpha = 0.01$.

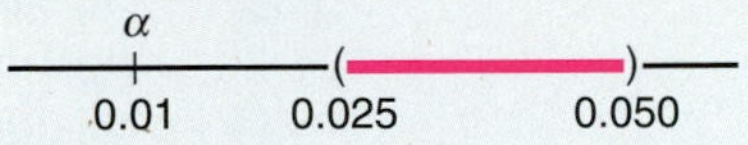

Fail to reject H_0.

(e) ***Interpret*** the conclusion in the context of the application.

➡ At the 1% level of significance, there is insufficient evidence to conclude that the variance of the radiation emitted by the machine is greater than that specified by government regulations. The evidence does not indicate that an adjustment is necessary at this time.

CONFIDENCE INTERVAL FOR σ^2

Confidence Interval

Sometimes it is important to have a *confidence interval* for the variance or standard deviation. Let us look at another example.

Mr. Wilson is a truck farmer in California who makes his living on a large single-vegetable crop of green beans. Because modern machinery is being used, the entire crop must be harvested at the same time. Therefore, it is important to plant a variety of green beans that mature all at once. This means that Mr. Wilson wants a small standard deviation between maturing times of individual plants. A seed company is trying to develop a new variety of green bean with a small standard deviation of maturing times. To test the new variety, Mr. Wilson planted 30 of the new seeds and carefully observed the number of days required for each plant to arrive at its peak of maturity. The maturing times for these plants had a sample standard deviation of $s = 3.4$ days. How can we find a 95% confidence interval for the population standard deviation of maturing times for this variety of green bean? The answer to this question is based on the following procedure.

PROCEDURE

HOW TO FIND A CONFIDENCE INTERVAL FOR σ^2 AND σ

Requirements

Let x be a random variable with a normal distribution and unknown population standard deviation σ. Take a random sample of size n from the x distribution and compute the sample standard deviation s.

Continued

Procedure

A **confidence interval for the population variance** σ^2 is

$$\frac{(n-1)s^2}{\chi_U^2} < \sigma^2 < \frac{(n-1)s^2}{\chi_L^2} \qquad (1)$$

and a **confidence interval for the population standard deviation** σ is

$$\sqrt{\frac{(n-1)s^2}{\chi_U^2}} < \sigma < \sqrt{\frac{(n-1)s^2}{\chi_L^2}}$$

where

c = confidence level $(0 < c < 1)$
n = sample size $(n \geq 2)$
χ_U^2 χ_U^2 = chi-square value from Table 7 of Appendix II using $d.f. = n - 1$ and right-tail area $= (1 - c)/2$
χ_L^2 χ_L^2 = chi-square value from Table 7 of Appendix II using $d.f. = n - 1$ and right-tail area $= (1 + c)/2$

FIGURE 10-12

Area Representing a *c* Confidence Level on a Chi-Square Distribution with *d.f.* $= n - 1$

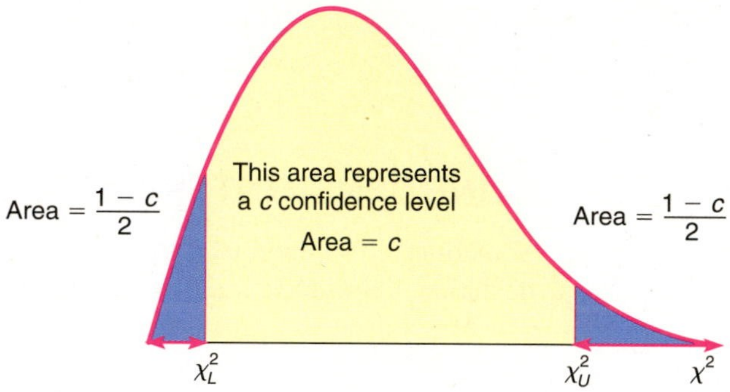

From Figure 10-12, we see that a *c* confidence level on a chi-square distribution with equal probability in each tail does not center the middle of the corresponding interval under the peak of the curve. This is to be expected because a chi-square curve is skewed to the right.

COMMENT Note that the method of computing confidence intervals for variances is different from the method of computing confidence intervals for means or proportions as studied in Chapter 7. Confidence intervals for σ^2 do not involve a maximal error of estimate *E*. Rather, the endpoints of the confidence interval are computed directly using the sample statistic s^2, the sample size, and the critical values.

Now let us finish our example regarding the variance of maturing times for green beans.

EXAMPLE 6

CONFIDENCE INTERVALS FOR σ^2 AND σ

A random sample of $n = 30$ green bean plants has a sample standard deviation of $s = 3.4$ days for maturity. Find a 95% confidence interval for the population variance σ^2. Assume the distribution of maturity times is normal.

Iain Frazer/Shutterstock.com

SOLUTION: To find the confidence interval, we use the following values:

$c = 0.95$	confidence level
$n = 30$	sample size
$d.f. = n - 1 = 30 - 1 = 29$	degrees of freedom
$s = 3.4$	sample standard deviation

To find the value of χ^2_U, we use Table 7 of Appendix II with $d.f. = 29$ and right-tail area $= (1 - c)/2 = (1 - 0.95)/2 = 0.025$. From Table 7, we get

$$\chi^2_U = 45.72$$

To find the value of χ^2_L, we use Table 7 of Appendix II with $d.f. = 29$ and right-tail area $= (1 + c)/2 = (1 + 0.95)/2 = 0.975$. From Table 7, we get

$$\chi^2_L = 16.05$$

Formula (1) tells us that our desired 95% confidence interval for σ^2 is

$$\frac{(n-1)s^2}{\chi^2_U} < \sigma^2 < \frac{(n-1)s^2}{\chi^2_L}$$

$$\frac{(30-1)(3.4)^2}{45.72} < \sigma^2 < \frac{(30-1)(3.4)^2}{16.05}$$

$$7.33 < \sigma^2 < 20.89$$

To find a 95% confidence interval for σ, we simply take square roots; therefore, a 95% confidence interval for σ is

$$\sqrt{7.33} < \sigma < \sqrt{20.89}$$

$$2.71 < \sigma < 4.57$$

GUIDED EXERCISE 9 | CONFIDENCE INTERVALS FOR σ^2 AND σ

A few miles off the Kona coast of the island of Hawaii, a research vessel lies anchored. This ship makes electrical energy from the solar temperature differential of (warm) surface water versus (cool) deep water. The basic idea is that the warm water is flushed over coils to vaporize a special fluid. The vapor is under pressure and drives electrical turbines. Then some electricity is used to pump up cold water to cool the vapor back to a liquid, and the process is repeated. Even though some electricity is used to pump up the cold water, there is plenty left to supply a moderate-sized Hawaiian town. The subtropic sun always warms up surface water to a reliable temperature, but ocean currents can change the temperature of the deep, cooler water. If the deep-water temperature is too variable, the power plant cannot operate efficiently or possibly cannot operate at all. To estimate the variability of deep ocean water temperatures, a random sample of 25 near-bottom readings gave a sample standard deviation of 7.3°C.

Find a 99% confidence interval for the variance σ^2 and standard deviation σ of the deep-water temperatures. Assume deep-water temperatures are normally distributed.

(a) Determine the following values: $c =$ _____; $n =$ _____; $d.f =$ _____; $s =$ _____.

→ $c = 0.99$; $n = 25$; $d.f. = 24$; $s = 7.3$

(b) What is the value of χ^2_U? of χ^2_L?

→ We use Table 7 of Appendix II with $d.f. = 24$.

For χ^2_U, right-tail area $= (1 - 0.99)/2 = 0.005$

$$\chi^2_U = 45.56$$

For χ^2_L, right-tail area $= (1 + 0.99)/2 = 0.995$

$$\chi^2_L = 9.89$$

Continued

GUIDED EXERCISE 9 *continued*

(c) Find a 99% confidence interval for σ^2.

$$\frac{(n-1)s^2}{\chi^2_U} < \sigma^2 < \frac{(n-1)s^2}{\chi^2_L}$$

$$\frac{(24)(7.3)^2}{45.56} < \sigma^2 < \frac{24(7.3)^2}{9.89}$$

$$28.07 < \sigma^2 < 129.32$$

(d) Find a 99% confidence interval for σ.

$$\sqrt{28.07} < \sqrt{\sigma^2} < \sqrt{129.32}$$

$$5.30 < \sigma < 11.37$$

VIEWPOINT Adoption—A Good Choice!

Cuckoos are birds that are known to lay their eggs in the nests of other (host) birds. The host birds then hatch the eggs and adopt the cuckoo chicks as their own. Birds such as the meadow pipit, tree pipit, hedge sparrow, robin, and wren have all played host to cuckoo eggs and adopted their chicks. L. H. C. Tippett (1902–1985) was a pioneer in the field of statistical quality control who collected data on cuckoo eggs found in the nests of other birds. For more information and data from Tippett's study, visit the Brase/Brase statistics site at **http://www.cengage.com/statistics/brase11e** *and find a link to DASL, the Carnegie Mellon University Data and Story Library. Find Biology under Data Subjects, and then select the Cuckoo Egg Length Data file.*

SECTION 10.3 PROBLEMS

1. *Statistical Literacy* Does the x distribution need to be normal in order to use the chi-square distribution to test the variance? Is it acceptable to use the chi-square distribution to test the variance if the x distribution is simply mound-shaped and more or less symmetric?

2. *Critical Thinking* The x distribution must be normal in order to use a chi-square distribution to test the variance. What are some methods you can use to assess whether the x distribution is normal? *Hint:* See Chapter 6 and goodness-of-fit tests.

For Problems 3–11, please provide the following information.

(a) What is the level of significance? State the null and alternate hypotheses.
(b) Find the value of the chi-square statistic for the sample. What are the degrees of freedom? What assumptions are you making about the original distribution?
(c) Find or estimate the P-value of the sample test statistic.
(d) Based on your answers in parts (a) to (c), will you reject or fail to reject the null hypothesis of independence?
(e) *Interpret* your conclusion in the context of the application.
(f) Find the requested confidence interval for the population variance or population standard deviation. Interpret the results in the context of the application.

In each of the following problems, assume a normal population distribution.

3. *Archaeology: Chaco Canyon* The following problem is based on information from *Archaeological Surveys of Chaco Canyon, New Mexico,* by A. Hayes, D. Brugge, and W. Judge, University of New Mexico Press. A *transect* is an archaeological study area that is 1/5 mile wide and 1 mile long. A *site* in a transect is the location of a significant archaeological find. Let x represent the

number of sites per transect. In a section of Chaco Canyon, a large number of transects showed that x has a population variance $\sigma^2 = 42.3$. In a different section of Chaco Canyon, a random sample of 23 transects gave a sample variance $s^2 = 46.1$ for the number of sites per transect. Use a 5% level of significance to test the claim that the variance in the new section is greater than 42.3. Find a 95% confidence interval for the population variance.

4. *Sociology: Marriage* The following problem is based on information from an article by N. Keyfitz in the *American Journal of Sociology* (Vol. 53, pp. 470–480). Let x = age in years of a rural Quebec woman at the time of her first marriage. In the year 1941, the population variance of x was approximately $\sigma^2 = 5.1$. Suppose a recent study of age at first marriage for a random sample of 41 women in rural Quebec gave a sample variance $s^2 = 3.3$. Use a 5% level of significance to test the claim that the current variance is less than 5.1. Find a 90% confidence interval for the population variance.

5. *Mountain Climbing: Accidents* The following problem is based on information taken from *Accidents in North American Mountaineering* (jointly published by The American Alpine Club and The Alpine Club of Canada). Let x represent the number of mountain climbers killed each year. The long-term variance of x is approximately $\sigma^2 = 136.2$. Suppose that for the past 8 years, the variance has been $s^2 = 115.1$. Use a 1% level of significance to test the claim that the recent variance for number of mountain climber deaths is less than 136.2. Find a 90% confidence interval for the population variance.

6. *Professors: Salaries* The following problem is based on information taken from *Academe, Bulletin of the American Association of University Professors.* Let x represent the average annual salary of college and university professors (in thousands of dollars) in the United States. For all colleges and universities in the United States, the population variance of x is approximately $\sigma^2 = 47.1$. However, a random sample of 15 colleges and universities in Kansas showed that x has a sample variance $s^2 = 83.2$. Use a 5% level of significance to test the claim that the variance for colleges and universities in Kansas is greater than 47.1. Find a 95% confidence interval for the population variance.

7. *Medical: Clinical Test* A new kind of typhoid shot is being developed by a medical research team. The old typhoid shot was known to protect the population for a mean time of 36 months, with a standard deviation of 3 months. To test the time variability of the new shot, a random sample of 23 people were given the new shot. Regular blood tests showed that the sample standard deviation of protection times was 1.9 months. Using a 0.05 level of significance, test the claim that the new typhoid shot has a smaller variance of protection times. Find a 90% confidence interval for the population standard deviation.

8. *Veterinary Science: Tranquilizer* Jim Mead is a veterinarian who visits a Vermont farm to examine prize bulls. In order to examine a bull, Jim first gives the animal a tranquilizer shot. The effect of the shot is supposed to last an average of 65 minutes, and it usually does. However, Jim sometimes gets chased out of the pasture by a bull that recovers too soon, and other times he becomes worried about prize bulls that take too long to recover. By reading journals, Jim has found that the tranquilizer should have a mean duration time of 65 minutes, with a standard deviation of 15 minutes. A random sample of 10 of Jim's bulls had a mean tranquilized duration time of close to 65 minutes but a standard deviation of 24 minutes. At the 1% level of significance, is Jim justified in the claim that the variance is larger than that stated in his journal? Find a 95% confidence interval for the population standard deviation.

9. *Engineering: Jet Engines* The fan blades on commercial jet engines must be replaced when wear on these parts indicates too much variability to pass inspection. If a single fan blade broke during operation, it could severely

endanger a flight. A large engine contains thousands of fan blades, and safety regulations require that variability measurements on the population of all blades not exceed $\sigma^2 = 0.18\ \text{mm}^2$. An engine inspector took a random sample of 61 fan blades from an engine. She measured each blade and found a sample variance of $0.27\ \text{mm}^2$. Using a 0.01 level of significance, is the inspector justified in claiming that all the engine fan blades must be replaced? Find a 90% confidence interval for the population standard deviation.

10. *Law: Bar Exam* A factor in determining the usefulness of an examination as a measure of demonstrated ability is the amount of spread that occurs in the grades. If the spread or variation of examination scores is very small, it usually means that the examination was either too hard or too easy. However, if the variance of scores is moderately large, then there is a definite difference in scores between "better," "average," and "poorer" students. A group of attorneys in a Midwest state has been given the task of making up this year's bar examination for the state. The examination has 500 total possible points, and from the history of past examinations, it is known that a standard deviation of around 60 points is desirable. Of course, too large or too small a standard deviation is not good. The attorneys want to test their examination to see how good it is. A preliminary version of the examination (with slight modifications to protect the integrity of the real examination) is given to a random sample of 24 newly graduated law students. Their scores give a sample standard deviation of 72 points.
    (i) Using a 0.01 level of significance, test the claim that the population standard deviation for the new examination is 60 against the claim that the population standard deviation is different from 60.
    (ii) Find a 99% confidence interval for the population variance.
    (iii) Find a 99% confidence interval for the population standard deviation.

11. *Engineering: Solar Batteries* A set of solar batteries is used in a research satellite. The satellite can run on only one battery, but it runs best if more than one battery is used. The variance σ^2 of lifetimes of these batteries affects the useful lifetime of the satellite before it goes dead. If the variance is too small, all the batteries will tend to die at once. Why? If the variance is too large, the batteries are simply not dependable. Why? Engineers have determined that a variance of $\sigma^2 = 23$ months (squared) is most desirable for these batteries. A random sample of 22 batteries gave a sample variance of 14.3 months (squared).
    (i) Using a 0.05 level of significance, test the claim that $\sigma^2 = 23$ against the claim that σ^2 is different from 23.
    (ii) Find a 90% confidence interval for the population variance σ^2.
    (iii) Find a 90% confidence interval for the population standard deviation σ.

PART II: INFERENCES USING THE *F* DISTRIBUTION

OVERVIEW OF THE *F* DISTRIBUTION

The *F* probability distribution was first developed by the English statistician Sir Ronald Fisher (1890–1962). Fisher had a long and distinguished career in statistics, including research work at the agricultural station at Rothamsted (see Problems 5 and 6 at the end of Section 10.4). During his time there he developed the subjects of experimental design and ANOVA (see Sections 10.5 and 10.6).

The *F* distribution is a ratio of two independent chi-square random variables, each with its own degrees of freedom,

$d.f._N$ = degrees of freedom in the numerator
$d.f._D$ = degrees of freedom in the denominator

FIGURE 10-13

Typical F Distribution ($d.f._N = 4$, $d.f._D = 7$)

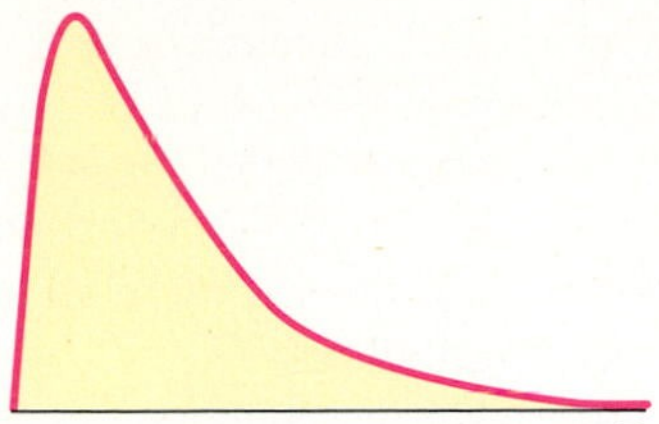

The F distribution depends on these *two* degrees of freedom, $d.f._N$ and $d.f._D$. Figure 10-13 shows a typical F distribution.

PROPERTIES OF THE *F* DISTRIBUTION

- The F distribution is not symmetrical. It is skewed to the right.
- Values of F are always greater than or equal to zero.
- A specific F distribution (see Table 8 of Appendix II) is determined from *two* degrees of freedom. These are called *degrees of freedom for the numerator* $d.f._N$ and *degrees of freedom for the denominator* $d.f._D$.
- Area under the entire F distribution is one.

The degrees of freedom used in the F distribution depend on the particular application. Table 8 of Appendix II shows areas in the right-tail of different F distributions according to the degrees of freedom in both the numerator, $d.f._N$, and the denominator, $d.f._D$.

Table 10-14 shows an excerpt from Table 8. Notice that $d.f._D$ are row headers. For each $d.f._D$, right-tail areas from 0.100 down to 0.001 are provided in the next column. Then, under column headers for $d.f._N$, values of F are given corresponding to $d.f._D$, the right-tail area, and $d.f._N$. For example, for $d.f._D = 2$, right-tail area $= 0.010$, and $d.f._N = 3$, the corresponding value of F is 99.17.

In this text we present three applications of the F distribution: testing two variances (Section 10.4), one-way ANOVA (Section 10.5), and two-way ANOVA (Section 10.6). Sections 10.4 and 10.5 are self-contained and can be studied independently. Section 10.5 should be studied before Section 10.6.

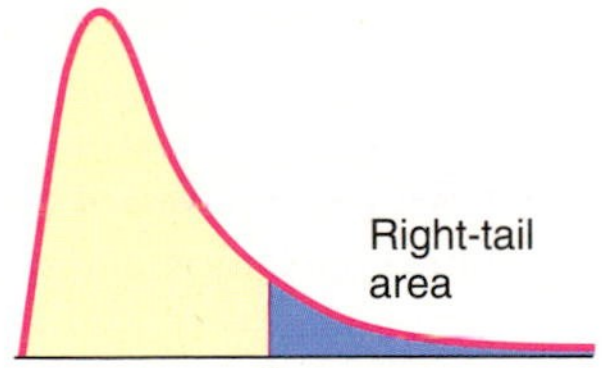

TABLE 10-14 **Excerpt from Table 8 (Appendix II): The *F* Distribution**

	Right-tail	Degrees of Freedom for Numerator $d.f._N$			
$d.f._D$	Area	1	2	3	4 . . .
⋮	⋮	⋮	⋮	⋮	⋮
	0.100	8.53	9.00	9.16	9.24
	0.050	18.51	19.00	19.16	19.25
✓2	0.025	38.51	39.00	39.17	39.25
	0.010	98.50	99.00	99.17	99.25
	0.001	998.50	999.00	999.17	9999.25

SECTION **10.4**

Testing Two Variances

FOCUS POINTS

- Set up a test for two variances σ_1^2 and σ_2^2.
- Use sample variances to compute the sample F statistic.
- Use the F distribution to estimate a P-value and conclude the test.

In this section, we present a method for testing two variances (or, equivalently, two standard deviations). We use *independent random samples* from two populations to test the claim that the population variances are equal. The concept of variation among data is very important, and there are many possible applications in science, industry, business administration, social science, and so on.

In Section 10.3, we tested a *single* variance. The main mathematical tool we used was the chi-square probability distribution. In this section, the main tool is the F probability distribution.

Let us begin by stating what we need to assume for a test of two population variances.

Basic requirements

- The two populations are independent of each other. Recall from Section 8.5 that two sampling distributions are *independent* if there is no relation whatsoever between specific values of the two distributions.
- The two populations each have a *normal* probability distribution. This is important because the test we will use is sensitive to changes away from normality.

Setup for test

Now that we know the basic requirements, let's consider the setup.

HOW TO SET UP THE TEST

STEP 1: Get Two Independent Random Samples, One from Each Population

We use the following notation:

Population I (larger s^2)	Population II (smaller s^2)
n_1 = sample size	n_2 = sample size
s_1^2 = sample variance	s_2^2 = sample variance
σ_1^2 = population variance	σ_2^2 = population variance

To simplify later discussion, we make the notational choice that

$$s_1^2 \geq s_2^2$$

This means that we *define* population I as the population with the *larger* (or equal, as the case may be) sample variance. This is only a notational convention and does not affect the general nature of the test.

STEP 2: Set Up the Hypotheses

Hypothesis tests about two variances

The null hypothesis will be that we have equal population variances.

$$H_0\colon \sigma_1^2 = \sigma_2^2$$

Reflecting on our notation setup, it makes sense to use an alternate hypothesis, either

$$H_1\colon \sigma_1^2 \neq \sigma_2^2 \quad \text{or} \quad H_1\colon \sigma_1^2 > \sigma_2^2$$

Notice that the test makes claims about variances. However, we can also use it for corresponding claims about standard deviations.

Hypotheses about Variances	Equivalent Hypotheses about Standard Deviations
$H_0\colon \sigma_1^2 = \sigma_2^2$	$H_0\colon \sigma_1 = \sigma_2$
$H_1\colon \sigma_1^2 \neq \sigma_2^2$	$H_1\colon \sigma_1 \neq \sigma_2$
$H_1\colon \sigma_1^2 > \sigma_2^2$	$H_1\colon \sigma_1 > \sigma_2$

STEP 3: Compute the Sample Test Statistic

Sample test statistic *F*

$$F = \frac{s_1^2}{s_2^2}$$

For two normally distributed populations with equal variances (H_0: $\sigma_1^2 = \sigma_2^2$), the sampling distribution we will use is the *F distribution* (see Table 8 of Appendix II).

The *F* distribution depends on *two* degrees of freedom.

For **tests of two variances,** it can be shown that

$d.f._N = n_1 - 1$ and $d.f._D = n_2 - 1$

STEP 4: Find (or Estimate) the *P*-value of the Sample Test Statistic

Use the *F* distribution (Table 8 of Appendix II) to find the *P*-value of the sample test statistic. You need to know the degrees of freedom for the numerator, $d.f._N = n_1 - 1$, and the degrees of freedom for the denominator, $d.f._D = n_2 - 1$. Find the block of entries with your $d.f._D$ as row header and your $d.f._N$ as column header. Within that block of values, find the position of the sample test statistic *F*. Then find the corresponding right-tail area. For instance, using Table 10-15 (Excerpt from Table 8), we see that for $d.f._D = 2$ and $d.f._N = 3$, sample $F = 55.2$ lies between 39.17 and 99.17, with corresponding right-tail areas of 0.025 and 0.010. The interval containing the *P*-value for $F = 55.2$ is $0.010 < P\text{-value} < 0.025$.

Table 10-16 gives a summary for computing the *P*-value for both right-tailed and two-tailed tests for two variances.

Now that we have steps 1 to 4 as an outline, let's look at a specific example.

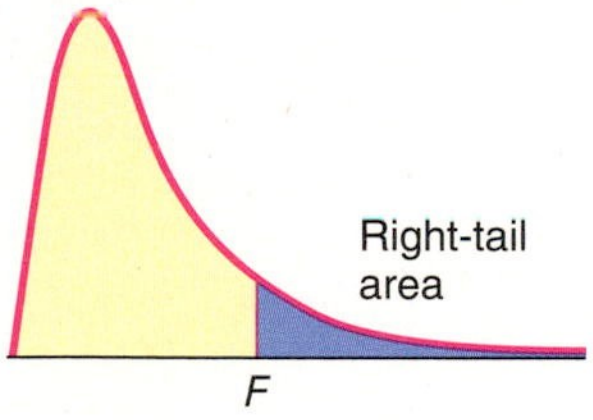

TABLE 10-15 Excerpt from Table 8 (Appendix II): The *F* Distribution

	Right-tail	Degrees of Freedom for Numerator $d.f._N$			
$d.f._D$	Area	1	2	3	4 ...
⋮	⋮	⋮	⋮	⋮	⋮
	0.100	8.53	9.00	9.16	9.24
	0.050	18.51	19.00	19.16	19.25
✓2	0.025	38.51	39.00	39.17	39.25
	0.010	98.50	99.00	99.17	99.25
	0.001	998.50	999.00	999.17	9999.25

TABLE 10-16 *P*-values for Testing Two Variances (Table 8, Appendix II)

(a) Right-tailed test

Since the *F*-distribution table gives right-tail probabilities, you can use the table directly to find or estimate the *P*-value.

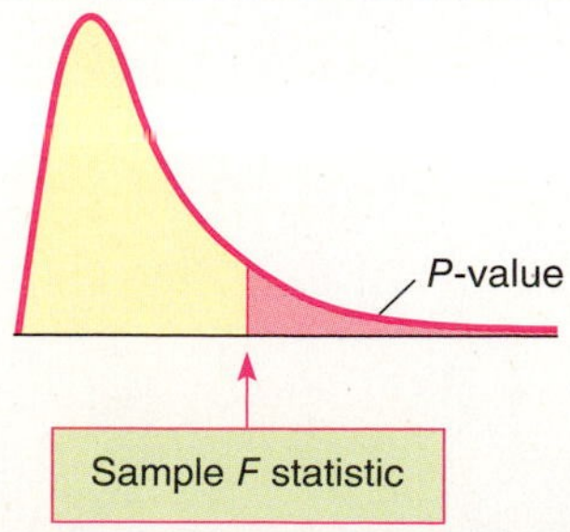

Continued

TABLE 10-16 ***P*-values for Testing Two Variances (Table 8, Appendix II)** *(continued)*

(b) Two-tailed test

Remember that the *P*-value is the probability of getting a test statistic as extreme as or more extreme than the test statistic computed from the sample. For a two-tailed test, we need to account for corresponding equal areas in *both* the upper and lower tails. This means that in each tail, we have an area of (P-value)/2. The total *P*-value is then

$$P\text{-value} = 2\left(\frac{P\text{-value}}{2}\right)$$

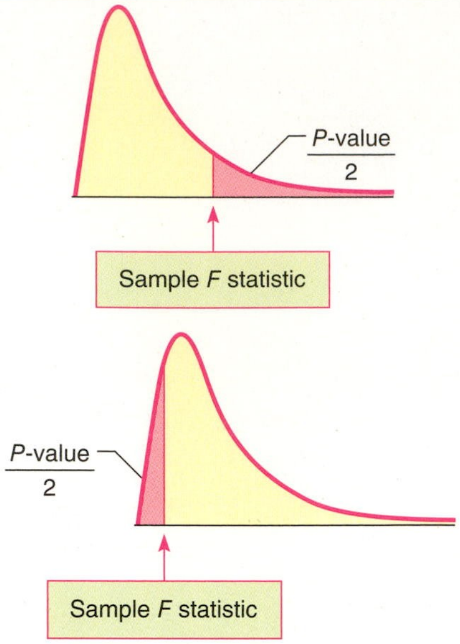

Be sure to choose the area in the appropriate tail (left or right) so that $\frac{P\text{-value}}{2} \le 0.5$.

EXAMPLE 7

TESTING TWO VARIANCES

Prehistoric Native Americans smoked pipes for ceremonial purposes. Most pipes were either carved-stone pipes or ceramic pipes made from clay. Clay pipes were easier to make, whereas stone pipes required careful drilling using hollow-core-bone drills and special stone reamers. An anthropologist claims that because clay pipes were easier to make, they show a greater variance in their construction. We want to test this claim using a 5% level of significance. Data for this example are taken from the Wind Mountain Archaeological Region (Source: *Mimbres Mogollon Archaeology* by A. I. Woosley and A. J. McIntyre, University of New Mexico Press). Assume the diameters of each type of pipe are normally distributed.

Ceramic Pipe Bowl Diameters (cm)

1.7	5.1	1.4	0.7	2.5	4.0
3.8	2.0	3.1	5.0	1.5	

Stone Pipe Bowl Diameters (cm)

1.6	2.1	3.1	1.4	2.2	2.1
2.6	3.2	3.4			

Richard A. Cooke/CORBIS

SOLUTION:

(a) *Check requirements* Assume that the pipe bowl diameters follow normal distributions and that the given data make up independent random samples of pipe measurements taken from archaeological excavations at Wind Mountain.

Use a calculator to verify the following:

Population I: Ceramic Pipes	Population II: Stone Pipes
$n_1 = 11$	$n_2 = 9$
$s_1^2 \approx 2.266$	$s_2^2 \approx 0.504$
σ_1^2 = population variance	σ_2^2 = population variance

Note: Because the sample variance for ceramic pipes (2.266) is larger than the sample variance for stone pipes (0.504), we designate population I as ceramic pipes.

(b) Set up the null and alternate hypotheses.

$$H_0\colon \sigma_1^2 = \sigma_2^2 \quad \text{(or the equivalent, } \sigma_1 = \sigma_2\text{)}$$
$$H_1\colon \sigma_1^2 > \sigma_2^2 \quad \text{(or the equivalent, } \sigma_1 > \sigma_2\text{)}$$

The null hypothesis states that there is no difference. The alternate hypothesis supports the anthropologist's claim that clay pipes have a larger variance.

(c) The sample test statistic is

$$F = \frac{s_1^2}{s_2^2} \approx \frac{2.266}{0.504} \approx 4.496$$

Now, if $\sigma_1^2 = \sigma_2^2$, then s_1^2 and s_2^2 also should be close in value. If this were the case, $F = s_1^2/s_2^2 \approx 1$. However, if $\sigma_1^2 > \sigma_2^2$, then we see that the sample statistic $F = s_1^2/s_2^2$ should be larger than 1.

(d) Find an interval containing the P-value for $F = 4.496$.
This is a right-tailed test (see Figure 10-14) with degrees of freedom

$$d.f._N = n_1 - 1 = 11 - 1 = 10 \quad \text{and} \quad d.f._D = n_2 - 1 = 9 - 1 = 8$$

The interval containing the P-value is

$$0.010 < P\text{-value} < 0.025$$

FIGURE 10-14

P-value

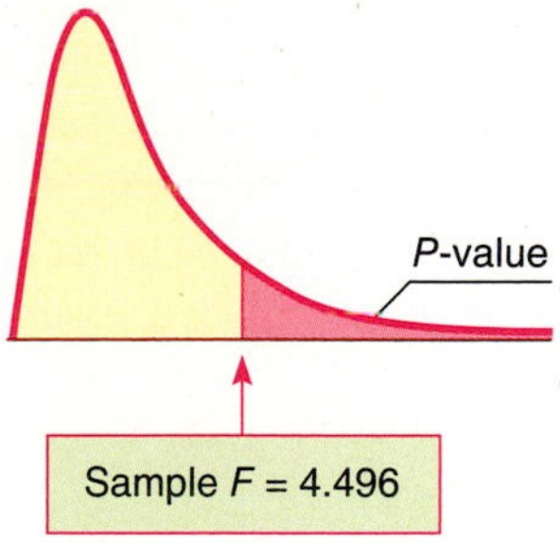

	Right-tail Area	$d.f._N$ 10
	0.100	2.54
$d.f._D$	0.050	3.35
✓ 8	0.025	4.30
	0.010	5.81
	0.001	11.54

(e) Conclude the test and *interpret* the results.
Since the P-value is less than $\alpha = 0.05$, we reject H_0.

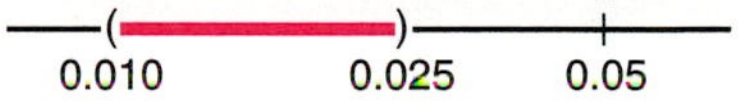

Technology gives P-value $\approx$ 0.0218. At the 5% level of significance, the evidence is sufficient to conclude that the variance for the ceramic pipes is larger.

We summarize the steps involved in testing two variances with the following procedure.

PROCEDURE

HOW TO TEST TWO VARIANCES σ_1^2 AND σ_2^2

Requirements

Assume that x_1 and x_2 are random variables that have *independent normal distributions* with unknown variances σ_1^2 and σ_2^2. Next, you need independent random samples of x_1 values and x_2 values, from which you compute sample variances s_1^2 and s_2^2. Use samples of sizes n_1 and n_2, respectively, with both samples of size at least 2. Without loss of generality, we may assume the notational setup is such that $s_1^2 \geq s_2^2$.

Procedure

1. Set the *level of significance* α. Use the *null hypothesis* H_0: $\sigma_1^2 = \sigma_2^2$. In the context of the problem, choose the *alternate hypothesis* to be H_1: $\sigma_1^2 > \sigma_2^2$ or H_1: $\sigma_1^2 \neq \sigma_2^2$.
2. Compute the *sample test statistic*
   $$F = \frac{s_1^2}{s_2^2}$$
   where $d.f._N = n_1 - 1$ (degrees of freedom numerator)
   $d.f._D = n_2 - 1$ (degrees of freedom denominator)
3. Use the *F* distribution and the type of test to find or estimate the *P-value.* Use Table 8 of Appendix II and the procedure shown in Table 10-16.
4. *Conclude* the test. If *P*-value $\leq \alpha$, then reject H_0. If *P*-value $> \alpha$, then do not reject H_0.
5. *Interpret your conclusion* in the context of the application.

GUIDED EXERCISE 10 | TESTING TWO VARIANCES

A large variance in blood chemistry components can result in health problems as the body attempts to return to equilibrium. J. B. O'Sullivan and C. M. Mahan conducted a study reported in the *American Journal of Clinical Nutrition* (Vol. 19, pp. 345–351) that concerned the glucose (blood sugar) levels of pregnant and nonpregnant women at Boston City Hospital. For both groups, a fasting (12-hour fast) blood glucose test was done. The following data are in units of milligrams of glucose per 100 milliliters of blood. Assume blood glucose levels for each group are normally distributed.

Glucose Test: Nonpregnant Women

73	61	104	75	85	65	62	98	92	106

Glucose Test: Pregnant Women

72	84	90	95	66	70	79	85

Medical researchers question if the variance of the glucose test results for nonpregnant women is *different* (either way) from the variance for pregnant women. Let's conduct a test using a 5% level of significance.

(a) ***Check requirements*** What assumptions must be made about the two populations and the samples?

→ The population measurements must follow independent normal distributions. The samples must be random samples from each population.

(b) Use a calculator to compute the sample variance for each data group. Then complete the following:

→ Recall that we choose our notation so that population I has the *larger* sample variance.

Continued

GUIDED EXERCISE 10 *continued*

Population I	Population II
n_1 = ____	n^2 = ____
s_1^2 = ____	s_2^2 = ____

Population I	Population II
$n_1 = 10$	$n_2 = 8$
$s_1^2 \approx 298.25$	$s_2^2 \approx 103.84$

(c) What is α? State the null and alternate hypotheses.

→ $\alpha = 0.05$; H_0: $\sigma_1^2 = \sigma_2^2$; H_1: $\sigma_1^2 \neq \sigma_2^2$

(d) Compute the sample test statistic F, $d.f._N$, and $d.f._D$.

→ $F = \frac{s_1^2}{s_2^2} \approx \frac{298.25}{103.84} \approx 2.87$

$d.f._N = n_1 - 1 = 10 - 1 = 9$

$d.f._D = n_2 - 1 = 8 - 1 = 7$

(e) Estimate the P-value.

→ Because this is a two-tailed test, we look up the area to the right of $F = 2.87$ and double it.

FIGURE 10-15 *P*-value

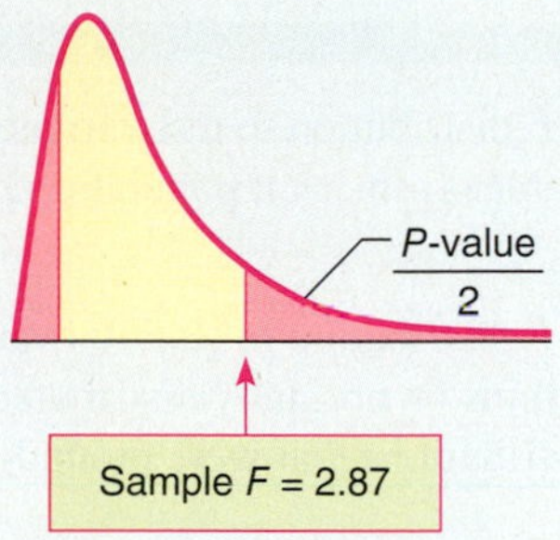

	Right-tail Area	$d.f._N$10
	0.100	2.72
$d.f._D$	0.050	3.68
✓ 7	0.025	4.82
	0.010	6.72
	0.001	14.33

$0.050 < \frac{P\text{-value}}{2} < 0.100$

$0.100 < P\text{-value} < 0.200$

(f) Conclude the test.

→ Since the P-value is greater than $\alpha = 0.05$, we do not reject H_0.

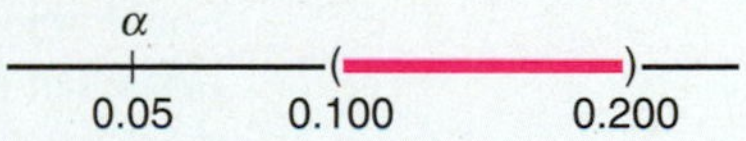

Technology gives P-value ≈ 0.1780.

(g) ***Interpret*** the results.

→ At the 5% level of significance, the evidence is insufficient to reject the claim of equal variances.

TECH NOTES The TI-84Plus/TI-83Plus/TI-*n*spire calculators support tests of two variances. Use **STAT**, **TESTS**, and option **D:2-SampFTest**. For results consistent with the notational convention that the larger variance goes in the numerator of the sample F statistic, put the data with the larger variance in **List1**.

Minitab Release 15 of Minitab supports tests of two variances. Use menu choices **Stat ➤ *Basic Statistics* ➤ 2 Variances.**

Variety is said to be the spice of life. However, in statistics, when we want to compare two populations, we will often need the assumption that the population variances are the same. As long as the two populations follow normal distributions, we can use the methods of this section and random samples from the populations to determine if the assumption of equal variances is reasonable at a given level of significance.

VIEWPOINT Mercury in Bass?

Largemouth bass were studied in 53 different lakes to examine factors that influence the level of mercury contamination. In many cases, the contamination was fairly low, except for older (trophy) fish, in which the mercury levels were much higher. Using information you have learned in this section, you can test variances of mercury contamination for different lakes and/or different regions. For more information, see the article by Lange, Royals, and Connor, "Transactions of the American Fisheries Society." To find this article, visit the Brase/Brase statistics site at **http://www.cengage.com/statistics/brase11e** *and find a link to DASL, the Carnegie Mellon University Data and Story Library. From the DASL site, link to Nature under Data Subjects, and select Mercury in the Bass Data file.*

SECTION 10.4 PROBLEMS

1. *Statistical Literacy* When using the F distribution to test variances from two populations, should the random variables from each population be independent or dependent?

2. *Statistical Literacy* When using the F distribution to test two variances, is it essential that each of the two populations be normally distributed? Would it be all right if the populations had distributions that were mound-shaped and more or less symmetrical?

3. *Statistical Literacy* In general, is the F distribution symmetrical? Can values of F be negative?

4. *Statistical Literacy* To use the F distribution, what degrees of freedom need to be calculated?

For Problems 5–12, please provide the following information.

(a) What is the level of significance? State the null and alternate hypotheses.
(b) Find the value of the sample F statistic. What are the degrees of freedom? What assumptions are you making about the original distribution?
(c) Find or estimate the P-value of the sample test statistic.
(d) Based on your answers in parts (a) to (c), will you reject or fail to reject the null hypothesis?
(e) *Interpret* your conclusion in the context of the application.

Assume that the data values in each problem come from independent populations and that each population follows a normal distribution.

5. *Agriculture: Wheat* Rothamsted Experimental Station (England) has studied wheat production since 1852. Each year, many small plots of equal size but different soil/fertilizer conditions are planted with wheat. At the end of the growing season, the yield (in pounds) of the wheat on the plot is measured. The following data are based on information taken from an article by G. A. Wiebe in the *Journal of Agricultural Research* (Vol. 50, pp. 331–357). For a random sample of years, one plot gave the following annual wheat production (in pounds):

4.15	4.21	4.27	3.55	3.50	3.79	4.09	4.42
3.89	3.87	4.12	3.09	4.86	2.90	5.01	3.39

Use a calculator to verify that, for this plot, the sample variance is $s^2 \approx 0.332$.

Another random sample of years for a second plot gave the following annual wheat production (in pounds):

4.03	3.77	3.49	3.76	3.61	3.72	4.13	4.01
3.59	4.29	3.78	3.19	3.84	3.91	3.66	4.35

Use a calculator to verify that the sample variance for this plot is $s^2 \approx 0.089$.

Test the claim that the population variance of annual wheat production for the first plot is larger than that for the second plot. Use a 1% level of significance.

6. *Agriculture: Wheat* Two plots at Rothamsted Experimental Station (see reference in Problem 5) were studied for production of wheat straw. For a random sample of years, the annual wheat straw production (in pounds) from one plot was as follows:

6.17	6.05	5.89	5.94	7.31	7.18
7.06	5.79	6.24	5.91	6.14	

Use a calculator to verify that, for the preceding data, $s^2 \approx 0.318$.

Another random sample of years for a second plot gave the following annual wheat straw production (in pounds):

6.85	7.71	8.23	6.01	7.22	5.58	5.47	5.86

Use a calculator to verify that, for these data, $s^2 \approx 1.078$.

Test the claim that there is a difference (either way) in the population variance of wheat straw production for these two plots. Use a 5% level of significance.

7. *Economics: Productivity* An economist wonders if corporate productivity in some countries is more *volatile* than that in other countries. One measure of a company's productivity is annual percentage yield based on total company assets. Data for this problem are based on information taken from *Forbes Top Companies*, edited by J. T. Davis. A random sample of leading companies in France gave the following percentage yields based on assets:

4.4	5.2	3.7	3.1	2.5	3.5	2.8	4.4	5.7	3.4	4.1
6.8	2.9	3.2	7.2	6.5	5.0	3.3	2.8	2.5	4.5	

Use a calculator to verify that $s^2 \approx 2.044$ for this sample of French companies.

Another random sample of leading companies in Germany gave the following percentage yields based on assets:

3.0	3.6	3.7	4.5	5.1	5.5	5.0	5.4	3.2
3.5	3.7	2.6	2.8	3.0	3.0	2.2	4.7	3.2

Use a calculator to verify that $s^2 \approx 1.038$ for this sample of German companies.

Test the claim that there is a difference (either way) in the population variance of percentage yields for leading companies in France and Germany. Use a 5% level of significance. How could your test conclusion relate to the economist's question regarding *volatility* (data spread) of corporate productivity of large companies in France compared with large companies in Germany?

8. *Economics: Productivity* A random sample of leading companies in South Korea gave the following percentage yields based on assets (see reference in Problem 7):

2.5	2.0	4.5	1.8	0.5	3.6	2.4
0.2	1.7	1.8	1.4	5.4	1.1	

Use a calculator to verify that $s^2 = 2.247$ for these South Korean companies.

Another random sample of leading companies in Sweden gave the following percentage yields based on assets:

2.3 3.2 3.6 1.2 3.6 2.8 2.3 3.5 2.8

Use a calculator to verify that $s^2 = 0.624$ for these Swedish companies.

Test the claim that the population variance of percentage yields on assets for South Korean companies is higher than that for companies in Sweden. Use a 5% level of significance. How could your test conclusion relate to an economist's question regarding *volatility* of corporate productivity of large companies in South Korea compared with that in Sweden?

9. *Investing: Mutual Funds* You don't need to be rich to buy a few shares in a mutual fund. The question is, "How *reliable* are mutual funds as investments?" That depends on the type of fund you buy. The following data are based on information taken from *Morningstar,* a mutual fund guide available in most libraries. A random sample of percentage annual returns for mutual funds holding stocks in aggressive-growth small companies is shown below.

−1.8	14.3	41.5	17.2	−16.8	4.4	32.6	−7.3	16.2	2.8	34.3
−10.6	8.4	−7.0	−2.3	−18.5	25.0	−9.8	−7.8	−24.6	22.8	

Use a calculator to verify that $s^2 \approx 348.43$ for the sample of aggressive-growth small company funds.

Another random sample of percentage annual returns for mutual funds holding value (i.e., market underpriced) stocks in large companies is shown below.

16.2	0.3	7.8	−1.6	−3.8	19.4	−2.5	15.9	32.6	22.1	3.4
−0.5	−8.3	25.8	−4.1	14.6	6.5	18.0	21.0	0.2	−1.6	

Use a calculator to verify that $s^2 \approx 137.31$ for value stocks in large companies.

Test the claim that the population variance for mutual funds holding aggressive-growth small stocks is larger than the population variance for mutual funds holding value stocks in large companies. Use a 5% level of significance. How could your test conclusion relate to the question of *reliability* of returns for each type of mutual fund?

10. *Investing: Mutual Funds* How *reliable* are mutual funds that invest in bonds? Again, this depends on the bond fund you buy (see reference in Problem 9). A random sample of annual percentage returns for mutual funds holding short-term U.S. government bonds is shown below.

4.6	4.7	1.9	9.3	−0.8	4.1	10.5
4.2	3.5	3.9	9.8	−1.2	7.3	

Use a calculator to verify that $s^2 \approx 13.59$ for the preceding data.

A random sample of annual percentage returns for mutual funds holding intermediate-term corporate bonds is shown below.

−0.8	3.6	20.2	7.8	−0.4	18.8	−3.4	10.5
8.0	−0.9	2.6	−6.5	14.9	8.2	18.8	14.2

Use a calculator to verify that $s^2 \approx 72.06$ for returns from mutual funds holding intermediate-term corporate bonds.

Use $\alpha = 0.05$ to test the claim that the population variance for annual percentage returns of mutual funds holding short-term government bonds is different from the population variance for mutual funds holding intermediate-term corporate bonds. How could your test conclusion relate to the question of *reliability* of returns for each type of mutual fund?

11. *Engineering: Fuel Injection* A new fuel injection system has been engineered for pickup trucks. The new system and the old system both produce about the same average miles per gallon. However, engineers question which system (old or new) will give better *consistency* in fuel consumption (miles per gallon) under a variety of driving conditions. A random sample of 31 trucks were fitted with the new fuel injection system and driven under different conditions. For these trucks, the sample variance of gasoline consumption was 58.4. Another random sample of 25 trucks were fitted with the old fuel injection system and driven under a variety of different conditions. For these trucks, the sample variance of gasoline consumption was 31.6. Test the claim that there is a difference in population variance of gasoline consumption for the two injection systems. Use a 5% level of significance. How could your test conclusion relate to the question regarding the *consistency* of fuel consumption for the two fuel injection systems?

12. *Engineering: Thermostats* A new thermostat has been engineered for the frozen food cases in large supermarkets. Both the old and the new thermostats hold temperatures at an average of 25°F. However, it is hoped that the new thermostat might be more *dependable* in the sense that it will hold temperatures closer to 25°F. One frozen food case was equipped with the new thermostat, and a random sample of 21 temperature readings gave a sample variance of 5.1. Another, similar frozen food case was equipped with the old thermostat, and a random sample of 16 temperature readings gave a sample variance of 12.8. Test the claim that the population variance of the old thermostat temperature readings is larger than that for the new thermostat. Use a 5% level of significance. How could your test conclusion relate to the question regarding the *dependability* of the temperature readings?

SECTION **10.5**

One-Way ANOVA: Comparing Several Sample Means

FOCUS POINTS

- Learn about the risk α of a type I error when we test several means at once.
- Learn about the notation and setup for a one-way ANOVA test.
- Compute mean squares between groups and within groups.
- Compute the sample F statistic.
- Use the F distribution to estimate a P-value and conclude the test.

In our past work, to determine the existence (or nonexistence) of a significant difference between population means, we restricted our attention to only two data groups representing the means in question. Many statistical applications in psychology, social science, business administration, and natural science involve many means and many data groups. Questions commonly asked are: Which of *several* alternative methods yields the best results in a particular setting? Which of *several* treatments leads to the highest incidence of patient recovery? Which of *several* teaching methods leads to greatest student retention? Which of *several* investment schemes leads to greatest economic gain?

Using our previous methods (Sections 8.4 and 8.5) of comparing only *two* means would require many tests of significance to answer the preceding questions. For example, even if we had only 5 variables, we would be required to perform 10 tests of significance in order to compare each variable to each of the other variables. If we had the time and patience, we could perform all 10 tests, but what about the risk of accepting a difference where there really is no difference (a type I error)? If the risk of a type I error on each test is $\alpha = 0.05$, then on 10 tests we expect the number of tests with a type I error to be 10(0.05), or 0.5 (see expected value, Section 5.3). This situation may not seem too serious to you, but remember that in a "real-world" problem and with the aid of a high-speed computer, a researcher may want to study the effect of 50 variables on the outcome of an experiment. Using a little mathematics, we can show that the study would require 1225 separate tests to check *each pair* of variables for a significant difference of means. At the $\alpha = 0.05$ level of significance for each test, we could expect (1225)(0.05), or 61.25, of the tests to have a type I error. In other words, these 61.25 tests would say that there are differences between means when there really are no differences.

Introduction

ANOVA

To avoid such problems, statisticians have developed a method called *analysis of variance* (abbreviated *ANOVA*). We will study single-factor analysis of variance (also called *one-way ANOVA*) in this section and two-way ANOVA in Section 10.6. With appropriate modification, methods of single-factor ANOVA generalize to n-dimensional ANOVA, but we leave that topic to more advanced studies.

EXAMPLE 8

ONE-WAY ANOVA TEST

Explanation of notation and formulas

A psychologist is studying the effects of dream deprivation on a person's anxiety level during waking hours. Brain waves, heart rate, and eye movements can be used to determine if a sleeping person is about to enter into a dream period. Three groups of subjects were randomly chosen from a large group of college students who volunteered to participate in the study. Group I subjects had their sleep interrupted four times each night, but never during or immediately before a dream. Group II subjects had their sleep interrupted four times also, but on two occasions they were wakened at the onset of a dream. Group III subjects were wakened four times, each time at the onset of a dream. This procedure was repeated for 10 nights, and each day all subjects were given a test to determine their levels of anxiety. The data in Table 10-17 record the total of the test scores for each person over the entire project. Higher totals mean higher anxiety levels.

From Table 10-17, we see that group I had $n_1 = 6$ subjects, group II had $n_2 = 7$ subjects, and group III had $n_3 = 5$ subjects. For each subject, the anxiety score (x value) and the square of the test score (x^2 value) are also shown. In addition, special sums are shown.

Tanya Constantine/Blend Images/Jupiter Images

TABLE 10-17 Dream Deprivation Study

Group I $n_1 = 6$ Subjects		Group II $n_2 = 7$ Subjects		Group III $n_3 = 5$ Subjects	
x_1	x_1^2	x_2	x_2^2	x_3	x_3^2
9	81	10	100	15	225
7	49	9	81	11	121
3	9	11	121	12	144
6	36	10	100	9	81
5	25	7	49	10	100
8	64	6	36		
		8	64		
$\Sigma x_1 = 38$	$\Sigma x_1^2 = 264$	$\Sigma x_2 = 61$	$\Sigma x_2^2 = 551$	$\Sigma x_3 = 57$	$\Sigma x_3^2 = 671$

$N = n_1 + n_2 + n_3 = 18$

$\Sigma x_{TOT} = \Sigma x_1 + \Sigma x_2 + \Sigma x_3 = 156$

$\Sigma x_{TOT}^2 = \Sigma x_1^2 + \Sigma x_2^2 + \Sigma x_3^2 = 1486$

We will outline the procedure for single-factor ANOVA in six steps. Each step will contain general methods and rationale appropriate to all single-factor ANOVA tests. As we proceed, we will use the data of Table 10-17 for a specific reference example.

Basic requirements of ANOVA

Our application of ANOVA has three basic requirements. In a general problem with k groups:

1. We require that each of our k groups of measurements is obtained from a population with a *normal* distribution.
2. Each group is randomly selected and is *independent* of all other groups. In particular, this means that we will not use the same subjects in more than one group and that the scores of one subject will not have an effect on the scores of another subject.
3. We assume that the variables from each group come from distributions with approximately the *same standard deviation.*

STEP 1: Determine the Null and Alternate Hypotheses

The purpose of an ANOVA test is to determine the existence (or nonexistence) of a statistically significant difference *among* the group means. In a general problem with k groups, we call the (population) mean of the first group μ_1, the population mean of the second group μ_2, and so forth. The null hypothesis is simply that *all* the group population means are the same. Since our basic requirements state that each of the k groups of measurements comes from normal, independent distributions with common standard deviation, the null hypothesis states that all the sample groups come from *one and the same* population. The alternate hypothesis is that *not all* the group population means are equal. Therefore, in a problem with k groups, we have

Hypotheses

HYPOTHESES FOR ONE-WAY ANOVA

H_0: $\mu_1 = \mu_2 = \cdots = \mu_k$
H_1: At least two of the means $\mu_1, \mu_2, \ldots, \mu_k$ are not equal.

Notice that the alternate hypothesis claims that *at least* two of the means are not equal. If more than two of the means are unequal, the alternate hypothesis is, of course, satisfied.

In our dream problem, we have $k = 3$; μ_1 is the population mean of group I, μ_2 is the population mean of group II, and μ_3 is the population mean of group III. Therefore,

H_0: $\mu_1 = \mu_2 = \mu_3$
H_1: At least two of the means μ_1, μ_2, μ_3 are not equal.

We will test the null hypothesis using an $\alpha = 0.05$ level of significance. Notice that only one test is being performed even though we have $k = 3$ groups and three corresponding means. Using ANOVA avoids the problem mentioned earlier of using multiple tests.

STEP 2: Find SS_{TOT}

Sum of squares, *SS*

The concept of *sum of squares* is very important in statistics. We used a sum of squares in Chapter 3 to compute the sample standard deviation and sample variance.

$$s = \sqrt{\frac{\Sigma(x - \bar{x})^2}{n - 1}} \quad \text{sample standard deviation}$$

$$s^2 = \frac{\Sigma(x - \bar{x})^2}{n - 1} \quad \text{sample variance}$$

The numerator of the sample variance is a special sum of squares that plays a central role in ANOVA. Since this numerator is so important, we give it the special name SS (for "sum of squares").

$$SS = \Sigma(x - \bar{x})^2 \tag{2}$$

Using some college algebra, it can be shown that the following, simpler formula is equivalent to Equation (2) and involves fewer calculations:

$$SS = \Sigma x^2 - \frac{(\Sigma x)^2}{n} \tag{3}$$

where n is the sample size.

In future references to SS, we will use Equation (3) because it is easier to use than Equation (2).

The **total sum of squares SS_{TOT}** can be found by using the entire collection of all data values in all groups:

$$SS_{TOT} = \Sigma x_{TOT}^2 - \frac{(\Sigma x_{TOT})^2}{N} \tag{4}$$

where $N = n_1 + n_2 + \cdots + n_k$ is the total sample size from all groups.

$$\Sigma x_{TOT} = \text{sum of all data} = \Sigma x_1 + \Sigma x_2 + \cdots + \Sigma x_k$$
$$\Sigma x_{TOT}^2 = \text{sum of all data squares} = \Sigma x_1^2 + \cdots + \Sigma x_k^2$$

Using the specific data given in Table 10-17 for the dream example, we have

$$k = 3 \quad \text{total number of groups}$$
$$N = n_1 + n_2 + n_3 = 6 + 7 + 5 = 18 \quad \text{total number of subjects}$$

$$\Sigma x_{TOT} = \text{total sum of } x \text{ values} = \Sigma x_1 + \Sigma x_2 + \Sigma x_3 = 38 + 61 + 57 = 156$$
$$\Sigma x_{TOT}^2 = \text{total sum of } x^2 \text{ values} = \Sigma x_1^2 + \Sigma x_2^2 + \Sigma x_3^2 = 264 + 551 + 671 = 1486$$

Therefore, using Equation (4), we have

$$SS_{TOT} = \Sigma x_{TOT}^2 - \frac{(\Sigma x_{TOT})^2}{N} = 1486 - \frac{(156)^2}{18} = 134$$

The numerator for the total variation for all groups in our dream example is $SS_{TOT} = 134$. What interpretation can we give to SS_{TOT}? If we let $\bar{x}_{TOT}$ be the mean of all x values for all groups, then

$$\text{Mean of all } x \text{ values} = \bar{x}_{TOT} = \frac{\Sigma x_{TOT}}{N}$$

Under the null hypothesis (that all groups come from the same normal distribution), $SS_{TOT} = \Sigma(x_{TOT} - \bar{x}_{TOT})^2$ represents the numerator of the sample variance for all groups. Therefore, SS_{TOT} represents total variability of the data. Total variability can occur in two ways:

SS_{BET}

SS_W

1. Scores may differ from one another because they belong to *different groups* with different means (recall that the alternate hypothesis states that the means are not all equal). This difference is called **between-group variability** and is denoted **SS_{BET}**.
2. Inherent differences unique to each subject and differences due to chance may cause a particular score to be different from the mean of its *own group*. This difference is called **within-group variability** and is denoted **SS_W**.

Because total variability SS_{TOT} is the sum of between-group variability SS_{BET} and within-group variability SS_W, we may write

$$SS_{TOT} = SS_{BET} + SS_W$$

As we will see, SS_{BET} and SS_W are going to help us decide whether or not to reject the null hypothesis. Therefore, our next two steps are to compute these two quantities.

STEP 3: Find SS_{BET}

Sum of squares between groups

Recall that $\bar{x}_{TOT}$ is the mean of all x values from all groups. Between-group variability (SS_{BET}) measures the variability of group means. Because different groups may have different numbers of subjects, we must "weight" the variability contribution from each group by the group size n_i.

$$SS_{BET} = \sum_{\text{all groups}} n_i(\bar{x}_i - \bar{x}_{TOT})^2$$

where n_i = sample size of group i
x_i = sample mean of group i
$\bar{x}_{TOT}$ = mean for values from all group

If we use algebraic manipulations, we can write the formula for SS_{BET} in the following computationally easier form:

SUM OF SQUARES BETWEEN GROUPS

$$SS_{BET} = \sum_{\text{all groups}} \left(\frac{(\Sigma x_i)^2}{n_i}\right) - \frac{(\Sigma x_{TOT})^2}{N} \quad (5)$$

where, as before, $N = n_1 + n_2 + \cdots + n_k$
Σx_i = sum of data in group i
Σx_{TOT} = sum of data from all groups

Using data from Table 10-17 for the dream example, we have

$$\begin{aligned} SS_{BET} &= \sum_{\text{all groups}} \left(\frac{(\Sigma x_i)^2}{n_i}\right) - \frac{(\Sigma x_{TOT})^2}{N} \\ &= \frac{(\Sigma x_1)^2}{n_1} + \frac{(\Sigma x_2)^2}{n_2} + \frac{(\Sigma x_3)^2}{n_3} - \frac{(\Sigma x_{TOT})^2}{N} \\ &= \frac{(38)^2}{6} + \frac{(61)^2}{7} + \frac{(57)^2}{5} - \frac{(156)^2}{18} \\ &= 70.038 \end{aligned}$$

Therefore, the numerator of the between-group variation is

$$SS_{BET} = 70.038$$

STEP 4: Find SS_W

Sum of squares within groups

We could find the value of SS_W by using the formula relating SS_{TOT} to SS_{BET} and SS_W and solving for SS_W:

$$SS_W = SS_{TOT} - SS_{BET}$$

However, we prefer to compute SS_W in a different way and to use the preceding formula as a check on our calculations.

SS_W is the numerator of the variation within groups. Inherent differences unique to each subject and differences due to chance create the variability assigned to SS_W. In a general problem with k groups, the variability within the ith group can be represented by

$$SS_i = \Sigma(x_i - \bar{x}_i)^2$$

or by the mathematically equivalent formula

$$SS_i = \Sigma x_i^2 - \frac{(\Sigma x_i)^2}{n_i} \tag{6}$$

Because SS_i represents the variation within the ith group and we are seeking SS_W, the variability within *all* groups, we simply add SS_i for all groups:

SUM OF SQUARES WITHIN GROUPS

$$SS_W = SS_1 + SS_2 + \cdots + SS_k \tag{7}$$

Using Equations (6) and (7) and the data of Table 10-17 with $k = 3$, we have

$$SS_1 = \Sigma x_1^2 - \frac{(\Sigma x_1)^2}{n_1} = 264 - \frac{(38)^2}{6} = 23.333$$

$$SS_2 = \Sigma x_2^2 - \frac{(\Sigma x_2)^2}{n_2} = 551 - \frac{(61)^2}{7} = 19.429$$

$$SS_3 = \Sigma x_3^2 - \frac{(\Sigma x_3)^2}{n_3} = 671 - \frac{(57)^2}{5} = 21.200$$

$$SS_W = SS_1 + SS_2 + SS_3 = 23.333 + 19.429 + 21.200 = 63.962$$

Let us check our calculation by using SS_{TOT} and SS_{BET}.

$$SS_{TOT} = SS_{BET} + SS_W$$

$$134 = 70.038 + 63.962 \quad \text{(from steps 2 and 3)}$$

We see that our calculation checks.

STEP 5: Find Variance Estimates (Mean Squares)

Mean squares

In steps 3 and 4, we found SS_{BET} and SS_W. Although these quantities represent variability between groups and within groups, they are not yet the variance estimates we need for our ANOVA test. You may recall our study of the Student's t distribution, in which we introduced the concept of degrees of freedom. Degrees of freedom represent the number of values that are free to vary once we have placed certain restrictions on our data. In ANOVA, there are two types of degrees of freedom: $d.f._{BET}$, representing the degrees of freedom between groups, and $d.f._W$, representing degrees of freedom within groups. A theoretical discussion beyond the scope of this text would show

DEGREES OF FREEDOM BETWEEN AND WITHIN GROUPS

$d.f._{BET} = k - 1$ where k is the number of groups

$d.f._W = N - k$ where N is the total sample size

(*Note:* $d.f._{BET} + d.f._W = N - 1$.)

The variance estimates we are looking for are designated as follows:

MS_{BET}

MS_W

MS_{BET}, the variance between groups (read "mean square between")

MS_W, the variance within groups (read "mean square within")

In the literature of ANOVA, the variances between and within groups are usually referred to as *mean squares between* and *within* groups, respectively. We will use the mean-square notation because it is used so commonly. However, remember that the notations MS_{BET} and MS_W both refer to *variances,* and you might occasionally see the variance notations S^2_{BET} and S^2_W used for these quantities. The formulas for the variances between and within samples follow the pattern of the basic formula for sample variance.

$$\text{Sample variance} = s^2 = \frac{\Sigma(x - \bar{x})^2}{n - 1} = \frac{SS}{n - 1}$$

However, instead of using $n - 1$ in the denominator for MS_{BET} and MS_W variances, we use their respective degrees of freedom.

$$\text{Mean square between} = MS_{BET} = \frac{SS_{BET}}{d.f._{BET}} = \frac{SS_{BET}}{k - 1}$$

$$\text{Mean square within} = MS_W = \frac{SS_W}{d.f._W} = \frac{SS_W}{N - k}$$

Using these two formulas and the data of Table 10-17, we find the mean squares within and between variances for the dream deprivation example:

$$MS_{BET} = \frac{SS_{BET}}{k - 1} = \frac{70.038}{3 - 1} = 35.019$$

$$MS_W = \frac{SS_W}{N - k} = \frac{63.962}{18 - 3} = 4.264$$

STEP 6: Find the *F* Ratio and Complete the ANOVA Test

The logic of our ANOVA test rests on the fact that one of the variances, MS_{BET}, *can* be influenced by population differences among means of the several groups, whereas the other variance, MS_W, *cannot* be so influenced. For instance, in the dream deprivation and anxiety study, the variance between groups MS_{BET} will be affected if any of the treatment groups has a population mean anxiety score that is *different* from that of any other group. On the other hand, the variance within groups MS_W compares anxiety scores of each treatment group to its own group anxiety mean, and the fact that group means might differ *does not* affect the MS_W value.

Recall that the null hypothesis claims that all the groups are samples from populations having the *same* (normal) distributions. The alternate hypothesis states that at least two of the sample groups come from populations with *different* (normal) distributions.

If the *null* hypothesis is *true,* MS_{BET} and MS_W should both estimate the *same* quantity. Therefore, if H_0 is true, the F ratio

***F* ratio**

Sample *F* statistic

SAMPLE *F* RATIO

$$F = \frac{MS_{BET}}{MS_W}$$

The *F* ratio is the *sample test statistic F* for ANOVA tests.

should be approximately 1, and variations away from 1 should occur only because of sampling errors. The variance within groups MS_W is a good estimate of the overall population variance, but the variance between groups MS_{BET} consists of the population variance *plus* an additional variance stemming from the differences between samples. Therefore, if the *null* hypothesis is *false,* MS_{BET} will be larger than MS_W, and the F ratio will tend to be *larger* than 1.

The decision of whether or not to reject the null hypothesis is determined by the relative size of the F ratio. Table 8 of Appendix II gives F values.

For our example about dreams, the computed F ratio is

$$F = \frac{MS_{BET}}{MS_W} = \frac{35.019}{4.264} = 8.213$$

ANOVA test is right-tailed

Because large F values tend to discredit the null hypothesis, we use a *right-tailed test* with the *F distribution.* To find (or estimate) the P-value for the sample F statistic, we use the F-distribution table, Table 8 of Appendix II. The table requires us to know *degrees of freedom for the numerator* and *degrees of freedom for the denominator.*

Degrees of freedom for sample test statistic *F*

DEGREES OF FREEDOM FOR SAMPLE TEST STATISTIC *F* IN ONE-WAY ANOVA

Degrees of freedom numerator $= d.f._N = d.f._{BET} = k - 1$

Degrees of freedom denominator $= d.f._D = d.f._W = N - k$

where k = number of groups

N = total sample size across all groups

For our example about dreams,

$$d.f._N = k - 1 = 3 - 1 = 2 \qquad d.f._D = N - k = 18 - 3 = 15$$

Finding the *P*-value

Let's use the F-distribution table (Table 8, Appendix II) to find the P-value of the sample statistic $F = 8.213$. The P-value is a *right-tail area,* as shown in Figure 10-16. In Table 8, look in the block headed by column $d.f._N = 2$ and row $d.f._D = 15$. For convenience, the entries are shown in Table 10-18 (Excerpt from Table 8). We see that the sample $F = 8.213$ falls between the entries 6.36 and 11.34, with corresponding right-tail areas 0.010 and 0.001.

FIGURE 10-16

P-value

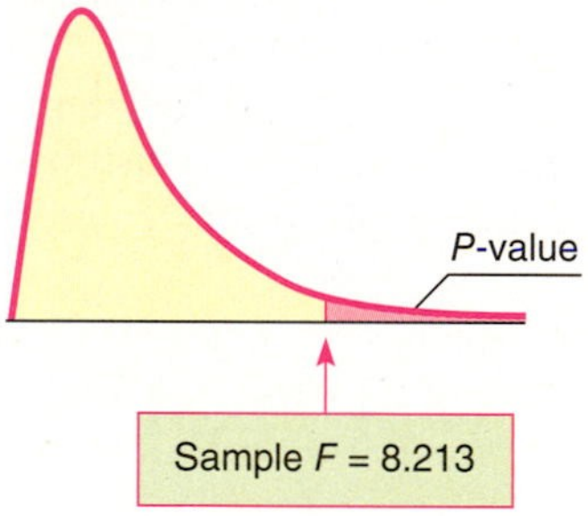

TABLE 10-18 Excerpt from Table 8, Appendix II

	Right-tail Area	$d.f._N$ 2
	0.100	2.70
$d.f._D$	0.050	3.68
✓ 15	0.025	4.77
	0.010	6.36
	0.001	11.34

Test conclusion

The P-value is in the interval $0.001 < P\text{-value} < 0.010$. Since $\alpha = 0.05$, we see that the P-value is less than α and we reject H_0.

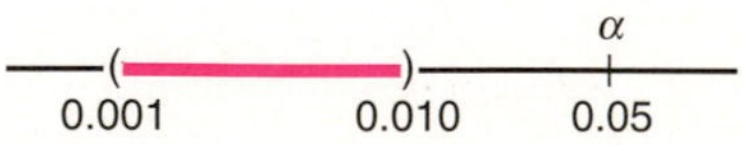

At the 5% level of significance, we reject H_0 and conclude that not all the means are equal. The amount of dream deprivation *does* make a difference in mean anxiety level. *Note:* Technology gives P-value ≈ 0.0039.

This completes our single-factor ANOVA test. Before we consider another example, let's summarize the main points.

PROCEDURE

HOW TO CONSTRUCT A ONE-WAY ANOVA TEST

Requirements

You need k independent data groups, with each group belonging to a normal distribution and all groups having (approximately) the same standard deviation. N is the total number of data values across all groups.

Procedure

1. Set the *level of significance* α and the *hypotheses*

   H_0: $\mu_1 = \mu_2 = \cdots = \mu_k$
   H_1: not all of $\mu_1, \mu_2, \ldots, \mu_k$, are equal

   where μ_i is the population mean of group i.
2. Compute the *sample test statistic* F using the following steps or appropriate technology.

   (a) $SS_{TOT} = \Sigma x_{TOT}^2 - \dfrac{(\Sigma x_{TOT})^2}{N}$

   where Σx_{TOT} is the sum of all data elements from all groups
   Σx_{TOT}^2 is the sum of all data elements squared from all groups
   N is the total sample size

   (b) $SS_{TOT} = SS_{BET} + SS_W$

   where $SS_{BET} = \sum_{\text{all groups}} \left(\dfrac{(\Sigma x_i)^2}{n_i} \right) - \dfrac{(\Sigma x_{TOT})^2}{N}$

   n_i is the number of data elements in group i
   Σx_i is the sum of the data elements in group i

   $SS_W = \sum_{\text{all groups}} \left(\Sigma x_i^2 - \dfrac{(\Sigma x_i)^2}{n_i} \right)$

   (c) $MS_{BET} = \dfrac{SS_{BET}}{d.f._{BET}}$ where $d.f._{BET} = k - 1$

   $MS_W = \dfrac{SS_W}{d.f._W}$ where $d.f._W = N - k$

   (d) $F = \dfrac{MS_{BET}}{MS_W}$ with $d.f._N = k - 1$ and $d.f._D = N - k$

   Because an ANOVA test requires a number of calculations, we recommend that you summarize your results in a table such as Table 10-19. This is the type of table that is often generated by computer software.
3. Find (or estimate) the *P-value* using the F distribution (Table 8, Appendix II). The test is a *right-tailed* test.
4. *Conclude* the test. If P-value $\leq \alpha$, then reject H_0. If P-value $> \alpha$, then do not reject H_0.
5. *Interpret your conclusion* in the context of the application.

Summary table for ANOVA

TABLE 10-19 Summary of ANOVA Results

Source of Variation	Sum of Squares	Degrees of Freedom	Mean Square (Variance)	F Ratio	P-value	Test Decision
			Basic Model			
Between groups	SS_{BET}	$d.f._{BET}$	MS_{BET}	$\frac{MS_{BET}}{MS_W}$	From table	Reject H_0 or fail to reject H_0
Within groups	SS_W	$d.f._W$	MS_W			
Total	SS_{TOT}	$N - 1$				
			Summary of ANOVA Results from Dream Experiment (Example 8)			
Between groups	70.038	2	35.019	8.213	< 0.010	Reject H_0
Within groups	63.962	15	4.264			
Total	134	17				

GUIDED EXERCISE 11 ONE-WAY ANOVA TEST

A psychologist is studying pattern-recognition skills under four laboratory settings. In each setting, a fourth-grade child is given a pattern-recognition test with 10 patterns to identify. In setting I, the child is given *praise* for each correct answer and no comment about wrong answers. In setting II, the child is given *criticism* for each wrong answer and no comment about correct answers. In setting III, the child is given no praise or criticism, but the observer expresses *interest* in what the child is doing. In setting IV, the observer remains *silent* in an adjacent room watching the child through a one-way mirror. A random sample of fourth-grade children was used, and each child participated in the test only once. The test scores (number correct) for each group follow. (See Table 10-20.)

(a) Fill in the missing entries of Table 10-20.

$\Sigma x_{TOT} =$ ______

$\Sigma x^2_{TOT} =$ ______

$N =$ ______

$k =$ ______

$$\Sigma x_{TOT} = \Sigma x_1 + \Sigma x_2 + \Sigma x_3 + \Sigma x_4$$
$$= 41 + 14 + 38 + 28 = 121$$
$$\Sigma x^2_{TOT} = \Sigma x_1^2 + \Sigma x_2^2 + \Sigma x_3^2 + \Sigma x_4^2$$
$$= 339 + 54 + 264 + 168 = 825$$
$$N = n_1 + n_2 + n_3 + n_4 = 5 + 4 + 6 + 5 = 20$$
$$k = 4 \text{ groups}$$

(b) What assumptions are we making about the data to apply a single-factor ANOVA test?

Because each of the groups comes from independent random samples (no child was tested twice), we need assume only that each group of data came from a normal distribution, and that all the groups came from distributions with about the same standard deviation.

TABLE 10-20 Pattern-Recognition Experiment

Group I (Praise) $n_1 = 5$		Group II (Criticism) $n_2 = 4$		Group III (Interest) $n_3 = 6$		Group IV (Silence) $n_4 = 5$	
x_1	x_1^2	x_2	x_2^2	x_3	x_3^2	x_4	x_4^2
9	81	2	4	9	81	5	25
8	64	5	25	3	9	7	49
8	64	4	16	7	49	3	9
9	81	3	9	8	64	6	36
7	49			5	25	7	49
				6	36		
$\Sigma x_1 = 41$		$\Sigma x_2 = 14$		$\Sigma x_3 = 38$		$\Sigma x_4 = 28$	
$\Sigma x_1^2 = 339$		$\Sigma x_2^2 = 54$		$\Sigma x_3^2 = 264$		$\Sigma x_4^2 = 168$	
$\Sigma x_{TOT} =$ ____		$\Sigma x^2_{TOT} =$ ____		$N =$ ____		$k =$ ____	

Continued

GUIDED EXERCISE 11 *continued*

(c) What are the null and alternate hypotheses?

$H_0: \mu_1 = \mu_2 = \mu_3 = \mu_4$

In words, all the groups have the same population mean, and this hypothesis, together with the basic assumptions of part (b), states that all the groups come from the same population.

H_1: not all the means $\mu_1, \mu_2, \mu_3, \mu_4$ are equal.

In words, not all the groups have the same population mean, so at least one group did not come from the same population as the others.

(d) Find the value of SS_{TOT}.

$$SS_{TOT} = \Sigma x_{TOT}^2 - \frac{(\Sigma x_{TOT})^2}{N} = 825 - \frac{(121)^2}{20} = 92.950$$

(e) Find SS_{BET}.

$$SS_{BET} = \sum_{\text{all groups}} \left(\frac{(\Sigma x_i)^2}{n_i} \right) - \frac{(\Sigma x_{TOT})^2}{N}$$

$$= \frac{(41)^2}{5} + \frac{(14)^2}{4} + \frac{(38)^2}{6} + \frac{(28)^2}{5} - \frac{(121)^2}{20} = 50.617$$

(f) Find SS_W and check your calculations using the formula

$SS_{TOT} = SS_{BET} + SS_W$

$$SS_W = \sum_{\text{all groups}} \left(\Sigma x_i^2 - \frac{(\Sigma x_i)^2}{n_i} \right)$$

$$SS_W = SS_1 + SS_2 + SS_3 + SS_4$$

$$SS_1 = \Sigma x_1^2 - \frac{(\Sigma x_1)^2}{n_1} = 339 - \frac{(41)^2}{5} = 2.800$$

$$SS_2 = \Sigma x_2^2 - \frac{(\Sigma x_2)^2}{n_2} = 54 - \frac{(14)^2}{4} = 5.000$$

$$SS_3 = \Sigma x_3^2 - \frac{(\Sigma x_3)^2}{n_3} = 264 - \frac{(38)^2}{6} \approx 23.333$$

$$SS_4 = \Sigma x_4^2 - \frac{(\Sigma x_4)^2}{n_4} = 168 - \frac{(28)^2}{5} = 11.200$$

$$SS_W = 42.333$$

Check: $SS_{TOT} = SS_{BET} + SS_W$

$92.950 = 50.617 + 42.333$ checks

(g) Find $d.f._{BET}$ and $d.f._W$.

$d.f._{BET} = k - 1 = 4 - 1 = 3$

$d.f._W = N - k = 20 - 4 = 16$

Check: $N - 1 = d.f._{BET} + d.f._W$

$20 - 1 = 3 + 16$ checks

Continued

GUIDED EXERCISE 11 *continued*

(h) Find the mean squares MS_{BET} and MS_W.

$$MS_{BET} = \frac{SS_{BET}}{d.f._{BET}} = \frac{50.617}{3} \approx 16.872$$

$$MS_W = \frac{SS_W}{d.f._W} = \frac{42.333}{16} \approx 2.646$$

(i) Find the F ratio (sample test statistic F).

$$F = \frac{MS_{BET}}{MS_W} = \frac{16.872}{2.646} \approx 6.376$$

(j) Estimate the P-value for the sample $F = 6.376$. (Use Table 8 of Appendix II.)

$d.f._N = d.f._{BET} = k - 1 = 3$
$d.f._D = d.f._W = N - k = 16$

FIGURE 10-17 *P*-value

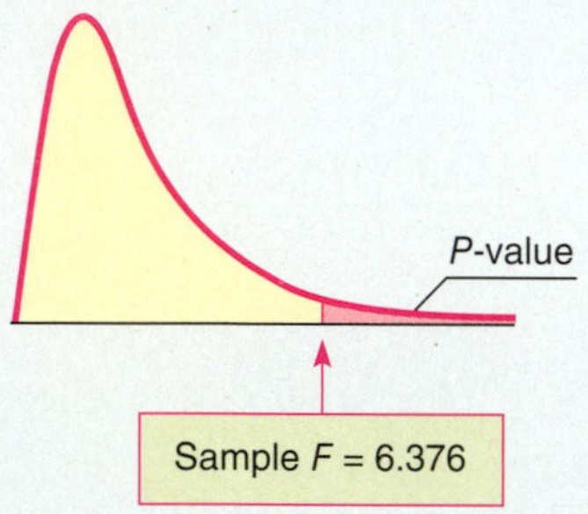

	Right-tail Area	$d.f._N$ 3
	0.100	2.46
$d.f._D$	0.050	3.24
✓ 16	0.025	4.08
	0.010	5.29
	0.001	9.01

$0.001 < P\text{-value} < 0.010$

(k) Conclude the test using a 1% level of significance. Does the test indicate that we should reject or fail to reject the null hypothesis? Explain.

$\alpha = 0.01$. Technology gives P-value ≈ 0.0048.

Since the P-value is less than 0.01, we reject H_0 and conclude that there is a significant difference in population means among the four groups. The laboratory setting *does* affect the mean scores.

(l) Make a summary table of this ANOVA test.

See Table 10-21.

TABLE 10-21 **Summary of ANOVA Results for Pattern-Recognition Experiment**

Source of Variation	Sum of Squares	Degrees of Freedom	Mean Square (Variance)	F Ratio	P-value	Test Decision
Between groups	50.617	3	16.872	6.376	< 0.01	Reject H_0
Within groups	42.333	16	2.646			
Total	92.950	19				

TECH NOTES After you understand the process of ANOVA, technology tools offer valuable assistance in performing one-way ANOVA. The TI-84Plus/TI-83Plus/TI-*n*spire calculators, Excel 2010, and Minitab all support one-way ANOVA. Both the TI-84Plus/TI-83Plus/TI-*n*spire calculators and Minitab use the terminology

Factor for Between Groups
Error for Within Groups

In all the technologies, enter the data for each group in separate columns. The displays show results for Guided Exercise 11.

TI-84Plus/TI-83Plus/TI-*n*spire (with TI-84Plus keypad) Use **STAT, TESTS,** and option **F:ANOVA** and enter the lists containing the data.

```
One-way ANOVA
 F=6.376902887
 p=.0047646422
 Factor
  df=3
  SS=50.6166667
↓ MS=16.8722222
```

```
One-way ANOVA
↑  MS=16.8722222
 Error
  df=16
  SS=42.3333333
  MS=2.64583333
 Sxp=1.62660177
```

Excel 2010 Enter the data for each group in separate columns. On the home screen, click the **Data** tab. Then in the Analysis group, click **Data Analysis**. In the dialogue box, select **Anova: Single Factor.**

Anova: Single Factor						
SUMMARY						
Groups	*Count*	*Sum*	*Average*	*Variance*		
x1	5	41	8.2	0.7		
x2	4	14	3.5	1.666666667		
x3	6	38	6.333333333	4.666666667		
x4	5	28	5.6	2.8		
ANOVA						
Source of Variation	*SS*	*df*	*MS*	*F*	*P-value*	*F crit*
Between Groups	50.61666667	3	16.87222222	6.376902887	0.004764642	5.292235983
Within Groups	42.33333333	16	2.645833333			
Total	92.95	19				

Minitab Enter the data for each group in separate columns. Use **Stat ➤ ANOVA ➤ Oneway(unstacked).**

```
One-way Analysis of Variance
Analysis of Variance
Source    DF      SS       MS       F       P
Factor     3   50.62    16.87    6.38   0.005
Error     16   42.33     2.65
Total     19   92.95
                                Individual 95% CIs For Mean
                             Based on Pooled StDev
Level      N    Mean    StDev   ---+---------+---------+---------+---
x1         5   8.200    0.837                        (-----*-----)
x2         4   3.500    1.291   (------*------)
x3         6   6.333    2.160                (-----*-----)
x4         5   5.600    1.673            (------*------)
                             ---+---------+---------+---------+---
Pooled StDev = 1.627             2.5       5.0       7.5      10.0
```

VIEWPOINT

Gear Guide!

How much does that backpack cost? Do brand names really make any difference in cost? One way to answer such a question is to construct a one-way ANOVA test to study list prices for various backpacks with similar features but different brand names. For more information, see Backpacker Magazine, *Vol. 25, issue 157.*

SECTION 10.5 PROBLEMS

In each problem, assume that the distributions are normal and have approximately the same population standard deviation.

In each problem, please provide the following information.

(a) What is the level of significance? State the null and alternate hypotheses.

(b) Find SS_{TOT}, SS_{BET}, and SS_W and check that $SS_{TOT} = SS_{BET} + SS_W$. Find $d.f._{BET}$, $d.f._W$, MS_{BET}, and MS_W. Find the value of the sample test statistic F (F ratio). What are the degrees of freedom?

(c) Find (or estimate) the P-value of the sample test statistic.

(d) Based on your answers in parts (a) to (c), will you reject or fail to reject the null hypothesis?

(e) *Interpret* your conclusion in the context of the application.

(f) Make a summary table for your ANOVA test.

1. *Archaeology: Ceramics* Wind Mountain is an archaeological study area located in southwestern New Mexico. Pot sherds are broken pieces of prehistoric Native American clay vessels. One type of painted ceramic vessel is called *Mimbres classic black-on-white*. At three different sites, the number of such sherds was counted in local dwelling excavations (Source: Based on information from *Mimbres Mogollon Archaeology* by A. I. Woosley and A. J. McIntyre, University of New Mexico Press).

Site I	Site II	Site III
61	25	12
34	18	36
25	54	69
12	67	27
79		18
55		14
20		

Shall we reject or not reject the claim that there is no difference in population mean Mimbres classic black-on-white sherd counts for the three sites? Use a 1% level of significance.

2. *Archaeology: Ceramics* Another type of painted ceramic vessel is called *three-circle red-on-white* (see reference in Problem 1). At four different sites in the Wind Mountain archaeological region, the number of such sherds was counted in local dwelling excavations.

Site I	Site II	Site III	Site IV
17	18	32	13
23	4	19	19
6	33	18	14
19	8	43	34
11	25		12
	16		15

Shall we reject or not reject the claim that there is no difference in the population mean three-circle red-on-white sherd counts for the four sites? Use a 5% level of significance.

3. *Economics: Profits per Employee* How productive are U.S. workers? One way to answer this question is to study annual profits per employee. A random sample of companies in computers (I), aerospace (II), heavy equipment (III), and broadcasting (IV) gave the following data regarding annual profits per employee (units in thousands of dollars) (Source: *Forbes Top Companies,* edited by J. T. Davis, John Wiley and Sons).

I	**II**	**III**	**IV**
27.8	13.3	22.3	17.1
23.8	9.9	20.9	16.9
14.1	11.7	7.2	14.3
8.8	8.6	12.8	15.2
11.9	6.6	7.0	10.1
	19.3		9.0

Shall we reject or not reject the claim that there is no difference in population mean annual profits per employee in each of the four types of companies? Use a 5% level of significance.

4. *Economics: Profits per Employee* A random sample of companies in electric utilities (I), financial services (II), and food processing (III) gave the following information regarding annual profits per employee (units in thousands of dollars). (See reference in Problem 3.)

I	II	III
49.1	55.6	39.0
43.4	25.0	37.3
32.9	41.3	10.8
27.8	29.9	32.5
38.3	39.5	15.8
36.1		42.6
20.2		

Shall we reject or not reject the claim that there is no difference in population mean annual profits per employee in each of the three types of companies? Use a 1% level of significance.

5. *Ecology: Deer* Where are the deer? Random samples of square-kilometer plots were taken in different ecological locations of Mesa Verde National Park. The deer counts per square kilometer were recorded and are shown in the following table (Source: *The Mule Deer of Mesa Verde National Park,* edited by G. W. Mierau and J. L. Schmidt, Mesa Verde Museum Association).

Mountain Brush	**Sagebrush Grassland**	**Pinon Juniper**
30	20	5
29	58	7
20	18	4
29	22	9

Shall we reject or accept the claim that there is no difference in the mean number of deer per square kilometer in these different ecological locations? Use a 5% level of significance.

6. *Ecology: Vegetation* Wild irises are beautiful flowers found throughout the United States, Canada, and northern Europe. This problem concerns the length of the sepal (leaf-like part covering the flower) of different species of wild iris. Data are based on information taken from an article by R. A. Fisher in *Annals of Eugenics* (Vol. 7, part 2, pp. 179–188). Measurements of sepal length in centimeters from random samples of *Iris setosa* (I), *Iris versicolor* (II), and *Iris virginica* (III) are as follows:

I	II	III
5.4	5.5	6.3
4.9	6.5	5.8
5.0	6.3	4.9
5.4	4.9	7.2
4.4	5.2	5.7
5.8	6.7	6.4
5.7	5.5	
	6.1	

Shall we reject or not reject the claim that there are no differences among the population means of sepal length for the different species of iris? Use a 5% level of significance.

7. *Insurance: Sales* An executive at the home office of Big Rock Life Insurance is considering three branch managers as candidates for promotion to vice president. The branch reports include records showing sales volume for each salesperson in the branch (in hundreds of thousands of dollars). A random sample of these records was selected for salespersons in each branch. All three branches are located in cities in which per capita income is the same. The executive wishes to compare these samples to see if there is a significant difference in performance of salespersons in the three different branches. If so, the information will be used to determine which of the managers to promote.

Branch Managed by Adams	Branch Managed by McDale	Branch Managed by Vasquez
7.2	8.8	6.9
6.4	10.7	8.7
10.1	11.1	10.5
11.0	9.8	11.4
9.9		
10.6		

Use an $\alpha = 0.01$ level of significance. Shall we reject or not reject the claim that there are no differences among the performances of the salespersons in the different branches?

8. *Ecology: Pollution* The quantity of dissolved oxygen is a measure of water pollution in lakes, rivers, and streams. Water samples were taken at four different locations in a river in an effort to determine if water pollution varied from location to location. Location I was 500 meters above an industrial plant water discharge point and near the shore. Location II was 200 meters above the discharge point and in midstream. Location III was 50 meters downstream from the discharge point and near the shore. Location IV was 200 meters downstream from the discharge point and in midstream. The following table shows the results. Lower dissolved oxygen readings mean more pollution. Because of the difficulty in getting midstream samples, ecology students collecting the data had fewer of these samples. Use an $\alpha = 0.05$ level of significance. Do we reject or not reject the claim that the quantity of dissolved oxygen does not vary from one location to another?

Location I	Location II	Location III	Location IV
7.3	6.6	4.2	4.4
6.9	7.1	5.9	5.1
7.5	7.7	4.9	6.2
6.8	8.0	5.1	
6.2		4.5	

9. *Sociology: Ethnic Groups* A sociologist studying New York City ethnic groups wants to determine if there is a difference in income for immigrants from four different countries during their first year in the city. She obtained the data in the following table from a random sample of immigrants from these countries (incomes in thousands of dollars). Use a 0.05 level of significance to test the claim that there is no difference in the earnings of immigrants from the four different countries.

Country I	Country II	Country III	Country IV
12.7	8.3	20.3	17.2
9.2	17.2	16.6	8.8
10.9	19.1	22.7	14.7
8.9	10.3	25.2	21.3
16.4		19.9	19.8

SECTION **10.6**

Introduction to Two-Way ANOVA

FOCUS POINTS

- Learn the notation and setup for two-way ANOVA tests.
- Learn about the three main types of deviations and how they break into additional effects.
- Use mean-square values to compute different sample *F* statistics.
- Use the *F* distribution to estimate *P*-values and conclude the test.
- Summarize experimental design features using a completely randomized design flow chart.

Suppose that Friendly Bank is interested in average customer satisfaction regarding the issue of obtaining bank balances and a summary of recent account transactions. Friendly Bank uses two systems, the first being a completely automated voice mail information system requiring customers to enter account numbers and passwords using the telephone keypad, and the second being the use of bank tellers or bank representatives to give the account information personally to customers. In addition, Friendly Bank wants to learn if average customer satisfaction is the same regardless of the time of day of contact. Three times of day are under study: morning, afternoon, and evening.

Friendly Bank could do two studies: one regarding average customer satisfaction with regard to type of contact (automated or bank representative) and one regarding average customer satisfaction with regard to time of day. The first study could be done using a difference-of-means test because there are only two types of contact being studied. The second study could be accomplished using one-way ANOVA.

Two-way ANOVA

However, Friendly Bank could use just *one* study and the technique of *two-way analysis of variance* (known as *two-way ANOVA*) to simultaneously study average customer satisfaction with regard to the variable type of contact and the variable time of day, and *also* with regard to the *interaction* between the two variables. An interaction is present if, for instance, the difference in average customer satisfaction regarding type of contact is much more pronounced during the evening hours than, say, during the afternoon hours or the morning hours.

Interaction

Factor
Level

Let's begin our study of two-way ANOVA by considering the organization of data appropriate to two-way ANOVA. Two-way ANOVA involves *two* variables. These variables are called *factors*. The *levels* of a factor are the different values the factor can assume. Example 9 demonstrates the use of this terminology for the information Friendly Bank is seeking.

EXAMPLE 9

FACTORS AND LEVELS

For the Friendly Bank study discussed earlier, identify the factors and their levels, and create a table displaying the information.

SOLUTION: There are two factors. Call factor 1 *time of day*. This factor has three levels: morning, afternoon, and evening. Factor 2 is *type of contact*. This factor has two levels: automated contact and personal contact through a bank representative. Table 10-22 shows how the information regarding customer satisfaction can be organized with respect to the two factors.

TABLE 10-22 **Table for Recording Average Customer Response**

Factor 1: Time of Day	Factor 2: Type of Contact	
	Automated	Bank Representative
Morning	Morning Automated	Morning Bank Representative
Afternoon	Afternoon Automated	Afternoon Bank Representative
Evening	Evening Automated	Evening Bank Representative

Cell

When we look at Table 10-22, we see six contact–time-of-day combinations. Each such combination is called a *cell* in the table. The number of cells in any two-way ANOVA data table equals the product of the number of levels of the row factor times the number of levels of the column factor. In the case illustrated by Table 10-22, we see that the number of cells is 3×2, or 6.

Basic requirements of two-way ANOVA

Just as for one-way ANOVA, our application of two-way ANOVA has some basic requirements:

1. The measurements in each cell of a two-way ANOVA model are assumed to be drawn from a population with a normal distribution.
2. The measurements in each cell of a two-way ANOVA model are assumed to come from distributions with approximately the same variance.
3. The measurements in each cell come from *independent* random samples.
4. There are the *same number of measurements* in each cell.

PROCEDURE TO CONDUCT A TWO-WAY ANOVA TEST (MORE THAN ONE MEASUREMENT PER CELL)

We will outline the procedure for two-way ANOVA in five steps. Each step will contain general methods and rationale appropriate to all two-way ANOVA tests with more than one data value in each cell. As we proceed, we will see how the method is applied to the Friendly Bank study.

Total mean, $\bar{\bar{x}}$

Let's assume that Friendly Bank has taken random samples of customers fitting the criteria of each of the six cells described in Table 10-22. This means that a random sample of four customers fitting the morning-automated cell were surveyed. Another random sample of four customers fitting the afternoon-automated cell were surveyed, and so on. The bank measured customer satisfaction on a scale of 0 to 10 (10 representing highest customer satisfaction). The data appear in Table 10-23. Table 10-23 also shows cell means, row means, column means, and the *total mean* $\bar{\bar{x}}$ computed for all 24 data values. We will use these means as we conduct the two-way ANOVA test.

Overview

As in any statistical test, the first task is to establish the hypotheses for the test. Then, as in one-way ANOVA, the F distribution is used to determine the test conclusion. To compute the sample F value for a given null hypothesis, many of the same kinds of computations are done as are done in one-way ANOVA. In particular, we will use degrees of freedom $d.f. = N - 1$ (where N is the total sample size) allocated among the row factor, the column factor, the interaction, and the error (corresponding to "within groups" of one-way ANOVA). We look at the sum of squares SS (which measures variation) for the row factor, the column factor, the interaction, and the error. Then we compute the mean square MS for each category by taking the SS value and dividing by the corresponding degrees of freedom. Finally, we compute the sample F statistic for each factor and for the interaction by dividing the appropriate MS value by the MS value of the error.

TABLE 10-23 **Customer Satisfaction at Friendly Bank**

Factor 1: Time of Day	Factor 2: Type of Contact		
	Automated	Bank Representative	Row Means
Morning	6, 5, 8, 4 $\bar{x} = 5.75$	8, 7, 9, 9 $\bar{x} = 8.25$	Row 1 $\bar{x} = 7.00$
Afternoon	3, 5, 6, 5 $\bar{x} = 4.75$	9, 10, 6, 8 $\bar{x} = 8.25$	Row 2 $\bar{x} = 6.50$
Evening	5, 5, 7, 5 $\bar{x} = 5.50$	9, 10, 10, 9 $\bar{x} = 9.50$	Row 3 $\bar{x} = 7.50$
Column means	Column 1 $\bar{x} = 5.33$	Column 2 $\bar{x} = 8.67$	Total $\bar{\bar{x}} = 7.00$

STEP 1: Establish the Hypotheses

Because we have two factors, we have hypotheses regarding each of the factors separately (called *main effects*) and then hypotheses regarding the interaction between the factors.

These three sets of hypotheses are

HYPOTHESES

1. H_0: There is no difference in population means among the levels of the row factor.
   H_1: At least two population means are different among the levels of the row factor.
2. H_0: There is no difference in population means among the levels of the column factor.
   H_1: At least two population means are different among the levels of the column factor.
3. H_0: There is no interaction between the factors.
   H_1: There is an interaction between the factors.

In the case of Friendly Bank, the hypotheses regarding the main effects are

H_0: There is no difference in population mean satisfaction depending on time of contact.
H_1: At least two population mean satisfaction measures are different depending on time of contact.
H_0: There is no difference in population mean satisfaction between the two types of customer contact.
H_1: There is a difference in population mean satisfaction between the two types of customer contact.

The hypotheses regarding interaction between factors are

H_0: There is no interaction between type of contact and time of contact.
H_1: There is an interaction between type of contact and time of contact.

STEP 2: Compute Sum of Squares (*SS*) Values

The calculations for the SS values are usually done on a computer. The main questions are whether population means differ according to the factors or the interaction of the factors. As we look at the Friendly Bank data in Table 10-23, we see that sample averages for customer satisfaction differ not only in each cell but also across the rows and across the columns. In addition, the total sample mean (designated $\bar{\bar{x}}$) differs from almost all the means. We know that different samples of the same size from the same population certainly can have different sample means. We need to decide if the differences are simply due to chance (sampling error) or are occurring because the samples have been taken from different populations with means that are not the same.

The tools we use to analyze the differences among the data values, the cell means, the row means, the column means, and the total mean are similar to those we used in Section 10.5 for one-way ANOVA. In particular, we first examine deviations of various measurements from the total mean $\bar{\bar{x}}$, and then we compute the sum of the squares *SS*.

There are basically three types of deviations:

Total deviation	**= Treatment deviation**	**+ Error deviation**
Compare each data value with the total mean $\bar{\bar{x}}$, $(x - \bar{\bar{x}})$.	For each data value, compare the mean of each cell with the total mean $\bar{\bar{x}}$, (cell $\bar{x} - \bar{\bar{x}}$).	Compare each data value with the mean of its cell, (x − cell $\bar{x}$).

The treatment deviation breaks down further as

Treatment deviation	**= Deviation for main effect of factor 1**	**+ Deviation for main effect of factor 2**	**+ Deviation for interaction**
For each data value, (cell $\bar{x} - \bar{\bar{x}}$).	For each data value, compare the row mean with the total mean $\bar{\bar{x}}$, (row $\bar{x} - \bar{\bar{x}}$).	For each data value, compare the column mean with the total mean $\bar{\bar{x}}$, (column $\bar{x} - \bar{\bar{x}}$).	For each data value, (cell $\bar{x}$ − corresponding row $\bar{x}$ − corresponding column $\bar{x} + \bar{\bar{x}}$).

The deviations for each data value, row mean, column mean, or cell mean are then *squared and totaled over all the data*. This results in sums of squares, or variations. The *treatment variations* correspond to *between-group variations* of one-way

ANOVA. The *error variation* corresponds to the *within-group variation* of one-way ANOVA.

$$\begin{array}{ccccc} \textbf{Total varition} & = & \textbf{Treatment variation} & + & \textbf{Error variation} \\ \downarrow & & \downarrow & & \downarrow \\ \sum_{\text{all data}} (x - \bar{\bar{x}})^2 & = & \sum_{\text{all data}} (\text{cell } \bar{x} - \bar{\bar{x}})^2 & + & \sum_{\text{all data}} (x - \text{cell } \bar{x})^2 \\ \downarrow & & \downarrow & & \downarrow \\ SS_{TOT} & = & SS_{TR} & + & SS_E \end{array}$$

where

$$\begin{array}{ccccccc} \textbf{Treatment variation} & = & \textbf{Factor 1 variation} & + & \textbf{Factor 2 variation} & + & \textbf{Interaction variation} \\ \downarrow & & \downarrow & & \downarrow & & \downarrow \\ SS_{TR} & = & SS_{F1} & + & SS_{F2} & + & SS_{F1 \times F2} \\ \downarrow & & \downarrow & & \downarrow & & \downarrow \\ \sum_{\text{all data}} (\text{cell } \bar{x} - \bar{\bar{x}})^2 & = & \sum_{\text{all data}} (\text{row } \bar{x} - \bar{\bar{x}})^2 & + & \sum_{\text{all data}} (\text{col } \bar{x} - \bar{\bar{x}})^2 & + & \sum_{\text{all data}} (\text{cell } \bar{x} - \text{row } \bar{x} - \text{col } \bar{x} + \bar{\bar{x}})^2 \end{array}$$

The actual calculation of all the required SS values is quite time-consuming. In most cases, computer packages are used to obtain the results. For the Friendly Bank data, the following table is a Minitab printout giving the sum of squares *SS* for the type-of-contact factor, the time-of-day factor, the interaction between time and type of contact, and the error.

Minitab Printout for Customer Satisfaction at Friendly Bank

Analysis of Variance for Response

Source	DF	SS	MS	F	P
Time	2	4.00	2.00	1.24	0.313
Type	1	66.67	66.67	41.38	0.000
Interaction	2	2.33	1.17	0.72	0.498
Error	18	29.00	1.61		
Total	23	102.00			

We see that $SS_{\text{type}} = 66.77$, $SS_{\text{time}} = 4.00$, $SS_{\text{interaction}} = 2.33$, $SS_{\text{error}} = 29.00$, and $SS_{TOT} = 102$ (the total of the other four sums of squares).

STEP 3: Compute the Mean Square (*MS*) Values

The calculations for the MS values are usually done on a computer. Although the sum of squares computed in step 2 represents variation, we need to compute mean-square (*MS*) values for two-way ANOVA. As in one-way ANOVA, we compute *MS* values by dividing the *SS* values by respective degrees of freedom:

$$\text{Mean square } MS = \frac{\text{Corresponding sum of squares } SS}{\text{Respective degrees of freedom}}$$

For two-way ANOVA with more than one data value per cell, the degrees of freedom are

DEGREES OF FREEDOM

d.f. of row factor $= r - 1$

d.f. of column factor $= c - 1$

d.f. of total $= nrc - 1$

d.f. of interaction $= (r - 1)(c - 1)$

d.f. of error $= rc(n - 1)$

where r = number of rows, c = number of columns, and n = number of data values in one cell.

The Minitab table shows the degrees of freedom and the *MS* values for the main effect factors, the interaction, and the error for the Friendly Bank study.

STEP 4: Compute the Sample *F* Statistic for Each Factor and for the Interaction

Under the assumption of the respective null hypothesis, we have

$$\text{Sample } F \text{ for row factor} = \frac{MS \text{ for row factor}}{MS \text{ for error}}$$

with degrees of freedom numerator, $d.f._N = d.f.$ of row factor

degrees of freedom denominator, $d.f._D = d.f.$ of error

$$\text{Sample } F \text{ for column factor} = \frac{MS \text{ for column factor}}{MS \text{ for error}}$$

with degrees of freedom numerator, $d.f._N = d.f.$ of column factor

degrees of freedom denominator, $d.f._D = d.f.$ of error

$$\text{Sample } F \text{ for interaction} = \frac{MS \text{ for interaction}}{MS \text{ for error}}$$

with degrees of freedom numerator, $d.f._N = d.f.$ of interaction

degrees of freedom denominator, $d.f._D = d.f.$ of error

For the Friendly Bank study, the sample *F* values are

$$\text{Sample } F \text{ for time:} \quad F = \frac{MS_{\text{time}}}{MS_{\text{error}}} = \frac{2.00}{1.61} = 1.24$$

$$d.f._N = 2 \quad \text{and} \quad d.f._D = 18$$

$$\text{Sample } F \text{ for type of contact:} \quad F = \frac{MS_{\text{type}}}{MS_{\text{error}}} = \frac{66.67}{1.61} = 41.41$$

$$d.f._N = 1 \quad \text{and} \quad d.f._D = 18$$

$$\text{Sample } F \text{ for interaction:} \quad F = \frac{MS_{\text{interaction}}}{MS_{\text{error}}} = \frac{1.17}{1.61} = 0.73$$

$$d.f._N = 2 \quad \text{and} \quad d.f._D = 18$$

Due to rounding, the sample *F* values we just computed differ slightly from those shown in the Minitab printout.

STEP 5: Conclude the Test

As with one-way ANOVA, larger values of the sample *F* statistic discredit the null hypothesis that there is no difference in population means across a given factor. The smaller the area to the right of the sample *F* statistic, the more likely there is an actual difference in some population means across the different factors. Smaller areas to the right of the sample *F* for interaction indicate greater likelihood of interaction between factors. Consequently, the *P*-value of a sample *F* statistic is the area of the *F* distribution to the *right* of the sample *F* statistic. Figure 10-18 shows the *P*-value associated with a sample *F* statistic.

Finding the *P*-value

Most statistical computer software packages provide *P*-values for the sample test statistic. You can also use the *F* distribution (Table 8 of Appendix II) to estimate the *P*-value. Once you have the *P*-value, compare it to the preset level of significance α. If the *P*-value is less than or equal to α, then reject H_0. Otherwise, do not reject H_0.

Be sure to test for interaction between the factors *first*. If you *reject* the null hypothesis of no interaction, then you should *not* test for a difference of means in the levels of the row factors or for a difference of means in the levels of the column factors because the interaction of the factors makes interpretation of the results of the main effects more complicated. A more extensive study of two-way ANOVA beyond the scope of this book shows how to interpret the results of the test of the main factors when there is interaction. For our purposes, we will simply stop the analysis rather than draw misleading conclusions.

FIGURE 10-18

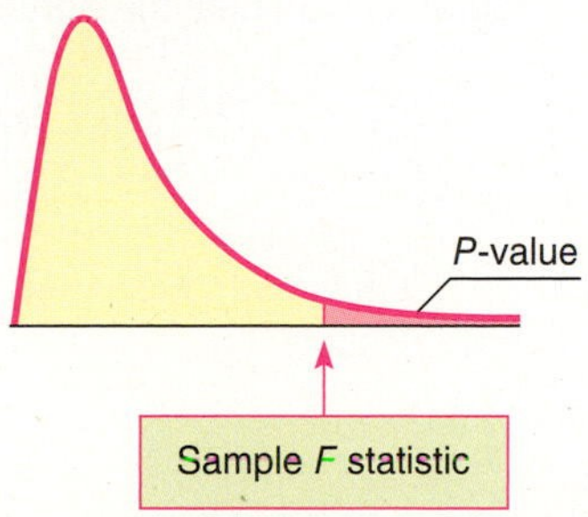

For two-way ANOVA, the *P*-value of the sample statistic is the area of the *F* distribution to the *right* of the sample *F* statistic.

In two-way ANOVA, test for *interaction* first. If you reject the null hypothesis of no interaction, then do not continue with any tests of differences of means among other factors (unless you know techniques more advanced than those presented in this section).

If the test for interaction between the factors indicates that there is no evidence of interaction, then proceed to test the hypotheses regarding the levels of the row factor and the hypotheses regarding the levels of the column factor.

For the Friendly Bank study, we proceed as follows:

1. First, we determine if there is any evidence of interaction between the factors. The sample test statistic for interaction is $F = 0.73$, with P-value ≈ 0.498. Since the P-value is greater than $\alpha = 0.05$, we do not reject H_0. There is no evidence of interaction. Because there is no evidence of interaction between the main effects of type of contact and time of day, we proceed to test each factor for a difference in population mean satisfaction among the respective levels of the factors.
2. Next, we determine if there is a difference in mean satisfaction according to type of contact. The sample test statistic for type of contact is $F = 41.41$, with P-value ≈ 0.000 (to three places after the decimal). Since the P-value is less than $\alpha = 0.05$, we reject H_0. At the 5% level of significance, we conclude that there is a difference in average customer satisfaction between contact with an automated system and contact with a bank representative.
3. Finally, we determine if there is a difference in mean satisfaction according to time of day. The sample test statistic for time of day is $F = 1.24$, with P-value ≈ 0.313. Because the P-value is greater than $\alpha = 0.05$, we do not reject H_0. We conclude that at the 5% level of significance, there is no evidence that population mean customer satisfaction is different according to time of day.

SPECIAL CASE: ONE OBSERVATION IN EACH CELL WITH NO INTERACTION

In the case where our data consist of only one value in each cell, there are no measures for sum of squares *SS* interaction or mean-square *MS* interaction, and we cannot test for interaction of factors using two-way ANOVA. If it *seems reasonable* (based on other information) to assume that there is *no* interaction between the factors, then we can use two-way ANOVA techniques to test for average response differences due to the main effects. In Guided Exercise 12, we look at two-way ANOVA applied to the special case of only one measurement per cell and no interactions.

GUIDED EXERCISE 12 | SPECIAL-CASE TWO-WAY ANOVA

Let's use two-way ANOVA to test if the average fat content (grams of fat per 3-oz serving) of potato chips is different according to the brand or according to which laboratory made the measurement. Use $\alpha = 0.05$. (See the following Minitab tables.)

Average Grams of Fat in a 3-oz Serving of Potato Chips

	Laboratory		
Brand	Lab I	Lab II	Lab III
Texas Chips	32.4	33.1	32.9
Great Chips	37.9	37.7	37.8
Chip Ooh	29.1	29.4	29.5

Minitab Printout for Potato Chip Data

Analysis of Variance for Fat

Source	DF	SS	MS	F	P
Brand	2	108.7022	54.3511	968.63	0.000
Lab	2	0.1422	0.0711	1.27	0.375
Error	4	0.2244	0.0561		
Total	8	109.0689			

(a) List the factors and the number of levels for each.

The factors are brand and laboratory. Each factor has three levels.

(b) Assuming there is no interaction, list the hypotheses for each factor.

For brand,
H_0: There is no difference in population mean fat by brands.
H_1: At least two brands have different population mean fat contents.
For laboratory,
H_0: There is no difference in mean fat content as measured by the labs.
H_1: At least two of the labs give different mean fat measurements.

(c) Calculate the sample *F* statistic for brands and compare it to the value given in the Minitab printout. Look at the *P*-value in the printout. What is your conclusion regarding average fat content among brands?

For brand,

$$\text{Sample } F = \frac{MS_{\text{brand}}}{MS_{\text{error}}} \approx \frac{54.35}{0.056} \approx 970$$

Using the Minitab printout, we see *P*-value $\approx$ 0.000 (to three places after the decimal). Using Table 8 of Appendix II, we see *P*-value < 0.001. Since the *P*-value is less than $\alpha = 0.05$, we reject H_0 and conclude that at the 5% level of significance, at least two of the brands have different mean fat contents.

(d) Calculate the sample *F* statistic for laboratories and compare it to the value given in the Minitab printout. What is your conclusion regarding average fat content as measured by the different laboratories?

For laboratories,

$$\text{Sample } F = \frac{MS_{\text{lab}}}{MS_{\text{error}}} \approx \frac{0.0711}{0.056} \approx 1.27$$

Using the Minitab printout, we see *P*-value $\approx$ 0.375. Using Table 8 of Appendix II, we see *P*-value > 0.100. Because the *P*-value is greater than $\alpha = 0.05$, we conclude that at the 5% level of significance, there is no evidence of differences in average measurements of fat content as determined by the different laboratories.

TECH NOTES The calculations involved in two-way ANOVA are usually done using statistical or spreadsheet software. Basic printouts from various software packages are similar. Specific instructions for using Excel 2010, Minitab, and SPSS are given in the Using Technology section at the end of the chapter.

EXPERIMENTAL DESIGN

In the preceding section and in this section, we have seen aspects of one-way and two-way ANOVA, respectively. Now let's take a brief look at some experimental design features that are appropriate for the use of these techniques.

Completely randomized design

For one-way ANOVA, we have one factor. Different levels for the factor form the treatment groups under study. In a *completely randomized design*, independent random samples of experimental subjects or objects are selected for each treatment group. For example, suppose a researcher wants to study the effects of different treatments for the condition of slightly high blood pressure. Three treatments are under study: diet, exercise, and medication. In a completely randomized design, the people participating in the experiment are *randomly* assigned to each treatment group. Table 10-24 shows the process.

TABLE 10-24 Completely Randomized Design Flow Chart

Subjects with slightly high blood pressure	→	Random assignment	→ Treatment 1: Diet → Treatment 2: Exercise → Treatment 3: Medication

For two-way ANOVA, there are *two* factors. When we *block* experimental subjects or objects together based on a similar characteristic that might affect responses to treatments, we have a *block design*. For example, suppose the researcher studying treatments for slightly high blood pressure believes that the age of subjects might affect the response to the three treatments. In such a case, blocks of subjects in specified age groups are used. The factor "age" is used to form blocks. Suppose age has three levels: under age 30, ages 31–50, and over age 50. The same number of subjects is assigned to each block. Then the subjects in each block are randomly assigned to the different treatments of diet, exercise, or medication. Table 10-25 shows the *randomized block design*.

Randomized block design

TABLE 10-25 Randomized Block Design Flow Chart

Blocks				Treatments	
Subjects with slightly high blood pressure	→ Under 30	→	Random assignment	→ Treatment 1:	Diet
				→ Treatment 2:	Exercise
				→ Treatment 3:	Medication
	→ Ages 31–50	→	Random assignment	→ Treatment 1:	Diet
				→ Treatment 2:	Exercise
				→ Treatment 3:	Medication
	→ Over 50	→	Random assignment	→ Treatment 1:	Diet
				→ Treatment 2:	Exercise
				→ Treatment 3:	Medication

Experimental design is an essential component of good statistical research. The design of experiments can be quite complicated, and if the experiment is complex, the services of a professional statistician may be required. The use of blocks helps the researcher account for some of the most important sources of variability among the experimental subjects or objects. Then, randomized assignments to different treatment groups help average out the effects of other remaining sources of variability. In this way, differences among the treatment groups are more likely to be caused by the treatments themselves rather than by other sources of variability.

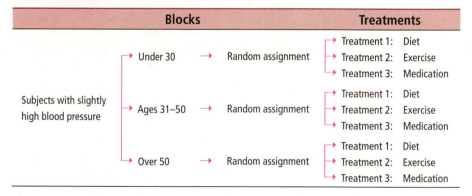

VIEWPOINT Who Watches Cable TV?

Consider the following claim: Average cable TV viewers are, generally speaking, as affluent as newspaper readers and better off than radio or magazine audiences. How do we know whether this claim is true? One way to answer such a question is to construct a two-way ANOVA test in which the rows represent household income levels and the columns represent cable TV viewers, news paper readers, magazine audiences, and radio listeners. The response variable is an index that represents the ratio of the specific medium compared with U.S. averages. For more information and data, see American Demographics *(Vol. 17, No. 6).*

SECTION 10.6 PROBLEMS

1. *Statistical Literacy: Physical Therapy* Does talking while walking slow you down? A study reported in the journal *Physical Therapy* (Vol. 72, No. 4) considered mean cadence (steps per minute) for subjects using no walking device, a standard walker, and a rolling walker. In addition, the cadence was measured when the subjects had to perform dual tasks. The second task was to respond vocally to a signal while walking. Cadence was measured for subjects who were just walking (using no device, a standard walker, or a rolling walker) and for subjects required to respond to a signal while walking. List the factors and the number of levels of each factor. How many cells are there in the data table?

2. *Statistical Literacy: Salary Survey* *Academe, Bulletin of the American Association of University Professors* (Vol. 83, No. 2) presents results of salary surveys (average salary) by rank of the faculty member (professor, associate, assistant, instructor) and by type of institution (public, private). List the factors and the number of levels of each factor. How many cells are there in the data table?

3. *Critical Thinking: Physical Therapy* For the study regarding mean cadence (see Problem 1), two-way ANOVA was used. Recall that the two factors were walking device (none, standard walker, rolling walker) and dual task (being required to respond vocally to a signal or no dual task required). Results of two-way ANOVA showed that there was no evidence of interaction between the factors. However, according to the article, "The ANOVA conducted on the cadence data revealed a main effect of walking device." When the hypothesis regarding no difference in mean cadence according to which, if any, walking device was used, the sample F was 30.94, with $d.f._N = 2$ and $d.f._D = 18$. Further, the P-value for the result was reported to be less than 0.01. From this information, what is the conclusion regarding any difference in mean cadence according to the factor "walking device used"?

4. *Education: Media Usage* In a study of media usage versus education level (*American Demographics,* Vol. 17, No. 6), an index was used to measure media usage, where a measurement of 100 represents the U.S. average. Values above 100 represent above-average media usage.

	Media				
Education Level	**Cable Network**	**Prime-Time TV**	**Radio**	**Newspaper**	**Magazine**
Less than high school	80	112	87	76	85
High school graduate	103	105	100	99	101
Some college	107	94	106	105	107
College graduate	108	90	106	116	108

(a) List the factors and the number of levels of each factor.
(b) Assume there is no interaction between the factors. Use two-way ANOVA and the following Minitab printout to determine if there is a difference in population mean index based on education. Use $\alpha = 0.05$.
(c) Determine if there is a difference in population mean index based on media. Use $\alpha = 0.05$.

```
Minitab Printout for Media/Education Data

Analysis of Variance for Index
Source    DF      SS      MS       F        P
Edu        3     961     320    2.96    0.075
Media      4       5       1    0.01    1.000
Error     12    1299     108
Total     19    2264
```

5. *Income: Media Usage* In the same study described in Problem 4, media usage versus household income also was considered. The media usage indices for the various media and income levels follow.

	Media				
Income Level	**Cable Network**	**Prime-Time TV**	**Radio**	**Newspaper**	**Magazine**
Less than $20,000	78	112	89	80	91
$20,000–$39,999	97	105	100	97	100
$40,000–$74,999	113	92	106	111	105
$75,000 or more	121	94	107	121	105

Source: American Demographics by Staff. Copyright 1995 by CRAIN COMMUNICATIONS INC. Reproduced with permission of CRAIN COMMUNICATIONS INC in the formats Textbook and Other book via Copyright Clearance Center.

(a) List the factors and the number of levels of each factor.
(b) Assume there is no interaction between the factors. Use two-way ANOVA and the following Minitab printout to determine if there is a difference in population mean index based on income. Use $\alpha = 0.05$.
(c) Determine if there is a difference in population mean index based on media. Use $\alpha = 0.05$.

```
Minitab Printout for Media/Income Data

Analysis of Variance for Index
Source    DF      SS      MS       F        P
Income     3    1078     359    2.77    0.088
Media      4      15       4    0.03    0.998
Error     12    1558     130
Total     19    2651
```

6. *Gender: Grade Point Average* Does college grade point average (GPA) depend on gender? Does it depend on class (freshman, sophomore, junior, senior)? In a study, the following GPA data were obtained for random samples of college students in each of the cells.

	Class							
Gender	**Freshman**		**Sophomore**		**Junior**		**Senior**	
Male	2.8	2.1	2.5	2.3	3.1	2.9	3.8	3.6
	2.7	3.0	2.9	3.5	3.2	3.8	3.5	3.1
Female	2.3	2.9	2.6	2.4	2.6	3.6	3.2	3.5
	3.5	3.9	3.3	3.6	3.3	3.7	3.8	3.6

(a) List the factors and the number of levels of each factor.

(b) Use two-way ANOVA and the following Minitab printout to determine if there is any evidence of interaction between the two factors at a level of significance of 0.05.

Minitab Printout of GPA Based on Gender and Class

Analysis of Variance for GPA

Source	DF	SS	MS	F	P
Gender	1	0.281	0.281	1.26	0.273
Class	3	2.226	0.742	3.32	0.037
Interaction	3	0.286	0.095	0.43	0.736
Error	24	5.365	0.224		
Total	31	8.159			

(c) If there is no evidence of interaction, use two-way ANOVA and the Minitab printout to determine if there is a difference in mean GPA based on class. Use $\alpha = 0.05$.

(d) If there is no evidence of interaction, use two-way ANOVA and the Minitab printout to determine if there is a difference in mean GPA based on gender. Use $\alpha = 0.05$.

7. *Experimental Design: Teaching Style* A researcher forms three blocks of students interested in taking a history course. The groups are based on grade point average (GPA). The first group consists of students with a GPA less than 2.5, the second group consists of students with a GPA between 2.5 and 3.1, and the last group consists of students with a GPA greater than 3.1. History courses are taught in three ways: traditional lecture, small-group collaborative method, and independent study. The researcher randomly assigns 10 students from each block to sections of history taught each of the three ways. Sections for each teaching style then have 10 students from each block. The researcher records the scores on a common course final examination administered to each student. Draw a flow chart showing the design of this experiment. Does the design fit the model for randomized block design?

CHAPTER REVIEW

SUMMARY

In this chapter, we introduced applications of two probability distributions: the chi-square distribution and the *F* distribution.

- The chi-square distribution is used for tests of independence or homogeneity, tests of goodness of fit, tests of variance σ^2, and tests to estimate a variance σ^2.
- The *F* distribution is used for tests of two variances, one-way ANOVA, and two-way ANOVA.

ANOVA tests are used to determine whether there are differences among means for several groups.

- If groups are based on the value of only one variable, we have one-way ANOVA.
- If groups are formed using two variables, we use two-way ANOVA to test for differences of means based on either variable or on an interaction between the variables.

IMPORTANT WORDS & SYMBOLS

Chi-square distribution overview 622

Section 10.1

Independence test 623
Observed frequency, O 623
Contingency table 623
Expected frequency, E 624
Row total 625
Column total 625
Degrees of freedom, $d.f. = (R - 1)(C - 1)$ for χ^2 distribution and tests of independence 627
Test of homogeneity 630
Multinomial experiments 633

Section 10.2

Goodness-of-fit test 639
Degrees of freedom $d.f.$ for χ^2 distribution and goodness-of-fit tests 640

Section 10.3

Hypotheses tests about σ^2 651
Confidence interval for σ^2 655
χ^2_U, χ^2_L 656

***F* distribution overview 660**

Section 10.4

Hypothesis tests about two variances 662

Section 10.5

ANOVA 672
Sum of squares, SS 673
SS_{TOT}, SS_{BET}, SS_W 674
MS_{BET}, MS_W 676
F ratio 677
Sample F statistic 677
Degrees of freedom for numerator for F distribution 678
Degrees of freedom for denominator for F distribution 678
Summary table for ANOVA 680

Section 10.6

Two-way ANOVA 687
Interaction 687
Factor 688
Level 688
Cell 688
Completely randomized design 695
Randomized block design 695

VIEWPOINT Movies and Money!

Young adults are the movie industry's best customers. However, going to the movies is expensive, which may explain why attendance rates increase with household income. Using what you have learned in this chapter, you can create appropriate chi-square tests to determine how good a fit exists between national percentage rates of attendance by household income and attendance rates in your demographic area. For more information and national data, see American Demographics *(Vol. 18, No. 12).*

CHAPTER REVIEW PROBLEMS

1. *Statistical Literacy* Of the following random variables, which have only non-negative values: z, t, chi-square, F?

2. *Statistical Literacy* Of the following probability distributions, which are always symmetric: normal, Student's t, chi-square, F?

3. *Critical Thinking* Suppose you took random samples from three distinct age groups. Through a survey, you determined how many respondents from each age group preferred to get news from TV, newspapers, the Internet, or another source (respondents could select only one mode). What type of test would be appropriate to determine if there is sufficient statistical evidence to claim that the proportions of each age group preferring the different modes of obtaining news are not the same? Select from tests of independence, homogeneity, goodness of fit, and ANOVA.

4. *Critical Thinking* Suppose you take a random sample from a normal population and you want to determine whether there is sufficient statistical evidence to claim that the population variance differs from a corresponding variance specified in a government contract. Which type of test is appropriate, a test of one variance or a test of two variances?

Before you solve Problems 5–14, first classify the problem as one of the following:

Chi-square test of independence or homogeneity
Chi-square goodness of fit
Chi-square for testing or estimating σ^2 or σ
F test for two variances
One-way ANOVA
Two-way ANOVA

Then, in each of the problems when a test is to be performed, do the following:

(i) Give the value of the level of significance. State the null and alternate hypotheses.
(ii) Find the sample test statistic.
(iii) Find or estimate the P-value of the sample test statistic.
(iv) Conclude the test.
(v) *Interpret* the conclusion in the context of the application.
(vi) In the case of one-way ANOVA, make a summary table.

5. *Sales: Packaging* The makers of Country Boy Corn Flakes are thinking about changing the packaging of the cereal in the hope of improving sales. In an experiment, five stores of similar size in the same region sold Country Boy Corn Flakes in different-shaped containers for 2 weeks. Total packages sold are given in the following table. Using a 0.05 level of significance, shall we reject or fail to reject the hypothesis that the mean sales are the same, no matter which box shape is used?

Cube	Cylinder	Pyramid	Rectangle
120	110	74	165
88	115	62	98
65	180	110	125
95	96	66	87
71	85	83	118

6. *Education: Exams* Professor Fair believes that extra time does not improve grades on exams. He randomly divided a group of 300 students into two groups and gave them all the same test. One group had exactly 1 hour in which to finish the test, and the other group could stay as long as desired. The results are shown in the following table. Test at the 0.01 level of significance that time to complete a test and test results are independent.

Time	A	B	C	F	Row Total
1 h	23	42	65	12	142
Unlimited	17	48	85	8	158
Column Total	40	90	150	20	300

7. *Tires: Blowouts* A consumer agency is investigating the blowout pressures of Soap Stone tires. A Soap Stone tire is said to blow out when it separates from the wheel rim due to impact forces usually caused by hitting a rock or a pothole in the road. A random sample of 30 Soap Stone tires were inflated to the recommended pressure, and then forces measured in foot-pounds were applied to each tire (1 foot-pound is the force of 1 pound dropped from a height of 1 foot). The customer complaint is that some Soap Stone tires blow out under small-impact forces, while other tires seem to be well made and don't have

this fault. For the 30 test tires, the sample standard deviation of blowout forces was 1353 foot-pounds.

(a) Soap Stone claims its tires will blow out at an average pressure of 20,000 foot-pounds, with a standard deviation of 1020 foot-pounds. The average blowout force is not in question, but the variability of blowout forces is in question. Using a 0.01 level of significance, test the claim that the variance of blowout pressures is more than Soap Stone claims it is.

(b) Find a 95% confidence interval for the variance of blowout pressures, using the information from the random sample.

8. *Computer Science: Data Processing* Anela is a computer scientist who is formulating a large and complicated program for a type of data processing. She has three ways of storing and retrieving data: CD, tape, and disk. As an experiment, she sets up her program in three different ways: one using CDs, one using tapes, and the other using disks. Then she makes four test runs of this type of data processing on each program. The time required to execute each program is shown in the following table (in minutes). Use a 0.01 level of significance to test the hypothesis that the mean processing time is the same for each method.

CD	Tape	Disks
8.7	7.2	7.0
9.3	9.1	6.4
7.9	7.5	9.8
8.0	7.7	8.2

9. *Teacher Ratings: Grades* Professor Stone complains that students' teacher ratings depend on the grade students receive. In other words, according to Professor Stone, a teacher who gives good grades gets good ratings, and a teacher who gives bad grades gets bad ratings. To test this claim, the Student Assembly took a random sample of 300 teacher ratings on which the students' grades for the course also were indicated. The results are given in the following table. Test the hypothesis that teacher ratings and student grades are independent at the 0.01 level of significance.

Rating	A	B	C	F (or withdrawal)	Row Total
Excellent	14	18	15	3	50
Average	25	35	75	15	150
Poor	21	27	40	12	100
Column Total	60	80	130	30	300

10. *Packaging: Corn Flakes* A machine that puts corn flakes into boxes is adjusted to put an average of 15 ounces into each box, with standard deviation of 0.25 ounce. If a random sample of 12 boxes gave a sample standard deviation of 0.38 ounce, do these data support the claim that the variance has increased and the machine needs to be brought back into adjustment? (Use a 0.01 level of significance.)

11. *Sociology: Age Distribution* A sociologist is studying the age of the population in Blue Valley. Ten years ago, the population was such that 20% were under 20 years old, 15% were in the 20- to 35-year-old bracket, 30% were between 36 and 50, 25% were between 51 and 65, and 10% were over 65. A study done this year used a random sample of 210 residents. This sample showed

Under 20	20–35	36–50	51–65	Over 65
26	27	69	68	20

At the 0.01 level of significance, has the age distribution of the population of Blue Valley changed?

12. *Engineering: Roller Bearings* Two processes for manufacturing large roller bearings are under study. In both cases, the diameters (in centimeters) are being examined. A random sample of 21 roller bearings from the old manufacturing process showed the sample variance of diameters to be $s^2 = 0.235$. Another random sample of 26 roller bearings from the new manufacturing process showed the sample variance of their diameters to be $s^2 = 0.128$. Use a 5% level of significance to test the claim that there is a difference (either way) in the population variances between the old and new manufacturing processes.

13. *Engineering: Light Bulbs* Two processes for manufacturing 60-watt light bulbs are under study. In both cases, the life (in hours) of the bulb before it burns out is being examined. A random sample of 18 light bulbs manufactured using the old process showed the sample variance of lifetimes to be $s^2 = 51.87$. Another random sample of 16 light bulbs manufactured using the new process showed the sample variance of the lifetimes to be $s^2 = 135.24$. Use a 5% level of significance to test the claim that the population variance of lifetimes for the new manufacturing process is larger than that of the old process.

14. *Advertising: Newspapers* Does the newspaper section in which an ad is placed make a difference in the average daily number of people responding to the ad? Is there a difference in average daily number of people responding to the ad if it is placed in the Sunday newspaper as compared with the Wednesday newspaper? Video Entertainment is a video club that sells videotapes featuring movies of all kinds—instructional videos, videos of TV specials, etc. To attract new customers, Video Entertainment ran ads in the Wednesday newspaper and in the Sunday newspaper offering six free videotapes or DVDs to new members. In addition, the ads were placed in the sports section, the entertainment section, and the business section of the local newspaper. Ads running in the different sections carried different promotion codes. Different codes also were used for Wednesday ads compared with Sunday ads. The numbers of people responding to the ads for days selected at random were recorded according to the promotion codes. The results follow:

	Section of Newspaper								
Day	**Sports**			**Entertainment**			**Business**		
Wed.	12	15	11	22	14	17	2	4	3
	12	15	20	22	18	12	5	0	1
Sun.	20	23	25	32	26	28	13	16	13
	33	15	17	31	25	41	15	14	10

(a) List the factors and the number of levels for each factor.

(b) Use the following Minitab printout and the *F* distribution table (Table 8 of Appendix II) to test for interaction between the variables. Use a 1% level of significance.

```
Minitab Printout for Mean Number of Responses per Day
Analysis of Variance for Response
Source        DF        SS        MS        F        P
Day            1    1024.0    1024.0    55.28    0.000
Section        2    1573.6     786.8    42.48    0.000
Interaction    2      38.0      19.0     1.03    0.371
Error         30     555.7      18.5
Total         35    3191.2
```

(c) If there is no evidence of interaction between the factors, test for a difference in mean number of daily responses for the levels of the day factor. Use $\alpha = 0.01$.

(d) If there is no evidence of interaction between the factors, test for a difference in mean number of daily responses for the levels of the section factor. Use $\alpha = 0.01$.

DATA HIGHLIGHTS: GROUP PROJECTS

Break into small groups and discuss the following topics. Organize a brief outline in which you summarize the main points of your group discussion.

Distribution of Drunk Driver Arrests by Age

Age	National Percentage	Number in Freemont County
16–17	3.7	8
18–24	18.9	35
25–29	12.9	23
30–34	10.3	19
35–39	8.5	12
40–44	7.9	14
45–49	8.0	16
50–54	7.9	13
55–59	6.8	10
60–64	5.7	9
65 and over	9.4	15
	100%	174

The *Statistical Abstract of the United States* reported information about the percentage of arrests of all drunk drivers according to age group. In the following table, the entry 3.7 in the first row means that in the entire United States, about 3.7% of all people arrested for drunk driving were in the age group 16–17 years. The Freemont County Sheriff's Office obtained data about the number of drunk drivers arrested in each age group over the past several years. In the following table, the entry 8 in the first row means that eight people in the age group 16–17 years were arrested for drunk driving in Freemont County.

Use a chi-square test with 5% level of significance to test the claim that the age distribution of drunk drivers arrested in Freemont County is the same as the national age distribution of drunk drivers arrested.

(a) State the null and alternate hypotheses.
(b) Find the value of the chi-square test statistic from the sample.
(c) Find the degrees of freedom and the *P*-value of the test statistic.
(d) Decide whether you should reject or not reject the null hypothesis.
(e) State your conclusion in the context of the problem.
(f) How could you gather data and conduct a similar test for the city or county in which you live? Explain.

LINKING CONCEPTS: WRITING PROJECTS

Discuss each of the following topics in class or review the topics on your own. Then write a brief but complete essay in which you summarize the main points. Please include formulas and graphs as appropriate.

1. In this chapter, you studied the chi-square distribution and three principal applications of the distribution.
   (a) Outline the basic ideas behind the chi-square test of independence. What is a contingency table? What are the null and alternate hypotheses? How is the test statistic constructed? What basic assumptions underlie this application of the chi-square distribution?
   (b) Outline the basic ideas behind the chi-square test of goodness of fit. What are the null and alternate hypotheses? How is the test statistic constructed? There are a number of direct similarities between tests of independence and tests of goodness of fit. Discuss and summarize these similarities.
   (c) Outline the basic ideas behind the chi-square method of testing and estimating a standard deviation. What basic assumptions underlie this process?
   (d) Outline the basic ideas behind the chi-square test of homogeneity. What are the null and alternate hypotheses? How is the test statistic constructed? What basic assumptions underlie the application of the chi-square distribution?
2. The *F* distribution is used to construct a one-way ANOVA test for comparing several sample means.
   (a) Outline the basic purpose of ANOVA. How does ANOVA avoid high risk due to multiple type I errors?
   (b) Outline the basic assumption for ANOVA.
   (c) What are the null and alternate hypotheses in an ANOVA test? If the test conclusion is to reject the null hypothesis, do we know which of the population means are different from each other?
   (d) What is the *F* distribution? How are the degrees of freedom for numerator and denominator determined?
   (e) What do we mean by a summary table of ANOVA results? What are the main components of such a table? How is the final decision made?

USING TECHNOLOGY

Application

Analysis of Variance (One-Way ANOVA)

The following data comprise a winter mildness/severity index for three European locations near 50° north latitude. For each decade, the number of unmistakably mild months minus the number of unmistakably severe months for December, January, and February is given.

Decade	Britain	Germany	Russia
1800	−2	−1	+1
1810	−2	−3	−1
1820	0	0	0
1830	−3	−2	−1
1840	−3	−2	+1
1850	−1	−2	+3
1860	+8	+6	+1
1870	0	0	−3
1880	−2	0	+1
1890	−3	−1	+1
1900	+2	0	+2
1910	+5	+6	+1
1920	+8	+6	+2
1930	+4	+4	+5
1940	+1	−1	−1
1950	0	+1	+2

Table is based on data from Exchanging Climate *by H. H. Lamb; copyright © 1966. Reprinted by permission of Routledge, UK.*

1. We wish to test the null hypothesis that the mean winter indices for Britain, Germany, and Russia are all equal against the alternate hypothesis that they are not all equal. Use a 5% level of significance.
2. What is the sum of squares between groups? Within groups? What is the sample F ratio? What is the P-value? Shall we reject or fail to reject the statement that the mean winter indices for these locations in Britain, Germany, and Russia are the same?
3. What is the smallest level of significance at which we could conclude that the mean winter indices for these locations are not all equal?

Technology Hints (One-Way ANOVA)

SPSS

Enter all the data in one column. In another column, use integers to designate the group to which each data value belongs. Use the menu selections **Analyze ➤ Compare Means ➤ One-Way ANOVA.** Move the column containing the data to the dependent list. Move the column containing group designation to the factor list.

Technology Hints (Two-Way ANOVA)

Excel 2010

Excel has two commands for two-way ANOVA, depending on how many data values are in each cell. On the home screen, click the **Data** tab. Then in the Analysis group, click **Data Analysis**. Then use **ANOVA: Two-Factor with Replication** if there are two or more sample measurements for each factor combination cell. Again, there must be the same number of data in each cell.

ANOVA: Two-Factor without Replication if there is only one data value in each factor combination cell.

Data entry is fairly straightforward. For example, look at the Excel spreadsheet for the data of Guided Exercise 12 regarding the fat content of different brands of potato chips as measured by different labs.

	A	B	C	D
1		**Lab 1**	**Lab 2**	**Lab 3**
2	Texas	32.4	33.1	32.9
3	Great	37.9	37.7	37.8
4	Chip Ooh	29.1	29.4	29.5

Minitab

For Minitab, all the data for the response variable (in this case, fat content) are entered into a single column. Create two more columns, one for the row number of the cell containing the data value and one for the column number of the cell containing the data value. For the potato chip example, the rows correspond to the brand and the columns to the lab doing the analysis. Use the menu choices **Stat ➤** ANOVA **➤ Two-Way.**

	C1	C2	C3
	Brand	Lab	Fat
1	1	1	32.4
2	1	2	33.1
3	1	3	32.9
4	2	1	37.9
5	2	2	37.7
6	2	3	37.8
7	3	1	29.1
8	3	2	29.4
9	3	3	29.5

SPSS

Data entry for SPSS is similar to that for Minitab. Enter all the data in one column. Use a separate column for each factor and use a label or integer to designate the group in the particular factor. Under Variable View, type appropriate labels for the columns of data. Use the menu selections **Analyze ➤ General Linear Model ➤ Univariate. . . .** In the dialogue box, the dependent variable is the quantity represented by the data. The factors are those found in each factor column. For the special case of only one datum per cell, click the Model button. Select Custom and move the desired factors into Model.

SPSS Data Editor

File Edit View Data Transform Analyze Graphs

	fat	lab	brand
1	32.40	1	Texas
2	37.90	1	Great
3	29.10	1	Chip
4	33.10	2	Texas
5	37.70	2	Great
6	29.40	2	Chip
7	32.90	3	Texas
8	37.80	3	Great
9	29.50	3	Chip

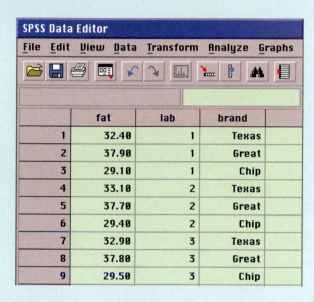

11

Business Wire/Getty Images Publicity/Getty Images

Photo Researchers / Alamy

Make everything as simple as possible, but no simpler.

—Albert Einstein

The brilliant German-born American physicist Albert Einstein (1879–1955) formulated the theory of relativity.

NONPARAMETRIC STATISTICS

James Shafer/PhotoEdit

PREVIEW QUESTIONS

What if you cannot make assumptions about a population distribution? Can you still use statistical methods? What are the advantages and disadvantages? (SECTION 11.1)

What are nonparametric tests? How do you handle a "before and after" situation? (SECTION 11.1)

If you can't make assumptions about the population and you have independent samples, how do you set up a nonparametric test? (SECTION 11.2)

Suppose you are interested only in rank data (ordinal-type data). If you have ordered pairs (x, y) of ranked data, is there a way to measure and test correlation? (SECTION 11.3)

Is a sequence random or is there a pattern associated with the sequence? (SECTION 11.4)

FOCUS PROBLEM

How Cold? Compared to What?

Juneau is the capital of Alaska. The terrain surrounding Juneau is very rugged, and storms that sweep across the Gulf of Alaska usually hit Juneau. However, Juneau is located in southern Alaska, near the ocean, and temperatures are often comparable with those found in the lower 48 states. Madison is the capital of Wisconsin. The city is located between two large lakes. The climate of Madison is described as the typical continental climate of interior North America. Consider the long-term average temperatures (in degrees Fahrenheit) paired by month for the two cities (Source: National Weather Bureau). Use a sign test with a 5% level of significance to test the claim that the overall temperature distribution of Madison is different (either way) from that of Juneau. (See Problem 12 of Section 11.1.)

Month	Madison	Juneau
January	17.5	22.2
February	21.1	27.3
March	31.5	31.9
April	46.1	38.4
May	57.0	46.4
June	67.0	52.8
July	71.3	55.5
August	69.8	54.1
September	60.7	49.0
October	51.0	41.5
November	35.7	32.0
December	22.8	26.9

Mediolmages / Photodisc / Getty Images

For online student resources, visit the Brase/Brase, *Understandable Statistics*, 11th edition web site at http://www.cengage.com/statistics/brase11e

SECTION **11.1**

The Sign Test for Matched Pairs

FOCUS POINTS

- State the criteria for setting up a matched pair sign test.
- Complete a matched pair sign test.
- Interpret the results in the context of the application.

Nonparametric statistics

There are many situations in which very little is known about the population from which samples are drawn. Therefore, we cannot make assumptions about the population distribution, such as assuming the distribution is normal or binomial. In this chapter, we will study methods that come under the heading of *nonparametric statistics*. These methods are called *nonparametric* because they require no assumptions about the population distributions from which samples are drawn. The obvious advantages of these tests are that they are quite general and (as we shall see) not difficult to apply. The disadvantages are that they tend to waste information and tend to result in acceptance of the null hypothesis more often than they should. As such, nonparametric tests are sometimes *less sensitive* than other tests.

Sign test

The easiest of all the nonparametric tests is probably the *sign test*. The sign test is used when we compare sample distributions from two populations that are *not independent*. This occurs when we measure the sample twice, as in "before and after" studies. The following example shows how the sign test is constructed and used:

Criteria for sign test

As part of their training, 15 police cadets took a special course on identification awareness. To determine how the course affects a cadet's ability to identify a suspect, the 15 cadets were first given an identification-awareness exam and then, after the course, were tested again. The police school would like to use the results of the two tests to see if the identification-awareness course *improves* a cadet's score. Table 11-1 gives the scores for each exam.

Science Source/Photo Researchers, Inc.

The sign of the difference is obtained by subtracting the precourse score from the postcourse score. If the difference is positive, we say that the sign of the difference is $+$, and if the difference is negative, we indicate it with $-$. No sign is indicated if the scores are identical; in essence, such scores are ignored when using the sign test. To use the sign test, we need to compute the *proportion x of plus signs* to all signs. We ignore the pairs with no difference of signs. This is demonstrated in Guided Exercise 1.

TABLE 11-1 Scores for 15 Police Cadets

Cadet	Postcourse Score	Precourse Score	Sign of Difference
1	93	76	+
2	70	72	−
3	81	75	+
4	65	68	−
5	79	65	+
6	54	54	No difference
7	94	88	+
8	91	81	+
9	77	65	+
10	65	57	+
11	95	86	+
12	89	87	+
13	78	78	No difference
14	80	77	+
15	76	76	No difference

GUIDED EXERCISE 1 | PROPORTION OF PLUS SIGNS

Look at Table 11-1 under the "Sign of Difference" column.

(a) How many plus signs do you see? → 10

(b) How many plus and minus signs do you see? → 12

(c) The *proportion of plus signs* is → $x = \frac{10}{12} = \frac{5}{6} \approx 0.833$

$$x = \frac{\text{Number of plus signs}}{\text{Total number of plus and minus signs}}$$

Use parts (a) and (b) to find x.

Null hypothesis

We observe that x is the sample proportion of plus signs, and we use p to represent the population proportion of plus signs (if *all* possible police cadets were tested).

The null hypothesis is

> H_0: $p = 0.5$ (the distributions of scores before and after the course are the same)

The null hypothesis states that the identification-awareness course does *not* affect the distribution of scores. Under the null hypothesis, we expect the number of plus signs and minus signs to be about equal. This means that the proportion of plus signs should be approximately 0.5.

Alternate hypothesis

The police department wants to see if the course *improves* a cadet's score. Therefore, the alternate hypothesis will be

> H_1: $p > 0.5$ (the distribution of scores after the course is shifted higher than the distribution before the course)

The alternate hypothesis states that the identification-awareness course tends to improve scores. This means that the proportion of plus signs should be greater than 0.5.

Sampling distribution

To test the null hypothesis H_0: $p = 0.5$ against the alternate hypothesis H_1: $p > 0.5$, we use methods of Section 8.3 for tests of proportions. As in Section 8.3, we will assume that all our samples are sufficiently large to permit a normal approximation to the binomial distribution. For most practical work, this will be the case if the total number of plus and minus signs is 12 or more ($n \geq 12$).

When the total number of plus and minus signs is 12 or more, the sample statistic x (proportion of plus signs) has a distribution that is approximately normal, with mean p and standard deviation $\sqrt{pq/n}$ where $q = 1 - p$. (See Section 6.6.)

Sample test statistic

Under the null hypothesis H_0: $p = 0.5$, we assume that the population proportion p of plus signs is 0.5. Therefore, the z value corresponding to the sample test statistic x is

$$z = \frac{x - p}{\sqrt{\frac{pq}{n}}} = \frac{x - 0.5}{\sqrt{\frac{(0.5)(0.5)}{n}}} = \frac{x - 0.5}{\sqrt{\frac{0.25}{n}}}$$

where n is the total number of plus and minus signs, x is the total number of plus signs divided by n, and $q = 1 - p$.

FIGURE 11-1

P-value

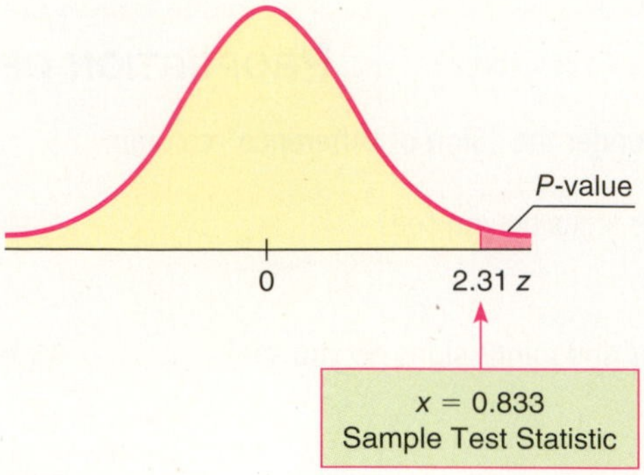

For the police cadet example, we found $x \approx 0.833$ in Guided Exercise 1. The value of n is 12. (Note that of the 15 cadets in the sample, 3 had no difference in precourse and postcourse test scores, so there are no signs for these 3.) The z value corresponding to $x = 0.833$ is then

$$z \approx \frac{0.833 - 0.5}{\sqrt{\frac{0.25}{12}}} \approx 2.31$$

P-value

We use the standard normal distribution table (Table 5 of Appendix II) to find *P*-values for the sign test. This table gives areas to the left of z. Recall from Section 8.2 that Table 5 of Appendix II can be used directly to find *P*-values of one-tailed tests. For *two-tailed* tests, we must *double* the value given in the table. To review the process of finding areas to the right or left of z using Table 5, see Section 6.2.

The alternate hypothesis for the police cadet example is $H_1: p > 0.5$. The *P*-value for the sample test statistic $z = 2.31$ is shown in Figure 11-1. For a right-tailed test, the *P*-value is the area to the right of the sample test statistic $z = 2.31$. From Table 5 of Appendix II, $P(z > 2.31) = 0.0104$.

Conclude the test and interpret the results

In our example, the police department wishes to use a 5% level of significance to test the claim that the identification-awareness course improves a cadet's score. Since the *P*-value of 0.0104 is less than $\alpha = 0.05$, we reject the null hypothesis H_0 that the course makes no difference. Instead, at the 5% level of significance, we say the results are significant. The evidence is sufficient to claim that the identification-awareness course improves cadets' scores.

The steps used to construct a sign test for matched pairs are summarized in the next procedure.

PROCEDURE

HOW TO CONSTRUCT A SIGN TEST FOR MATCHED PAIRS

Setup and Requirements

You first need a random sample of data pairs (A, B). Next, you take the differences $A - B$ and record the sign change for each difference: plus, minus, or no change. The number of data pairs should be large enough that the total number of plus and minus signs is at least 12. The sample proportion of plus signs is

$$x = \frac{\text{number of plus signs}}{\text{total number of plus and minus signs}}$$

Let p represent the population proportion of plus signs if the entire population of all possible data pairs (A, B) were to be used.

Procedure

1. Set the *level of significance* α. The *null hypothesis* is $H_0: p = 0.5$. In the context of the application, set the *alternate hypothesis:* $H_1: p > 0.5$, $H_1: p < 0.5$, or $H_1: p \neq 0.5$.

2. The *sample test statistic* is

$$z = \frac{x - 0.5}{\sqrt{\frac{0.25}{n}}}$$

where $n \geq 12$ is the total number of plus and minus signs.
3. Use the standard normal distribution and the type of test, one-tailed or two-tailed, to find the *P-value* corresponding to the test statistic.
4. *Conclude* the test. If P-value $\leq \alpha$, then reject H_0. If P-value $> \alpha$, then do not reject H_0.
5. *Interpret your conclusion* in the context of the application.

GUIDED EXERCISE 2 | SIGN TEST

Dr. Kick-a-poo's Traveling Circus made a stop at Middlebury, Vermont, where the doctor opened a booth and sold bottles of Dr. Kick-a-poo's Magic Gasoline Additive. The additive is supposed to increase gas mileage when used according to instructions. Twenty local people purchased bottles of the additive and used it according to instructions. These people carefully recorded their mileage with and without the additive. The results are shown in Table 11-2.

TABLE 11-2 **Mileage Before and After Kick-a-poo's Additive**

Car	With Additive	Without Additive	Sign of Difference
1	17.1	16.8	+
2	21.2	20.1	+
3	12.3	12.3	No difference (N.D.)
4	19.6	21.0	−
5	22.5	20.9	+
6	17.0	17.9	___
7	24.2	25.4	___
8	22.2	20.1	___
9	18.3	19.1	___
10	11.0	12.3	___
11	17.6	14.2	___
12	22.1	23.7	___
13	29.9	30.2	___
14	27.6	27.6	___
15	28.4	27.7	___
16	16.1	16.1	___
17	19.0	19.5	___
18	38.7	37.9	___
19	17.6	19.7	___
20	21.6	22.2	___

TABLE 11-3 **Completion of Table 11-2**

Car	Sign of Difference
6	−
7	−
8	+
9	−
10	−
11	+
12	−
13	−
14	N.D.
15	+
16	N.D.
17	−
18	+
19	−
20	−

Continued

GUIDED EXERCISE 2 *continued*

(a) In Table 11-2, complete the column headed "Sign of Difference." How many plus signs are there? How many total plus and minus signs are there? What is the value of x, the proportion of plus signs?

→ There are 7 plus signs and 17 total plus and minus signs. The proportion of plus signs is

$$x = \frac{7}{17} \approx 0.412$$

(b) Most people claim that the additive has no effect. Let's use a 0.05 level of significance to test this claim against the alternate hypothesis that the additive did have an effect (one way or the other). State the null and alternate hypotheses.

→ We use
H_0: $p = 0.5$ (mileage distributions are the same)
H_1: $p \neq 0.5$ (mileage distributions are different)

(c) Convert the sample x value, $x = 0.412$, to a z value.

→ To find the z value corresponding to $x = 0.412$, we use $n = 17$ (total number of signs).

$$z = \frac{x - 0.5}{\sqrt{0.25/n}} \approx \frac{0.412 - 0.5}{\sqrt{0.25/17}} \approx -0.73$$

(d) Find the corresponding P-value.

FIGURE 11-2 *P*-value

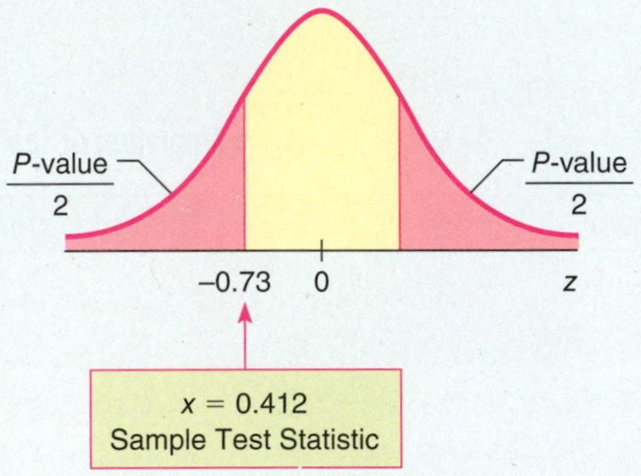

→ Table 5 of Appendix II gives the area to the left of $z = -0.73$.
$P(z < -0.73) = 0.2327$

Because this is a two-tailed test, the P-value is double this area.
P-value $= 2(0.2327) = 0.4654$

(e) Conclude the test.

→ For $\alpha = 0.05$, we see that the P-value $= 0.4654$ is greater than α. We fail to reject H_0.

(f) *Interpret* the results.

→ At the 5% level of significance, the data are not statistically significant, and we cannot reject the hypothesis that the mileage distribution is the same with or without the additive.

VIEWPOINT Yukon News

The Yukon News *featured an article titled "Resurgence of the Dreaded White Plague," about the resurgence of tuberculosis (TB) in the far north. TB, also known as the white plague, has been present in Canada since it was brought in by European immigrants in the 17th century. Although antibiotics are widely used today, the disease has never been eradicated. Canadian National Health data suggest that TB is spreading faster in the Yukon than elsewhere in Canada. Because of this, the Canadian government has established many new TB clinics in remote Yukon villages. Using what you have learned in this section and Canadian National Health data, can you think of a way to use a sign test to study the claim that in these villages, the rate of TB in the population dropped after the clinics were activated?*

SECTION 11.1 PROBLEMS

1. *Statistical Literacy* To apply the sign test, do you need independent or dependent (matched pair) data?

2. *Statistical Literacy* For the sign test of matched pairs, do pairs for which the difference in values is zero enter into any calculations?

For Problems 3–12, please provide the following information.
(a) What is the level of significance? State the null and alternate hypotheses.
(b) Compute the sample test statistic. What is the sampling distribution?
(c) Find the P-value of the sample test statistic.
(d) Conclude the test.
(e) *Interpret* the conclusion in the context of the application.

3. *Economic Growth: Asia* Asian economies impact some of the world's largest populations. The growth of an economy has a big influence on the everyday lives of ordinary people. Are Asian economies changing? A random sample of 15 Asian economies gave the following information about annual percentage growth rate (Reference: *Handbook of International Economic Statistics,* U.S. Government Documents).

Region	1	2	3	4	5	6	7	8
Modern Growth Rate %	4.0	2.3	7.8	2.8	0.7	5.1	2.9	4.2
Historic Growth Rate %	3.3	1.9	7.0	5.5	3.3	6.0	3.2	8.2

Region	9	10	11	12	13	14	15
Modern Growth Rate %	4.9	5.8	6.8	3.6	3.2	0.8	7.3
Historic Growth Rate %	6.4	7.2	6.1	1.5	1.0	2.1	5.1

Does this information indicate a change (either way) in the growth rate of Asian economies? Use a 5% level of significance.

4. *Debt: Developing Countries* Borrowing money may be necessary for business expansion. However, too much borrowed money can also mean trouble. Are developing countries tending to borrow more? A random sample of 20 developing countries gave the following information regarding foreign debt per capita (in U.S. dollars, inflation adjusted) (Reference: *Handbook of International Economic Statistics,* U.S. Government Documents).

Country	1	2	3	4	5	6	7	8	9	10
Modern Debt per Capita	179	157	129	125	91	80	31	25	29	85
Historic Debt per Capita	144	132	88	112	53	66	31	30	40	75

Country	11	12	13	14	15	16	17	18	19	20
Modern Debt per Capita	27	20	17	21	195	189	143	126	106	76
Historic Debt per Capita	21	19	15	24	104	150	142	118	117	79

Does this information indicate that foreign debt per capita is increasing in developing countries? Use a 1% level of significance.

5. *Education: Exams* A high school science teacher decided to give a series of lectures on current events. To determine if the lectures had any effect on student awareness of current events, an exam was given to the class before

the lectures, and a similar exam was given after the lectures. The scores follow. Use a 0.05 level of significance to test the claim that the lectures made no difference against the claim that the lectures did make some difference (one way or the other).

Student	1	2	3	4	5	6	7	8	9
After Lectures	107	115	120	78	83	56	71	89	77
Before Lectures	111	110	93	75	88	56	75	73	83

Student	10	11	12	13	14	15	16	17	18
After Lectures	44	119	130	91	99	96	83	100	118
Before Lectures	40	115	101	110	90	98	76	100	109

6. *Grain Yields: Feeding the World* With an ever-increasing world population, grain yields are extremely important. A random sample of 16 large grain-producing regions in the world gave the following information about grain production (in kg/hectare) (Reference: *Handbook of International Economic Statistics,* U.S. Government Documents).

Region	1	2	3	4	5	6	7	8
Modern Production	1610	2230	5270	6990	2010	4560	780	6510
Historic Production	1590	2360	5161	7170	1920	4760	660	6320

Region	9	10	11	12	13	14	15	16
Modern Production	2850	3550	1710	2050	2750	2550	6750	3670
Historic Production	2920	2440	1340	2180	3110	2070	7330	2980

Does this information indicate that modern grain production is higher? Use a 5% level of significance.

7. *Identical Twins: Reading Skills* To compare two elementary schools regarding teaching of reading skills, 12 sets of identical twins were used. In each case, one child was selected at random and sent to school A, and his or her twin was sent to school B. Near the end of fifth grade, an achievement test was given to each child. The results follow:

Twin Pair	1	2	3	4	5	6
School A	177	150	112	95	120	117
School B	86	135	115	110	116	84

Twin Pair	7	8	9	10	11	12
School A	86	111	110	142	125	89
School B	93	77	96	130	147	101

Use a 0.05 level of significance to test the hypothesis that the two schools have the same effectiveness in teaching reading skills against the alternate hypothesis that the schools are not equally effective.

8. *Incomes: Electricians and Carpenters* How do the average weekly incomes of electricians and carpenters compare? A random sample of 17 regions in the United States gave the following information about

average weekly income (in dollars) (Reference: U.S. Department of Labor, Bureau of Labor Statistics).

Region	1	2	3	4	5	6	7	8	9
Electricians	461	713	593	468	730	690	740	572	805
Carpenters	540	812	512	473	686	507	785	657	475

Region	10	11	12	13	14	15	16	17
Electricians	593	593	700	572	863	599	596	653
Carpenters	485	646	675	382	819	600	559	501

Does this information indicate a difference (either way) in the average weekly incomes of electricians compared to those of carpenters? Use a 5% level of significance.

9. *Quitting Smoking: Hypnosis* One program to help people stop smoking cigarettes uses the method of posthypnotic suggestion to remind subjects to avoid smoking. A random sample of 18 subjects agreed to test the program. All subjects counted the number of cigarettes they usually smoke a day; then they counted the number of cigarettes smoked the day after hypnosis. (*Note:* It usually takes several weeks for a subject to stop smoking completely, and the method does not work for everyone.) The results follow.

	Cigarettes Smoked per Day			Cigarettes Smoked per Day	
Subject	After Hypnosis	Before Hypnosis	Subject	After Hypnosis	Before Hypnosis
1	28	28	10	5	19
2	15	35	11	12	32
3	2	14	12	20	42
4	20	20	13	30	26
5	31	25	14	19	37
6	19	40	15	0	19
7	6	18	16	16	38
8	17	15	17	4	23
9	1	21	18	19	24

Using a 1% level of significance, test the claim that the number of cigarettes smoked per day was less after hypnosis.

10. *Incomes: Lawyers and Architects* How do the average weekly incomes of lawyers and architects compare? A random sample of 18 regions in the United States gave the following information about average weekly incomes (in dollars) (Reference: U.S. Department of Labor, Bureau of Labor Statistics).

Region	1	2	3	4	5	6	7	8	9
Lawyers	709	898	848	1041	1326	1165	1127	866	1033
Architects	859	936	887	1100	1378	1295	1039	888	1012

Region	10	11	12	13	14	15	16	17	18
Lawyers	718	835	1192	992	1138	920	1397	872	1142
Architects	794	900	1150	1038	1197	939	1124	911	1171

Does this information indicate that architects tend to have a larger average weekly income? Use $\alpha = 0.05$.

11. *High School Dropouts: Male versus Female* Is the high school dropout rate higher for males or females? A random sample of population regions gave the following information about percentage of 15- to 19-year-olds who are high school dropouts (Reference: *Statistical Abstract of the United States,* 121st Edition).

Region	1	2	3	4	5	6	7	8	9	10
Male	7.3	7.5	7.7	21.8	4.2	12.2	3.5	4.2	8.0	9.7
Female	7.5	6.4	6.0	20.0	2.6	5.2	3.1	4.9	12.1	10.8

Region	11	12	13	14	15	16	17	18	19	20
Male	14.1	3.6	3.6	4.0	5.2	6.9	15.6	6.3	8.0	6.5
Female	15.6	6.3	4.0	3.9	9.8	9.8	12.0	3.3	7.1	8.2

Does this information indicate that the dropout rates for males and females are different (either way)? Use $\alpha = 0.01$.

12. *Focus Problem: Meteorology* The Focus Problem at the beginning of this chapter asks you to use a sign test with a 5% level of significance to test the claim that the overall temperature distribution of Madison, Wisconsin, is different (either way) from that of Juneau, Alaska. The monthly average data (in °F) are as follows.

Month	Jan.	Feb.	March	April	May	June
Madison	17.5	21.1	31.5	46.1	57.0	67.0
Juneau	22.2	27.3	31.9	38.4	46.4	52.8

Month	July	Aug.	Sept.	Oct.	Nov.	Dec.
Madison	71.3	69.8	60.7	51.0	35.7	22.8
Juneau	55.5	54.1	49.0	41.5	32.0	26.9

What is your conclusion?

SECTION **11.2**

The Rank-Sum Test

FOCUS POINTS

- State the criteria for setting up a rank-sum test.
- Use the distribution of ranks to complete the test.
- Interpret the results in the context of the application.

The sign test is used when we have paired data values coming from dependent samples, as in "before and after" studies. However, if the data values are *not paired,* the sign test should *not* be used.

Criteria for rank-sum test

For the situation in which we draw *independent random samples* from two populations, there is another nonparametric method for testing the difference between sample means; it is called the *rank-sum test* (also called the *Mann–Whitney test*). The rank-sum test can be used when assumptions about *normal* populations are not satisfied. To fix our thoughts on a definite problem, let's consider the following example:

TABLE 11-4 **Decompression Times for 23 Navy Divers (in min)**

Group A (had pill)	41	56	64	42	50	70	44	57	63	65	52	
	Mean time = 54.91 min											
Group B (no pill)	66	43	72	62	55	80	74	75	77	78	47	60
	Mean time = 65.75 min											

When a scuba diver makes a deep dive, nitrogen builds up in the diver's blood. After returning to the surface, the diver must wait in a decompression chamber until the nitrogen level of the blood returns to normal. A physiologist working with the Navy has invented a pill that a diver takes 1 hour before diving. The pill is supposed to reduce the waiting time spent in the decompression chamber. Twenty-three Navy divers volunteered to help the physiologist determine if the pill has any effect. The divers were randomly divided into two groups: group A had 11 divers who took the pill, and group B had 12 divers who did not take the pill. All the divers worked the same length of time on a deep salvage operation and returned to the decompression chamber. A monitoring device in the decompression chamber measured the waiting time for each diver's nitrogen level to return to normal. These times are recorded in Table 11-4.

Rank the data

The means of our two samples are 54.91 and 65.75 minutes. We will use the rank-sum test to decide whether the difference between the means is significant. First, we arrange the two samples jointly in order of increasing time. To do this, we use the data of groups A and B as if they were one sample. The times (in minutes), groups, and ranks are shown in Table 11-5.

Group A occupies the ranks 1, 2, 4, 6, 7, 9, 10, 13, 14, 15, and 17, while group B occupies the ranks 3, 5, 8, 11, 12, 16, 18, 19, 20, 21, 22, and 23. We add up the ranks of the group with the *smaller* sample size, in this case, group A.

Sum the ranks of the smaller group

The sum of the ranks is denoted by R:

$$R = 1 + 2 + 4 + 6 + 7 + 9 + 10 + 13 + 14 + 15 + 17 = 98$$

Let n_1 be the size of the *smaller sample* and n_2 be the size of the *larger sample*. In the case of the divers, $n_1 = 11$ and $n_2 = 12$. So, R is the sum of the ranks from the smaller sample. If both samples are of the same size, then $n_1 = n_2$ and R is the sum of the ranks of either group (but not both groups).

Andrew G. Wood/Science Source/Photo Researchers, Inc.

TABLE 11-5 **Ranks for Decompression Time**

Time	Group	Rank	Time	Group	Rank
41	A	1	63	A	13
42	A	2	64	A	14
43	B	3	65	A	15
44	A	4	66	B	16
47	B	5	70	A	17
50	A	6	72	B	18
52	A	7	74	B	19
55	B	8	75	B	20
56	A	9	77	B	21
57	A	10	78	B	22
60	B	11	80	B	23
62	B	12			

Distribution of ranks

When both n_1 and n_2 are sufficiently large (each greater than 10), advanced mathematical statistics can be used to show that R is approximately normally distributed, with mean

$$\mu_R = \frac{n_1(n_1 + n_2 + 1)}{2}$$

and standard deviation

$$\sigma_R = \sqrt{\frac{n_1 n_2(n_1 + n_2 + 1)}{12}}$$

GUIDED EXERCISE 3 — MEAN AND STANDARD DEVIATION OF RANKS

For the Navy divers, compute μ_R and σ_R. (Recall that $n_1 = 11$ and $n_2 = 12$.)

$$\mu_R = \frac{n_1(n_1 + n_2 + 1)}{2} = \frac{11(11 + 12 + 1)}{2} = 132$$

$$\sigma_R = \sqrt{\frac{n_1 n_2(n_1 + n_2 + 1)}{12}}$$

$$= \sqrt{\frac{11 \cdot 12(11 + 12 + 1)}{12}} \approx 16.25$$

Sample test statistic

Since $n_1 = 11$ and $n_2 = 12$, the samples are large enough to assume that the rank R is approximately normally distributed. We convert the sample test statistic R to a z value using the following formula, with $R = 98$, $\mu_R = 132$, and $\sigma_R \approx 16.25$:

$$z = \frac{R - \mu_R}{\sigma_R} \approx \frac{98 - 132}{16.25} \approx -2.09$$

Hypotheses

When using the rank-sum test, the null hypothesis is that the distributions are the same, while the alternate hypothesis is that the distributions are different. In the case of the Navy divers, we have

H_0: Decompression time distributions are the same.
H_1: Decompression time distributions are different.

We'll test the decompression time distributions using level of significance 5%.

***P*-value**

To find the P-value of the sample test statistic $z = -2.09$, we use the normal distribution (Table 5 of Appendix II) and the fact that we have a two-tailed test. Figure 11-3 shows the P-value.

The area to the left of -2.09 is 0.0183. This is a two-tailed test, so

$$P\text{-value} = 2(0.0183) = 0.0366$$

Conclusion

Since the P-value is less than $\alpha = 0.05$, we reject H_0. At the 5% level of significance, we have sufficient evidence to conclude that the pill changes decompression times for divers.

The steps necessary for a rank-sum test are summarized by the procedure on the next page.

FIGURE 11-3

P-value

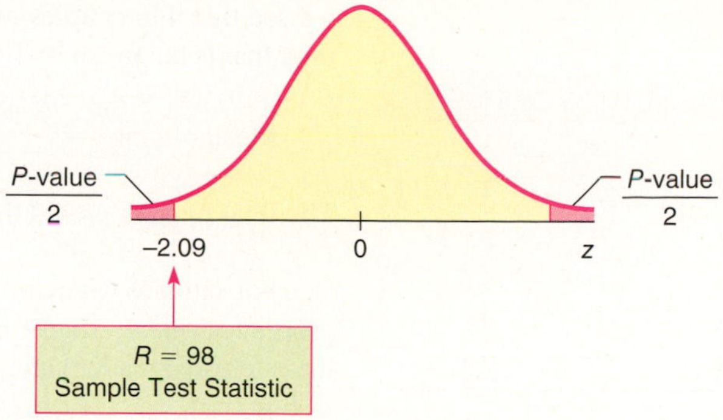

PROCEDURE

HOW TO CONSTRUCT A RANK-SUM TEST

Setup and Requirements

You first need independent random samples (both of size 11 or more) from two populations *A* and *B*. Let n_1 be the sample size of the *smaller* sample and let n_2 be the sample size of the larger sample. If the sample sizes are equal, then simply use the common value for n_1 and n_2. Next, you need to rank-order the data as if they were one big sample. Label each rank *A* or *B* according to the population from which it came. Let *R* be a random variable that represents the sum of ranks from the sample of size n_1. If $n_1 = n_2$, then *R* is the sum of ranks from either group (but not both).

Procedure

1. Set the *level of significance* α. The *null* and *alternate hypotheses* are

   H_0: The two samples come from populations with the same distribution (the two populations are identical).

   H_1: The two samples come from populations with different distributions (the populations differ in some way).

2. The *sample test statistic* is

$$z = \frac{R - \mu_R}{\sigma_R}$$

   where $R =$ sum of ranks from the sample of size n_1 (smaller sample),

$$\mu_R = \frac{n_1(n_1 + n_2 + 1)}{2}$$

$$\sigma_R = \sqrt{\frac{n_1 n_2(n_1 + n_2 + 1)}{12}}$$

   and $n_1 > 10$, $n_2 > 10$.

3. Use the standard normal distribution with a two-tailed test to find the *P-value* corresponding to the test statistic.
4. *Conclude* the test. If P-value $\leq \alpha$, then reject H_0. If P-value $> \alpha$, then do not reject H_0.
5. *Interpret your conclusion* in the context of the application.

Procedure for tied ranks

NOTE For the decompression time data, there were no ties for any rank. If a tie does occur, then each of the tied observations is given the *mean* of the ranks that it occupies. For example, if we rank the numbers

41 42 44 44 44 44

TABLE 11-6

Observation	Rank
41	1
42	2
44	4.5
44	4.5
44	4.5
44	4.5

we see that 44 occupies ranks 3, 4, 5, and 6. Therefore, we give each of the 44's a rank that is the mean of 3, 4, 5, and 6:

$$\text{Mean of ranks} = \frac{3 + 4 + 5 + 6}{4} = 4.5$$

The final ranking would then be that shown in Table 11-6.

For samples where n_1 or n_2 is less than 11, there are statistical tables that give appropriate critical values for the rank-sum test. Most libraries contain such tables, and the interested reader can find such information by looking under the *Mann–Whitney U Test*.

GUIDED EXERCISE 4 RANK-SUM TEST

A biologist is doing research on elk in their natural Colorado habitat. Two regions are under study, both having about the same amount of forage and natural cover. However, region A seems to have fewer predators than region B. To determine if there is a difference in elk life spans between the two regions, a sample of 11 mature elk from each region are tranquilized and have a tooth removed. A laboratory examination of the teeth reveals the ages of the elk. Results for each sample are given in Table 11-7. The biologist uses a 5% level of significance to test for a difference in life spans.

TABLE 11-7 **Ages of Elk**

Group A	4	10	11	2	2	3	9	4	12	6	6
Group B	7	3	8	4	8	5	6	4	2	4	3

(a) Fill in the remaining ranks of Table 11-8. Be sure to use the process of taking the mean of tied ranks.

TABLE 11-8 **Ranks of Elk**

Age	Group	Rank	Age	Group	Rank
2	A	2	5	B	12
2	A	2	6	A	—
2	B	2	6	A	—
3	A	5	6	B	—
3	B	5	7	B	—
3	B	5	8	B	—
4	A	9	8	B	—
4	A	9	9	A	—
4	B	9	10	A	—
4	B	9	11	A	—
4	B	9	12	A	—

Rank
12
14
14
14
16
17.5
17.5
19
20
21
22

(b) What is α? State the null and alternate hypotheses.

$\alpha = 0.05$

H_0: Distributions of life spans are the same.
H_1: Distributions of life spans are different.

Continued

GUIDED EXERCISE 3 *continued*

(c) Find μ_R, σ_R, and R. Convert R to a sample z statistic.

Since $n_1 = 11$ and $n_2 = 11$,

$$\mu_R = \frac{(11)(11 + 11 + 1)}{2} = 126.5$$

$$\sigma_R = \sqrt{\frac{11 \cdot 11(11 + 11 + 1)}{12}} \approx 15.23$$

Since $n_1 = n_2 = 11$, we can use the sum of the ranks of either the A group or the B group. Let's use the A group. The A group ranks are 2, 2, 5, 9, 9, 14, 14, 19, 20, 21, and 22. Therefore,

$$R = 2 + 2 + 5 + 9 + 9 + 14 + 14 + 19 + 20 + 21 + 22 = 137$$

$$z = \frac{R - \mu_R}{\sigma_R} = \frac{137 - 126.5}{15.23} \approx 0.69$$

(d) Find the *P*-value shown in Figure 11-4.

FIGURE 11-4 *P*-value

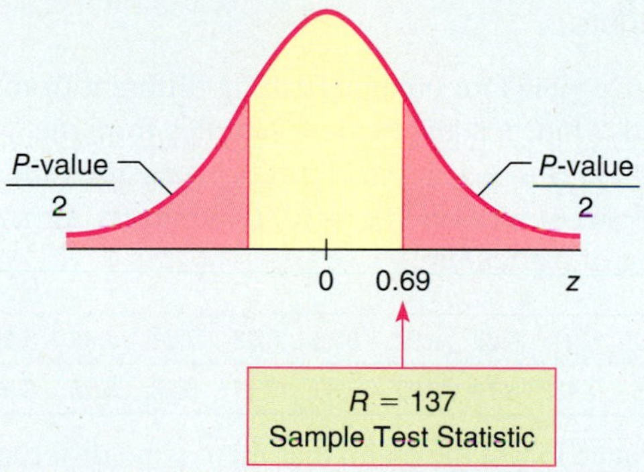

Using Table 5 of Appendix II, the area to the right of 0.69 is 0.2451. Since this is a two-tailed test,

$$P\text{-value} = 2(0.2451) = 0.4902$$

Comment: If we use the sum of ranks of group B, then $R_B = 116$ and $z = -0.69$. The *P*-value is again 0.4902, and we have the same conclusion.

(e) *Interpretation* What is the conclusion?

The *P*-value of 0.4902 is greater than $\alpha = 0.05$, so we do not reject H_0. The evidence does not support the claim that the age distribution of elk is different between the two regions.

VIEWPOINT Point Barrow, Alaska

Point Barrow is located very near the northernmost point of land in the United States. In 1935, Will Rogers (an American humorist, social critic, and philosopher) was killed with Wiley Post (a pioneer aviator) at a landing strip near Point Barrow. Since 1920, a weather station at the (now named) Wiley Post–Will Rogers Memorial Landing Strip has recorded daily high and low temperatures. From these readings, annual mean maximum and minimum temperatures have been computed. Is Point Barrow warming up, cooling down, or neither? Can you think of a way to gather data and construct a nonparametric test to investigate long-term temperature highs and lows at Point Barrow? For weather-related data, visit the Brase/Brase statistics site at **http://www.cengage.com/statistics/brase11e** *and find a link to the Geophysical Institute at the University of Alaska in Fairbanks. Then follow the links to Point Barrow.*

SECTION 11.2 PROBLEMS

1. *Statistical Literacy* When applying the rank-sum test, do you need independent or dependent samples?

2. *Statistical Literacy* If two or more data values are the same, how is the rank of each of the tied data computed?

For Problems 3–11, please provide the following information.
(a) What is the level of significance? State the null and alternate hypotheses.
(b) Compute the sample test statistic. What is the sampling distribution? What conditions are necessary to use this distribution?
(c) Find the P-value of the sample test statistic.
(d) Conclude the test.
(e) *Interpret* the conclusion in the context of the application.

3. *Agriculture: Lima Beans* Are yields for organic farming different from conventional farming yields? Independent random samples from method A (organic farming) and method B (conventional farming) gave the following information about yield of lima beans (in tons/acre) (Reference: *Agricultural Statistics,* U.S. Department of Agriculture).

Method A	1.83	2.34	1.61	1.99	1.78	2.01	2.12	1.15	1.41	1.95	1.25	
Method B	2.15	2.17	2.11	1.89	1.34	1.88	1.96	1.10	1.75	1.80	1.53	2.21

Use a 5% level of significance to test the hypothesis that there is no difference between the yield distributions.

4. *Agriculture: Sweet Corn* Are yields for organic farming different from conventional farming yields? Independent random samples from method A (organic farming) and method B (conventional farming) gave the following information about yield of sweet corn (in tons/acre) (Reference: *Agricultural Statistics,* U.S. Department of Agriculture).

Method A	6.88	6.86	7.12	5.91	6.80	6.92	6.25	6.98	7.21	7.33	5.85	6.72
Method B	5.71	6.93	7.05	7.15	6.79	6.87	6.45	7.34	5.68	6.78	6.95	

Use a 5% level of significance to test the claim that there is no difference between the yield distributions.

5. *Horse Trainer: Jumps* A horse trainer teaches horses to jump by using two methods of instruction. Horses being taught by method A have a lead horse that accompanies each jump. Horses being taught by method B have no lead horse. The table shows the number of training sessions required before each horse performed the jumps properly.

Method A	28	35	19	41	37	31	38	40	25	27	36	43
Method B	42	33	26	24	44	46	34	20	48	39	45	

Use a 5% level of significance to test the claim that there is no difference between the training session distributions.

6. *Violent Crime: FBI Report* Is the crime rate in New York different from the crime rate in New Jersey? Independent random samples from region A (cities in New York) and region B (cities in New Jersey) gave the following information about violent crime rate (number of violent crimes per 100,000 population) (Reference: U.S. Department of Justice, Federal Bureau of Investigation).

Region A	554	517	492	561	577	621	512	580	543	605	531	
Region B	475	419	505	575	395	433	521	388	375	411	586	415

Use a 5% level of significance to test the claim that there is no difference in the crime rate distributions of the two states.

7. *Psychology: Testing* A cognitive aptitude test consists of putting together a puzzle. Eleven people in group A took the test in a competitive setting (first and second to finish received a prize). Twelve people in group B took the test in a noncompetitive setting. The results follow (in minutes required to complete the puzzle).

Group A	7	12	10	15	22	17	18	13	8	16	11	
Group B	9	16	30	11	33	28	19	14	24	27	31	29

Use a 5% level of significance to test the claim that there is no difference in the distributions of time to complete the test.

8. *Psychology: Testing* A psychologist has developed a mental alertness test. She wishes to study the effects (if any) of type of food consumed on mental alertness. Twenty-one volunteers were randomly divided into two groups. Both groups were told to eat the amount they usually eat for lunch at noon. At 2:00 P.M., all subjects were given the alertness test. Group A had a low-fat lunch with no red meat, lots of vegetables, carbohydrates, and fiber. Group B had a high-fat lunch with red meat, vegetable oils, and low fiber. The only drink for both groups was water. The test scores are shown below.

Group A	76	93	52	81	68	79	88	90	67	85	60	
Group B	44	57	60	91	62	86	82	65	96	42	68	98

Use a 1% level of significance to test the claim that there is no difference in mental alertness distributions based on type of lunch.

9. *Lifestyles: Exercise* Is there a link between exercise and level of education? Independent random samples of adults from group A (college graduates) and group B (no high school diploma) gave the following information about percentage who exercise regularly (Reference: Center for Disease Control and Prevention).

A(%)	63.3	55.1	50.0	47.1	58.2	60.0	44.3	49.1	68.7	57.3	59.9	
B(%)	33.7	40.1	53.3	36.9	29.1	59.6	35.7	44.2	38.2	46.6	45.2	60.2

Use a 1% level of significance to test the claim that there is no difference in the exercise rate distributions according to education level.

10. *Doctor's Degree: Years of Study* Does the average length of time to earn a doctorate differ from one field to another? Independent random samples from large graduate schools gave the following averages for length of registered time (in years) from bachelor's degree to doctorate. Sample A was taken from the humanities field, and sample B from the social sciences field (Reference: *Education Statistics*, U.S. Department of Education).

Field A	8.9	8.3	7.2	6.4	8.0	7.5	7.1	6.0	9.2	8.7	7.5	
Field B	7.6	7.9	6.2	5.8	7.8	8.3	8.5	7.0	6.3	5.4	5.9	7.7

Use a 1% level of significance to test the claim that there is no difference in the distributions of time to complete a doctorate for the two fields.

11. *Education: Spelling* Twenty-two fourth-grade children were randomly divided into two groups. Group A was taught spelling by a phonetic method. Group B was taught spelling by a memorization method. At the

end of the fourth grade, all children were given a standard spelling exam. The scores are as follows.

Group A	77	95	83	69	85	92	61	79	87	93	65	78
Group B	62	90	70	81	63	75	80	72	82	94	65	79

Use a 1% level of significance to test the claim that there is no difference in the test score distributions based on instruction method.

SECTION **11.3**

Spearman Rank Correlation

FOCUS POINTS

- Learn about monotone relations and the Spearman rank correlation coefficient.
- Compute the Spearman correlation coefficient and conduct statistical tests for significance.
- Interpret the results in the context of the application.

Data given in ranked form (ordinal type) are different from data given in measurement form (interval or ratio type). For instance, if we compared the test performances of three students and, say, Elizabeth did the best, Joel did next best, and Sally did the worst, we are giving the information in ranked form. We cannot say how much better Elizabeth did than Sally or Joel, but we do know how the three scores compare. If the actual test scores for the three tests were given, we would have data in measurement form and could tell exactly how much better Elizabeth did than Joel or Sally. In Chapter 9, we studied linear correlation of data in measurement form. In this section, we will study correlation of data in ranked form.

As a specific example of a situation in which we might want to compare ranked data from two sources, consider the following. Hendricks College has a new faculty position in its political science department. A national search to fill this position has resulted in a large number of qualified candidates. The political science faculty reserves the right to make the final hiring decision. However, the faculty is interested in comparing its opinion with student opinion about the teaching ability of the candidates. A random sample of nine equally qualified candidates were asked to give a classroom presentation to a large class of students. Both faculty and students attended the lectures. At the end of each lecture, both faculty and students filled out a questionnaire about the teaching performance of the candidate. Based on these questionnaires, each candidate was given an overall rank from the faculty and an overall rank from the students. The results are shown in Table 11-9. Higher ranks mean better teaching performance.

TABLE 11-9 **Faculty and Student Ranks of Candidates**

Candidate	Faculty Rank	Student Rank
1	3	5
2	7	7
3	5	6
4	9	8
5	2	3
6	8	9
7	1	1
8	6	4
9	4	2

FIGURE 11-5

Examples of Monotone Relations

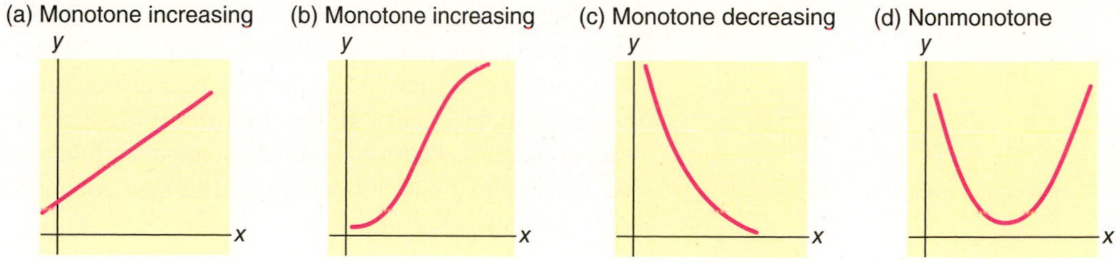

Using data in ranked form, we answer the following questions:

1. Do candidates getting higher ranks from faculty tend to get higher ranks from students?
2. Is there any relation between faculty rankings and student rankings?
3. Do candidates getting higher ranks from faculty tend to get lower ranks from students?

We will use the Spearman rank correlation to answer such questions. In the early 1900s, Charles Spearman of the University of London developed the techniques that now bear his name. The Spearman test of rank correlation requires us to use *ranked variables.* Because we are using only ranks, we cannot use the Spearman test to check for the existence of a linear relationship between the variables as we did with the Pearson correlation coefficient (Section 9.1). The Spearman test checks only for the existence of a *monotone* relationship between the variables. (See Figure 11-5.)

Monotone relationship

By a *monotone relationship** between variables x and y, we mean a relationship in which

1. as x increases, y also increases, or
2. as x increases, y decreases.

The relationship shown in Figure 11-5(d) is a nonmonotone relationship because as x increases, y at first decreases, but later starts to increase. Remember, for a relation to be monotone, as x increases, y must *always* increase or *always* decrease. In a nonmonotone relationship, as x increases, y sometimes increases and sometimes decreases or stays unchanged.

GUIDED EXERCISE 5 | MONOTONIC BEHAVIOR

Identify each of the relations in Figure 11-6 as monotone increasing, monotone decreasing, or nonmonotone.

FIGURE 11-6

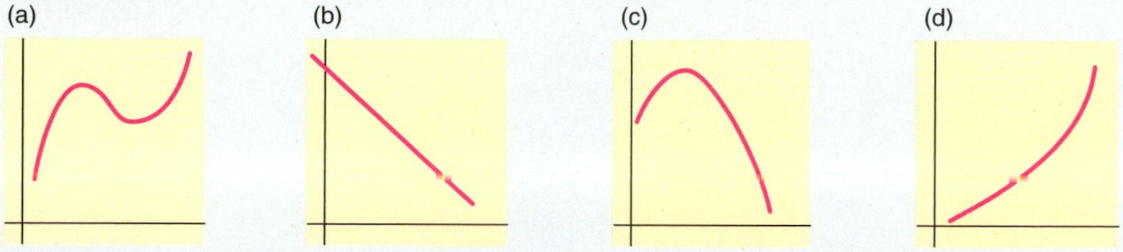

Answers: (a) nonmonotone, (b) monotone decreasing, (c) nonmonotone, (d) monotone increasing

*Some advanced texts call the monotone relationship we describe *strictly monotone*.

Before we can complete the solution of our problem about the political science department at Hendricks College, we need the following information.

Spearman rank correlation coefficient r_s

Suppose we have a sample of size n of randomly obtained ordered pairs (x, y), where both the x and y values are from *ranked variables*. If there are no ties in the ranks, then the Pearson product-moment correlation coefficient (Section 9.1) can be reduced to a simpler equation. The new equation produces the *Spearman rank correlation coefficient*, r_s.

SPEARMAN RANK CORRELATION COEFFICIENT

$$r_s = 1 - \frac{6\Sigma d^2}{n(n^2 - 1)} \quad \text{where } d = x - y$$

The Spearman rank correlation coefficient has the following properties.

PROPERTIES OF THE SPEARMAN RANK CORRELATION COEFFICIENT

1. $-1 \le r_s \le 1$. If $r_s = -1$, the relation between x and y is perfectly monotone decreasing. If $r_s = 0$, there is no monotone relation between x and y. If $r_s = 1$, the relation between x and y is perfectly monotone increasing. Values of r_s close to 1 or -1 indicate a strong tendency for x and y to have a monotone relationship (increasing or decreasing). Values of r_s close to 0 indicate a very weak (or perhaps nonexistent) monotone relationship.
2. The probability distribution of r_s depends on the sample size n. It is symmetric about $r_s = 0$. Table 9 of Appendix II gives critical values for certain specified one-tail and two-tail areas. Use of the table requires no assumptions that x and y are normally distributed variables. In addition, we make no assumption about the x and y relationship being linear.
3. The Spearman rank correlation coefficient r_s is the *sample* estimate for the *population* Spearman rank correlation coefficient ρ_s.

Population Spearman rank correlation coefficient ρ_s

Hypotheses

We construct a test of significance for the Spearman rank correlation coefficient in much the same way that we tested the Pearson correlation coefficient (Section 9.3). The null hypothesis states that there is no monotone relation between x and y (either increasing or decreasing).

H_0: $\rho_s = 0$

The alternate hypothesis is one of the following:

H_1: $\rho_s < 0$	H_1: $\rho_s > 0$	H_1: $\rho_s \ne 0$
(left-tailed)	(right-tailed)	(two-tailed)

A left-tailed alternate hypothesis claims there is a monotone-decreasing relation between x and y. A right-tailed alternate hypothesis claims there is a monotone-increasing relation between x and y, while a two-tailed alternate hypothesis claims there is a monotone relation (either increasing or decreasing) between x and y.

Figure 11-7 shows the type of test and corresponding P-value region.

FIGURE 11-7

Type of Test and *P*-value Region

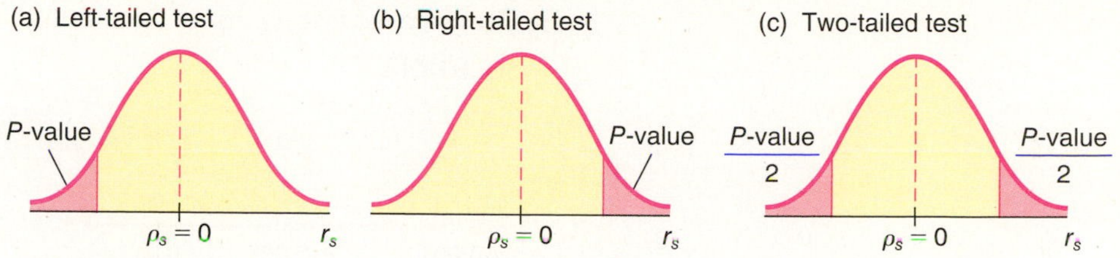

EXAMPLE 1

TESTING THE SPEARMAN RANK CORRELATION COEFFICIENT

Using the information about the Spearman rank correlation coefficient, let's finish our problem about the search for a new member of the political science department at Hendricks College. Our work is organized in Table 11-10, where the rankings given by students and faculty are listed for each of the nine candidates.

Zigy Kaluzny-Charles Thatcher/Stone/Getty Images

(a) Using a 1% level of significance, let's test the claim that the faculty and students tend to agree about a candidate's teaching ability. This means that the x and y variables should be monotone increasing (as x increases, y increases). Since ρ_s is the population Spearman rank correlation coefficient, we have

H_0: $\rho_s = 0$ (There is no monotone relation.)

H_1: $\rho_s > 0$ (There is a monotone-increasing relation.)

(b) Compute the sample test statistic.

SOLUTION: Since the sample size is $n = 9$, and from Table 11-10 we see that $\Sigma d^2 = 16$, the Spearman rank correlation coefficient is

$$r_s = 1 - \frac{6\Sigma d^2}{n(n^2 - 1)} = 1 - \frac{6(16)}{9(81 - 1)} \approx 0.867$$

(c) Find or estimate the *P*-value.

P-value

SOLUTION: To estimate the *P*-value for the sample test statistic $r_s = 0.867$, we use Table 9 of Appendix II. The sample size is $n = 9$ and the test is a one-tailed test. We find the location of the sample test statistic in row 9, and then read the corresponding one-tail area. From the Table 9, Appendix II excerpt, we see that the sample test statistic $r_s = 0.867$ falls between the entries 0.834 and 0.917 in the $n = 9$ row. These values correspond to *one-tail areas* between 0.005 and 0.001.

$$0.001 < P\text{-value} < 0.005$$

TABLE 11-10 Student and Faculty Ranks of Candidates and Calculations for the Spearman Rank Correlation Test

Candidate	Faculty Rank x	Student Rank y	$d = x - y$	d^2
1	3	5	−2	4
2	7	7	0	0
3	5	6	−1	1
4	9	8	1	1
5	2	3	−1	1
6	8	9	−1	1
7	1	1	0	0
8	6	4	2	4
9	4	2	2	4
				$\Sigma d^2 = 16$

(d) Conclude the test. *Interpret* the results.

SOLUTION:

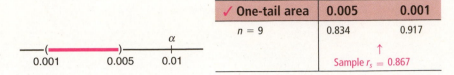

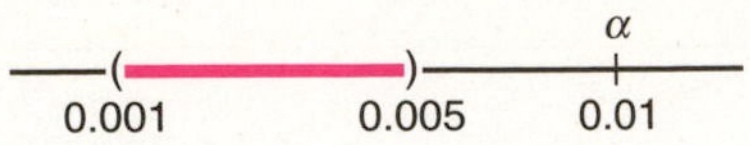

✓ One-tail area	0.005	0.001
$n = 9$	0.834	0.917
	↑ Sample $r_s = 0.867$	

Since the P-value is less than $\alpha = 0.01$, we reject H_0. At the 1% level of significance, we conclude that the relation between faculty and student ratings is monotone increasing. This means that faculty and students tend to rank the teaching performance of candidates in a similar way: Higher student ratings of a candidate correspond with higher faculty ratings of the same candidate.

The following procedure summarizes the steps involved in testing the population Spearman rank correlation coefficient.

PROCEDURE

HOW TO TEST THE SPEARMAN RANK CORRELATION COEFFICIENT ρ_S

Setup

You first need a random sample (of size n) of data pairs (x, y), where both the x and y values are *ranked* variables. Let ρ_s represent the population Spearman rank correlation coefficient, which is in theory computed from the population of all possible (x, y) data pairs.

Procedure

1. Set the *level of significance* α. The *null hypothesis* is H_0: $\rho_s = 0$. In the context of the application, choose the *alternate hypothesis* to be H_1: $\rho_s > 0$ or H_1: $\rho_s < 0$ or H_1: $\rho_s \neq 0$.
2. If there are no ties in the ranks, or if the number of ties is small compared to the number of data pairs n, then compute the *sample test statistic*

$$r_s = 1 - \frac{6\Sigma d^2}{n(n^2 - 1)}$$

where $d = x - y$ is the difference in ranks
$n =$ number of data pairs
and the sum is over all sample data pairs.

3. Use Table 9 of Appendix II to find or estimate the *P-value* corresponding to r_s and $n =$ number of data pairs.
4. *Conclude* the test. If P-value $\leq \alpha$, then reject H_0. If P-value $> \alpha$, then do not reject H_0.
5. *Interpret your conclusion* in the context of the application.

GUIDED EXERCISE 6 TESTING THE SPEARMAN RANK CORRELATION COEFFICIENT

Fishermen in the Adirondack Mountains are complaining that acid rain caused by air pollution is killing fish in their region. To research this claim, a team of biologists studied a random sample of 12 lakes in the region. For each lake, they measured the level of acidity of rain in the drainage leading into the lake and the density of fish in the lake (number of fish per acre-foot of water). They then did a ranking of $x =$ acidity and $y =$ density of fish. The results are shown in Table 11-11. Higher x ranks mean more acidity, and higher y ranks mean higher density of fish.

Continued

GUIDED EXERCISE 6 *continued*

TABLE 11-11 **Acid Rain and Density of Fish**

Lake	Acidity x	Fish Density y	$d = x - y$	d^2
1	5	8	-3	9
2	8	6	2	4
3	3	9	-6	36
4	2	12	-10	100
5	6	7	-1	1
6	1	10	-9	81
7	10	2	8	64
8	12	1	___	___
9	7	5	___	___
10	4	11	___	___
11	9	4	___	___
12	11	3	___	___
				$\Sigma d^2 =$ ___

(a) Complete the entries in the d and d^2 columns of Table 11-11, and find Σd^2.

Lake	x	y	d	d^2
8	12	1	11	121
9	7	5	2	4
10	4	11	-7	49
11	9	4	5	25
12	11	3	8	64
				$\Sigma d^2 = 558$

(b) Compute r_s.

$$r_s = 1 - \frac{6\Sigma d^2}{n(n^2 - 1)} = 1 - \frac{6(558)}{12(144 - 1)} \approx -0.951$$

(c) The fishermen claim that more acidity means lower density of fish. Does this claim state that x and y have a monotone-increasing relation, a monotone-decreasing relation, or no monotone relation?

The claim states that as x increases, y decreases, so the relation between x and y is monotone decreasing.

(d) To test the fishermen's claim, what should we use for the null hypothesis and for the alternate hypothesis? Use $\alpha = 0.01$.

H_0: $\rho_s = 0$ (no monotone relation)
H_1: $\rho_s < 0$ monotone-decreasing relation)

(e) Find or estimate the P-value of the sample test statistic $r_s = -0.951$.

Use Table 9 of Appendix II. There are $n = 12$ data pairs. The sample statistic r_s is negative. Because the r_s distribution is symmetric about 0, we look up the corresponding

Continued

GUIDED EXERCISE 6 *continued*

positive value 0.951 in the row headed by $n = 12$. Use one-tail areas, since this is a left-tailed test.

✓ One-tail area	0.001
$n = 12$	0.826
	↑
	$-r_s = 0.951$

As positive r_s values increase, corresponding right-tail areas decrease. Therefore, P-value < 0.001

(f) Use $\alpha = 0.01$ and conclude the test.

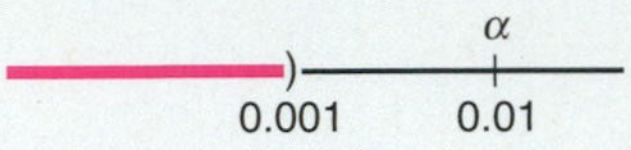

Since the P-value is less than $\alpha = 0.01$, we reject H_0 and conclude that there is a monotone-decreasing relationship between the acidity of the water and the number of fish.

(g) *Interpretation* Do the data support the claim that higher acidity means fewer fish?

At the 1% level of significance, we conclude that higher acidity means fewer fish.

Ties of ranks

If ties occur in the assignment of ranks, we follow the usual method of averaging tied ranks. This method was discussed in Section 11.2 (The Rank-Sum Test). The next example illustrates the method.

COMMENT Technically, the use of the given formula for r_s requires that there be no ties in rank. However, if the number of ties in rank is small relative to the number of ranks, the formula can be used with quite a bit of reliability.

EXAMPLE 2

TIED RANKS

Do people who smoke more tend to drink more cups of coffee? The following data were obtained from a random sample of $n = 10$ cigarette smokers who also drink coffee.

Person	Cigarettes per Day	Cups of Coffee per Day
1	8	4
2	15	7
3	20	10
4	5	3
5	22	9
6	15	5
7	15	8
8	25	11
9	30	18
10	35	18

(a) To use the Spearman rank correlation test, we need to rank the data. It does not matter if we rank from smallest to largest or from largest to smallest. The only requirement is that we be consistent in our rankings. Let us rank from smallest to largest.

First, we rank the data for each variable as though there were no ties; then we average the ties as shown in Tables 11-12 and 11-13.

(b) Using 0.01 as the level of significance, we test the claim that x and y have a monotone-increasing relationship. In other words, we test the claim that people who tend to smoke more tend to drink more cups of coffee (Table 11-14).

H_0: $\rho_s = 0$ (There is no montone relation.)

H_1: $\rho_s > 0$ (Right-tailed test)

(c) Next, we compute the observed sample test statistic r_s using the results shown in Table 11-14.

$$r_s = 1 - \frac{6\Sigma d^2}{n(n^2 - 1)} = 1 - \frac{6(4.5)}{10(100 - 1)} \approx 0.973$$

(d) Find or estimate the P-value for the sample test statistic $r_s = 0.973$.
We use Table 9 of Appendix II to estimate the P-value. Using $n = 10$ and a one-tailed test, we see that $r_s = 0.973$ is to the right of the entry 0.879. Therefore, the P-value is smaller than 0.001.

✓ One-tail area	0.001
$n = 9$	0.879
	↑ Sample $r_s = 0.973$

TABLE 11-12 Rankings of Cigarettes Smoked per Day

Person	Cigarettes per Day		Rank		Average Rank x	
4	5		1		1	
1	8		2		2	
2	15	Ties	3	Average rank is 4.	4	Use the average rank for tied data.
6	15		4		4	
7	15		5		4	
3	20		6		6	
5	22		7		7	
8	25		8		8	
9	30		9		9	
10	35		10		10	

TABLE 11-13 Rankings of Cups of Coffee per Day

Person	Cups of Coffee per Day		Rank		Average Rank y	
4	3		1		1	
1	4		2		2	
6	5		3		3	
2	7		4		4	
7	8		5		5	
5	9		6		6	
3	10		7		7	
8	11		8		8	
9	18	Ties	9	Average rank is 9.5.	9.5	Use the average rank for tied data.
10	18		10		9.5	

TABLE 11-14 Ranks to Be Used for a Spearman Rank Correlation Test

Person	Cigarette Rank x	Coffee Rank y	$d = x - y$	d^2
1	2	2	0	0
2	4	4	0	0
3	6	7	−1	1
4	1	1	0	0
5	7	6	1	1
6	4	3	1	1
7	4	5	−1	1
8	8	8	0	0
9	9	9.5	−0.5	0.25
10	10	9.5	0.5	0.25
				$\Sigma d^2 = 4.5$

(e) Conclude the test and *interpret* the results.

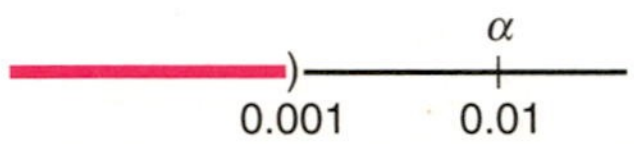

Since the P-value is less than $\alpha = 0.01$, we reject H_0. At the 1% level of significance, it appears that there is a monotone-increasing relationship between the number of cigarettes smoked and the amount of coffee consumed. People who smoke more cigarettes tend to drink more coffee.

VIEWPOINT Rug Rats!

When do babies start to crawl? Janette Benson, in her article "Infant Behavior and Development," claims that crawling age is related to temperature during the month in which babies first try to crawl. To find a data file for this subject, visit the Brase/Brase statistics site at **http://www.cengage.com/statistics/brase11e** *and find the link to DASL, the Carnegie Mellon University Data and Story Library. Then look under Psychology in the Data Subjects and select the Crawling Datafile. Can you think of a way to gather data and construct a nonparametric test to study this claim?*

SECTION 11.3 PROBLEMS

1. *Statistical Literacy* For data pairs (x, y), if y always increases as x increases, is the relationship monotone increasing, monotone decreasing, or nonmonotone?

2. *Statistical Literacy* Consider the Spearman rank correlation coefficient r_s for data pairs (x, y). What is the monotone relationship, if any, between x and y implied by a value of
(a) $r_s = 0$?
(b) r_s close to 1?
(c) r_s close to -1?

For Problems 3–11, please provide the following information.
(a) What is the level of significance? State the null and alternate hypotheses.
(b) Compute the sample test statistic.
(c) Find or estimate the P-value of the sample test statistic.
(d) Conclude the test.
(e) *Interpret* the conclusion in the context of the application.

3. *Training Program: Sales* A data-processing company has a training program for new salespeople. After completing the training program, each trainee is ranked by his or her instructor. After a year of sales, the same class of trainees is again ranked by a company supervisor according to net value of the contracts they have acquired for the company. The results for a random sample of 11 salespeople trained in the previous year follow, where x is rank in training class and y is rank in sales after 1 year. Lower ranks mean higher standing in class and higher net sales.

Person	1	2	3	4	5	6	7	8	9	10	11
x rank	6	8	11	2	5	7	3	9	1	10	4
y rank	4	9	10	1	6	7	8	11	3	5	2

Using a 0.05 level of significance, test the claim that the relation between x and y is monotone (either increasing or decreasing).

4. *Economics: Stocks* As an economics class project, Debbie studied a random sample of 14 stocks. For each of these stocks, she found the cost per share (in dollars) and ranked each of the stocks according to cost. After 3 months, she found the earnings per share for each stock (in dollars). Again, Debbie ranked each of the stocks according to earnings. The way Debbie ranked, higher ranks mean higher cost and higher earnings. The results follow, where x is the rank in cost and y is the rank in earnings.

Stock	1	2	3	4	5	6	7	8	9	10	11	12	13	14
x rank	5	2	4	7	11	8	12	3	13	14	10	1	9	6
y rank	5	13	1	10	7	3	14	6	4	12	8	2	11	9

Using a 0.01 level of significance, test the claim that there is a monotone relation, either way, between the ranks of cost and earnings.

5. *Psychology: Rat Colonies* A psychology professor is studying the relation between overcrowding and violent behavior in a rat colony. Eight colonies with different degrees of overcrowding are being studied. By using a television monitor, lab assistants record incidents of violence. Each colony has been ranked for crowding and violence. A rank of 1 means most crowded or most violent. The results for the eight colonies are given in the following table, with x being the population density rank and y the violence rank.

Colony	1	2	3	4	5	6	7	8
x rank	3	5	6	1	8	7	4	2
y rank	1	3	5	2	8	6	4	7

Using a 0.05 level of significance, test the claim that lower crowding ranks mean lower violence ranks (i.e., the variables have a monotone-increasing relationship).

6. *FBI Report: Murder and Arson* Is there a relation between murder and arson? A random sample of 15 Midwest cities (over 10,000 population) gave the following information about annual number of murder and arson cases (Reference: Federal Bureau of Investigation, U.S. Department of Justice).

City	1	2	3	4	5	6	7	8	9	10	11	12	13	14	15
Murder	12	7	25	4	10	15	9	8	11	18	23	19	21	17	6
Arson	62	12	153	2	63	93	31	29	47	131	175	129	162	115	4

(i) Rank-order murder using 1 as the largest data value. Also rank-order arson using 1 as the largest data value. Then construct a table of ranks to be used for a Spearman rank correlation test.

(ii) Use a 1% level of significance to test the claim that there is a monotone-increasing relationship between the ranks of murder and arson.

7. *Psychology: Testing* An army psychologist gave a random sample of seven soldiers a test to measure sense of humor and another test to measure aggressiveness. Higher scores mean greater sense of humor or more aggressiveness.

Soldier	1	2	3	4	5	6	7
Score on humor test	60	85	78	90	93	45	51
Score on aggressiveness test	78	42	68	53	62	50	76

(i) Ranking the data with rank 1 for highest score on a test, make a table of ranks to be used in a Spearman rank correlation test.

(ii) Using a 0.05 level of significance, test the claim that rank in humor has a monotone-decreasing relation to rank in aggressiveness.

8. *FBI Report: Child Abuse and Runaway Children* Is there a relation between incidents of child abuse and number of runaway children? A random sample of 15 cities (over 10,000 population) gave the following information about the number of reported incidents of child abuse and the number of runaway children (Reference: Federal Bureau of Investigation, U.S. Department of Justice).

City	1	2	3	4	5	6	7	8	9	10	11	12	13	14	15
Abuse cases	49	74	87	10	26	119	35	13	89	45	53	22	65	38	29
Runaways	382	510	581	163	210	791	275	153	491	351	402	209	410	312	210

(i) Rank-order abuse using 1 as the largest data value. Also rank-order runaways using 1 as the largest data value. Then construct a table of ranks to be used for a Spearman rank correlation test.

(ii) Use a 1% level of significance to test the claim that there is a monotone-increasing relationship between the ranks of incidents of abuse and number of runaway children.

9. *Demographics: Police and Fire Protection* Is there a relation between police protection and fire protection? A random sample of large population areas gave the following information about the number of local police and the number of local firefighters (units in thousands) (Reference: *Statistical Abstract of the United States*).

Area	1	2	3	4	5	6	7	8	9	10	11	12	13
Police	11.1	6.6	8.5	4.2	3.5	2.8	5.9	7.9	2.9	18.0	9.7	7.4	1.8
Firefighters	5.5	2.4	4.5	1.6	1.7	1.0	1.7	5.1	1.3	12.6	2.1	3.1	0.6

(i) Rank-order police using 1 as the largest data value. Also rank-order firefighters using 1 as the largest data value. Then construct a table of ranks to be used for a Spearman rank correlation test.

(ii) Use a 5% level of significance to test the claim that there is a monotone relationship (either way) between the ranks of number of police and number of firefighters.

10. *Ecology: Wetlands* Turbid water is muddy or cloudy water. Sunlight is necessary for most life forms; thus turbid water is considered a threat to wetland ecosystems. Passive filtration systems are commonly used to

reduce turbidity in wetlands. Suspended solids are measured in mg/L. Is there a relation between input and output turbidity for a passive filtration system and, if so, is it statistically significant? At a wetlands environment in Illinois, the inlet and outlet turbidity of a passive filtration system have been measured. A random sample of measurements is shown below (Reference: *EPA Wetland Case Studies*).

Reading	1	2	3	4	5	6	7	8	9	10	11	12
Inlet (mg/L)	8.0	7.1	24.2	47.7	50.1	63.9	66.0	15.1	37.2	93.1	53.7	73.3
Outlet (mg/L)	2.4	3.6	4.5	14.9	7.4	7.4	6.7	3.6	5.9	8.2	6.2	18.1

(i) Rank-order the inlet readings using 1 as the largest data value. Also rank-order the outlet readings using 1 as the largest data value. Then construct a table of ranks to be used for a Spearman rank correlation test.

(ii) Use a 1% level of significance to test the claim that there is a monotone relation ship (either way) between the ranks of the inlet readings and outlet readings.

11. *Insurance: Sales* Big Rock Insurance Company did a study of per capita income and volume of insurance sales in eight Midwest cities. The volume of sales in each city was ranked, with 1 being the largest volume. The per capita income was rounded to the nearest thousand dollars.

City	1	2	3	4	5	6	7	8
Rank of insurance sales volume	6	7	1	8	3	2	5	4
Per capita income in $1000	17	18	19	11	16	20	15	19

(i) Using a rank of 1 for the highest per capita income, make a table of ranks to be used for a Spearman rank correlation test.

(ii) Using a 0.01 level of significance, test the claim that there is a monotone relation (either way) between rank of sales volume and rank of per capita income.

SECTION **11.4**

Runs Test for Randomness

FOCUS POINTS

- Test a sequence of *symbols* for randomness.
- Test a sequence of *numbers* for randomness about the median.

Astronomers have made an extensive study of galaxies that are $\pm 16°$ above and below the celestial equator. Of special interest is the flux, or change in radio signals, that originates from large electromagnetic disturbances deep in space. The flux units (10^{-26} watts/m^2/Hz) are very small. However, modern radio astronomy can detect and analyze these signals using large antennas (Reference: *Journal of Astrophysics,* Vol. 148, pp. 321–365).

A very important question is the following: Are changes in flux simply random, or is there some kind of nonrandom pattern? Let us use the symbol S to represent a strong or moderate flux and the symbol W to represent a faint or weak flux. Astronomers have received the following signals in order of occurrence.

S S W W W S W W S S S W W W S S W W W S S

Is there a statistical test to help us decide whether or not this sequence of radio signals is random? Well, we're glad you asked, because that is the topic of this section.

We consider applications in which *two* symbols are used (e.g., S or W). Applications using more than two symbols are left to specialized studies in mathematical combinatorics.

Sequence

Run

A **sequence** is an *ordered set* of consecutive symbols.
A **run** is a sequence of one or more occurrences of the *same* symbol.
n_1 = number of times the first symbol occurs in a sequence
n_2 = number of times the second symbol occurs in a sequence
R is a random variable that represents the **number of runs in a sequence.**

EXAMPLE 3

Basic Terminology

In this example, we use the symbols S and W, where S is the first symbol and W is the second symbol, to demonstrate sequences and runs. Identify the runs.

(a) S S W W W is a sequence.

SOLUTION: Table 11-15 shows the sequence of runs. There are $R = 2$ runs in the sequence. The first symbol S occurs $n_1 = 2$ times. The second symbol W occurs $n_2 = 3$ times.

TABLE 11-15 Runs

Run 1	Run 2
S S	W W W

(b) S S W W W S W W S S S S W is a sequence.

SOLUTION: The sequence of runs are shown in Table 11-16. There are $R = 6$ runs in the sequence. The first symbol S occurs $n_1 = 7$ times. The second symbol W occurs $n_2 = 6$ times.

TABLE 11-16 Runs

Run 1	Run 2	Run 3	Run 4	Run 5	Run 6
S S	W W W	S	W W	S S S S	W

Hypotheses

To test a sequence of two symbols for randomness, we use the following hypotheses.

Hypotheses for runs test for randomness

Hypotheses for runs test for randomness
H_0: The symbols are randomly mixed in the sequence.
H_1: The symbols are not randomly mixed in the sequence.

The decision procedure will reject H_0 if either R is too small (too few runs) or R is too large (too many runs).

Sample test statistic

The number of runs R is a *sample test statistic* with its own sampling distribution. Table 10 of Appendix II gives critical values of R for a significance level $\alpha = 0.05$. There are two parameters associated with R. They are n_1 and n_2, the numbers of times the first and second symbols appear in the sequence, respectively. If either $n_1 > 20$ or $n_2 > 20$, you can apply the normal approximation to construct the test. This will be discussed in Problems 11 and 12 at the end of this section. For now, we assume that $n_1 \leq 20$ and $n_2 \leq 20$.

Critical values

For each pair of n_1 and n_2 values, Table 10 of Appendix II provides two critical values: a smaller value denoted c_1 and a larger value denoted c_2. These two values are used to decide whether or not to reject the null hypothesis H_0 that the symbols are randomly mixed in the sequence.

DECISION PROCESS WHEN $n_1 \le 20$ AND $n_2 \le 20$

Use Table 10 of Appendix II with n_1 and n_2 to find the critical values c_1 and c_2. At the $\alpha = 5\%$ level of significance, use the following decision process, where R is the number of runs: If either $R \le c_1$ (too few runs) or $R \ge c_2$ (too many runs), then *reject* H_0. Otherwise, *do not reject* H_0.

COMMENT If either n_1 or n_2 is larger than 20, a normal approximation can be used. See Problems 11 and 12 at the end of this section.

Let's apply this decision process to the astronomy example regarding the sequence of strong and weak electromagnetic radio signals coming from a distant galaxy.

EXAMPLE 4

RUNS TEST

Recall that our astronomers had received the following sequence of electromagnetic signals, where S represents a strong flux and W represents a weak flux.

S S W W W S W W S S S W W W S S W W W S S

Is this a random sequence or not? Use a 5% level of significance.

(a) What is the level of significance α? State the null and alternate hypotheses.

SOLUTION: $\alpha = 0.05$

H_0: The symbols S and W are randomly mixed in the sequence.

H_1: The symbols S and W are not randomly mixed in the sequence.

(b) Find the sample test statistic R and the parameters n_1 and n_2.

SOLUTION: We break the sequence according to runs.

Run 1	Run 2	Run 3	Run 4	Run 5	Run 6	Run 7	Run 8	Run 9
SS	WWW	S	WW	SSS	WWW	SS	WWW	SS

We see that there are $n_1 = 10$ S symbols and $n_2 = 11$ W symbols. The number of runs is $R = 9$.

(c) Use Table 10 of Appendix II to find the critical values c_1 and c_2.

SOLUTION: Since $n_1 = 10$ and $n_2 = 11$, then $c_1 = 6$ and $c_2 = 17$.

(d) Conclude the test.

SOLUTION:

$R \le 6$	✓ $7 \le R \le 16$	$R \ge 17$
Reject H_0.	✓ Fail to reject H_0.	Reject H_0.

Since $R = 9$, we fail to reject H_0 at the 5% level of significance.

(e) *Interpret* the conclusion in the context of the problem.

SOLUTION: At the 5% level of significance, there is insufficient evidence to conclude that the sequence of electromagnetic signals is not random.

Eastcott-Momatiuk/The Image Works

Randomness about the median

An important application of the runs test is to help us decide if a sequence of numbers is a random sequence about the median. This is done using the *median* of the sequence of numbers. The process is explained in the next example.

Billy Calzada/ZUMA Press/Newscom

EXAMPLE 5

Runs test about the median

Silver iodide seeding of summer clouds was done over the Santa Catalina mountains of Arizona. Of great importance is the direction of the wind during the seeding process. A sequence of consecutive days gave the following compass readings for wind direction at seeding level at 5 A.M. (0° represents true north) (Reference: *Proceedings of the National Academy of Science,* Vol. 68, pp. 649–652).

174	160	175	288	195	140	124	219	197	184
183	224	33	49	175	74	103	166	27	302
61	72	93	172						

We will test this sequence for randomness above and below the median using a 5% level of significance.

Part I: Adjust the sequence so that it has only two symbols, A and B.

SOLUTION: First rank-order the data and find the median (see Section 3.1). Doing this, we find the median to be 169. Next, give each data value in the original sequence the label A if it is *above* the median and the label B if it is *below* the median. Using the original sequence, we get

A | B | AAA | BB | AAAAA | BB | A | BBBB | A | BBB | A

We see that

$n_1 = 12$(number of A's) $n_2 = 12$(number of B's) $R = 11$(number of runs)

Note: In this example, none of the data values actually equals the median. If a data value *equals the median,* we put neither A nor B in the sequence. This eliminates from the sequence any data values that equal the median.

Part II: Test the sequence of A and B symbols for randomness.

(a) What is the level of significance α? State the null and alternate hypotheses.

SOLUTION: $\alpha = 0.05$

H_0: The symbols A and B are randomly mixed in the sequence.

H_1: The symbols A and B are not randomly mixed in the sequence.

(b) Find the sample test statistic R and the parameters n_1 and n_2.

SOLUTION: As shown in Part I, for the sequence of A's and B's,

$n_1 = 12;\ n_2 = 12;\ R = 11$

(c) Use Table 10 of Appendix II to find the critical values c_1 and c_2.

SOLUTION: Since $n_1 = 12$ and $n_2 = 12$, we find $c_1 = 7$ and $c_2 = 19$.

(d) Conclude the test.

SOLUTION:

$R \le 7$	✓ $8 \le R \le 18$	$R \ge 19$
Reject H_0.	✓ Fail to reject H_0.	Reject H_0.

Since $R = 11$, we fail to reject H_0 at the 5% level of significance.

(e) *Interpret* the conclusion in the context of the problem.

SOLUTION: At the 5% level of significance, there is insufficient evidence to conclude that the sequence of wind directions above and below the median direction is not random.

PROCEDURE

HOW TO CONSTRUCT A RUNS TEST FOR RANDOMNESS

Setup

You need a sequence (ordered set) consisting of two symbols. If your sequence consists of measurements of some type, then convert it to a sequence of two symbols in the following way:

(a) Find the median of the entries in the sequence.
(b) Label an entry A if it is above the median and B if it is below the median. If an entry equals the median, then put neither A nor B in the sequence.

Now you have a sequence with two symbols.

Let n_1 = number of times the first symbol occurs in the sequence.
n_2 = number of times the second symbol occurs in the sequence.
Note: Either symbol can be called the "first" symbol.
Let R = number of runs in the sequence.

Procedure

1. The *level of significance* is $\alpha = 0.05$. The *null and alternate hypotheses* are:
   H_0: The two symbols are randomly mixed in the sequence.
   H_1: The two symbols are not randomly mixed in the sequence.
2. The *sample test statistic* is the number of runs R.
3. Use Table 10, Appendix II, with parameters n_1 and n_2 to find the *lower and upper critical values* c_1 and c_2.
4. Use the *critical values* c_1 and c_2 in the following *decision process*.

$R \le c_1$	$c_1 + 1 \le R \le c_2 - 1$	$R \le c_2$
Reject H_0.	Fail to reject H_0.	Reject H_0.

5. *Interpret your conclusion* in the context of the application.

   Note: If your original sequence consisted of measurements (not just symbols), it is important to remember that you are testing for randomness about the median of these measurements. In any case, you are testing for randomness regarding a mix of two symbols in a given sequence.

Problem 11 describes how to use a normal approximation for the sample test statistic. Problem 12 gives additional practice.

COMMENT In many applications, $n_1 \le 20$ and $n_2 \le 20$. What happens if either $n_1 > 20$ or $n_2 > 20$? In this case, you can use the normal approximation, which is presented in Problems 11 and 12 at the end of this section.

GUIDED EXERCISE 7 | RUNS TEST FOR RANDOMNESS OF TWO SYMBOLS

The majority party of the U.S. Senate for each year from 1973 to 2003 is shown below, where D and R represent Democrat and Republican, respectively (Reference: *Statistical Abstract of the United States*).

D D D D R R R D D D D R R R R D D R

Test the sequence for randomness. Use a 5% level of significance.

(a) What is α? State the null and alternate hypotheses.

→ $\alpha = 0.05$

H_0: The two symbols are randomly mixed.
H_1: The two symbols are not randomly mixed.

Continued

GUIDED EXERCISE 7 *continued*

(b) Block the sequence into runs. Find the values of n_1, n_2, and R.

→ DDDD | RRR | DDDD | RRRR | DD | R

Letting D be the first symbol, we have

$n_1 = 10$; $n_2 = 8$; $R = 6$

(c) Use Table 10 of Appendix II to find the critical values c_1 and c_2.

→ Lower critical value $c_1 = 5$
Upper critical value $c_2 = 15$

(d) Using critical values, do you reject or fail to reject H_0?

→

$R \le 5$	✓ $6 \le R \le 14$	$R \ge 15$
Reject H_0.	✓ Fail to reject H_0.	Reject H_0.

Since $R = 6$, we fail to reject H_0.

(e) ***Interpret*** the conclusion in the context of the application.

→ The sequence of party control of the U.S. Senate appears to be random. At the 5% level of significance, the evidence is insufficient to reject H_0, that the sequence is random.

GUIDED EXERCISE 8 | RUNS TEST FOR RANDOMNESS ABOUT THE MEDIAN

The national percentage distribution of burglaries is shown by month, starting in January (Reference: *FBI Crime Report,* U.S. Department of Justice).

7.8 6.7 7.6 7.7 8.3 8.2 9.0 9.1 8.6 9.3 8.8 8.9

Test the sequence for randomness about the median. Use a 5% level of significance.

(a) What is α? State the null and alternate hypotheses.

→ $\alpha = 0.05$

H_0: The sequence of values above and below the median is random.

H_1: The sequence of values above and below the median is not random.

(b) Find the median. Assign the symbol A to values above the median and the symbol B to values below the median. Next block the sequence of A's and B's into runs. Find n_1, n_2, and R.

→ First order the numbers. Then find the median.
Median = 8.45. The original sequence translates to

BBBBBB | AAAAAA |

$n_1 = 6$; $n_2 = 6$; $R = 2$

(c) Use Table 10 of Appendix II to find the critical values c_1 and c_2.

→ Lower critical value $c_1 = 3$
Upper critical value $c_2 = 11$

(d) Using the critical values, do you reject or fail to reject H_0?

→

✓ $R \le 3$	$4 \le R \le 10$	$R \le 11$
✓ Reject H_0.	Fail to reject H_0.	Reject H_0.

Since $R = 2$, we reject H_0.

(e) ***Interpret*** the conclusion in the context of the application.

→ At the 5% level of significance, there is sufficient evidence to claim that the sequence of burglaries is not random about the median. It appears that from January to June, there tend to be fewer burglaries.

TECH NOTES **Minitab** Enter your sequence of numbers in a column. Use the menu choices **Stat ➤ Nonparametrics ➤ Runs.** In the dialogue box, select the column containing the sequence. The default is to test the sequence for randomness above and below the mean. Otherwise, you can test for randomness above and below any other value, such as the median.

SECTION 11.4 PROBLEMS

1. *Statistical Literacy* To apply a runs test for randomness as described in this section to a sequence of symbols, how many different symbols are required?

2. *Statistical Literacy* Suppose your data consist of a sequence of numbers. To apply a runs test for randomness about the median, what process do you use to convert the numbers into two distinct symbols?

For Problems 3–10, please provide the following information.
(a) What is the level of significance? State the null and alternate hypotheses.
(b) Find the sample test statistic R, the number of runs.
(c) Find the upper and lower critical values in Table 10 of Appendix II.
(d) Conclude the test.
(e) *Interpret* the conclusion in the context of the application.

3. *Presidents: Party Affiliation* For each successive presidential term from Teddy Roosevelt to George W. Bush (first term), the party affiliation controlling the White House is shown below, where R designates Republican and D designates Democrat (Reference: *The New York Times Almanac*).

   R R R D D R R D D D D D R D R R D R R R D D R

   Historical Note: In cases in which a president died in office or resigned, the period during which the vice president finished the term is not counted as a new term. Test the sequence for randomness. Use $\alpha = 0.05$.

4. *Congress: Party Affiliation* The majority party of the U.S. House of Representatives for each year from 1973 to 2003 is shown below, where D and R represent Democrat and Republican, respectively (Reference: *Statistical Abstract of the United States*).

   D D D D D D D D D D D R R R R R R R

   Test the sequence for randomness. Use $\alpha = 0.05$.

5. *Cloud Seeding: Arizona* Researchers experimenting with cloud seeding in Arizona want a random sequence of days for their experiments (Reference: *Proceedings of the National Academy of Science,* Vol. 68, pp. 649–652). Suppose they have the following itinerary for consecutive days, where S indicates a day for cloud seeding and N indicates a day for no cloud seeding.

   S S S N S N S S S S N N S N S S S N N S S S S

   Test this sequence for randomness. Use $\alpha = 0.05$.

6. *Astronomy: Earth's Rotation* Changes in the earth's rotation are exceedingly small. However, a very long-term trend could be important. (Reference: *Journal of Astronomy,* Vol. 57, pp. 125–146). Let I represent an increase and D a decrease in the rate of the earth's rotation. The following sequence represents historical increases and decreases measured every consecutive fifth year.

   D D D D D I I I D D D D D I I I I I I I I I I D I I I I I

   Test the sequence for randomness. Use $\alpha = 0.05$.

7. *Random Walk: Stocks* Many economists and financial experts claim that the *price level* of a stock or bond is not random; rather, the *price changes* tend to follow a random sequence over time. The following data represent annual percentage returns on Vanguard Total Stock Index for a sequence of recent years. This fund represents nearly all publicly traded U.S. stocks (Reference: *Morningstar Mutual Fund Analysis*).

   10.4 10.6 −0.2 35.8 21.0 31.0 23.3 23.8 −10.6
   −11.0 −21.0 12.8

   (i) Convert this sequence of numbers to a sequence of symbols A and B, where A indicates a value above the median and B a value below the median.
   (ii) Test the sequence for randomness about the median. Use $\alpha = 0.05$.

8. *Random Walk: Bonds* The following data represent annual percentage returns on Vanguard Total Bond Index for a sequence of recent years. This fund represents nearly all publicly traded U.S. bonds (Reference: *Morningstar Mutual Fund Analysis*).

   7.1 9.7 −2.7 18.2 3.6 9.4 8.6 −0.8 11.4 8.4 8.3 0.8

   (i) Convert this sequence of numbers to a sequence of symbols A and B, where A indicates a value above the median and B a value below the median.
   (ii) Test the sequence for randomness about the median. Use $\alpha = 0.05$.

9. *Civil Engineering: Soil Profiles* Sand and clay studies were conducted at the West Side Field Station of the University of California (Reference: Professor D. R. Nielsen, University of California, Davis). Twelve consecutive depths, each about 15 cm deep, were studied and the following percentages of sand in the soil were recorded.

   19.0 27.0 30.0 24.3 33.2 27.5 24.2 18.0 16.2 8.3 1.0 0.0

   (i) Convert this sequence of numbers to a sequence of symbols A and B, where A indicates a value above the median and B a value below the median.
   (ii) Test the sequence for randomness about the median. Use $\alpha = 0.05$.

10. *Civil Engineering: Soil Profiles* Sand and clay studies were conducted at the West Side Field Station of the University of California (Reference: Professor D. R. Nielsen, University of California, Davis). Twelve consecutive depths, each about 15 cm deep, were studied and the following percentages of clay in the soil were recorded.

    47.4 43.4 48.4 42.6 41.4 40.7 46.4 44.8 36.5 35.7 33.7 42.6

    (i) Convert this sequence of numbers to a sequence of symbols A and B, where A indicates a value above the median and B a value below the median.
    (ii) Test the sequence for randomness about the median. Use $\alpha = 0.05$.

11. *Expand Your Knowledge: Either $n_1 > 20$ or $n_2 > 20$* For each successive presidential term from Franklin Pierce (the 14th president, elected in 1853) to George W. Bush (43rd president), the party affiliation controlling the White House is shown below, where R designates Republican and D designates Democrat (Reference: *The New York Times Almanac*).

    Historical Note: We start this sequence with the 14th president because earlier presidents belonged to political parties such as the Federalist or Wigg (not Democratic or Republican) party. In cases in which a president died in office or resigned, the period during which the vice president finished the term is not counted as a new term. The one exception is the case in which Lincoln (a Republican) was assassinated and the vice president, Johnson (a Democrat), finished the term.

D D R R D R R R R D R D R R R R D D R R

D D D D D R R D D R R D R R R D D R

Test the sequence for randomness at the 5% level of significance. Use the following outline.

(a) State the null and alternate hypotheses.

(b) Find the number of runs R, n_1, and n_2. Let n_1 = number of Republicans and n_2 = number of Democrats.

(c) In this case, $n_1 = 21$, so we cannot use Table 10 of Appendix II to find the critical values. Whenever either n_1 or n_2 exceeds 20, the number of runs R has a distribution that is approximately normal, with

$$\mu_R = \frac{2n_1n_2}{n_1 + n_2} + 1 \quad \text{and} \quad \sigma_R = \sqrt{\frac{(2n_1n_2)(2n_1n_2 - n_1 - n_2)}{(n_1 + n_2)^2(n_1 + n_2 - 1)}}$$

We convert the number of runs R to a z value, and then use the normal distribution to find the critical values. Convert the sample test statistic R to z using the formula

$$z = \frac{R - \mu_R}{\sigma_R}$$

(d) The critical values of a normal distribution for a two-tailed test with level of significance $\alpha = 0.05$ are -1.96 and 1.96 (see Table 5(c) of Appendix II). Reject H_0 if the sample test statistic $z \leq -1.96$ or if the sample test statistic $z \geq 1.96$. Otherwise, do not reject H_0.

Sample $z \leq -1.96$	**$-1.96 <$ Sample $z < 1.96$**	**Sample $z \geq 1.96$**
Reject H_0.	Fail to reject H_0.	Reject H_0.

Using this decision process, do you reject or fail to reject H_0 at the 5% level of significance? What is the P-value for this two-tailed test? At the 5% level of significance, do you reach the same conclusion using the P-value that you reach using critical values? Explain.

(e) *Interpret* your results in the context of the application.

12. *Expand Your Knowledge: Either $n_1 > 20$ or $n_2 > 20$* Professor Cornish studied rainfall cycles and sunspot cycles (Reference: *Australian Journal of Physics*, Vol. 7, pp. 334–346). Part of the data include amount of rain (in mm) for 6-day intervals. The following data give rain amounts for consecutive 6-day intervals at Adelaide, South Australia.

6	29	6	0	68	0	0	2	23	5	18	0	50	163
64	72	26	0	0	3	8	142	108	3	90	43	2	5
0	21	2	57	117	51	3	157	43	20	14	40	0	23
18	73	25	61	114	38	31	72	54	38	9	1	17	0
13	6	2	0	1	5	9	11						

Verify that the median is 17.5.

(a) Convert this sequence of numbers to a sequence of symbols A and B, where A indicates a value above the median and B a value below the median.

(b) Test the sequence for randomness about the median at the 5% level of significance. Use the large sample theory outlined in Problem 11.

CHAPTER REVIEW

SUMMARY

When we cannot assume that data come from a normal, binomial, or Student's t distribution, we can employ tests that make no assumptions about data distribution. Such tests are called nonparametric tests. We studied four widely used tests: the sign test, the rank-sum test, the Spearman rank correlation coefficient test, and the runs test for randomness. Nonparametric tests have both advantages and disadvantages:

- Advantages of nonparametric tests
  No requirements concerning the distributions of populations under investigation.
  Easy to use.
- Disadvantages of nonparametric tests
  Waste information.
  Are less sensitive.

It is usually good advice to use standard tests when possible, keeping nonparametric tests for situations wherein assumptions about the data distribution cannot be made.

IMPORTANT WORDS & SYMBOLS

Section 11.1

Nonparametric statistics 708
Sign test 708

Section 11.2

Rank-sum test 716

Section 11.3

Monotone relationship 725
Spearman rank correlation coefficient r_s 726
Population Spearman rank correlation coefficient ρ_s 726

Section 11.4

Runs test for randomness 735
Sequence 736
Run 736

VIEWPOINT Lending a Hand

Whom would you ask for help if you were sick? in need of money? upset with your spouse? depressed? Consider the following claims: People look to sisters for emotional help and brothers for physical help. After that, people look to parents, clergy, or friends. Can you think of nonparametric tests to study such claims? For more information, see American Demographics, *Vol. 18, No. 8.*

CHAPTER REVIEW PROBLEMS

1. *Statistical Literacy* For nonparametric tests, what assumptions, if any, need to be made concerning the distributions of the populations under investigation?

2. *Critical Thinking* Suppose you want to test whether there is a difference in means in a matched pair, "before and after" situation. If you know that the populations under investigation are at least mound-shaped and symmetrical and you have a large sample, is it better to use the parametric paired differences test or the nonparametric sign test for matched pairs? Explain.

For Problems 3–10, please provide the following information.
(a) State the test used.
(b) Give α. State the null and alternate hypotheses.
(c) Find the sample test statistic.

(d) For the sign test, rank-sum test, and Spearman correlation coefficient test, find the P-value of the sample test statistic. For the runs test of randomness, find the critical values from Table 10 of Appendix II.

(e) Conclude the test and *interpret* the results in the context of the application.

3. *Chemistry: Lubricant* In the production of synthetic motor lubricant from coal, a new catalyst has been discovered that seems to affect the viscosity index of the lubricant. In an experiment consisting of 23 production runs, 11 used the new catalyst and 12 did not. After each production run, the viscosity index of the lubricant was determined to be as follows.

With catalyst	1.6	3.2	2.9	4.4	3.7	2.5	1.1	1.8	3.8	4.2	4.1	
Without catalyst	3.9	4.6	1.5	2.2	2.8	3.6	2.4	3.3	1.9	4.0	3.5	3.1

The two samples are independent. Use a 0.05 level of significance to test the null hypothesis that the viscosity index is unchanged by the catalyst against the alternate hypothesis that the viscosity index has changed.

4. *Self-Improvement: Memory* Professor Adams wrote a book called *Improving Your Memory*. The professor claims that if you follow the program outlined in the book, your memory will definitely improve. Fifteen people took the professor's course, in which the book and its program were used. On the first day of class, everyone took a memory exam; and on the last day, everyone took a similar exam. The paired scores for each person follow.

Last exam	225	120	115	275	85	76	114	200	99	135	170	110	216	280	78
First exam	175	110	115	200	60	85	160	190	70	110	140	10	190	200	92

Use a 0.05 level of significance to test the null hypothesis that the scores are the same whether or not people have taken the course against the alternate hypothesis that the scores of people who have taken the course are higher.

5. *Sales: Paint* A chain of hardware stores is trying to sell more paint by mailing pamphlets describing the paint. In 15 communities containing one of these hardware stores, the paint sales (in dollars) were recorded for the months before and after the ads were sent out. The paired results for each store follow.

Sales after	610	150	790	288	715	465	280	640	500	118	265	365	93	217	280
Sales before	460	216	640	250	685	430	220	470	370	118	117	360	93	291	430

Use a 0.01 level of significance to test the null hypothesis that the advertising had no effect on sales against the alternate hypothesis that it improved sales.

6. *Dogs: Obedience School* An obedience school for dogs experimented with two methods of training. One method involved rewards (food, praise); the other involved no rewards. The dogs were randomly placed into two independent groups of 11 each. The number of sessions required to train each of 22 dogs follows.

With rewards	12	17	15	10	16	20	9	23	8	14	10
No rewards	19	22	11	18	13	25	24	28	21	20	21

Use a 0.05 level of significance to test the hypothesis that the number of sessions was the same for the two groups against the alternate hypothesis that the number of sessions was not the same.

7. *Training Program: Fast Food* At McDouglas Hamburger stands, each employee must undergo a training program before he or she is assigned. A group of nine people went through the training program and were assigned to work at the Teton Park McDouglas Hamburger stand. Rankings in performance after the training program and after one month on the job are shown (a rank of 1 is for best performance).

Employee	1	2	3	4	5	6	7	8	9
Rank, training program	8	9	7	3	6	4	1	2	5
Rank on job	9	8	6	7	5	1	3	4	2

Using a 0.05 level of significance, test the claim that there is a monotone-increasing relation between rank from the training program and rank in performance on the job.

8. *Cooking School: Chocolate Mousse* Two expert French chefs judged chocolate mousse made by students in a Paris cooking school. Each chef ranked the best chocolate mousse as 1.

Student	1	2	3	4	5
Rank by Chef Pierre	4	2	3	1	5
Rank by Chef André	4	1	2	3	5

Use a 0.10 level of significance to test the claim that there is a monotone relation (either way) between ranks given by Chef Pierre and by Chef André.

9. *Education: True–False Questions* Dr. Gill wants to arrange the answers to a true–false exam in random order. The answers in order of occurrence are shown below.

T T T T F T T F F T T T T F F F F F T T T T T T

Test the sequence for randomness using $\alpha = 0.05$.

10. *Agriculture: Wheat* For the past 16 years, the yields of wheat (in tons) grown on a plot at Rothamsted Experimental Station (England) are shown below. The sequence is by year.

3.8 1.9 0.6 1.7 2.0 3.5 3.0 1.4 2.7 2.3 2.6 2.1
2.4 2.7 1.8 1.9

Use level of significance 5% to test for randomness about the median.

DATA HIGHLIGHTS: GROUP PROJECTS

Break into small groups and discuss the following topics. Organize a brief outline in which you summarize the main points of your group discussion.

In the world of business and economics, to what extent do assets determine profits? Do the big companies with large assets always make more profits? Is there a rank correlation between assets and profits? The following table is based on information taken from

Company	Asset Rank	Profit Rank
Pepsico	4	2
McDonald's	1	1
Aramark	6	4
Darden Restaurants	7	5
Flagstar	11	11
VIAD	10	8
Wendy's International	2	3
Host Marriott Services	9	10
Brinker International	5	7
Shoney's	3	6
Food Maker	8	9

Fortune (Vol. 135, No. 8). A rank of 1 means highest profits or highest assets. The companies are food service companies.

(a) Compute the Spearman rank correlation coefficient for these data.

(b) Using a 5% level of significance, test the claim that there is a monotone-increasing relation between the ranks of earnings and growth.

(c) Decide whether you should reject or not reject the null hypothesis. Interpret your conclusion in the context of the problem.

(d) As an investor, what are some other features of food companies that you might be interested in ranking? Identify any such features that you think might have a monotone relation.

LINKING CONCEPTS: WRITING PROJECTS

Discuss each of the following topics in class or review the topics on your own. Then write a brief but complete essay in which you summarize the main points. Please include formulas and graphs as appropriate.

1. (a) What do we mean by the term *nonparametric statistics*? What do we mean by the term *parametric statistics*? How do nonparametric methods differ from the methods we studied earlier?

   (b) What are the advantages of nonparametric statistical methods? How can they be used in problems to which other methods we have learned would not apply?

   (c) Are there disadvantages to nonparametric statistical methods? What do we mean when we say that nonparametric methods tend to waste information? Why do we say that nonparametric methods are not as *sensitive* as parametric methods?

   (d) List three random variables from ordinary experience to which you think nonparametric methods would definitely apply and the application of parametric methods would be questionable.

2. Outline the basic logic and ideas behind the sign test. Describe how the binomial probability distribution was used in the construction of the sign test. What assumptions must be made about the sign test? Why is the sign test so extremely general in its possible applications? Why is it a special test for "before and after" studies?

3. Outline the basic logic and ideas behind the rank-sum test. Under what conditions would you use the rank-sum test and *not* the sign test? What assumptions must be made in order to use the rank-sum test? List two advantages the rank-sum test has that the methods of Section 8.5 do not have. List some advantages the methods of Section 8.5 have that the rank-sum test does not have.

4. What do we mean by a monotone relationship between two variables x and y? What do we mean by ranked variables? Give a graphic example of two variables x and y that have a monotone relationship but do *not* have a linear relationship. Does the Spearman test check for a monotone relationship or a linear relationship? Under what conditions does the Pearson product-moment correlation coefficient reduce to the Spearman rank correlation coefficient? Summarize the basic logic and ideas behind the test for Spearman rank correlation. List variables x and y from daily experience for which you think a strong Spearman rank correlation coefficient exists even though the variables are *not* linearly related.

5. What do we mean by a runs test for randomness? What is a run in a sequence? How can we test for randomness about the median? Why is this an important concept? List at least three applications from your own experience.

CUMULATIVE REVIEW PROBLEMS

CHAPTER 10-11

1. *Goodness-of-Fit Test: Rare Events* This cumulative review problem uses material from Chapters 3, 5, and 10. Recall that the Poisson distribution deals with rare events. Death from the kick of a horse is a rare event, even in the Prussian army. The following data are a classic example of a Poisson application to rare events. A reproduction of the original data can be found in C. P. Winsor, *Human Biology*, Vol. 19, pp. 154–161. The data represent the number of deaths from the kick of a horse per army corps per year for 10 Prussian army corps for 20 years (1875–1894). Let x represent the number of deaths and f the frequency of x deaths.

x	0	1	2	3 or more
f	109	65	22	4

(a) First, we fit the data to a Poisson distribution (see Section 5.4).

Poission distribution: $P(x) = \dfrac{e^{-\lambda}\lambda^x}{x!}$

where $\lambda \approx \bar{x}$ (sample mean of x values)
From our study of weighted averages (see Section 3.1),

$$\bar{x} = \frac{\Sigma xf}{\Sigma f}$$

Verify that $\bar{x} \approx 0.61$ *Hint:* For the category 3 or more, use 3.

(b) Now we have $P(x) = \dfrac{e^{-0.61}(0.61)^x}{x!}$ for $x = 0, 1, 2, 3 \ldots$

Find $P(0)$, $P(1)$, $P(2)$, and $P(3 \le x)$. Round to three places after the decimal.

(c) The total number of observations is $\Sigma f = 200$. For a given x, the expected frequency of x deaths is $200P(x)$. The following table gives the observed frequencies O and the expected frequencies $E = 200P(x)$.

x	$O = f$	$E = 200P(x)$
0	109	$200(0.543) = 108.6$
1	65	$200(0.331) = 66.2$
2	22	$200(0.101) = 20.2$
3 or more	4	$200(0.025) = 5$

Compute $\chi^2 = \Sigma \dfrac{(O - E)^2}{E}$

(d) State the null and alternate hypotheses for a chi-square goodness-of-fit test. Set the level of significance to be $\alpha = 0.01$. Find the P-value for a goodness-of-fit test. Interpret your conclusion in the context of this application. Is there reason to believe that the Poisson distribution fits the raw data provided by the Prussian army? Explain.

2. *Test of Independence: Agriculture* Three types of fertilizer were used on 132 identical plots of maize. Each plot was harvested and the yield (in kg) was recorded (Reference: Caribbean Agricultural Research and Development Institute).

	Type of Fertilizer			
Yield (kg)	**I**	**II**	**III**	**Row Total**
0–2.9	12	10	15	37
3.0–5.9	18	21	11	50
6.0–8.9	16	19	10	45
Column Total	46	50	36	132

Use a 5% level of significance to test the hypothesis that type of fertilizer and yield of maize are independent. Interpret the results.

3. *Testing and Estimating Variances: Iris* Random samples of two species of iris gave the following petal lengths (in cm) (Reference: R. A. Fisher, *Annals of Eugenics*, Vol. 7).

x_1, *Iris virginica*	5.1	5.9	4.5	4.9	5.7	4.8	5.8	6.4	5.6	5.9
x_2, *Iris versicolor*	4.5	4.8	4.7	5.0	3.8	5.1	4.4	4.2		

(a) Use a 5% level of significance to test the claim that the population standard deviation of x_1 is larger than 0.55.
(b) Find a 90% confidence interval for the population standard deviation of x_1.
(c) Use a 1% level of significance to test the claim that the population variance of x_1 is larger than that of x_2. Interpret the results.

4. *Sign Test: Wind Direction* The following data are paired by date. Let x and y be random variables representing wind direction at 5 A.M. and 5 P.M., respectively (units are degrees on a compass, with 0° representing true north). The readings were taken at seeding level in a cloud seeding experiment. (Reference: *Proceedings of the National Academy of Science*, Vol. 68, pp. 649–652.) A random sample of days gave the following information.

x	177	140	197	224	49	175	257	72	172
y	142	142	217	125	53	245	218	35	147

x	214	265	110	193	180	190	94	8	93
y	205	218	100	170	245	117	140	99	60

Use the sign test with a 5% level of significance to test the claim that the distributions of wind directions at 5 A.M. and 5 P.M. are different. Interpret the results.

5. *Rank-Sum Test: Apple Trees* Commercial apple trees usually consist of two parts grafted together. The upper part, or graft, determines the character of the fruit, while the root stock determines the size of the tree. (Reference: East Malling Research Station, England.) The following data are from two root stocks A and B. The data represent total extension growth (in meters) of the grafts after 4 years.

Stock A	2.81	2.26	1.94	2.37	3.11	2.58	2.74	2.10	3.41	2.94	2.88
Stock B	2.52	3.02	2.86	2.91	2.78	2.71	1.96	2.44	2.13	1.58	2.77

Use a 1% level of significance and the rank-sum test to test the claim that the distributions of growths are different for root stocks A and B. Interpret the results.

Eastcott-Momatiuk/The Image Works

6. *Spearman Rank Correlation: Calcium Tests* Random collections of nine different solutions of a calcium compound were given to two laboratories A and B. Each laboratory measured the calcium content (in mmol per liter) and reported the results. The data are paired by calcium compound (Reference: *Journal of Clinical Chemistry and Clinical Biochemistry*, Vol. 19, pp. 395–426).

Compound	1	2	3	4	5	6	7	8	9
Lab A	13.33	15.79	14.78	11.29	12.59	9.65	8.69	10.06	11.58
Lab B	13.17	15.72	14.66	11.47	12.65	9.60	8.75	10.25	11.56

(a) Rank-order the data using 1 for the lowest calcium reading. Make a table of ranks to be used in a Spearman rank correlation test.
(b) Use a 5% level of significance to test for a monotone relation (either way) between ranks. Interpret the results.

7. *Runs Test for Randomness: Sunspots* The January mean number of sunspots is recorded for a sequence of recent Januaries (Reference: *International Astronomical Union Quarterly Bulletin on Solar Activity*).

57.9	38.7	19.8	15.3	17.5	28.2
110.9	121.8	104.4	111.5	9.13	61.5
43.4	27.6	18.9	8.1	16.4	51.9

Use level of significance 5% to test for randomness about the median. Interpret the results.

Part I

Bayes's Theorem

The Reverend Thomas Bayes (1702–1761) was an English mathematician who discovered an important relation for conditional probabilities. This relation is referred to as *Bayes's rule* or *Bayes's theorem*. It uses conditional probabilities to adjust calculations so that we can accommodate new, relevant information. We will restrict our attention to a special case of Bayes's theorem in which an event B is partitioned into only *two* mutually exclusive events (see Figure AI-1). The general formula is a bit complicated but is a straightforward extension of the basic ideas we will present here. Most advanced texts contain such an extension.

Note: We use the following compact notation in the statement of Bayes's theorem:

Notation	Meaning
A^c	complement of A; *not* A
$P(B\|A)$	probability of event B, *given* event A; $P(B$, *given* $A)$
$P(B\|A^c)$	probability of event B, *given* the complement of A; $P(B$, *given not* $A)$

We will use Figure AI-1 to motivate Bayes's theorem. Let A and B be events in a sample space that have probabilities not equal to 0 or 1. Let A^c be the complement of A.

Here is Bayes's theorem: $$P(A|B) = \frac{P(B|A)P(A)}{P(B|A)P(A) + P(B|A^c)P(A^c)} \quad (1)$$

Overview of Bayes's Theorem

Suppose we have an event A and we calculate $P(A)$, the unconditional probability of A standing by itself. Now suppose we have a "new" event B and we know the probability of B given that A occurs $P(B|A)$, as well as the probability of B given that A does not occur $P(B|A^c)$. Where does such an event B come from? The event B can be constructed in many possible ways. For example, B can be constructed as the result of a consulting service, a testing procedure, or a sorting activity. In the examples and problems, you will find more ways to construct such an event B.

FIGURE AI-1

A Typical Setup for Bayes's Theorem

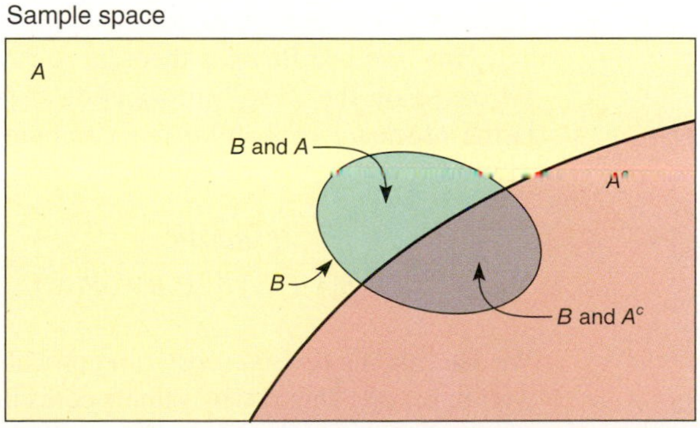

How can we use this "new" information concerning the event B to adjust our calculation of the probability of event A, given B? That is, how can we make our calculation of the probability of A more realistic by including information about the event B? The answer is that we will use Equation (1) of Bayes's theorem.

Let's look at some examples that use Equation (1) of Bayes's theorem. We are grateful to personal friends in the oil and natural gas business in Colorado who provided the basic information in the following example.

EXAMPLE 1

BAYES'S THEOREM

A geologist has examined seismic data and other geologic formations in the vicinity of a proposed site for an oil well. Based on this information, the geologist reports a 65% chance of finding oil. The oil company decides to go ahead and start drilling. As the drilling progresses, sample cores are taken from the well and studied by the geologist. These sample cores have a history of predicting oil *when there is oil* about 85% of the time. However, about 6% of the time the sample cores will predict oil *when there is no oil*. (Note that these probabilities need not add up to 1.) Our geologist is delighted because the sample cores predict oil for this well.

Use the "new" information from the sample cores to revise the geologist's original probability that the well will hit oil. What is the new probability?

SOLUTION: To use Bayes's theorem, we need to identify the events A and B. Then we need to find $P(A)$, $P(A^c)$, $P(B|A)$, and $P(B|A^c)$. From the description of the problem, we have

A is the event that the well strikes oil.

A^c is the event that the well is dry (no oil).

B is the event that the core samples indicate oil.

Again, from the description, we have

$$P(A) = 0.65, \qquad \text{so} \qquad P(A^c) = 1 - 0.65 = 0.35$$

These are our *prior* (before new information) probabilities. New information comes from the sample cores. Probabilities associated with the new information are

$$P(B|A) = 0.85$$

This is the probability that core samples indicate oil when there actually is oil.

$$P(B|A^c) = 0.06$$

This is the probability that core samples indicate oil when there is no oil (dry well).

Now we use Bayes's theorem to revise the probability that the well will hit oil based on the "new" information from core samples. The revised probability is the *posterior* probability we compute that uses the new information from the sample cores:

$$P(A|B) = \frac{P(B|A)P(A)}{P(B|A)P(A) + P(B|A^c)P(A^c)} = \frac{(0.85)(0.65)}{(0.85)(0.65) + (0.06)(0.35)} = 0.9634$$

We see that the revised (*posterior*) probability indicates about a 96% chance for the well to hit oil. This is why sample cores that are good can attract money in the form of venture capital (for independent drillers) on a big, expensive well.

GUIDED EXERCISE 1 | BAYES'S THEOREM

The Anasazi were prehistoric pueblo people who lived in what is now the southwestern United States. Mesa Verde, Pecos Pueblo, and Chaco Canyon are beautiful national parks and monuments, but long ago they were home to many Anasazi. In prehistoric times, there were several Anasazi migrations, until finally their pueblo homes were completely abandoned. The delightful book *Proceedings of the Anasazi Symposium, 1981*, published by Mesa Verde Museum Association, contains a very interesting discussion about methods anthropologists use to (approximately) date Anasazi objects. There are two popular ways. One is to compare environmental data to other objects of known dates. The other is radioactive carbon dating.

Carbon dating has some variability in its accuracy, depending on how far back in time the age estimate goes and also on the condition of the specimen itself. Suppose experience has shown that the carbon method is correct 75% of the time it is used on an object from a known (given) time period. However, there is a 10% chance that the carbon method will predict that an object is from a certain period even when we already know the object is not from that period.

Using environmental data, an anthropologist reported the probability to be 40% that a fossilized deer bone bracelet was from a certain Anasazi migration period. Then, as a follow-up study, the carbon method also indicated that the bracelet was from this migration period. How can the anthropologist adjust her estimated probability to include the "new" information from the carbon dating?

(a) To use Bayes's theorem, we must identify the events A and B. From the description of the problem, what are A and B?

➡ A is the event that the bracelet is from the given migration period. B is the event that carbon dating indicates that the bracelet is from the given migration period.

(b) Find $P(A)$, $P(A^c)$, $P(B|A)$, and $P(B|A^c)$.

➡ From the description,

$$P(A) = 0.40$$
$$P(A^c) = 0.60$$
$$P(B|A) = 0.75$$
$$P(B|A^c) = 0.10$$

(c) Compute $P(A|B)$, and explain the meaning of this number.

➡ Using Bayes's theorem and the results of part (b), we have

$$P(A|B) = \frac{P(B|A)P(A)}{P(B|A)P(A) + P(B|A^c)P(A^c)}$$

$$= \frac{(0.75)(0.40)}{(0.75)(0.40) + (0.10)(0.60)} = 0.8333$$

The prior (before carbon dating) probability was only 40%. However, the carbon dating enabled us to revise this probability to 83%. Thus, we are about 83% sure that the bracelet came from the given migration period. Perhaps additional research at the site will uncover more information to which Bayes's theorem could be applied again.

The next example is a classic application of Bayes's theorem. Suppose we are faced with two competing hypotheses. Each hypothesis claims to explain the same phenomenon; however, only one hypothesis can be correct. Which hypothesis should we accept? This situation occurs in the natural sciences, the social sciences, medicine, finance, and many other areas of life. Bayes's theorem will help us compute the probabilities that one or the other hypothesis is correct. Then what do we do? Well, the great mathematician and philosopher René Descartes can guide us. Descartes once said, "When it is not in our power to determine what is true, we ought to follow what is most probable." Just knowing probabilities does not allow us with absolute certainty to choose the correct hypothesis, but it does permit us to identify which hypothesis is *most likely* to be correct.

EXAMPLE 2

COMPETING HYPOTHESES

A large hospital uses two medical labs for blood work, biopsies, throat cultures, and other medical tests. Lab I does 60% of the reports. The other 40% of the reports are done by Lab II. Based on long experience, it is known that about 10% of the reports from Lab I contain errors and that about 7% of the reports from Lab II contain errors. The hospital recently received a lab report that, through additional medical work, was revealed to be incorrect. One hypothesis is that the report with the mistake came from Lab I. The competing hypothesis is that the report with the mistake came from Lab II. Which lab do you suspect is the culprit? Why?

SOLUTION: Let's use the following notation.

$$A = \text{event report is from Lab I}$$
$$A^c = \text{event report is from Lab II}$$
$$B = \text{event report contains a mistake}$$

From the information given,

$$P(A) = 0.60 \qquad P(A^c) = 0.40$$
$$P(B|A) = 0.10 \qquad P(B|A^c) = 0.07$$

The probability that the report is from Lab I *given* we have a mistake is $P(A|B)$. Using Bayes's theorem, we get

$$\begin{aligned} P(A|B) &= \frac{P(B|A)P(A)}{P(B|A)P(A) + P(B|A^c)P(A^c)} \\ &= \frac{(0.10)(0.60)}{(0.10)(0.60) + (0.07)(0.40)} \\ &= \frac{0.06}{0.088} \approx 0.682 \approx 68\% \end{aligned}$$

So, the probability is about 68% that Lab I supplied the report with the error. It follows that the probability is about 100% − 68% = 32% that the erroneous report came from Lab II.

PROBLEM

BAYES'S THEOREM APPLIED TO QUALITY CONTROL

A company that makes steel bolts knows from long experience that about 12% of its bolts are defective. If the company simply ships all bolts that it produces, then 12% of the shipment the customer receives will be defective. To decrease the percentage of defective bolts shipped to customers, an electronic scanner is installed. The scanner is positioned over the production line and is supposed to pick out the good bolts. However, the scanner itself is not perfect. To test the scanner, a large number of (pretested) "good" bolts were run under the scanner, and it accepted 90% of the bolts as good.

Continued

Then a large number of (pretested) defective bolts were run under the scanner, and it accepted 3% of these as good bolts.

(a) If the company does not use the scanner, what percentage of a shipment is expected to be good? What percentage is expected to be defective?

(b) The scanner itself makes mistakes, and the company is questioning the value of using it. Suppose the company does use the scanner and ships only what the scanner passes as "good" bolts. In this case, what percentage of the shipment is expected to be good? What percentage is expected to be defective?

Partial Answer

To solve this problem, we use Bayes's theorem. The result of using the scanner is a dramatic improvement in the quality of the shipped product. If the scanner is not used, only 88% of the shipped bolts will be good. However, if the scanner is used and only the bolts it passes as good are shipped, then 99.6% of the shipment is expected to be good. Even though the scanner itself makes a considerable number of mistakes, it is definitely worth using. Not only does it increase the quality of a shipment, the bolts it rejects can also be recycled into new bolts.

Part II

The Hypergeometric Probability Distribution

In Chapter 5, we examined the binomial distribution. The binomial probability distribution assumes *independent trials*. If the trials are constructed by drawing samples from a population, then we have two possibilities: We sample either *with replacement* or *without replacement*. If we draw random samples with replacement, the trials can be taken to be independent. If we draw random samples without replacement and the population is very large, then it is reasonable to say that the trials are approximately independent. In this case, we go ahead and use the binomial distribution. However, if the population is relatively small and we draw samples without replacement, the assumption of independent trials is not valid, and we should not use the binomial distribution.

The *hypergeometric distribution* is a probability distribution of a random variable that has two outcomes when sampling is done *without replacement*.

Consider the following notational setup (see Figure AI-2). Suppose we have a population with only *two* distinct types of objects. Such a population might be made up of females and males, students and faculty, residents and nonresidents, defective and nondefective items, and so on. For simplicity of reference, let us call one type of object (your choice) "success" and the other "failure." Let's use the letter a to

FIGURE A1-2

Notational Setup for Hypergeometric Distribution

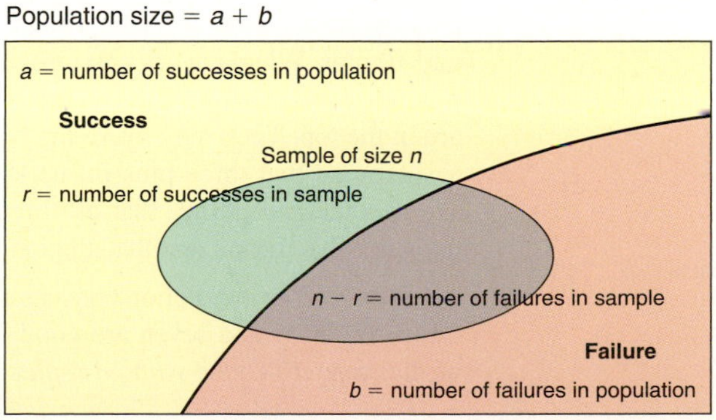

designate the number of successes in the population and the letter b to designate the number of failures in the population. Thus, the total population size is $a + b$. Next, we draw a random sample (without replacement) of size n from this population. Let r be the number of successes in this sample. Then $n - r$ is the number of failures in the sample. The hypergeometric distribution gives us the probability of r successes in the sample of size n.

Recall from Section 4.3 that the number of combinations of k objects taken j at a time can be computed as

$$C_{k,j} = \frac{k!}{j!(k-j)!}$$

Using the notation of Figure AI-2 and the formula for combinations, the hypergeometric distribution can be calculated.

HYPERGEOMETRIC DISTRIBUTION

Given that a population has two distinct types of objects, success and failure,
a counts the number of successes in the population.
b counts the number of failures in the population.
For a random sample of size *n* taken *without* replacement from this population, the probability $P(r)$ of getting *r* successes in the *sample* is

$$P(r) = \frac{C_{a,r}C_{b,n-r}}{C_{(a+b),n}} \quad (2)$$

The expected value and standard deviation are

$$\mu = \frac{na}{a+b} \text{ and } \sigma = \left(\sqrt{n\left(\frac{a}{a+b}\right)\left(\frac{b}{a+b}\right)\left(\frac{a+b-n}{a+b-1}\right)}\right)$$

EXAMPLE 3

HYPERGEOMETRIC DISTRIBUTION

A section of an Interstate 95 bridge across the Mianus River in Connecticut collapsed suddenly on the morning of June 28, 1983. (See *To Engineer Is Human: The Role of Failure in Successful Designs* by Henry Petroski.) Three people were killed when their vehicles fell off the bridge. It was determined that the collapse was caused by the failure of a metal hanger design that left a section of the bridge with no support when something went wrong with the pins. Subsequent inspection revealed many cracked pins and hangers in bridges across the United States.

FIGURE A1-3

Steel Hanger Design for Bridge Support

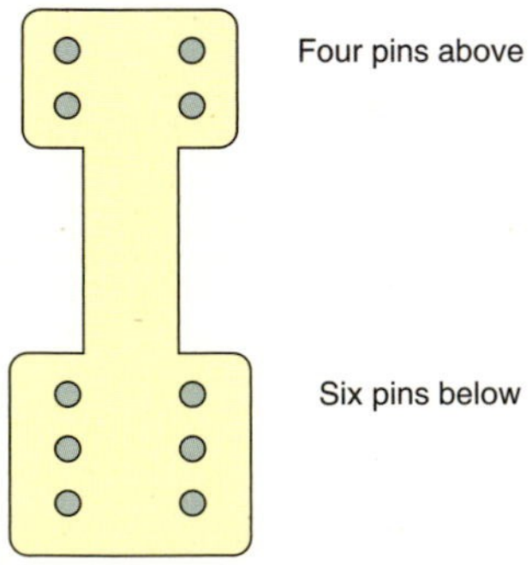

(a) Suppose a hanger design uses four pins in the upper part and six pins in the lower part, as shown in Figure AI-3. The hangers come in a kit consisting of the hanger and 10 pins. When a work crew installs a hanger, they start with the top part and randomly select a pin, which is put into place. This is repeated until all four pins are in the top. Then they finish the lower part.

Assume that three pins in the kit are faulty. The other seven are all right. What is the probability that all three faulty pins get put in the top part of the hanger? This means that the support is held up, in effect, by only one good pin.

SOLUTION: The population consists of 10 pins identical in appearance. However, three are faulty and seven are good. The sampling of four pins for the top part of the hanger is done *without replacement*. Since we are interested in the faulty pins, let us label them "success" (only a convenient label). Using the notation of Figure AI-2 and the hypergeometric distribution, we have

a = number of successes in the population (bad pins) = 3
b = number of failures in the population (good pins) = 7
n = sample size (number of pins put in top) = 4
r = number of successes in sample (number of bad pins in top) = 3

The hypergeometric distribution applies because the population is relatively small (10 pins) and sampling is done without replacement. By Equation (2), we compute $P(r)$:

$$P(r) = \frac{C_{a,r}C_{b,n-r}}{C_{(a+b),n}}$$

Using the preceding information about a, b, n, and r, we get

$$P(r = 3) = \frac{C_{3,3}C_{7,1}}{C_{10,4}}$$

Using the formula for $C_{k,j}$, Table 2 of Appendix II, or the combinations key on a calculator, we get

$$P(r = 3) = \frac{1 \cdot 7}{210} = 0.0333$$

We see that there is a better than 3.3% chance of getting three out of four bad pins in the top part of the hanger.

(b) Suppose that all the hanger kits are like the one described in part (a). On a long bridge that uses 200 such hangers, how many do you expect are held up by only one good pin? How might this affect the safety of the bridge?

SOLUTION: We would expect

$$200(0.0333) \approx 6.7$$

That is, between six and seven hangers are expected to be held up by only one good pin. As time goes on, this pin will corrode and show signs of wear as the bridge vibrates. With only one good pin, there is much less margin of safety.

Professor Petroski discusses the bridge on I-95 across the Mianus River in his book mentioned earlier. He points out that this dramatic accidental collapse resulted in better quality control (for hangers and pins) as well as better overall design of bridges. In addition to this, the government has greatly increased programs for maintenance and inspection of bridges.

GUIDED EXERCISE 2 | HYPERGEOMETRIC DISTRIBUTION

The biology club weekend outing has two groups. One group with seven people will camp at Diamond Lake. The other group with 10 people will camp at Arapahoe Pass. Seventeen duffels were prepacked by the outing committee, but six of these had the tents accidentally left out of the duffel. The group going to Diamond Lake picked up their duffels at random from the collection and started off on the trail. The group going to Arapahoe Pass used the remaining duffels. What is the probability that all six duffels without tents were picked up by the group going to Diamond Lake?

(a) What is success? Are the duffels selected with or without replacement? Which probability distribution applies?

→ Success is taking a duffel without a tent. The duffels are selected without replacement. The hypergeometric distribution applies.

Continued

GUIDED EXERCISE 2 *continued*

(b) Use the hypergeometric distribution to compute the probability of $r = 6$ successes in the sample of seven people going to Diamond Lake.

To use the hypergeometric distribution, we need to know the values of

a = number of successes in population = 6
b = number of failures in population = 11
n = sample size = 7, since seven people are going to Diamond Lake
r = number of successes in sample = 6

Then, $P(r = 6) = \dfrac{C_{6,6}C_{11,1}}{C_{17,7}} = \dfrac{1 \cdot 11}{19448} = 0.0006$

The probability that all six duffels without tents are taken by the seven hikers to Diamond Lake is 0.0006.

TABLE 1 Random Numbers

92630	78240	19267	95457	53497	23894	37708	79862	76471	66418
79445	78735	71549	44843	26104	67318	00701	34986	66751	99723
59654	71966	27386	50004	05358	94031	29281	18544	52429	06080
31524	49587	76612	39789	13537	48086	59483	60680	84675	53014
06348	76938	90379	51392	55887	71015	09209	79157	24440	30244
28703	51709	94456	48396	73780	06436	86641	69239	57662	80181
68108	89266	94730	95761	75023	48464	65544	96583	18911	16391
99938	90704	93621	66330	33393	95261	95349	51769	91616	33238
91543	73196	34449	63513	83834	99411	58826	40456	69268	48562
42103	02781	73920	56297	72678	12249	25270	36678	21313	75767
17138	27584	25296	28387	51350	61664	37893	05363	44143	42677
28297	14280	54524	21618	95320	38174	60579	08089	94999	78460
09331	56712	51333	06289	75345	08811	82711	57392	25252	30333
31295	04204	93712	51287	05754	79396	87399	51773	33075	97061
36146	15560	27592	42089	99281	59640	15221	96079	09961	05371
29553	18432	13630	05529	02791	81017	49027	79031	50912	09399
23501	22642	63081	08191	89420	67800	55137	54707	32945	64522
57888	85846	67967	07835	11314	01545	48535	17142	08552	67457
55336	71264	88472	04334	63919	36394	11196	92470	70543	29776
10087	10072	55980	64688	68239	20461	89381	93809	00796	95945
34101	81277	66090	88872	37818	72142	67140	50785	21380	16703
53362	44940	60430	22834	14130	96593	23298	56203	92671	15925
82975	66158	84731	19436	55790	69229	28661	13675	99318	76873
54827	84673	22898	08094	14326	87038	42892	21127	30712	48489
25464	59098	27436	89421	80754	89924	19097	67737	80368	08795
67609	60214	41475	84950	40133	02546	09570	45682	50165	15609
44921	70924	61295	51137	47596	86735	35561	76649	18217	63446
33170	30972	98130	95828	49786	13301	36081	80761	33985	68621
84687	85445	06208	17654	51333	02878	35010	67578	61574	20749
71886	56450	36567	09395	96951	35507	17555	35212	69106	01679
00475	02224	74722	14721	40215	21351	08596	45625	83981	63748
25993	38881	68361	59560	41274	69742	40703	37993	03435	18873
92882	53178	99195	93803	56985	53089	15305	50522	55900	43026

Continued

TABLE 1 *continued*

25138	26810	07093	15677	60688	04410	24505	37890	67186	62829
84631	71882	12991	83028	82484	90339	91950	74579	03539	90122
34003	92326	12793	61453	48121	74271	28363	66561	75220	35908
53775	45749	05734	86169	42762	70175	97310	73894	88606	19994
59316	97885	72807	54966	60859	11932	35265	71601	55577	67715
20479	66557	50705	26999	09854	52591	14063	30214	19890	19292
86180	84931	25455	26044	02227	52015	21820	50599	51671	65411
21451	68001	72710	40261	61281	13172	63819	48970	51732	54113
98062	68375	80089	24135	72355	95428	11808	29740	81644	86610
01788	64429	14430	94575	75153	94576	61393	96192	03227	32258
62465	04841	43272	68702	01274	05437	22953	18946	99053	41690
94324	31089	84159	92933	99989	89500	91586	02802	69471	68274
05797	43984	21575	09908	70221	19791	51578	36432	33494	79888
10395	14289	52185	09721	25789	38562	54794	04897	59012	89251
35177	56986	25549	59730	64718	52630	31100	62384	49483	11409
25633	89619	75882	98256	02126	72099	57183	55887	09320	73463
16464	48280	94254	45777	45150	68865	11382	11782	22695	41988

Source: MILLION RANDOM DIGITS WITH 100,000 NORMAL DEVIATES. By Staff. Copyright 1983 by RAND CORPORATION. Reproduced with permission of RAND CORPORATION in the formats Textbook and Other Book via Copyright Clearance Center.

TABLE 2 **Binomial Coefficients $C_{n,r}$**

n \ r	0	1	2	3	4	5	6	7	8	9	10
1	1	1									
2	1	2	1								
3	1	3	3	1							
4	1	4	6	4	1						
5	1	5	10	10	5	1					
6	1	6	15	20	15	6	1				
7	1	7	21	35	35	21	7	1			
8	1	8	28	56	70	56	28	8	1		
9	1	9	36	84	126	126	84	36	9	1	
10	1	10	45	120	210	252	210	120	45	10	1
11	1	11	55	165	330	462	462	330	165	55	11
12	1	12	66	220	495	792	924	792	495	220	66
13	1	13	78	286	715	1,287	1,716	1,716	1,287	715	286
14	1	14	91	364	1,001	2,002	3,003	3,432	3,003	2,002	1,001
15	1	15	105	455	1,365	3,003	5,005	6,435	6,435	5,005	3,003
16	1	16	120	560	1,820	4,368	8,008	11,440	12,870	11,440	8,008
17	1	17	136	680	2,380	6,188	12,376	19,448	24,310	24,310	19,448
18	1	18	153	816	3,060	8,568	18,564	31,824	43,758	48,620	43,758
19	1	19	171	969	3,876	11,628	27,132	50,388	75,582	92,378	92,378
20	1	20	190	1,140	4,845	15,504	38,760	77,520	125,970	167,960	184,756

TABLE 3 **Binomial Probability Distribution $C_{n,r}p^rq^{n-r}$**

This table shows the probability of r successes in n independent trials, each with probability of success p.

		p																		
n	r	.01	.05	.10	.15	.20	.25	.30	.35	.40	.45	.50	.55	.60	.65	.70	.75	.80	.85	.90	.95
2	0	.980	.902	.810	.723	.640	.563	.490	.423	.360	.303	.250	.203	.160	.123	.090	.063	.040	.023	.010	.002
	1	.020	.095	.180	.255	.320	.375	.420	.455	.480	.495	.500	.495	.480	.455	.420	.375	.320	.255	.180	.095
	2	.000	.002	.010	.023	.040	.063	.090	.123	.160	.203	.250	.303	.360	.423	.490	.563	.640	.723	.810	.902
3	0	.970	.857	.729	.614	.512	.422	.343	.275	.216	.166	.125	.091	.064	.043	.027	.016	.008	.003	.001	.000
	1	.029	.135	.243	.325	.384	.422	.441	.444	.432	.408	.375	.334	.288	.239	.189	.141	.096	.057	.027	.007
	2	.000	.007	.027	.057	.096	.141	.189	.239	.288	.334	.375	.408	.432	.444	.441	.422	.384	.325	.243	.135
	3	.000	.000	.001	.003	.008	.016	.027	.043	.064	.091	.125	.166	.216	.275	.343	.422	.512	.614	.729	.857
4	0	.961	.815	.656	.522	.410	.316	.240	.179	.130	.092	.062	.041	.026	.015	.008	.004	.002	.001	.000	.000
	1	.039	.171	.292	.368	.410	.422	.412	.384	.346	.300	.250	.200	.154	.112	.076	.047	.026	.011	.004	.000
	2	.001	.014	.049	.098	.154	.211	.265	.311	.346	.368	.375	.368	.346	.311	.265	.211	.154	.098	.049	.014
	3	.000	.000	.004	.011	.026	.047	.076	.112	.154	.200	.250	.300	.346	.384	.412	.422	.410	.368	.292	.171
	4	.000	.000	.000	.001	.002	.004	.008	.015	.026	.041	.062	.092	.130	.179	.240	.316	.410	.522	.656	.815
5	0	.951	.774	.590	.444	.328	.237	.168	.116	.078	.050	.031	.019	.010	.005	.002	.001	.000	.000	.000	.000
	1	.048	.204	.328	.392	.410	.396	.360	.312	.259	.206	.156	.113	.077	.049	.028	.015	.006	.002	.000	.000
	2	.001	.021	.073	.138	.205	.264	.309	.336	.346	.337	.312	.276	.230	.181	.132	.088	.051	.024	.008	.001
	3	.000	.001	.008	.024	.051	.088	.132	.181	.230	.276	.312	.337	.346	.336	.309	.264	.205	.138	.073	.021
	4	.000	.000	.000	.002	.006	.015	.028	.049	.077	.113	.156	.206	.259	.312	.360	.396	.410	.392	.328	.204
	5	.000	.000	.000	.000	.000	.001	.002	.005	.010	.019	.031	.050	.078	.116	.168	.237	.328	.444	.590	.774
6	0	.941	.735	.531	.377	.262	.178	.118	.075	.047	.028	.016	.008	.004	.002	.001	.000	.000	.000	.000	.000
	1	.057	.232	.354	.399	.393	.356	.303	.244	.187	.136	.094	.061	.037	.020	.010	.004	.002	.000	.000	.000
	2	.001	.031	.098	.176	.246	.297	.324	.328	.311	.278	.234	.186	.138	.095	.060	.033	.015	.006	.001	.000
	3	.000	.002	.015	.042	.082	.132	.185	.236	.276	.303	.312	.303	.276	.236	.185	.132	.082	.042	.015	.002
	4	.000	.000	.001	.006	.015	.033	.060	.095	.138	.186	.234	.278	.311	.328	.324	.297	.246	.176	.098	.031
	5	.000	.000	.000	.000	.002	.004	.010	.020	.037	.061	.094	.136	.187	.244	.303	.356	.393	.399	.354	.232
	6	.000	.000	.000	.000	.000	.000	.001	.002	.004	.008	.016	.028	.047	.075	.118	.178	.262	.377	.531	.735
7	0	.932	.698	.478	.321	.210	.133	.082	.049	.028	.015	.008	.004	.002	.001	.000	.000	.000	.000	.000	.000
	1	.066	.257	.372	.396	.367	.311	.247	.185	.131	.087	.055	.032	.017	.008	.004	.001	.000	.000	.000	.000
	2	.002	.041	.124	.210	.275	.311	.318	.299	.261	.214	.164	.117	.077	.047	.025	.012	.004	.001	.000	.000
	3	.000	.004	.023	.062	.115	.173	.227	.268	.290	.292	.273	.239	.194	.144	.097	.058	.029	.011	.003	.000
	4	.000	.000	.003	.011	.029	.058	.097	.144	.194	.239	.273	.292	.290	.268	.227	.173	.115	.062	.023	.004

Continued

TABLE 3 *continued*

n	r	p = .01	.05	.10	.15	.20	.25	.30	.35	.40	.45	.50	.55	.60	.65	.70	.75	.80	.85	.90	.95
7	5	.000	.000	.000	.001	.004	.012	.025	.047	.077	.117	.164	.214	.261	.299	.318	.311	.275	.210	.124	.041
	6	.000	.000	.000	.000	.000	.001	.004	.008	.017	.032	.055	.087	.131	.185	.247	.311	.367	.396	.372	.257
	7	.000	.000	.000	.000	.000	.000	.000	.001	.002	.004	.008	.015	.028	.049	.082	.133	.210	.321	.478	.698
8	0	.923	.663	.430	.272	.168	.100	.058	.032	.017	.008	.004	.002	.001	.000	.000	.000	.000	.000	.000	.000
	1	.075	.279	.383	.385	.336	.267	.198	.137	.090	.055	.031	.016	.008	.003	.001	.000	.000	.000	.000	.000
	2	.003	.051	.149	.238	.294	.311	.296	.259	.209	.157	.109	.070	.041	.022	.010	.004	.001	.000	.000	.000
	3	.000	.005	.033	.084	.147	.208	.254	.279	.279	.257	.219	.172	.124	.081	.047	.023	.009	.003	.000	.000
	4	.000	.000	.005	.018	.046	.087	.136	.188	.232	.263	.273	.263	.232	.188	.136	.087	.046	.018	.005	.000
	5	.000	.000	.000	.003	.009	.023	.047	.081	.124	.172	.219	.257	.279	.279	.254	.208	.147	.084	.033	.005
	6	.000	.000	.000	.000	.001	.004	.010	.022	.041	.070	.109	.157	.209	.259	.296	.311	.294	.238	.149	.051
	7	.000	.000	.000	.000	.000	.000	.001	.003	.008	.016	.031	.055	.090	.137	.198	.267	.336	.385	.383	.279
	8	.000	.000	.000	.000	.000	.000	.000	.000	.001	.002	.004	.008	.017	.032	.058	.100	.168	.272	.430	.663
9	0	.914	.630	.387	.232	.134	.075	.040	.021	.010	.005	.002	.001	.000	.000	.000	.000	.000	.000	.000	.000
	1	.083	.299	.387	.368	.302	.225	.156	.100	.060	.034	.018	.008	.004	.001	.000	.000	.000	.000	.000	.000
	2	.003	.063	.172	.260	.302	.300	.267	.216	.161	.111	.070	.041	.021	.010	.004	.001	.000	.000	.000	.000
	3	.000	.008	.045	.107	.176	.234	.267	.272	.251	.212	.164	.116	.074	.042	.021	.009	.003	.001	.000	.000
	4	.000	.001	.007	.028	.066	.117	.172	.219	.251	.260	.246	.213	.167	.118	.074	.039	.017	.005	.001	.000
	5	.000	.000	.001	.005	.017	.039	.074	.118	.167	.213	.246	.260	.251	.219	.172	.117	.066	.028	.007	.001
	6	.000	.000	.000	.001	.003	.009	.021	.042	.074	.116	.164	.212	.251	.272	.267	.234	.176	.107	.045	.008
	7	.000	.000	.000	.000	.000	.001	.004	.010	.021	.041	.070	.111	.161	.216	.267	.300	.302	.260	.172	.063
	8	.000	.000	.000	.000	.000	.000	.000	.001	.004	.008	.018	.034	.060	.100	.156	.225	.302	.368	.387	.299
	9	.000	.000	.000	.000	.000	.000	.000	.000	.000	.001	.002	.005	.010	.021	.040	.075	.134	.232	.387	.630
10	0	.904	.599	.349	.197	.107	.056	.028	.014	.006	.003	.001	.000	.000	.000	.000	.000	.000	.000	.000	.000
	1	.091	.315	.387	.347	.268	.188	.121	.072	.040	.021	.010	.004	.002	.000	.000	.000	.000	.000	.000	.000
	2	.004	.075	.194	.276	.302	.282	.233	.176	.121	.076	.044	.023	.011	.004	.001	.000	.000	.000	.000	.000
	3	.000	.010	.057	.130	.201	.250	.267	.252	.215	.166	.117	.075	.042	.021	.009	.003	.001	.000	.000	.000
	4	.000	.001	.011	.040	.088	.146	.200	.238	.251	.238	.205	.160	.111	.069	.037	.016	.006	.001	.000	.000
	5	.000	.000	.001	.008	.026	.058	.103	.154	.201	.234	.246	.234	.201	.154	.103	.058	.026	.008	.001	.000
	6	.000	.000	.000	.001	.006	.016	.037	.069	.111	.160	.205	.238	.251	.238	.200	.146	.088	.040	.011	.001
	7	.000	.000	.000	.000	.001	.003	.009	.021	.042	.075	.117	.166	.215	.252	.267	.250	.201	.130	.057	.010

10	8	.000	.000	.000	.000	.000	.000	.001	.004	.011	.023	.044	.076	.121	.176	.233	.282	.302	.276	.194	.075
	9	.000	.000	.000	.000	.000	.000	.000	.000	.002	.004	.010	.021	.040	.072	.121	.188	.268	.347	.387	.315
	10	.000	.000	.000	.000	.000	.000	.000	.000	.000	.000	.001	.003	.006	.014	.028	.056	.107	.197	.349	.599
11	0	.895	.569	.314	.167	.086	.042	.020	.009	.004	.001	.000	.000	.000	.000	.000	.000	.000	.000	.000	.000
	1	.099	.329	.384	.325	.236	.155	.093	.052	.027	.013	.005	.002	.001	.000	.000	.000	.000	.000	.000	.000
	2	.005	.087	.213	.287	.295	.258	.200	.140	.089	.051	.027	.013	.005	.002	.001	.000	.000	.000	.000	.000
	3	.000	.014	.071	.152	.221	.258	.257	.225	.177	.126	.081	.046	.023	.010	.004	.001	.000	.000	.000	.000
	4	.000	.001	.016	.054	.111	.172	.220	.243	.236	.206	.161	.113	.070	.038	.017	.006	.002	.000	.000	.000
	5	.000	.000	.002	.013	.039	.080	.132	.183	.221	.236	.226	.193	.147	.099	.057	.027	.010	.002	.000	.000
	6	.000	.000	.000	.002	.010	.027	.057	.099	.147	.193	.226	.236	.221	.183	.132	.080	.039	.013	.002	.000
	7	.000	.000	.000	.000	.002	.006	.017	.038	.070	.113	.161	.206	.236	.243	.220	.172	.111	.054	.016	.001
	8	.000	.000	.000	.000	.000	.001	.004	.010	.023	.046	.081	.126	.177	.225	.257	.258	.221	.152	.071	.014
	9	.000	.000	.000	.000	.000	.000	.001	.002	.005	.013	.027	.051	.089	.140	.200	.258	.295	.287	.213	.087
	10	.000	.000	.000	.000	.000	.000	.000	.000	.001	.002	.005	.013	.027	.052	.093	.155	.236	.325	.384	.329
	11	.000	.000	.000	.000	.000	.000	.000	.000	.000	.000	.000	.001	.004	.009	.020	.042	.086	.167	.314	.569
12	0	.886	.540	.282	.142	.069	.032	.014	.006	.002	.001	.000	.000	.000	.000	.000	.000	.000	.000	.000	.000
	1	.107	.341	.377	.301	.206	.127	.071	.037	.017	.008	.003	.001	.000	.000	.000	.000	.000	.000	.000	.000
	2	.006	.099	.230	.292	.283	.232	.168	.109	.064	.034	.016	.007	.002	.001	.000	.000	.000	.000	.000	.000
	3	.000	.017	.085	.172	.236	.258	.240	.195	.142	.092	.054	.028	.012	.005	.001	.000	.000	.000	.000	.000
	4	.000	.002	.021	.068	.133	.194	.231	.237	.213	.170	.121	.076	.042	.020	.008	.002	.001	.000	.000	.000
	5	.000	.000	.004	.019	.053	.103	.158	.204	.227	.223	.193	.149	.101	.059	.029	.011	.003	.001	.000	.000
	6	.000	.000	.000	.004	.016	.040	.079	.128	.177	.212	.226	.212	.177	.128	.079	.040	.016	.004	.000	.000
	7	.000	.000	.000	.001	.003	.011	.029	.059	.101	.149	.193	.223	.227	.204	.158	.103	.053	.019	.004	.000
	8	.000	.000	.000	.000	.001	.002	.008	.020	.042	.076	.121	.170	.213	.237	.231	.194	.133	.068	.021	.002
	9	.000	.000	.000	.000	.000	.000	.001	.005	.012	.028	.054	.092	.142	.195	.240	.258	.236	.172	.085	.017
	10	.000	.000	.000	.000	.000	.000	.000	.001	.002	.007	.016	.034	.064	.109	.168	.232	.283	.292	.230	.099
	11	.000	.000	.000	.000	.000	.000	.000	.000	.000	.001	.003	.008	.017	.037	.071	.127	.206	.301	.377	.341
	12	.000	.000	.000	.000	.000	.000	.000	.000	.000	.000	.000	.001	.002	.006	.014	.032	.069	.142	.282	.540
15	0	.860	.463	.206	.087	.035	.013	.005	.002	.000	.000	.000	.000	.000	.000	.000	.000	.000	.000	.000	.000
	1	.130	.366	.343	.231	.132	.067	.031	.013	.005	.002	.000	.000	.000	.000	.000	.000	.000	.000	.000	.000
	2	.009	.135	.267	.286	.231	.156	.092	.048	.022	.009	.003	.001	.000	.000	.000	.000	.000	.000	.000	.000
	3	.000	.031	.129	.218	.250	.225	.170	.111	.063	.032	.014	.005	.002	.000	.000	.000	.000	.000	.000	.000
	4	.000	.005	.043	.116	.188	.225	.219	.179	.127	.078	.042	.019	.007	.002	.001	.000	.000	.000	.000	.000
	5	.000	.001	.010	.045	.103	.165	.206	.212	.186	.140	.092	.051	.024	.010	.003	.001	.000	.000	.000	.000
	6	.000	.000	.002	.013	.043	.092	.147	.191	.207	.191	.153	.105	.061	.030	.012	.003	.001	.000	.000	.000

Continued

TABLE 3 *continued*

		p																		
n	r	.01	.05	.10	.15	.20	.25	.30	.35	.40	.45	.50	.55	.60	.65	.70	.75	.80	.85	.90	.95
15	7	.000	.000	.000	.003	.014	.039	.081	.132	.177	.201	.196	.165	.118	.071	.035	.013	.003	.001	.000	.000
	8	.000	.000	.000	.001	.003	.013	.035	.071	.118	.165	.196	.201	.177	.132	.081	.039	.014	.003	.000	.000
	9	.000	.000	.000	.000	.001	.003	.012	.030	.061	.105	.153	.191	.207	.191	.147	.092	.043	.013	.002	.000
	10	.000	.000	.000	.000	.000	.001	.003	.010	.024	.051	.092	.140	.186	.212	.206	.165	.103	.045	.010	.001
	11	.000	.000	.000	.000	.000	.000	.001	.002	.007	.019	.042	.078	.127	.179	.219	.225	.188	.116	.043	.005
	12	.000	.000	.000	.000	.000	.000	.000	.000	.002	.005	.014	.032	.063	.111	.170	.225	.250	.218	.129	.031
	13	.000	.000	.000	.000	.000	.000	.000	.000	.000	.001	.003	.009	.022	.048	.092	.156	.231	.286	.267	.135
	14	.000	.000	.000	.000	.000	.000	.000	.000	.000	.000	.000	.002	.005	.013	.031	.067	.132	.231	.343	.366
	15	.000	.000	.000	.000	.000	.000	.000	.000	.000	.000	.000	.000	.000	.002	.005	.013	.035	.087	.206	.463
16	0	.851	.440	.185	.074	.028	.010	.003	.001	.000	.000	.000	.000	.000	.000	.000	.000	.000	.000	.000	.000
	1	.138	.371	.329	.210	.113	.053	.023	.009	.003	.001	.000	.000	.000	.000	.000	.000	.000	.000	.000	.000
	2	.010	.146	.275	.277	.211	.134	.073	.035	.015	.006	.002	.001	.000	.000	.000	.000	.000	.000	.000	.000
	3	.000	.036	.142	.229	.246	.208	.146	.089	.047	.022	.009	.003	.001	.000	.000	.000	.000	.000	.000	.000
	4	.000	.006	.051	.131	.200	.225	.204	.155	.101	.057	.028	.011	.004	.001	.000	.000	.000	.000	.000	.000
	5	.000	.001	.014	.056	.120	.180	.210	.201	.162	.112	.067	.034	.014	.005	.001	.000	.000	.000	.000	.000
	6	.000	.000	.003	.018	.055	.110	.165	.198	.198	.168	.122	.075	.039	.017	.006	.001	.000	.000	.000	.000
	7	.000	.000	.000	.005	.020	.052	.101	.152	.189	.197	.175	.132	.084	.044	.019	.006	.001	.000	.000	.000
	8	.000	.000	.000	.001	.006	.020	.049	.092	.142	.181	.196	.181	.142	.092	.049	.020	.006	.001	.000	.000
	9	.000	.000	.000	.000	.001	.006	.019	.044	.084	.132	.175	.197	.189	.152	.101	.052	.020	.005	.000	.000
	10	.000	.000	.000	.000	.000	.001	.006	.017	.039	.075	.122	.168	.198	.198	.165	.110	.055	.018	.003	.000
	11	.000	.000	.000	.000	.000	.000	.001	.005	.014	.034	.067	.112	.162	.201	.210	.180	.120	.056	.014	.001
	12	.000	.000	.000	.000	.000	.000	.000	.001	.004	.011	.028	.057	.101	.155	.204	.225	.200	.131	.051	.006
	13	.000	.000	.000	.000	.000	.000	.000	.000	.001	.003	.009	.022	.047	.089	.146	.208	.246	.229	.142	.036
	14	.000	.000	.000	.000	.000	.000	.000	.000	.000	.001	.002	.006	.015	.035	.073	.134	.211	.277	.275	.146
	15	.000	.000	.000	.000	.000	.000	.000	.000	.000	.000	.000	.001	.003	.009	.023	.053	.113	.210	.329	.371
	16	.000	.000	.000	.000	.000	.000	.000	.000	.000	.000	.000	.000	.000	.001	.003	.010	.028	.074	.185	.440
20	0	.818	.358	.122	.039	.012	.003	.001	.000	.000	.000	.000	.000	.000	.000	.000	.000	.000	.000	.000	.000
	1	.165	.377	.270	.137	.058	.021	.007	.002	.000	.000	.000	.000	.000	.000	.000	.000	.000	.000	.000	.000
	2	.016	.189	.285	.229	.137	.067	.028	.010	.003	.001	.000	.000	.000	.000	.000	.000	.000	.000	.000	.000
	3	.001	.060	.190	.243	.205	.134	.072	.032	.012	.004	.001	.000	.000	.000	.000	.000	.000	.000	.000	.000

20	4	.000	.013	.090	.182	.218	.190	.130	.074	.035	.014	.005	.001	.000	.000	.000	.000	.000	.000	.000	.000
	5	.000	.002	.032	.103	.175	.202	.179	.127	.075	.036	.015	.005	.001	.000	.000	.000	.000	.000	.000	.000
	6	.000	.000	.009	.045	.109	.169	.192	.171	.124	.075	.036	.015	.005	.001	.000	.000	.000	.000	.000	.000
	7	.000	.000	.002	.016	.055	.112	.164	.184	.166	.122	.074	.037	.015	.005	.001	.000	.000	.000	.000	.000
	8	.000	.000	.000	.005	.022	.061	.114	.161	.180	.162	.120	.073	.035	.014	.004	.001	.000	.000	.000	.000
	9	.000	.000	.000	.001	.007	.027	.065	.116	.160	.177	.160	.119	.071	.034	.012	.003	.000	.000	.000	.000
	10	.000	.000	.000	.000	.002	.010	.031	.069	.117	.159	.176	.159	.117	.069	.031	.010	.002	.000	.000	.000
	11	.000	.000	.000	.000	.000	.003	.012	.034	.071	.119	.160	.177	.160	.116	.065	.027	.007	.001	.000	.000
	12	.000	.000	.000	.000	.000	.001	.004	.014	.035	.073	.120	.162	.180	.161	.114	.061	.022	.005	.000	.000
	13	.000	.000	.000	.000	.000	.000	.001	.005	.015	.037	.074	.122	.166	.184	.164	.112	.055	.016	.002	.000
	14	.000	.000	.000	.000	.000	.000	.000	.001	.005	.015	.037	.075	.124	.171	.192	.169	.109	.045	.009	.000
	15	.000	.000	.000	.000	.000	.000	.000	.000	.001	.005	.015	.036	.075	.127	.179	.202	.175	.103	.032	.002
	16	.000	.000	.000	.000	.000	.000	.000	.000	.000	.001	.005	.014	.035	.074	.130	.190	.218	.182	.090	.013
	17	.000	.000	.000	.000	.000	.000	.000	.000	.000	.000	.001	.004	.012	.032	.072	.134	.205	.243	.190	.060
	18	.000	.000	.000	.000	.000	.000	.000	.000	.000	.000	.000	.001	.003	.010	.028	.067	.137	.229	.285	.189
	19	.000	.000	.000	.000	.000	.000	.000	.000	.000	.000	.000	.000	.000	.002	.007	.021	.058	.137	.270	.377
	20	.000	.000	.000	.000	.000	.000	.000	.000	.000	.000	.000	.000	.000	.000	.001	.003	.012	.039	.122	.358

TABLE 4 **Poisson Probability Distribution**

For a given value of λ, entry indicates the probability of obtaining a specified value of r.

r	λ = 0.1	0.2	0.3	0.4	0.5	0.6	0.7	0.8	0.9	1.0
0	.9048	.8187	.7408	.6703	.6065	.5488	.4966	.4493	.4066	.3679
1	.0905	.1637	.2222	.2681	.3033	.3293	.3476	.3595	.3659	.3679
2	.0045	.0164	.0333	.0536	.0758	.0988	.1217	.1438	.1647	.1839
3	.0002	.0011	.0033	.0072	.0126	.0198	.0284	.0383	.0494	.0613
4	.0000	.0001	.0003	.0007	.0016	.0030	.0050	.0077	.0111	.0153
5	.0000	.0000	.0000	.0001	.0002	.0004	.0007	.0012	.0020	.0031
6	.0000	.0000	.0000	.0000	.0000	.0000	.0001	.0002	.0003	.0005
7	.0000	.0000	.0000	.0000	.0000	.0000	.0000	.0000	.0000	.0001

r	λ = 1.1	1.2	1.3	1.4	1.5	1.6	1.7	1.8	1.9	2.0
0	.3329	.3012	.2725	.2466	.2231	.2019	.1827	.1653	.1496	.1353
1	.3662	.3614	.3543	.3452	.3347	.3230	.3106	.2975	.2842	.2707
2	.2014	.2169	.2303	.2417	.2510	.2584	.2640	.2678	.2700	.2707
3	.0738	.0867	.0998	.1128	.1255	.1378	.1496	.1607	.1710	.1804
4	.0203	.0260	.0324	.0395	.0471	.0551	.0636	.0723	.0812	.0902
5	.0045	.0062	.0084	.0111	.0141	.0176	.0216	.0260	.0309	.0361
6	.0008	.0012	.0018	.0026	.0035	.0047	.0061	.0078	.0098	.0120
7	.0001	.0002	.0003	.0005	.0008	.0011	.0015	.0020	.0027	.0034
8	.0000	.0000	.0001	.0001	.0001	.0002	.0003	.0005	.0006	.0009
9	.0000	.0000	.0000	.0000	.0000	.0000	.0001	.0001	.0001	.0002

r	λ = 2.1	2.2	2.3	2.4	2.5	2.6	2.7	2.8	2.9	3.0
0	.1225	.1108	.1003	.0907	.0821	.0743	.0672	.0608	.0550	.0498
1	.2572	.2438	.2306	.2177	.2052	.1931	.1815	.1703	.1596	.1494
2	.2700	.2681	.2652	.2613	.2565	.2510	.2450	.2384	.2314	.2240
3	.1890	.1966	.2033	.2090	.2138	.2176	.2205	.2225	.2237	.2240
4	.0992	.1082	.1169	.1254	.1336	.1414	.1488	.1557	.1622	.1680
5	.0417	.0476	.0538	.0602	.0668	.0735	.0804	.0872	.0940	.1008
6	.0146	.0174	.0206	.0241	.0278	.0319	.0362	.0407	.0455	.0504
7	.0044	.0055	.0068	.0083	.0099	.0118	.0139	.0163	.0188	.0216
8	.0011	.0015	.0019	.0025	.0031	.0038	.0047	.0057	.0068	.0081
9	.0003	.0004	.0005	.0007	.0009	.0011	.0014	.0018	.0022	.0027
10	.0001	.0001	.0001	.0002	.0002	.0003	.0004	.0005	.0006	.0008
11	.0000	.0000	.0000	.0000	.0000	.0001	.0001	.0001	.0002	.0002
12	.0000	.0000	.0000	.0000	.0000	.0000	.0000	.0000	.0000	.0001

TABLE 4 *continued*

	λ									
r	**3.1**	**3.2**	**3.3**	**3.4**	**3.5**	**3.6**	**3.7**	**3.8**	**3.9**	**4.0**
0	.0450	.0408	.0369	.0334	.0302	.0273	.0247	.0224	.0202	.0183
1	.1397	.1304	.1217	.1135	.1057	.0984	.0915	.0850	.0789	.0733
2	.2165	.2087	.2008	.1929	.1850	.1771	.1692	.1615	.1539	.1465
3	.2237	.2226	.2209	.2186	.2158	.2125	.2087	.2046	.2001	.1954
4	.1734	.1781	.1823	.1858	.1888	.1912	.1931	.1944	.1951	.1954
5	.1075	.1140	.1203	.1264	.1322	.1377	.1429	.1477	.1522	.1563
6	.0555	.0608	.0662	.0716	.0771	.0826	.0881	.0936	.0989	.1042
7	.0246	.2078	.0312	.0348	.0385	.0425	.0466	.0508	.0551	.0595
8	.0095	.0111	.0129	.0148	.0169	.0191	.0215	.0241	.0269	.0298
9	.0033	.0040	.0047	.0056	.0066	.0076	.0089	.0102	.0116	.0132
10	.0010	.0013	.0016	.0019	.0023	.0028	.0033	.0039	.0045	.0053
11	.0003	.0004	.0005	.0006	.0007	.0009	.0011	.0013	.0016	.0019
12	.0001	.0001	.0001	.0002	.0002	.0003	.0003	.0004	.0005	.0006
13	.0000	.0000	.0000	.0000	.0001	.0001	.0001	.0001	.0002	.0002
14	.0000	.0000	.0000	.0000	.0000	.0000	.0000	.0000	.0000	.0001

	λ									
r	**4.1**	**4.2**	**4.3**	**4.4**	**4.5**	**4.6**	**4.7**	**4.8**	**4.9**	**5.0**
0	.0166	.0150	.0136	.0123	.0111	.0101	.0091	.0082	.0074	.0067
1	.0679	.0630	.0583	.0540	.0500	.0462	.0427	.0395	.0365	.0337
2	.1393	.1323	.1254	.1188	.1125	.1063	.1005	.0948	.0894	.0842
3	.1904	.1852	.1798	.1743	.1687	.1631	.1574	.1517	.1460	.1404
4	.1951	.1944	.1933	.1917	.1898	.1875	.1849	.1820	.1789	.1755
5	.1600	.1633	.1662	.1687	.1708	.1725	.1738	.1747	.1753	.1755
6	.1093	.1143	.1191	.1237	.1281	.1323	.1362	.1398	.1432	.1462
7	.0640	.0686	.0732	.0778	.0824	.0869	.0914	.0959	.1002	.1044
8	.0328	.0360	.0393	.0428	.0463	.0500	.0537	.0575	.0614	.0653
9	.0150	.0168	.0188	.0209	.0232	.0255	.0280	.0307	.0334	.0363
10	.0061	.0071	.0081	.0092	.0104	.0118	.0132	.0147	.0164	.0181
11	.0023	.0027	.0032	.0037	.0043	.0049	.0056	.0064	.0073	.0082
12	.0008	.0009	.0011	.0014	.0016	.0019	.0022	.0026	.0030	.0034
13	.0002	.0003	.0004	.0005	.0006	.0007	.0008	.0009	.0011	.0013
14	.0001	.0001	.0001	.0001	.0002	.0002	.0003	.0003	.0004	.0005
15	.0000	.0000	.0000	.0000	.0001	.0001	.0001	.0001	.0001	.0002

Continued

TABLE 4 *continued*

	λ									
r	**5.1**	**5.2**	**5.3**	**5.4**	**5.5**	**5.6**	**5.7**	**5.8**	**5.9**	**6.0**
0	.0061	.0055	.0050	.0045	.0041	.0037	.0033	.0030	.0027	.0025
1	.0311	.0287	.0265	.0244	.0225	.0207	.0191	.0176	.0162	.0149
2	.0793	.0746	.0701	.0659	.0618	.0580	.0544	.0509	.0477	.0446
3	.1348	.1293	.1239	.1185	.1133	.1082	.1033	.0985	.0938	.0892
4	.1719	.1681	.1641	.1600	.1558	.1515	.1472	.1428	.1383	.1339
5	.1753	.1748	.1740	.1728	.1714	.1697	.1678	.1656	.1632	.1606
6	.1490	.1515	.1537	.1555	.1571	.1584	.1594	.1601	.1605	.1606
7	.1086	.1125	.1163	.1200	.1234	.1267	.1298	.1326	.1353	.1377
8	.0692	.0731	.0771	.0810	.0849	.0887	.0925	.0962	.0998	.1033
9	.0392	.0423	.0454	.0486	.0519	.0552	.0586	.0620	.0654	.0688
10	.0200	.0220	.0241	.0262	.0285	.0309	.0334	.0359	.0386	.0413
11	.0093	.0104	.0116	.0129	.0143	.0157	.0173	.0190	.0207	.0225
12	.0039	.0045	.0051	.0058	.0065	.0073	.0082	.0092	.0102	.0113
13	.0015	.0018	.0021	.0024	.0028	.0032	.0036	.0041	.0046	.0052
14	.0006	.0007	.0008	.0009	.0011	.0013	.0015	.0017	.0019	.0022
15	.0002	.0002	.0003	.0003	.0004	.0005	.0006	.0007	.0008	.0009
16	.0001	.0001	.0001	.0001	.0001	.0002	.0002	.0002	.0003	.0003
17	.0000	.0000	.0000	.0000	.0000	.0000	.0001	.0001	.0001	.0001

	λ									
r	**6.1**	**6.2**	**6.3**	**6.4**	**6.5**	**6.6**	**6.7**	**6.8**	**6.9**	**7.0**
0	.0022	.0020	.0018	.0017	.0015	.0014	.0012	.0011	.0010	.0009
1	.0137	.0126	.0116	.0106	.0098	.0090	.0082	.0076	.0070	.0064
2	.0417	.0390	.0364	.0340	.0318	.0296	.0276	.0258	.0240	.0223
3	.0848	.0806	.0765	.0726	.0688	.0652	.0617	.0584	.0552	.0521
4	.1294	.1249	.1205	.1162	.1118	.1076	.1034	.0992	.0952	.0912
5	.1579	.1549	.1519	.1487	.1454	.1420	.1385	.1349	.1314	.1277
6	.1605	.1601	.1595	.1586	.1575	.1562	.1546	.1529	.1511	.1490
7	.1399	.1418	.1435	.1450	.1462	.1472	.1480	.1486	.1489	.1490
8	.1066	.1099	.1130	.1160	.1188	.1215	.1240	.1263	.1284	.1304
9	.0723	.0757	.0791	.0825	.0858	.0891	.0923	.0954	.0985	.1014
10	.0441	.0469	.0498	.0528	.0558	.0588	.0618	.0649	.0679	.0710
11	.0245	.0265	.0285	.0307	.0330	.0353	.0377	.0401	.0426	.0452
12	.0124	.0137	.0150	.0164	.0179	.0194	.0210	.0227	.0245	.0264
13	.0058	.0065	.0073	.0081	.0089	.0098	.0108	.0119	.0130	.0142
14	.0025	.0029	.0033	.0037	.0041	.0046	.0052	.0058	.0064	.0071
15	.0010	.0012	.0014	.0016	.0018	.0020	.0023	.0026	.0029	.0033
16	.0004	.0005	.0005	.0006	.0007	.0008	.0010	.0011	.0013	.0014
17	.0001	.0002	.0002	.0002	.0003	.0003	.0004	.0004	.0005	.0006
18	.0000	.0001	.0001	.0001	.0001	.0001	.0001	.0002	.0002	.0002
19	.0000	.0000	.0000	.0000	.0000	.0000	.0000	.0001	.0001	.0001

TABLE 4 *continued*

	λ									
r	7.1	7.2	7.3	7.4	7.5	7.6	7.7	7.8	7.9	8.0
0	.0008	.0007	.0007	.0006	.0006	.0005	.0005	.0004	.0004	.0003
1	.0059	.0054	.0049	.0045	.0041	.0038	.0035	.0032	.0029	.0027
2	.0208	.0194	.0180	.0167	.0156	.0145	.0134	.0125	.0116	.0107
3	.0492	.0464	.0438	.0413	.0389	.0366	.0345	.0324	.0305	.0286
4	.0874	.0836	.0799	.0764	.0729	.0696	.0663	.0632	.0602	.0573
5	.1241	.1204	.1167	.1130	.1094	.1057	.1021	.0986	.0951	.0916
6	.1468	.1445	.1420	.1394	.1367	.1339	.1311	.1282	.1252	.1221
7	.1489	.1486	.1481	.1474	.1465	.1454	.1442	.1428	.1413	.1396
8	.1321	.1337	.1351	.1363	.1373	.1382	.1388	.1392	.1395	.1396
9	.1042	.1070	.1096	.1121	.1144	.1167	.1187	.1207	.1224	.1241
10	.0740	.0770	.0800	.0829	.0858	.0887	.0914	.0941	.0967	.0993
11	.0478	.0504	.0531	.0558	.0585	.0613	.0640	.0667	.0695	.0722
12	.0283	.0303	.0323	.0344	.0366	.0388	.0411	.0434	.0457	.0481
13	.0154	.0168	.0181	.0196	.0211	.0227	.0243	.0260	.0278	.0296
14	.0078	.0086	.0095	.0104	.0113	.0123	.0134	.0145	.0157	.0169
15	.0037	.0041	.0046	.0051	.0057	.0062	.0069	.0075	.0083	.0090
16	.0016	.0019	.0021	.0024	.0026	.0030	.0033	.0037	.0041	.0045
17	.0007	.0008	.0009	.0010	.0012	.0013	.0015	.0017	.0019	.0021
18	.0003	.0003	.0004	.0004	.0005	.0006	.0006	.0007	.0008	.0009
19	.0001	.0001	.0001	.0002	.0002	.0002	.0003	.0003	.0003	.0004
20	.0000	.0000	.0001	.0001	.0001	.0001	.0001	.0001	.0001	.0002
21	.0000	.0000	.0000	.0000	.0000	.0000	.0000	.0000	.0001	.0001

	λ									
r	8.1	8.2	8.3	8.4	8.5	8.6	8.7	8.8	8.9	9.0
0	.0003	.0003	.0002	.0002	.0002	.0002	.0002	.0002	.0001	.0001
1	.0025	.0023	.0021	.0019	.0017	.0016	.0014	.0013	.0012	.0011
2	.0100	.0092	.0086	.0079	.0074	.0068	.0063	.0058	.0054	.0050
3	.0269	.0252	.0237	.0222	.0208	.0195	.0183	.0171	.0160	.0150
4	.0544	.0517	.0491	.0466	.0443	.0420	.0398	.0377	.0357	.0337
5	.0882	.0849	.0816	.0784	.0752	.0722	.0692	.0663	.0635	.0607
6	.1191	.1160	.1128	.1097	.1066	.1034	.1003	.0972	.0941	.0911
7	.1378	.1358	.1338	.1317	.1294	.1271	.1247	.1222	.1197	.1171
8	.1395	.1392	.1388	.1382	.1375	.1366	.1356	.1344	.1332	.1318
9	.1256	.1269	.1280	.1290	.1299	.1306	.1311	.1315	.1317	.1318
10	.1017	.1040	.1063	.1084	.1104	.1123	.1140	.1157	.1172	.1186
11	.0749	.0776	.0802	.0828	.0853	.0878	.0902	.0925	.0948	.0970
12	.0505	.0530	.0555	.0579	.0604	.0629	.0654	.0679	.0703	.0728
13	.0315	.0334	.0354	.0374	.0395	.0416	.0438	.0459	.0481	.0504
14	.0182	.0196	.0210	.0225	.0240	.0256	.0272	.0289	.0306	.0324
15	.0098	.0107	.0116	.0126	.0136	.0147	.0158	.0169	.0182	.0194

Continued

TABLE 4 *continued*

	λ									
r	**8.1**	**8.2**	**8.3**	**8.4**	**8.5**	**8.6**	**8.7**	**8.8**	**8.9**	**9.0**
16	.0050	.0055	.0060	.0066	.0072	.0079	.0086	.0093	.0101	.0109
17	.0024	.0026	.0029	.0033	.0036	.0040	.0044	.0048	.0053	.0058
18	.0011	.0012	.0014	.0015	.0017	.0019	.0021	.0024	.0026	.0029
19	.0005	.0005	.0006	.0007	.0008	.0009	.0010	.0011	.0012	.0014
20	.0002	.0002	.0002	.0003	.0003	.0004	.0004	.0005	.0005	.0006
21	.0001	.0001	.0001	.0001	.0001	.0002	.0002	.0002	.0002	.0003
22	.0000	.0000	.0000	.0000	.0001	.0001	.0001	.0001	.0001	.0001

	λ									
r	**9.1**	**9.2**	**9.3**	**9.4**	**9.5**	**9.6**	**9.7**	**9.8**	**9.9**	**10**
0	.0001	.0001	.0001	.0001	.0001	.0001	.0001	.0001	.0001	.0000
1	.0010	.0009	.0009	.0008	.0007	.0007	.0006	.0005	.0005	.0005
2	.0046	.0043	.0040	.0037	.0034	.0031	.0029	.0027	.0025	.0023
3	.0140	.0131	.0123	.0115	.0107	.0100	.0093	.0087	.0081	.0076
4	.0319	.0302	.0285	.0269	.0254	.0240	.0226	.0213	.0201	.0189
5	.0581	.0555	.0530	.0506	.0483	.0460	.0439	.0418	.0398	.0378
6	.0881	.0851	.0822	.0793	.0764	.0736	.0709	.0682	.0656	.0631
7	.1145	.1118	.1091	.1064	.1037	.1010	.0982	.0955	.0928	.0901
8	.1302	.1286	.1269	.1251	.1232	.1212	.1191	.1170	.1148	.1126
9	.1317	.1315	.1311	.1306	.1300	.1293	.1284	.1274	.1263	.1251
10	.1198	.1210	.1219	.1228	.1235	.1241	.1245	.1249	.1250	.1251
11	.0991	.1012	.1031	.1049	.1067	.1083	.1098	.1112	.1125	.1137
12	.0752	.0776	.0799	.0822	.0844	.0866	.0888	.0908	.0928	.0948
13	.0526	.0549	.0572	.0594	.0617	.0640	.0662	.0685	.0707	.0729
14	.0342	.0361	.0380	.0399	.0419	.0439	.0459	.0479	.0500	.0521
15	.0208	.0221	.0235	.0250	.0265	.0281	.0297	.0313	.0330	.0347
16	.0118	.0127	.0137	.0147	.0157	.0168	.0180	.0192	.0204	.0217
17	.0063	.0069	.0075	.0081	.0088	.0095	.0103	.0111	.0119	.0128
18	.0032	.0035	.0039	.0042	.0046	.0051	.0055	.0060	.0065	.0071
19	.0015	.0017	.0019	.0021	.0023	.0026	.0028	.0031	.0034	.0037
20	.0007	.0008	.0009	.0010	.0011	.0012	.0014	.0015	.0017	.0019
21	.0003	.0003	.0004	.0004	.0005	.0006	.0006	.0007	.0008	.0009
22	.0001	.0001	.0002	.0002	.0002	.0002	.0003	.0003	.0004	.0004
23	.0000	.0001	.0001	.0001	.0001	.0001	.0001	.0001	.0002	.0002
24	.0000	.0000	.0000	.0000	.0000	.0000	.0000	.0001	.0001	.0001

TABLE 4 ***continued***

	λ									
r	**11**	**12**	**13**	**14**	**15**	**16**	**17**	**18**	**19**	**20**
0	.0000	.0000	.0000	.0000	.0000	.0000	.0000	.0000	.0000	.0000
1	.0002	.0001	.0000	.0000	.0000	.0000	.0000	.0000	.0000	.0000
2	.0010	.0004	.0002	.0001	.0000	.0000	.0000	.0000	.0000	.0000
3	.0037	.0018	.0008	.0004	.0002	.0001	.0000	.0000	.0000	.0000
4	.0102	.0053	.0027	.0013	.0006	.0003	.0001	.0001	.0000	.0000
5	.0224	.0127	.0070	.0037	.0019	.0010	.0005	.0002	.0001	.0001
6	.0411	.0255	.0152	.0087	.0048	.0026	.0014	.0007	.0004	.0002
7	.0646	.0437	.0281	.0174	.0104	.0060	.0034	.0018	.0010	.0005
8	.0888	.0655	.0457	.0304	.0194	.0120	.0072	.0042	.0024	.0013
9	.1085	.0874	.0661	.0473	.0324	.0213	.0135	.0083	.0050	.0029
10	.1194	.1048	.0859	.0663	.0486	.0341	.0230	.0150	.0095	.0058
11	.1194	.1144	.1015	.0844	.0663	.0496	.0355	.0245	.0164	.0106
12	.1094	.1144	.1099	.0984	.0829	.0661	.0504	.0368	.0259	.0176
13	.0926	.1056	.1099	.1060	.0956	.0814	.0658	.0509	.0378	.0271
14	.0728	.0905	.1021	.1060	.1024	.0930	.0800	.0655	.0514	.0387
15	.0534	.0724	.0885	.0989	.1024	.0992	.0906	.0786	.0650	.0516
16	.0367	.0543	.0719	.0866	.0960	.0992	.0963	.0884	.0772	.0646
17	.0237	.0383	.0550	.0713	.0847	.0934	.0963	.0936	.0863	.0760
18	.0145	.0256	.0397	.0554	.0706	.0830	.0909	.0936	.0911	.0844
19	.0084	.0161	.0272	.0409	.0557	.0699	.0814	.0887	.0911	.0888
20	.0046	.0097	.0177	.0286	.0418	.0559	.0692	.0798	.0866	.0888
21	.0024	.0055	.0109	.0191	.0299	.0426	.0560	.0684	.0783	.0846
22	.0012	.0030	.0065	.0121	.0204	.0310	.0433	.0560	.0676	.0769
23	.0006	.0016	.0037	.0074	.0133	.0216	.0320	.0438	.0559	.0669
24	.0003	.0008	.0020	.0043	.0083	.0144	.0226	.0328	.0442	.0557
25	.0001	.0004	.0010	.0024	.0050	.0092	.0154	.0237	.0336	.0446
26	.0000	.0002	.0005	.0013	.0029	.0057	.0101	.0164	.0246	.0343
27	.0000	.0001	.0002	.0007	.0016	.0034	.0063	.0109	.0173	.0254
28	.0000	.0000	.0001	.0003	.0009	.0019	.0038	.0070	.0117	.0181
29	.0000	.0000	.0001	.0002	.0004	.0011	.0023	.0044	.0077	.0125
30	.0000	.0000	.0000	.0001	.0002	.0006	.0013	.0026	.0049	.0083
31	.0000	.0000	.0000	.0000	.0001	.0003	.0007	.0015	.0030	.0054
32	.0000	.0000	.0000	.0000	.0001	.0001	.0004	.0009	.0018	.0034
33	.0000	.0000	.0000	.0000	.0000	.0001	.0002	.0005	.0010	.0020
34	.0000	.0000	.0000	.0000	.0000	.0000	.0001	.0002	.0006	.0012
35	.0000	.0000	.0000	.0000	.0000	.0000	.0000	.0001	.0003	.0007
36	.0000	.0000	.0000	.0000	.0000	.0000	.0000	.0001	.0002	.0004
37	.0000	.0000	.0000	.0000	.0000	.0000	.0000	.0000	.0001	.0002
38	.0000	.0000	.0000	.0000	.0000	.0000	.0000	.0000	.0000	.0001
39	.0000	.0000	.0000	.0000	.0000	.0000	.0000	.0000	.0000	.0001

Source: Biometricka, June 1964, The χ^2 Distribution, H. L. Herter (Table 7). Used by permission of Oxford University Press.

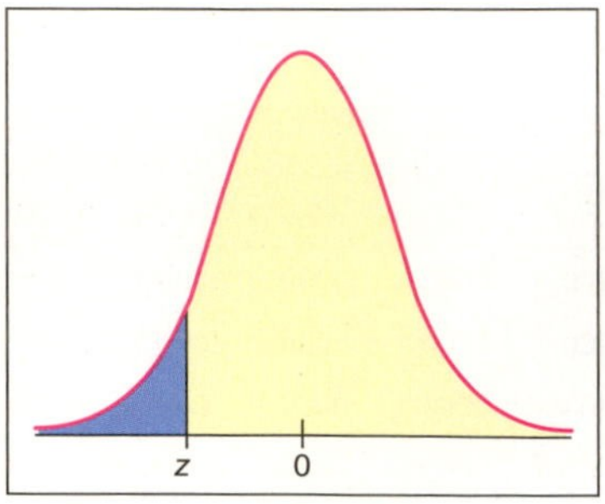

The table entry for *z* is the area to the left of *z*.

TABLE 5 Areas of a Standard Normal Distribution

(a) Table of Areas to the Left of *z*

z	.00	.01	.02	.03	.04	.05	.06	.07	.08	.09
−3.4	.0003	.0003	.0003	.0003	.0003	.0003	.0003	.0003	.0003	.0002
−3.3	.0005	.0005	.0005	.0004	.0004	.0004	.0004	.0004	.0004	.0003
−3.2	.0007	.0007	.0006	.0006	.0006	.0006	.0006	.0005	.0005	.0005
−3.1	.0010	.0009	.0009	.0009	.0008	.0008	.0008	.0008	.0007	.0007
−3.0	.0013	.0013	.0013	.0012	.0012	.0011	.0011	.0011	.0010	.0010
−2.9	.0019	.0018	.0018	.0017	.0016	.0016	.0015	.0015	.0014	.0014
−2.8	.0026	.0025	.0024	.0023	.0023	.0022	.0021	.0021	.0020	.0019
−2.7	.0035	.0034	.0033	.0032	.0031	.0030	.0029	.0028	.0027	.0026
−2.6	.0047	.0045	.0044	.0043	.0041	.0040	.0039	.0038	.0037	.0036
−2.5	.0062	.0060	.0059	.0057	.0055	.0054	.0052	.0051	.0049	.0048
−2.4	.0082	.0080	.0078	.0075	.0073	.0071	.0069	.0068	.0066	.0064
−2.3	.0107	.0104	.0102	.0099	.0096	.0094	.0091	.0089	.0087	.0084
−2.2	.0139	.0136	.0132	.0129	.0125	.0122	.0119	.0116	.0113	.0110
−2.1	.0179	.0174	.0170	.0166	.0162	.0158	.0154	.0150	.0146	.0143
−2.0	.0228	.0222	.0217	.0212	.0207	.0202	.0197	.0192	.0188	.0183
−1.9	.0287	.0281	.0274	.0268	.0262	.0256	.0250	.0244	.0239	.0233
−1.8	.0359	.0351	.0344	.0336	.0329	.0322	.0314	.0307	.0301	.0294
−1.7	.0446	.0436	.0427	.0418	.0409	.0401	.0392	.0384	.0375	.0367
−1.6	.0548	.0537	.0526	.0516	.0505	.0495	.0485	.0475	.0465	.0455
−1.5	.0668	.0655	.0643	.0630	.0618	.0606	.0594	.0582	.0571	.0559
−1.4	.0808	.0793	.0778	.0764	.0749	.0735	.0721	.0708	.0694	.0681
−1.3	.0968	.0951	.0934	.0918	.0901	.0885	.0869	.0853	.0838	.0823
−1.2	.1151	.1131	.1112	.1093	.1075	.1056	.1038	.1020	.1003	.0985
−1.1	.1357	.1335	.1314	.1292	.1271	.1251	.1230	.1210	.1190	.1170
−1.0	.1587	.1562	.1539	.1515	.1492	.1469	.1446	.1423	.1401	.1379
−0.9	.1841	.1814	.1788	.1762	.1736	.1711	.1685	.1660	.1635	.1611
−0.8	.2119	.2090	.2061	.2033	.2005	.1977	.1949	.1922	.1894	.1867
−0.7	.2420	.2389	.2358	.2327	.2296	.2266	.2236	.2206	.2177	.2148
−0.6	.2743	.2709	.2676	.2643	.2611	.2578	.2546	.2514	.2483	.2451
−0.5	.3085	.3050	.3015	.2981	.2946	.2912	.2877	.2843	.2810	.2776
−0.4	.3446	.3409	.3372	.3336	.3300	.3264	.3228	.3192	.3156	.3121
−0.3	.3821	.3783	.3745	.3707	.3669	.3632	.3594	.3557	.3520	.3483
−0.2	.4207	.4168	.4129	.4090	.4052	.4013	.3974	.3936	.3897	.3859
−0.1	.4602	.4562	.4522	.4483	.4443	.4404	.4364	.4325	.4286	.4247
−0.0	.5000	.4960	.4920	.4880	.4840	.4801	.4761	.4721	.4681	.4641

For values of *z* less than −3.49, use 0.000 to approximate the area.

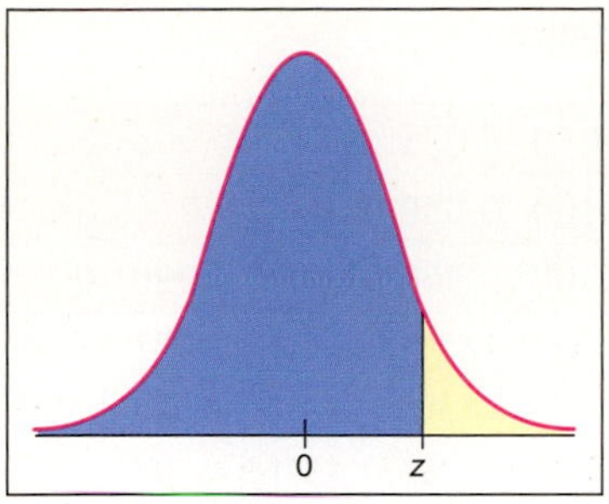

The table entry for z is the area to the left of z.

TABLE 5a *continued*

z	.00	.01	.02	.03	.04	.05	.06	.07	.08	.09
0.0	.5000	.5040	.5080	.5120	.5160	.5199	.5239	.5279	.5319	.5359
0.1	.5398	.5438	.5478	.5517	.5557	.5596	.5636	.5675	.5714	.5753
0.2	.5793	.5832	.5871	.5910	.5948	.5987	.6026	.6064	.6103	.6141
0.3	.6179	.6217	.6255	.6293	.6331	.6368	.6406	.6443	.6480	.6517
0.4	.6554	.6591	.6628	.6664	.6700	.6736	.6772	.6808	.6844	.6879
0.5	.6915	.6950	.6985	.7019	.7054	.7088	.7123	.7157	.7190	.7224
0.6	.7257	.7291	.7324	.7357	.7389	.7422	.7454	.7486	.7517	.7549
0.7	.7580	.7611	.7642	.7673	.7704	.7734	.7764	.7794	.7823	.7852
0.8	.7881	.7910	.7939	.7967	.7995	.8023	.8051	.8078	.8106	.8133
0.9	.8159	.8186	.8212	.8238	.8264	.8289	.8315	.8340	.8365	.8389
1.0	.8413	.8438	.8461	.8485	.8508	.8531	.8554	.8577	.8599	.8621
1.1	.8643	.8665	.8686	.8708	.8729	.8749	.8770	.8790	.8810	.8830
1.2	.8849	.8869	.8888	.8907	.8925	.8944	.8962	.8980	.8997	.9015
1.3	.9032	.9049	.9066	.9082	.9099	.9115	.9131	.9147	.9162	.9177
1.4	.9192	.9207	.9222	.9236	.9251	.9265	.9279	.9292	.9306	.9319
1.5	.9332	.9345	.9357	.9370	.9382	.9394	.9406	.9418	.9429	.9441
1.6	.9452	.9463	.9474	.9484	.9495	.9505	.9515	.9525	.9535	.9545
1.7	.9554	.9564	.9573	.9582	.9591	.9599	.9608	.9616	.9625	.9633
1.8	.9641	.9649	.9656	.9664	.9671	.9678	.9686	.9693	.9699	.9706
1.9	.9713	.9719	.9726	.9732	.9738	.9744	.9750	.9756	.9761	.9767
2.0	.9772	.9778	.9783	.9788	.9793	.9798	.9803	.9808	.9812	.9817
2.1	.9821	.9826	.9830	.9834	.9838	.9842	.9846	.9850	.9854	.9857
2.2	.9861	.9864	.9868	.9871	.9875	.9878	.9881	.9884	.9887	.9890
2.3	.9893	.9896	.9898	.9901	.9904	.9906	.9909	.9911	.9913	.9916
2.4	.9918	.9920	.9922	.9925	.9927	.9929	.9931	.9932	.9934	.9936
2.5	.9938	.9940	.9941	.9943	.9945	.9946	.9948	.9949	.9951	.9952
2.6	.9953	.9955	.9956	.9957	.9959	.9960	.9961	.9962	.9963	.9964
2.7	.9965	.9966	.9967	.9968	.9969	.9970	.9971	.9972	.9973	.9974
2.8	.9974	.9975	.9976	.9977	.9977	.9978	.9979	.9979	.9980	.9981
2.9	.9981	.9982	.9982	.9983	.9984	.9984	.9985	.9985	.9986	.9986
3.0	.9987	.9987	.9987	.9988	.9988	.9989	.9989	.9989	.9990	.9990
3.1	.9990	.9991	.9991	.9991	.9992	.9992	.9992	.9992	.9993	.9993
3.2	.9993	.9993	.9994	.9994	.9994	.9994	.9994	.9995	.9995	.9995
3.3	.9995	.9995	.9995	.9996	.9996	.9996	.9996	.9996	.9996	.9997
3.4	.9997	.9997	.9997	.9997	.9997	.9997	.9997	.9997	.9997	.9998

For z values greater than 3.49, use 1.000 to approximate the area.

TABLE 5 *continued*

(b) Confidence Interval Critical Values z_c	
Level of Confidence c	**Critical Value z_c**
0.70, or 70%	1.04
0.75, or 75%	1.15
0.80, or 80%	1.28
0.85, or 85%	1.44
0.90, or 90%	1.645
0.95, or 95%	1.96
0.98, or 98%	2.33
0.99, or 99%	2.58

TABLE 5 *continued*

(c) Hypothesis Testing, Critical Values z_0		
Level of Significance	$\alpha = 0.05$	$\alpha = 0.01$
Critical value z_0 for a left-tailed test	−1.645	−2.33
Critical value z_0 for a right-tailed test	1.645	2.33
Critical values $\pm z_0$ for a two-tailed test	±1.96	±2.58

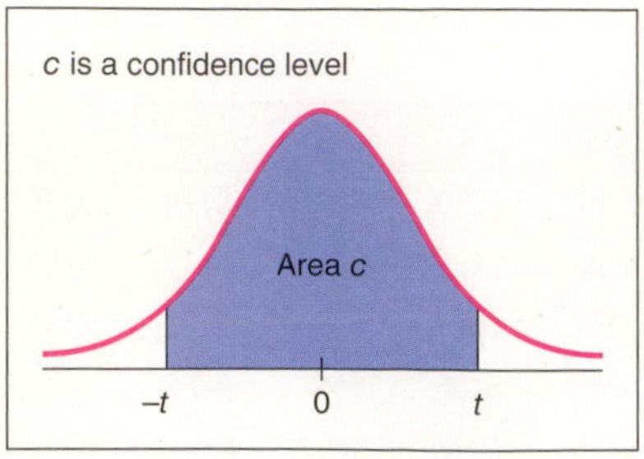

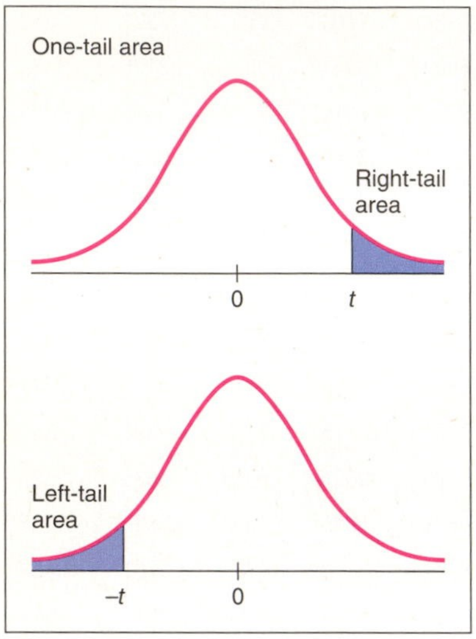

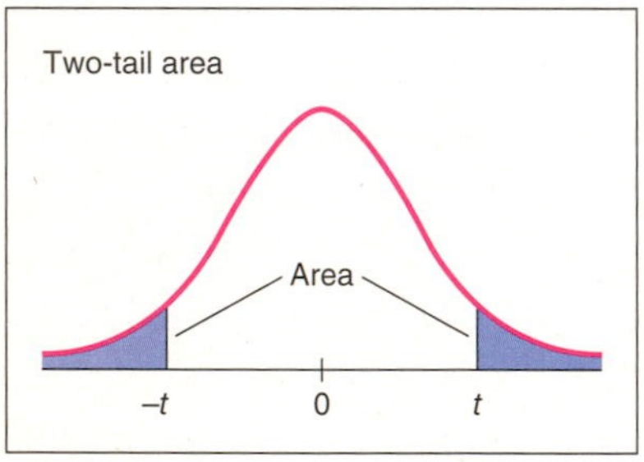

TABLE 6 Critical Values for Student's *t* Distribution

one-tail area	0.250	0.125	0.100	0.075	0.050	0.025	0.010	0.005	0.0005
two-tail area	**0.500**	**0.250**	**0.200**	**0.150**	**0.100**	**0.050**	**0.020**	**0.010**	**0.0010**
d.f.* \ *c	**0.500**	**0.750**	**0.800**	**0.850**	**0.900**	**0.950**	**0.980**	**0.990**	**0.999**
1	1.000	2.414	3.078	4.165	6.314	12.706	31.821	63.657	636.619
2	0.816	1.604	1.886	2.282	2.920	4.303	6.965	9.925	31.599
3	0.765	1.423	1.638	1.924	2.353	3.182	4.541	5.841	12.924
4	0.741	1.344	1.533	1.778	2.132	2.776	3.747	4.604	8.610
5	0.727	1.301	1.476	1.699	2.015	2.571	3.365	4.032	6.869
6	0.718	1.273	1.440	1.650	1.943	2.447	3.143	3.707	5.959
7	0.711	1.254	1.415	1.617	1.895	2.365	2.998	3.499	5.408
8	0.706	1.240	1.397	1.592	1.860	2.306	2.896	3.355	5.041
9	0.703	1.230	1.383	1.574	1.833	2.262	2.821	3.250	4.781
10	0.700	1.221	1.372	1.559	1.812	2.228	2.764	3.169	4.587
11	0.697	1.214	1.363	1.548	1.796	2.201	2.718	3.106	4.437
12	0.695	1.209	1.356	1.538	1.782	2.179	2.681	3.055	4.318
13	0.694	1.204	1.350	1.530	1.771	2.160	2.650	3.012	4.221
14	0.692	1.200	1.345	1.523	1.761	2.145	2.624	2.977	4.140
15	0.691	1.197	1.341	1.517	1.753	2.131	2.602	2.947	4.073
16	0.690	1.194	1.337	1.512	1.746	2.120	2.583	2.921	4.015
17	0.689	1.191	1.333	1.508	1.740	2.110	2.567	2.898	3.965
18	0.688	1.189	1.330	1.504	1.734	2.101	2.552	2.878	3.922
19	0.688	1.187	1.328	1.500	1.729	2.093	2.539	2.861	3.883
20	0.687	1.185	1.325	1.497	1.725	2.086	2.528	2.845	3.850
21	0.686	1.183	1.323	1.494	1.721	2.080	2.518	2.831	3.819
22	0.686	1.182	1.321	1.492	1.717	2.074	2.508	2.819	3.792
23	0.685	1.180	1.319	1.489	1.714	2.069	2.500	2.807	3.768
24	0.685	1.179	1.318	1.487	1.711	2.064	2.492	2.797	3.745
25	0.684	1.178	1.316	1.485	1.708	2.060	2.485	2.787	3.725
26	0.684	1.177	1.315	1.483	1.706	2.056	2.479	2.779	3.707
27	0.684	1.176	1.314	1.482	1.703	2.052	2.473	2.771	3.690
28	0.683	1.175	1.313	1.480	1.701	2.048	2.467	2.763	3.674
29	0.683	1.174	1.311	1.479	1.699	2.045	2.462	2.756	3.659
30	0.683	1.173	1.310	1.477	1.697	2.042	2.457	2.750	3.646
35	0.682	1.170	1.306	1.472	1.690	2.030	2.438	2.724	3.591
40	0.681	1.167	1.303	1.468	1.684	2.021	2.423	2.704	3.551
45	0.680	1.165	1.301	1.465	1.679	2.014	2.412	2.690	3.520
50	0.679	1.164	1.299	1.462	1.676	2.009	2.403	2.678	3.496
60	0.679	1.162	1.296	1.458	1.671	2.000	2.390	2.660	3.460
70	0.678	1.160	1.294	1.456	1.667	1.994	2.381	2.648	3.435
80	0.678	1.159	1.292	1.453	1.664	1.990	2.374	2.639	3.416
100	0.677	1.157	1.290	1.451	1.660	1.984	2.364	2.626	3.390
500	0.675	1.152	1.283	1.442	1.648	1.965	2.334	2.586	3.310
1000	0.675	1.151	1.282	1.441	1.646	1.962	2.330	2.581	3.300
∞	0.674	1.150	1.282	1.440	1.645	1.960	2.326	2.576	3.291

For degrees of freedom *d.f.* not in the table, use the closest *d.f.* that is *smaller.*

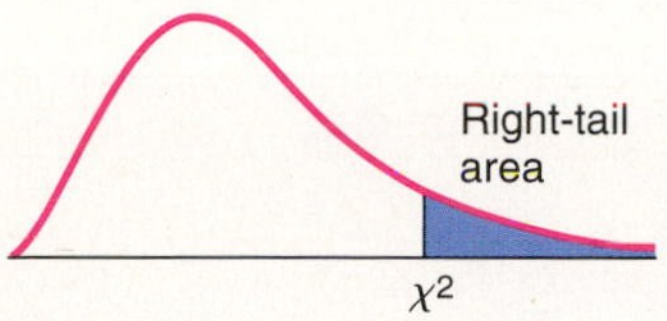

For *d.f.* ≥ 3

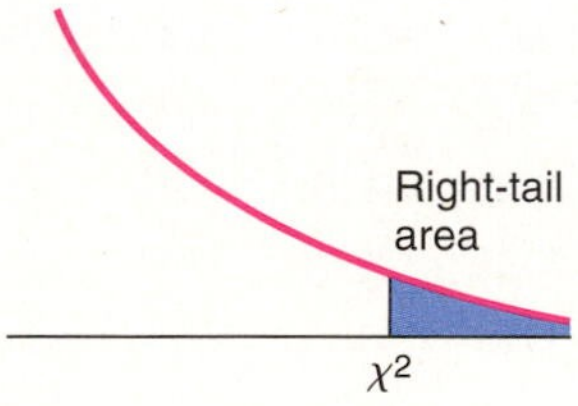

For *d.f.* = 1 or 2

TABLE 7 **The χ^2 Distribution**

	Right-tail Area									
d.f.	.995	.990	.975	.950	.900	.100	.050	.025	.010	.005
1	0.0^4393	0.0^3157	0.0^3982	0.0^2393	0.0158	2.71	3.84	5.02	6.63	7.88
2	0.0100	0.0201	0.0506	0.103	0.211	4.61	5.99	7.38	9.21	10.60
3	0.072	0.115	0.216	0.352	0.584	6.25	7.81	9.35	11.34	12.84
4	0.207	0.297	0.484	0.711	1.064	7.78	9.49	11.14	13.28	14.86
5	0.412	0.554	0.831	1.145	1.61	9.24	11.07	12.83	15.09	16.75
6	0.676	0.872	1.24	1.64	2.20	10.64	12.59	14.45	16.81	18.55
7	0.989	1.24	1.69	2.17	2.83	12.02	14.07	16.01	18.48	20.28
8	1.34	1.65	2.18	2.73	3.49	13.36	15.51	17.53	20.09	21.96
9	1.73	2.09	2.70	3.33	4.17	14.68	16.92	19.02	21.67	23.59
10	2.16	2.56	3.25	3.94	4.87	15.99	18.31	20.48	23.21	25.19
11	2.60	3.05	3.82	4.57	5.58	17.28	19.68	21.92	24.72	26.76
12	3.07	3.57	4.40	5.23	6.30	18.55	21.03	23.34	26.22	28.30
13	3.57	4.11	5.01	5.89	7.04	19.81	22.36	24.74	27.69	29.82
14	4.07	4.66	5.63	6.57	7.79	21.06	23.68	26.12	29.14	31.32
15	4.60	5.23	6.26	7.26	8.55	22.31	25.00	27.49	30.58	32.80
16	5.14	5.81	6.91	7.96	9.31	23.54	26.30	28.85	32.00	34.27
17	5.70	6.41	7.56	8.67	10.09	24.77	27.59	30.19	33.41	35.72
18	6.26	7.01	8.23	9.39	10.86	25.99	28.87	31.53	34.81	37.16
19	6.84	7.63	8.91	10.12	11.65	27.20	30.14	32.85	36.19	38.58
20	7.43	8.26	9.59	10.85	12.44	28.41	31.41	34.17	37.57	40.00
21	8.03	8.90	10.28	11.59	13.24	29.62	32.67	35.48	38.93	41.40
22	8.64	9.54	10.98	12.34	14.04	30.81	33.92	36.78	40.29	42.80
23	9.26	10.20	11.69	13.09	14.85	32.01	35.17	38.08	41.64	44.18
24	9.89	10.86	12.40	13.85	15.66	33.20	36.42	39.36	42.98	45.56
25	10.52	11.52	13.12	14.61	16.47	34.38	37.65	40.65	44.31	46.93
26	11.16	12.20	13.84	15.38	17.29	35.56	38.89	41.92	45.64	48.29
27	11.81	12.88	14.57	16.15	18.11	36.74	40.11	43.19	46.96	49.64
28	12.46	13.56	15.31	16.93	18.94	37.92	41.34	44.46	48.28	50.99
29	13.21	14.26	16.05	17.71	19.77	39.09	42.56	45.72	49.59	52.34
30	13.79	14.95	16.79	18.49	20.60	40.26	43.77	46.98	50.89	53.67
40	20.71	22.16	24.43	26.51	29.05	51.80	55.76	59.34	63.69	66.77
50	27.99	29.71	32.36	34.76	37.69	63.17	67.50	71.42	76.15	79.49
60	35.53	37.48	40.48	43.19	46.46	74.40	79.08	83.30	88.38	91.95
70	43.28	45.44	48.76	51.74	55.33	85.53	90.53	95.02	100.4	104.2
80	51.17	53.54	57.15	60.39	64.28	96.58	101.9	106.6	112.3	116.3
90	59.20	61.75	65.65	69.13	73.29	107.6	113.1	118.1	124.1	128.3
100	67.33	70.06	74.22	77.93	82.36	118.5	124.3	129.6	135.8	140.2

Source: Biometricka, June 1964, The χ^2 Distribution, H. L. Herter (Table 7). Used by permission of Oxford University Press.

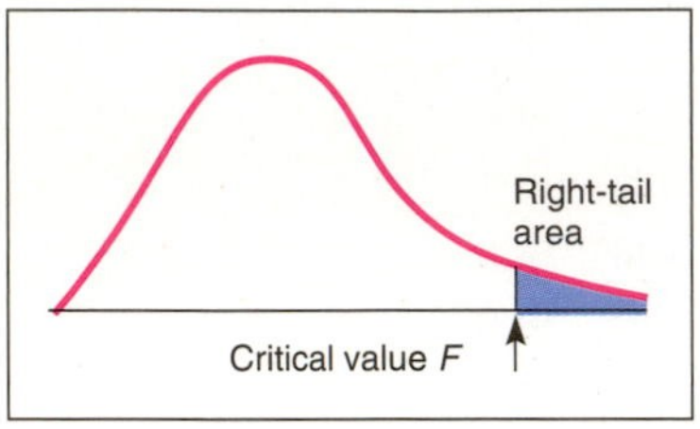

TABLE 8 Critical Values for *F* Distribution

Degrees of Freedom Denominator, $d.f._D$	Right-tail Area	Degrees of Freedom Numerator, $d.f._N$: 1	2	3	4	5	6	7	8	9
	0.100	39.86	49.50	53.59	55.83	57.24	58.20	58.91	59.44	59.86
	0.050	161.45	199.50	215.71	224.58	230.16	233.99	236.77	238.88	240.54
1	0.025	647.79	799.50	864.16	899.58	921.85	937.11	948.22	956.66	963.28
	0.010	4052.2	4999.5	5403.4	5624.6	5763.6	5859.0	5928.4	5981.1	6022.5
	0.001	405284	500000	540379	562500	576405	585937	592873	598144	602284
	0.100	8.53	9.00	9.16	9.24	9.29	9.33	9.35	9.37	9.38
	0.050	18.51	19.00	19.16	19.25	19.30	19.33	19.35	19.37	19.38
2	0.025	38.51	39.00	39.17	39.25	39.30	39.33	39.36	39.37	39.39
	0.010	98.50	99.00	99.17	99.25	99.30	99.33	99.36	99.37	99.39
	0.001	998.50	999.00	999.17	999.25	999.30	999.33	999.36	999.37	999.39
	0.100	5.54	5.46	5.39	5.34	5.31	5.28	5.27	5.25	5.24
	0.050	10.13	9.55	9.28	9.12	9.01	8.94	8.89	8.85	8.81
3	0.025	17.44	16.04	15.44	15.10	14.88	14.73	14.62	14.54	14.47
	0.010	34.12	30.82	29.46	28.71	28.24	27.91	27.67	27.49	27.35
	0.001	167.03	148.50	141.11	137.10	134.58	132.85	131.58	130.62	129.86
	0.100	4.54	4.32	4.19	4.11	4.05	4.01	3.98	3.95	3.94
	0.050	7.71	6.94	6.59	6.39	6.26	6.16	6.09	6.04	6.00
4	0.025	12.22	10.65	9.98	9.60	9.36	9.20	9.07	8.98	8.90
	0.010	21.20	18.00	16.69	15.98	15.52	15.21	14.98	14.80	14.66
	0.001	74.14	61.25	56.18	53.44	51.71	50.53	49.66	49.00	48.47
	0.100	4.06	3.78	3.62	3.52	3.45	3.40	3.37	3.34	3.32
	0.050	6.61	5.79	5.41	5.19	5.05	4.95	4.88	4.82	4.77
5	0.025	10.01	8.43	7.76	7.39	7.15	6.98	6.85	6.76	6.68
	0.010	16.26	13.27	12.06	11.39	10.97	10.67	10.46	10.29	10.16
	0.001	47.18	37.12	33.20	31.09	29.75	28.83	28.16	27.65	27.24
	0.100	3.78	3.46	3.29	3.18	3.11	3.05	3.01	2.98	2.96
	0.050	5.99	5.14	4.76	4.53	4.39	4.28	4.21	4.15	4.10
6	0.025	8.81	7.26	6.60	6.23	5.99	5.82	5.70	5.60	5.52
	0.010	13.75	10.92	9.78	9.15	8.75	8.47	8.26	8.10	7.98
	0.001	35.51	27.00	23.70	21.92	20.80	20.03	19.46	19.03	18.69
	0.100	3.59	3.26	3.07	2.96	2.88	2.83	2.78	2.75	2.72
	0.050	5.59	4.74	4.35	4.12	3.97	3.87	3.79	3.73	3.68
7	0.025	8.07	6.54	5.89	5.52	5.29	5.12	4.99	4.90	4.82
	0.010	12.25	9.55	8.45	7.85	7.46	7.19	6.99	6.84	6.72
	0.001	29.25	21.69	18.77	17.20	16.21	15.52	15.02	14.63	14.33
	0.100	3.46	3.11	2.92	2.81	2.73	2.67	2.62	2.59	2.56
	0.050	5.32	4.46	4.07	3.84	3.69	3.58	3.50	3.44	3.39
8	0.025	7.57	6.06	5.42	5.05	4.82	4.65	4.53	4.43	4.36
	0.010	11.26	8.65	7.59	7.01	6.63	6.37	6.18	6.03	5.91
	0.001	25.41	18.49	15.83	14.39	13.48	12.86	12.40	12.05	11.77

TABLE 8 *continued*

Degrees of Freedom Denominator, $d.f._D$	Right-tail Area	Degrees of Freedom Numerator, $d.f._N$ 10	12	15	20	25	30	40	50	60	120	1000
	0.100	60.19	60.71	61.22	61.74	62.05	62.26	62.53	62.69	62.79	63.06	63.30
	0.050	241.88	243.91	245.95	248.01	249.26	250.10	251.14	251.77	252.20	253.25	254.19
1	0.025	968.63	976.71	984.87	993.10	998.08	1001.4	1005.6	1008.1	1009.8	1014.0	1017.7
	0.010	6055.8	6106.3	6157.3	6208.7	6239.8	6260.6	6286.8	6302.5	6313.0	6339.4	6362.7
	0.001	605621	610668	615764	620908	624017	626099	628712	630285	631337	633972	636301
	0.100	9.39	9.41	9.42	9.44	9.45	9.46	9.47	9.47	9.47	9.48	9.49
	0.050	19.40	19.41	19.43	19.45	19.46	19.46	19.47	19.48	19.48	19.49	19.49
2	0.025	39.40	39.41	39.43	39.45	39.46	39.46	39.47	39.48	39.48	39.49	39.50
	0.010	99.40	99.42	99.43	99.45	99.46	99.47	99.47	99.48	99.48	99.49	99.50
	0.001	999.40	999.42	999.43	999.45	999.46	999.47	999.47	999.48	999.48	999.49	999.50
	0.100	5.23	5.22	5.20	5.18	5.17	5.17	5.16	5.15	5.15	5.14	5.13
	0.050	8.79	8.74	8.70	8.66	8.63	8.62	8.59	8.58	8.57	8.55	8.53
3	0.025	14.42	14.34	14.25	14.17	14.12	14.08	14.04	14.01	13.99	13.95	13.91
	0.010	27.23	27.05	26.87	26.69	26.58	26.50	26.41	26.35	26.32	26.22	26.14
	0.001	129.25	128.32	127.37	126.42	125.84	125.45	124.96	124.66	124.47	123.97	123.53
	0.100	3.92	3.90	3.87	3.84	3.83	3.82	3.80	3.80	3.79	3.78	3.76
	0.050	5.96	5.91	5.86	5.80	5.77	5.75	5.72	5.70	5.69	5.66	5.63
4	0.025	8.84	8.75	8.66	8.56	8.50	8.46	8.41	8.38	8.36	8.31	8.26
	0.010	14.55	14.37	14.20	14.02	13.91	13.84	13.75	13.69	13.65	13.56	13.47
	0.001	48.05	47.41	46.76	46.10	45.70	45.43	45.09	44.88	44.75	44.40	44.09
	0.100	3.30	3.27	3.24	3.21	3.19	3.17	3.16	3.15	3.14	3.12	3.11
	0.050	4.74	4.68	4.62	4.56	4.52	4.50	4.46	4.44	4.43	4.40	4.37
5	0.025	6.62	6.52	6.43	6.33	6.27	6.23	6.18	6.14	6.12	6.07	6.02
	0.010	10.05	9.89	9.72	9.55	9.45	9.38	9.29	9.24	9.20	9.11	9.03
	0.001	26.92	26.42	25.91	25.39	25.08	24.87	24.60	24.44	24.33	24.06	23.82
	0.100	2.94	2.90	2.87	2.84	2.81	2.80	2.78	2.77	2.76	2.74	2.72
	0.050	4.06	4.00	3.94	3.87	3.83	3.81	3.77	3.75	3.74	3.70	3.67
6	0.025	5.46	5.37	5.27	5.17	5.11	5.07	5.01	4.98	4.96	4.90	4.86
	0.010	7.87	7.72	7.56	7.40	7.30	7.23	7.14	7.09	7.06	6.97	6.89
	0.001	18.41	17.99	17.56	17.12	16.85	16.67	16.44	16.31	16.21	15.98	15.77
	0.100	2.70	2.67	2.63	2.59	2.57	2.56	2.54	2.52	2.51	2.49	2.47
	0.050	3.64	3.57	3.51	3.44	3.40	3.38	3.34	3.32	3.30	3.27	3.23
7	0.025	4.76	4.67	4.57	4.47	4.40	4.36	4.31	4.28	4.25	4.20	4.15
	0.010	6.62	6.47	6.31	6.16	6.06	5.99	5.91	5.86	5.82	5.74	5.66
	0.001	14.08	13.71	13.32	12.93	12.69	12.53	12.33	12.20	12.12	11.91	11.72
	0.100	2.54	2.50	2.46	2.42	2.40	2.38	2.36	2.35	2.34	2.32	2.30
	0.050	3.35	3.28	3.22	3.15	3.11	3.08	3.04	3.02	3.01	2.97	2.93
8	0.025	4.30	4.20	4.10	4.00	3.94	3.89	3.84	3.81	3.78	3.73	3.68
	0.010	5.81	5.67	5.52	5.36	5.26	5.20	5.12	5.07	5.03	4.95	4.87
	0.001	11.54	11.19	10.84	10.48	10.26	10.11	9.92	9.80	9.73	9.53	9.36

Continued

TABLE 8 *continued*

Degrees of Freedom Denominator, $d.f._D$	Right-tail Area	Degrees of Freedom Numerator, $d.f._N$ 1	2	3	4	5	6	7	8	9
	0.100	3.36	3.01	2.81	2.69	2.61	2.55	2.51	2.47	2.44
	0.050	5.12	4.26	3.86	3.63	3.48	3.37	3.29	3.23	3.18
9	0.025	7.21	5.71	5.08	4.72	4.48	4.32	4.20	4.10	4.03
	0.010	10.56	8.02	6.99	6.42	6.06	5.80	5.61	5.47	5.35
	0.001	22.86	16.39	13.90	12.56	11.71	11.13	10.70	10.37	10.11
	0.100	3.29	2.92	2.73	2.61	2.52	2.46	2.41	2.38	2.35
	0.050	4.96	4.10	3.71	3.48	3.33	3.22	3.14	3.07	3.02
10	0.025	6.94	5.46	4.83	4.47	4.24	4.07	3.95	3.85	3.78
	0.010	10.04	7.56	6.55	5.99	5.64	5.39	5.20	5.06	4.94
	0.001	21.04	14.91	12.55	11.28	10.48	9.93	9.52	9.20	8.96
	0.100	3.23	2.86	2.66	2.54	2.45	2.39	2.34	2.30	2.27
	0.050	4.84	3.98	3.59	3.36	3.20	3.09	3.01	2.95	2.90
11	0.025	6.72	5.26	4.63	4.28	4.04	3.88	3.76	3.66	3.59
	0.010	9.65	7.21	6.22	5.67	5.32	5.07	4.89	4.74	4.63
	0.001	19.69	13.81	11.56	10.35	9.58	9.05	8.66	8.35	8.12
	0.100	3.18	2.81	2.61	2.48	2.39	2.33	2.28	2.24	2.21
	0.050	4.75	3.89	3.49	3.26	3.11	3.00	2.91	2.85	2.80
12	0.025	6.55	5.10	4.47	4.12	3.89	3.73	3.61	3.51	3.44
	0.010	9.33	6.93	5.95	5.41	5.06	4.82	4.64	4.50	4.39
	0.001	18.64	12.97	10.80	9.63	8.89	8.38	8.00	7.71	7.48
	0.100	3.14	2.76	2.56	2.43	2.35	2.28	2.23	2.20	2.16
	0.050	4.67	3.81	3.41	3.18	3.03	2.92	2.83	2.77	2.71
13	0.025	6.41	4.97	4.35	4.00	3.77	3.60	3.48	3.39	3.31
	0.010	9.07	6.70	5.74	5.21	4.86	4.62	4.44	4.30	4.19
	0.001	17.82	12.31	10.21	9.07	8.35	7.86	7.49	7.21	6.98
	0.100	3.10	2.73	2.52	2.39	2.31	2.24	2.19	2.15	2.12
	0.050	4.60	3.74	3.34	3.11	2.96	2.85	2.76	2.70	2.65
14	0.025	6.30	4.86	4.24	3.89	3.66	3.50	3.38	3.29	3.21
	0.010	8.86	6.51	5.56	5.04	4.69	4.46	4.28	4.14	4.03
	0.001	17.14	11.78	9.73	8.62	7.92	7.44	7.08	6.80	6.58
	0.100	3.07	2.70	2.49	2.36	2.27	2.21	2.16	2.12	2.09
	0.050	4.54	3.68	3.29	3.06	2.90	2.79	2.71	2.64	2.59
15	0.025	6.20	4.77	4.15	3.80	3.58	3.41	3.29	3.20	3.12
	0.010	8.68	6.36	5.42	4.89	4.56	4.32	4.14	4.00	3.89
	0.001	16.59	11.34	9.34	8.25	7.57	7.09	6.74	6.47	6.26
	0.100	3.05	2.67	2.46	2.33	2.24	2.18	2.13	2.09	2.06
	0.050	4.49	3.63	3.24	3.01	2.85	2.74	2.66	2.59	2.54
16	0.025	6.12	4.69	4.08	3.73	3.50	3.34	3.22	3.12	3.05
	0.010	8.53	6.23	5.29	4.77	4.44	4.20	4.03	3.89	3.78
	0.001	16.12	10.97	9.01	7.94	7.27	6.80	6.46	6.19	5.98

TABLE 8 *continued*

Degrees of Freedom Denominator, $d.f._D$	Right-tail Area	Degrees of Freedom Numerator, $d.f._N$ 10	12	15	20	25	30	40	50	60	120	1000
9	0.100	2.42	2.38	2.34	2.30	2.27	2.25	2.23	2.22	2.21	2.18	2.16
	0.050	3.14	3.07	3.01	2.94	2.89	2.86	2.83	2.80	2.79	2.75	2.71
	0.025	3.96	3.87	3.77	3.67	3.60	3.56	3.51	3.47	3.45	3.39	3.34
	0.010	5.26	5.11	4.96	4.81	4.71	4.65	4.57	4.52	4.48	4.40	4.32
	0.001	9.89	9.57	9.24	8.90	8.69	8.55	8.37	8.26	8.19	8.00	7.84
10	0.100	2.32	2.28	2.24	2.20	2.17	2.16	2.13	2.12	2.11	2.08	2.06
	0.050	2.98	2.91	2.85	2.77	2.73	2.70	2.66	2.64	2.62	2.58	2.54
	0.025	3.72	3.62	3.52	3.42	3.35	3.31	3.26	3.22	3.20	3.14	3.09
	0.010	4.85	4.71	4.56	4.41	4.31	4.25	4.17	4.12	4.08	4.00	3.92
	0.001	8.75	8.45	8.13	7.80	7.60	7.47	7.30	7.19	7.12	6.94	6.78
11	0.100	2.25	2.21	2.17	2.12	2.10	2.08	2.05	2.04	2.03	2.00	1.98
	0.050	2.85	2.79	2.72	2.65	2.60	2.57	2.53	2.51	2.49	2.45	2.41
	0.025	3.53	3.43	3.33	3.23	3.16	3.12	3.06	3.03	3.00	2.94	2.89
	0.010	4.54	4.40	4.25	4.10	4.01	3.94	3.86	3.81	3.78	3.69	3.61
	0.001	7.92	7.63	7.32	7.01	6.81	6.68	6.52	6.42	6.35	6.18	6.02
12	0.100	2.19	2.15	2.10	2.06	2.03	2.01	1.99	1.97	1.96	1.93	1.91
	0.050	2.75	2.69	2.62	2.54	2.50	2.47	2.43	2.40	2.38	2.34	2.30
	0.025	3.37	3.28	3.18	3.07	3.01	2.96	2.91	2.87	2.85	2.79	2.73
	0.010	4.30	4.16	4.01	3.86	3.76	3.70	3.62	3.57	3.54	3.45	3.37
	0.001	7.29	7.00	6.71	6.40	6.22	6.09	5.93	5.83	5.76	5.59	5.44
13	0.100	2.14	2.10	2.05	2.01	1.98	1.96	1.93	1.92	1.90	1.88	1.85
	0.050	2.67	2.60	2.53	2.46	2.41	2.38	2.34	2.31	2.30	2.25	2.21
	0.025	3.25	3.15	3.05	2.95	2.88	2.84	2.78	2.74	2.72	2.66	2.60
	0.010	4.10	3.96	3.82	3.66	3.57	3.51	3.43	3.38	3.34	3.25	3.18
	0.001	6.80	6.52	6.23	5.93	5.75	5.63	5.47	5.37	5.30	5.14	4.99
14	0.100	2.10	2.05	2.01	1.96	1.93	1.91	1.89	1.87	1.86	1.83	1.80
	0.050	2.60	2.53	2.46	2.39	2.34	2.31	2.27	2.24	2.22	2.18	2.14
	0.025	3.15	3.05	2.95	2.84	2.78	2.73	2.67	2.64	2.61	2.55	2.50
	0.010	3.94	3.80	3.66	3.51	3.41	3.35	3.27	3.22	3.18	3.09	3.02
	0.001	6.40	6.13	5.85	5.56	5.38	5.25	5.10	5.00	4.94	4.77	4.62
15	0.100	2.06	2.02	1.97	1.92	1.89	1.87	1.85	1.83	1.82	1.79	1.76
	0.050	2.54	2.48	2.40	2.33	2.28	2.25	2.20	2.18	2.16	2.11	2.07
	0.025	3.06	2.96	2.86	2.76	2.69	2.64	2.59	2.55	2.52	2.46	2.40
	0.010	3.80	3.67	3.52	3.37	3.28	3.21	3.13	3.08	3.05	2.96	2.88
	0.001	6.08	5.81	5.54	5.25	5.07	4.95	4.80	4.70	4.64	4.47	4.33
16	0.100	2.03	1.99	1.94	1.89	1.86	1.84	1.81	1.79	1.78	1.75	1.72
	0.050	2.49	2.42	2.35	2.28	2.23	2.19	2.15	2.12	2.11	2.06	2.02
	0.025	2.99	2.89	2.79	2.68	2.61	2.57	2.51	2.47	2.45	2.38	2.32
	0.010	3.69	3.55	3.41	3.26	3.16	3.10	3.02	2.97	2.93	2.84	2.76
	0.001	5.81	5.55	5.27	4.99	4.82	4.70	4.54	4.45	4.39	4.23	4.08

Continued

TABLE 8 *continued*

Degrees of Freedom Denominator, $d.f._D$	Right-tail Area	Degrees of Freedom Numerator, $d.f._N$ 1	2	3	4	5	6	7	8	9
17	0.100	3.03	2.64	2.44	2.31	2.22	2.15	2.10	2.06	2.03
	0.050	4.45	3.59	3.20	2.96	2.81	2.70	2.61	2.55	2.49
	0.025	6.04	4.62	4.01	3.66	3.44	3.28	3.16	3.06	2.98
	0.010	8.40	6.11	5.19	4.67	4.34	4.10	3.93	3.79	3.68
	0.001	15.72	10.66	8.73	7.68	7.02	6.56	6.22	5.96	5.75
18	0.100	3.01	2.62	2.42	2.29	2.20	2.13	2.08	2.04	2.00
	0.050	4.41	3.55	3.16	2.93	2.77	2.66	2.58	2.51	2.46
	0.025	5.98	4.56	3.95	3.61	3.38	3.22	3.10	3.01	2.93
	0.010	8.29	6.01	5.09	4.58	4.25	4.01	3.84	3.71	3.60
	0.001	15.38	10.39	8.49	7.46	6.81	6.35	6.02	5.76	5.56
19	0.100	2.99	2.61	2.40	2.27	2.18	2.11	2.06	2.02	1.98
	0.050	4.38	3.52	3.13	2.90	2.74	2.63	2.54	2.48	2.42
	0.025	5.92	4.51	3.90	3.56	3.33	3.17	3.05	2.96	2.88
	0.010	8.18	5.93	5.01	4.50	4.17	3.94	3.77	3.63	3.52
	0.001	15.08	10.16	8.28	7.27	6.62	6.18	5.85	5.59	5.39
20	0.100	2.97	2.59	2.38	2.25	2.16	2.09	2.04	2.00	1.96
	0.050	4.35	3.49	3.10	2.87	2.71	2.60	2.51	2.45	2.39
	0.025	5.87	4.46	3.86	3.51	3.29	3.13	3.01	2.91	2.84
	0.010	8.10	5.85	4.94	4.43	4.10	3.87	3.70	3.56	3.46
	0.001	14.82	9.95	8.10	7.10	6.46	6.02	5.69	5.44	5.24
21	0.100	2.96	2.57	2.36	2.23	2.14	2.08	2.02	1.98	1.95
	0.050	4.32	3.47	3.07	2.84	2.68	2.57	2.49	2.42	2.37
	0.025	5.83	4.42	3.82	3.48	3.25	3.09	2.97	2.87	2.80
	0.010	8.02	5.78	4.87	4.37	4.04	3.81	3.64	3.51	3.40
	0.001	14.59	9.77	7.94	6.95	6.32	5.88	5.56	5.31	5.11
22	0.100	2.95	2.56	2.35	2.22	2.13	2.06	2.01	1.97	1.93
	0.050	4.30	3.44	3.05	2.82	2.66	2.55	2.46	2.40	2.34
	0.025	5.79	4.38	3.78	3.44	3.22	3.05	2.93	2.84	2.76
	0.010	7.95	5.72	4.82	4.31	3.99	3.76	3.59	3.45	3.35
	0.001	14.38	9.61	7.80	6.81	6.19	5.76	5.44	5.19	4.99
23	0.100	2.94	2.55	2.34	2.21	2.11	2.05	1.99	1.95	1.92
	0.050	4.28	3.42	3.03	2.80	2.64	2.53	2.44	2.37	2.32
	0.025	5.75	4.35	3.75	3.41	3.18	3.02	2.90	2.81	2.73
	0.010	7.88	5.66	4.76	4.26	3.94	3.71	3.54	3.41	3.30
	0.001	14.20	9.47	7.67	6.70	6.08	5.65	5.33	5.09	4.89
24	0.100	2.93	2.54	2.33	2.19	2.10	2.04	1.98	1.94	1.91
	0.050	4.26	3.40	3.01	2.78	2.62	2.51	2.42	2.36	2.30
	0.025	5.72	4.32	3.72	3.38	3.15	2.99	2.87	2.78	2.70
	0.010	7.82	5.61	4.72	4.22	3.90	3.67	3.50	3.36	3.26
	0.001	14.03	9.34	7.55	6.59	5.98	5.55	5.23	4.99	4.80

TABLE 8 *continued*

Degrees of Freedom Denominator, $d.f._D$	Right-tail Area	Degrees of Freedom Numerator, $d.f._N$										
		10	12	15	20	25	30	40	50	60	120	1000
17	0.100	2.00	1.96	1.91	1.86	1.83	1.81	1.78	1.76	1.75	1.72	1.69
	0.050	2.45	2.38	2.31	2.23	2.18	2.15	2.10	2.08	2.06	2.01	1.97
	0.025	2.92	2.82	2.72	2.62	2.55	2.50	2.44	2.41	2.38	2.32	2.26
	0.010	3.59	3.46	3.31	3.16	3.07	3.00	2.92	2.87	2.83	2.75	2.66
	0.001	5.58	5.32	5.05	4.78	4.60	4.48	4.33	4.24	4.18	4.02	3.87
18	0.100	1.98	1.93	1.89	1.84	1.80	1.78	1.75	1.74	1.72	1.69	1.66
	0.050	2.41	2.34	2.27	2.19	2.14	2.11	2.06	2.04	2.02	1.97	1.92
	0.025	2.87	2.77	2.67	2.56	2.49	2.44	2.38	2.35	2.32	2.26	2.20
	0.010	3.51	3.37	3.23	3.08	2.98	2.92	2.84	2.78	2.75	2.66	2.58
	0.001	5.39	5.13	4.87	4.59	4.42	4.30	4.15	4.06	4.00	3.84	3.69
19	0.100	1.96	1.91	1.86	1.81	1.78	1.76	1.73	1.71	1.70	1.67	1.64
	0.050	2.38	2.31	2.23	2.16	2.11	2.07	2.03	2.00	1.98	1.93	1.88
	0.025	2.82	2.72	2.62	2.51	2.44	2.39	2.33	2.30	2.27	2.20	2.14
	0.010	3.43	3.30	3.15	3.00	2.91	2.84	2.76	2.71	2.67	2.58	2.50
	0.001	5.22	4.97	4.70	4.43	4.26	4.14	3.99	3.90	3.84	3.68	3.53
20	0.100	1.94	1.89	1.84	1.79	1.76	1.74	1.71	1.69	1.68	1.64	1.61
	0.050	2.35	2.28	2.20	2.12	2.07	2.04	1.99	1.97	1.95	1.90	1.85
	0.025	2.77	2.68	2.57	2.46	2.40	2.35	2.29	2.25	2.22	2.16	2.09
	0.010	3.37	3.23	3.09	2.94	2.84	2.78	2.69	2.64	2.61	2.52	2.43
	0.001	5.08	4.82	4.56	4.29	4.12	4.00	3.86	3.77	3.70	3.54	3.40
21	0.100	1.92	1.87	1.83	1.78	1.74	1.72	1.69	1.67	1.66	1.62	1.59
	0.050	2.32	2.25	2.18	2.10	2.05	2.01	1.96	1.94	1.92	1.87	1.82
	0.025	2.73	2.64	2.53	2.42	2.36	2.31	2.25	2.21	2.18	2.11	2.05
	0.010	3.31	3.17	3.03	2.88	2.79	2.72	2.64	2.58	2.55	2.46	2.37
	0.001	4.95	4.70	4.44	4.17	4.00	3.88	3.74	3.64	3.58	3.42	3.28
22	0.100	1.90	1.86	1.81	1.76	1.73	1.70	1.67	1.65	1.64	1.60	1.57
	0.050	2.30	2.23	2.15	2.07	2.02	1.98	1.94	1.91	1.89	1.84	1.79
	0.025	2.70	2.60	2.50	2.39	2.32	2.27	2.21	2.17	2.14	2.08	2.01
	0.010	3.26	3.12	2.98	2.83	2.73	2.67	2.58	2.53	2.50	2.40	2.32
	0.001	4.83	4.58	4.33	4.06	3.89	3.78	3.63	3.54	3.48	3.32	3.17
23	0.100	1.89	1.84	1.80	1.74	1.71	1.69	1.66	1.64	1.62	1.59	1.55
	0.050	2.27	2.20	2.13	2.05	2.00	1.96	1.91	1.88	1.86	1.81	1.76
	0.025	2.67	2.57	2.47	2.36	2.29	2.24	2.18	2.14	2.11	2.04	1.98
	0.010	3.21	3.07	2.93	2.78	2.69	2.62	2.54	2.48	2.45	2.35	2.27
	0.001	4.73	4.48	4.23	3.96	3.79	3.68	3.53	3.44	3.38	3.22	3.08
24	0.100	1.88	1.83	1.78	1.73	1.70	1.67	1.64	1.62	1.61	1.57	1.54
	0.050	2.25	2.18	2.11	2.03	1.97	1.94	1.89	1.86	1.84	1.79	1.74
	0.025	2.64	2.54	2.44	2.33	2.26	2.21	2.15	2.11	2.08	2.01	1.94
	0.010	3.17	3.03	2.89	2.74	2.64	2.58	2.49	2.44	2.40	2.31	2.22
	0.001	4.64	4.39	4.14	3.87	3.71	3.59	3.45	3.36	3.29	3.14	2.99

Continued

TABLE 8 *continued*

Degrees of Freedom Denominator, $d.f._D$	Right-tail Area	Degrees of Freedom Numerator, $d.f._N$ 1	2	3	4	5	6	7	8	9
	0.100	2.92	2.53	2.32	2.18	2.09	2.02	1.97	1.93	1.89
	0.050	4.24	3.39	2.99	2.76	2.60	2.49	2.40	2.34	2.28
25	0.025	5.69	4.29	3.69	3.35	3.13	2.97	2.85	2.75	2.68
	0.010	7.77	5.57	4.68	4.18	3.85	3.63	3.46	3.32	3.22
	0.001	13.88	9.22	7.45	6.49	5.89	5.46	5.15	4.91	4.71
	0.100	2.91	2.52	2.31	2.17	2.08	2.01	1.96	1.92	1.88
	0.050	4.23	3.37	2.98	2.74	2.59	2.47	2.39	2.32	2.27
26	0.025	5.66	4.27	3.67	3.33	3.10	2.94	2.82	2.73	2.65
	0.010	7.72	5.53	4.64	4.14	3.82	3.59	3.42	3.29	3.18
	0.001	13.74	9.12	7.36	6.41	5.80	5.38	5.07	4.83	4.64
	0.100	2.90	2.51	2.30	2.17	2.07	2.00	1.95	1.91	1.87
	0.050	4.21	3.35	2.96	2.73	2.57	2.46	2.37	2.31	2.25
27	0.025	5.63	4.24	3.65	3.31	3.08	2.92	2.80	2.71	2.63
	0.010	7.68	5.49	4.60	4.11	3.78	3.56	3.39	3.26	3.15
	0.001	13.61	9.02	7.27	6.33	5.73	5.31	5.00	4.76	4.57
	0.100	2.89	2.50	2.29	2.16	2.06	2.00	1.94	1.90	1.87
	0.050	4.20	3.34	2.95	2.71	2.56	2.45	2.36	2.29	2.24
28	0.025	5.61	4.22	3.63	3.29	3.06	2.90	2.78	2.69	2.61
	0.010	7.64	5.45	4.57	4.07	3.75	3.53	3.36	3.23	3.12
	0.001	13.50	8.93	7.19	6.25	5.66	5.24	4.93	4.69	4.50
	0.100	2.89	2.50	2.28	2.15	2.06	1.99	1.93	1.89	1.86
	0.050	4.18	3.33	2.93	2.70	2.55	2.43	2.35	2.28	2.22
29	0.025	5.59	4.20	3.61	3.27	3.04	2.88	2.76	2.67	2.59
	0.010	7.60	5.42	4.54	4.04	3.73	3.50	3.33	3.20	3.09
	0.001	13.39	8.85	7.12	6.19	5.59	5.18	4.87	4.64	4.45
	0.100	2.88	2.49	2.28	2.14	2.05	1.98	1.93	1.88	1.85
	0.050	4.17	3.32	2.92	2.69	2.53	2.42	2.33	2.27	2.21
30	0.025	5.57	4.18	3.59	3.25	3.03	2.87	2.75	2.65	2.57
	0.010	7.56	5.39	4.51	4.02	3.70	3.47	3.30	3.17	3.07
	0.001	13.29	8.77	7.05	6.12	5.53	5.12	4.82	4.58	4.39
	0.100	2.84	2.44	2.23	2.09	2.00	1.93	1.87	1.83	1.79
	0.050	4.08	3.23	2.84	2.61	2.45	2.34	2.25	2.18	2.12
40	0.025	5.42	4.05	3.46	3.13	2.90	2.74	2.62	2.53	2.45
	0.010	7.31	5.18	4.31	3.83	3.51	3.29	3.12	2.99	2.89
	0.001	12.61	8.25	6.59	5.70	5.13	4.73	4.44	4.21	4.02
	0.100	2.81	2.41	2.20	2.06	1.97	1.90	1.84	1.80	1.76
	0.050	4.03	3.18	2.79	2.56	2.40	2.29	2.20	2.13	2.07
50	0.025	5.34	3.97	3.39	3.05	2.83	2.67	2.55	2.46	2.38
	0.010	7.17	5.06	4.20	3.72	3.41	3.19	3.02	2.89	2.78
	0.001	12.22	7.96	6.34	5.46	4.90	4.51	4.22	4.00	3.82

TABLE 8 *continued*

Degrees of Freedom Denominator, $d.f._D$	Right-tail Area	Degrees of Freedom Numerator, $d.f._N$ 10	12	15	20	25	30	40	50	60	120	1000
	0.100	1.87	1.82	1.77	1.72	1.68	1.66	1.63	1.61	1.59	1.56	1.52
	0.050	2.24	2.16	2.09	2.01	1.96	1.92	1.87	1.84	1.82	1.77	1.72
25	0.025	2.61	2.51	2.41	2.30	2.23	2.18	2.12	2.08	2.05	1.98	1.91
	0.010	3.13	2.99	2.85	2.70	2.60	2.54	2.45	2.40	2.36	2.27	2.18
	0.001	4.56	4.31	4.06	3.79	3.63	3.52	3.37	3.28	3.22	3.06	2.91
	0.100	1.86	1.81	1.76	1.71	1.67	1.65	1.61	1.59	1.58	1.54	1.51
	0.050	2.22	2.15	2.07	1.99	1.94	1.90	1.85	1.82	1.80	1.75	1.70
26	0.025	2.59	2.49	2.39	2.28	2.21	2.16	2.09	2.05	2.03	1.95	1.89
	0.010	3.09	2.96	2.81	2.66	2.57	2.50	2.42	2.36	2.33	2.23	2.14
	0.001	4.48	4.24	3.99	3.72	3.56	3.44	3.30	3.21	3.15	2.99	2.84
	0.100	1.85	1.80	1.75	1.70	1.66	1.64	1.60	1.58	1.57	1.53	1.50
	0.050	2.20	2.13	2.06	1.97	1.92	1.88	1.84	1.81	1.79	1.73	1.68
27	0.025	2.57	2.47	2.36	2.25	2.18	2.13	2.07	2.03	2.00	1.93	1.86
	0.010	3.06	2.93	2.78	2.63	2.54	2.47	2.38	2.33	2.29	2.20	2.11
	0.001	4.41	4.17	3.92	3.66	3.49	3.38	3.23	3.14	3.08	2.92	2.78
	0.100	1.84	1.79	1.74	1.69	1.65	1.63	1.59	1.57	1.56	1.52	1.48
	0.050	2.19	2.12	2.04	1.96	1.91	1.87	1.82	1.79	1.77	1.71	1.66
28	0.025	2.55	2.45	2.34	2.23	2.16	2.11	2.05	2.01	1.98	1.91	1.84
	0.010	3.03	2.90	2.75	2.60	2.51	2.44	2.35	2.30	2.26	2.17	2.08
	0.001	4.35	4.11	3.86	3.60	3.43	3.32	3.18	3.09	3.02	2.86	2.72
	0.100	1.83	1.78	1.73	1.68	1.64	1.62	1.58	1.56	1.55	1.51	1.47
	0.050	2.18	2.10	2.03	1.94	1.89	1.85	1.81	1.77	1.75	1.70	1.65
29	0.025	2.53	2.43	2.32	2.21	2.14	2.09	2.03	1.99	1.96	1.89	1.82
	0.010	3.00	2.87	2.73	2.57	2.48	2.41	2.33	2.27	2.23	2.14	2.05
	0.001	4.29	4.05	3.80	3.54	3.38	3.27	3.12	3.03	2.97	2.81	2.66
	0.100	1.82	1.77	1.72	1.67	1.63	1.61	1.57	1.55	1.54	1.50	1.46
	0.050	2.16	2.09	2.01	1.93	1.88	1.84	1.79	1.76	1.74	1.68	1.63
30	0.025	2.51	2.41	2.31	2.20	2.12	2.07	2.01	1.97	1.94	1.87	1.80
	0.010	2.98	2.84	2.70	2.55	2.45	2.39	2.30	2.25	2.21	2.11	2.02
	0.001	4.24	4.00	3.75	3.49	3.33	3.22	3.07	2.98	2.92	2.76	2.61
	0.100	1.76	1.71	1.66	1.61	1.57	1.54	1.51	1.48	1.47	1.42	1.38
	0.050	2.08	2.00	1.92	1.84	1.78	1.74	1.69	1.66	1.64	1.58	1.52
40	0.025	2.39	2.29	2.18	2.07	1.99	1.94	1.88	1.83	1.80	1.72	1.65
	0.010	2.80	2.66	2.52	2.37	2.27	2.20	2.11	2.06	2.02	1.92	1.82
	0.001	3.87	3.64	3.40	3.14	2.98	2.87	2.73	2.64	2.57	2.41	2.25
	0.100	1.73	1.68	1.63	1.57	1.53	1.50	1.46	1.44	1.42	1.38	1.33
	0.050	2.03	1.95	1.87	1.78	1.73	1.69	1.63	1.60	1.58	1.51	1.45
50	0.025	2.32	2.22	2.11	1.99	1.92	1.87	1.80	1.75	1.72	1.64	1.56
	0.010	2.70	2.56	2.42	2.27	2.17	2.10	2.01	1.95	1.91	1.80	1.70
	0.001	3.67	3.44	3.20	2.95	2.79	2.68	2.53	2.44	2.38	2.21	2.05

Continued

TABLE 8 *continued*

Degrees of Freedom Denominator, $d.f._D$	Right-tail Area	Degrees of Freedom Numerator, $d.f._N$ 1	2	3	4	5	6	7	8	9
	0.100	2.79	2.39	2.18	2.04	1.95	1.87	1.82	1.77	1.74
	0.050	4.00	3.15	2.76	2.53	2.37	2.25	2.17	2.10	2.04
60	0.025	5.29	3.93	3.34	3.01	2.79	2.63	2.51	2.41	2.33
	0.010	7.08	4.98	4.13	3.65	3.34	3.12	2.95	2.82	2.72
	0.001	11.97	7.77	6.17	5.31	4.76	4.37	4.09	3.86	3.69
	0.100	2.76	2.36	2.14	2.00	1.91	1.83	1.78	1.73	1.69
	0.050	3.94	3.09	2.70	2.46	2.31	2.19	2.10	2.03	1.97
100	0.025	5.18	3.83	3.25	2.92	2.70	2.54	2.42	2.32	2.24
	0.010	6.90	4.82	3.98	3.51	3.21	2.99	2.82	2.69	2.59
	0.001	11.50	7.41	5.86	5.02	4.48	4.11	3.83	3.61	3.44
	0.100	2.73	2.33	2.11	1.97	1.88	1.80	1.75	1.70	1.66
	0.050	3.89	3.04	2.65	2.42	2.26	2.14	2.06	1.98	1.93
200	0.025	5.10	3.76	3.18	2.85	2.63	2.47	2.35	2.26	2.18
	0.010	6.76	4.71	3.88	3.41	3.11	2.89	2.73	2.60	2.50
	0.001	11.15	7.15	5.63	4.81	4.29	3.92	3.65	3.43	3.26
	0.100	2.71	2.31	2.09	1.95	1.85	1.78	1.72	1.68	1.64
	0.050	3.85	3.00	2.61	2.38	2.22	2.11	2.02	1.95	1.89
1000	0.025	5.04	3.70	3.13	2.80	2.58	2.42	2.30	2.20	2.13
	0.010	6.66	4.63	3.80	3.34	3.04	2.82	2.66	2.53	2.43
	0.001	10.89	6.96	5.46	4.65	4.14	3.78	3.51	3.30	3.13

TABLE 8 *continued*

Degrees of Freedom Denominator, $d.f._D$	Right-tail Area	Degrees of Freedom Numerator, $d.f._N$ 10	12	15	20	25	30	40	50	60	120	1000
	0.100	1.71	1.66	1.60	1.54	1.50	1.48	1.44	1.41	1.40	1.35	1.30
	0.050	1.99	1.92	1.84	1.75	1.69	1.65	1.59	1.56	1.53	1.47	1.40
60	0.025	2.27	2.17	2.06	1.94	1.87	1.82	1.74	1.70	1.67	1.58	1.49
	0.010	2.63	2.50	2.35	2.20	2.10	2.03	1.94	1.88	1.84	1.73	1.62
	0.001	3.54	3.32	3.08	2.83	2.67	2.55	2.41	2.32	2.25	2.08	1.92
	0.100	1.66	1.61	1.56	1.49	1.45	1.42	1.38	1.35	1.34	1.28	1.22
	0.050	1.93	1.85	1.77	1.68	1.62	1.57	1.52	1.48	1.45	1.38	1.30
100	0.025	2.18	2.08	1.97	1.85	1.77	1.71	1.64	1.59	1.56	1.46	1.36
	0.010	2.50	2.37	2.22	2.07	1.97	1.89	1.80	1.74	1.69	1.57	1.45
	0.001	3.30	3.07	2.84	2.59	2.43	2.32	2.17	2.08	2.01	1.83	1.64
	0.100	1.63	1.58	1.52	1.46	1.41	1.38	1.34	1.31	1.29	1.23	1.16
	0.050	1.88	1.80	1.72	1.62	1.56	1.52	1.46	1.41	1.39	1.30	1.21
200	0.025	2.11	2.01	1.90	1.78	1.70	1.64	1.56	1.51	1.47	1.37	1.25
	0.010	2.41	2.27	2.13	1.97	1.87	1.79	1.69	1.63	1.58	1.45	1.30
	0.001	3.12	2.90	2.67	2.42	2.26	2.15	2.00	1.90	1.83	1.64	1.43
	0.100	1.61	1.55	1.49	1.43	1.38	1.35	1.30	1.27	1.25	1.18	1.08
	0.050	1.84	1.76	1.68	1.58	1.52	1.47	1.41	1.36	1.33	1.24	1.11
1000	0.025	2.06	1.96	1.85	1.72	1.64	1.58	1.50	1.45	1.41	1.29	1.13
	0.010	2.34	2.20	2.06	1.90	1.79	1.72	1.61	1.54	1.50	1.35	1.16
	0.001	2.99	2.77	2.54	2.30	2.14	2.02	1.87	1.77	1.69	1.49	1.22

Source: From Biometrika, Tables of Statisticans, Vol. I; Critical Values for *F* Distribution. (Table 8). Reprinted by permission of Oxford University Press.

TABLE 9 Critical Values for Spearman Rank Correlation, r_s

For a right- (left-) tailed test, use the positive (negative) critical value found in the table under One-tail Area. For a two-tailed test, use both the positive and the negative of the critical value found in the table under Two-tail Area; n = number of pairs.

	One-tail Area			
	0.05	0.025	0.005	0.001
	Two-tail Area			
n	0.10	0.05	0.01	0.002
5	0.900	1.000		
6	0.829	0.886	1.000	
7	0.715	0.786	0.929	1.000
8	0.620	0.715	0.881	0.953
9	0.600	0.700	0.834	0.917
10	0.564	0.649	0.794	0.879
11	0.537	0.619	0.764	0.855
12	0.504	0.588	0.735	0.826
13	0.484	0.561	0.704	0.797
14	0.464	0.539	0.680	0.772
15	0.447	0.522	0.658	0.750
16	0.430	0.503	0.636	0.730
17	0.415	0.488	0.618	0.711
18	0.402	0.474	0.600	0.693
19	0.392	0.460	0.585	0.676
20	0.381	0.447	0.570	0.661
21	0.371	0.437	0.556	0.647
22	0.361	0.426	0.544	0.633
23	0.353	0.417	0.532	0.620
24	0.345	0.407	0.521	0.608
25	0.337	0.399	0.511	0.597
26	0.331	0.391	0.501	0.587
27	0.325	0.383	0.493	0.577
28	0.319	0.376	0.484	0.567
29	0.312	0.369	0.475	0.558
30	0.307	0.363	0.467	0.549

Source: From G. J. Glasser and R. F. Winter, "Critical Values of the Coefficient of Rank Correlation for Testing the Hypothesis of Independence," *Biometrika, 48*, 444 (1961). Reprinted by permission of Oxford University Press.

TABLE 10 Critical Values for Number of Runs R (Level of significance $\alpha = 0.05$)

Value of n_1	Value of n_2: 2	3	4	5	6	7	8	9	10	11	12	13	14	15	16	17	18	19	20
2	1	1	1	1	1	1	1	1	1	1	2	2	2	2	2	2	2	2	2
	6	6	6	6	6	6	6	6	6	6	6	6	6	6	6	6	6	6	6
3	1	1	1	1	2	2	2	2	2	2	2	2	2	3	3	3	3	3	3
	6	8	8	8	8	8	8	8	8	8	8	8	8	8	8	8	8	8	8
4	1	1	1	2	2	2	3	3	3	3	3	3	3	3	4	4	4	4	4
	6	8	9	9	9	10	10	10	10	10	10	10	10	10	10	10	10	10	10
5	1	1	2	2	3	3	3	3	3	4	4	4	4	4	4	4	5	5	5
	6	8	9	10	10	11	11	12	12	12	12	12	12	12	12	12	12	12	12
6	1	2	2	3	3	3	3	4	4	4	4	5	5	5	5	5	5	6	6
	6	8	9	10	11	12	12	13	13	13	13	14	14	14	14	14	14	14	14
7	1	2	2	3	3	3	4	4	5	5	5	5	5	6	6	6	6	6	6
	6	8	10	11	12	13	13	14	14	14	14	15	15	15	16	16	16	16	16
8	1	2	3	3	3	4	4	5	5	5	6	6	6	6	6	7	7	7	7
	6	8	10	11	12	13	14	14	15	15	16	16	16	16	17	17	17	17	17
9	1	2	3	3	4	4	5	5	5	6	6	6	7	7	7	7	8	8	8
	6	8	10	12	13	14	14	15	16	16	16	17	17	18	18	18	18	18	18
10	1	2	3	3	4	5	5	5	6	6	7	7	7	7	8	8	8	8	9
	6	8	10	12	13	14	15	16	16	17	17	18	18	18	19	19	19	20	20
11	1	2	3	4	4	5	5	6	6	7	7	7	8	8	8	9	9	9	9
	6	8	10	12	13	14	15	16	17	17	18	19	19	19	20	20	20	21	21
12	2	2	3	4	4	5	6	6	7	7	7	8	8	8	9	9	9	10	10
	6	8	10	12	13	14	16	16	17	18	19	19	20	20	21	21	21	22	22
13	2	2	3	4	5	5	6	6	7	7	8	8	9	9	9	10	10	10	10
	6	8	10	12	14	15	16	17	18	19	19	20	20	21	21	22	22	23	23
14	2	2	3	4	5	5	6	7	7	8	8	9	9	9	10	10	10	11	11
	6	8	10	12	14	15	16	17	18	19	20	20	21	22	22	23	23	23	24
15	2	3	3	4	5	6	6	7	7	8	8	9	9	10	10	11	11	11	12
	6	8	10	12	14	15	16	18	18	19	20	21	22	22	23	23	24	24	25
16	2	3	4	4	5	6	6	7	8	8	9	9	10	10	11	11	11	12	12
	6	8	10	12	14	16	17	18	19	20	21	21	22	23	23	24	25	25	25
17	2	3	4	4	5	6	7	7	8	9	9	10	10	11	11	11	12	12	13
	6	8	10	12	14	16	17	18	19	20	21	22	23	23	24	25	25	26	26
18	2	3	4	5	5	6	7	8	8	9	9	10	10	11	11	12	12	13	13
	6	8	10	12	14	16	17	18	19	20	21	22	23	24	25	25	26	26	27
19	2	3	4	5	6	6	7	8	8	9	10	10	11	11	12	12	13	13	13
	6	8	10	12	14	16	17	18	20	21	22	23	23	24	25	26	26	27	27
20	2	3	4	5	6	6	7	8	9	9	10	10	11	12	12	13	13	13	14
	6	8	10	12	14	16	17	18	20	21	22	23	24	25	25	26	27	27	28

Answers and Key Steps

TO ODD-NUMBERED PROBLEMS

CHAPTER 1

Section 1.1

1. An individual is a member of the population of interest. A variable is an aspect of an individual subject or object being measured.
3. A parameter is a numerical measurement describing data from a population. A statistic is a numerical measurement describing data from a sample.
5. (a) Nominal level. There is no apparent order relationship among responses.
(b) Ordinal level. There is an increasing relationship from worst to best level of service. The interval between service levels is not meaningful, nor are ratios.
7. (a) Response regarding meal ordered at fast-food restaurants. (b) Qualitative. (c) Responses for *all* adult fast-food customers in the U.S.
9. (a) Nitrogen concentration (mg nitrogen/L water). (b) Quantitative. (c) Nitrogen concentration (mg nitrogen/L water) in the entire lake.
11. (a) Ratio. (b) Interval. (c) Nominal. (d) Ordinal. (e) Ratio. (f) Ratio.
13. (a) Nominal. (b) Ratio. (c) Interval. (d) Ordinal. (e) Ratio. (f) Interval.
15. Answers vary.
(a) For example: Use pounds. Round weights to the nearest pound. Since backpacks might weigh as much as 30 pounds, you might use a high-quality bathroom scale. (b) Some students may not allow you to weigh their backpacks for privacy reasons, etc. (c) Possibly. Some students may want to impress you with the heaviness of their backpacks, or they may be embarrassed about the "junk" they have stowed inside and thus may clean out their backpacks.

Section 1.2

1. In a stratified sample, random samples from each stratum are included. In a cluster sample, the clusters to be included are selected at random and then all members of each selected cluster are included.
3. The advice is wrong. A sampling error accounts only for the difference in results based on the use of a sample rather than of the entire population.
5. No, even though the sample is random, some students younger than 18 or older than 20 may not have been included in the sample.
7. (a) Stratified. (b) No, because each pooled sample would have 100 season ticket holders for men's basketball games and 100 for women's basketball games. Samples with, say, 125 ticket holders for men's basketball games and 75 for women's games are not possible.
9. Use a random-number table to select four distinct numbers corresponding to people in your class. (a) Reasons may vary. For instance, the first four students may make a special effort to get to class on time. (b) Reasons may vary. For instance, four students who come in late might all be nursing students enrolled in an anatomy and physiology class that meets the hour before in a faraway building. They may be more motivated than other students to complete a degree requirement. (c) Reasons may vary. For instance, four students sitting in the back row might be less inclined to participate in class discussions. (d) Reasons may vary. For instance, the tallest students might all be male.
11. Answers vary. Use groups of two digits.
13. Select a starting place in the table and group the digits in groups of four. Scan the table by rows and include the first six groups with numbers between 0001 and 8615.
15. (a) Yes, when a die is rolled several times, the same number may appear more than once. Outcome on the fourth roll is 2. (b) No, for a fair die, the outcomes are random.
17. Since there are five possible outcomes for each question, read single digits from a random-number table. Select a starting place and proceed until you have 10 digits from 1 to 5. Repetition is required. The correct answer for each question will be the letter choice corresponding to the digit chosen for that question.
19. (a) Simple random sample. (b) Cluster sample. (c) Convenience sample. (d) Systematic sample. (e) Stratified sample.

Section 1.3

1. Answers vary. People with higher incomes are more likely to have high-speed Internet access and to spend more time online. People with high-speed Internet access might spend less time watching TV news or programming. People with higher incomes might have less time to spend watching TV because of access to other entertainment venues.
3. No, respondents do not constitute a random sample from the community for several reasons, for instance, the sample frame includes only those at a farmer's market, Jill might not have approached people with large dogs or those who were busy, and participation was voluntary. Jill's T-shirt may have influenced respondents.

5. (a) No, those ages 18–29 in 2006 became ages 20–31 in 2008.
   (b) 1977 to 1988 (inclusive).
7. (a) Observational study. (b) Experiment.
   (c) Experiment. (d) Observational study.
9. (a) Use random selection to pick 10 calves to inoculate; test all calves; no placebo. (b) Use random selection to pick 9 schools to visit; survey all schools; no placebo. (c) Use random selection to pick 40 volunteers for skin patch with drug; survey all volunteers; placebo used.
11. Based on the information given, Scheme A is best because it blocks all plots bordering the river together and all plots not bordering the river together. The blocks of Scheme B do not seem to differ from each other.

Chapter 1 Review

1. Because of the requirement that each number appear only once in any row, column, or box, it would be very inefficient to use a random-number table to select the numbers. It's better to simply look at existing numbers, list possibilities that meet the requirement, and eliminate numbers that don't work.
3. (a) Stratified. (b) Students on your campus with work-study jobs. (c) Hours scheduled; quantitative; ratio. (d) Rating of applicability of work experience to future employment; qualitative; ordinal. (e) Statistic. (f) 60%; The people choosing not to respond may have some characteristics, such as not working many hours, that would bias the study. (g) No. The sample frame is restricted to one campus.
5. Assign digits so that 3 out of the 10 digits 0 through 9 correspond to the answer "Yes" and 7 of the digits correspond to the answer "No." One assignment is digits 0, 1, and 2 correspond to "Yes," while digits 3, 4, 5, 6, 7, 8, and 9 correspond to "No." Starting with line 1, block 1 of Table 1, this assignment of digits gives the sequence No, Yes, No, No, Yes, No, No.
7. (a) Observational study. (b) Experiment.
9. Possible directions on survey questions: Give height in inches, give age as of last birthday, give GPA to one decimal place, and so forth. Think about the types of responses you wish to have on each question.
11. (a) Experiment, since a treatment is imposed on one colony. (b) The control group receives normal daylight/darkness conditions. The treatment group has light 24 hours per day. (c) The number of fireflies living at the end of 72 hours. (d) Ratio.

CHAPTER 2

Section 2.1

1. Class limits are possible data values. Class limits specify the span of data values that fall within a class. Class boundaries are not possible data values; rather, they are values halfway between the upper class limit of one class and the lower class limit of the next.
3. The classes overlap so that some data values, such as 20, fall within two classes.
5. Class width = 9; class limits: 20–28, 29–37, 38–46, 47–55, 56–64, 65–73, 74–82.
7. (a) Answers vary. Skewed right, if you hope most of the waiting times are low, with only a few times at the higher end of the distribution of waiting times.
   (b) A bimodal distribution might reflect the fact that when there are lots of customers, most of the waiting times are longer, especially since the lines are likely to be long. On the other hand, when there are fewer customers, the lines are short or almost nonexistent, and most of the waiting times are briefer.
9. (a) Yes
   (b) Histogram of Highway mpg

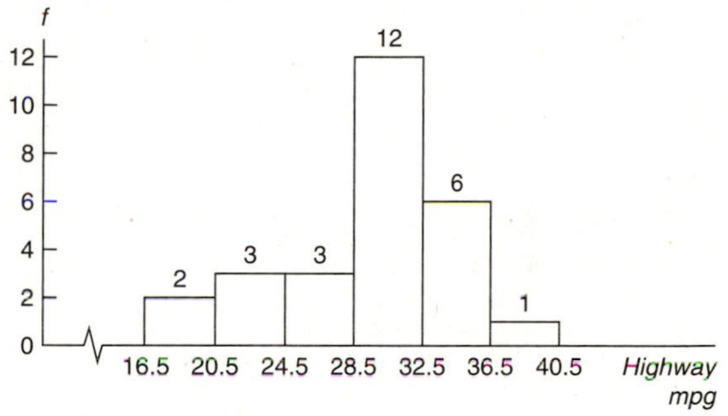

11. (a) The range of data seem to fall from 7 and 13, with the bulk of the data between 8 and 12.
    (b) All three histograms are somewhat mound-shaped, with the top of the mound between 9.5 and 10.5 In all three histograms, the bulk of the data fall between 8 and 12.
13. (a) Since there are 50 data values, each cumulative frequency was divided by 50 and converted to a percent. (b) 35. (c) 6. (d) 2%.
15. (a) Class width = 25.
    (b)

Class Limits	Class Boundaries	Midpoint	Frequency	Relative Frequency	Cumulative Frequency
236–260	235.5–260.5	248	4	0.07	4
261–285	260.5–285.5	273	9	0.16	13
286–310	285.5–310.5	298	25	0.44	38
311–335	310.5–335.5	323	16	0.28	54
336–360	335.5–360.5	348	3	0.05	57

(c, d) Hours to Complete the Iditarod—Histogram, Relative-Frequency Histogram

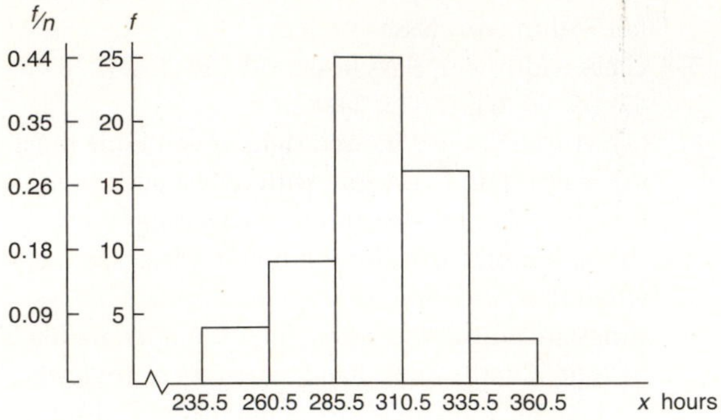

(e) Approximately mound-shaped symmetrical.

(f) Hours to Complete the Iditarod—Ogive

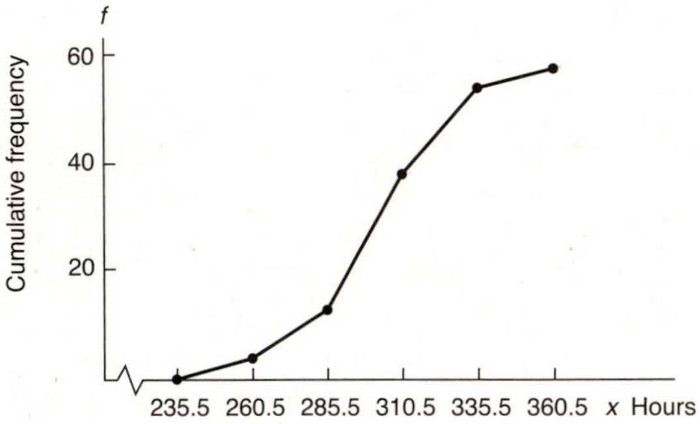

17. (a) Class width = 12.

(b)

Class Limits	Class Boundaries	Midpoint	Frequency	Relative Frequency	Cumulative Frequency
1–12	0.5–12.5	6.5	6	0.14	6
13–24	12.5–24.5	18.5	10	0.24	16
25–36	24.5–36.5	30.5	5	0.12	21
37–48	36.5–48.5	42.5	13	0.31	34
49–60	48.5–60.5	54.5	8	0.19	42

(c, d) Months Before Tumor Recurrence—Histogram, Relative-Frequency Histogram

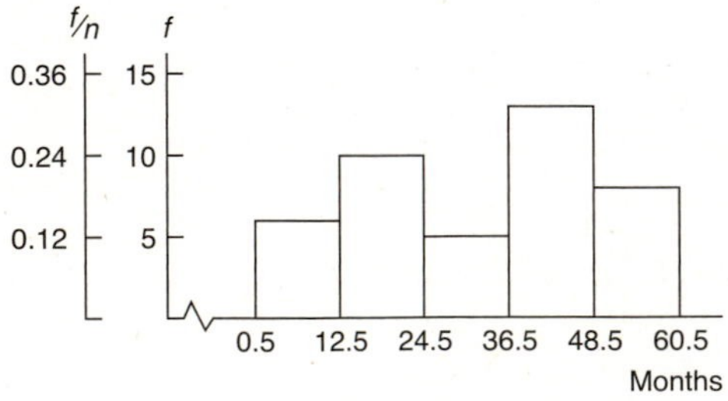

(e) Somewhat bimodal.

(f) Months Before Tumor Recurrence—Ogive

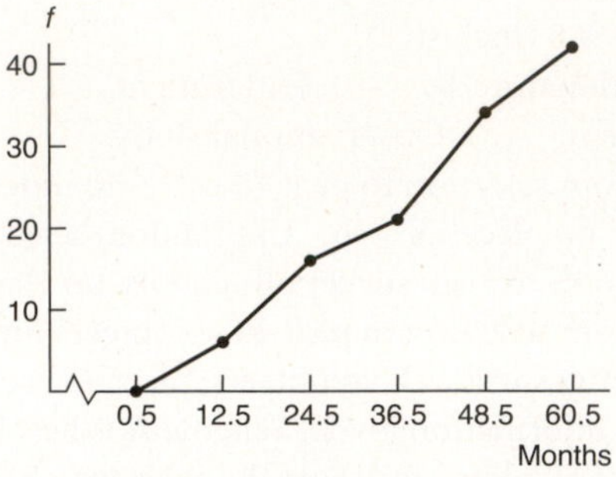

19. (a) Class width = 9.

(b)

Class Limits	Class Boundaries	Midpoint	Frequency	Relative Frequency	Cumulative Frequency
10–18	9.5–18.5	14	6	0.11	6
19–27	18.5–27.5	23	26	0.47	32
28–36	27.5–36.5	32	20	0.36	52
37–45	36.5–45.5	41	1	0.02	53
46–54	45.5–54.5	50	2	0.04	55

(c, d) Fuel Consumption (mpg)—Histogram, Relative-Frequency Histogram

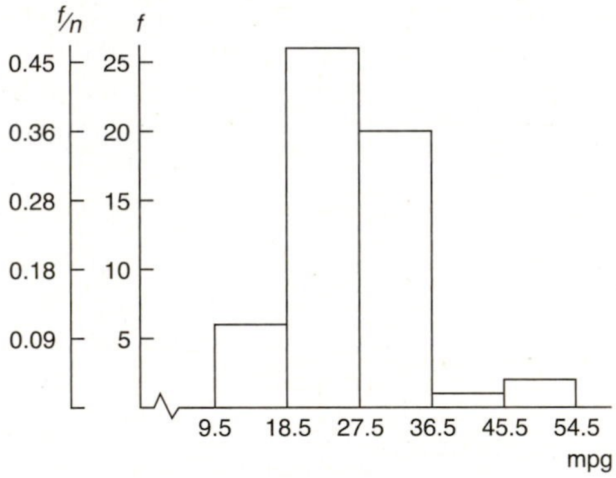

(e) Skewed slightly right.

(f) Fuel Consumption (mpg)—Ogive

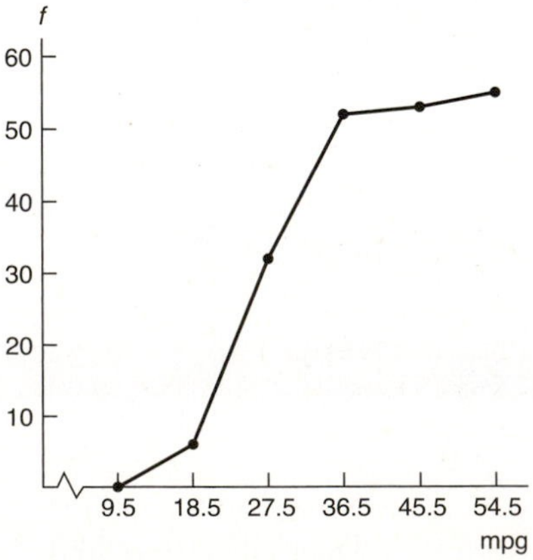

21. (a) Clear the decimals.
(b, c) Class width = 0.40.

Class Limits	Boundaries	Midpoint	Frequency
0.46–0.85	0.455–0.855	0.655	4
0.86–1.25	0.855–1.255	1.055	5
1.26–1.65	1.255–1.655	1.455	10
1.66–2.05	1.655–2.055	1.855	5
2.06–2.45	2.055–2.455	2.255	5
2.46–2.85	2.455–2.855	2.655	3

(c) Tonnes of Wheat—Histogram

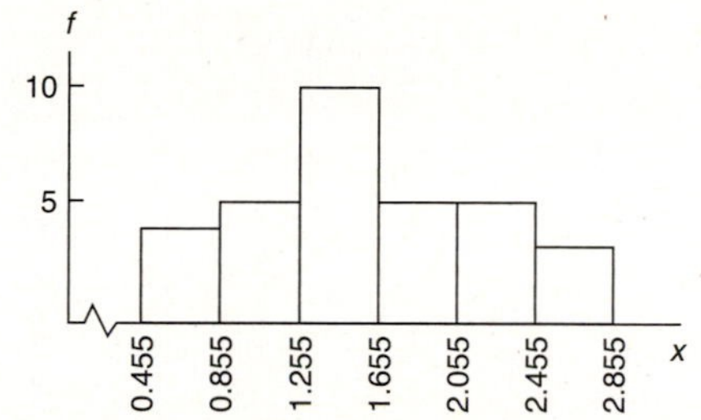

23. (a) One. (b) 5/51 or 9.8%. (c) Interval from 650 to 750.

25. Dotplot for Months Before Tumor Recurrence

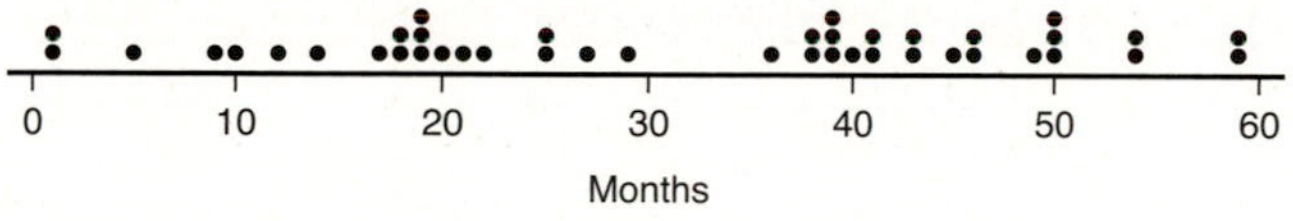

Section 2.2

1. (a) Yes, the percentages total more than 100%.
(b) No, in a circle graph the percentages must total 100% (within rounding error).
(c) Yes, the graph is organized in order from most frequently selected reason to least.

3. Pareto chart, because it shows the items in order of importance to the greatest number of employees.

5. Highest Level of Education and Average Annual Household Income (in thousands of dollars).

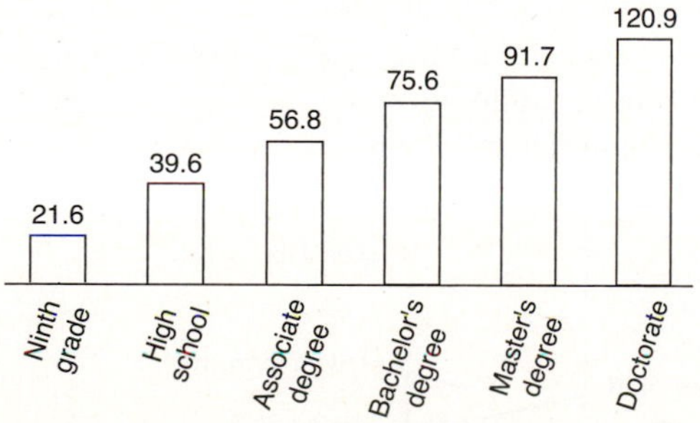

7. Annual Harvest (1000 Metric Tons)—Pareto Chart

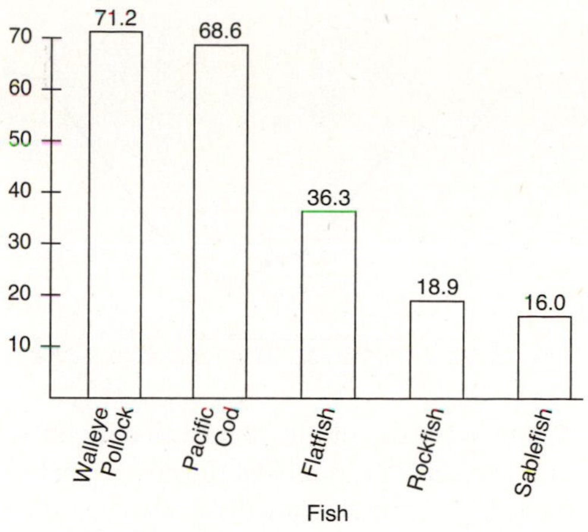

9. Where We Hide the Mess

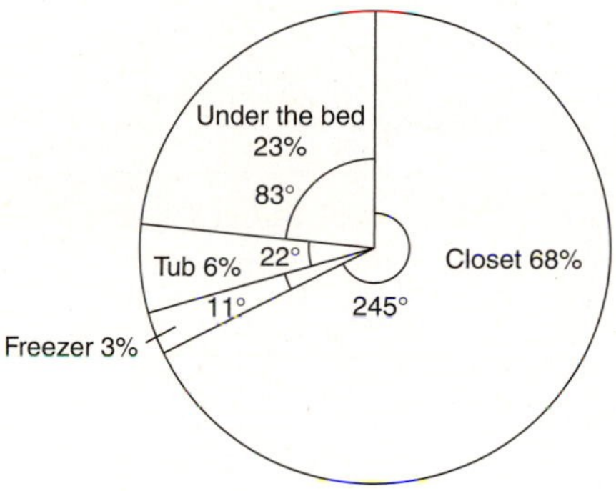

11. (a) Hawaii Crime Rate per 100,000 Population

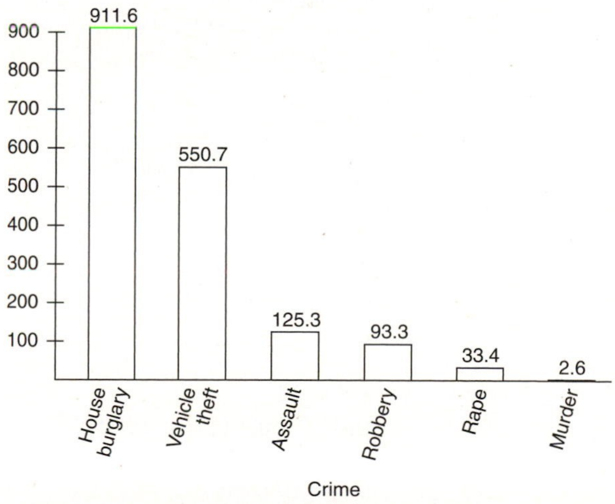

(b) A circle graph is not appropriate because the data do not reflect all types of crime. Also, the same person may have been the victim of more than one crime.

13.　Elevation of Pyramid Lake Surface—Time Plot

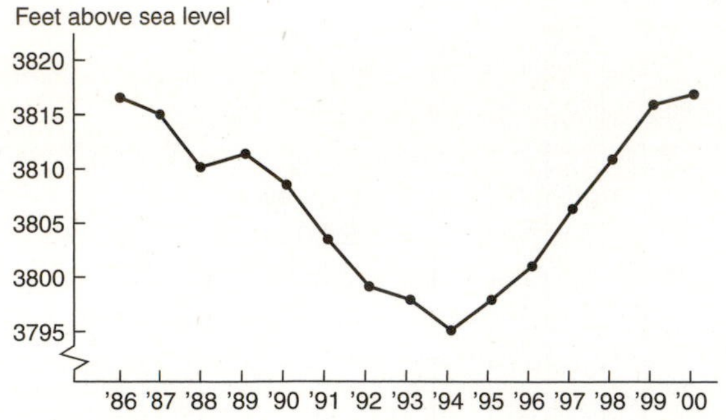

15.　(a) The size of the donut hole. Make all donuts exactly the same size, with the radius of the respective holes the same as well. Data labels showing percentages for each response would also be useful. (b) College graduates have a higher frequency of "no" response than those having only high school or less.

Section 2.3

1.　(a) Longevity of Cowboys

4	I	7 = 47 years
4		7
5		2788
6		16688
7		02233567
8		44456679
9		011237

(b) Yes, certainly these cowboys lived long lives.

3.　Average Length of Hospital Stay

5	I	2 = 5.2 years
5		235567
6		0246677888899
7		00000011122233334455668
8		457
9		469
10		03
11		1

The distribution is skewed right.

5.　(a) Minutes Beyond 2 Hours (1961–1980)

0	I	9 = 9 minutes past 2 hours
0		99
1		002334
1		55667889
2		0233

(b) Minutes Beyond 2 Hours (1981–2000)

0	I	7 = 7 minutes past 2 hours
0		777888899999999
1		00114

(c) In more recent years, the winning times have been closer to 2 hours, with all the times between 7 and 14 minutes over 2 hours. In the earlier period, more than half the times were more than 2 hours and 14 minutes.

7.　Milligrams of Tar per Cigarette

1	I	0 = 1.0 mg tar		
1		0	11	4
2			12	048
3			13	7
4		15	14	159
5			15	0128
6			16	06
7		38	17	0
8		068		
9		0		
10			29	8

The value 29.8 may be an outlier.

9.　Milligrams of Nicotine per Cigarette

0	I	1 = 0.1 milligram
0		144
0		566677788999
1		000000012
1		
2		0

Chapter 2 Review

1.　(a) Bar graph, Pareto chart, pie chart.　(b) All.
3.　Any large gaps between bars or stems with leaves at the beginning or end of the data set might indicate that the extreme data values are outliers.
5.　(a) Yes, with lines used instead of bars. However, because of the perspective nature of the drawing, the lengths of the bars do not represent the mileages. Thus, the scale for each bar changes.　(b) Yes. The scale does not change, and the viewer is not distracted by the graphic of the highway.
7.　Problems with Tax Returns

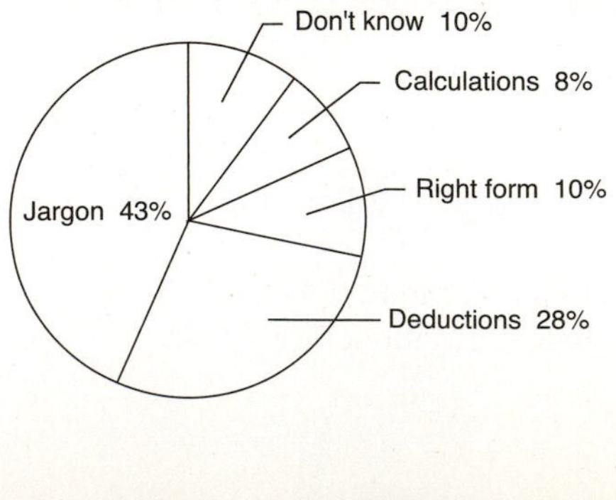

9. (a) Class width = 11.

Class Limits	Class Boundaries	Midpoint	Frequency	Relative Frequency	Cumulative Frequency
69–79	68.5–79.5	74	2	0.03	2
80–90	79.5–90.5	85	3	0.05	5
91–101	90.5–101.5	96	8	0.13	13
102–112	101.5–112.5	107	19	0.32	32
113–123	112.5–123.5	118	22	0.37	54
124–134	123.5–134.5	129	3	0.05	57
135–145	134.5–145.5	140	3	0.05	60

(b, c) Trunk Circumference (mm)—Histogram, Relative-Frequency Histogram

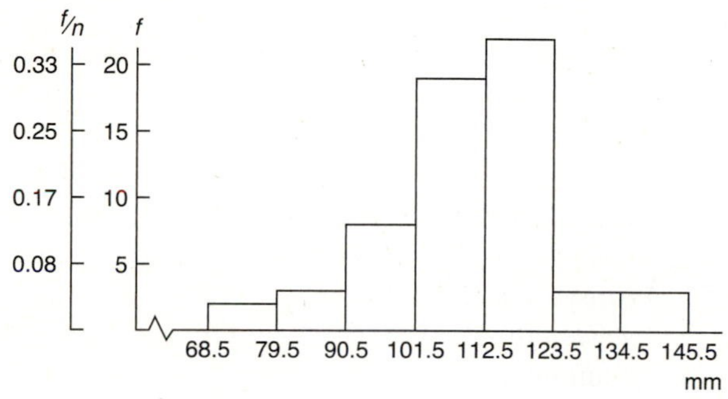

(d) Skewed slightly left.

(e) Trunk Circumference (mm)—Ogive

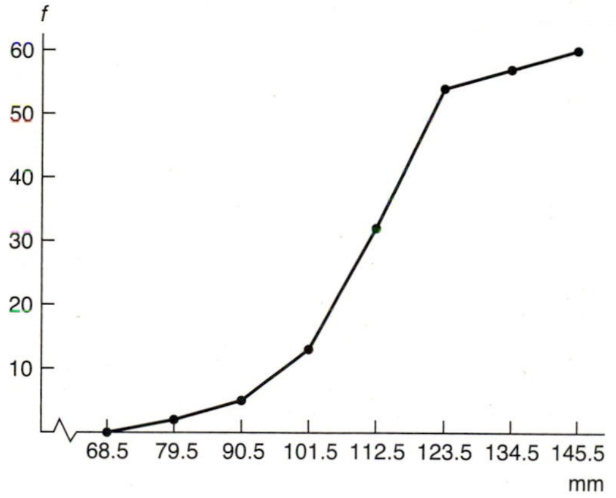

11. (a) 1240s had 40 data values. (b) 75. (c) From 1203 to 1212. Little if any repairs or new construction.

CHAPTER 3

Section 3.1

1. Median; mode; mean.
3. First add up all the data values, then divide by the number of data.
5. $\bar{x} = 5$; median = 6; mode = 2.
7. $\bar{x} = 5$; median = 5.5; mode = 2.
9. (a) No, the sum of the data does not change. (b) No, changing extreme data values does not change the median (c) Yes, depending on which data value occurs most frequently after the data are changed.
11. Mean, median, and mode are approximately equal.
13. (a) Mode = 5; median = 4; mean = 3.8. (b) Mode. (c) Mean, median, and mode. (d) Mode, median.
15. The supervisor has a legitimate concern because at least half the clients rated the employee below satisfactory. From the information given, it seems that this employee is very inconsistent in her performance.
17. (a) Mode = 2; median = 3; mean = 4.6. (b) Mode = 10; median = 15; mean = 23. (c) Corresponding values are 5 times the original averages. In general, multiplying each data value by a constant c results in the mode, median, and mean changing by a factor of c. (d) Mode = 177.8 cm; median = 172.72 cm; mean = 180.34 cm.
19. $\bar{x} \approx 167.3$°F; median = 171°F; mode = 178°F.
21. (a) $\bar{x} \approx 3.27$; median = 3; mode = 3. (b) $\bar{x} \approx 4.21$; median = 2; mode = 1. (c) Lower Canyon mean is greater; median and mode are less. (d) Trimmed mean = 3.75 and is closer to Upper Canyon mean.
23. (a) $\bar{x}$ = \$136.15; median = \$66.50; mode = \$60. (b) 5% trimmed mean ≈ \$121.28; yes, but still higher than the median. (c) Median. The low and high prices would be useful.
25. 23.
27. $\Sigma wx = 85$; $\Sigma w = 10$; weighted average = 8.5.
29. Approx. 66.67 mph.

Section 3.2

1. Mean.
3. Yes. For the sample standard deviation s, the sum $\Sigma(x - \bar{x})^2$ is divided by $n - 1$, where n is the sample size. For the population standard deviation σ, the sum $\Sigma(x - \mu)^2$ is divided by N, where N is the population size.
5. (a) Range is 4. (b) $s \approx 1.58$. (c) $\sigma \approx 1.41$.
7. For a data set in which not all data values are equal, σ is less than s. The reason is that to compute σ, we divide the sum of the squares by n, and to compute s we divide by the smaller number $n - 1$.
9. (a) (i), (ii), (iii). (b) The data change between data sets (i) and (ii) increased the sum of squared differences $\Sigma(x - \bar{x})^2$ by 10, whereas the data change between data sets (ii) and (iii) increased the sum of squared differences $\Sigma(x - \bar{x})^2$ by only 6.
11. (a) $s \approx 3.6$. (b) $s \approx 18.0$. (c) When each data value is multiplied by 5, the standard deviation is five times greater than that of the original data set. In general, multiplying each data value by the same constant c results in the standard deviation being $|c|$ times as large. (d) No. Multiply 3.1 miles by 1.6 kilometers/mile to obtain $s \approx 4.96$ kilometers.

13. (a) 15. (b) Use a calculator. (c) 37; 6.08. (d) 37; 6.08. (e) $\sigma^2 \approx 29.59$; $\sigma \approx 5.44$.

15. (a) $CV = 10\%$. (b) 14 to 26.

17. (a) 7.87. (b) Use a calculator. (c) $\bar{x} \approx 1.24$; $s^2 \approx 1.78$; $s \approx 1.33$. (d) $CV \approx 107\%$. The standard deviation of the time to failure is just slightly larger than the average time.

19. (a) Use a calculator. (b) $\bar{x} = 49$; $s^2 \approx 687.49$; $s \approx 26.22$. (c) $\bar{y} = 44.8$; $s^2 \approx 508.50$; $s \approx 22.55$.
(d) Mallard nests, $CV \approx 53.5\%$; Canada goose nests, $CV \approx 50.3\%$. The CV gives the ratio of the standard deviation to the mean; the CV for mallard nests is slightly higher.

21. Since $CV = s/\bar{x}$, then $s = CV(\bar{x})$; $s = 0.033$.

23. Midpoints: 25.5, 35.5, 45.5; $\bar{x} \approx 35.80$; $s^2 \approx 61.1$; $s \approx 7.82$.

25. Midpoints: 10.55, 14.55, 18.55, 22.55, 26.55; $\bar{x} \approx 15.6$; $s^2 \approx 23.4$; $s \approx 4.8$.

29. (a) $n_1 \approx 59$; $n_2 \approx 23$; $n_3 \approx 168$.
(b) $\mu \approx 7.46$.

Section 3.3

1. 82% or more of the scores were at or below Angela's score; 18% or fewer of the scores were above Angela's score.

3. No, the score 82 might have a percentile rank less than 70.

5. (a) Low = 2; $Q_1 = 5$; median = 7; $Q_3 = 8.5$; high = 10. (b) $IQR = 3.5$. (c) Box-and-Whisker Plot

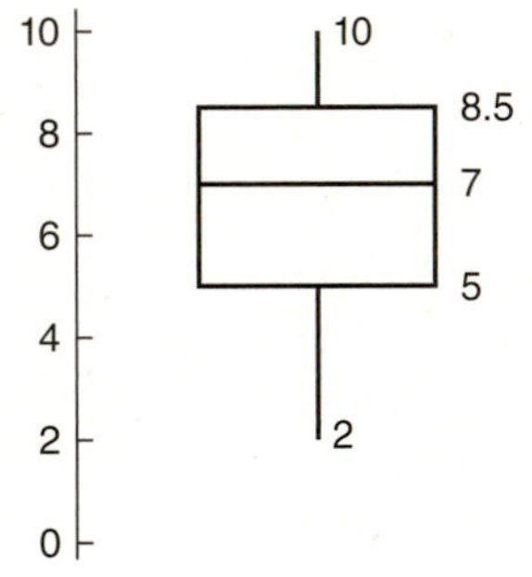

7. Low = 2; $Q_1 = 9.5$; median = 23; $Q_3 = 28.5$; high = 42; $IQR = 19$.
Nurses' Length of Employment (months)

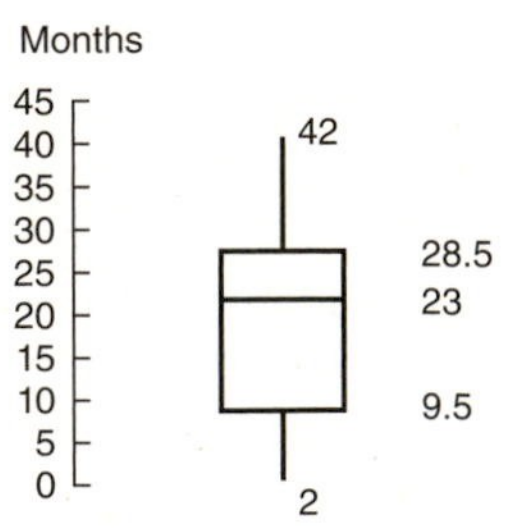

9. (a) Low = 17; $Q_1 = 22$; median = 24; $Q_3 = 27$; high = 38; $IQR = 5$. (b) Third quartile, since it is between the median and Q_3.
Bachelor's Degree Percentage by State

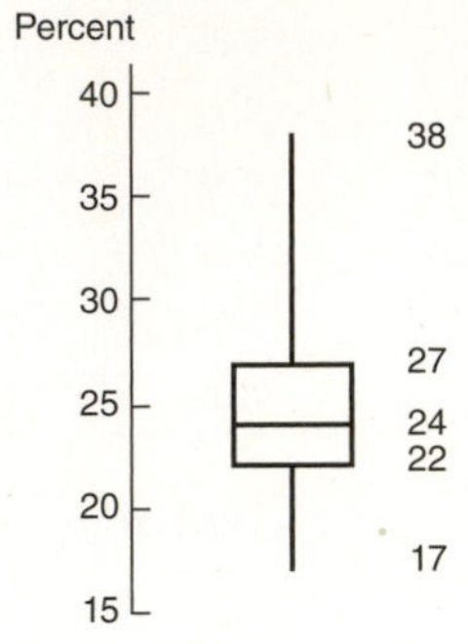

11. (a) California has the lowest premium. Pennsylvania has the highest. (b) Pennsylvania has the highest median premium. (c) California has the smallest range. Texas has the smallest interquartile range. (d) Part (a) is the five-number summary for Texas. It has the smallest *IQR*. Part (b) is the five-number summary for Pennsylvania. It has the largest minimum. Part (c) is the five-number summary for California. It has the lowest minimum.

Chapter 3 Review

1. (a) Variance and standard deviation. (b) Box-and-whisker plot.

3. (a) For both data sets, mean = 20 and range = 24.
(b) The C1 distribution seems more symmetric because the mean and median are equal, and the median is in the center of the interquartile range. In the C2 distribution, the mean is less than the median.
(c) The C1 distribution has a larger interquartile range that is symmetric around the median. The C2 distribution has a very compressed interquartile range with the median equal to Q_3.

5. (a) Low = 31; $Q_1 = 40$; median = 45; $Q_3 = 52.5$; high = 68; $IQR = 12.5$.
Percentage of Democratic Vote by County

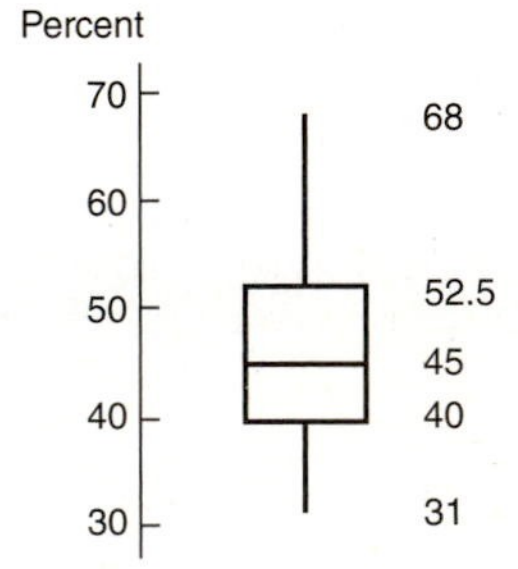

(b) Class width = 8.

Class	Midpoint	f
31–38	34.5	11
39–46	42.5	24
47–54	50.5	15
55–62	58.5	7
63–70	66.5	3

$\bar{x} \approx 46.1$; $s \approx 8.64$; 28.82 to 63.38.

(c) $\bar{x} \approx 46.15$; $s \approx 8.63$.

7. Mean weight = 156.25 pounds.
9. No.
   (b) \$37,091 to \$71,091.
   (c) Public 4-year \$5,105 to \$8,689; private 4-year \$19,854 to \$34,417; 15%; 50%.
11. $\Sigma w = 16$, $\Sigma wx = 121$, average = 7.56.

CUMULATIVE REVIEW PROBLEMS

Chapters 1–3

1. (a) Median, percentile. (b) Mean, variance, standard deviation.
2. (a) Gap between first bar and rest of bars or between last bar and rest of bars. (b) Large gap between data on far-left or far-right side and rest of data. (c) Several empty stems after stem including lowest values or before stem including highest values. (d) Data beyond fences placed at $Q_1 - 1.5(IQR)$ and $Q_3 + 1.5(IQR)$.
3. (a) Same. (b) Set B has a higher mean. (c) Set B has a higher standard deviation. (d) Set B has a much longer whisker beyond Q_3.
4. (a) Set A, because 86 is the relatively higher score, since a larger percentage of scores fall below it.
   (b) Set B, because 86 is more standard deviations above the mean.
5. Assign consecutive numbers to all the wells in the study region. Then use a random-number table, computer, or calculator to select 102 values that are less than or equal to the highest number assigned to a well in the study region. The sample consists of the wells with numbers corresponding to those selected.
6. Ratio.
7.

7	0 represents a pH level of 7.0
7	0 0 0 0 0 0 0 0 1 1 1 1 1 1 1 1 1
7	2 2 2 2 2 2 2 2 2 2 2 3 3 3 3 3 3 3 3 3 3
7	4 4 4 4 4 4 4 4 4 5 5 5 5 5 5 5
7	6 6 6 6 6 6 6 6 6 7 7 7 7 7 7
7	8 8 8 8 8 9 9 9 9 9
8	0 1 1 1 1 1 1 1
8	2 2 2 2 2 2 2
8	4 5
8	6 7
8	8 8

8. Clear the decimals. Then the highest value is 88 and the lowest is 70. The class width for the whole numbers is 4. For the actual data, the class width is 0.4.

Class Limits	Class Boundaries	Midpoint	Frequency	Relative Frequency
7.0–7.3	6.95–7.35	7.15	39	0.38
7.4–7.7	7.35–7.75	7.55	32	0.31
7.8–8.1	7.75–8.15	7.95	18	0.18
8.2–8.5	8.15–8.55	8.35	9	0.09
8.6–8.9	8.55–8.95	8.75	4	0.04

Levels of pH in West Texas Wells—Histogram, Relative-Frequency Histogram

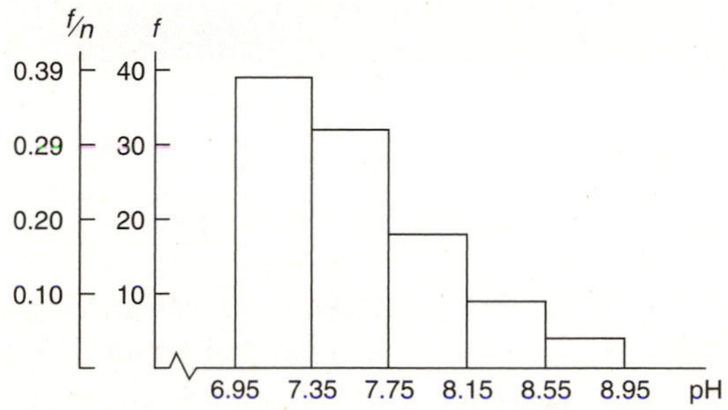

9. Levels of pH in West Texas Wells—Ogive

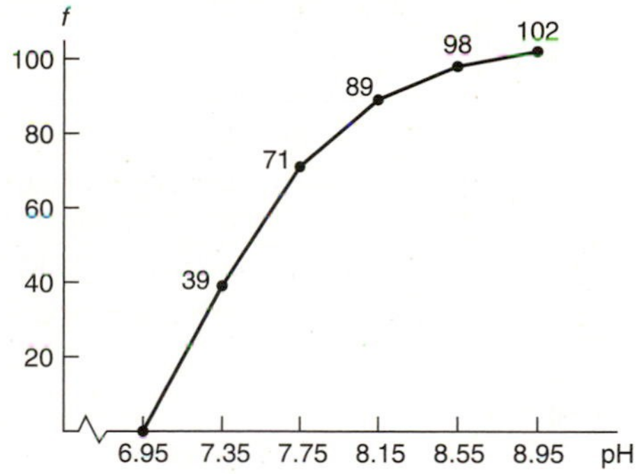

10. Range = 1.8; $\bar{x} \approx 7.58$; median = 7.5; mode = 7.3.
11. (a) Use a calculator or computer.
    (b) $s^2 \approx 0.20$; $s \approx 0.45$; $CV \approx 5.9\%$.
12. 6.68 to 8.48.
13. Levels of pH in West Texas Wells

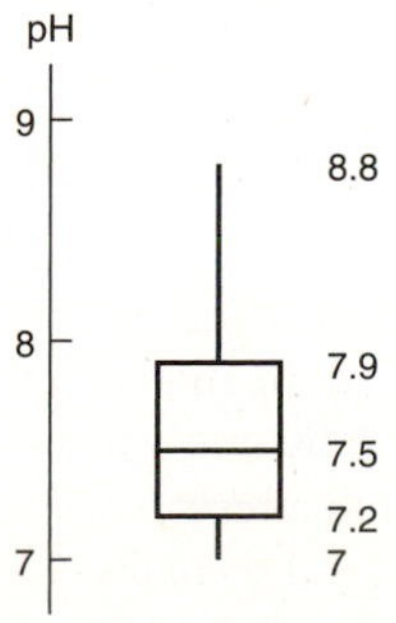

$IQR = 0.7$.

14. Skewed right. Lower values are more common.
15. 89%; 50%.
16. No, there are no gaps in the plot, but only 6 out of 102, or about 6%, have pH levels at or above 8.4. Eight wells are neutral.
17. Half the wells have pH levels between 7.2 and 7.9. The data are skewed toward the high values, with the upper half of the pH levels spread out more than the lower half. The upper half ranges between 7.5 and 8.8, while the lower half is clustered between 7 and 7.5.
18. The report should emphasize the relatively low mean, median, and mode, and the fact that half the wells have a pH level less than 7.5. The data are clustered at the low end of the range.

CHAPTER 4

Section 4.1

1. Equally likely outcomes, relative frequency, intuition.
3. (a) 1. (b) 0.
5. No, the probability was stated for drivers in the age range from 18 to 24. We have no information for other age groups. Other age groups may not behave the same way as the 18- to 24-year-olds.
7. $627/1010 \approx 0.62$.
9. Although the probability is high that you will make money, it is not completely certain that you will. In fact, there is a small chance that you could lose your entire investment. If you can afford to lose all of the investment, it might be worthwhile to invest, because there is a high chance of doubling your money.
11. (a) MMM MMF MFM MFF FMM FMF FFM FFF. (b) $P(\text{MMM}) = 1/8$. $P(\text{at least one female}) = 1 - P(\text{MMM}) = 7/8$.
13. No. The probability of heads on the second toss is 0.50 regardless of the outcome on the first toss.
15. Answers vary. Probability as a relative frequency. One concern is whether the students in the class are more or less adept at wiggling their ears than people in the general population.
17. (a) $P(0) = 15/375$; $P(1) = 71/375$; $P(2) = 124/375$; $P(3) = 131/375$; $P(4) = 34/375$. (b) Yes, the listed numbers of similar preferences form the sample space.
19. (a) $P(\text{best idea 6 A.M.–12 noon}) = 290/966 \approx 0.30$; $P(\text{best idea 12 noon–6 P.M.}) = 135/966 \approx 0.14$; $P(\text{best idea 6 P.M.–12 midnight}) = 319/966 \approx 0.33$; $P(\text{best idea 12 midnight–6 A.M.}) = 222/966 \approx 0.23$. (b) The probabilities add up to 1. They should add up to 1 (within rounding errors), provided the intervals do not overlap and each inventor chose only one interval. The sample space is the set of four time intervals.
21. (b) $P(\text{success}) = 2/17 \approx 0.118$. (c) $P(\text{make shot}) = 3/8$ or 0.375.
23. (a) $P(\text{enter if walks by}) = 58/127 \approx 0.46$. (b) $P(\text{buy if entered}) = 25/58 \approx 0.43$. (c) $P(\text{walk in and buy}) = 25/127 \approx 0.20$. (d) $P(\text{not buy}) = 1 - P(\text{buy}) \approx 1 - 0.43 = 0.57$.

Section 4.2

1. No. By definition, mutually exclusive events cannot occur together.
3. (a) 0.7. (b) 0.6.
5. (a) 0.08. (b) 0.04.
7. (a) 0.15. (b) 0.55.
9. $P(A \textit{ and } B)$ is the probability that both events A and B occur. It cannot exceed the probability that either event occurs. When the assigned probabilities are used to get $P(A \mid B)$, the result exceeds 1.
11. (a) Because the events are mutually exclusive, A cannot occur if B occurred. $P(A \mid B) = 0$. (b) Because $P(A \mid B) \neq P(A)$, the events A and B are not independent.
13. (a) $P(A \textit{ and } B)$. (b) $P(B \mid A)$. (c) $P(A^c \mid B)$. (d) $P(A \textit{ or } B)$. (e) $P(B^c \textit{ or } A)$.
15. (a) 0.2; yes. (b) 0.4; yes. (c) $1.0 - 0.2 = 0.8$.
17. (a) Yes. (b) $P(5 \text{ on green } \textit{and } 3 \text{ on red}) = P(5) \cdot P(3) = (1/6)(1/6) = 1/36 \approx 0.028$. (c) $P(3 \text{ on green } \textit{and } 5 \text{ on red}) = P(3) \cdot P(5) = (1/6)(1/6) = 1/36 \approx 0.028$. (d) $P((5 \text{ on green } \textit{and } 3 \text{ on red}) \textit{ or } (3 \text{ on green } \textit{and } 5 \text{ on red})) = (1/36) + (1/36) = 1/18 \approx 0.056$.
19. (a) $P(\text{sum of } 6) = P(1 \textit{ and } 5) + P(2 \textit{ and } 4) + P(3 \textit{ and } 3) + P(4 \textit{ and } 2) + P(5 \textit{ and } 1) = (1/36) + (1/36) + (1/36) + (1/36) + (1/36) = 5/36$. (b) $P(\text{sum of } 4) = P(1 \textit{ and } 3) + P(2 \textit{ and } 2) + P(3 \textit{ and } 1) = (1/36) + (1/36) + (1/36) = 3/36$ or $1/12$. (c) $P(\text{sum of 6 } \textit{or} \text{ sum of } 4) = P(\text{sum of } 6) + P(\text{sum of } 4) = (5/36) + (3/36) = 8/36$ or $2/9$; yes.
21. (a) No, after the first draw the sample space becomes smaller and probabilities for events on the second draw change. (b) $P(\text{Ace on 1st } \textit{and} \text{ King on 2nd}) = P(\text{Ace}) \cdot P(\text{King} \mid \text{Ace}) = (4/52)(4/51) = 4/663$. (c) $P(\text{King on 1st } \textit{and} \text{ Ace on 2nd}) = P(\text{King}) \cdot P(\text{Ace} \mid \text{King}) = (4/52)(4/51) = 4/663$. (d) $P(\text{Ace } \textit{and} \text{ King in either order}) = P(\text{Ace on 1st } \textit{and} \text{ King on 2nd}) + P(\text{King on 1st } \textit{and} \text{ Ace on 2nd}) = (4/663) + (4/663) = 8/663$.
23. (a) Yes, replacement of the card restores the sample space and all probabilities for the second draw remain unchanged regardless of the outcome of the first card. (b) $P(\text{Ace on 1st } \textit{and} \text{ King on 2nd}) = P(\text{Ace}) \cdot P(\text{King}) = (4/52)(4/52) = 1/169$. (c) $P(\text{King on 1st } \textit{and} \text{ Ace on 2nd}) = P(\text{King}) \cdot P(\text{Ace}) = (4/52)(4/52) = 1/169$. (d) $P(\text{Ace } \textit{and} \text{ King in either order}) = P(\text{Ace on 1st } \textit{and} \text{ King on 2nd}) + P(\text{King on 1st } \textit{and} \text{ Ace on 2nd}) = (1/169) + (1/169) = 2/169$.
25. (a) $P(6 \text{ years old } \textit{or} \text{ older}) = P(6\text{–}9) + P(10\text{–}12) + P(13 \text{ and over}) = 0.27 + 0.14 + 0.22 = 0.63$. (b) $P(12 \text{ years old } \textit{or} \text{ younger}) = P(2 \text{ and under}) + P(3\text{–}5) + P(6\text{–}9) + P(10\text{–}12) = 0.15 + 0.22 + 0.27 + 0.14 = 0.78$. (c) $P(\text{between 6 and 12}) = P(6\text{–}9) +$

$P(10\text{–}12) = 0.27 + 0.14 = 0.41$. (d) $P(\text{between 3 and 9}) = P(3\text{–}5) + P(6\text{–}9) = 0.22 + 0.27 = 0.49$. The category 13 and over contains far more ages than the group 10–12. It is not surprising that more toys are purchased for this group, since there are more children in this group.

27. The information from James Burke can be viewed as conditional probabilities. $P(\text{reports lie} \mid \text{person is lying}) = 0.72$ and $P(\text{reports lie} \mid \text{person is not lying}) = 0.07$. (a) $P(\text{person is not lying}) = 0.90$; $P(\text{person is not lying } \textit{and} \text{ polygraph reports lie}) = P(\text{person is not lying}) \times P(\text{reports lie} \mid \text{person not lying}) = (0.90)(0.07) = 0.063$ or 6.3%. (b) $P(\text{person is lying}) = 0.10$; $P(\text{person is lying } \textit{and} \text{ polygraph reports lie}) = P(\text{person is lying}) \times P(\text{reports lie} \mid \text{person is lying}) = (0.10)(0.72) = 0.072$ or 7.2%. (c) $P(\text{person is not lying}) = 0.5$; $P(\text{person is lying}) = 0.5$; $P(\text{person is not lying } \textit{and} \text{ polygraph reports lie}) = P(\text{person is not lying}) \times P(\text{reports lie} \mid \text{person not lying}) = (0.50)(0.07) = 0.035$ or 3.5%. $P(\text{person is lying } \textit{and} \text{ polygraph reports lie}) = P(\text{person is lying}) \times P(\text{reports lie} \mid \text{person is lying}) = (0.50)(0.72) = 0.36$ or 36%. (d) $P(\text{person is not lying}) = 0.15$; $P(\text{person is lying}) = 0.85$; $P(\text{person is not lying } \textit{and} \text{ polygraph reports lie}) = P(\text{person is not lying}) \times P(\text{reports lie} \mid \text{person is not lying}) = (0.15)(0.07) = 0.0105$ or 1.05%. $P(\text{person is lying } \textit{and} \text{ polygraph reports lie}) = P(\text{person is lying}) \times P(\text{reports lie} \mid \text{person is lying}) = (0.85)(0.72) = 0.612$ or 61.2%.

29. (a) 686/1160; 270/580; 416/580. (b) No. (c) 270/1160; 416/1160. (d) 474/1160; 310/580. (e) No. (f) 686/1160 + 580/1160 − 270/1160 = 996/1160.

31. (a) 72/154. (b) 82/154. (c) 79/116. (d) 37/116. (e) 72/270. (f) 82/270.

33. (a) $P(A) = 0.65$. (b) $P(B) = 0.71$. (c) $P(B \mid A) = 0.87$. (d) $P(A \textit{ and } B) = P(A) \cdot P(B \mid A) = (0.65)(0.87) \approx 0.57$. (e) $P(A \textit{ or } B) = P(A) + P(B) - P(A \textit{ and } B) \approx 0.65 + 0.71 - 0.57 = 0.79$. (f) $P(\text{not close}) = P(\text{profit 1st year } \textit{or} \text{ profit 2nd year}) = P(A \textit{ or } B) \approx 0.79$; $P(\text{close}) = 1 - P(\text{not close}) \approx 1 - 0.79 = 0.21$.

35. (a) $P(\text{TB } \textit{and} \text{ positive}) = P(\text{TB})P(\text{positive} \mid \text{TB}) = (0.04)(0.82) \approx 0.033$. (b) $P(\text{does not have TB}) = 1 - P(\text{TB}) = 1 - 0.04 = 0.96$. (c) $P(\text{no TB } \textit{and} \text{ positive}) = P(\text{no TB})P(\text{positive} \mid \text{no TB}) = (0.96)(0.09) \approx 0.086$.

37. True. A^c consists of all events not in A.

39. False. If event A^c has occurred, then event A cannot occur.

41. True. $P(A \textit{ and } B) = P(B) \cdot P(A \mid B)$. Since $0 < P(B) < 1$, the product $P(B) \cdot P(A \mid B) \leq P(A \mid B)$.

43. True. All the outcomes in event A *and* B are also in event A.

45. True. All the outcomes in event A^c *and* B^c are also in event A^c.

47. False. See Problem 11.

49. True. Since $P(A \textit{ and } B) = P(A) \cdot P(B) = 0$, either $P(A) = 0$ or $P(B) = 0$.

51. True. All simple events of the sample space under the condition "given B" are included in either the event A or the disjoint event A^c

Section 4.3

1. The permutations rule counts the number of different *arrangements* of r items out of n distinct items, whereas the combinations rule counts only the *number* of groups of r items out of n distinct items. The number of permutations is larger than the number of combinations.

3. (a) Use the combinations rule, since only the items in the group and not their arrangement is of concern. (b) Use the permutations rule, since the number of arrangements within each group is of interest.

5. (a) Outcomes for Flipping a Coin Three Times

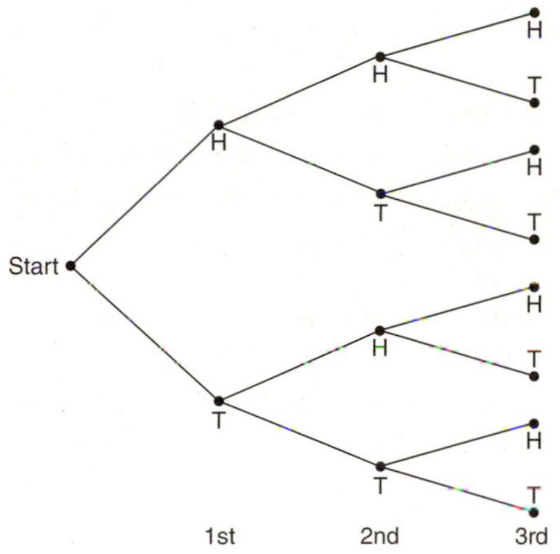

(b) 3. (c) 3/8.

7. (a) Outcomes for Drawing Two Balls (without replacement)

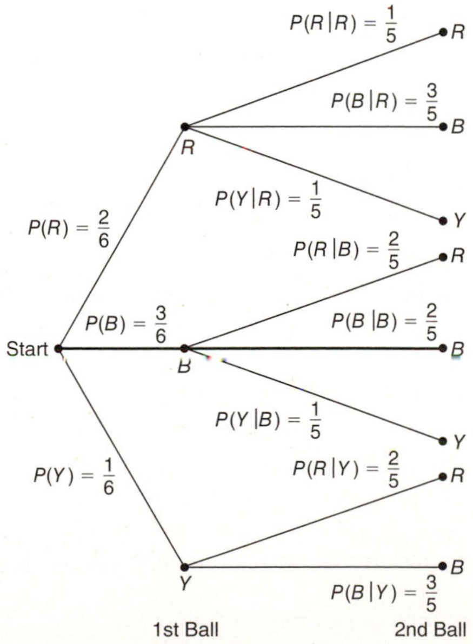

(b) $P(R \text{ and } R) = 2/6 \cdot 1/5 = 1/15$.
$P(R \text{ 1st and } B \text{ 2nd}) = 2/6 \cdot 3/5 = 1/5$.
$P(R \text{ 1st and } Y \text{ 2nd}) = 2/6 \cdot 1/5 = 1/15$.
$P(B \text{ 1st and } R \text{ 2nd}) = 3/6 \cdot 2/5 = 1/5$.
$P(B \text{ 1st and } B \text{ 2nd}) = 3/6 \cdot 2/5 = 1/5$.
$P(B \text{ 1st and } Y \text{ 2nd}) = 3/6 \cdot 1/5 = 1/10$.
$P(Y \text{ 1st and } R \text{ 2nd}) = 1/6 \cdot 2/5 = 1/15$.
$P(Y \text{ 1st and } B \text{ 2nd}) = 1/6 \cdot 3/5 = 1/10$.

9. $4 \cdot 3 \cdot 2 \cdot 1 = 24$ sequences.
11. $4 \cdot 3 \cdot 3 = 36$.
13. $P_{5,2} = (5!/3!) = 5 \cdot 4 = 20$.
15. $P_{7,7} = (7!/0!) = 7! = 5040$.
17. $C_{5,2} = (5!/(2!3!)) = 10$.
19. $C_{7,7} = (7!/(7!0!)) = 1$.
21. $P_{15,3} = 2730$.
23. $5 \cdot 4 \cdot 3 = 60$.
25. $C_{15,5} = (15!/(5!10!)) = 3003$.
27. (a) $C_{12,6} = (12!/(6!6!)) = 924$.
(b) $C_{7,6} = (7!/(6!1!)) = 7$. (c) $7/924 \approx 0.008$.

Chapter 4 Review

1. (a) The individual does not own a cell phone. (b) The individual owns a cell phone as well as a laptop computer. (c) The individual owns either a cell phone or a laptop computer, and maybe both. (d) The individual owns a cell phone, given he or she owns a laptop computer. (e) The individual owns a laptop computer, given he or she owns a cell phone.
3. For independent events A and B, $P(A) = P(A \mid B)$.
5. (a) $P(\text{offer job 1 } and \text{ offer job 2}) = 0.56$. The probability of getting offers for both jobs is less than the probability of getting each individual job offer.
(b) $P(\text{offer job 1 } or \text{ offer job 2}) = 0.94$. The probability of getting at least one of the job offers is greater than the probability of getting each individual job offer. It seems worthwhile to apply for both jobs since the probability is high of getting at least one offer.
7. (a) No. You need to know that the events are independent or you need to know the value of $P(A \mid B)$ or $P(B \mid A)$. (b) Yes. For independent events, $P(A \text{ and } B) = P(A) \cdot P(B)$.
9. $P(\text{asked}) = 24\%$; $P(\text{received} \mid \text{asked}) = 45\%$; $P(\text{asked } and \text{ received}) = (0.24)(0.45) = 10.8\%$.
11. (a) Drop a fixed number of tacks and count how many land flat side down. Then form the ratio of the number landing flat side down to the total number dropped.
(b) Up, down. (c) $P(\text{up}) = 160/500 = 0.32$; $P(\text{down}) = 340/500 = 0.68$.
13. (a)

Outcomes x	2	3	4	5	6
$P(x)$	0.028	0.056	0.083	0.111	0.139

x	7	8	9	10	11	12
$P(x)$	0.167	0.139	0.111	0.083	0.056	0.028

15. $C_{8,2} = (8!/(2!6!)) = (8 \cdot 7/2) = 28$.
17. $4 \cdot 4 \cdot 4 \cdot 4 \cdot 4 = 1024$ choices; $P(\text{all correct}) = 1/1024 \approx 0.00098$.
19. $10 \cdot 10 \cdot 10 = 1000$.

CHAPTER 5

Section 5.1

1. (a) Discrete. (b) Continuous. (c) Continuous. (d) Discrete. (e) Continuous.
3. (a) Yes. (b) No; probabilities total more than 1.
5. No, even though the outcomes in the sample space are the same, the individual Probabilities may differ in a way that produces the same μ but a different standard deviation.
7. Expected value ≈ 0.9. $\sigma \approx 0.6245$.
9. (a) Yes, 7 of the 10 digits represent "making a basket."
(b) Let S represent "making a basket" and F represent "missing the shot." $F, F, S, S, S, F, F, F, S, S$.
(c) Yes. Again, 7 of the 10 digits represent "making a basket." $S, S, S, S, S, S, S, S, S, S$.
11. (a) Yes, events are distinct and probabilities total 1.
(b) Income Distribution ($1000)

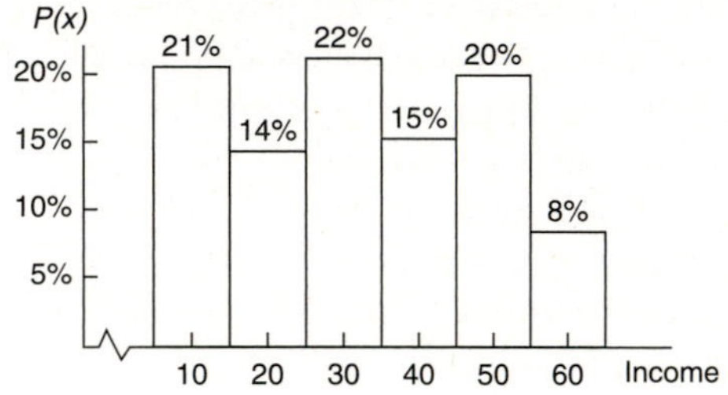

(c) 32.3 thousand dollars. (d) 16.12 thousand dollars.
13. (a) Number of Fish Caught in a 6-Hour Period at Pyramid Lake, Nevada

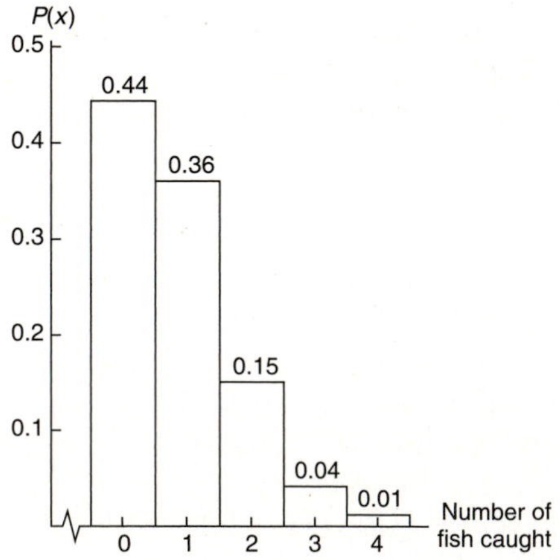

(b) 0.56. (c) 0.20. (d) 0.82. (e) 0.899.
15. (a) 15/719; 704/719. (b) $0.73; $14.27.
17. (a) 0.01191; $595.50. (b) $646; $698; $751.50; $806.50; $3497.50 total. (c) $4197.50. (d) $1502.50.

19. (a) $\mu_W = 1.5$; $\sigma_W^2 = 208$; $\sigma_W \approx 14.4$.
(b) $\mu_W = 107.5$; $\sigma_W^2 = 52$; $\sigma_W \approx 7.2$.
(c) $\mu_L = 90$; $\sigma_L^2 = 92.16$; $\sigma_L \approx 9.6$.
(d) $\mu_L = 90$; $\sigma_L^2 = 57.76$; $\sigma_L \approx 7.6$.

21. (a) $\mu_W = 50.2$; $\sigma_W^2 = 66.125$; $\sigma_W \approx 8.13$.
(b) The means are the same. (c) The standard deviation for two policies is smaller. (d) As we include more policies, the coefficients in W decrease, resulting in smaller σ_W^2 and σ_W. For instance, for three policies, $W = (\mu_1 + \mu_2 + \mu_3)/3 \approx 0.33\mu_1 + 0.33\mu_2 + 0.33\mu_3$ and $\sigma_W^2 \approx (0.33)^2\sigma_1^2 + (0.33)^2\sigma_2^2 + (0.33)^2\sigma_3^2$. Yes, the risk appears to decrease by a factor of $1/\sqrt{n}$.

23. (a) Essay.
(b) Square the standard deviations and substitute the results in the formula to evaluate c_1, c_2, c_3.
(c) $\sigma_w \approx 5.608$

Section 5.2

1. The random variable measures the number of successes out of n trials. This text uses the letter r for the random variable.

3. Two outcomes, success or failure.

5. Any monitor failure might endanger patient safely, so you should be concerned about the probability of *at least* one failure, not just exactly one failure.

7. (a) No. A binomial probability model applies to only two outcomes per trial. (b) Yes. Assign outcome A to "success" and outcomes B and C to "failure." $p = 0.40$.

9. (a) A trial consists of looking at the class status of a student enrolled in introductory statistics. Two outcomes are "freshman" and "not freshman." Success is freshman status; failure is any other class status. $P(\text{success}) = 0.40$. (b) Trials are not independent. With a population of only 30 students, in 5 trials without replacement, the probability of success rounded to the nearest hundredth changes for the later trials. Use the hypergeometric distribution for this situation.

11. (a) 0.082. (b) 0.918.

13. (a) 0.000. (b) Yes, the probability of 0 or 1 success is 0.000 to three places after the decimal. It would be a very rare event to get fewer than 2 successes when the probability of success on a single trial is so high.

15. A trial is one flip of a fair quarter. Success = coin shows heads. Failure = coin shows tails. $n = 3$; $p = 0.5$; $q = 0.5$. (a) $P(r = 3 \text{ heads}) = C_{3,3}p^3q^0 = 1(0.5)^3(0.5)^0 = 0.125$. To find this value in Table 3 of Appendix II, use the group in which $n = 3$, the column headed by $p = 0.5$, and the row headed by $r = 3$. (b) $P(r = 2 \text{ heads}) = C_{3,2}p^2q^1 = 3(0.5)^2(0.5)^1 = 0.375$. To find this value in Table 3 of Appendix II, use the group in which $n = 3$, the column headed by $p = 0.5$, and the row headed by $r = 2$. (c) $P(r \text{ is 2 or more}) = P(r = 2 \text{ heads}) + P(r = 3 \text{ heads}) = 0.375 + 0.125 = 0.500$. (d) The probability of getting three tails when you toss a coin three times is the same as getting zero heads. Therefore, $P(3 \text{ tails}) = P(r = 0 \text{ heads}) = C_{3,0}p^0q^3 = 1(0.5)^0(0.5)^3 = 0.125$. To find this value in Table 3 of Appendix II, use the group in which $n = 3$, the column headed by $p = 0.5$, and the row headed by $r = 0$.

17. A trial is recording the gender of one wolf. Success = male. Failure = female. $n = 12$; $p = 0.55$; $q = 0.45$.
(a) $P(r \geq 6) = 0.740$. Six or more females means $12 - 6 = 6$ or fewer males; $P(r \leq 6) = 0.473$. Fewer than four females means more than $12 - 4 = 8$ males; $P(r > 8) = 0.135$. (b) A trial is recording the gender of one wolf. Success = male. Failure = female. $n = 12$; $p = 0.70$; $q = 0.30$. $P(r \geq 6) = 0.961$; $P(r \leq 6) = 0.117$; $P(r > 8) = 0.493$.

19. A trial consists of a woman's response regarding her mother-in-law. Success = dislike. Failure = like. $n = 6$; $p = 0.90$; $q = 0.10$. (a) $P(r = 6) = 0.531$. (b) $P(r = 0) = 0.000$ (to three digits). (c) $P(r \geq 4) = P(r = 4) + P(r = 5) + P(r = 6) = 0.098 + 0.354 + 0.531 = 0.983$. (d) $P(r \leq 3) = 1 - P(r \geq 4) \approx 1 - 0.983 = 0.017$ or 0.016 directly from table.

21. A trial is taking a polygraph exam. Success = pass. Failure = fail. $n = 9$; $p = 0.85$; $q = 0.15$.
(a) $P(r = 9) = 0.232$. (b) $P(r \geq 5) = P(r = 5) + P(r = 6) + P(r = 7) + P(r = 8) + P(r = 9) = 0.028 + 0.107 + 0.260 + 0.368 + 0.232 = 0.995$. (c) $P(r \leq 4) = 1 - P(r \geq 5) \approx 1 - 0.995 = 0.005$ or 0.006 directly from table. (d) $P(r = 0) = 0.000$ (to three digits).

23. (a) A trial consists of using the Myers–Briggs instrument to determine if a person in marketing is an extrovert. Success = extrovert. Failure = not extrovert. $n = 15$; $p = 0.75$; $q = 0.25$. $P(r \geq 10) = 0.851$; $P(r \geq 5) = 0.999$; $P(r = 15) = 0.013$. (b) A trial consists of using the Myers–Briggs instrument to determine if a computer programmer is an introvert. Success = introvert. Failure = not introvert. $n = 5$; $p = 0.60$; $q = 0.40$. $P(r = 0) = 0.010$; $P(r \geq 3) = 0.683$; $P(r = 5) = 0.078$.

25. $n = 8$; $p = 0.53$; $q = 0.47$. (a) 0.812515; yes, truncated at five digits. (b) 0.187486; 0.18749; yes, rounded to five digits.

27. (a) They are the same. (b) They are the same. (c) $r = 1$. (d) The column headed by $p = 0.80$.

29. (a) $n = 8$; $p = 0.65$; $P(6 \leq r \mid 4 \leq r) = P(6 \leq r)/P(4 \leq r) = 0.428/0.895 \approx 0.478$. (b) $n = 10$; $p = 0.65$; $P(8 \leq r \mid 6 \leq r) = P(8 \leq r)/P(6 \leq r) = 0.262/0.752 \approx 0.348$. (c) Essay. (d) Use event $A = 6 \leq r$ and event $B = 4 \leq r$ in the formula.

31. (a) $P(r_1 = 3, r_2 = 2, r_3 = 1) = \frac{6!}{3!2!1!}(0.5)^3(0.3)^2(0.2)^1 = 0.135$. (b) $P(r_1 = 4, r_2 = 2, r_3 = 0) = \frac{6!}{4!2!0!}(0.5)^4(0.3)^2(0.2)^0 \approx 0.084$.

Section 5.3

1. The average number of successes.

3. (a) $\mu = 1.6$; $\sigma \approx 1.13$. (b) Yes, 5 successes is more than 2.5σ above the expected value. $P(r \geq 5) = 0.010$.

5. (a) Yes, 120 is more than 2.5 standard deviations above the expected value. (b) Yes, 40 is less than 2.5 standard deviations below the expected value. (c) No, 70 to 90 successes is within 2.5 standard deviations of the expected value.

7. (a) Binomial Distribution
The distribution is symmetrical.

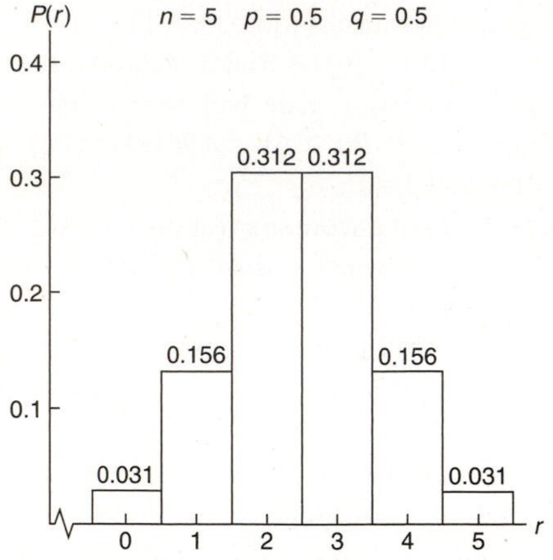

(b) Binomial Distribution
The distribution is skewed right.

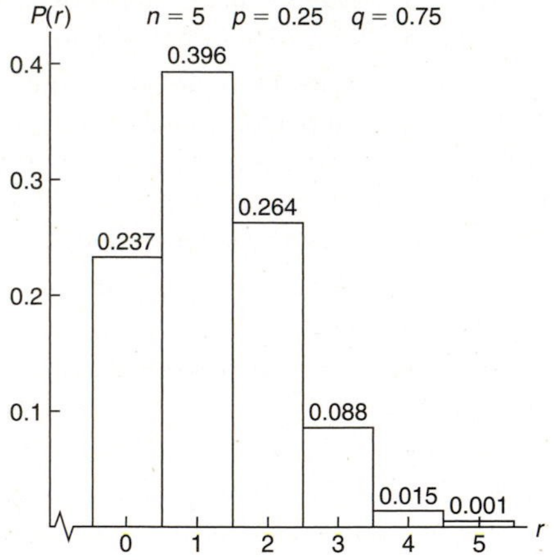

(c) Binomial Distribution
The distribution is skewed left.

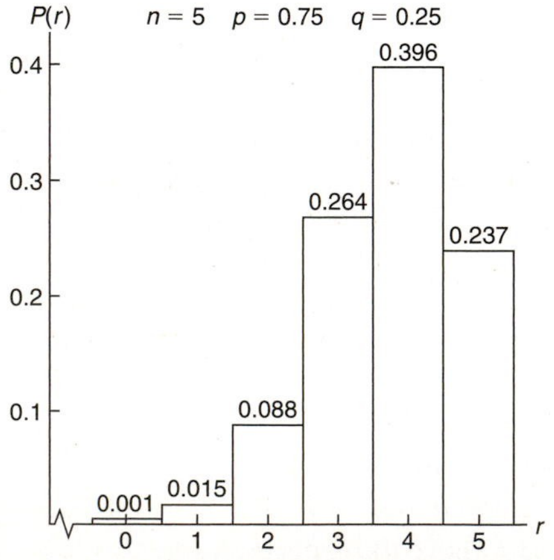

(d) The distributions are mirror images of one another.
(e) The distribution would be skewed left for $p = 0.73$ because the more likely numbers of successes are to the right of the middle.

9. (a) Skewed left.
(b) $\mu = 8.5$.
(c) Very low; the expected number of successes in 10 trials is 8.5 and p is so high it would unusual to have so few successes in 10 trials.
(d) Very high; the expected number of successes in 10 trials is 8.5 and p is so high it would be common to have 8 or more successes in 10 trials.

11. (a) Households with Children Under 2 That Buy Photo Gear

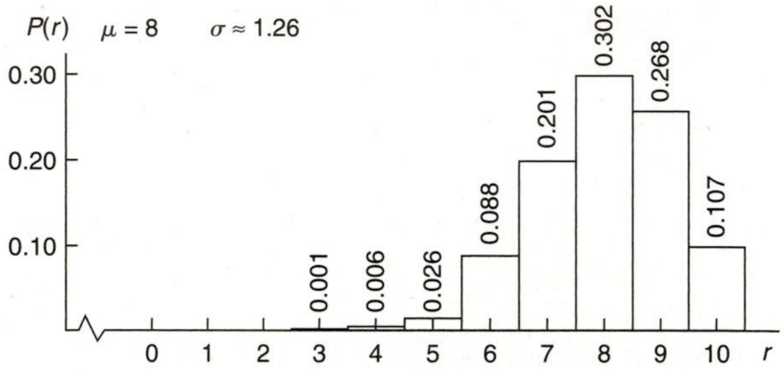

(b) Households with No Children Under 21 That Buy Photo Gear

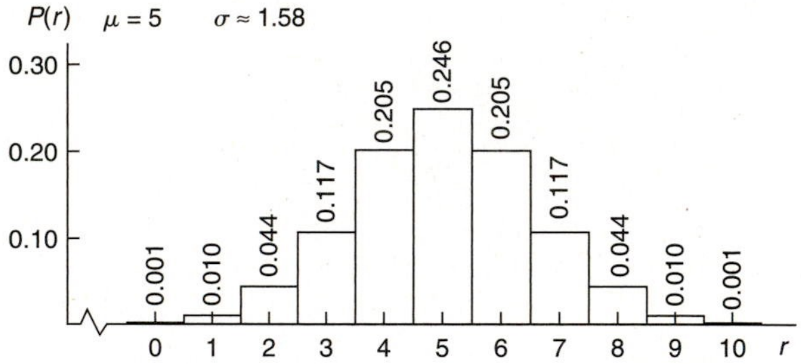

(c) Yes. Adults with children seem to buy more photo gear.

13. (a) Binomial Distribution for Number of Addresses Found

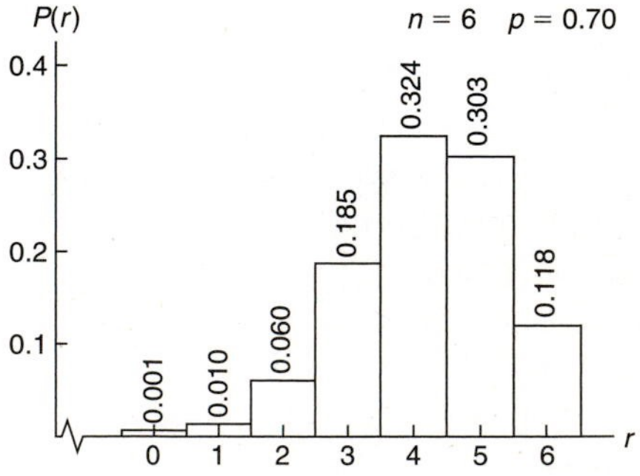

(b) $\mu = 4.2$; $\sigma \approx 1.122$. (c) $n = 5$. Note that $n = 5$ gives $P(r \geq 2) = 0.97$.

15. (a) Binomial Distribution for Number of Illiterate People

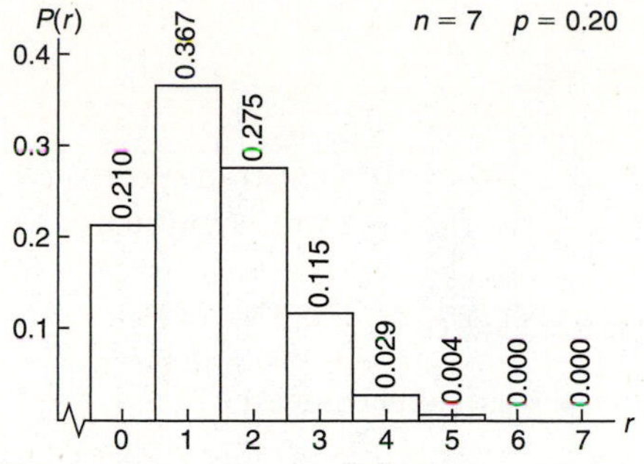

(b) $\mu = 1.4$; $\sigma \approx 1.058$. (c) $n = 12$. Note that $n = 12$ gives $P(r \geq 7) = 0.98$, where success = literate and $p = 0.80$.

17. (a) Binomial Distribution for Number of Gullible Consumers

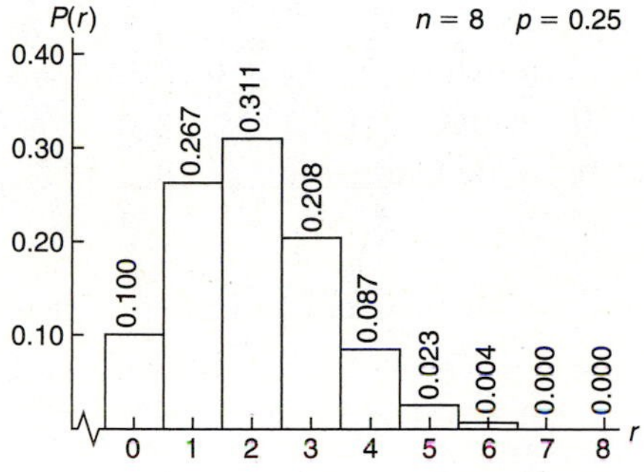

(b) $\mu = 2$; $\sigma \approx 1.225$. (c) $n = 16$. Note that $n = 16$ gives $P(r \geq 1) = 0.99$.

19. (a) $P(r = 0) = 0.004$; $P(r = 1) = 0.047$; $P(r = 2) = 0.211$; $P(r = 3) = 0.422$; $P(r = 4) = 0.316$. (b) Binomial Distribution for Number of Parolees Who Do Not Become Repeat Offenders

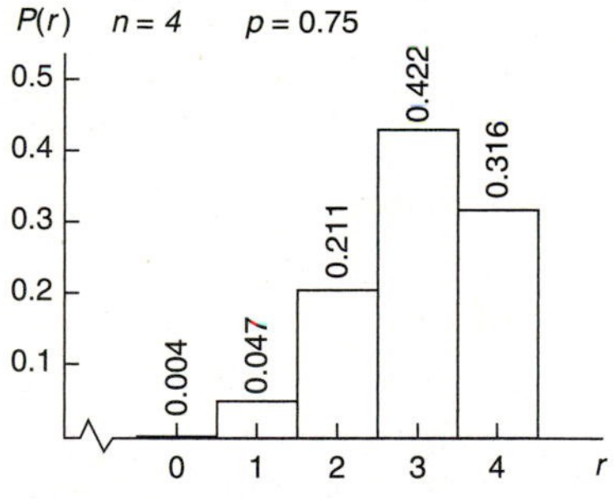

(c) $\mu = 3$; $\sigma \approx 0.866$. (d) $n = 7$. Note that $n = 7$ gives $P(r \geq 3) = 0.987$.

21. $n = 12$; $p = 0.25$ do not serve; $p = 0.75$ serve. (a) $P(r = 12 \text{ serve}) = 0.032$. (b) $P(r \geq 6 \text{ do not serve}) = 0.053$. (c) For serving, $\mu = 9$; $\sigma = 1.50$. (d) To be at least 95.9% sure that 12 are available to serve, call 20.

23. $n = 6$; $p = 0.80$ do not solve; $p = 0.20$ solve. (a) $P(r = 6 \text{ not solved}) = 0.262$. (b) $P(r \geq 1 \text{ solved}) = 0.738$. (c) For solving crime, $\mu = 1.2$; $\sigma \approx 0.98$. (d) To be 90% sure of solving one or more crimes, investigate $n = 11$ crimes.

25. (a) $P(r = 7 \text{ guilty in U.S.}) = 0.028$; $P(r = 7 \text{ guilty in Japan}) = 0.698$. (b) For guilty in Japan, $\mu = 6.65$; $\sigma \approx 0.58$; for guilty in U.S., $\mu = 4.2$; $\sigma \approx 1.30$. (c) To be 99% sure of at least two guilty convictions in the U.S., look at $n = 8$ trials. To be 99% sure of at least two guilty convictions in Japan, look at $n = 3$ trials.

27. (a) 9. (b) 10.

Section 5.4

1. Geometric distribution.
3. No, $n = 50$ is not large enough.
5. 0.144.
7. $\lambda = 8$; 0.1396.
9. (a) $p = 0.77$; $P(n) = (0.77)(0.23)^{n-1}$. (b) $P(1) = 0.77$. (c) $P(2) = 0.1771$. (d) $P(3 \text{ or more tries}) = 1 - P(1) - P(2) = 0.0529$. (e) 1.29, or 1.
11. (a) $P(n) = (0.80)(0.20)^{n-1}$. (b) $P(1) = 0.8$; $P(2) = 0.16$; $P(3) = 0.032$. (c) $P(n \geq 4) = 1 - P(1) - P(2) - P(3) = 1 - 0.8 - 0.16 - 0.032 = 0.008$. (d) $P(n) = (0.04)(0.96)^{n-1}$; $P(1) = 0.04$; $P(2) = 0.0384$; $P(3) = 0.0369$; $P(n \geq 4) = 0.8847$.
13. (a) $P(n) = (0.30)(0.70)^{n-1}$. (b) $P(3) = 0.147$. (c) $P(n > 3) = 1 - P(1) - P(2) - P(3) = 1 - 0.300 - 0.210 - 0.147 = 0.343$. (d) 3.33, or 3.
15. (a) $\lambda = (1.7/10) \times (3/3) = 5.1$ per 30-minute interval; $P(r) = e^{-5.1}(5.1)^r/r!$. (b) Using Table 4 of Appendix II with $\lambda = 5.1$, we find $P(4) = 0.1719$; $P(5) = 0.1753$; $P(6) = 0.1490$. (c) $P(r \geq 4) = 1 - P(0) - P(1) - P(2) - P(3) = 1 - 0.0061 - 0.0311 - 0.0793 - 0.1348 = 0.7487$. (d) $P(r < 4) = 1 - P(r \geq 4) = 1 - 0.7487 = 0.2513$.
17. (a) Births and deaths occur somewhat rarely in a group of 1000 people in a given year. For 1000 people, $\lambda = 16$ births; $\lambda = 8$ deaths. (b) By Table 4 of Appendix II, $P(10 \text{ births}) = 0.0341$; $P(10 \text{ deaths}) = 0.0993$; $P(16 \text{ births}) = 0.0992$; $P(16 \text{ deaths}) = 0.0045$. (c) $\lambda(\text{births}) = (16/1000) \times (1500/1500) = 24$ per 1500 people. $\lambda(\text{deaths}) = (8/1000) \times (1500/1500) = 12$ per 1500 people. By the table, $P(10 \text{ deaths}) = 0.1048$; $P(16 \text{ deaths}) = 0.0543$. Since $\lambda = 24$ is not in the table, use the formula for $P(r)$ to find $P(10 \text{ births}) = 0.00066$; $P(16 \text{ births}) = 0.02186$. (d) $\lambda(\text{births}) = (16/1000) \times (750/750) = 12$ per 750 people. $\lambda(\text{deaths}) = (8/1000) \times (750/750) = 6$ per 750 people. By Table 4 of Appendix II, $P(10 \text{ births}) = 0.1048$; $P(10 \text{ deaths}) = 0.0413$; $P(16 \text{ births}) = 0.0543$; $P(16 \text{ deaths}) = 0.0003$.
19. (a) The Poisson distribution is a good choice for r because gale-force winds occur rather rarely. The occurrences are usually independent. (b) For interval of 108 hours, $\lambda = (1/60) \times (108/108) = 1.8$ per 108 hours. Using Table 4 of Appendix II, we find that $P(2) = 0.2678$; $P(3) = 0.1607$; $P(4) = 0.0723$; $P(r < 2) = P(0) + P(1) = 0.1653 + 0.2975 = 0.4628$.

(c) For interval of 180 hours, $\lambda = (1/60) \times (180/180) =$ 3 per 180 hours. Table 4 of Appendix II gives $P(3) = 0.2240$; $P(4) = 0.1680$; $P(5) = 0.1008$; $P(r < 3) = P(0) + P(1) + P(2) = 0.0498 + 0.1494 + 0.2240 = 0.4232$.

21. (a) The sales of large buildings are rare events. It is reasonable to assume that they are independent. The variable r = number of sales in a fixed time interval. (b) For a 60-day period, $\lambda = (8/275) \times (60/60) = 1.7$ per 60 days. By Table 4 of Appendix II, $P(0) = 0.1827$; $P(1) = 0.3106$; $P(r \geq 2) = 1 - P(0) - P(1) = 0.5067$. (c) For a 90-day period, $\lambda = (8/275) \times (90/90) = 2.6$ per 90 days. By Table 4 of Appendix II, $P(0) = 0.0743$; $P(2) = 0.2510$; $P(r \geq 3) = 1 - P(0) - P(1) - P(2) = 1 - 0.0743 - 0.1931 - 0.2510 = 0.4816$.

23. (a) The problem satisfies the conditions for a binomial experiment with small $p = 0.0018$ and large $n = 1000$. $np = 1.8$, which is less than 10, so the Poisson approximation to the binomial distribution would be a good choice. $\lambda = np = 1.8$. (b) By Table 4, Appendix II, $P(0) = 0.1653$. (c) $P(r > 1) = 1 - P(0) - P(1) = 1 - 0.1653 - 0.2975 = 0.5372$. (d) $P(r > 2) = 1 - P(0) - P(1) - P(2) = 1 - 0.1653 - 0.2975 - 0.2678 = 0.2694$. (e) $P(r > 3) = 1 - P(0) - P(1) - P(2) - P(3) = 1 - 0.1653 - 0.2975 - 0.2678 - 0.1607 = 0.1087$.

25. (a) The problem satisfies the conditions for a binomial experiment with n large, $n = 175$, and p small. $np = (175)(0.005) = 0.875 < 10$. The Poisson distribution would be a good approximation to the binomial. $n = 175$; $p = 0.005$; $\lambda = np \approx 0.9$. (b) By Table 4 of Appendix II, $P(0) = 0.4066$. (c) $P(r \geq 1) = 1 - P(0) = 0.5934$. (d) $P(r \geq 2) = 1 - P(0) - P(1) = 0.2275$.

27. (a) $n = 100$; $p = 0.02$; $r = 2$; $P(2) = C_{100,2}(0.02)^2 (0.98)^{98} \approx 0.2734$. (b) $\lambda = np = 2$; $P(2) = [e^{-2}(2)^2]/2! \approx 0.2\ 707$. (c) The approximation is correct to two decimal places. (d) $n = 100$; $p = 0.02$; $r = 3$. By the formula for the binomial distribution, $P(3) \approx 0.1823$. By the Poisson approximation, $P(3) \approx 0.1804$. The approximation is correct to two decimal places.

29. (a) $\lambda \approx 3.4$. (b) $P(r \geq 4 \mid r \geq 2) = P(r \geq 4)/P(r \geq 2) \approx 0.4416/0.8531 \approx 0.5176$. (c) $P(r < 6 \mid r \geq 3) = P(3 \leq r < 6)/P(r \geq 3) \approx 0.5308/0.6602 \approx 0.8040$.

31. (a) $P(n) = C_{n-1,11}(0.80^{12})(0.20^{n-12})$. (b) $P(12) \approx 0.0687$; $P(13) \approx 0.1649$; $P(14) \approx 0.2144$. (c) 0.4480. (d) 0.5520. (e) $\mu = 15$; $\sigma \approx 1.94$. Susan can expect to get the bonus if she makes 15 contacts, with a standard deviation of about 2 contacts.

Chapter 5 Review

1. A description of all distinct possible values of a random variable x, with a probability assignment $P(x)$ for each value or range of values. $0 \leq P(x) \leq 1$ and $\Sigma P(x) = 1$.

3. (a) Yes. $\mu = 2$ and $\sigma \approx 1.3$. Numbers of successes above 5.25 are unusual. (b) No. It would be unusual to get more than five questions correct.

5. (a) 38; 11.6.
(b) Duration of Leases in Months

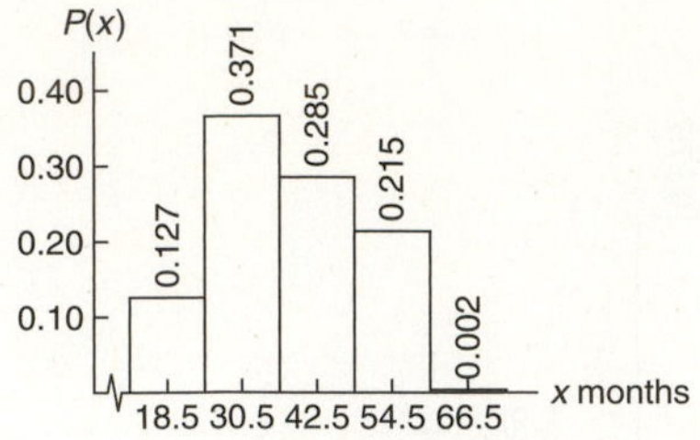

7. (a) Number of Claimants Under 25

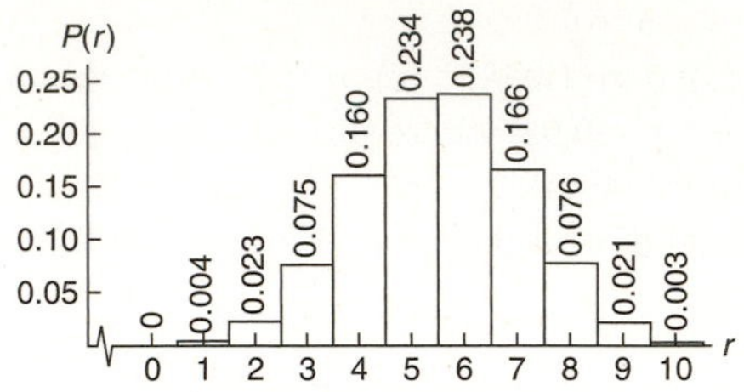

(b) $P(r \geq 6) = 0.504$. (c) $\mu = 5.5$; $\sigma \approx 1.57$.

9. (a) 0.039. (b) 0.403. (c) 8.

11. (a) Number of Good Grapefruit

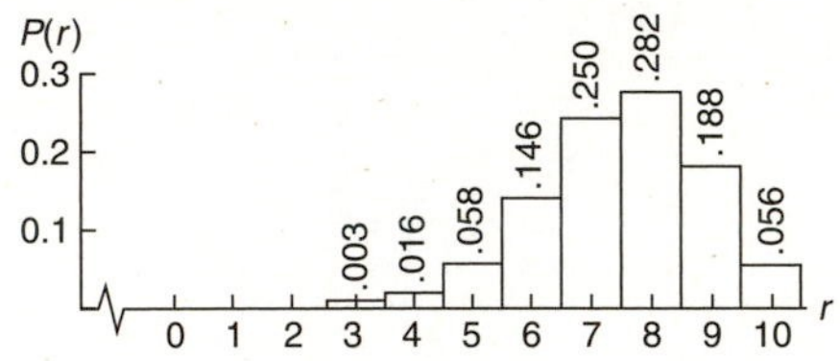

(b) 0.244, 0.999. (c) 7.5. (d) 1.37.

13. $P(r \leq 2) = 0.000$ (to three digits). The data seem to indicate that the percent favoring the increase in fees is less than 85%.

15. (a) Coughs are a relatively rare occurrence. It is reasonable to assume that they are independent events, and the variable is the number of coughs in a fixed time interval. (b) $\lambda = 11$ coughs per minute; $P(r \leq 3) = P(0) + P(1) + P(2) + P(3) = 0.000 + 0.002 + 0.0010 + 0.0037 = 0.0049$. (c) $\lambda = (11/1) \times (0.5/0.5) = 5.5$ coughs per 30-second period. $P(r \geq 3) = 1 - P(0) - P(1) - P(2) = 1 - 0.0041 - 0.0225 - 0.0618 = 0.9116$.

17. The loan-default problem satisfies the conditions for a binomial experiment. Moreover, p is small, n is large, and $np < 10$. Use of the Poisson approximation to the binomial distribution is appropriate. $n = 300$; $p = 1/350 \approx 0.0029$; and $\lambda = np \approx 300(0.0029) = 0.86 \approx 0.9$; $P(r \geq 2) = 1 - P(0) - P(1) = 1 - 0.4066 - 0.3659 = 0.2275$.

19. (a) Use the geometric distribution with $p = 0.5$. $P(n = 2) = (0.5)(0.5) = 0.25$. As long as you toss the coin at least twice, it does not matter how many more times you toss it. To get the first head on the second toss, you must get a tail on the first and a head on the second. (b) $P(n = 4) = (0.5)(0.5)^3 = 0.0625$; $P(n > 4) = 1 - P(1) - P(2) - P(3) - P(4) = 1 - 0.5 - 0.5^2 - 0.5^3 - 0.5^4 = 0.0625$.

CHAPTER 6

Section 6.1

1. (a) No, it's skewed. (b) No, it crosses the horizontal axis. (c) No, it has three peaks. (d) No, the curve is not smooth.
3. Figure 6-12 has the larger standard deviation. The mean of Figure 6-12 is $\mu = 10$. The mean of Figure 6-13 is $\mu = 4$.
5. (a) 50%. (b) 68%. (c) 99.7%.
7. (a) 50%. (b) 50%. (c) 68%. (d) 95%.
9. (a) From 1207 to 1279. (b) From 1171 to 1315. (c) From 1135 to 1351.
11. (a) From 1.70 mA to 4.60 mA. (b) From 0.25 mA to 6.05 mA.
13. (a) Tri-County Bank Monthly Loan Request—First Year (thousands of dollars)

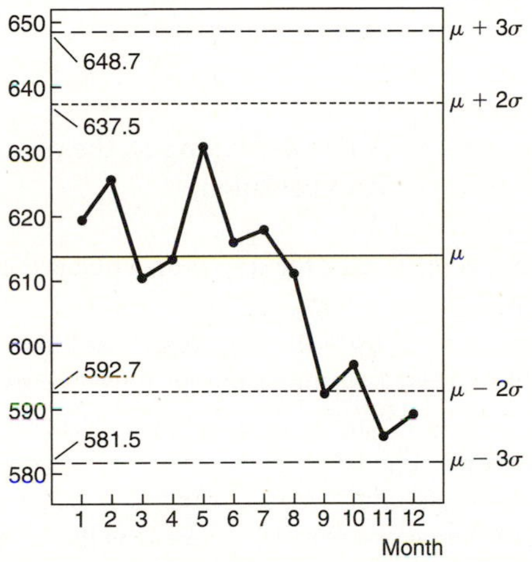

The process is out of control with a type III warning signal, since two of three consecutive points are more than 2 standard deviations below the mean. The trend is down.

(b) Tri-County Bank Monthly Loan Requests—Second Year (thousands of dollars)

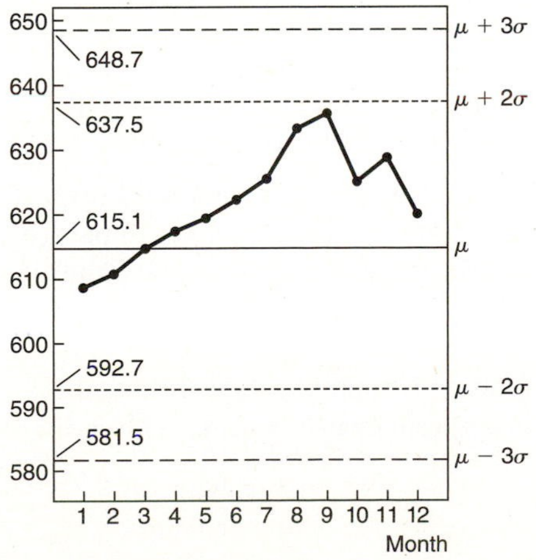

The process shows warning signal II, a run of nine consecutive points above the mean. The economy is probably heating up.

15. Visibility Standard Index

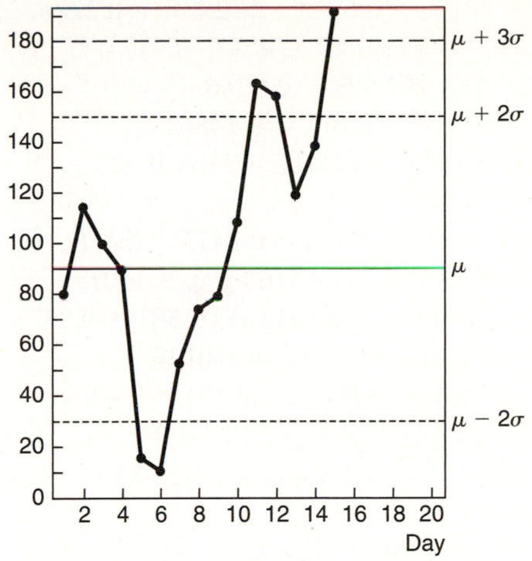

There is one point above $\mu + 3\sigma$. Thus control signal I indicates "out of control." Control signal III is present. There are two consecutive points below $\mu - 2\sigma$ and two consecutive points above $\mu + 2\sigma$. The out-of-control signals that cause the most concern are those above the mean. Special pollution regulations may be appropriate for those periods.

17. (a) 0.8000. (b) 0.7000. (c) 0.5000. (d) $\mu = 0$; $\sigma \approx 0.0289$. Since $\mu = 0$, the measurements are unbiased.
19. (a) 0.4493. (b) 0.8454. (c) 0.1857. (d) 120.71.

Section 6.2

1. The number of standard deviations from the mean.
3. 0.
5. (a) −1. (b) 2.4. (c) 20. (d) 36.5.
7. They are the same, since both are 1 standard deviation below the mean.
9. (a) Robert, Juan, and Linda each scored above the mean. (b) Joel scored on the mean. (c) Susan and Jan scored below the mean. (d) Robert, 172; Juan, 184; Susan, 110; Joel, 150; Jan, 134; Linda, 182.
11. (a) $-1.00 < z$. (b) $z < -2.00$. (c) $-2.67 < z < 2.33$. (d) $x < 4.4$. (e) $5.2 < x$. (f) $4.1 < x < 4.5$. (g) A red blood cell count of 5.9 or higher corresponds to a standard z score of 3.67. Practically no data values occur this far above the mean. Such a count would be considered unusually high for a healthy female.

13. 0.5000. 15. 0.0934. 17. 0.6736. 19. 0.0643.
21. 0.8888. 23. 0.4993. 25. 0.8953. 27. 0.3471.
29. 0.0306. 31. 0.5000. 33. 0.4483. 35. 0.8849.
37. 0.0885. 39. 0.8849. 41. 0.8808. 43. 0.3226.
45. 0.4474. 47. 0.2939. 49. 0.6704.

Section 6.3

1. 0.50.
3. Negative.
5. $P(3 \le x \le 6) = P(-0.50 \le z \le 1.00) = 0.5328$.
7. $P(50 \le x \le 70) = P(0.67 \le z \le 2.00) = 0.2286$.
9. $P(8 \le x \le 12) = P(-2.19 \le z \le -0.94) = 0.1593$.
11. $P(x \ge 30) = P(z \ge 2.94) = 0.0016$.
13. $P(x \ge 90) = P(z \ge -0.67) = 0.7486$.
15. -1.555. 17. 0.13. 19. 1.41. 21. -0.92.
23. ± 2.33.
25. (a) $P(x > 60) = P(z > -1) = 0.8413$. (b) $P(x < 110) = P(z < 1) = 0.8413$. (c) $P(60 \le x \le 110) = P(-1.00 \le z \le 1.00) = 0.8413 - 0.1587 = 0.6826$. (d) $P(x > 140) = P(z > 2.20) = 0.0139$.
27. (a) $P(x < 3.0 \text{ mm}) = P(z < -2.33) = 0.0099$. (b) $P(x > 7.0 \text{ mm}) = P(z > 2.11) = 0.0174$. (c) $P(3.0 \text{ mm} < x < 7.0 \text{ mm}) = P(-2.33 < z < 2.11) = 0.9727$.
29. (a) $P(x < 36 \text{ months}) = P(z < -1.13) = 0.1292$. The company will replace 13% of its batteries. (b) $P(z < z_0) = 10\%$ for $z_0 = -1.28$; $x = -1.28(8) + 45 = 34.76$. Guarantee the batteries for 35 months.
31. (a) According to the empirical rule, about 95% of the data lies between $\mu - 2\sigma$ and $\mu + 2\sigma$. Since this interval is 4σ wide, we have $4\sigma \approx 6$ years, so $\sigma \approx 1.5$ years. (b) $P(x > 5) = P(z > -2.00) = 0.9772$. (c) $P(x < 10) = P(z < 1.33) = 0.9082$. (d) $P(z < z_0) = 0.10$ for $z_0 = -1.28$; $x = -1.28(1.5) + 8 = 6.08$ years. Guarantee the TVs for about 6.1 years.
33. (a) $\sigma \approx 12$ beats/minute. (b) $P(x < 25) = P(z < -1.75) = 0.0401$. (c) $P(x > 60) = P(z > 1.17) = 0.1210$. (d) $P(25 \le x \le 60) = P(-1.75 \le z \le 1.17) = 0.8389$. (e) $P(z \le z_0) = 0.90$ for $z_0 = 1.28$; $x = 1.28(12) + 46 = 61.36$ beats/minute. A heart rate of 61 beats/minute corresponds to the 90% cutoff point of the distribution.
35. (a) $P(z \ge z_0) = 0.99$ for $z_0 = -2.33$; $x = -2.33(3.7) + 90 \approx 81.38$ months. Guarantee the microchips for 81 months. (b) $P(x \le 84) = P(z \le -1.62) = 0.0526$. (c) Expected loss $= (50{,}000{,}000)(0.0526) = \$2{,}630{,}000$. (d) Profit $= \$370{,}000$.
37. (a) $z = 1.28$; $x \approx 4.9$ hours. (b) $z = -1.04$; $x \approx 2.9$ hours. (c) Yes; work and/or school schedules may be different on Saturday.
39. (a) In general, $P(A \mid B) = P(A \text{ and } B)/P(B)$; $P(x > 20) = P(z > 0.50) = 0.3085$; $P(x > 15) = P(z > -0.75) = 0.7734$; $P(x > 20 \mid x > 15) = 0.3989$. (b) $P(x > 25) = P(z > 1.75) = 0.0401$; $P(x > 18) = P(z > 0.00) = 0.5000$; $P(x > 25 \mid x > 18) = 0.0802$. (c) Use event $A = x > 20$ and event $B = x > 15$ in the formula.

Section 6.4

1. A set of measurements or counts either existing or conceptual. For example, the population of ages of all people in Colorado; the population of weights of all students in your school; the population count of all antelope in Wyoming.
3. A numerical descriptive measure of a population, such as μ, the population mean; σ, the population standard deviation; or σ^2, the population variance.
5. A statistical inference is a conclusion about the value of a population parameter. We will do both estimation and testing.
7. They help us visualize the sampling distribution through tables and graphs that approximately represent the sampling distribution.
9. We studied the sampling distribution of mean trout lengths based on samples of size 5. Other such sampling distributions abound.

Section 6.5

Note: Answers may differ slightly depending on the number of digits carried in the standard deviation.

1. The standard deviation.
3. $\bar{x}$ is an unbiased estimator for μ; $\hat{p}$ is an unbiased estimator for p.
5. (a) Normal; $\mu_{\bar{x}} = 8$; $\sigma_{\bar{x}} = 2$. (b) 0.50. (c) 0.3085. (d) No, about 30% of all such samples have means exceeding 9.
7. (a) 30 or more. (b) No.
9. The second. The standard error of the first is $\sigma/10$, while that of the second is $\sigma/15$, where σ is the standard deviation of the original x distribution.
11. (a) $\mu_{\bar{x}} = 15$; $\sigma_{\bar{x}} = 2.0$; $P(15 \le \bar{x} \le 17) = P(0 \le z \le 1.00) = 0.3413$. (b) $\mu_{\bar{x}} = 15$; $\sigma_{\bar{x}} = 1.75$; $P(15 \le \bar{x} \le 17) = P(0 \le z \le 1.14) = 0.3729$. (c) The standard deviation is smaller in part (b) because of the larger sample size. Therefore, the distribution about $\mu_{\bar{x}}$ is narrower in part (b).
13. (a) $P(x < 74.5) = P(z < -0.63) = 0.2643$. (b) $P(\bar{x} < 74.5) = P(z < -2.79) = 0.0026$. (c) No. If the weight of coal in only one car were less than 74.5 tons, we could not conclude that the loader is out of adjustment. If the mean weight of coal for a sample of 20 cars were less than 74.5 tons, we would suspect that the loader is malfunctioning. As we see in part (b), the probability of this happening is very low if the loader is correctly adjusted.
15. (a) $P(x < 40) = P(z < -1.80) = 0.0359$. (b) Since the x distribution is approximately normal, the $\bar{x}$ distribution is approximately normal, with mean 85 and

standard deviation 17.678. $P(\bar{x} < 40) = P(z < -2.55) = 0.0054$. (c) $P(\bar{x} < 40) = P(z < -3.12) = 0.0009$. (d) $P(\bar{x} < 40) = P(z < -4.02) < 0.0002$. (e) Yes; if the average value based on five tests were less than 40, the patient is almost certain to have excess insulin.

17. (a) $P(x < 54) = P(z < -1.27) = 0.1020$. (b) The expected number undernourished is 2200(0.1020), or about 224. (c) $P(\bar{x} \le 60) = P(z \le -2.99) = 0.0014$. (d) $P(\bar{x} < 64.2) = P(z < 1.20) = 0.8849$. Since the sample average is above the mean, it is quite unlikely that the doe population is undernourished.
19. (a) Since x itself represents a sample mean return based on a large (random) sample of stocks, x has a distribution that is approximately normal (central limit theorem). (b) $P(1\% \le \bar{x} \le 2\%) = P(-1.63 \le z \le 1.09) = 0.8105$. (c) $P(1\% \le \bar{x} \le 2\%) = P(-3.27 \le z \le 2.18) = 0.9849$. (d) Yes. The standard deviation decreases as the sample size increases. (e) $P(\bar{x} < 1\%) = P(z < -3.27) = 0.0005$. This is very unlikely if $\mu = 1.6\%$. One would suspect that μ has slipped below 1.6%.
21. (a) The total checkout time for 30 customers is the sum of the checkout times for each individual customer. Thus, $w = x_1 + x_2 + \cdots + x_{30}$, and the probability that the total checkout time for the next 30 customers is less than 90 is $P(w < 90)$. (b) $w < 90$ is equivalent to $x_1 + x_2 + \cdots + x_{30} < 90$. Divide both sides by 30 to get $\bar{x} < 3$ for samples of size 30. Therefore, $P(w < 90) = P(\bar{x} < 3)$. (c) By the central limit theorem, $\bar{x}$ is approximately normal, with $\mu_{\bar{x}} = 2.7$ minutes and $\sigma_{\bar{x}} = 0.1095$ minute. (d) $P(\bar{x} < 3) = P(z < 2.74) = 0.9969$.
23. (a) $P(w > 90) = P(\bar{x} > 18) = P(z > 0.68) = 0.2483$. (b) $P(w < 80) = P(\bar{x} < 16) = P(z < -0.68) = 0.2483$. (c) $P(80 < w < 90) = P(16 < \bar{x} < 18) = P(-0.68 < z < 0.68) = 0.5034$.

Section 6.6

1. $np > 5$ and $nq > 5$, where $q = 1 - p$.
3. (a) Yes, both $np > 5$ and $nq > 5$. (b) $\mu = 20$; $\sigma \approx 3.162$. (c) $r \ge 23$ corresponds to $x \ge 22.5$. (d) $P(r \ge 23) \approx P(x \ge 22.5) \approx P(z \ge 0.79) \approx 0.2148$. (e) No, the probability that this will occur is about 21%.
5. No, $np = 4.3$ and does not satisfy the criterion that $np > 5$.

Note: Answers may differ slightly depending on how many digits are carried in the computation of the standard deviation and z.

7. $np > 5$; $nq > 5$. (a) $P(r \ge 50) = P(x \ge 49.5) = P(z \ge -27.53) \approx 1$, or almost certain. (b) $P(r \ge 50) = P(x \ge 49.5) = P(z \ge 7.78) \approx 0$, or almost impossible for a random sample.
9. $np > 5$; $nq > 5$. (a) $P(r \ge 15) = P(x \ge 14.5) = P(z \ge -2.35) = 0.9906$. (b) $P(r \ge 30) = P(x \ge 29.5) = P(z \ge 0.62) = 0.2676$. (c) $P(25 \le r \le 35) = P(24.5 \le x \le 35.5) = P(-0.37 \le z \le 1.81) = 0.6092$. (d) $P(r > 40) = P(r \ge 41) = P(x \ge 40.5) = P(z \ge 2.80) = 0.0026$.
11. $np > 5$; $nq > 5$. (a) $P(r \ge 47) = P(x \ge 46.5) = P(z \ge -1.94) = 0.9738$. (b) $P(r \le 58) = P(x \le 58.5) = P(z \le 1.75) = 0.9599$. In parts (c) and (d), let r be the number of products that succeed, and use $p = 1 - 0.80 = 0.20$. (c) $P(r \ge 15) = P(x \ge 14.5) = P(z \ge 0.40) = 0.3446$. (d) $P(r < 10) = P(r \le 9) = P(x \le 9.5) = P(z \le -1.14) = 0.1271$.
13. $np > 5$; $nq > 5$. (a) $P(r > 180) = P(x \ge 180.5) = P(z > -1.11) = 0.8665$. (b) $P(r < 200) = P(x \le 199.5) = P(z \le 1.07) = 0.8577$. (c) P(take sample and buy product) $= P$(take sample) $\cdot$ P(buy | take sample) $= 0.222$. (d) $P(60 \le r \le 80) = P(59.5 \le x \le 80.5) = P(-1.47 \le z \le 1.37) = 0.8439$.
15. $np > 5$; $nq > 5$. (a) 0.94. (b) $P(r \le 255)$. (c) $P(r \le 255) = P(x \le 255.5) = P(z \le 1.16) = 0.8770$.
17. $np > 5$ and $nq > 5$.
19. Yes, since the mean of the approximate sampling distribution is $\mu_{\hat{p}} = p$.
21. (a) Yes, both np and nq exceed 5. $\mu_{\hat{p}} = 0.23$; $\sigma_{\hat{p}} \approx 0.042$. (b) No, $np = 4.6$ and does not exceed 5.

Chapter 6 Review

1. Normal probability distributions are distributions of continuous random variables. They are symmetric about the mean and bell-shaped. Most of the data fall within 3 standard deviations of the mean. The mean and median are the same.
3. No.
5. The points lie close to a straight line.
7. $\sigma_{\bar{x}} = \sigma/\sqrt{n}$.
9. (a) A normal distribution. (b) The mean μ of the x distribution. (c) $\sigma/\sqrt{n}$, where σ is the standard deviation of the x distribution. (d) They will both be approximately normal with the same mean, but the standard deviations will be $\sigma/\sqrt{50}$ and $\sigma/\sqrt{100}$, respectively.
11. (a) 0.9821. (b) 0.3156. (c) 0.2977.
13. 1.645.
15. (a) 0.8665. (b) 0.7330.
17. (a) 0.0166. (b) 0.975.
19. (a) 0.9772. (b) 17.3 hours.
21. (a) $P(x \ge 40) = P(z \ge 0.71) = 0.2389$. (b) $P(\bar{x} \ge 40) = P(z \ge 2.14) = 0.0162$.
23. $P(98 \le \bar{x} \le 102) = P(-1.33 \le z \le 1.33) = 0.8164$.
25. (a) Yes, np and nq both exceed 5. (b) $\mu_{\hat{p}} = 0.4$, $\sigma_{\hat{p}} \approx 0.1$

CUMULATIVE REVIEW PROBLEMS

1. The specified ranges of readings are disjoint and cover all possible readings.
2. Essay.
3. Yes; the events constitute the entire sample space.

4. (a) 0.85. (b) 0.70. (c) 0.70. (d) 0.30. (e) 0.15. (f) 0.75. (g) 0.30. (h) 0.05.
5. 0.17
6.

x	$P(x)$
5	0.25
15	0.45
25	0.15
35	0.10
45	0.05

$\mu \approx 17.5$; $\sigma \approx 10.9$.
7. (a) $p = 0.10$. (b) $\mu = 1.2$; $\sigma \approx 1.04$. (c) 0.718. (d) 0.889.
8. (a) 0.05. (b) $P(n) = (0.05)(0.95)^{n-1}$; $n \geq 1$. (c) 0.81.
9. (a) Yes; since $n = 100$ and $np = 5$, the criteria $n \geq 100$ and $np < 10$ are satisfied. $\lambda = 5$. (b) 0.7622. (c) 0.0680.
10. (a) Yes; both np and nq exceed 5. (b) 0.9925. (c) np is too large ($np > 10$) and n is too small ($n < 100$).
11. (a) $\sigma \approx 1.7$. (b) 0.1314. (c) 0.1075.
12. Essay based on material from Chapter 6 and Section 1.2.
13. (a) Because of the large sample size, the central limit theorem describes the $\bar{x}$ distribution (approximately). (b) $P(\bar{x} \leq 6820) = P(z \leq -2.75) = 0.0030$. (c) The probability that the average white blood cell count for 50 healthy adults is as low as or lower than 6820 is very small, 0.0030. Based on this result, it would be reasonable to gather additional facts.
14. (a) Yes, both np and nq exceed 5. (b) $\mu_{\hat{p}} = p = 0.45$; $\sigma_{\hat{p}} \approx 0.09$
15. Essay.

CHAPTER 7

Section 7.1

1. True. By definition, critical values z_c are values such that c% of the area under the normal curve falls between $-z_c$ and z_c.
3. True. By definition, the margin of error is the magnitude of the difference between $\bar{x}$ and μ.
5. False. The maximal margin of error is $E = z_c \dfrac{\sigma}{\sqrt{n}}$. As the sample size n increases, the maximal error decreases, resulting in a shorter confidence interval for μ.
7. False. The maximal error of estimate E controls the length of the confidence interval regardless of the value of $\bar{x}$.
9. μ is either in the interval 10.1 to 12.2 or not. Therefore, the probability that μ is in this interval is either 0 or 1, not 0.95.
11. (a) Yes, the x distribution is normal and σ is known so the $\bar{x}$ distribution is also normal. (b) 47.53 to 52.47. (c) You are 90% confident that the confidence interval computed is one that contains μ.
13. (a) 217. (b) Yes, by the central limit theorem.
15. (a) 3.04 gm to 3.26 gm; 0.11 gm. (b) Distribution of weights is normal with known σ. (c) There is an 80% chance that the confidence interval is one of the intervals that contain the population average weight of Allen's hummingbirds in this region. (d) $n = 28$.
17. (a) 34.62 ml/kg to 40.38 ml/kg; 2.88 ml/kg. (b) The sample size is large (30 or more) and σ is known. (c) There is a 99% chance that the confidence interval is one of the intervals that contain the population average blood plasma level for male firefighters. (d) $n = 60$.
19. (a) 125.7 to 151.3 larceny cases; 12.8 larceny cases. (b) 123.3 to 153.7 larceny cases; 15.2 larceny cases. (c) 118.4 to 158.6 larceny cases; 20.1 larceny cases. (d) Yes. (e) Yes.
21. (a) 26.64 to 33.36; 3.36. (b) 27.65 to 32.35; 2.35. (c) 28.43 to 31.57; 1.57. (d) Yes. (e) Yes.
23. (a) The mean rounds to the value given. (b) Using the rounded value of part (a), the 75% interval is from 34.19 thousand to 37.81 thousand. (c) Yes; 30 thousand dollars is below the lower bound of the 75% confidence interval. We can say with 75% confidence that the mean lies between 34.19 thousand and 37.81 thousand. (d) Yes; 40 thousand is above the upper bound of the 75% confidence interval. (e) 33.41 thousand to 38.59 thousand. We can say with 90% confidence that the mean lies between 33.4 thousand and 38.6 thousand dollars. 30 thousand is below the lower bound and 40 thousand is above the upper bound.
25. (a) 92.5°C to 101.5°C. (b) The balloon will go up.

Section 7.2

1. 2.110.
3. 1.721.
5. $t = 0$.
7. $n = 10$, with $d.f. = 9$.
9. Shorter. For $d.f. = 40$, z_c is less than t_c and the resulting margin of error E is smaller.
11. (a) Yes, since x has a mound-shaped distribution. (b) 9.12 to 10.88. (c) There is a 90% chance that the confidence interval you computed is one of the confidence intervals that contain μ.
13. (a) The mean and standard deviation round to the values given. (b) Using the rounded values for the mean and standard deviation given in part (a), the interval is from 1249 to 1295. (c) We are 90% confident that the computed interval is one that contains the population mean for the tree-ring date.
15. (a) Use a calculator. (b) 74.7 pounds to 107.3 pounds. (c) We are 75% confident that the computed interval is one that contains the population mean weight of adult mountain lions in the region.
17. (a) The mean and standard deviation round to the given values. (b) 8.41 to 11.49. (c) Since all values in the 99.9% confidence interval are above 6, we can be almost certain that this patient no longer has a calcium deficiency.

19. (a) Boxplots differ in length of interquartile box, location of median, and length of whiskers. The boxplots come from different samples. (b) Yes; yes; for 95% confidence intervals, we expect about 95% of the samples to generate intervals that contain the mean of the population.
21. (a) The mean and standard deviation round to the given values. (b) 21.6 to 28.8. (c) 19.4 to 31.0. (d) Using both confidence intervals, we can say that the P/E for Bank One is well below the population average. The P/E for AT&T Wireless is well above the population average. The P/E for Disney is within both confidence intervals. It appears that the P/E for Disney is close to the population average P/E. (e) By the central limit theorem, when n is large, the $\bar{x}$ distribution is approximately normal. In general, $n \geq 30$ is considered large.
23. (a) $d.f. = 30$; 43.59 to 46.82; 43.26 to 47.14; 42.58 to 47.81. (b) 43.63 to 46.77; 43.33 to 47.07; 42.74 to 47.66. (c) Yes; the respective intervals based on the Student's t distribution are slightly longer. (d) For Student's t, $d.f. = 80$; 44.22 to 46.18; 44.03 to 46.37; 43.65 to 46.75. For standard normal, 44.23 to 46.17; 44.05 to 46.35; 43.68 to 46.72. The intervals using the t distribution are still slightly longer than the corresponding intervals using the standard normal distribution. However, with a larger sample size, the differences between the two methods are less pronounced.

Section 7.3

1. $\hat{p} = r/n$.
3. (a) No. (b) The difference between $\hat{p}$ and p. In other words, the margin of error is the difference between results based on a random sample and results based on a population.
5. No, Jerry does not have a random sample of all laptops. In fact, he does not even have a random sample of laptops from the computer science class. Also, because all the laptops he tested for spyware are those of students from the same computer class, it could be that students shared software with classmates and spread the infection among the laptops owned by the students of the class.
7. (a) $n\hat{p} = 30$ and $n\hat{q} = 70$, so both products exceed 5. Also, the trials are binomial trials. (b) 0.225 to 0.375. (c) You are 90% confident that the confidence interval you computed is one of the intervals that contain p.
9. (a) 73. (b) 97.
11. (a) $\hat{p} = 39/62 \approx 0.6290$. (b) 0.51 to 0.75. If this experiment were repeated many times, about 95% of the intervals would contain p. (c) Both np and nq are greater than 5. If either is less than 5, the normal curve will not necessarily give a good approximation to the binomial.
13. (a) $\hat{p} = 1619/5222 \approx 0.3100$. (b) 0.29 to 0.33. If we repeat the survey with many different samples of 5222 dwellings, about 99% of the intervals will contain p. (c) Both np and nq are greater than 5. If either is less than 5, the normal curve will not necessarily give a good approximation to the binomial.
15. (a) $\hat{p} = 0.5420$. (b) 0.53 to 0.56. (c) Yes. Both np and nq are greater than 5.
17. (a) $\hat{p} = 0.0304$. (b) 0.02 to 0.05. (c) Yes. Both np and nq are greater than 5.
19. (a) $\hat{p} = 0.8603$. (b) 0.84 to 0.89. (c) A recent study shows that 86% of women shoppers remained loyal to their favorite supermarket last year. The margin of error was 2.5 percentage points.
21. (a) $\hat{p} = 0.25$. (b) 0.22 to 0.28. (c) A survey of 1000 large corporations has shown that 25% will choose a nonsmoking job candidate over an equally qualified smoker. The margin of error was 2.7%.
23. (a) Estimate a proportion; 208. (b) 68.
25. (a) Estimate a proportion; 666. (b) 662.
27. (a) $1/4 - (p - 1/2)^2 = 1/4 - (p^2 - p + 1/4) = -p^2 + p = p(1 - p)$. (b) Since $(p - 1/2)^2 \geq 0$, then $1/4 - (p - 1/2)^2 \leq 1/4$ because we are subtracting $(p - 1/2)^2$ from 1/4.

Section 7.4

1. Two random samples are independent if sample data drawn from one population are completely unrelated to the selection of sample data from the other population.
3. Josh's, because the critical value t_c is smaller based on larger $d.f.$; Kendra's, because her value for t_c is larger.
5. $\mu_1 < \mu_2$.
7. (a) Normal distribution by Theorem 7.1 and the fact that the samples are independent and the population standard deviations are known. (b) $E \approx 1.717$; interval from -3.717 to -0.283. (c) Student's t distribution with $d.f. = 19$, based on the fact that the original distributions are normal and the samples are independent. (d) $t_{0.90} = 1.729$; $E \approx 1.805$; interval from -3.805 to -0.195. (e) $d.f. \approx 42.85$; interval from -3.755 to -0.245. (f) Since the 90% confidence interval contains all negative values, you can be 90% confident that μ_1 is less than μ_2.
9. (a) Yes, $n_1\hat{p}_1$, $n_1\hat{q}_1$, $n_2\hat{p}_2$, $n_2\hat{q}_2$ all exceed 5. (b) $\hat{\sigma} \approx 0.0943$; $E \approx 0.155$; -0.205 to 0.105. (c) No, the 90% confidence interval contains both negative and positive values.
11. (a) Use a calculator. (b) $d.f. \approx 11$; $E \approx 129.9$; interval from -121.3 to 138.5 ppm. (c) Because the interval contains both positive and negative numbers, we cannot say at the 90% confidence level that one region is more interesting than the other. (d) Student's t because σ_1 and σ_2 are unknown.
13. (a) Use a calculator. (b) $d.f. \approx 15$; $E \approx 5.42$; interval from 12.64% to 23.48% foreign revenue. (c) Because the interval contains only positive values, we can say at the 85% confidence level that technology companies have

a higher population mean percentage foreign revenue. (d) Student's t because σ_1 and σ_2 are unknown.

15. (a) Use a calculator. (b) $d.f. \approx 39$; to use Table 6, round down to $d.f. \approx 35$; $E \approx 0.125$; interval from -0.399 to -0.149 feet. (c) Since the interval contains all negative numbers, it seems that at the 90% confidence level the population mean height of pro football players is less than that of pro basketball players. (d) Student's t distribution because σ_1 and σ_2 are unknown. Both samples are large, so no assumptions about the original distributions are needed.
17. (a) Yes, the sample sizes, number of successes, and number of failures are sufficiently large. (b) $\hat{\sigma} \approx 0.0232$; $E = 0.0599$; the interval is from 0.67 to 0.79. (c) The confidence interval contains values that are all positive, so we can be 99% sure that $p_1 > p_2$.
19. (a) Normal distribution since the sample sizes are sufficiently large and both σ_1 and σ_2 are known. (b) $E = 0.3201$; the interval is from -9.12 to -8.48. (c) The interval consists of negative values only. At the 99% confidence level, we can conclude that $\mu_1 < \mu_2$.
21. (a) Yes, the sample sizes, number of successes, and number of failures are sufficiently large. (b) $\hat{p}_1 = 0.3095$; $\hat{p}_2 = 0.1184$; $\hat{\sigma} = 0.0413$; interval from 0.085 to 0.297. (c) The interval contains numbers that are all positive. A greater proportion of hogans exist in Fort Defiance.
23. (a) Use a calculator. (b) Student's t distribution because the population standard deviations are unknown. In addition, since the original distributions may not be normal, the sample sizes are too small. (c) $d.f. \approx 9$; $E \approx 5.3$; 3.7 to 14.3 pounds. (d) Interval contains all positive values. At the 85% confidence level, it appears that the population mean weight of gray wolves in Chihuahua is greater than that of gray wolves in Durango.
25. (a) -1.35 to 2.39. (b) 0.06 to 3.86. (c) -0.61 to 3.49. (d) At the 85% confidence level, we can say that the mean index of self-esteem based on competence is greater than the mean index of self-esteem based on physical attractiveness. We cannot conclude that there is a difference between the mean index of self-esteem based on competence and that based on social acceptance. We also cannot conclude that there is a difference in the mean indices based on social acceptance and physical attractiveness.
27. (a) Based on the same data, a 99% confidence interval is longer than a 95% confidence interval. Therefore, if the 95% confidence interval has both positive and negative values, so will the 99% confidence interval. However, for the same data, a 90% confidence interval is shorter than a 95% confidence interval. The 90% confidence interval might contain only positive or only negative values even if the 95% interval contains both. (b) Based on the same data, a 99% confidence interval is longer than a 95% confidence interval. Even if the 95% confidence interval contains values that are all positive, the longer 99% interval could contain both positive and negative values. Since, for the same data, a 90% confidence interval is shorter than a 95% confidence interval, if the 95% confidence interval contains only positive values, so will the 90% confidence interval.
29. (a) $n = 896.1$, or 897 couples in each sample. (b) $n = 768.3$, or 769 couples in each sample.
31. (a) Pooled standard deviation $s \approx 8.6836$; interval from 3.9 to 14.1. (b) The pooled standard deviation method has a shorter interval and a larger $d.f.$

Chapter 7 Review

1. See text.
3. (a) No, the probability that μ is in the interval is either 0 or 1. (b) Yes, 99% confidence intervals are constructed in such a way that 99% of all such confidence intervals based on random samples of the designated size will contain μ.
5. Interval for a mean; 176.91 to 180.49.
7. Interval for a mean. (a) Use a calculator. (b) 64.1 to 84.3.
9. Interval for a proportion; 0.50 to 0.54.
11. Interval for a proportion. (a) $\hat{p} = 0.4072$. (b) 0.333 to 0.482.
13. Difference of means. (a) Use a calculator. (b) $d.f. \approx 71$; to use Table 6, round down to $d.f. \approx 70$; $E \approx 0.83$; interval from -0.06 to 1.6. (c) Because the interval contains both positive and negative values, we cannot conclude at the 95% confidence level that there is any difference in soil water content between the two fields. (d) Student's t distribution because σ_1 and σ_2 are unknown. Both samples are large, so no assumptions about the original distributions are needed.
15. Difference of means. (a) $d.f. \approx 17$; $E \approx 2.5$; interval from 5.5 to 10.5 pounds. (b) Yes, the interval contains values that are all positive. At the 75% level of confidence, it appears that the average weight of adult male wolves from the Northwest Territories is greater.
17. Difference of proportions. (a) $\hat{p}_1 = 0.8495$; $\hat{p}_2 = 0.8916$; -0.1409 to 0.0567. (b) The interval contains both negative and positive numbers. We do not detect a difference in the proportions at the 95% confidence level.
19. (a) $P(A_1 < \mu_1 < B_1$ *and* $A_2 < \mu_2 < B_2) = (0.80)(0.80) = 0.64$. The complement of the event $A_1 < \mu_1 < B_1$ *and* $A_2 < \mu_2 < B_2$ is that either μ_1 is not in the first interval or μ_2 is not in the second interval, or both. Thus, $P(\text{at least one interval fails}) = 1 - P(A_1 < \mu_1 < B_1$ *and* $A_2 < \mu_2 < B_2) = 1 - 0.64 = 0.36$. (b) Suppose $P(A_1 < \mu_1 < B_1) = c$ and $P(A_2 < \mu_2 < B_2) = c$. If we want the probability that both hold to be 90%, and if

x_1 and x_2 are independent, then $P(A_1 < \mu_1 < B_1$ *and* $A_2 < \mu_2 < B_2) = 0.90$ means $P(A_1 < \mu_1 < B_1) \cdot P(A_2 < \mu_2 < B_2) = 0.90$, so $c^2 = 0.90$, or $c = 0.9487$. (c) In order to have a high probability of success for the whole project, the probability that each component will perform as specified must be significantly higher.

CHAPTER 8

Section 8.1

1. See text.
3. No, if we fail to reject the null hypothesis, we have not proved it beyond all doubt. We have failed only to find sufficient evidence to reject it.
5. Level of significance; α; type I.
7. Fail to reject H_0
9. 0.0184.
11. (a) H_0: $\mu = 40$. (b) H_1: $\mu \neq 40$. (c) H_1: $\mu > 40$. (d) H_1: $\mu < 40$.
13. (a) Yes, because x has a normal distribution. (b) $z \approx 1.12$. (c) 0.2628. (d) Fail to reject H_0 because P-value $> \alpha$.
15. (a) H_0: $\mu = 60$ kg. (b) H_1: $\mu < 60$ kg. (c) H_1: $\mu > 60$ kg. (d) H_1: $\mu \neq 60$ kg. (e) For part (b), the P-value area region is on the left. For part (c), the P-value area is on the right. For part (d), the P-value area is on both sides of the mean.
17. (a) H_0: $\mu = 16.4$ feet. (b) H_1: $\mu > 16.4$ feet. (c) H_1: $\mu < 16.4$ feet. (d) H_1: $\mu \neq 16.4$ feet. (e) For part (b), the P-value area is on the right. For part (c), the P-value area is on the left. For part (d), the P-value area is on both sides of the mean.
19. (a) $\alpha = 0.01$; H_0: $\mu = 4.7\%$; H_1: $\mu > 4.7\%$; right-tailed. (b) Normal; $\bar{x} = 5.38$; $z \approx 0.90$. (c) P-value ≈ 0.1841; on standard normal curve, shade area to the right of 0.90. (d) P-value of $0.1841 > 0.01$ for α; fail to reject H_0. (e) Insufficient evidence at the 0.01 level to reject claim that average yield for bank stocks equals average yield for all stocks.
21. (a) $\alpha = 0.01$; H_0: $\mu = 4.55$ grams; H_1: $\mu < 4.55$ grams; left-tailed. (b) Normal; $\bar{x} = 3.75$ grams; $z \approx -2.80$. (c) P-value ≈ 0.0026; on standard normal curve, shade area to the left of -2.80. (d) P-value of $0.0026 \leq 0.01$ for α; reject H_0. (e) The sample evidence is sufficient at the 0.01 level to justify rejecting H_0. It seems that the hummingbirds in the Grand Canyon region have a lower average weight.
23. (a) $\alpha = 0.01$; H_0: $\mu = 11\%$; H_1: $\mu \neq 11\%$; two-tailed. (b) Normal; $\bar{x} = 12.5\%$; $z = 1.20$. (c) P-value $= 2(0.1151) = 0.2302$; on standard normal curve, shade areas to the right of 1.20 and to the left of -1.20. (d) P-value of $0.2302 > 0.01$ for α; fail to reject H_0. (e) There is insufficient evidence at the 0.01 level to reject H_0. It seems that the average hail damage to wheat crops in Weld County matches the national average.

Section 8.2

1. The P-value for a two-tailed test of μ is twice that for a one-tailed test, based on the same sample data and null hypothesis.
3. $d.f. = n - 1$.
5. Yes. When P-value < 0.01, it is also true that P-value < 0.05.
7. (a) $0.010 < P$-value < 0.020; technology gives P-value ≈ 0.0150. (b) $0.005 < P$-value < 0.010; technology gives P-value ≈ 0.0075.
9. (a) Yes, since the original distribution is mound-shaped and symmetric and σ is unknown; $d.f. = 24$. (b) H_0: $\mu = 9.5$; H_1: $\mu \neq 9.5$. (c) $t \approx 1.250$. (d) $0.200 < P$-value < 0.250; technology gives P-value ≈ 0.2234. (e) Fail to reject H_0 because the entire interval containing the P-value > 0.05 for α. (f) The sample evidence is insufficient at the 0.05 level to reject H_0.
11. (a) $\alpha = 0.01$; H_0: $\mu = 16.4$ feet; H_1: $\mu > 16.4$ feet. (b) Normal; $z \approx 1.54$. (c) P-value ≈ 0.068; on standard normal curve, shade area to the right of $z \approx 1.54$. (d) P-value of $0.0618 > 0.01$ for α; fail to reject H_0. (e) At the 1% level, there is insufficient evidence to say that the average storm level is increasing.
13. (a) $\alpha = 0.01$; H_0: $\mu = 1.75$ years; H_1: $\mu > 1.75$ years. (b) Student's t, $d.f. = 45$; $t \approx 2.481$. (c) $0.005 < P$-value < 0.010; on t graph, shade area to the right of 2.481. From TI-84, P-value ≈ 0.0084. (d) Entire P-interval ≤ 0.01 for α; reject H_0. (e) At the 1% level of significance, the sample data indicate that the average age of the Minnesota region coyotes is higher than 1.75 years.
15. (a) $\alpha = 0.05$; H_0: $\mu = 19.4$; H_1: $\mu \neq 19.4$ (b) Student's t, $d.f. = 35$; $t \approx -1.731$. (c) $0.050 < P$-value < 0.100; on t graph, shade area to the right of 1.731 and to the left of -1.731. From TI-84, P-value ≈ 0.0923. (d) P-value interval > 0.05 for α; fail to reject H_0. (e) At the 5% level of significance, the sample evidence does not support rejecting the claim that the average P/E of socially responsible funds is different from that of the S&P stock index.
17. i. Use a calculator. Rounded values are used in part ii. ii. (a) $\alpha = 0.05$; H_0: $\mu = 4.8$; H_1: $\mu < 4.8$. (b) Student's t, $d.f. = 5$; $t \approx -3.499$. (c) $0.005 < P$-value < 0.010; on t graph, shade area to the left of -3.499. From TI-84, P-value ≈ 0.0086. (d) P-value interval ≤ 0.05 for α; reject H_0. (e) At the 5% level of significance, sample evidence supports the claim that the average RBC count for this patient is less than 4.8.
19. i. Use a calculator. Rounded values are used in part ii. ii. (a) $\alpha = 0.01$; H_0: $\mu = 67$; H_1: $\mu \neq 67$. (b) Student's t, $d.f. = 15$; $t \approx -1.962$. (c) $0.050 < P$-value < 0.100; on t graph, shade area to the right of 1.962 and to the left of -1.962. From TI-84, P-value ≈ 0.0686. (d) P-value interval > 0.01; fail to reject H_0. (e) At the 1% level of significance, the sample evidence does not support a claim that the average thickness of slab avalanches in Vail is different from that in Canada.

21. i. Use a calculator. Rounded values are used in part ii. ii. (a) $\alpha = 0.05$; H_0: $\mu = 8.8$; H_1: $\mu \neq 8.8$. (b) Student's t, $d.f. = 13$; $t \approx -1.337$. (c) $0.200 < P\text{-value} < 0.250$; on t graph, shade area to the right of 1.337 and to the left of -1.337. From TI-84, P-value ≈ 0.2042. (d) P-value interval > 0.05; fail to reject H_0. (e) At the 5% level of significance, we cannot conclude that the average catch is different from 8.8 fish per day.

23. (a) The P-value of a one-tailed test is smaller. For a two-tailed test, the P-value is doubled because it includes the area in both tails. (b) Yes; the P-value of a one-tailed test is smaller, so it might be smaller than α, whereas the P-value of a corresponding two-tailed test may be larger than α. (c) Yes; if the two-tailed P-value is less than α, the smaller one-tail area is also less than α. (d) Yes, the conclusions can be different. The conclusion based on the two-tailed test is more conservative in the sense that the sample data must be more extreme (differ more from H_0) in order to reject H_0.

25. (a) For $\alpha = 0.01$, confidence level $c = 0.99$; interval from 20.28 to 23.72; hypothesized $\mu = 20$ is not in the interval; reject H_0. (b) H_0: $\mu = 20$; H_1: $\mu \neq 20$; $z = 3.000$; P-value ≈ 0.0026; P-value of $0.0026 \leq 0.01$ for α; reject H_0; conclusions are the same.

27. Critical value $z_0 = 2.33$; critical region is values to the right of 2.33; since the sample statistic $z = 1.54$ is not in the critical region, fail to reject H_0. At the 1% level, there is insufficient evidence to say that the average storm level is increasing. Conclusion is same as with P-value method.

29. Critical value is $t_0 = 2.412$ for one-tailed test with $d.f. = 45$; critical region is values to the right of 2.412. Since the sample test statistic $t = 2.481$ is in the critical region, reject H_0. At the 1% level, the sample data indicate that the average age of Minnesota region coyotes is higher than 1.75 years. Conclusion is same as with P-value method.

Section 8.3

1. For the conditions $np > 5$ and $nq > 5$, use the value of p from H_0. Note that $q = 1 - p$.

3. Yes. The corresponding P-value for a one-tailed test is half that for a two-tailed test, so the P-value of the one-tailed test is also less than 0.01.

5. (a) Yes, np and nq are both greater than 5. (b) H_0: $p = 0.50$; H_1: $p \neq 0.50$. (c) $\hat{p} = 0.40$; $z \approx -1.10$. (d) 0.2714. (e) Fail to reject H_0 because P-value of $0.2714 > 0.05$ for α. (f) The sample $\hat{p}$ value based on 30 trials is not sufficiently different from 0.50 to justify rejecting H_0 for $\alpha = 0.05$.

7. i. (a) $\alpha = 0.01$; H_0: $p = 0.301$; H_1: $p < 0.301$. (b) Standard normal; yes, $np \approx 64.7 > 5$ and $nq \approx 150.3 > 5$; $\hat{p} \approx 0.214$; $z \approx -2.78$. (c) P-value $\approx$ 0.0027; on standard normal curve, shade area to the left of -2.78. (d) P-value of $0.0027 \leq 0.01$ for α; reject H_0. (e) At the 1% level of significance, the sample data indicate that the population proportion of numbers with a leading "1" in the revenue file is less than 0.301, predicted by Benford's Law.
ii. Yes; the revenue data file seems to include more numbers with higher first nonzero digits than Benford's Law predicts.
iii. We have not proved H_0 to be false. However, because our sample data led us to reject H_0 and to conclude that there are too few numbers with a leading digit of 1, more investigation is merited.

9. (a) $\alpha = 0.01$; H_0: $p = 0.70$; H_1: $p \neq 0.70$. (b) Standard normal; $\hat{p} = 0.75$; $z \approx 0.62$. (c) P-value = 2(0.2676) = 0.5352; on standard normal curve, shade areas to the right of 0.62 and to the left of -0.62. (d) P-value of $0.5352 > 0.01$ for α; fail to reject H_0. (e) At the 1% level of significance, we cannot say that the population proportion of arrests of males aged 15 to 34 in Rock Springs is different from 70%.

11. (a) $\alpha = 0.01$; H_0: $p = 0.77$; H_1: $p < 0.77$. (b) Standard normal; $\hat{p} \approx 0.5556$; $z \approx -2.65$. (c) P-value ≈ 0.004; on standard normal curve, shade area to the left of -2.65. (d) P-value of $0.004 \leq 0.01$ for α; reject H_0. (e) At the 1% level of significance, the data show that the population proportion of driver fatalities related to alcohol is less than 77% in Kit Carson County.

13. (a) $\alpha = 0.01$; H_0: $p = 0.50$; H_1: $p < 0.50$. (b) Standard normal; $\hat{p} \approx 0.2941$; $z \approx -2.40$. (c) P-value = 0.0082; on standard normal curve, shade region to the left of -2.40. (d) P-value of $0.0082 \leq 0.01$ for α; reject H_0. (e) At the 1% level of significance, the data indicate that the population proportion of female wolves is now less than 50% in the region.

15. (a) $\alpha = 0.01$; H_0: $p = 0.261$; H_1: $p \neq 0.261$. (b) Standard normal; $\hat{p} \approx 0.1924$; $z \approx -2.78$. (c) P-value = 2(0.0027) = 0.0054; on standard normal curve, shade area to the right of 2.78 and to the left of -2.78. (d) P-value of $0.0054 \leq 0.01$ for α; reject H_0. (e) At the 1% level of significance, the sample data indicate that the population proportion of the five-syllable sequence is different from that of Plato's *Republic*.

17. (a) $\alpha = 0.01$; H_0: $p = 0.47$; H_1: $p > 0.47$. (b) Standard normal; $\hat{p} \approx 0.4871$; $z \approx 1.09$. (c) P-value = 0.1379; on standard normal curve, shade area to the right of 1.09. (d) P-value of $0.1379 > 0.01$ for α; fail to reject H_0. (e) At the 1% level of significance, there is insufficient evidence to support the claim that the population proportion of customers loyal to Chevrolet is more than 47%.

19. (a) $\alpha = 0.05$; H_0: $p = 0.092$; H_1: $p > 0.092$. (b) Standard normal; $\hat{p} \approx 0.1480$; $z \approx 2.71$. (c) P-value = 0.0034; on standard normal curve, shade region to the right of 2.71. (d) P-value of $0.0034 \leq 0.05$ for α; reject H_0. (e) At the 5% level of significance, the data indicate that the population proportion of students with hypertension during final exams week is higher than 9.2%.

21. (a) $\alpha = 0.01$; $H_0: p = 0.82$; $H_1: p \neq 0.82$. (b) Standard normal; $\hat{p} \approx 0.7671$; $z \approx -1.18$. (c) P-value $= 2(0.1190) = 0.2380$; on standard normal curve, shade area to the right of 1.18 and to the left of -1.18.
(d) P-value of $0.2380 > 0.01$ for α; fail to reject H_0.
(e) At the 1% level of significance, the evidence is insufficient to indicate that the population proportion of extroverts among college student government leaders is different from 82%.

23. Critical value is $z_0 = -2.33$. The critical region consists of values less than -2.33. The sample test statistic $z = -2.65$ is in the critical region, so we reject H_0. This result is consistent with the P-value conclusion.

Section 8.4

1. Paired data are dependent.
3. $H_0: \mu_d = 0$; that is, the mean of the differences is 0, so there is no difference.
5. $d.f. = n - 1$.
7. (a) Yes. The sample size is sufficiently large. Student's t with $d.f. = 35$. (b) $H_0: \mu_d = 0$; $H_1: \mu_d \neq 0$. (c) $t = 2.400$ with $d.f. = 35$. (d) $0.020 < P$-value < 0.050. TI-84 gives P-value ≈ 0.0218. (e) Reject H_0 since the entire interval containing the P-value < 0.05 for α.
(f) At the 5% level of significance and for a sample size of 36, the sample mean of the differences is sufficiently different from 0 that we conclude the population mean of the differences is not zero.
9. (a) $\alpha = 0.05$; $H_0: \mu_d = 0$; $H_1: \mu_d \neq 0$. (b) Student's t, $d.f. = 7$; $\bar{d} \approx 2.25$; $t \approx 0.818$. (c) $0.250 < P$-value < 0.500; on t graph, shade area to the left of -0.818 and to the right of 0.818. From TI-84, P-value ≈ 0.4402.
(d) P-value interval > 0.05 for α; fail to reject H_0.
(e) At the 5% level of significance, the evidence is insufficient to claim a difference in population mean percentage increases for corporate revenue and CEO salary.
11. (a) $\alpha = 0.01$; $H_0: \mu_d = 0$; $H_1: \mu_d > 0$. (b) Student's t, $d.f. = 4$; $\bar{d} \approx 12.6$; $t \approx 1.243$. (c) $0.125 < P$-value < 0.250; on t graph, shade area to the right of 1.243. From TI-84, P-value ≈ 0.1408. (d) P-value interval > 0.01 for α; fail to reject H_0. (e) At the 1% level of significance, the evidence is insufficient to claim that the average peak wind gusts are higher in January.
13. (a) $\alpha = 0.05$; $H_0: \mu_d = 0$; $H_1: \mu_d > 0$. (b) Student's t, $d.f. = 7$; $\bar{d} \approx 6.125$; $t \approx 1.762$. (c) $0.050 < P$-value < 0.075; on t graph, shade area to the right of 1.762. From TI-84, P-value ≈ 0.0607. (d) P-value interval > 0.05 for α; fail to reject H_0. (e) At the 5% level of significance, the evidence is insufficient to indicate that the population average percentage of male wolves in winter is higher.
15. (a) $\alpha = 0.05$; $H_0: \mu_d = 0$; $H_1: \mu_d > 0$. (b) Student's t, $d.f. = 7$; $\bar{d} \approx 6.0$; $t \approx 0.788$. (c) $0.125 < P$-value < 0.250; on t graph, shade area to the right of 0.788. From TI-84, P-value ≈ 0.2282. (d) P-value interval > 0.05 for α; fail to reject H_0. (e) At the 5% level of significance, the evidence is insufficient to show that the population mean number of inhabited houses is greater than that of hogans.
17. i. Use a calculator. Nonrounded results are used in part ii.
ii. (a) $\alpha = 0.05$; $H_0: \mu_d = 0$; $H_1: \mu_d > 0$. (b) Student's t, $d.f. = 35$; $\bar{d} \approx 2.472$; $t \approx 1.223$. (c) $0.100 < P$-value < 0.125; on t graph, shade area to the right of 1.223. From TI-84, P-value ≈ 0.1147. (d) P-value interval > 0.05 for α; fail to reject H_0. (e) At the 5% level of significance, the evidence is insufficient to claim that the population mean cost of living index for housing is higher than that for groceries.
19. (a) $\alpha = 0.05$; $H_0: \mu_d = 0$; $H_1: \mu_d > 0$. (b) Student's t, $d.f. = 8$; $\bar{d} = 2.0$; $t \approx 1.333$. (c) $0.100 < P$-value < 0.125; on t graph, shade area to the right of 1.333. From TI-84, P-value ≈ 0.1096. (d) P-value interval > 0.05 for α; fail to reject H_0. (e) At the 5% level of significance, the evidence is insufficient to claim that the population score on the last round is higher than that on the first.
21. (a) $\alpha = 0.05$; $H_0: \mu_d = 0$; $H_1: \mu_d > 0$. (b) Student's t, $d.f. = 7$; $\bar{d} \approx 0.775$; $t \approx 2.080$. (c) $0.025 < P$-value < 0.050; on t graph, shade area to the right of 2.080. From TI-84, P-value ≈ 0.0380. (d) P-value interval ≤ 0.05 for α; reject H_0. (e) At the 5% level of significance, the evidence is sufficient to claim that the population mean time for rats receiving larger rewards to climb the ladder is less.
23. For a two-tailed test with $\alpha = 0.05$ and $d.f. = 7$, the critical values are $\pm t_0 = \pm 2.365$. The sample test statistic $t = 0.818$ is between -2.365 and 2.365, so we do not reject H_0. This conclusion is the same as that reached by the P-value method.

Section 8.5

1. (a) H_0 says that the population means are equal.
(b) $$z = \frac{\bar{x}_1 - \bar{x}_2}{\sqrt{\frac{\sigma_1^2}{n_1} + \frac{\sigma_2^2}{n_2}}}.$$
(c) $$t = \frac{\bar{x}_1 - \bar{x}_2}{\sqrt{\frac{s_1^2}{n_1} + \frac{s_2^2}{n_2}}}$$ with $d.f.$ = smaller sample size - 1 or $d.f.$ is from Satterthwaite's formula.
3. $H_0: \mu_1 = \mu_2$ or $H_0: \mu_1 - \mu_2 = 0$.
5. $$\bar{p} = \frac{r_1 + r_2}{n_1 + n_2}.$$
7. $H_1: \mu_1 > \mu_2$; $H_1: \mu_1 - \mu_2 > 0$.
9. (a) Student's t with $d.f. = 48$. Samples are independent, population standard deviations are not known, and sample sizes are sufficiently large. (b) $H_0: \mu_1 = \mu_2$; $H_1: \mu_1 \neq \mu_2$. (c) $\bar{x}_1 - \bar{x}_2 = -2$; $t \approx -3.037$.
(d) $0.0010 < P$-value < 0.010 (using $d.f. = 45$ and Table 6). TI-84 gives P-value ≈ 0.0030 with

$d.f. \approx 110.96$. (e) Because the entire interval containing the P-value < 0.01 for α, reject H_0. (f) At the 1% level of significance, the sample evidence is sufficiently strong to reject H_0 and conclude that the population means are different.

11. (a) Standard normal. Samples are independent, population standard deviations are known, and sample sizes are sufficiently large. (b) $H_0: \mu_1 = \mu_2$; $H_1: \mu_1 \neq \mu_2$. (c) $\bar{x}_1 - \bar{x}_2 = -2$; $z \approx -3.04$. (d) 0.0024. (e) P-value $0.0024 < 0.01$ for α, reject H_0. (f) At the 1% level of significance, the sample evidence is sufficiently strong to reject H_0 and conclude that the population means are different.

13. (a) $\bar{p} \approx 0.657$. (b) Standard normal distribution because $n_1\bar{p}$, $n_1\bar{q}$, $n_2\bar{p}$, $n_2\bar{q}$ are each greater than 5. (c) $H_0: p_1 = p_2$; $H_1: p_1 \neq p_2$ (d) $\hat{p}_1 - \hat{p}_2 = -0.1$; $z \approx -1.38$. (e) P-value ≈ 0.1676. (f) Since P-value of $0.1676 \geq 0.05$ for α, fail to reject H_0. (g) At the 5% level of significance, the difference between the sample probabilities of success for the two binomial experiments is too small to justify rejecting the hypothesis that the probabilities are equal.

15. (a) $\alpha = 0.01$; $H_0: \mu_1 = \mu_2$; $H_1: \mu_1 > \mu_2$. (b) Standard normal; $\bar{x}_1 - \bar{x}_2 = 0.7$; $z \approx 2.57$. (c) P-value $= P(z > 2.57) \approx 0.0051$; on standard normal curve, shade area to the right of 2.57. (d) P-value of $0.0051 \leq 0.01$ for α; reject H_0. (e) At the 1% level of significance, the evidence is sufficient to indicate that the population mean REM sleep time for children is more than that for adults.

17. (a) $\alpha = 0.05$; $H_0: \mu_1 = \mu_2$; $H_1: \mu_1 \neq \mu_2$. (b) Standard normal; $\bar{x}_1 - \bar{x}_2 = 0.6$; $z \approx 2.16$. (c) P-value $= 2P(z > 2.16) \approx 2(0.0154) = 0.0308$; on standard normal curve, shade area to the right of 2.16 and to the left of -2.16. (d) P-value of $0.0308 \leq 0.05$ for α; reject H_0. (e) At the 5% level of significance, the evidence is sufficient to show that there is a difference between mean responses regarding preference for camping or fishing.

19. i. Use rounded results to compute t.
 ii. (a) $\alpha = 0.01$; $H_0: \mu_1 = \mu_2$; $H_1: \mu_1 < \mu_2$. (b) Student's t, $d.f. = 9$; $\bar{x}_1 - \bar{x}_2 = -0.36$; $t \approx -0.965$. (c) $0.125 < P$-value < 0.250; on t graph, shade area to the left of -0.965. From TI-84, $d.f. \approx 19.96$; P-value ≈ 0.1731. (d) P-value interval > 0.01 for α; do not reject H_0. (e) At the 1% level of significance, the evidence is insufficient to indicate that the violent crime rate in the Rocky Mountain region is higher than that in New England.

21. (a) $\alpha = 0.05$; $H_0: \mu_1 = \mu_2$; $H_1: \mu_1 \neq \mu_2$. (b) Student's t, $d.f. = 29$; $\bar{x}_1 - \bar{x}_2 = -9.7$; $t \approx -0.751$. (c) $0.250 < P$-value < 0.500; on t graph, shade area to the right of 0.751 and to the left of -0.751. From TI-84, $d.f. \approx 57.92$; P-value ≈ 0.4556. (d) P-value interval > 0.05 for α; do not reject H_0. (e) At the 5% level of significance, the evidence is insufficient to indicate that there is a difference between the control and experimental groups in the mean score on the vocabulary portion of the test.

23. i. Use rounded results to compute t.
 ii. (a) $\alpha = 0.05$; $H_0: \mu_1 = \mu_2$; $H_1: \mu_1 \neq \mu_2$. (b) Student's t, $d.f. = 14$; $\bar{x}_1 - \bar{x}_2 = 0.82$; $t \approx 0.869$. (c) $0.250 < P$-value < 0.500; on t graph, shade area to the right of 0.869 and to the left of -0.869. From TI-84, $d.f. \approx 28.81$; P-value ≈ 0.3940. (d) P-value interval > 0.05 for α; do not reject H_0. (e) At the 5% level of significance, the evidence is insufficient to indicate that there is a difference in the mean number of cases of fox rabies between the two regions.

25. i. Use rounded results to compute t.
 ii. (a) $\alpha = 0.05$; $H_0: \mu_1 = \mu_2$; $H_1: \mu_1 \neq \mu_2$. (b) Student's t, $d.f. = 6$; $\bar{x}_1 - \bar{x}_2 = -1.64$; $t \approx -1.041$. (c) $0.250 < P$-value < 0.500; on t graph, shade area to the right of 1.041 and to the left of -1.041. From TI-84, $d.f. \approx 12.28$; P-value ≈ 0.3179. (d) P-value interval > 0.05 for α; do not reject H_0. (e) At the 5% level of significance, the evidence is insufficient to indicate that the mean time lost due to hot tempers is different from that lost due to technical workers' attitudes.

27. (a) $d.f. = 19.96$ (Some software will truncate this to 19.) (b) $d.f. = 9$; the convention of using the smaller of $n_1 - 1$ and $n_2 - 1$ leads to a $d.f.$ that is always less than or equal to that computed by Satterthwaite's formula.

29. (a) $\alpha = 0.05$; $H_0: p_1 = p_2$; $H_1: p_1 \neq p_2$. (b) Standard normal; $\bar{p} \approx 0.2911$; $\hat{p}_1 - \hat{p}_2 \approx -0.052$; $z \approx -1.13$. (c) P-value $\approx 2P(z < -1.13) \approx 2(0.1292) = 0.2584$ on standard normal curve, shade area to the right of 1.13 and to the left of -1.13. (d) P-value of $0.2584 > 0.05$ for α; fail to reject H_0. (e) At the 5% level of significance, there is insufficient evidence to conclude that the population proportion of women favoring more tax dollars for the arts is different from the proportion of men.

31. (a) $\alpha = 0.01$; $H_0: p_1 = p_2$; $H_1: p_1 \neq p_2$. (b) Standard normal; $\bar{p} \approx 0.0676$; $\hat{p}_1 - \hat{p}_2 \approx 0.0237$; $z \approx 0.79$. (c) P-value $\approx 2P(z > 0.79) \approx 2(0.2148) = 0.4296$; on standard normal curve, shade area to the right of 0.79 and to the left of -0.79. (d) P-value of $0.4296 > 0.01$ for α; fail to reject H_0. (e) At the 1% level of significance, there is insufficient evidence to conclude that the population proportion of high school dropouts on Oahu is different from that of Sweetwater County.

33. (a) $\alpha = 0.01$; $H_0: p_1 = p_2$; $H_1: p_1 < p_2$. (b) Standard normal; $\bar{p} = 0.42$; $\hat{p}_1 - \hat{p}_2 = -0.10$; $z \approx -1.43$. (c) P-value $\approx P(z < -1.43) \approx 0.0764$; on standard normal curve, shade area to the left of -1.43. (d) P-value of $0.0764 > 0.01$ for α; fail to reject H_0. (e) At the 1% level of significance, there is insufficient

evidence to conclude that the population proportion of adults who believe in extraterrestrials and who attended college is higher than the proportion who believe in extraterrestrials but did not attend college.

35. (a) $\alpha = 0.05$; $H_0: p_1 = p_2$; $H_1: p_1 < p_2$. (b) Standard normal; $\bar{p} \approx 0.2189$; $\hat{p}_1 - \hat{p}_2 \approx -0.074$; $z \approx -2.04$. (c) P-value $\approx P(z < -2.04) \approx 0.0207$; on standard normal curve, shade area to the left of -2.04. (d) P-value of $0.0207 \le 0.05$ for α; reject H_0. (e) At the 5% level of significance, there is sufficient evidence to conclude that the population proportion of trusting people in Chicago is higher for the older group.

37. $H_0: \mu_1 = \mu_2$; $H_1: \mu_1 < \mu_2$; for $d.f. = 9$, $\alpha = 0.01$ in the *one-tail area* row, the critical value is $t_0 = -2.821$; sample test statistic $t = -0.965$ is not in the critical region; fail to reject H_0. This result is consistent with that obtained by the P-value method.

Chapter 8 Review

1. Look at the original x distribution. If it is normal or $n \ge 30$, and σ is known, use the standard normal distribution. If the x distribution is mound-shaped or $n \ge 30$, and σ is unknown, use the Student's t distribution. The $d.f.$ is determined by the application.

3. A larger sample size increases the $|z|$ or $|t|$ value of the sample test statistic.

5. Single mean. (a) $\alpha = 0.05$; $H_0: \mu = 11.1$; $H_1: \mu \ne 11.1$. (b) Standard normal; $z = -3.00$. (c) P-value $= 0.0026$; on standard normal curve, shade area to the right of 3.00 and to the left of -3.00. (d) P-value of $0.0026 \le 0.05$ for α; reject H_0. (e) At the 5% level of significance, the evidence is sufficient to say that the miles driven per vehicle in Chicago is different from the national average.

7. Single mean. (a) $\alpha = 0.01$; $H_0: \mu = 0.8$; $H_1: \mu > 0.8$. (b) Student's t, $d.f. = 8$; $t \approx 4.390$. (c) $0.0005 < P$-value < 0.005; on t graph, shade area to the right of 4.390. From TI-84, P-value ≈ 0.0012. (d) P-value interval ≤ 0.01 for α; reject H_0. (e) At the 1% level of significance, the evidence is sufficient to say that the Toylot claim of 0.8 A is too low.

9. Single proportion. (a) $\alpha = 0.01$; $H_0: p = 0.60$; $H_1: p < 0.60$. (b) Standard normal; $z = -3.01$. (c) P-value $= 0.0013$; on standard normal curve, shade area to the left of -3.01. (d) P-value of $0.0013 \le 0.01$ for α; reject H_0. (e) At the 1% level of significance, the evidence is sufficient to show that the mortality rate has dropped.

11. Single mean. (a) $\alpha = 0.01$; $H_0: \mu = 40$; $H_1: \mu > 40$. (b) Standard normal; $z = 3.34$. (c) P-value $= 0.0004$; on standard normal curve, shade area to the right of 3.34. (d) P-value of $0.0004 \le 0.01$ for α; reject H_0. (e) At the 1% level of significance, the evidence is sufficient to say that the population average number of matches is larger than 40.

13. Difference of means. (a) $\alpha = 0.05$; $H_0: \mu_1 = \mu_2$; $H_1: \mu_1 \ne \mu_2$. (b) Student's t, $d.f. = 50$; $\bar{x}_1 - \bar{x}_2 = 0.3$ cm; $t \approx 1.808$. (c) $0.050 < P$-value < 0.100; on t graph, shade area to the right of 1.808 and to the left of -1.808. From TI-84, $d.f. \approx 100.27$, P-value ≈ 0.0735. (d) P-value interval > 0.05 for α; do not reject H_0. (e) At the 5% level of significance, the evidence is insufficient to indicate a difference in population mean length between the two types of projectile points.

15. Single mean. (a) $\alpha = 0.05$; $H_0: \mu = 7$ oz; $H_1: \mu \ne 7$ oz. (b) Student's t, $d.f. = 7$; $t \approx 1.697$. (c) $0.100 < P$-value < 0.150; on t graph, shade area to the right of 1.697 and to the left of -1.697. From TI-84, P-value ≈ 0.1335. (d) P-value interval > 0.05 for α; do not reject H_0. (e) At the 5% level of significance, the evidence is insufficient to show that the population mean amount of coffee per cup is different from 7 oz.

17. Paired difference test. (a) $\alpha = 0.05$; $H_0: \mu_d = 0$; $H_1: \mu_d < 0$. (b) Student's t, $d.f. = 4$; $\bar{d} \approx -4.94$; $t = -2.832$. (c) $0.010 < P$-value < 0.025; on t graph, shade area to the left of -2.832. From TI-84, P-value ≈ 0.0236. (d) P-value interval ≤ 0.05 for α; reject H_0. (e) At the 5% level of significance, there is sufficient evidence to claim that the population average net sales improved.

CHAPTER 9

Section 9.1

1. Explanatory variable is placed along horizontal axis, usually x axis. Response variable is placed along vertical axis, usually y axis.

3. Decreases.

5. (a) Moderate. (b) None. (c) High.

7. (a) No. (b) Increasing population might be a lurking variable causing both variables to increase.

9. (a) No. (b) One lurking variable responsible for average annual income increases is inflation. Better training might be a lurking variable responsible for shorter times to run the mile.

11. The correlation coefficient is moderate and negative. It suggests that as gasoline prices increase, consumption decreases, and the relationship is moderately linear. It is risky to apply these results to gasoline prices much higher than \$5.30 per gallon. It could be that many of the discretionary and technical means of reducing consumption have already been applied, so consumers cannot reduce their consumption much more.

13. (a) Ages and Average Weights of Shetland Ponies

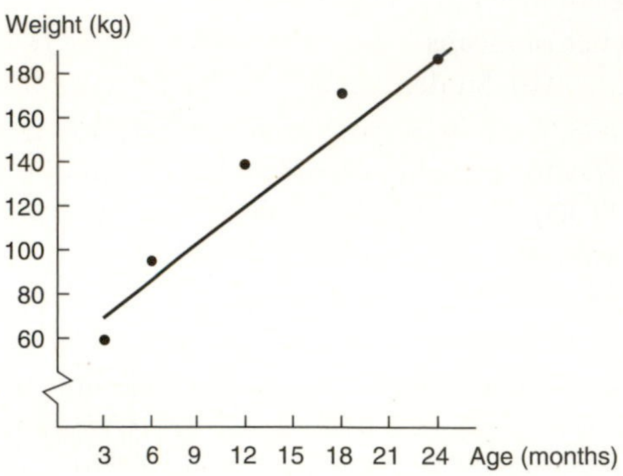

Line slopes upward.
(b) Strong; positive. (c) $r \approx 0.972$; increase.

15. (a) Lowest Barometric Pressure and Maximum Wind Speed for Tropical Cyclones

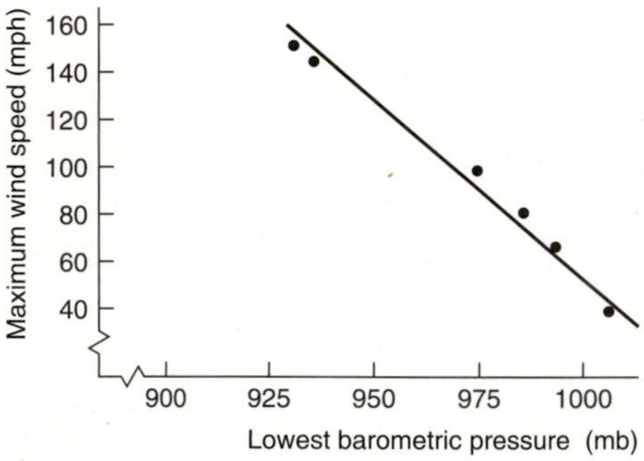

Line slopes downward.
(b) Strong; negative. (c) $r \approx -0.990$; decrease.

17. (a) Batting Average and Home Run Percentage

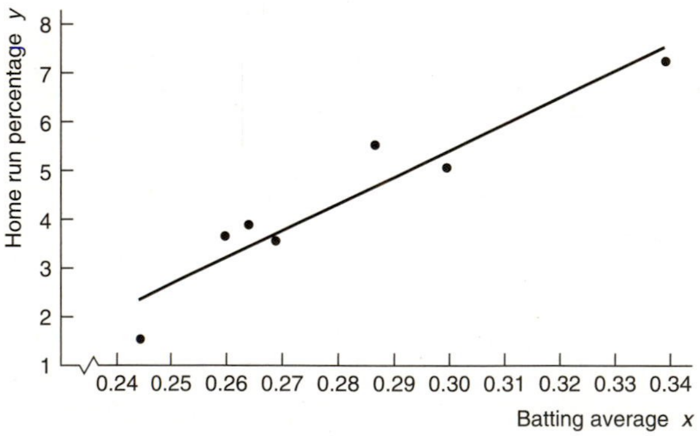

Line slopes upward.
(b) High; positive. (c) $r \approx 0.948$; increase.

19. (a) Unit Length on y Same as That on x

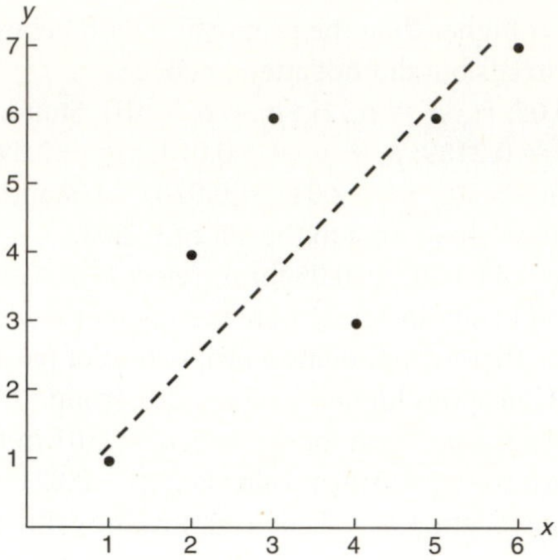

(b) Unit Length on y Twice That on x

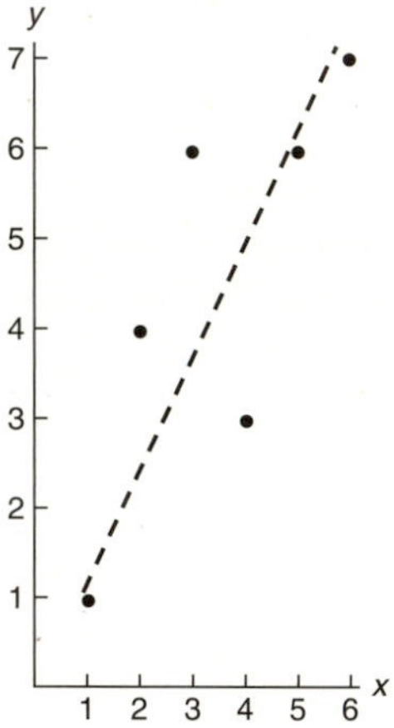

(c) Unit Length on y Half That on x

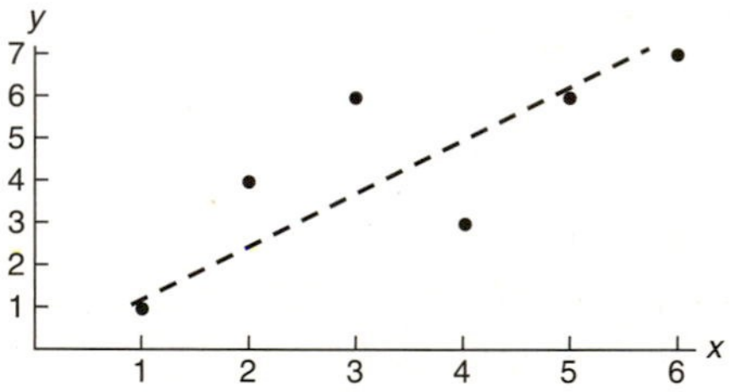

(d) The line in part (b) appears steeper than the line in part (a), whereas the line in part (c) appears flatter than the line in part (a). The slopes actually are all the same, but the lines look different because of the change in unit lengths on the y and x axes.

21. (a) $r \approx 0.972$ with $n = 5$ is significant for $\alpha = 0.05$. For this α, we conclude that age and weight of Shetland ponies are correlated. (b) $r \approx -0.990$ with $n = 6$ is significant for $\alpha = 0.01$. For this α, we conclude that lowest barometric pressure reading and maximum wind speed for cyclones are correlated.

23. (a) Average Hours Lost per Person versus Average Fuel Wasted per Person in Traffic Delays

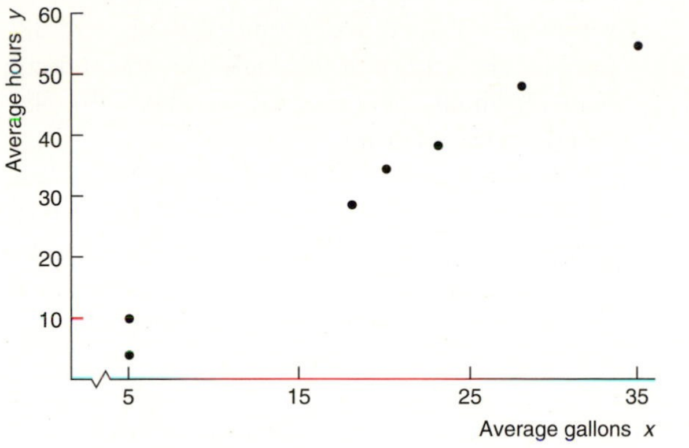

$r \approx 0.991$.

(b) For variables based on averages, $\bar{x} = 19.25$ hr; $s_x \approx 10.33$ hr; $\bar{y} = 31.13$ gal; $s_y \approx 17.76$ gal. For variables based on single individuals, $\bar{x} = 20.13$ hr; $s_x \approx 13.84$ hr; $\bar{y} = 31.87$ gal; $s_y \approx 25.18$ gal. Dividing by larger numbers results in a smaller value.

(c) Hours Lost per Person versus Fuel Wasted per Person in Traffic Delays

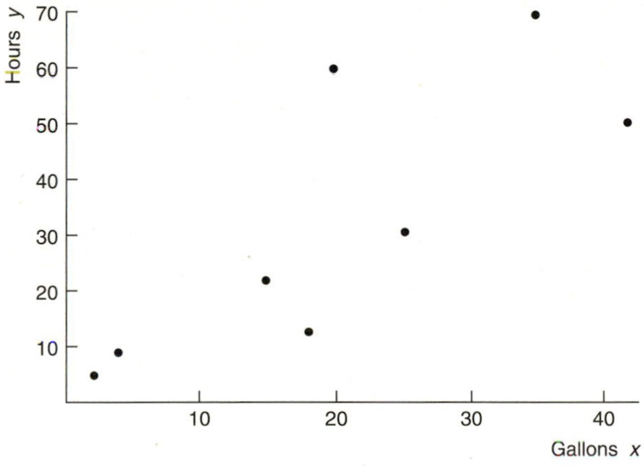

$r \approx 0.794$.

(d) Yes; by the central limit theorem, the $\bar{x}$ distribution has a smaller standard deviation than the corresponding x distribution.

Section 9.2

1. $b = -2$. When x changes by 1 unit, y decreases by 2 units.
3. Extrapolating Extrapolating beyond the range of the data is dangerous because the relationship pattern might change.
5. (a) $\hat{y} \approx 318.16 - 30.878x$. (b) About 31 fewer frost-free days. (c) $r \approx -0.981$. Note that if the slope is negative, r is also negative. (d) 96.3% of variation explained and 3.7% unexplained.
7. (a) Total Number of Jobs and Number of Entry-Level Jobs (Units in 100's)

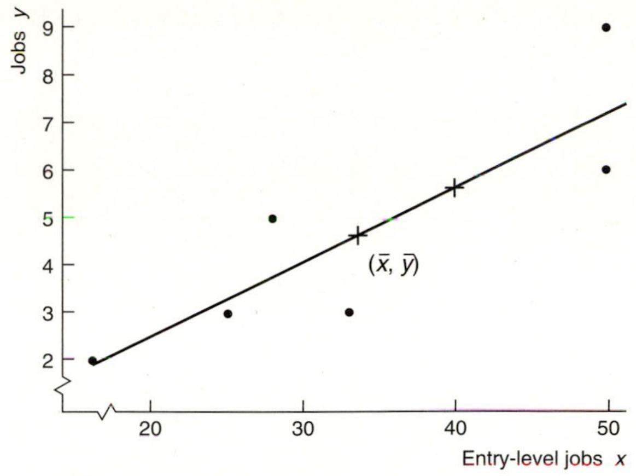

(b) Use a calculator. (c) $\bar{x} \approx 33.67$ jobs; $\bar{y} \approx 4.67$ entry-level jobs; $a \approx -0.748$; $b \approx 0.161$; $\hat{y} \approx -0.748 + 0.161x$ (d) See figure in part (a). (e) $r^2 \approx 0.740$; 74.0% of variation explained and 26.0% unexplained. (f) 5.69 jobs.

9. (a) Weight of Cars and Gasoline Mileage

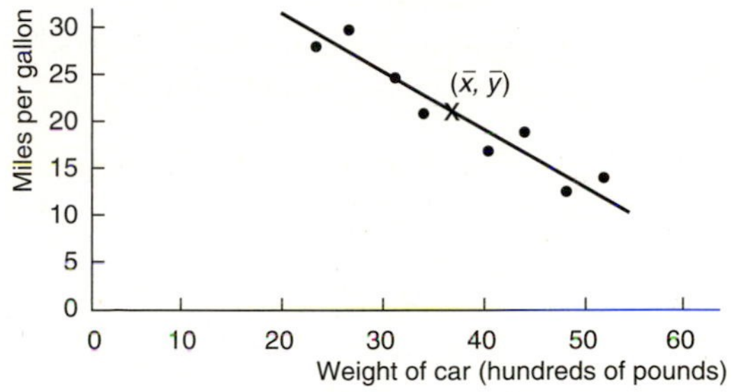

(b) Use a calculator. (c) $\bar{x} \approx 37.375$; $\bar{y} \approx 20.875$ mpg; $a \approx 43.326$; $b \approx -0.6007$; $\hat{y} \approx 43.326 - 0.6007x$.
(d) See figure in part (a). (e) $r^2 \approx 0.895$; 89.5% of variation explained and 10.5% unexplained.
(f) 20.5 mpg.

11. (a) Age and Percentage of Fatal Accidents Due to Speeding

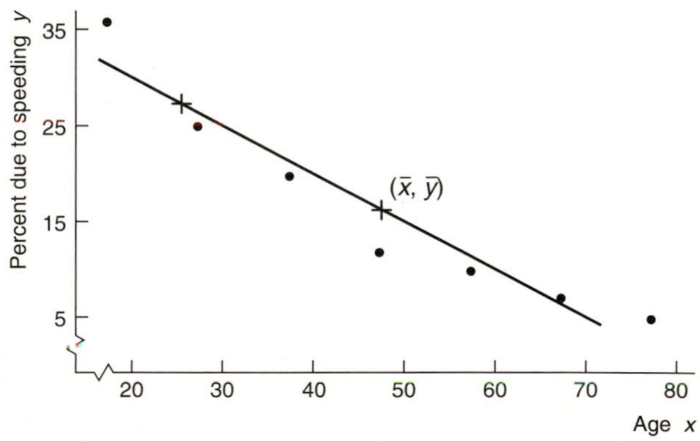

(b) Use a calculator. (c) $\bar{x} \approx 47$ years; $\bar{y} \approx 16.43\%$; $a \approx 39.761$; $b \approx -0.496$; $\hat{y} \approx 39.761 - 0.496x$.
(d) See figure in part (a). (e) $r^2 \approx 0.920$; 92.0% of variation explained and 8.0% unexplained. (f) 27.36%.

13. (a) Per Capita Income ($1000) and M.D.s per 10,000 Residents

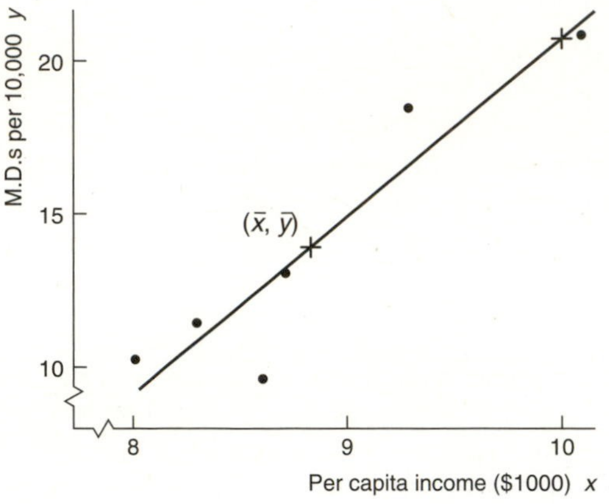

(b) Use a calculator. (c) $\bar{x} = \$8.83$; $\bar{y} \approx 13.95$ M.D.s; $a \approx -36.898$; $b \approx 5.756$; $\hat{y} \approx -36.898 + 5.756x$.

(d) See figure in part (a). (e) $r^2 \approx 0.872$; 87.2% of variation explained, 12.8% unexplained. (f) 20.7 M.D.s per 10,000 residents.

15. (a) Percentage of 16- to 19-Year-Olds Not in School and Violent Crime Rate per 1000 Residents

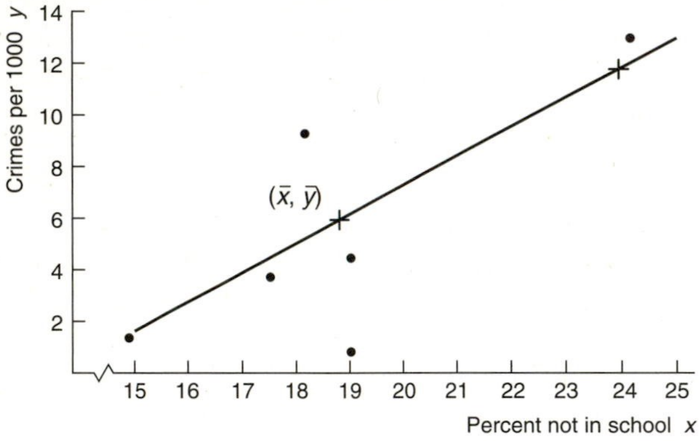

(b) Use a calculator. (c) $\bar{x} = 18.8\%$; $\bar{y} = 5.4$; $a \approx -17.204$; $b \approx 1.202$; $\bar{y} \approx -17.204 + 1.202x$. (d) See figure in part (a). (e) $r^2 \approx 0.584$; 58.4% of variation explained, 41.6% unexplained. (f) 11.6 crimes per 1000 residents.

17. (a) Elevation of Archaeological Sites and Percentage of Unidentified Artifacts

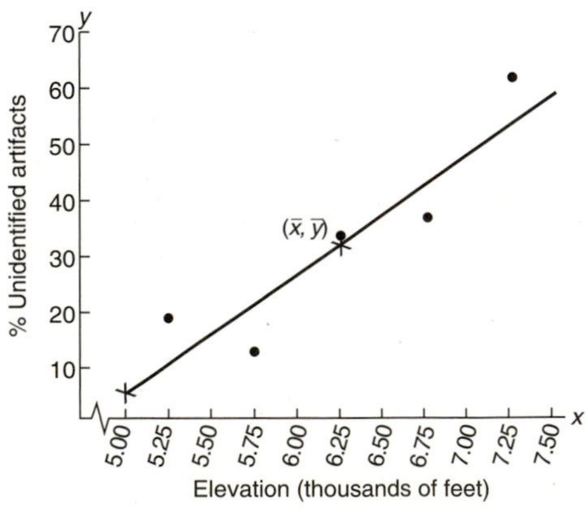

(b) Use a calculator. (c) $\bar{x} = 6.25$; $\bar{y} = 32.8$; $a = -104.7$; $b = 22$; $\hat{y} = -104.7 + 22x$.

(d) See figure in part (a). (e) $r^2 \approx 0.833$; 83.3% of variation explained, 16.7% unexplained. (f) 38.3.

19. (a) Yes. The pattern of residuals appears randomly scattered about the horizontal line at 0. (b) No. There do not appear to be any outliers.

21. (a) Result checks. (b) Result checks. (c) Yes. (d) The equation $x = 0.718y - 0.047$ does not match part (b). (e) No. The least-squares equation changes depending on which variable is the explanatory variable and which is the response variable.

23. (a) Model with (x, y) Data Pairs

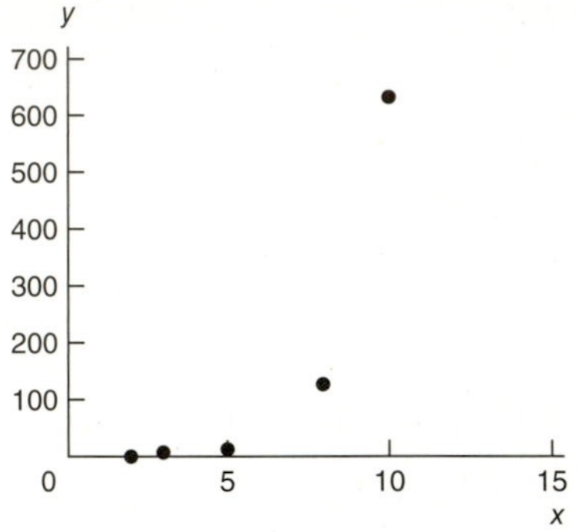

(b) Model with (x, y') Data Pairs

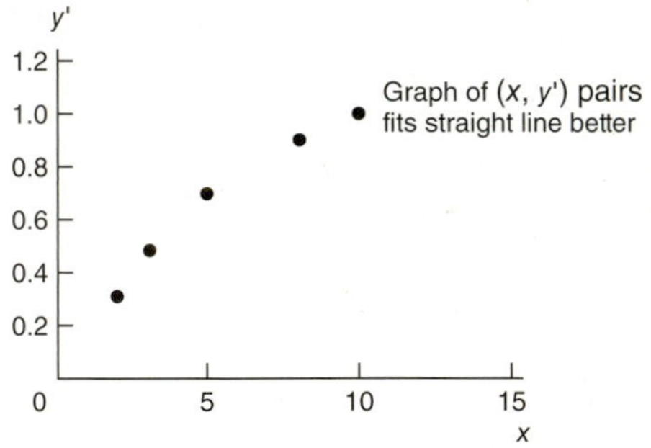

(c) $y' \approx -0.421 + 0.318x$; $r \approx 0.998$.
(d) $\alpha \approx 0.371$; $\beta \approx 2.080$; $y \approx 0.371(2.080)^x$.

25. (a) Model with (x', y') Data Pairs

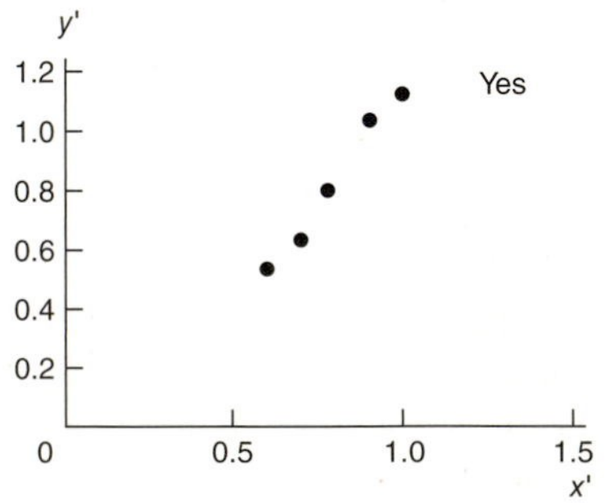

(b) $y' \approx -0.451 + 1.600x'$; $r \approx 0.991$.
(c) $\alpha \approx 0.354$; $\beta \approx 1.600$; $y \approx 0.354x^{1.600}$.

Section 9.3

1. ρ (Greek letter rho).
3. As x becomes further away from $\bar{x}$, the confidence interval for the predicted y becomes longer.
5. (a) Diameter. (b) $a = -0.223$; $b = 0.7848$; $\hat{y} = -0.223 + 0.7848x$. (c) P-value of b is 0.001. H_0: $\beta = 0$; H_1: $\beta \neq 0$. Since P-value < 0.01, reject H_0 and conclude that the slope is not zero. (d) $r \approx 0.896$. Yes. P-value is 0.001, so we reject H_0 for $\alpha = 0.01$.
7. (a) Use a calculator. (b) $\alpha = 0.05$; H_0: $\rho = 0$; H_1: $\rho > 0$; sample $t \approx 2.522$; $d.f. = 4$; $0.025 < P\text{-value} < 0.050$; reject H_0. There seems to be a positive correlation between x and y. From TI-84, P-value ≈ 0.0326. (c) Use a calculator. (d) 45.36%. (e) Interval from 39.05 to 51.67. (f) $\alpha = 0.05$; H_0: $\beta = 0$; H_1: $\beta > 0$; sample $t \approx 2.522$; $d.f. = 4$; $0.025 < P\text{-value} < 0.050$; reject H_0. There seems to be a positive slope between x and y. From TI-84, P-value ≈ 0.0326. (g) Interval from 0.064 to 0.760. For every percentage increase in successful free throws, the percentage of successful field goals increases by an amount between 0.06 and 0.76.
9. (a) Use a calculator. (b) $\alpha = 0.01$; H_0: $\rho = 0$; H_1: $\rho < 0$; sample $t \approx -10.06$; $d.f. = 5$; P-value < 0.0005; reject H_0. The sample evidence supports a negative correlation. From TI-84, P-value ≈ 0.00008. (c) Use a calculator. (d) 2.39 hours. (e) Interval from 2.12 to 2.66 hours. (f) $\alpha = 0.01$; H_0: $\beta = 0$; H_1: $\beta < 0$; sample $t \approx -10.06$; $d.f. = 5$; P-value < 0.0005; reject H_0. The sample evidence supports a negative slope. From TI-84, P-value ≈ 0.00008. (g) Interval from -0.065 to -0.044. For every additional meter of depth, the optimal time decreases by between 0.04 and 0.07 hour.
11. (a) Use a calculator. (b) $\alpha = 0.01$; H_0: $\rho = 0$; H_1: $\rho > 0$; sample $t \approx 6.534$; $d.f. = 4$; $0.0005 < P\text{-value} < 0.005$; reject H_0. The sample evidence supports a positive correlation. From TI-84, P-value ≈ 0.0014. (c) Use a calculator. (d) \$12.577 thousand. (e) Interval from 12.247 to 12.907 (thousand dollars). (f) $\alpha = 0.01$; H_0: $\beta = 0$; H_1: $\beta > 0$; sample $t \approx 6.534$; $d.f. = 4$; $0.0005 < P\text{-value} < 0.005$; reject H_0. The sample evidence supports a positive slope. From TI-84, P-value ≈ 0.0014. (g) Interval from 0.436 to 1.080. For every \$1000 increase in list price, the dealer price increase is between \$436 and \$1080 higher.
13. (a) H_0: $\rho = 0$; H_1: $\rho \neq 0$; $d.f. = 4$; sample $t = 4.129$; $0.01 < P\text{-value} < 0.02$; do not reject H_0; r is not significant at the 0.01 level of significance. (b) H_0: $\rho = 0$; H_1: $\rho \neq 0$; $d.f. = 8$; sample $t = 5.840$; P-value < 0.001; reject H_0; r is significant at the 0.01 level of significance. (c) As n increases, the t value corresponding to r also increases, resulting in a smaller P-value.
15. (a) $\hat{y} \approx 1.9938 + 0.9165x$; when $x = 5.8$, $\hat{y} \approx 7.3095$. (b) $r \approx 0.9815$, $r^2 \approx 0.9633$; H_0: $\rho = 0$, H_1: $\rho > 0$; P-value ≈ 0.000044; for $\alpha = 0.1$, reject H_0. The data support a positve correlation and indicate a predictable original time series from one week to the next.
17. (b) $\hat{y} \approx 4.1415 + 0.9785x$; when $x = \$42$, $\hat{y} \approx \$45.24$. (c) $r \approx 0.9668$, $r^2 \approx 9.347$; H_0: $\rho = 0$, H_1: $\rho > 0$; P-value ≈ 0.00008; for $\alpha = 00.1$, reject H_0. The data support a positve correlation and indicate a predictable original time series from one week to the next.

Section 9.4

1. (a) Response variable is x_1. Explanatory variables are x_2, x_3, x_4. (b) 1.6 is the constant term; 3.5 is the coefficient of x_2; -7.9 is the coefficient of x_3; and 2.0 is the coefficient of x_4. (c) $x_1 = 10.7$. (d) 3.5 units; 7 units; -14 units. (e) $d.f. = 8$; $t = 1.860$; 2.72 to 4.28. (f) $\alpha = 0.05$; H_0: $\beta_2 = 0$; H_1: $\beta_2 \neq 0$; $d.f. = 8$; $t = 8.35$; P-value < 0.001; reject H_0.
3. (a) $CVx_1 \approx 9.08$; $CVx_2 \approx 14.59$; $CVx_3 \approx 8.88$; x_2 has greatest spread; x_3 has smallest. (b) $r^2x_1x_2 \approx 0.958$; $r^2x_1x_3 \approx 0.942$; $r^2x_2x_3 \approx 0.895$; x_2; yes; 95.8%; 94.2%. (c) 97.7%. (d) $x_1 = 30.99 + 0.861x_2 + 0.335x_3$; 3.35; 8.61. (e) $\alpha = 0.05$; H_0: coefficient $= 0$; H_1: coefficient $\neq 0$; $d.f. = 8$; for β_2, $t = 3.47$ with P-value $= 0.008$; for β_3, $t = 2.56$ with P-value $= 0.034$; reject H_0 for each coefficient and conclude that the coefficients of x_2 and x_3 are not zero. (f) $d.f. = 8$; $t = 1.86$; C.I. for β_2 is 0.40 to 1.32; C.I. for β_3 is 0.09 to 0.58. (g) 153.9; 148.3 to 159.4.
5. (a) $CVx_1 \approx 39.64$; $CVx_2 \approx 44.45$; $CVx_3 \approx 50.62$; $CVx_4 \approx 52.15$; x_4; x_1 has a small CV because we divide by a large mean. (b) $r^2x_1x_2 \approx 0.842$; $r^2x_1x_3 \approx 0.865$; $r^2x_1x_4 \approx 0.225$; $r^2x_2x_3 \approx 0.624$; $r^2x_2x_4 \approx 0.184$; $r^2x_3x_4 \approx 0.089$; x_3; 84.2%. (c) 96.7%. (d) $x_1 = 7.68 + 3.66x_2 + 7.62x_3 + 0.83x_4$; 7.62 million dollars. (e) $\alpha = 0.05$; H_0: coefficient $= 0$; H_1: coefficient $\neq 0$; $d.f. = 6$; for β_2, $t = 3.28$ with P-value $= 0.017$; for β_3, $t = 4.60$ with P-value $= 0.004$; for β_4, $t = 1.54$ with P-value $= 0.175$; reject H_0 for β_2 and β_3 and conclude that the coefficients of x_2 and x_3 are not zero. For β_4, fail to reject H_0 and conclude that the coefficient of x_4 could be zero. (f) $d.f. = 6$; $t = 1.943$; C.I. for β_2 is 1.49 to 5.83; C.I. for β_3 is 4.40 to 10.84; C.I. for β_4 is -0.22 to 1.88. (g) 91.95; 77.6 to 106.3. (h) 5.63; 4.21 to 7.04.
7. Depends on data.

Chapter 9 Review

1. r will be close to 0.
3. Results are more reliable for interpolation.
5. (a) Age and Mortality Rate for Bighorn Sheep

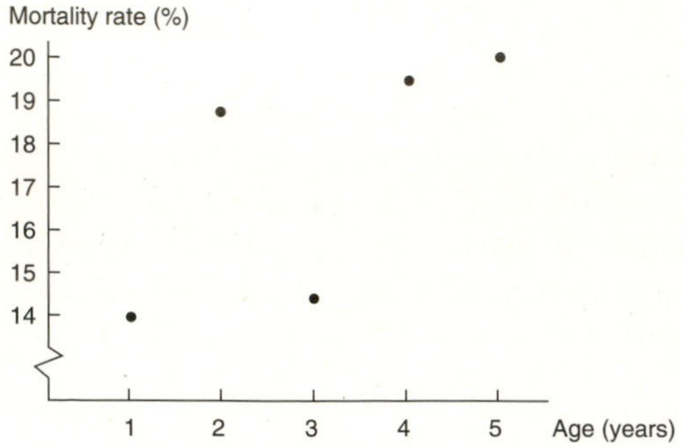

(b) $\bar{x} = 3$; $\bar{y} \approx 17.38$; $b \approx 1.27$; $\hat{y} \approx 13.57 + 1.27x$. (c) $r \approx 0.685$; $r^2 \approx 0.469$. (d) $\alpha = 0.01$; H_0: $\rho = 0$; H_1: $\rho > 0$; $d.f. = 3$; $t = 1.627$; $0.100 < P\text{-value} < 0.125$; do not reject H_0. There does not seem to be a positive correlation between age and mortality rate of bighorn sheep. From TI-84, P-value ≈ 0.1011. (e) No. Based on these limited data, predictions from the least-squares line model might be misleading. There appear to be other lurking variables that affect the mortality rate of sheep in different age groups.

7. (a) Weight of 1-Year-Old versus Weight of Adult

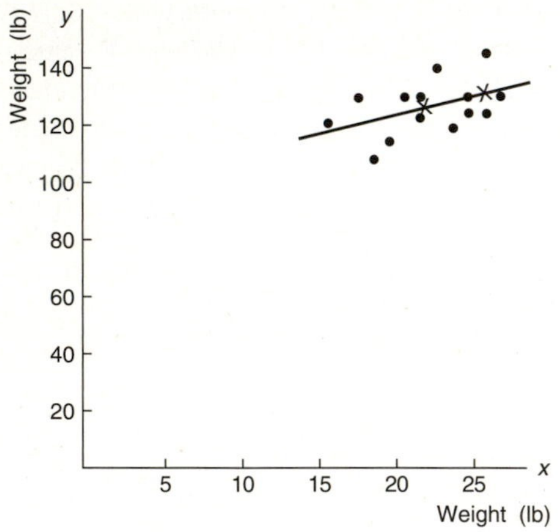

(b) $\bar{x} \approx 21.43$; $\bar{y} \approx 126.79$; $b \approx 1.285$; $\hat{y} \approx 99.25 + 1.285x$. (c) $r \approx 0.468$; $r^2 \approx 0.219$; 21.9% explained. (d) $\alpha = 0.01$; H_0: $\rho = 0$; H_1: $\rho > 0$; $d.f. = 12$; $t = 1.835$; $0.025 < P\text{-value} < 0.050$; do not reject H_0. At the 1% level of significance, there does not seem to be a positive correlation between weight of baby and weight of adult. From TI-84, P-value ≈ 0.0457. (e) 124.95 pounds. However, since r is not significant, this prediction may not be useful. Other lurking variables seem to have an effect on adult weight. (f) Use a calculator. (g) 105.91 to 143.99 pounds. (h) $\alpha = 0.01$; H_0: $\beta = 0$; H_1: $\beta > 0$; $d.f. = 12$; $t = 1.835$; $0.025 < P\text{-value} < 0.050$; do not reject H_0. At the 1% level of significance, there does not seem to be a positive slope between weight of baby x and weight of adult y. From TI-84, P-value ≈ 0.0457. (i) 0.347 to 2.223. At the 80% confidence level, we can say that for each additional pound a female infant weighs at 1 year, the female's adult weight changes by 0.35 to 2.22 pounds.

9. (a) Weight of Mail versus Number of Employees Required

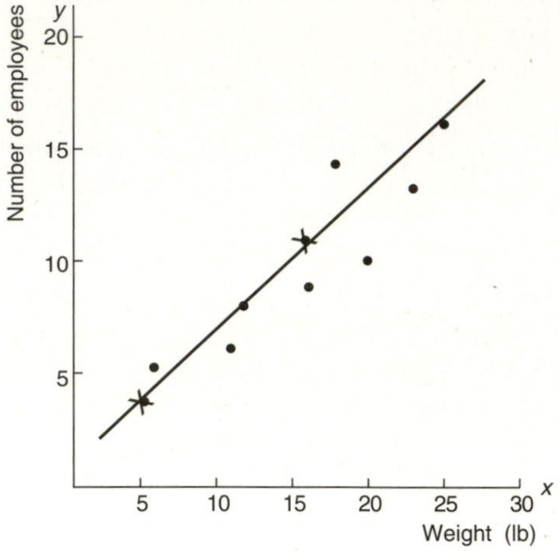

(b) $\bar{x} \approx 16.38$; $\bar{y} \approx 10.13$; $b \approx 0.554$; $\hat{y} \approx 1.051 + 0.554x$. (c) $r \approx 0.913$; $r^2 \approx 0.833$; 83.3% explained. (d) $\alpha = 0.01$; H_0: $\rho = 0$; H_1: $\rho > 0$; $d.f. = 6$; $t = 5.467$; $0.0005 < P\text{-value} < 0.005$; reject H_0. At the 1% level of significance, there is sufficient evidence to show a positive correlation between pounds of mail and number of employees required to process the mail. From TI-84, P-value ≈ 0.0008. (e) 9.36. (f) Use a calculator. (g) 4.86 to 13.86. (h) $\alpha = 0.01$; H_0: $\beta = 0$; H_1: $\beta > 0$; $d.f. = 6$; $t = 5.467$; $0.0005 < P\text{-value} < 0.005$; reject H_0. At the 1% level of significance, there is sufficient evidence to show a positive slope between pounds of mail x and number of employees required to process the mail y. From TI-84, P-value ≈ 0.0008. (i) 0.408 to 0.700. At the 80% confidence level, we can say that for each additional pound of mail, between 0.4 and 0.7 additional employees are needed.

CUMULATIVE REVIEW PROBLEMS

1. (a) i. $\alpha = 0.01$; H_0: $\mu = 2.0\ \mu g/L$; H_1: $\mu > 2.0\ \mu g/L$.
   ii. Standard normal; $z = 2.53$.
   iii. P-value ≈ 0.0057; on standard normal curve, shade area to the right of 2.53.
   iv. P-value of $0.0057 \leq 0.01$ for α; reject H_0.
   v. At the 1% level of significance, the evidence is sufficient to say that the population mean discharge level of lead is higher.

   (b) $2.13\ \mu g/L$ to $2.99\ \mu g/L$. (c) $n = 48$.
2. (a) Use rounded results to compute t in part (b).
   (b) i. $\alpha = 0.05$; H_0: $\mu = 10\%$; H_1: $\mu > 10\%$.
   ii. Student's t, $d.f. = 11$; $t \approx 1.248$.
   iii. $0.100 < P\text{-value} < 0.125$; on t graph, shade area to the right of 1.248. From TI-84, P-value ≈ 0.1190.
   iv. P-value interval > 0.05 for α; fail to reject H_0.

v. At the 5% level of significance, the evidence does not indicate that the patient is asymptomatic.

(c) 9.27% to 11.71%.

3. (a) i. $\alpha = 0.05$; H_0: $p = 0.10$; H_1: $p \neq 0.10$; yes, $np > 5$ and $nq > 5$; necessary to use normal approximation to the binominal.

ii. Standard normal; $\hat{p} \approx 0.147$; $z = 1.29$.

iii. P-value $= 2P(z > 1.29) \approx 0.1970$; on standard normal curve, shade area to the right of 1.29 and to the left of -1.29.

iv. P-value of $0.1970 > 0.05$ for α; fail to reject H_0.

v. At the 5% level of significance, the data do not indicate any difference from the national average for the population proportion of crime victims.

(b) 0.063 to 0.231. (c) From sample, $p \approx \hat{p} \approx 0.147$; $n = 193$.

4. (a) i. $\alpha = 0.05$; H_0: $\mu_d = 0$; H_1: $\mu_d \neq 0$.

ii. Student's t, $d.f. = 6$; $\bar{d} \approx -0.0039$, $t \approx -0.771$.

iii. $0.250 < P$-value < 0.500; on t graph, shade area to the right of 0.771 and to the left of -0.771. From TI-84, P-value ≈ 0.4699.

iv. P-value interval > 0.05 for α; fail to reject H_0.

v. At the 5% level of significance, the evidence does not show a population mean difference in phosphorous reduction between the two methods.

5. (a) i. $\alpha = 0.05$; H_0: $\mu_1 = \mu_2$; H_1: $\mu_1 \neq \mu_2$.

ii. Student's t, $d.f. = 15$; $t \approx 1.952$.

iii. $0.050 < P$-value < 0.100; on t graph, shade area to the right of 1.952 and to the left of -1.952. From TI-84, P-value ≈ 0.0609.

iv. P-value interval > 0.05 for α; fail to reject H_0.

v. At the 5% level of significance, the evidence does not show any difference in the population mean proportion of on-time arrivals in summer versus winter.

(b) -0.43% to 9.835%. (c) x_1 and x_2 distributions are approximately normal (mound-shaped and symmetric).

6. (a) i. $\alpha = 0.05$; H_0: $p_1 = p_2$; H_1: $p_1 > p_2$.

ii. Standard normal; $\hat{p}_1 \approx 0.242$; $\hat{p}_2 \approx 0.207$; $\bar{p} \approx 0.2246$; $z \approx 0.58$.

iii. P-value ≈ 0.2810; on standard normal curve, shade area to the right of 0.58.

iv. P-value interval > 0.05 for α; fail to reject H_0.

v. At the 5% level of significance, the evidence does not indicate that the population proportion of single men who go out dancing occasionally differs from the proportion of single women who do so.

Since $n_1\bar{p}$, $n_1\bar{q}$, $n_2\bar{p}$, and $n_2\bar{q}$ are all greater than 5, the normal approximation to the binomial is justified. (b) -0.065 to 0.139.

7. (a) Essay. (b) Outline of study.

8. Answers vary.

9. (a) Blood Glucose Level

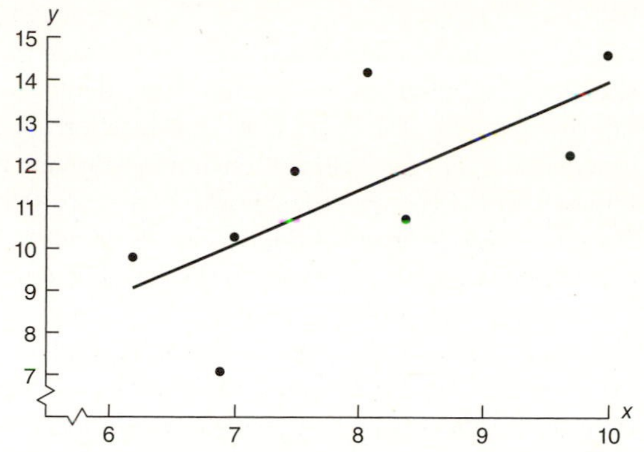

(b) $\hat{y} \approx 1.135 + 1.279x$. (c) $r \approx 0.700$; $r^2 \approx 0.490$; 49% of the variance in y is explained by the model and the variance in x. (d) 12.65; 9.64 to 15.66. (e) $\alpha = 0.01$; H_0: $\rho = 0$; H_1: $\rho \neq 0$; $r \approx 0.700$ with $t \approx 2.40$; $d.f. = 6$; $0.05 < P$-value < 0.10; do not reject H_0. At the 1% level of significance, the evidence is insufficient to conclude that there is a linear correlation. (f) $S_e \approx 1.901$; $t_c = 1.645$; 0.40 to 2.16.

CHAPTER 10

Section 10.1

1. Skewed right.

3. Right-tailed test.

5. Take random samples from each of the 4 age groups and record the number of people in each age group who recycle each of the 3 product types. Make a contingency table with age groups as labels for rows (or columns) and products as labels for columns (or rows).

7. (a) $d.f. = 6$; $0.005 < P$-value < 0.01. At the 1% level of significance, we reject H_0 since the P-value is less than 0.01. At the 1% level of significance, we conclude that the age groups differ in the proportions of who recycles each of the specified products.

(b) No. All he can say is that the 4 age groups differ in the proportions of those recycling each specified product. For this study, he cannot determine how the age groups differ regarding the proportions of those recycling the listed products.

9. (a) $\alpha = 0.05$; H_0: Myers–Briggs preference and profession are independent; H_1: Myers–Briggs preference and profession are not independent. (b) $\chi^2 = 8.649$; $d.f. = 2$. (c) $0.010 < P$-value < 0.025. From TI-84, P-value ≈ 0.0132. (d) Reject H_0. (e) At the 5% level of significance, there is sufficient evidence to conclude that Myers–Briggs preference and profession are not independent.

11. (a) $\alpha = 0.01$; H_0: Site type and pottery type are independent; H_1: Site type and pottery type are not independent. (b) $\chi^2 = 0.5552$; $d.f. = 4$. (c) $0.950 <$ P-value < 0.975. From TI-84, P-value ≈ 0.9679.

(d) Do not reject H_0. (e) At the 1% level of significance, there is insufficient evidence to conclude that site type and pottery type are not independent.

13. (a) $\alpha = 0.05$; H_0: Age distribution and location are independent; H_1: Age distribution and location are not independent. (b) $\chi^2 = 0.6704$; $d.f. = 4$. (c) $0.950 <$ P-value < 0.975. From TI-84, P-value ≈ 0.9549. (d) Do not reject H_0. (e) At the 5% level of significance, there is insufficient evidence to conclude that age distribution and location are not independent.

15. (a) $\alpha = 0.05$; H_0: Age of young adult and movie preference are independent; H_1: Age of young adult and movie preference are not independent. (b) $\chi^2 = 3.6230$; $d.f. = 4$. (c) $0.100 <$ P-value < 0.900. From TI-84, P-value ≈ 0.4594. (d) Do not reject H_0. (e) At the 5% level of significance, there is insufficient evidence to conclude that age of young adult and movie preference are not independent.

17. (a) $\alpha = 0.05$; H_0: Stone tool construction material and site are independent; H_1: Stone tool construction material and site are not independent. (b) $\chi^2 = 11.15$; $d.f. = 3$. (c) $0.010 <$ P-value < 0.025. From TI-84, P-value ≈ 0.0110. (d) Reject H_0. (e) At the 5% level of significance, there is sufficient evidence to conclude that stone tool construction material and site are not independent.

19. (i) Communication Preference by Percentage of Age Group

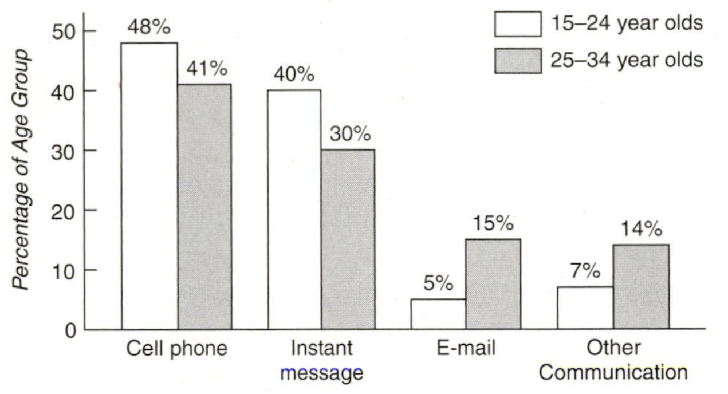

(ii) (a) H_0: The proportions of the different age groups having each communication preference are the same. H_1: The proportions of the different age groups having each communication preference are not the same. (b) $\chi^2 = 9.312$; $d.f. = 3$. (c) $0.025 <$ P-value < 0.050. From TI-84, P-value ≈ 0.0254. (d) Reject H_0. (e) At the 5% level of significance, there is sufficient evidence to conclude that the two age groups do not have the same proportions of communications preferences.

Section 10.2

1. $d.f.$ = number of categories − 1.

3. The greater the differences between the observed frequencies and the expected frequencies, the higher the sample χ^2 value. Greater χ^2 values lead to the conclusion that the differences between expected and observed frequencies are too large to be explained by chance alone.

5. (a) $\alpha = 0.05$; H_0: The distributions are the same; H_1: The distributions are different. (b) Sample $\chi^2 = 11.788$; $d.f. = 3$. (c) $0.005 <$ P-value < 0.010. (d) Reject H_0. (e) At the 5% level of significance, the evidence is sufficient to conclude that the age distribution of the Red Lake Village population does not fit the age distribution of the general Canadian population.

7. (a) $\alpha = 0.01$; H_0: The distributions are the same; H_1: The distributions are different. (b) Sample $\chi^2 = 0.1984$; $d.f. = 4$. (c) P-value > 0.995. (Note that as the χ^2 values decrease, the area in the right tail increases, so $\chi^2 < 0.207$ means that the corresponding P-value > 0.995.) (d) Do not reject H_0. (e) At the 1% level of significance, the evidence is insufficient to conclude that the regional distribution of raw materials does not fit the distribution at the current excavation site.

9. (i) Answers vary. (ii) (a) $\alpha = 0.01$; H_0: The distributions are the same; H_1: The distributions are different. (b) Sample $\chi^2 = 1.5693$; $d.f. = 5$. (c) $0.900 <$ P-value < 0.950. (d) Do not reject H_0. (e) At the 1% level of significance, the evidence is insufficient to conclude that the average daily July temperature does not follow a normal distribution.

11. (a) $\alpha = 0.05$; H_0: The distributions are the same; H_1: The distributions are different. (b) Sample $\chi^2 = 9.333$; $d.f. = 3$. (c) $0.025 <$ P-value < 0.050. (d) Reject H_0. (e) At the 5% level of significance, the evidence is sufficient to conclude that the current fish distribution is different than it was 5 years ago.

13. (a) $\alpha = 0.01$; H_0: The distributions are the same; H_1: The distributions are different. (b) Sample $\chi^2 = 13.70$; $d.f. = 5$. (c) $0.010 <$ P-value < 0.025. (d) Do not reject H_0. (e) At the 1% level of significance, the evidence is insufficient to conclude that the census ethnic origin distribution and the ethnic origin distribution of city residents are different.

15. (a) $\alpha = 0.01$; H_0: The distributions are the same; H_1: The distributions are different. (b) Sample $\chi^2 = 3.559$; $d.f. = 8$. (c) $0.100 <$ P-value < 0.900. (d) Do not reject H_0. (e) At the 1% level of significance, the evidence is insufficient to conclude that the distribution of first nonzero digits in the accounting file does not follow Benford's Law.

17. (a) $P(0) \approx 0.179$; $P(1) \approx 0.308$; $P(2) \approx 0.265$; $P(3) \approx 0.152$; $P(r \geq 4) \approx 0.096$. (b) For $r = 0$, $E \approx 16.11$; for $r = 1$, $E \approx 27.72$; for $r = 2$, $E \approx 23.85$; for $r = 3$, $E \approx 13.68$; for $r \geq 4$, $E \approx 8.64$. (c) $\chi^2 \approx 12.55$ with $d.f. = 4$. (d) $\alpha = 0.01$; H_0: The Poisson distribution fits; H_1: The Poisson distribution does not fit; $0.01 <$ P-value < 0.025; do not reject H_0. At the 1% level of significance, we cannot say that the Poisson distribution does not fit the sample data.

Section 10.3

1. Yes. No, the chi-square test of variance requires that the x distribution be a normal distribution.
3. (a) $\alpha = 0.05$; $H_0: \sigma^2 = 42.3$; $H_1: \sigma^2 > 42.3$. (b) $\chi^2 \approx 23.98$; $d.f. = 22$. (c) $0.100 < P\text{-value} < 0.900$. (d) Do not reject H_0. (e) At the 5% level of significance, there is insufficient evidence to conclude that the variance is greater in the new section. (f) $\chi^2_U = 36.78$; $\chi^2_L = 10.98$. Interval for σ^2 is from 27.57 to 92.37.
5. (a) $\alpha = 0.01$; $H_0: \sigma^2 = 136.2$; $H_1: \sigma^2 < 136.2$. (b) $\chi^2 \approx 5.92$; $d.f. = 7$. (c) Right-tailed area between 0.900 and 0.100; $0.100 < P\text{-value} < 0.900$. (d) Do not reject H_0. (e) At the 1% level of significance, there is insufficient evidence to conclude that the variance for number of mountain climber deaths is less than 136.2. (f) $\chi^2_U = 14.07$; $\chi^2_L = 2.17$. Interval for σ^2 is from 57.26 to 371.29.
7. (a) $\alpha = 0.05$; $H_0: \sigma^2 = 9$; $H_1: \sigma^2 < 9$. (b) $\chi^2 \approx 8.82$; $d.f. = 22$. (c) Right-tail area is between 0.995 and 0.990; $0.005 < P\text{-value} < 0.010$. (d) Reject H_0. (e) At the 5% level of significance, there is sufficient evidence to conclude that the variance of protection times for the new typhoid shot is less than 9. (f) $\chi^2_U = 33.92$; $\chi^2_L = 12.34$. Interval for σ is from 1.53 to 2.54.
9. (a) $\alpha = 0.01$; $H_0: \sigma^2 = 0.18$; $H_1: \sigma^2 > 0.18$. (b) $\chi^2 = 90$; $d.f. = 60$. (c) $0.005 < P\text{-value} < 0.010$. (d) Reject H_0. (e) At the 1% level of significance, there is sufficient evidence to conclude that the variance of measurements for the fan blades is higher than the specified amount. The inspector is justified in claiming that the blades must be replaced. (f) $\chi^2_U = 79.08$; $\chi^2_L = 43.19$. Interval for σ is from 0.45 mm to 0.61 mm.
11. (i) (a) $\alpha = 0.05$; $H_0: \sigma^2 = 23$; $H_1: \sigma^2 \neq 23$. (b) $\chi^2 \approx 13.06$; $d.f. = 21$. (c) The area to the left of $\chi^2 = 13.06$ is less than 50%, so we double the left-tail area to find the P-value for the two-tailed test. Right-tail area is between 0.950 and 0.900. Subtracting each value from 1, we find that the left-tail area is between 0.050 and 0.100. Doubling the left-tail area for a two-tailed test gives $0.100 < P\text{-value} < 0.200$. (d) Do not reject H_0. (e) At the 5% level of significance, there is insufficient evidence to conclude that the variance of battery lifetimes is different from 23. (ii) $\chi^2_U = 32.67$; $\chi^2_L = 11.59$. Interval for σ^2 is from 9.19 to 25.91. (iii) Interval for σ is from 3.03 to 5.09.

Section 10.4

1. Independent.
3. F distributions are not symmetrical. Values of the F distribution are all nonnegative.
5. (a) $\alpha = 0.01$; population 1 is annual production from the first plot; $H_0: \sigma_1^2 = \sigma_2^2$; $H_1: \sigma_1^2 > \sigma_2^2$;. (b) $F \approx 3.73$; $d.f._N = 15$; $d.f._D = 15$. (c) $0.001 < P\text{-value} < 0.010$. From TI-84, $P\text{-value} \approx 0.0075$. (d) Reject H_0. (e) At the 1% level of significance, there is sufficient evidence to show that the variance in annual wheat production of the first plot is greater than that of the second plot.
7. (a) $\alpha = 0.05$; population 1 has data from France; $H_0: \sigma_1^2 = \sigma_2^2$; $H_1: \sigma_1^2 \neq \sigma_2^2$. (b) $F \approx 1.97$; $d.f._N = 20$; $d.f._D = 17$. (c) $0.050 < \text{right-tail area} < 0.100$; $0.100 < P\text{-value} < 0.200$. From TI-84, $P\text{-value} \approx 0.1631$. (d) Do not reject H_0. (e) At the 5% level of significance, there is insufficient evidence to show that the variance in corporate productivity of large companies in France and of those in Germany differ. Volatility of corporate productivity does not appear to differ.
9. (a) $\alpha = 0.05$; population 1 has data from aggressive-growth companies; $H_0: \sigma_1^2 = \sigma_2^2$; $H_1: \sigma_1^2 > \sigma_2^2$. (b) $F \approx 2.54$; $d.f._N = 20$; $d.f._D = 20$. (c) $0.010 < P\text{-value} < 0.025$. From TI-84, $P\text{-value} \approx 0.0216$. (d) Reject H_0. (e) At the 5% level of significance, there is sufficient evidence to show that the variance in percentage annual returns for funds holding aggressive-growth small stocks is larger than that for funds holding value stocks.
11. (a) $\alpha = 0.05$; population 1 has data from the new system; $H_0: \sigma_1^2 = \sigma_2^2$; $H_1: \sigma_1^2 \neq \sigma_2^2$. (b) $F \approx 1.85$; $d.f._N = 30$; $d.f._D = 24$. (c) $0.050 < \text{right-tail area} < 0.100$; $0.100 < P\text{-value} < 0.200$. From TI-84, $P\text{-value} \approx 0.1266$. (d) Do not reject H_0. (e) At the 5% level of significance, there is insufficient evidence to show that the variance in gasoline consumption for the two injection systems is different.

Section 10.5

1. (a) $\alpha = 0.01$; $H_0: \mu_1 = \mu_2 = \mu_3$; H_1: Not all the means are equal.
(b–f)

Source of Variation	Sum of Squares	Degrees of Freedom	MS	F Ratio	P-value	Test Decision
Between groups	520.280	2	260.14	0.48	> 0.100	Do not reject H_0
Within groups	7544.190	14	538.87			
Total	8064.470	16				

From TI-84, $P\text{-value} \approx 0.6270$.

3. (a) $\alpha = 0.05$; H_0: $\mu_1 = \mu_2 = \mu_3 = \mu_4$; H_1: Not all the means are equal.
(b–f)

Source of Variation	Sum of Squares	Degrees of Freedom	*MS*	*F* Ratio	*P*-value	Test Decision
Between groups	89.637	3	29.879	0.846	> 0.100	Do not reject H_0
Within groups	635.827	18	35.324			
Total	725.464	21				

From TI-84, *P*-value $\approx$ 0.4867.

5. (a) $\alpha = 0.05$; H_0: $\mu_1 = \mu_2 = \mu_3$; H_1: Not all the means are equal.
(b–f)

Source of Variation	Sum of Squares	Degrees of Freedom	*MS*	*F* Ratio	*P*-value	Test Decision
Between groups	1303.167	2	651.58	5.005	between	Reject H_0
Within groups	1171.750	9	130.19		0.025 and 0.050	
Total	2474.917	11				

From TI-84, *P*-value $\approx$ 0.0346.

7. (a) $\alpha = 0.01$; H_0: $\mu_1 = \mu_2 = \mu_3$; H_1: Not all the means are equal.
(b–f)

Source of Variation	Sum of Squares	Degrees of Freedom	*MS*	*F* Ratio	*P*-value	Test Decision
Between groups	2.042	2	1.021	0.336	> 0.100	Do not reject H_0
Within groups	33.428	11	3.039			
Total	35.470	13				

From TI-84, *P*-value $\approx$ 0.7217.

9. (a) $\alpha = 0.05$; H_0: $\mu_1 = \mu_2 = \mu_3 = \mu_4$; H_1: Not all the means are equal.
(b–f)

Source of Variation	Sum of Squares	Degrees of Freedom	*MS*	*F* Ratio	*P*-value	Test Decision
Between groups	238.225	3	79.408	4.611	between	Reject H_0
Within groups	258.340	15	17.223		0.010 and 0.025	
Total	496.565	18				

From TI-84, *P*-value $\approx$ 0.0177.

Section 10.6

1. Two factors; walking device with 3 levels and task with 2 levels; data table has 6 cells.
3. Since the *P*-value is less than 0.01, there is a significant difference in mean cadence according to the factor "walking device used."
5. (a) Two factors: income with 4 levels and media type with 5 levels. (b) $\alpha = 0.05$; For income level, H_0: There is no difference in population mean index based on income level; H_1: At least two income levels have different population mean indices; $F_{\text{income}} \approx 2.77$ with *P*-value $\approx$ 0.088. At the 5% level of significance, do not reject H_0. The data do not indicate any differences in population mean index according to income level. (c) $\alpha = 0.05$; For media, H_0: There is no difference in population mean index according to media type; H_1: At least two media types have different population mean indices; $F_{\text{media}} \approx 0.03$ with *P*-value $\approx$ 0.998. At the 5% level of significance, do not reject H_0. The data do not indicate any differences in population mean index according to media type.

7. Randomized Block Design

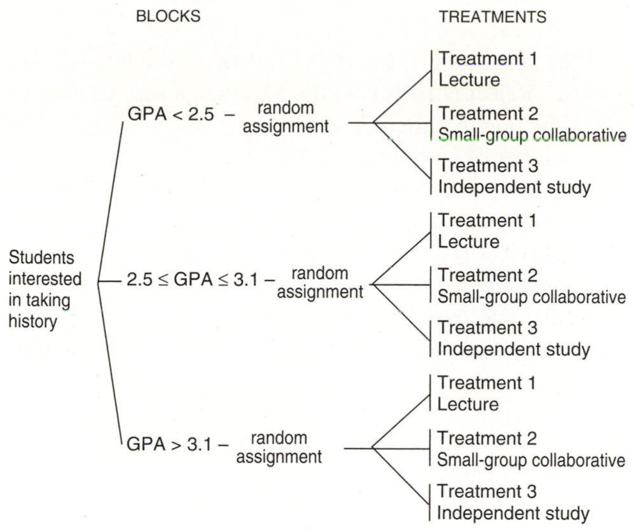

Yes, the design fits the model for randomized block design.

Chapter 10 Review

1. Chi-square, F.
3. Test of homogeneity.
5. One-way ANOVA. $\alpha = 0.05$; H_0: $\mu_1 = \mu_2 = \mu_3 = \mu_4$; H_1: Not all the means are equal.

Source of Variation	Sum of Squares	Degrees of Freedom	*MS*
Between groups	6149.75	3	2049.917
Within groups	12,454.80	16	778.425
Total	18,604.55	19	

F Ratio	P-value	Test Decision
2.633	between 0.050 and 0.100	Do not reject H_0

From TI-84, P-value ≈ 0.0854.

7. (a) Chi-square test of σ^2. (i) $\alpha = 0.01$; H_0: $\sigma^2 = 1{,}040{,}400$; H_1: $\sigma^2 > 1{,}040{,}400$. (ii) $\chi^2 \approx 51.03$; $d.f. = 29$. (iii) $0.005 < P\text{-value} < 0.010$. (iv) Reject H_0. (v) At the 1% level of significance, there is sufficient evidence to conclude that the variance is greater than claimed. (b) $\chi^2_U = 45.72$; $\chi^2_L = 16.05$; $1{,}161{,}147.4 < \sigma^2 < 3{,}307{,}642.4$.
9. Chi-square test of independence. (i) $\alpha = 0.01$; H_0: Student grade and teacher rating are independent; H_1: Student grade and teacher rating are not independent. (ii) $\chi^2 \approx 9.80$; $d.f. = 6$. (iii) $0.100 < P\text{-value} < 0.900$. From TI-84, P-value ≈ 0.1337. (iv) Do not reject H_0. (v) At the 1% level of significance, there is insufficient evidence to claim that student grade and teacher rating are not independent.
11. Chi-square test of goodness of fit. (i) $\alpha = 0.01$; H_0: The distributions are the same; H_1: The distributions are different. (ii) $\chi^2 \approx 11.93$; $d.f. = 4$. (iii) $0.010 < P\text{-value} < 0.025$. (iv) Do not reject H_0. (v) At the 1% level of significance, there is insufficient evidence to claim that the age distribution of the population of Blue Valley has changed.
13. F test for two variances. (i) $\alpha = 0.05$; H_0; $\sigma_1^2 = \sigma_2^2$; H_1: $\sigma_1^2 > \sigma_2^2$. (ii) $F \approx 2.61$; $d.f._N = 15$; $d.f._D = 17$. (iii) $0.025 < P\text{-value} < 0.050$. From TI-84, P-value ≈ 0.0302. (iv) Reject H_0. (v) At the 5% level of significance, there is sufficient evidence to show that the variance for the lifetimes of bulbs manufactured using the new process is larger than that for bulbs made by the old process.

CHAPTER 11

Section 11.1

1. Dependent (matched pairs).
3. (a) $\alpha = 0.05$; H_0: Distributions are the same; H_1: Distributions are different. (b) $x = 7/15 \approx 0.4667$; $z \approx -0.26$. (c) P-value $= 2(0.3974) = 0.7948$. (d) Do not reject H_0. (e) At the 5% level of significance, the data are not significant. The evidence is insufficient to conclude that the economic growth rates are different.
5. (a) $\alpha = 0.05$; H_0: Distributions are the same; H_1: Distributions are different. (b) $x = 10/16 = 0.625$; $z = 1.00$. (c) P-value $= 2(0.1587) = 0.3174$. (d) Do not reject H_0. (e) At the 5% level of significance, the data are not significant. The evidence is insufficient to conclude that the lectures had any effect on student awareness of current events.
7. (a) $\alpha = 0.05$; H_0: Distributions are the same; H_1: Distributions are different. (b) $x = 7/12 \approx 0.5833$; $z \approx 0.58$. (c) P-value $= 2(0.2810) = 0.5620$. (d) Do not reject H_0. (e) At the 5% level of significance, the data are not significant. The evidence is insufficient to conclude that the schools are not equally effective.
9. (a) $\alpha = 0.01$; H_0: Distributions are the same; H_1: Distribution after hypnosis is lower. (b) $x = 3/16 = 0.1875$; $z \approx -2.50$. (c) P-value $= 0.0062$. (d) Reject H_0. (e) At the 1% level of significance, the data are significant. The evidence is sufficient to conclude that the number of cigarettes smoked per day was less after hypnosis.
11. (a) $\alpha = 0.01$; H_0: Distributions are the same; H_1: Distributions are different. (b) $x = 10/20 = 0.5000$; $z = 0$. (c) P-value $= 2(0.5000) = 1$. (d) Do not reject H_0. (e) At the 1% level of significance, the data are not significant. The evidence is insufficient to conclude that the distribution of dropout rates is different for males and females.

Section 11.2

1. Independent.
3. (a) $\alpha = 0.05$; H_0: Distributions are the same; H_1: Distributions are different. (b) $R_A = 126$; $\mu_R = 132$; $\sigma_R \approx 16.25$; $z \approx -0.37$. (c) P-value $\approx 2(0.3557) = 0.7114$. (d) Do not reject H_0. (e) At the 5% level of significance, the evidence is insufficient to conclude that the yield distributions for organic and conventional farming methods are different.
5. (a) $\alpha = 0.05$; H_0: Distributions are the same; H_1: Distributions are different. (b) $R_B = 148$; $\mu_R = 132$; $\sigma_R \approx 16.25$; $z \approx 0.98$. (c) P-value $\approx 2(0.1635) = 0.3270$. (d) Do not reject H_0. (e) At the 5% level of significance, the evidence is insufficient to conclude that the distributions of the training sessions are different.
7. (a) $\alpha = 0.05$; H_0: Distributions are the same; H_1: Distributions are different. (b) $R_A = 92$; $\mu_R = 132$; $\sigma_R \approx 16.25$; $z \approx -2.46$. (c) P-value $\approx 2(0.0069) = 0.0138$. (d) Reject H_0. (e) At the 5% level of significance, the evidence is sufficient to conclude that the completion time distributions for the two settings are different.
9. (a) $\alpha = 0.01$; H_0: Distributions are the same; H_1: Distributions are different. (b) $R_A = 176$; $\mu_R = 132$; $\sigma_R \approx 16.25$; $z \approx 2.71$. (c) P-value $\approx 2(0.0034) = 0.0068$. (d) Reject H_0. (e) At the 1% level of significance, the evidence is sufficient to conclude that the distributions showing percentage of exercisers differ by education level.
11. (a) $\alpha = 0.01$; H_0: Distributions are the same; H_1: Distributions are different. (b) $R_A = 166$; $\mu_R = 150$; $\sigma_R \approx 17.32$; $z \approx 0.92$. (c) P-value $\approx 2(0.1788) = 0.3576$. (d) Do not reject H_0. (e) At the 1% level of significance, the evidence is insufficient to conclude that the distributions of test scores differ according to instruction method.

Section 11.3

1. Monotone increasing.
3. (a) $\alpha = 0.05$; H_0: $\rho_s = 0$; H_1: $\rho_s \neq 0$. (b) $r_s \approx 0.682$. (c) $n = 11$; $0.01 < P$-value < 0.05. (d) Reject H_0. (e) At the 5% level of significance, we conclude that there is a monotone relationship (either increasing or decreasing) between rank in training class and rank in sales.
5. (a) $\alpha = 0.05$; H_0: $\rho_s = 0$; H_1: $\rho_s > 0$. (b) $r_s \approx 0.571$. (c) $n = 8$; P-value > 0.05. (d) Do not reject H_0. (e) At the 5% level of significance, there is insufficient evidence to indicate a monotone-increasing relationship between crowding and violence.
7. (ii) (a) $\alpha = 0.05$; H_0: $\rho_s = 0$; H_1: $\rho_s < 0$. (b) $r_s \approx -0.214$. (c) $n = 7$; P-value > 0.05. (d) Do not reject H_0. (e) At the 5% level of significance, the evidence is insufficient to conclude that there is a monotone-decreasing relationship between the ranks of humor and aggressiveness.
9. (ii) (a) $\alpha = 0.05$; H_0: $\rho_s = 0$; H_1: $\rho_s \neq 0$. (b) $r_s \approx 0.930$. (c) $n = 13$; P-value < 0.002. (d) Reject H_0. (e) At the 5% level of significance, we conclude that there is a monotone relationship between number of firefighters and number of police.
11. (ii) (a) $\alpha = 0.01$; H_0: $\rho_s = 0$; H_1: $\rho_s \neq 0$. (b) $r_s \approx 0.661$. (c) $n = 8$; $0.05 < P$-value < 0.10. (d) Do not reject H_0. (e) At the 1% level of significance, we conclude that there is insufficient evidence to reject the null hypothesis of no monotone relationship between rank of insurance sales and rank of per capita income.

Section 11.4

1. Exactly two.
3. (a) $\alpha = 0.05$; H_0: The symbols are randomly mixed in the sequence; H_1: The symbols are not randomly mixed in the sequence. (b) $R = 11$. (c) $n_1 = 12$; $n_2 = 11$; $c_1 = 7$; $c_2 = 18$. (d) Do not reject H_0. (e) At the 5% level of significance, the evidence is insufficient to conclude that the sequence of presidential party affiliations is not random.
5. (a) $\alpha = 0.05$; H_0: The symbols are randomly mixed in the sequence; H_1: The symbols are not randomly mixed in the sequence. (b) $R = 11$. (c) $n_1 = 16$; $n_2 = 7$; $c_1 = 6$; $c_2 = 16$. (d) Do not reject H_0. (e) At the 5% level of significance, the evidence is insufficient to conclude that the sequence of days for seeding and not seeding is not random.
7. (i) Median $= 11.7$; BBBAAAAABBBA. (ii) (a) $\alpha = 0.05$; H_0: The numbers are randomly mixed about the median; H_1: The numbers are not randomly mixed about the median. (b) $R = 4$. (c) $n_1 = 6$; $n_2 = 6$; $c_1 = 3$; $c_2 = 11$. (d) Do not reject H_0. (e) At the 5% level of significance, the evidence is insufficient to conclude that the sequence of returns is not random about the median.
9. (i) Median $= 21.6$; BAAAAAABBBBB. (ii) (a) $\alpha = 0.05$; H_0: The numbers are randomly mixed about the median; H_1: The numbers are not randomly mixed about the median. (b) $R = 3$. (c) $n_1 = 6$; $n_2 = 6$; $c_1 = 3$; $c_2 = 11$. (d) Reject H_0. (e) At the 5% level of significance, we can conclude that the sequence of percentages of sand in the soil at successive depths is not random about the median.
11. (a) H_0: The symbols are randomly mixed in the sequence. H_1: The symbols are not randomly mixed in the sequence. (b) $n_1 = 21$; $n_2 = 17$; $R = 18$. (c) $\mu_R \approx 19.80$; $\sigma_R \approx 3.01$; $z \approx -0.60$. (d) Since $-1.96 < z < 1.96$, do not reject H_0; P-value $\approx 2(0.2743) = 0.5486$; at the 5% level of significance, the P-value also tells us not to reject H_0. (e) At the 5% level of significance, the evidence is insufficient to reject the null hypothesis of a random sequence of Democratic and Republican presidential terms.

Chapter 11 Review

1. No assumptions about population distributions are required.
3. (a) Rank-sum test. (b) $\alpha = 0.05$; H_0: Distributions are the same; H_1: Distributions are different. (c) $R_A = 134$; $\mu_R = 132$; $\sigma_R \approx 16.25$; $z \approx 0.12$. (d) P-value $= 2(0.4522) = 0.9044$. (e) Do not reject H_0. At the 5% level of significance, there is insufficient evidence to conclude that the viscosity index distribution has changed with use of the catalyst.
5. (a) Sign test. (b) $\alpha = 0.01$; H_0: Distributions are the same; H_1: Distribution after ads is higher. (c) $x = 0.77$; $z = 1.95$. (d) P-value $= 0.0256$. (e) Do not reject H_0. At the 1% level of significance, the evidence is insufficient to claim that the distribution is higher after the ads.
7. (a) Spearman rank correlation coefficient test. (b) $\alpha = 0.05$; H_0: $\rho = 0$; H_1: $\rho > 0$. (c) $r_s \approx 0.617$. (d) $n = 9$; $0.025 < P\text{-value} < 0.05$. (e) Reject H_0. At the 5% level of significance, we conclude that there is a monotone-increasing relation between the ranks for the training program and the ranks on the job.
9. (a) Runs test for randomness. (b) $\alpha = 0.05$; H_0: The symbols are randomly mixed in the sequence; H_1: The symbols are not randomly mixed in the sequence. (c) $R = 7$. (d) $n_1 = 16$; $n_2 = 9$; $c_1 = 7$; $c_2 = 18$. (e) Reject H_0. At the 5% level of significance, we can conclude that the sequence of answers is not random.

CUMULATIVE REVIEW PROBLEMS

1. (a) Use a calculator. (b) $P(0) \approx 0.543$; $P(1) \approx 0.331$; $P(2) \approx 0.101$; $P(3) \approx 0.025$. (c) 0.3836; $d.f. = 3$. (d) $\alpha = 0.01$; H_0: The distributions are the same; H_1: The distributions are different; $\chi^2 \approx 0.3836$; $0.900 < P\text{-value} < 0.950$; do not reject H_0. At the 1% level of significance, the evidence is insufficient to claim that the distribution does not fit the Poisson distribution.
2. $\alpha = 0.05$; H_0: Yield and fertilizer type are independent; H_1: Yield and fertilizer type are not independent; $\chi^2 \approx 5.005$; $d.f. = 4$; $0.100 < P\text{-value} < 0.900$; do not reject H_0. At the 5% level of significance, the evidence is insufficient to conclude that fertilizer type and yield are not independent.
3. (a) $\alpha = 0.05$; H_0: $\sigma = 0.55$; H_1: $\sigma > 0.55$; $s \approx 0.602$; $d.f. = 9$; $\chi^2 \approx 10.78$; $0.100 < P\text{-value} < 0.900$; do not reject H_0. At the 5% level of significance, there is insufficient evidence to conclude that the standard deviation of petal lengths is greater than 0.55. (b) Interval from 0.44 to 0.99. (c) $\alpha = 0.01$; H_0: $\sigma_1^2 = \sigma_2^2$; H_1: $\sigma_1^2 > \sigma_2^2$; $F \approx 1.95$; $d.f._N = 9$, $d.f._D = 7$; P-value > 0.100; do not reject H_0. At the 1% level of significance, the evidence is insufficient to conclude that the variance of the petal lengths for *Iris virginica* is greater than that for *Iris versicolor*.
4. $\alpha = 0.05$; H_0: $p = 0.5$ (wind direction distributions are the same); H_1: $p \neq 0.5$ (wind direction distributions are different); $x = 11/18$; $z \approx 0.94$; P-value $= 2(0.1736) = 0.3472$; do not reject H_0. At the 5% level of significance, the evidence is insufficient to conclude that the wind direction distributions are different.
5. $\alpha = 0.01$; H_0: Growth distributions are the same; H_1: Growth distributions are different; $\mu_R = 126.5$; $\sigma_R \approx 15.23$; $R_A = 135$; $z \approx 0.56$; P-value $= 2(0.2877) = 0.5754$; do not reject H_0. At the 1% level of significance, the evidence is insufficient to conclude that the growth distributions are different for the two root stocks.
6. (b) $\alpha = 0.05$; H_0: $\rho_s = 0$; H_1: $\rho_s \neq 0$; $r_s = 1$; P-value < 0.002; reject H_0. At the 5% level of significance, we can say that there is a monotone relationship between the calcium contents as measured by the labs.
7. Median $= 33.45$; AABBBBAAAABAABBBBA; $\alpha = 0.05$; H_0: Numbers are random about the median; H_1: Numbers are not random about the median; $R = 7$; $n_1 = n_2 = 9$; $c_1 = 5$; $c_2 = 15$; do not reject H_0. At the 5% level of significance, there is insufficient evidence to conclude that the sunspot activity about the median is not random.

ANSWERS TO SELECTED
Even-Numbered Problems

CHAPTER 2

Even-numbered answers not included here appear in the margins of the chapters, next to the problems.

Section 2.1

10. (a) Employee Salaries—Histogram

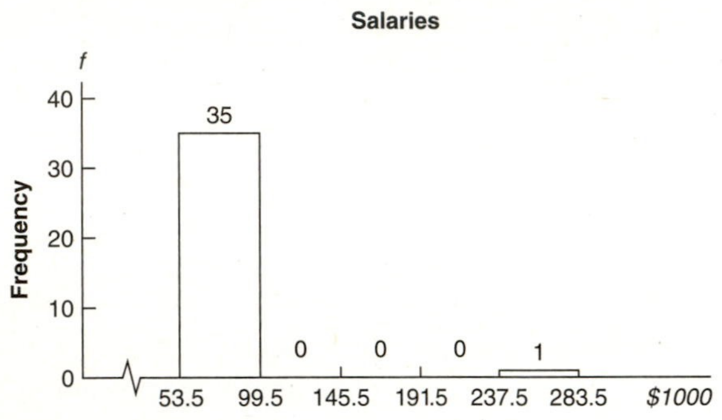

(c) Employee Salaries—Histogram

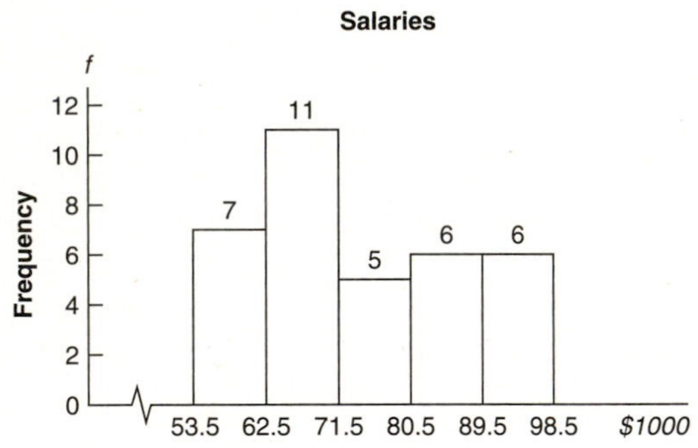

16. (a) Class width = 11.

(b)

Class Limits	Class Boundaries	Midpoint	Frequency	Relative Frequency	Cumulative Frequency
45–55	44.5–55.5	50	3	0.04	3
56–66	55.5–66.5	61	7	0.10	10
67–77	66.5–77.5	72	22	0.31	32
78–88	77.5–88.5	83	26	0.37	58
89–99	88.5–99.5	94	9	0.13	67
100–110	99.5–110.5	105	3	0.04	70

(c, d) Glucose Level (mg/100 ml)—Histogram, Relative-Frequency Histogram

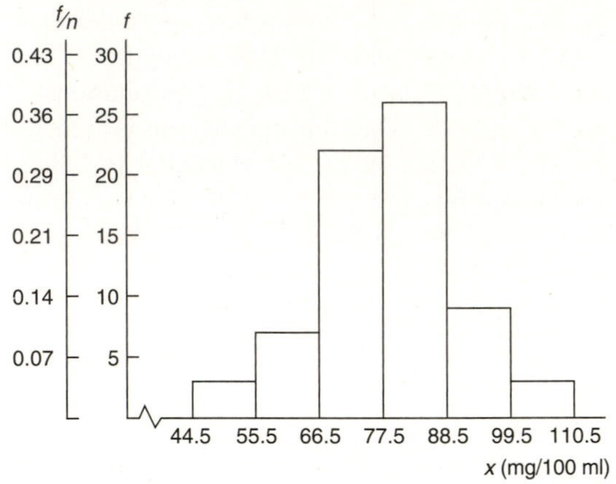

(f) Glucose Level (mg/100 ml)—Ogive

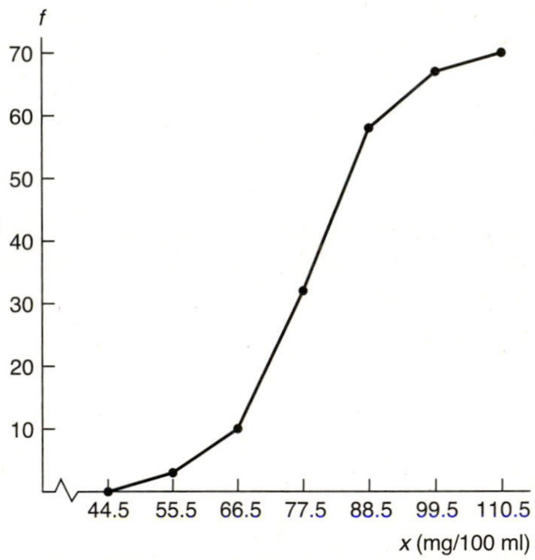

18. (a) Class width = 28.

(b)

Class Limits	Class Boundaries	Midpoint	Frequency	Relative Frequency	Cumulative Frequency
10–37	9.5–37.5	23.5	7	0.10	7
38–65	37.5–65.5	51.5	25	0.34	32
66–93	65.5–93.5	79.5	26	0.36	58
94–121	93.5–121.5	107.5	9	0.12	67
122–149	121.5–149.5	135.5	5	0.07	72
150–177	149.5–177.5	163.5	0	0.00	72
178–205	177.5–205.5	191.5	1	0.01	73

(c, d) Depth of Artifacts (cm)—Histogram, Relative-Frequency Histogram

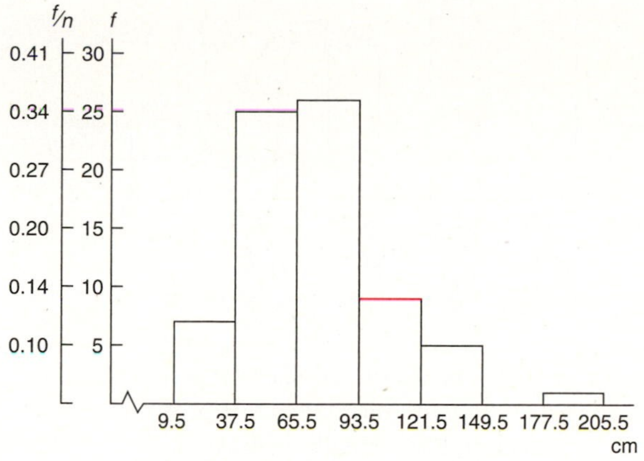

(f) Depth of Artifacts (cm)—Ogive

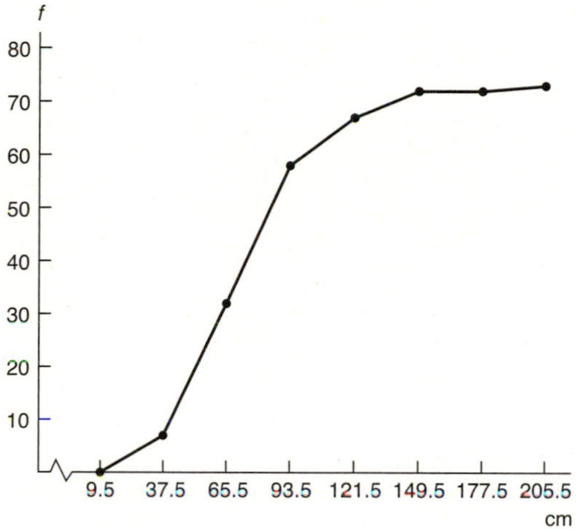

20. (a) Class width = 6.
(b)

Words of Three Syllables or More

Class Limits	Class Boundaries	Midpoint	Frequency	Relative Frequency	Cumulative Frequency
0–5	−0.5–5.5	2.5	13	0.24	13
6–11	5.5–11.5	8.5	15	0.27	28
12–17	11.5–17.5	14.5	11	0.20	39
18–23	17.5–23.5	20.5	3	0.05	42
24–29	23.5–29.5	26.5	6	0.11	48
30–35	29.5–35.5	32.5	4	0.07	52
36–41	35.5–41.5	38.5	2	0.04	54
42–47	41.5–47.5	44.5	1	0.02	55

(c, d) Words of Three Syllables or More—Histogram, Relative-Frequency Histogram

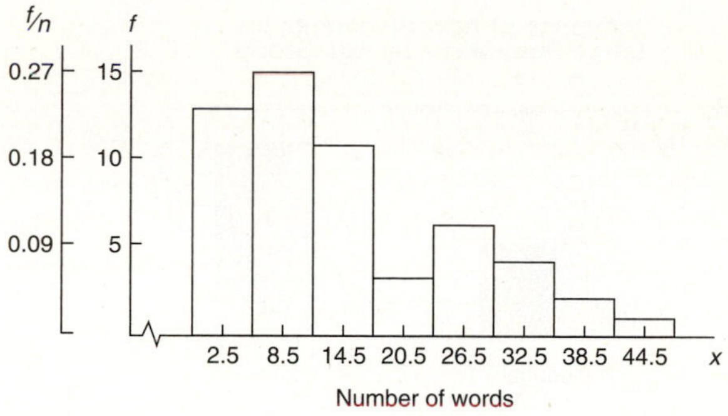

(f) Ogive for Words of Three Syllables or More

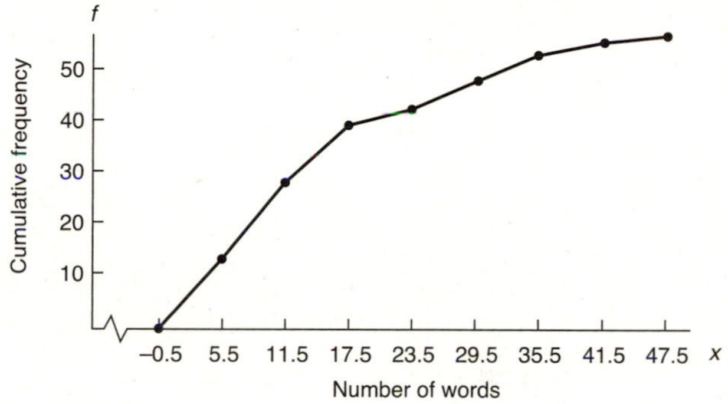

22. (b)

Baseball Batting Averages (class width = 0.043)

Class Limits	Class Boundaries	Midpoint	Frequency
0.107–0.149	0.1065–0.1495	0.128	3
0.150–0.192	0.1495–0.1925	0.171	4
0.193–0.235	0.1925–0.2355	0.214	3
0.236–0.278	0.2355–0.2785	0.257	10
0.279–0.321	0.2785–0.3215	0.3	6

(b, c) Baseball Batting Averages—Histogram

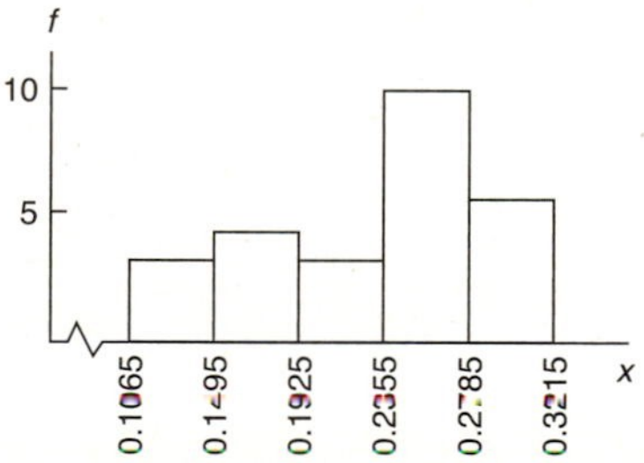

24. Dotplot for Iditarod Finish Time (in hours)

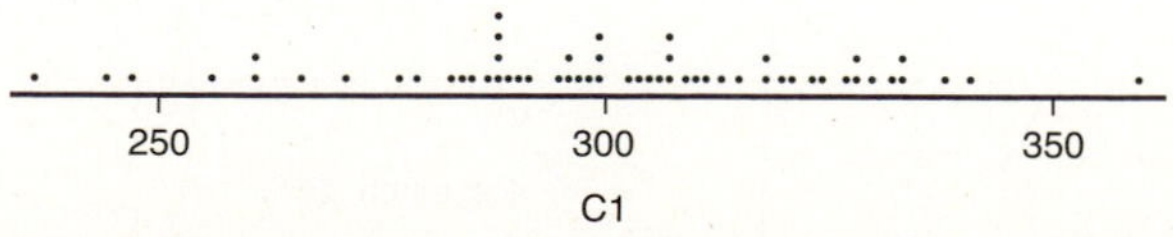

Section 2.2

6. (b)

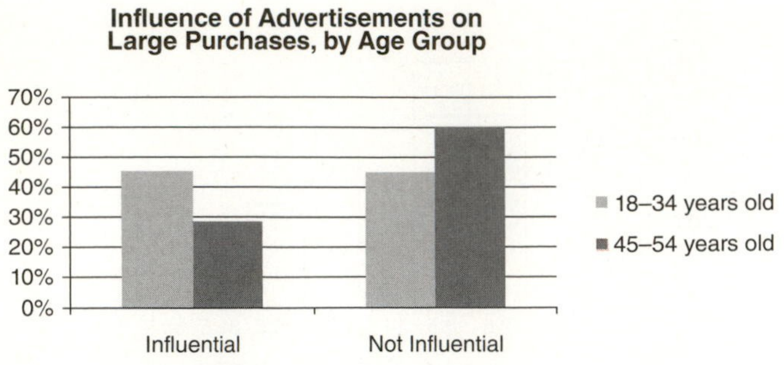

8. (a) Number of Spearheads—Pareto Chart

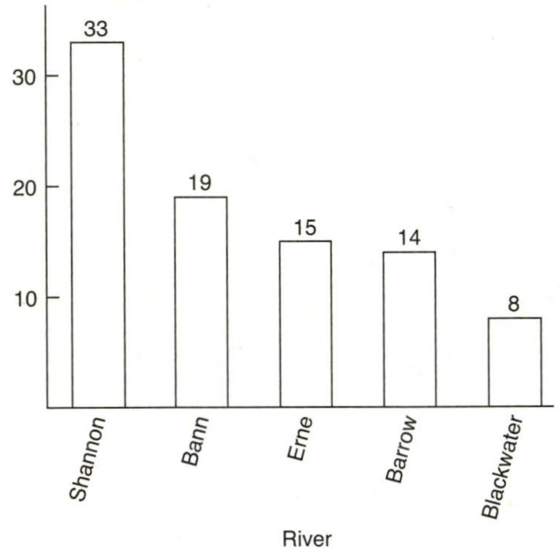

(b) Number of Spearheads—Circle Graph

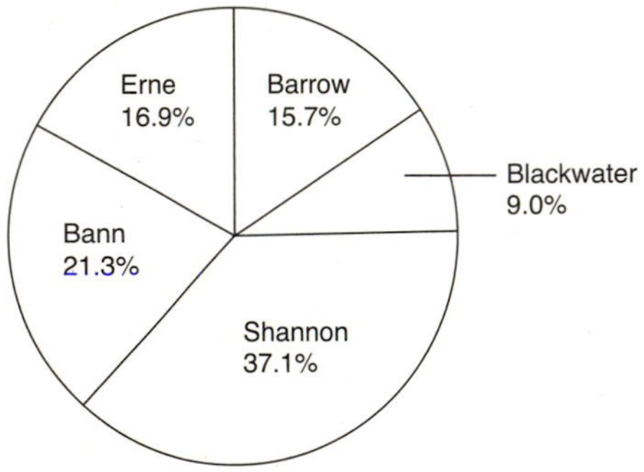

10. How College Professors Spend Their Time—Circle Graph

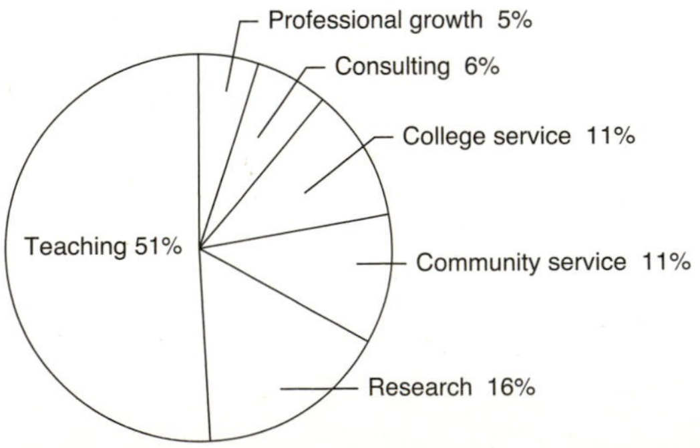

12. Driving Problems—Pareto Chart

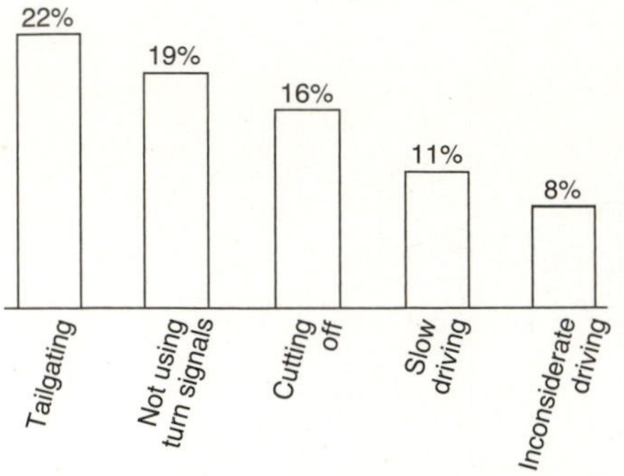

No. The total is not 100%, and it is not clear if respondents could mark more than one problem.

14. Changes in Boys' Height with Age—Time-Series Graph

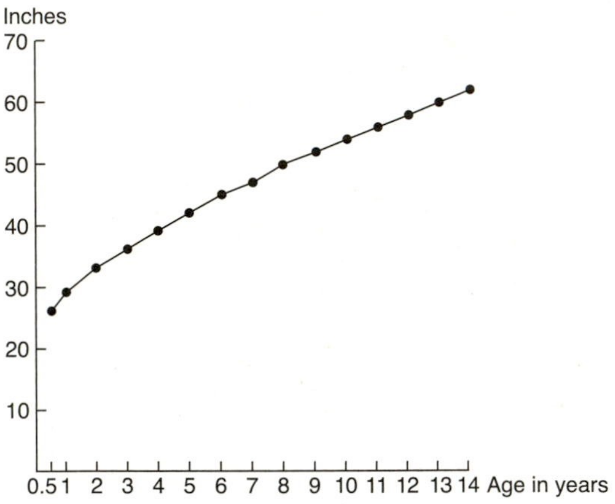

Chapter 2 Review

8. (a)

Age of DUI Arrests

1	6 = 16 years
1	6 8
2	0 1 1 2 2 2 3 4 4 5 6 6 6 7 7 7 9
3	0 0 1 1 2 3 4 4 5 5 6 7 8 9
4	0 0 1 3 5 6 7 7 9 9
5	1 3 5 6 8
6	3 4

(b) Class width = 7.

Age Distribution of DUI Arrests

Class Limits	Class Boundaries	Midpoint	Frequency	Relative Frequency	Cumulative Frequency
16–22	15.5–22.5	19	8	0.16	8
23–29	22.5–29.5	26	11	0.22	19
30–36	29.5–36.5	33	11	0.22	30
37–43	36.5–43.5	40	7	0.14	37
44–50	43.5–50.5	47	6	0.12	43
51–57	50.5–57.5	54	4	0.08	47
58–64	57.5–64.5	61	3	0.06	50

(c) Age Distribution of DUI Arrests—Histogram

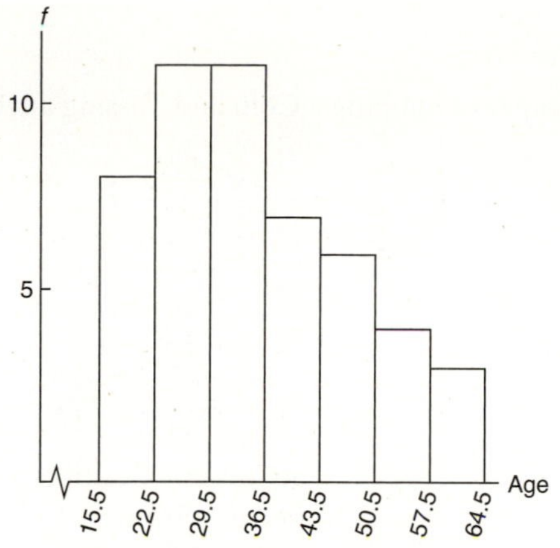

10. (a) Distribution of Civil Justice Caseloads Involving Businesses—Pareto Chart

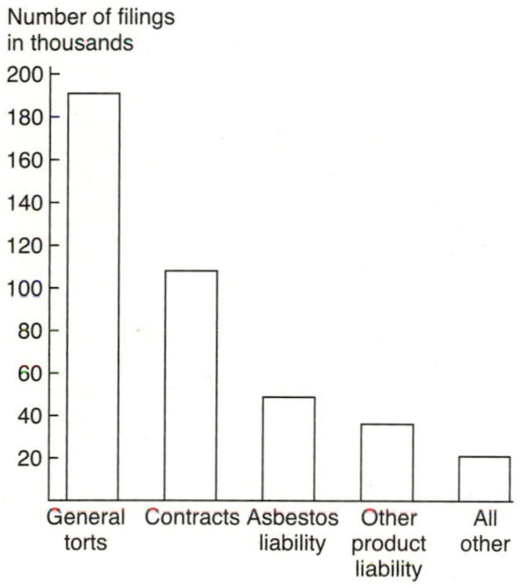

(b) Distribution of Civil Justice Caseloads Involving Businesses—Pie Chart

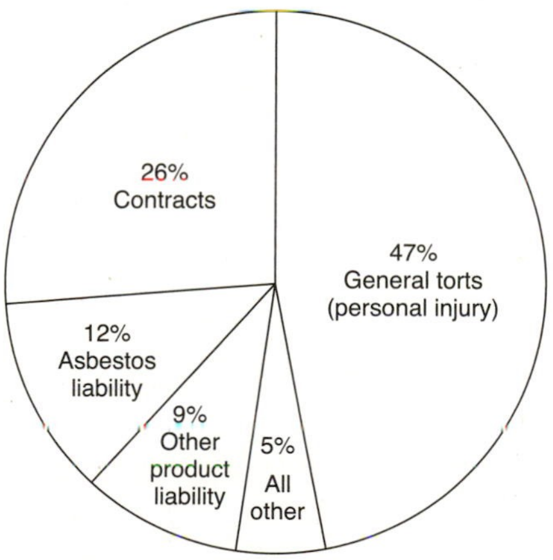

CHAPTER 3

Section 3.3

6. (c) Box-and-Whisker Plot

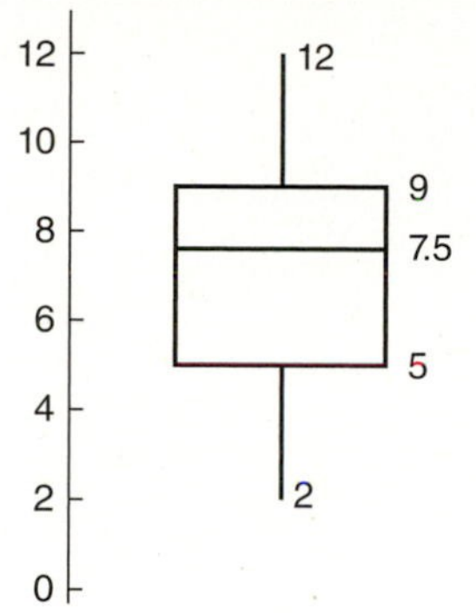

8. (a) Low = 3; $Q_1 = 16$; median = 23; $Q_3 = 30$; high = 72; $IQR = 14$.
Clerical Staff Length of Employment (months)

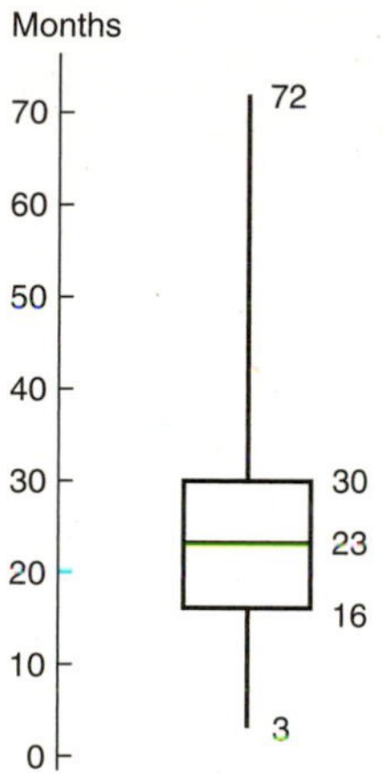

10. (a) Low = 5; $Q_1 = 9$; median = 10; $Q_3 = 12$; high = 15; $IQR = 3$.
(b) First quartile, since it is below Q_1.
High School Dropout Percentage by State

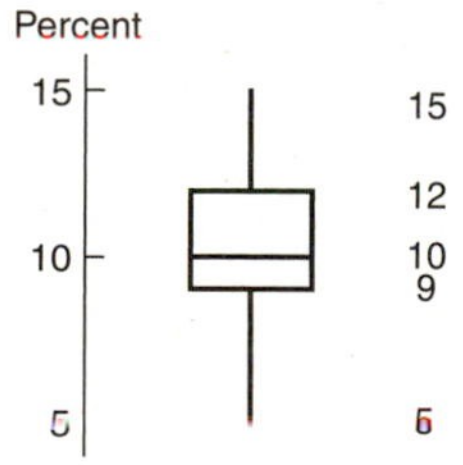

12. (a) Low value = 4; $Q_1 = 61.5$; median = 65.5; $Q_3 = 71.5$; high value = 80.
(b) $IQR = 10$.
(c) Lower limit = 46.5; upper limit = 86.5.

(d) Yes, the value 4 is below the lower limit and is probably an error. Our guess is that one of the students is 4 feet tall and listed height in feet instead of inches. There are no values above the upper limit.
Students' Heights (inches)

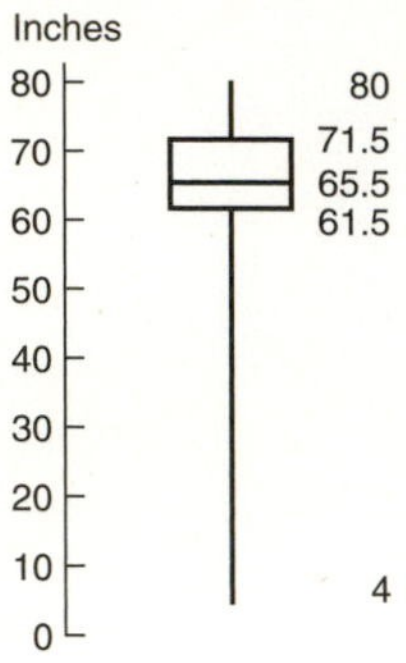

Chapter 3 Review

8. (a) Low = 7.8; Q_1 = 14.2; (kilograms) median = 20.25; Q_3 = 23.8; high = 29.5.
(b) IQR = 9.6 kilograms.
(d) Yes, the lower half shows slightly more spread.
Maize Harvest

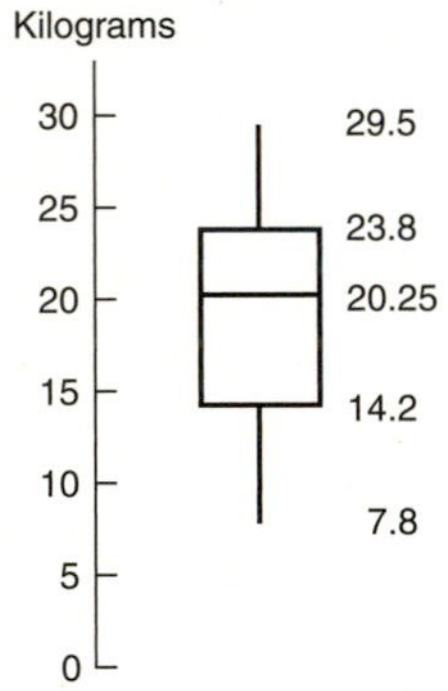

10. (a) Low = 6; Q_1 = 10; median = 11; Q_3 = 13; high = 16; IQR = 3.
Soil Water Content

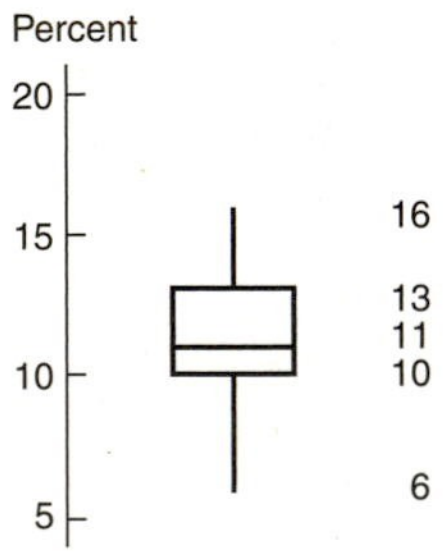

CHAPTER 4

Section 4.3

6. (a) Outcomes of Flipping a Coin and Tossing a Die

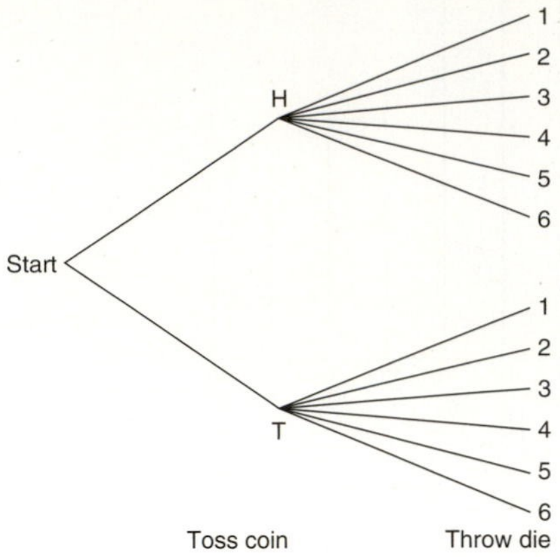

8. (a) Outcomes of Three Multiple-Choice Questions

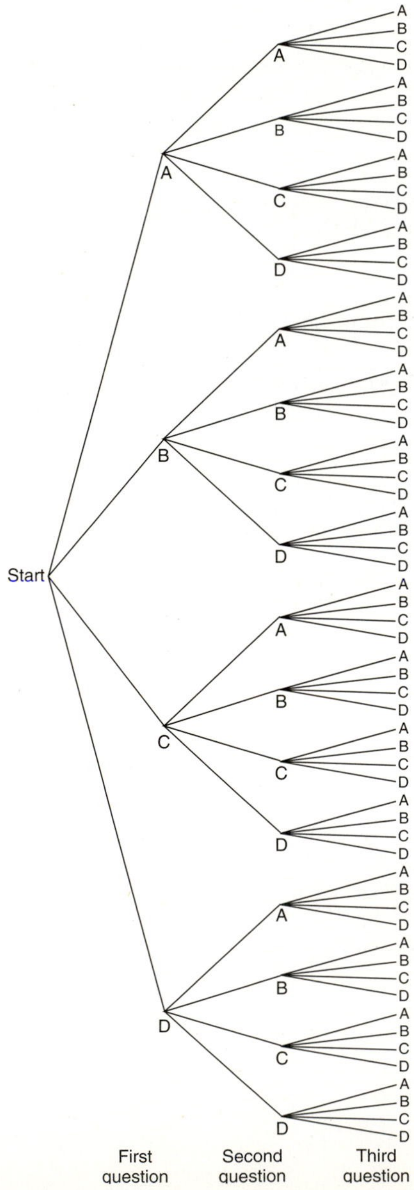

Chapter 4 Review

18. Ways to Satisfy Literature, Social Science, and Philosophy Requirements

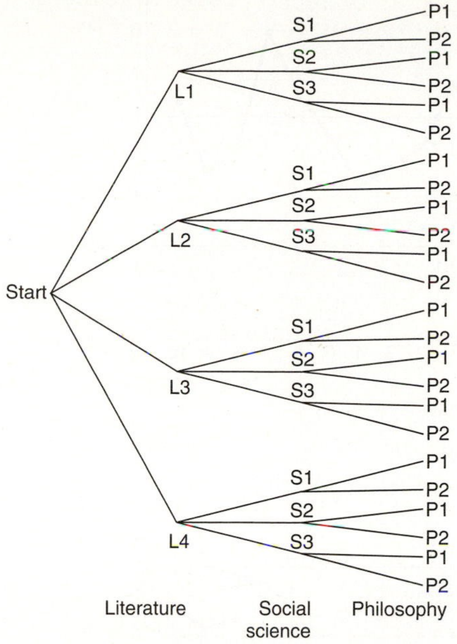

CHAPTER 5

Section 5.1

10. (b) Age of Promotion-Sensitive Shoppers

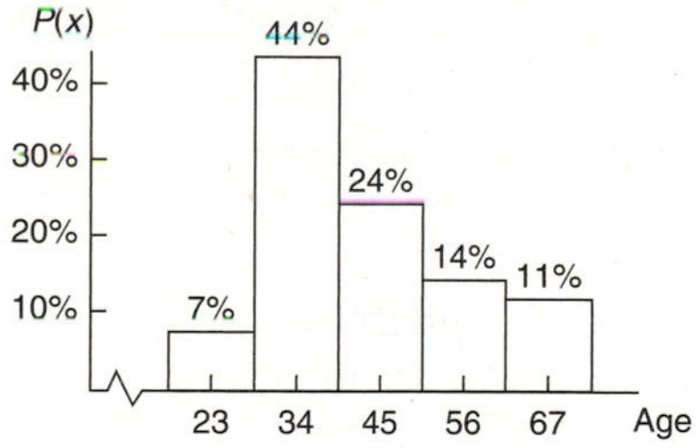

12. (b) Age of Nurses

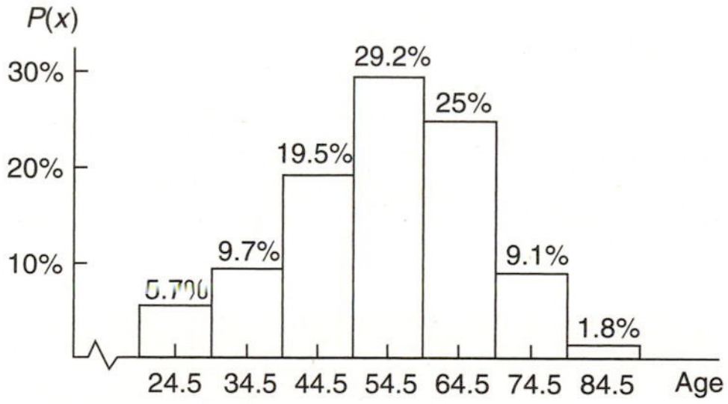

Section 5.3

12. (a) Binomial Distribution for Number of Defective Syringes

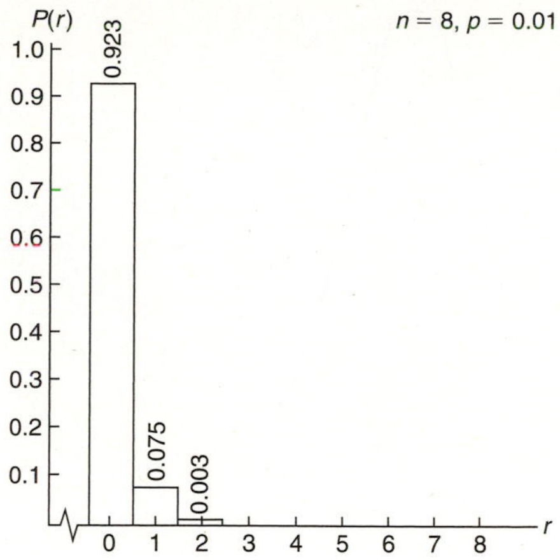

14. (a) Binomial Distribution for Number of Automobile Damage Claims by People Under Age 25

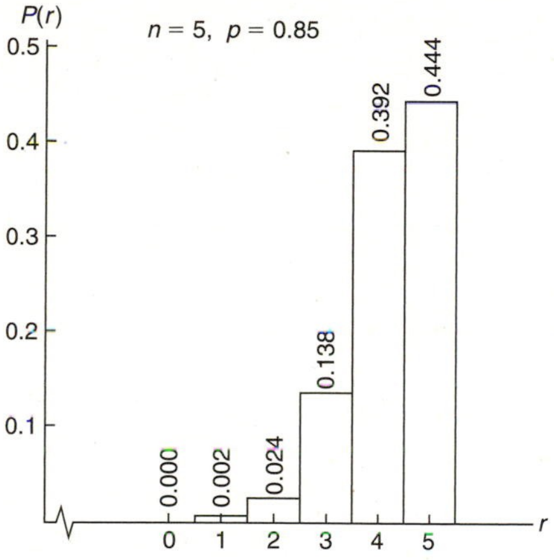

16. (a) Binomial Distribution for Drivers Who Tailgate

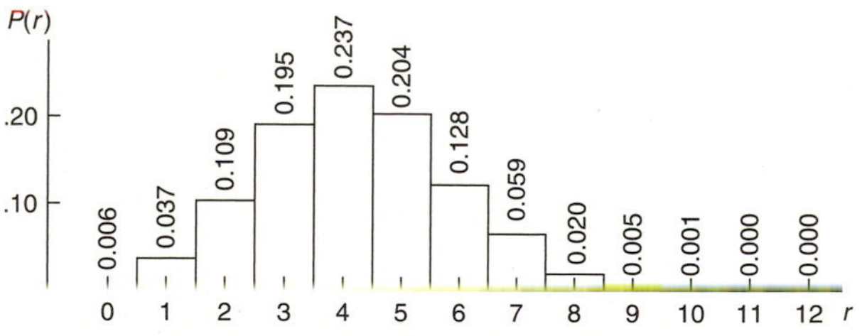

CHAPTER 6

Section 6.1

4. (a) Normal Curve

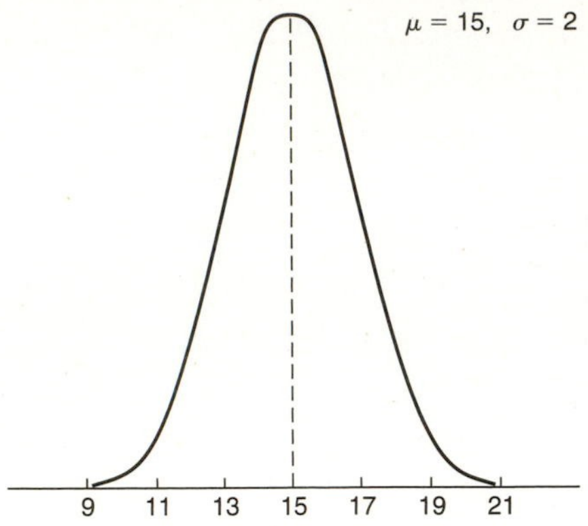

(b) Normal Curve

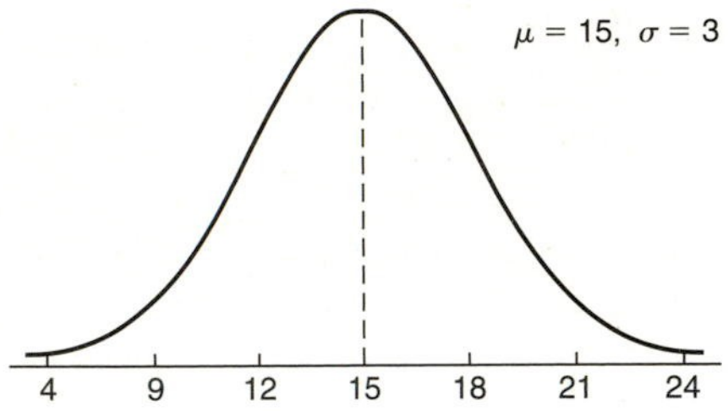

(c) Normal Curve

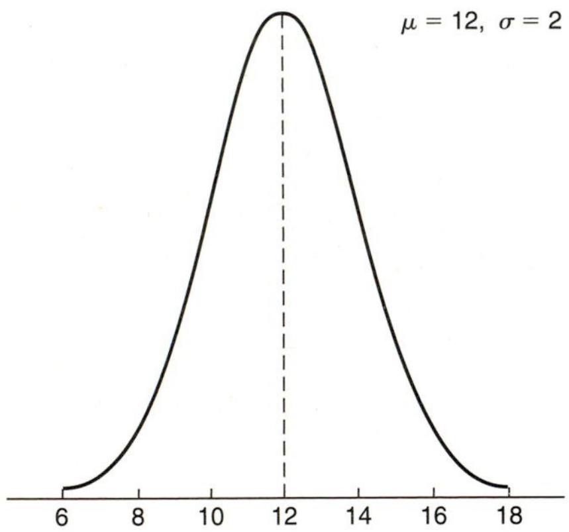

(d) Normal Curve

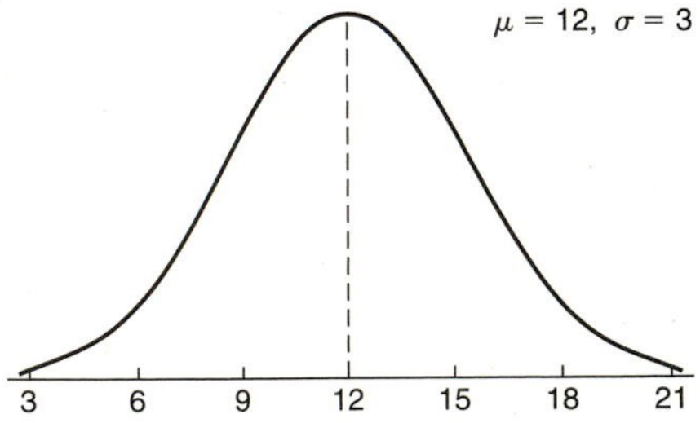

12. (a) Visitors Treated Each Day by YPMS (first period)

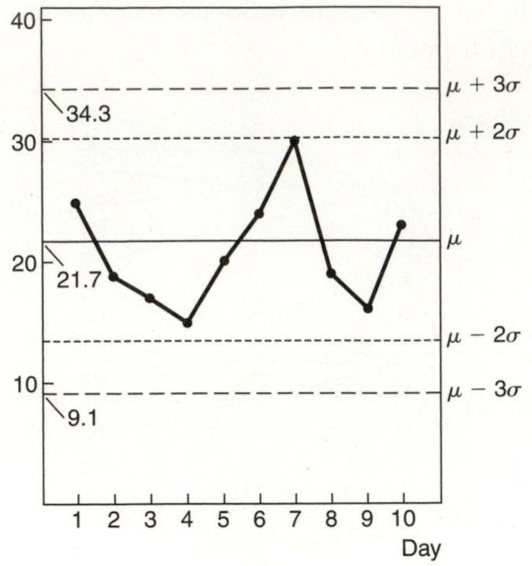

In control.

(b) Visitors Treated Each Day by YPMS (second period)

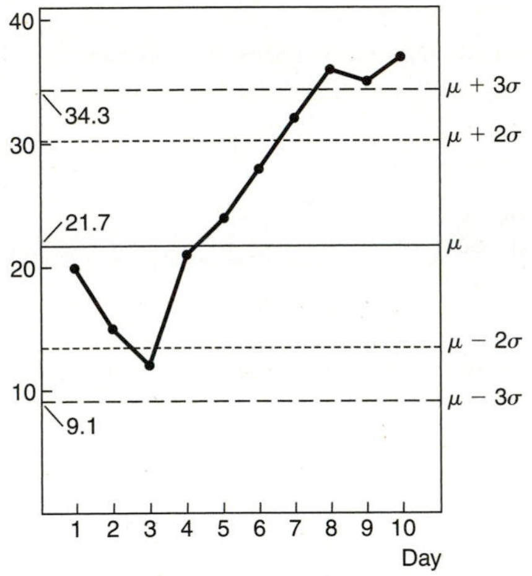

Out-of-control signals I and III are present.

14. (a) Number of Rooms Rented (first period)

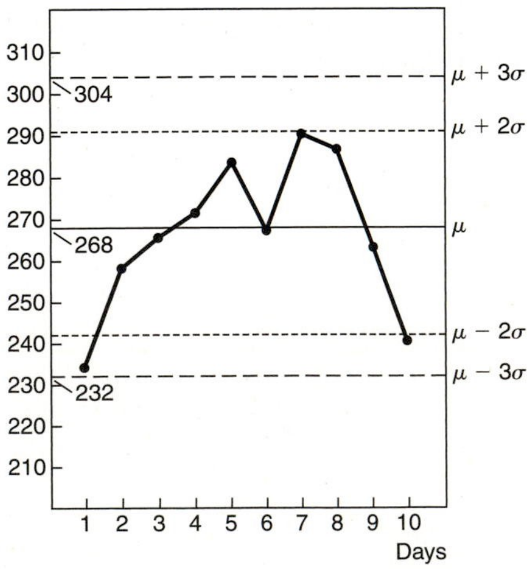

In control.

(b) Number of Rooms Rented (second period)

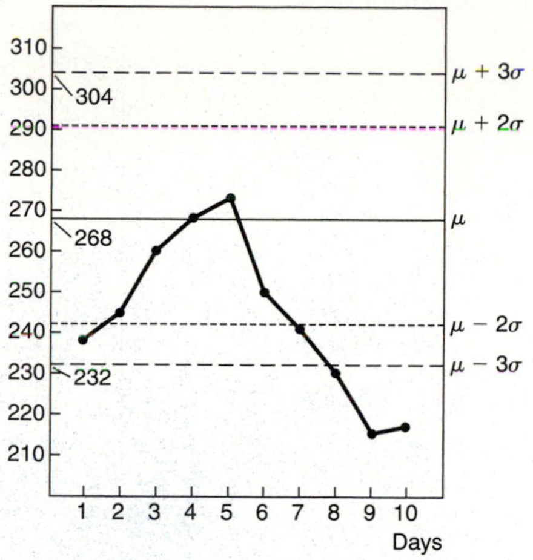

Out-of-control signals I and III are present.

Chapter 6 Review

20. (a) Hydraulic Pressure in Main Cylinder of Landing Gear of Airplanes (psi)—First Data Set

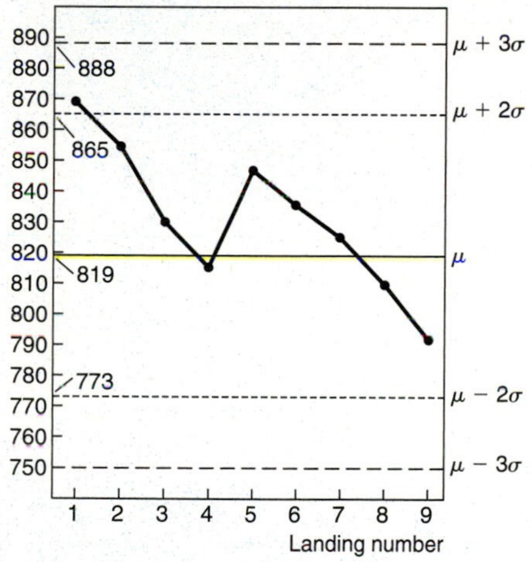

In control.

(b) Hydraulic Pressure in Main Cylinder of Landing Gear of Airplanes (psi)—Second Data Set

Out of control signals I and III are present.

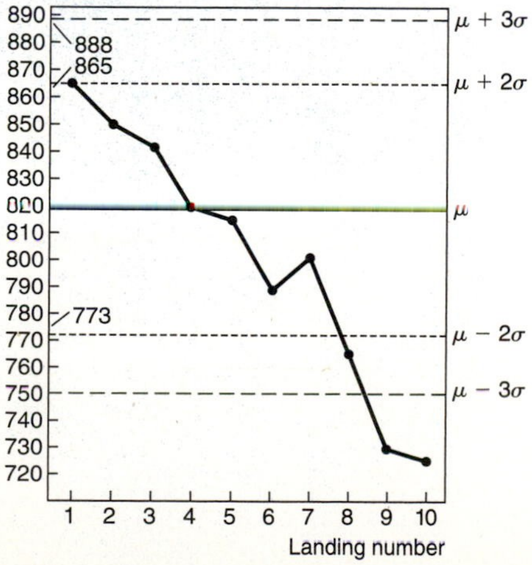

CHAPTER 9

Section 9.1

14. (a) Group Health Insurance Plans: Average Number of Employees versus Administrative Costs as a Percentage of Claims

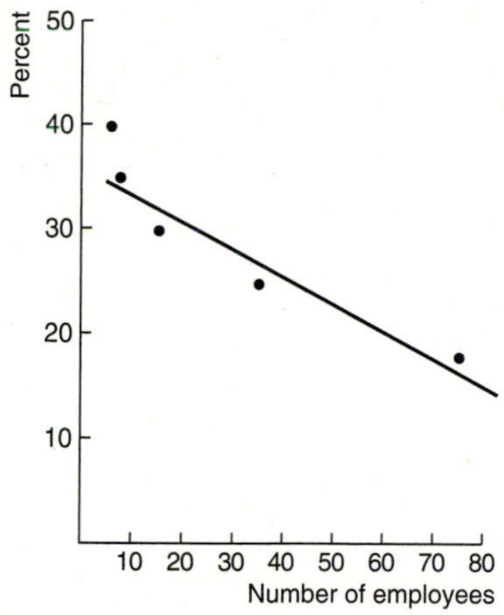

16. (a) Magnitude (Richter Scale) and Depth (km) of Earthquakes

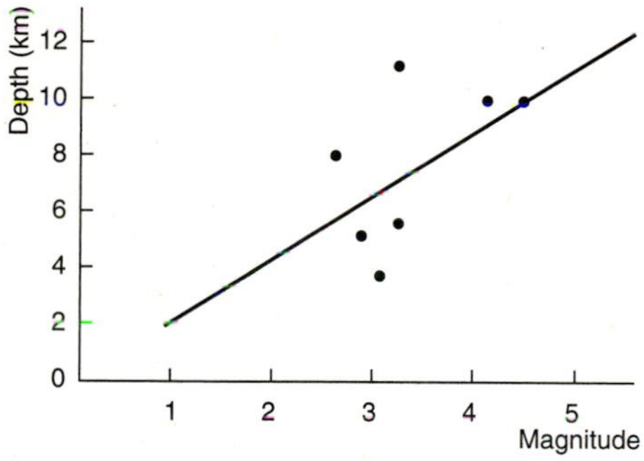

18. (a) Student Enrollment (in thousands) versus Number of Burglaries

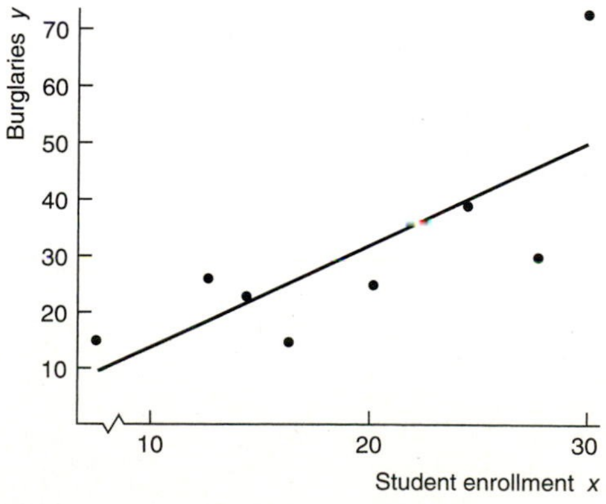

Section 9.2

8. (a) Age and Weight of Healthy Calves

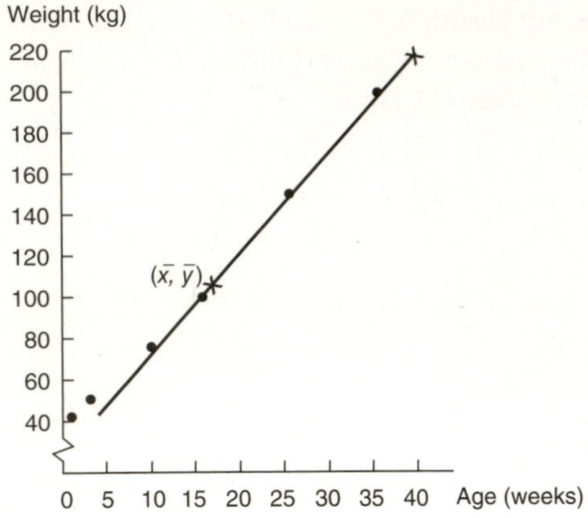

10. (a) Fouls and Basketball Wins

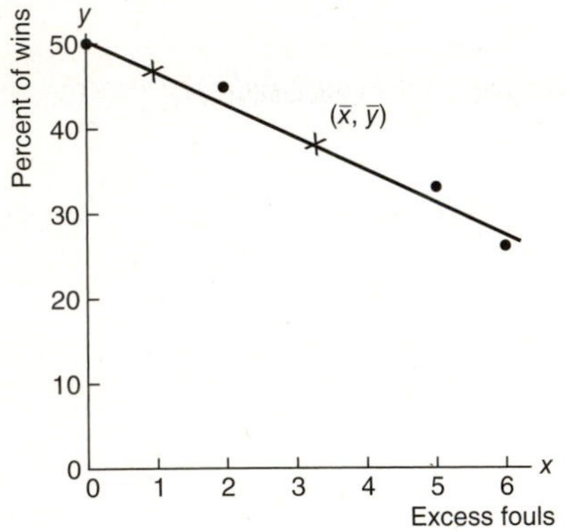

12. (a) Age and Percentage of Fatal Accidents Due to Failure to Yield

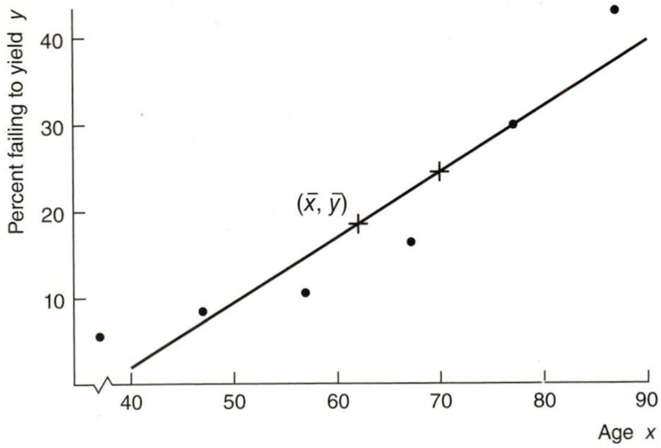

14. (a) Percent Change in Rate of Violent Crime and Percent Change in Rate of Imprisonment in U.S. Population

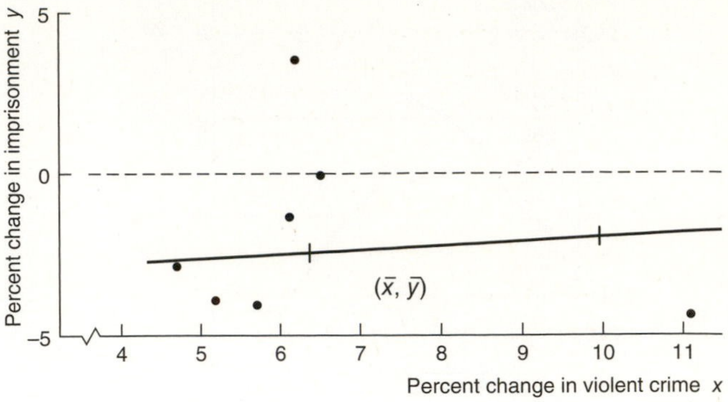

16. (a) Number of Research Programs and Mean Number of Patents per Program

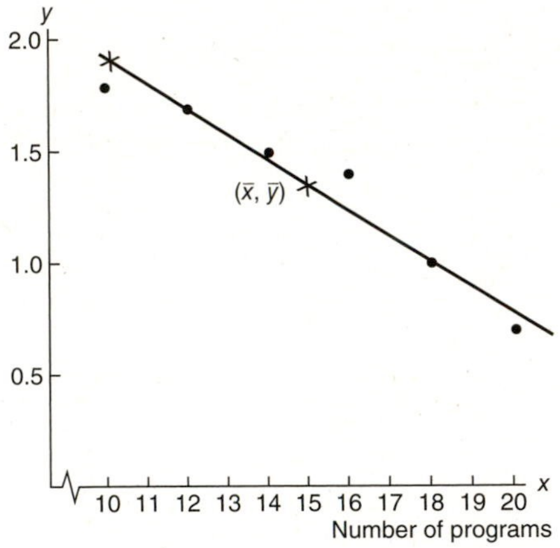

18. (a) Chirps per Second and Temperature (°F)

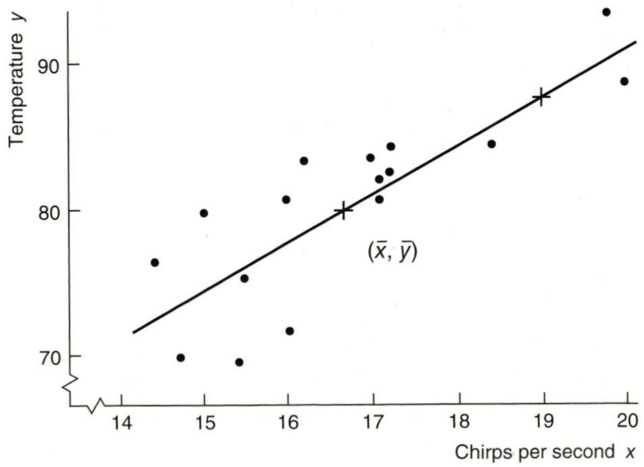

20. (a) Residuals: 2.9; 2.1; −0.1; −2.1; −0.5; −2.3; −1.9; 1.9.
Residual Plot

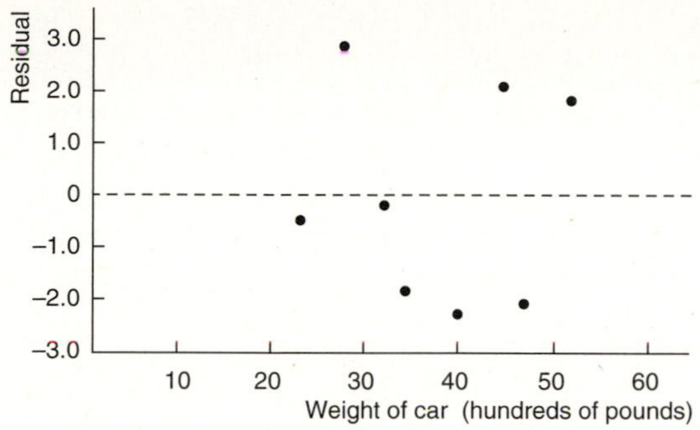

24. (a) Model with $(x'y')$ Data Pairs

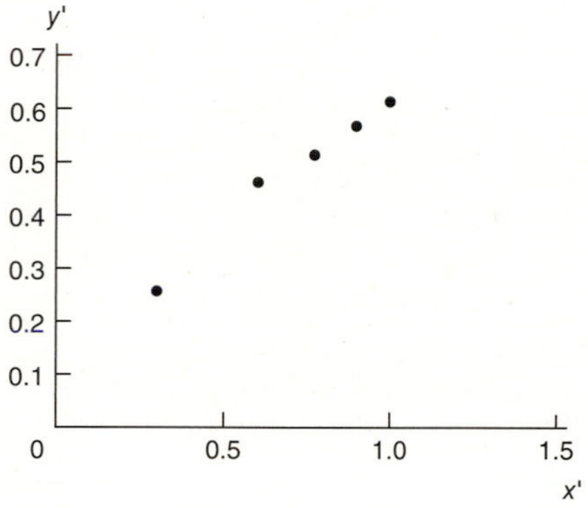

Chapter 9 Review

6. (a) Annual Salary (thousands) and Number of Job Changes

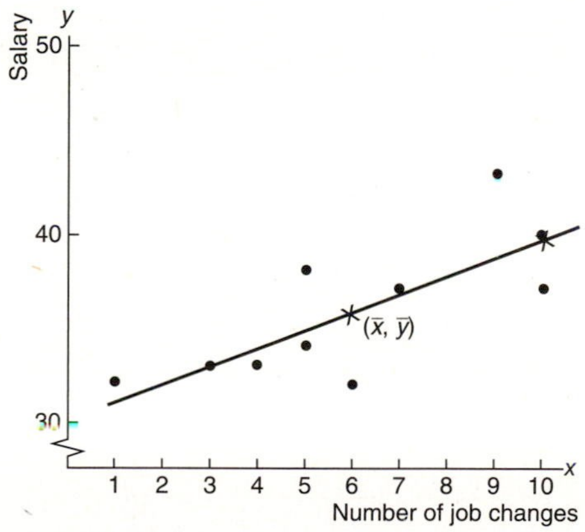

8. (a) Number of Insurance Sales and Number of Visits

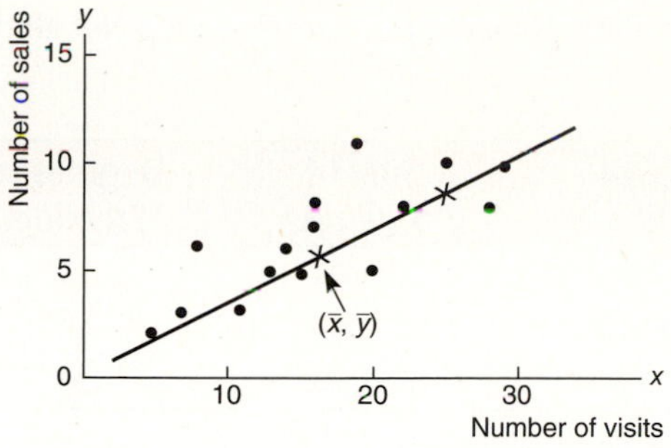

10. (a) Percent Population Change and Crime Rate

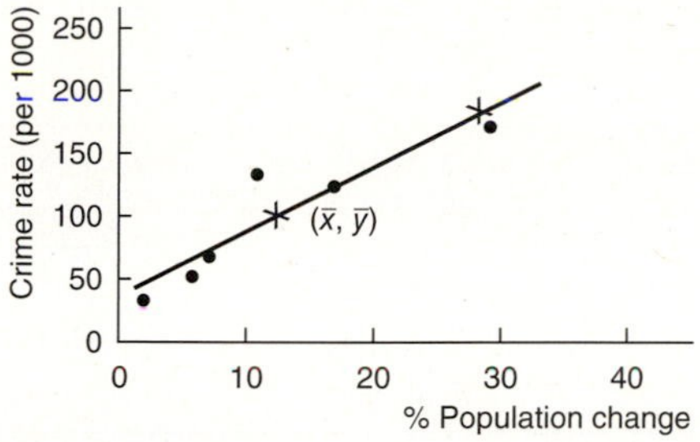

CHAPTER 10

Section 10.1

14. (i) Percentage of Each Party Spending Designated Amount

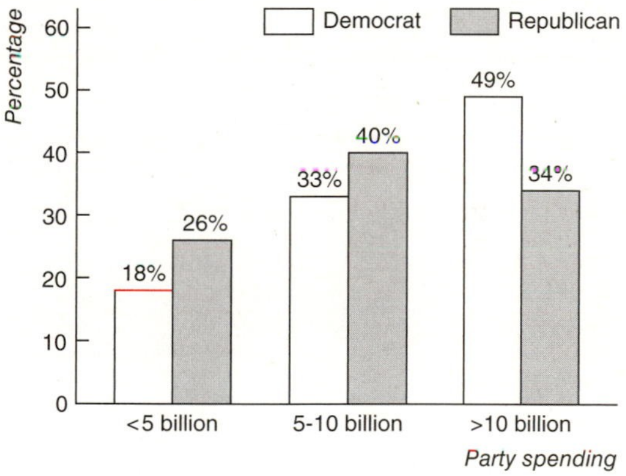

Section 10.5

2. (a) $\alpha = 0.05$; H_0: $\mu_1 = \mu_2 = \mu_3 = \mu_4$; H_1: Not all the means are equal.
(b–f)

Source of Variation	Sum of Squares	Degrees of Freedom	Mean Square	*F* Ratio	*P*-value	Test Decision
Between groups	421.033	3	140.344	1.573	> 0.100	Do not reject H_0
Within groups	1516.967	17	89.233			
Total	1938.000	20				

From TI-84, P-value ≈ 0.2327.

4. (a) $\alpha = 0.01$; H_0: $\mu_1 = \mu_2 = \mu_3$; H_1: Not all the means are equal.
(b–f)

Source of Variation	Sum of Squares	Degrees of Freedom	Mean Square	*F* Ratio	*P*-value	Test Decision
Between groups	215.680	2	107.840	0.816	> 0.100	Do not reject H_0
Within groups	1981.725	15	132.115			
Total	2197.405	17				

From TI-84, P-value ≈ 0.4608.

6. (a) $\alpha = 0.05$; H_0: $\mu_1 = \mu_2 = \mu_3$; H_1: Not all the means are equal.
(b–f)

Source of Variation	Sum of Squares	Degrees of Freedom	Mean Square	*F* Ratio	*P*-value	Test Decision
Between groups	2.441	2	1.2207	2.95	between	Do not reject H_0
Within groups	7.448	18	0.4138		0.050 and 0.100	
Total	9.890	20				

From TI-84, P-value ≈ 0.0779.

8. (a) $\alpha = 0.05$; H_0: $\mu_1 = \mu_2 = \mu_3 = \mu_4$; H_1: Not all the means are equal.
(b–f)

Source of Variation	Sum of Squares	Degrees of Freedom	Mean Square	*F* Ratio	*P*-value	Test Decision
Between groups	18.965	3	6.322	14.910	< 0.001	Reject H_0
Within groups	5.517	13	0.424			
Total	24.482	16				

From TI-84, P-value ≈ 0.0002.

Chapter 10 Review

8. One-way ANOVA. H_0: $\mu_1 = \mu_2 = \mu_3$; H_1: Not all the means are equal.

Source of Variation	Sum of Squares	Degrees of Freedom	Mean Square	*F* Ratio	*P*-value	Test Decision
Between groups	1.002	2	0.501	0.443	> 0.100	Fail to reject H_0
Within groups	10.165	9	1.129			
Total	11.167	11				

TI-84 gives P-value ≈ 0.6651.

Index

FREQUENTLY USED FORMULAS

n = sample size N = population size f = frequency

Chapter 2

Class width $= \dfrac{\text{high} - \text{low}}{\text{number of classes}}$ (increase to next integer)

Class midpoint $= \dfrac{\text{upper limit} + \text{lower limit}}{2}$

Lower boundary = lower boundary of previous class + class width

Chapter 3

Sample mean $\bar{x} = \dfrac{\Sigma x}{n}$

Population mean $\mu = \dfrac{\Sigma x}{N}$

Weighted average $= \dfrac{\Sigma xw}{\Sigma w}$

Range = largest data value − smallest data value

Sample standard deviation $s = \sqrt{\dfrac{\Sigma(x - \bar{x})^2}{n - 1}}$

Computation formula $s = \sqrt{\dfrac{\Sigma x^2 - (\Sigma x)^2/n}{n - 1}}$

Population standard deviation $\sigma = \sqrt{\dfrac{\Sigma(x - \mu)^2}{N}}$

Sample variance s^2

Population variance σ^2

Sample coefficient of variation $CV = \dfrac{s}{\bar{x}} \cdot 100$

Sample mean for grouped data $\bar{x} = \dfrac{\Sigma xf}{n}$

Sample standard deviation for grouped data

$$s = \sqrt{\frac{\Sigma(x - \bar{x})^2 f}{n - 1}} = \sqrt{\frac{\Sigma x^2 f - (\Sigma xf)^2/n}{n - 1}}$$

Chapter 4

Probability of the complement of event A
$P(A^c) = 1 - P(A)$

Multiplication rule for independent events
$P(A \text{ and } B) = P(A) \cdot P(B)$

General multiplication rules
$P(A \text{ and } B) = P(A) \cdot P(B|A)$
$P(A \text{ and } B) = P(B) \cdot P(A|B)$

Addition rule for mutually exclusive events
$P(A \text{ or } B) = P(A) + P(B)$

General addition rule
$P(A \text{ or } B) = P(A) + P(B) - P(A \text{ and } B)$

Permutation rule $P_{n,r} = \dfrac{n!}{(n - r)!}$

Combination rule $C_{n,r} = \dfrac{n!}{r!(n - r)!}$

Chapter 5

Mean of a discrete probability distribution $\mu = \Sigma xP(x)$

Standard deviation of a discrete probability distribution

$\sigma = \sqrt{\Sigma(x - \mu)^2 P(x)}$

Given $L = a + bx$

$\mu_L = a + b\mu$

$\sigma_L = |b|\sigma$

Given $W = ax_1 + bx_2$ (x_1 and x_2 independent)

$\mu_W = a\mu_1 + b\mu_2$

$\sigma_W = \sqrt{a^2\sigma_1^2 + b^2\sigma_2^2}$

For Binomial Distributions

r = number of successes; p = probability of success;

$q = 1 - p$

Binomial probability distribution $P(r) = C_{n,r}\, p^r q^{n-r}$

Mean $\mu = np$

Standard deviation $\sigma = \sqrt{npq}$

Geometric Probability Distribution

n = number of trial on which first success occurs

$P(n) = p(1 - p)^{n-1}$

Poisson Probability Distribution

r = number of successes

λ = mean number of successes over given interval

$P(r) = \dfrac{e^{-\lambda}\lambda^r}{r!}$

Chapter 6

Raw score $x = z\sigma + \mu$ Standard score $z = \dfrac{x - \mu}{\sigma}$

Mean of $\bar{x}$ distribution $\mu_{\bar{x}} = \mu$

Standard deviation of $\bar{x}$ distribution $\sigma_{\bar{x}} = \dfrac{\sigma}{\sqrt{n}}$

Standard score for $\bar{x}$ $z = \dfrac{\bar{x} - \mu}{\sigma/\sqrt{n}}$

Mean of $\hat{p}$ distribution $\mu_{\hat{p}} = p$

Standard deviation of $\hat{p}$ distribution $\sigma_{\hat{p}} = \sqrt{\dfrac{pq}{n}}$; $q = 1 - p$

Chapter 7

Confidence Interval

for μ

$\bar{x} - E < \mu < \bar{x} + E$

where $E = z_c \dfrac{\sigma}{\sqrt{n}}$ when σ is known

$E = t_c \dfrac{s}{\sqrt{n}}$ when σ is unknown

with $d.f. = n - 1$

for p ($np > 5$ and $n(1 - p) > 5$)

$\hat{p} - E < p < \hat{p} + E$

where $E = z_c \sqrt{\dfrac{\hat{p}(1 - \hat{p})}{n}}$

$\hat{p} = \dfrac{r}{n}$

for $\mu_1 - \mu_2$ (independent samples)

$(\bar{x}_1 - \bar{x}_2) - E < \mu_1 - \mu_2 < (\bar{x}_1 - \bar{x}_2) + E$

where $E = z_c \sqrt{\dfrac{\sigma_1^2}{n_1} + \dfrac{\sigma_2^2}{n_2}}$ when σ_1 and σ_2 are known

$E = t_c \sqrt{\dfrac{s_1^2}{n_1} + \dfrac{s_2^2}{n_2}}$ when σ_1 or σ_2 is unknown

with $d.f.$ = smaller of $n_1 - 1$ and $n_2 - 1$

(*Note*: Software uses Satterthwaite's approximation for degrees of freedom $d.f.$)

for difference of proportions $p_1 - p_2$

$(\hat{p}_1 - \hat{p}_2) - E < p_1 - p_2 < (\hat{p}_1 - \hat{p}_2) + E$

where $E = z_c \sqrt{\dfrac{\hat{p}_1\hat{q}_1}{n_1} + \dfrac{\hat{p}_2\hat{q}_2}{n_2}}$

$\hat{p}_1 = r_1/n_1$; $\hat{p}_2 = r_2/n_2$

$\hat{q}_1 = 1 - \hat{p}_1$; $\hat{q}_2 = 1 - \hat{p}_2$

Sample Size for Estimating

means $n = \left(\dfrac{z_c\sigma}{E}\right)^2$

proportions

$n = p(1 - p)\left(\dfrac{z_c}{E}\right)^2$ with preliminary estimate for p

$n = \dfrac{1}{4}\left(\dfrac{z_c}{E}\right)^2$ without preliminary estimate for p

Chapter 8

Sample Test Statistics for Tests of Hypotheses

for μ (σ k0nown) $z = \dfrac{\bar{x} - \mu}{\sigma/\sqrt{n}}$

for μ (σ unknown) $t = \dfrac{\bar{x} - \mu}{s/\sqrt{n}}$; $d.f. = n - 1$

for p ($np > 5$ and $nq > 5$) $z = \dfrac{\hat{p} - p}{\sqrt{pq/n}}$

where $q = 1 - p$; $\hat{p} = r/n$

for paired differences d $t = \dfrac{\bar{d} - \mu_d}{s_d/\sqrt{n}}$; $d.f. = n - 1$

for difference of means, σ_1 and σ_2 known

$$z = \frac{\bar{x}_1 - \bar{x}_2}{\sqrt{\dfrac{\sigma_1^2}{n_1} + \dfrac{\sigma_2^2}{n_2}}}$$

for difference of means, σ_1 or σ_2 unknown

$$t = \frac{\bar{x}_1 - \bar{x}_2}{\sqrt{\dfrac{s_1^2}{n_1} + \dfrac{s_2^2}{n_2}}}$$

$d.f.$ = smaller of $n_1 - 1$ and $n_2 - 1$

(*Note*: Software uses Satterthwaite's approximation for degrees of freedom $d.f.$)

for difference of proportions

$$z = \frac{\hat{p}_1 - \hat{p}_2}{\sqrt{\dfrac{\bar{p}\,\bar{q}}{n_1} + \dfrac{\bar{p}\,\bar{q}}{n_2}}}$$

where $\bar{p} = \dfrac{r_1 + r_2}{n_1 + n_2}$ and $\bar{q} = 1 - \bar{p}$

$\hat{p}_1 = r_1/n_1$; $\hat{p}_2 = r_2/n_2$

Chapter 9

Regression and Correlation

Pearson product-moment correlation coefficient

$$r = \frac{n\sum xy - (\sum x)(\sum y)}{\sqrt{n\sum x^2 - (\sum x)^2}\sqrt{n\sum y^2 - (\sum y)^2}}$$

Least-squares line $\hat{y} = a + bx$

where $b = \dfrac{n\sum xy - (\sum x)(\sum y)}{n\sum x^2 - (\sum x)^2}$

$a = \bar{y} - b\bar{x}$

Coefficient of determination $= r^2$

Sample test statistic for r

$t = \dfrac{r\sqrt{n - 2}}{\sqrt{1 - r^2}}$ with $d.f. = n - 2$

Standard error of estimate $S_e = \sqrt{\dfrac{\sum y^2 - a\sum y - b\sum xy}{n - 2}}$

Confidence interval for y

$\hat{y} - E < y < \hat{y} + E$

where $E = t_c S_e \sqrt{1 + \dfrac{1}{n} + \dfrac{n(x - \bar{x})^2}{n\sum x^2 - (\sum x)^2}}$

with $d.f. = n - 2$